Register Now to Improve Your Grade!

MasteringBIOLOGY

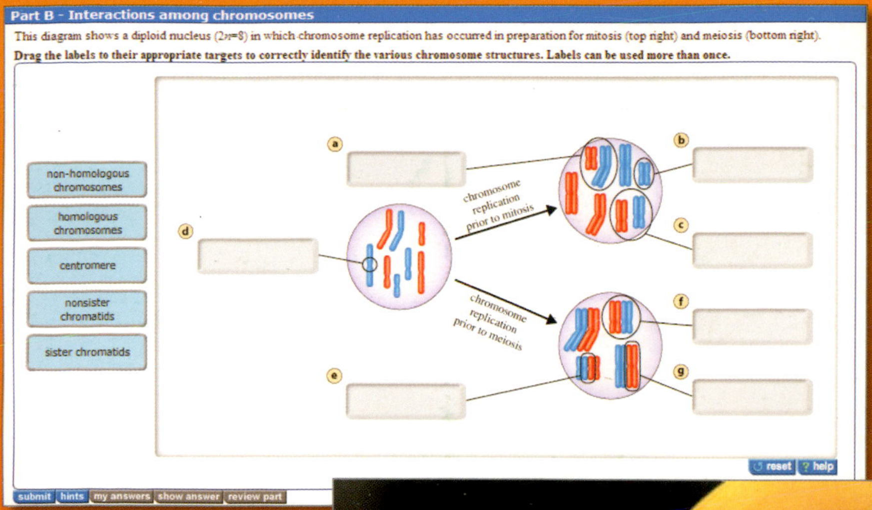

- Access your textbook online 24/7
- Take practice quizzes to prepare for tests
- Grasp difficult concepts with 3-D BioFlix® animations, engaging activities, and more

www.masteringbiology.com

STUDENTS

To Register Using the Student Access Kit

Your textbook may have been packaged with a **Mastering Student Access Kit**. This kit contains your access code to this valuable website.

To Purchase Access Online

If your textbook was not packaged with a **Mastering Student Access Kit**, you can purchase access online using a major credit card or PayPal account.

INSTRUCTORS

To Request Access Online

To request access online, go to www.masteringbiology.com and click New Instructors. Please contact your sales representative for more information.

To the Student: How to Use This Book

Find important information in this book by focusing on the Gold Thread.

UNIT 1

THE MOLECULES OF LIFE

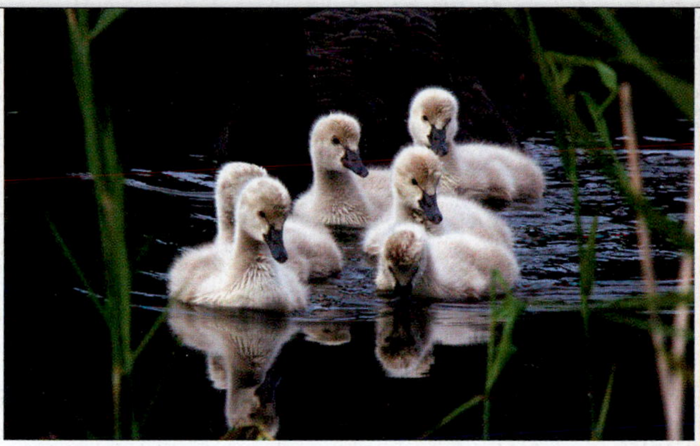

These black swan chicks are being introduced to the world by their mother, who trails close behind them. Likewise, this chapter introduces the key ideas that launched biological science as a discipline.

Biology and the Tree of Life 1

In essence, biological science is a search for ideas and observations that unify our understanding of the diversity of life, from bacteria living in rocks a mile underground to hedgehogs and humans. Chapter 1 is an introduction to this search.

The goals of this chapter are to introduce the nature of life and explore how biologists go about studying it. The chapter also introduces themes that will resonate throughout this book: (1) analyzing how organisms work at the molecular level, (2) understanding organisms in terms of their evolutionary history, and (3) helping you learn to think like a biologist.

Let's begin with what may be the most fundamental question of all: What is life?

KEY CONCEPTS

- Organisms obtain and use energy, are made up of cells, process information, replicate, and as populations evolve.
- The cell theory proposes that all organisms are made of cells and that all cells come from preexisting cells.
- The theory of evolution by natural selection maintains that species change through time because individuals with certain heritable traits produce more offspring than other individuals do.
- A phylogenetic tree is a graphical representation of the evolutionary relationships between species. These relationships can be estimated by analyzing similarities and differences in traits. Species that share distinctive traits are closely related and are placed close to each other on the tree of life.
- Biologists ask questions, generate hypotheses to answer them, and design experiments that test the predictions made by competing hypotheses.

1.1 What Does It Mean to Say That Something Is Alive?

An organism is a life-form—a living entity made up of one or more cells. Although there is no simple definition of life that is endorsed by all biologists, most agree that organisms share a suite of five fundamental characteristics.

- *Energy* To stay alive and reproduce, organisms have to acquire and use energy. To give just two examples: plants absorb sunlight; animals ingest food.
- *Cells* Organisms are made up of membrane-bound units called cells. A cell's membrane regulates the passage of materials between exterior and interior spaces.
- *Information* Organisms process hereditary or genetic information, encoded in units called genes, along with information they acquire from the environment. Right now cells throughout your body are using genetic information to make the molecules that keep you alive; your eyes and brain are decoding information on this page that will help you learn some biology.

✓ When you see this checkmark, stop and test yourself. Answers are available in Appendix B.

Key Concepts

Start with Key Concepts on the first page of every chapter. Read these gold key points first to familiarize yourself with the chapter's big ideas.

MORE! Bulleted Lists

Take note of bulleted lists that "chunk" information and ideas. This will help you manage the information that you are learning in the course.

Gold Highlighting

Watch for important information highlighted in gold. Gold highlighting is always a signal to slow down and pay special attention.

Gold Key

Material related to Key Concepts will be signaled with a gold key.

The best way to succeed on exams...

As you read the text and view the figures, practice with the Blue Thread.

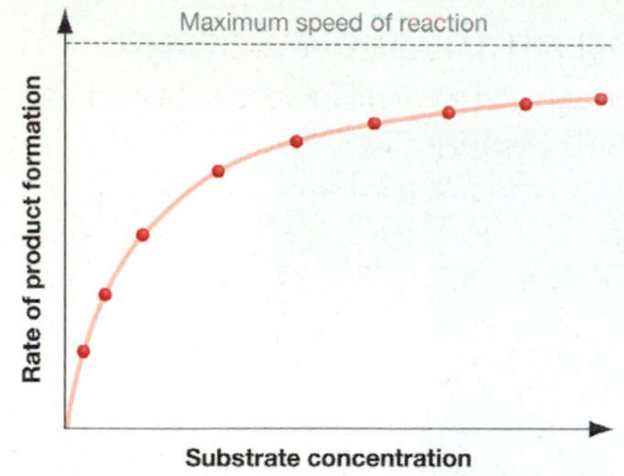

FIGURE 3.22 Kinetics of an Enzyme-Catalyzed Reaction. The general shape of this curve is characteristic of enzyme-catalyzed reactions.

✔**QUESTION** Explain which part of the graph represents where (1) the reaction rate is most sensitive to changes in substrate concentration and (2) most or all of the active sites present are occupied.

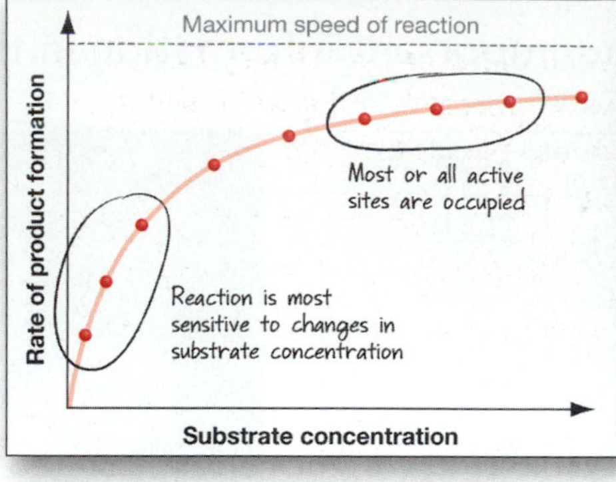

Blue Thread Questions

Many figures include Blue Thread Questions or Exercises to help you check your understanding of the material they present.

Drawing Exercises

Some Blue Thread Questions contain artwork from the textbook that you will be asked to draw on or modify.

NEW! Suggested Answers

Suggested answers for the Blue Thread Questions and Exercises are provided in Appendix B.

"You Should be Able To" Exercises

Text passages flagged with blue type and the words "you should be able to" offer exercises on concepts that professors and students have identified as most difficult. These are the topics most students struggle with on exams.

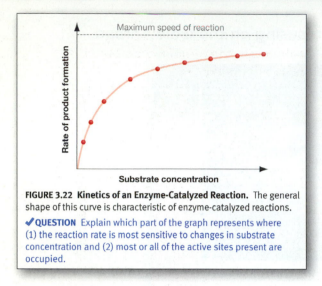

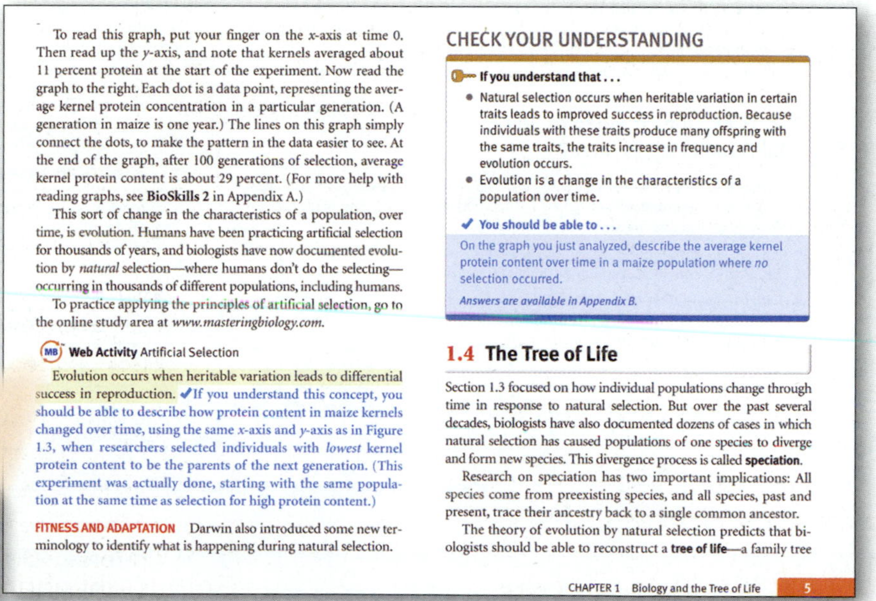

To read this graph, put your finger on the *x*-axis at time 0. Then read up the *y*-axis, and note that kernels averaged about 11 percent protein at the start of the experiment. Now read the graph to the right. Each dot is a data point, representing the average kernel protein concentration in a particular generation. (A generation in maize is one year.) The lines on this graph simply connect the dots, to make the pattern in the data easier to see. At the end of the graph, after 100 generations of selection, average kernel protein content is about 29 percent. (For more help with reading graphs, see **BioSkills 2** in Appendix A.)

This sort of change in the characteristics of a population, over time, is evolution. Humans have been practicing artificial selection for thousands of years, and biologists have now documented evolution by *natural* selection—where humans don't do the selecting—occurring in thousands of different populations, including humans.

To practice applying the principles of artificial selection, go to the online study area at *www.masteringbiology.com*.

MB **Web Activity** Artificial Selection

Evolution occurs when heritable variation leads to differential success in reproduction. ✔If you understand this concept, you should be able to describe how protein content in maize kernels changed over time, using the same *x*-axis and *y*-axis as in Figure 1.3, when researchers selected individuals with *lowest* kernel protein content to be the parents of the next generation. (This experiment was actually done, starting with the same population at the same time as selection for high protein content.)

FITNESS AND ADAPTATION Darwin also introduced some new terminology to identify what is happening during natural selection.

CHECK YOUR UNDERSTANDING

If you understand that . . .

- Natural selection occurs when heritable variation in certain traits leads to improved success in reproduction. Because individuals with these traits produce many offspring with the same traits, the traits increase in frequency and evolution occurs.
- Evolution is a change in the characteristics of a population over time.

✔ **You should be able to . . .**

On the graph you just analyzed, describe the average kernel protein content over time in a maize population where *no* selection occurred.

Answers are available in Appendix B.

1.4 The Tree of Life

Section 1.3 focused on how individual populations change through time in response to natural selection. But over the past several decades, biologists have also documented dozens of cases in which natural selection has caused populations of one species to diverge and form new species. This divergence process is called **speciation**.

Research on speciation has two important implications: All species come from preexisting species, and all species, past and present, trace their ancestry back to a single common ancestor.

The theory of evolution by natural selection predicts that biologists should be able to reconstruct a **tree of life**—a family tree

CHAPTER 1 Biology and the Tree of Life 5

Check Your Understanding

The blue half of the Check Your Understanding boxes asks you to do something with the information in the top half. If you can't complete these exercises, go back and re-read that section of the chapter.

MasteringBIOLOGY

Make Learning Part of the Grade®

It's possible that your professor will include these Blue Thread Questions in a graded assignment at www.masteringbiology.com.

...is to practice. Here's how.

NEW! Bulleted Summary of Key Concepts

The succinct Summary of Key Concepts reviews important concepts in short, manageable bullet points.

Blue Thread Exercises

End-of-chapter Blue Thread Exercises help you review the major themes of the chapter and synthesize information.

CHAPTER 6 REVIEW *For media, go to the study area at www.masteringbiology.com*

Summary of Key Concepts

Phospholipids are amphipathic molecules—they have a hydrophilic region and a hydrophobic region. In solution, phospholipids spontaneously form bilayers that are selectively permeable—meaning that only certain substances cross them readily.

- The plasma membrane forms a physical barrier between the internal and external environment—often between life and nonlife.
- The basic structure of plasma membranes is created by a phospholipid bilayer.

solutes to the region of lower water concentration and higher solute concentration.

- Osmosis is a passive process driven by an increase in entropy.

✔You should be able to imagine a beaker with solutions separated by a plasma membrane, and then predict what will happen after addition of a solute to one side if the solute (1) crosses the membrane readily or (2) is incapable of crossing the membrane.

Web Activity Diffusion and Osmosis

MasteringBIOLOGY®

Make Learning Part of the Grade®

Visit www.masteringbiology.com for practice quizzes, 3-D animations, the eText, and more.

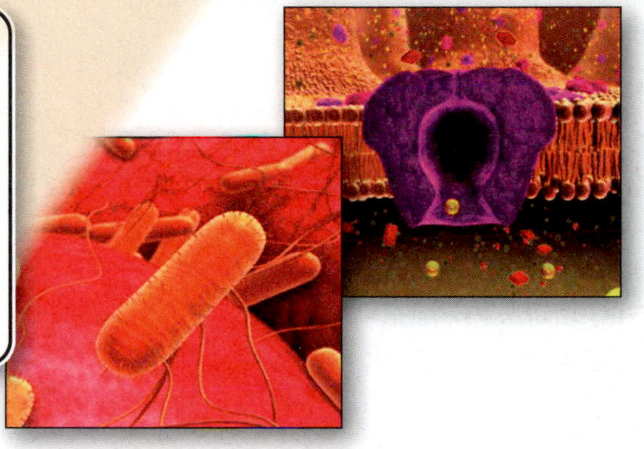

BioFlix™

BioFlix™ 3-D animations are included in MasteringBiology's Study Area and are available as automatically graded assignments.

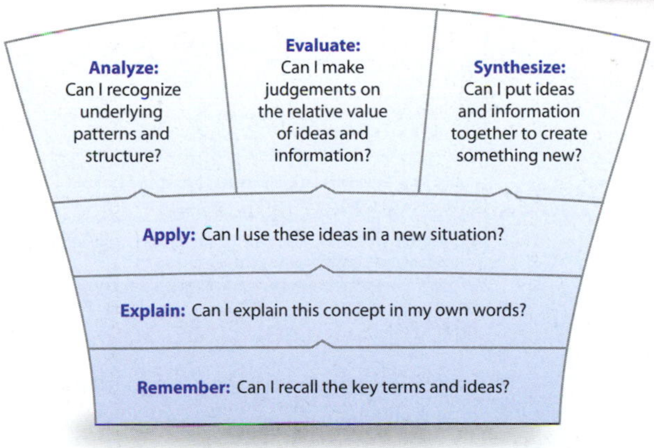

Analyze: Can I recognize underlying patterns and structure?
Evaluate: Can I make judgements on the relative value of ideas and information?
Synthesize: Can I put ideas and information together to create something new?
Apply: Can I use these ideas in a new situation?
Explain: Can I explain this concept in my own words?
Remember: Can I recall the key terms and ideas?

Bloom's Taxonomy

Bloom's Taxonomy categorizes six levels of learning competency. The Blue Thread Questions and Exercises in the textbook test on the higher levels of the scale—Explain, Apply, Analyze, Evaluate, and Synthesize—to help you develop critical thinking skills and prepare you for exams.

Steps to Understanding

End of Chapter questions are scaled along Bloom's Taxonomy.

✔TEST YOUR KNOWLEDGE

Begin by testing your knowledge of new facts.

✔TEST YOUR UNDERSTANDING

Once you're confident in your knowledge of the material, demonstrate your understanding by answering the Test Your Understanding questions.

✔APPLYING CONCEPTS TO NEW SITUATIONS

Challenge yourself even further by applying your understanding of the concepts to new situations.

Learn to think like a scientist. Here's how.

A unique emphasis on the process of scientific discovery and experimental design teaches you how to think like a scientist as you learn fundamental biology concepts.

Experiment Boxes

Study Experiment Boxes to help you understand how experiments are designed and give you practice interpreting data.

MasteringBIOLOGY
Make Learning Part of the Grade®
www.masteringbiology.com

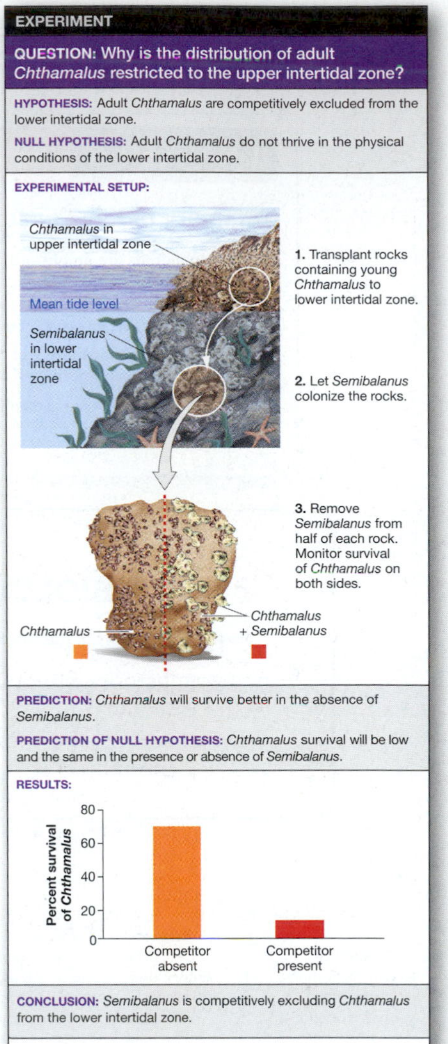

EXPERIMENT
QUESTION: Why is the distribution of adult *Chthamalus* restricted to the upper intertidal zone?
HYPOTHESIS: Adult *Chthamalus* are competitively excluded from the lower intertidal zone.
NULL HYPOTHESIS: Adult *Chthamalus* do not thrive in the physical conditions of the lower intertidal zone.
EXPERIMENTAL SETUP:
Chthamalus in upper intertidal zone
Mean tide level
Semibalanus in lower intertidal zone
1. Transplant rocks containing young *Chthamalus* to lower intertidal zone.
2. Let *Semibalanus* colonize the rocks.
3. Remove *Semibalanus* from half of each rock. Monitor survival of *Chthamalus* on both sides.
Chthamalus
Chthamalus + *Semibalanus*
PREDICTION: *Chthamalus* will survive better in the absence of *Semibalanus*.
PREDICTION OF NULL HYPOTHESIS: *Chthamalus* survival will be low and the same in the presence or absence of *Semibalanus*.
RESULTS:
Percent survival of *Chthamalus*
80
60
40
20
0
Competitor absent
Competitor present
CONCLUSION: *Semibalanus* is competitively excluding *Chthamalus* from the lower intertidal zone.

FIGURE 53.6 Experimental Evidence for Competitive Exclusion.

SOURCE: Connell, J. H. 1961. The influence of interspecific competition and other factors on the distribution of the barnacle *Chthamalus stellatus*. *Ecology* 42: 710–723.

✔QUESTION Why was it important to carry out both treatments on the same rock? Why not use separate rocks?

NEW! Experimental Inquiry Tutorials

Experimental Inquiry Tutorials based on some of biology's most seminal experiments can be found on www.masteringbiology.com. Your instructor may assign these. They will give you practice analyzing the experimental design and data, and help you understand reasoning that led scientists from the data they collected to their conclusions.

Some of the topics include:

- The Process of Science
- Engelmann's Photosynthesis and Wavelengths of Light
- Morgan's Cross with White-Eyed Males
- Meselson-Stahl's Semiconservative Replication
- Steinhardt et al and Hafner et al's Polyspermy
- Grant's Changes in Finch Beak Size
- Went's Phototropism and Auxin Distribution
- Coleman's Obesity Gene
- Connell's Competition in Barnacles
- Bormann, Likens et al's Nutrient Cycling in Hubbard Brook Forest

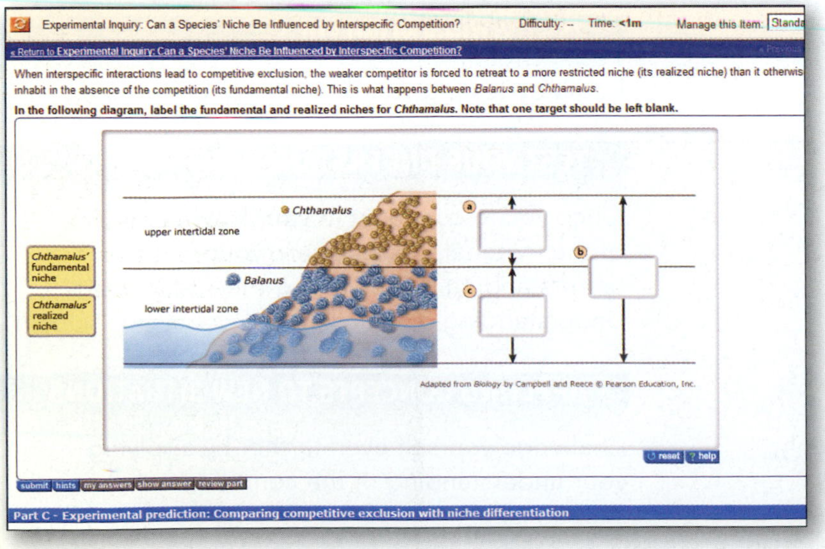

inhabit in the absence of the competition (its fundamental niche). This is what happens between *Balanus* and *Chthamalus*.

NEW! Source Citations

Each Experiment Box now cites the original research paper, encouraging you to extend your learning by exploring the primary literature.

NEW! Experiment Box Questions

Each Experiment Box now includes a question that asks students to analyze the design of the experiment.

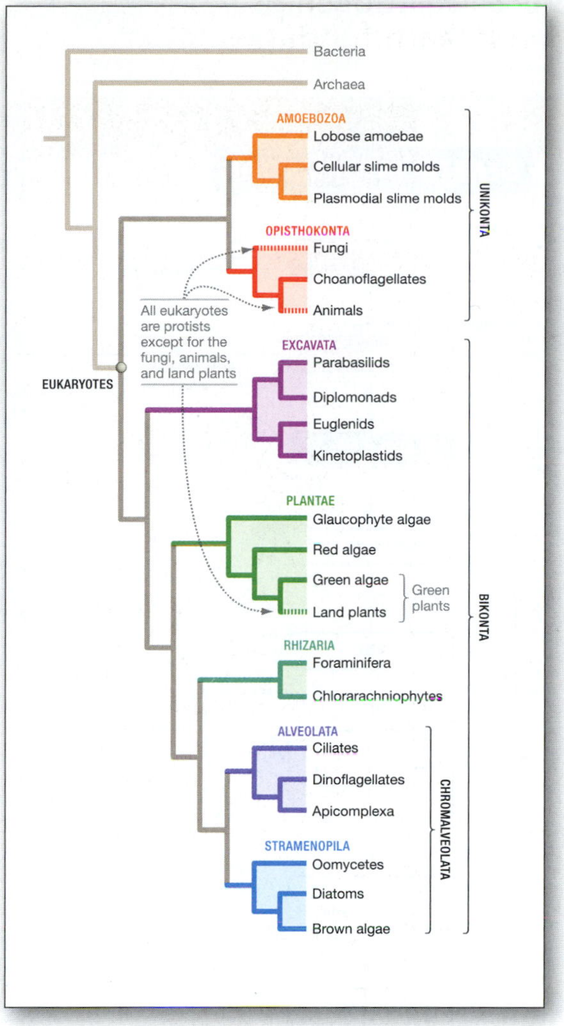

NEW! Redesigned Phylogenetic Trees

Practice "tree thinking" using these newly redesigned phylogenetic trees. Their U-shaped, top-to-bottom format is consistent with the way such trees are most commonly depicted in the scientific literature.

Expanded BioSkills Appendix

BIOSKILLS

Build skills that will be important to your success in future courses. At relevant points in the text, you'll find references to the expanded BioSkills Appendix that will help you learn and practice the following foundational skills:

- NEW! The Metric System
- Reading Graphs
- Reading a Phylogenetic Tree
- NEW! Some Common Latin and Greek Roots Used in Biology
- Using Statistical Tests and Interpreting Standard Error Bars
- Reading Chemical Structures
- Using Logarithms
- Making Concept Maps
- Separating and Visualizing Molecules
- Biological Imaging: Microscopy and X-Ray Crystallography
- NEW! Separating Cell Components by Centrifugation
- NEW! Cell Culture Methods
- Combining Probabilities
- NEW! Model Organisms

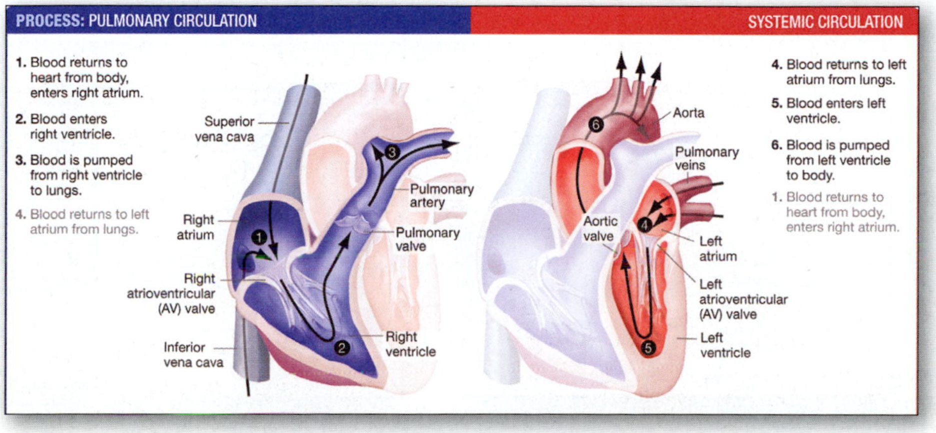

Informative Figures

Think through complex biological processes with figures that clearly define concepts.

Keep sight of the big picture.

Concept maps help you to keep sight of "big picture" relationships among biological concepts.

NEW! Big Picture Concept Maps

Four remarkable Big Picture concept maps help you synthesize information across the chapters on energy, genetics, evolution, and ecology.

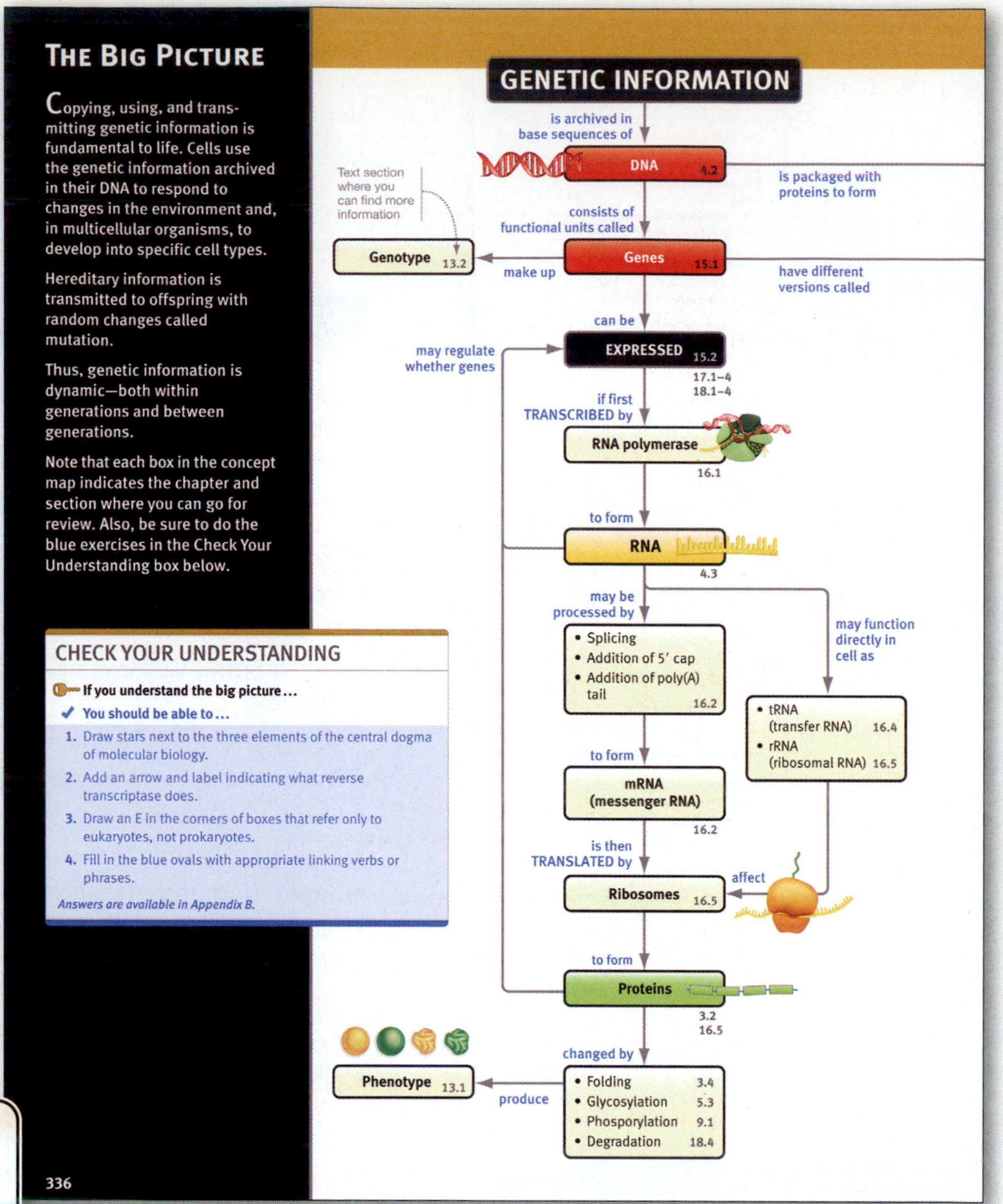

THE BIG PICTURE

Copying, using, and transmitting genetic information is fundamental to life. Cells use the genetic information archived in their DNA to respond to changes in the environment and, in multicellular organisms, to develop into specific cell types.

Hereditary information is transmitted to offspring with random changes called mutation.

Thus, genetic information is dynamic—both within generations and between generations.

Note that each box in the concept map indicates the chapter and section where you can go for review. Also, be sure to do the blue exercises in the Check Your Understanding box below.

CHECK YOUR UNDERSTANDING

If you understand the big picture . . .

You should be able to . . .

1. Draw stars next to the three elements of the central dogma of molecular biology.
2. Add an arrow and label indicating what reverse transcriptase does.
3. Draw an E in the corners of boxes that refer only to eukaryotes, not prokaryotes.
4. Fill in the blue ovals with appropriate linking verbs or phrases.

Answers are available in Appendix B.

336

Check Your Understanding

Check your understanding of these big picture relationships by answering the Blue Thread Questions.

Your professor may assign interactive Big Picture concept map exercises at www.masteringbiology.com.

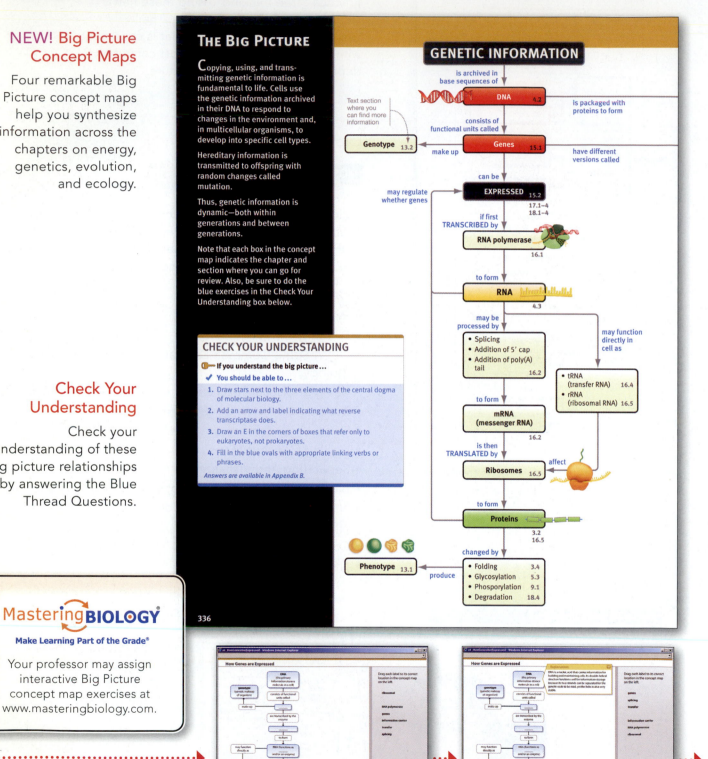

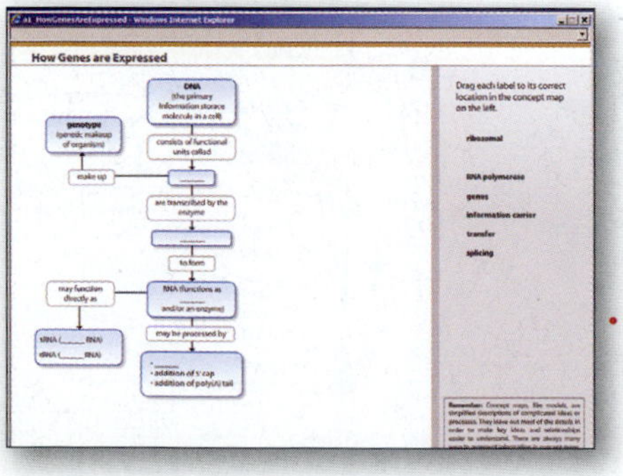

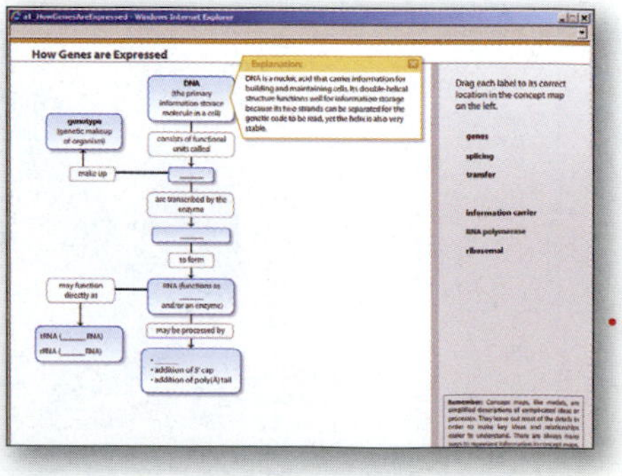

More Big Picture activities are available in the study area at www.masteringbiology.com

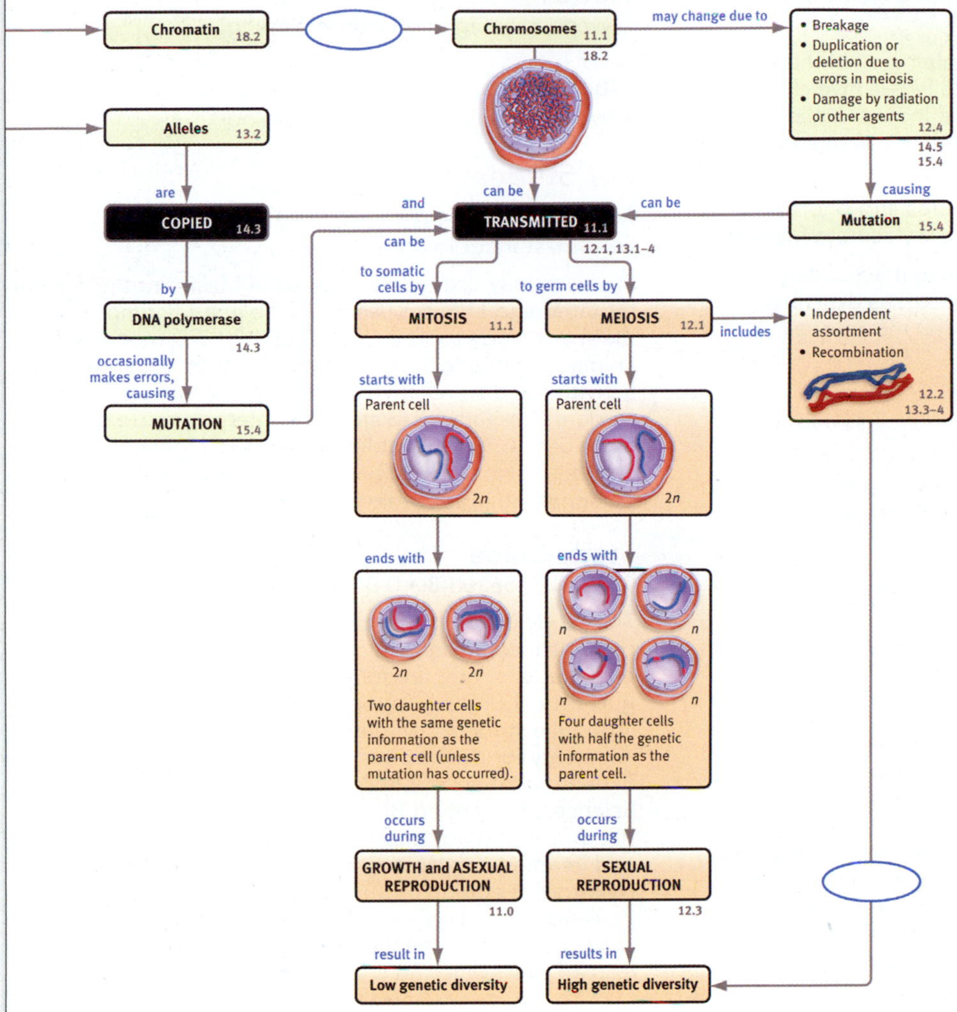
Chromatin 18.2
Chromosomes 11.1
18.2
may change due to
• Breakage
• Duplication or deletion due to errors in meiosis
• Damage by radiation or other agents
12.4
14.5
15.4
causing
Mutation 15.4
can be
Alleles 13.2
are
COPIED 14.3
and
can be
TRANSMITTED 11.1
can be
12.1, 13.1–4
by
DNA polymerase
14.3
occasionally makes errors, causing
MUTATION 15.4
to somatic cells by
to germ cells by
MITOSIS 11.1
MEIOSIS 12.1
includes
• Independent assortment
• Recombination
12.2
13.3–4
starts with
Parent cell
2n
ends with
2n
2n
Two daughter cells with the same genetic information as the parent cell (unless mutation has occurred).
n
Four daughter cells with half the genetic information as the parent cell.
occurs during
GROWTH and ASEXUAL REPRODUCTION
11.0
SEXUAL REPRODUCTION
12.3
result in
results in
Low genetic diversity
High genetic diversity

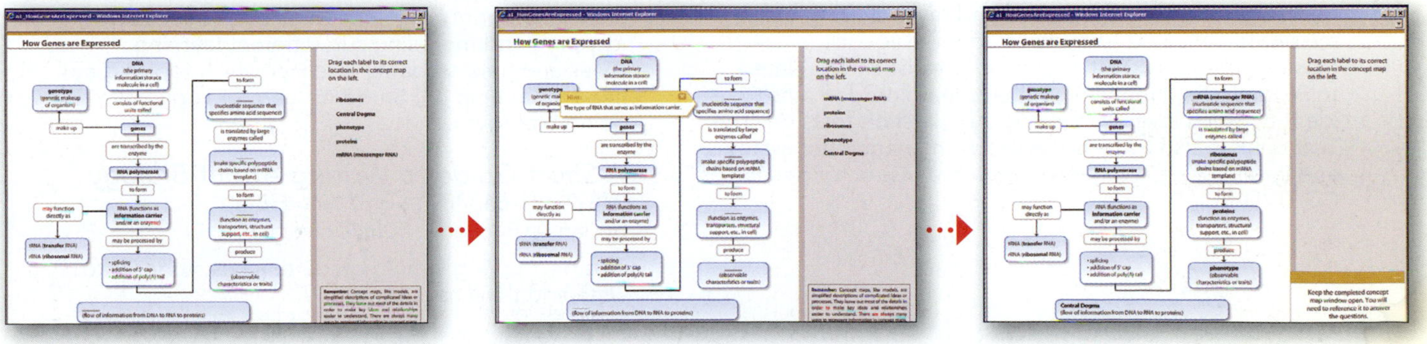
How Genes are Expressed
How Genes are Expressed
How Genes are Expressed

Teaching and Learning Resources

For Instructors

Instructor Resource CD/DVD-ROM

978-0-321-61351-6 • 0-321-61351-1

Everything instructors need for lectures is in one place, including video segments that demonstrate how to incorporate active-learning techniques into your own classroom. The Instructor Resource CD/DVD-ROM includes:

- PowerPoint® Lecture Tools containing all of the figures and photos, which have editable labels; five clicker questions per chapter; pre-made lecture outlines containing select images from the text with embedded animations
- JPEG images of all textbook figures and photos including printer-ready transparency acetate masters
- Over 300 animations and videos that accurately depict complex topics and dynamic processes described in the book; five new BioFlix™ 3-D movie-quality animations
- Instructor's Guides for *Biological Science* and *Practicing Biology* are available as well as a list of all primary literature citations
- The full test bank for *Biological Science* and the *Active Learning Workshop DVD* also come included with the IR-DVD

MasteringBiology® with Pearson eText

www.masteringbiology.com

Assign dynamic homework into your course with automatic grading and adaptive tutoring. Choose from a wide variety of stimulating activities, including visually stunning and scientifically accurate tutorials, ranking questions, 3-D animations, and test bank questions. The powerful gradebook compiles all your favorite teaching diagnostics—the hardest concept, class grade distribution, which students are spending the most or the least time on homework—with the click of a button. Instructors are empowered to customize the eText for themselves and students, including highlighting key text, annotating with comments, adding weblinks, and hiding chapters.

TestGen®

978-0-321-60533-7 • 0-321-60533-0

All of the exam questions in the test bank have been rigorously peer reviewed and revised using the metadata collected from real student usage in MasteringBiology. Test questions have been correlated to Bloom's Taxonomy of cognitive learning domains to identify which level of learning the question tests on. The Test Bank is also available in course management systems and in Microsoft® Word format on the Instructor Resources CD/DVD-ROM.

Course Management Options

CourseCompass™

www.pearsonhighered.com/elearning

This course management system contains preloaded content such as testing and assessment question pools.

WebCT

www.pearsonhighered.com/elearning

Blackboard

www.pearsonhighered.com/elearning

For Students

MasteringBiology with Pearson eText

www.masteringbiology.com

Students may use the Study Area in MasteringBiology for targeted and efficient use of valuable study time. Some of the many study tools include BioFlix™ 3-D movie-quality animations that focus on the toughest topics, engaging activities and cumulative chapter quizzes that help students prepare for exams. The interactive eText is available 24/7 and enables students to highlight text, add their own study notes, and review their Instructor's personalized notes at their convenience.

Study Guide

978-0-321-56168-8 • 0-321-56168-6

The Study Guide presents a breakdown of key biological concepts, difficult topics, and quizzes to help students prepare for exams. Unique to this study guide are four introductory, stand-alone chapters that introduce students to foundational ideas and skills necessary for classroom success: Introduction to Experimentation and Research in the Biological Sciences, Presenting Biological Data, Understanding Patterns in Biology and Improving Study Techniques, and Reading and Writing to Understand Biology. New to this edition of the Study Guide are "Looking Forward" and "Looking Back" sections that help students make connections across the chapters instead of viewing them as discrete entities.

Practicing Biology: A Student Workbook

978-0-321-61264-9 • 0-321-61264-7

This workbook focuses on key ideas, principles, and concepts that are fundamental to understanding biology. A variety of hands-on activities such as mapping and modeling suit different learning styles and help students discover which topics they need more help on. Students learn biology by doing biology.

A Short Guide to Writing About Biology

978-0-321-66838-7 • 0-321-66838-3

by Jan A. Pechenik, Tufts University

This best-selling writing guide teaches students to write and think as biologists.

BIOLOGICAL SCIENCE

VOLUME 2 Evolution, Diversity, & Ecology

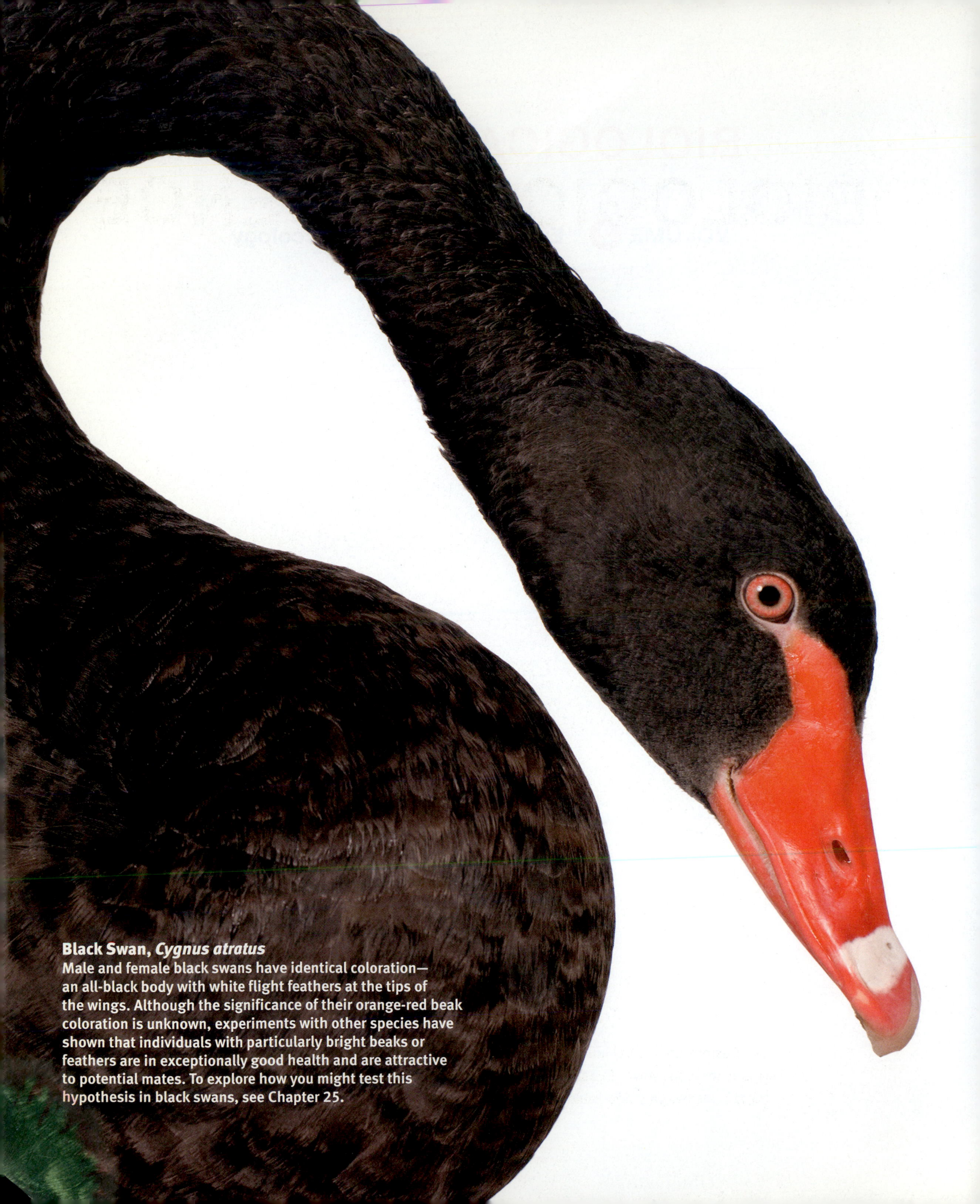

Black Swan, *Cygnus atratus*
Male and female black swans have identical coloration—an all-black body with white flight feathers at the tips of the wings. Although the significance of their orange-red beak coloration is unknown, experiments with other species have shown that individuals with particularly bright beaks or feathers are in exceptionally good health and are attractive to potential mates. To explore how you might test this hypothesis in black swans, see Chapter 25.

BIOLOGICAL SCIENCE

FOURTH EDITION

SCOTT FREEMAN
University of Washington

Benjamin Cummings
Boston Columbus Indianapolis New York San Francisco Upper Saddle River
Amsterdam Cape Town Dubai London Madrid Milan Munich Paris Montréal Toronto
Delhi Mexico City São Paulo Sydney Hong Kong Seoul Singapore Taipei Tokyo

VP, Editor-in-Chief, Biology: Beth Wilbur
Acquisitions Editor: Becky Ruden
Executive Director of Development, Biology: Deborah Gale
Editorial Project Manager: Sonia DiVittorio
Development Editors: Alice Fugate, Moira Lerner-Nelson, William O'Neal, Susan Teahan
Art Editor: Kelly Murphy
Assistant Editor: Brady Golden
Editorial Assistant: Leslie Allen
Senior Media Producer: Laura Tommasi
Director of Editorial Content, Mastering Biology: Tania Mlawer
Developmental Editor, Mastering Biology: Sarah Jensen
Director of Marketing: Christy Lawrence
Executive Marketing Manager: Lauren Harp
Director of Production, Science: Erin Gregg
Managing Editor, Biology: Michael Early
Production Supervisor: Lori Newman
Media Production Supervisor: James Bruce
Supplements Production Supervisor: Jane Brundage
Production Management and Composition: S4Carlisle Publishing Services
Design Manager and Interior Designer: Marilyn Perry
Cover Designer: Riezebos Holzbaur Design Group
Illustrators: Kim Quillin, Imagineering Media Services
Photo Researcher: Maureen Spuhler
Manufacturing Buyer: Michael Penne
Cover Printer: Phoenix Color Corp.
Printer and Binder: Courier, Kendalville

Cover Photo Credits: Black Swan—*Cygnus atratus,* © Eric Isselée/Fotolia (front cover); Black swan among white swans, Hokkaido, Japan, North-East Asia © Keren Su/Getty Images, Inc. (back cover)

Credits and acknowledgments borrowed from other sources and reproduced with permission in this textbook appear on the appropriate page within the text or beginning on page C:1 of the backmatter.

Library of Congress Cataloging-in-Publication Data

Freeman, Scott
Biological science / Scott Freeman.—4th ed.
p. cm.
Includes index.
ISBN 978-0-32-159820-2 (student ed.)—ISBN 978-0-32-159819-6 (professional copy)—ISBN 978-0-32-161347-9 (v. 1 : the cell, genetics, and development—ISBN 978-0-32-160530-6 (v. 2 : evolution, diversity, and ecology)—ISBN 978-0-32-157676-7 (v. 3 : how plants and animals work)
1. Biology—Textbooks. I. Title.

QH308. 2. F73 2011
570—dc22 2009047825

ISBN 10: 0-32-159820-2; ISBN 13: 978-0-32-159820-2 (Student edition)
ISBN 10: 0-32-159819-9; ISBN 13: 978-0-32-159819-6 (Professional copy)
ISBN 10: 0-32-161347-3; ISBN 13: 978-0-32-161347-9 (Volume 1)
ISBN 10: 0-32-160530-6; ISBN 13: 978-0-32-160530-6 (Volume 2)
ISBN 10: 0-32-157676-4; ISBN 13: 978-0-32-157676-7 (Volume 3)

1 2 3 4 5 6 7 8 9 10—CRK—14 13 12 11 10

Benjamin Cummings
is an imprint of

PEARSON

www.pearsonhighered.com

Brief Contents: Volume 2

Detailed Contents

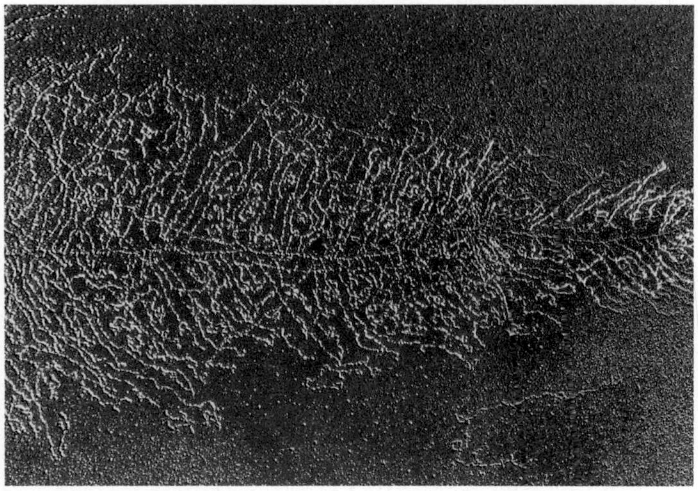

About the Author

SCOTT FREEMAN received his Ph.D. in Zoology from the University of Washington and was subsequently awarded an Alfred P. Sloan Postdoctoral Fellowship in Molecular Evolution at Princeton University. His current research focuses on the scholarship of teaching and learning—specifically **(1)** how active learning and peer teaching techniques increase student learning and improve performance in introductory biology, and **(2)** how the levels of exam questions vary among introductory biology courses, standardized postgraduate entrance exams, and professional school courses. He has also done research in evolutionary biology on topics ranging from nest parasitism to the molecular systematics of the blackbird family. Scott teaches introductory biology for majors at the University of Washington and is coauthor, with Jon Herron, of the standard-setting undergraduate text *Evolutionary Analysis*.

Unit Advisors

Twelve cherished colleagues guided the revision process by synthesizing reviews, providing citations for recent high-impact publications, and drawing on their extensive teaching experience and subject-matter expertise to advise Scott on hundreds of questions, ranging from what to include to which analogies might communicate best to students. It is hard to overstate just how critical these people were to this revision. Through their own teaching and research and their work on this book, they are having a profound effect on how biology is taught.

Jason Flores, University of North Carolina, Charlotte (Unit 1)

Lisa Elfring, University of Arizona, Tucson (Unit 1)

Suzanne Simon-Westendorf, Ohio University (Unit 2)

Gregory Podgorski, Utah State University (Unit 3)

Kathleen Marrs, Indiana University–Purdue University, Indianapolis (Units 4, 6, and 7)

Jon Monroe, James Madison University (Units 4, 6, and 7)

Warren Burggren, University of North Texas (Units 4 and 8)

Joan Sharp, Simon Fraser University (Units 5 and 6)

Michael Black, California Polytechnic State University, San Luis Obispo (Units 6 and 8)

Kathleen Hunt, University of Portland (Units 6 and 9)

Emily Taylor, California Polytechnic State University, San Luis Obispo (Unit 8)

Fred Wasserman, Boston University (Unit 9)

Illustrator

KIM QUILLIN combines expertise in biology and information design to create lucid visual representations of biological principles. She received her B.A. in Biology at Oberlin College and her Ph.D. in Integrative Biology from the University of California, Berkeley (as a National Science Foundation Graduate Fellow), and taught undergraduate biology at both schools. Students and instructors alike have praised Kim's illustration programs for *Biological Science,* as well as *Biology: A Guide to the Natural World* by David Krogh and *Biology: Science for Life* by Colleen Belk and Virginia Borden, for their success in the visual communication of biology. Kim is a lecturer in the Department of Biological Sciences at Salisbury University.

Preface to Instructors

This book is a response to calls—from the National Academy of Sciences, the Howard Hughes Medical Institute and American Association of Medical Colleges, and the National Science Foundation—for changes in the way introductory biology is taught. Reports like *Biology 2010, Scientific Foundations for Future Physicians,* and *Vision and Change* are asking that introductory students not only learn the language of biology and understand fundamental concepts, but begin to apply those concepts in new situations, analyze experimental design, synthesize results, and evaluate hypotheses and data.

I wrote this book for instructors who embrace this challenge—who want to help their students learn how to think like a biologist. The essence of higher education is to promote higher-order thinking. Our job is to help students understand biological science at all six levels of Bloom's taxonomy of learning.

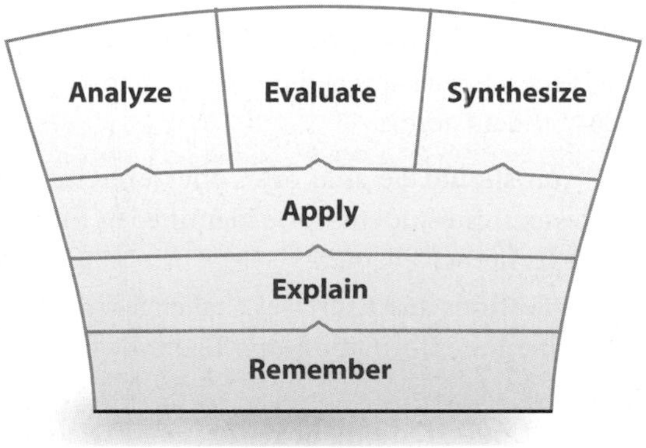

Bloom's Taxonomy. An annotated version of this graphic can be found in "To the Student: How to Use This Book" at the front of this book.

The Evolution of a Textbook

Evolution can be extremely fast in populations with short generation times and high mutation rates. Biology textbooks are no exception. Generation times have to be short because the pace of research in biology and student learning is so fast. This book, in particular, evolves quickly because it incorporates so many new ideas with each edition. Some of these "alleles" are novel mutations, but most arrive via lateral transfer—from advisors, reviewers, friends, students, and the literature.

In comparing the first edition of *Biological Science* with the fourth, the thing that jumps out is what educational researchers call "scaffolding"—tools that help professors and students teach and learn at the level demanded by the NAS, HHMI, AMC, and NSF. It's essential to have high expectations of our students, but we also need to provide the help and practice they need to meet those expectations.

What's New in This Edition

This revision was about making the book a better teaching and learning tool. To help students manage the mass of information and ideas that is contemporary biology, I broke long paragraphs into shorter paragraphs, made liberal use of numbered lists and bulleted lists to "chunk" information and ideas, and broke out dozens of new sections and subsections. I could almost hear my mother's voice: "Take small bites, and chew them well." I also made the book over 100 pages shorter by following Thoreau's dictum to "Simplify, simplify, simplify"—focusing on the most critical concepts that an introductory student needs to master.

In addition, the book team and I came up with a long list of new or expanded features.

- **The Big Picture** These new, two-page spreads are meant to help students see the forest for the trees. They are concept maps that focus on particularly critical areas—Energy, Genetic Information, Evolution, and Ecology. Each synthesizes content and concepts from an array of chapters and includes exercises for students to complete. You'll recognize these pages readily—their edges are colored black (for example, see The Big Picture: Energy on pages 192–193). In addition, the book's MasteringBiology® website has ten new concept map activities based on Big Picture content that will allow you to explore the concepts and their connections with your students during lecture.
- **BioSkills** *Biology 2010, Scientific Foundations,* and *Vision and Change* all place a premium on skills—the ability to read a graph, interpret an equation, understand the bands on a gel. The third edition of *Biological Science* introduced a series of appendixes focused on key skills for introductory biology students. Instructors and students found them extraordinarily helpful. New in this edition are BioSkills on using the metric system, common Latin and Greek roots, techniques for isolating and visualizing cell components, cell and tissue culture methods, and model organisms. BioSkills are located in Appendix A.
- **Answer Key** New to the Fourth Edition are suggested answers to all questions and exercises in the textbook. Students asked us to make this important change between editions to make the book a more complete study tool. The answer key will allow them to self-check their understanding while reading, and when reviewing for exams. Answers are in Appendix B.
- **Experiment Boxes** This text's hallmark has always been its emphasis on experimental evidence—on teaching how we know what we know. In the second edition, key experiments were converted to a boxed format so students could easily

navigate through the logic of the question, hypothesis, and test. In this edition, I added a new question to every experiment box to encourage students to analyze some aspect of the experiment's design.

- **Art Program** Recent research shows that students are more likely to interpret phylogenetic trees correctly if the trees are designed with U-shaped branches instead of Y-shaped branches. We responded by redesigning every phylogenetic tree in the text. To make other subject areas more accessible to visual learners, we enlarged figures, replaced hundreds of photos with clearer images, and strove to streamline labels and graphics across the board. (More on improvements to the art program below.)
- **MasteringBiology Quizzes** MasteringBiology gives students round-the-clock access to quizzes. We developed 550 new assignable questions based on the book's "Blue Thread" questions (more on the "Blue Thread" and its evolution below). We also developed 1100 new quiz questions, along with a cumulative practice test to simulate what a real exam might be like. To help students keep up with their reading, we created 55 new reading quizzes—one for each chapter—that you can assign through Mastering Biology.
- **MasteringBiology Experimental Inquiry Tutorials** The call to teach students about the process of science has never been louder. In response, a team lead by Tom Owens of Cornell University developed 10 new interactive tutorials on classic scientific experiments—ranging from Meselson–Stahl on DNA replication to the Grants' work on Galápagos finches and Connell's work on competition. Students who use these interactive tutorials should be better prepared to think critically about experimental design and evaluate the wider implications of the data—preparing them to do the work of real scientists in the future.
- **MasteringBiology BioFlix Animations and Tutorials** BioFlix™ are movie-quality, 3-D animations available on MasteringBiology. They focus on the most difficult core topics and are accompanied by in-depth, online tutorials that provide hints and feedback to help guide student learning. Thirteen BioFlix were available with the third edition of *Biological Science.* Five new BioFlix 3-D animations and tutorials have been developed for this edition—on mechanisms of evolution, homeostasis, gas exchange, population ecology, and the carbon cycle.

Changes to Gold Thread Scaffolding

The third edition introduced a dramatically expanded set of tools designed to help with a chronic problem for novice learners: picking out important information. Novices highlight every line in the text and try to memorize everything mentioned in lecture; experts instinctively home in on the key unifying ideas.

For students to make the novice-to-expert transition, we have to help them with features like:

1. **Key concepts** that are declared at the start of each chapter, highlighted with a key icon within the chapter, and reviewed at the end of the chapter.
2. **In-text highlighting**, in gold, that directs their attention to particularly important ideas.
3. **Check Your Understanding boxes**, at the end of key sections, with a bulleted list of key points.
4. **Summary tables** that pull information together in a compact format that is easy to review and synthesize.

The book team and I scrubbed every aspect of the gold thread—reviewing, revising, and re-revising. The third edition was a coming-out party for the gold thread; in the Fourth Edition we wanted it to dance.

Changes to Blue Thread Scaffolding

Each edition of this text has added tools to help students with metacognition—understanding what they do and don't understand. Novices like to receive information passively, and easily persuade themselves that they know what's going on. Experts are skeptical—they want to solve some problems before they're convinced that they know and understand an idea.

In the third edition, we formalized the metacognitive tools in *Biological Science* as a "Blue Thread" set of questions; in this edition, we revised each question and put answers in the back of the book for easy student access.

1. **In-text "You should be able to's"** offer exercises on topics that professors and students have identified as the most difficult concepts in each chapter.
2. **Caption Questions and Exercises** challenge students to critically examine the information in a figure or table—not just absorb it.
3. **Check Your Understanding boxes** present two to three tasks that students should be able to complete in order to demonstrate a mastery of summarized key ideas.
4. **Chapter Summaries** include "You should be able to" problems or exercises related to each of the key concepts declared in the gold thread.
5. **End-of-Chapter Questions** are organized around Bloom's taxonomy of learning, so students can test their understanding at the knowledge, comprehension, and application levels.

The fundamental idea is that if students really understand a piece of information or a concept, they should be able to do something with it. How do you get to Carnegie Hall? Practice.

As students mature as biologists-in-training and start taking upper-division courses, most or all of this scaffolding can disappear. By the time our students are juniors and seniors, they should have enough expertise to construct a high-level understanding on their own. But if a well-designed scaffold isn't there to get them started in their first and second years, when they are novices, most will flounder. We have to help them learn how to become good students.

Supporting Visual Learners

Figures can help students, especially visual learners, at all levels of Bloom's taxonomy—not only to understand and remember the material, but also to exercise higher levels of critical thinking. The overall goal of the Fourth Edition art revision was to hone the figures for accessibility to help novice learners recognize and engage with important visual information. In addition to re-designing the previously mentioned phylogenetic trees, Kim Quillin led the effort to enhance virtually every other aspect of the visual-teaching program.

- **Art and Photos** Kim enlarged art and photographs in figures throughout the book to increase clarity by making details physically easier to see. She also reduced the amount of detail in labels and graphics to simplify, simplify, simplify.
- **Color Use** Kim continues to use color strategically to draw attention to important parts of the figures. In this revision, she boosted color contrast in many figures to make the art more vibrant and the details easier to see.
- **Molecular Icons** Kim redesigned many molecular icons to simplify their shapes. The overall contours are based on molecular coordinates, when available, to accurately represent size and geometry, but she smoothed the textures for a simpler appearance—one that is more memorable and pleasing.
- **Molecular Models** New molecular models have been introduced to help students visualize structure-function relationships. In Chapter 5, for example, redesigned 2-D line drawings of sugars are now paired with 3-D ball-and-stick models.
- **"Pointers"** The Fourth Edition figures still use pointer annotations as a "whisper in the ear" to guide students in interpreting figures, but Kim has replaced the hand with an arrow to be more precise.

Serving a Community of Teachers

I love students, but I love teachers even more. There is nothing I like better than to sit in a workshop with a bunch of biology instructors, steal some good ideas, and come home to try them out—in some cases in a framework where I can collect data and test the hypothesis that the new approaches are improving student learning.

Research on biology education is gathering momentum, trying to catch up on the trail blazed by physics education researchers, bringing the same level of rigor to our classrooms that we bring to our lab benches and field sites. I try to bring the spirit and practice of evidence-based teaching into this textbook, and welcome your comments, suggestions, and questions.

Thank you for considering this text, and for your work on behalf of your students. We have the best jobs in the world.

SCOTT FREEMAN
University of Washington

Content Highlights of the Fourth Edition

As discussed in the preface, a major focus of this revision is to enhance the pedagogical utility of *Biological Science*. Another major goal is to ensure that the content reflects the current state of science and is accurate. In addition, every chapter has been rigorously evaluated for discussions that, in the previous edition, may have been too complex or overly detailed. As a result of this scrutiny, certain sections in every chapter have been simplified, content has been pruned judiciously, and the approach to certain topics has been re-envisioned to enhance student comprehension. In this section, some of the key content improvements to the textbook are highlighted.

Unit 1 The Molecules of Life

Chapter 1 A new experiment on ant navigation and discussions of tree-based naming systems and artificial selection in maize has been added. Coverage is expanded on the definition of life and the nature of science and religion.

Chapter 2 The descriptions of bond angles and the geometry of simple molecules are simplified. Added is a discussion on the hot-start hypothesis as well as a new Key Concept on the nature of chemical energy.

Chapter 3 This chapter has been streamlined by eliminating discussion of optical isomers/chirality and reducing coverage of enzyme kinetics and reaction rates.

Chapter 4 The discussion of RNA is expanded to include recently discovered roles for RNAs in cells. Also added is a new summary table (**Table 4.1**) comparing DNA and RNA structure.

Chapter 5 A stronger emphasis on the link between electronegativity of atoms and potential energy in C—C, C—H, and C—O bonds is developed. New ball-and-stick models are added to clarify the differences in location and orientation of functional groups.

Chapter 6 Coverage of secondary active transport has been expanded. Also included in this chapter is current research on the "first cell" and a discussion of non-random distribution of membrane proteins and phospholipids.

Unit 2 Cell Structure and Function

Chapter 7 New research on bacterial cell structure has been included, and a more explicit connection between lysosomes and the endomembrane system is emphasized. Centrifugation is moved to **BioSkills 11** in Appendix A.

Chapter 8 New sections on quorum sensing in bacteria and cross-talk among signal-transduction pathways have been added.

Chapter 9 The discussions of mitochondrial structure, ATP yield from glucose oxidation, and the role of GDP in the citric acid cycle have been updated. The introductory section on cellular respiration has been simplified.

Chapter 10 A new section on regulation (inhibition) has been added. The sections on C_4 and CAM photosynthesis now emphasize the role of these pathways in increasing CO_2 concentrations versus water conservation.

Chapter 11 Micrographs have been added to the phases of mitosis figure (**Figure 11.5**). The discussion on the role of activated MPF has been updated to include the triggering M phase of the cell cycle. Animal-cell culture methods are moved to **BioSkills 12** in Appendix A.

Unit 3 Gene Structure and Expression

Chapter 12 The discussions of recombination rates and aneuploidy rates in humans are updated. New micrographs have been added to the phases of meiosis figure (**Figure 12.7**).

Chapter 13 The linkage discussion and notation in fly crosses have been simplified. Sex-linkage is moved to the Mendelian section (Section 13.4 The Chromosome Theory of Inheritance), and mapping is now covered in Box 13.1 Quantitative Methods: Linkage. A new summary table (**Table 13.3**) presenting basic vocabulary used in Mendelian genetics has been added.

Chapter 14 A new space-filling model of DNA has been added to **Figure 14.4**.

Chapter 15 Discussions on mutation in the melanocortin receptor (link to mouse-coat-color camouflage) and karyotypes of cancerous cells have been added.

Chapter 16 The sections on transcription in bacteria and eukaryotes are now combined. The structure of the translation initiation complex in bacteria has been updated to reflect current science; snRNAs have been added to the discussion of RNA splicing.

Chapter 17 The chapter was streamlined with the removal of discussions of DNA fingerprinting and the structure of the operator and DNA-binding proteins. Treatment of catabolite repression/positive control has been trimmed.

Chapter 18 Included in this chapter is a new summary table (**Table 18.1**) comparing control of gene expression in bacteria and eukaryotes. Also added are discussions on ubiquitination and protein degradation, the importance of epigenetic inheritance (chromosome structure), and the histone code hypothesis.

Chapter 19 Southern/Northern/Western blots have moved to **BioSkills 9** in Appendix A. The discussions on golden rice, the impact of GM crops, and SNP association studies for human diseases have been updated with the most recent research. Notes on "next generation" sequencing technologies have been included.

Chapter 20 Human health applications now emphasize the use of genomics and microarrays to study cancer. Several datasets are updated, including sequencing database totals. New notes on miRNA genes, metagenomics, and the definition of the gene have been added.

Unit 4 Developmental Biology

Chapter 21 The discussions of *bicoid* and regulatory gene cascades are simplified. New material on auxin as a master regulator in early development and the importance of apoptosis has been added.

Chapter 22 The discussion about sea urchin fertilization and variation has been streamlined.

Chapter 23 A new section introducing basic concepts in angiosperm gametogenesis is added.

Unit 5 Evolutionary Processes and Patterns

Chapter 24 A section on the internal consistency of diverse data as evidence for evolution, including a new phylogeny and timeline of whale evolution, has been added. **Figure 24.6**, depicting the evolution of the Galápagos mockingbird, and long-term data on ground finches (**Figure 24.17**) are updated to reflect the most current science. There is a new graph on the evolution of drug resistance in pathogenic bacteria (**Figure 24.14**).

Chapter 25 The genetic drift example has changed from breeding in a small population on Pitcairn Island to coin flips simulating mating in a single couple (using data from the author's classroom). The prairie lupine gene flow example is replaced by recent work on an island population of *Parus major*. Notes on balancing selection and interactions among evolutionary forces have been included.

Chapter 26 The speciation-by-vicariance example has been changed from ratites to snapping shrimp, and the sympatric speciation example featuring soapberry bugs has been changed to apple/hawthorn flies.

Chapter 27 The sections on adaptive radiation and mass extinction have been completely reorganized. A new hypothesis for the cause of the Cambrian explosion is included, and detail on the "new genes, new bodies" hypothesis has been removed. Presentation of "Life's Timeline" has been significantly overhauled (see **Figures 27.8, 27.9,** and **27.10**).

Unit 6 The Diversification of Life

The model organisms have been moved to **BioSkills 14** in Appendix A. Phylogenetic trees have been redrawn to reflect a horizontal orientation with U-shaped branches for easier comprehension.

Chapter 28 New information on mechanisms of pathogenicity is added. Extensive updates include new notes on archaeon-eukaryote polymerases, the discovery of extensive biomass in the marine subfloor, an archaeon associated with a human disease, discovery of N-fixation and nitrification in archaea, and bacteriorhodopsin's role in phototrophy.

Chapter 29 A stronger emphasis on endosymbiosis as a theme in protist diversification has been threaded throughout this chapter.

Chapter 30 New content on green algae as a grade and on convergence in vascular tissue in mosses-vascular plants and gnetophytes-angiosperms has been added.

Chapter 31 The dynamic nature of mycelia, the importance of glomalin in soil, the role of mating types, and the discovery of "multigenomic" asexual glomales all have new supporting material.

Chapter 32 The treatment of embryonic tissues, developmental patterns, the coelom, and body symmetry have been updated to reflect the latest scientific thinking. A shift in emphasis to the origin of the neuron and cephalization has been implemented.

Chapter 33 New commentary on the independent transitions to land as well as a clarified discussion on the nature of the ecdysozoan-lophotrochozoan split are included. The discussion of annelids is updated to reflect recent results.

Chapter 34 The coverage of the echinoderm endoskeleton has been expanded and a phylogeny of early tetrapods has been added to the fin-to-limb transition figure (**Figure 34.16**). New data have been incorporated in the evolution-of-fishes timeline (**Figure 34.11**). The treatment of hagfish-lamprey, evolution-of-the-jaw (**Figure 34.14**) and *H. sapiens* migration (**Figure 34.40**) also include the most recent data available. The emphasis on the adaptive significance of the amniotic egg has changed from watertightness to increased size and support. Emphasis in the discussion of viviparity has changed to the adaptive advantage of embryo portability and temperature control. The recent analysis of *Ardipithecus ramidus* as the first hominin, with data on estimated body mass and braincase volume, has been included.

Chapter 35 The material on HIV phylogeny has been moved to the section on emerging viruses.

Unit 7 How Plants Work

Chapter 36 Surface area-to-volume ratios have been added as a theme in root and shoot systems. New information on contractile roots in *Ficus* and bulbs is incorporated into this chapter.

Chapter 37 New content on aquaporins and the transmembrane route to root xylem has been added, and coverage of why air has such low water pressure potential has been expanded.

Chapter 38 The description of nitrogen fixation has been clarified.

Chapter 39 **Figure 39.8** on the acid-growth hypothesis has been redesigned, and the discussion of polar auxin transport is simplified. New commentary on the role of brassinosteroids in growth regulation and on "talking trees" is included. The coverage of the receptors for GA, auxin, ABA, and brassinosteroids, and MeSA's role in the SAR has been updated with the most current research. Plant-tissue culture methods have been moved to **BioSkills 12** in Appendix A.

Chapter 40 Comments on day-length sensing and on pollination syndromes are new to this chapter.

Unit 8 How Animals Work

Chapter 41 New details on tissue types (especially connective tissue) have been incorporated. The discussion of thermoregulation has been completely reorganized for a more logical flow.

Chapter 42 The sections on the shark rectal gland and the mammalian loop of Henle have been revised to improve focus.

Chapter 43 A description of incomplete digestive systems is now included, and coverage of comparative aspects of digestive tract structure and function has been expanded.

Chapter 44 Information on the types of circulatory systems and types of blood vessels has been consolidated. Details on surface tension and lung elasticity have been removed while new content on countercurrent exchange in fish gills has been added.

Chapter 45 The chapter and section introductions have been rewritten to introduce a comparative context and to make the neuron-to-systems chapter organization more transparent. New content on interspecific variation in nervous systems has been added.

Chapter 46 The chapter has been shortened and its focus sharpened by the removal of nonessential information.

Chapter 47 New material on EPO abuse in athletes has been included.

Chapter 48 The section on sperm competition includes new data from experiments on seed beetles.

Chapter 49 The discussion of the V regions of BCRs and antibodies and recombination in BCR/TCR genes has been simplified. New content on autoimmune disorders and diseases associated with immunosuppression, allergies, and immunodeficiency diseases has been added. The discussion of vaccination has been expanded.

Unit 9 Ecology

Chapter 50 New information on the importance of nutrient availability in aquatic ecosystems, with details on lake turnover and ocean upwelling, is included. A new section on the Wallace line has also been added.

Chapter 51 The content in this chapter has been completely reorganized to increase cohesiveness. It is presented as a series of questions in behavioral ecology, with each question addressed at the proximate and ultimate levels with separate case studies. Material on modes of learning, innate behavior, bat-moth interactions, sex change in wrasses, and acoustic and visual signaling in red-winged blackbirds has been trimmed or dropped. New content on child abuse in humans is added.

Chapter 52 Discussion of the hare–lynx-cycle field experiment has been reorganized for clarity, with new supporting "Results" data added to accompanying **Figure 52.12**.

Chapter 53 New content has been added on species richness and resistance of communities to invasion, the use of predators or parasites as biocontrol agents, and character displacement in finches. The discussion of succession in Glacier Bay is reorganized and simplified. The discussion of alternative hypotheses to explain the latitudinal gradient in species richness has been expanded and clarified.

Chapter 54 The chapter was rewritten and reorganized to sharpen its focus on human impacts. Sections on trophic cascades and biomagnification have been added, as have recent data on human appropriation of NPP, sources of nutrient gain and loss, and the impact of ocean acidification on coral growth.

Chapter 55 New content on the impact of global climate change and a new section on ways to preserve biodiversity are now included. Two new boxes on quantitative methods have been added: one on estimating species numbers and species losses and the other on population viability analysis.

Acknowledgments

Reviewers

The peer review system is the key to quality and clarity in science publishing. In addition to providing a filter, the investment that respected individuals make in vetting the material—catching errors or inconsistencies and making suggestions to improve the presentation—gives authors, editors, and readers confidence that what they are publishing and reading meets rigorous professional standards.

Peer review plays the same role in textbook publishing. The time and care that this book's reviewers have invested is a tribute to their professional integrity, their scholarship, and their concern for the quality of teaching. Virtually every paragraph in this edition has been revised and improved based on insights from the following individuals.

Ann Aguanno, *Marymount Manhattan College*
Adrienne Alaie, *City University of New York, Hunter College*
John Alcock, *Arizona State University*
Sylvester Allred, *Northern Arizona University*
Suzanne Alonzo, *Yale University*
Dan Ardia, *Franklin and Marshall College*
Peter Armbruster, *Georgetown University*
David Asch, *University of Pennsylvania*
Andrea Aspbury, *Texas State University, San Marcos*
Nicanor Austriaco, *Providence College*
Mitchell Balish, *Miami University*
Elizabeth Balko, *State University of New York, Oswego*
Ralston Bartholomew, *Warren County Community College*
Christine Barton, *Centre College*
Robert Bauman, *Amarillo College*
Wayne Becker, *University of Wisconsin, Madison*
Peter Berget, *Carnegie Mellon University*
Ethan Bier, *University of California, San Diego*
Michael Black, *California Polytechnic State University, San Luis Obispo*
Anthony Bledsoe, *University of Pittsburgh*
James Bottesch, *Brevard Community College*
Scott Bowling, *Auburn University*
Robert Boyd, *Auburn University*
Ronald Breaker, *Yale University*
Diane Bridge, *Elizabethtown College*
Andrew Brower, *Middle Tennessee State University*
Rebecca Brown, *College of Marin*
Mark Browning, *Purdue University*
Arthur Buikema, *Virginia Polytechnic Institute and State University*
Warren Burggren, *University of North Texas*
Jennifer Carbrey, *Duke University*
Dale Casamatta, *University of North Florida*
Gregory Chandler, *University of North Carolina, Wilmington*
Deborah Chapman, *University of Pittsburgh*
Curt Coffman, *Vincennes University*
Patricia Colberg, *University of Wyoming*
Kathleen Cornely, *Providence College*
Elizabeth Cowles, *Eastern Connecticut State University*
Jason Curtis, *Purdue University, North Central*
Karen Curto, *University of Pittsburgh*
Farahad Dastoor, *University of Maine*
Robin Lee Davies, *Sweet Briar College*
Charles Delwiche, *University of Maryland, College Park*
Jean DeSaix, *University of North Carolina, Chapel Hill*
Hudson DeYoe, *University of Texas, Pan American*
Sunethra Dharmasari, *Texas State University, San Marcos*
Lisa Elfring, *University of Arizona, Tucson*
Eric Engstrom, *College of William and Mary*
Jean Everett, *College of Charleston*
Brent Ewers, *University of Wyoming*
Andrew Fabich, *Tennessee Temple University*
Gary Firestone, *University of California, Berkeley*
Ryan Fisher, *Salem State College*
David Fitch, *New York University*
Jason Flores, *University of North Carolina, Charlotte*
Andrew Forbes, *University of Notre Dame*
Edward Freeman, *St. John Fisher College*
Caitlin Gabor, *Texas State University, San Marcos*
George Gilchrist, *College of William and Mary*
Lynda Goff, *University of California, Santa Cruz*
Elliot Goldstein, *Arizona State University*
Andrew Goliszek, *North Carolina Agricultural & Technical State University*
Margaret Goodman, *Wittenberg University*
Joyce Gordon, *University of British Columbia*
Steve Gorsich, *Central Michigan University*
Susan Michele Green, *Texas State University, San Marcos*
Paul Greenwood, *Colby College*
Stanley Guffey, *University of Tennessee, Knoxville*
Wendy Hanna-Rose, *Pennsylvania State University*
Kiki Harbitz, *Gustavus Adolphus College*
Jeffrey Hardin, *University of Wisconsin, Madison*
Jana Henson, *Georgetown College*
Helen Hess, *College of the Atlantic*
Tracey Hickox, *University of Illinois, Urbana-Champaign*
Sara Hoot, *University of Wisconsin, Milwaukee*
Laurie Host, *Harford Community College*
Kelly Howe, *University of New Mexico, Valencia*
Kathleen Hunt, *University of Portland*
Christine Janis, *Brown University*
Eric Jellen, *Brigham Young University*
Warren Johnson, *University of Wisconsin, Green Bay*
Cindy Johnson-Groh, *Gustavus Adolphus College*
Greg M. Kelly, *University of Western Ontario*
Paul King, *Massasoit Community College*
Joel Kingsolver, *University of North Carolina, Chapel Hill*
Roger Koeppe, *Oklahoma State University*
David Kooyman, *Brigham Young University*

Jocelyn Krebs, *University of Alaska, Anchorage*
John Krenetsky, *Metro State College of Denver*
Patrick Krug, *California State University, Los Angeles*
Holly Kupfer, *Central Piedmont Community College*
Mary Rose Lamb, *University of Puget Sound*
John Lammert, *Gustavus Adolphus College*
Hans Landel, *Edmonds Community College*
Dominic Lannutti, *El Paso Community College*
Janet Lanza, *University of Arkansas, Little Rock*
Georgia Lind, *Kingsborough Community College*
Debra Linton, *Central Michigan University*
Curtis Loer, *University of California, San Diego*
Frank Logiudice, *University of Central Florida*
Barbara Lom, *Davidson College*
James Maller, *University of Colorado, Denver*
James Manser, *Harvey Mudd College (retired)*
Kathleen Marrs, *Indiana University–Purdue University, Indianapolis*
Jennifer Martin, *University of Colorado, Boulder*
Kathy Martin-Troy, *Central Connecticut State University*
John H. McDonald, *University of Delaware*
Robert McLean, *Texas State University, San Marcos*
Michael Meighan, *University of California, Berkeley*
Mariana Melo, *North Essex Community College*
Philip Meneely, *Haverford College*
Dennis Minchella, *Purdue University*
Alan Molumby, *University of Illinois, Chicago*
Vertigo Moody, *Santa Fe Community College*
Elizabeth Morgan, *Lonestar College Kingswood*
Nancy Morvillo, *Florida Southern College*
Mike Muller, *University of Illinois, Chicago*
Dennis Nyberg, *University of Illinois, Chicago*
Amanda N. Orenstein, *Centenary College of New Jersey*
Stephanie Scher Pandolfi, *Michigan State University*
Lisa Parks, *North Carolina State University*
Robert Paul, *Kennesaw State University*
Andrew Pease, *Stevenson University*
Andrew Pekosz, *Johns Hopkins University*
Nancy Pelaez, *Purdue University*
Roger Persell, *City University of New York, Hunter College*
John Peters, *College of Charleston*
Debra Pires, *University of California, Los Angeles*
Gregory Podgorski, *Utah State University*
Robert Podolsky, *College of Charleston*
Therese Poole, *Georgia State University*
Harvey Pough, *Rochester Institute of Technology*
Vanessa Quinn, *Purdue University North Central*
Stephanie Randell, *McLennan Community College*
Clifford Ross, *University of North Florida*
Michael Rutledge, *Middle Tennessee State University*
James Ryan, *Hobart and William Smith Colleges*
Margaret Saha, *College of William and Mary*
Mark Sandheinrich, *University of Wisconsin, La Crosse*
Terry Saropoulos, *Vanier College*
Thomas Sasek, *University of Louisiana, Monroe*
Jon Scales, *Midwestern State University*
Gregory Schmaltz, *Great Basin College*
Oswald Schmitz, *Yale University*
Joan Sharp, *Simon Fraser University*
Michele Shuster, *New Mexico State University*
Suzanne Simon-Westendorf, *Ohio State University*
David Skelly, *Yale University*
Meredith Somerville-Norris, *University of North Carolina, Charlotte*
Sally Sommers-Smith, *Boston University*
Eric Stavney, *Pierce College*
Scott Steinmaus, *California Polytechnic State University, San Luis Obispo*
Janet Steven, *Sweet Briar College*
Kirk Stowe, *University of South Carolina*
Christine Strand, *California Polytechnic State University, San Luis Obispo*
Cynthia Surmacz, *Bloomsburg University*
Jackie Swanik, *North Carolina Central University*
Jerilyn Swann, *Maryville College*
Brad Swanson, *Central Michigan University*
Emily Taylor, *California Polytechnic State University, San Luis Obispo*
Eric Thobaben, *Carroll University*
Ken Thomas, *North Essex Community College*
Briana Timmerman, *University of South Carolina*
Alexandru M. F. Tomescu, *Humboldt State University*
James Traniello, *Boston University*
William Velhagen, *New York University*
Sara Via, *University of Maryland, College Park*
Susan Waaland, *University of Washington*
Frederick Wasserman, *Boston University*
Elizabeth Weiss-Kuziel, *University of Texas, Austin*
Jason Wiles, *Syracuse University*
Kelly P. Williams, *Virginia Polytechnic Institute and State University*
Elizabeth Willott, *University of Arizona, Tucson*
Clifford Wilson, *Kennedy King College*
Charles Wimpee, *University of Wisconsin, Milwaukee*
Leslie Wooten-Blanks, *College of Charleston*
Todd Yetter, *University of the Cumberlands*

Correspondents

One of the most enjoyable interactions I have as a textbook author is correspondence or conversations with researchers and teachers who take the time and trouble to contact me to discuss an issue with the book, or who respond to my queries about a particular data set or study. I'm always amazed and heartened by the generosity of these individuals. They care, deeply.

Julie Aires, *Florida Community College, Jacksonville*
Göran Arnqvist, *Uppsala University*
Terry Bidleman, *Environment Canada*
Brian Buchwitz, *University of Washington*
Helaine Burstein, *Ohio University*
Curt Coffman, *Vincennes University*
Mark Cooper, *University of Washington*
Scott Creel, *Montana State University*
Laura DiCaprio, *Ohio University*
Mary Durant, *Lone Star College*
Victoria Finnerty, *Emory University*
Kathleen Foltz, *University of California, Santa Barbara*
Kathy Gillen, *Kenyon College*
Peter Grant, *Princeton University*
Rosemary Grant, *Princeton University*
Takato Imaizumi, *University of Washington*
Kathleen Janech, *College of Charleston*
Paul King, *Massosoit Community College*

John Lammert, *Gustavus Adolphus College*
Curtis Loer, *University of San Diego*
Scott Meissner, *Cornell University*
Philip Meneely, *Haverford College*
Diarmaid Ó Foighil, *University of Michigan*
John Roth, *University of California, Davis*
Brian Spohn, *Florida Community College, Jacksonville*
Fayla Schwartz, *Everett Community College*
Elizabeth Willott, *University of Arizona, Tucson*
Dan Wulff, *University at Albany, State University of New York*
Glenn Yasuda, *Seattle University*

Supplements Contributors

Instructors depend on an impressive array of support materials—in print and online—to design and deliver their courses. The student experience would be much weaker without the study guide, test bank, activities, animations, quizzes, and tutorials written by the following individuals.

Brian Bagatto, *University of Akron*
Michael Black, *California Polytechnic State University, San Luis Obispo*
Jay L. Brewster, *Pepperdine University*
Warren Burggren, *University of North Texas*
Patricia Colberg, *University of Wyoming*
Clarissa Dirks, *Evergreen State College*
Lisa Elfring, *University of Arizona, Tucson*
Brent Ewers, *University of Wyoming*
Miriam Ferzli, *North Carolina State University*
Cheryl Frederick, *University of Washington*
Cindee Giffen, *University of Wisconsin, Madison*
Kathy M. Gillen, *Kenyon College*
Mary Catherine Hager
Christopher Harendza, *Montgomery County Community College*
Laurel Hester, *University of South Carolina*
Jean Heitz, *University of Wisconsin, Madison*
Tracey Hickox, *University of Illinois, Urbana-Champaign*
Kathleen Hunt, *University of Portland*
Barbara Lom, *Davidson College*
Jennifer Nauen, *University of Delaware*
Chris Pagliarulo, *University of Arizona, Tucson*
Stephanie Scher Pandolfi, *Michigan State University*
Debra Pires, *University of California, Los Angeles*
Gregory Podgorski, *Utah State University*
Carol Pollock, *University of British Columbia*
Jessica Poulin, *University at Buffalo, the State University of New York*
Vanessa Quinn, *Purdue University North Central*
Eric Ribbens, *Western Illinois University*
Joan Sharp, *Simon Fraser University*
Suzanne Simon-Westendorf, *Ohio University*
Emily Taylor, *California Polytechnic State University, San Luis Obispo*
Fred Wasserman, *Boston University*
Cindy White, *University of Northern Colorado*

Book Team

People who watch film credits won't be surprised to learn how many talented people are involved in making a textbook happen. Ruth Steyn provided incisive comments on the earliest drafts of the revised manuscript, which were then implemented by Development Editors Alice Fugate, Moira Lerner-Nelson, Bill O'Neal, and Susan Teahan; the final version of the text was copyedited by Michael Rossa and proofread by Pete Shanks. This editorial team was directed by Executive Director of Development Deborah Gale.

As in the first three editions, the book's figures were designed by Dr. Kim Quillin. Kim's artistic sensibilities, scientific training, and teaching talent are what make the figures in this book sing. If imitation really is the sincerest form of flattery, then Kim should be blushing—we've started seeing her style copied in textbook figures and scientific illustrations by other illustrator/designers. Kim's designs were rendered by Imagineering Media Services; the final art manuscript was vetted by Art Editor Kelly Murphy and rendered art expertly proofread by Frank Purcell. Maureen Spuhler did a thorough review of the photography program and researched the hundreds of images new to the Fourth Edition.

The book's clean, elegant design is the brainchild of Marilyn Perry; the text and art were set in Marilyn's design by S4Carlisle Publishing Services. The book's production was supervised by Lori Newman and Mike Early.

The extensive supplements program was managed by Assistant Editor Brady Golden. All of the individuals I've mentioned—and more—were supported by Editorial Assistant Leslie Allen.

Creating MasteringBiology® tutorials and activities requires a talented team of people playing many different roles. Media content development was overseen by Tania Mlawer and Sarah Jensen who benefited from the program expertise of Caroline Power and Karen Sheh. Julia Henderson and Laura Tommasi worked together as Media Producers with the leadership of Senior Media Producer Deb Greco. Lauren Fogel (VP, Director, Media Development), Stacy Treco (VP, Director, Media Product Strategy), and Laura ensured that the complete media program that accompanies the Fourth Edition, including MasteringBiology, will meet the needs of the students and professors who use our offerings.

Pearson's talented Sales Reps, who listen to professors, advise the editorial staff, and get the book in students' hands, are supported by tireless Executive Marketing Manager Lauren Harp and Director of Marketing Christy Lawrence. The marketing materials that support the outreach effort were produced by Lillian Carr and her colleagues in Pearson's MarCom group.

The vision and resources required to run this entire enterprise are the responsibility of Vice President and Editor-in-Chief Beth Wilbur, Senior Vice President and Editorial Director Frank Ruggirello, President of Pearson Science Paul Corey, and President of Pearson Science & Math Linda Davis.

Finally, the driving forces behind this edition were Project Manager Sonia DiVittorio and Acquisitions Editor Becky Ruden. Sonia's razor-sharp mind, depth of heart, and dedication to excellence shine in every page. Becky's energy and belief in this book surmounted obstacles that would have felled any other editor. I thank the two of them from the bottom of my heart.

Media Guide

Students who purchase a new copy of the text receive free access to MasteringBiology® *(www.masteringbiology.com)*, which contains valuable videos, animations, and practice quizzes to help students learn and prepare for exams.

THE BIG PICTURE New to the Fourth Edition, The Big Pictures are interactive concept maps based on four overarching topics in biology that help students synthesize information across broad concepts and not get lost in the details.

Energy (Chapters 9 and 10)

- How Photosynthesis Yields Sugar
- How Cellular Respiration Yields ATP
- How Photosynthesis Relates to Cellular Respiration

Genetic Information (Chapters 12–18)

- How Genes Are Expressed
- How Genetic Information Is Copied and Transmitted
- How Genetic Information Changes

Evolution (Chapters 24–27)

- How Species Evolve
- How Species Form the Tree of Life

Ecology (Chapters 50–55)

- How Organisms Interact in Their Environment
- How Energy and Nutrients Flow through Ecosystems

BIOFLIX™ BioFlix are 3-D movie-quality animations with carefully constructed student tutorials, labeled slide shows, study sheets, and quizzes which bring biology to life.

WEB ACTIVITIES Web Activities help students learn biological concepts via simple, cartoon-style animations and contain pre-quizzes and post-quizzes to test student's understanding of biology's dynamic processes and concepts.

DISCOVERY VIDEOS Brief videos from the Discovery Channel on 29 different biology topics are available for student viewing along with a corresponding video quiz.

VIDEOS Additional molecular and microscopy videos provide vivid images of processes of the cell.

BIOSKILLS BioSkills (in Appendix A) provide background on key skills and techniques for introductory biology students. New to the Fourth Edition are online questions that give students practice building their skill set.

GRAPHIT! Graphing tutorials show students how to plot, interpret, and critically evaluate real data.

Chapter 1

- An Introduction to Graphing

Chapter 50

- Animal Food Production Efficiency and Food Policy
- Atmospheric CO_2 and Temperature Changes

Chapter 52

- Age Pyramids and Population Growth

Chapter 53

- Species Area Effect and Island Biogeography

Chapter 55

- Forestation Change
- Global Fisheries and Overfishing
- Municipal Solid Waste Trends in the U.S.
- Global Freshwater Resources
- Prospects for Renewable Energy
- Global Soil Degradation

WORD STUDY TOOLS New to the Fourth Edition are Latin and Greek root word flash cards to help students practice the language of biology. In addition, an audio glossary provides correct pronunciation to help students learn key terms introduced in the book.

CUMULATIVE TEST Every chapter offers 20 Practice Test questions that students can pool from different chapters into a Cumulative Test to simulate a practice exam.

RSS FEEDS Real Simple Syndication directly links breaking news from four important sources: National Public Radio, *Scientific American*, *Science Daily News*, and *BioScience*. Current articles reinforce the dynamic nature of science in our daily lives.

eTEXT The eText of *Biological Science*, Fourth Edition, is available online 24/7 for students' convenience. New annotation, highlighting, and bookmarking tools allow students to personalize the material for efficient review.

MEDIA AT A GLANCE

	BIOFLIX	WEB ACTIVITIES	DISCOVERY VIDEOS	VIDEOS	BIOSKILLS
1 Biology and the Tree of Life		Artificial Selection; Introduction to Experimental Design	Cells; Charles Darwin; Early Life; Bacteria		The Metric System; Reading Graphs; Reading a Phylogenetic Tree; Some Common Latin and Greek Roots Used in Biology
Unit 1 The Molecules of Life					
2 Water and Carbon: The Chemical Basis of Life		The Properties of Water			Reading Chemical Structures; Using Logarithms; Making Concept Maps; Reading Graphs
3 Protein Structure and Function		Condensation and Hydrolysis Reactions; Activation Energy and Enzymes		An Idealized Alpha Helix (A); An Idealized Alpha Helix (B); An Idealized Beta-Pleated Sheet (A); An Idealized Beta-Pleated Sheet (B)	
4 Nucleic Acids and the RNA World		Structure of RNA and DNA		Stick Model of DNA; Surface Model of DNA	Separating and Visualizing Molecules; Biological Imaging: Microscopy and X-Ray Crystallography
5 An Introduction to Carbohydrates		Carbohydrate Structure and Function			
6 Lipids, Membranes, and the First Cells	Membrane Transport	Diffusion and Osmosis; Membrane Transport Proteins		Space-filling Model of Cholesterol; Stick Model of Cholesterol; Space-filling Model of Phosphatidylcholine; Stick Model of a Phosphatidylcholine	Biological Imaging: Microscopy and X-Ray Crystallography; Separating and Visualizing Molecules
Unit 2 Cell Structure and Function					
7 Inside the Cell	Tour of an Animal Cell; Tour of a Plant Cell	Transport into the Nucleus; A Pulse-Chase Experiment	Bacteria; Cells	Confocal vs. Standard Fluorescence Microscopy; Cytoplasmic Streaming; Crawling Amoeba	Separating Cell Components by Centrifugation; Biological Imaging: Microscopy and X-Ray Crystallography; Separating and Visualizing Molecules
8 Cell-Cell Interactions			Cells	Connexon Structure	Separating and Visualizing Molecules
9 Cellular Respiration and Fermentation	Cellular Respiration	Redox Reactions; Glucose Metabolism		Space-filling Model of ATP (adenosine triphosphate); Stick Model of ATP (adenosine triphosphate)	
10 Photosynthesis	Photosynthesis	Chemiosmosis; Photosynthesis; Strategies for Carbon Fixation	Space Plants	Space-filling Model of Chlorophyll	
11 The Cell Cycle	Mitosis	The Phases of Mitosis; Four Phases of the Cell Cycle	Fighting Cancer	Mitosis	Separating and Visualizing Molecules; Cell and Tissue Culture Methods

	BIOFLIX	WEB ACTIVITIES	DISCOVERY VIDEOS	VIDEOS	BIOSKILLS
Unit 3 Gene Structure and Expression					
12 Meiosis	Meiosis	Meiosis; Mistakes in Meiosis			Combining Probabilities; Using Statistical Tests and Interpreting Standard Error Bars
13 Mendel and the Gene		Mendel's Experiments; The Principle of Independent Assortment			Model Organisms; Combining Probabilities; Reading Graphs
14 DNA and the Gene: Synthesis and Repair	DNA Replication	DNA Synthesis			Separating Cell Components by Centrifugation; Cell and Tissue Culture Methods; Using Logarithms; Reading Graphs
15 How Genes Work		The One-Gene One-Enzyme Hypothesis; The Triplet Nature of the Genetic Code			
16 Transcription, RNA Processing, and Translation	Protein Synthesis	RNA Synthesis; Synthesizing Proteins		A Stick-and-Ribbon Rendering of a tRNA	
17 Control of Gene Expression in Bacteria		The *lac* Operon		Cartoon Model of the *lac* Repressor from *E. coli*	
18 Control of Gene Expression in Eukaryotes		Transcription Initiation in Eukaryotes		Cartoon Model of the DNA-Binding Portion of TATA-Binding Protein Interacting with DNA; Cartoon Model of the GAL4 Transcription Factor from the yeast *S. cerevisiae*	Biological Imaging: Microscopy and X-Ray Crystallography; Separating and Visualizing Molecules
19 Analyzing and Engineering Genes		Producing Human Growth Hormone; The Polymerase Chain Reaction	Transgenics	Cartoon Model of the BamH1a Endonuclease	Separating and Visualizing Molecules
20 Genomics		Human Genome Sequencing Strategies	DNA Forensics		Model Organisms; Using Logarithms
Unit 4 Developmental Biology					
21 Principles of Development		Early Pattern Formation in *Drosophila*	Cloning	A Cartoon and Stick Model of the Homeodomain of the *Engrailed* Protein from *Drosophila* Interacting with DNA	Model Organisms; Cell and Tissue Culture Methods
22 An Introduction to Animal Development		Early Stages of Animal Development			
23 An Introduction to Plant Development					Model Organisms
Unit 5 Evolutionary Processes and Patterns					
24 Evolution by Natural Selection		Natural Selection for Antibiotic Resistance	Charles Darwin; Antibiotics		Reading a Phylogenetic Tree; Model Organisms; Reading Graphs

MEDIA AT A GLANCE *(continued)*

	BIOFLIX	WEB ACTIVITIES	DISCOVERY VIDEOS	VIDEOS	BIOSKILLS
25 Evolutionary Processes	Mechanisms of Evolution	The Hardy-Weinberg Principle; Three Modes of Natural Selection			Combining Probabilities; Using Statistical Tests and Interpreting Standard Error Bars; Reading Graphs
26 Speciation		Allopatric Speciation; Speciation by Changes in Ploidy			Reading a Phylogenetic Tree
27 Phylogenies and the History of Life		Adaptive Radiation	Mass Extinctions		Reading a Phylogenetic Tree
Unit 6 The Diversification of Life					
28 Bacteria and Archaea		The Tree of Life	Bacteria; Antibiotics; Molds; Tasty Bacteria; Early Life		Reading a Phylogenetic Tree; Model Organisms
29 Protists		Alternation of Generations in a Protist		A Crawling Amoeba	Biological Imaging: Microscopy and X-Ray Crystallography; Model Organisms
30 Green Algae and Land Plants		Plant Evolution and the PhylogeneticTree	Colored Cotton; Space Plants; Plant Pollination		
31 Fungi		Life Cycle of a Mushroom	Leafcutter Ants; Molds		
32 An Introduction to Animals		The Architecture of Animals	Leafcutter Ants; Invertebrates		
33 Protostome Animals		Protostome Diversity			Model Organisms
34 Deuterostome Animals		Deuterostome Diversity	Invertebrates		
35 Viruses		The HIV Replicative Cycle	Vaccines; Emerging Diseases		Biological Imaging: Microscopy and X-Ray Crystallography; Separating and Visualizing Molecules
Unit 7 How Plants Work					
36 Plant Form and Function		Plant Growth			
37 Water and Sugar Transport in Plants	Water Transport in Plants	Solute Transport in Plants		Plasmolysis of Plant Cells	
38 Plant Nutrition		Soil Formation and Nutrient Uptake			
39 Plant Sensory Systems, Signals, and Responses		Sensing Light; Plant Hormones; Plant Defenses			Cell and Tissue Culture Methods
40 Plant Reproduction		Reproduction in Flowering Plants; Fruit Structure and Development	Colored Cotton; Plant Pollination		
Unit 8 How Animals Work					
41 Animal Form and Function		Surface Area/Volume Relationships; Homeostasis	Blood; Human Body		Using Logarithms
42 Water and Electrolyte Balance in Animals		The Mammalian Kidney			

MEDIA AT A GLANCE *(continued)*

	BIOFLIX	WEB ACTIVITIES	DISCOVERY VIDEOS	VIDEOS	BIOSKILLS
43 Animal Nutrition	Homeostasis: Regulating Blood Sugar	The Digestion and Absorption of Food; Understanding Diabetes Mellitus	Nutrition		Biological Imaging: Microscopy and X-Ray Crystallography; Separating and Visualizing Molecules
44 Gas Exchange and Circulation	Gas Exchange	Gas Exchange in the Lungs and Tissues; The Human Heart	Blood		
45 Electrical Signals in Animals	How Neurons Work; How Synapses Work	Membrane Potentials; Action Potentials	Novelty Gene; Teen Brains	The Acetylcholine Receptor	Using Logarithms
46 Animal Sensory Systems and Movement	Muscle Contraction	The Vertebrate Eye; Structure and Contraction of Muscle Fibers	Muscles & Bones	The Acetylcholine Receptor	
47 Chemical Signals in Animals		Endocrine System Anatomy; Hormone Actions on Target Cells	Endocrine System	Cartoon Model of the DNA Binding Motif of a Zinc Finger Transcription Factor Binding to DNA	Separating Cell Components by Centrifugation
48 Animal Reproduction		Human Gametogenesis; Human Reproduction			Using Logarithms; Reading a Phylogenetic Tree
49 The Immune System in Animals		The Inflammatory Response; The Adaptive Immune Response	Emerging Diseases; Vaccines	Chemotaxis of a Neutrophil	
Unit 9 Ecology					
50 An Introduction to Ecology		Tropical Atmospheric Circulation	Rain Forests; Trees; Introduced Species		
51 Behavioral Ecology		Homing Behavior in Digger Wasps	Novelty Gene		
52 Population Ecology	Population Ecology	Modeling Population Growth; Human Population Growth and Regulation			
53 Community Ecology		Life Cycle of a Malaria Parasite; Succession	Leafcutter Ants		
54 Ecosystems	The Carbon Cycle	The Global Carbon Cycle			
55 Biodiversity and Conservation Biology		Habitat Fragmentation	Rain Forests		Using Logarithms

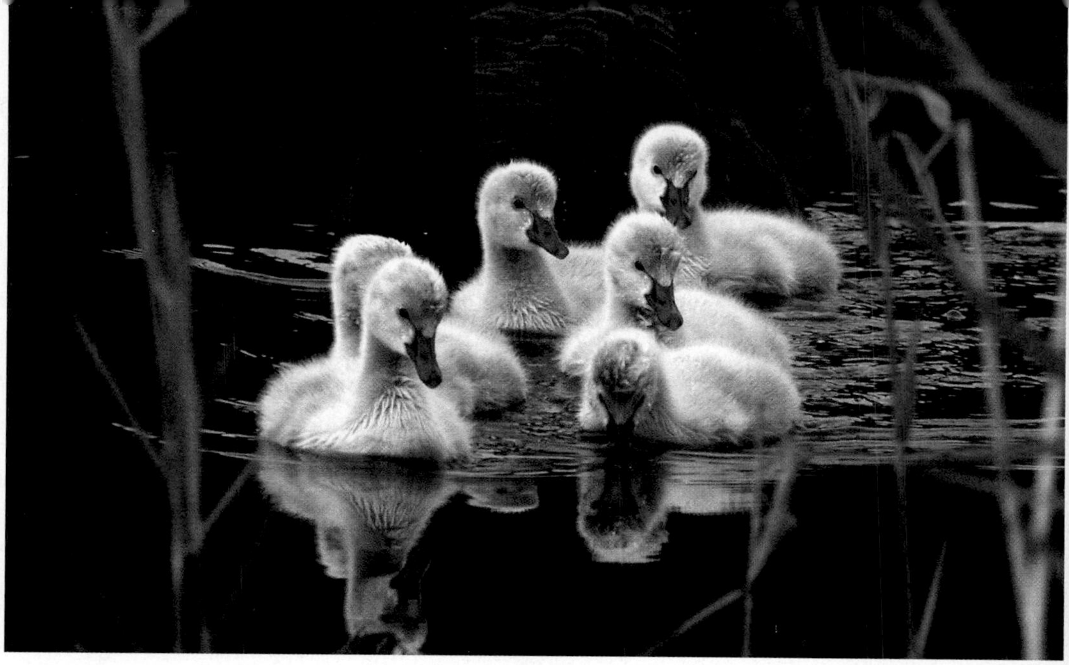

These black swan chicks are being introduced to the world by their mother, who trails close behind them. Likewise, this chapter introduces the key ideas that launched biological science as a discipline.

Biology and the Tree of Life 1

In essence, biological science is a search for ideas and observations that unify our understanding of the diversity of life, from bacteria living in rocks a mile underground to hedgehogs and humans. Chapter 1 is an introduction to this search.

The goals of this chapter are to introduce the nature of life and explore how biologists go about studying it. The chapter also introduces themes that will resonate throughout this book: (1) analyzing how organisms work at the molecular level, (2) understanding organisms in terms of their evolutionary history, and (3) helping you learn to think like a biologist.

Let's begin with what may be the most fundamental question of all: What is life?

KEY CONCEPTS

- Organisms obtain and use energy, are made up of cells, process information, replicate, and as populations evolve.
- The cell theory proposes that all organisms are made of cells and that all cells come from preexisting cells.
- The theory of evolution by natural selection maintains that species change through time because individuals with certain heritable traits produce more offspring than other individuals do.
- A phylogenetic tree is a graphical representation of the evolutionary relationships between species. These relationships can be estimated by analyzing similarities and differences in traits. Species that share distinctive traits are closely related and are placed close to each other on the tree of life.
- Biologists ask questions, generate hypotheses to answer them, and design experiments that test the predictions made by competing hypotheses.

1.1 What Does It Mean to Say That Something Is Alive?

An **organism** is a life-form—a living entity made up of one or more cells. Although there is no simple definition of life that is endorsed by all biologists, most agree that organisms share a suite of five fundamental characteristics.

- *Energy* To stay alive and reproduce, organisms have to acquire and use energy. To give just two examples: plants absorb sunlight; animals ingest food.
- *Cells* Organisms are made up of membrane-bound units called cells. A cell's membrane regulates the passage of materials between exterior and interior spaces.
- *Information* Organisms process hereditary or genetic information, encoded in units called genes, along with information they acquire from the environment. Right now cells throughout your body are using genetic information to make the molecules that keep you alive; your eyes and brain are decoding information on this page that will help you learn some biology.

✔ When you see this checkmark, stop and test yourself. Answers are available in Appendix B.

- *Replication* One of the great biologists of the twentieth century, François Jacob, said that the "dream of a bacterium is to become two bacteria." Almost everything an organism does contributes to one goal: replicating itself.
- *Evolution* Organisms are the product of evolution, and their populations continue to evolve.

You can think of this text as one long exploration of these five traits. Here's to life!

1.2 The Cell Theory

Two of the greatest unifying ideas in all of science laid the groundwork for modern biology: the cell theory and the theory of evolution by natural selection. Formally, scientists define a **theory** as an explanation for a very general class of phenomena or observations. The cell theory and theory of evolution address fundamental questions: What are organisms made of? Where do they come from?

When these concepts emerged in the mid-1800s, they revolutionized the way biologists think about the world. They established two of the five attributes of life: Organisms are cellular, and their populations change over time.

Neither insight came easily, however. The cell theory, for example, emerged after some 200 years of work. In 1665 Robert Hooke used a crude microscope to examine the structure of cork (a bark tissue) from an oak tree. The instrument magnified objects to just 30× (30 times) their normal size, but it allowed Hooke to see something extraordinary. In the cork he observed small, pore-like compartments that were invisible to the naked eye. These structures came to be called cells.

Soon after Hooke published his results, Anton van Leeuwenhoek succeeded in developing much more powerful microscopes, some capable of magnifications up to 300×. With these instruments, Leeuwenhoek inspected samples of pond water and made the first observations of human blood cells, and of sperm cells, shown in **Figure 1.1**.

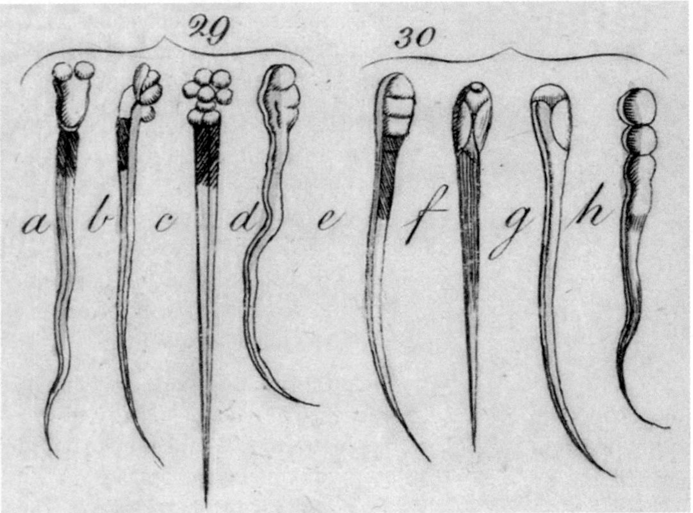

FIGURE 1.1 First Images of Cells. This drawing, by Anton van Leeuwenhoek, shows sperm cells.

In the 1670s a researcher who was studying the leaves and stems of plants with a microscope concluded that these large, complex structures are composed of many individual cells. By the early 1800s enough data had accumulated for a biologist to claim that *all* organisms consist of cells.

Are *All* Organisms Made of Cells?

The smallest organisms known today are bacteria that are barely 80 nanometers wide, or 80 *billionths* of a meter. (See **BioSkills 1** in Appendix A to review the metric system and its prefixes.[1]) It would take 12,500 of these organisms lined up end to end to span a millimeter. This is the distance between the smallest hash marks on a metric ruler.

In contrast, sequoia trees can be over 100 meters tall. This is the equivalent of a 20-story building. Bacteria and sequoias are composed of the same fundamental building block, however—the cell. Bacteria consist of a single cell; sequoias are made up of many cells.

Today, a **cell** is defined as a highly organized compartment that is bounded by a thin, flexible structure called a plasma membrane and that contains concentrated chemicals in an aqueous (watery) solution. The chemical reactions that sustain life take place inside cells. Most cells are also capable of reproducing by dividing—in effect, by making a copy of themselves.

The realization that all organisms are made of cells was fundamentally important, but it formed only the first part of the cell theory. In addition to understanding what organisms are made of, scientists wanted to understand how cells come to be.

Where Do Cells Come From?

Most scientific theories have two components: The first describes a pattern in the natural world; the second identifies a mechanism or process that is responsible for creating that pattern. Hooke and his fellow scientists articulated the pattern component of the cell theory. In 1858 Rudolph Virchow added the process component by stating that all cells arise from preexisting cells.

The complete **cell theory** can be stated as follows: All organisms are made of cells, and all cells come from preexisting cells.

TWO HYPOTHESES The cell theory was a direct challenge to the prevailing explanation of where cells come from, called spontaneous generation. At the time, most biologists believed that organisms arise spontaneously under certain conditions. For example, the bacteria and fungi that spoil foods such as milk and wine were thought to appear in these nutrient-rich media of their own accord—springing to life from nonliving materials. Spontaneous generation was a **hypothesis**: a proposed explanation.

The all-cells-from-cells hypothesis, in contrast, maintained that cells do not spring to life spontaneously but are produced only when preexisting cells grow and divide.

Biologists usually use the word theory to refer to proposed explanations for broad patterns in nature and prefer hypothesis to refer to explanations for more tightly focused questions.

[1]BioSkills are located in the first appendix at the back of the book. They focus on general skills that you'll use throughout this course. More than a few students have found them to be a life-saver. Please use them!

AN EXPERIMENT TO SETTLE THE QUESTION Soon after the all-cells-from-cells hypothesis appeared in print, Louis Pasteur set out to test its predictions experimentally. A **prediction** is something that can be measured and that must be correct if a hypothesis is valid.

Pasteur wanted to determine whether microorganisms could arise spontaneously in a nutrient broth or whether they appear only when a broth is exposed to a source of preexisting cells. To address the question, he created two treatment groups: a broth that was not exposed to a source of preexisting cells and a broth that was.

The spontaneous generation hypothesis predicted that cells would appear in both treatment groups. The all-cells-from-cells hypothesis predicted that cells would appear only in the treatment exposed to a source of preexisting cells.

Figure 1.2 shows Pasteur's experimental setup. Note that the two treatments are identical in every respect but one. Both used glass flasks filled with the same amount of the same nutrient broth. Both were boiled for the same amount of time to kill any existing organisms such as bacteria or fungi. But because the flask pictured in Figure 1.2a had a straight neck, it was exposed to preexisting cells after sterilization by the heat treatment. These preexisting cells are the bacteria and fungi that cling to dust particles in the air. They could drop into the nutrient broth because the neck of the flask was straight.

In contrast, the flask drawn in Figure 1.2b had a long swan neck. Pasteur knew that water would condense in the crook of the swan neck after the boiling treatment and that this pool of water

EXPERIMENT

QUESTION: Do cells arise spontaneously or from other cells?

SPONTANEOUS GENERATION HYPOTHESIS: Cells arise spontaneously from nonliving materials.

ALL-CELLS-FROM-CELLS HYPOTHESIS: Cells are produced only when preexisting cells grow and divide.

(a) Pasteur experiment with straight-necked flask:

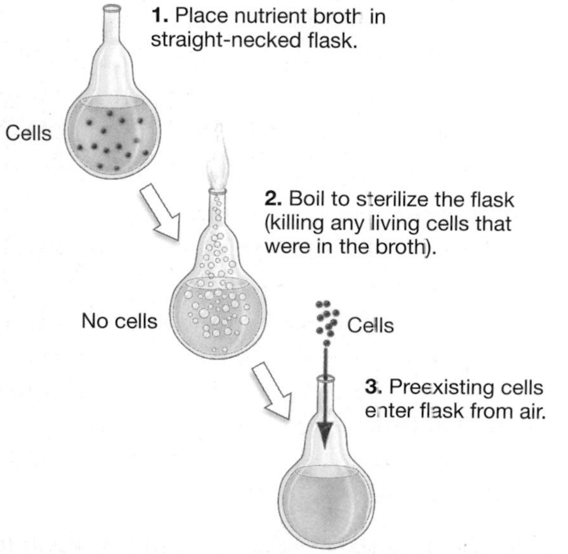

1. Place nutrient broth in straight-necked flask.
Cells
2. Boil to sterilize the flask (killing any living cells that were in the broth).
No cells
Cells
3. Preexisting cells enter flask from air.

(b) Pasteur experiment with swan-necked flask:

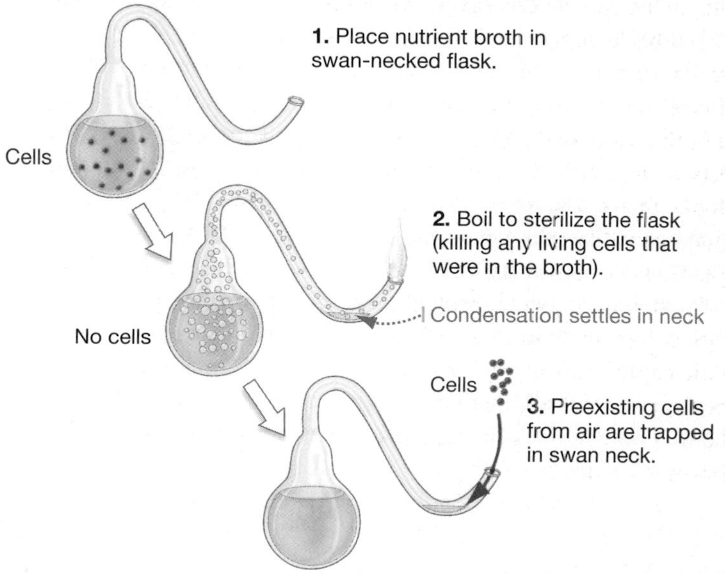

1. Place nutrient broth in swan-necked flask.
Cells
2. Boil to sterilize the flask (killing any living cells that were in the broth).
Condensation settles in neck
No cells
Cells
3. Preexisting cells from air are trapped in swan neck.

(a)	(b)
PREDICTION OF SPONTANEOUS GENERATION HYPOTHESIS: Cells will appear in broth.	**PREDICTION OF SPONTANEOUS GENERATION HYPOTHESIS:** Cells will appear in broth.
PREDICTION OF ALL-CELLS-FROM-CELLS HYPOTHESIS: Cells will appear in broth.	**PREDICTION OF ALL-CELLS-FROM-CELLS HYPOTHESIS:** Cells will not appear in broth.

RESULTS:

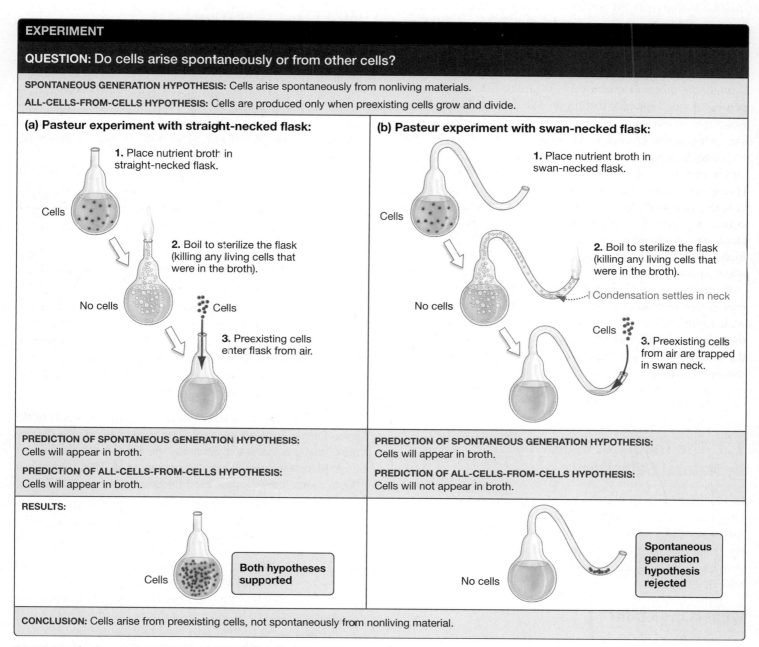

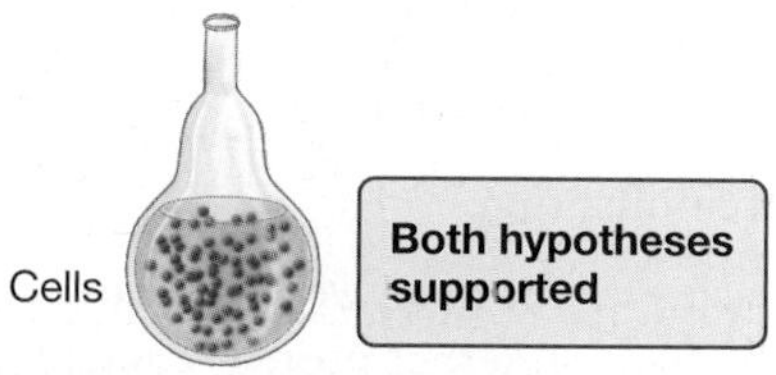

Cells
Both hypotheses supported

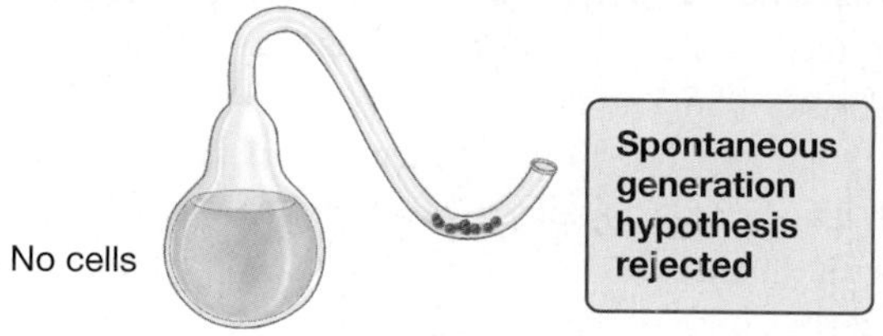

No cells
Spontaneous generation hypothesis rejected

CONCLUSION: Cells arise from preexisting cells, not spontaneously from nonliving material.

FIGURE 1.2 The Spontaneous Generation and All-Cells-from-Cells Hypotheses Were Tested Experimentally.

✔**QUESTION** What problem would arise if Pasteur had (1) put different types of broth in the two treatments, (2) heated them for different lengths of time, or (3) used a ceramic flask for one treatment and a glass flask for the other?

would trap any bacteria or fungi that entered on dust particles. Thus, the contents of the swan-necked flask was isolated from any source of preexisting cells even though still open to the air.

Pasteur's experimental setup was effective because there was only one difference between the two treatments and because that difference was the factor being tested—in this case, a broth's exposure to preexisting cells.

ONE HYPOTHESIS SUPPORTED And Pasteur's results? As Figure 1.2 shows, the treatment exposed to preexisting cells quickly filled with bacteria and fungi. This observation was important because it showed that the heat sterilization step had not altered the nutrient broth's capacity to support growth.

The treatment in the swan-necked flask remained sterile, however. Even when the broth was left standing for months, no organisms appeared in it. This result was inconsistent with the hypothesis of spontaneous generation.

Because Pasteur's data were so conclusive—meaning that there was no other reasonable explanation for them—the results persuaded most biologists that the all-cells-from-cells hypothesis was correct. However, Chapters 2–6 will show that biologists now have evidence that life did arise from nonlife early in Earth's history, through a process called chemical evolution.

The success of the cell theory's process component had an important implication: If all cells come from preexisting cells, it follows that all individuals in an isolated population of single-celled organisms are related by common ancestry. Similarly, in you and other multicellular individuals, all the cells present are descended from preexisting cells, tracing back to a fertilized egg. A fertilized egg is a cell created by the fusion of sperm and egg—cells that formed in individuals of the previous generation. In this way, all the cells in a multicellular organism are connected by common ancestry.

The second great founding idea in biology is similar, in spirit, to the cell theory. It also happened to be published the same year as the all-cells-from-cells hypothesis. This was the realization, made independently by Charles Darwin and Alfred Russel Wallace, that all species—all distinct, identifiable types of organisms—are connected by common ancestry.

1.3 The Theory of Evolution by Natural Selection

In 1858 short papers written separately by Darwin and Wallace were read to a small group of scientists attending a meeting of the Linnean Society of London. A year later, Darwin published a book that expanded on the idea summarized in those brief papers. The book was called *The Origin of Species.* The first edition sold out in a day.

What Is Evolution?

Like the cell theory, the theory of evolution by natural selection has a pattern and a process component. Darwin and Wallace's theory made two important claims concerning patterns that exist in the natural world.

- Species are related by common ancestry. This contrasted with the prevailing view in science at the time, which was that species represent independent entities created separately by a divine being.
- In contrast to the accepted view that species remain unchanged through time, Darwin and Wallace proposed that the characteristics of species can be modified from generation to generation. Darwin called this process "descent with modification."

Evolution is a change in the characteristics of a population over time. It means that species are not independent and unchanging entities, but are related to one another and can change through time.

What Is Natural Selection?

This pattern component of the theory of evolution was actually not original to Darwin and Wallace. Several scientists had already come to the same conclusions about the relationships between species. The great insight by Darwin and Wallace was in proposing a process, called **natural selection**, that explains *how* evolution occurs.

TWO CONDITIONS OF NATURAL SELECTION Natural selection occurs whenever two conditions are met.

1. Individuals within a population vary in characteristics that are **heritable**—meaning, traits that can be passed on to offspring. A **population** is defined as a group of individuals of the same species living in the same area at the same time.
2. In a particular environment, certain versions of these heritable traits help individuals survive better or reproduce more than do other versions.

If certain heritable traits lead to increased success in producing offspring, then those traits become more common in the population over time. In this way, the population's characteristics change as a result of natural selection acting on individuals. This is a key insight: Natural selection acts on individuals, but evolutionary change occurs in populations.

SELECTION ON MAIZE AS AN EXAMPLE To clarify how natural selection works, consider an example of **artificial selection**—changes in populations that occur when *humans* select certain individuals to produce the most offspring. Beginning in 1896, researchers began a long-term selection experiment on maize (corn).

1. In the original population, the percentage of protein in maize kernels was variable among individuals. Kernel protein content is a heritable trait—parents tend to pass the trait on to their offspring.
2. Each year for many years, researchers chose individuals with the highest kernel protein content to be the parents of the next generation. In this environment, individuals with high kernel protein content produced more offspring than individuals with low kernel protein content.

Figure 1.3 shows the results. Note that this graph plots generation number on the *x*-axis, starting from the first generation

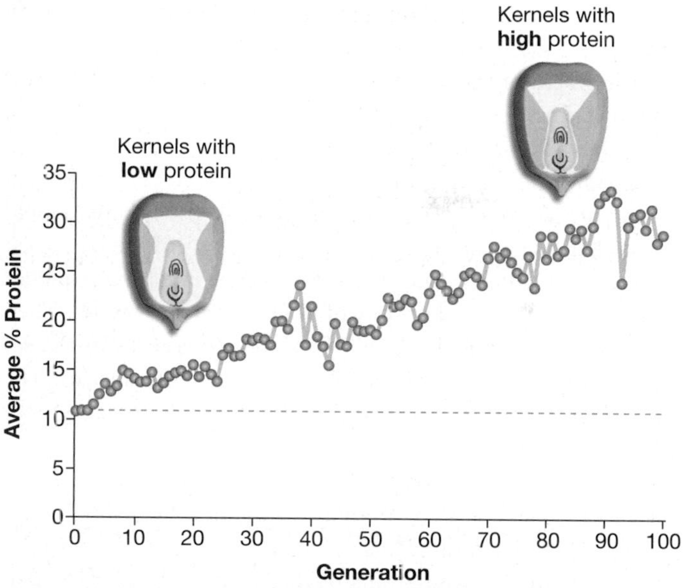

FIGURE 1.3 Response to Selection for High Kernel Protein Content in Maize. The average protein content goes down in a few years because of poor growing conditions or chance changes in how the many genes responsible for this trait interact.

("0") and continuing for 100 generations. The average percentage of protein in a kernel among individuals in this population is plotted on the y-axis.

To read this graph, put your finger on the x-axis at time 0. Then read up the y-axis, and note that kernels averaged about 11 percent protein at the start of the experiment. Now read the graph to the right. Each dot is a data point, representing the average kernel protein concentration in a particular generation. (A generation in maize is one year.) The lines on this graph simply connect the dots, to make the pattern in the data easier to see. At the end of the graph, after 100 generations of selection, average kernel protein content is about 29 percent. (For more help with reading graphs, see **BioSkills 2** in Appendix A.)

This sort of change in the characteristics of a population, over time, is evolution. Humans have been practicing artificial selection for thousands of years, and biologists have now documented evolution by *natural* selection—where humans don't do the selecting—occurring in thousands of different populations, including humans.

To practice applying the principles of artificial selection, go to the online study area at *www.masteringbiology.com*.

(MB) **Web Activity** Artificial Selection

Evolution occurs when heritable variation leads to differential success in reproduction. ✔If you understand this concept, you should be able to describe how protein content in maize kernels changed over time, using the same x-axis and y-axis as in Figure 1.3, when researchers selected individuals with *lowest* kernel protein content to be the parents of the next generation. (This experiment was actually done, starting with the same population at the same time as selection for high protein content.)

FITNESS AND ADAPTATION Darwin also introduced some new terminology to identify what is happening during natural selection.

- In everyday English, fitness means health and well-being. But in biology, **fitness** means the ability of an individual to produce offspring. Individuals with high fitness produce many surviving offspring.
- In everyday English, adaptation means that an individual is adjusting and changing to function in new circumstances. But in biology, an **adaptation** is a trait that increases the fitness of an individual in a particular environment.

Once again, consider kernel protein content in maize: In the environment of the experiment graphed in Figure 1.3, individuals with high kernel protein content produced more offspring and had higher fitness than individuals with lower kernel protein content. In this population and this environment, high kernel protein content was an adaptation that allowed certain individuals to thrive.

Note that during this process, the amount of protein in the kernels of any individual maize plant did not change within its lifetime—the change occurred in the characteristics of the population over time.

Together, the cell theory and the theory of evolution provided the young science of biology with two central, unifying ideas:

1. The cell is the fundamental structural unit in all organisms.
2. All species are related by common ancestry and have changed over time in response to natural selection.

CHECK YOUR UNDERSTANDING

If you understand that . . .

- Natural selection occurs when heritable variation in certain traits leads to improved success in reproduction. Because individuals with these traits produce many offspring with the same traits, the traits increase in frequency and evolution occurs.
- Evolution is a change in the characteristics of a population over time.

✔ **You should be able to . . .**

On the graph you just analyzed, describe the average kernel protein content over time in a maize population where *no* selection occurred.

Answers are available in Appendix B.

1.4 The Tree of Life

Section 1.3 focused on how individual populations change through time in response to natural selection. But over the past several decades, biologists have also documented dozens of cases in which natural selection has caused populations of one species to diverge and form new species. This divergence process is called **speciation**.

Research on speciation has two important implications: All species come from preexisting species, and all species, past and present, trace their ancestry back to a single common ancestor.

The theory of evolution by natural selection predicts that biologists should be able to reconstruct a **tree of life**—a family tree of organisms. If life on Earth arose just once, then such a diagram would describe the genealogical relationships between species with a single, ancestral species at its base.

Has this task been accomplished? If the tree of life exists, what does it look like?

Using Molecules to Understand the Tree of Life

One of the great breakthroughs in research on the tree of life occurred when Carl Woese (pronounced *woes*) and colleagues began analyzing the chemical components of organisms, as a way to understand their evolutionary relationships. Their goal was to understand the **phylogeny** of all organisms—their actual genealogical relationships. Translated literally, phylogeny means "tribe-source."

To understand which organisms are closely versus distantly related, Woese and co-workers needed to study a molecule that is found in all organisms. The molecule they selected is called small subunit ribosomal RNA (rRNA). It is an essential part of the machinery that all cells use to grow and reproduce.

Although rRNA is a large and complex molecule, its underlying structure is simple. The rRNA molecule is made up of sequences of four smaller chemical components called ribonucleotides. These ribonucleotides are symbolized by the letters A, U, C, and G. In rRNA, ribonucleotides are connected to one another linearly, like boxcars of a freight train (**Figure 1.4**).

ANALYZING rRNA Why might rRNA be useful for understanding the relationships between organisms? The answer is that the ribonucleotide sequence in rRNA is a trait that can change during the course of evolution. Although rRNA performs the same function in all organisms, the sequence of ribonucleotide building blocks in this molecule is not identical among species.

In land plants, for example, the molecule might start with the sequence A-U-A-U-C-G-A-G. In green algae, which are closely related to land plants, the same section of the molecule might contain A-U-A-U-**G**-G-A-G. But in brown algae, which are not closely related to green algae or to land plants, the same part of the molecule might consist of A-**A**-A-U-**G**-G-A-**C**.

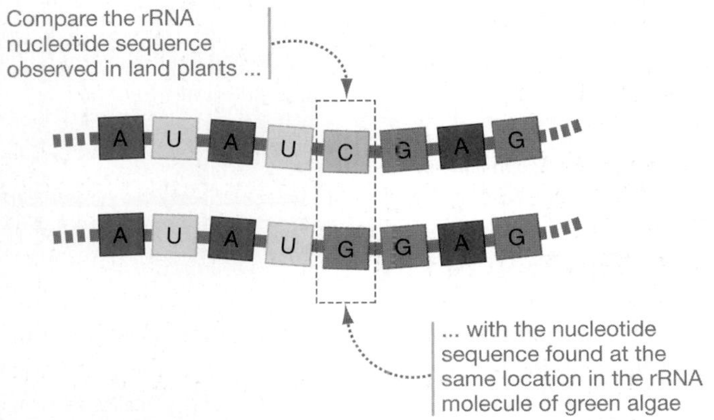
Compare the rRNA nucleotide sequence observed in land plants ...
A
U
A
U
C
G
A
G
A
U
A
U
G
G
A
G
... with the nucleotide sequence found at the same location in the rRNA molecule of green algae

FIGURE 1.4 RNA Molecules Are Made Up of Smaller Molecules. The complete small subunit rRNA molecule contains about 2000 ribonucleotides; just 8 are shown in this comparison.

✔**QUESTION** Suppose that in the same portion of rRNA, molds and other fungi have the sequence A-U-A-U-G-G-A-C. According to these data, are fungi more closely related to green algae or to land plants? Explain your logic.

The research program that Woese and co-workers pursued was based on a simple premise: If the theory of evolution is correct, then rRNA sequences should be very similar in closely related organisms but less similar in organisms that are less closely related. Species that are part of the same evolutionary lineage, like the plants, should share certain changes in rRNA that no other species have.

To test this premise, the researchers determined the sequence of ribonucleotides in the rRNA of a wide array of species. Then they considered what the similarities and differences in the sequences implied about relationships between the species. The goal was to produce a diagram that described the phylogeny of the organisms in the study.

A diagram that depicts evolutionary history in this way is called a phylogenetic tree. Just as a family tree shows relationships between individuals, a phylogenetic tree shows relationships between species. On a phylogenetic tree, branches that share a recent common ancestor represent species that are closely related; branches that don't share recent common ancestors represent species that are more distantly related.

THE TREE OF LIFE ESTIMATED FROM AN ARRAY OF GENES To construct a phylogenetic tree, researchers use a computer to find the arrangement of branches that is most consistent with the similarities and differences observed in the data.

Although the initial work was based only on the sequences of ribonucleotides observed in rRNA, biologists now use data sets that include sequences from a wide array of genes. **Figure 1.5** shows a recent tree produced by comparing these sequences. Because this tree includes such a diverse array of species, it is often called the universal tree, or the tree of life. (For help in learning how to read a phylogenetic tree, see **BioSkills 3** in Appendix A.)

The tree of life implied by rRNA and other genetic data established that there are three fundamental groups or lineages of organisms: (1) the Bacteria, (2) the Archaea, and (3) the Eukarya. In all **eukaryotes**, cells have a prominent component called the nucleus (**Figure 1.6a**). Translated literally, the word eukaryotes means "true kernel." Because the vast majority of bacterial and archaeal cells lack a nucleus, they are referred to as **prokaryotes** (literally, "before kernel"; see **Figure 1.6b**). The vast majority of bacteria and archaea are unicellular ("one-celled"); many eukaryotes are multicellular ("many-celled").

When these results were first published, biologists were astonished. For example:

- Prior to Woese's work and follow-up studies, biologists thought that the most fundamental division among organisms was between prokaryotes and eukaryotes. The Archaea were virtually unknown—much less recognized as a major and highly distinctive branch on the tree of life.
- Fungi were thought to be closely related to plants. Instead, they are actually much more closely related to animals.
- Traditional approaches for classifying organisms—including the system of five kingdoms divided into various classes, or-

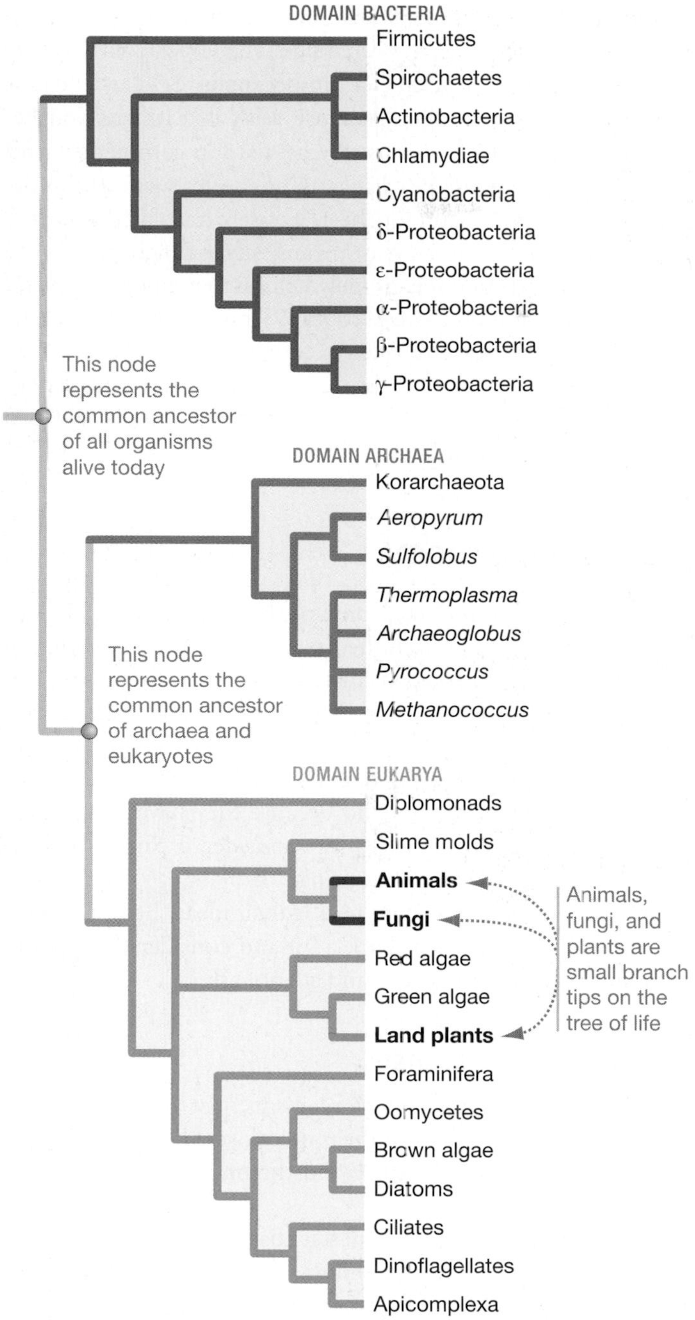

FIGURE 1.5 The Tree of Life. "Universal tree" estimated from a large amount of gene sequence data. The three domains of life revealed by the analysis are labeled. Common names are given for most lineages in the domains Bacteria and Eukarya. Genus names are given for members of the domain Archaea, because most of these organisms have no common names.

ders, and families that you may have learned in high school—are inaccurate in many cases, because they do not reflect the actual evolutionary history of the organisms involved.

THE TREE OF LIFE IS A WORK IN PROGRESS Just as researching your family tree can help you understand who you are and where you came from, so the tree of life helps biologists understand the relationships between organisms and the history of species. The discovery of the Archaea and the accurate placement of lineages

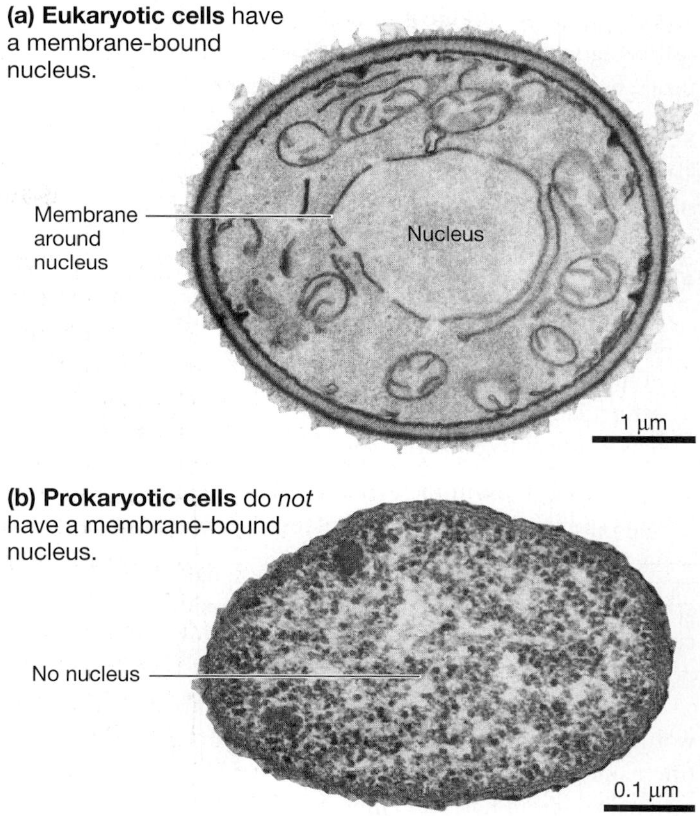

FIGURE 1.6 Eukaryotes and Prokaryotes.

✔**QUESTION** How many times larger is the eukaryotic cell in this figure than the prokaryotic cell? (Hint: study the scale bars.)

such as the fungi qualify as exciting breakthroughs in our understanding of evolutionary history and life's diversity.

Work on the tree of life continues at a furious pace, however, and the location of certain branches on the tree is hotly debated. As databases expand and as techniques for analyzing data improve, the shape of the tree of life presented in Figure 1.5 will undoubtedly change. Our understanding of the tree of life, like our understanding of every other topic in biological science, is dynamic.

How Should We Name Branches on the Tree of Life?

In science, the effort to name and classify organisms is called **taxonomy**. Any named group is called a **taxon** (plural: **taxa**). Currently, biologists are working to create a taxonomy, or naming system, that accurately reflects the phylogeny of organisms.

For example, Woese proposed a new taxonomic category called the **domain**. The three domains of life are the Bacteria, Archaea, and Eukarya.

Biologists often use the term **phylum** (plural: **phyla**) to refer to major lineages within each domain. Although the designation is somewhat arbitrary, each phylum is considered a major branch on the tree of life. For example, within the lineage called animals, biologists name about 35 phyla—each of which is distinguished by distinctive aspects of its body structure as well as by distinctive gene sequences. The mollusks (clams, squid, octopuses) constitute a phylum, as do chordates (the vertebrates and their close relatives).

Because the tree of life is so new, though, naming systems are still being worked out. One thing that hasn't changed for centuries, however, is the naming system for individual species.

SCIENTIFIC (SPECIES) NAMES In 1735, a Swedish botanist named Carolus Linnaeus established a system for naming species that is still in use today. Linnaeus created a two-part name unique to each type of organism.

- The first part indicates the organism's **genus** (plural: **genera**). A genus is made up of a closely related group of species. For example, Linnaeus put humans in the genus *Homo*. Although humans are the only living species in this genus, at least five extinct organisms, all of which walked upright and made extensive use of tools, were later also assigned to *Homo*.
- The second term in the two-part name identifies the organism's species. Linnaeus gave humans the species name *sapiens*.

An organism's genus and species designation is called its **scientific name** or Latin name. Scientific names are always italicized. Genus names are always capitalized, but species names are not—for instance, *Homo sapiens*.

Scientific names are based on Latin or Greek word roots or on words "Latinized" from other languages. Linnaeus gave a scientific name to every species then known, and also Latinized his own name—from Karl von Linné to Carolus Linnaeus.

Linnaeus maintained that different types of organisms should not be given the same genus and species names. Other species may be assigned to the genus *Homo*, and members of other genera may be named *sapiens*, but only humans are named *Homo sapiens*. Each scientific name is unique.

SCIENTIFIC NAMES ARE OFTEN DESCRIPTIVE Scientific names and terms are often based on Latin or Greek word roots that are descriptive. For example, *Homo sapiens* is derived from the Latin *homo* for "man" and *sapiens* for "wise" or "knowing." The yeast that bakers use to produce bread and that brewers use to brew beer is called *Saccharomyces cerevisiae*. The Greek root *saccharo* means "sugar," and *myces* refers to a fungus. *Saccharomyces* is aptly named "sugar fungus" because yeast is a fungus and because the domesticated strains of yeast used in commercial baking and brewing are often fed sugar. The species name of this organism, *cerevisiae*, is Latin for "beer." Loosely translated, then, the scientific name of brewer's yeast means "sugar-fungus for beer."

Most biologists find it extremely helpful to memorize some of the common Latin and Greek roots. To aid you in this process, new terms in this text are often accompanied by a reference to their Latin or Greek word roots in parentheses, and a glossary of common root words with translations and examples is provided in **BioSkills 4** in Appendix A.

CHECK YOUR UNDERSTANDING

If you understand that . . .

- A phylogenetic tree shows the evolutionary relationships between species.
- To infer where species belong on a phylogenetic tree, biologists examine the characteristics of the species involved. Closely related species should have similar characteristics, while less closely related species should be less similar.

✔ You should be able to . . .

Examine the following sequences and draw a phylogenetic tree showing the relationships between species A, B, and C that these data imply:

Species A: A A C T A G C G C G A T
Species B: A A C T A G C G C C A T
Species C: T T C T A G C G G T A T

Answers are available in Appendix B.

1.5 Doing Biology

This chapter has introduced some of the great ideas in biology. The development of the cell theory and the theory of evolution by natural selection provided cornerstones when the science was young; the tree of life is a relatively recent insight that has revolutionized our understanding of life's diversity.

These theories are considered great because they explain fundamental aspects of nature, and because they have consistently been shown to be correct. They are considered correct because they have withstood extensive testing.

How do biologists go about testing their ideas? Before answering this question, let's step back a bit and consider the types of questions that researchers can and cannot ask.

The Nature of Science

Biologists ask questions about organisms, just as physicists and chemists ask questions about the physical world or geologists ask questions about Earth's history and the ongoing processes that shape landforms.

No matter what their field, all scientists ask questions that can be answered by measuring things—by collecting data. Conversely, scientists cannot address questions that can't be answered by measuring things.

This distinction is important. It is at the root of continuing controversies about teaching evolution in publicly funded schools. In the United States and in Turkey, in particular, some Christian and Islamic leaders have been particularly successful in pushing their claim that evolution and religious faith are in conflict. Even though the theory of evolution is considered one of the most successful and best-substantiated ideas in the history of science, they object to teaching it.

The vast majority of biologists and religious leaders reject this claim; they see no conflict between evolution and religious faith. Their view is that science and religion are compatible because they address different types of questions.

- Science is about formulating hypotheses and finding evidence that supports or conflicts with those hypotheses.

- Religious faith addresses questions that cannot be answered by data. The questions addressed by the world's great religions focus on why we exist and how we should live.

Both types of questions are seen as legitimate and important.

So how do biologists go about answering questions? Let's consider two issues currently being addressed by researchers.

Why Do Giraffes Have Long Necks? An Introduction to Hypothesis Testing

If you were asked why giraffes have long necks, you might say that long necks enable giraffes to reach food that is unavailable to other mammals. This hypothesis is expressed in African folktales and has traditionally been accepted by many biologists. The food competition hypothesis is so plausible, in fact, that for decades no one thought to test it.

In the mid-1990s, however, Robert Simmons and Lue Scheepers assembled data suggesting that the food competition hypothesis is only part of the story. Their analysis supports an alternative hypothesis—that long necks allow giraffes to use their heads as effective weapons for battering their opponents.

How did biologists test the food competition hypothesis? What data support their alternative explanation? Before attempting to answer these questions, it's important to recognize that hypothesis testing is a two-step process:

1. State the hypothesis as precisely as possible and list the predictions it makes.
2. Design an observational or experimental study that is capable of testing those predictions.

If the predictions are accurate, the hypothesis is supported. If the predictions are not met, then researchers do further tests, modify the original hypothesis, or search for alternative explanations.

THE FOOD COMPETITION HYPOTHESIS: PREDICTIONS AND TESTS

The food competition hypothesis claims that giraffes compete for food with other species of mammals. When food is scarce, as it is during the dry season, giraffes with longer necks can reach food that is unavailable to other species and to giraffes with shorter necks. As a result, the longest-necked individuals in a giraffe population survive better and produce more young than do shorter-necked individuals, and average neck length of the population increases with each generation.

To use the terms introduced earlier, long necks are adaptations that increase the fitness of individual giraffes during competition for food. This type of natural selection has gone on so long that the population has become extremely long necked.

The food competition hypothesis makes several explicit predictions. For example, the food competition hypothesis predicts that

1. neck length is variable among giraffes;
2. neck length in giraffes is heritable; and
3. giraffes feed high in trees, especially during the dry season, when food is scarce and the threat of starvation is high.

The first prediction is correct. Studies in zoos and natural populations confirm that neck length is variable among individuals.

The researchers were unable to test the second prediction, however, because they studied giraffes in a natural population and were unable to do breeding experiments. As a result, they simply had to accept this prediction as an assumption. In general, though, biologists prefer to test every assumption behind a hypothesis.

What about the prediction regarding feeding high in trees? According to Simmons and Scheepers, this is where the food competition hypothesis breaks down.

Consider, for example, data collected by a different research team on the amount of time that giraffes spend feeding in vegetation of different heights. **Figure 1.7a** plots the height of vegetation versus the percentage of bites taken by a giraffe, for males and for females from the same population in Kenya. The dashed line on each graph indicates the average height of a male or female in this population.

Note that the average height of a giraffe in this population is much greater than the height where most feeding takes place. In

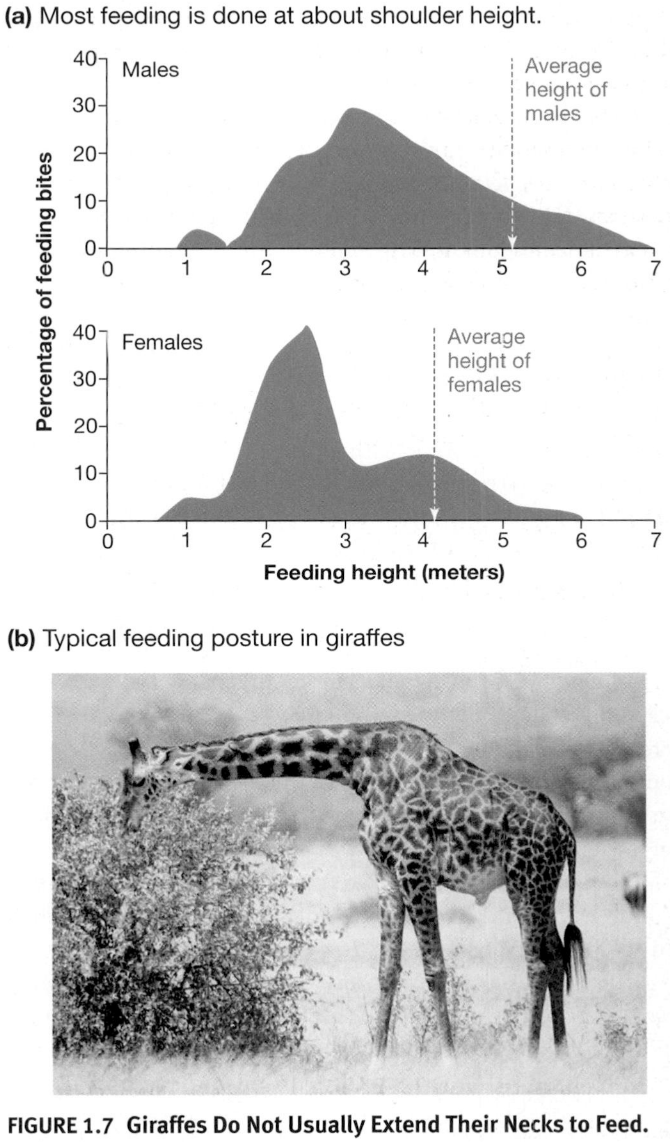

FIGURE 1.7 Giraffes Do Not Usually Extend Their Necks to Feed.

this population, both male and female giraffes spend most of their feeding time eating vegetation that averages just 60 percent of their full height. Studies on other populations of giraffes, during both the wet and dry seasons, are consistent with these data. Giraffes usually feed with their necks bent (**Figure 1.7b**).

These data cast doubt on the food competition hypothesis, because one of its predictions does not appear to hold. Biologists have not abandoned this hypothesis completely, though, because feeding high in trees may be particularly valuable during extreme droughts, when a giraffe's ability to reach leaves far above the ground could mean the difference between life and death. Still, Simmons and Scheepers have offered an alternative explanation for why giraffes have long necks. The new hypothesis is based on the mating system of giraffes.

THE SEXUAL COMPETITION HYPOTHESIS: PREDICTIONS AND TESTS Giraffes have an unusual mating system. Breeding occurs year round rather than seasonally. To determine when females are coming into estrus or "heat" and are thus receptive to mating, the males nuzzle the rumps of females. In response, the females urinate into the males' mouths. The males then tip their heads back and pull their lips to and fro, as if tasting the liquid. Biologists who have witnessed this behavior have proposed that the males taste the females' urine to detect whether estrus has begun.

Once a female giraffe enters estrus, males fight among themselves for the opportunity to mate. Combat is spectacular. The bulls stand next to one another, swing their necks, and strike thunderous blows with their heads. Researchers have seen males knocked unconscious for 20 minutes after being hit and have cataloged numerous instances in which the loser died. Giraffes are the only animals known to fight in this way.

These observations inspired a new explanation for why giraffes have long necks. The sexual competition hypothesis is based on the idea that longer-necked giraffes are able to strike harder blows during combat than can shorter-necked giraffes. In engineering terms, longer necks provide a longer "moment arm." A long moment arm increases the force of the impact. (Think about the type of sledgehammer you'd use to bash down a concrete wall—one with a short handle or one with a long handle?)

The idea here is that longer-necked males should win more fights and, as a result, father more offspring than shorter-necked males do. If neck length in giraffes is inherited, then the average neck length in the population should increase over time. Under the sexual competition hypothesis, long necks are adaptations that increase the fitness of males during competition for females.

Although several studies have shown that long-necked males are more successful in fighting and that the winners of fights gain access to estrous females, the question of why giraffes have long necks is not closed. With the data collected to date, most biologists would probably concede that the food competition hypothesis needs further testing and refinement and that the sexual selection hypothesis appears promising. It could also be true that both hypotheses are correct. For our purposes, the important take-home message is that all hypotheses must be tested rigorously.

In many cases in biological science, testing hypotheses rigorously involves experimentation. Experimenting on giraffes is difficult. But in the case study considered next, biologists were able to test an interesting hypothesis experimentally.

How Do Ants Navigate? An Introduction to Experimental Design

Experiments are a powerful scientific tool because they allow researchers to test the effect of a single, well-defined factor on a particular phenomenon. Because experiments testing the effect of neck length on food and sexual competition in giraffes haven't been done yet, let's consider a different question: When ants leave their nest to search for food, how do they find their way back?

The Saharan desert ant lives in colonies and makes a living by scavenging the dead carcasses of insects. Individuals leave the burrow and wander about searching for food at midday, when temperatures at the surface can reach 60°C (140°F) and predators are hiding from the heat.

Foraging trips can take the ants hundreds of meters—an impressive distance when you consider that these animals are only about a centimeter long. But when an ant returns, it doesn't follow the same long, wandering route it took on its way away from the nest. Instead, individuals return in a straight line. How do they do this?

THE PEDOMETER HYPOTHESIS Early work on navigation in desert ants showed that they use the Sun's position as a compass—meaning that they always know the approximate direction of the nest relative to the Sun. But how do they know how far to go?

After experiments had shown that the ants do not use landmarks to navigate, Matthias Wittlinger and co-workers set out to test a novel idea. The biologists proposed that Saharan desert ants know how far they are from the nest by integrating information from leg movements.

According to this pedometer hypothesis, the ants always know how far they are from the nest because they track the number of steps they have taken and their stride length. The idea is that they can make a beeline back to the burrow because they integrate information on the angles they have traveled *and* the distance they have gone—based on step number and stride length.

TESTING THE HYPOTHESIS To test their idea, Wittlinger's group allowed ants to walk from a nest to a feeder through a channel—a distance of 10 m. Then they caught ants at the feeder and created three test groups, each with 25 individuals:

- *Stumps* By cutting the lower legs of some individuals off, they created ants with shorter-than-normal legs.
- *Normal* Some individuals were left alone, meaning that they had normal leg length.
- *Stilts* By gluing pig bristles onto each leg, the biologists created ants with longer-than-normal legs.

Next they put the ants in a different channel and recorded how far they traveled before starting the characteristic set of 180° turns that ants make when they are looking for the nest hole. To see the data they collected, look at the graph on the left side of the "Results" section in **Figure 1.8**.

- *Stumps* The ants with stumps stopped short, by about 5 m, before starting to look for the nest opening.
- *Normal* The normal ants walked the correct distance—about 10 m.
- *Stilts* The ants with stilts walked about 5 m too far before starting to look for the nest opening.

To check the validity of this result, they put the test ants back in the nest and recaptured them one to several days later, when they had walked to the feeder on their stumps, normal legs, or stilts. Now when the ants were put into the other channel to "walk back," they all traveled the correct distance—10 m—before starting to look for the nest (see the graph on the right side of the "Results" section in Figure 1.8).

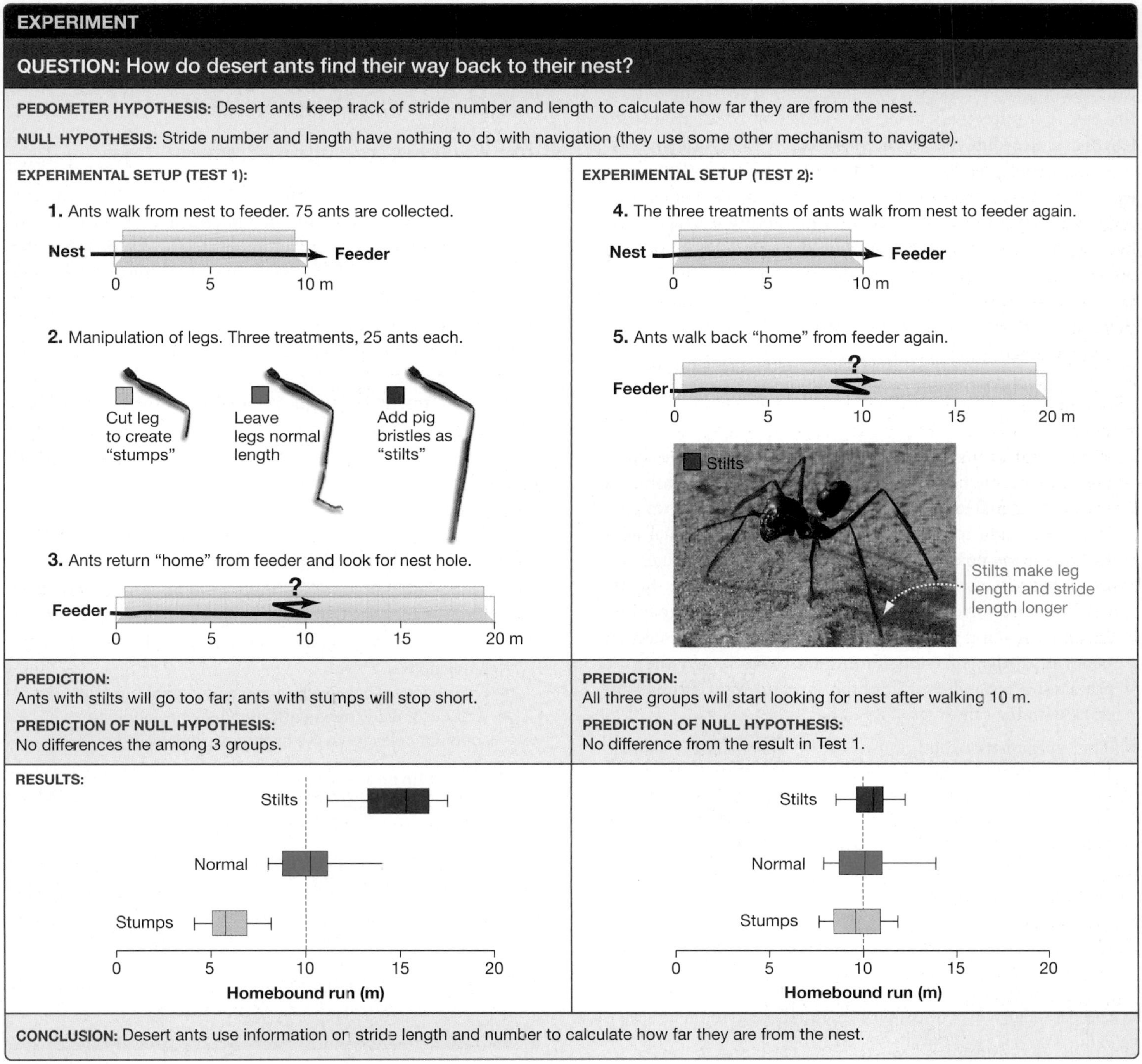

EXPERIMENT
QUESTION: How do desert ants find their way back to their nest?
PEDOMETER HYPOTHESIS: Desert ants keep track of stride number and length to calculate how far they are from the nest.
NULL HYPOTHESIS: Stride number and length have nothing to do with navigation (they use some other mechanism to navigate).
EXPERIMENTAL SETUP (TEST 1):
1. Ants walk from nest to feeder. 75 ants are collected.
Nest
Feeder
0
5
10 m
2. Manipulation of legs. Three treatments, 25 ants each.
Cut leg to create "stumps"
Leave legs normal length
Add pig bristles as "stilts"
3. Ants return "home" from feeder and look for nest hole.
Feeder
?
0
5
10
15
20 m
EXPERIMENTAL SETUP (TEST 2):
4. The three treatments of ants walk from nest to feeder again.
Nest
Feeder
0
5
10 m
5. Ants walk back "home" from feeder again.
Feeder
?
0
5
10
15
20 m
Stilts
Stilts make leg length and stride length longer
PREDICTION:
Ants with stilts will go too far; ants with stumps will stop short.
PREDICTION OF NULL HYPOTHESIS:
No differences the among 3 groups.
PREDICTION:
All three groups will start looking for nest after walking 10 m.
PREDICTION OF NULL HYPOTHESIS:
No difference from the result in Test 1.
RESULTS:
Stilts
Normal
Stumps
0
5
10
15
20
Homebound run (m)
Stilts
Normal
Stumps
0
5
10
15
20
Homebound run (m)
CONCLUSION: Desert ants use information on stride length and number to calculate how far they are from the nest.

FIGURE 1.8 An Experimental Test: Do Desert Ants Use a "Pedometer"?

SOURCE: Wittlinger, M., R. Wehner, and H. Wolf. 2006. The ant odometer: Stepping on stilts and stumps. *Science* 312: 1965–1967.

✔**QUESTION** How would you interpret the experiment if the researchers had used just one ant in each group instead of 25?

The graphs in the "Results" display "box-and-whisker" plots. Each box indicates the range of distances where 50 percent of the ants stopped to search for the nest; the whiskers indicate the range where 95 percent of the ants stopped to search. The vertical line inside each box indicates the median—meaning that half the ants stopped above this distance and half below. For more details on how biologists report averages and indicate the variability and uncertainty in data, see **BioSkills 5** in Appendix A.

INTERPRETING THE RESULTS The pedometer hypothesis predicts that an ant's ability to walk home depends on the number and length of steps taken on its outbound trip. Recall that a prediction specifies what we should observe if a hypothesis is correct. Good scientific hypotheses make testable predictions—predictions that can be supported or rejected by collecting and analyzing data. In this case, the researchers tested the prediction by altering stride length and recording the distance traveled on the return trip.

If the pedometer hypothesis is wrong, however, then stride length and step number should have no effect on the ability of an ant to get back to its nest. This latter possibility is called a **null hypothesis**. A null hypothesis specifies what we should observe when the hypothesis being tested isn't correct. Under the null hypothesis in this experiment, all the ants should have walked 10 m in the first test before they started looking for their nest.

IMPORTANT CHARACTERISTICS OF GOOD EXPERIMENTAL DESIGN In relation to designing effective experiments, this study illustrates several important points:

- It is critical to include **control** groups. A control checks for factors, other than the one being tested, that might influence the experiment's outcome. In this case, there were two controls. Including a normal, unmanipulated individual controlled for the possibility that switching the individuals to a new channel altered their behavior. In addition, the researchers had to control for the possibility that the manipulation itself—and not the change in leg length—affected the behavior of the stilts and stumps ants. This is why they did the second test, where the outbound and return runs were done with the same legs.
- The experimental conditions must be as constant or equivalent as possible. The investigators used ants of the same species, from the same nest, at the same time of day, under the same humidity and temperature conditions, at the same feeders, in the same channels. Controlling all the variables except one—leg length in this case—is crucial because it eliminates alternative explanations for the results.
- Repeating the test is essential. It is almost universally true that larger sample sizes in experiments are better. By testing many individuals, the amount of distortion or "noise" in the data caused by unusual individuals or circumstances is reduced.

✔You should be able to explain: (1) What issue would arise if in the first test, the normal individual had not walked 10 m on the return trip before looking for the nest; and (2) What you would conclude if the stilts and stumps ants had not navigated normally during the second test.

From the outcomes of these experiments, the researchers concluded that desert ants use stride length and number to measure how far they are from the nest. They interpreted their results as strong support for the pedometer hypothesis.

Biologists practice evidence-based decision-making. They ask questions about how organisms work, pose hypotheses to answer those questions, and use experimental or observational evidence to decide which hypotheses are correct.

To review the principles of experimental design, go to the study area at *www.masteringbiology.com.*

(MB) **Web Activity** Introduction to Experimental Design

The data on giraffes and ants are a taste of things to come. In this text you will encounter hypotheses and experiments on questions ranging from how water gets to the top of 100-meter-tall sequoia trees to why the bacterium that causes tuberculosis has become resistant to antibiotics. As you work through this book, you'll get lots of practice thinking about hypotheses and predictions, analyzing the nature of control treatments, and interpreting graphs.

A commitment to tough-minded hypothesis testing and sound experimental design is a hallmark of biological science. Understanding their value is an important first step in becoming a biologist.

CHECK YOUR UNDERSTANDING

If you understand that . . .

- Hypotheses are proposed explanations that make testable predictions.
- Predictions are observable outcomes of particular conditions.
- Well-designed experiments alter just one condition—a condition relevant to the hypothesis being tested.

✔ **You should be able to . . .**

Design an experiment to test the hypothesis that desert ants feed during the hottest part of the day because it allows them to avoid being eaten by lizards. Then answer the following questions about your experimental design:

1. How does the presence of a control group in your experiment allow you to test the null hypothesis?
2. How are experimental conditions controlled or standardized in a way that precludes alternative explanations of the data?

Answers are available in Appendix B.

CHAPTER 1 REVIEW

For media, go to the study area at www.masteringbiology.com

Summary of Key Concepts

Organisms obtain and use energy, are made up of cells, process information, replicate, and as populations evolve.

- There is no single, well-accepted definition of life. Instead, biologists point to five characteristics that organisms share.

✔You should be able to explain why the cells in a dead organism are different from the cells in a live organism.

The cell theory proposes that all organisms are made of cells and that all cells come from preexisting cells.

- The cell theory identified the fundamental structural unit common to all life.

✔You should be able to describe the evidence that supported the pattern and the process components of the cell theory.

The theory of evolution by natural selection maintains that species change through time because individuals with certain heritable traits produce more offspring than other individuals do.

- The theory of evolution states that all organisms are related by common ancestry.
- Natural selection is a well-tested explanation for why species change through time and why they are so well adapted to their habitats.

✔You should be able to explain why the average protein content of seeds in a natural population of a grass species would increase over time, if seeds with higher protein content survive better and grow into individuals that produce many seeds with high protein content when they mature.

MB **Web Activity** Artificial Selection

A phylogenetic tree is a graphical representation of the evolutionary relationships between species. These relationships can be estimated by analyzing similarities and differences in traits. Species that share distinctive traits are closely related and are placed close to each other on the tree of life.

- The cell theory and the theory of evolution predict that all organisms are part of a genealogy of species, and that all species trace their ancestry back to a single common ancestor.
- To reconstruct this phylogeny, biologists have analyzed the sequence of components in rRNA and other molecules found in all cells.
- A tree of life, based on similarities and differences in these molecules has three major lineages: the Bacteria, Archaea, and Eukarya.

✔You should be able to explain how biologists can determine whether newly discovered species are members of the Bacteria, Archaea, or Eukarya by analyzing their rRNA or other molecules.

Biologists ask questions, generate hypotheses to answer them, and design experiments that test the predictions made by competing hypotheses.

- Biology is a hypothesis-driven, experimental science.

✔You should be able to explain (1) the relationship between a hypothesis and a prediction and (2) why experiments are convincing ways to test predictions.

MB **Web Activity** Introduction to Experimental Design

Questions

✔TEST YOUR KNOWLEDGE

Answers are available in Appendix B

1. Anton van Leeuwenhoek made an important contribution to the development of the cell theory. How?
   - **a.** He articulated the pattern component of the theory—that all organisms are made of cells.
   - **b.** He articulated the process component of the theory—that all cells come from preexisting cells.
   - **c.** He invented the first microscope and saw the first cell.
   - **d.** He invented more powerful microscopes and was the first to describe the diversity of cells.
2. What does it mean to say that experimental conditions are controlled?
   - **a.** The test groups consist of the same individuals.
   - **b.** The null hypothesis is correct.
   - **c.** There is no difference in outcome between the control and experimental treatment.
   - **d.** Physical conditions are identical for all groups tested.
3. What does the term *evolution* mean?
   - **a.** The strongest individuals produce the most offspring.
   - **b.** The characteristics of an individual change through the course of its life, in response to natural selection.
   - **c.** The characteristics of populations change through time.
   - **d.** The characteristics of species become more complex over time.
4. What does it mean to say that a characteristic of an organism is heritable?
   - **a.** The characteristic evolves.
   - **b.** The characteristic can be passed on to offspring.
   - **c.** The characteristic is advantageous to the organism.
   - **d.** The characteristic does not vary in the population.
5. In biology, to what does the term *fitness* refer?
   - **a.** The degree of training and muscle mass an individual has, relative to others in the same population
   - **b.** An individual's slimness, relative to others in the same population
   - **c.** The longevity of a particular individual
   - **d.** An individual's ability to survive and reproduce
6. Could *both* the food competition hypothesis and the sexual selection hypothesis explain why giraffes have long necks? Why or why not?
   - **a.** No. In science, only one hypothesis can be correct.
   - **b.** No. Observations have shown that the food competition hypothesis cannot be correct.
   - **c.** Yes. Long necks could be advantageous for more than one reason.
   - **d.** Yes. All giraffes have been shown to feed at the highest possible height and fight for mates.

✔ TEST YOUR UNDERSTANDING

Answers are available in Appendix B

1. What would researchers have to demonstrate to convince you that they had discovered life on another planet?
2. It was once thought that the deepest split between life-forms was between two groups: prokaryotes and eukaryotes. Draw and label a phylogenetic tree that represents this hypothesis. Then draw and label a phylogenetic tree that shows the actual relationships between the three domains of organisms.
3. Why was it important for Linnaeus to establish the rule that only one type of organism can have a particular genus and species name?
4. What does it mean to say that a species is adapted to a particular habitat?
5. Explain how selection occurs during natural selection. What is selected, and why?
6. The following two statements explain the logic behind the use of molecular sequence data to estimate evolutionary relationships:

   "If the theory of evolution is true, then rRNA sequences should be very similar in closely related organisms but less similar in organisms that are less closely related."

   "On a phylogenetic tree, branches that share a recent common ancestor represent species that are closely related; branches that don't share recent common ancestors represent species that are more distantly related."

   Is the logic of these statements sound? Why or why not?

✔ APPLYING CONCEPTS TO NEW SITUATIONS

Answers are available in Appendix B

1. A scientific theory is a set of propositions that defines and explains some aspect of the world. This definition contrasts sharply with the everyday usage of the word theory, which often carries meanings such as "speculation" or "guess." Explain the difference between the two definitions, using the cell theory and the theory of evolution by natural selection as examples.
2. Turn back to the tree of life shown in Figure 1.5. Note that Bacteria and Archaea are prokaryotes, while Eukarya are eukaryotes. On the simplified tree below, draw an arrow that points to the branch where the structure called the nucleus originated. Explain your reasoning.

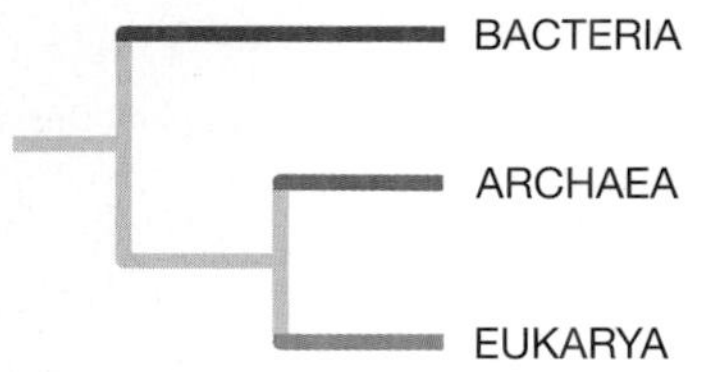

3. The proponents of the cell theory could not "prove" that it was correct in the sense of providing incontrovertible evidence that all organisms are made up of cells. They could state only that all organisms examined to date were made of cells. Why was it reasonable for them to conclude that the theory was valid?
4. Some humans have heritable traits that make them resistant to infection by HIV. In areas of the world where HIV infection rates are high, are human populations evolving? Explain your logic.

Natural selection acts on individuals in populations such as these sea stars, but only populations evolve. One of Darwin's greatest contributions to science was the introduction of population thinking to the theory of evolution.

24 Evolution by Natural Selection

KEY CONCEPTS

- Populations and species evolve, meaning that their heritable characteristics change through time. Evolution is change in allele frequencies over time.
- Evolution by natural selection occurs when individuals with certain alleles produce the most surviving offspring in a population. An adaptation is a genetically based trait that increases an individual's ability to produce offspring in a particular environment.
- Evolution by natural selection is not progressive, and it does not change the characteristics of the individuals that are selected—it changes only the characteristics of the population. Animals do not do things for the good of the species, and not all traits are adaptive. All adaptations are constrained by trade-offs and genetic and historical factors.

This chapter is about one of the great ideas in science. The theory of evolution by natural selection, formulated independently by Charles Darwin and Alfred Russel Wallace, explains how organisms have come to be adapted to environments ranging from arctic tundra to tropical wet forest. It revealed one of the five attributes of life introduced in Chapter 1: Populations of organisms evolve—meaning that they change through time.

As an example of a revolutionary breakthrough in our understanding of the world, the theory of evolution by natural selection ranks alongside Copernicus's theory of the Sun as the center of our solar system, Newton's laws of motion and theory of gravitation, the germ theory of disease, the theory of plate tectonics, and Einstein's general theory of relativity. These theories are the foundation stones of modern science; all are accepted on the basis of overwhelming evidence.

Evolution by natural selection is one of the best-supported and most important theories in the history of scientific research. But like most scientific breakthroughs, this one did not come easily. When Darwin published his theory in 1859 in a book called *On the Origin of Species by Means of Natural Selection*, it unleashed a firestorm of protest throughout Europe. At that time, the leading explanation for the diversity of organisms was a theory called special creation.

To understand the contrast between the theory of special creation and the theory of evolution by natural selection more thoroughly, recall from Chapter 1 that scientific theories usually have two components: a pattern and a process.

- The pattern component is a statement that summarizes a series of observations about the natural world. The pattern component is about facts—about how things *are* in nature.
- The process component is a mechanism that produces that pattern or set of observations.

✔ When you see this checkmark, stop and test yourself. Answers are available in Appendix B.

The pattern component of the theory of special creation held that: (1) All species are independent, in the sense of being unrelated to each other; (2) life on Earth is young—perhaps just 6000 years old; and (3) species are immutable, or incapable of change. The process that explained this pattern was the instantaneous and independent creation of living organisms by a supernatural being.

Darwin's ideas were radically different. What are the pattern and process components of the theory of evolution by natural selection?

24.1 The Evolution of Evolutionary Thought

People often use the word revolutionary to describe the theory of evolution by natural selection. Revolutions overturn things—they replace an existing entity with something new and often radically different. A political revolution removes the ruling class or group and replaces it with another. The industrial revolution replaced small shops for manufacturing goods by hand with huge, mechanized assembly lines.

A scientific revolution, in contrast, overturns an existing idea about how nature works and replaces it with another, radically different, idea. The idea that Darwin and Wallace overturned—that species were specially, not naturally, created—had dominated thinking about the nature of organisms for over 2000 years.

Plato and Typological Thinking

The Greek philosopher Plato claimed that every organism was an example of a perfect essence, or type, created by God, and that these types were unchanging. Plato acknowledged that the individual organisms present on Earth might deviate slightly from the perfect type, but he said this deviation was similar to seeing the perfect type in a shadow on a wall. The key to understanding life, in Plato's mind, was to ignore the shadows and focus on understanding each type of unchanging, perfect essence.

Today, philosophers and biologists refer to ideas like this as typological thinking. Typological thinking is based on the idea that species are unchanging types and that variations within species are unimportant or even misleading. Typological thinking also occurs in the Bible's book of Genesis, where God creates one of each type of organism.

Aristotle and the Great Chain of Being

Not long after Plato developed his ideas, Aristotle ordered the types of organisms known at the time into a linear scheme called the great chain of being, also called the scale of nature (**Figure 24.1**). Aristotle proposed that species were organized into a sequence based on increased size and complexity, with humans at the top. He also claimed that the characteristics of species were fixed—they did not change through time.

In the 1700s Aristotle's ideas were still popular in scientific and religious circles. The central claims were that (1) species are fixed types, and (2) some species are higher—in the sense of being more complex or "better"—than others.

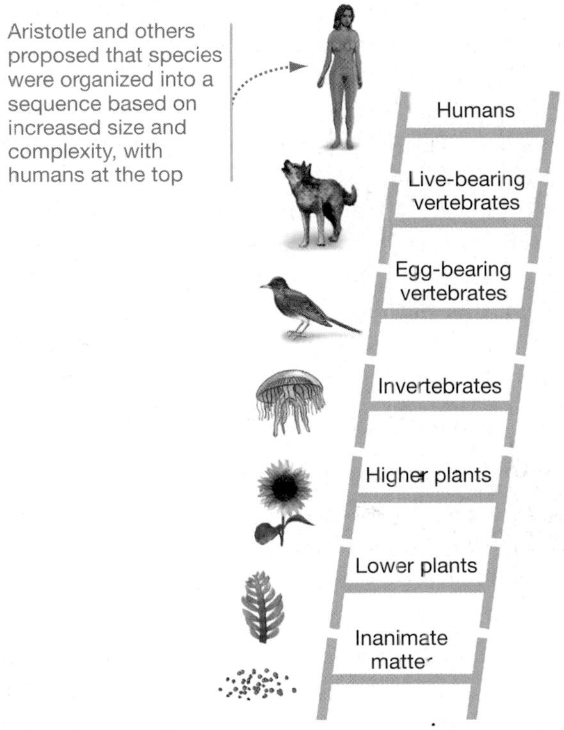

Aristotle and others proposed that species were organized into a sequence based on increased size and complexity, with humans at the top
Humans
Live-bearing vertebrates
Egg-bearing vertebrates
Invertebrates
Higher plants
Lower plants
Inanimate matter

FIGURE 24.1 The Great Chain of Being, or Scale of Nature

Lamarck and the Idea of Evolution as Change through Time

Typological thinking eventually began to break down, however. In 1809 the biologist Jean-Baptiste de Lamarck proposed a formal theory of evolution—that species are not static but change through time. The pattern component of Lamarck's theory was initially based on the great chain of being, however.

When he started his work on evolution, Lamarck claimed that simple organisms originate at the base of the chain by spontaneous generation (see Chapter 1) and then evolve by moving up the chain over time. Thus, Lamarckian evolution is progressive in the sense of always producing larger and more complex, or "better," species. To capture this point, biologists often say that Lamarck turned the ladder of life into an escalator.

Lamarck also contended that species change through time via the inheritance of acquired characters. The idea here is that an individual's phenotype changes as they develop in response to challenges posed by the environment, and they pass on these phenotypic changes to offspring. A classic Lamarckian scenario is that giraffes develop long necks as they stretch to reach leaves high in treetops, and they then produce offspring with elongated necks.

Darwin and Wallace and Evolution by Natural Selection

As his thinking matured, Lamarck eventually abandoned his linear and progressive view of life. Darwin and Wallace concurred. What is more important, they emphasized that the process responsible for change through time—evolution—occurs because traits vary among the individuals in a population, and because individuals with certain traits leave more offspring than others do. A **population** consists of individuals of the same species that are living in the same area at the same time.

Darwin and Wallace's proposal was a radical break from the typological thinking that had dominated scientific thought since Plato. Darwin claimed that instead of being unimportant or an illusion, variation among individuals in a population was the key to understanding the nature of species. Biologists refer to this view as **population thinking**.

The theory of evolution by natural selection was revolutionary for several reasons:

1. It overturned the idea that species are static and unchanging.
2. It replaced typological thinking with population thinking.
3. It was scientific. It proposed a mechanism that could account for change through time and made predictions that could be tested through observation and experimentation.

Plato and his followers emphasized the existence of fixed types; evolution by natural selection is all about change and diversity.

Now the questions are: What evidence backs the claim that species are not fixed types? What data convince biologists that the theory of evolution by natural selection is correct?

24.2 The Pattern of Evolution: Have Species Changed through Time?

In *On the Origin of Species*, Darwin repeatedly described evolution as **descent with modification**. He meant that species that lived in the past are the ancestors of the species existing today, and that species and their descendant species change through time. Descendant species are modified.

This view was a radical departure from the pattern of independently created and immutable species embodied in Plato's work and in the theory of special creation. In essence, the pattern component of the theory of evolution by natural selection makes two statements about the nature of species:

1. Species change through time.
2. Species are related by common ancestry.

Let's consider the evidence for each of these claims in turn.

Evidence for Change through Time

When Darwin began his work, biologists and geologists had just begun to assemble and interpret the fossil record. A **fossil** is any trace of an organism that lived in the past. These traces range from bones and branches to shells, tracks or impressions, and dung (**Figure 24.2**). The **fossil record** consists of all the fossils that have been found and described in the scientific literature.

Why did data in the fossil record support the hypothesis that species have changed through time? And what data from **extant species**—those living today—support the claim that they are modified forms of ancestral species?

THE VASTNESS OF GEOLOGIC TIME Initially, fossils were organized according to their relative ages. This was possible because geologists had created the **geologic time scale**: a sequence of named intervals called eons, eras, and periods that represented the major events in Earth history (see Chapter 27).

The geologic time scale, in turn, was based on a series of principles derived from observations about rock formation. **Sedimentary rocks**, for example, form from sand or mud or other materials deposited at locations such as beaches or river mouths. Sedimentary rocks, along with rocks derived from volcanic ash or lava, are known to form in layers—with younger layers deposited on top of older layers. In a similar vein, rock deposits that contained boulders were inferred to be younger than the rocks that the boulder material originated from.

As the geologic record was being established, then, researchers placed fossils in a younger-to-older sequence, based on their relative position in layers of sedimentary rock. They also realized that vast amounts of time were required to form the thick layers of sedimentary rock that they were studying, because erosion and deposition of sediments are such slow processes.

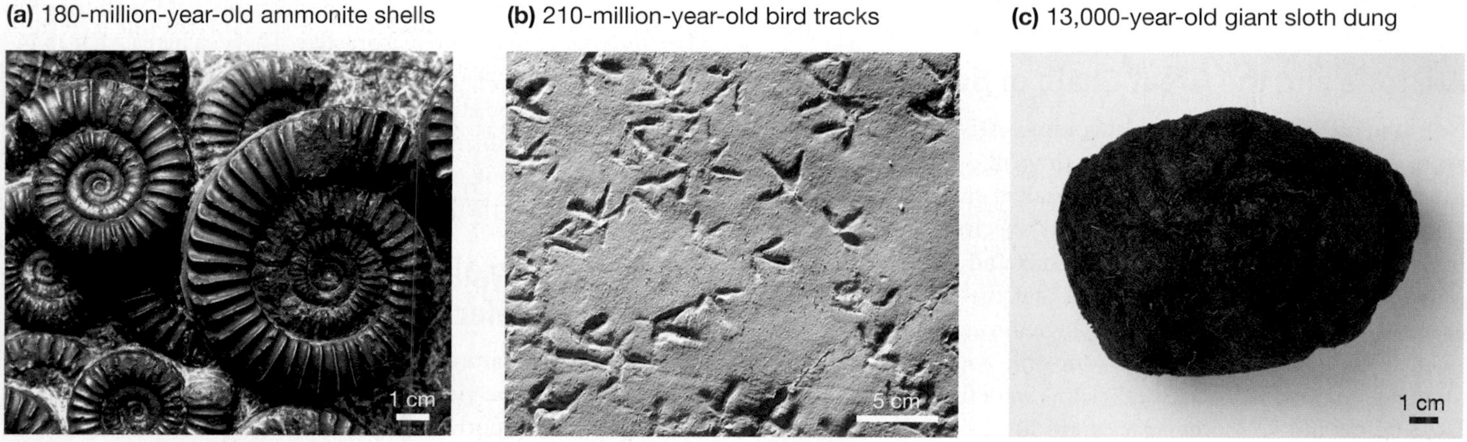

(a) 180-million-year-old ammonite shells
(b) 210-million-year-old bird tracks
(c) 13,000-year-old giant sloth dung
1 cm
5 cm
1 cm

FIGURE 24.2 A Fossil Is *Any* Trace of an Organism That Lived in the Past. In addition to **(a)** body parts such as shells or bones or branches, fossils may consist of **(b)** tracks or impressions, or even **(c)** pieces of dung.

This was an important insight. The geologic record indicated that the Earth was much, much older than the 6000 years claimed by proponents of the theory of special creation.

After the discovery of radioactivity in the late 1800s, researchers realized that radioactive decay—the steady rate at which unstable or "parent" atoms are converted into more stable "daughter" atoms—furnished a way to assign absolute ages, in years, to the relative ages in the geologic time scale.

Radiometric dating is based on three pieces of information:

1. Observed decay rates of parent to daughter atoms.
2. The ratio of parent to daughter atoms present in newly formed rocks—such as the amount of uranium atoms versus lead atoms when uranium-containing molten rock first cools. (Uranium decays to form lead.)
3. The ratio of parent to daughter atoms present in a particular rock sample.

Combining information from these two ratios with information on the decay rate allows researchers to estimate how long ago a rock formed. According to data from radiometric dating, Earth is about 4.6 billion years old, and the earliest signs of life appear in rocks that formed 3.4–3.8 billion years ago.

Data from relative and absolute dating techniques agree: Life on Earth is ancient. There has been a great deal of time for change through time to occur.

EXTINCTION CHANGES THE SPECIES PRESENT OVER TIME In the early nineteenth century, researchers began discovering fossil bones, leaves, and shells that were unlike structures from any known animal or plant. At first, many scientists insisted that living examples of these species would be found in unexplored regions of the globe. But as research continued and the number and diversity of fossil collections grew, the argument became less and less plausible.

The issue was finally settled when Baron Georges Cuvier published a detailed analysis of an **extinct** species—that is, a species that no longer exists—called the Irish "elk" in 1812. Scientists accepted the fact of extinction because this gigantic deer was judged to be too large to have escaped discovery and too distinctive to be classified as a large-bodied population of an existing species (**Figure 24.3**).

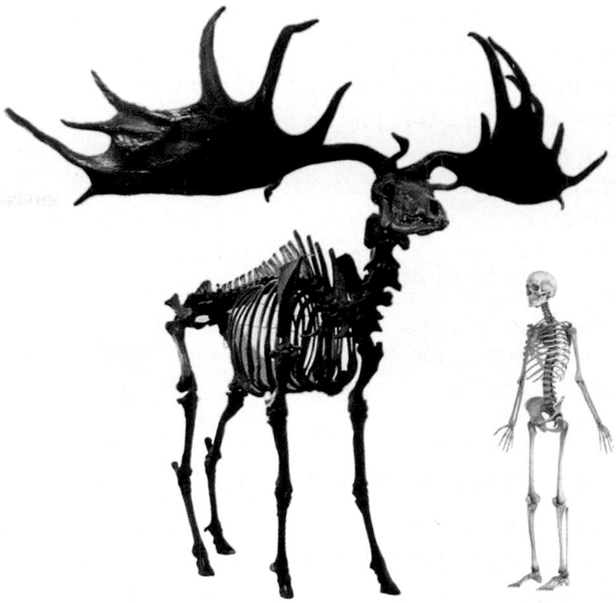

FIGURE 24.3 Evidence of Extinction. The skeleton of the Irish "elk" dwarfs a human. Scientists agreed that the deer was too large and unique to be overlooked if it were alive; it must have gone extinct.

Advocates of the theory of special creation argued that the fossil species were victims of the flood at the time of Noah. Darwin, in contrast, interpreted extinct forms as evidence that species are not static, immutable entities, unchanged since the moment of special creation. His reasoning was that if species have gone extinct, then the array of species living on Earth has changed through time.

Recent analyses of the fossil record suggest that over 99 percent of all the species that have ever lived are now extinct. The data also indicate that species have gone extinct continuously throughout Earth's history—not just in one or even a few catastrophic events.

TRANSITIONAL FEATURES LINK OLDER AND YOUNGER SPECIES Long before Darwin published his theory, researchers reported striking resemblances between the fossils found in the rocks underlying certain regions and the living species found in the same geographic areas. The pattern was so widespread that it became known as the "law of succession." The general observation was that extinct species in the fossil record were succeeded, in the same region, by similar species.

Early in the nineteenth century, the pattern was simply reported and not interpreted. But later, Darwin pointed out that it provided strong evidence in favor of the hypothesis that species had changed through time. His idea was that the extinct forms and living forms were related—that they represented ancestors and descendants.

As the fossil record improved, researchers discovered species with characteristics that broadened the scope of the law of succession. A **transitional feature** is a trait in a fossil species that is intermediate between those of older and younger species. For example, intensive work over the past several decades has yielded fossils that document a gradual change over time from aquatic animals that had fins to terrestrial animals that had limbs (**Figure 24.4** on page 418). Over a period of about 25 million years, the fins of species similar to today's lungfish changed into limbs similar to those found in today's amphibians, reptiles, and mammals—a group called the tetrapods (literally, "four-footed").

These observations support the hypothesis that an ancestral lungfish-like species began living on land, and that their descendants became more and more like today's tetrapods in appearance and lifestyle. Lungfish and tetrapod species have clearly changed through time.

Similar sequences of transitional features document changes that led to the evolution of feathers and flight in birds, stomata and vascular tissue in plants, upright posture, flattened faces, and large brains in humans, jaws in vertebrates (animals with

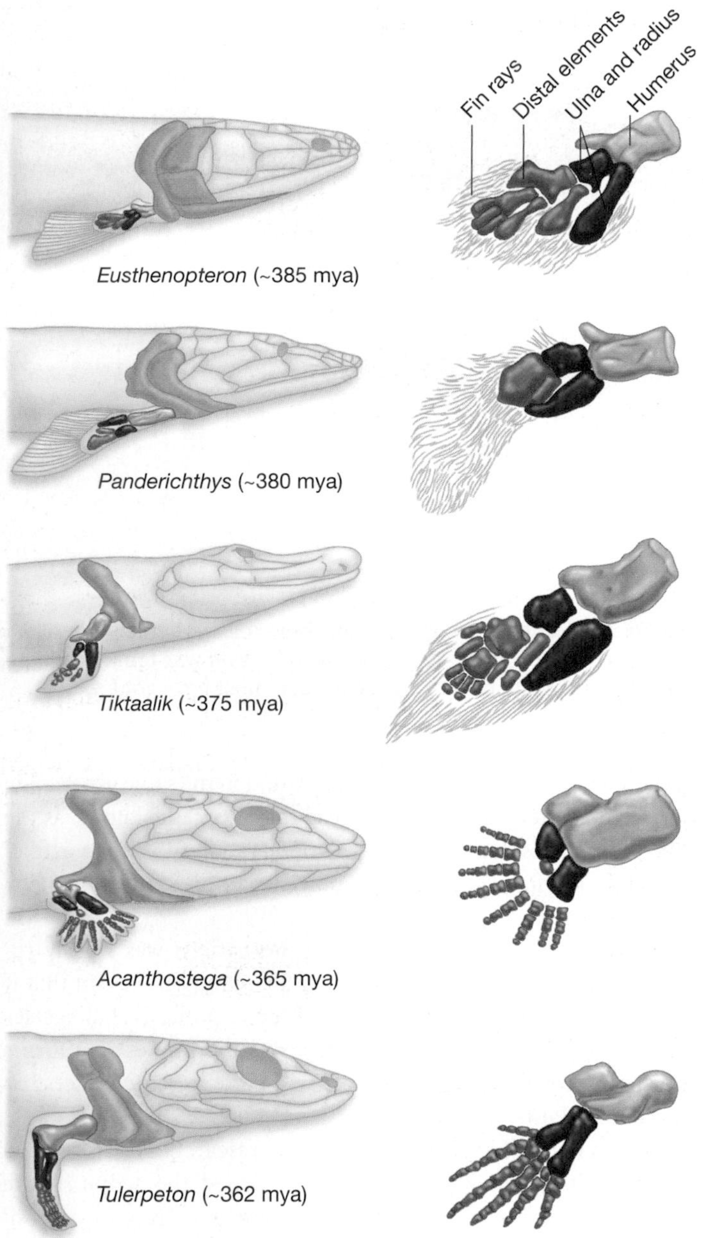

FIGURE 24.4 Transitional Features during the Evolution of the Tetrapod Limb. Fossil species similar to today's lungfish and tetrapods have fin and limb bones that are transitional features. *Eusthenopteron* was aquatic; *Tulerpeton* was probably semi-aquatic (mya = million years ago).

✔**QUESTION** How would observations of transitional features be explained under the theory of special creation?

backbones), the loss of limbs in snakes, and other traits. Data like these are consistent with predictions from the theory of evolution: If the traits observed in more recent species evolved from traits in more ancient species, then intermediate forms are expected to occur in the appropriate time sequence.

The fossil record provides compelling evidence that species have evolved. What data from extant forms supports the hypothesis that the characteristics of species change through time?

VESTIGIAL TRAITS ARE EVIDENCE OF CHANGE THROUGH TIME Darwin was the first to provide a widely accepted interpretation of vestigial traits. A **vestigial trait** is a reduced or incompletely developed structure that has no function or reduced function, but is clearly similar to functioning organs or structures in closely related species.

Biologists have documented thousands of examples of vestigial traits.

- The genomes of humans and other organisms contain hundreds of pseudogenes—the functionless DNA sequences introduced in Chapter 20.
- Bowhead whales and rubber boas have tiny hip and leg bones that do not help them swim or slither.
- Ostriches and kiwis have reduced wings and cannot fly.
- Blind cave-dwelling fish still have eye sockets.
- Even though marsupial mammals give birth to live young, an eggshell forms briefly early in their development; in some species, newborns have a non-functioning "egg tooth" similar to those seen in birds and reptiles.
- Monkeys and many other primates have long tails; but our coccyx, illustrated in **Figure 24.5a**, is too small to help us maintain balance or grab tree limbs for support.
- Many mammals, including primates, are able to erect their hair when they are cold or excited. But our sparse fur does little to keep us warm, and goose bumps are largely ineffective in signaling our emotional state (**Figure 24.5b**).

The existence of vestigial traits is inconsistent with the theory of special creation, which maintains that species were perfectly designed by a supernatural being and that the characteristics of species are static. Instead, vestigial traits are evidence that the characteristics of species have changed over time.

CURRENT EXAMPLES OF CHANGE THROUGH TIME Biologists have documented hundreds of contemporary populations that are changing in response to changes in their environment. Bacteria have evolved resistance to drugs; insects have evolved resistance to pesticides; weedy plants have evolved resistance to herbicides. Section 24.4 provides a detailed analysis of research on two examples of evolution in action.

To summarize, change through time continues and can be measured directly. Evidence from the fossil record and living species indicates that life is ancient, that species have changed through the course of Earth's history, and that species continue to change. The take-home message is that species are dynamic—not static, unchanging, and fixed types, as claimed by Plato, Aristotle, and the theory of special creation.

Evidence of Descent from a Common Ancestor

Data from the fossil record and contemporary species refute the hypothesis that species are immutable. What about the claim that species were created independently—meaning that they are unrelated to each other?

(a) The human tailbone is a vestigial trait.

(b) Goose bumps are a vestigial trait.

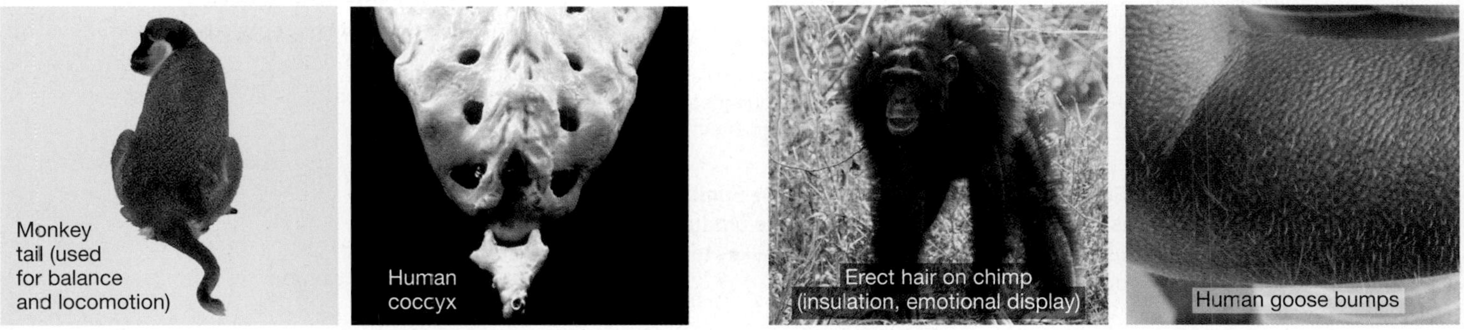

FIGURE 24.5 Vestigial Traits Are Reduced Versions of Traits in Other Species. **(a)** The tailbone and **(b)** goose bumps are human traits that have reduced function. They are similar to larger, fully functional structures in other species.

✔**QUESTION** How would observations of vestigial traits be explained if evolution occurred via inheritance of acquired characters?

SIMILAR SPECIES ARE FOUND IN THE SAME GEOGRAPHIC AREA Charles Darwin began to realize that species are related by common ancestry, just as individuals within a family are, during a five-year voyage he took aboard the English naval ship HMS *Beagle.* While fulfilling its mission to explore and map the coast of South America, the *Beagle* spent considerable time in the Galápagos Islands off the coast of present-day Ecuador. Darwin had taken over the role of ship's naturalist and as the first scientist to study the area, gathered extensive collections of the plants and animals found in these islands. Among the birds he collected were what came to be known as the Galápagos mockingbirds, pictured in **Figure 24.6a**.

Several years after Darwin returned to England, a biologist pointed out that the mockingbirds collected on different islands

(a) Pattern: Although the Galápagos mockingbirds are extremely similar, distinct species are found on different islands.

(b) Recent data support Darwin's hypothesis that the Galápagos mockingbirds share a common ancestor.

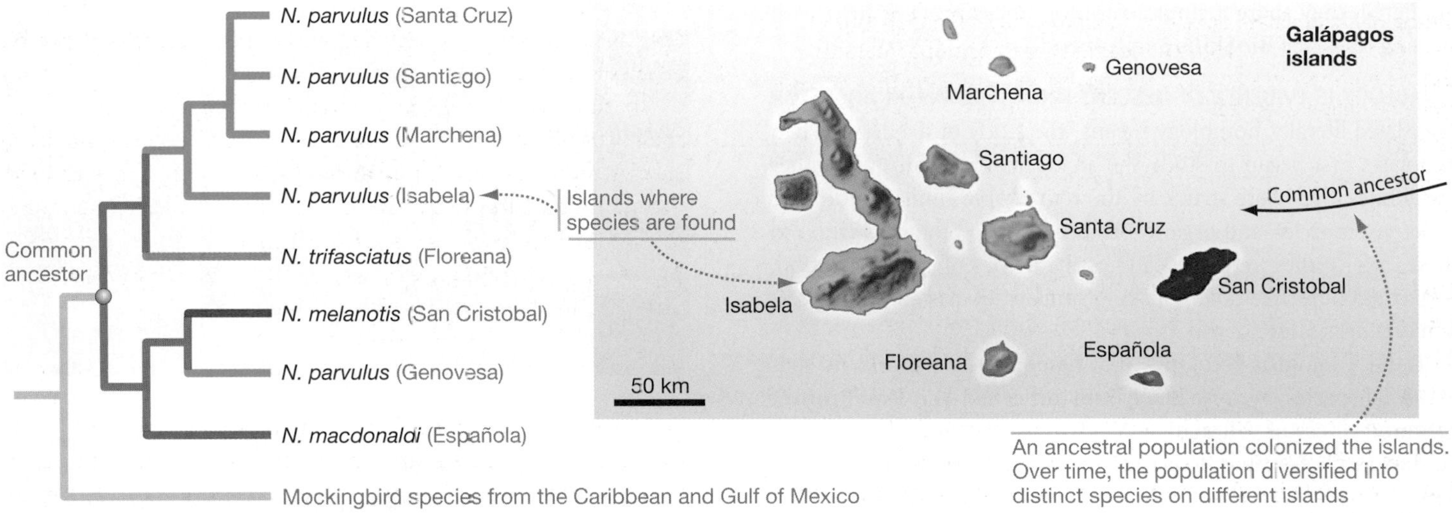

FIGURE 24.6 Close Relationships among Island Forms Argue for Shared Ancestry.

Gene:	Amino acid sequence (single-letter abbreviations):
Aniridia (Human)	LQRNRTSFTQEQIEALEKEFERTHYPDVFARERLAAKIDLPEARIQVWFSNRRAKWRREE
eyeless (Fruit fly)	LQRNRTSFTNDQIDSLEKEFERTHYPDVFARERLAGKIGLPEARIQVWFSNRRAKWRREE

Only six of the 60 amino acids in these sequences are different. The two sequences are 90% identical

FIGURE 24.7 Genetic Homology: Genes from Different Species May Be Similar in DNA Sequence or Other Attributes. Amino acid sequences from a portion of the *Aniridia* gene product found in humans are almost identical to those found in the *Drosophila eyeless* gene product. For a key to the single-letter abbreviations used for the amino acids, see Figure 3.3.

were distinct species, based on differences in coloration and beak size and shape. This struck Darwin as remarkable. Why would species that inhabit neighboring islands be so similar, yet clearly distinct? This turns out to be a widespread pattern: In island groups across the globe, it is routine to find similar but distinct species on neighboring islands.

Darwin realized that this pattern—puzzling when examined as a product of special creation—made perfect sense when interpreted in the context of evolution, or descent with modification. He proposed that the mockingbirds were similar because they had descended from the same common ancestor. Instead of being created independently, he proposed that mockingbird populations that colonized different islands had changed through time and formed new species (**Figure 24.6b**).

Recent analyses of DNA sequences in these mockingbirds support Darwin's hypothesis. Specifically, the molecular data are consistent with the prediction that the mockingbirds are part of a **phylogeny**—a family tree of populations or species.

Using techniques that will be introduced in Chapter 27, researchers have placed the mockingbirds on a phylogenetic tree—a branching diagram that describes the ancestor–descendant relationships among species. A phylogenetic tree is similar to a genealogy describing the ancestor–descendant relationships among individual humans. As Figure 24.6b shows, the Galápagos mockingbirds are each others' closest living relatives. As Darwin predicted, they share a single common ancestor. (For help with reading trees, see **BioSkills 3** in Appendix A.)

HOMOLOGY IS EVIDENCE OF DESCENT FROM A COMMON ANCESTOR Translated literally, homology means "the study of likeness." When biologists first began to study the anatomy of humans and other vertebrates, they were struck by the remarkable similarity of their skeletons, muscles, and organs. But because the biologists who did these early studies were advocates of the theory of special creation, they could not explain why striking similarities existed among certain organisms but not others.

Today, biologists recognize that **homology** is a similarity that exists in species because they both inherited the trait from a common ancestor. Human hair and dog fur are homologous. Humans have hair and dogs have hair because they share a common ancestor—an early mammal species—that had hair.

Homology can be recognized and studied at three levels:

- **Genetic homology** occurs in DNA sequences. For example, the *eyeless* gene in fruit flies and the *Aniridia* gene in humans are so similar that their protein products are 90 percent identical in amino acid sequence (**Figure 24.7**). Both genes act in determining where eyes will develop—even though fruit flies have a compound eye with many lenses and humans have a camera eye with a single lens.
- **Developmental homology** is recognized in embryos. For example, early chick, human, and cat embryos have tails and structures called gill pouches (**Figure 24.8**). Later, gill pouches are lost in all three species and tails are lost in humans. But in fish, the embryonic gill pouches stay intact and give rise to functioning gills in adults. To explain this observation, biologists hypothesize that gill pouches and tails exist in chicks, humans, and cats because they existed in the fishlike species that was the common ancestor of today's vertebrates. Embryonic gill pouches are a vestigial trait in chicks, humans, and cats; embryonic tails are a vestigial trait in humans.
- **Structural homology** is a similarity in adult **morphology**, or form. A classic example is the common structural plan observed in the limbs of vertebrates (**Figure 24.9**). In Darwin's

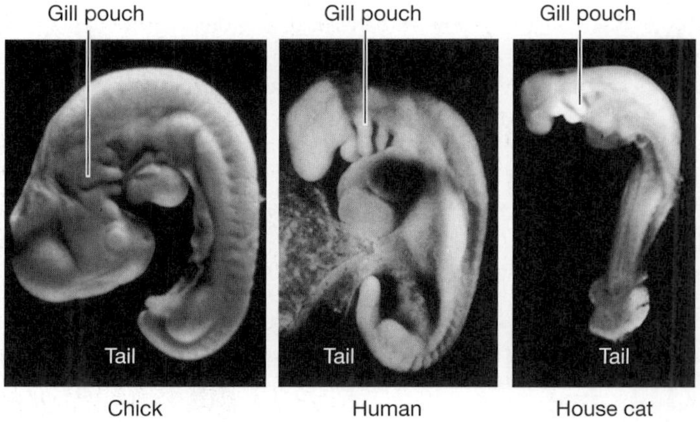

FIGURE 24.8 Developmental Homology: Structures That Appear Early in Development Are Similar. The early embryonic stages of a chick, a human, and a cat, showing a strong resemblance.

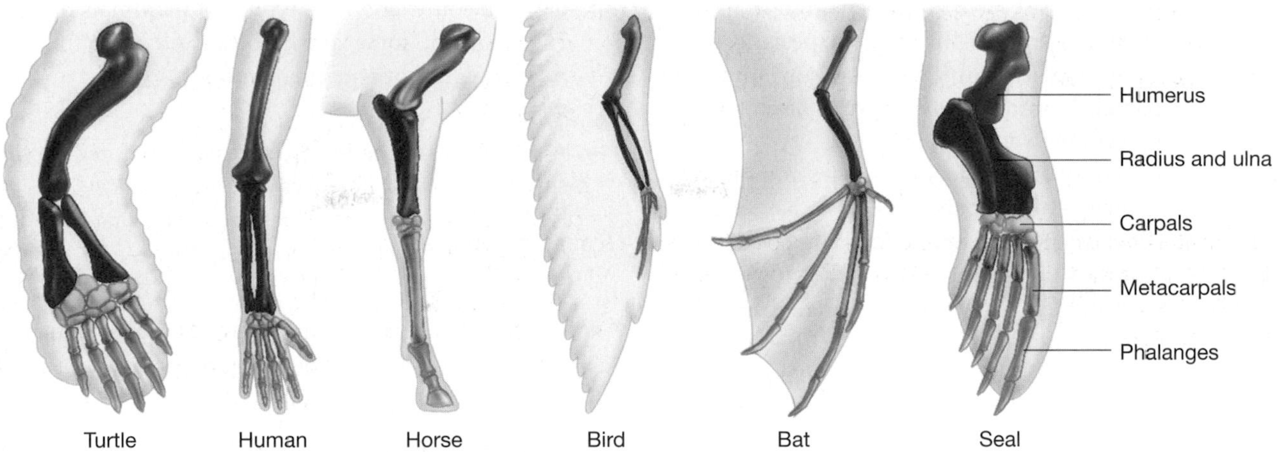

FIGURE 24.9 Structural Homology: Limbs with Different Functions Have the Same Underlying Structure. Even though their function varies, all vertebrate limbs are modifications of the same number and arrangement of bones. Darwin interpreted structural homologies like these as a product of descent with modification. (These limbs are not drawn to scale.)

own words, "What could be more curious than that the hand of a man, formed for grasping, that of a mole for digging, the leg of the horse, the paddle of the porpoise, and the wing of the bat, should all be constructed on the same pattern, and should include the same bones, in the same relative positions?" An engineer would never use the same underlying structure to design a grasping tool, a digging implement, a walking device, a propeller, and a wing. Instead, the structural homology exists because mammals evolved from the lungfish-like ancestor in Figure 24.4, which had the same general arrangement of bones in its fins.

The three levels of homology interact. Genetic homologies cause the developmental homologies observed in embryos, which then lead to the structural homologies recognized in adults. Perhaps the most fundamental of all homologies is the genetic code. With a few minor exceptions, all organisms use the same rules for transferring the information coded in DNA into proteins (see Chapter 15).

In some cases, hypotheses about homology can be tested experimentally. For example, researchers **(1)** isolated genes from mice or squid that were thought to be homologous to the fruit fly *eyeless* gene, **(2)** inserted the mouse or squid gene into fruit fly embryos, **(3)** stimulated expression of the foreign gene in locations that normally give rise to appendages, and **(4)** observed formation of eyes on legs and antennae (**Figure 24.10**). The inserted genes' function was identical to the function of *eyeless*. This result was strong evidence that the fruit fly, mouse, and squid genes are homologous, as predicted from their sequence similarity.

Homology is a key concept in contemporary biology:

- Chemicals that are cancer-causing in humans can be identified by testing their effects on mutation rates in bacteria, because the molecular machinery responsible for copying and repairing DNA is homologous in all organisms (see Chapter 14).

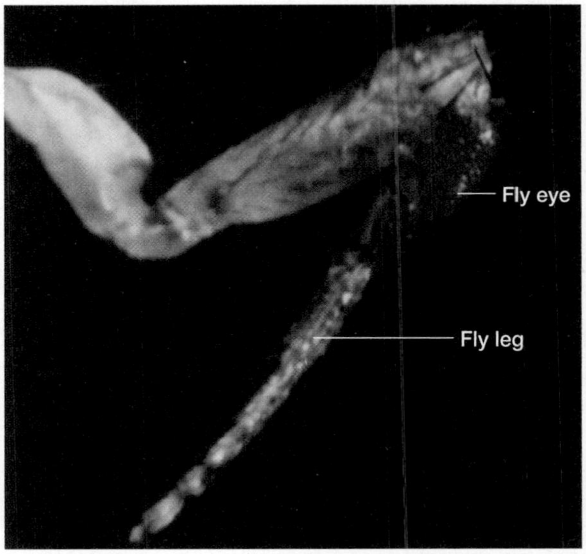

FIGURE 24.10 Evidence for Homology: A Mouse Gene Expressed in Fruit Flies. As an embryo, this fruit fly received a mouse gene that signals where eyes should form. A fruit fly eye formed in the location where the mouse gene was expressed.

- Drugs intended for human use can be tested on mice or rabbits if the molecules targeted by the drugs are homologous.
- Unknown sequences in the human, rice, or other genomes can be identified if they are homologous to known sequences in yeast, fruit flies, *Arabidopsis thaliana*, or other well-studied model organisms (see Chapter 20 and **BioSkills 14** in Appendix A).

The theory of evolution by natural selection predicts that homologies will occur. If species were created independently of one another, as the theory of special creation claims, these types of similarities would not occur.

CURRENT EXAMPLES Biologists have documented dozens of contemporary populations that are undergoing speciation—a process that results in one species splitting into two or more descendant species. Chapter 26 introduces what may be the best-studied example of speciation in action.

In most cases, the identity of the ancestral species and the descendant species is known—meaning that biologists have established a direct link between ancestral and descendant species. In addition, the reason for the splitting event is usually known.

The contemporary examples of new species being formed are powerful evidence that species living today are the descendants of species that lived in the past. They support the claim that all organisms are related by descent from a common ancestor.

Evolution's "Internal Consistency"—the Importance of Independent Datasets

Biologists draw upon data from several sources to challenge the hypothesis that species are immutable and were created independently. The data support the idea that species have descended, with modification, from a common ancestor. **Table 24.1** summarizes this evidence.

Perhaps the most powerful evidence for any scientific theory, including evolution by natural selection, is what scientists call "internal consistency." This is the observation that data from independent sources agree in supporting predictions made by a theory.

As an example, consider the evolution of whales and dolphins—a group called the cetaceans.

- The fossil record contains a series of species that are clearly identified as cetaceans, on the basis of unusual ear bones found only in this group. Some of the species have the long legs and compact bodies typical of mammals that live primarily on land; some are limbless and have the streamlined bodies typical of aquatic mammals; some have intermediate features.
- A phylogeny of the fossil cetaceans, estimated on the basis of similarities and differences in morphological traits other than limbs and overall body shape, indicates that a gradual transition occurred between terrestrial forms and aquatic, whale-like forms (**Figure 24.11**).
- Relative dating, based on the positions of sedimentary rocks where the fossils were found, agrees with the order of species indicated in the phylogeny.
- Absolute dating, based on analyses of radioactive atoms in rocks in or near the layers where the fossils were found, also agrees with the order of species indicated in the phylogeny.
- A phylogeny of living whales and dolphins, estimated from similarities and differences in DNA sequences, indicates that hippos—which spend much of their time in shallow water—are the closest living relative of cetaceans. This observation supports the hypothesis that whales and hippos shared a common ancestor that was semi-aquatic.
- Some whales have vestigial hip and limb bones as adults, and some dolphin embryos have vestigial hindlimb buds—outgrowths where legs form in other mammals.

The general message here is that many independent lines of evidence converge on the same conclusion: Whales gradually evolved from a terrestrial ancestor, over the span of about 12 million years.

As you evaluate the evidence supporting the pattern component of the theory of evolution, though, it's important to recognize that no single observation or experiment instantly "proved" the fact of evolution and swept aside belief in special creation. Rather, data from many different sources are much more consistent with evolution than with special creation. Descent with modification is a more successful and powerful scientific theory because it explains observations—such as vestigial traits and the close relationships among species on neighboring islands—that special creation does not.

What about the process component of the theory of evolution by natural selection? If the limbs of bats and humans were not created independently and recently, how did they come to be?

TABLE 24.1 **Evidence for Evolution**

Prediction 1: Species Are Not Static, but Change through Time
• Most species have gone extinct.
• Fossil (extinct) species frequently resemble living species found in the same area.
• Transitional features document change in traits through time.
• Vestigial traits are common.
• The characteristics of populations can be observed changing today.
Prediction 2: Species Are Related, Not Independent
• Closely related species often live in the same geographic area.
• Homologous traits are common and are recognized at three levels:
1. genetic (gene structure and the genetic code)
2. developmental (embryonic structures and processes)
3. structural (morphological traits in adults)
• The formation of new species, from preexisting species, can be observed today.

24.3 The Process of Evolution: How Does Natural Selection Work?

Darwin's greatest contribution did not lie in recognizing the fact of evolution. Lamarck and other researchers had already proposed evolution as a pattern in nature long before Darwin began his work. Instead, Darwin's crucial insight lay in recognizing a process, called **natural selection**, that could explain the pattern of descent with modification.

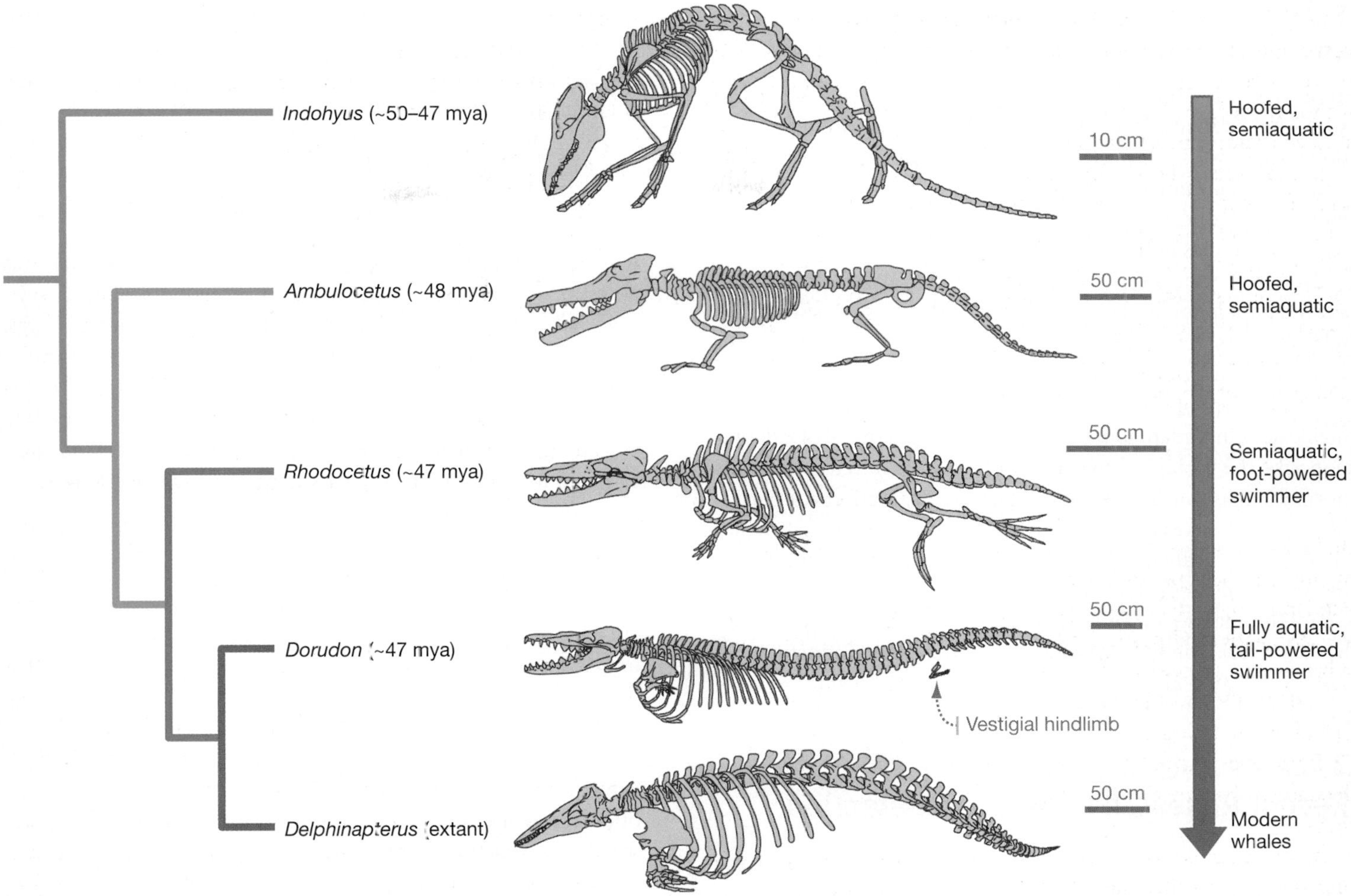

FIGURE 24.11 Data on Evolution from Independent Sources Are Consistent. This phylogeny of fossil cetaceans is consistent with data from relative dating, absolute dating, and phylogenies estimated from molecular traits in living species—all agree that whales evolved from terrestrial ancestors that were related to today's hippos.

Darwin's Four Postulates

In his original formulation, Darwin broke the process of evolution by natural selection into four simple postulates—meaning, steps in a logical sequence:

1. The individual organisms that make up a population vary in the traits they possess, such as their size and shape.
2. Some of the trait differences are heritable, meaning that they are passed on to offspring genetically. For example, tall parents may tend to have tall offspring.
3. In each generation, many more offspring are produced than can possibly survive. Thus, only some individuals in the population survive long enough to produce offspring; and among the individuals that produce offspring, some will produce more than others.
4. The subset of individuals that survive best and produce the most offspring is not a random sample of the population. Instead, individuals with certain heritable traits are more likely to survive and reproduce. Natural selection occurs when individuals with certain characteristics produce more offspring than do individuals without those characteristics. The individuals are selected naturally—meaning, by the environment.

Because the selected traits are passed on to offspring, the frequency of the selected traits increases from one generation to the next. Evolution is simply the outcome of these four steps. Each of the steps, and evolution, is observable. Evolution is defined as a change in allele frequencies in a population over time.

In studying these steps, you should realize that variation among individuals is essential if evolution is to occur. You should also recognize that Darwin had to introduce population thinking into biology because it is populations that change over time when evolution occurs. To come up with these postulates and understand their consequences, Darwin had to think in a revolutionary way.

Today, biologists usually condense Darwin's four postulates into a two-part statement that communicates the essence of evolution by natural selection more forcefully: Evolution by natural selection occurs when (1) heritable variation leads to (2) differential reproductive success.

The Biological Definitions of Fitness and Adaptation

To explain the process of natural selection, Darwin referred to successful individuals as "more fit" than other individuals. In doing so, he gave the word fitness a definition different from its everyday English usage. **Biological fitness** is the ability of an individual to produce surviving offspring, relative to that ability in other individuals in the population.

Note that fitness is a measurable quantity. When researchers study a population in the lab or in the field, they can estimate the relative fitness of individuals by counting the number of offspring each individual produces and comparing the data.

The concept of fitness, in turn, provides a compact way of formally defining adaptation. The biological meaning of adaptation, like the biological meaning of fitness, is different from its normal English usage. In biology, an adaptation is a heritable trait that increases the fitness of an individual in a particular environment relative to individuals lacking the trait. Adaptations increase fitness—the ability to produce offspring.

To summarize, evolution by natural selection occurs when heritable variation in traits leads to differential reproductive success among individuals.

24.4 Evolution in Action: Recent Research on Natural Selection

The theory of evolution by natural selection is testable. If the theory is correct, biologists should be able to test the validity of each of Darwin's postulates—documenting heritable variation and differential reproductive success in a wide array of natural populations.

This section summarizes two examples in which evolution by natural selection is being observed in nature. Literally hundreds of other case studies are available, involving a wide variety of traits and organisms. To begin, let's explore the evolution of drug resistance: one of the great challenges facing today's biomedical researchers and physicians.

Case Study 1: How Did *Mycobacterium tuberculosis* Become Resistant to Antibiotics?

Mycobacterium tuberculosis, the bacterium that causes **tuberculosis**, or TB, has long been a scourge of humankind. In Europe and the U.S., TB was once as great a public health issue as cancer is now. It receded in importance during the early 1900s, though, for two reasons:

1. Advances in nutrition made people better able to fight off most *M. tuberculosis* infections quickly.
2. The development of antibiotics such as rifampin allowed physicians to stop even advanced infections.

In the late 1980s, however, rates of *M. tuberculosis* infection surged in many countries, and in 1993 the World Health Organization (WHO) declared TB a global health emergency. Physicians were particularly alarmed because the strains of *M. tuberculosis* responsible for the increase were largely or completely resistant to rifampin and other antibiotics that were once extremely effective.

How and why did the evolution of drug resistance occur? The case of a single patient—a young man who lived in Baltimore—will illustrate what is happening all over the world.

A PATIENT HISTORY The story begins when the individual was admitted to the hospital with fever and coughing. Chest X-rays, followed by bacterial cultures of fluid ejected from the lungs, showed that he had an active TB infection. He was given several antibiotics for 6 weeks, followed by twice-weekly doses of rifampin and isoniazid for an additional 33 weeks. Ten months after therapy started, bacterial cultures from his chest fluid indicated no *M. tuberculosis* cells. His chest X-rays were also normal. The antibiotics seemed to have cleared the infection.

Just two months after the TB tests proved normal, however, the young man was readmitted to the hospital with a fever, severe cough, and labored breathing. Despite being treated with a variety of antibiotics, including rifampin, he died of respiratory failure 10 days later. Samples of material from his lungs showed that *M. tuberculosis* was again growing actively there. But this time the bacterial cells were completely resistant to rifampin.

Drug-resistant bacteria had killed this patient. Where did they come from? Is it possible that a strain that was resistant to antibiotic treatment evolved *within* him? To answer this question, a research team analyzed DNA from the drug-resistant strain and compared it with stored DNA from *M. tuberculosis* cells that had been isolated a year earlier from the same patient. After examining extensive stretches from each genome, the biologists were able to find only one difference: a point mutation in a gene called *rpoB*.

A MUTATION IN A BACTERIAL GENE CONFERS RESISTANCE The *rpoB* gene codes for a component of the enzyme RNA polymerase. Recall from Chapter 16 that RNA polymerase transcribes DNA to mRNA, and that a point mutation is a single base change in DNA (see Chapter 15). In this case, the mutation in the bacterial gene changed a cytosine to a thymine, altering the normal codon TCG to a mutant one, TTG (**Figure 24.12**). As a result, the bacterial RNA polymerase produced by the drug-resistant strain had leucine instead of serine at the 153rd amino acid in the polypeptide chain.

This result is meaningful. Rifampin, the drug that was being used to treat the patient, works by binding to the RNA polymerase of *M. tuberculosis*. When the drug enters an *M. tuberculosis* cell and binds to RNA polymerase, it interferes with transcription. If sufficient quantities of rifampin are present for long enough and if the drug binds tightly, bacterial cells don't make proteins efficiently and they produce few offspring. But apparently the substitution of a leucine for a serine prevents rifampin from binding efficiently. Consequently, cells with the C $\rightarrow$ T mutation continue to produce offspring efficiently even in the presence of the drug.

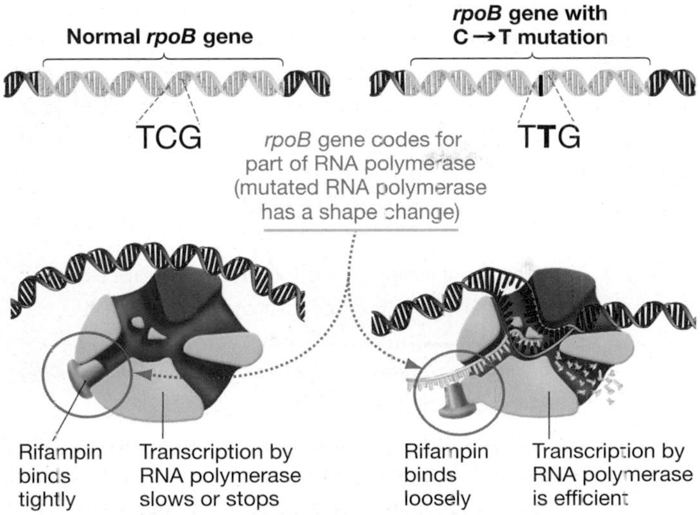

Normal rpoB gene
rpoB gene with C → T mutation
TCG
TTG
rpoB gene codes for part of RNA polymerase (mutated RNA polymerase has a shape change)
Rifampin binds tightly
Transcription by RNA polymerase slows or stops
Rifampin binds loosely
Transcription by RNA polymerase is efficient

FIGURE 24.12 Mutation in *rpoB* gene Confers Drug Resistance. The drug rifampin cannot bind tightly to RNA polymerase in *M. tuberculosis* cells with the C → T mutation in the *rpoB* gene. As a result, the drug-resistant bacteria continue to reproduce efficiently, even in the presence of the drug.

These results suggest that a chain of events led to this patient's death (**Figure 24.13**).

1. By chance, one or a few of the cells present in the patient, before drug therapy, started, happened to have an *rpoB* gene with the C → T mutation. Under normal conditions, mutant forms of RNA polymerase do not work as well as the more common form, so cells with the C → T mutation would not produce many offspring and would stay at low frequency—even while the overall population grew to the point of inducing symptoms that sent the young man to the hospital.
2. Therapy with rifampin began. In response, cells in the population with normal RNA polymerase began to grow much more slowly or to die outright. As a result, the overall bacterial population declined in size so drastically that the patient appeared to be cured—his symptoms began to disappear.
3. Cells with the C → T mutation had an advantage in the new environment. They began to grow more rapidly than the normal cells and continued to increase in number after therapy ended. Eventually the *M. tuberculosis* population regained its former abundance, and the patient's symptoms reappeared.
4. Drug-resistant cells now dominated the population, so the second round of rifampin therapy was futile.

Note that in most individuals, the immune system is able to eliminate the few bacteria that remain at step 2. This individual had AIDS, however, so his immune system was damaged.

✔If you understand these concepts, you should be able to explain: (1) Why the relapse in step 3 occurred, and (2) whether a family member or health care worker who got TB from this patient at step 3 or step 4 would respond to drug therapy.

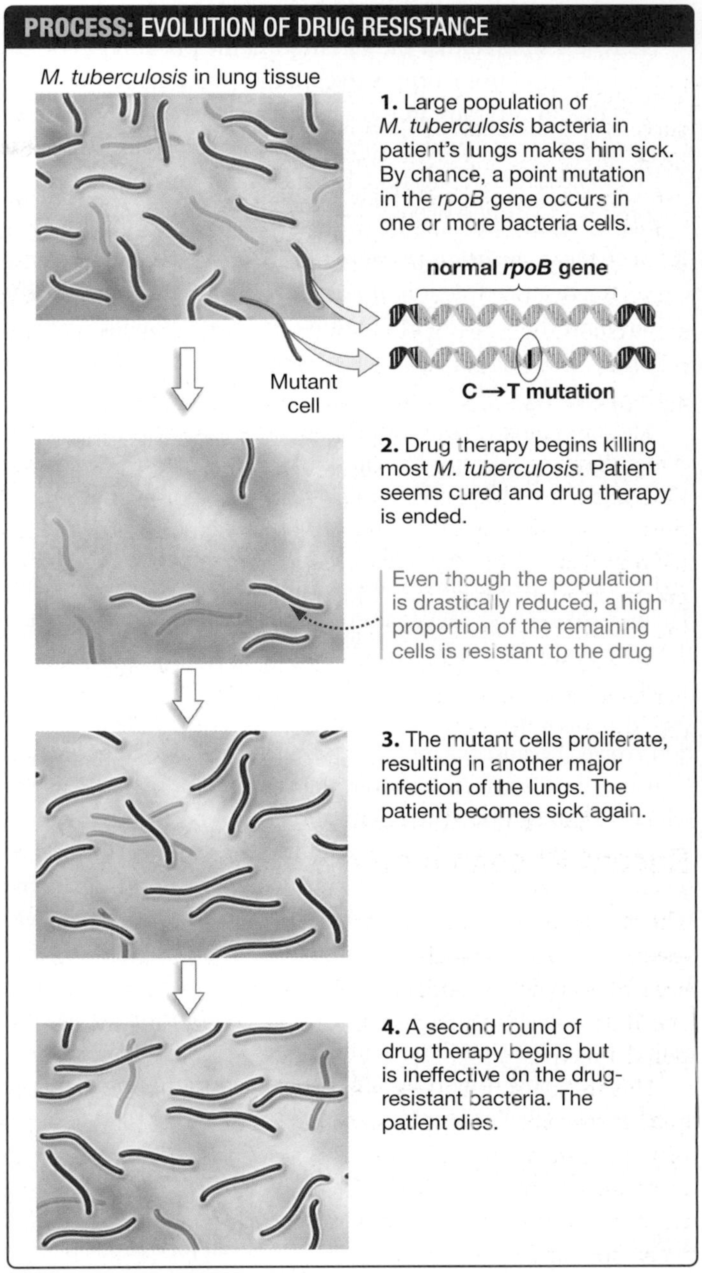

PROCESS: EVOLUTION OF DRUG RESISTANCE
M. tuberculosis in lung tissue
1. Large population of M. tuberculosis bacteria in patient's lungs makes him sick. By chance, a point mutation in the rpoB gene occurs in one or more bacteria cells.
normal rpoB gene
Mutant cell
C → T mutation
2. Drug therapy begins killing most M. tuberculosis. Patient seems cured and drug therapy is ended.
Even though the population is drastically reduced, a high proportion of the remaining cells is resistant to the drug
3. The mutant cells proliferate, resulting in another major infection of the lungs. The patient becomes sick again.
4. A second round of drug therapy begins but is ineffective on the drug-resistant bacteria. The patient dies.

FIGURE 24.13 Alleles That Confer Drug Resistance Increase in Frequency When Drugs Are Used.

TESTING DARWIN'S POSTULATES Does the sequence of events illustrated in Figure 24.13 mean that evolution by natural selection occurred? One way of answering this question is to review Darwin's four postulates and test whether each was verified:

1. ***Did variation exist in the population?*** The answer is yes. Due to mutation, both resistant and nonresistant strains of TB were present prior to administration of the drug. Most *M. tuberculosis* populations, in fact, exhibit variation for the trait; studies on cultured *M. tuberculosis* show that a mutation conferring resistance to rifampin is present in one out of every 10^7 to 10^8 cells.

2. ***Was this variation heritable?*** The answer is yes. The researchers showed that the variation in the phenotypes of the two strains—from drug susceptibility to drug resistance—was due to variation in their genotypes. Because the mutant *rpoB* gene is passed on to daughter cells when a *Mycobacterium* replicates, the allele and the phenotype it produces—drug resistance—are passed on to offspring.
3. ***Was there variation in reproductive success?*** The answer is yes. Only a tiny fraction of *M. tuberculosis* cells in the patient survived the first round of antibiotics long enough to reproduce. Most cells died and left no or almost no offspring.
4. ***Did selection occur?*** The answer is yes. When rifampin was present, certain cells—those with the drug-resistant allele—had higher reproductive success than cells with the normal allele.

M. tuberculosis individuals with the mutant *rpoB* gene had higher fitness in an environment where rifampin was present. The mutant allele produces a protein that is an adaptation when the cell's environment contains the antibiotic.

This study verified all four postulates and confirmed that evolution by natural selection had occurred. The *M. tuberculosis* population evolved because the mutant *rpoB* allele increased in frequency.

It is critical to note, however, that the individual cells themselves did not evolve. When natural selection occurred, the individual cells did not change through time; they simply survived or died, or produced more or fewer offspring. This is a fundamentally important point: Natural selection acts on individuals, because individuals experience differential reproductive success. But only populations evolve. Allele frequencies change in populations, not in individuals. Understanding evolution by natural selection requires population thinking.

To review how drug resistance evolves, go to the study area at *www.masteringbiology.com.*

(MB) **Web Activity** Natural Selection for Antibiotic Resistance

A WIDESPREAD PROBLEM The events reviewed for a single patient have occurred many times in other patients. Recent surveys indicate that drug-resistant strains now account for about 10 percent of the *M. tuberculosis*–causing infections throughout the world.

Unfortunately, the emergence of drug resistance in TB is far from unusual. Resistance to a wide variety of insecticides, fungicides, antibiotics, antiviral drugs, and herbicides has evolved in hundreds of insects, fungi, bacteria, viruses, and plants. In every case, evolution has occurred because individuals with the heritable ability to resist some chemical compound were present in the original population. As the susceptible individuals die from the pesticide, herbicide, or drug, the resistance alleles increase in frequency.

To drive home the prevalence of evolution in response to drugs and other human-induced changes in the environment, consider the data in **Figure 24.14**. The graph shows changes through time in the percentage of infections, in intensive care units in the United States, caused by strains of the bacterium *Staphylococcus aureus* that are resistant to the antibiotic vancomycin. Most of these *S. aureus* cells are also resistant to methicillin and other antibiotics as well—a phenomenon known as multi-drug resistance. In some cases, physicians have no effective antibiotics available to treat these infections.

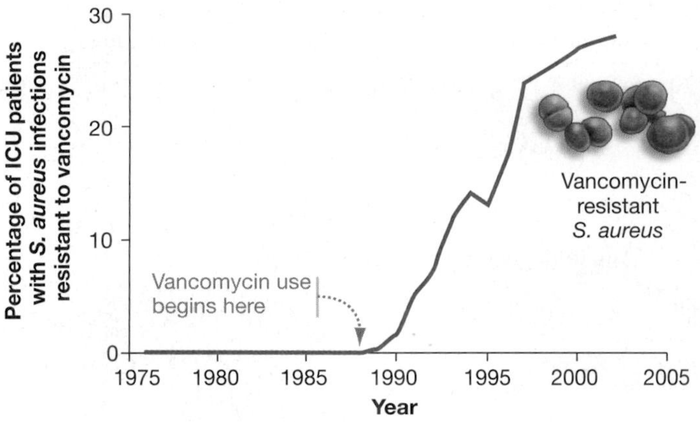

FIGURE 24.14 Trends in Infections Due to Antibiotic-resistant Bacteria. These data are from large hospitals in the USA. The line indicates changes in the percentage of *S. aureus* infections—acquired in the hospitals—that are resistant to the antibiotic vancomycin.

Case Study 2: Why Are Beak Size, Beak Shape, and Body Size Changing in Galápagos Finches?

Can biologists study evolution in response to natural environmental change—when humans are not involved? The answer is yes. As an example, consider research led by Peter and Rosemary Grant. These biologists have been investigating changes in beak size, beak shape, and body size that have occurred in finches native to the Galápagos Islands.

The medium ground finch makes its living by eating seeds. Finches crack seeds with their beaks. For well over three decades, the population of medium ground finches on Isle Daphne Major of the Galápagos has been studied intensively by the Grants' team (**Figure 24.15a**). Because Daphne Major is small—about the size of 80 football fields (see **Figure 24.15b**)—the researchers have been able to catch, weigh, and measure all individuals and mark each one with a unique combination of colored leg bands.

Early studies of the finch population established that beak size and shape and body size vary among individuals, and that beak morphology and body size are heritable. Stated another way, parents with particularly deep beaks tend to have offspring with deep beaks. Large parents also tend to have large offspring. Beak size and shape and body size are traits with heritable variation.

SELECTION DURING DROUGHT CONDITIONS Not long after the team began to study the population, a dramatic selection event occurred. In the annual wet season of 1977, Daphne Major re-

(a) Medium ground finches (*Geospiza fortis*)

(b) Isle Daphne Major of the Galápagos

FIGURE 24.15 Studying Evolution-in-Action on the Galápagos. The Galápagos Islands are the tops of undersea volcanoes.

ceived just 24 mm of rain instead of the 130 mm that normally falls. During the drought, few plants were able to produce seeds, and 84 percent (about 660 individuals) of the medium ground finch population disappeared.

Two observations support the hypothesis that most or all of these individuals died of starvation:

- The researchers found a total of 38 dead birds, and all were emaciated.
- None of the missing individuals were spotted on nearby islands, and none reappeared once the drought had ended and food supplies returned to normal.

The research team realized that the die-off was a **natural experiment**. Instead of comparing groups created by direct manipulation under controlled conditions, natural experiments allow researchers to compare treatment groups created by an unplanned change in conditions. In this case, the Grants' team could test whether natural selection occurred by comparing the population before and after the drought.

Were the survivors different from nonsurvivors? The histograms in **Figure 24.16** show the distribution of beak sizes in the population before and after the drought. (For more on how histograms are constructed, see **BioSkills 2** in Appendix A.) On average, survivors tended to have much deeper beaks than did the birds that died.

EXPERIMENT

QUESTION: Did natural selection on ground finches occur when the environment changed?

HYPOTHESIS: Beak characteristics changed in response to a drought.

NULL HYPOTHESIS: No changes in beak characteristics occurred in response to a drought.

EXPERIMENTAL SETUP:

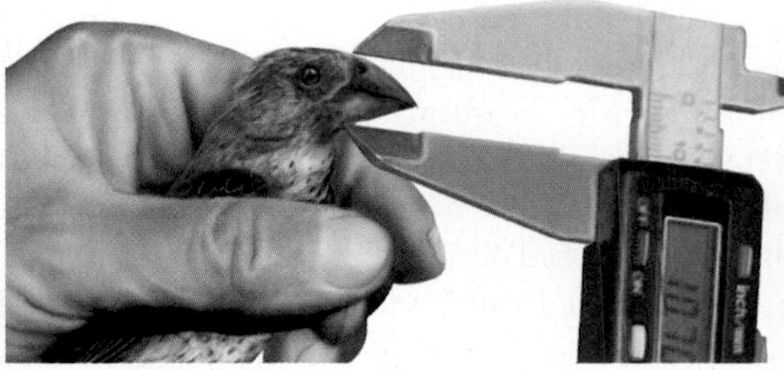

Weigh and measure all birds in the population before and after the drought.

PREDICTION:

PREDICTION OF NULL HYPOTHESIS:

RESULTS:

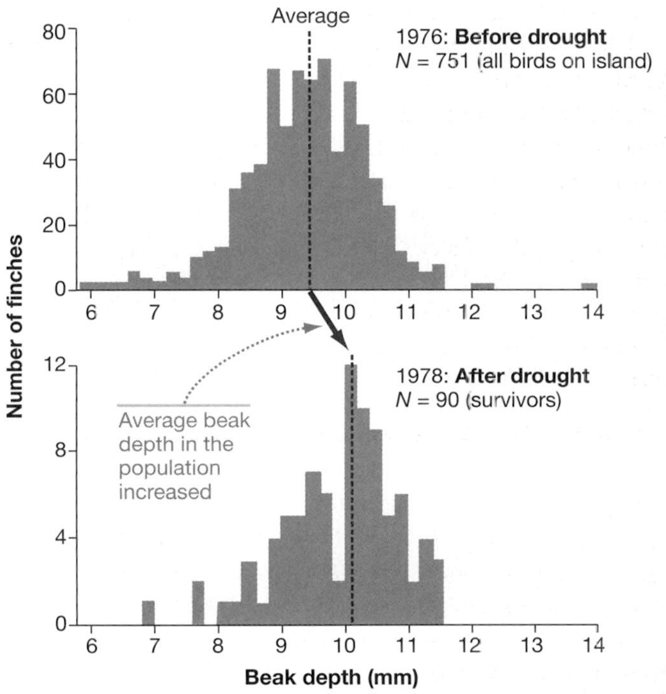

CONCLUSION: Natural selection occurred. The characteristics of the population have changed.

FIGURE 24.16 A Natural Experiment: Changes in a Medium Ground Finch Population in Response to a Change in the Environment (a Drought). The results show the distribution of beak depth in the population of medium ground finches on Daphne Major before and after the drought of 1977. N is the sample size.

SOURCE: Boag, P. T. and P. R. Grant. 1981. Intense natural selection in a population of Darwin's finches (Geospizinae) in the Galápagos. *Science* 214: 82-85.

✔**EXERCISE** Fill in the predictions made by the two hypotheses.

This was an important finding, because the type of seeds available to the finches had changed dramatically as the drought continued. At the drought's peak, most seed sources were absent and the tough fruits of a plant called *Tribulus cistoides* served as the finches' primary food source. These fruits are so difficult to crack that they are ignored in years when food supplies are normal. The group hypothesized that individuals with particularly large and deep beaks were more likely to crack these fruits efficiently enough to survive.

At this point, the Grants had shown that natural selection led to an increase in average beak depth in the population. When breeding resumed in 1978, the offspring that were produced had beaks that were half a millimeter deeper, on average, than those in the population that existed before the drought. This result confirmed that evolution had occurred.

In only one generation, natural selection led to a measurable change in the characteristics of the population. Alleles that led to the development of deep beaks had increased in frequency in the population. Large, deep beaks were an adaptation for cracking large fruits and seeds.

CONTINUED CHANGES IN THE ENVIRONMENT, CONTINUED SELECTION, CONTINUED EVOLUTION In 1983, the environment on the Galápagos Islands changed again. Over a seven-month period, a total of 1359 mm of rain fell. Plant growth was luxuriant, and finches fed primarily on small, soft seeds that were being produced in abundance. During this interval, small individuals with small, pointed beaks had exceptionally high reproductive success—meaning that they had higher fitness. As a result, the characteristics of the population changed again. Alleles associated with small, pointed beaks increased in frequency.

Over subsequent decades, the Grants have documented continued evolution in response to continued changes in the environment. **Figure 24.17** documents changes that have occurred in average body size, beak size, and beak shape over 35 years. From 1972 to 2006, average beak size declined. Beak shape also changed dramatically—on average, finch beaks got much pointier. In addition, average body size got smaller.

Long-term studies such as this have been powerful, because they have succeeded in documenting natural selection in response to changes in the environment.

WHICH GENES ARE UNDER SELECTION? Characteristics like beak size, beak shape, and overall body size are polygenic, meaning that many genes—each one exerting a relatively small effect—influence the trait (see Chapter 13). Because many genes are involved, it can be difficult for researchers to know exactly which alleles are changing in frequency when polygenic traits evolve.

To explore which medium ground finch genes might be under selection, researchers in Clifford Tabin's lab began studying beak development in an array of Galápagos finch species. More specifically, they looked for variation in the pattern of expression of cell-cell signals that had already been identified as important in the development of chicken beaks. The hope was that homologous genes might affect beak development in finches.

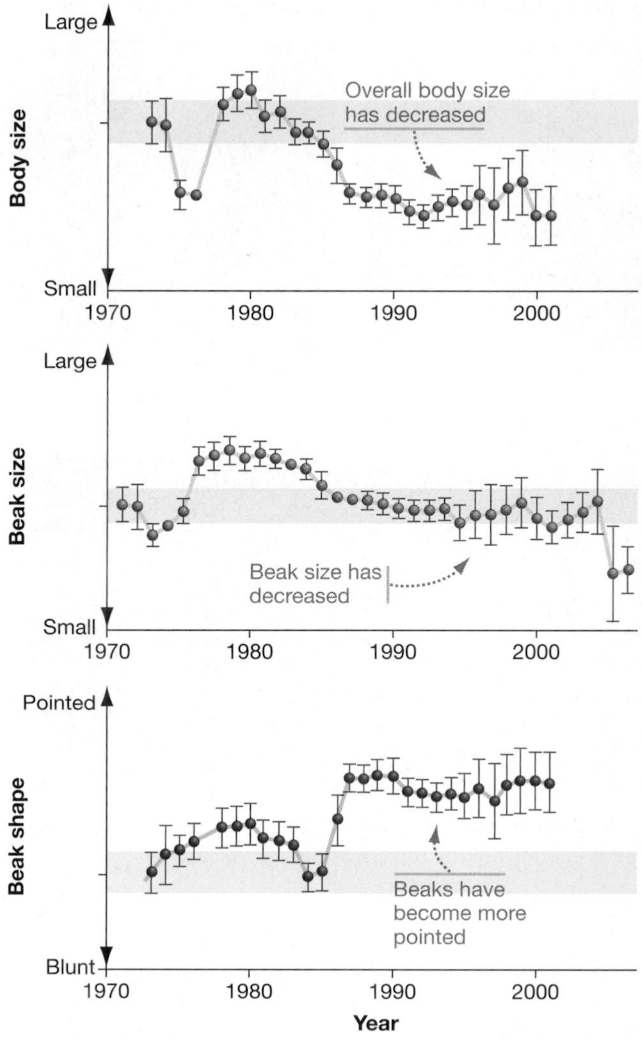

FIGURE 24.17 Body Size, Beak Size, and Beak Shape in Finches Changed over a 35-Year Interval.

The researchers struck pay dirt when they did in situ hybridizations—a technique featured in Chapter 21—showing where a cell-cell signal gene called *Bmp4* is expressed.

- There is a strong correlation between the amount of *Bmp4* expression when beaks are developing in young Galápagos finches and the width and depth of adult beaks (**Figure 24.18**).
- When the researchers experimentally increased *Bmp4* expression in young chickens, they found that beaks got wider and deeper than normal.

Similar experiments suggest that variation in alleles for a molecule called calmodulin, which is involved in calcium signaling during development, affects beak length.

Based on these data, biologists suspect that alleles associated with *Bmp4* and calmodulin expression may be under selection in the population of medium ground finches that the Grants are studying. If so, the research community will have made a direct connection between natural selection on phenotypes and evolutionary change in genotypes.

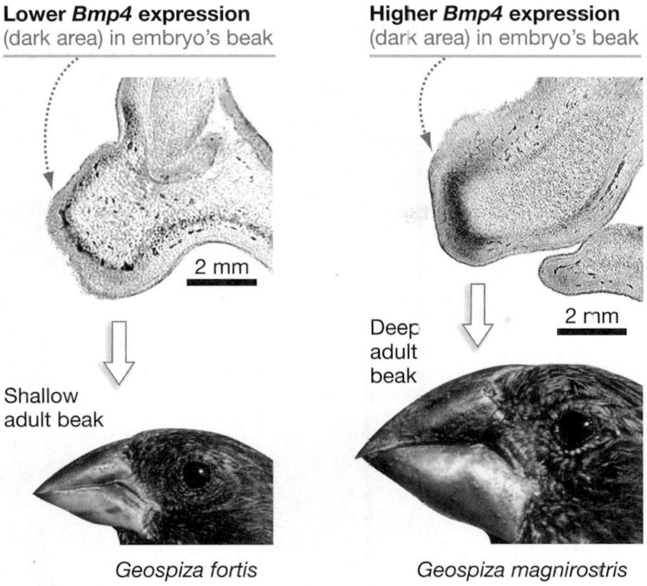

FIGURE 24.18 Changes in *Bmp4* Expression Change Beak Depth and Width. These micrographs are in situ hybridizations (see Chapter 21) showing the location and extent of *Bmp4* expression in young *Geospiza fortis* and *G. magnirostris*. In these and four other species that were investigated, the amount of Bmp4 protein produced correlates with the depth and width of the adult beak.

CHECK YOUR UNDERSTANDING

If you understand that . . .

- If individuals with certain alleles produce the most offspring in a population, then those alleles increase in frequency over time. Evolution—a change in allele frequencies—results from this process (natural selection on heritable variation).

✓ You should be able to . . .

1. List Darwin's four postulates, and indicate which are related to heritable variation and which are related to differential reproductive success.
2. Explain how data on beak size and shape and body size of Galápagos finch populations provide examples of heritable variation and differential reproductive success.

Answers are available in Appendix B.

24.5 Common Misconceptions about Natural Selection and Adaptation

Evolution by natural selection is a simple process—just the logical outcome of some straightforward postulates. Ironically, it can be extremely difficult to understand.

Research has shown that evolution by natural selection is often misunderstood. To help clarify how the process works, let's consider some of the more common misconceptions about natural selection in light of data on drug resistance in the TB bacterium and changes in finch populations.

Selection Acts on Individuals, but Evolutionary Change Occurs in Populations

Perhaps the most important point to clarify about natural selection is that during the process, individuals do not change—only the population does. During the drought, the beaks of individual finches did not become deeper. Rather, the average beak depth in the population increased over time, because deep-beaked individuals produced more offspring than shallow-beaked individuals did. Natural selection acted on individuals, but the evolutionary change occurred in the characteristics of the population.

In the same way, individual bacterial cells did not change when rifampin was introduced to their environment. Each *M. tuberculosis* cell had the same polymerase alleles all its life. But because the mutant allele increased in frequency over time, the characteristics of the bacterial population changed.

This point should make sense, given that evolution is defined as changes in allele frequencies. An individual's allele frequencies cannot change over time—it has the alleles it was born with all its life.

A CONTRAST WITH "LAMARCKIAN" INHERITANCE There is a sharp contrast between evolution by natural selection and evolution by the inheritance of acquired characters—the hypothesis promoted by Jean-Baptiste de Lamarck. If you recall, Lamarck proposed that (1) individuals change in response to challenges posed by the environment, and (2) the changed traits are then passed on to offspring. The key claim is that the important evolutionary changes occur in individuals.

In contrast, Darwin realized that individuals do not change when they are selected. Instead, they simply produce more offspring than other individuals do. When this happens, alleles found in the selected individuals become more frequent in the population.

Darwin was correct: there is no mechanism that makes it possible for natural selection to change the nature of an allele inside an individual. An individual's heritable characteristics don't change when natural selection occurs. Natural selection just sorts existing variants—it doesn't change them.

ACCLIMATION IS *NOT* ADAPTATION The issue of change in individuals is tricky because individuals often *do* change in response to changes in the environment. For example, wood frogs native to northern North America are exposed to extremely cold temperatures as they overwinter. When ice begins to form in their skin, their bodies begin producing a sort of natural antifreeze—molecules that protect their tissues from being damaged by the ice crystals. These individuals are changing in response to a change in temperature.[1] You may have observed changes in your own body as you got accustomed to living at high elevation or in a particularly hot or cold environment.

[1]In some species of frogs, so much extracellular fluid freezes during cold snaps that individuals appear to be frozen solid. Their hearts also stop beating. When temperatures warm in the spring, their hearts start beating again, their tissues thaw, and they resume normal activities.

Biologists use the term **acclimation** to describe changes in an individual's phenotype that occur in response to changes in environmental conditions. The key is to realize that phenotypic changes due to acclimation are not passed on to offspring, because no alleles have changed in composition. As a result, acclimation does not cause evolution. ✔**If you understand this concept, you should be able to (1) explain the difference between the biological definition of adaptation and its use in everyday English, and (2) explain the difference between acclimation and adaptation.**

Evolution Is Not Goal Directed

It is tempting to think that evolution by natural selection is goal directed. For example, you might hear a fellow student say that *M. tuberculosis* cells "wanted" or "needed" the mutant, drug-resistant allele so that they could survive and continue to reproduce in an environment that included rifampin. This does not happen. The mutation that created the mutant allele occurred randomly, due to an error during DNA synthesis, and it just happened to be advantageous when the environment changed.

Stated another way, the mutation that conferred resistance did not occur because of the presence of the drug. It just happened. Every mutation is equally likely to occur in every environment. There is no mechanism that makes it possible for the environment to direct which mistakes DNA polymerase makes when it copies genes. Adaptations do not occur because organisms want or need them.

EVOLUTION IS NOT PROGRESSIVE It is often tempting to think that evolution by natural selection is progressive—meaning organisms have gotten "better" over time. (In this context, *better* usually means bigger, stronger, or more complex.) It is true that the groups appearing later in the fossil record are often more morphologically complex or "advanced" than closely related groups that appeared earlier. Flowering plants are considered more complex than mosses, and most biologists would agree that the morphology of mammals is more complex than that of the first vertebrates in the fossil record. But there is nothing predetermined or absolute about this tendency.

In fact, complex traits are routinely lost or simplified over time as a result of evolution by natural selection. You've already analyzed evidence on limb loss in snakes (Chapter 21) and whales (this chapter). Chapter 20 presented data indicating that parasitic bacteria often lose large portions of their genomes.

Populations that become parasitic are particularly prone to loss of complex traits. Tapeworms, for example, lack a mouth and digestive system. As parasites that live in the intestines of humans and other mammals, they simply absorb nutrients directly from their environment, across their plasma membranes. But tapeworms evolved from species with a sophisticated digestive tract. Tapeworms lost their digestive tract as a result of evolution by natural selection, but they did not become bigger or stronger or more complex.

THERE IS NO SUCH THING AS A HIGHER OR LOWER ORGANISM The nonprogressive nature of evolution by natural selection contrasts

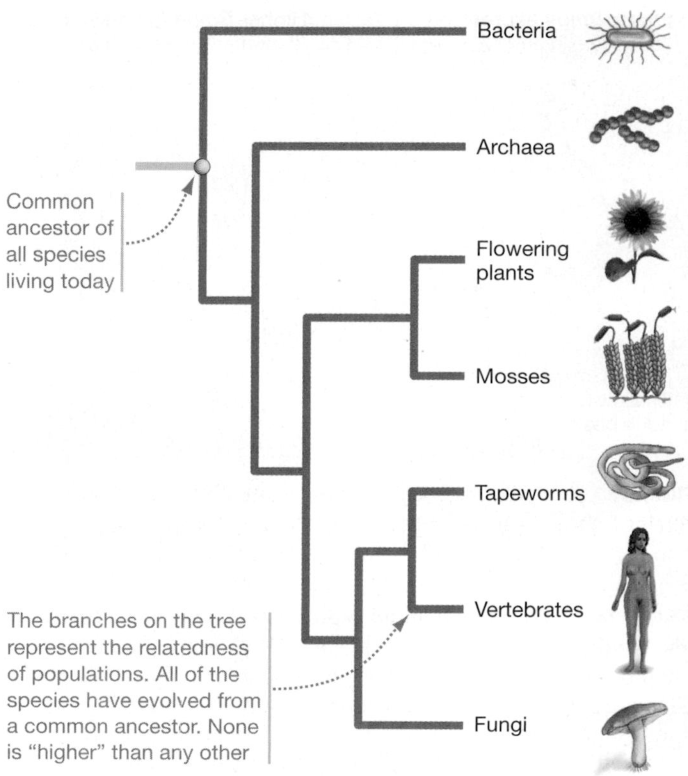

FIGURE 24.19 Evolution Produces a Tree of Life, Not a Progressive Ladder of Life. Under evolution by natural selection, species are related by common ancestry and all have evolved through time. (Not all branches of the tree of life are shown.)

sharply with Lamarck's conception of the evolutionary process, in which organisms progress over time to higher and higher levels on a chain of being.

Under Aristotle's and Lamarck's hypothesis, it is sensible to refer to "higher" and "lower" organisms. But under evolution by natural selection, there is no such thing as a higher or lower organism (**Figure 24.19**). Mosses may be a more ancient group than flowering plants, but neither group is higher or lower than the other. Mosses simply have a different suite of adaptations than do flowering plants, so they thrive in different types of environments. A human is no higher than its tapeworm parasite; each is well adapted to its environment.

All populations have evolved by natural selection based on their ability to gather resources and produce offspring. All organisms are adapted to their environment, and are related by common ancestry.

Natural selection is not goal directed or progressive. It simply favors individuals that happen to be better adapted to the environment existing at the time.

Organisms Do Not Act for the Good of the Species

Consider the widely circulated story that rodents called lemmings sacrifice themselves for the good of their species. The story claims that when lemming populations are high, overgrazing is so exten-

FIGURE 24.20 Self-Sacrificing Behavior Cannot Evolve if "Selfish" Alleles Exist. Suppose that most individuals in a lemming population had an allele that led to self-sacrificing behavior, or even suicide, for "the good of the species." Now suppose that this individual had alleles that prevented self-sacrifice. Which alleles would increase in frequency in the population over time?

sive that the entire species is threatened with starvation and extinction. In response, some individuals throw themselves into the sea and drown. This lowers the overall population size and allows the vegetation to recover enough to save the species. Even though individuals suffer, the good-of-the-species hypothesis maintains that the behavior evolved because the group benefits.

The lemming suicide story is false. Although lemmings do disperse from areas of high population density in order to find habitats with higher food availability, they do not throw themselves into the sea.

To understand why this type of self-sacrificing behavior does not occur, suppose that certain alleles predispose lemmings to sacrifice themselves for others. But consider what happens if alleles exist that prevent this type of behavior—what biologists call a "selfish" allele (**Figure 24.20**). Individuals with self-sacrificing alleles die and do not produce offspring. But individuals with selfish, cheater alleles survive and produce offspring. As a result, selfish alleles increase in frequency while self-sacrificing alleles decrease in frequency. Thus, it is not possible for individuals to sacrifice themselves for the good of the species.

No instance of purely self-sacrificing behavior—where the individual received no fitness benefit in return—has ever been recorded in nature. Chapter 52 provides additional details on this point.

Limitations of Natural Selection

Although organisms are often exquisitely adapted to their environment, adaptation is far from perfect. A long list of circumstances limits the effectiveness of natural selection; only a few of the most important are discussed here.

NON-ADAPTIVE TRAITS Vestigial traits such as the human coccyx (tailbone), goose bumps, and appendix do not increase the fitness of individuals with those traits. The structures are not adaptive. They exist simply because they were present in the ancestral population.

Vestigial traits are not the only types of structures with no function. Some adult traits exist as holdovers from structures that appear early in development. For example, human males have rudimentary mammary glands. The structures are not adaptive. They exist only because nipples form in the human embryo before sex hormones begin directing the development of male organs instead of female organs.

Perhaps the best example of nonadaptive traits involves evolutionary changes in DNA sequences. Recall from Chapter 16 that mutation may change a base in the third position of a codon without changing the amino acid sequence of the protein encoded by that gene. Changes such as these are said to be silent. They occur because of the redundancy of the genetic code (see Chapter 16). Silent changes in DNA sequences are extremely common. But because they don't change the phenotype, they can't be acted on by natural selection.

The general point here is that not all traits are adaptive. Evolution by natural selection does not lead to "perfection." Besides carrying an array of traits that have no function, the adaptations that organisms have are constrained in a variety of important ways.

GENETIC CONSTRAINTS The Grants' team analyzed data on the characteristics of finches that survived the 1977 drought, and the team made an interesting observation: Although individuals with deep beaks survived better than individuals with shallow beaks, birds with particularly narrow beaks survived better than individuals with wider beaks.

This observation made sense because finches crack *Tribulus* fruits by twisting them. Narrow beaks concentrate the twisting force more efficiently than wider beaks, so they are especially useful for cracking the fruits. But narrower beaks did not evolve in the population.

To explain why, the biologists noted that parents with deep beaks tend to have offspring with beaks that are both deep and wide. This is a common pattern. Many alleles that affect body size have an effect on all aspects of size—not just one structure or dimension. As a result, selection for increased beak depth overrode selection for narrow beaks, even though a deep and narrow beak would have been more advantageous.

The general point here is that selection was not able to optimize all aspects of a trait. In the case of the finches, wider beaks were not the best possible beak shape for individuals living in an arid habitat. Wider beaks evolved anyway, due to a type of constraint called a **genetic correlation**. Genetic correlations occur because of pleiotropy (see Chapter 13)—in which a single allele affects multiple traits. In this case, selection on alleles for one trait (increased beak depth) caused a correlated, though suboptimal, increase in another trait (beak width).

Genetic correlations are not the only genetic constraint on adaptation. Lack of genetic variation is also important. Consider that salamanders have the ability to regrow severed limbs. Some eels and sharks can sense electric fields. Birds can sense magnetic fields and see ultraviolet light. Even though it is possible that these traits would confer increased reproductive success in humans, they do not exist—because the requisite genes are lacking.

FITNESS TRADE-OFFS In everyday English, the term trade-off refers to a compromise between competing goals. It is difficult to design a car that is both large and fuel efficient, a bicycle that is both rugged and light, or a plane that is both fast and maneuverable.

In nature, selection occurs in the context of fitness trade-offs. A **fitness trade-off** is a compromise between traits, in terms of how those traits perform in the environment. During the drought in the Galápagos, for example, medium ground finches with large bodies had an advantage because they won fights over the few remaining sources of seeds. But individuals with large bodies also require large amounts of food to maintain their mass; they also tend to be slower and less nimble than smaller individuals. When food is short, large individuals are more prone to starvation. Even if large size is advantageous in an environment, there is always counteracting selection that prevents individuals from getting even bigger.

Biologists have documented trade-offs between the size of eggs or seeds that an individual makes and the number of offspring it can produce, between rapid growth and long life span, and between bright coloration and tendency to attract predators. The message of this research is simple: Because selection acts on many traits at once, every adaptation is a compromise.

HISTORICAL CONSTRAINTS In addition to being constrained by genetic correlations, lack of genetic variation, and fitness trade-offs, adaptations are constrained by history. The reason is simple: All traits have evolved from previously existing traits.

Natural selection acts on structures that originally had a very different function. For example, the tiny hammer, anvil, and stirrup bones found in your middle ear evolved from bones that were part of the jaw and braincase in the ancestors of mammals. These bones now function in the transmission and amplification of sound from your outer ear to your inner ear. Biologists routinely interpret these bones as adaptations that improve your ability to hear airborne sounds. But are the bones a "perfect" solution to the problem of transmitting sound from the outside of the ear to the inside? The answer is no. They are the best solution possible, given an important historical constraint. Other vertebrates have different structures involved in transmitting sound to the ear. In at least some cases, those structures may be more efficient than our hammer, anvil, and stirrup.

To summarize, not all traits are adaptive, and even adaptive traits are constrained by genetic and historical factors. In addition, natural selection is not the only process that causes evolutionary change. Chapter 25 introduces three other processes that change allele frequencies over time. Compared with natural selection, these processes have very different consequences.

CHECK YOUR UNDERSTANDING

If you understand that...

- Selection by drugs on the TB bacterium and changes in seed availability to finches in the Galápagos are well-studied examples of evolution by natural selection.
- Evolution by natural selection is simple in concept but widely misunderstood.

✓ You should be able to...

1. Explain why individuals do not change when natural selection occurs.
2. Explain why trade-offs and genetic and historical constraints prevent adaptations from being "perfect."

Answers are available in Appendix B.

CHAPTER 24 REVIEW

For media, go to the study area at www.masteringbiology.com

Summary of Key Concepts

Populations and species evolve, meaning that their heritable characteristics change through time. Evolution is change in allele frequencies over time.

- Data on (1) the resemblance of modern to fossil forms; (2) transitional features in fossils; (3) the fact of extinction; (4) the presence of vestigial traits; and (5) change in contemporary populations are inconsistent with the claim that species have remained unchanged through time.
- Data on (1) the geographic proximity of closely related species such as the Galápagos mockingbirds; (2) the existence of structural, developmental, and genetic homologies; and (3) contemporary formation of new species are inconsistent with the theory that species were formed instantaneously, recently, and independently by a divine being.
- Evidence for evolution is internally consistent—meaning that data from several independent sources is mutually reinforcing.

✓You should be able to predict how changes in *Mycobacterium* populations would be explained under the theory of special creation and under evolution by inheritance of acquired characters.

Evolution by natural selection occurs when individuals with certain alleles produce the most surviving offspring in a population. An adaptation is a genetically based trait that increases an individual's ability to produce offspring in a particular environment.

- Evolution by natural selection occurs whenever genetically based differences among individuals lead to differences in their ability to reproduce—when heritable variation leads to differential reproductive success.
- Alleles or traits that increase the reproductive success of an individual are said to increase the individual's fitness.

- A trait that leads to higher fitness, relative to individuals without the trait, is an adaptation.
- If a particular allele increases fitness and leads to adaptation, the allele will increase in frequency in the population.
- Evolution is an outcome of natural selection. Evolution by natural selection has been confirmed by a wide variety of studies and has long been considered to be the central organizing principle of biology.

✔You should be able to explain the difference between the biological and everyday English definitions of fitness.

(MB) **Web Activity** Natural Selection for Antibiotic Resistance

Evolution by natural selection is not progressive, and it does not change the characteristics of the individuals that are selected—it changes only the characteristics of the population. Animals do not do things for the good of the species, and not all traits are adaptive. All adaptations are constrained by trade-offs and genetic and historical factors.

- Individuals that are naturally selected are not changed by the process—they simply produce more offspring than other individuals do.
- Traits that increase in frequency under natural selection do so because they improve fitness, not because they are necessarily larger or more complex.
- Because natural selection acts only on existing traits, it does not lead to perfection.

✔You should be able to discuss how an adaptation—such as the large brains of *Homo sapiens* or the ability of falcons to fly very fast—is constrained.

Questions

✔TEST YOUR KNOWLEDGE

Answers are available in Appendix B

1. How can biological fitness be estimated?
   a. Document how long different individuals in a population survive.
   b. Count the number of offspring produced by different individuals in a population.
   c. Determine which individuals are strongest.
   d. Determine which phenotype is the most common one in a given population.
2. Why are some traits considered vestigial?
   a. They improve the fitness of an individual who bears them, compared with the fitness of individuals without those traits.
   b. They change in response to environmental influences.
   c. They existed long ago.
   d. They are reduced in size, complexity, and function compared with traits in related species.
3. What is an adaptation?
   a. a trait that improves the fitness of its bearer, compared with individuals without the trait
   b. a trait that changes in response to environmental influences within the individual's lifetime
   c. an ancestral trait—one that was modified to form the trait observed today
   d. the ability to produce offspring
4. Why does the presence of extinct forms and transitional features in the fossil record support the pattern component of the theory of evolution by natural selection?
   a. It supports the hypothesis that individuals change over time.
   b. It supports the hypothesis that weaker species are eliminated by natural selection.
   c. It supports the hypothesis that species evolve to become more complex and better adapted over time.
   d. It supports the hypothesis that species have changed through time.
5. Why are homologous traits similar?
   a. They are derived from a common ancestor.
   b. They are derived from different ancestors.
   c. They result from convergent evolution.
   d. Their appearance, structure, or development is similar.
6. According to data presented in this chapter, which of the following statements is correct?
   a. When individuals change in response to challenges from the environment, their altered traits are passed on to offspring.
   b. Species are created independently of each other and do not change over time.
   c. Populations—not individuals—change when natural selection occurs.
   d. The Earth is young, and most of today's landforms were created during the floods at the time of Noah.

✔TEST YOUR UNDERSTANDING

Answers are available in Appendix B

1. Compare and contrast the theory of evolution by natural selection and the theory of special creation and evolution by inheritance of acquired characters. What testable predictions does each make?
2. Some biologists encapsulate evolution by natural selection with the phrase "mutation proposes, selection disposes." Mutation is a process that creates heritable variation. Explain what the phrase means.
3. Review the section on the evolution of drug resistance in *Mycobacterium tuberculosis*.
   - What evidence do researchers have that a drug-resistant strain evolved in the patient analyzed in their study, instead of having been transmitted from another infected individual?
   - If the antibiotic rifampin were banned, would the mutant *rpoB* gene have lower or higher fitness in the new environment? Would strains carrying the mutation continue to increase in frequency in *M. tuberculosis* populations?

4. Compare and contrast typological thinking with population thinking. Why was Darwin's emphasis on the importance of variation among individuals so crucial to his theory, and why was it a revolutionary idea in Western science?
5. The evidence supporting the pattern component of the theory of evolution can be criticized on the grounds that it is indirect. For example, no one has directly observed the formation of a vestigial trait over time. Is indirect evidence for a scientific theory legitimate? Why or why not?
6. Why isn't evolution by natural selection progressive? Why don't the biggest and strongest individuals in a population always produce the most offspring?

✓ APPLYING CONCEPTS TO NEW SITUATIONS

Answers are available in Appendix B

1. The geneticist James Crow wrote that successful scientific theories have the following characteristics: (1) They explain otherwise puzzling observations; (2) they provide connections between otherwise disparate observations; (3) they make predictions that can be tested; and (4) they are heuristic, meaning that they open up new avenues of theory and experimentation. Crow added two other elements that he considered important on a personal, emotional level: (5) They should be elegant, in the sense of being simple and powerful; and (6) they should have an element of surprise. How well does the theory of evolution by natural selection fulfill these six criteria?
2. The average height of humans has increased steadily for the past 100 years in industrialized nations. This trait has clearly changed over time. Most physicians and human geneticists hypothesize that the change is due to better nutrition and a reduced incidence of disease. Has human height evolved?
3. Genome sequencing projects may dramatically affect how biologists analyze evolutionary changes in quantitative traits. For example, suppose that the genomes of many living humans are sequenced and that genomes could be sequenced from many people who lived 100 years ago. (That might be possible with preserved tissue.) If 20 genes have been shown to influence height, how could you use the sequence data from these genes to test the hypothesis that human height has evolved in response to natural selection?
4. In some human populations, individuals tan in response to exposure to sunlight. Tanning is an acclimation response to a short-term change in the environment. It is adaptive because it prevents sunburn; it may also prevent a vitamin called folate from being destroyed by sunlight. The ability to tan varies among individuals in these populations, however. Is the ability to tan an acclimation or an adaptation? Explain your logic.

This chapter is part of The Big Picture. See how on pages 494–495.

A male raggiana bird of paradise, left, displays for a female, right. His long, colorful feathers and dramatic behavior result from sexual selection—a process introduced in this chapter.

Evolutionary Processes 25

Chapter 24 defined evolution as a change in allele frequencies. One of the key concepts from that chapter was that even though natural selection acts on individuals, evolutionary change occurs in **populations**. A population is a group of individuals from the same species that live in the same area and regularly interbreed.

Natural selection is not the only process that causes evolution, however. There are actually four mechanisms that shift allele frequencies in populations:

1. *Natural selection* increases the frequency of certain alleles—the ones that contribute to reproductive success in a particular environment.
2. *Genetic drift* causes allele frequencies to change randomly. In some cases, drift may cause alleles that decrease fitness to increase in frequency.
3. *Gene flow* occurs when individuals leave one population, join another, and breed. Allele frequencies may change when gene flow occurs, because arriving individuals introduce alleles to their new population and departing individuals remove alleles from their old population.
4. *Mutation* modifies allele frequencies by continually introducing new alleles. The alleles created by mutation may be beneficial or detrimental or have no effect on fitness.

This chapter has two fundamental messages: Natural selection is not the only agent responsible for evolution, and each of the four evolutionary processes has different consequences. Natural selection is the only mechanism that acting alone can result in adaptation. Mutation, gene flow, and genetic drift do not favor certain alleles over others. Mutation and drift introduce a nonadaptive component into evolution.

Let's take a closer look at the four evolutionary processes by examining a null hypothesis—what happens to allele frequencies when the evolutionary mechanisms are *not* operating.

KEY CONCEPTS

- The Hardy-Weinberg principle acts as a null hypothesis when researchers want to test whether evolution or nonrandom mating is occurring at a particular gene.
- Each of the four evolutionary mechanisms has different consequences. Only natural selection produces adaptation. Genetic drift causes random fluctuations in allele frequencies. Gene flow equalizes allele frequencies between populations. Mutation introduces new alleles.
- Inbreeding changes genotype frequencies but does not change allele frequencies.
- Sexual selection leads to the evolution of traits that help individuals attract mates. It is usually stronger on males than on females.

✔ When you see this checkmark, stop and test yourself. Answers are available in Appendix B.

25.1 Analyzing Change in Allele Frequencies: The Hardy-Weinberg Principle

To study how the four evolutionary processes affect populations, biologists take a three-pronged approach.

1. They create mathematical models that predict the fate of alleles over time under various conditions.
2. They collect data to test predictions made by the models.
3. They apply the results to solve problems in human genetics, conservation of endangered species, or other fields.

This research strategy began in 1908, when G. H. Hardy and Wilhelm Weinberg each published a major result independently. At the time, it was commonly believed that changes in allele frequency occur simply as a result of sexual reproduction—meiosis followed by the random fusion of gametes (eggs and sperm) to form offspring. Some biologists claimed that dominant alleles inevitably increase in frequency when gametes combine at random. Others predicted that two alleles of the same gene inevitably reach a frequency of 0.5.

To test these hypotheses, Hardy and Weinberg analyzed what happens to the frequencies of alleles when many individuals in a population mate and produce offspring. Instead of thinking about the consequences of a mating between two parents with a specific pair of genotypes, as we did with Punnett squares in Chapter 13, Hardy and Weinberg wanted to know what happened in an entire population, when *all* of the individuals—and thus all possible genotypes—bred. Like Darwin, Hardy and Weinberg were engaged in population thinking.

The Gene Pool Concept

To analyze the consequences of matings among all of the individuals in a population, Hardy and Weinberg invented a novel approach: They imagined that all of the gametes produced in each generation go into a single group called the **gene pool** and then combine at random to form offspring. Something very much like this happens in species like clams and sea stars and sea urchins, which release their gametes into the water, where they mix randomly with gametes from other individuals in the population and combine to form zygotes.

To determine which genotypes would be present in the next generation and in what frequency, Hardy and Weinberg simply had to calculate what happened when two gametes were plucked at random out of the gene pool, many times, and each of these gamete pairs was then combined to form offspring. These calculations would predict the genotypes of the offspring that would be produced, as well as the frequency of each genotype.

Deriving the Hardy-Weinberg Principle

Hardy and Weinberg began by analyzing the simplest situation possible—that just two alleles of a particular gene exist in a population. Let's call these alleles A_1 and A_2. We'll use p to symbolize the frequency of A_1 alleles in the gene pool and q to symbolize the frequency of A_2 alleles in the same gene pool. Because there are only two alleles, the two frequencies must add up to 1; that is, $p + q = 1$. Now follow the steps in **Figure 25.1**:

Step 1 Although p and q can have any value between 0 and 1, let's suppose that the initial frequency of A_1 is 0.7 and that of A_2 is 0.3.

Step 2 In this gene pool, 70 percent of the gametes carry A_1 and 30 percent carry A_2.

Step 3 Each time a gamete is involved in forming an offspring, there is a 70 percent chance that it carries A_1 and a 30 percent chance that it carries A_2. In general, there is a p chance that it carries A_1 and a q chance that it carries A_2.

Step 4 Because only two alleles are present, three genotypes are possible in the offspring generation: A_1A_1, A_1A_2, and A_2A_2. What will the frequency of these three genotypes be? According to the logic of Hardy's and Weinberg's result:

- The frequency of the A_1A_1 genotype is p^2.
- The frequency of the A_1A_2 genotype is $2pq$.
- The frequency of the A_2A_2 genotype is q^2.

The genotype frequencies in the offspring generation must add up to 1, which means that $p^2 + 2pq + q^2 = 1$. In our numerical example, $0.49 + 0.42 + 0.09 = 1$.

The last two steps in the figure show how the frequencies of alleles A_1 and A_2 are calculated from these genotype frequencies:

Step 5 The easiest way to calculate the allele frequencies in the offspring is to imagine that they form gametes that go into a gene pool. All of the gametes from A_1A_1 individuals carry A_1, so 49 percent (q^2) of the gametes in the gene pool will carry A_1. But half of the gametes from A_1A_2 will also carry A_1, so an additional $\frac{1}{2}(0.42) = 0.21$ (this is $\frac{1}{2} \times 2pq = pq$) gametes in the gene pool will carry A_1, for a total of $0.49 + 0.21 = 0.70$ or $q^2 + pq = q(q + p) = q$. Use the same logic to figure out the frequency of A_2.

Step 6 In our example, the frequency of allele A_1 in the offspring generation is still 0.7 and the frequency of allele A_2 is still 0.3. Thus, the frequency of allele A_1 is still p and the frequency of allele A_2 is still q.

No allele frequency change occurred. Even if A_1 is dominant to A_2, it does not increase in frequency. And there is no trend toward both alleles reaching a frequency of 0.5. This result is called the **Hardy-Weinberg principle**.

Figure 25.2 illustrates the same result a little differently. The figure uses a Punnett square to predict the outcome of random mating—meaning, random combinations of all gametes in a population. (Recall that in Chapter 13 you used Punnett squares to predict the outcome of a mating between two individuals.) The outcome of this analysis is the same as in Figure 25.1.

The Hardy-Weinberg principle makes two fundamental claims:

1. If the frequencies of alleles A_1 and A_2 in a population are given by p and q, then the frequencies of genotypes A_1A_1, A_1A_2, and A_2A_2 will be given by p^2, $2pq$, and q^2 for generation after generation.

PROCESS: A NUMERICAL EXAMPLE OF THE HARDY-WEINBERG PRINCIPLE

Parental allele frequencies: Allele A_1 $p = 0.7$ Allele A_2 $q = 0.3$

1. Allele frequencies in parents: Happen to be 0.7 and 0.3 in this example.

Gene pool

A_2 A_1 A_2 A_1 A_1 A_1 A_2 A_1 A_1 A_1 A_1 A_1 A_1
A_2 A_2 A_1 A_1 A_2 A_1 A_1 A_1 A_1 A_2 A_2 A_1
A_1 A_1 A_1 A_1 A_1 A_2 A_2 A_1 A_1 A_2 A_1 A_2 A_1 A_1

2. Resulting gene pool (possible gametes): 70% carry A_1, 30% carry A_2.

Offspring genotypes

A_1 A_1
$0.7 \times 0.7 = 0.49$
$p \times p = p^2$

A_1 A_2
$0.7 \times 0.3 = 0.21$
$p \times q = pq$

A_2 A_1
$0.3 \times 0.7 = 0.21$
$q \times p = pq$

A_2 A_2
$0.3 \times 0.3 = 0.09$
$q \times q = q^2$

$0.21 + 0.21 = 0.42$

3. Random pairing of gametes to produce offspring: Chance of getting picked for a pairing—meaning that gemetes combine to form an offspring—is 70% for A_1 and a 30% for A_2.

A_1A_1 genotype frequency is $p^2 = 0.49$

A_1A_2 genotype frequency is $2pq = 0.42$

A_2A_2 genotype frequency is $q^2 = 0.09$

4. Three possible offspring genotypes: A_1A_1, A_1A_2, and A_2A_2. Calculate the frequencies.

Offspring allele frequencies

49% A_1A_1 offspring. All will contribute A_1 alleles to the new gene pool.

42% A_1A_2 offspring. Alleles contributed to the nex: gene pool will be half A_1, half A_2.

9% A_2A_2 offspring. All will contribute A_2 alleles to the next gene pool.

5. Offspring allele frequencies: Calculate frequencies of A_1 and A_2 alleles that will enter next gene pool.

$p = 0.49 + \frac{1}{2}(0.42) = 0.7$
$p =$ frequency of allele A_1

$q = \frac{1}{2}(0.42) + 0.09 = 0.3$
$q =$ frequency of allele A_2

6. Allele frequencies have not changed from parents to offspring. Evolution has not occurred.

Under the conditions modeled here, allele frequencies do not change and genotype frequencies are given by $p^2 : 2pq : q^2$.

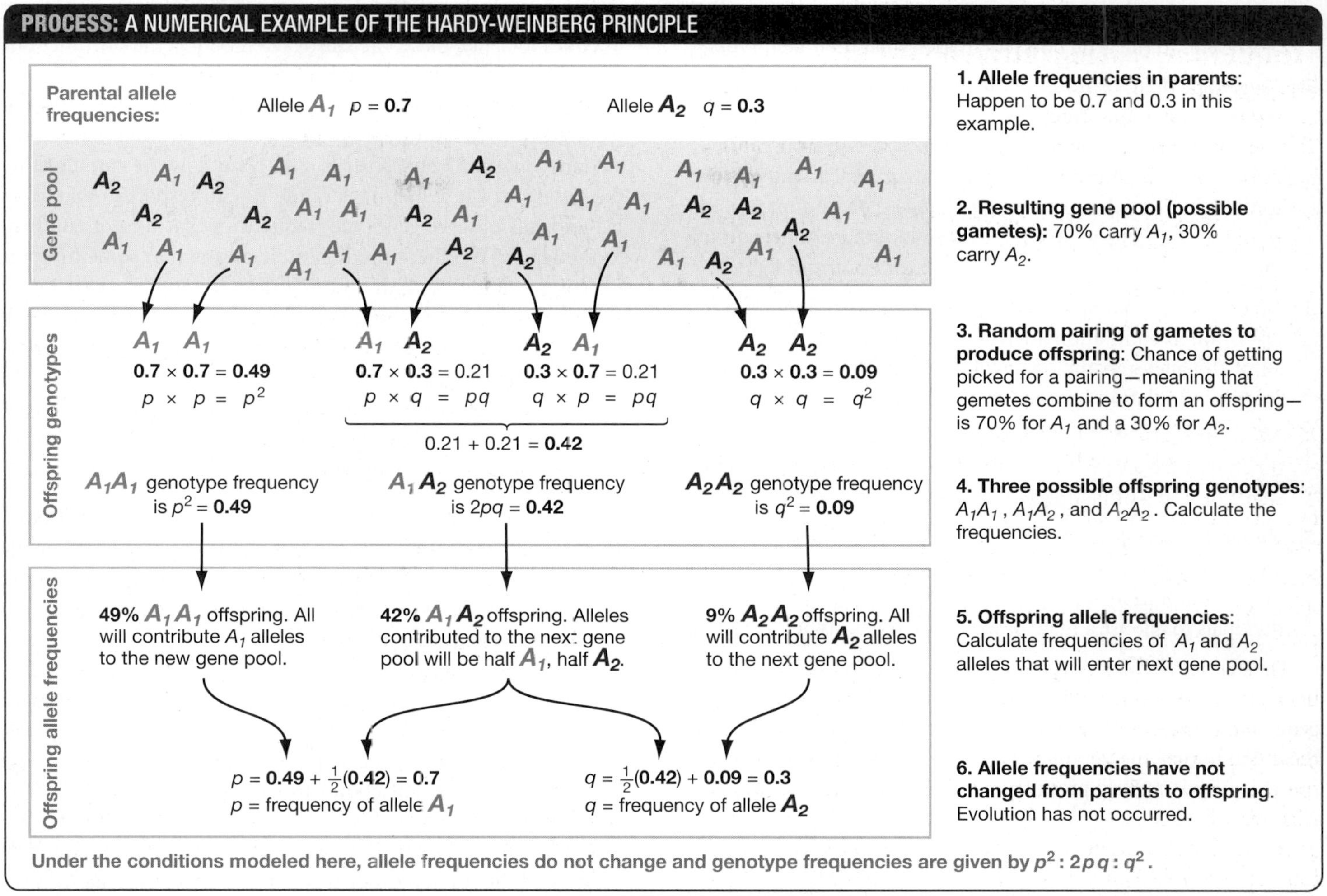

FIGURE 25.1 Deriving the Hardy-Weinberg Principle. To understand the logic behind calculating the frequency of A_1A_2 genotypes in step 4, see **BioSkills 13** in Appendix A.

Allele frequencies in parental generation:

$A1$ $p = 0.7$ A_2 $q = 0.3$

All eggs in gene pool

A_1 $p = 0.7$ A_2 $q = 0.3$

All sperm in gene pool

A_1 $p = 0.7$
A_1A_1 $p^2 = 0.49$
A_1A_2 $pq = 0.21$

A_2 $q = 0.3$
A_2A_1 $pq = 0.21$
A_2A_2 $q^2 = 0.09$

Allele frequencies have no: changed

Genotype frequencies in offspring generation:
A_1A_1 $p^2 = 0.49$
A_1A_2 $2pq = 0.42$
A_2A_2 $q^2 = 0.09$

Allele frequencies in offspring generation:
A_1 $p = 0.49 + \frac{1}{2}(0.42) = 0.70$
A_2 $q = \frac{1}{2}(0.42) + 0.09 = 0.30$

2. When alleles are transmitted via meiosis and random combination of gametes, their frequencies do not change over time. For evolution to occur, some other factor or factors must come into play.

What are these other factors?

The Hardy-Weinberg Model Makes Important Assumptions

The Hardy-Weinberg model is based on important assumptions about how populations and alleles behave. Specifically, for a population to conform to the Hardy-Weinberg principle, none of the four mechanisms of evolution can be acting on the population. In addition, the model assumes that mating is random with respect to the gene in question. Thus, here are the five assumptions that must be met:

1. *No natural selection at the gene in question* In step 2 of Figure 25.1, the model assumed that all members of the

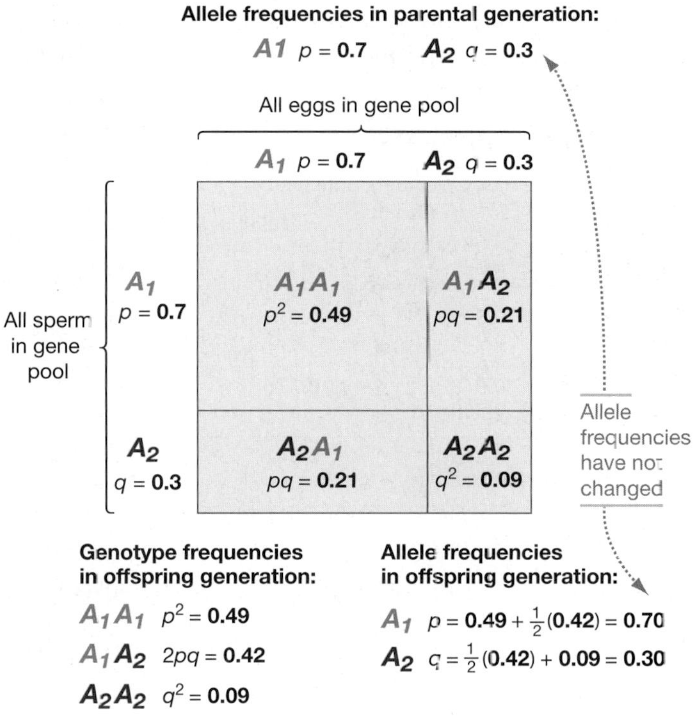

FIGURE 25.2 A Punnett Square Illustrates the Hardy-Weinberg Principle.

parental generation survived and contributed equal numbers of gametes to the gene pool, no matter what their genotype.

2. ***No genetic drift, or random allele frequency changes, affecting the gene in question*** We avoided this type of allele frequency change in step 4 of Figure 25.1 by assuming that we drew alleles in their exact frequencies p and q, and not at some different values caused by chance. For example, allele A_1 did not "get lucky" and get drawn more than 70 percent of the time. No random changes due to luck occurred.
3. ***No gene flow*** No new alleles were added by immigration or lost through emigration anywhere in Figure 25.1. As a result, all of the alleles in the offspring population came from the original population's gene pool.
4. ***No mutation*** We didn't consider that new A_1s or A_2s or other, new alleles might be introduced into the gene pool in step 2 or step 5 of Figure 25.1.
5. ***Random mating with respect to the gene in question*** We enforced this condition by picking gametes from the gene pool at random in step 3 of Figure 25.1. We did not allow individuals to choose a mate based on their genotype.

The Hardy-Weinberg principle tells us what to expect if selection, genetic drift, gene flow, and mutation are not affecting a gene, *and* if mating is random with respect to that gene. Under these conditions, the genotypes A_1A_1, A_1A_2, and A_2A_2 should be in the Hardy-Weinberg proportions p^2, $2pq$, and q^2, and no evolution will occur.

How Does the Hardy-Weinberg Principle Serve as a Null Hypothesis?

Recall from Chapter 1 that a null hypothesis predicts there are no differences among the treatment groups in an experiment. Biologists often want to test whether natural selection is acting on a particular gene, nonrandom mating is occurring, or one of the other evolutionary mechanisms is at work. In addressing questions like these, the Hardy-Weinberg principle functions as a null hypothesis.

Given a set of allele frequencies, the Hardy-Weinberg Principle predicts what genotype frequencies will be when natural selection, mutation, genetic drift, and gene flow are not affecting the gene; and when mating is random with respect to that gene. If biologists observe genotype frequencies that do not conform to the Hardy-Weinberg prediction, it means that something interesting is going on: Either nonrandom mating is occurring, or allele frequencies are changing for some reason. Further research is needed to determine which of the five Hardy-Weinberg conditions is being violated.

Let's consider two examples to illustrate how the Hardy-Weinberg principle is used as a null hypothesis: MN blood types and *HLA* genes, both in humans.

CASE STUDY 1: ARE MN BLOOD TYPES IN HUMANS IN HARDY-WEINBERG PROPORTIONS? One of the first genes that geneticists could analyze in natural populations was the MN blood group of humans. Most human populations have two alleles, designated *M* and *N*, at this gene.

Because the *MN* gene codes for a protein found on the surface of red blood cells—with the M allele coding for the M version and the N allele coding for the N version—researchers could determine whether individuals are *MM*, *MN*, or *NN* by treating blood samples with antibodies to each protein (this technique was first introduced in Chapter 8). The *M* and *N* alleles are codominant—meaning that heterozygotes have both M and N versions of the protein on their red blood cells (see Chapter 13).

To estimate the frequency of each genotype in a population, geneticists obtain data from a large number of individuals and then divide the number of individuals with each genotype by the total number of individuals in the sample.

Table 25.1 shows *MN* genotype frequencies for populations from throughout the world and illustrates how observed genotype

TABLE 25.1 **The MN Blood Group of Humans: Observed and Expected Genotype Frequencies**

The expected genotype frequencies are calculated from the observed allele frequencies, using the Hardy-Weinberg principle.

		Genotype Frequencies			Allele Frequencies	
Population and Location	**Data Type**	***MM***	***MN***	***NN***	***M***	***N***
Inuit (Greenland)	Observed	0.835	0.156	0.009	0.913	0.087
	Expected	0.834	0.159	0.008		
Native Americans (U.S.)	Observed	0.600	0.351	0.049	0.776	0.224
	Expected	0.602	0.348	0.050		
Caucasians (U.S.)	Observed	0.292	0.494	0.213	0.540	0.460
	Expected	0.290	0.497	0.212		
Aborigines (Australia)	Observed	0.025	0.304	0.672	0.178	0.825
	Expected	0.031	0.290	0.679		
Ainu (Japan)	Observed (Step 1 →)	0.179	0.502	0.319 (Step 2 →)		
	Expected (← Step 3)					

✔**EXERCISE** Fill in the values for observed allele frequencies and expected genotype frequencies for the Ainu people of Japan.

frequencies are compared with the genotype frequencies expected if the Hardy-Weinberg principle holds. The analysis is based on the following steps:

Step 1 Estimate genotype frequencies by observation—in this case, by testing many blood samples for the *M* and *N* alleles. These frequencies are given in the rows labeled "observed" in Table 25.1.

Step 2 Calculate observed allele frequencies from the observed genotype frequencies. In this case, the frequency of the *M* allele is the frequency of *MM* homozygotes plus half the frequency of *MN* heterozygotes; the frequency of the *N* allele is the frequency of *NN* homozygotes plus half the frequency of *MN* heterozygotes. (You can review the logic behind this calculation in steps 5 and 6 of Figure 25.1.)

Step 3 Use the observed allele frequencies to calculate the genotypes expected according to the Hardy-Weinberg principle. Under the null hypothesis of no evolution and random mating, the expected genotype frequencies are $p^2 : 2pq : q^2$.

Step 4 Compare the observed and expected values. Researchers use statistical tests to determine whether the differences between the observed and expected genotype frequencies are small enough to be due to chance or large enough to reject the null hypothesis of no evolution and random mating.

Although using statistical testing is beyond the scope of this text (see **BioSkills 5** in Appendix A for a brief introduction to the topic), you should be able to inspect the numbers and comment on them. In these populations, for example, the observed and expected *MN* genotype frequencies are almost identical. (A statistical test shows that the small differences observed are probably due to chance.) For every population surveyed, genotypes at the *MN* gene are in Hardy-Weinberg proportions. As a result, biologists conclude that the assumptions of the Hardy-Weinberg model are valid for this locus.

The results imply that when these data were collected, the *M* and *N* alleles in these populations were not being affected by the four evolutionary mechanisms and that mating was random with respect to this gene—meaning that humans were not choosing mates on the basis of their *MN* genotype.

Before moving on, however, it is important to note that a study such as this does not mean that the *MN* gene has never been under selection or subject to nonrandom mating or genetic drift. Even if selection has been very strong for many generations, one generation of no evolutionary forces and of random mating will result in genotype frequencies that conform to Hardy-Weinberg expectations.

The Hardy-Weinberg principle is used to test the hypothesis that currently no evolution is occurring at a particular gene and that in the previous generation, mating was random with respect to the gene in question.

CASE STUDY 2: ARE *HLA* GENES IN HUMANS IN HARDY-WEINBERG EQUILIBRIUM? A research team collected data on the genotypes of 122 individuals from the Havasupai tribe native to Arizona. These biologists were studying two genes that are important in the functioning of the human immune system. More specifically, the genes that they analyzed code for proteins that help immune system cells recognize and destroy invading bacteria and viruses.

Previous work had shown that different alleles exist at both the *HLA-A* and *HLA-B* genes, and that the alleles at each gene code for proteins that recognize proteins from slightly different disease-causing organisms. Like the *M* and *N* alleles, *HLA* alleles are codominant.

As a result, the research group hypothesized that individuals who are heterozygous at one or both of these genes may have a strong fitness advantage. The logic is that heterozygous people have a wider variety of HLA proteins, so their immune systems can recognize and destroy more types of bacteria and viruses. They should be healthier and have more offspring than homozygous people do.

To test this hypothesis, the researchers used their data on observed genotype frequencies to determine the frequency of each allele present. When they used these allele frequencies to calculate the expected number of each genotype according to the Hardy-Weinberg principle, they found the observed and expected values reported in **Table 25.2**.

When you inspect these data, notice that there are many more heterozygotes and many fewer homozygotes than expected under Hardy-Weinberg conditions. Statistical tests show it is extremely unlikely that the difference between the observed and expected numbers could occur purely by chance.

These results supported the team's prediction and indicated that one of the assumptions behind the Hardy-Weinberg principle was being violated. But which assumption? The researchers argued that mutation, migration, and drift are negligible in this case and offered two competing explanations for their data:

1. ***Mating may not be random with respect to the HLA genotype.*** Specifically, people may subconsciously prefer mates with *HLA* genotypes unlike their own and thus produce an excess of heterozygous offspring. This hypothesis is plausible. For example, experiments have shown that college students can distinguish each others' genotypes at genes related to *HLA* on the basis of body odor. Individuals in this study were more attracted to the smell of genotypes unlike their own. If this is true among the Havasupai, then nonrandom mating

TABLE 25.2 ***HLA* Genes of Humans: Observed and Expected Genotypes**

The expected numbers of homozygous and heterozygous genotypes are calculated from observed allele frequencies, according to the Hardy-Weinberg principle.

		Genotype counts (n = 122)	
Gene	**Data Type**	**Homozygotes**	**Heterozygotes**
HLA-A	Observed	38	84
	Expected	48	74
HLA-B	Observed	21	101
	Expected	30	92

SOURCE: T. Markow et al., *HLA* polymorphism in the Havasupai: Evidence for balancing selection. *American Journal of Human Genetics* 53 (1993): 943–952. Published by the University of Chicago Press.

would lead to an excess of heterozygotes compared with the proportion expected under Hardy-Weinberg.

2. ***Heterozygous individuals may have higher fitness.*** This hypothesis is supported by data collected by a different research team, who studied the Hutterite people living in South Dakota. In the Hutterite population, married women who have the same *HLA*-related alleles as their husbands have more trouble getting pregnant and experience higher rates of spontaneous abortion than do women with *HLA*-related alleles different from those of their husbands. The data suggest that homozygous fetuses have lower fitness than do fetuses heterozygous at these genes. If this were true among the Havasupai, selection would lead to an excess of heterozygotes relative to Hardy-Weinberg expectations.

Which explanation is correct? It is possible that both are. But the fact is, no one knows. Using the Hardy-Weinberg principle as a null hypothesis allowed biologists to detect an interesting pattern in a natural population. Research continues on the question of why the pattern exists.

To review how Hardy and Weinberg derived their principle and how it is used as a null hypothesis, go to the study area at *www.masteringbiology.com.*

MB **Web Activity** The Hardy-Weinberg Principle

CHECK YOUR UNDERSTANDING

If you understand that . . .

- The Hardy-Weinberg principle functions as a null hypothesis when researchers test whether nonrandom mating or evolution is occurring at a particular gene.

✔ You should be able to . . .

Analyze whether a gene suspected of causing hypertension in humans is in Hardy-Weinberg proportions, and comment on why or why not. In one study, the observed genotype frequencies were A_1A_1 0.574; A_1A_2 0.339; A_2A_2 0.087. (Note: The sample size in this study was so large that a difference of 3 percent or more in any of the observed versus expected frequencies indicated a statistically significant difference—meaning, a difference that was not due to chance.)

Answers are available in Appendix B.

25.2 Types of Natural Selection

Natural selection occurs when individuals with certain phenotypes produce more offspring than individuals with other phenotypes do. If certain alleles are associated with the favored phenotypes, they increase in frequency while other alleles decrease in frequency. The result is evolution. To use the language introduced in Chapter 24, evolution by natural selection occurs when heritable variation leads to differential success in survival and reproduction.

Although you should have a solid understanding of why evolution by natural selection occurs, it is important to recognize that natural selection occurs in a wide variety of patterns. Each of these patterns has different causes and consequences.

When biologists analyze the consequences of different patterns of selection, they often focus on **genetic variation**—the number and relative frequency of alleles that are present in a particular population. The reason is simple: Lack of genetic variation in a population is usually a bad thing.

To understand why this is so, recall from Chapter 24 that selection can occur only if heritable variation exists in a population. If genetic variation is low and the environment changes—perhaps due to the emergence of a new disease-causing virus, a rapid change in climate, or a reduction in the availability of a particular food source—it is unlikely that any alleles will be present that have high fitness under the new conditions. As a result, the average fitness of the population will decline. If the environmental change is severe enough, the population may even be faced with extinction.

Let's examine some of the different types of natural selection with this question in mind: How do they affect the level of genetic variation in the population?

Directional Selection

According to the data introduced in Chapter 24, natural selection has increased the frequency of drug-resistant strains of the tuberculosis bacterium and caused changes in beak shape and body size in medium ground finches. This type of natural selection is called **directional selection**, because the average phenotype of the populations changed in one direction.

DIRECTIONAL SELECTION TENDS TO REDUCE GENETIC VARIATION The top graph in **Figure 25.3a** plots the value of a trait on the *x*-axis and the number of individuals with a particular value of that trait is plotted on the *y*-axis. (In a histogram like this, the *y*-axis could also plot the frequency of individuals with a particular trait value—see **BioSkills 2** in Appendix A.) Note that the trait in question has a bell-shaped, normal distribution in this hypothetical population. Recall from Chapter 13 that when many different genes influence a trait, the distribution of phenotypes in the population tends to form a bell-shaped curve like this.

The middle and bottom graphs in the figure show what happens when directional selection acts on this trait. Note that in cases like this, directional selection is acting on many different genes at once. In contrast, selection on drug resistance in the TB bacterium was acting on a single gene.

Directional selection tends to reduce the genetic diversity of populations. If directional selection continues over time, the favored alleles will eventually approach a frequency of 1.0 while disadvantageous alleles will approach a frequency of 0.0. Alleles that reach a frequency of 1.0 are said to be fixed; those that reach a frequency of 0.0 are said to be lost. When disadvantageous alleles decline in frequency, **purifying selection** is said to occur.

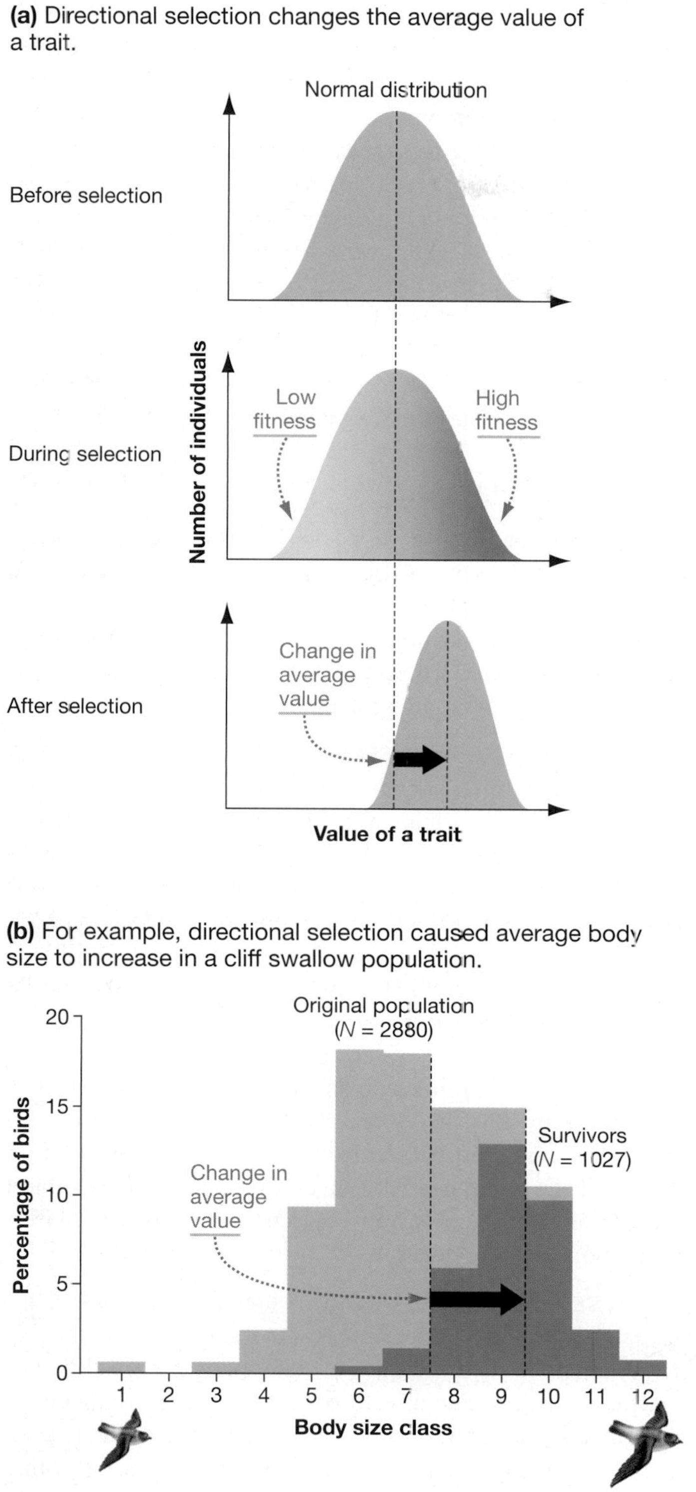

FIGURE 25.3 Directional Selection. (a) When directional selection acts on traits that have a normal distribution, individuals at one end of the distribution experience poor reproductive success. **(b)** The light green histogram shows the distribution of overall body size in cliff swallows prior to an extended cold snap that killed many individuals. (The various body-size classes were calculated from measurements of wing length, tail length, leg length, and beak size.) The dark green histogram shows the size distribution of individuals from the same population that survived the cold spell. Here *N* indicates the sample size.

Fixation and loss may not always occur under directional selection, however. To appreciate why, consider recent data on the body size of swallows.

DIRECTIONAL SELECTION ON BODY SIZE IN CLIFF SWALLOWS In 1996 a population of cliff swallows native to the Great Plains of North America endured a six-day period of exceptionally cold, rainy weather. Cliff swallows feed by catching mosquitoes and other insects in flight. Insects disappeared during this cold snap, however, and the biologists recovered the bodies of 1853 swallows that died of starvation.

As soon as the weather improved, the researchers caught and measured the body size of 1027 survivors from the same population. As the histograms in **Figure 25.3b** show, survivors were much larger on average than the birds that died.

Directional selection, favoring large body size, had occurred. To explain this observation, the investigators suggest that larger birds survived because they had larger fat stores and did not get as cold as the smaller birds. As a result, the larger birds were less likely to die of exposure to cold and more likely to avoid starvation until the weather warmed up and insects were again available.

If the differences in body size among individuals were due in part to differences in their genotypes—meaning that heritable variation in body size existed—then the population evolved. If so, and if this type of directional selection continued, then alleles that contribute to small body size would quickly decline in frequency in the cliff swallow population.

It is not clear that this will be the case, however, because directional selection is rarely constant throughout a species' range and through time. By examining weather records, the researchers established that cold spells as severe as the one that occurred in 1996 are rare.

COUNTERVAILING SELECTION AND FITNESS TRADE-OFFS Research on other swallow species suggests that smaller birds are more maneuverable in flight and thus more efficient when they feed. If so, then selection for feeding efficiency could counteract selection by cold weather. When this is the case, individuals with intermediate body size should be favored.

In swallows and many other species, it is common to find that one cause of directional selection on a trait is counterbalanced by a different factor that causes selection in the opposite direction. This concept, known as a fitness trade-off, was introduced in Chapter 24. In such cases, the optimal phenotype is intermediate. The same pattern can result from a different pattern of natural selection, called stabilizing selection.

Stabilizing Selection

When cliff swallows were exposed to cold weather, selection greatly reduced one extreme in the range of phenotypes and resulted in a directional change in the average characteristics of the population. But selection can also reduce both extremes in a

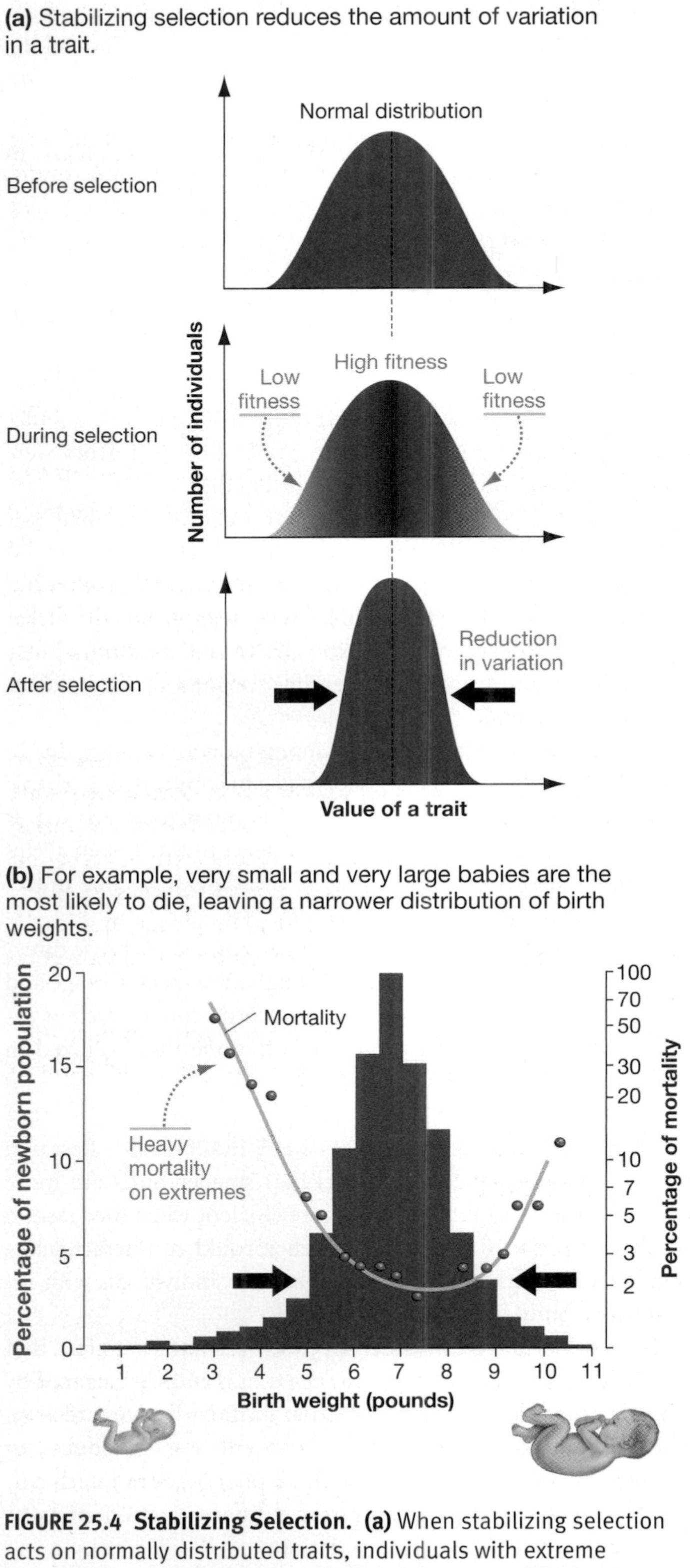

FIGURE 25.4 Stabilizing Selection. (a) When stabilizing selection acts on normally distributed traits, individuals with extreme phenotypes experience poor reproductive success. **(b)** Histogram showing the percentages of newborns with various birth weights on the left-hand axis. The purple dots indicate the percentage of newborns in each weight class that died, plotted on the logarithmic scale shown on the right.

population, as illustrated in **Figure 25.4a**. This pattern of selection is called **stabilizing selection**. It has two important consequences: There is no change in the average value of a trait over time, and genetic variation in the population is reduced.

Figure 25.4b shows a classical data set in humans illustrating stabilizing selection. Biologists who analyzed birth weights and mortality in 13,730 babies born in British hospitals in the 1950s found that babies of average size (slightly over 7 pounds) survived best. Mortality was high for very small babies and very large babies. This is persuasive evidence that birth weight was under strong stabilizing selection in this population. Alleles associated with high birth weight or low birth weight were subject to purifying selection, and alleles associated with intermediate birth weight increased in frequency.

Disruptive Selection

Disruptive selection has the opposite effect of stabilizing selection. Instead of favoring phenotypes near the average value and eliminating extreme phenotypes, it eliminates phenotypes near the average value and favors extreme phenotypes (**Figure 25.5a**). When disruptive selection occurs, the overall amount of genetic variation in the population is maintained.

DISRUPTIVE SELECTION ON BEAK SIZE IN BLACK-BELLIED SEEDCRACKERS Recent research has shown that disruptive selection is responsible for the striking distribution of bills of black-bellied seedcrackers (**Figure 25.5b**). The data plotted in the graph show that individuals with either very short or very long beaks survive best and that birds with intermediate phenotypes are at a disadvantage.

In this case, the agent that causes natural selection is food. At a study site in south-central Cameroon, West Africa, a researcher found that only two sizes of seed are available to the seedcrackers: large and small. Birds with small beaks crack and eat small seeds efficiently. Birds with large beaks handle large seeds efficiently. But birds with intermediate beaks have trouble with both, so alleles associated with medium-sized beaks are subject to purifying selection. Disruptive selection maintains high overall variation in this population.

DISRUPTIVE SELECTION CAN LEAD TO FORMATION OF NEW SPECIES Disruptive selection is important because it sometimes plays a part in speciation, or the formation of new species. If small-beaked seedcrackers began mating with other small-beaked individuals, their offspring would tend to be small-beaked and would feed on small seeds. Similarly, if large-beaked individuals chose only other large-beaked individuals as mates, they would tend to produce large-beaked offspring that would feed on large seeds.

In this way, selection would result in two distinct populations. Under some conditions, the populations may eventually form two new species. The process of species formation, based on disruptive selection and other mechanisms, is explored in detail in Chapter 26.

Balancing Selection

Directional selection, stabilizing selection, and disruptive selection describe how natural selection can act on polygenic traits in

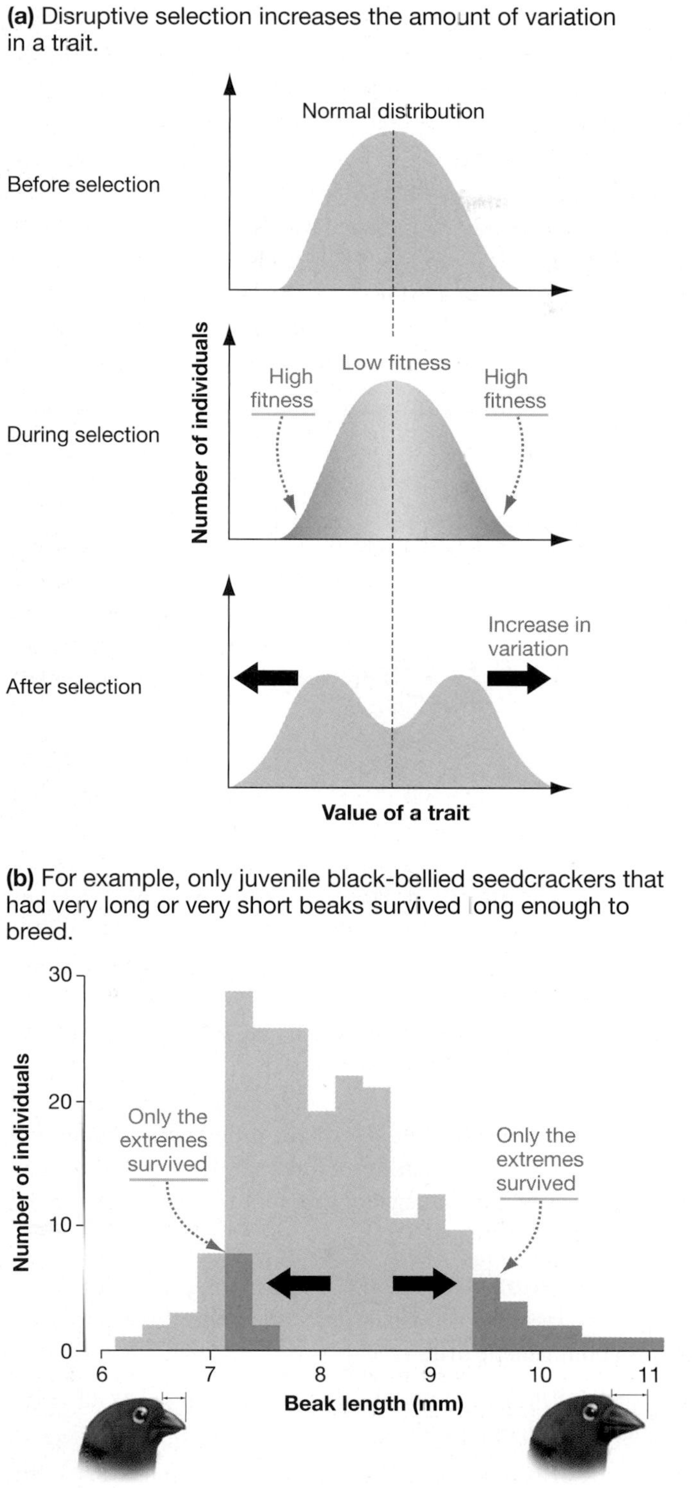

FIGURE 25.5 Disruptive Selection. (a) When disruptive selection occurs on traits with a normal distribution, individuals with extreme phenotypes experience high reproductive success. **(b)** Histogram showing the distribution of beak length in a population of black-bellied seedcrackers. The light orange bars represent all juveniles; the dark orange bars, juveniles that survived to adulthood.

a single generation or episode. To review how they work, go to the study area at *www.masteringbiology.com*.

(MB) **Web Activity** Three Modes of Natural Selection

It's important to recognize, though, that they are not the only patterns of selection. For example, the second explanation for the data in Table 25.2 is a pattern of natural selection called **heterozygote advantage**. When selection operates in this way, heterozygous individuals have higher fitness than homozygous individuals do. The consequence of this pattern is that genetic variation is maintained in populations.

Heterozygote advantage is one mechanism responsible for a more general phenomenon known as **balancing selection**. When balancing selection occurs, no single allele has a distinct advantage and increases in frequency. Instead, there is a balance among several alleles in terms of their fitness and frequency. Balancing selection also occurs when:

1. The environment varies over time or in different geographic areas occupied by a population—meaning that certain alleles are favored by natural selection at different times or in different places. As a result, overall genetic variation in the population is maintained or increased.
2. Certain alleles are favored when they are rare, but not when they are common—a pattern known as **frequency-dependent selection**. For example, rare alleles responsible for coloration in guppies are favored because predators learn to recognize common color patterns. Alleles for common colors get eliminated; alleles for rare colors increase in frequency. As a result, overall genetic variation in the population is maintained or increased.

No matter how natural selection occurs, though, its most fundamental attribute is the same: It increases fitness and leads to adaptation.

25.3 Genetic Drift

Natural selection is not random. It is directed by the environment and results in adaptation. Genetic drift, in contrast, is undirected and random.

Genetic drift is defined as any change in allele frequencies in a population that is due to chance. The process is aptly named, because it causes allele frequencies to drift up and down randomly over time. When drift occurs, allele frequencies change due to blind luck—what is formally known as **sampling error.** Drift occurs in every population, in every generation.

Simulation Studies of Genetic Drift

To understand why genetic drift occurs, imagine a couple marooned on a deserted island. Suppose that at gene *A*, the wife's genotype is A_TA_H and the husband is also A_TA_H. In this population, the two alleles are each at a frequency of 0.5.

Now suppose that the couple produce five children over their lifetime. Half of the eggs produced by the wife carry allele A_T and half carry allele A_H. Likewise, half of the sperm produced by the husband carry allele A_T and half carry allele A_H. To simulate which sperm and which egg happen to combine to produce each of the five offspring, you can flip a coin for each sperm and each egg, with tails standing for allele A_T and heads standing for allele A_H.

The following coin flips were done by a pair of students in a recent biology class:

	Sperm	Egg	Genotype
First offspring	A_H	A_H	A_HA_H
Second offspring	A_T	A_T	A_TA_T
Third offspring	A_T	A_H	A_HA_T
Fourth offspring	A_H	A_H	A_HA_H
Fifth offspring	A_T	A_H	A_HA_T

When the parents die, there are a total of 10 alleles in the population. But note that the allele frequencies have changed. In this generation, six of the 10 alleles (60 percent) are A_H; four of the 10 alleles (40 percent) are A_T. Evolution—a change in allele frequencies in a population—occurred due to genetic drift.

Instead of each allele being sampled in exactly its original frequency when offspring formed, as the Hardy-Weinberg principle assumes, a chance sampling error occurred. Allele A_H got lucky; allele A_T was unlucky.

COMPUTER SIMULATIONS **Figure 25.6** shows what happens when a computer simulates the same process of random combinations in gametes over time. The program that generated the graphs combines the alleles in a gene pool at random to create an offspring generation, calculates the allele frequencies in the offspring generation, and uses those allele frequencies to create a new gene pool.

In this example, the process was continued for 100 generations. The *x*-axis on each graph plots time in generations; the *y*-axis plots the frequency of one of the two alleles present at the *A* gene in a hypothetical population.

The top graph shows eight replicates of this process with a population size of 4; the bottom graph shows eight replicates with a population of 400. Notice (**1**) the striking differences between the effects of drift in the small versus large population and (**2**) the consequences for genetic variation when alleles drift to fixation or loss.

Given enough time, drift can be an important factor even in large populations. To drive this point home, consider two types of alleles that were introduced in earlier chapters and that have no effect on fitness. Recall from Chapter 15 that alleles containing silent mutations, usually in the third position of a codon, do not change the gene product. As a result, most of these alleles have little or no effect on the phenotype. Yet these alleles routinely drift to high frequency or even fixation over time. Similarly, recall from Chapter 20 that pseudogenes do not code for a product. Although their presence does not affect an individual's phenotype, dozens of pseudogenes in the human genome have reached fixation, due to drift.

✔If you understand genetic drift, you should be able to examine the MN blood group genotype frequencies in Table 25.1 and describe how drift could explain differences in genotype frequencies among populations. Note that there are no data indicating a selective advantage for different MN genotypes in different environments.

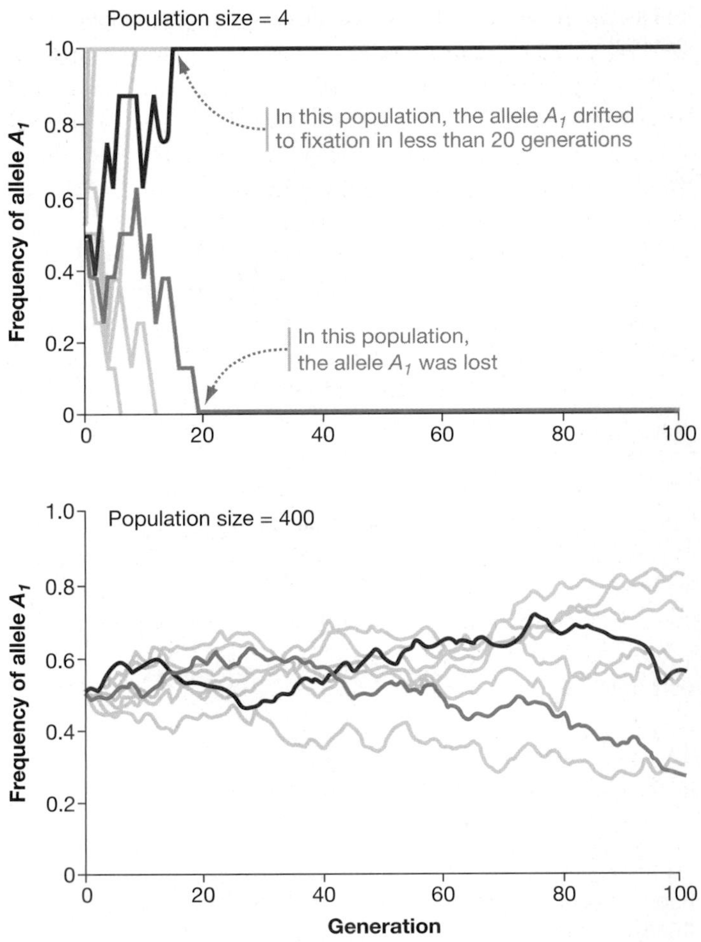

FIGURE 25.6 Genetic Drift Is More Pronounced in Small Populations than Large Populations.

✔EXERCISE Draw graphs predicting what these graphs would look like for a population size of 4000.

KEY POINTS ABOUT GENETIC DRIFT The matings that were simulated in a class and on a computer illustrate three important points about genetic drift:

- Genetic drift is random with respect to fitness. The allele frequency changes it produces are not adaptive.
- Genetic drift is most pronounced in small populations. In the computer simulation, allele frequencies changed much less in the large population than the small population. And if the couple on the deserted island had produced 50 children instead of five, it is almost certain that allele frequencies in the next generation would have been much closer to 0.5.
- Over time, genetic drift can lead to the random loss or fixation of alleles. In the computer simulation with a population of 4, it took at most 20 generations for one allele to be fixed or lost. When random loss or fixation occurs, genetic variation in the population declines.

The importance of drift in small populations is a particular concern for conservation biologists, because many populations are being drastically reduced in size by habitat destruction and other human activities. Small populations that occupy nature reserves or zoos are particularly susceptible to genetic drift. If drift leads to a loss of genetic diversity, it could darken the already bleak outlook for some endangered species.

Experimental Studies of Genetic Drift

Research on genetic drift began with theoretical work in the 1930s and 1940s, which used mathematical models to predict the effect of genetic drift on allele frequencies and genetic variation. In the mid-1950s, Warwick Kerr and Sewall Wright did an experiment to show how drift works in practice.

Kerr and Wright started with a large laboratory population of fruit flies that contained a **genetic marker**—a specific allele that causes a distinctive phenotype. In this case, the marker was the morphology of bristles. Fruit flies have bristles on their bodies that can be either straight or bent (**Figure 25.7a**). This difference in bristle phenotype depends on a single gene. Kerr and Wright's lab population contained just two alleles—normal (straight) and "forked" (bent).

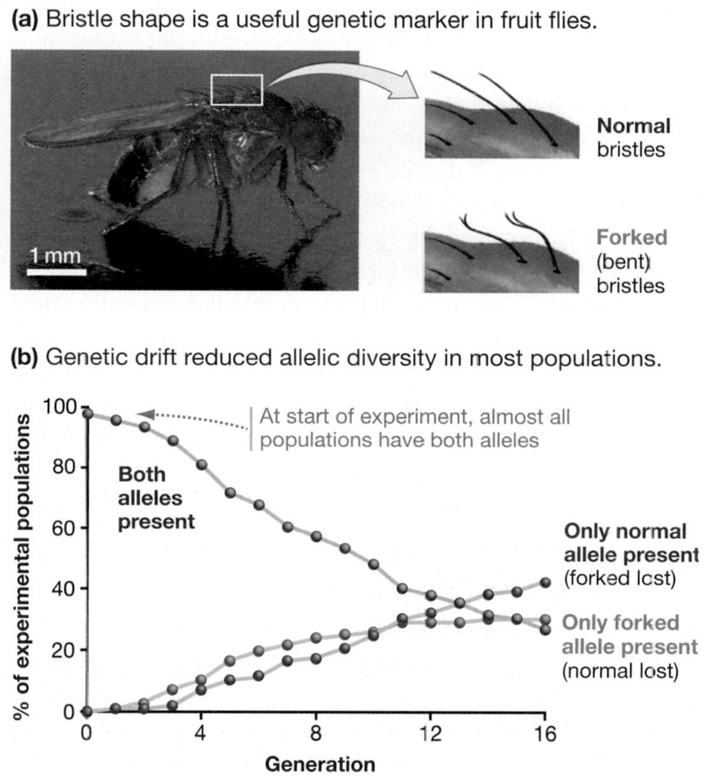

FIGURE 25.7 An Experiment on the Effects of Genetic Drift in Small Populations. At the start of an experiment on fruit flies, the frequencies of the wild-type (normal) bristle alleles and forked-bristle alleles were each 50 percent in all of 96 separate populations. Population size was kept at 8 individuals for 16 generations. By the end of the experiment, 73 percent of the 96 populations had lost either the normal or the forked allele; 26 percent of the populations still had both alleles. The only evolutionary force acting in these populations was genetic drift.

- The researchers set up 96 cages in their lab.
- They placed four adult females and four adult males of the fruit fly *Drosophila melanogaster* in each.

They chose individual flies to begin these experimental populations so that the frequency of the normal and forked alleles in each of the 96 starting populations was 0.5. The two alleles do not affect the fitness of flies in the lab environment, so Kerr and Wright could be confident that if changes in the frequency of normal and forked phenotypes occurred, they would not be due to natural selection.

- After these first-generation adults bred, Kerr and Wright reared their offspring.
- In the offspring (F_1) generation, they randomly chose four males and four females—meaning that they simply grabbed individuals without regard to whether their bristles were normal or forked—from each of the 96 offspring populations and allowed them to breed and produce the next generation.
- They repeated this procedure until all 96 populations had undergone a total of 16 generations.

During the entire course of the experiment, no migration from one population to another occurred. Previous studies had shown that mutations from normal to forked bristles (and forked to normal) are rare. Thus, the only evolutionary process operating during the experiment was genetic drift. It was as if random accidents claimed the lives of all but eight individuals in each generation, so that only eight bred.

Their result? After 16 generations, the 96 populations fell into three groups. Forked bristles were found on all of the individuals in 29 of the experimental populations. Due to drift, the forked allele had been fixed in these 29 populations and the normal allele had been lost (**Figure 25.7b**). In 41 other populations, however, the opposite was true: All individuals had normal bristles. In these populations, the forked allele had been lost due to chance. Both alleles were still present in 26 of the populations.

The message of the study is startling: In 73 percent of the experimental populations (70 out of the 96), genetic drift had reduced allelic diversity at this gene to zero.

As predicted, genetic drift decreased genetic variation within populations and increased genetic differences between populations. Is drift important in natural populations as well?

What Causes Genetic Drift in Natural Populations?

The sampling process that occurs during fertilization occurs in every population in every generation in every species that reproduces sexually. Similarly, accidents that remove individuals at random occur in every population in every generation.

It is important to realize, though, that because drift is caused by sampling error, it can occur by *any* process or event that involves sampling—not just the sampling of gametes that occurs during fertilization or the loss of unlucky individuals due to accidents. Let's consider two such examples, called founder effects and bottlenecks.

FOUNDER EFFECTS ON THE GREEN IGUANAS OF ANGUILLA When a group of individuals immigrates to a new geographic area and establishes a new population, a founder event is said to occur. If the group is small enough, the allele frequencies in the new population are almost guaranteed to be different from those in the source population—meaning, the population in the place from which the group emigrated—due to sampling error. A change in allele frequencies that occurs when a new population is established is called a **founder effect** (**Figure 25.8a**).

In 1995, fishermen on the island of Anguilla in the Caribbean witnessed a founder event involving green iguanas. A few weeks after two major hurricanes swept through the region, a large raft composed of downed logs tangled with other debris floated onto a beach on Anguilla. The fishermen noticed green iguanas on the raft and several on shore. Because green iguanas had not previously been found on Anguilla, the fishermen notified biologists. The researchers were able to document that at least 15 individuals had arrived; two years later they were able to confirm that at least some of the individuals were breeding. A new population had formed.

During this founder event, it is extremely unlikely that allele frequencies in the new Anguilla population of green iguanas exactly matched those of the source population, thought to be on the islands of Guadeloupe.

Colonization events like these have been the major source of populations that occupy islands all over the world, as well as island-like habitats such as mountain meadows, caves, and ponds.

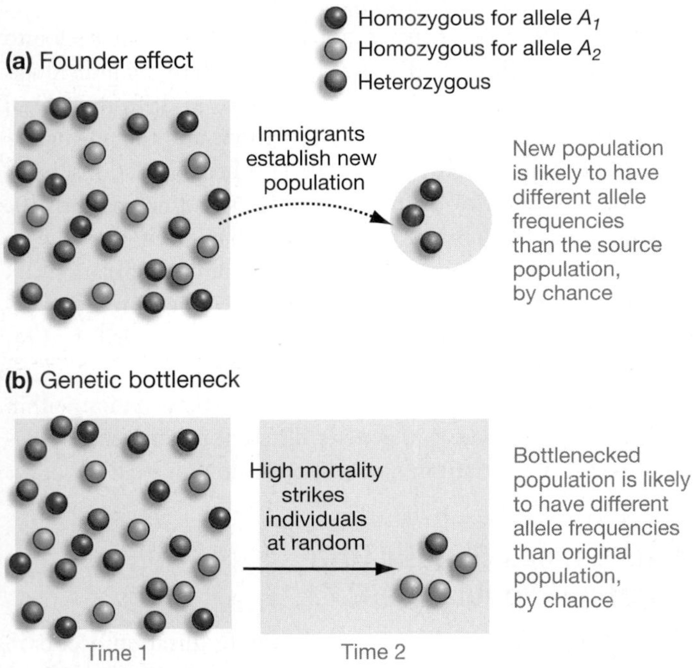

Homozygous for allele A_1
Homozygous for allele A_2
Heterozygous
(a) Founder effect
Immigrants establish new population
New population is likely to have different allele frequencies than the source population, by chance
(b) Genetic bottleneck
High mortality strikes individuals at random
Bottlenecked population is likely to have different allele frequencies than original population, by chance
Time 1
Time 2

FIGURE 25.8 Two Causes of Genetic Drift in Natural Populations. The smaller the new population, the higher the likelihood that genetic drift will result not only in differences in allele frequencies, but also loss of alleles.

✔**EXERCISE** The original population in (a) consists of 9 A_1A_1, 11 A_1A_2, and 7 A_2A_2 individuals (this hypothetical population is very small for simplicity). Compare the frequencies of the A_1 allele in the original and new populations.

Each time a founder event occurs, a founder effect is likely to accompany it, changing allele frequencies through genetic drift.

GENETIC BOTTLENECK ON PINGELAP ATOLL If a large population experiences a sudden reduction in size, a population bottleneck is said to occur. The term comes from the metaphor of a few individuals passing through the neck of a bottle, by chance. Disease outbreaks, natural catastrophes such as floods or fires or storms, or other events can cause population bottlenecks.

Genetic bottlenecks follow population bottlenecks, just as founder effects follow founder events. A **genetic bottleneck** is a sudden reduction in the number of alleles in a population. Drift occurs during genetic bottlenecks and causes a change in allele frequencies. **Figure 25.8b** provides a hypothetical example; for an example from a natural population consider the humans who occupy Pingelap Atoll in the South Pacific.

On this island, only about 20 people out of a population of several thousand managed to survive the effects of a typhoon and a subsequent famine that occurred around 1775. The survivors apparently included at least one individual who carried a loss-of-function allele at a gene called *CNGB 3*, which codes for a protein involved in color vision.

The *CNGB 3* allele is recessive, and when it is homozygous it causes a serious vision deficit called achromatopsia. The condition is extremely rare in most populations, with the frequency of the *CNGB 3* allele estimated to be under 1.0 percent.

In the population that survived the Pingelap Atoll disaster, however, the loss-of-function allele was at a frequency of about 1/40, or 2.5 percent. If the allele was at the typical frequency of 1.0 percent or less prior to the population bottleneck, then a large frequency change occurred during the bottleneck, due to drift.

In today's population on Pingelap Atoll, over 1 in 20 people is afflicted with achromatopsia, and the allele which causes the affliction is at a frequency of well over 20 percent. Because it is extremely unlikely that the loss-of-function allele is favored by directional selection or heterozygote advantage, researchers hypothesize that the frequency of the allele in this small population has continued to increase over the past 230 years due to drift.

CHECK YOUR UNDERSTANDING

If you understand that . . .

- Genetic drift occurs any time allele frequencies change due to chance.
- Drift violates the assumptions of the Hardy-Weinberg principle and occurs during many different types of events, including random fusion of gametes at fertilization, founder events, and population bottlenecks.

✔ **You should be able to . . .**

1. Explain why drift leads to a random loss or fixation of alleles.
2. Explain why drift is particularly important as an evolutionary force in small populations.

Answers are available in Appendix B.

25.4 Gene Flow

Gene flow is the movement of alleles from one population to another. It occurs when individuals leave one population, join another, and breed.

As an evolutionary mechanism, gene flow usually has one outcome: It equalizes allele frequencies between the source population and the recipient population. When alleles move from one population to another, the populations tend to become more alike. To capture this point, biologists say that gene flow homogenizes allele frequencies among populations (**Figure 25.9**).

Gene Flow in Natural Populations

Theoretical work quickly established how gene flow should affect populations. Studying gene flow in natural populations has been more challenging, however. To see how researchers are documenting gene flow's effects, consider a population of birds called great tits, native to Western Europe.

On the island of Vlieland off the coast of the Netherlands, great tits breed in two sets of woodlands. One lies to the west; one to the east. In both areas, mated pairs nest in boxes that have been maintained by biologists since 1955. For many decades, all of the great tits hatched on the island have been marked with leg bands.

Recently, biologists analyzed data on nesting that took place between 1975 and 1995 and found that individuals hatched in the eastern woodlands laid smaller clutches—groups of eggs that are incubated together—than individuals hatched in the western woodlands. Because the difference occurred whether the birds actually bred in the eastern or western habitats, the researchers inferred that it was due to a genetic difference between the eastern and western subpopulations.

Further, the data indicated that females hatched in the eastern woodlands survive much better than females hatched in the western habitats. The eastern birds appear to be much better adapted to the island environment.

Why aren't individuals from the western population as well-adapted to local conditions? The answer is gene flow. In the western woodlands, 43 percent of individuals breeding for the first time are unbanded—meaning that they just arrived from the mainland (**Figure 25.10a**). But in the eastern woodlands, only 13 percent of the individuals breeding for the first time are immigrants. Gene flow is over three times as high in the west as it is in the east.

To see the impact of gene flow on this population, study the data graphed in **Figure 25.10b**. The *x*-axis plots the number of grandparents that a female has that were hatched on the island. This is a measure of how local an individual's ancestry is. The *y*-axis plots the average chance that a female survives from one year to the next. This is a measure of fitness. Note that if a female has 0–2 grandparents from the island, they have a 30–35% chance of surviving from year to year. But if they have 3 or 4 grandparents who were local, their chance of surviving jumps to about 45 percent. Females with more island-born grandparents have higher fitness.

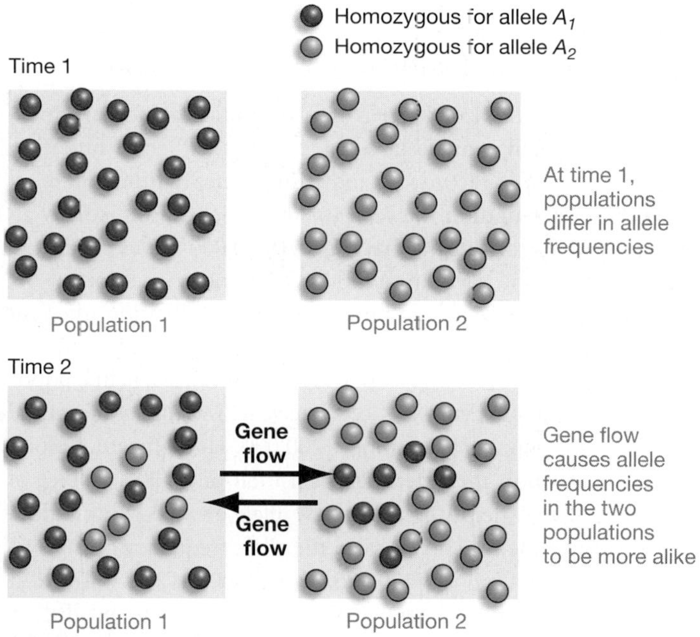

FIGURE 25.9 Gene Flow Makes Allele Frequencies More Similar between Populations. Gene flow can occur in one or both directions. If gene flow continues and no selection occurs, allele frequencies will eventually be identical.

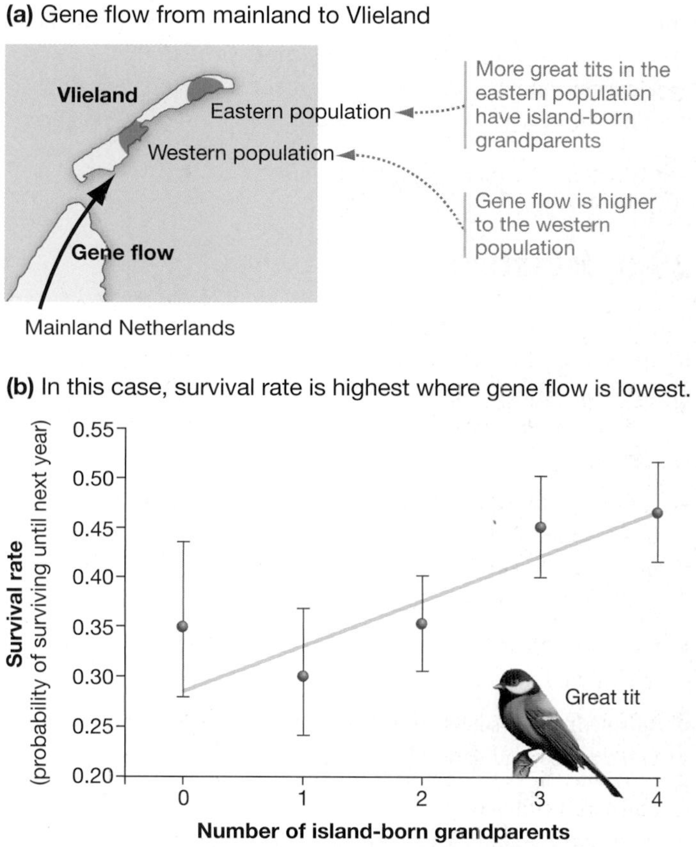

FIGURE 25.10 Gene Flow Reduces Fitness in a Population of Great Tits. In this study, the fewer immigrant grandparents an island bird has, the higher its survival rate.

How Does Gene Flow Affect Fitness?

In the case of great tits on Vlieland, immigrants bring in alleles that have relatively low fitness in the island environment. Natural selection and gene flow are working in opposition: Selection favors small clutch sizes and other traits that affect survival in the island environment; gene flow constantly introduces alleles that have high fitness in mainland habitats but low fitness on the island. The homogenizing effect of gene flow is bad for island birds.

It is not true, though, that gene flow always reduces fitness. If a population has lost alleles due to genetic drift, then the arrival of new alleles via gene flow should increase genetic diversity. If increased genetic diversity results in better resistance to infections by bacteria or viruses or other parasites, gene flow would increase the average fitness of individuals. Chapter 55 provides an example of how gene flow aided an endangered population.

Gene flow is random with respect to fitness—the arrival or departure of alleles can increase or decrease average fitness, depending on the situation. But in every case, a movement of alleles between populations tends to reduce genetic differences between them.

This latter generalization is particularly important in our own species right now. Large numbers of people from Africa, the Middle East, Mexico, Central America, and Asia are immigrating to the countries of the European Union and the United States. Because individuals from different cultural and ethnic groups are intermarrying frequently and having offspring, allele frequencies in human populations are becoming more similar.

To review how natural selection, genetic drift, and gene flow affect populations, go to the study area at *www.masteringbiology.com.*

(MB) **BioFlix™** Mechanisms of Evolution

25.5 Mutation

To appreciate the role of mutation as an evolutionary force, let's return to one of the central questions that biologists ask about an evolutionary mechanism: How does it affect genetic variation in a population? To appreciate why this is an important question, recall that:

- Gene flow increases genetic diversity in a recipient population if new alleles arrive with immigrating individuals. But gene flow may decrease genetic variation in the source population if alleles leave with emigrating individuals.
- Genetic drift tends to decrease genetic diversity over time, as alleles are randomly lost or fixed.
- Most forms of selection favor certain alleles and lead to a decrease in overall genetic variation.

If most of the evolutionary mechanisms lead to a loss of genetic diversity over time, what restores it? In particular, where do entirely new alleles come from? The answer to both of these questions is **mutation**.

As Chapter 15 noted, mutations occur when DNA polymerase makes an error as it copies a DNA molecule, resulting in a change in the sequence of deoxyribonucleotides. If a mutation occurs in a stretch of DNA that codes for a protein, the changed codon may result in a polypeptide with a novel amino acid sequence. Mutations also occur at the level of chromosomes—when the number of chromosomes changes due to errors in meiosis or when chromosome segments detach and change position, due to damage.

Because errors and chromosome damage are inevitable, mutation constantly introduces new alleles into populations in every generation. Mutation is an evolutionary mechanism that increases genetic diversity in populations.

Despite the fact that it consistently leads to an increase in genetic diversity in a population, mutation is random with respect to the affected allele's impact on the fitness of the individual. Changes in the makeup of chromosomes or in specific DNA sequences do not occur in ways that tend to increase fitness or decrease fitness. Mutation just happens.

But because most organisms are well adapted to their current habitat, random changes in genes usually result in products that do not work as well as the alleles that currently exist. Stated another way, most mutations in sequences that code for a functional protein or RNA result in **deleterious** alleles—alleles that lower fitness. Deleterious alleles tend to be eliminated by purifying selection.

On rare occasions, however, mutation in these types of sequences produces a **beneficial** allele—an allele that allows individuals to produce more offspring. Beneficial alleles should increase in frequency in the population due to natural selection.

Because mutation produces new alleles, it can in principle change the frequencies of alleles through time. But does mutation occur often enough to make it an important factor in changing allele frequencies? The short answer is no.

Mutation as an Evolutionary Mechanism

To understand why mutation is not a significant mechanism of evolutionary change by itself, consider that the highest mutation rates that have been recorded at individual genes in humans are on the order of 1 mutation in every 10,000 gametes produced by an individual. This rate means that for every 10,000 alleles produced, on average one will have a mutation at the gene in question.

When two gametes combine to form an offspring, then, at most about 1 in every 5000 offspring will carry a mutation at a particular gene. Now suppose that 195,000 humans live in a population; that 5000 offspring are born one year; and that at the end of that year, the population numbers 200,000. Humans are diploid, so in a population this size, there is a total of 400,000 copies of each gene. Only one of them is a new allele created by mutation, however. Over the course of a year, the allele frequency change introduced by mutation is 1/400,000, or 0.0000025 (2.5×10^{-6}). At this rate, it would take 4000 years for mutation to produce a change in allele frequency of 1 percent.

These calculations support the conclusion that mutation does little to change allele frequencies on its own. Although mutation can be a significant evolutionary force in bacteria and archaea, which have extremely short generation times, mutation in eukaryotes rarely causes a change from the genotype frequencies expected under the Hardy-Weinberg principle. As an evolutionary mechanism, mutation is slow compared with selection, genetic drift, and gene flow.

But mutation still plays a role in evolution. Humans have over 20,000 genes, so each individual carries at least 40,000 alleles. Although the rate of mutation per allele may be very low, the total number of alleles is high. Multiplying the estimated number of genes in a human by the average mutation rate per gene suggests that an average person contains about 1.1 new alleles created by mutation.

Mutation introduces new alleles into every individual in every population in every generation. And in species that undergo sexual reproduction, meiosis and genetic recombination create variation in terms of the allele combinations present in each individual. Even if selection and drift are eliminating genetic diversity, mutation renews it.

Experimental Studies of Mutation

To get a better appreciation for how mutation affects evolution, consider an experiment conducted by Richard Lenski and colleagues. The experiment was designed to evaluate the role that mutation plays in many genes over many generations.

EXPERIMENTAL EVOLUTION Lenski's group focused on *Escherichia coli*, a bacterium that is a common resident of the human intestine. To begin, they set up a large series of populations, each founded with a single cell (see **Figure 25.11**).

To track the evolution of the experimental populations over time, the researchers transferred a small number of cells from each of the populations into a new batch of the same growth medium, under the same light and temperature conditions, every day for over four years. In this way, each population grew continuously.

Over the course of the experiment, the researchers estimated that each population underwent a total of 10,000 generations. This is the equivalent of over 200,000 years of human evolution. In addition, the biologists saved a sample of cells from each population at regular intervals and stored them in a freezer. Because frozen *E. coli* cells resume growth when they are thawed, the frozen cells archived individuals that existed over the 10,000 generation time interval.

The strain of *E. coli* used in the experiment is completely asexual and reproduces by cell division. Thus, mutation was the only source of genetic variation in these populations. Although no

EXPERIMENT

QUESTION: How does average fitness in a population change over time?

HYPOTHESIS: Average fitness increases over time.

NULL HYPOTHESIS: Average fitness does not increase over time.

EXPERIMENTAL SETUP:

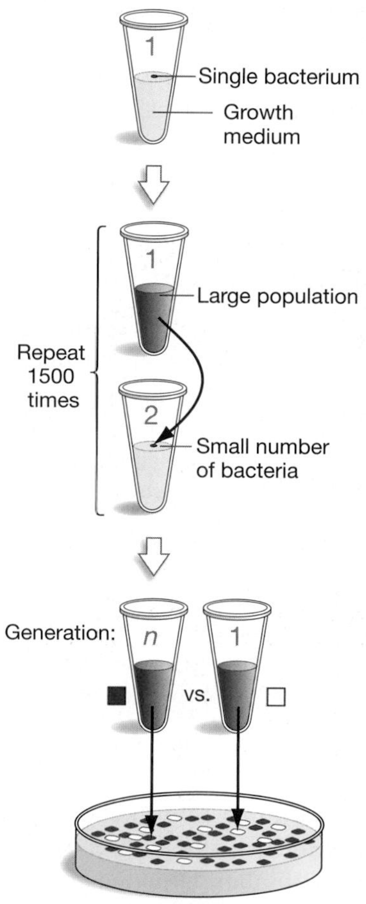

1. Place 10 mL of identical growth medium into many replicate tubes with one bacterium in each.

2. Incubate overnight. Average population in each tube is now 5×10^8 cells.

3. Remove 0.1 mL from each tube and move to 10 mL of fresh medium. Freeze remaining cells for later analysis.

4. Competition experiment. Put an equal number of cells from generation 1 and a later generation (*n*) in fresh growth medium.

5. Incubate overnight and count the cells (color indicator distinguishes generations). Which are more numerous?

PREDICTION: Descendant populations have higher average fitness. (There will be more individuals from descendant than ancestral populations on the plates.)

PREDICTION OF NULL HYPOTHESIS: There will be no difference in fitness between descendant and ancestral populations.

RESULTS:

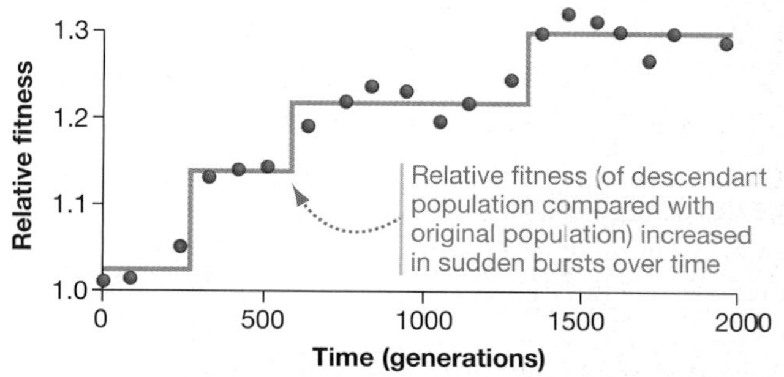

CONCLUSION: Descendant populations have higher fitness than do ancestral populations.

FIGURE 25.11 Testing Changes in Fitness over Time.

✔**QUESTION** The authors of this study claim that the frozen cells from each generation form a fossil record, except that the fossils can be brought back to life. Is their analogy valid?

SOURCE: Elena, S. F., V. S. Cooper, and R. E. Lenski. 1996. Punctuated evolution caused by selection of rare beneficial mutations. *Science* 272: 1802–1804.

SUMMARY TABLE 25.3 **Evolutionary Mechanisms**

	Definition and Notes	Effect on Genetic Variation	Effect on Average Fitness
Natural selection	Certain alleles are favored	Can lead to maintenance, increase, or reduction of genetic variation	Can produce adaptation
Genetic drift	Random changes in allele frequencies; most important in small populations	Tends to reduce genetic variation via loss or fixation of alleles	Random with respect to fitness; usually reduces average fitness
Gene flow	Movement of alleles between populations; reduces differences between populations	May increase genetic variation by introducing new alleles; may decrease it by removing alleles	Random with respect to fitness; may increase or decrease average fitness by introducing high- or low-fitness alleles
Mutation	Production of new alleles	Increases genetic variation by introducing new alleles	Random with respect to fitness; most mutations in coding sequences lower fitness

gene flow occurred, both selection and genetic drift were operating in each population.

Were cells from the older and newer generations of each population different? Lenski's group used competition experiments to address this question. Taking advantage of a neutral indicator that can distinguish two populations by color, they grew cells from two different generations on the same plate and compared their growth rates. The populations of cells that were more numerous had grown the fastest, meaning that they were better adapted to the experimental environment.

In this way the researchers could measure the fitness of descendant populations relative to ancestral populations. If relative fitness was greater than 1, it meant that recent-generation cells outnumbered older-generation cells when the competition was over.

FITNESS INCREASED IN FITS AND STARTS The data from the competition experiments are graphed in the "Results" section of Figure 25.11. Notice that relative fitness increased dramatically—almost 30 percent—over time. But notice also that fitness increased in fits and starts. This pattern is emphasized by the solid line on the graph, which represents a mathematical function fitted to the data points.

What caused this stair-step pattern? Lenski's group hypothesizes that genetic drift was relatively unimportant in this experiment because population sizes were so large. Instead, they propose that each jump was caused by a novel mutation that conferred a fitness benefit. Their interpretation is that cells that happened to have the beneficial mutation grew rapidly and came to dominate the population.

After a beneficial mutation occurred, the fitness of the population stabilized—sometimes for hundreds of generations—until another random but beneficial mutation occurred and produced another jump in fitness.

TAKE-HOME MESSAGES The data on mutation rates and experimental evolution in *E. coli* reinforce fundamental messages about mutation's role in evolution.

1. Mutation is the ultimate source of genetic variation. Crossing over and independent assortment shuffle existing alleles into new combinations, but only mutation creates new alleles.
2. If mutation did not occur, evolution would eventually stop. Recall that natural selection and genetic drift tend to eliminate alleles. Without mutation, eventually there would be no variation for selection and drift to act on. Without mutation, there would be no fitness increases observed in the *E. coli* experiment.
3. Mutation alone is usually inconsequential in changing allele frequencies at a particular gene. When considered across the genome and when combined with natural selection, however, it becomes an important evolutionary mechanism—it is the basis of fitness increases like those documented in the *E. coli* populations.

Table 25.3 summarizes these points and similar conclusions about the four evolutionary mechanisms. Each of the four evolutionary forces has different consequences for allele frequencies.

If one or more of these processes affects a gene, then genotypes will not be in Hardy-Weinberg proportions. But we've yet to consider the effects of another assumption in the Hardy-Weinberg model—that mating takes place at random with respect to the gene in question. What happens when the random mating assumption is violated?

25.6 Nonrandom Mating

In the Hardy-Weinberg model, gametes were picked from the gene pool at random and paired to create offspring genotypes. In nature, however, matings between individuals may not be random with respect to the gene in question. Even in species like clams that simply broadcast their gametes into the surrounding water, gametes from individuals who live close to each other are more likely to combine than gametes from individuals that live farther apart. In insects, vertebrates, and many other animals, females don't mate at random but actively choose certain males.

Two mechanisms that violate the Hardy-Weinberg assumption of random mating have been studied intensively. One is the phenomenon known as inbreeding; the other is the process called sexual selection.

Inbreeding

Mating between relatives is called **inbreeding**. By definition, relatives share a recent common ancestor. Individuals that inbreed are likely to share alleles they inherited from their common ancestor.

HOW DOES INBREEDING AFFECT ALLELE FREQUENCIES AND GENOTYPE FREQUENCIES? To understand how inbreeding affects populations, let's follow the fate of alleles and genotypes when inbreeding occurs. Focus again on a single locus with two alleles, A_1 and A_2, and suppose that these alleles initially have equal frequencies of 0.5.

Now imagine that the gametes produced by individuals in the population don't go into a gene pool. Instead, individuals self-fertilize. Many flowering plants, for example, contain both male and female organs and routinely self-pollinate. Self-fertilization, or selfing, is the most extreme form of inbreeding.

As **Figure 25.12a** shows, homozygous parents that self-fertilize produce all homozygous offspring. Heterozygous parents, in contrast, produce homozygous and heterozygous offspring in a 1:2:1 ratio.

Figure 25.12b shows the outcome for the population as a whole. In this figure, the width of the boxes represents the frequency of the three genotypes, which start out at the Hardy-Weinberg ratio of $p^2 : 2pq : q^2$. Notice that the homozygous proportion of the population increases each generation, while the heterozygous proportion is halved. At the end of the four generations illustrated, heterozygotes are rare. The same outcomes occur, more slowly, with less-extreme forms of inbreeding.

This simple exercise demonstrates two fundamental points about inbreeding:

1. Inbreeding increases homozygosity. In essence, inbreeding takes alleles from heterozygotes and puts them into homozygotes.
2. Inbreeding does not cause evolution, because allele frequencies do not change in the population as a whole.

Inbreeding and other forms of nonrandom mating change genotype frequencies—not allele frequencies. ✔**If you understand these concepts, you should be able to predict how observed genotype frequencies should differ from those expected under the Hardy-Weinberg principle, when inbreeding is occurring.**

WHY DOES INBREEDING DEPRESSION OCCUR? Inbreeding has attracted a great deal of attention because of a phenomenon called inbreeding depression. **Inbreeding depression** is a decline in average fitness that takes place when homozygosity increases and heterozygosity decreases in a population. Inbreeding depression results from two causes:

1. *Many recessive alleles represent loss-of-function mutations.* Because these alleles are usually rare, there are normally very few homozygous recessive individuals in a population. Instead, most loss-of-function alleles exist in heterozygous individuals. The alleles have little or no effect when they occur in heterozygotes, because one normal allele usually produces enough functional protein to support a normal phenotype. But inbreeding increases the frequency of homozygous recessive individuals. Loss-of-function mutations are usually deleterious or even lethal when they are homozygous.

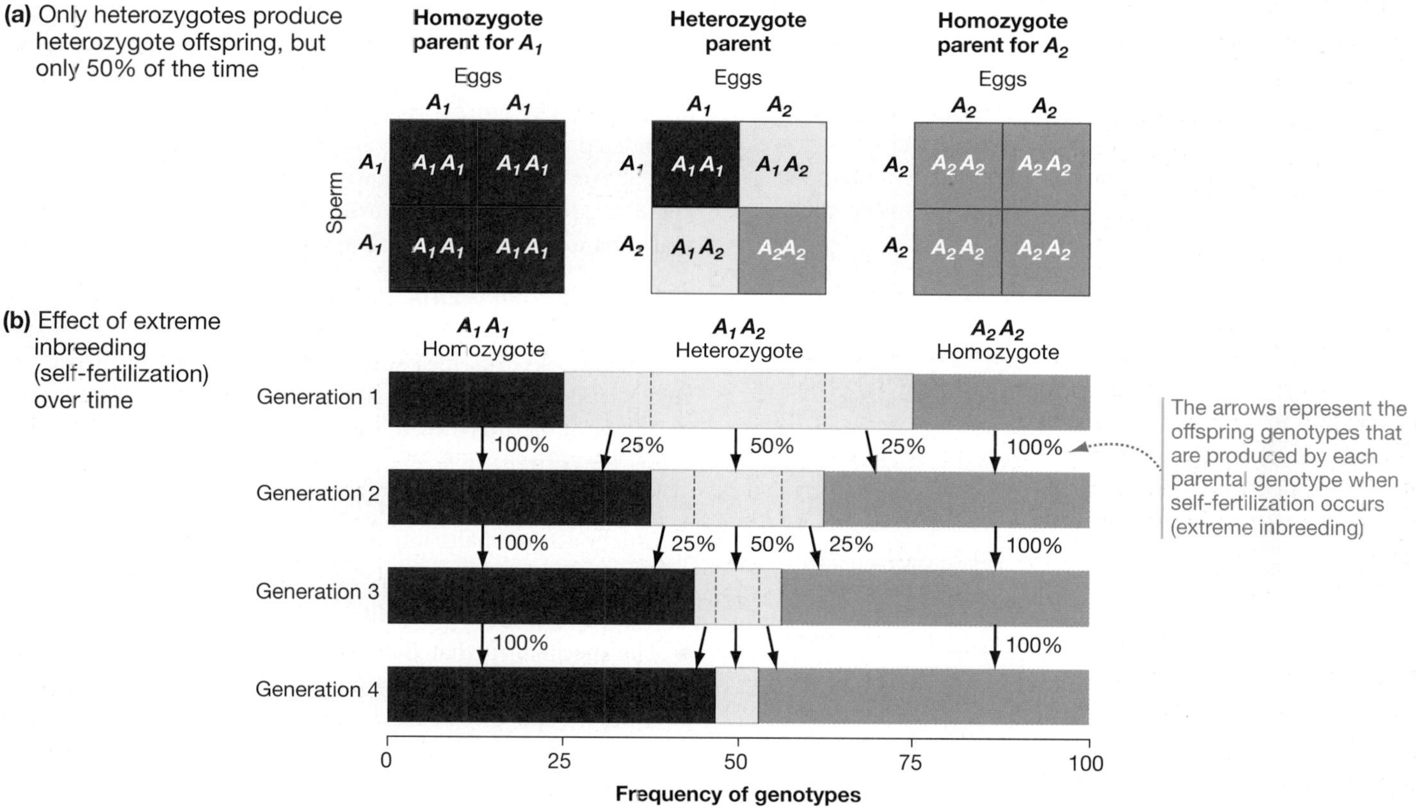

FIGURE 25.12 Inbreeding Increases Homozygosity and Decreases Heterozygosity. (a) Heterozygous parents produce homozygous and heterozygous offspring in a 1:2:1 ratio. **(b)** The width of the boxes corresponds to the frequency of each genotype.

2. ***Many genes—especially those involved in fighting disease—are under intense selection for heterozygote advantage.*** If an individual is homozygous at these genes, then fitness declines.

The upshot here is that the offspring of inbred matings are expected to have lower fitness than the progeny of outcrossed matings. This prediction has been verified in a wide variety of species. **Figure 25.13** shows data from controlled matings between *Lobelia cardinalis* individuals, with the fitness of offspring plotted against their age. Note that the two sets of data points compare the fitnesses of offspring from inbred and non-inbred matings. **Table 25.4** provides data on inbreeding depression in an array of human populations. Because inbreeding has such deleterious consequences in humans, it is not surprising that many contemporary human societies have laws forbidding marriages between individuals who are related as first cousins or closer.

The trickiest point to grasp about inbreeding is that even though it does not cause evolution directly—because it does not change allele frequencies—it can speed the rate of evolutionary change. More specifically, it increases the rate at which purifying selection eliminates recessive deleterious alleles from a population.

A moment's thought should convince you why this is so. Deleterious recessives are usually very rare in populations, because they lower fitness. Rare alleles are usually found in heterozygotes because it is more likely for individuals to have one copy of a rare allele than two. When no inbreeding is occurring, the recessive deleterious alleles found in heterozygotes cannot be eliminated by natural selection. But when inbreeding occurs, more recessive deleterious alleles are found in homozygotes and are quickly eliminated by selection.

✔**If you understand inbreeding depression, you should be able to explain (1) why inbreeding helps "purge" recessive deleterious alleles, and (2) why it does not occur in species like garden peas, where self-fertilization has occurred routinely for many generations and most or all deleterious recessive alleles have been eliminated.**

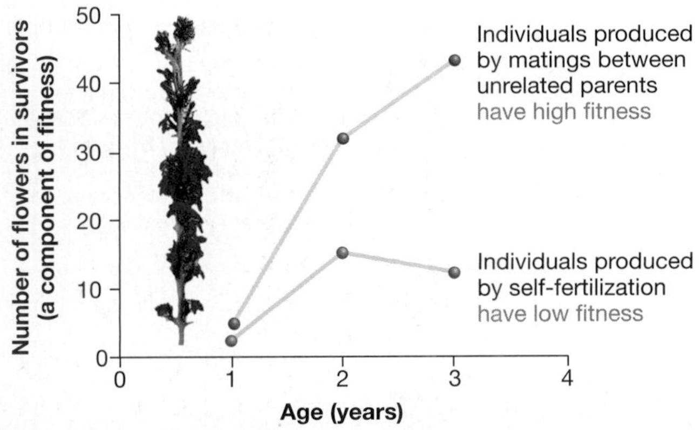

FIGURE 25.13 Inbreeding Depression Occurs in *Lobelia cardinalis*. Inbreeding depression is the fitness difference between non-inbred and inbred individuals.

✔**EXERCISE** Does inbreeding depression increase with age in this species or remain constant throughout life?

TABLE 25.4 **Inbreeding Reduces Fitness in Humans**

The percentages reported here give the mortality rate of children produced by first-cousin marriages versus marriages between nonrelatives. In every study, children of first-cousin marriages have a higher mortality rate.

		Deaths (%)	
Age of Children	**Time Period**	**Children of First Cousins**	**Children of Nonrelatives**
Under 20 years (U.S.)	18th–19th century	17.0	12.0
Under 10 years (U.S.)	1920–1956	8.1	2.4
At or before birth (France)	1919–1950	9.3	3.9
Children (France)	1919–1950	14.0	10.0
Under 1 year (Japan)	1948–1954	5.8	3.5
1–8 years (Japan)	1948–1954	4.6	1.5

SOURCE: From *Principles of Human Genetics*, by Curt Stern: © 1973 by W. H. Freeman. Used with permission.

Sexual Selection

If female peacocks choose males with the longest and most iridescent tails as mates, then nonrandom mating is occurring. Due to this type of nonrandom mating, the frequency of alleles that contribute to long, iridescent tails will increase in the population.

Charles Darwin was the first biologist to recognize that selection based on success in courtship is a mechanism of evolutionary change. The process is called **sexual selection**, and it can be considered a special case of natural selection. Sexual selection occurs when individuals within a population differ in their ability to attract mates. It favors individuals with heritable traits that enhance their ability to obtain mates.

THEORY: THE FUNDAMENTAL ASYMMETRY OF SEX In 1948 A. J. Bateman contributed a fundamental insight about how sexual selection works. His idea was elaborated by Robert Trivers in 1972. The Bateman-Trivers theory contains two elements: a claim about a pattern in the natural world and a process that causes the pattern.

- The pattern component of their theory is that sexual selection usually acts on males much more strongly than on females. As a result, traits that attract members of the opposite sex are much more highly elaborated in males.
- The mechanism that Bateman and Trivers proposed to explain this pattern can be summarized with a quip: "Eggs are expensive, but sperm are cheap." That is, the energetic cost of creating a large egg is enormous, whereas a sperm contains few energetic resources.

In most species, females invest much more in their offspring than do males. This phenomenon is called the fundamental asymmetry of sex. It is characteristic of almost all sexual species and has two important consequences:

1. Because eggs are large and energetically expensive, females produce relatively few young over the course of a lifetime. A female's fitness is limited not by the ability to find a mate but primarily by her ability to gain the resources needed to produce more eggs and healthier young.
2. Sperm are so simple to produce that a male can father an almost limitless number of offspring. Thus, a male's fitness is limited not by the ability to acquire the resources needed to produce sperm but by the number of females he can mate with.

PREDICTIONS OF THE BATEMAN-TRIVERS THEORY The theory of sexual selection makes strong predictions.

- If females invest a great deal in each offspring, then they should protect that investment by being choosy about their mates. Conversely, if males invest little in each offspring, then they should be willing to mate with almost any female.
- If there are an equal number of males and females in the population, and if males are trying to mate with any female possible, then males will compete with each other for mates.

Stated another way, there should be two broad types of sexual selection: female choice and male-male competition. In addition:

- If male fitness is limited by access to mates, then any allele that increases a male's attractiveness to females or success in male-male competition should increase rapidly in the population. Thus, sexual selection should act more strongly on males than on females.

Stated another way, traits that evolve due to sexual selection—meaning traits that are useful only in courtship or in competition for mates—should be found primarily in males.

Do data from experimental or observational studies agree with these predictions? Let's consider each of them in turn.

FEMALE CHOICE FOR "GOOD ALLELES" If females are choosy about which males they mate with, what criteria do females use to make their choice? Recent experiments have shown that in several bird species, females prefer to mate with males that are well fed and in good health. These experiments were motivated by three key observations:

1. In many bird species, the existence of colorful feathers or a colorful beak is due to the presence of the red and yellow pigments called carotenoids.
2. Carotenoids protect tissues and stimulate the immune system to fight disease more effectively.
3. Animals cannot synthesize their own carotenoids, but plants can. To obtain carotenoids, animals have to eat carotenoid-rich plant tissues.

These observations suggest that the healthiest and best-nourished birds in a population have the most colorful beaks and feathers. Sick birds have dull coloration because they are using all of their carotenoids to stimulate their immune system. Poorly fed birds have dull coloration because they have few carotenoids available. By choosing a colorful male as the father of her offspring, a female is likely to have offspring with alleles that will help the offspring fight disease effectively and feed efficiently.

To test the hypothesis that females prefer to mate with colorful males, a team of researchers experimented with zebra finches (**Figure 25.14a**). They fed one group of male zebra finches a diet that was heavily supplemented with carotenoids, and they fed a second group of male zebra finches (the control group) a diet that was similar in every way except for the additional carotenoids.

To control for other differences between the groups, the researchers used individuals that had been raised and maintained in environments that were as similar as possible. In addition, they identified pairs of brothers, and randomly assigned one brother to the control group and one brother to the treatment group. Thus, there should have been no pronounced genetic differences between individuals in the treatments.

As predicted, the males eating the carotenoid-supplemented diet developed more colorful beaks than did the males fed the carotenoid-poor diet (**Figure 25.14b**). When given a choice of mating with either of the two brothers, 9 out of the 10 females tested preferred the more-colorful male. These results are strong evidence that females of this species are choosy about their mates and that they prefer to mate with healthy, well-fed males.

Enough experiments have been done on other bird species to support a general conclusion: Colorful beaks and feathers, along with songs and dances and other types of courtship displays, carry the message "I'm healthy and well fed because I have good alleles. Mate with me."

FEMALE CHOICE FOR PATERNAL CARE Choosing "good alleles" is not the entire story in sexual selection via female choice. In many species, females prefer to mate with males that care for young or that provide the resources required to produce eggs.

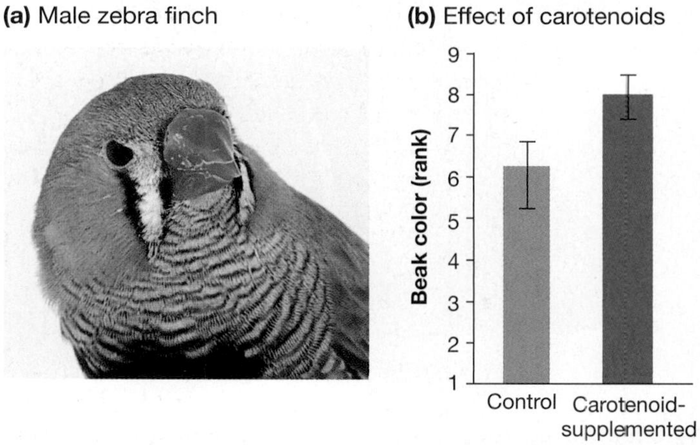

FIGURE 25.14 If Male Zebra Finches Are Fed Carotenoids, Their Beaks Get More Colorful. In this experiment, beak color was ranked on a scale where higher numbers indicate more intense orange color.

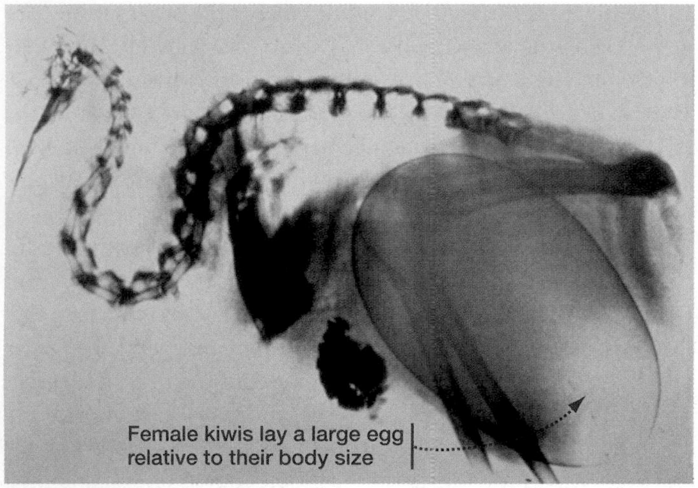

FIGURE 25.15 In Many Species, Females Make a Large Investment in Each Offspring. X-ray of a female kiwi, ready to lay an egg.

Brown kiwi females make an enormous initial investment in their offspring—their eggs routinely represent over 15 percent of the mother's total body weight (**Figure 25.15**)—but choose to mate with males that take over all of the incubation and other care of the offspring. It is common to find that female fish prefer to mate with males that protect a nest site and care for the eggs until they hatch. In humans and many species of birds, males provide food, protection, and other resources required for rearing young.

To summarize, females may choose mates on the basis of (1) physical characteristics that signal male genetic quality, (2) resources or parental care provided by males, or (3) both. In some species, however, females do not have the luxury of choosing a male. Instead, competition among males is the primary cause of sexual selection.

MALE-MALE COMPETITION As an example of research on how males compete for mates, consider data from a long-term study of a northern elephant seal population breeding on Año Nuevo Island, off the coast of California.

Elephant seals feed on marine fish and spend most of the year in the water. But when females are ready to mate and give birth, they haul themselves out of the water onto land. Females prefer to give birth on islands, where newborn pups are protected from terrestrial and marine predators. Because elephant seals have flippers that are ill suited for walking, females can haul themselves out of the water only on the few beaches that have gentle slopes. As a result, large numbers of females congregate in tiny areas to breed.

Male elephant seals establish territories on breeding beaches by fighting (**Figure 25.16a**). A **territory** is an area that is actively defended and that provides exclusive use by the owner. Males

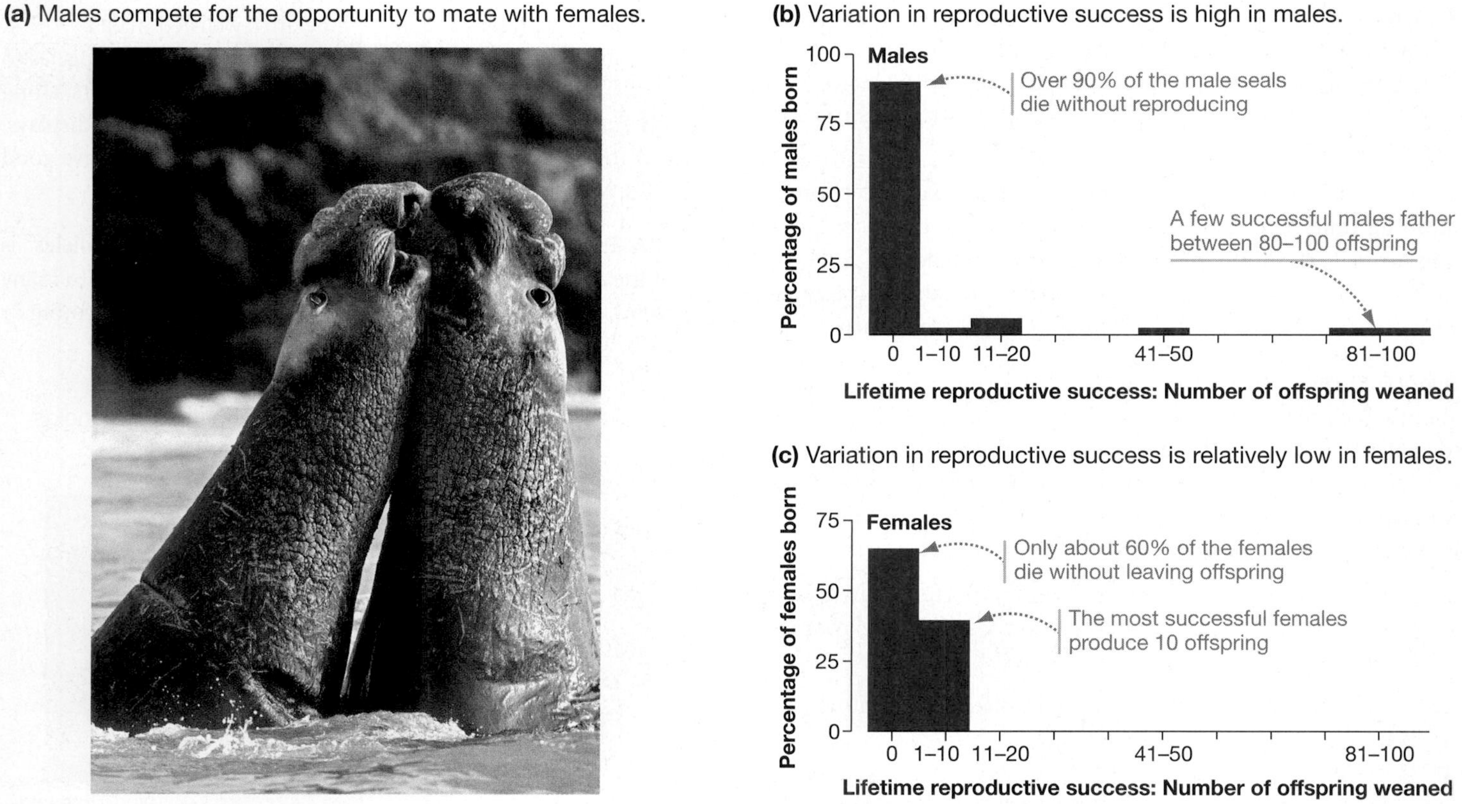

FIGURE 25.16 Intense Sexual Selection on Male Elephant Seals. The histograms show that variation in lifetime reproductive success is even higher **(b)** in male northern elephant seals than it is **(c)** in females.

✔**QUESTION** Consider an allele that increases reproductive success in elephant seal males versus an allele that increases reproductive success in females. Which allele will increase in frequency faster, and why?

that win battles with other males monopolize matings with the females residing in their territories. Females don't choose among males—they simply mate with the winning male. Males that lose battles are relegated to territories with few females or are excluded from the beach. Fights are essentially slugging contests and are usually won by the larger male. The males stand face to face, bite each other, and land blows with their heads.

Based on these observations, it is not surprising that male northern elephant seals frequently weigh 2700 kg (5940 lbs) and are over four times more massive, on average, than females. The logic here runs as follows:

- Males that own beaches with large congregations of females father large numbers of offspring. Males that lose fights father few or no offspring.
- The alleles of territory-owning males rapidly increase in frequency in the population.
- If the ability to win fights and produce offspring is determined primarily by body size, then alleles for large body size have a significant fitness advantage, leading to the evolution of large male size. The fitness advantage is due to sexual selection.

Figure 25.16b provides evidence for intense sexual selection in males. Biologists have marked many of the individuals in the seal population on Año Nuevo to track the lifetime reproductive success of a large number of individuals. The *x*-axis indicates fitness, plotted as 0 offspring produced over a lifetime, 1–10, 11–20, and so on. The *y*-axis indicates the percentage of males in the population that achieved each category of offspring production. As the data show, in this population a few males father a large number of offspring, while most males father few or none.

Among females, variation in reproductive success is also high; but it is much lower than in males (**Figure 25.16c**). In this species, most sexual selection is driven by male-male competition rather than female choice.

WHAT ARE THE CONSEQUENCES OF SEXUAL SELECTION? In elephant seals and most other animals studied, most females that survive to adulthood get a mate. In contrast, many males do not. Because sexual selection tends to be much more intense in males than females, males tend to have many more traits that function only in courtship or male-male competition. Stated another way, sexually selected traits often differ sharply between the sexes.

Sexual dimorphism (literally, "two-forms") refers to any trait that differs between males and females. **Figure 25.17** illustrates sexually dimorphic traits. They range from weapons that males use to fight over females, such as antlers and horns, to the elaborate ornamentation and behavior used in courtship displays. Humans are sexually dimorphic in size, distribution of body hair, and many other traits.

Like inbreeding and other forms of nonrandom mating, sexual selection violates the assumptions of the Hardy-Weinberg principle. Unlike inbreeding, however, it causes certain alleles to increase or decrease in frequency and results in evolution.

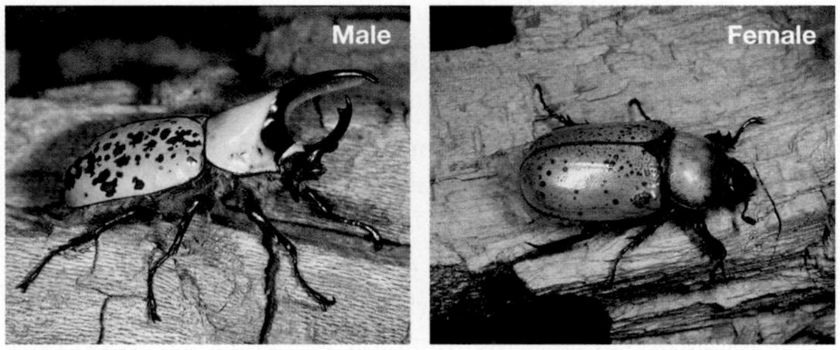

Beetle. During breeding season, males of the beetle *Dynastes granti* use their elongated horns to fight over females.

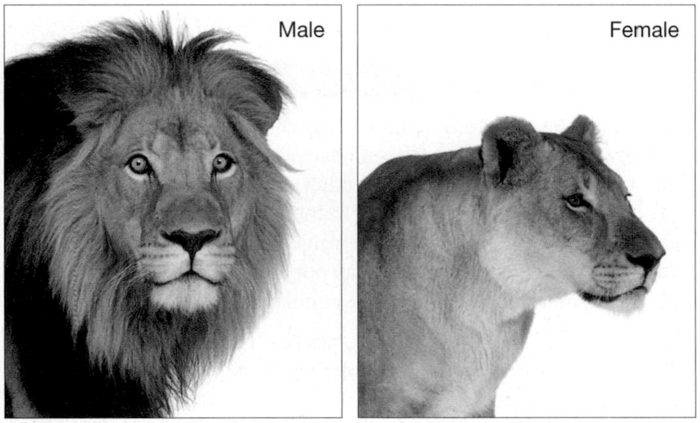

Lion. Male lions are larger than female lions and have an elaborate ruff of fur called a mane.

FIGURE 25.17 Sexually Selected Traits Are Used to Compete for Mates. Males often have exaggerated traits that they use in fighting or courtship. In many species, females lack these traits.

CHECK YOUR UNDERSTANDING

If you understand that . . .

- Inbreeding and sexual selection are forms of nonrandom mating.
- Inbreeding is mating between relatives. It is not an evolutionary process, because it changes genotype frequencies—not allele frequencies. Inbreeding increases homozygosity and may lead to inbreeding depression.
- Sexual selection is based on differential success in obtaining mates. It causes evolution by increasing the frequency of alleles associated with successful courtship.

✓ You should be able to . . .

1. Define the fundamental asymmetry of sex.
2. Explain why males are usually the sex with exaggerated traits used in courtship.

Answers are available in Appendix B.

Summary of Key Concepts

The Hardy-Weinberg principle acts as a null hypothesis when researchers want to test whether evolution or nonrandom mating is occurring at a particular gene.

- The Hardy-Weinberg principle can serve as a null hypothesis in evolutionary studies because it predicts what genotype and allele frequencies are expected to be if mating is random with respect to the gene in question and none of the four evolutionary processes is operating on that gene.

✔You should be able to predict how genotype frequencies differ from Hardy-Weinberg proportions under directional selection, when alleles are codominant.

MB **Web Activity** The Hardy-Weinberg Principle

Each of the four evolutionary mechanisms has different consequences. Only natural selection produces adaptation. Genetic drift causes random fluctuations in allele frequencies. Gene flow equalizes allele frequencies between populations. Mutation introduces new alleles.

- Directional selection favors phenotypes at one end of a distribution; stabilizing selection eliminates phenotypes with extreme characteristics. Both decrease allelic diversity in populations.
- Genetic drift causes random changes in allele frequency, is particularly important in small populations, and tends to reduce overall genetic diversity.
- Gene flow homogenizes allele frequencies among populations and can serve as an important source of new alleles.
- Mutation is too infrequent to be a major cause of allele frequency change, but is the only mechanism that creates new alleles.

✔You should be able to consider how the evolutionary forces affect the management of endangered species. What effect will drift have? Predict how selection will affect allele frequencies in populations that are being maintained in captivity.

MB **Web Activity** Three Modes of Natural Selection, **BioFlix™** Mechanisms of Evolution

Inbreeding changes genotype frequencies but does not change allele frequencies.

- Inbreeding, or mating among relatives, is a form of nonrandom mating.
- Inbreeding does not change allele frequencies, so it is not an evolutionary mechanism.
- Inbreeding changes genotype frequencies. It leads to an increase in homozygosity and a decrease in heterozygosity.
- These patterns can accelerate natural selection and can cause inbreeding depression.

✔You should be able to predict how extensive inbreeding during the 1700s and 1800s affected the royal families of Europe.

Sexual selection leads to the evolution of traits that help individuals attract mates. It is usually stronger on males than on females.

- Sexual selection is a form of nonrandom mating.
- Sexual selection occurs when certain traits help males succeed in contests over mates or when certain traits are attractive to females.
- Sexual selection is responsible for the evolution of phenotypic differences between males and females.

✔You should be able to predict whether males or females are the most brightly colored in birds called red-necked phalaropes. In this species, males make a much larger investment in offspring than females do because they build the nest and incubate the eggs.

Questions

✔TEST YOUR KNOWLEDGE

Answers are available in Appendix B

1. Why isn't inbreeding considered an evolutionary mechanism?
   a. It does not change genotype frequencies.
   b. It does not change allele frequencies.
   c. It does not occur often enough to be important in evolution.
   d. It does not violate the assumptions of the Hardy-Weinberg principle.
2. Why is genetic drift aptly named?
   a. It causes allele frequencies to drift up or down randomly.
   b. It is the ultimate source of genetic variability.
   c. It is an especially important mechanism in small populations.
   d. It occurs when populations drift into new habitats.
3. How do sexual selection and inbreeding differ?
   a. Unlike inbreeding, sexual selection changes allele frequencies and affects only genes involved in attracting mates.
   b. Unlike sexual selection, inbreeding changes allele frequencies and involves any mating between relatives—not just self-fertilization.
   c. Unlike inbreeding, sexual selection results from the random fusion of gametes during fertilization. It is particularly important in small populations, where few mates are available.
   d. Inbreeding occurs only in small populations, while sexual selection can occur in any size population.
4. What does it mean when an allele reaches "fixation"?
   a. It is eliminated from the population.
   b. It has a frequency of 1.0.
   c. It is dominant to all other alleles.
   d. It is adaptively advantageous.

5. In what sense is the Hardy-Weinberg principle a null hypothesis, similar to the control treatment in an experiment?
   a. It defines what genotype frequencies should be if nonrandom mating is occurring.
   b. Expected genotype frequencies can be calculated from observed allele frequencies and then compared with the observed genotype frequencies.
   c. It defines what genotype frequencies should be if natural selection, genetic drift, gene flow, or mutation is occurring *and* if mating is random.
   d. It defines what genotype frequencies should be if evolutionary mechanisms are *not* occurring.

6. Mutation is the ultimate source of genetic variability. Why is this statement correct?
   a. DNA polymerase (the enzyme that copies DNA) is remarkably accurate.
   b. "Mutation proposes and selection disposes."
   c. Mutation is the only source of new alleles.
   d. Mutation occurs in response to natural selection. It generates the alleles that are required for a population to adapt to a particular habitat.

✓ TEST YOUR UNDERSTANDING

Answers are available in Appendix B

1. Consider selection, drift, gene flow, mutation, and inbreeding. Which of the five processes decreases genetic variation? Which increases it?
2. Directional selection can lead to the fixation of favored alleles. When this occurs, genetic variation is zero and evolution stops. Explain why this rarely occurs.
3. Why does sexual selection often lead to sexual dimorphism?
4. In what sense does the Hardy-Weinberg principle function as a null hypothesis?
5. Explain why small populations become inbred.
6. How can allele frequencies change under stabilizing selection, even if the average phenotype in the population does not?

✓ APPLYING CONCEPTS TO NEW SITUATIONS

Answers are available in Appendix B

1. In humans, albinism is caused by loss-of-function mutations in genes involved in the synthesis of melanin, the dark pigment in skin. Only people homozygous for a loss-of-function allele have the relevant phenotype. In Americans of northern European ancestry, albino individuals are present at a frequency of about 1 in 10,000 (or 0.0001). Knowing this genotype frequency and assuming that genotypes are in Hardy-Weinberg proportions, we can calculate the frequency of the loss-of-function alleles. If we let p_2 stand for this frequency, we know that $p_2^2 = 0.0001$; therefore $p_2 = \sqrt{0.0001} = 0.01$. By subtraction, the frequency of normal alleles is 0.99. Under the Hardy-Weinberg principle, what is the frequency of "carriers"—or people who are heterozygous for this condition? Your answer estimates the percentage of Caucasians in the United States who carry an allele for albinism.
2. Conservation managers frequently use gene flow, in the form of transporting individuals or releasing captive-bred young, to counteract the effects of drift on small, endangered populations. Explain how gene flow can also mitigate the effects of inbreeding.
3. Suppose you were studying several species of human. In one, males never lifted a finger to help females raise children. In another, males did just as much parental care as females except for actually carrying the baby during pregnancy. How does the fundamental asymmetry of sex compare in the two species? How would you expect sexual dimorphism to compare between the two species?
4. You are a conservation biologist charged with creating a recovery plan for an endangered species of turtle. The turtle's habitat has been fragmented into small, isolated but protected areas by suburbanization and highway construction. Some evidence indicates that certain turtle populations are adapted to marshes that are normal, whereas others are adapted to acidic wetlands or salty habitats. Further, some turtle populations number less than 25 breeding adults, making genetic drift and inbreeding a major concern. In creating a recovery plan, the tools at your disposal are captive breeding, the capture and transfer of adults to create gene flow, or the creation of habitat corridors between wetlands to make migration possible. How would you use gene flow to help this species?

This chapter is part of The Big Picture. See how on pages 494–495.

UNIT 5

EVOLUTIONARY PROCESSES AND PATTERNS

New sunflower species have formed recently in southwestern North America. This chapter explains how.

26 Speciation

KEY CONCEPTS

- Speciation occurs when populations of the same species become genetically isolated by lack of gene flow and then diverge from each other due to selection, genetic drift, or mutation.
- Populations can be recognized as distinct species if they are reproductively isolated from each other, if they have distinct morphological characteristics, or if they form independent branches on a phylogenetic tree.
- Populations can become genetically isolated from each other if they occupy different geographic areas, if they use different habitats or resources within the same area, or if one population is polyploid and cannot breed with the other.
- When populations that have diverged come back into contact, they may fuse, continue to diverge, stay partially differentiated, or have offspring that form a new species.

Although Darwin called his masterwork *On the Origin of Species by Means of Natural Selection*, he actually had little to say about how new species arise. Instead, his data and analyses focused on the process of natural selection and the changes that occur within populations over time. He spent much less time considering changes that occur *between* populations.

Since Darwin, however, biologists have realized that populations of the same species may diverge from each other when they are isolated in terms of gene flow. Recall from Chapter 25 that gene flow makes allele frequencies more similar among populations. If gene flow ends, allele frequencies in isolated populations are free to diverge—meaning that the populations begin to evolve independently of each other. For example, when a new mutation creates an allele that changes the phenotype of individuals in one population, there is no longer any way for that allele to appear in the other population. If mutation, selection, and genetic drift cause isolated populations to diverge sufficiently, distinct types, or species, form—that is, the process of speciation takes place.

Speciation is a splitting event that creates two or more distinct species from a single ancestral group (**Figure 26.1**). When speciation is complete, a new branch has been added to the tree of life.

In essence, then, speciation results from genetic isolation and genetic divergence. Genetic isolation results from lack of gene flow, and divergence occurs because selection, genetic drift, and mutation proceed independently in the isolated populations. How does genetic isolation come about? And how do selection, drift, and mutation cause divergence?

26.1 How Are Species Defined and Identified?

Like the Galápagos finches in Figure 26.1, species are distinct from one another in appearance, behavior, habitat use, or other traits. These characteristics differ among species because their genetic characteristics differ. Genetic distinctions occur because

✔ When you see this checkmark, stop and test yourself. Answers are available in Appendix B.

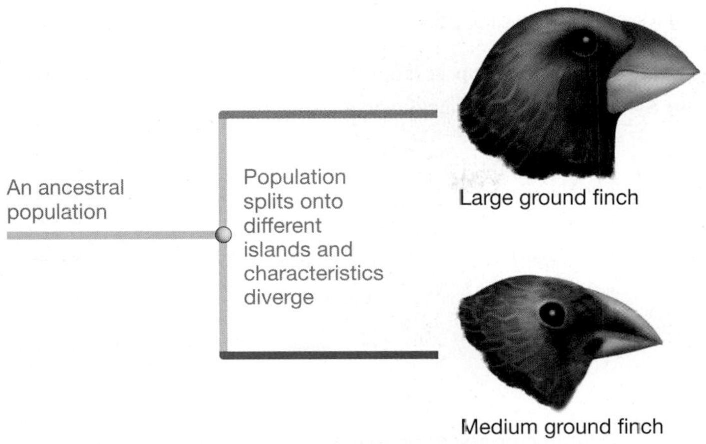

FIGURE 26.1 Speciation Creates Evolutionarily Independent Populations. The large ground finch and medium ground finch are derived from the same ancestral population. This ancestral population split into two populations isolated by lack of gene flow. Because the populations began evolving independently, over time they acquired the distinctive characteristics observed today.

mutation, selection, and drift act on each species independently of what is happening in other populations.

Formally, then, a **species** is defined as an evolutionarily independent population or group of populations. Even though this definition sounds straightforward, it can be exceedingly difficult to put into practice.

How can evolutionarily independent populations be identified in the field and in the fossil record? There is no single, universal answer. Even though biologists agree on the definition of a species, they frequently have to use different sets of criteria to identify them. Three criteria for identifying species are in common use: (1) the biological species concept, (2) the morphospecies concept, and (3) the phylogenetic species concept.

The Biological Species Concept

According to the **biological species concept**, the critical criterion for identifying species is reproductive isolation. This is a logical yardstick because no gene flow occurs between populations that are reproductively isolated from each other. Specifically, if two different populations do not interbreed in nature, or if they fail to produce viable and fertile offspring when matings take place, then they are considered distinct species. Groups that naturally or potentially interbreed, and that are reproductively isolated from other groups, belong to the same species. Biologists can be confident that reproductively isolated populations are evolutionarily independent.

Reproductive isolation can result from a wide variety of events and processes. To organize the various mechanisms that stop gene flow between populations, biologists distinguish

- **prezygotic** (literally, "before-zygote") **isolation**, which prevents individuals of different species from mating; and
- **postzygotic** ("after-zygote") **isolation**, in which the offspring of matings between members of different species do not survive or reproduce.

Table 26.1 outlines some of the more important mechanisms of prezygotic and postzygotic isolation.

TABLE 26.1 **Mechanisms of Reproductive Isolation**

	Process	Example
Prezygotic Isolation		
Temporal	Populations are isolated because they breed at different times.	Bishop pines and Monterey pines release their pollen at different times of the year.
Habitat	Populations are isolated because they breed in different habitats.	Parasites that begin to exploit new host species are isolated from their original population.
Behavioral	Populations do not interbreed because their courtship displays differ.	To attract male fireflies, female fireflies give a species-specific sequence of flashes.
Gametic barrier	Matings fail because eggs and sperm are incompatible.	In sea urchins, a protein called bindin allows sperm to penetrate eggs. Differences in the amino acid sequence of bindin cause matings to fail between closely related populations.
Mechanical	Matings fail because male and female reproductive structures are incompatible.	In alpine skypilots (a flowering plant), the length of the floral tube varies. Bees can pollinate in populations with short tubes, but only hummingbirds can pollinate in populations with long tubes.
Postzygotic Isolation		
Hybrid viability	Hybrid offspring do not develop normally and die as embryos.	When ring-necked doves mate with rock doves, less than 6 percent of eggs hatch.
Hybrid sterility	Hybrid offspring mature but are sterile as adults.	Eastern meadowlarks and western meadowlarks are almost identical morphologically, but hybrid offspring are largely infertile.

Although the biological species concept has a strong theoretical foundation, it has disadvantages. The criterion of reproductive isolation cannot be evaluated in fossils or in species that reproduce asexually. In addition, it is difficult to apply when closely related populations do not happen to overlap with each other geographically. In this case, biologists are left to guess whether interbreeding and gene flow would occur if the populations happened to come into contact.

The Morphospecies Concept

How do biologists identify species when the criterion of reproductive isolation cannot be applied? Under the **morphospecies** ("form-species") **concept**, researchers identify evolutionarily independent lineages by differences in size, shape, or other morphological features. The logic behind the morphospecies concept is that distinguishing features are most likely to arise if populations are independent and isolated from gene flow.

The morphospecies concept is compelling simply because it is so widely applicable. It is a useful criterion when biologists have no data on the extent of gene flow, and it is equally applicable to sexual, asexual, or fossil species. Its disadvantages are that: (1) it cannot identify **cryptic species**, which differ in traits other than morphology, such as the meadowlarks in Table 26.1, and (2) the morphological features used to distinguish species are subjective. In the worst case, different researchers working on the same populations disagree on the characters that distinguish species. Disagreements like these often end in a stalemate, because no independent criteria exist for resolving the conflict.

The Phylogenetic Species Concept

The **phylogenetic species concept** is a recent addition to the tools available for identifying evolutionarily independent lineages and is based on reconstructing the evolutionary history of populations. Proponents of this approach argue that it is widely applicable and precise.

The reasoning behind the phylogenetic species concept begins with Darwin's claim that all species are related by common ancestry. In modern terms, Darwin was suggesting that all species form a monophyletic ("one-tribe") group—the tree of life.

A **monophyletic group**, also called a **clade** or **lineage**, consists of an ancestral population, all of its descendants, and *only* those descendants. On any given evolutionary tree (whether the tree of life, or any smaller part of the tree of life), there are many monophyletic groups (**Figure 26.2a**).

Monophyletic groups, in turn, are identified by traits called synapomorphies ("unique-forms"). A **synapomorphy** is a trait found in certain groups of organisms that exists in no others. It is a homologous trait—one inherited from a common ancestor—that is unique to certain populations or lineages. Fur and lactation, for example, are synapomorphies that identify mammals as a monophyletic group. The placenta is a synapomorphy that distinguishes the placental mammals as a monophyletic group within mammals, distinct from the monophyletic group called marsupials.

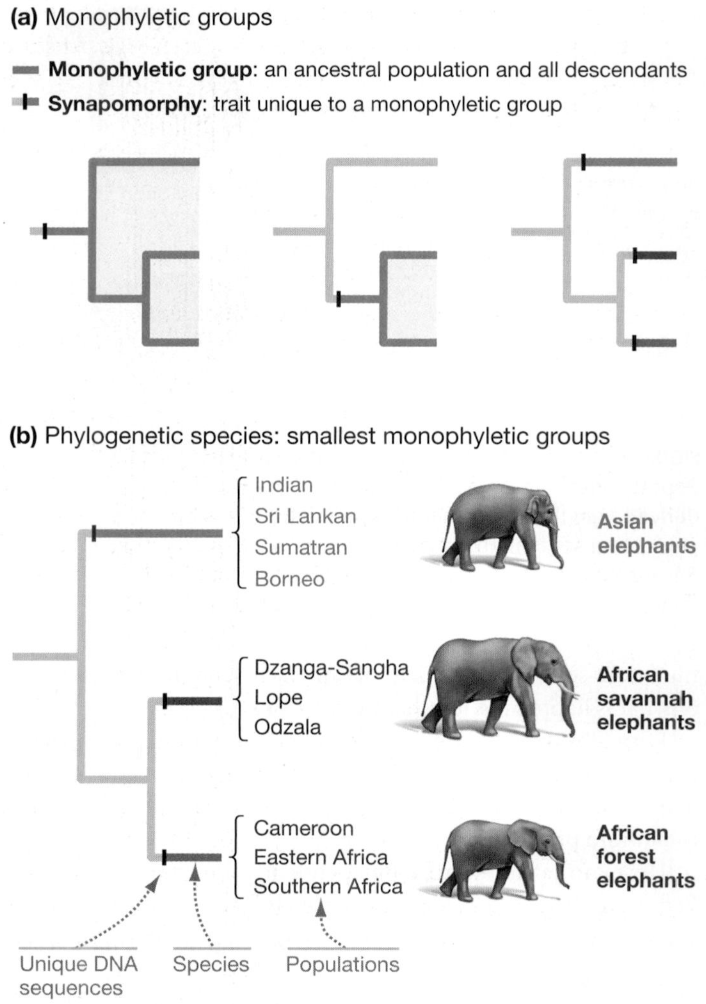

FIGURE 26.2 The Phylogenetic Species Concept Is Based on Monophyletic Groups. Although there are several isolated populations of elephants on two continents, the phylogenetic species concept identifies no more or less than three species of elephants. The populations within species are too similar to form separate branches on the tree. This tree was created by comparing DNA sequences. Thus, the synapomorphies in this case are DNA sequences unique to each species.

Like other homologous traits, synapomorphies can be identified at the genetic, developmental, or structural level. In many cases, researchers use DNA sequence data to identify synapomorphies and estimate phylogenetic trees. Chapter 27 explores the data and logic that biologists use to reconstruct phylogenies, and **BioSkills 3** in Appendix A provides more help on how to read phylogenetic trees.

Under the phylogenetic species concept, species are defined as the smallest monophyletic groups on the tree of life. Phylogenetic species are made up of populations that share one or more unique synapomorphies.

The tree in **Figure 26.2b**, for example, is based on DNA sequence data from an array of elephant populations; but many of the populations do not have unique synapomorphies—they are not distinctive enough to represent separate branches. The phylogenetic species on the tree are labeled Asian elephants, African forest elephants, and African savanna elephants. These are the smallest monophyletic groups on the tree.

The names at some of the tips on this tree (Cameroon, Eastern Africa, Southern Africa, etc.) represent populations within species. These populations may be separated geographically, but their characteristics are so similar that they do not form independent branches on the tree. They are simply part of the same monophyletic group containing other populations.

The phylogenetic species concept has two distinct advantages: (1) It can be applied to any population (fossil, asexual, or sexual), and (2) it is logical because different species have different synapomorphies only if they are isolated from gene flow and have evolved independently. The approach has a distinct disadvantage, however: Carefully estimated phylogenies are available only for a tiny (though growing) subset of populations on the tree of life.

Critics of this approach also point out that it would probably lead to recognition of many more species than either the morphospecies or biological species concept. Proponents counter that, far from being a disadvantage, the recognition of increased numbers of species might better reflect the extent of life's diversity.

In actual practice, researchers use all three species concepts to identify evolutionarily independent populations in nature. The concepts are summarized in **Table 26.2**. Conflicts have occurred, however, when different species concepts are applied to the real world. To appreciate this point, consider the case of the dusky seaside sparrow.

Species Definitions in Action: The Case of the Dusky Seaside Sparrow

Seaside sparrows live in salt marshes along the Atlantic and Gulf Coasts of the United States. The scientific name of this species is *Ammodramus maritimus*. (Recall from Chapter 1 that scientific names consist of a genus name followed by a species name.)

Using the morphospecies concept, researchers had traditionally named a variety of seaside sparrow "subspecies." **Subspecies** are populations that live in discrete geographic areas and have distinguishing features, such as coloration or calls, but are not considered distinct enough to be called separate morphospecies.

Because salt marshes are often destroyed for agriculture or oceanfront housing, by the late 1960s biologists began to be concerned about the future of some seaside sparrow populations. A subspecies called the dusky seaside sparrow (*A. m. nigrescens*) was in particular trouble; by 1980 only six individuals from this population remained. All were males.

At this point government and private conservation agencies sprang into action under the auspices of the Endangered Species Act, a law whose goal is to prevent the extinction of species. The law uses the biological species concept to identify species and calls for the rescue of endangered species through active management.

Because current populations of seaside sparrows are physically isolated from one another, and because young seaside sparrows tend to breed near where they hatched, researchers believed that little to no gene flow occurred among populations. Under the biological species concept and morphospecies concept, there may be as many as six species of seaside sparrow (**Figure 26.3a** on page 462). The dusky seaside sparrow subspecies became a priority for conservation efforts because it was reproductively isolated.

To launch the rescue program, the remaining male dusky seaside sparrows were taken into captivity and bred with females from a nearby subspecies: *A. m. peninsulae*. Officials planned to use these hybrid offspring as breeding stock for a reintroduction program. The goal was to preserve as much genetic diversity as possible by reestablishing a healthy population of dusky-like birds. The plan was thrown into turmoil, however, when a different group of biologists estimated the phylogeny of the seaside sparrows by comparing gene sequences.

The tree in **Figure 26.3b** shows that seaside sparrows represent just two distinct monophyletic groups: one native to the Atlantic Coast and the other native to the Gulf Coast. Under the phylogenetic species concept, only two species of seaside sparrow exist. Far from being an important, reproductively isolated population, the phylogeny showed that the dusky sparrow is part of

SUMMARY TABLE 26.2 **Species Concepts**

	Criterion for Identifying Populations as Species	Advantages	Disadvantages
Biological	Reproductive isolation between populations (they don't breed and don't produce viable offspring)	Reproductive isolation = evolutionary independence	Not applicable to asexual or fossil species; difficult to assess if populations do not overlap geographically.
Morphospecies	Morphologically distinct populations	Widely applicable	Subjective (researchers often disagree about how much or what kinds of morphological distinction indicate speciation); misses cryptic species.
Phylogenetic	Smallest monophyletic group on phylogenetic tree	Widely applicable; based on testable criteria	Relatively few well-estimated phylogenies are currently available.

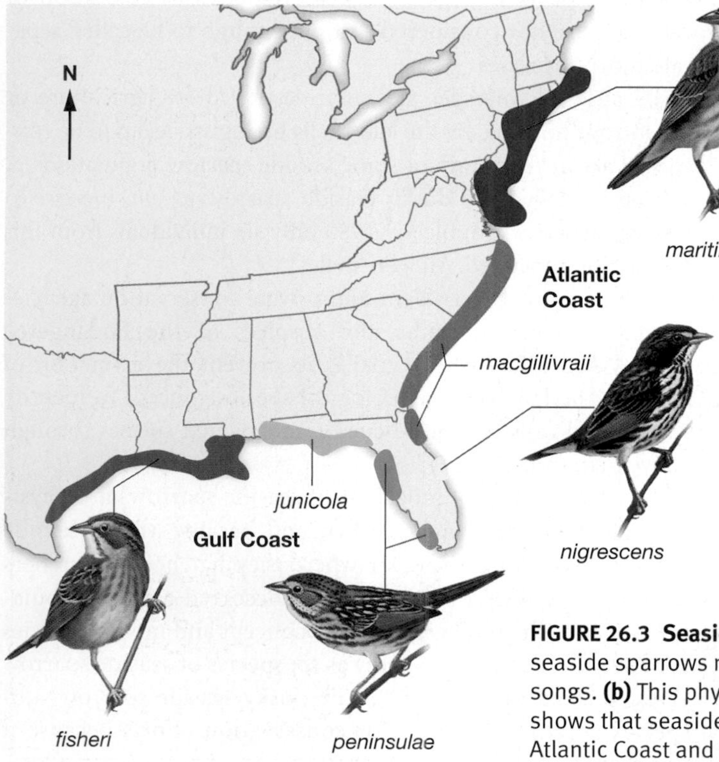

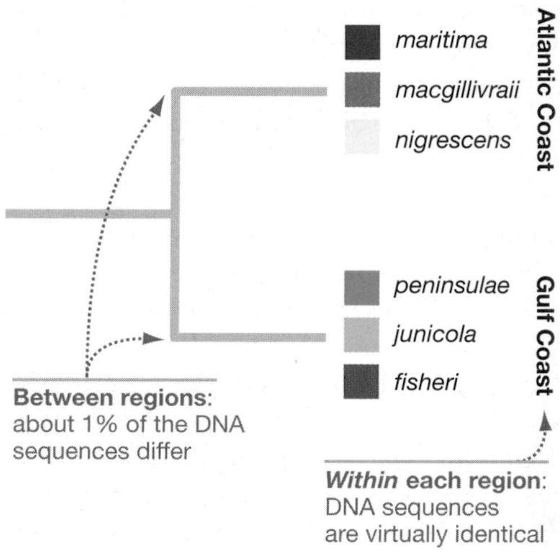

FIGURE 26.3 Seaside Sparrows Form Two Monophyletic Groups. (a) The subspecies of seaside sparrows named on this map are distinguished by their distinctive coloration and songs. **(b)** This phylogenetic tree was constructed by comparing DNA sequences. The tree shows that seaside sparrows represent two distinct monophyletic groups, one native to the Atlantic Coast and the other native to the Gulf Coast.

✓**QUESTION** If you were a conservation biologist and could save only two subspecies of seaside sparrows from extinction, would you choose two subspecies from the Atlantic Coast, two from the Gulf Coast, or one from the Atlantic and one from the Gulf? Explain why.

the same monophyletic group that includes the other Atlantic Coast sparrows.

Further, officials had unwittingly crossed the dusky males (*A. m. nigrescens* from the Atlantic Coast) with females from the Gulf Coast lineage. Because the goal of the conservation effort was to preserve existing genetic diversity, this was the wrong population to use.

The researchers who did the phylogenetic analysis maintained that the biological and morphospecies concepts had misled a well-intentioned conservation program. Under the phylogenetic species concept, they claimed that officials should have allowed the dusky sparrow to go extinct and then concentrated their efforts on simply preserving one or more populations from each coast. In this way, the two monophyletic groups of sparrows—and the most genetic diversity—would be preserved.

Under the morphospecies concept, however, officials did the right thing by preserving distinct types. They argue that dusky seaside sparrows had distinctive, heritable traits like coloration and songs that are now lost forever.

When conservation funding is scarce, life-and-death decisions like these are crucial. Now our task is to consider an even more fundamental question: How do isolation and divergence produce the event called speciation?

CHECK YOUR UNDERSTANDING

If you understand that . . .

- Species are evolutionarily independent because no gene flow occurs between them and other species.
- Biologists use an array of criteria to identify evolutionarily independent groups.

✓ **You should be able to . . .**

1. Explain why the criteria invoked by the biological, morphological, and phylogenetic species concepts allow biologists to identify evolutionarily independent groups.
2. Describe the disadvantages of the biological, morphological, and phylogenetic species concepts.

Answers are available in Appendix B.

26.2 Isolation and Divergence in Allopatry

Speciation begins when gene flow between populations is reduced or eliminated. Genetic isolation happens routinely when populations become physically separated. Physical isolation, in turn, occurs in one of two ways: dispersal or vicariance.

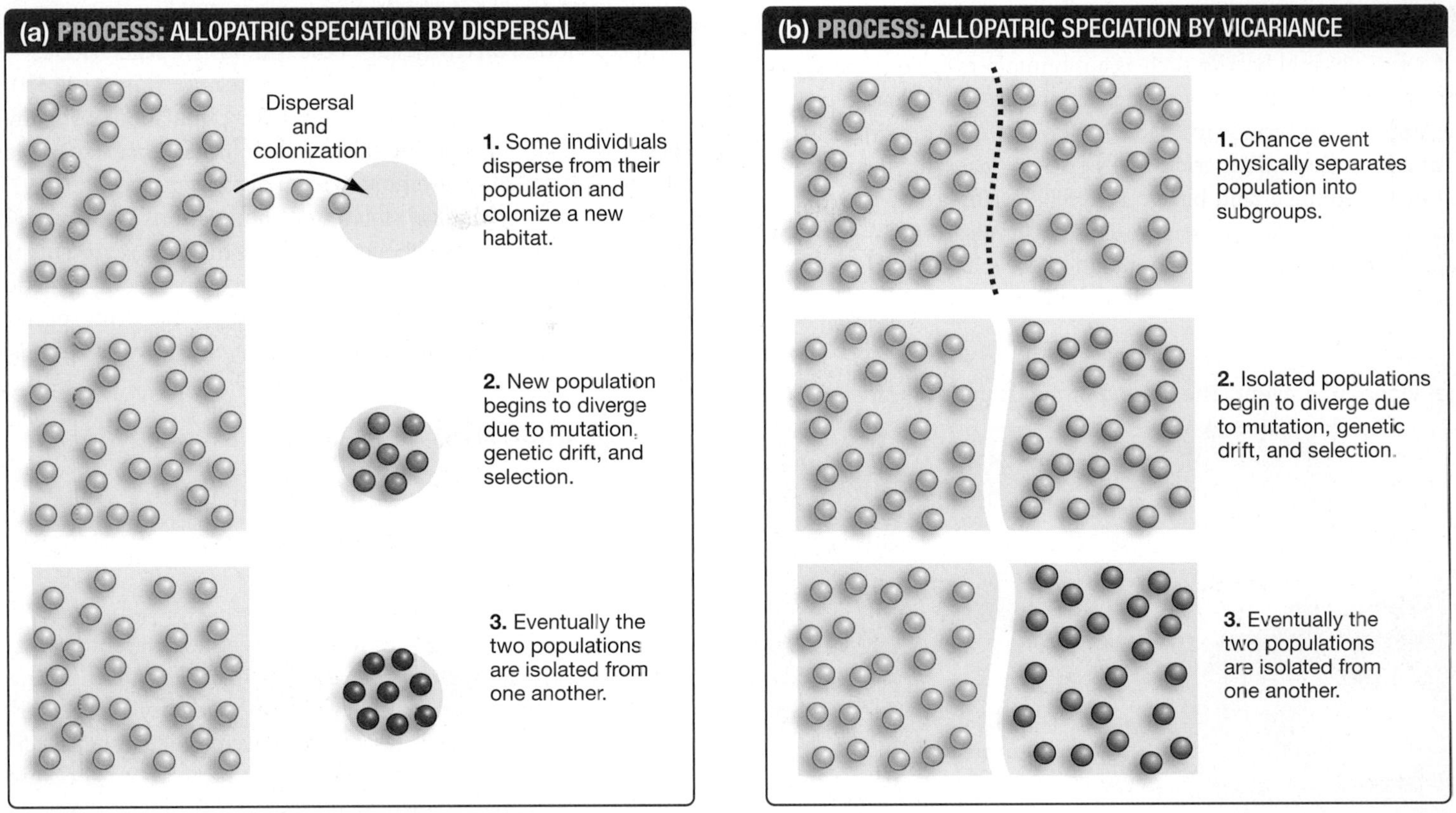

FIGURE 26.4 Allopatric Speciation Begins via Dispersal or Vicariance. (a) When dispersal occurs, colonists establish a new population in a novel location. **(b)** In vicariance, a widespread population becomes fragmented into isolated subgroups.

As **Figure 26.4a** illustrates, a population can disperse to a new habitat, colonize it, and found a new population. Alternatively, a new physical barrier can split a widespread population into two or more subgroups that are physically isolated from each other (**Figure 26.4b**). A physical splitting of habitat is called **vicariance**.

Speciation that begins with physical isolation via either dispersal or vicariance is known as **allopatric** ("different-home-land") **speciation**. Populations that live in different areas are said to be in **allopatry**.

The case studies that follow address two questions: How do colonization and range-splitting events occur? Answering this question takes us into the field of **biogeography**—the study of how species and populations are distributed geographically. Once populations are physically isolated, how do genetic drift and selection produce divergence?

Dispersal and Colonization Isolate Populations

Peter Grant and Rosemary Grant witnessed a colonization event while working in the Galápagos Islands off the coast of South America. Recall from Chapter 24 that the Grants have been studying medium ground finches on the island of Daphne Major since 1971. In 1982 five members of a new species, called the large ground finch, arrived and began nesting. These colonists had apparently dispersed from a population that lived on a nearby island in the Galápagos. Because finches normally stay on the same island year-round, the colonists represented a new population, allopatric with their source population.

Could this colonization event lead to speciation? To evaluate this question, Grant and Grant caught, weighed, and measured most of the parents and offspring produced on Daphne Major over the succeeding 12 years. When they compared these data with measurements of large ground finches in other populations, they discovered that the average beak size in the new population was much larger.

Two evolutionary processes could be responsible for the change in beak size:

1. Genetic drift produced a colonizing population that happened to have particularly large beaks relative to the source population.
2. Natural selection in the new environment could favor alleles associated with large beaks.

The general message here is that the characteristics of a colonizing population are likely to be different from the characteristics of the source population due to chance—that is, genetic drift. Subsequent natural selection may extend the rapid divergence that begins with genetic drift.

Colonization, followed by genetic drift and natural selection, is thought to be responsible for speciation in Galápagos finches and many other island groups.

Vicariance Isolates Populations

If a new physical barrier such as a mountain range or river splits the geographic range of a species, vicariance has taken place. An example of vicariance in speciation comes from the Isthmus of Panama in Central America. Geologists estimate that the isthmus closed, forming a land bridge between North and South America, about 3 million years ago. This was a vicariance event that split Caribbean and Pacific populations of many marine species, including snapping shrimp (**Figure 26.5a**).

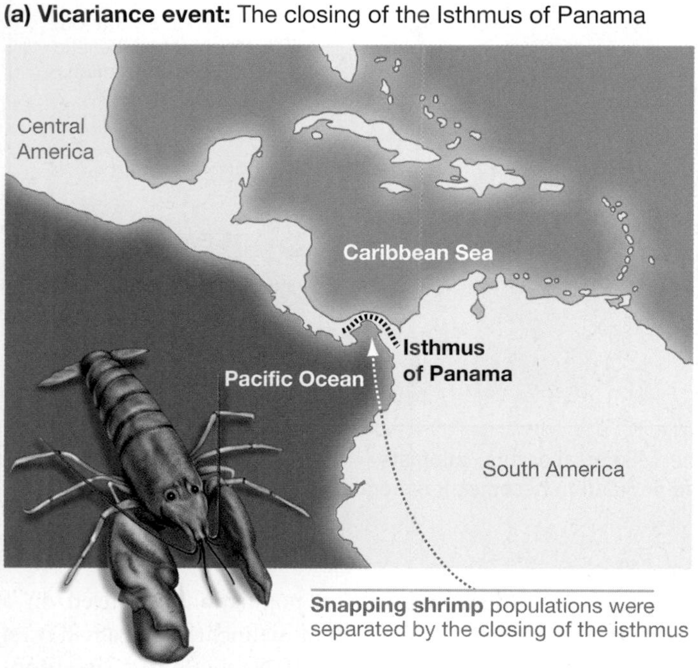

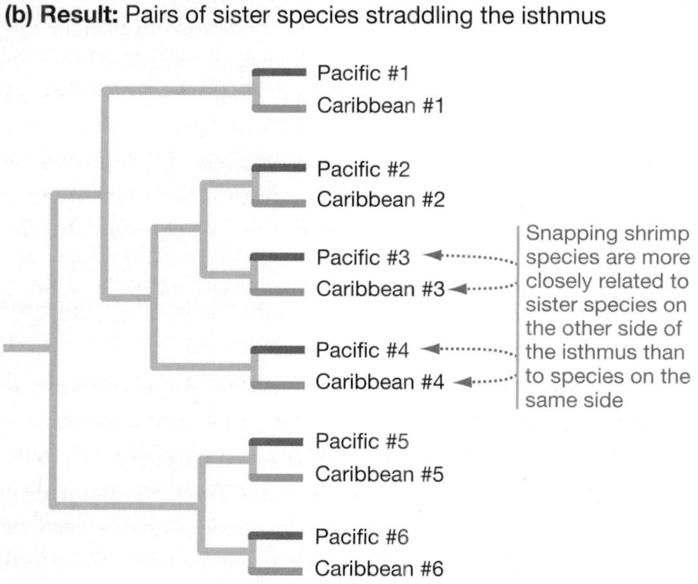

FIGURE 26.5 Evidence for Speciation by Vicariance in Snapping Shrimp. The most logical interpretation of this phylogenetic tree is that 6 species of snapping shrimp split into 12 species when the Isthmus of Panama formed and separated the Caribbean Sea from the Pacific Ocean.

A phylogenetic tree of the snapping shrimp shows that many of the species found on either side of the Isthmus are **sister species**—meaning that they are each others' closest relative (**Figure 26.5b**). This is exactly the pattern predicted if the populations of many species were split in two, with the populations on either side of the Isthmus subsequently diverging to form distinct species.

To summarize, physical isolation of populations via dispersal or vicariance produces genetic isolation—the first requirement of speciation. When genetic isolation is accompanied by genetic divergence due to mutation, selection, and genetic drift, speciation results. To review these concepts, go to the study area at *www.masteringbiology.com*:

MB **Web Activity** Allopatric Speciation

26.3 Isolation and Divergence in Sympatry

When populations or species live in the same geographic area, or at least close enough to one another to make interbreeding possible, biologists say that they live in **sympatry** ("together-homeland"). Traditionally, researchers have predicted that speciation could not occur among sympatric populations, because gene flow is possible (**Figure 26.6**). The prediction was that gene

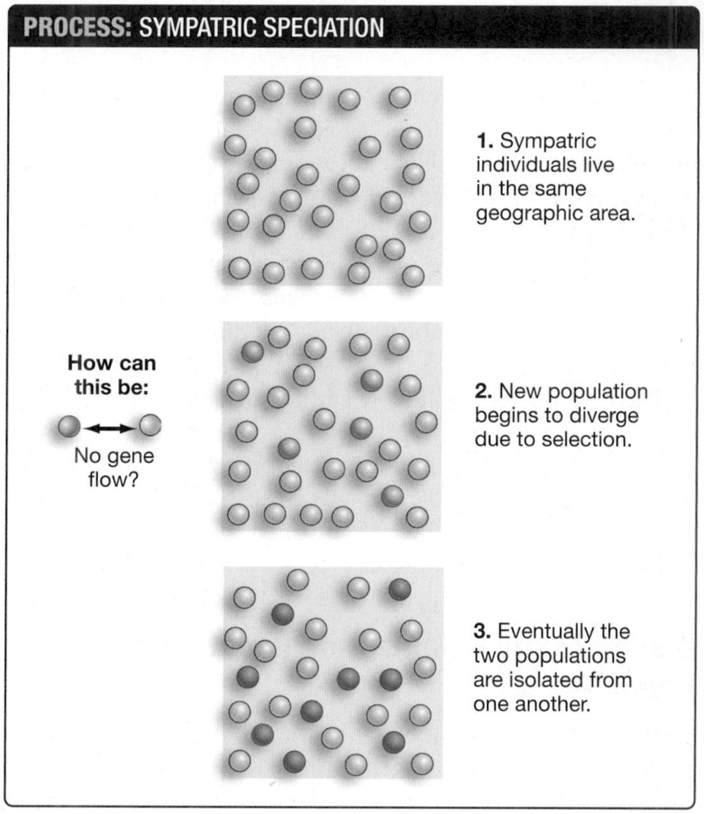

FIGURE 26.6 Sympatric Speciation Has Long Perplexed Researchers. Recent studies have documented several mechanisms that reduce gene flow and result in sympatric speciation.

flow would easily overwhelm any differences among populations created by genetic drift and natural selection. As Chapter 25 showed, gene flow can homogenize gene frequencies even between populations that are allopatric, such as an island population close to a population on a continent.

In general, gene flow overwhelms the diversifying force of natural selection and prevents speciation. Is this always the case?

Can Natural Selection Cause Speciation Even When Gene Flow Is Possible?

Recently, several well-documented examples have upset the traditional view that **sympatric speciation**—speciation that occurs even though gene flow is possible—is rare or nonexistent. These studies are fueling a growing awareness that under certain circumstances, natural selection that causes populations to diverge can overcome gene flow and cause speciation.

The key realization is that even though sympatric populations are not physically isolated, they may be isolated by preferences for different habitats. As an example, let's consider research on speciation in apple maggot flies.

After apple maggot flies court and mate on apple fruits, females lay an egg inside. After the egg hatches, the larva eats and grows as the fruit ripens and drops to the ground. The larva leaves, burrows into the ground, and pupates—meaning that it secretes a protective case and undergoes metamorphosis (see Chapter 32). Individuals emerge as adults the following spring, starting the cycle anew.

Apple trees were introduced to North America from Europe less than 300 years ago, however. Where did apple maggot flies come from?

Phylogenetic trees, estimated from synapomorphies in DNA sequence data, indicate that apple maggot flies are extremely closely related to hawthorn flies, which are native to North America. Hawthorn flies lay their eggs in hawthorn fruits.

Hawthorn trees and apple trees often grow almost side by side. But by following marked individuals in the field, biologists have determined that only about 6 percent of the total matings observed are between apple flies and hawthorn flies. The data in **Figure 26.7** show why. The bars on the graphs indicate the percentage of apple flies (top) or hawthorn flies (bottom) that land on a surface containing scents from apple, hawthorn, both apple and hawthorn, or neither apple nor hawthorn, in laboratory tests. Note that:

- Apple flies respond most strongly to apple scents; hawthorn flies respond most strongly to hawthorn scents.
- In both types of flies, there is no difference in the response to a mix of both scents and no scent at all.
- Apple flies avoid hawthorn scent, and hawthorn flies avoid apple scent—both types respond less to the "wrong" scent than to no scent at all.

Other experiments have (1) established that a fly's ability to discriminate scents has a genetic basis—meaning that apple flies and hawthorn flies have different alleles associated with attraction to fruit; (2) identified the specific odor receptor cells responsible for the difference in scent response, and (3) shown that hybrid individuals cannot discriminate between apple and hawthorn fruit.

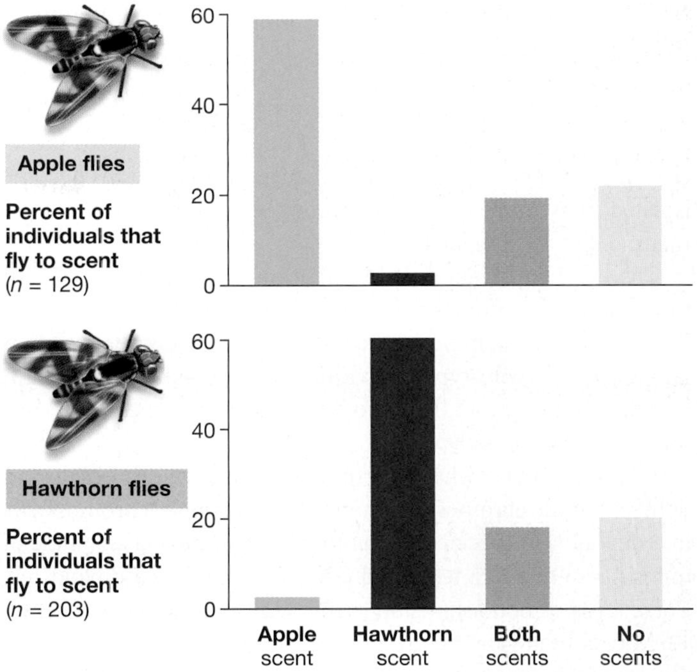

FIGURE 26.7 Disruptive Selection on Fruit Preference in Flies. Each fly was tested with four types of scent, one at a time, in a lab setting.

✔**QUESTION** To promote divergence, why is it important for apple maggot flies to mate on their food plants?

The upshot is that apple flies mate on apples and hawthorn flies mate on hawthorn fruits. Disruptive selection—a process introduced in Chapter 25—is occurring. The two populations are becoming genetically isolated.

Although they are not yet separate species on the basis of the biological, morphological, or phylogenetic species concepts, apple flies and hawthorn flies are diverging. They are currently in the process of becoming distinct species.

✔**If you understand the speciation process occurring in these flies, you should be able to predict why natural selection would favor divergence based on two observations: (1) the sugars and other molecules in apple fruit and hawthorn fruit are different, and (2) apple fruits drop much earlier than hawthorn fruits—meaning that apple larvae develop in warmer temperatures and endure a longer winter before emerging.**

Although the apple maggot fly's story might seem localized and specific, the events may be common. Biologists currently estimate that over 3 million insect species exist. Most of these species are associated with specific host plants. Based on the data from apple maggot flies, it is reasonable to hypothesize that switching host plants has been a major trigger for speciation throughout the course of insect evolution.

How Can Polyploidy Lead to Speciation?

Based on the theory and data reviewed thus far, it is clear that gene flow, genetic drift, and natural selection play important roles in speciation. Can the fourth evolutionary process—mutation—influence speciation as well?

At first glance, the answer to this question would appear to be no. Chapter 25 emphasized that even though mutation is the ultimate source of genetic variation in populations, it is an inefficient mechanism of evolutionary change. If populations become isolated, it is unlikely that mutation, on its own, could cause them to diverge appreciably.

There is a particular type of mutation, though, that turns out to be extremely important in speciation—particularly in plants. The key is that the mutation reduces gene flow between mutant and normal, or wild-type, individuals. It does so because mutant individuals have more than two sets of chromosomes. This condition is known as **polyploidy**.

Polyploidy occurs when an error in meiosis or mitosis results in a doubling of the chromosome number. For example, chromosomes in a diploid ($2n$) species may fail to pull apart during anaphase of mitosis, resulting in a tetraploid cell ($4n$) instead of a diploid cell. These types of nondisjunction events were introduced in Chapter 12. Mutations are any change in an organism's DNA present; polyploidization is a massive mutation that affects entire chromosomes.

To understand why polyploid individuals are genetically isolated from wild-type individuals, consider what happens when that tetraploid cell undergoes meiosis to form gametes and mates with a diploid individual (**Figure 26.8**).

- By meiosis, diploid individuals produce haploid gametes and tetraploid individuals produce diploid gametes. These gametes unite to form a triploid ($3n$) zygote.
- Even if this offspring develops normally and reaches sexual maturity, its three homologous chromosomes cannot synapse and separate correctly during meiosis. Thus, they are distributed to daughter cells unevenly. Virtually all of the gametes produced by the triploid individual end up with an uneven number of chromosomes.
- Because almost all of its gametes contain a dysfunctional set of chromosomes, the triploid individual is virtually sterile.

Tetraploid and diploid individuals rarely produce fertile offspring when they mate. As a result, tetraploid and diploid populations are reproductively isolated.

How do the polyploid individuals involved in speciation form? There are two general mechanisms:

1. **Autopolyploid** ("same-many-form") individuals are produced when a mutation results in a doubling of chromosome number and the chromosomes all come from the same species.
2. **Allopolyploid** ("different-many-form") individuals are created when parents that belong to different species mate and produce an offspring where chromosome number doubles. Allopolyploid individuals have chromosome sets from different species.

Let's consider specific examples to illustrate how speciation by polyploidy occurs.

AUTOPOLYPLOIDY Although autopolyploidy is thought to be much less common than allopolyploidy, biologists recently documented autopolyploidy in the maidenhair fern. This plant inhabits woodlands across North America. During the normal life cycle of a fern, individuals alternate between a haploid (n) stage and a diploid ($2n$) stage.

Biologists initially set out to do a routine survey of allelic diversity in a population of these ferns. They happened to be examining individuals in the haploid stage and found several individuals that had *two* versions of each gene instead of just one. These individuals were diploid even though they had the "haploid" growth form. The biologists followed these individuals through their life cycle and confirmed that when the ferns mated, they produced offspring that were tetraploid ($4n$). The researchers had stumbled upon polyploid mutants within a normal population.

To follow up on the observation, researchers located the parent of the mutant individuals. The parent turned out to have a defect in meiosis that caused non-disjunction of chromosomes. Instead of producing normal, haploid cells as a result of meiosis,

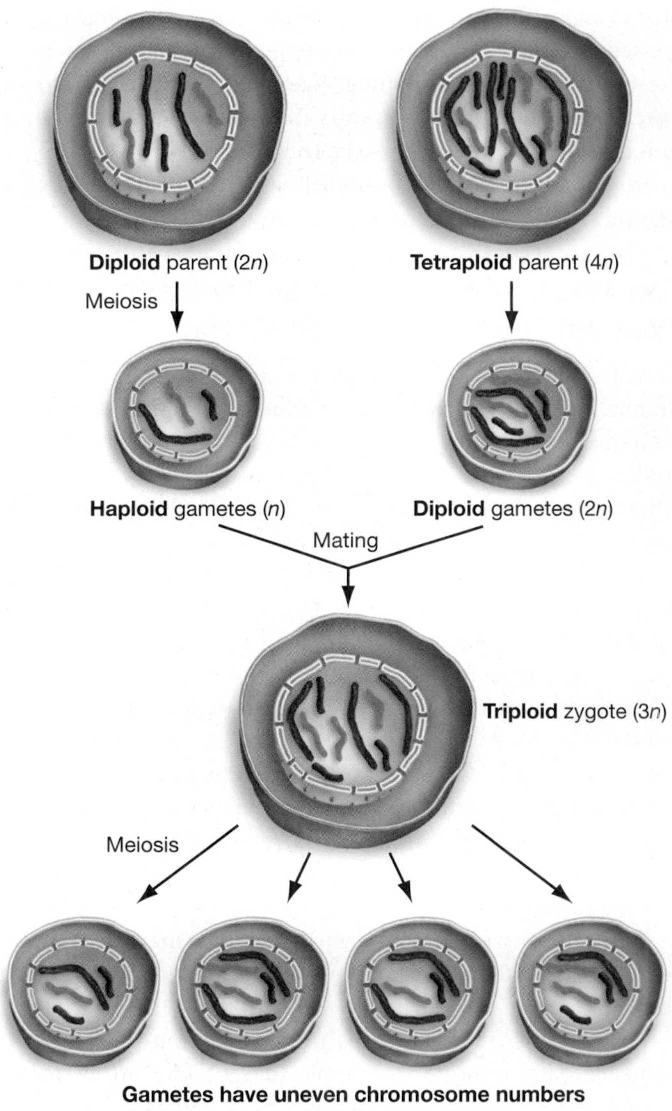

FIGURE 26.8 Polyploidy Can Lead to Reproductive Isolation. The mating diagrammed here illustrates why tetraploid individuals are reproductively isolated from diploid individuals.

the mutant individual produced diploid cells. These diploid cells eventually led to the production of diploid gametes.

Because maidenhair ferns can self-fertilize, the diploid gametes could combine to form tetraploid offspring. The tetraploid offspring could then self-fertilize or mate with their tetraploid parent or each other. If the process continued, a polyploid population of maidenhair ferns would be established.

Polyploid individuals are genetically isolated from the original diploid population and thus evolutionarily independent, because tetraploid individuals can breed with other tetraploids but not with diploids. If genetic drift and selection then caused the two populations to diverge, speciation would be under way.

✓If you understand how autopolyploidy works, you should be able to create a scenario explaining how the process gave rise to a tetraploid grape with extra-large fruit, from a diploid population with smaller fruit. (You've probably seen both types of fruit in the supermarket.)

This autopolyploidy study documented the critical first step in speciation—the establishment of genetic isolation. In this population of maidenhair ferns, as in apple maggot flies, speciation is under way right before our eyes.

ALLOPOLYPLOIDY New tetraploid species may be created when two diploid species hybridize. **Figure 26.9a** shows how. The top three drawings show a diploid offspring forming from a mating between two different species. Because the offspring has chromosomes that do not pair normally during meiosis, it is sterile. But if a mutation occurs that doubles the chromosome number in this individual prior to or during meiosis, then each chromosome gains a homolog. When these homologs synapse, meiosis can proceed, and diploid gametes are produced. When diploid gametes fuse during self-fertilization, a tetraploid individual results.

Exactly this chain of events occurred after three European species of weedy plants in the genus *Tragopogon* were introduced to western North America in the early 1900s. In 1950 a biologist described the first of two new tetraploid species that have been discovered. Based on an analysis of their chromosomes, both were clearly the descendants of matings between the introduced diploids. Follow-up work has shown that at least one of the new tetraploid species is expanding its geographic range (**Figure 26.9b**).

✓If you understand how alloploidy works, you should be able to create a scenario explaining how a cross between a tetraploid population called Emmer wheat and a wild, diploid wheat gave rise to the hexaploid bread wheat grown throughout the world today.

WHY IS SPECIATION BY POLYPLOIDY SO COMMON IN PLANTS? The claim that speciation by polyploidization has been particularly important in plants is backed by the observation that many diploid species have close relatives that are polyploid. Three

(a) If chromosome doubling occurs, allopolyploid offspring can be fertile and form new species.

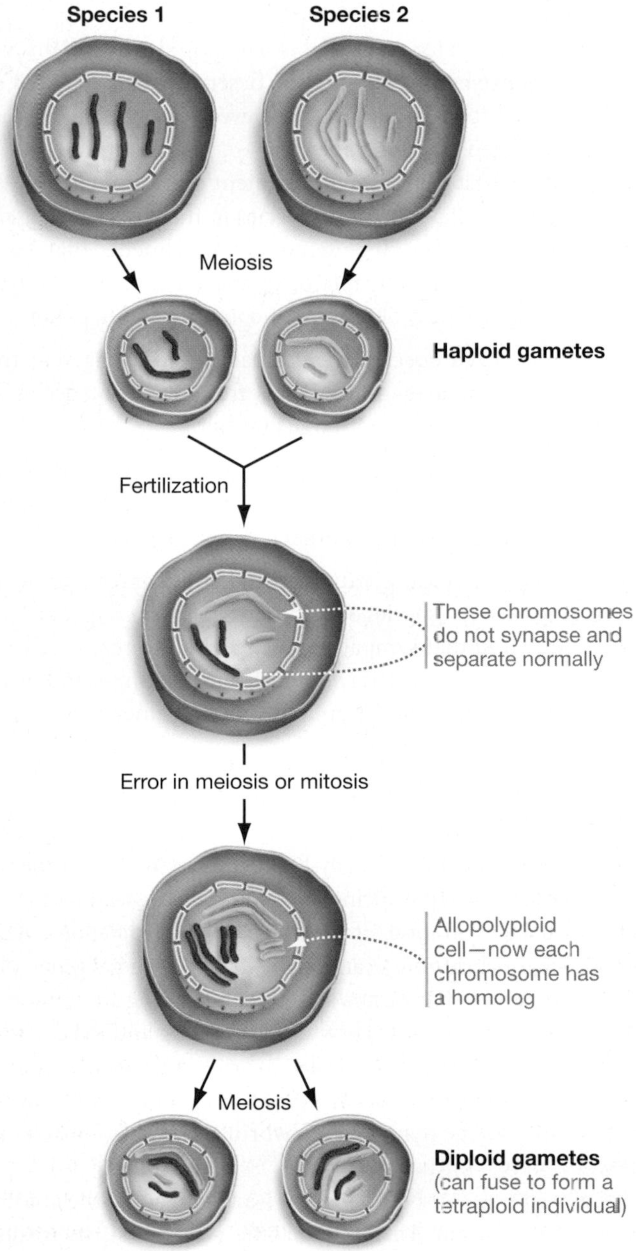

(b) An allopolyploid species that formed recently.

One of the **diploid species** introduced to North America (*Tragopogon dubius*)

New tetraploid species resulting from allopolyploidy (*Tragopogon mirus*)

FIGURE 26.9 Allopolyploids Can Form New Species.

properties of plants have been noteworthy in making this mode of speciation possible:

1. In plants, somatic cells that have undergone many rounds of mitosis can undergo meiosis and produce gametes. If sister chromatids separate during anaphase of one of these mitotic divisions but do not migrate to opposite poles, the result can be a tetraploid daughter cell that later undergoes meiosis to form diploid gametes.
2. The ability of some plant species to self-fertilize makes it possible for diploid gametes to fuse and create genetically isolated tetraploid populations.
3. Hybridization between plant species is common, creating opportunities for speciation via formation of allopolyploids.

To summarize, speciation by polyploidization is driven by chromosome-level mutations and occurs in sympatry (**Table 26.3** on page 471). Compared to the gradual process of speciation by geographic isolation or by disruptive selection in sympatry, speciation by polyploidy is virtually instantaneous. It is fast, sympatric, and common. Before moving on, be sure to go to the study area at *www.masteringbiology.com* and review how polyploidization can trigger speciation.

(MB) **Web Activity** Speciation by Changes in Ploidy

CHECK YOUR UNDERSTANDING

If you understand that . . .

- Speciation occurs when populations become isolated genetically and then diverge due to selection, genetic drift, or mutation.

✔ **You should be able to . . .**

Give an example of an event that can lead to the genetic isolation of populations, and explain why selection and drift would cause the populations to diverge.

Answers are available in Appendix B.

26.4 What Happens When Isolated Populations Come into Contact?

Suppose two populations that have been isolated come into contact again. If divergence has taken place and if divergence has affected when, where, or how individuals in the populations mate, then it is unlikely that interbreeding will take place. In cases such as this, prezygotic isolation exists. When it does, mating between the populations is rare, gene flow is minimal, and the populations continue to diverge.

But what if prezygotic isolation does not exist, and the populations begin interbreeding? The simplest outcome is that the populations fuse over time, as gene flow erases any distinctions between them. Several other possibilities exist, however. Let's explore three of them: reinforcement, hybrid zones, and speciation by hybridization.

Reinforcement

If two populations have diverged extensively and are distinct genetically, it is reasonable to expect that their hybrid offspring will have lower fitness than their parents. The logic here is that if populations are well-adapted to different habitats, then hybrid offspring will not be well-adapted to either habitat. If the two populations have diverged enough genetically, hybrid offspring also may fail to develop normally or may be infertile.

When postzygotic isolation occurs, there should be strong selection against interbreeding because hybrid offspring represent a wasted effort on the part of parents. Individuals that do not interbreed, due to a different courtship ritual or pollination system or other form of prezygotic isolation, should be favored because they produce more viable offspring.

Natural selection for traits that isolate populations in this way is called **reinforcement**. The name is descriptive because the selected traits reinforce differences that evolved while the populations were isolated from one another.

Some of the best data on reinforcement come from laboratory studies of closely related fruit fly species in the genus *Drosophila*. Researchers analyzed a large series of experiments that tested whether members of closely related fly species are willing to mate with one another. The biologists found an interesting pattern:

- If closely related species are sympatric—meaning that they live in the same area—individuals from the two species are seldom willing to mate with one another.
- If the species are allopatric—meaning that they live in different areas—then individuals from the two species are often willing to mate with one another.

The pattern is logical because natural selection can act to reduce mating between species only if their ranges overlap. Thus, it is reasonable to find that sympatric species exhibit prezygotic isolation but that allopatric species do not. There is a long-standing debate, however, over just how important reinforcement is in groups other than the genus *Drosophila*.

Hybrid Zones

Hybrid offspring are not always dysfunctional. In some cases they are capable of mating and producing offspring and have features that are intermediate between those of the two parental populations. When this is the case, hybrid zones can form. A **hybrid zone** is a geographic area where interbreeding occurs and hybrid offspring are common.

Depending on the fitness of hybrid offspring and the extent of breeding between parental species, hybrid zones can be narrow or wide, and long or short lived. As an example of how researchers analyze the dynamics of hybrid zones, let's consider recent work on two bird species.

Townsend's warblers and hermit warblers live in the coniferous forests of North America's Pacific Northwest. In western

(a) Hybrids have intermediate characteristics.

(b) Hybrids inherit species-specific mtDNA sequences from their mothers.

Townsend's warbler ■

Townsend's-hermit hybrid ▨

Hermit warbler ▦

N

Pacific Ocean

Individuals that look like Townsend's warblers but have hermit mtDNA

- ● All individuals have Townsend's mtDNA
- ◑ Some individuals have Townsend's mtDNA, others have hermit mtDNA
- ○ All individuals have hermit mtDNA

Present hybrid zone where two ranges meet {

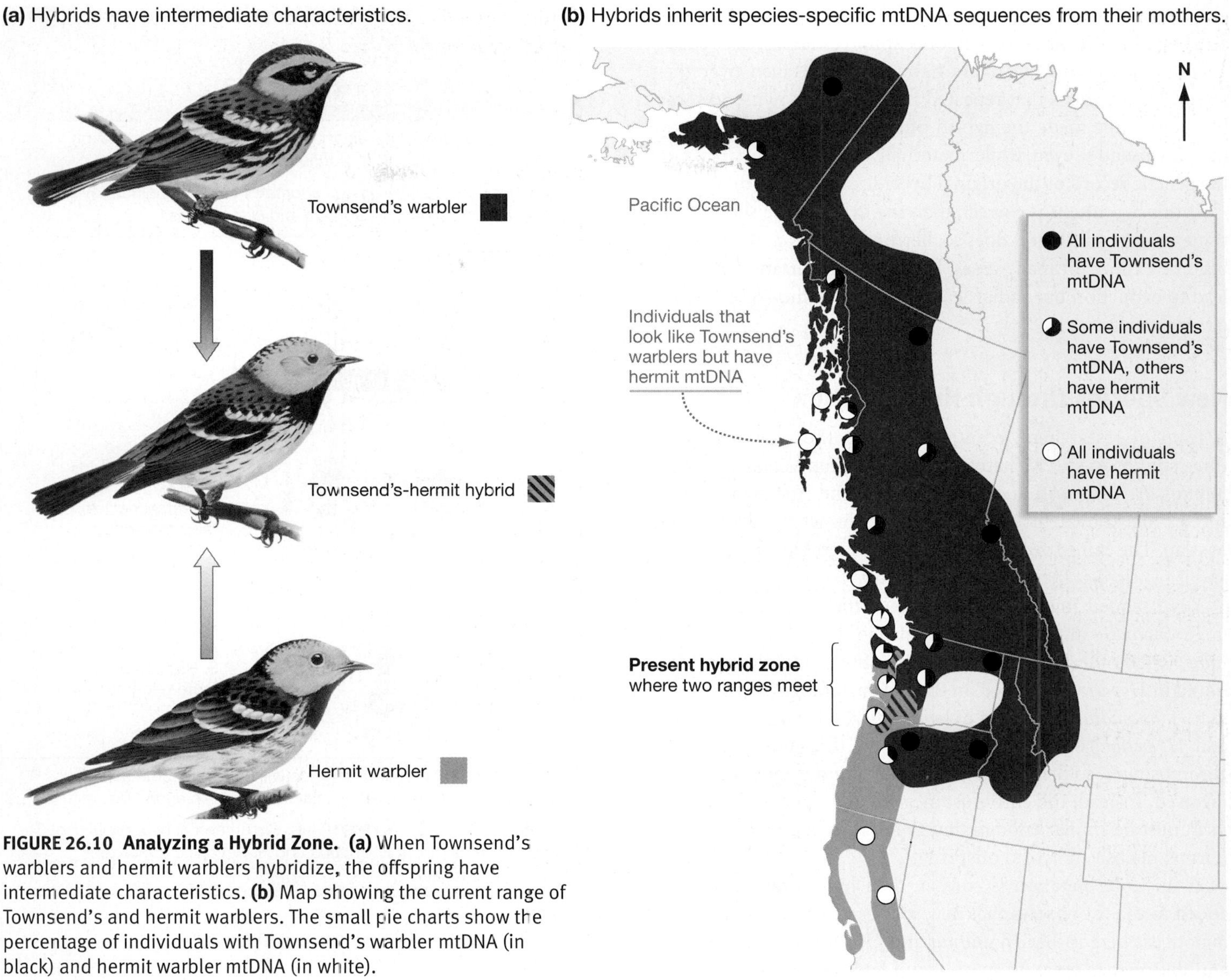

FIGURE 26.10 Analyzing a Hybrid Zone. (a) When Townsend's warblers and hermit warblers hybridize, the offspring have intermediate characteristics. **(b)** Map showing the current range of Townsend's and hermit warblers. The small pie charts show the percentage of individuals with Townsend's warbler mtDNA (in black) and hermit warbler mtDNA (in white).

Washington State, where their ranges overlap, the two species hybridize extensively. As **Figure 26.10a** shows, hybrid offspring have characteristics that are intermediate relative to the two parental species.

To explore the dynamics of this hybrid zone, a team of biologists examined gene sequences in the mitochondrial DNA (mtDNA) of a large number of Townsend's, hermit, and hybrid warblers collected from forests throughout the region. The team found that each of the parental species has certain species-specific mtDNA sequences. This result allowed the researchers to infer how hybridization was occurring.

To grasp the reasoning here, it is critical to realize that mtDNA is maternally inherited in most animals and plants. If a hybrid individual has Townsend's mtDNA, its mother had to be a Townsend's warbler while its father had to be a hermit warbler. In this way, identifying mtDNA types allowed the research team

to infer whether Townsend's females were mating with hermit males, or vice versa, or both.

Their data presented a clear pattern: Most hybrids form when Townsend's males mate with hermit warbler females. One of the investigators followed up on this result with experiments showing that Townsend's males are extremely aggressive in establishing territories and that they readily attack hermit warbler males. The data suggest that Townsend's males invade hermit territories, drive off the hermit males, and mate with hermit females.

The team also found something completely unexpected. When they analyzed the distribution of mtDNA types along the Pacific Coast and in the northern Rocky Mountains, they found that many Townsend's warblers actually had hermit mtDNA. **Figure 26.10b** shows that in some regions—such as the larger islands off the coast of British Columbia—*all* of the warblers had hermit mtDNA, even though they looked like full-blooded Townsend's warblers.

How could this be? To explain the result, the team hypothesized that hermit warblers were once found as far north as Alaska and that Townsend's warblers have gradually taken over their range. Their logic is that repeated mating with Townsend's warblers over time made the hybrid offspring look more and more like Townsend's, even while maternally inherited mtDNA kept the genetic record of the original hybridization event intact.

If this hypothesis is correct, then the hybrid zone should continue moving south. If it does so, hermit warblers may eventually become extinct. In many cases, however, hybridization does not lead to extinction but rather leads to the opposite—the creation of new species.

New Species through Hybridization

A team of researchers recently examined the relationships of three sunflower species native to the American West: *Helianthus annuus*, *H. petiolaris*, and *H. anomalus*. The first two of these species are known to hybridize in regions where their ranges overlap. The third species, *H. anomalus*, resembles these hybrids and is pictured at the start of this chapter.

DNA sequencing studies have shown that some gene regions in *H. anomalus* are remarkably similar to those found in *H. annuus*, while other gene sequences are almost identical to those found in *H. petiolaris*. Based on these data, biologists hypothesize that *H. anomalus* originated in hybridization between *H. annuus* and *H. petiolaris*. All three species have the same number of chromosomes, so neither allopolyploidy nor autopolyploidy was involved. Instead, the chromosomes of *H. annuus* and *H. petiolaris* must be similar enough that they can synapse and undergo normal meiosis in hybrid offspring.

If this interpretation is correct, then hybridization must be added to the list of ways that new species can form. The specific hypothesis here is that *H. annuus* and *H. petiolaris* were isolated and diverged as separate species, and later began interbreeding. The hybrid offspring created a third, new species that had unique combinations of alleles from each parental species and therefore different characteristics.

This hypothesis is supported by the observation that *H. anomalus* grows in much drier habitats than either of the parental species—suggesting that a unique combination of alleles allowed *H. anomalus* to thrive in certain habitats.

Biologists set out to test the hybridization hypothesis by trying to re-create the speciation event experimentally (**Figure 26.11**).

Step 1 They mated individuals from the two parental species and raised the offspring in a greenhouse.

Step 2 When these hybrid individuals were mature, the researchers either mated the plants to other hybrid individuals or "backcrossed" them to individuals from one of the parental species.

Step 3 This breeding program continued for four more generations before the experiment ended. Ultimately, the experimental lines were backcrossed twice, and they were mated to other hybrid offspring three times.

EXPERIMENT

QUESTION: Can new species arise by hybridization between existing species?

HYPOTHESIS: *Helianthus anomalus* originated by hybridization between *H. annuus* and *H. petiolaris*.

NULL HYPOTHESIS: *Helianthus anomalus* did not originate by hybridization between *H. annuus* and *H. petiolaris*.

EXPERIMENTAL SETUP:

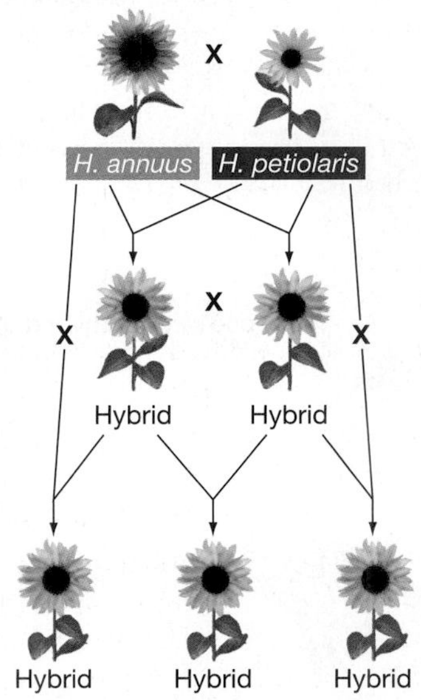

1. Mate *H. annuus* and *H. petiolaris* and raise offspring.

2. Mate F_1 hybrids or backcross F_1s to parental species; raise offspring.

3. Repeat for four more generations.

PREDICTION: Experimental hybrids will have the same mix of *H. annuus* and *H. petiolaris* genes as natural *H. anomalus*.

PREDICTION OF NULL HYPOTHESIS: Experimental hybrids will not have the same mix of *H. annuus* and *H. petiolaris* genes as natural *H. anomalus*.

RESULTS:
DNA comparison of a chromosomal region called S:

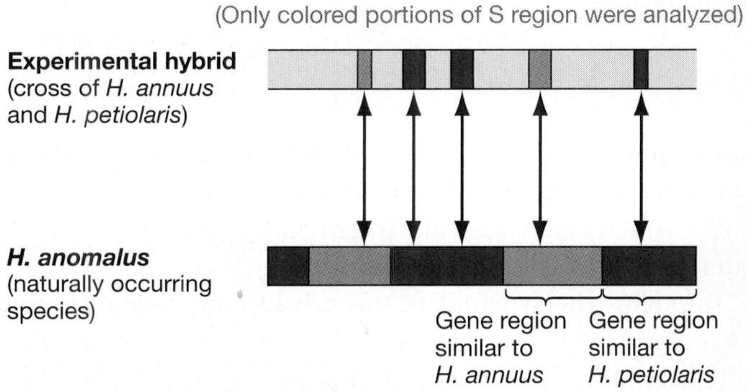

CONCLUSION: New species may arise via hybridization between existing species.

FIGURE 26.11 Experimental Evidence That New Species Can Originate in Hybridization Events.

SOURCE: Rieseberg, L. H., B. Sinervo, et al. 1996. Role of gene interactions in hybrid speciation: Evidence from ancient and experimental hybrids. *Science* 272: 741–745.

✔**QUESTION** Why it is valid to use experiments with living organisms—like this one—to infer what happened during historical events?

SUMMARY TABLE 26.3 **Mechanisms of Sympatric Speciation**

	Process	Example
Disruptive selection	Genetic divergence is caused by natural selection for different habitats or resources. Must be accompanied by some mechanism of genetic isolation.	Apple maggot flies that are adapted to breed on apple versus hawthorn fruits mate on those fruits, so little or no interbreeding occurs.
Polyploidization	Genetic isolation is created by formation of polyploid individuals (mutants with more than two sets of chromosomes) that can breed only with each other.	Particularly common in plants because frequent hybridization occurs between species and many mitotic divisions occur prior to meiosis.
• *Autopolyploidy*	Autopolyploids gain duplicate chromosome sets from the same species due to chromosome doubling during mitosis or meiosis.	A maidenhair fern individual became tetraploid due to an error in meiosis.
• *Allopolyploidy*	Allopolyploids gain duplicate chromosome sets from different species due to a hybridization event, followed by chromosome doubling.	*Tragopogon* species introduced to North America hybridized and formed offspring that became tetraploid and formed new species.

The goal of these crosses was to simulate matings that might have occurred naturally.

The experimental hybrids looked like the natural hybrid species, but did they resemble them genetically? To answer this question, the research team constructed genetic maps of each population, using a large series of genetic markers similar to the types of markers introduced in Chapter 19 and Chapter 20. Because each parental population had a large number of unique markers in their genomes, the research team hoped to identify which genes found in the experimental hybrids came from which parental species.

Some of their data are diagrammed in the Results section of Figure 26.11. The bottom bar in the illustration represents a region called S in the genome of the naturally occurring species *Helianthus anomalus*. As the legend indicates, this region contains three sections of sequences that are also found in *H. petiolaris* (indicated with the color red) and two that are also found in *H. annuus* (indicated with the color orange). The top bar shows the composition of this same region in the genome of the experimental hybrid lines.

The key observation here is that the genetic composition of the synthesized hybrids matches that of the naturally occurring hybrid species. In effect, the researchers had succeeded in re-creating a speciation event. Their results provide strong support for the hybridization hypothesis for the origin of *H. anomalus*.

Secondary contact of two populations can produce a dynamic range of possible outcomes: fusion of the populations, reinforcement of divergence, founding of stable hybrid zones, extinction of one population, or the creation of new species. **Table 26.4** summarizes the outcomes of secondary contact.

SUMMARY TABLE 26.4 **Possible Outcomes of Secondary Contact between Populations**

	Process	Example
Fusion of the populations	The two populations freely interbreed.	Occurs whenever populations of the same species come into contact.
Reinforcement of divergence	If hybrid offspring have low fitness, natural selection favors the evolution of traits that prevent interbreeding between the populations.	Appears to be common in fruit fly species that occupy the same geographic areas.
Hybrid zone formation	There is a well-defined geographic area where hybridization occurs. This area may move over time or be stable.	Many stable hybrid zones have been described; the hybrid zone between hermit and Townsend's warblers appears to have moved over time.
Extinction of one population	If one population or species is a better competitor for shared resources, then the poorer competitor may be driven to extinction.	Townsend's warblers may be driving hermit warblers to extinction.
Creation of new species	If the combination of genes in hybrid offspring allows them to occupy distinct habitats or use novel resources, they may form a new species.	Hybridization between sunflowers gave rise to a new species with unique characteristics.

CHAPTER 26 REVIEW

For media, go to the study area at www.masteringbiology.com

Summary of Key Concepts

Speciation occurs when populations of the same species become genetically isolated by lack of gene flow and then diverge from each other due to selection, genetic drift, or mutation.

- Speciation is a splitting event in which one lineage gives rise to two or more independent descendant lineages.
- Speciation begins when populations become genetically isolated—meaning that gene flow does not occur.
- Once populations are isolated, natural selection, genetic drift, and mutation act on them independently. As a result, the populations diverge over time.
- Eventually, the populations become so different that they can be recognized as distinct species—evolutionarily independent populations.

✔You should be able to design an experiment that would, given enough time, result in the production of two species from a single ancestral population.

Populations can be recognized as distinct species if they are reproductively isolated from each other, if they have distinct morphological characteristics, or if they form independent branches on a phylogenetic tree.

- Researchers use several criteria to test whether populations represent distinct species.
- The biological species concept focuses on the degree of hybridization between species to determine whether gene flow is occurring.
- The morphospecies concept infers that speciation has occurred if populations have distinctive morphological traits.
- The phylogenetic species concept identifies species as the smallest monophyletic groups on the tree of life.

✔You should be able to explain whether human populations would be considered separate species under the biological, morphological, and phylogenetic species concepts.

Populations can become genetically isolated from each other if they occupy different geographic areas, if they use different habitats or resources within the same area, or if one population is polyploid and cannot breed with the other.

- Speciation often begins when small groups of individuals colonize a new habitat or when a large, continuous population becomes fragmented into isolated habitats.
- When populations are sympatric, speciation can occur when disruptive selection favors individuals that breed in different habitats.
- Mutations that produce polyploidy can trigger rapid speciation in sympatry because they lead to reproductive isolation between diploid and tetraploid populations.

✔You should be able to evaluate whether your experiment on speciation (see above) represents a case of dispersal, vicariance, different habitat use, or polyploidy.

MB **Web Activity** Allopatric Speciation, **Web Activity** Speciation by Changes in Ploidy

When populations that have diverged come back into contact, they may fuse, continue to diverge, stay partially differentiated, or have offspring that form a new species.

- If gene flow occurs, populations that have diverged may fuse into a single species.
- If prezygotic isolation exists, populations that come back into contact will probably continue to diverge.
- Secondary contact can lead to reinforcement—the evolution of mechanisms that prevent hybridization.
- Gene flow between different species can lead to the formation of hybrid zones that move over time or are stable.
- In some cases, hybridization between species can create new species with unique combinations of traits.

✔You should be able to predict how a hybrid zone will change over time when hybrid offspring have higher fitness than the parental populations.

Questions

✔TEST YOUR KNOWLEDGE

Answers are available in Appendix B

1. What distinguishes a morphospecies?
   a. It has distinctive characteristics, such as size, shape, or coloration.
   b. It represents a distinct twig in a phylogeny of populations.
   c. It is reproductively isolated from other species.
   d. It is a fossil from a distinct time in Earth history.

2. When does vicariance occur?
   a. Small populations coalesce into one large population.
   b. A population is fragmented into isolated subpopulations.
   c. Individuals colonize a novel habitat.
   d. Individuals disperse and found a new population.

3. Why is "reinforcement" an appropriate name for the concept that natural selection should favor divergence and genetic isolation if populations experience postzygotic isolation?
   a. Selection should reinforce high fitness for hybrid offspring.
   b. Selection should reinforce the fact that they are "good species" under the morphological species concept.
   c. Selection acts because hybrid offspring do not develop at all or are sterile when mature.
   d. It reinforces selection for divergence that began when the species were geographically isolated.

4. The biological species concept can be applied only to which of the following groups?
   a. bird species living today
   b. dinosaurs
   c. bacteria
   d. archaea
5. Why are genetic isolation and genetic divergence occurring in apple maggot flies, even though populations occupy the same geographic area?
   a. Different populations feed and mate on different types of fruit.
   b. One population is tetraploid; others are diploid.
   c. The introduction of a nonnative host plant caused vicariance.
   d. Responses to scents have changed due to disruptive selection.
6. When the ranges of different species meet, a stable "hybrid zone" occupied by hybrid individuals may form. How is this possible?
   a. Hybrid individuals may have intermediate characteristics that are advantageous in a given region.
   b. Hybrid individuals are always allopolyploid and are thus unable to mate with either of the original species.
   c. Hybrid individuals may have reduced fitness and thus be strongly selected against.
   d. One species has a selective advantage, so as hybridization continues, the other species will go extinct.

✓ TEST YOUR UNDERSTANDING

Answers are available in Appendix B

1. Which studies in this chapter represent direct observation of speciation? Which are indirect studies of historical events?
2. In the case of the seaside sparrow, how did the species identified by the biological species concept, the morphospecies concept, and the phylogenetic species concept conflict?
3. Explain why genetic drift occurs during colonization events. Explain why natural selection occurs after colonization events.
4. Explain how isolation and divergence are occurring in apple maggot flies. Of the four evolutionary processes (mutation, gene flow, drift, and selection), which two are most important in causing this event?
5. Unlike animal gametes, plant reproductive cells do not differentiate until late in life. Explain why plants are much more likely to produce diploid gametes and polyploid offspring than are animals.
6. Individuals from closely related fruit fly species are less likely to mate if the related species are sympatric versus allopatric. Why is this considered strong evidence for reinforcement?

✓ APPLYING CONCEPTS TO NEW SITUATIONS

Answers are available in Appendix B

1. A large amount of gene flow is now occurring among human populations due to intermarriage among people from different ethnic groups and regions of the world. Is this phenomenon increasing or decreasing racial differences in our species? Explain.
2. Humans have introduced thousands of species to new locations around the globe, forming geographically isolated populations. Few, if any, of these colonization events have resulted in speciation. Use these data to evaluate the hypothesis that founder events trigger speciation.
3. A friend says that apple maggot flies prefer apple fruit scents because they need to, in order to survive. Another agrees and adds that the flies acquire the ability to distinguish the apple scents by spending time on the fruit, and that's why their offspring prefer apples. What's wrong with these statements?
4. All over the world, natural habitats are being fragmented into tiny islands as suburbs, ranches, and farms expand. Explain why this fragmentation process could lead to extinction. Then explain how it could lead to speciation.

This chapter is part of The Big Picture. See how on pages 494–495.

A fossilized trilobite. The last trilobites disappeared during a mass extinction event analyzed in Section 27.4.

27 Phylogenies and the History of Life

KEY CONCEPTS

- Phylogenetic trees document the evolutionary relationships among organisms and are estimated from data.
- The fossil record provides physical evidence of organisms that lived in the past.
- Adaptive radiations are a major pattern in the history of life. They are instances of rapid diversification associated with new ecological opportunities and new morphological innovations.
- Mass extinctions have occurred repeatedly throughout the history of life. They are environmental catastrophes that rapidly eliminate most of the species alive.

This chapter is about time and change. More specifically, it's about vast amounts of time and profound change in organisms. Both of these topics can be difficult for humans to grasp. Our lifetimes are measured in decades, and our knowledge of history is usually measured in centuries or millennia. But this chapter analyzes events that occurred over millions and even billions of years. A million years is completely beyond our experience and almost beyond our imagination.

It takes practice to get comfortable analyzing the profound changes that occur in organisms over deep time. To help you get started, the chapter begins by introducing the two major analytical tools that biologists use to reconstruct the history of life: phylogenetic trees and the fossil record. The remaining two sections explore episodes called adaptive radiations, which produce large numbers of highly diverse new species, and mass extinctions, which wipe out large numbers of species.

27.1 Tools for Studying History: Phylogenetic Trees

The evolutionary history of a group of organisms is called its **phylogeny**. Phylogenies are usually summarized and depicted in the form of a phylogenetic tree. A **phylogenetic tree** shows the ancestor-descendant relationships among populations or species, and clarifies who is related to whom. In a phylogenetic tree:

- a **branch** represents a population through time;
- the point where two branches diverge, called a **node** (or fork), represents the point in time when an ancestral species split into two or more descendant species; and
- a **tip** (or terminal node), the endpoint of a branch, represents a group (a species or larger taxon) that is living today or ended in extinction.

✔ When you see this checkmark, stop and test yourself. Answers are available in Appendix B.

Evolutionary trees have been introduced at various points in earlier chapters—often with just enough information to help you understand a new concept. **BioSkills 3** in Appendix A introduces the parts of a phylogenetic tree and how to read one. (It would be a *very* good idea to review **BioSkills 3**, right now.) Here let's focus on how biologists go about building them.

How Do Researchers Estimate Phylogenies?

Phylogenetic trees are an extremely effective way of summarizing data on the evolutionary history of a group of organisms. But like any other pattern or measurement in nature, from the average height of a person in a particular human population to the speed of a passing airplane, the genealogical relationships among species cannot be known with absolute certainty. Instead, the relationships depicted in an evolutionary tree are estimated from data.

To infer the historical relationships among species, researchers analyze the species' morphological or genetic characteristics, or both. For example, to reconstruct relationships among fossil species of humans, scientists analyze aspects of tooth, jaw, and skull structure. To reconstruct relationships among contemporary human populations, investigators usually compare the sequences of bases in a particular gene.

The fundamental idea in phylogeny inference is that closely related species should share many of their characteristics, while distantly related species should share fewer characteristics. But there are two general strategies for using data to estimate trees: the phenetic approach and the cladistic approach.

The **phenetic approach** to estimating trees is based on computing a statistic that summarizes the overall similarity among populations, based on the data. For example, researchers might use gene sequences to compute an overall "genetic distance" between two populations. A genetic distance summarizes the average percentage of bases in a DNA sequence that differ between two populations. A computer program then builds a tree that clusters the most similar populations and places more-divergent populations on more-distant branches.

The **cladistic approach** to inferring trees is based on the realization that relationships among species can be reconstructed by identifying shared derived characters in the species being studied—the synapomorphies introduced in Chapter 26. Recall that a synapomorphy is a trait that certain groups of organisms have that exists in no others. Synapomorphies allow biologists to recognize monophyletic groups—also called clades or lineages. Synapomorphies are characteristics that are shared because they are derived from traits that existed in their common ancestor.

Figure 27.1 illustrates the logic behind a cladistic analysis. When the ancestral population at the left of the figure splits into two descendant lineages at node A, each descendant group begins evolving independently and acquires unique traits. These traits are derived from their common ancestor via mutation, selection, and genetic drift.

An **ancestral trait** is a characteristic that existed in an ancestor; a **derived trait** is one that is a modified form of the ancestral trait,

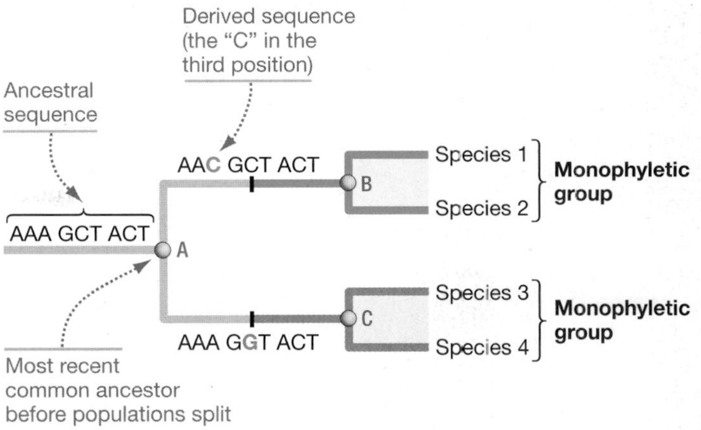

FIGURE 27.1 Synapomorphies Identify Monophyletic Groups. In this example, synapomorphies are changes in a DNA sequence.

found in a descendant. It's important to recognize that ancestral and derived traits are relative. If you are comparing mammals with the fossil forms called mammal-like reptiles, then fur and lactation are derived traits. But if you are comparing whales and humans, then fur and lactation are ancestral traits.

As an example, the traits used in a cladistic analysis might be a particular region of DNA that changes as follows:

AAA GCT ACT	ancestral population
AAC GCT ACT	a descendant population
AAA GGT ACT	another descendant population

When the two lineages themselves split at nodes B and C, the species that result share the derived characteristics. In this way, the ancestor at node B and species 1 and 2 can be recognized as a monophyletic group. Similarly, the ancestor at node C and species 3 and 4 can be recognized as a different monophyletic group. When many such traits have been measured, a computer program can be used to identify which traits are unique to each monophyletic group and then place the groups in a tree in the correct relationship to each other.

How Can Biologists Distinguish Homology from Homoplasy?

Although the logic behind phenetic and cladistic analyses is elegant, problems arise. The issue is that traits can be similar in two species not because those traits were present in a common ancestor, but because similar traits evolved independently in two distantly related groups.

In the example given in Figure 27.1, it is possible that species 2 is not at all closely related to species 1. Its ancestors may have had the sequence TAT GGT AGT, which happened to change to AAC GCT ACT due to mutation, selection, and drift that took place independently of the changes that took place in the ancestors of species 1.

WHAT IS HOMOPLASY? **Homology** (literally, "same-source") occurs when traits are similar due to shared ancestry; **homoplasy**

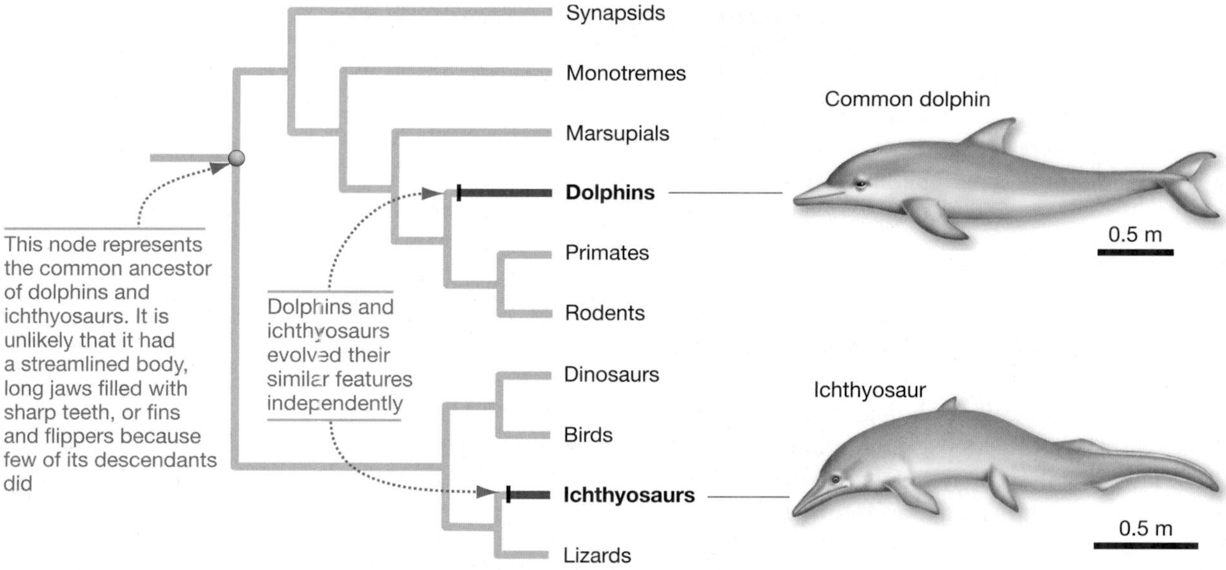

FIGURE 27.2 Homoplasy: Traits Are Similar but Not Inherited from a Common Ancestor. Dolphins and ichthyosaurs look similar but are not closely related—dolphins are mammals; ichthyosaurs are reptiles.

("same-form") occurs when traits are similar for reasons other than common ancestry. For example, the aquatic reptiles called ichthyosaurs were strikingly similar to modern dolphins (**Figure 27.2**). Both are large marine animals with streamlined bodies and large dorsal fins. Both chase down fish and capture them between elongated jaws filled with dagger-like teeth. But no one would argue that ichthyosaurs and dolphins are similar because the traits they share existed in a common ancestor. As the phylogeny in Figure 27.2 shows, analyses of other traits show that ichthyosaurs are reptiles whereas dolphins are mammals.

Based on these data, it is logical to argue that the similarities between ichthyosaurs and dolphins result from a common cause of homoplasy called convergent evolution. **Convergent evolution** occurs when natural selection favors similar solutions to the problems posed by a similar way of making a living. But convergent traits, or what biologists once called analogous traits, do not occur in the common ancestor of the similar species. Streamlined bodies and elongated jaws filled with sharp teeth are adaptations that help *any* species—whether it is a reptile or a mammal—chase down fish in open water.

EVIDENCE FOR HOMOLOGY In many cases, homology and homoplasy are much more difficult to distinguish than in the ichthyosaur and dolphin example. How do biologists recognize homology in such cases? As an example, consider the *Hox* genes of insects and vertebrates introduced in Chapter 21. Even though insects and vertebrates last shared a common ancestor some 600–700 million years ago, biologists argue that their *Hox* genes are derived from the same ancestral sequences. There are several lines of evidence to support this hypothesis:

- Groups of *Hox* genes are organized on chromosomes in a similar way. **Figure 27.3** shows that *Hox* genes in both insects and vertebrates are found in gene complexes, with similar genes found adjacent to one another on the chromosome. Recall from Chapter 20 that genes with these characteristics are called gene families. The organization of the gene families is nearly identical among insect and vertebrate species.
- All of the *Hox* genes share a 180-base-pair sequence called the homeobox. The portion of a protein encoded by the homeobox (introduced in Chapter 23) is almost identical in insects and vertebrates and has a similar function. It binds to DNA and regulates the expression of other genes.
- The products of the *Hox* genes have similar functions (see Chapter 21): identifying the locations of cells in embryos. They are also expressed in similar patterns in time and space.

In addition, many other animals, on lineages that branched off between insects and mammals, have similar genes. This is a crucial observation: If similar traits found in distantly related lineages are indeed similar due to common ancestry, then similar traits should be found in many intervening lineages on the tree of life—because all of the species in question inherited the trait from the same common ancestor.

Now suppose that a researcher set out to infer a phylogenetic tree from a set of morphological traits or DNA sequences. Without already having a tree in hand, it can be difficult or impossible to tell which traits are homologous and qualify as synapomorphies. Just as any data set contains unavoidable errors, or "noise," the data sets used to infer phylogenies inevitably include homoplasy.

USING PARSIMONY TO MINIMIZE THE IMPACT OF "NOISE" IN THE DATA To reduce the chance that homoplasy will lead to erroneous conclusions about which species are most closely related, biologists who are using cladistic approaches invoke the logical principle of **parsimony**. Under parsimony, the most likely explanation or pattern is the one that implies the least amount of change.

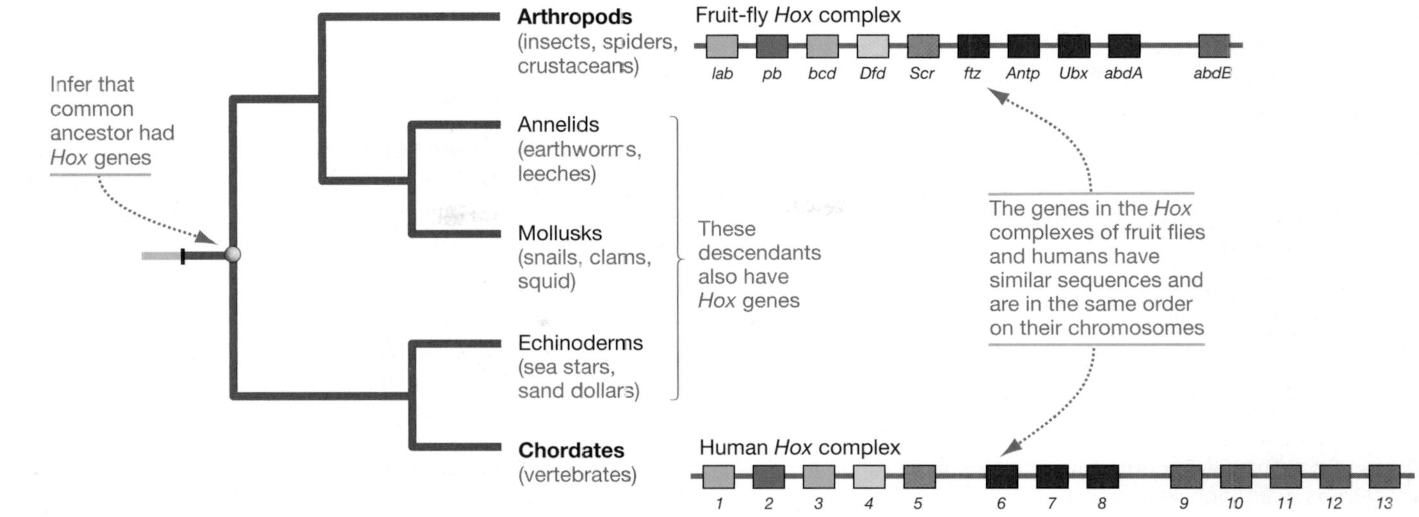

FIGURE 27.3 Homology: Similarities Are Inherited from a Common Ancestor. All of the animal groups illustrated on this phylogeny have *Hox* complexes that are similar to those illustrated for fruit flies and humans.

To implement a parsimony analysis, a computer compares the branching patterns that are theoretically possible and counts the number of changes in DNA sequences required to produce each pattern. For example, the tree in **Figure 27.4a** requires two changes in base sequence; the tree in **Figure 27.4b** requires four.

Convergent evolution and other causes of homoplasy should be rare compared with similarity due to shared descent. Thus, the tree that implies the fewest overall evolutionary changes should be the one that most accurately reflects what really happened during evolution.

If the branching pattern in Figure 27.4a is the most parsimonious of all possible trees, then biologists conclude that it is the most likely representation of actual phylogeny based on the data in hand.

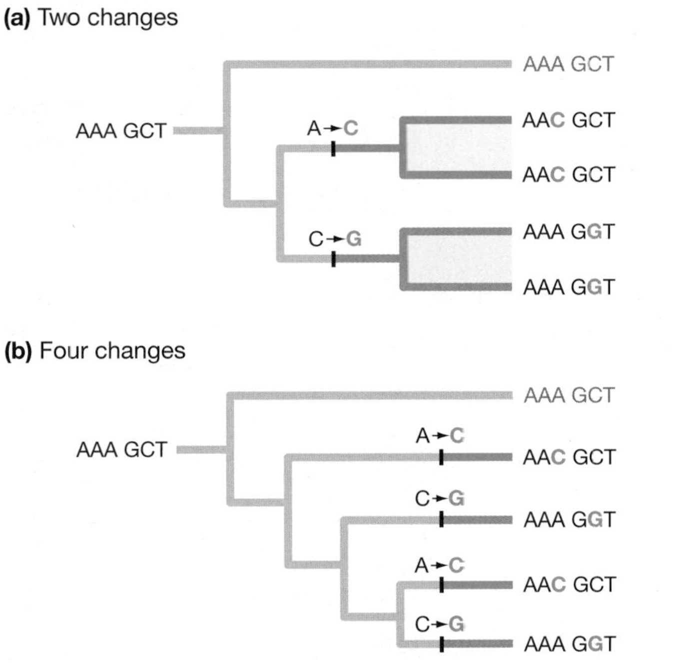

FIGURE 27.4 Parsimony Is One Method for Choosing among the Many Possible Trees. If you were using DNA sequence data to infer the phylogeny of five species, many different trees are possible. (Only two are shown here.) Under parsimony, the best tree is the one that requires the fewest changes to explain the sequences observed in the five species.

Whale Evolution: A Case History

As an example of how a cladistic approach works, consider the evolutionary relationships of the lineage of mammals called the Artiodactyla and the whales. Chapter 24 introduced the early evolution of whales—specifically how data from the fossil record support the hypothesis that whales evolved from terrestrial mammals. That chapter claimed that hippos, which are semi-aquatic, are the closest living relative of today's whales. What data back that claim?

A PHYLOGENY BASED ON MORPHOLOGICAL TRAITS Hippos, along with cows, deer, pigs, and camels, are artiodactyls. Members of this group have hooves and an even number of toes. They also share another feature: the unusual pulley shape of an ankle bone called the astragalus. Along with having feet with hooves and an even number of toes, the shape of the astragalus is a synapomorphy that identifies the artiodactyls as a monophyletic group.

These data support the tree shown in **Figure 27.5a** on page 478. Note that whales do not have an astragalus and are shown as an **outgroup** on this tree—that is, a species or group that is closely related to the monophyletic group but not part of it. It is logical to map the gain of the pulley-shaped astragalus in the ancestral population marked by a black bar in Figure 27.5a, because all of the descendants of that ancestor have the trait, but members of the outgroup do not.

A PHYLOGENY BASED ON DNA SEQUENCE DATA When researchers began comparing DNA sequences of artiodactyls and other species of mammals, however, the data showed that whales share many similarities with hippos. These results supported the tree shown in **Figure 27.5b**.

Here, it is still logical to map the evolution of the pulley-shaped astragalus at the same ancestor as in Figure 27.5a. But to recognize that the trait had to have been lost in an ancestor of today's whales—which do not have a pulley-shaped astragalus—the tree contains another black bar mapping the trait loss.

The tree supported by the DNA data conflicts with the tree supported by morphological data because it implies that the pulley-shaped astragalus evolved in artiodactyls and then was lost during whale evolution. The tree in Figure 27.5b implies two changes in the astragalus (a gain *and* a loss of the astragalus), while the tree in Figure 27.5a implies just one (a gain only). In terms of the evolution of the astragalus, the "whale + hippo" tree is less parsimonious than the "whales-are-not-artiodactyls" tree.

A PHYLOGENY BASED ON TRANSPOSABLE ELEMENTS The conflict between the data sets was resolved when researchers analyzed the distribution of the parasitic gene sequences called **SINEs (short interspersed nuclear elements)**, which occasionally insert themselves into the genomes of mammals. SINEs are transposable elements, similar to the LINEs introduced in Chapter 20.

As the data in **Figure 27.5c** show, whales and hippos share several types of SINES that are not found in other groups.

(a) The astragalus is a synapomorphy that identifies artiodactyls as a monophyletic group.

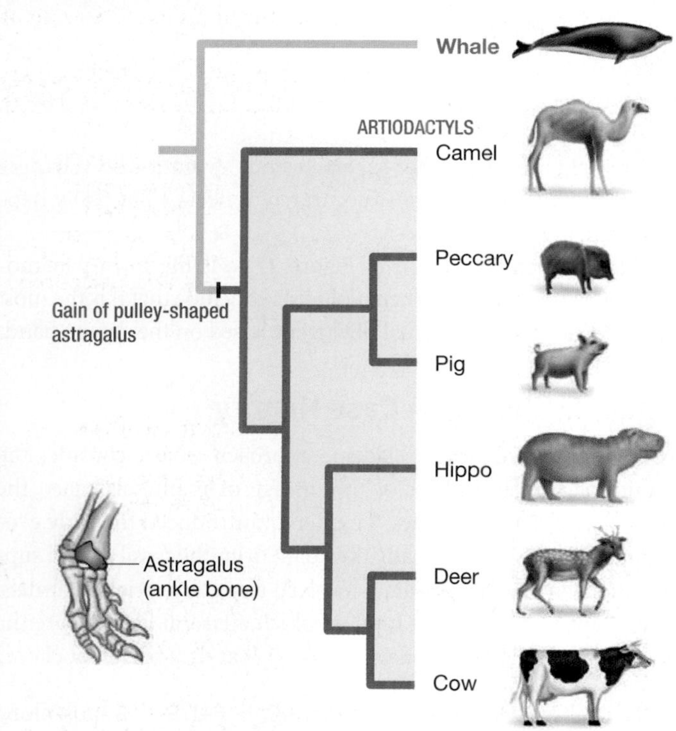

(b) If whales are related to hippos, then two changes occurred in the astragalus.

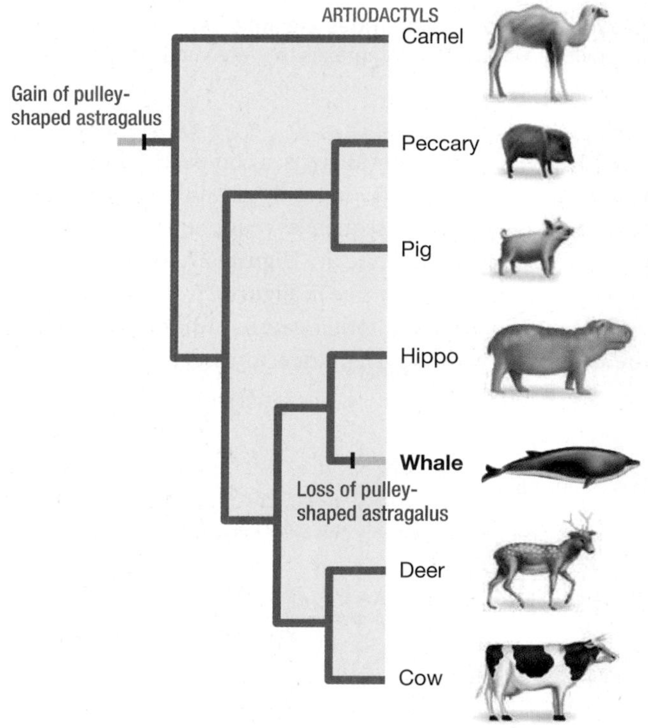

(c) Data on the presence and absence of SINE genes support the close relationship between whales and hippos.

Locus	1	2	3	4	5	6	7	8	9	10	11	12	13	14	15	16	17	18	19	20
Cow	0	0	0	0	0	0	0	1	1	1	1	1	1	1	1	1	1	1	0	0
Deer	0	0	0	0	0	0	0	1	?	1	1	1	1	1	1	?	1	1	0	0
Whale	1	1	1	1	1	1	1	0	?	1	0	1	1	0	0	0	?	1	0	0
Hippo	0	?	0	1	1	1	1	0	1	1	0	1	1	0	0	0	?	1	0	0
Pig	0	0	0	?	0	0	0	0	?	0	0	0	?	?	0	0	0	1	1	1
Peccary	?	?	?	?	?	?	?	?	?	?	?	?	?	?	?	?	?	?	1	1
Camel	0	0	0	0	0	0	0	0	0	0	0	0	0	0	0	0	0	0	0	0

1 = gene present
0 = gene absent
? = still undetermined

Whales and hippos share four *unique* SINE genes (4, 5, 6, and 7)

FIGURE 27.5 Evidence That Whales and Hippos Form a Monophyletic Group. Based on a parsimony analysis of the pulley-shaped astragalus, biologists favored **(a)** a tree excluding whales from the Artiodactyla over **(b)** the hypothesis that whales are artiodactyls and are closely related to hippos. **(c)** Data on the presence and absence of SINE genes support the tree in part (b).

✔**EXERCISE** The presence of SINE genes 4, 5, 6, 7 identifies hippos and whales as part of a monophyletic group. The presence of SINE genes 8, 11, 14, 15, 17 identifies deer and cows as part of a monophyletic group. SINE genes 10, 12, 13 identify hippos, whales, deer, and cows as part of a monophyletic group. Map the origin of these SINE genes on the tree in part (b). Where did SINE genes 19 and 20 first insert themselves into the genomes of artiodactyls?

- Whales and hippos share the SINEs numbered 4, 5, 6, and 7.
- Other SINE genes are present in some artiodactyls but not in others.
- Camels have no SINE genes at all.

To explain these data, biologists hypothesize that no SINEs were present in the population that is ancestral to all of the species in the study. Then, after the branching event that led to the split between the camels and all the other artiodactyls, different SINEs became inserted into the genomes of descendant populations.

If this hypothesis is correct, then the presence of a particular SINE represents a derived character. Because whales and hippos share four of these derived characters, it is logical to conclude that these animals are closely related.

✓If you understand this concept, you should be able to explain why SINES numbered 4–7 are synapomorphies that identify whales and hippos as a monophyletic group, and why the similarity in these SINES is unlikely to represent homoplasy.

Based on these data, most biologists accepted the phylogeny shown in Figure 27.5b as the most accurate estimate of evolutionary history. According to this phylogeny, whales are artiodactyls and share a relatively recent common ancestor with hippos. This observation inspired the hypothesis that both whales and dolphins are descended from a population of artiodactyls that spent most of their time feeding in shallow water, much as hippos do today.

In 2001, the discovery of the fossil artiodactyls featured in Figure 24.11 supported this hypothesis in spectacular fashion. These fossil species were clearly related to whales—they have an unusual ear bone found only in whales—and yet had a pulley-shaped astragalus.

The combination of DNA sequence data and data from the fossil record has clarified how a particularly interesting group of mammals evolved. What else do fossils have to say?

CHECK YOUR UNDERSTANDING

If you understand that . . .

- Phylogenies can be estimated by finding synapomorphies that identify monophyletic groups.

✓ You should be able to . . .

Explain whether the following traits represent homoplasy or homology: hair in humans and whales; extensive hair loss in humans and whales; limbs in humans and whales; social behavior in certain whales (e.g., dolphins) and humans.

Answers are available in Appendix B.

27.2 Tools for Studying History: The Fossil Record

Phylogenetic analyses are powerful ways to infer the order in which events occurred during evolution and to understand how particular groups of species are related. But only the fossil record provides direct evidence about what organisms that lived in the past looked like, where they lived, and when they existed.

A **fossil** is a piece of physical evidence from an organism that lived in the past. The **fossil record** is the total collection of fossils that have been found throughout the world. The fossil record is housed in thousands of private and public collections.

Let's review how fossils form, analyze the strengths and weaknesses of the fossil record, and then summarize major events that have taken place in life's approximately 3.5-billion-year history. (Although most evidence suggests that life originated 3.4–3.6 billion years ago, research continues.)

How Do Fossils Form?

Most of the processes that form fossils begin when part or all of an organism is buried in ash, sand, mud, or some other type of sediment. **Figure 27.6** illustrates the leaves of a tree falling onto a patch of mud, where they are buried by soil and debris before they decay. Pollen and seeds settle into the muck at the bottom of

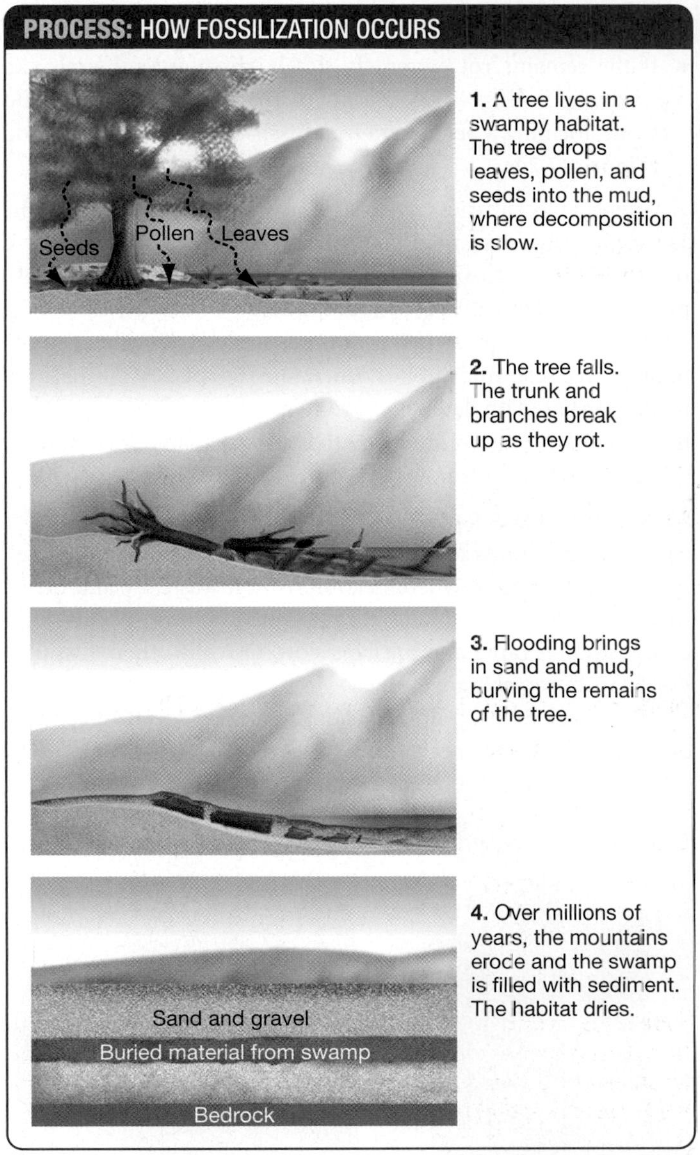

PROCESS: HOW FOSSILIZATION OCCURS
Seeds
Pollen
Leaves
1. A tree lives in a swampy habitat. The tree drops leaves, pollen, and seeds into the mud, where decomposition is slow.
2. The tree falls. The trunk and branches break up as they rot.
3. Flooding brings in sand and mud, burying the remains of the tree.
4. Over millions of years, the mountains erode and the swamp is filled with sediment. The habitat dries.
Sand and gravel
Buried material from swamp
Bedrock

FIGURE 27.6 Fossilization Preserves Traces of Organisms That Lived in the Past. Fossilization occurs most readily when the remains of an organism are buried in sediments, where decay is slow.

the swamp, where decomposition is slow. The stagnant water is too acidic and too oxygen-poor to support large populations of bacteria and fungi, so much of this material is buried intact before it decomposes. The trunk and branches that sit above the water line rot fairly quickly, but as pieces break off they, too, sink to the bottom and are buried.

PRESERVATION AFTER BURIAL Once burial occurs, several things can happen.

- If decomposition does not occur, the organic remains can be preserved intact—like the fossil pollen in **Figure 27.7a**.
- If sediments accumulate on top of the material and become cemented into rocks such as mudstone or shale, the sediments' weight can compress the organic material below into a thin, carbonaceous film. This happened to the leaf in **Figure 27.7b**.
- If the remains decompose *after* they are buried—as did the branch in **Figure 27.7c**—the hole that remains can fill with dissolved minerals and faithfully create a **cast** of the remains.
- If the remains rot extremely slowly, dissolved minerals can gradually infiltrate the interior of the cells and harden into stone, forming a permineralized fossil, such as petrified wood (**Figure 27.7d**).

After many centuries have passed, fossils can be exposed at the surface by erosion, a road cut, quarrying, or other processes. If researchers find a fossil, they can prepare it for study by painstakingly clearing away the surrounding rock.

If the species represented is new, researchers describe its morphology in a scientific publication, name the species, estimate the fossil's age based on dates assigned to nearby rock layers, and add the specimen to a collection so that it is available for study by other researchers. It is now part of the fossil record.

FOSSILIZATION IS A RARE EVENT The scenario just presented is based on conditions that are ideal for fossilization: The tree fell into an environment where decomposition was slow and burial was rapid. In most habitats the opposite situation occurs—decomposition is rapid and burial is slow. In reality, then, fossilization is an extremely rare event.

To appreciate this point, consider that there are 10 specimens of the first bird to appear in the fossil record, *Archaeopteryx*. All were found at the same site in Germany where limestone is quarried for printmaking (the bird's specific name is *lithographica*). If you accept an estimate that crow-sized birds native to wetland habitats in northern Europe would have a population size of around 10,000 and a life span of 10 years, and if you accept the current estimate that the species existed for about 2 million years, then you can calculate that about 2 billion *Archaeopteryx* lived. But as far as researchers currently know, only 1 out of every 200,000,000 individuals fossilized. For this species, the odds of becoming a fossil were almost 40 times worse than your odds are of winning the grand prize in a state lottery.

Limitations of the Fossil Record

Before looking at how the fossil record is used to answer questions about the history of life, it is essential to review the nature of this archive and recognize several features.

HABITAT BIAS Because burial in sediments is so crucial to fossilization, there is a strong habitat bias in the database. Organisms that live in areas where sediments are actively being deposited—including beaches, mudflats, and swamps—are much more likely to form fossils than are organisms that live in other habitats.

Within these habitats, burrowing organisms such as clams are already underground—pre-buried—at death and are therefore much more likely to fossilize. Organisms that live aboveground in dry forests, grasslands, and deserts are much less likely to fossilize.

TAXONOMIC AND TISSUE BIAS Slow decay is almost always essential to fossilization, so organisms with hard parts such as bones or shells are most likely to leave fossil evidence. This requirement introduces a strong taxonomic bias into the record. Clams, snails, and other organisms with hard parts have a much higher tendency to be preserved than do worms.

A similar bias exists for tissues within organisms. For instance, pollen grains are encased in a tough outer coat that resists decay, so they fossilize much more readily than do flowers. Shark teeth are abundant in the fossil record; but shark skeletal elements, which are made of cartilage, are almost nonexistent.

TEMPORAL BIAS Recent fossils are much more common than ancient fossils. This causes a temporal bias in the fossil record.

(a) Intact fossil (pollen) **(b)** Compression fossil (leaf) **(c)** Cast fossil (bark) **(d)** Permineralized fossil (trunk)

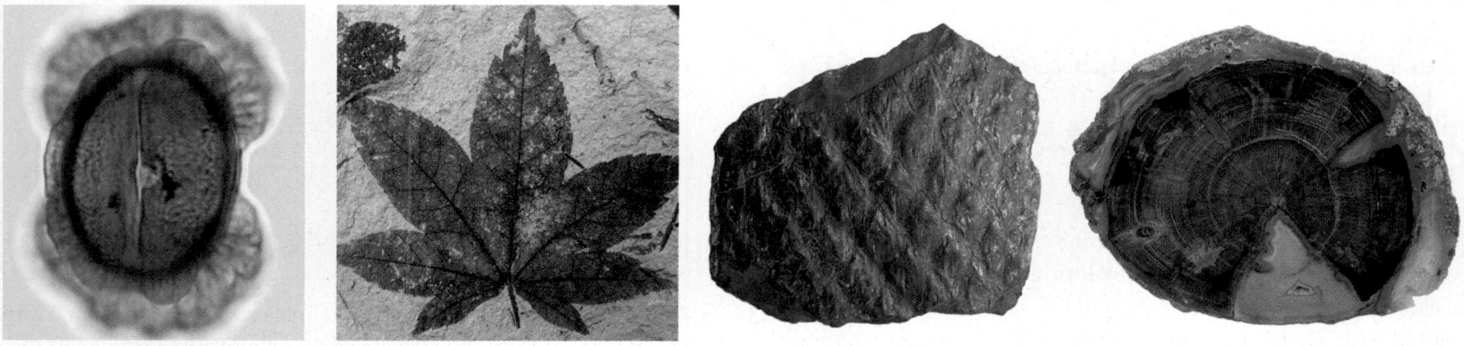

FIGURE 27.7 Fossils Are Formed in Several Ways. Different preservation processes give rise to different types of fossils.

To understand why, consider that when two of Earth's tectonic plates converge, the edge of one plate usually sinks beneath the other plate. The rocks composing the edge of the descending plate are either melted or radically altered by the increased heat and pressure they encounter as they move downward into Earth's interior. These alterations obliterate any fossils in the rock.

In addition, fossil-bearing rocks on land are constantly being broken apart and destroyed by wind and water erosion. The older a fossil is, the more likely it is to be demolished.

ABUNDANCE BIAS The fossil record has an abundance bias; it is weighted toward common species. Organisms that are abundant, widespread, and present on Earth for long periods of time leave evidence much more often than do species that are rare, local, or ephemeral.

To summarize, the fossil database represents a highly nonrandom sample of the past. **Paleontologists**—scientists who study fossils—recognize that they are limited to asking questions about tiny and scattered segments on the tree of life.

And yet, as this chapter shows, the record is a scientific treasure trove. Analyzing fossils is the only way scientists have of examining the physical appearance of extinct forms and inferring how they lived. The fossil record is like an ancient library, filled with volumes that give us glimpses of what life was like millions of years before humans appeared.

Life's Time Line

The best data available indicate that the Earth started to form about 4.6 billion years ago, and that life began around 3.5 billion years ago. To organize the tremendous sweep of time between then and now, researchers divide Earth history into segments called eons, eras, and periods.

Originally, geologists used distinctive rock formations or fossilized organisms to identify the boundaries between named time intervals. Later, researchers were able to use radiometric dating to assign absolute dates—expressed as years before the present—to events in the fossil record. Radiometric dating is based on the well-studied decay rates of certain radioactive isotopes (see Chapter 24). By dating rocks near fossils, researchers can also assign an absolute age to many of the species in the fossil record.

To summarize the history of life, researchers create time lines that record key "evolutionary firsts"—the appearance of important new lineages or innovations. It's important to recognize, though, that the times assigned to these first appearances underestimate when lineages appeared and events occurred. The reason is simple: It is possible for a particular species or lineage to exist for millions of years before leaving fossil evidence. The fossil record and efforts to date fossils are constantly improving, so time lines are always a work in progress.

PRECAMBRIAN **Figure 27.8** is a time line for the interval between the formation of Earth about 4.6 billion years ago and the

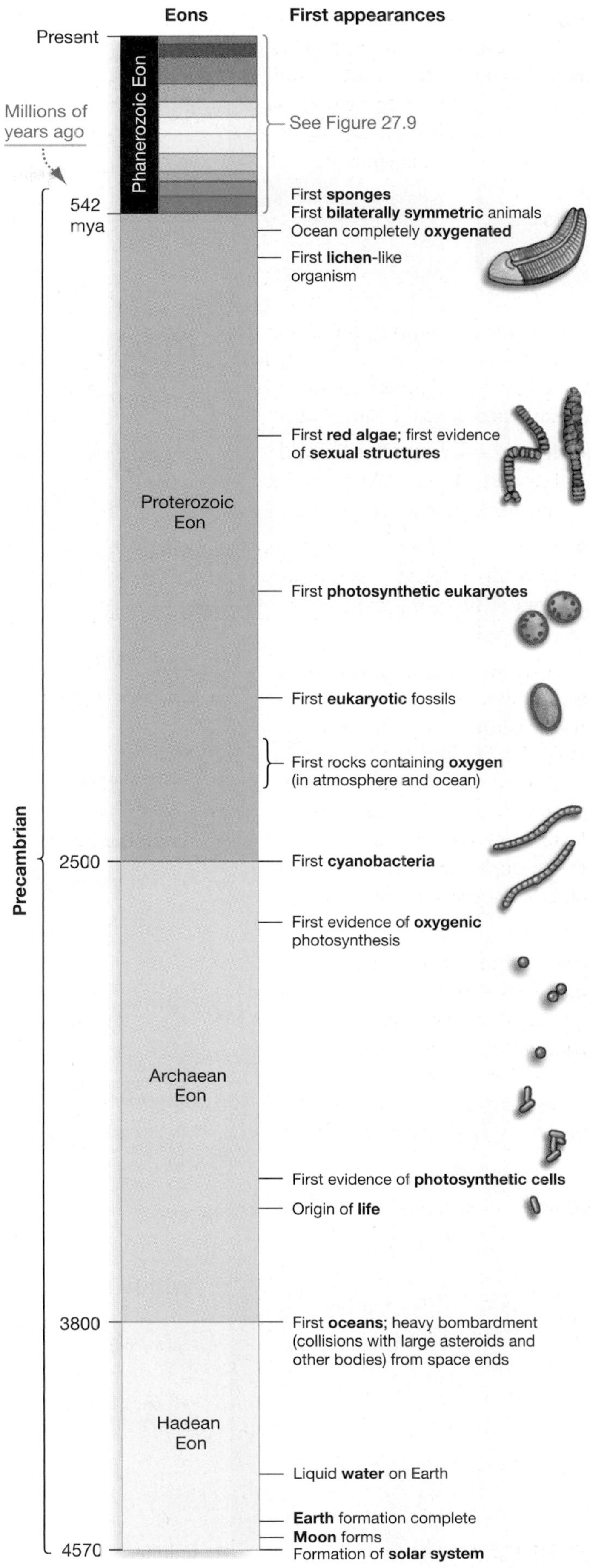

FIGURE 27.8 Major Events of the Precambrian. Life, photosynthesis, and the oxygen atmosphere all originated in the Precambrian.

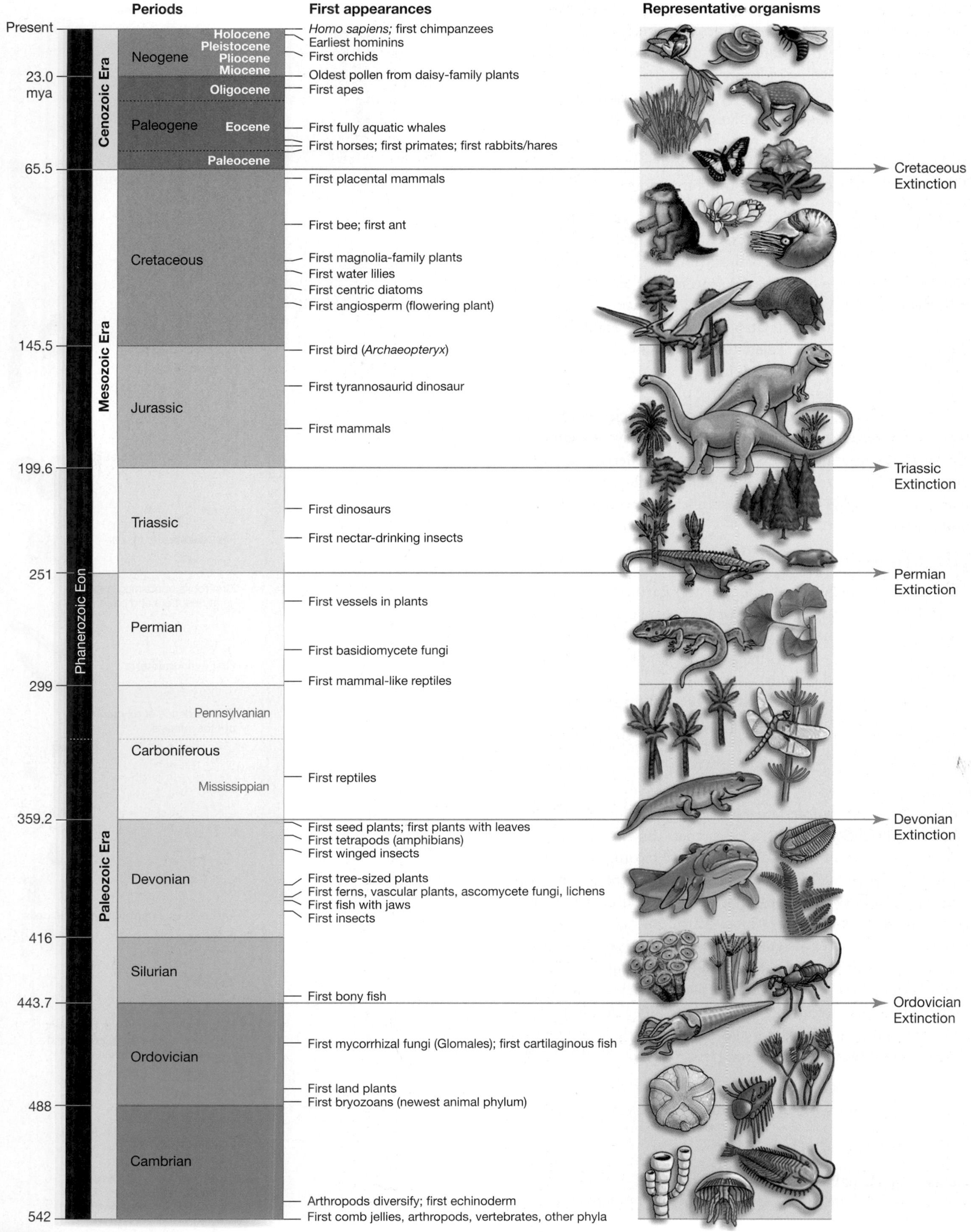
Periods
First appearances
Representative organisms
Present
23.0 mya
65.5
145.5
199.6
251
299
359.2
416
443.7
488
542
Phanerozoic Eon
Cenozoic Era
Mesozoic Era
Paleozoic Era
Neogene
Holocene
Pleistocene
Pliocene
Miocene
Oligocene
Paleogene
Eocene
Paleocene
Cretaceous
Jurassic
Triassic
Permian
Pennsylvanian
Carboniferous
Mississippian
Devonian
Silurian
Ordovician
Cambrian
Homo sapiens; first chimpanzees
Earliest hominins
First orchids
Oldest pollen from daisy-family plants
First apes
First fully aquatic whales
First horses; first primates; first rabbits/hares
First placental mammals
First bee; first ant
First magnolia-family plants
First water lilies
First centric diatoms
First angiosperm (flowering plant)
First bird (Archaeopteryx)
First tyrannosaurid dinosaur
First mammals
First dinosaurs
First nectar-drinking insects
First vessels in plants
First basidiomycete fungi
First mammal-like reptiles
First reptiles
First seed plants; first plants with leaves
First tetrapods (amphibians)
First winged insects
First tree-sized plants
First ferns, vascular plants, ascomycete fungi, lichens
First fish with jaws
First insects
First bony fish
First mycorrhizal fungi (Glomales); first cartilaginous fish
First land plants
First bryozoans (newest animal phylum)
Arthropods diversify; first echinoderm
First comb jellies, arthropods, vertebrates, other phyla
Cretaceous Extinction
Triassic Extinction
Permian Extinction
Devonian Extinction
Ordovician Extinction

FIGURE 27.9 Major Events of the Phanerozoic Eon (at left). The Phanerozoic began with the initial diversification of animals, continued with the evolution and early diversification of land plants and fungi, and includes the subsequent movement of animals to land. A total of five mass extinctions occurred during the Eon.

appearance of most animal groups about 542 million years ago (abbreviated mya). The entire interval is called the **Precambrian**; it is divided into the Hadean, Archaean, and Proterozoic eons.

The important things to note about the Precambrian are that:

- life was exclusively unicellular for most of Earth's history, and
- oxygen was virtually absent from the oceans and atmosphere for almost 2 billion years after the origin of life.

PHANEROZOIC EON The interval between 542 mya and the present is called the Phanerozoic eon and is divided into three eras (**Figure 27.9**). Each of these eras is further divided into intervals called periods.

1. The **Paleozoic** ("ancient life") **era** begins with the appearance of most major animal lineages and ends with the obliteration of almost all multicellular life-forms at the end of the Permian period. The Paleozoic saw the origin and initial diversification of the animals, land plants, and fungi, as well as the appearance of land animals.
2. The **Mesozoic** ("middle life") **era** begins with the end-Permian extinction events and ends with the extinction of the dinosaurs and other groups at the boundary between the Cretaceous period and Paleogene period. In terrestrial environments of the Mesozoic, gymnosperms were the most important plants and dinosaurs were the most important vertebrates.
3. The **Cenozoic** ("recent life") **era** is divided into the Paleogene period and the Neogene period. On land, angiosperms were the most important plants and mammals were the most important vertebrates. Events that occur today are considered to be part of the Cenozoic era.

CHANGES IN THE OCEANS AND CONTINENTS The changes in environments and life-forms that are recorded in Figures 27.8 and 27.9 took place in the context of radical changes in climate and in the extent and location of Earth's oceans and continents. Earth's crust is broken into enormous plates that are in constant motion, driven by heat rising from the planet's core.

Figure 27.10 summarizes the most recent data on how the positions of the continents and oceans have changed through time, starting at the beginning of the Phanerozoic Eon. The figure also includes notes on the prevailing climate. Over the past 542 million

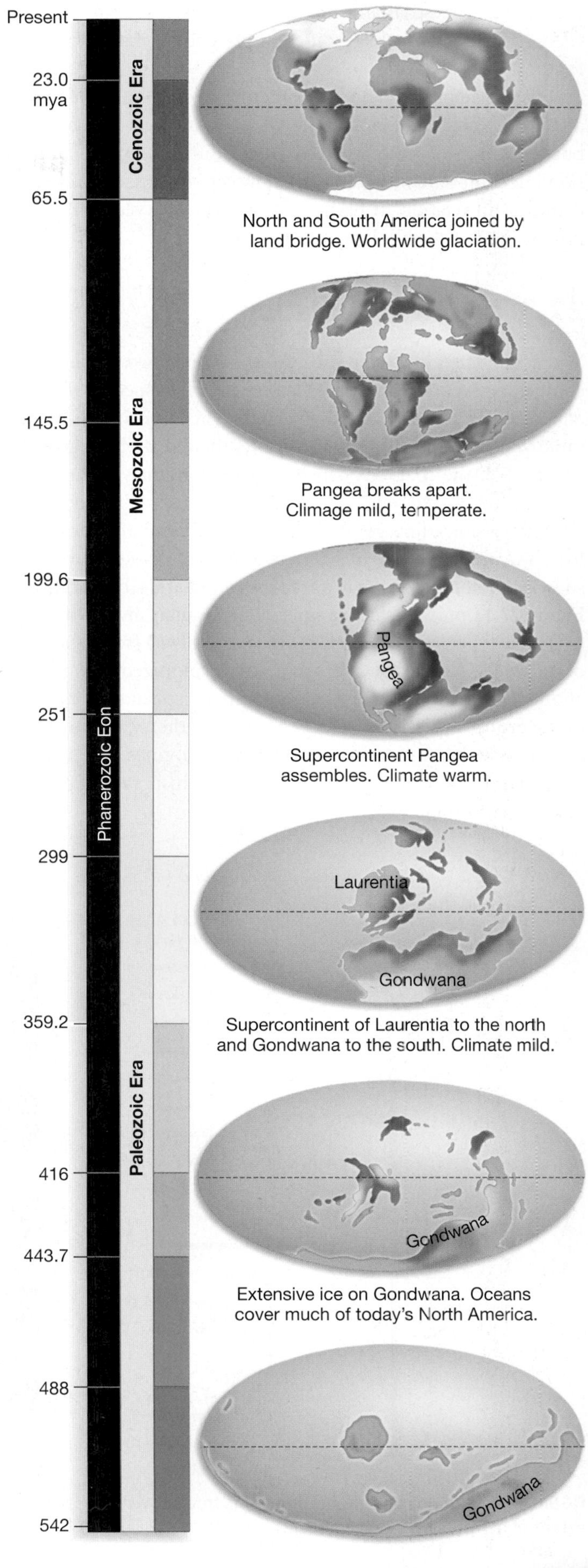

FIGURE 27.10 Continental Positions during the Phanerozoic (at right). Dramatic changes in the extent and position of the continents took place during the Phanerozoic. These changes affected the total amount of land area, the relative amounts of land in the tropics versus northern latitudes, and the nature of ocean currents.

years alone, terrestrial and marine landforms and climates have changed radically. Earth is a different place today than it has been for most of its history.

Taken together, the fundamental message in time-line data is a story of constant change. The changes are well documented, but the sweep of time involved is still difficult for the human mind to comprehend. A semester can seem long to a college student; but in relation to Earth's history, 100,000 years or even a million years is the blink of an eye.

27.3 Adaptive Radiation

When biologists consider the history of life, two of the most compelling events to study are (1) periods when species originate and diversify rapidly, and (2) periods when species go extinct rapidly. Let's focus on dramatic events that create biological diversity first; the chapter's concluding section analyzes how that diversity gets wiped out.

When a single lineage produces many descendant species that live in a wide diversity of habitats and use a wide array of resources, biologists say that an **adaptive radiation** has occurred. **Figure 27.11** provides an example: the Hawaiian silverswords.

The 30 species in this plant lineage evolved from a species of tarweed, native to California, that colonized the islands about 5 million years ago. Compared to speciation rates in other groups, this is extremely rapid. Further, today's silverswords vary from mosslike mat-formers to vines to shrubs to trees. They live in habitats ranging from lush rain forests to austere lava flows.

The Hawaiian silverswords fulfill the three hallmarks of an adaptive radiation: (1) they are a monophyletic group, (2) they speciated rapidly, and (3) they diversified ecologically—meaning, in terms of the resources they use and the habitats they occupy. Biologists use the term **niche** (pronounced *nitch*) to describe the range of resources that a species can use and the range of conditions that it can tolerate. Silverswords occupy a wide array of niches.

Why Do Adaptive Radiations Occur?

Adaptive radiations are a major pattern in the history of life. But why do some lineages diversify rapidly, while others do not?

Two general mechanisms can trigger adaptive radiations: new resources, and new ways to exploit resources. Let's consider each in turn.

ECOLOGICAL OPPORTUNITY Ecological opportunity—meaning the availability of new or novel types of resources—has driven a wide array of adaptive radiations. For example, biologists explain the diversification of silverswords by hypothesizing that few other flowering plant species were present on the Hawaiian Islands 5 million years ago. With few competitors, the descendants of the colonizing tarweed were able to grow in a wide range of habitats. Over time, some became specialized for growth on dry or wet sites; others evolved the different growth forms illustrated in Figure 27.11.

The same type of ecological opportunity was hypothesized to explain the adaptive radiation of *Anolis* lizards on islands in the

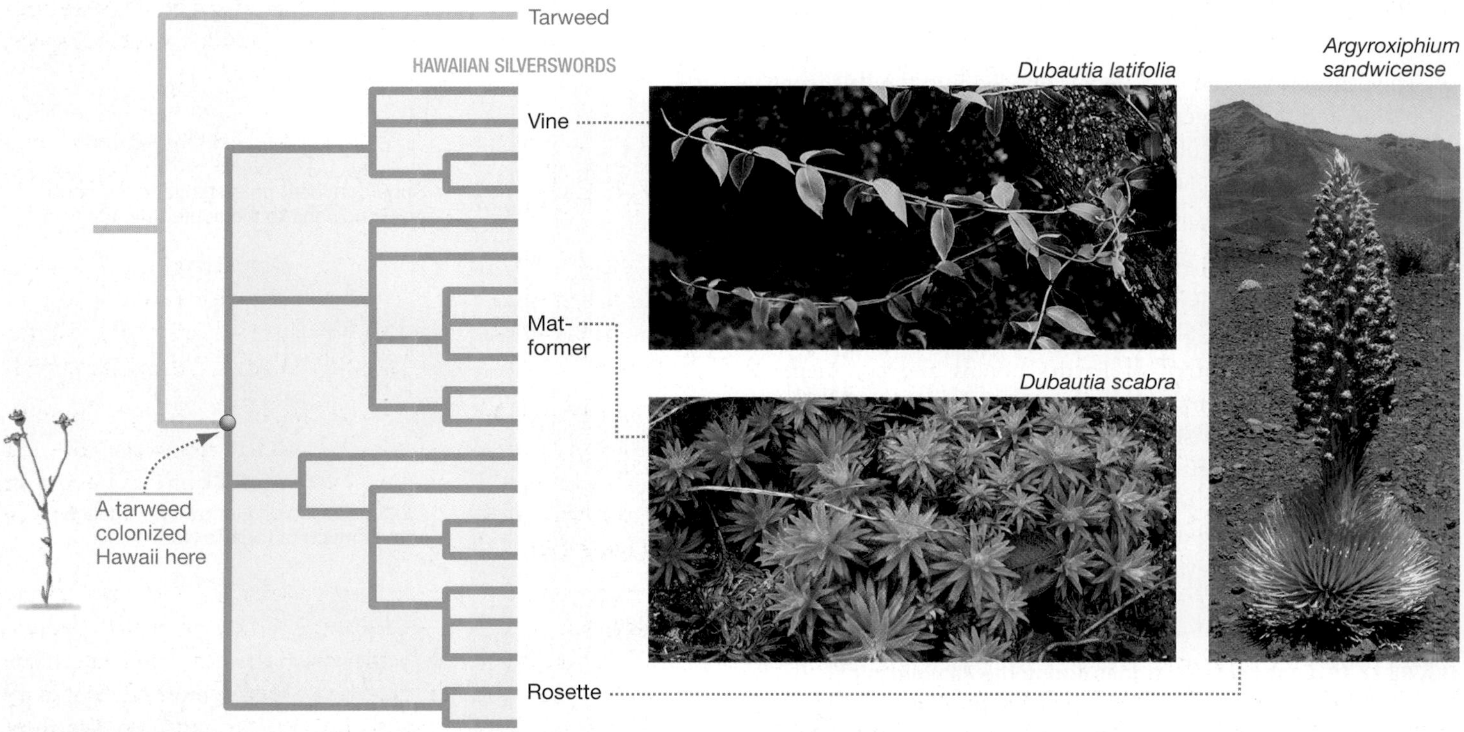

FIGURE 27.11 An Adaptive Radiation. This tree shows a subset of the Hawaiian silverswords, illustrating the extent of morphological divergence.

(a) Short-legged lizard species spend most of their time on the twigs of trees and bushes.

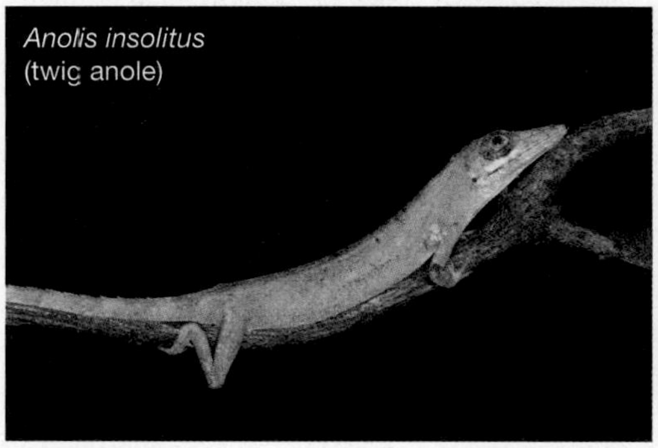

(b) Long-legged lizard species live on tree trunks and the ground.

FIGURE 27.12 Adaptive Radiations of *Anolis* Lizards. (a, b) Species of *Anolis* lizards vary in leg length and tail length. **(c)** Evolutionary relationships among lizard species on the islands of Hispaniola and Jamaica. The initial colonist species was different on these islands; but in terms of how they look and where they live, a similar suite of four species evolved.

(c) The same adaptive radiation of *Anolis* has occurred on different islands, starting from different types of colonists.

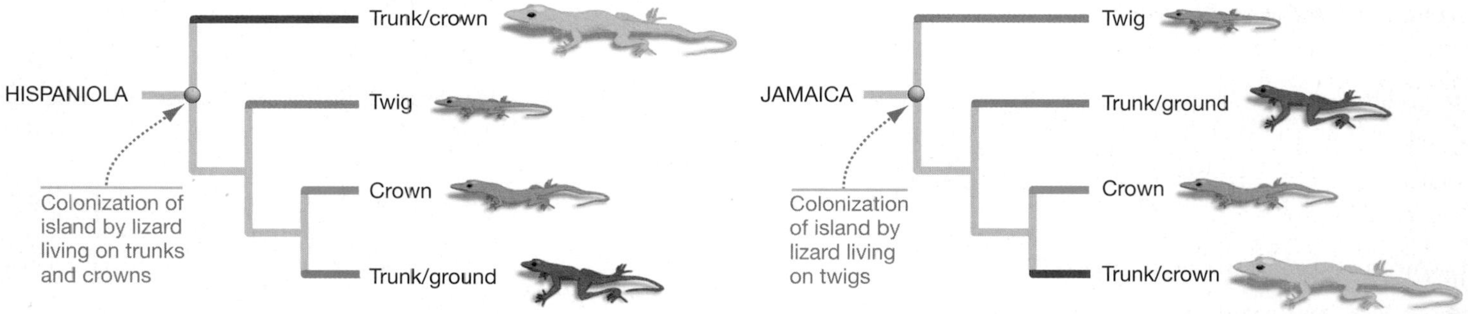

Caribbean. The lineage includes 150 species. They thrive in a wide array of habitats and have diverse body sizes and shapes. And in most cases, a lizard's size and shape are correlated with the habitat it occupies. For example:

- Species that live on tree twigs have short legs and tails that allow them to move efficiently on narrow surfaces (**Figure 27.12a**).
- Species that spend most of their time clinging to broad tree trunks or running along the ground have long legs and tails, making them fast and agile on broad surfaces (**Figure 27.12b**).

Most islands in the Caribbean have a distinct suite of lizard species. And in most cases, each island has a species that lives only in the twigs, the ground, or other distinctive habitats. The classical explanation for this pattern was that a mini-radiation occurred on each island: An original colonizing group encountered no competitors, and diversified in a way that led to efficient use of the available resources by a group of descendant species.

To test this hypothesis rigorously, a group of biologists estimated the phylogeny of *Anolis* from DNA sequence data. The results supported a key claim of the ecological opportunity hypothesis: The lizards on each island were monophyletic.

The critical data, though, are shown in **Figure 27.12c**. These phylogenetic trees, for species found on two different islands, are typical. The key observation is that the original colonist on each island was specialized for a different niche. The initial species on Hispaniola lived on the trunks and crowns of trees, while the original colonist on Jamaica occupied twigs.

From different evolutionary starting points, then, an adaptive radiation filled the same niches on both islands. This is exactly what the ecological opportunity hypothesis predicts. In *Anolis* lizards in the Caribbean, a series of mini-radiations were triggered when colonists arrived on an island that had food, space, and other resources, but lacked competitors.

✔If you understand this concept, you should be able to generate a hypothesis to explain why tarweeds in California and *Anolis* lizards on the mainland are not particularly species-rich or ecologically diverse.

MORPHOLOGICAL INNOVATION The evolution of a key morphological trait—one that allowed descendants to live in new areas, exploit new sources of food, or move in new ways—triggered many of the important diversification events in the history of life. For example:

- The evolution of wings, three pairs of legs, complex mouthparts, and a strong external skeleton helped insects move and feed efficiently. Today insects are the most diverse lineage on Earth, with perhaps well over 3 million species in existence (**Figure 27.13a** on page 486).
- Flowers are a unique reproductive structure that helped trigger the diversification of angiosperms (flowering plants). Because flowers are particularly efficient at attracting pollinators, the evolution of the flower made angiosperms more efficient in reproduction. Today angiosperms are far and away the most species-rich lineage of land plants. Over 250,000 species are known (**Figure 27.13b** on page 486).

(a) Insect body plan (b) Angiosperm flowers (c) Cichlids "throat jaws" (d) Dinosaur feathers

FIGURE 27.13 Some Adaptive Radiations Are Associated with Morphological Innovations.
[Part (c) ©Don P. Northup, www.africancichlidphotos.com]

- Cichlids are a lineage of fish that evolved a unique set of jaws in their throat. These second jaws make food processing extremely efficient. Different species have throat jaws specialized for crushing snail shells, shredding tissue from other fish, or mashing bits of algae. Over 300 species of cichlid live in Africa's Lake Victoria alone (**Figure 27.13c**).
- Feathers and wings gave some dinosaurs the ability to fly. Feathers originally evolved for display or insulation; later they were used in gliding and in powered flight. The dinosaur in **Figure 27.13d**, *Archaeopteryx*, was covered with complex feathers and could probably fly, at least short distances. Today the lineage called birds contains about 10,000 species, with representatives that live in virtually every habitat on the planet.

In each case, the evolution of new morphological features is hypothesized to have supported rapid speciation and ecological divergence.

To review fundamental concepts about adaptive radiation, go to the study area at *www.masteringbiology.com.*

(MB) **Web Activity** Adaptive Radiation

Then you'll be ready to analyze what may be the most spectacular adaptive radiation in the history of life.

The Cambrian Explosion

Almost all life-forms were unicellular for almost 3 *billion* years after the origin of life. The exceptions were several lineages of small multicellular algae, which show up in the fossil record about 1 billion years ago. Then, about 565 million years ago, the first animals—sponges, jellyfish, and perhaps simple worms—appear in the fossil record. A mere 50 million years later, virtually every major group of animals had appeared.

In a relatively short time, creatures with shells, exoskeletons, internal skeletons, legs, heads, tails, eyes, antennae, jaw-like mandibles, segmented bodies, muscles, and brains had evolved. It was arguably the most spectacular period of evolutionary change in the history of life. Because much of it occurred during the Cambrian period, this adaptive radiation is called the **Cambrian explosion**.

All organisms were small for 3 billion years, then life got big in just 50 million years—1/60th of the total time life had existed. Before asking how this happened, let's explore the fossilized evidence for what happened.

THE DOUSHANTUO, EDIACARAN, AND CAMBRIAN FOSSILS The Cambrian explosion is documented by three major fossil assemblages that record the state of animal life at 570 mya, at 565 to 542 mya, and at 525 to 515 mya. As **Figure 27.14a** indicates, the species collected from each of these intervals are referred to respectively as the Doushantuo fossils (from the Doushantuo formation in China), Ediacaran fossils (from Ediacara Hills, Australia), and Burgess Shale fossils (from British Columbia, Canada). Fossils from the Ediacaran interval and Burgess Shale interval have now been found at localities throughout the world.

Each fossil assemblage records a distinctive **fauna**—or collection of animal species (**Figure 27.14b**):

- *Doushantuo microfossils* include tiny sponges, less than 1 mm across, and clusters of cells interpreted as animal embryos because they resemble the stages of cleavage observed in today's animals (see Chapter 22). These creatures—the first animals on Earth—probably made their living by filtering organic debris from the water.
- *Ediacaran faunas* include sponges, jellyfish, and comb jellies as well as fossilized burrows, tracks, and other traces from unidentified animals. As the scale bars in Figure 27.14 show, these organisms were small; none have shells, limbs, heads, mouths, or feeding appendages. It is likely that Ediacaran animals simply filtered or absorbed organic material as they burrowed in sediments, sat immobile on the seafloor, or floated in the water.
- *Burgess Shale faunas* are among the most sensational additions ever made to the fossil record. Sponges, jellyfish, and comb jel-

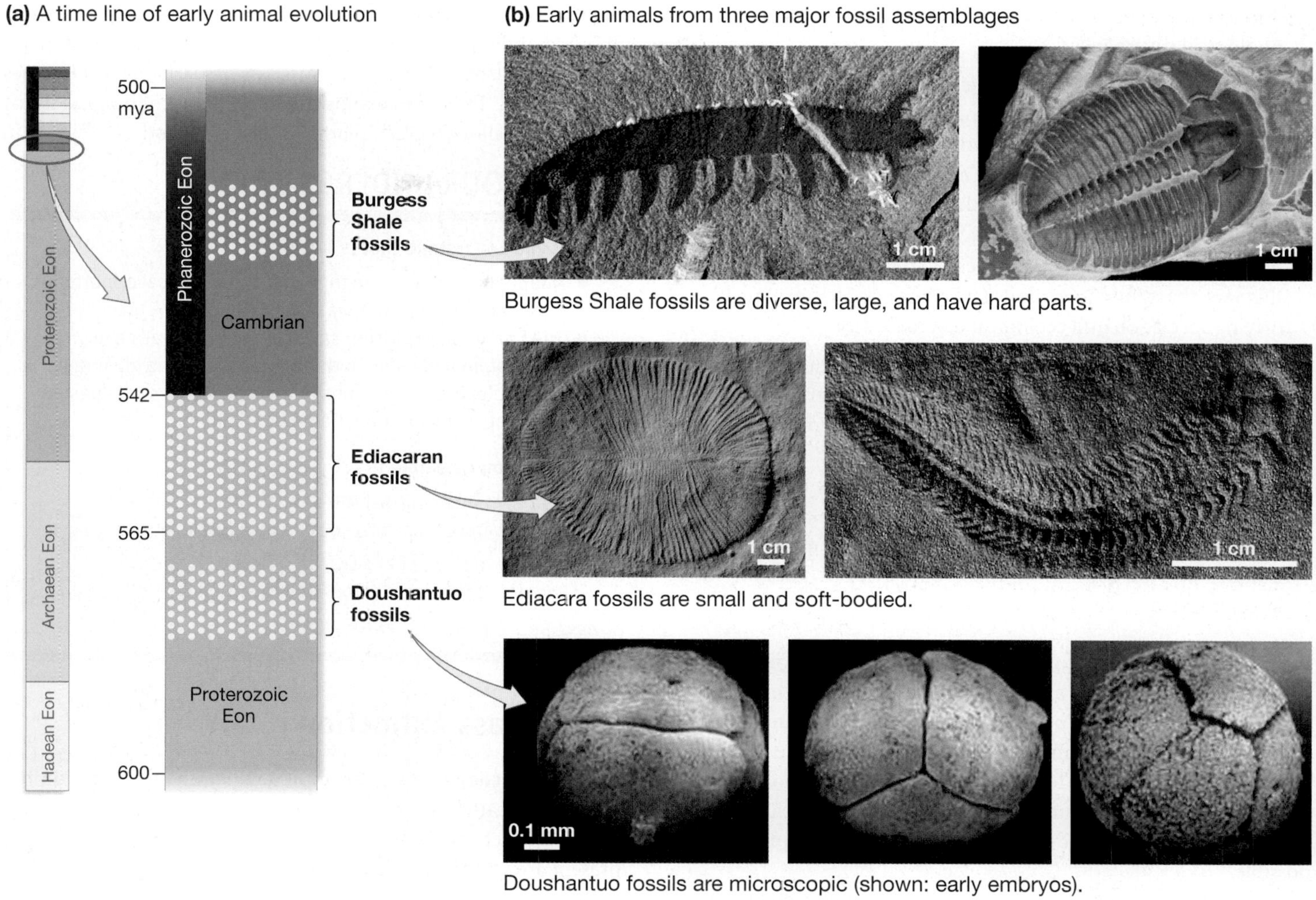

FIGURE 27.14 Fossils Document the Cambrian Explosion. The origin of animals and their diversification during the Cambrian explosion is documented by three major fossil assemblages.

lies are abundant in these rocks; but so are arthropods and mollusks. Today, the arthropods include the spiders, insects, and crustaceans (crabs, shrimp, and lobsters); mollusks include the clams, mussels, squid, and octopi. Echinoderms (sea stars and sea urchins), several types of worm and wormlike creatures, and even vertebrates—in short, virtually every major animal lineage—are found in these fossil faunas. A tremendous increase in the size and morphological complexity of animals occurred, accompanied by diversification in how they made a living. Species in this fauna had eyes, mouths, limbs, and shells. They swam, burrowed, walked, ran, slithered, clung, or floated; there were predators, scavengers, filter feeders, and grazers. The diversification created and filled many of the ecological niches found in marine habitats today.

The Cambrian explosion still echoes. Animals that fill today's teeming tide pools, beaches, and mudflats trace their ancestry to species preserved in the Burgess Shale.

WHAT TRIGGERED THE CAMBRIAN EXPLOSION? The Doushantuo, Ediacaran, and Burgess Shale faunas document what happened during the Cambrian explosion. Now the question is: *How* did all this speciation, morphological change, and ecological diversification come about?

To answer this question, biologists point to an array of datasets and hypotheses:

- *Higher oxygen levels* By analyzing the composition of rocks formed during the Proterozoic, geologists have established that oxygen levels gradually rose in the atmosphere and ocean. Increased oxygen levels make aerobic respiration (see Chapter 9) more efficient; increased aerobic respiration is required to support larger bodies and more active movements. Some biologists suggest that oxygen levels reached a critical threshold at the start of the Cambrian explosion, making the evolution of big, mobile animals possible.
- *The evolution of predation* Prior to the Cambrian explosion, animals made their living by eating organic material that settled on the seafloor or filtering cells and debris from the water. But Cambrian fossils include shelled animals with holes in the shells—evidence that a predator drilled through and ate the animal inside. When predators evolved, they exerted selection pressure for shells, hard exoskeletons, rapid movement, and other adaptations for

prey to defend themselves. This selection would drive morphological divergence.

- *New niches beget more new niches* The vast majority of Doushantuo and Ediacaran animals lived on the ocean floor—in what biologists call benthic habitats. Once animals could move off this substrate, they could exploit algae and other resources that were available above the ocean floor. The presence of animals at an array of depths created selection pressure for the evolution of species that could eat them. In this way, the ability to exploit new niches created new niches, driving speciation and ecological diversification.
- *New genes, new bodies* **Figure 27.15** shows the *Hox* genes found in some major animal groups. As Chapter 21 pointed out, *Hox* genes play a key role in organizing the development of the animal body by signaling where cells are in the embryo. The key observation is that the earliest animals in the fossil record had few or no *Hox* genes; most groups that appear later have more. The idea here is that mutations increased the number of *Hox* genes in animals and made it possible for larger, more complex bodies to evolve.

It's important to recognize that most or all of these hypotheses could be correct. They are not mutually exclusive. If increased oxygen levels made larger bodies and more rapid movement possible, then animals could move into new habitats off the seafloor and become large enough to eat smaller animals. Selection would favor individuals with mutations in *Hox* genes and other DNA sequences that make the development of a large, complex body possible.

About 100 million years after the Cambrian explosion, a similar adaptive radiation occurred after the first plants—the descendants of green algae—adapted to life on land. In the span of about 56 million years, an array of growth forms and most major lineages of plants appear in the fossil record. Chapter 30 features more on this "Devonian explosion" and the morphological innovations that allowed green plants to thrive on land.

CHECK YOUR UNDERSTANDING

If you understand that . . .

- Adaptive radiations are triggered by ecological opportunity and morphological innovation.
- The Cambrian explosion saw the rise of virtually every major animal lineage, and the evolution of a wide array of morphological innovations and food-getting strategies, in the relatively short time frame of 50 million years.

✓ You should be able to . . .

1. Compare and contrast the animals documented in the Doushantuo, Ediacara, and Burgess Shale faunas.
2. Explain the role of ecological opportunity and morphological innovation in the Cambrian explosion.

Answers are available in Appendix B.

27.4 Mass Extinction

A **mass extinction** refers to the rapid extinction of a large number of lineages scattered throughout the tree of life. More specifically, a mass extinction occurs when at least 60 percent of the species present are wiped out within 1 million years.

Mass extinction events are evolutionary hurricanes. They buffet the tree of life, snapping twigs and breaking branches. They

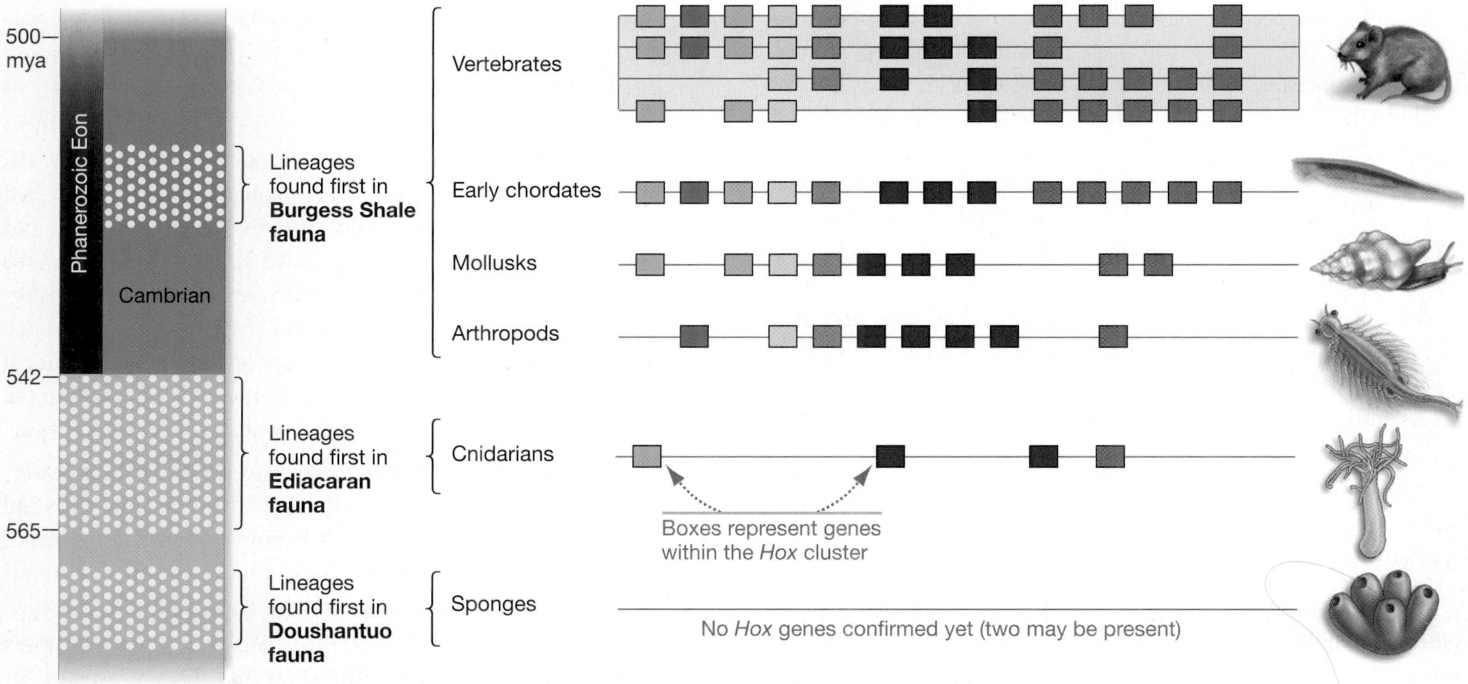

FIGURE 27.15 *Hox* Genes May Have Been Important in Animal Diversification.

are catastrophic episodes that wipe out huge numbers of species and lineages in a short time, giving the tree of life a drastic pruning. They are the polar opposite of adaptive radiation.

Before analyzing two of the best-studied mass extinctions in the fossil record, let's step back and ask how they differ from normal extinction events.

How Do Mass Extinctions Differ From Background Extinctions?

Mass extinction events are distinguished from background extinctions. **Background extinction** refers to the lower, average rate of extinction observed when a mass extinction is not occurring.

Although there is no hard-and-fast rule for distinguishing between the two extinction rates, paleontologists traditionally recognize and study five mass extinction events. **Figure 27.16**, for example, plots the percentage of plant and animal lineages called families that died out during each stage in the geologic time scale since the Cambrian explosion. Five spikes in the graph—denoting a large number of extinctions within a short time—are drawn in red. These are referred to as "The Big Five."

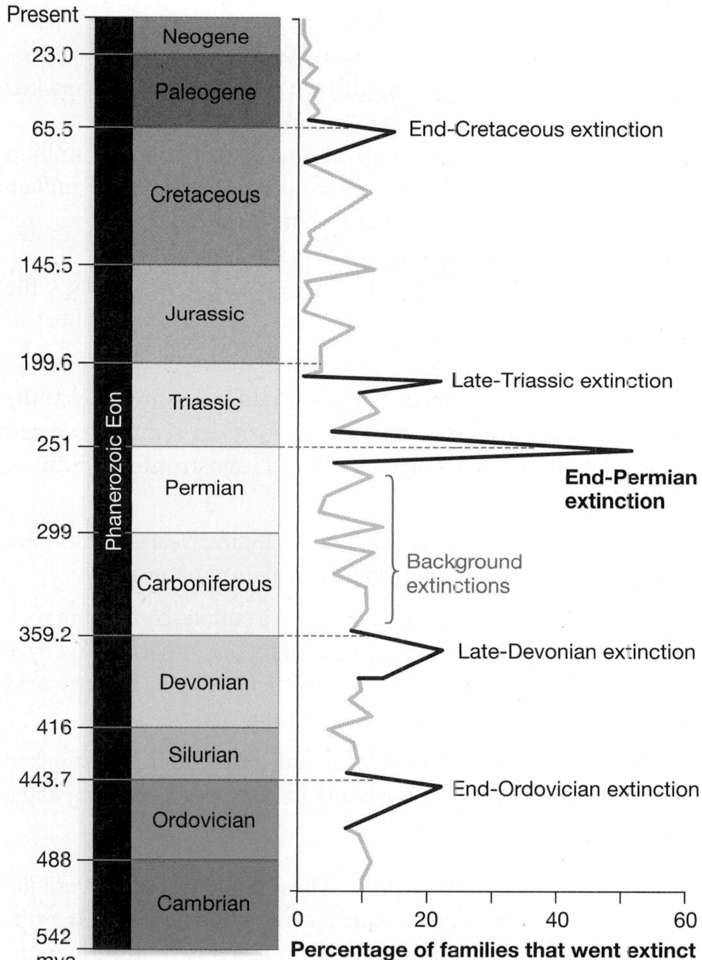

FIGURE 27.16 The Big Five Mass Extinction Events. This graph shows the percentage of lineages called families that went extinct over each interval in the fossil record since the Cambrian explosion. Over 50 percent of families and 90 percent of species went extinct during the end-Permian extinction.

✔**QUESTION** Which extinction event ended the era of the dinosaurs 65 million years ago?

Biologists are interested in distinguishing between background and mass extinctions because these events have contrasting causes and effects.

- *Background extinctions* are thought to occur when normal environmental change, emerging diseases, or competition with other species reduces certain populations to zero.
- *Mass extinctions* result from extraordinary, sudden, and temporary changes in the environment. During a mass extinction, species do not die out because individuals are poorly adapted to normal or gradually changing environmental conditions. Rather, species die out from exposure to exceptionally harsh, short-term conditions—such as huge volcanic eruptions or catastrophic sea-level changes.

In a general sense, background extinctions are thought to result primarily from natural selection. Mass extinctions, in contrast, function like genetic drift. The extinctions they cause are largely random with respect to the fitness of individuals under normal conditions.

To drive these points home, consider a mass extinction event that nearly uprooted the tree of life entirely. The end-Permian extinction, which occurred about 251 million years ago, came close to ending multicellular life on Earth.

The End-Permian Extinction

The end-Permian has been called the Mother of Mass Extinctions. To appreciate the scale of what happened, imagine that you took a walk along a seashore and identified 100 different species of algae and animals living on the beach and tide pools and the shallow water offshore. Now imagine that you snapped your fingers and 90 of those species disappeared forever. Only 10 species are left. An area that was teeming with diverse forms of life would look barren.

This is what happened, all over the world, during the end-Permian extinction. The event was a catastrophe of almost unimaginable proportions. On a personal level, it would be like nine of your ten best friends dying.

Although biologists have long appreciated the scale of the end-Permian extinction, research on its causes is ongoing. Consider the following:

- Flood basalts are outpourings of molten rock that flow across the Earth's surface. The largest flood basalts on Earth, called the Siberian traps, occurred during the end-Permian. They added enormous quantities of heat, CO_2, and sulfur dioxide to the atmosphere. The CO_2 led to intense global warming, and sulfur dioxide reacted with water to form sulfuric acid, which is toxic to most organisms.
- Rocks that formed during the interval indicate that the oceans became completely or largely anoxic—meaning that they lacked oxygen. These conditions are fatal to organisms that rely on aerobic respiration.

- There is convincing evidence that sea level dropped dramatically during the extinction event, reducing the amount of habitat available for marine organisms.
- Terrestrial animals may have been restricted to small patches of low-elevation habitats, due to low oxygen concentrations and high CO_2 levels in the atmosphere.

In short, both marine and terrestrial environments deteriorated dramatically for organisms that depend on oxygen to live. A prominent researcher has captured this point by naming the suite of killing mechanisms the "world went to hell hypothesis."

What biologists don't understand is *why* the environment changed so radically, and so quickly. For example, no one has yet been able to establish a convincing connection between the eruption of the Siberian traps and the other, more global, environmental changes that occurred.

The cause of the end-Permian extinction may be the most important unsolved question in research on the history of life. The cause of the dinosaur's demise, in contrast, is settled.

What Killed the Dinosaurs?

The end-Cretaceous extinction of 65 million years ago is as satisfying a murder mystery as you could hope for, but the butler didn't do it. The **impact hypothesis** for the extinction of the dinosaurs, first put forth in the early 1970s, proposed that an asteroid struck Earth and snuffed out an estimated 60–80 percent of the multicellular species alive.

EVIDENCE FOR THE IMPACT HYPOTHESIS The impact hypothesis was intensely controversial at first. As researchers set out to test its predictions, however, support began to grow:

- Worldwide, sedimentary rocks that formed at the Cretaceous-Paleogene (K-P)[1] boundary were found to contain extremely high quantities of the element iridium. Iridium is vanishingly rare in Earth rocks, but it is an abundant component of asteroids and meteorites (**Figure 27.17a**).
- Shocked quartz and microtektites are minerals that are found only at documented meteorite impact sites (**Figure 27.17b**). Shocked quartz forms when shock waves from an asteroid impact alter the structure of sand grains. Microtektites form when minerals are melted at an impact site and then cool and resolidify. In Haiti and an array of other locations, both shocked quartz and microtektites have been discovered in abundance in rock layers dated to 65 million years ago.
- A crater the size of Sicily was found just off the northwest coast of Mexico's Yucatán peninsula (**Figure 27.17c**). Microtektites are abundant in sediments from the crater's walls, and the crater dates to the K-P boundary.

Taken together, these data provided conclusive evidence in favor of the impact hypothesis. Researchers agree that the mystery is solved.

[1]Geologists use *K* to abbreviate Cretaceous, because *C* refers to the Cambrian period.

NATURE OF THE IMPACT The shape of the Yucatán crater and the distribution of shocked quartz and microtektites dated to 65 mya indicates that the asteroid hit Earth at an angle and splashed material over much of southeastern North America. Based on currently available data, astronomers and paleontologists estimate that the asteroid was about 10 km across.

To get a sense of the event's scale, consider that Mt. Everest is about 10 km above sea level and that planes cruise at an altitude of about 10 km. Imagine Earth being hit by a rock the size of Mt. Everest, or a rock that would fill the space between you and a jet in the sky.

To understand the impact's consequences, consider the results of the Tunguska event. On June 30, 1908, a piece of a comet about 30 m across and about 2 megatons in mass exploded at least 5 km above Earth's surface near the Tunguska River in Siberia. The explosion released about 1000 times as much energy as the atomic bomb that destroyed Hiroshima in World War II. The event incinerated vegetation over hundreds of square kilometers, leveled trees over thousands of square kilometers, and significantly increased dust levels across the Northern Hemisphere. A deafening blast was heard 500 km (300 mi) away; people standing 60 km away were thrown to the ground or knocked unconscious.

These events pale in comparison to what happened 65 million years ago. The energy the K-P asteroid unleashed was 37 *million* times greater than the Tunguska event. And the energy was released at the surface—not 5 km above the Earth.

According to both computer models and geologic data, the consequences of the K-P asteroid strike were nothing short of devastating.

- A tremendous fireball of hot gas would have spread from the impact site; large soot and ash deposits in sediments dated to 65 million years ago testify to catastrophic wildfires, worldwide.
- The largest tsunami in the last 3.5 billion years would have disrupted ocean sediments and circulation patterns.
- The impact site itself is underlain by a sulfate-containing rock called anhydrite. The SO_4^{2-} released by the impact would have reacted with water in the atmosphere to form sulfuric acid (H_2SO_4), triggering extensive acid rain.
- Massive quantities of dust, ash, and soot would have blocked the Sun for long periods, leading to rapid global cooling and a crash in plant productivity.

SELECTIVITY OF THE EXTINCTIONS The asteroid impact did not kill indiscriminately. Perhaps by chance, certain lineages escaped virtually unscathed while others vanished. Among vertebrates, for example, the dinosaurs, pterosaurs (flying reptiles), and all of the large-bodied marine reptiles (mosasaurs, ichthyosaurs, and plesiosaurs) expired; mammals, crocodilians, amphibians, and turtles survived.

Why? Answering this question has sparked intense debate. For years the leading hypothesis was that the K-P extinction event was size selective. The logic here was that the extended darkness and cold would affect large organisms disproportionately, be-

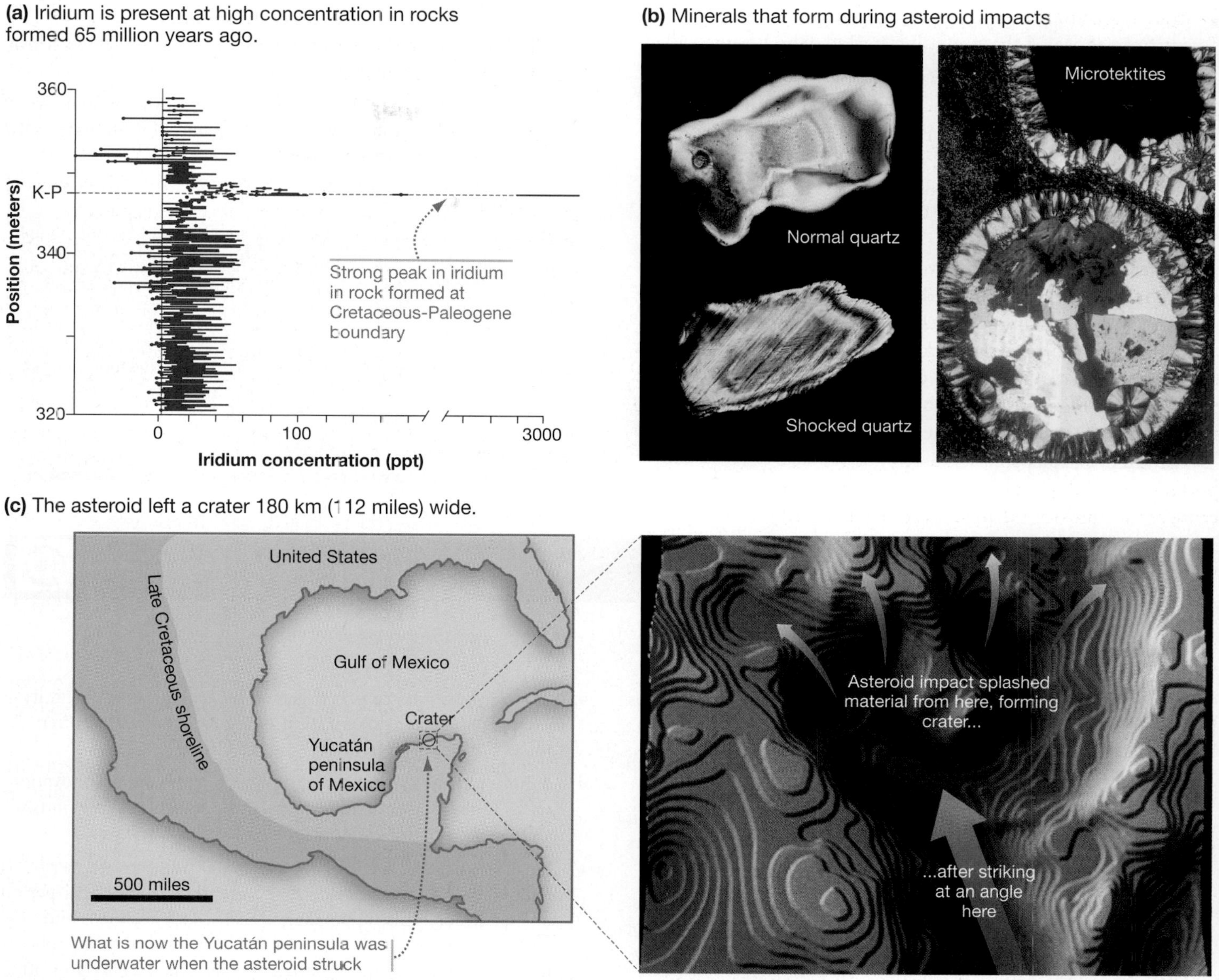
(a) Iridium is present at high concentration in rocks formed 65 million years ago.
360
K-P
340
320
Position (meters)
0
100
3000
Iridium concentration (ppt)
Strong peak in iridium in rock formed at Cretaceous-Paleogene boundary
(b) Minerals that form during asteroid impacts
Normal quartz
Shocked quartz
Microtektites
(c) The asteroid left a crater 180 km (112 miles) wide.
United States
Late Cretaceous shoreline
Gulf of Mexico
Crater
Yucatán peninsula of Mexico
500 miles
What is now the Yucatán peninsula was underwater when the asteroid struck
Asteroid impact splashed material from here, forming crater...
...after striking at an angle here

FIGURE 27.17 Evidence of an Asteroid Impact 65 Million Years Ago. **(a)** The concentration of iridium in rocks that formed on either side of the Cretaceous-Paleogene (K-P) boundary, in parts per trillion (ppt). On the *y*-axis, the position in meters refers to the location of rocks on either side of the rocks dated to the K-P boundary. Note that the *x*-axis is broken—otherwise the value of 3000 ppt at the K-P boundary would be far off the page. **(b)** Normal quartz grains are markedly different from shocked quartz grains. The striations in the shocked quartz are caused by a sudden increase in pressure. Microtektites (right) are tiny glass particles formed when minerals melt at an impact site and then recrystallize. **(c)** Geologists have identified the walls of a crater off the northwest coast of the Yucatán Peninsula, dated to 65 million years ago.

cause they require more food than do small organisms. But extensive data on the survival and extinction of marine clams and snails have shown no hint of size selectivity, and small-bodied and juvenile dinosaurs perished along with large-bodied and adult forms.

One hypothesis currently being tested is that organisms that were capable of inactivity for long periods—by hibernating or resting as long-lived seeds or spores—were able to survive the catastrophe. But this aspect of the mystery is still unsolved.

RECOVERY FROM THE EXTINCTION After the K-P extinction, fern fronds and fern spores dominate the plant fossil record from North America and Australia. These data suggest that extensive stands of ferns replaced diverse assemblages of cone-bearing and flowering plants after the impact. The fundamental message here is that terrestrial ecosystems around the world were radically simplified. In marine environments, some invertebrate groups do not exhibit normal levels of species diversity in the fossil record until 4–8 million years past the K-P boundary. Recovery was slow.

The organisms present in the Paleogene were markedly different from those of the preceding period. The lineage called Mammalia, which had consisted largely of rat-sized predators and scavengers in the heyday of the dinosaurs, exploded after the impact and took the place of the dinosaurs. Within 10–15 million years, all of the major mammalian orders observed today had appeared—from pigs to primates. Why? A major branch on the tree of life had disappeared. With competitors removed, mammals flourished.

The change in the terrestrial vertebrate fauna was not due to a competitive superiority conferred by adaptations such as fur and lactation. Rather, it was due to a chance event: a once-in-a-billion-years collision with a massive rock from outer space.

The diversification of lineages across the tree of life—including mammals—is the subject of Unit 6. A mass extinction that is currently occurring, and pruning the tree of life before our very eyes, is analyzed in Chapter 55.

CHECK YOUR UNDERSTANDING

If you understand that . . .

- Mass extinctions have occurred repeatedly throughout the history of life, and are due to cataclysmic, short-term changes in the environment.

✔ **You should be able to . . .**

1. Evaluate the strength of the evidence supporting the impact hypothesis for the extinction of the dinosaurs, and identify which evidence is strongest. (In science, evidence is considered strong if it cannot be explained by alternative hypotheses.)
2. Explain why mass extinctions wipe out species more or less randomly—much the way genetic drift affects changes in allele frequencies.

Answers are available in Appendix B.

CHAPTER 27 REVIEW

For media, go to the study area at www.masteringbiology.com

Summary of Key Concepts

Phylogenetic trees document the evolutionary relationships among organisms and are estimated from data.

- Phylogenetic trees document the genealogical relationships among species and identify the order in which events occurred.
- Phylogenetic trees can be estimated by grouping species based on overall similarity in traits or by analyzing shared, derived characters (synapomorphies) that identify monophyletic groups.
- To minimize the impact of homoplasy when inferring phylogenies, researchers use parsimony or other approaches to decide which of the many trees that are possible is most likely to reflect actual evolutionary history.

✔**You should be able to explain why upright posture and bipedalism (walking on two legs) is a synapomorphy that defines a monophyletic group called hominins, which includes humans.**

The fossil record provides physical evidence of organisms that lived in the past.

- The fossil record is the only direct source of data about what extinct organisms looked like and where they lived.
- The fossil record is critical to understanding the history of life, but it is biased: common and recent species that burrow and that have hard parts are most likely to be present in the record.

✔**You should be able to explain why teeth and pollen grains are the most common types of vertebrate and plant fossils, respectively.**

Adaptive radiations are a major pattern in the history of life. They are instances of rapid diversification associated with new ecological opportunities and new morphological innovations.

- Speciation events and morphological change occur rapidly during an adaptive radiation, as a single lineage diversifies into a wide variety of ecological roles.
- Adaptive radiations can be triggered by ecological opportunity—for example, by the colonization of a new habitat that offers resources but lacks competitors.
- Adaptive radiations can be triggered by morphological innovations such as limbs, flowers, and feathers, if the structure allows individuals to exploit resources more efficiently or in a new way.
- The diversification of animals over a 50-million-year period is perhaps the best-studied adaptive radiation. During this period the first heads, tails, appendages, shells, exoskeletons, and segmented bodies evolved. The Cambrian explosion was probably caused by increases in oxygen and expansion of *Hox* and other gene sequences, which made the evolution of large, motile bodies possible.

✔**You should be able to explain why the evolution of animals that could move throughout marine environments—not just in benthic habitats—created an ecological opportunity.**

MB **Web Activity** Adaptive Radiation

Mass extinctions have occurred repeatedly throughout the history of life. They are environmental catastrophes that rapidly eliminate most of the species alive.

- Mass extinctions have altered the course of evolutionary history at least five times.
- They prune the tree of life more or less randomly and have marked the end of several prominent lineages and the rise of new branches.
- The end-Permian extinction killed over 90 percent of existing species; its underlying cause is unknown.
- The end-Cretaceous extinction killed 60 to 80 percent of existing species and was caused by an asteroid impact.

- After the devastation of a mass extinction, the fossil record indicates that it can take 10–15 million years for ecosystems to recover their former levels of diversity.

✔You should be able to evaluate whether environmental changes caused by humans are eliminating species in a random manner, as opposed to those species being poorly adapted to the environment.

Questions

✔TEST YOUR KNOWLEDGE

Answers are available in Appendix B

1. Choose the best definition of a fossil.
   a. any trace of an organism that has been converted into rock
   b. a bone, tooth, shell, or other hard part of an organism that has been preserved
   c. any trace of an organism that lived in the past
   d. the process that leads to preservation of any body part from an organism that lived in the past

2. What is the difference between a branch and a node on a phylogenetic tree?
   a. A branch is a population through time; a node is where a population splits into two independent populations.
   b. A branch represents a common ancestor; a node is a species.
   c. A branch is a lineage; a node is any named taxonomic group.
   d. A branch is a population through time; a node is the common ancestor or all species or lineages present on a particular tree.

3. Which of the following best characterizes an adaptive radiation?
   a. Speciation occurs extremely rapidly, and descendant populations occupy a large geographic area.
   b. A single lineage diversifies rapidly, and descendant populations occupy many habitats and ecological roles.
   c. Natural selection is particularly intense, because disruptive selection occurs.
   d. Species recover after a mass extinction.

4. Which of the following is most accurate?
   a. Mass extinctions are due to asteroid impacts; background extinctions may have a wide variety of causes.
   b. Mass extinctions focus on particularly prominent groups, such as dinosaurs; background extinctions affect species from throughout the tree of life.
   c. Only five mass extinctions have occurred, but hundreds of background extinctions have occurred.
   d. Mass extinctions extinguish groups rapidly and randomly; background extinctions are slower and often result from natural selection.

5. **BioSkills 3** in Appendix A recommends a "one-snip test" to identify monophyletic groups—meaning that if you cut any branch on a tree, everything that "falls off" is a monophyletic group. Why is this valid?
   a. Monophyletic groups are nested on a tree—meaning that they are hierarchical.
   b. Monophyletic groups can also be called clades or lineages.
   c. Species are the smallest monophyletic groups on the tree of life.
   d. One snip gets an ancestor and all of its descendants.

6. What is homoplasy?
   a. similarity not due to homology
   b. similarity not due to inheritance from a common ancestor
   c. "noise" in the data sets used to infer phylogenies
   d. all of the above

✔TEST YOUR UNDERSTANDING

Answers are available in Appendix B

1. The text claims that the fossil record is biased in several ways. What are these biases? If the database is biased, is it still an effective tool to use in studying the diversification of life? Explain.

2. The initial diversification of animals took place over some 50 million years, at the start of the Cambrian period. Why is the diversification called an "explosion"?

3. Why is parsimony usually a reliable way to minimize the effect of homoplasy in assessing which phylogenetic tree is most accurate? Why was parsimony misleading in the case of the astragalus during evolution of artiodactyls?

4. Give an example of an adaptive radiation that occurred after a colonization event and after a morphological innovation. In each case, provide a hypothesis to explain why the adaptive radiation occurred.

5. What is the "world went to hell" hypothesis for the end-Permian extinction. What underlying mechanism was responsible?

6. Why are monophyletic groups identified by shared, derived traits?

✔APPLYING CONCEPTS TO NEW SITUATIONS

Answers are available in Appendix B

1. Suppose that the dying wish of a famous eccentric was that his remains be fossilized. His family has come to you for expert advice. What steps would you recommend to maximize the chances that his wish will be fulfilled?

2. Using data from the molecular phylogenies presented in this section and data on the fossil record of whales presented in Chapter 24, summarize how whales evolved from the common ancestor they share with today's hippos.

3. Some researchers contend that the end-Permian extinction event was also caused by an impact with a large extraterrestrial object. List the evidence, ordered from least convincing to most convincing, that you would like to see before you accept this hypothesis. Explain your rankings.

4. One of the "triggers" proposed for the Cambrian explosion is a dramatic rise in oxygen concentrations that occurred in the oceans about 800 mya. Review material in Chapter 9 on how oxygen compares with other atoms or compounds as an electron acceptor during cellular respiration (look near the end of Section 9.6, where aerobic and anaerobic respiration are compared). Then explain the logic behind the "oxygen-trigger" hypothesis for the Cambrian explosion.

THE BIG PICTURE

Geneticist and evolutionary biologist Theodosius Dobzhansky said that "Nothing in biology makes sense except in the light of evolution." Use this concept map to study how ideas introduced in Unit 5 fit together.

The key is to connect the four evolutionary processes that work at the level of populations—natural selection, genetic drift, mutation, and gene flow—to processes, events, and outcomes at higher levels of organization: speciation, adaptive radiation, mass extinction, and the tree of life.

It's all about changes in allele frequencies. Over time, small changes that occur between populations lead to large changes that distinguish major lineages on the tree of life.

Note that each box in the concept map indicates the chapter and section where you can go for review. Also, be sure to do the blue exercises in the Check Your Understanding box below.

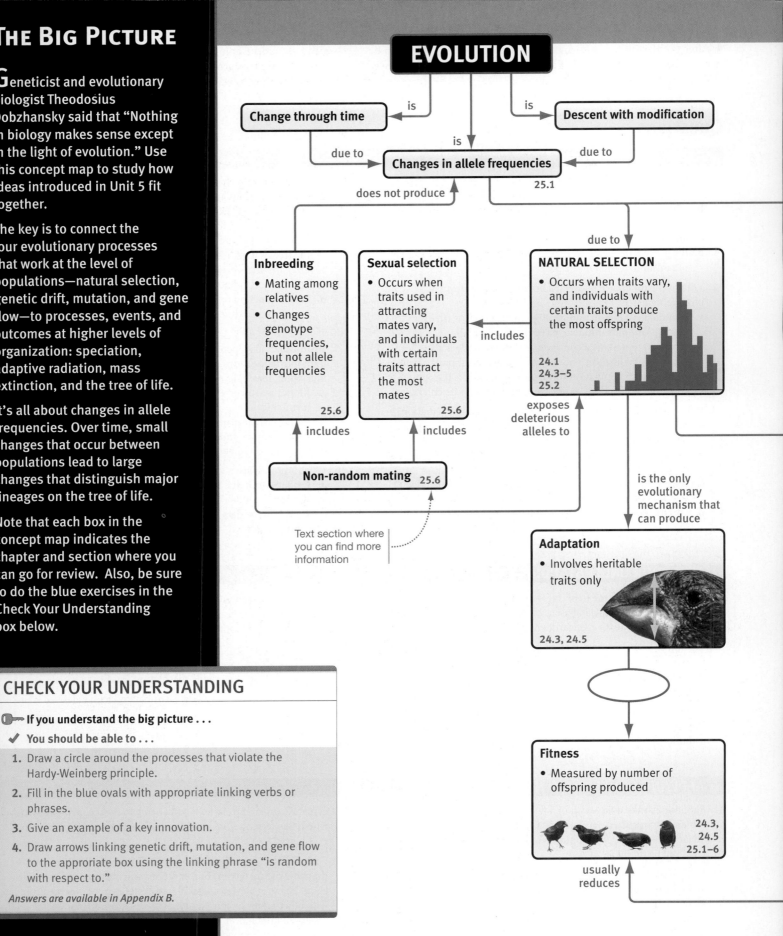

CHECK YOUR UNDERSTANDING

If you understand the big picture . . .

✓ You should be able to . . .

1. Draw a circle around the processes that violate the Hardy-Weinberg principle.
2. Fill in the blue ovals with appropriate linking verbs or phrases.
3. Give an example of a key innovation.
4. Draw arrows linking genetic drift, mutation, and gene flow to the approriate box using the linking phrase "is random with respect to."

Answers are available in Appendix B.

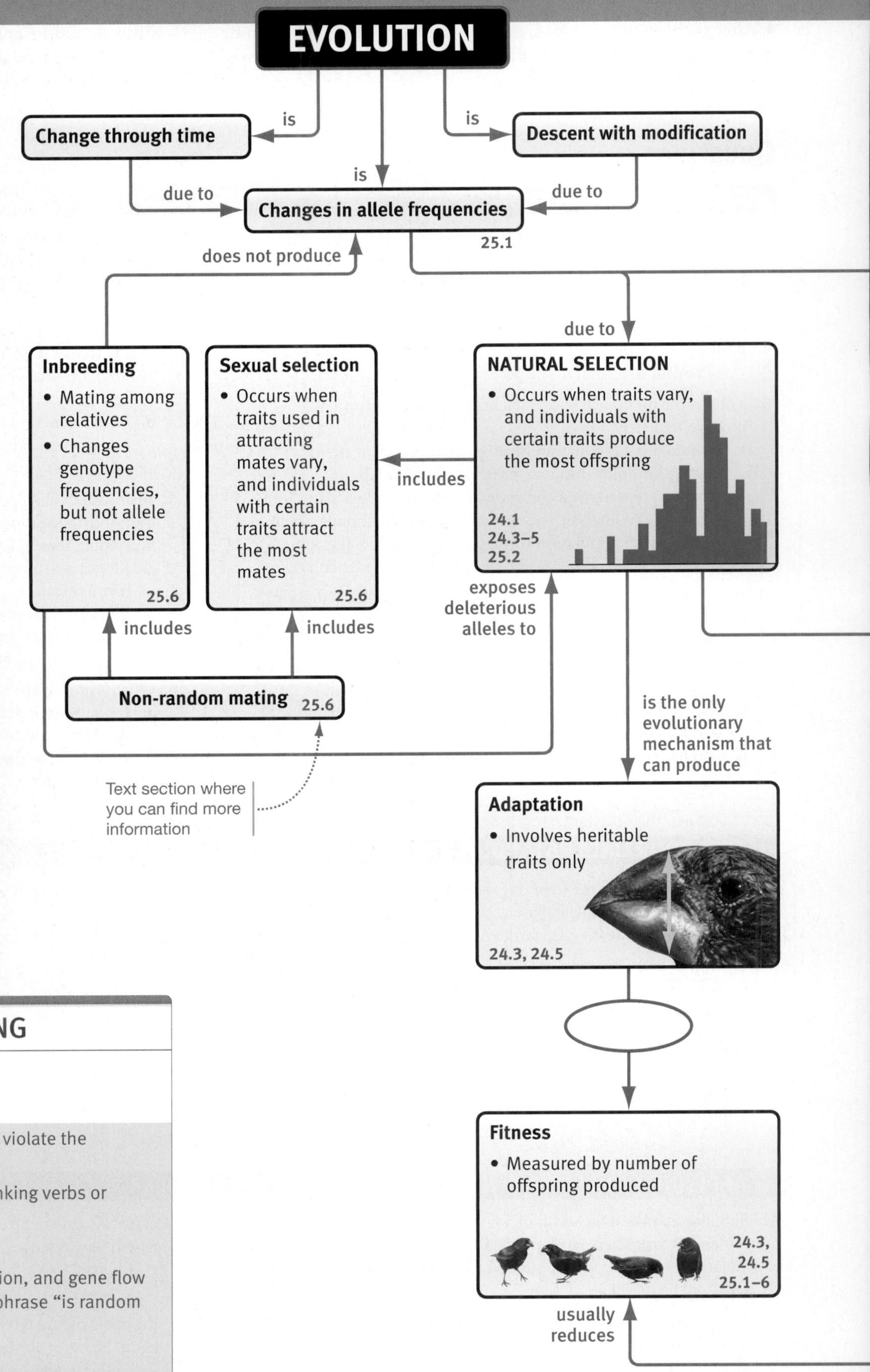

due to

GENETIC DRIFT
- Changes in allele frequencies due entirely to chance
- Especially important in small populations

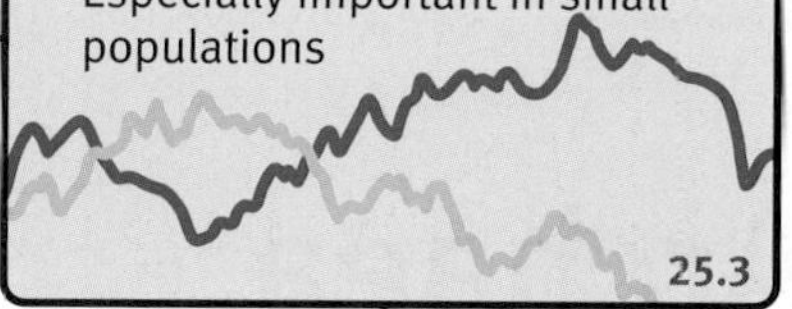

25.3

due to

MUTATION
- Random changes in DNA
- Creates new alleles
- Occurs in every individual in every generation, at low frequency

AT TAT TGT CTA GTA CCC CA

15.4, 25.5

due to

GENE FLOW
- Occurs when individuals move between populations
- Homogenizes allele frequencies between populations

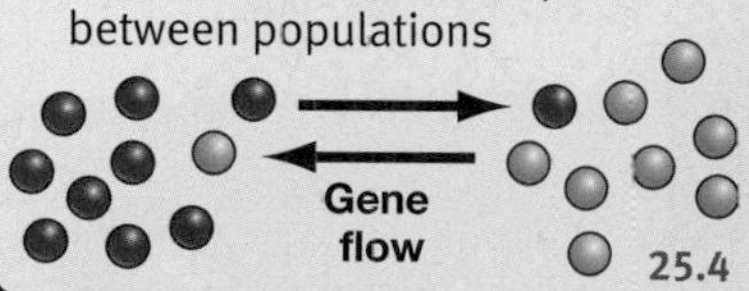

25.4

produces divergence required for

produces divergence required for

produces divergence required for

due to lack of

SPECIATION

Results from:

1. Genetic isolation, followed by
2. Genetic divergence

26.2–4

creates new branches on

form smallest possible tips on

The TREE OF LIFE
- Describes the evolutionary relationships among species

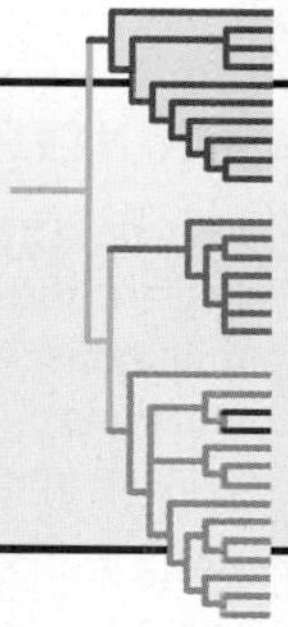

1.3, 27.1

"prune"

forms new

Species

Evolutionarily independent units in nature, identified by:

1. Reproductive isolation, and/or
2. Phylogenetic analysis, and/or
3. Morphological differences

26.1

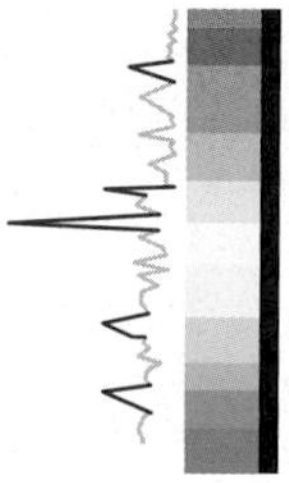

MASS EXTINCTIONS
- 60% of species are lost in less than 1 million years
- 5 events in the past 542 million years
- Is analogous to genetic drift

27.4

may occur after

with

Synamorphies
- Traits that are unique to a single lineage (found in some species but not others)
- Arise in a common ancestor

26.1
27.1

that may be

Key innovations
- Traits that allow species to exploit resources in a new way or use new habitats

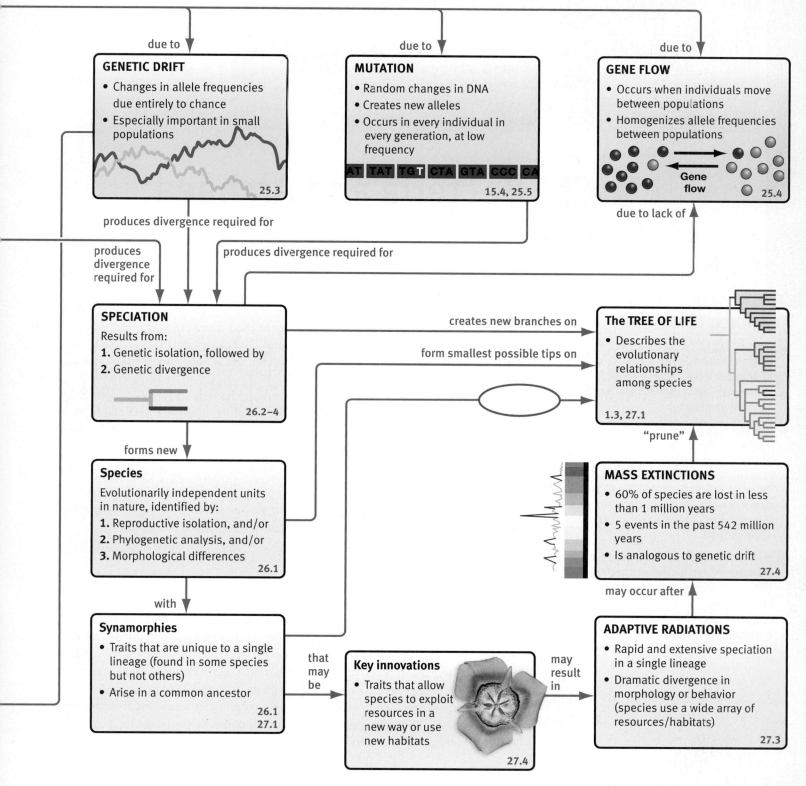

27.4

may result in

ADAPTIVE RADIATIONS
- Rapid and extensive speciation in a single lineage
- Dramatic divergence in morphology or behavior (species use a wide array of resources/habitats)

27.3

UNIT 6

THE DIVERSIFICATION OF LIFE

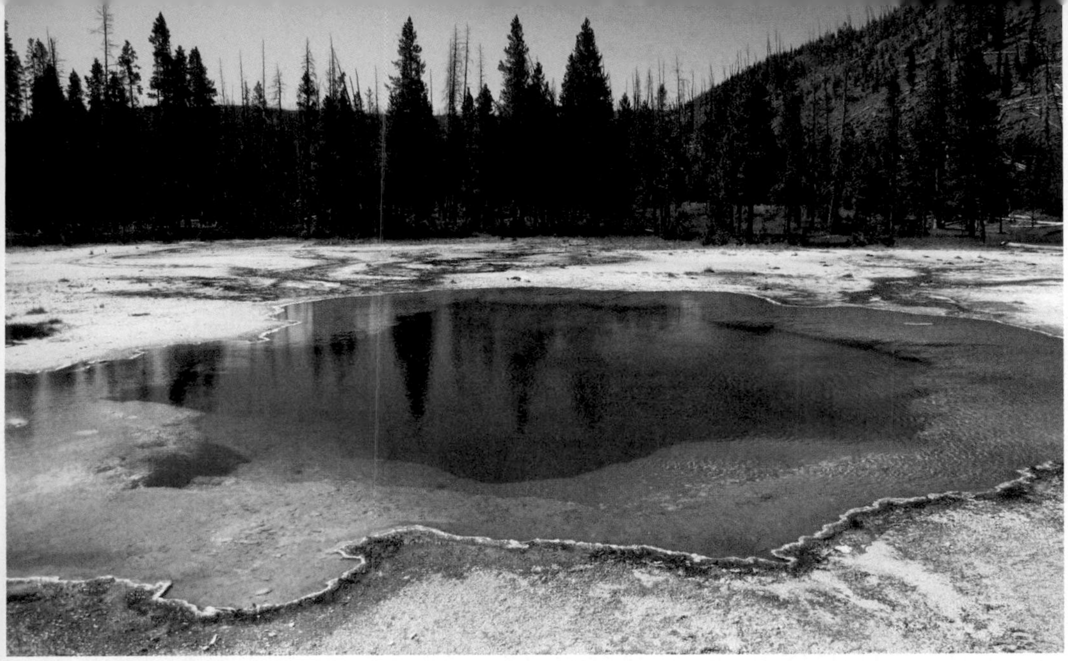

Although this hot spring looks devoid of life, it is actually teeming with billions of bacterial and archaeal cells.

28 Bacteria and Archaea

KEY CONCEPTS

- Bacteria and archaea may be tiny, but they are ancient, diverse, abundant, and ubiquitous. A few bacteria cause infectious disease; some are effective at cleaning up pollution.
- As a group, bacteria and archaea live in virtually every habitat known and use remarkably diverse types of compounds in cellular respiration and fermentation. Although cell size is usually small and overall morphology is relatively simple, the chemistry they can do is extremely sophisticated.
- Bacteria and archaea play a key role in ecosystems. Photosynthetic bacteria were responsible for the evolution of the oxygen atmosphere; bacteria and archaea cycle nutrients through terrestrial and aquatic environments.

Bacteria and Archaea (usually pronounced *ar-KAY-ah*) form two of the three largest branches on the tree of life (**Figure 28.1**). The third major branch, or domain, consists of eukaryotes and is called the Eukarya. Virtually all members of the Bacteria and Archaea domains are unicellular, and all are prokaryotic—meaning that they lack a membrane-bound nucleus.

Although their relatively simple morphology makes bacteria and archaea appear similar to the untrained eye, they are strikingly different at the molecular level (**Table 28.1**). Organisms in the Bacteria and Archaea domains are distinguished by the types of molecules that make up their plasma membranes and cell walls:

- **Bacteria** have a unique compound called peptidoglycan in their cell walls (see Chapter 5).
- **Archaea** have unique phospholipids in their plasma membranes—the hydrocarbon tails of the phospholipids contain isoprene (see Chapter 6).

If you were unicellular, bacteria and archaea would look as different to you as mammals and fish do now.

In addition, the machinery that bacteria and archaea use to process genetic information is strikingly different. More specifically, the DNA polymerases, RNA polymerases, transcription-initiation proteins, and ribosomes found in Archaea and Eukarya are distinct from those found in Bacteria and similar to each other. These differences have practical consequences: Antibiotics that poison bacterial ribosomes do not affect the ribosomes of archaea or eukaryotes. If all ribosomes were identical, these antibiotics would kill you along with the bacterial species that was supposed to be targeted.

As an introductory biology student, though, you probably know less about bacteria and archaea than about any other group on the tree of life. Before taking this course, it's likely that you'd never even heard of the Archaea.

✓ When you see this checkmark, stop and test yourself. Answers are available in Appendix B.

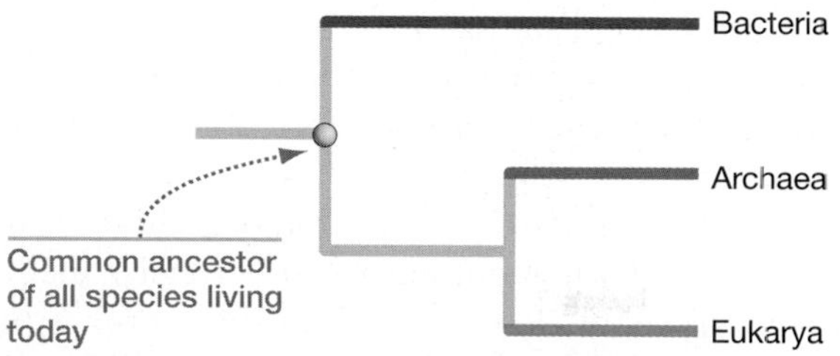

FIGURE 28.1 Bacteria, Archaea, and Eukarya Are the Three Domains of Life. Archaea are more closely related to eukaryotes than they are to bacteria.

✔**QUESTION** Was the common ancestor of all species living today prokaryotic or eukaryotic? Explain your reasoning.

This chapter's goal is to change this state of affairs, and convince you that even though bacteria and archaea are tiny, they have an enormous impact on you and the planet in general. By the time you finish reading the chapter, you should understand why a researcher summed up the bacteria and archaea by claiming, "They run this joint."

28.1 Why Do Biologists Study Bacteria and Archaea?

Biologists study bacteria and archaea for the same reasons they study any organisms. First, they are intrinsically fascinating. Discoveries such as finding bacterial cells living a kilometer underground or in 95°C hot springs keep biologists awake at night, staring at the ceiling. They can't wait to get into the lab in the morning and figure out how those cells are staying alive.

Second, there are practical benefits to understanding the species that share the planet with us. Understanding bacteria and archaea is particularly important—both in terms of understanding life on Earth and improving human health and welfare.

Biological Impact

The lineages in the domains Bacteria and Archaea are ancient, diverse, abundant, and ubiquitous. The oldest fossils of any type found to date are 3.5-billion-year-old carbon-rich deposits derived from bacteria. Because eukaryotes do not appear in the

SUMMARY TABLE 28.1 **Characteristics of Bacteria, Archaea, and Eukarya**

	Bacteria	Archaea	Eukarya
Nuclear envelope? (see Chapter 7)	No	No	Yes
Circular chromosome?	Yes (but linear in some species)	Yes	No
DNA associated with histone proteins? (see Chapter 18)	No	Yes	Yes
Organelles present? (see Chapter 7)	Some in limited number of species	None described to date	Extensive in number and density
Flagella present? (see Chapter 7)	Yes—spin like propeller	Yes; spin like bacterial flagella, but distinctive in molecular composition	Yes, but undulate back and forth and have completely different molecular composition compared with bacteria and archaea
Unicellular or multicellular?	Almost all unicellular	All unicellular	Many multicellular
Gene exchange between individuals? (see Chapter 12)	Transfer from viruses (transduction), environment (transformation), or direct contact from another cell (conjugation)	Transfer from viruses (transduction), environment (transformation), or direct contact from another cell (conjugation)	Sexual reproduction in some species—fusion of haploid genomes. Transduction and transformation can also occur.
Structure of lipids in plasma membrane (see Chapter 6)	Glycerol bonded to straight-chain fatty acids via ester linkage	Glycerol bonded to branched fatty acids (synthesized from isoprene subunits) via ether linkage	Glycerol bonded to straight-chain membrane fatty acids via ester linkage
Cell-wall material (see Chapter 5, Chapter 8)	Almost all include peptidoglycan, which contains muramic acid	Varies, but no peptidoglycan and no muramic acid	When present, usually made of cellulose or chitin
Information processing: DNA synthesis, transcription, and translation machinery (see Chapter 15, Chapter 16)	One relatively simple RNA polymerase; translation begins with formylmethionine; translation poisoned by several antibiotics that do not affect archaea or eukaryotes	DNA polymerase is eukaryote-like; one relatively complex RNA polymerase; eukaryote-like basal transcription complex; translation begins with methionine	Several relatively complex RNA polymerases; translation begins with methionine

✔**EXERCISE** Using the data in this table, add labeled marks to Figure 28.1 indicating where the following traits evolved: peptidoglycan in cell wall, archaeal-type plasma membrane, archaeal and eukaryote-type ribosomes/DNA polymerase/transcription machinery, nuclear envelope.

fossil record until 1.75 billion years ago, biologists infer that prokaryotes were the only form of life on Earth for at least 1.7 billion years.

Just how many bacteria and archaea are alive today? Although a mere 5000 species have been formally named and described to date—most by the morphological species concept introduced in Chapter 26—it is virtually certain that millions exist. Consider that over 400 species of prokaryotes are living in your small intestine right now, with another 128 in your stomach lining. About 500 species live in your mouth, of which just 300 have been described and named. Norman Pace points out that there may be tens of millions of different insect species but notes, "If we squeeze out any one of these insects and examine its contents under the microscope, we find hundreds or thousands of distinct microbial species." Most of these **microbes** (microscopic organisms) are bacteria or archaea. Virtually all are unnamed and undescribed. If you want to discover and name new species, then study bacteria or archaea.

ABUNDANCE In addition to recognizing how diverse bacteria and archaea are in terms of numbers of species, it's critical to appreciate their abundance.

- The approximately 10^{13} (10 trillion) cells in your body are vastly outnumbered by the bacterial and archaeal cells living on and in you. An estimated 10^{12} bacterial cells live on your skin, and an additional 10^{14} bacterial and archaeal cells occupy your stomach and intestines. You are a walking, talking habitat—one that is teeming with bacteria and archaea.
- A mere teaspoon of good-quality soil contains *billions* of microbial cells, most of which are bacteria and archaea.
- In sheer numbers, species in a lineage called the Group I marine archaea may be the most successful organisms on Earth. Biologists routinely find these cells at concentrations of 10,000 to 100,000 individuals per milliliter in most of the world's oceans. At these concentrations, a drop of seawater contains a population equivalent to that of a large human city. Yet this lineage was first described in the early 1990s.
- Recent research has found enormous numbers of bacterial and especially archaeal cells in rocks and sediments as much as 1600 meters underneath the world's oceans. Although recently discovered, the bacteria and archaea living under the ocean may make up 10 percent of the world's total mass of living material.
- Biologists estimate the total number of individual bacteria and archaea alive today at over 5×10^{30}. If they were lined up end to end, they would make a chain longer than the Milky Way galaxy. These cells contain 50 percent of all the carbon and 90 percent of all the nitrogen and phosphorus found in organisms.

In terms of the total volume of living material on our planet, bacteria and archaea are dominant life-forms.

HABITAT DIVERSITY Bacteria and archaea are found almost everywhere. They live in environments as unusual as oxygen-free mud, hot springs, and salt flats. In seawater they are found from the surface to depths of 10,000 m, at temperatures ranging from near 0°C in Antarctic sea ice to over 121°C near submarine volcanoes.

Although there are far more prokaryotes than eukaryotes, much more is known about eukaryotic diversity than about prokaryotic diversity. Researchers who study prokaryotic diversity are exploring one of the most wide-open frontiers in all of science. So little is known about the extent of these domains that recent collecting expeditions have turned up entirely new **phyla** (singular: **phylum**). These are names given to major lineages within each domain. To a biologist, this achievement is equivalent to the sudden discovery of a new group of eukaryotes as distinctive as flowering plants or animals with backbones.

The physical world has been explored and mapped, and many of the larger plants and animals are named. But in **microbiology**—the study of organisms that can be seen only with the aid of a microscope—this is an age of exploration and discovery.

Medical Importance

The first paper documenting that an archaeon was associated with a human disease—a dental condition called periodontitis—was published in 2004. But biologists have been studying disease-causing bacteria for over a century.

Of the hundreds or thousands of bacterial species living in and on your body, a tiny fraction can disrupt normal body functions enough to cause illness. Bacteria that cause disease are said to be **pathogenic** (literally, "disease-producing"). Pathogenic bacteria have been responsible for some of the most devastating epidemics in human history.

Table 28.2 lists some of the bacteria that cause illness in humans. The important things to note are that:

- Pathogenic forms come from several different lineages in the domain Bacteria.
- Pathogenic bacteria tend to affect tissues at the entry points to the body, such as wounds or pores in the skin, the respiratory and gastrointestinal tracts, and the urogenital canal.

KOCH'S POSTULATES Robert Koch was the first person to establish a link between a particular species of bacterium and a specific disease. When Koch began his work on the nature of disease in the late 1800s, microscopists had confirmed the existence of the particle-like organisms we now call bacteria, and Louis Pasteur had shown that bacteria and other microorganisms are responsible for spoiling milk, wine, broth, and other foods. Koch hypothesized that bacteria might also be responsible for causing infectious diseases, which spread by being passed from an infected individual to an uninfected individual.

Koch set out to test this hypothesis by identifying the organism that causes anthrax. Anthrax is a disease of cattle and other

TABLE 28.2 **Some Bacteria That Cause Illness in Humans**

Lineage	Species	Tissues Affected	Disease
Firmicutes	*Clostridium botulinum*	Gastrointestinal tract, nervous system	Food poisoning (botulism)
	Clostridium tetani	Wounds, nervous system	Tetanus
	Staphylococcus aureus	Skin, urogenital canal	Acne, boils, impetigo, toxic shock syndrome
	Streptococcus pneumoniae	Respiratory tract	Pneumonia
	Streptococcus pyogenes	Respiratory tract	Strep throat, scarlet fever
Spirochaetes	*Borrelia burgdorferi*	Skin and nerves	Lyme disease
	Treponema pallidum	Urogenital canal	Syphilis
Actinomycetes	*Mycobacterium leprae*	Skin and nerves	Leprosy
	Mycobacterium tuberculosis	Respiratory tract	Tuberculosis
	Propionibacterium acnes	Skin	Acne
Chlamydiales	*Chlamydia trachomatis*	Urogenital canal	Genital tract infection
ε-Proteobacteria	*Helicobacter pylori*	Stomach	Ulcer
β-Proteobacteria	*Neisseria gonorrhoeae*	Urogenital canal	Gonorrhea
γ-Proteobacteria	*Haemophilus influenzae*	Ear canal, nervous system	Ear infections, meningitis
	Pseudomonas aeruginosa	Urogenital canal, eyes, ear canal	Infections of eye, ear, urinary tract
	Salmonella enterica	Gastrointestinal tract	Food poisoning
	Vibrio parahaemolyticus	Gastrointestinal tract	Food poisoning
	Yersinia pestis	Lymph and blood	Plague

grazing mammals that can result in fatal blood poisoning. The disease also occurs infrequently in humans and mice.

To establish a causative link between a specific microbe and a specific disease, Koch proposed that four criteria had to be met:

1. The microbe must be present in individuals suffering from the disease and absent from healthy individuals. By careful microscopy, Koch was able to show that the bacterium *Bacillus anthracis* was always present in the blood of cattle suffering from anthrax, but absent from healthy individuals.
2. The organism must be isolated and grown in a pure culture away from the host organism. Koch was able to grow pure colonies of *B. anthracis* in glass dishes on a nutrient medium, using gelatin as a substrate.
3. If organisms from the pure culture are injected into a healthy experimental animal, the disease symptoms should appear. Koch demonstrated this crucial causative link in mice injected with *B. anthracis*. The symptoms of anthrax infection appeared, and then the infected mice died.
4. The organism should be isolated from the diseased experimental animal, again grown in pure culture, and demonstrated by its size, shape, and color to be the same as the original organism. Koch did this by purifying *B. anthracis* from the blood of diseased experimental mice.

These criteria, now called **Koch's postulates**, are still used to confirm a causative link between new diseases and a suspected infectious agent.

THE GERM THEORY Koch's experimental results also became the basis for the **germ theory of disease**.

- The pattern component of this theory is that certain diseases are infectious—meaning that they can be passed from person to person.
- The process responsible for this pattern is the transmission and growth of certain bacteria and viruses.

(Viruses are acellular particles that parasitize cells and are analyzed in detail in Chapter 35.)

The germ theory of disease laid the foundation for modern medicine. Initially its greatest impact was on sanitation—efforts to prevent transmission of pathogenic bacteria. During the American civil war, for example, it was common for surgeons to sharpen their scalpels on their shoe leather, after walking in horse manure. During that conflict, records indicate that more soldiers died of dysentery and typhoid fever, contracted from drinking water contaminated with human feces, than from wounds in battle.

Fortunately, improvements in sanitation and nutrition have caused dramatic reductions in mortality rates due to infectious diseases in the industrialized countries. For example, the graph

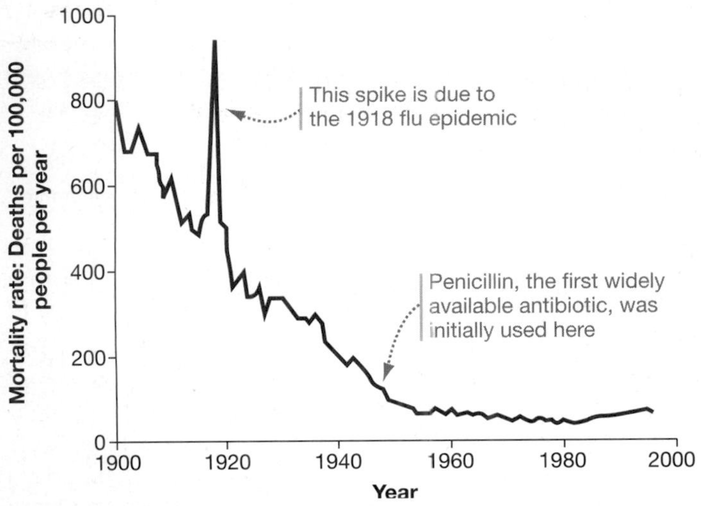

FIGURE 28.2 Deaths Due to Bacterial Infections Have Declined Dramatically in Some Countries. These data are from a country that industrialized in the late 1800s (the United States).

✔**QUESTION** Why did the death rate due to infectious disease drop to low levels, long before antibiotics were available?

in **Figure 28.2** shows the annual death rate due to infectious diseases—meaning, bacterial or viral infections—in the United States between 1900 and the late 1990s. Except for a large spike in the death rate due to a devastating flu epidemic in 1918, the death rate dropped steadily and dramatically during this period. The vast majority of this decline occurred long before antibiotics were introduced.

WHAT MAKES SOME BACTERIAL CELLS PATHOGENIC? Virulence, or the ability to cause disease, is a heritable trait that varies among individuals in a population. Most *Escherichia coli*, for example, are harmless inhabitants of the gastrointestinal tract of humans and other mammals. But some *E. coli* cells cause potentially fatal food poisoning.

What makes some cells of the same species pathogenic, while others are harmless? Biologists have answered this question for *E. coli* by sequencing the entire genome of a harmless lab strain and the pathogenic strain called O157. The genome of the virulent strain is much larger, and contains genes for proteins that allow the cells to adhere to host cells and secrete toxins that disrupt host cells. The diarrhea that results is advantageous to the bacterial cells—they are shed into the environment. If sanitation is poor, the virulent cells may infect many new hosts.

Similar types of studies are identifying the genes responsible for virulence in a wide array of pathogenic bacteria.

THE PAST, PRESENT, AND FUTURE OF ANTIBIOTICS **Antibiotics** are molecules that kill bacteria. They are produced naturally by a wide array of soil-dwelling bacteria and fungi. In these environments, antibiotics are hypothesized to be a weapon that helps cells reduce competition for nutrients and other resources.

The discovery of antibiotics by biomedical researchers in 1928, their development over subsequent decades, and widespread use starting in the late 1940s gave physicians effective tools to combat many bacterial infections.

Unfortunately, extensive use of antibiotics in the late twentieth century in clinics and animal feed led to the evolution of drug-resistant strains of pathogenic bacteria (see Chapter 24). One recent study found that there are now soil-dwelling bacteria in natural environments that—far from being killed by antibiotics—actually use them as food.

Coping with antibiotic resistance in pathogenic bacteria has become a great challenge of modern medicine. Some researchers even claim that we may be entering the "post-antibiotic era" in medicine.

Role in Bioremediation

Only a tiny proportion of bacteria and archaea cause disease in humans or other organisms. In the vast majority of cases, bacteria and archaea have no direct impact on humans or are beneficial. For example, researchers are using bacteria and archaea to clean up sites polluted with organic solvents—an effort called **bioremediation.**

Throughout the industrialized world some of the most serious pollutants in soils, rivers, and ponds consist of organic compounds that were originally used as solvents or fuels but leaked or were spilled into the environment. Most of these compounds are highly hydrophobic. Because they do not dissolve in water, they tend to accumulate in sediments. If the compounds are subsequently ingested by burrowing worms or clams or other organisms, they can be passed along to fish, insects, humans, birds, and other species.

At moderate to high concentrations, these pollutants are toxic to eukaryotes. Petroleum from oil spills and compounds that contain ring structures and chlorine atoms, such as the family of compounds called dioxins, are particularly notorious because of their toxicity to humans.

At sites like these, bioremediation is often based on complementary strategies:

- *Fertilizing contaminated sites to encourage the growth of existing bacteria and archaea that degrade toxic compounds.* After several recent oil spills, researchers added nitrogen to affected sites as a fertilizer, but left nearby beaches untreated as controls. Dramatic increases occurred in the growth of bacteria and archaea that use hydrocarbons in cellular respiration, probably because the cells used the added nitrogen to synthesize enzymes and other key compounds. In at least some cases, the fertilized sediments cleaned up much faster than the unfertilized sites (**Figure 28.3**).
- *"Seeding," or adding, specific species of bacteria and archaea to contaminated sites.* Seeding shows promise of alleviating pollution in some situations. For example, researchers have recently discovered bacteria that are able to render certain chlorinated, ring-containing compounds harmless. Instead of being poisoned by the pollutants, these bacteria use the chlorinated compounds as electron acceptors during cellular respiration. In at least some cases, the by-product is dechlorinated and nontoxic to humans and other eukaryotes.

FIGURE 28.3 Bacteria and Archaea Can Play a Role in Cleaning Up Pollution. On the left is a rocky coast that was polluted by an oil spill but fertilized to promote the growth of oil-eating bacteria. The portion of the beach on the right was untreated.

To follow up on these discoveries, researchers are now growing the bacteria in quantity and testing them in the field, to test the hypothesis that seeding can speed the rate of decomposition in contaminated sediments. Initial reports suggest that seeding may help clean up at least some polluted sites.

Extremophiles

Bacteria or archaea that live in high-salt, high-temperature, low-temperature, or high-pressure habitats are known as **extremophiles** ("extreme-lovers"). Studying them has been extraordinarily fruitful for understanding the tree of life, developing industrial applications, and exploring the structure and function of enzymes.

As an example of these habitats, consider hot springs at the bottom of the ocean, where water as hot as 300°C emerges and mixes with 4°C seawater. At locations like these, archaea are abundant forms of life. Researchers recently discovered an archaeon that grows so close to these hot springs that its surroundings are at 121°C—a record for life at high temperature. This organism can live and grow in water that is heated past its boiling point (100°C) and at pressures that would instantly destroy a human being. Other recently discovered bacteria and archaea can grow at:

- a pH less than 1.0;
- seawater at a depth of over 2500 m; and
- anoxic ("no-oxygen") habitats deep in the Mediterranean Sea, where the water is virtually saturated with salt.

Extremophiles have become a hot area of research. The genomes of a wide array of extremophiles have been sequenced, and expeditions regularly seek to characterize new species. Why?

- *Origin of life* Based on models of conditions that prevailed early in Earth's history, it appears likely that the first forms of life lived at high temperature and pressure in environments that lacked oxygen—conditions that we would call extreme. Thus, understanding extremophiles may help explain how life on Earth began.
- *Extraterrestrial life?* In a similar vein, many astrobiologists ("space-biologists") use extremophiles as model organisms in the search for extraterrestrial life. The idea is that if bacteria and archaea can thrive in extreme habitats on Earth, it is possible that cells might be found in similar environments on other planets or moons of planets.
- *Commercial applications* Because enzymes that function at extreme temperatures and pressures are useful in many industrial processes, extremophiles are of commercial interest as well. Chapter 19 introduced *Taq* polymerase—a DNA polymerase that is stable up to 95°C. Recall that *Taq* polymerase is used to run the polymerase chain reaction (PCR) in research and commercial settings. This enzyme was isolated from a bacterium called *Thermus aquaticus* ("hot water"), which was discovered in hot springs in Yellowstone National Park.

Bacteria and archaea may be small, but they thrive in an amazing range of conditions.

28.2 How Do Biologists Study Bacteria and Archaea?

Our understanding of the domains Bacteria and Archaea is advancing more rapidly right now than at any time during the past 100 years—and perhaps faster than our understanding of any other lineages on the tree of life.

As an introduction to the domains Bacteria and Archaea, let's examine a few of the techniques that biologists use to answer questions about them. Some of these research strategies have been used since bacteria were first discovered; some were invented less than 10 years ago.

Using Enrichment Cultures

Which species of bacteria and archaea are present at a particular location, and what do they use as food? To answer questions like these, biologists rely heavily on their ability to culture organisms in the lab. Of the 5000 species of bacteria and archaea that have been described to date, almost all were discovered when they were isolated from natural habitats and grown under controlled conditions in the laboratory.

One classical technique for isolating new types of bacteria and archaea is called **enrichment culture**. Enrichment cultures are based on establishing a specified set of growing conditions—temperature, lighting, substrate, types of available food, and so on. Cells that thrive under the specified conditions increase in numbers enough to be isolated and studied in detail.

To appreciate how this strategy works in practice, consider research on bacteria that live deep below Earth's surface. One study began with samples of rock and fluid from drilling operations in Virginia and Colorado. The samples came from sedimentary

EXPERIMENT

QUESTION: Can bacteria live a mile below Earth's surface?

HYPOTHESIS: Bacteria are capable of cellular respiration deep below Earth's surface by using H_2 as an electron donor and Fe^{3+} as an electron acceptor.

NULL HYPOTHESIS: Bacteria from this environment are not capable of using H_2 as an electron donor and Fe^{3+} as an electron acceptor.

EXPERIMENTAL SETUP:

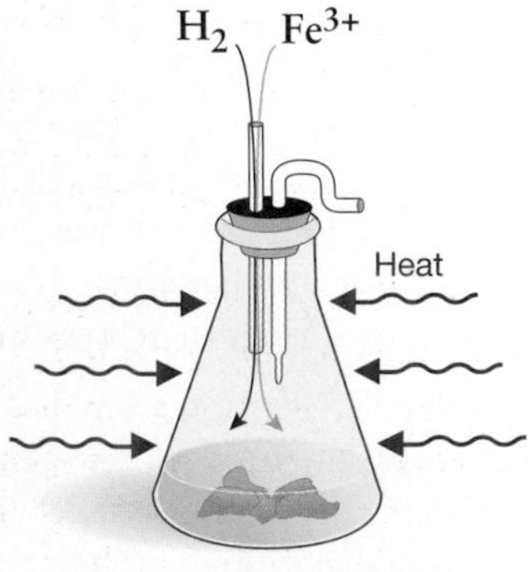

1. Prepare enrichment culture abundant in H_2 and Fe^{3+}; raise temperatures above 45°C.

2. Add rock and fluid samples extracted from drilling operations at depths of about 1000 m below Earth's surface.

PREDICTION: Black, magnetic grains of magnetite (Fe_3O_4) will accumulate because Fe^{3+} is reduced by growing cells and shed as waste product. Cells will be visible.

PREDICTION OF NULL HYPOTHESIS: No magnetite will appear. No cells will grow.

RESULTS: Cells are visible, and magnetite is detectable.

CONCLUSION: At least one bacterial species that can live deep below Earth's surface grew in this enrichment culture. Different culture conditions might result in the enrichment of different species present in the same sample.

FIGURE 28.4 Enrichment Cultures Isolate Large Populations of Cells That Grow under Specific Conditions.

✔**QUESTION** Suppose no organisms had grown in this culture. Explain why the lack of growth would be strong evidence or weak evidence on the question of whether organisms live a mile below the Earth's surface.

SOURCE: Liu, S. V., et al. 1997. Thermophilic Fe(III)-reducing bacteria from the deep subsurface: the evolutionary implications. *Science* 277: 1106–1109.

rocks at depths ranging from 860 to 2800 meters below the surface, where temperatures are between 42°C and 85°C. The questions posed in the study were simple: Is anything alive down there? If so, what do the organisms use to fuel cellular respiration?

The research team hypothesized that if organisms were living deep below the surface of the Earth, the cells might use hydrogen molecules (H_2) as an electron donor and the ferric ion (Fe^{3+}) as an electron acceptor (**Figure 28.4**). (Recall from Chapter 9 that most eukaryotes use sugars as electron donors and use oxygen as an electron acceptor during cellular respiration.) Fe^{3+} is the oxidized form of iron, and it is abundant in the rocks the biologists collected from great depths. It exists at great depths below the surface in the form of ferric oxyhydroxide. The researchers predicted that if an organism in the samples reduced the ferric ions during cellular respiration, then a black, oxidized, and magnetic mineral called magnetite (Fe_3O_4) would start appearing in the cultures as a by-product of cellular respiration.

What did their enrichment cultures produce? In some cultures, a black compound began to appear within a week. Using a variety of tests, the biologists confirmed that the black substance was indeed magnetite. As the "Results" section of Figure 28.4 shows, microscopy revealed the organisms themselves—previously undiscovered bacteria. Because they grow only when incubated at between 45°C and 75°C, these organisms are considered **thermophiles** ("heat-lovers"). The discovery was spectacular—it was one of the first studies demonstrating that Earth's crust is teeming with organisms to depths of over a mile below the surface. Enrichment culture continues to be a productive way to isolate and characterize new species of bacteria and archaea.

Using Direct Sequencing

Researchers estimate that of all the bacteria and archaea living today, less than 1 percent have been grown in culture. To augment research based on enrichment cultures, researchers are employing a technique called direct sequencing. **Direct sequencing** is a strategy for documenting the presence of bacteria and archaea that cannot be grown in culture. It is based on identifying phylogenetic species—populations that have enough distinctive characteristics to represent an independent branch on an evolutionary tree (see Chapter 26).

Direct sequencing allows biologists to identify and characterize organisms that have never been seen. The technique has revealed huge new branches on the tree of life and produced revolutionary data on the habitats where archaea are found.

Figure 28.5 outlines the steps performed in a direct sequencing study. The essence of the approach is to use the polymerase chain reaction, introduced in Chapter 19, to isolate specific genes from the sample. After sequencing these genes, biologists compare the data with sequences in existing databases. If the sequences are markedly different, the sample probably contains previously undiscovered organisms.

Direct sequencing studies have produced new and sometimes startling results. For two decades after the discovery of the Archaea, for example, researchers thought that these organisms could be conveniently grouped into just four categories:

- extreme **halophiles** ("salt-lovers") that live in salt lakes, salt ponds, and salty soils;
- **sulfate reducers**—cells that produce hydrogen sulfide (H_2S) as a by-product of cellular respiration (H_2S may be familiar because it smells like rotten eggs);
- **methanogens** produce methane (natural gas; CH_4) as a by-product of cellular respiration; and
- extreme **thermophiles** grow best at temperatures above 80°C.

Direct sequencing studies threw this generalization out the window, revealing archaea in habitats as diverse as rice paddies and the Arctic Ocean, and entirely new lineages and metabolic capabilities within the Archaea.

Direct sequencing has also revealed previously unrecognized diversity in the bacteria and archaea found on and in the human body. In combination with environmental genomics—a technique based on sequencing all or most of the genes found in a particular habitat (see Chapter 20)—direct sequencing is revolutionizing our understanding of bacterial and archaeal diversity.

Evaluating Molecular Phylogenies

To put data from enrichment culture and direct sequencing studies into context, biologists depend on the accurate placement of species on phylogenetic trees. Recall from Chapter 1, Chapter 27, and **BioSkills 3** in Appendix A that phylogenetic trees illustrate the evolutionary relationships among species and lineages. They are a pictorial summary of which species are more closely or distantly related to others.

Some of the most useful phylogenetic trees for the Bacteria and the Archaea have been based on studies of the RNA molecules found in the small subunit of ribosomes, or what biologists call 16S and 18S RNA. (See Chapter 16 for more information on the structure and function of ribosomes.) In the late 1960s Carl Woese and colleagues began a massive effort to determine and compare the base sequences of 16S and 18S RNA molecules from a wide array of species. The result of their analysis was the **universal tree**, or the **tree of life**, illustrated in Figure 28.1. To review the results of their work, go to the study area at *www.masteringbiology.com.*

(MB) **Web Activity** The Tree of Life

Woese's tree is now considered a classic result. Prior to its publication, biologists thought that the major division among organisms was between prokaryotes and eukaryotes.

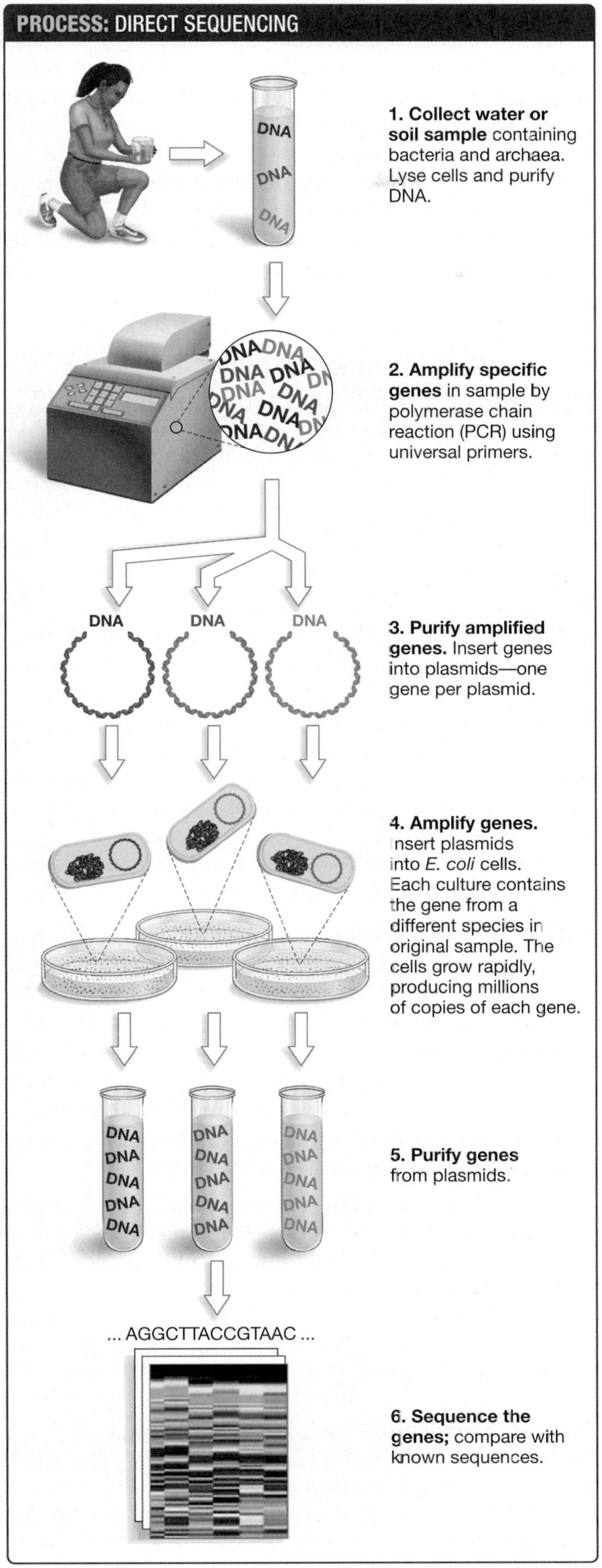

PROCESS: DIRECT SEQUENCING
1. Collect water or soil sample containing bacteria and archaea. Lyse cells and purify DNA.
2. Amplify specific genes in sample by polymerase chain reaction (PCR) using universal primers.
3. Purify amplified genes. Insert genes into plasmids—one gene per plasmid.
4. Amplify genes. Insert plasmids into *E. coli* cells. Each culture contains the gene from a different species in original sample. The cells grow rapidly, producing millions of copies of each gene.
5. Purify genes from plasmids.
6. Sequence the genes; compare with known sequences.
DNA
... AGGCTTACCGTAAC ...

FIGURE 28.5 Direct Sequencing Allows Researchers to Identify Species That Have Never Been Seen. Direct sequencing isolates specific genes from organisms in a sample. The polymerase chain reaction generates many copies of the genes from each species, so they can be sequenced.

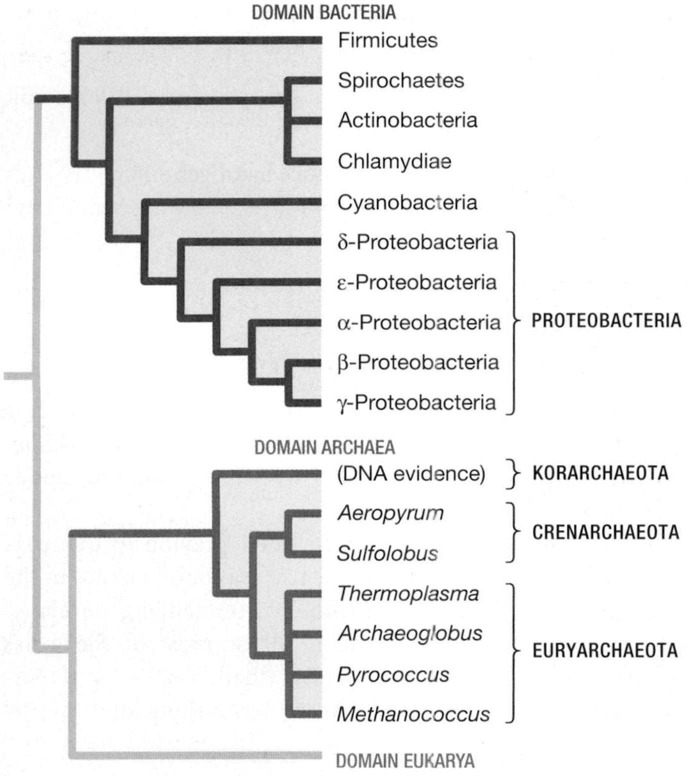

FIGURE 28.6 Phylogeny of Some Major Lineages in Bacteria and Archaea.

But based on data from the ribosomal RNA molecule, the major divisions of life-forms are actually the Bacteria, Archaea, and Eukarya. Follow-up work documented that Bacteria were the first of the three lineages to diverge from the common ancestor of all living organisms—meaning that the Archaea and Eukarya are more closely related to each other than they are to the Bacteria.

CHECK YOUR UNDERSTANDING

If you understand that . . .

- Enrichment cultures isolate cells that grow in response to specific conditions. They create an abundant sample of bacteria that thrive under particular conditions, allowing further study.
- Direct sequencing is based on isolating DNA from samples taken directly from the environment, purifying and sequencing specific genes, and then analyzing where those DNA sequences are found on the phylogenetic tree of Bacteria and Archaea.

✔ You should be able to . . .

1. Design an enrichment culture that would isolate species that could be used to clean up oil spills.
2. Outline a study designed to identify the bacterial and archaeal species present in a soil sample near the biology building on your campus.

Answers are available in Appendix B.

More recent analyses of morphological and molecular characteristics have succeeded in identifying a large series of monophyletic groups within the domains. Recall from Chapter 27 that a **monophyletic group** consists of an ancestral population and all of its descendants. Monophyletic groups can also be called clades or lineages.

The phylogenetic tree in **Figure 28.6** summarizes recent results but is still considered highly provisional. Work on molecular phylogenies continues at a brisk pace. Section 28.4 will explore some of these lineages in detail, but for now let's turn to the question of how all this diversification took place.

28.3 What Themes Occur in the Diversification of Bacteria and Archaea?

Initially, the diversity of bacteria and archaea can seem almost overwhelming. To make sense of the variation among lineages and species, biologists focus on two themes in diversification: morphology and metabolism. Regarding metabolism, the key question is which molecules are used as food. Bacteria and archaea are capable of living in a wide array of environments because they vary in cell structure and in how they make a living.

Morphological Diversity

Because we humans are so large, it is hard for us to appreciate the morphological diversity that exists among bacteria and archaea. To us, they all look small and similar. But at the scale of a bacterium or archaean, different species are wildly diverse in morphology.

SIZE, SHAPE, AND MOTILITY To appreciate how diverse these organisms are in terms of morphology, consider bacteria alone:

- *Size* Bacterial cells range in size from the smallest of all free-living cells—bacteria called mycoplasmas with volumes as small as 0.15 μm^3—to the largest bacterium known, *Thiomargarita namibiensis*, with volumes as large as 200×10^6 μm^3. Over a billion *Mycoplasma* cells could fit inside an individual *Thiomargarita* (**Figure 28.7a**). Comparing *Thiomargarita* and *E. coli* cell size is like comparing a blue whale to a newborn mouse.
- *Shape* Bacterial cells range in shape from filaments, spheres, rods, and chains to spirals (**Figure 28.7b**).
- *Motility* Many bacterial cells are motile, with swimming movements powered by flagella. Some cells can swim 10 or more body lengths per second—much faster than any human can sprint. Gliding movement, which allows cells to creep along a surface, also occurs in several groups, though the molecular mechanism responsible for this form of motility is still unknown (**Figure 28.7c**).

CELL WALL COMPOSITION For single-celled organisms, the composition of the plasma membrane and cell wall are particularly important. The introduction to this chapter highlighted the dramatic differences between the plasma membranes and cell walls of bacteria versus archaea.

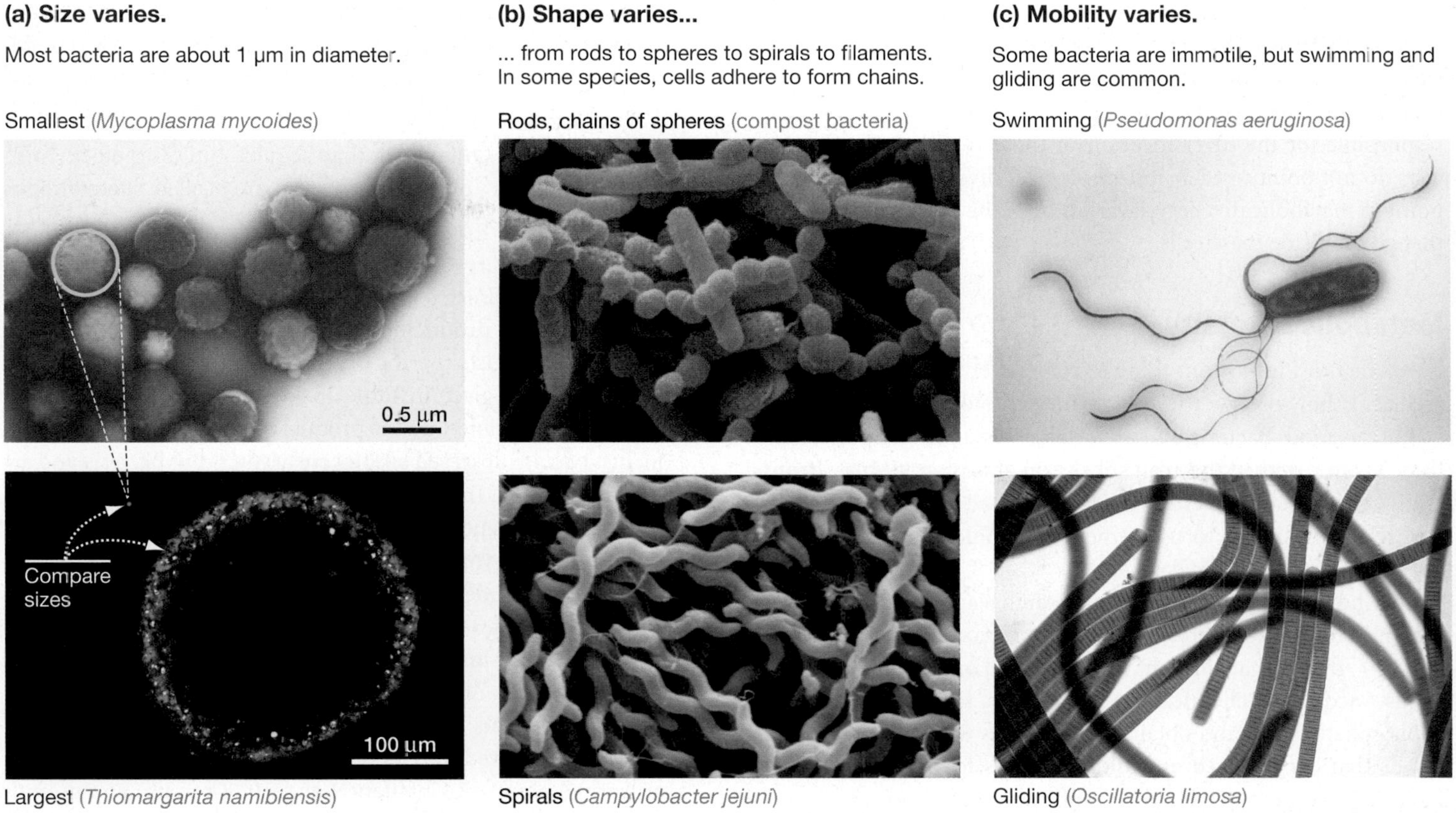

FIGURE 28.7 Morphological Diversity among Bacteria Is Extensive. Some of the cells in these micrographs have been colorized to make them more visible.

Within bacteria, biologists distinguish two general types of cell wall using a dyeing system called the **Gram stain**. As **Figure 28.8a** shows, Gram-positive cells look purple but Gram-negative cells look pink.

At the molecular level, cells that are **Gram-positive** have a plasma membrane surrounded by a cell wall with extensive peptidoglycan (**Figure 28.8b**). You might recall from Chapter 5 that peptidoglycan is a complex substance composed of carbohydrate strands that are cross-linked by short chains of amino acids. Cells that are **Gram-negative**, in contrast, have a plasma membrane surrounded by a cell wall that has two components—a thin gelatinous layer containing peptidoglycan and an outer phospholipid bilayer (**Figure 28.8c**).

Analyzing cell cultures with the Gram stain can be an important preliminary step in treating bacterial infections. Because they contain so much peptidoglycan, Gram-positive cells may respond to treatment by penicillin-like drugs that disrupt peptidoglycan synthesis. Gram-negative cells, in contrast, are more likely to be affected by erythromycin or other types of drugs that poison bacterial ribosomes.

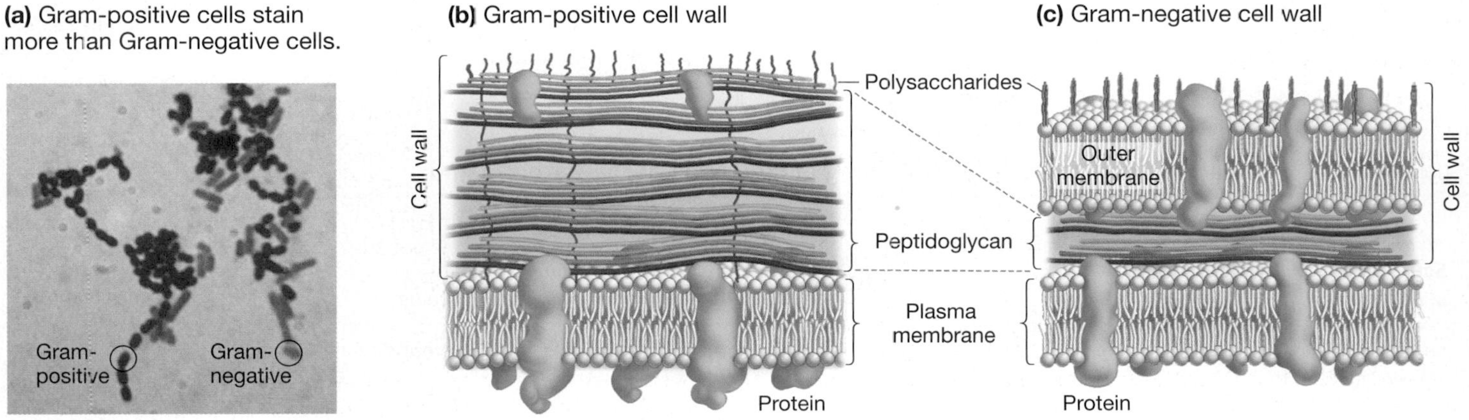

FIGURE 28.8 Gram Staining Distinguishes Two Types of Cell Walls in Bacteria. Cells with extensive peptidoglycan retain a large amount of stain and look purple; others retain little stain and look pink, as can be seen in part (a).

To summarize, members of the Bacteria and the Archaea are remarkably diverse in their overall size, shape, and motility, as well as in the composition of their cell walls and plasma membranes. But when asked to name the innovations that were most responsible for the diversification of these two domains, biologists do not point to their morphological diversity. Instead, they point to metabolic diversity—variation in the chemical reactions that go on inside these cells.

Metabolic Diversity

The most important thing to remember about bacteria and archaea is how diverse they are in the types of compounds they can use as food. Bacteria and archaea are the masters of metabolism. Taken together, they can subsist on almost anything—from hydrogen molecules to crude oil. Bacteria and archaea look small and relatively simple to us in their morphology, but their biochemical capabilities are dazzling.

Just how varied are bacteria and archaea when it comes to making a living? To appreciate the answer, recall from Chapters 9 and 10 that organisms have two fundamental nutritional needs—acquiring chemical energy in the form of adenosine triphosphate (ATP) and obtaining molecules with carbon-carbon bonds that can be used as building blocks for the synthesis of fatty acids, proteins, DNA, RNA, and other large, complex compounds required by the cell.

Bacteria and archaea produce ATP in three ways:

1. **Phototrophs** ("light-feeders") use light energy to promote electrons to the top of electron transport chains. ATP is produced by photophosphorylation (see Chapter 10).
2. **Chemoorganotrophs** oxidize organic molecules with high potential energy, such as sugars. ATP may be produced by cellular respiration—with sugars serving as electron donors—or via fermentation pathways (see Chapter 9).
3. **Chemolithotrophs** ("rock-feeders") oxidize inorganic molecules with high potential energy, such as ammonia (NH_3) or methane (CH_4). ATP is produced by cellular respiration, with inorganic compounds serving as the electron donor.

Bacteria and archaea fulfill their second nutritional need—obtaining building block compounds with carbon-carbon bonds—in two ways:

1. By synthesizing their own from simple starting materials such as CO_2 and CH_4. Organisms that manufacture their own building-block compounds are termed **autotrophs** ("self-feeders").
2. By absorbing ready-to-use organic compounds from their environment. Organisms that acquire building-block compounds from other organisms are called **heterotrophs** ("other-feeders").

Because there are three distinct ways of producing ATP and two general mechanisms for obtaining carbon, there are a total of six methods for producing ATP and obtaining carbon. The names that biologists use for organisms that use these six "feeding strategies" are given in **Table 28.3**.

Of the six possible ways of producing ATP and obtaining carbon, just two are observed among eukaryotes. But bacteria and archaea do them all. In addition, certain species can switch among modes of living, depending on environmental conditions. In their metabolism, eukaryotes are simple compared with bacteria and archaea. ✓**If you understand the essence of metabolic diversity in bacteria and archaea, you should be able to match the six example species described in Table 28.4 to the appropriate category in** Table 28.3.

What makes this remarkable diversity possible? Bacteria and archaea have evolved dozens of variations on the basic processes you learned about in Chapters 9 and 10. They use compounds with high potential energy to produce ATP via cellular respiration (electron transport chains) or fermentation, they use light to produce high-energy electrons, and they reduce carbon from CO_2 or other sources to produce sugars or other building-block molecules with carbon-carbon bonds.

The story of bacteria and archaea can be boiled down to two sentences: The basic chemistry required for photosynthesis, cellular respiration, and fermentation originated in these lineages. Then the evolution of variations on each of these processes allowed prokaryotes to diversify into millions of species that occupy diverse habitats. Let's take a closer look.

PRODUCING ATP THROUGH CELLULAR RESPIRATION: VARIATION IN ELECTRON DONORS AND ACCEPTORS Millions of bacterial, archaeal, and eukaryotic species—including animals and plants—are organotrophs. These organisms obtain the energy required to make ATP by breaking down organic compounds such as sugars, starch, or fatty acids.

SUMMARY TABLE 28.3 **Six General Methods for Obtaining Energy and Carbon-Carbon Bonds**

	Source of Energy (for synthesis of ATP)		
Source of C–C Bonds (for synthesis of complex organic compounds)	Phototrophs: from sunlight	Chemoorganotrophs: from organic molecules	Chemolithotrophs: from inorganic molecules
Autotrophs: self-synthesized from CO_2, CH_4, or other simple molecules	photoautotrophs	chemoorganoautotrophs	chemolitho[auto]trophs
Heterotrophs: from molecules produced by other organisms	photoheterotrophs	chemoorganoheterotrophs	chemolithotrophic heterotrophs

TABLE 28.4 **Six Examples of Metabolic Diversity**

Example	How ATP Is Produced	How Building-Block Molecules Are Synthesized
Cyanobacteria	via photosynthesis	from CO_2 via the Calvin cycle
Clostridium aceticum	fermentation of glucose	from CO_2 via reactions called the acetyl-CoA pathway
Nitrifying bacteria (e.g., *Nitrosomonas* sp.)	via cellular respiration, using ammonia (NH_3) as an electron donor	from CO_2 via the Calvin cycle
Heliobacteria	via photosynthesis	absorb carbon-containing building-block molecules from the environment
Escherichia coli	fermentation of organic compounds or cellular respiration, using organic compounds as electron donors	absorb carbon-containing building-block molecules from the environment
Beggiatoa	via cellular respiration, using hydrogen sulfide (H_2S) as an electron donor	absorb carbon-containing building-block molecules from the environment

As Chapter 9 showed, cellular enzymes can strip electrons from organic molecules that have high potential energy and then transfer these high-energy electrons to the electron carriers NADH and $FADH_2$. These compounds feed electrons to an electron transport chain (ETC), where electrons are stepped down from a high-energy state to a low-energy state (**Figure 28.9a**).

The energy that is released allows components of the ETC to generate a proton gradient across the plasma membrane (**Figure 28.9b**). The resulting flow of protons through the enzyme ATP synthase results in the production of ATP, via the process called chemiosmosis.

The essence of this process, called **cellular respiration**, is that a molecule with high potential energy serves as an original electron donor and is oxidized, while a molecule with low potential energy serves as a final electron acceptor and becomes reduced. Much of the potential energy difference between the electron donor and electron acceptor is transformed into chemical energy in the form of ATP.

Chapter 9 focused on how eukaryotes perform cellular respiration. In eukaryotes:

- Organic compounds with high potential energy—often glucose—serve as the original electron donor. When cellular respiration is complete, glucose is completely oxidized to CO_2, which is given off as a by-product.
- Oxygen is the final electron acceptor, and water is also produced as a by-product.

Many bacteria and archaea also rely on these molecules.

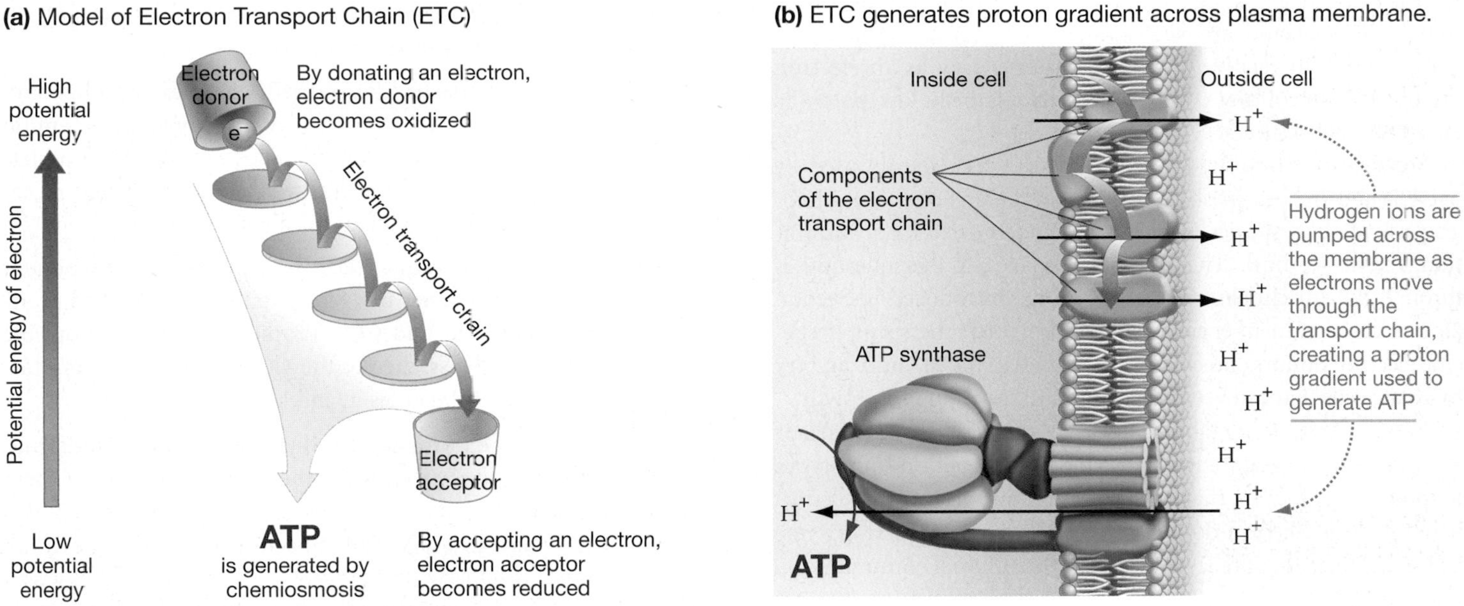

FIGURE 28.9 Cellular Respiration Is Based on Electron Transport Chains.

✔**EXERCISE** In part (a), add the chemical formula for a specific electron donor, electron acceptor, and reduced by-product for a species of bacteria or archaea. Then write in the electron donor, electron acceptor, and reduced by-product observed in humans.

TABLE 28.5 **Some Electron Donors and Acceptors Used by Bacteria and Archaea**

Electron Donor	Electron Acceptor	By-Products: From Electron Donor	By-Products: From Electron Acceptor	Category*
Sugars	O_2	CO_2	H_2O	Organotrophs
H_2 or organic compounds	SO_4^{2-}	H_2O or CO	H_2S or S^{2-}	Sulfate reducers
H_2	CO_2	H_2O	CH_4	Methanogens
CH_4	O_2	CO_2	H_2O	Methanotrophs
S^{2-} or H_2S	O_2	SO_4^{2-}	H_2O	Sulfur bacteria
Organic compounds	Fe^{3+}	CO_2	Fe^{2+}	Iron reducers
NH_3	O_2	NO_2^-	H_2O	Nitrifiers
Organic compounds	NO_3^-	CO_2	N_2O, NO, or N_2	Denitrifiers (or nitrate reducers)
NO_2^-	O_2	NO_3^-	H_2O	Nitrosifiers

*The name biologists use to identify species that use a particular metabolic strategy.

✔**QUESTION** Explain why the terms organotrophs, sulfate reducers, and methanogens are appropriate. (The word root *–gen* means source or origin; *–troph* refers to feeding.)

It is common, however, to find bacteria and archaea that employ an electron donor other than sugars and an electron acceptor other than oxygen during cellular respiration. These species produce by-products other than carbon dioxide and water (**Table 28.5**). In some bacteria and archaea:

- Inorganic ions or molecules with high potential energy serve as electron donors. The substances used as electron donors range from hydrogen molecules (H_2) and hydrogen sulfide (H_2S) to ammonia (NH_3) and methane (CH_4).
- Compounds with relatively low potential energy—including sulfate (SO_4^{2-}), nitrate (NO_3^-), carbon dioxide (CO_2), or ferric ions (Fe^{3+})—act as electron acceptors.

It is only a slight exaggeration to claim that researchers have found bacterial and archaeal species that can use almost any compound with relatively high potential energy as an electron donor and almost any compound with relatively low potential energy as an electron acceptor.

Because the electron donors and electron acceptors used by bacteria and archaea are so diverse, one of the first questions biologists ask about a species is whether it undergoes cellular respiration and if so, how. The best way to answer this question is through the enrichment culture technique introduced in Section 28.2. Recall that in an enrichment culture, researchers supply specific electron donors and electron acceptors in the medium and try to isolate cells that can use those compounds to support growth.

The remarkable metabolic diversity of bacteria and archaea explains why they play such a key role in cleaning up some types of pollution. Species that use organic solvents or petroleum-based fuels as electron donors or electron acceptors may excrete waste products that are less toxic than the original compounds.

PRODUCING ATP VIA FERMENTATION: VARIATION IN SUBSTRATES Chapter 9 introduced **fermentation** as a strategy for making ATP that does not involve electron transport chains. In fermentation, no outside electron acceptor is used.

Because fermentation is a much less efficient way to make ATP compared with cellular respiration, in many species it occurs as an alternative metabolic strategy when no electron acceptors are available to make cellular respiration possible. In other species, fermentation does not occur at all. But in many bacteria and archaea, fermentation is the only way that cells make ATP.

Although the presentation in Chapter 9 focused on how glucose is fermented to ethanol or lactic acid, some bacteria and archaea are capable of using other organic compounds as the starting point for fermentation. Bacteria and archaea that produce ATP via fermentation are still classified as organotrophs, but they are much more diverse in the substrates used. For example:

- The bacterium *Clostridium aceticum* can ferment ethanol, acetate, and fatty acids as well as glucose.
- Other species of *Clostridium* ferment complex carbohydrates (including cellulose or starch), proteins, purines, or amino acids. Species that ferment amino acids produce by-products with names such as cadaverine and putrescine. These molecules are responsible for the odor of rotting flesh.
- Other bacteria can ferment lactose, a prominent component of milk. In some species this fermentation has two end products: propionic acid and CO_2. Propionic acid is responsible for the taste of Swiss cheese; the CO_2 produced during fermentation creates the holes in cheese.

The diversity of enzymatic pathways observed in bacterial and archaeal fermentations extends the metabolic repertoire of these organisms. The diversity of substrates that are fermented also supports the claim that as a group, bacteria and archaea can use virtually any molecule with relatively high potential energy as a source of high-energy electrons for producing ATP.

PRODUCING ATP VIA PHOTOSYNTHESIS: VARIATION IN ELECTRON SOURCES AND PIGMENTS Instead of using molecules as a source of high-energy electrons, phototrophs pursue a radically different strategy: **photosynthesis**. Among bacteria and archaea, photosynthesis can happen in one of three ways:

- Light activates a pigment called bacteriorhodopsin, which uses the absorbed energy to transport protons across a membrane. The resulting flow of protons drives the synthesis of ATP via chemiosmosis (see Chapter 9).
- A recently discovered bacterium that lives near hydrothermal vents on the ocean floor performs photosynthesis not by absorbing light, but by absorbing geothermal radiation.
- Pigments that absorb light raise electrons to high-energy states. As these electrons are stepped down to lower energy states by electron transport chains, the energy released is used to generate ATP.

Chapter 10 introduced an important feature of this last mode of photosynthesis: The process requires a source of electrons. Recall that in cyanobacteria and plants the required electrons come from water. When these organisms "split" water molecules apart to obtain electrons, they generate oxygen as a by-product. Species that use water as a source of electrons for photosynthesis are said to complete **oxygenic** photosynthesis.

In contrast, many phototrophic bacteria use a molecule other than water as the source of electrons. In many cases, the electron donor is hydrogen sulfide (H_2S); a few species can use the ion known as ferrous iron (Fe^{2+}). Instead of producing oxygen as a by-product of photosynthesis, these cells produce elemental sulfur (S) or the ferric ion (Fe^{3+}). They are said to complete **anoxygenic** photosynthesis. They live in habitats where oxygen is rare.

Chapter 10 also introduced the photosynthetic pigments found in plants and explored the light-absorbing properties of chlorophylls *a* and *b*. Cyanobacteria have these two pigments. But researchers have isolated seven additional chlorophylls from bacterial phototrophs. Each major group of photosynthetic bacteria has one or more of these distinctive chlorophylls, and each type of chlorophyll absorbs light best at a different wavelength. As a result, a diverse array of photosynthetic bacteria can live in the same habitat without competing for light.

OBTAINING BUILDING-BLOCK COMPOUNDS: VARIATION IN PATHWAYS FOR FIXING CARBON In addition to acquiring energy, organisms must obtain building-block molecules that contain carbon-carbon bonds. Chapters 9 and 10 introduced the two mechanisms that organisms use to procure usable carbon—either making their own or getting it from other organisms. Autotrophs make their own building-block compounds; heterotrophs don't.

In many autotrophs, including cyanobacteria and plants, the enzymes of the Calvin cycle transform carbon dioxide (CO_2) to organic molecules that can be used in synthesizing cell material. The carbon atom in CO_2 is reduced during the process and is said to be "fixed." Animals and fungi, in contrast, obtain carbon from living plants or animals or by absorbing the organic compounds released as dead tissues decay.

Bacteria and archaea pursue these same two strategies. Some interesting twists occur among bacterial and archaeal autotrophs, however. Not all of them use the Calvin cycle to make building-block molecules, and not all start with CO_2 as a source of carbon atoms. For example:

- Several groups of bacteria fix CO_2 using pathways other than the Calvin cycle. Three of these distinctive pathways have been discovered to date.
- Some proteobacteria are called **methanotrophs** ("methane-eaters") because they use methane (CH_4) as their carbon source. (They also use CH_4 as an electron donor in cellular respiration.) Methanotrophs process CH_4 into more complex organic compounds via one of two enzymatic pathways, depending on the species.
- Some bacteria can use carbon monoxide (CO) or methanol (CH_3OH) as a starting material.

These observations drive home an important message from this chapter: Compared with eukaryotes, the metabolic capabilities of bacteria and archaea are remarkably sophisticated and complex.

Ecological Diversity and Global Change

The metabolic diversity observed among bacteria and archaea explains why these organisms can thrive in such a wide array of habitats.

- The array of electron donors, electron acceptors, and fermentation substrates exploited by bacteria and archaea allows the heterotrophic species to live just about anywhere.
- The evolution of three distinct types of photosynthesis—based on bacteriorhodopsin, geothermal energy, or pigments that donate high-energy elections to ETCs—extends the types of habitats that can support phototrophs. In addition, the remarkable diversity of bacterial chlorophylls allows photosynthetic species with different absorption spectra to live together without competing for light.

The complex chemistry that these cells carry out, combined with their numerical abundance, has made them potent forces for global change throughout Earth's history. Bacteria and archaea have altered the chemical composition of the oceans, atmosphere, and terrestrial environments for billions of years. They continue to do so today.

THE OXYGEN REVOLUTION Today, oxygen represents almost 21 percent of the molecules in Earth's atmosphere. But researchers who study the composition of the atmosphere are virtually certain that no free molecular oxygen (O_2) existed for the first 2.3 billion years of Earth's existence. This conclusion is based on two observations:

1. There was no plausible source of oxygen at the time the planet formed.
2. The oldest Earth rocks indicate that the first signs of oxygen appear many years after the formation of the planet. Even then, virtually all of the oxygen present reacted immediately with iron atoms to produce iron oxides, such as hematite (Fe_2O_3) and magnetite (Fe_3O_4).

Early in Earth's history, the atmosphere was dominated by nitrogen and carbon dioxide. Where did the oxygen we breathe come from? The answer is cyanobacteria.

FIGURE 28.10 Cyanobacteria Were the First Organisms to Perform Oxygenic Photosynthesis.

Cyanobacteria are a lineage of photosynthetic bacteria. According to the fossil record, species of cyanobacteria first became numerous in the oceans about 2.7–2.55 billion years ago. Their appearance was momentous because cyanobacteria were the first organisms to perform oxygenic ("oxygen-producing") photosynthesis (**Figure 28.10**).

The fossil record and geological record indicate that oxygen concentrations in the oceans and atmosphere began to increase 2.3–2.1 billion years ago. Once oxygen was common in the oceans, cells could begin to use it as the final electron acceptor during cellular respiration. **Aerobic** respiration was now a possibility. Prior to this, organisms had to use compounds other than oxygen as a final electron acceptor—only **anaerobic** respiration was possible.

The evolution of aerobic respiration was a crucial event in the history of life. Because oxygen is extremely electronegative, it is an efficient electron acceptor. Much more energy is released as electrons move through electron transport chains with oxygen as the ultimate acceptor than is released with other substances as the electron acceptor.

To drive this point home, study the graph in **Figure 28.11**. Note that the vertical axis plots free energy changes; the graph shows the energy released when glucose is oxidized with fumarate, nitrate, or oxygen as the final electron acceptor. Once oxygen was available, then, cells could produce much more ATP for each electron donated by NADH or $FADH_2$. As a result, the rate of energy production could rise dramatically.

To summarize, data indicate that cyanobacteria were responsible for a fundamental change in Earth's atmosphere—a high concentration of oxygen. Never before, or since, have organisms done so much to alter the nature of our planet.

NITROGEN FIXATION AND THE NITROGEN CYCLE In many environments, fertilizing forests or grasslands with nitrogen results in increased growth. Researchers infer from these results that plant growth is limited by the availability of nitrogen.

Organisms must have nitrogen to synthesize proteins and nucleic acids. Although molecular nitrogen (N_2) is extremely abundant in the atmosphere, most organisms cannot use it. To incorporate nitrogen atoms into amino acids and nucleotides, all

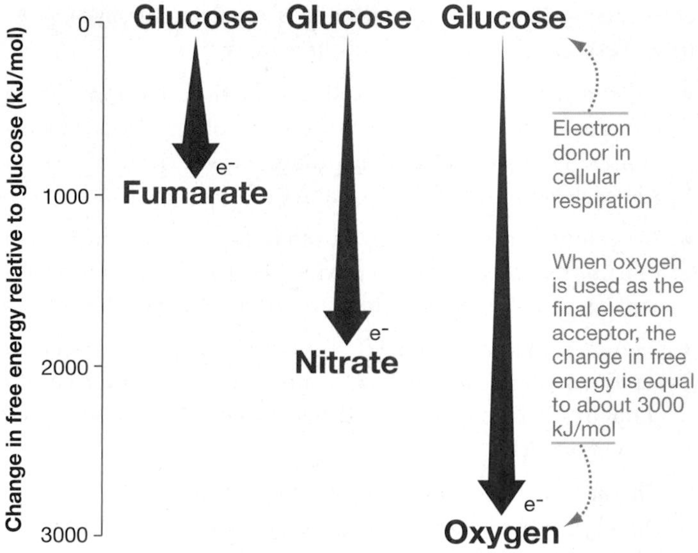

FIGURE 28.11 Cellular Respiration Can Produce More Energy When Oxygen Is the Final Electron Acceptor. Because oxygen has such high electronegativity, the potential energy of the electrons used in cellular respiration is much lower when oxygen is the final acceptor compared to other molecules or ions—resulting in the release of a larger amount of free energy.

✔**QUESTION** Which organisms grow faster—those using aerobic respiration or those using anaerobic respiration? Explain your reasoning.

eukaryotes and many bacteria and archaea have to obtain N in a form such as ammonia (NH_3) or nitrate (NO_3^-).

Certain bacteria and archaea are the only species that are capable of converting molecular nitrogen to ammonia. The steps in the process, called **nitrogen fixation**, are complex and highly endergonic reduction-oxidation (redox) reactions (see Chapter 9). The enzymes required to accomplish nitrogen fixation are found only in selected bacterial and archaeal lineages.

- Some cyanobacteria that live in surface waters of the ocean or in association with water plants are capable of fixing nitrogen.
- In terrestrial environments, nitrogen-fixing bacteria live in close association with plants—often taking up residence in special root structures called nodules (**Figure 28.12**).

If bacteria and archaea could not fix nitrogen, it is virtually certain that only a tiny fraction of life on Earth would exist today. Large or multicellular organisms would probably be rare to nonexistent, because too little nitrogen would be available to make large quantities of proteins and build a large body.

Nitrogen-fixation is only the beginning of the story, however. A quick glance back at Table 28.5 should convince you that bacteria and archaea use a wide array of nitrogen-containing compounds as electron donors and electron acceptors during cellular respiration.

To understand why this is important, consider that the nitrite (NO_2^-) that some bacteria produce as a by-product of respiration does not build up in the environment. Instead, it is used as an electron acceptor by other species and converted to molecular nitrate (NO_3^-). Nitrate, in turn, is converted to molecular nitrogen (N_2) by yet another suite of bacterial and archaeal species. In

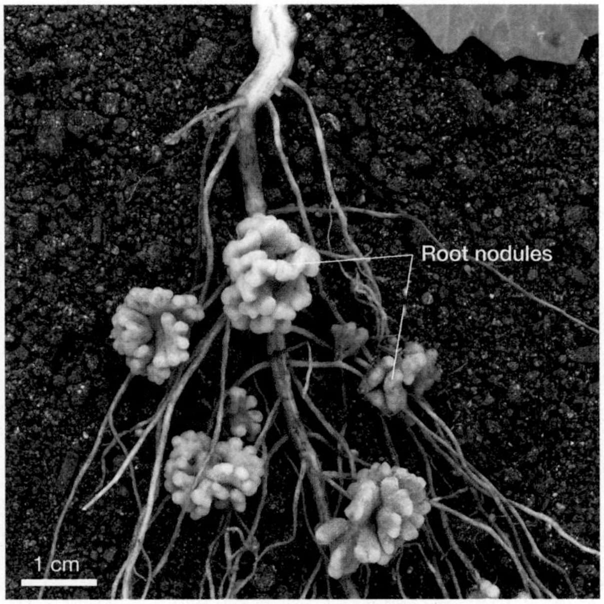

FIGURE 28.12 Root nodules form a protective structure for bacteria that fix nitrogen.

this way, bacteria and archaea are responsible for driving the movement of nitrogen atoms through ecosystems around the globe (**Figure 28.13**).

Similar types of interactions occur with molecules that contain phosphorus, sulfur, and carbon. In this way, bacteria and archaea play a key role in the cycling of nitrogen and other nutrients.

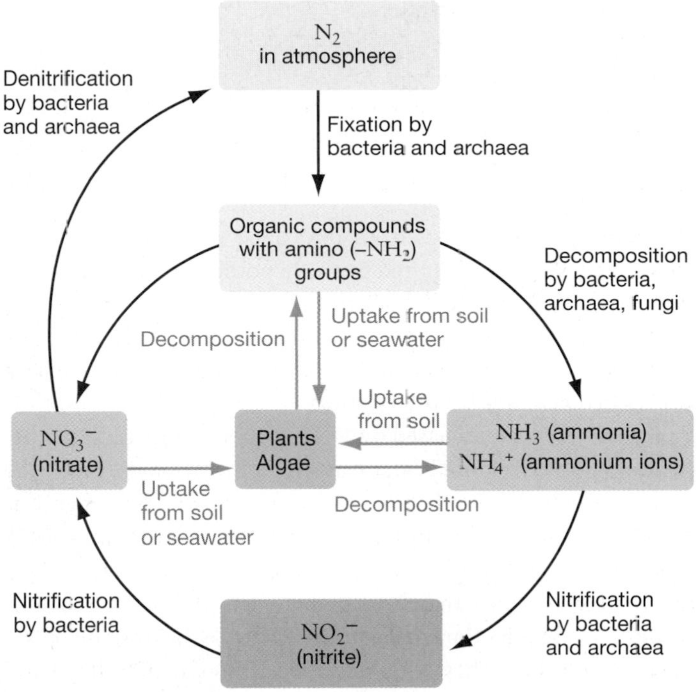

FIGURE 28.13 Bacteria and Archaea Drive the Movement of Nitrogen Atoms through Ecosystems. Nitrogen atoms cycle in different molecular forms. In addition to the conversions shown here, some bacteria can also convert nitrate to ammonium ions; others combine ammonium and nitrate ions to form nitrogen gas (N_2).

NITRATE POLLUTION Corn, rice, wheat, and many other crop plants do not live in association with nitrogen-fixing bacteria. To increase yields of these crops, farmers use fertilizers that are high in nitrogen. In parts of the world, massive additions of nitrogen in the form of ammonia are causing serious pollution problems.

Figure 28.14 shows why. When ammonia is added to a cornfield—in midwestern North America, for example—much of it never reaches the growing corn plants. Instead, a significant

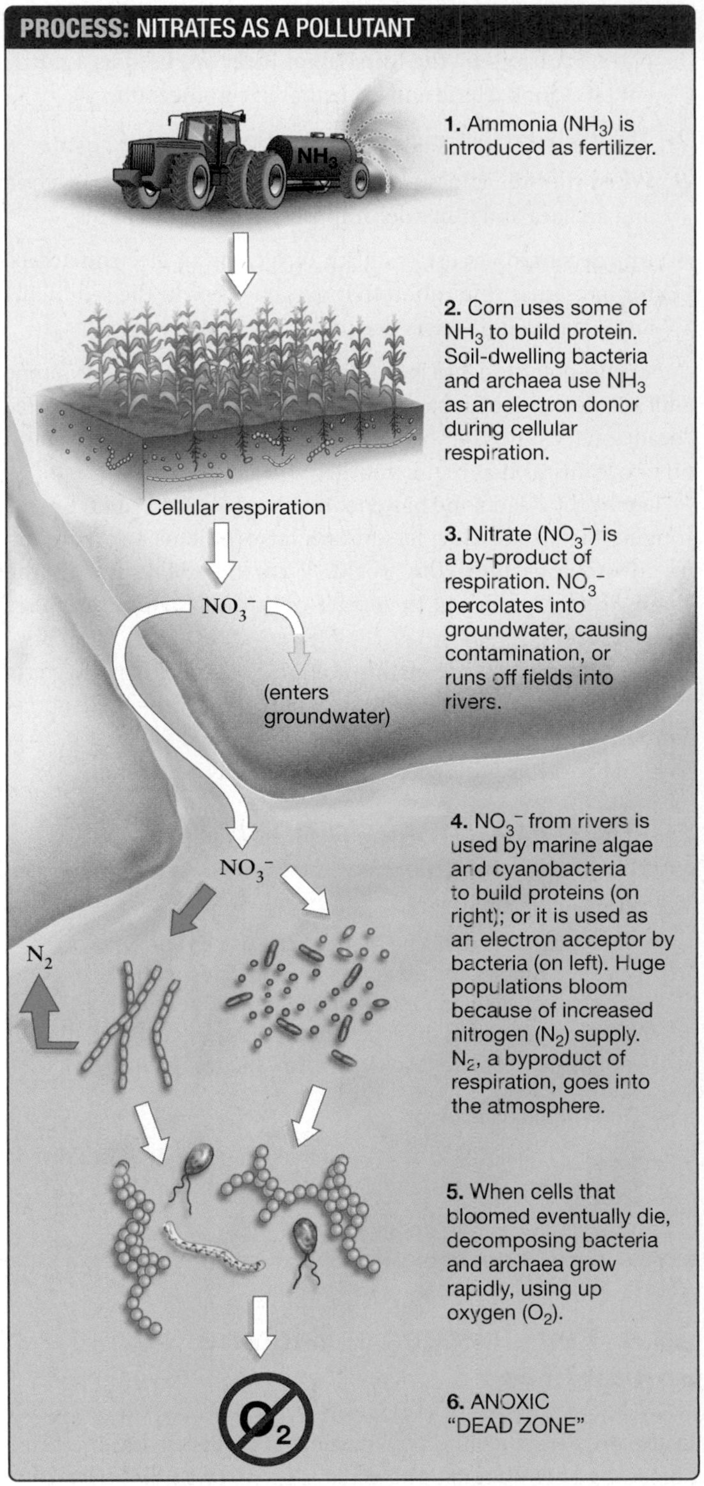

FIGURE 28.14 Nitrates Act as a Pollutant in Aquatic Ecosystems.

fraction of the ammonia molecules is used as food by bacteria in the soil. Bacteria that use ammonia as an electron donor to fuel cellular respiration release nitrite (NO_2^-) as a waste product. Other bacteria use nitrite as an electron donor and release nitrate (NO_3^-). Nitrate molecules are extremely soluble in water and tend to be washed out of soils into groundwater or streams. From there they eventually reach the ocean, where they can cause pollution.

To understand why nitrates can pollute the oceans, consider the Gulf of Mexico:

1. Nitrates carried by the Mississippi River are used as a nutrient by cyanobacteria and algae that live in the Gulf.
2. These cells explode in numbers in response.
3. When they die and sink to the bottom of the Gulf, bacteria and archaea and other decomposers use them as food.
4. The decomposers use so much oxygen as an electron acceptor in cellular respiration that oxygen levels in the sediments and even in Gulf waters decline.

Nitrate pollution has been so severe that large areas in the Gulf of Mexico are anoxic (lacking in oxygen). The oxygen-free "dead zone" in the Gulf of Mexico is devoid of fish, shrimp, and other organisms that require oxygen.

Lately, the dead zone has encompassed about 18,000 km^2—roughly the size of New Jersey. Similar problems are cropping up in other parts of the world. Virtually every link in the chain of events leading to nitrate pollution involves bacteria and archaea.

The general message of this section is simple: Bacteria and Archaea may be small in size, but because of their abundance, ubiquity, and ability to do sophisticated chemistry, they have an enormous influence on the global environment.

CHECK YOUR UNDERSTANDING

If you understand that . . .

- As a group, Bacteria and Archaea can use a wide array of electron donors and acceptors in cellular respiration, a diverse set of compounds in fermentation, perform non-oxygenic as well as oxygenic photosynthesis, and fix carbon from several different sources via a variety of pathways.

✓ You should be able to . . .

Defend the claim that in terms of metabolism, bacteria and archaea are much more sophisticated than eukaryotes.

Answers are available in Appendix B.

28.4 Key Lineages of Bacteria and Archaea

In the decades since the phylogenetic tree identifying the three domains of life was first published, dozens of studies have confirmed the result. It is now well established that all organisms alive today belong to one of the three domains, and that archaea and eukaryotes are more closely related to each other than either group is to bacteria.

Although the relationships among the major lineages within Bacteria and Archaea are still uncertain in some cases, many of the lineages themselves are well studied. Let's survey the attributes of species from selected major lineages within the Bacteria and Archaea, with an emphasis on themes explored earlier in the chapter: their morphological and metabolic diversity, their impacts on humans, and their importance to other species and to the environment.

Bacteria

The name *bacteria* comes from the Greek root *bacter*, meaning "rod" or "staff." The name was inspired by the first bacteria to be seen under a microscope, which were rod shaped. But as the following descriptions indicate, bacterial cells come in a wide variety of shapes.

Biologists who study bacterial diversity currently recognize at least 16 major lineages, or phyla, within the domain. Some of these lineages were recognized by distinctive morphological characteristics and others by phylogenetic analyses of gene sequence data. The lineages reviewed here are just a sampling of bacterial diversity.

- Bacteria > Firmicutes
- Bacteria > Spirochaetes (Spirochetes)
- Bacteria > Actinobacteria
- Bacteria > Chlamydiae
- Bacteria > Cyanobacteria
- Bacteria > Proteobacteria

Archaea

The name *archaea* comes from the Greek root *archae*, for "ancient." The name was inspired by the hypothesis that this is a particularly ancient group, which turned out to be incorrect. Also incorrect was the initial hypothesis that archaeans are restricted to hot springs, salt ponds, and other extreme habitats. Archaea live in virtually every habitat known.

Recent phylogenies based on DNA sequence data indicate that the domain is composed of at least three major phyla, called the Crenarchaeota, Euryarchaeota, and Korarchaeota. Although it is clear that these three groups are highly differentiated at the DNA sequence level, biologists are still searching for shared, derived morphological traits that help define each lineage as a monophyletic group. The Korarchaeota are known only from direct sequencing studies. They have never been grown in culture and almost nothing is known about them.

- Archaea > Crenarchaeota
- Archaea > Euryarchaeota

Bacteria > Firmicutes

The Firmicutes have also been called "low-GC Gram positives" because their cell walls react positively to the Gram stain and because they have a relatively low percentage of guanine and cytosine (G and C) in their DNA. In some species, G and C represent less than 25 percent of the bases present. There are over 1100 species. ✔**You should be able to mark the origin of the Gram-positive cell wall and low-GC genome on Figure 28.6 (only Firmicutes have a low-GC genome; Actinobacteria are the only other Gram-positive lineage).**

Morphological diversity Most are rod shaped or spherical. Some of the spherical species form chains or tetrads (groups of four cells). A few form a durable resting stage called a spore. One subgroup lacks cell walls entirely; another synthesizes a cell wall made of cellulose.

Metabolic diversity Some species can fix nitrogen; some perform non-oxygenic photosynthesis. Others make all of their ATP via various fermentation pathways; still others perform cellular respiration, using hydrogen gas (H_2) as an electron donor.

Human and ecological impacts Recent direct sequencing studies have shown that members of this lineage are extremely common in the human gut. Species in this group cause a variety of diseases, including anthrax, botulism, tetanus, walking pneumonia, boils, gangrene, and strep throat. *Bacillus thuringiensis* produces a toxin that is one of the most important insecticides currently used in farming. Species in the lactic acid bacteria group are used to ferment milk products into yogurt or cheese (**Figure 28.15**). Species in this group are important components of soil, where they speed the decomposition of dead plants, animals, and fungi.

Lactobacillus bulgaricus (rods) and *Streptococcus thermophilus*

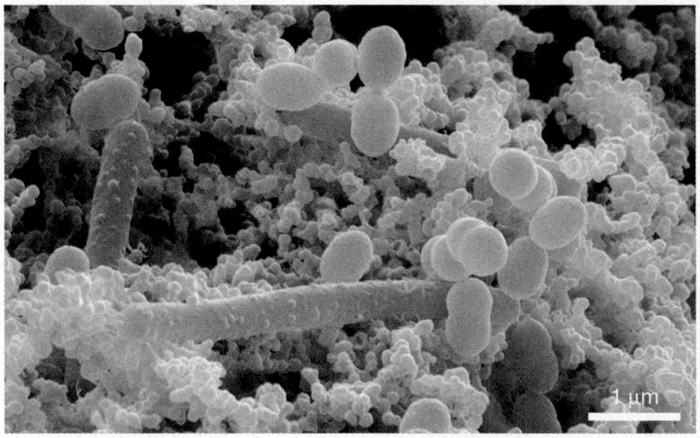

FIGURE 28.15 Firmicutes in Yogurt. (The cells in this micrograph have been colorized—see **BioSkills 10** in Appendix A.)

Bacteria > Spirochaetes (Spirochetes)

The spirochetes are one of the smaller bacterial phyla in terms of numbers of species: only 13 genera and a total of 62 species have been described to date. Recent analyses suggest that the spirochete lineage branched near the base of the bacterial phylogenetic tree.

Morphological diversity Spirochetes are distinguished by their unique corkscrew shape and flagella (**Figure 28.16**). Instead of extending into the water surrounding the cell, spirochete flagella are contained within a structure called the outer sheath, which surrounds the cell. When the flagella beat, the cell lashes back and forth and swims forward. ✔**You should be able to mark the origin of the spirochete flagellum on Figure 28.6.**

Metabolic diversity Most spirochetes manufacture ATP via fermentation. The substrate used in fermentation varies among species and may consist of sugars, amino acids, starch, or the pectin found in plant cell walls. A spirochete that lives only in the hindgut of termites can fix nitrogen.

Treponema pallidum

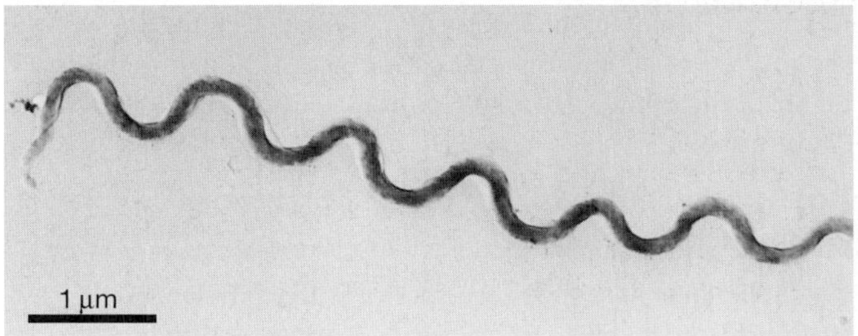

FIGURE 28.16 Spirochetes Are Corkscrew-Shaped Cells Inside an Outer Sheath.

Human and ecological impacts The sexually transmitted disease syphilis is caused by a spirochete. So is Lyme disease, which is transmitted to humans by deer ticks. Spirochetes are extremely common in freshwater and marine habitats; many live only under anaerobic conditions.

Bacteria > Actinobacteria

Actinobacteria are sometimes called the "high-GC Gram positives" because (1) their cell-wall material appears purple when treated with the Gram stain—meaning that they have a peptidoglycan-rich cell wall and lack an outer membrane—and (2) their DNA contains a relatively high percentage of guanine and cytosine. In some species, G and C represent over 75 percent of the bases present. Over 1100 species have been described to date (**Figure 28.17**). **✔You should be able to mark the origin of the high-GC genome in Actinobacteria on Figure 28.6.**

Morphological diversity Cell shape varies from rods to filaments. Many of the soil-dwelling species are found as chains of cells that form extensive branching filaments called **mycelia**.

Metabolic diversity Many are heterotrophs that use an array of organic compounds as electron donors and oxygen as an electron acceptor. There are a handful of parasitic species. Like other parasites, they get most of their nutrition from host organisms.

Human and ecological impacts Over 500 distinct antibiotics have been isolated from species in the genus *Streptomyces*; 60 of these—including streptomycin, neomycin, tetracycline, and erythromycin—are now actively prescribed to treat diseases in humans or domestic livestock. Tuberculosis and leprosy are caused by members of the Actinobacteria. One actinobacterium species is critical to the manufacture of Swiss cheese. Species in the genus *Streptomyces* and *Arthrobacter* are abundant in soil and are vital as decomposers of dead plant and animal material. Some species in these genera live in association with plant roots and fix nitrogen; others can break down toxins such as herbicides, nicotine, and caffeine.

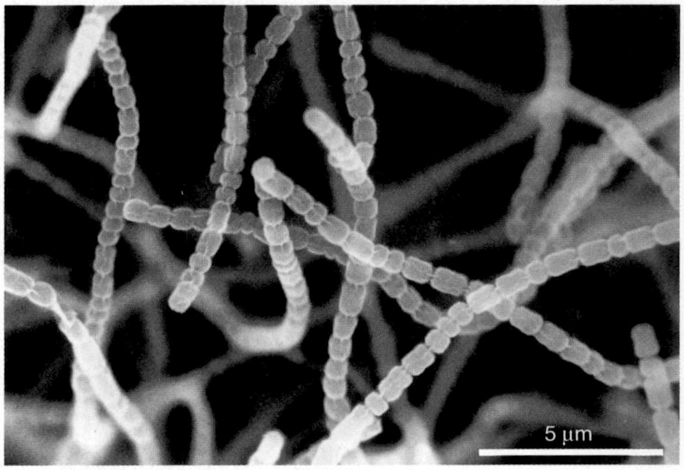

FIGURE 28.17 A *Streptomyces* Species That Produces the Antibiotic Streptomycin.

Bacteria > Chlamydiae

In terms of numbers of species living today, Chlamydiae may be the smallest of all major bacterial lineages. Although the group is highly distinct phylogenetically, only 13 species are known. All are Gram-negative.

Morphological diversity Chlamydiae are spherical. They are tiny, even by bacterial standards.

Metabolic diversity All known species in this phylum live as parasites *inside* host cells and are termed **endosymbionts** ("inside-together-living"). Chlamydiae acquire almost all of their nutrition from their hosts. In **Figure 28.18**, the chlamydiae have been colored red; the animal cells that they live in are colored brown. **✔You should be able to mark the origin of the endosymbiotic lifestyle in this lineage on Figure 28.6.** (The endosymbiotic lifestyle has also arisen in other bacterial lineages, independently of Chlamydiae.)

Human and ecological impacts *Chlamydia trachomatis* infections are the most common cause of blindness in humans. When the same organism is transmitted from person to person via sexual intercourse, it can cause serious urogenital tract infections. One species causes epidemics of a pneumonia-like disease in birds.

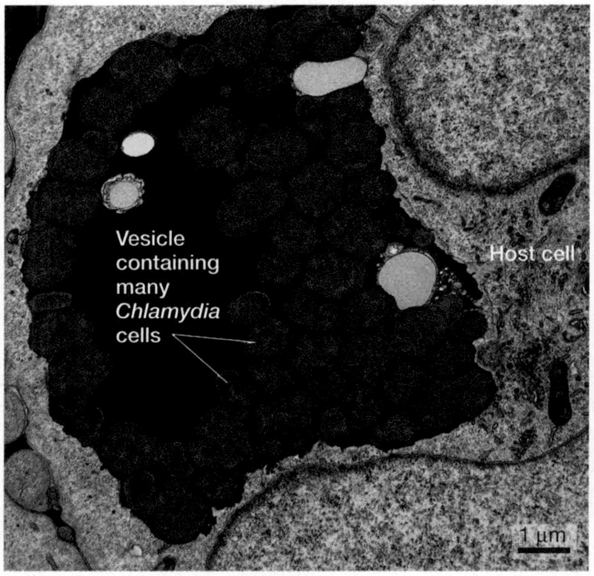

FIGURE 28.18 Chlamydiae Live Only inside Animal Cells.

Bacteria > Cyanobacteria

The cyanobacteria were formerly known as the "blue-green algae"—even though algae are eukaryotes. Only about 80 species of cyanobacteria have been described to date, but they are among the most abundant organisms on Earth. In terms of total mass, cyanobacteria dominate the surface waters in many marine and freshwater environments.

Morphological diversity Cyanobacteria may be found as independent cells, in chains that form filaments (**Figure 28.19**), or in the loose aggregations of individual cells called colonies. The shape of colonies varies from flat sheets to ball-like clusters of cells.

Metabolic diversity All perform oxygenic photosynthesis; many can also fix nitrogen. Because cyanobacteria can synthesize virtually every molecule they need, they can be grown in culture media that contain only CO_2, N_2, H_2O, and a few mineral nutrients. ✔You should be able to mark the origin of oxygenic photosynthesis on Figure 28.6.

Human and ecological impacts If cyanobacteria are present in high numbers, their waste products can make drinking water smell bad. Some species release molecules called microcystins that are toxic to plants and animals. Cyanobacteria were responsible for the origin of the oxygen atmosphere on Earth. Today they still produce much of the oxygen and nitrogen and many of the organic compounds that feed other organisms in freshwater and marine environments. A few species live in association with fungi, forming lichens.

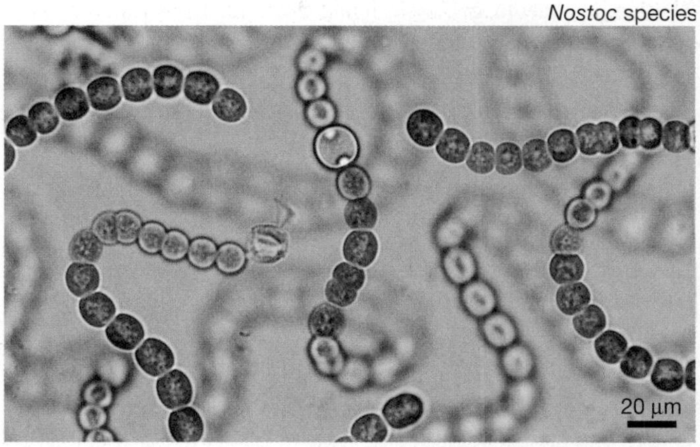

FIGURE 28.19 Cyanobacteria Contain Chlorophyll and Are Green.

Bacteria > Proteobacteria

The approximately 1200 species of proteobacteria form five major subgroups, designated by the Greek letters α (alpha), β (beta), γ (gamma), δ (delta), and ϵ (epsilon). Because they are so diverse in their morphology and metabolism, the lineage is named after the Greek god Proteus, who could assume many shapes.

Morphological diversity Proteobacterial cells can be rods, spheres, or spirals. Some form stalks (**Figure 28.20a**). Some are motile. In one group, cells may move together to form colonies, which then transform into the specialized cell aggregate shown in **Figure 28.20b**. This structure is known as a **fruiting body**. Cells that are surrounded by a durable coating are produced at the tips of fruiting bodies. These spores sit until conditions improve, and then they resume growth.

Metabolic diversity Proteobacteria make a living in virtually every way known to bacteria—except that none perform oxygenic photosynthesis. Various species may perform cellular respiration by using organic compounds, nitrite, methane, hydrogen gas, sulfur, or ammonia as electron donors and oxygen, sulfate, or sulfur as an electron acceptor. Some perform non-oxygenic photosynthesis.

Human and ecological impacts *Escherichia coli* may be the best-studied of all organisms and is a key species in biotechnology (see Chapter 19 and **BioSkills 14** in Appendix A). Pathogenic proteobacteria cause Legionnaire's disease, cholera, food poisoning, dysentery, gonorrhea, Rocky Mountain spotted fever, typhus, ulcers, and diarrhea. *Wolbachia* infections are common in insects and are often transmitted from mothers to offspring via eggs. Biologists use *Agrobacterium* cells to transfer new genes into crop plants. Certain acid-loving species of proteobacteria are used in the production of vinegars. Species in the genus *Rhizobium* (α-proteobacteria) live in association with plant roots and fix nitrogen. The bdellovibrios are a group in the δ-proteobacteria that are predators—they drill into other proteobacterial cells and digest them. Because some species use nitrogen-containing compounds as electron acceptors, proteobacteria are critical players in the cycling of nitrogen atoms through terrestrial and aquatic ecosystems.

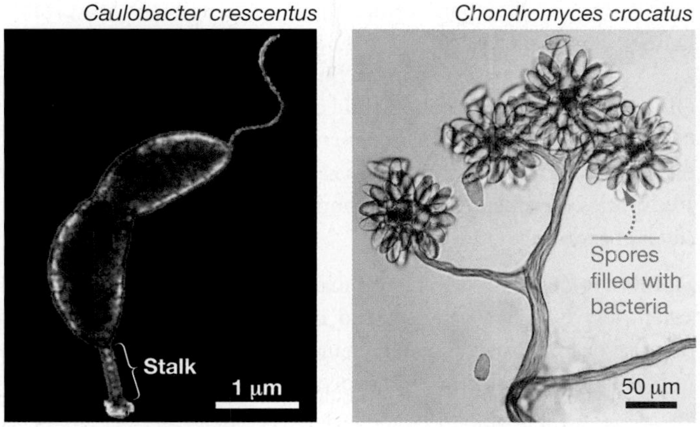

FIGURE 28.20 Some Proteobacteria Grow on Stalks or Form Fruiting Bodies. The stalked bacterium (a) has been colorized.

Archaea > Crenarchaeota

The Crenarchaeota got their name because they are considered similar to the oldest archaeans; the word root *cren-* refers to a source or fount. Although only 37 species have been named to date, it is virtually certain that thousands are yet to be discovered.

Morphological diversity Crenarchaeota cells can be shaped like filaments, rods, discs, or spheres. One species that lives in extremely hot habitats has a tough cell wall consisting solely of glycoprotein.

Metabolic diversity Depending on the species, cellular respiration can involve organic compounds, sulfur, hydrogen gas, ammonia, or Fe^{2+} ions as electron donors and oxygen, nitrate, sulfate, sulfur, carbon dioxide, or Fe^{3+} ions as electron acceptors. Some species make ATP exclusively through fermentation pathways.

Human and ecological impacts Crenarchaeota have yet to be used in the manufacture of commercial products. In certain extremely hot, high-pressure, cold, or acidic environments, crenarchaeota may be the only life-form present (**Figure 28.21**). Acid-loving species thrive in habitats with pH 1–5; some species are found in ocean sediments at depths ranging from 2500 to 4000 m below the surface.

Sulfolobus species
0.5 μm

FIGURE 28.21 Some Crenarchaeota Live in Sulfur-Rich Hot Springs. The cells in the micrograph have been colorized to make them more visible.

Archaea > Euryarchaeota

The Euryarchaeota are aptly named, because the word root *eury-* means "broad." Members of this phylum live in every conceivable habitat. Some species are adapted to high-salt habitats with pH 11.5—almost as basic as household ammonia (**Figure 28.22**). Other species are adapted to acidic conditions with a pH as low as 0. Species in the genus *Methanopyrus* live near hot springs called black smokers that are 2000 m (over 1 mile) below sea level. About 170 species have been identified thus far, and more are being discovered each year.

Morphological diversity Euryarchaeota cells can be spherical, filamentous, rod shaped, disc shaped, or spiral. Rod-shaped cells may be short or long or arranged in chains. Spherical cells can be found in ball-like aggregations. Some species have several flagella. Some species lack a cell wall; others have a cell wall composed entirely of glycoproteins.

Metabolic diversity The group includes a variety of methane-producing species. These methanogens can use up to 11 different organic compounds as electron acceptors during cellular respiration; all produce CH_4 as a by-product of respiration. In other species of Euryarchaeota, cellular respiration is based on hydrogen gas or Fe^{2+} ions as electron donors and nitrate or sulfate as electron acceptors. Species that live in high-salt environments can use the molecule retinal—which is responsible for light reception in your eyes—to capture light energy and perform photosynthesis.

Human and ecological impacts Species in the genus *Ferroplasma* live in piles of waste rock near abandoned mines. As a by-product of metabolism, they produce acids that drain into streams and pollute them. Methanogens live in the soils of swamps and the guts of mammals (including yours). They are responsible for adding about 2 billion tons of methane to the atmosphere each year. A methanogen in this phylum was also recently implicated in gum disease.

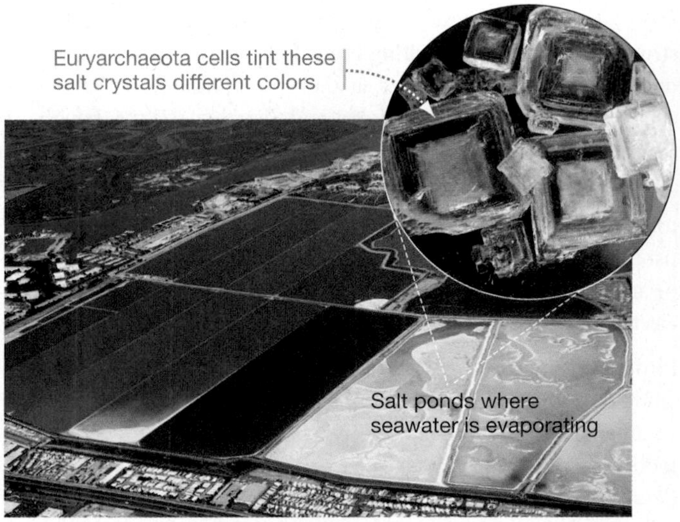
Euryarchaeota cells tint these salt crystals different colors
Salt ponds where seawater is evaporating

FIGURE 28.22 Some Euryarchaeota Live in High-Salt Habitats.

CHAPTER 28 REVIEW

For media, go to the study area at www.masteringbiology.com

Summary of Key Concepts

Bacteria and archaea may be tiny, but they are ancient, diverse, abundant, and ubiquitous. A few bacteria cause infectious disease; some are effective at cleaning up pollution.

- Bacteria and archaea are the most abundant organisms on Earth and are found in every habitat that has been sampled.
- Bacteria cause some of the most dangerous human diseases, including plague, syphilis, botulism, cholera, and tuberculosis.
- Enrichment cultures are used to grow large numbers of bacterial or archaeal cells that thrive under specified conditions.
- Biologists can study bacteria and archaea that cannot be cultured by extracting DNA directly from an environment, sequencing target genes, and using the data to place the organisms present on the tree of life.

✔You should be able to explain the difference between a bacterium, an archaeon, and a eukaryote.

MB **Web Activity** The Tree of Life

As a group, bacteria and archaea live in virtually every habitat known and use remarkably diverse types of compounds in cellular respiration and fermentation. Although cell size is usually small and overall morphology is relatively simple, the chemistry they can do is extremely sophisticated.

- Metabolic diversity and complexity are the hallmarks of the bacteria and archaea, just as morphological diversity and complexity are the hallmarks of the eukaryotes.
- Among bacteria and archaea, a wide array of inorganic or organic compounds with high potential energy can serve as electron donors in cellular respiration, and a wide variety of inorganic or organic molecules with low potential energy may serve as electron acceptors. Dozens of distinct organic compounds are fermented.
- Photosynthesis is widespread in bacteria. In cyanobacteria, water is used as a source of electrons and oxygen gas is generated as a by-product. But in other species, the electron excited by photon capture comes from a source other than water, and no oxygen is produced.
- To acquire building-block molecules containing carbon-carbon bonds, some species use the enzymes of the Calvin cycle to reduce CO_2. But three other biochemical pathways found in bacteria and archaea can also reduce simple organic compounds to sugars or carbohydrates.

✔You should be able to explain why species that release H_2S as a by-product and that use H_2S as an electron donor often live side by side.

Bacteria and archaea play a key role in ecosystems. Photosynthetic bacteria were responsible for the evolution of the oxygen atmosphere; bacteria and archaea cycle nutrients through terrestrial and aquatic environments.

- Because of their metabolic diversity, bacteria and archaea play a large role in global change and nutrient cycling.
- Nitrogen-fixing species provide nitrogen in forms that can be used by many other species, including plants and animals.

✔You should be able to explain what the composition of the atmosphere and what the nitrogen cycle would be like if bacteria and archaea did not exist.

Questions

✔TEST YOUR KNOWLEDGE

Answers are available in Appendix B

1. How do molecules that function as electron donors and those that function as electron acceptors differ?
   a. Electron donors are almost always organic molecules; electron acceptors are always inorganic.
   b. Electron donors are almost always inorganic molecules; electron acceptors are always organic.
   c. Electron donors have relatively high potential energy; electron acceptors have relatively low potential energy.
   d. Electron donors have relatively low potential energy; electron acceptors have relatively high potential energy.
2. What do some photosynthetic bacteria use as a source of electrons instead of water?
   a. oxygen (O_2)
   b. hydrogen sulfide (H_2S)
   c. organic compounds (e.g., CH_3COO^-)
   d. nitrate (NO_3^-)
3. What is distinctive about the chlorophylls found in different photosynthetic bacteria?
   a. their membranes
   b. their role in acquiring energy
   c. their role in carbon fixation
   d. their absorption spectra
4. What are organisms called that use inorganic compounds as electron donors in cellular respiration?
   a. phototrophs
   b. heterotrophs
   c. organotrophs
   d. lithotrophs

5. What has direct sequencing allowed researchers to do for the first time?
   a. sample organisms from an environment and grow them under defined conditions in the lab
   b. sample organisms from an environment and sequence the entire genomes present
   c. study organisms that cannot be cultured (grown in the lab)
   d. identify important morphological differences among species

6. Koch's postulates outline the requirements for which of the following?
   a. showing that an organism is autotrophic
   b. showing that a bacterium's cell wall lacks an outer membrane and consists primarily of peptidoglycan
   c. showing that an organism causes a particular disease
   d. showing that an organism can use a particular electron donor and electron acceptor

✔ TEST YOUR UNDERSTANDING

Answers are available in Appendix B

1. Biologists often use the term energy source as a synonym for "electron donor." Why?
2. The text claims that the tremendous ecological diversity of bacteria and archaea is possible because of their impressive metabolic diversity. Do you agree with this statement? Why or why not?
3. How is it possible for direct sequencing to identify species that no one has ever seen?
4. The text claims that the evolution of an oxygen atmosphere paved the way for increasingly efficient cellular respiration and higher growth rates in organisms. Explain.
5. Look back at Table 28.5 and note that the by-products of respiration in some organisms are used as electron donors or acceptors by other organisms. In the table, draw lines between the dual-use molecules listed in the "Electron Donor," "Electron Acceptor," and "By-Products" columns.
6. Explain the statement, "Prokaryotes are a paraphyletic group."

✔ APPLYING CONCEPTS TO NEW SITUATIONS

Answers are available in Appendix B

1. The researchers who observed that magnetite was produced by bacterial cultures from the deep subsurface carried out a follow-up experiment. These biologists treated some of the cultures with a drug that poisons the enzymes involved in electron transport chains. In cultures where the drug was present, no more magnetite was produced. Does this result support or undermine their hypothesis that the bacteria in the cultures perform cellular respiration? Explain your reasoning.
2. *Streptococcus mutans* obtains energy by oxidizing sucrose. This bacterium is abundant in the mouths of Western European and North American children and is a prominent cause of cavities. The organism is virtually absent in children from East Africa, where tooth decay is rare. Propose a hypothesis to explain this observation. Outline the design of a study that would test your hypothesis.
3. Suppose that you've been hired by a firm interested in using bacteria to clean up organic solvents found in toxic waste dumps. Your new employer is particularly interested in finding cells that are capable of breaking a molecule called benzene into less toxic compounds. Where would you go to look for bacteria that can metabolize benzene as an energy or carbon source? How would you design an enrichment culture capable of isolating benzene-metabolizing species?
4. Would you predict that disease-causing bacteria, such as those listed in Table 28.2, obtain energy from light, organic molecules, or inorganic molecules? Explain your answer.

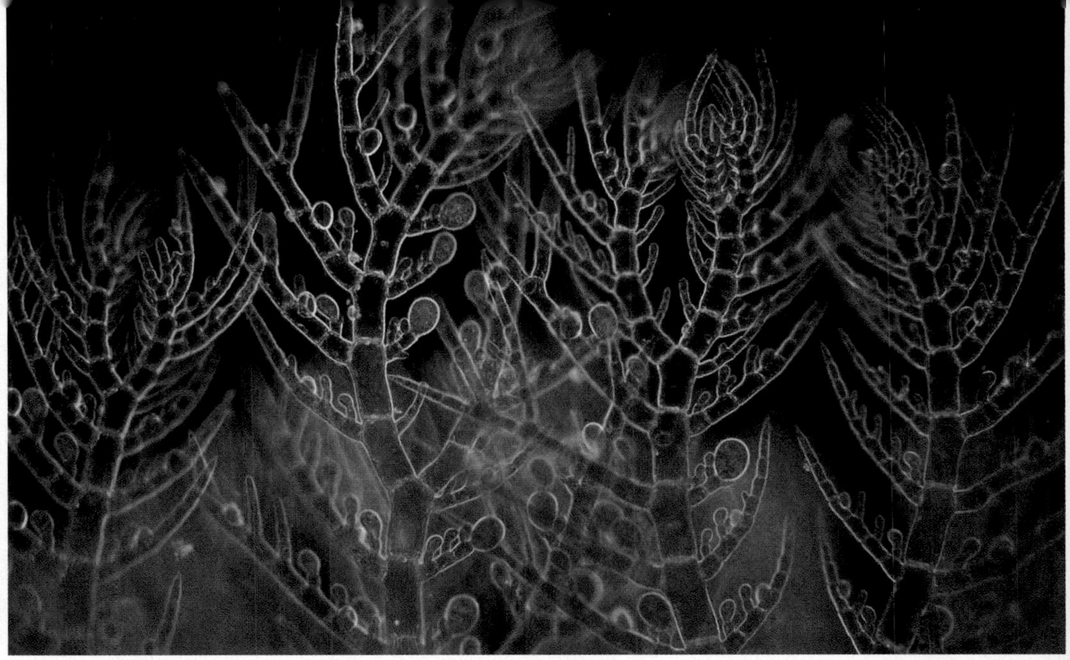

The red algae shown here live attached to rocks in shallow ocean waters. The algae and other species featured in this chapter are particularly abundant in the world's oceans.

Protists 29

This chapter introduces the third domain on the tree of life: the **Eukarya**. Eukaryotes range from single-celled organisms that are the size of bacteria to sequoia trees and blue whales. The largest and most morphologically complex organisms on the tree of life—algae, plants, fungi, and animals—are eukaryotes.

Although eukaryotes are astonishingly diverse, they share fundamental features that distinguish them from bacteria and archaea:

- The nuclear envelope is a synapomorphy that defines the Eukarya.
- Compared to bacteria and archaea, most eukaryotic cells are large, have many more organelles, and have a much more extensive system of structural proteins called the cytoskeleton.
- Multicellularity is rare in bacteria and unknown in archaea, but has evolved multiple times in eukaryotes.
- Bacteria and archaea reproduce asexually by fission; many eukaryotes reproduce asexually via mitosis.
- Many eukaryotes undergo meiosis and reproduce sexually.

One of this chapter's fundamental goals is to explore how these morphological innovations—features like the nuclear envelope and organelles—evolved. Another goal is to analyze how morphological innovations allowed eukaryotes to pursue novel ways of performing basic life tasks such as feeding, moving, and reproducing.

In introducing the Eukarya, this chapter focuses on an informal grab bag of lineages known as the protists. The term **protist** refers to all eukaryotes that are not green plants, fungi, or animals. Protist lineages are colored orange in **Figure 29.1** on page 520.

Protists do not make up a monophyletic group. Instead, they refer to a **paraphyletic group**—they represent some, but not all, of the descendants of a single common ancestor. To use the vocabulary introduced in Chapter 27, no synapomorphies

KEY CONCEPTS

- Protists are a paraphyletic grouping that includes all eukaryotes except the land plants, fungi, and animals. Biologists study them because they are important medically, ecologically, and evolutionarily.
- Key morphological innovations occurred as protists diversified: the nuclear envelope, multicellularity, and an array of structures that function in support and protection. In addition, the mitochondrion and chloroplast arose by endosymbiosis.
- Protists vary widely in terms of how they obtain food. Many species are photosynthetic; while others obtain carbon compounds by ingesting food or parasitizing other organisms.
- Protists vary widely in terms of how they reproduce. Sexual reproduction evolved in protists, and many protist species can reproduce both sexually and asexually.

✔ When you see this checkmark, stop and test yourself. Answers are available in Appendix B.

define the protists. There is no trait that is found in protists but no other organisms.

By definition, then, the protists are a diverse lot. The common feature among protists is that they tend to live in environments where they are surrounded by water (**Figure 29.2**). Most plants, fungi, and animals are terrestrial, but protists are found in wet soils, aquatic habitats, or the bodies of other organisms—including, perhaps, you.

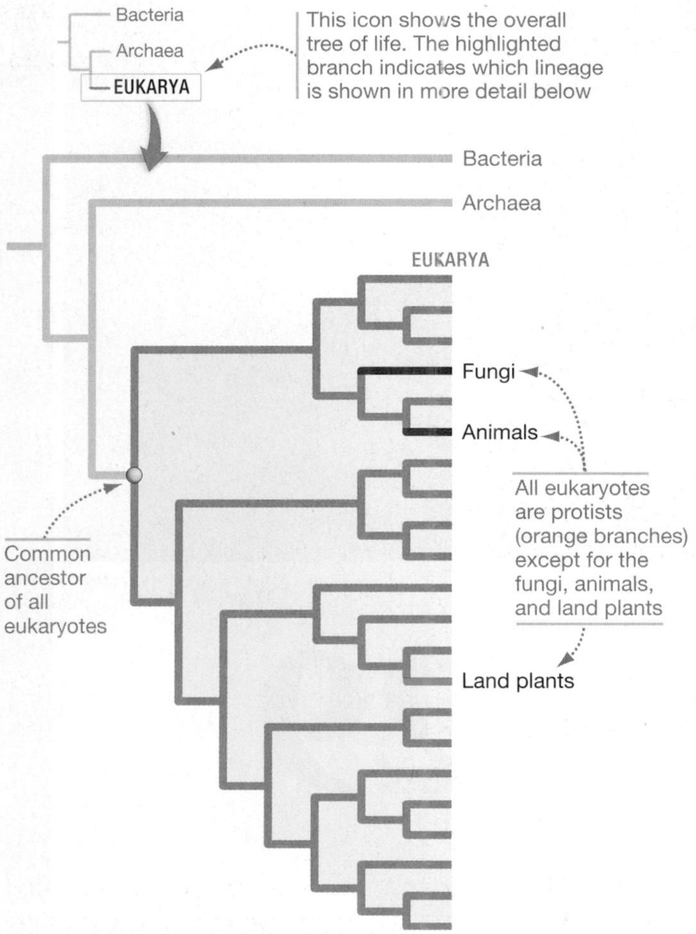

This icon shows the overall tree of life. The highlighted branch indicates which lineage is shown in more detail below

All eukaryotes are protists (orange branches) except for the fungi, animals, and land plants

FIGURE 29.1 Protists Are Paraphyletic. The group called protists includes some, but not all, descendants of a single common ancestor.

29.1 Why Do Biologists Study Protists?

Biologists study protists for three reasons, in addition to their intrinsic interest: (1) they are important medically, (2) they are important ecologically, and (3) they are critical to understanding the evolution of plants, fungi, and animals. The remainder of the chapter will focus on why protists are interesting in their own right and how they evolved; here let's consider their impact on the environment and human health.

Impacts on Human Health and Welfare

The most spectacular crop failure in history, the Irish potato famine, was caused by a protist. In 1845 most of the 3 million acres that had been planted to grow potatoes in Ireland became infested with *Phytophthora infestans*—a parasite that belongs to a lineage of protists called Oomycota (pronounced *oo-oh-my-COTE-ah*). Potato tubers that were infected with *P. infestans* rotted in the fields or in storage.

As a result of crop failures in Ireland for two consecutive years, an estimated 1 million people out of a population of less

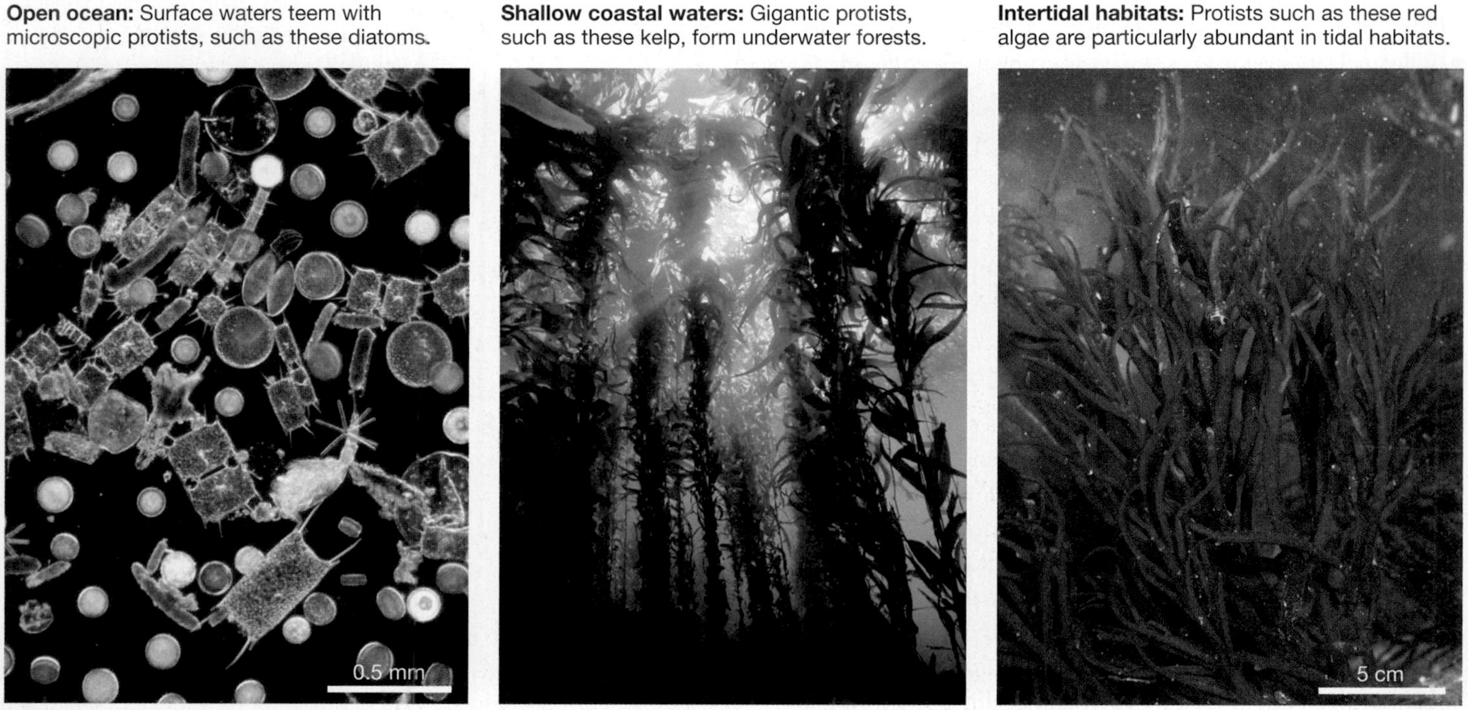

Open ocean: Surface waters teem with microscopic protists, such as these diatoms.
Shallow coastal waters: Gigantic protists, such as these kelp, form underwater forests.
Intertidal habitats: Protists such as these red algae are particularly abundant in tidal habitats.
0.5 mm
5 cm

FIGURE 29.2 Protists Are Particularly Abundant in Aquatic Environments.

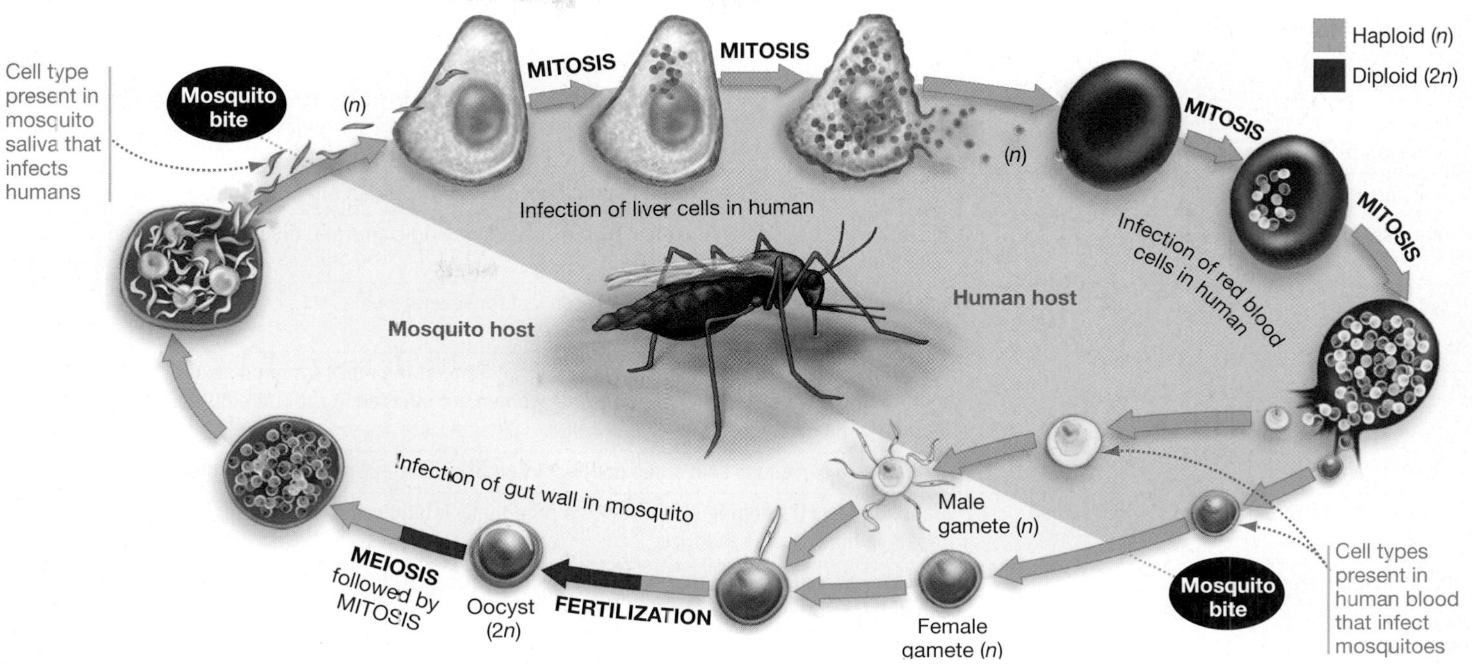

FIGURE 29.3 ***Plasmodium*** **Lives in Mosquitoes and in Humans, where It Causes Malaria.** Over the course of its life cycle, *Plasmodium falciparum* develops into a series of distinct cell types. Each type is specialized for infecting a different host cell in mosquitoes or humans. In humans, it infects and kills liver cells and red blood cells, contributing to anemia and high fever. In mosquitoes, the protist lives in the gut and salivary glands.

than 9 million died of starvation or starvation-related illnesses. Several million others emigrated. Many people of Irish heritage living in North America, New Zealand, and Australia trace their ancestry to relatives who left Ireland to evade the famine. As devastating as the potato famine was, however, it does not begin to approach the misery caused by the protist *Plasmodium.*

MALARIA Physicians and public health officials point to three major infectious diseases that are currently afflicting large numbers of people worldwide: tuberculosis, HIV, and malaria. Tuberculosis is caused by a bacterium and was introduced in Chapter 24; HIV is caused by a virus and is analyzed in Chapter 35. Malaria is caused by a protist—specifically, by species in the eukaryotic lineage called Apicomplexa.

Malaria ranks as the world's most chronic public health problem. In India alone, over 30 million people each year suffer from debilitating fevers caused by malaria. At least 300 million people worldwide are sickened by it each year, and over 1 million die from the disease annually. The toll is equivalent to eight 747s, loaded with passengers, crashing every day. Most of the dead are children of preschool age.

Four species of the protist *Plasmodium* are capable of parasitizing humans. Infections start when *Plasmodium* cells enter a person's bloodstream during a mosquito bite. As **Figure 29.3** shows, *Plasmodium* initially infects liver cells; later, some of the *Plasmodium* cells change into a distinctive cell type that infects the host's red blood cells as well. The *Plasmodium* cells multiply inside the host cells, which are killed as parasite cells exit to infect additional liver cells or red blood cells.

If infected red blood cells are transferred to a mosquito during a bite, they differentiate to form gametes. Inside the mosquito, gametes fuse to form a diploid cell called an oocyst, which undergoes meiosis. The haploid cells that result from meiosis can infect a human when the mosquito bites again.

Although *Plasmodium* is arguably the best studied of all protists, researchers have still not been able to devise effective and sustainable measures to control it.

- Natural selection has favored mosquito strains that are resistant to the insecticides that have been sprayed in their breeding habitats, in attempts to control malaria's spread.
- *Plasmodium* has evolved resistance to most of the drugs used to control its growth in infected people.
- Efforts to develop a vaccine against *Plasmodium* have been fruitless to date, in part because the parasite evolves so quickly (flu virus and HIV pose similar problems; see Chapter 49).

Unfortunately, malaria is not the only important human disease caused by protists. **Table 29.1** on page 522 lists protists that have been the cause of human suffering and economic losses.

HARMFUL ALGAL BLOOMS When a unicellular species experiences rapid population growth and reaches high densities in an aquatic environment, it is said to "bloom." Unfortunately, a handful of the many protist species involved in blooms can be harmful.

Harmful algal blooms are usually due to photosynthetic protists called dinoflagellates. Certain dinoflagellates synthesize toxins to protect themselves from predation by small animals called copepods. Because toxin-producing dinoflagellates have high

TABLE 29.1 **Human Health Problems Caused by Protists**

Species	Disease
Four species of *Plasmodium*, primarily *P. falciparum* and *P. vivax*	Malaria has the potential to affect 40 percent of the world's total population.
Toxoplasma	Toxoplasmosis may cause eye and brain damage in infants and in AIDS patients.
Many species of dinoflagellates	Toxins released during harmful algal blooms accumulate in clams and mussels and poison people if eaten.
Giardia	Diarrhea due to giardiasis (beaver fever) can last for several weeks.
Trichomonas	Trichomoniasis is a reproductive tract infection and one of the most common sexually transmitted diseases. About 2 million young women are infected in the United States each year; some of them become infertile.
Leishmania	Leishmaniasis can cause skin sores or affect internal organs—particularly the spleen and liver.
Trypanosoma gambiense and *T. rhodesiense*	Trypanosomiasis ("sleeping sickness") is a potentially fatal disease transmitted through bites from tsetse flies. Occurs in Africa.
Trypanosoma cruzi	Chagas disease affects 16–18 million people and causes 50,000 deaths annually, primarily in South and Central America.
Entamoeba histolytica	Amoebic dysentery results from severe infections.
Phytophthora infestans	An outbreak of this protist wiped out potato crops in Ireland in 1845–1847, causing famine.

concentrations of accessory pigments called xanthophylls, their blooms can sometimes discolor seawater (**Figure 29.4**).

Algal blooms can be harmful to people because clams and other shellfish filter photosynthetic protists out of the water as food. During a bloom, high levels of toxins can build up in the flesh of these shellfish. Typically, the shellfish themselves are not harmed. But if a person eats contaminated shellfish, several types of poisoning can result.

Paralytic shellfish poisoning, for example, occurs when people eat shellfish that have fed heavily on protists that synthesize poisons called saxitoxins. Saxitoxins block ion channels that have to open for electrical signals to travel through nerve cells (see Chapter 45). In humans, high dosages of saxitoxins cause unpleasant symptoms such as prickling sensations in the mouth or even life-threatening symptoms such as muscle weakness and paralysis.

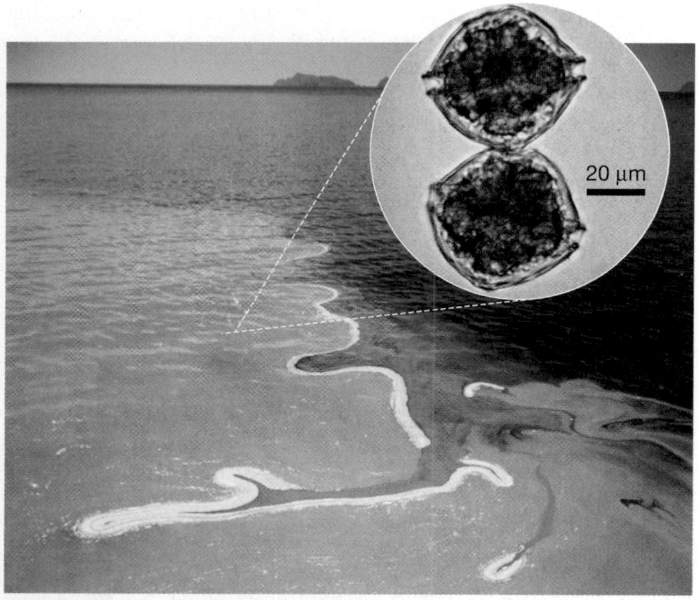

FIGURE 29.4 Harmful Algal Blooms Are Caused by Dinoflagellates.

Ecological Importance of Protists

As a whole, the protists represent just 10 percent of the total number of named eukaryote species. Although the species diversity of protists may be relatively low, their abundance is extraordinarily high. The number of individual protists found in some habitats is astonishing.

- One milliliter of pond water can contain well over 500 single-celled protists that swim with the aid of flagella.
- Under certain conditions, dinoflagellates can reach concentrations of 60 million cells per liter of seawater.

Why is this important?

PROTISTS PLAY A KEY ROLE IN AQUATIC FOOD CHAINS Photosynthetic protists take in carbon dioxide from the atmosphere and reduce, or "fix," it to form sugars or other organic compounds with high potential energy (Chapter 10). Photosynthesis transforms some of the energy in sunlight into chemical energy that organisms can use to grow and produce offspring.

Species that produce chemical energy in this way are called **primary producers**. Diatoms, for example, are photosynthetic protists that rank among the leading primary producers in the oceans, simply because they are so abundant. Primary production in the ocean represents almost half of the total carbon dioxide that is fixed on Earth.

Diatoms and other small organisms that live near the surface of oceans or lakes are called **plankton**. The sugars and other organic compounds produced by photosynthetic plankton are the basis of food chains in freshwater and marine environments.

A **food chain** describes nutritional relationships among organisms, and thus how chemical energy flows within ecosystems. In this case, photosynthetic protists and other primary producers are eaten by primary consumers, many of which are protists. Primary consumers are eaten by fish, shellfish, and other secondary consumers, which in turn are eaten by tertiary consumers—whales, squid, and large fish (such as tuna).

Many of the species at the base of food chains in aquatic environments are protists. Without protists, most food chains in freshwater and marine habitats would collapse.

COULD PROTISTS HELP REDUCE GLOBAL WARMING? Carbon dioxide levels in the atmosphere are increasing rapidly because humans are burning fossil fuels and forests (see Chapter 54). Carbon dioxide traps heat that is radiating from Earth, so high CO_2 levels in the atmosphere contribute to global warming—an issue that many observers consider today's most pressing environmental problem.

The carbon atoms in carbon dioxide molecules move to organisms in the soil or the ocean and then back to the atmosphere, in what researchers call the **global carbon cycle**. To reduce global warming, researchers are trying to figure out ways to decrease carbon dioxide concentrations in the atmosphere and increase the amount of carbon stored in terrestrial and marine environments.

To understand how this might be done, consider the marine carbon cycle diagrammed in **Figure 29.5**. The cycle starts when CO_2 from the atmosphere dissolves in water and is taken up by primary producers called phytoplankton and converted to organic matter. The phytoplankton are eaten by primary consumers, die and are consumed by decomposers or scavengers, or die and sink to the bottom of the ocean. There they may enter one of two long-lived repositories:

1. ***Sedimentary rocks*** Several lineages of protists have shells made of calcium carbonate ($CaCO_3$). When these shells rain down from the ocean surface and settle in layers at the bottom, the deposits that result are compacted by the weight of the water and by sediments accumulating above them. Eventually the deposits turn into rock. The limestone used to build the pyramids of Egypt consists of protist shells.
2. ***Petroleum*** Although the process of petroleum (oil) formation is not well understood, it begins with accumulations of dead bacteria, archaea, and protists at the bottom of the ocean.

Recent experiments have shown that the carbon cycle speeds up when habitats in the middle of the ocean are fertilized with iron. Iron is a critical component of the electron transport chains responsible for photosynthesis and respiration, but it is in particularly short supply in the open ocean. After iron is added to ocean waters, it is not uncommon to see populations of protists and other primary producers increase by a factor of 10.

Some researchers hypothesize that when these blooms occur, the amount of carbon that rains down into carbon sinks in the form of shells and dead cells may increase. If so, then fertilizing the ocean to promote blooms might be an effective way to reduce CO_2 concentrations in the atmosphere.

The effectiveness of iron fertilization is hotly debated, however. Fertilizing the ocean with iron might lead to large accumulations of dead organic matter and the formation of anaerobic dead zones like those described in Chapter 28. But if further research shows that iron fertilization is safe and effective, it could be added to the list of approaches to reduce global warming.

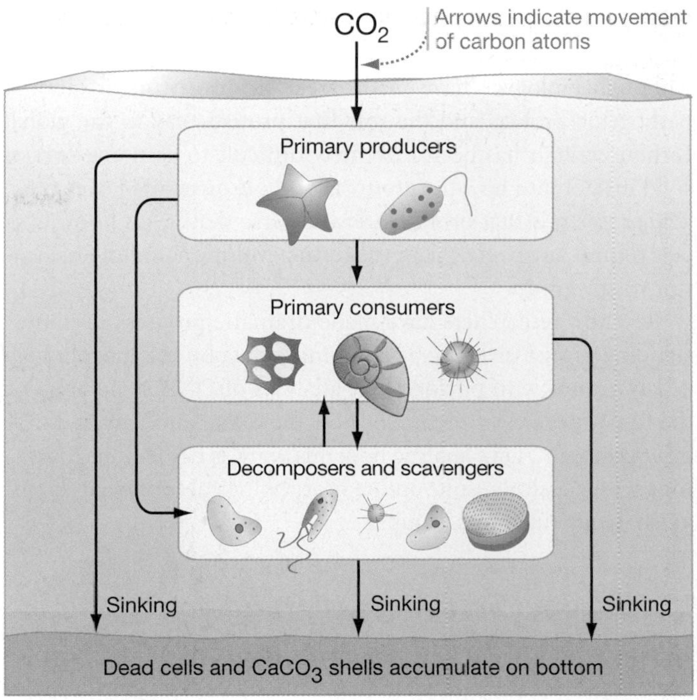

FIGURE 29.5 Protists Play a Key Role in the Marine Carbon Cycle. At the surface, carbon atoms tend to shuttle quickly among organisms. But if carbon atoms sink to the bottom of the ocean in the form of shells or dead cells, they may be locked up for long periods in carbon sinks. (The bottom of the ocean may be miles below the surface.)

CHECK YOUR UNDERSTANDING

If you understand that . . .

- Malaria is caused by a protist that lives in mosquitoes and in humans in different parts of its life cycle.
- Harmful algal blooms are caused by protists that produce a toxin as a defense against predation.
- Protists are key primary producers in aquatic environments. As a result, they play a key role in the global carbon cycle.

You should be able to . . .

1. Explain why public health workers are promoting the use of insecticide-treated sleeping nets as a way of reducing malaria.
2. Make a flowchart showing the chain of events that would start with massive iron fertilization and end with large deposits of carbon-containing compounds on the ocean floor.

Answers are available in Appendix B.

29.2 How Do Biologists Study Protists?

Although biologists have made great strides in understanding pathogenic protists and the role that protists play in the global carbon cycle, it has been extremely difficult to gain any sort of solid insight into how the group as a whole diversified over time. The problem is that protists are so diverse that it has been difficult to find any overall patterns in the evolution and diversification of the group.

Recently, researchers have made dramatic progress in understanding protist diversity by combining data on the morphology of key groups with phylogenetic analyses of DNA sequence. For the first time, a clear picture of how the Eukarya diversified may be within sight. Let's analyze how this work is being done, beginning with classical results on the morphological traits that distinguish major eukaryote groups.

Microscopy: Studying Cell Structure

Using light microscopy (see **BioSkills 10** in Appendix A), biologists were able to identify and name many of the protist species known today. When transmission electron microscopes became available, a major breakthrough in understanding protist diversity occurred: Detailed studies of cell structure revealed that protists could be grouped according to characteristic overall form, according to organelles with distinctive features, or both.

For example, both light and electron microscopy confirmed that the species that caused the Irish potato famine has reproductive cells with an unusual type of **flagellum**. Flagella are organelles that project from the cell and whip back and forth to produce swimming movements (Chapter 7). In reproductive cells of this species, one of the two flagella present has tiny, hollow, hairlike projections. Biologists noted that kelp and other forms of brown algae also have cells with this type of flagellum.

To make sense of these results, researchers interpreted these types of distinctive morphological features as **synapomorphies**—shared, derived traits that distinguish major monophyletic groups (see Chapter 27). Species that have a flagellum with hollow, hairlike projections became known as stramenopiles (literally, "straw-hairs"); the hairs typically have three branches at the tip (**Figure 29.6**).

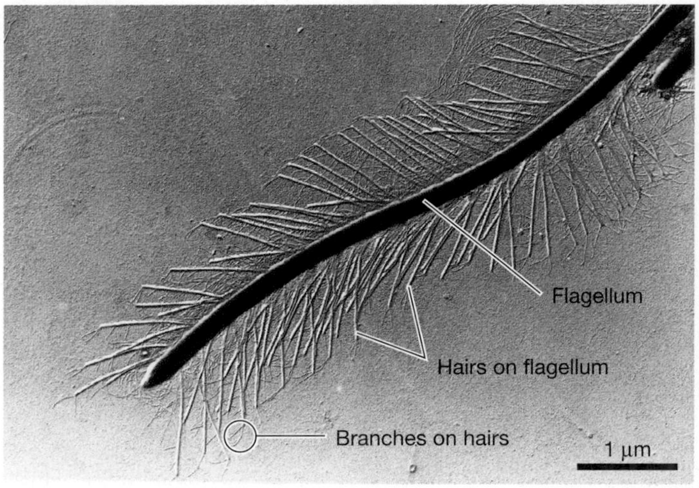

FIGURE 29.6 Species in the Lineage Called Stramenopiles Have a Distinctive Flagellum. The unusual, hollow "hairs" that decorate the flagella of stramenopiles often have three branches at the tip.

In recognizing this group, investigators hypothesized that because an ancestor had evolved a distinctive flagellum, all or most of its descendants also had this trait. The qualifier most is important, because it is not unusual for certain subgroups to lose particular traits over the course of evolution, much as humans are gradually losing fur and tailbones.

Eventually, seven major groups of eukaryotes came to be identified on the basis of diagnostic morphological characteristics. These groups and the synapomorphies that identify them are listed in **Table 29.2**. Note that in almost every case, the synapomorphies that define eukaryotic lineages are changes in structures that protect or support the cell or that influence the

SUMMARY TABLE 29.2 **Major Lineages of Eukaryotes**

Lineage	Distinguishing Morphological Features (synapomorphies)
Amoebozoa	Cells lack cell walls. When portions of the cell extend outward to move the cell, they form large lobes.
Opisthokonta	Reproductive cells have a single flagellum at their base. The cristae inside mitochondria are flat, not tube shaped as in other eukaryotes. (This lineage includes protists as well as the fungi and the animals. Fungi and animals are discussed in detail in Chapters 31 through 34.)
Excavata	Cells have a pronounced "feeding groove" where prey or organic debris is ingested. No functioning mitochondria are present, although genes derived from mitochondria are found in the nucleus.
Plantae	Cells have chloroplasts with a double membrane.
Rhizaria	Cells lack cell walls, although some produce an elaborate shell-like covering. When portions of the cell extend outward to move the cell, they are slender in shape.
Alveolata	Cells have sac-like structures called alveoli that form a continuous layer just under the plasma membrane. Alveoli are thought to provide support.
Stramenopila	If flagella are present, cells usually have two—one of which is covered with hairlike projections.

organism's ability to move or feed. The plants, fungi, and animals analyzed in Chapters 30 through 34 represent subgroups within two of the seven major eukaryotic lineages.

Evaluating Molecular Phylogenies

When researchers began using DNA sequence data to estimate the evolutionary relationships among eukaryotes, the analyses suggested that the seven groups identified on the basis of distinctive morphological characteristics were indeed monophyletic. This was important support for the hypothesis that the distinctive morphological features were shared, derived characters that existed in a common ancestor of each lineage.

The phylogenetic tree in **Figure 29.7** is the current best estimate of the group's evolutionary history. As you read this tree, note that:

- The Amoebozoa and the Opisthokonta—which include fungi and animals—form a monophyletic group called the Unikonta.
- The other five major lineages form a monophyletic group called the Bikonta.
- The Alveolata and Stramenopila form a monophyletic group called the Chromalveolata.

Understanding where the root or base of the tree lies has been more problematic. The latest data suggest that once the first eukaryotic organism evolved, the major split was between unikonts and bikonts—meaning, between eukaryotes that have one flagellum versus two flagella. This finding is tentative and controversial, however, given the data analyzed to date. Researchers continue to work on the issue of placing the root of the Eukarya.

Discovering New Lineages via Direct Sequencing

The effort to refine the phylogeny of the Eukarya is ongoing. But of all the research frontiers in eukaryotic diversity, the most exciting may be the one based on the technique called direct sequencing.

You might recall from Chapter 28 that **direct sequencing** is based on sampling soil or water, analyzing the DNA sequence of specific genes in the sample, and using the data to place the organisms in the sample on a phylogenetic tree. Direct sequencing led to the discovery of previously unknown but major lineages of Archaea. To the amazement of biologists all over the world, the same thing happened when researchers used direct sequencing to survey eukaryotes.

The first direct sequencing studies that focused on eukaryotes were published in 2001. One study sampled organisms at depths from 250 to 3000 m below the surface in waters off Antarctica; another focused on cells at depths of 75 m in the Pacific Ocean, near the equator. Both studies detected a wide array of species that were new to science.

Investigators who examined the samples under the microscope were astonished to find that many of the newly discovered eukaryotes were tiny—from 0.2 μm to 5 μm in diameter. Subsequent work has confirmed the existence of protists that are less than 0.2 μm in diameter. These cells overlap in size with bacteria, which typically range from 0.5 μm to 2 μm in diameter.

Eukaryotic cells are much more variable in size than previously imagined. A whole new world of tiny protists has just been discovered.

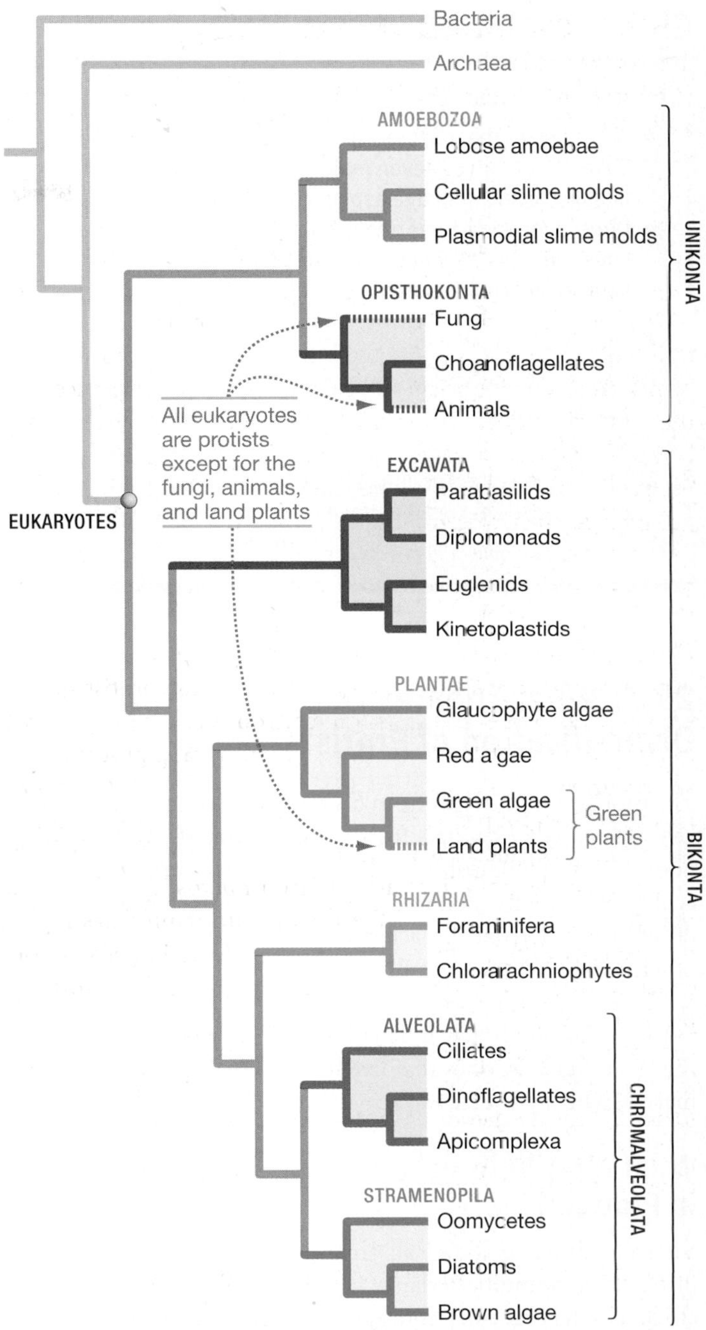

FIGURE 29.7 Phylogenetic Analyses Have Identified Seven Major Lineages of Eukaryotes. This tree shows selected subgroups from the seven major lineages discussed in this chapter. Many other lineages of protists have been identified. Note that kinetoplastids are not discussed in detail elsewhere in the chapter, but include the pathogens *Trypanosoma* and *Leishmania* described in Table 29.1.

CHECK YOUR UNDERSTANDING

If you understand that . . .

- According to the most recent analyses, the domain Eukarya comprises seven major lineages. Members of each lineage have distinctive aspects of cell structure.
- Direct sequencing has allowed investigators to recognize large numbers of previously undescribed eukaryotes, some of which are extremely small.

✔ You should be able to . . .

1. Explain why opisthokonts, alveolates, and stramenopiles got their names. (The root *opistho* refers to the back of a cell; the root *kont* refers to a flagellum.)
2. Explain why direct sequencing studies allow researchers to characterize species that have never been seen before.

Answers are available in Appendix B.

29.3 What Themes Occur in the Diversification of Protists?

The protists range in size from bacteria-sized single cells to giant kelp. They live in habitats from the open oceans to dank forest floors. They are almost bewildering in their morphological and ecological diversity. Because they are a paraphyletic group, they do not share derived characteristics that set them apart from all other lineages on the tree of life.

Fortunately, one general theme helps tie protists together. Once an important new innovation arose in protists, it triggered the evolution of species that live in a wide array of habitats and make a living in diverse ways.

What Morphological Innovations Evolved in Protists?

Virtually all bacteria and all archaea are unicellular. Given the distribution of multicellularity in eukaryotes, it is logical to conclude that the first eukaryote was also a single-celled organism.

Further, all eukaryotes alive today have (1) a nucleus and endomembrane system, (2) mitochondria or genes that are normally found in mitochondria, and (3) a cytoskeleton. Based on the distribution of cell walls in living eukaryotes, though, it is likely that the first eukaryotes lacked this feature.

Based on these observations, biologists hypothesize that the earliest eukaryotes were probably single-celled organisms with a nucleus and endomembrane system, mitochondria, and a cytoskeleton, but no cell wall.

It is also likely that the first eukaryotic cells swam using a novel type of flagellum. Eukaryotic flagella are completely different structures from bacterial flagella and evolved independently. The eukaryotic flagellum is made up of microtubules, and dynein is the major motor protein. An undulating motion occurs as dynein molecules walk down microtubules (see Chapter 7). The flagella of bacteria and archaea, in contrast, are composed primarily of a protein called flagellin. Instead of undulating, these flagella rotate to produce movement.

✔If you understand the synapomorphies that identify the eukaryotes as a monophyletic group, you should be able to map the origin of the nuclear envelope and the eukaryotic flagellum on Figure 29.7. Once you've done that, let's consider how several of these key new morphological features arose and influenced the subsequent diversification of protists, beginning with the trait that defines the Eukarya.

THE NUCLEAR ENVELOPE The leading hypothesis to explain the origin of the nuclear envelope is based on infoldings of the plasma membrane. As the drawings in **Figure 29.8** show, a stepwise process could give rise to small infoldings that were elaborated by mutation and natural selection over time, with the infolding eventually becoming detached from the plasma membrane. Note that the infoldings would have given rise to the nuclear envelope and the endoplasmic reticulum (ER) together.

Two lines of evidence support this hypothesis: Infoldings of the plasma membrane occur in some bacteria living today, and the nuclear envelope and ER of today's eukaryotes are continuous (see Chapter 7). **✔If you understand the infolding hypothesis, you should be able to explain why these observations support it.**

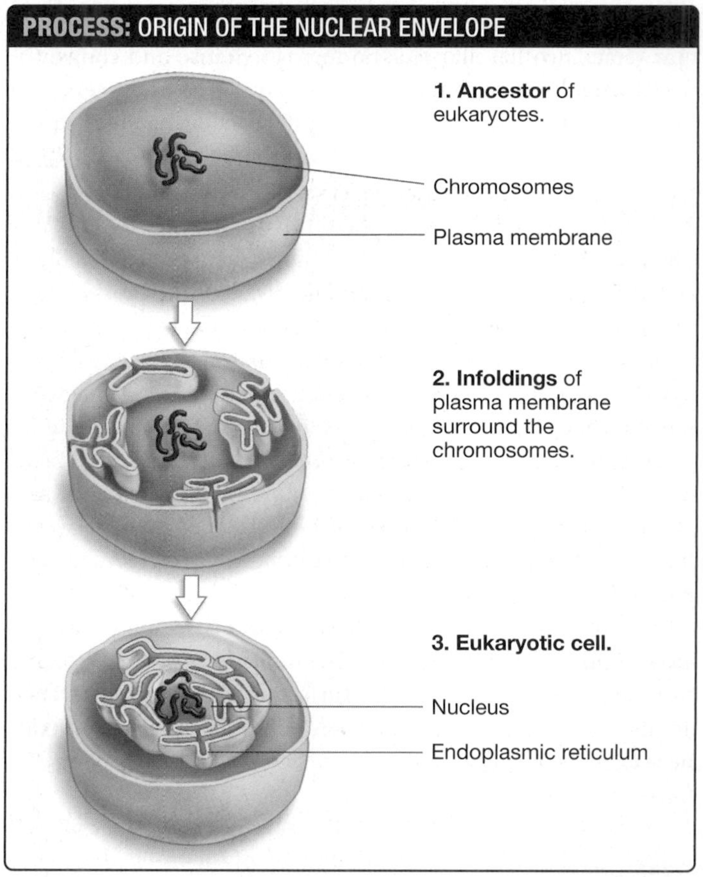

FIGURE 29.8 A Hypothesis for the Origin of the Nuclear Envelope. Infoldings of the plasma membrane, analogous to those shown here, have been observed in bacteria living today.

According to current thinking, the evolution of the nuclear envelope was advantageous because it separated transcription and translation. Recall from Chapter 16 that RNA transcripts are processed inside the nucleus but translated outside the nucleus. In bacteria and archaea, transcription and translation occur together.

Once a simple nuclear envelope was in place, alternative splicing and other forms of RNA processing could occur—giving the early eukaryotes a novel way to control gene expression (see Chapter 18). The take-home message here is that an important morphological innovation gave the early eukaryotes a new way to manage and process genetic information.

Once a nucleus had evolved, it underwent diversification. In some cases, unique types of nuclei are associated with the founding of important lineages of protists.

- Ciliates have a diploid micronucleus that is involved only in reproduction and a polyploid macronucleus where transcription occurs.
- Diplomonads have two nuclei that look identical; it is not known how they interact.
- In foraminifera, red algae, and plasmodial slime molds, certain cells may contain many nuclei.
- Dinoflagellates have chromosomes that lack histones and attach to the nuclear envelope.

In each case, the distinctive structure of the nucleus is a synapomorphy that allows us to recognize these lineages as distinct monophyletic groups.

ENDOSYMBIOSIS AND THE ORIGIN OF THE MITOCHONDRION Mitochondria are organelles that generate ATP using pyruvate as an electron donor and oxygen as the ultimate electron acceptor (see Chapter 9).

In 1981 Lynn Margulis expanded on a radical hypothesis—first proposed in the nineteenth century—to explain the origin of mitochondria. The **endosymbiosis theory** proposes that mitochondria originated when a bacterial cell took up residence inside a eukaryote about 2 billion years ago.

The theory's name was inspired by the Greek word roots *endo*, *sym*, and *bio* (literally, "inside-together-living"). **Symbiosis** is said to occur when individuals of two different species live in physical contact; **endosymbiosis** occurs when an organism of one species lives inside an organism of another species.

In its current form, the endosymbiosis theory proposes that mitochondria evolved through a series of steps, beginning with a eukaryotic cell that was capable only of anaerobic fermentation—meaning it could not use oxygen as an electron acceptor in cellular respiration (**Figure 29.9**):

1. The eukaryotic cell used its cytoskeletal elements to surround and engulf smaller prey.
2. Instead of fusing with a lysosome and being digested (see Chapter 7), an engulfed bacterium began to live inside a eukaryotic cell.

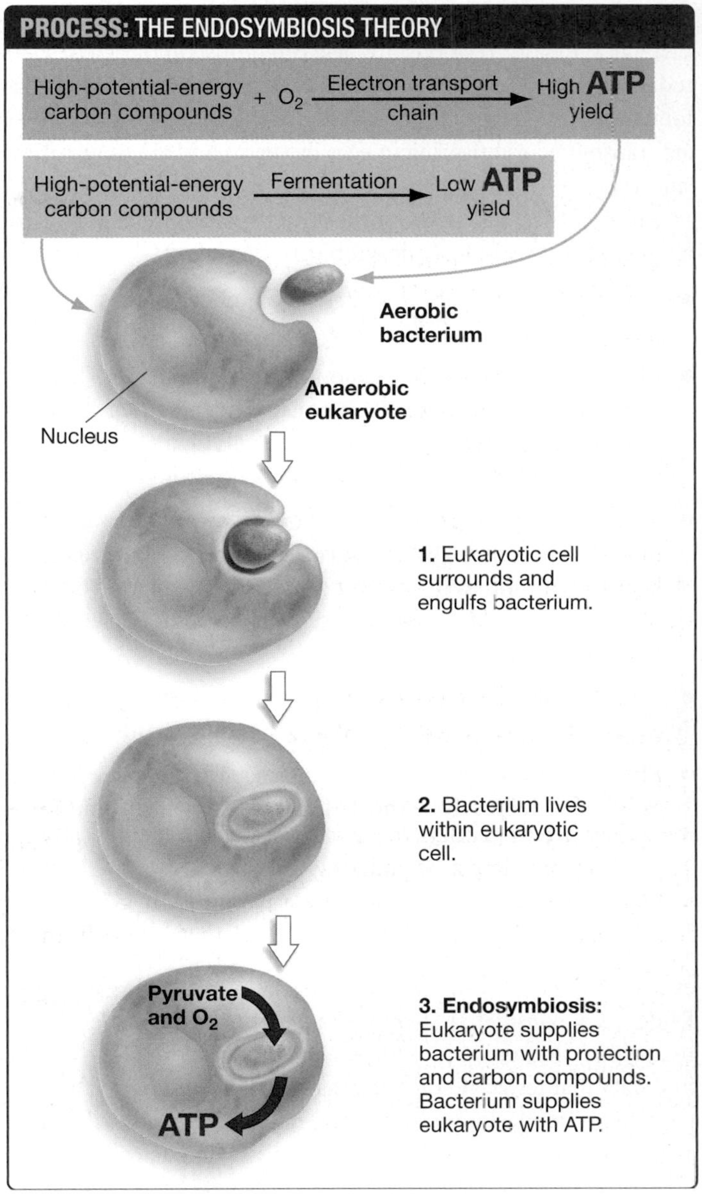

FIGURE 29.9 Proposed Initial Steps in the Evolution of the Mitochondrion.

✔**QUESTION** According to this hypothesis, how many membranes should surround a mitochondrion? Explain your logic.

3. The engulfed cell survived by absorbing carbon molecules with high potential energy from the eukaryotic cell and oxidizing them, using oxygen as a final electron acceptor.

The relationship between the host and the engulfed cell was presumed to be stable because a mutual advantage existed between them: The host supplied the bacterium with protection and carbon compounds from its other prey, while the bacterium produced much more ATP than the host cell could synthesize on its own. Cells that can use oxygen during cellular respiration are able to produce much more ATP than are cells that cannot (see Chapter 28).

When Margulis first began promoting the theory, it met with a storm of criticism—largely because it seemed slightly preposterous.

But gradually biologists began to examine it rigorously. For example, endosymbiotic relationships between protists and bacteria exist today. Among the α-proteobacteria alone, three major groups are found *only* inside eukaryotic cells. In many cases, the bacterial cells are transmitted to offspring in eggs or sperm and are required for survival.

Several observations about the structure of mitochondria are also consistent with the endosymbiosis theory:

- Mitochondria are about the size of an average α-proteobacterium.
- Mitochondria replicate by fission, as do bacterial cells. The duplication of mitochondria takes place independently of division by the host cell. When eukaryotic cells divide, each daughter cell receives some of the many mitochondria present.
- Mitochondria have their own ribosomes and manufacture their own proteins. Mitochondrial ribosomes closely resemble bacterial ribosomes in size and composition and are poisoned by antibiotics such as streptomycin that inhibit bacterial, but not eukaryotic, ribosomes.
- Mitochondria have double membranes, consistent with the engulfing mechanism of origin illustrated in Figure 29.9.
- Mitochondria have their own genomes, which are organized as circular molecules—much like a bacterial chromosome. Mitochondrial genes code for the enzymes needed to replicate and transcribe the mitochondrial genome.

Although these data are impressive, they are only consistent with the endosymbiosis theory. Stated another way, they do not exclude other explanations. This is a general principle in science: Evidence is considered strong when it cannot be explained by reasonable alternative hypotheses. In this case, the key was to find data that tested predictions made by Margulis's idea against predictions made by an alternative theory: that mitochondria evolved within eukaryotic cells, separately from bacteria.

A breakthrough occurred when researchers realized that according to the "within-eukaryotes" theory, the genes found in mitochondria are derived from nuclear genes in ancestral eukaryotes. Margulis's theory, in contrast, proposed that the genes found in mitochondria were bacterial in origin.

These predictions were tested by studies on the phylogenetic relationships of mitochondrial genes. Specifically, researchers compared gene sequences isolated from the nuclear DNA of eukaryotes, mitochondrial DNA from eukaryotes, and DNA from several species of bacteria. Exactly as the endosymbiosis theory predicted, mitochondrial gene sequences are much more closely related to the sequences from the α-proteobacteria than to sequences from the nuclear DNA of eukaryotes.

The result diagrammed in **Figure 29.10** was considered overwhelming evidence that the mitochondrial genome came from an α-proteobacterium rather than from a eukaryote. The endosymbiosis theory was the only reasonable explanation for the data. The results were a stunning vindication of a theory that had once been intensely controversial. Mitochondria evolved via endosymbiosis.

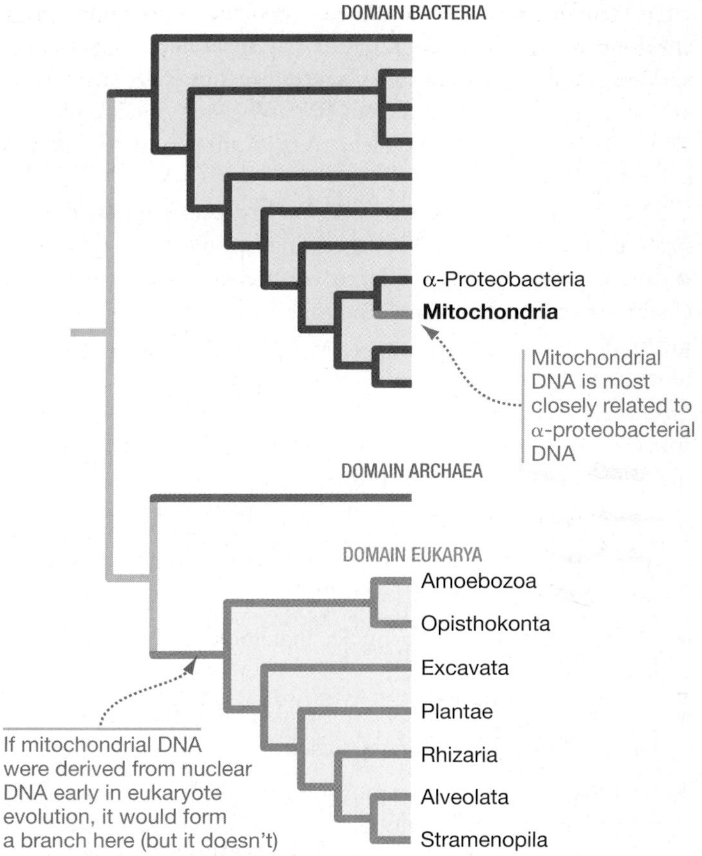

FIGURE 29.10 Phylogenetic Data Support the Endosymbiosis Theory.

✔If you understand the endosymbiosis theory and the evidence for it, you should be able to describe how the chloroplast—the organelle where photosynthesis takes place in eukaryotes—could arise via endosymbiosis.

Once the mitochondrion was present in the ancestor of today's protists, it underwent diversification.

- In Diplomonads and Parabasalids, the organelle was lost entirely or is now a vestigial trait.
- In Euglenida and related species, the sac-like cristae inside the mitochondrion—where the electron transport chain and ATP synthase are located—are disc-shaped.
- In the Opisthokonta, the cristae are flattened.

Presumably, variation in the presence or structure of mitochondria reflects variation in how ATP is produced by protists. In these cases, the mitochondrion's absence or structure qualifies as a synapomorphy that identifies a monophyletic group. The mitochondrion was a morphological innovation that subsequently diversified.

STRUCTURES FOR SUPPORT AND PROTECTION As far as is known, the basic structure of the cytoskeleton does not vary much among protists. In contrast, other structures that provide support and protection for the cell vary a great deal.

Many protists have cell walls outside their plasma membrane; others have hard external structures called a **shell**; others have rigid structures inside the plasma membrane. In many cases, these novel structures represent synapomorphies that identify monophyletic groups among protists. For example:

- Diatoms are surrounded by a glass-like, silicon-dioxide cell wall (**Figure 29.11a**). The cell wall is made up of two pieces that fit together in a box-and-lid arrangement, like the petri plates you may have seen in lab.
- Dinoflagelletes have a cell wall made up of cellulose plates (**Figure 29.11b**).
- Within Foraminifera, some lineages secrete an intricate, chambered calcium carbonate shell (**Figure 29.11c**).
- Members of other Foraminifera lineages, and some amoebae, cover themselves with tiny pebbles.
- The parabasalids have a unique internal support rod, consisting of cross-linked microtubules that run the length of the cell.
- The euglenids have a collection of protein strips located just under the plasma membrane. The strips are supported by microtubules and stiffen the cell.
- The alveolates have distinctive sac-like structures called alveoli, located just under the plasma membrane, that help stiffen the cell.

In many cases, the diversification of protists has been associated with the evolution of innovative structures for support and protection.

MULTICELLULARITY Multicellular individuals contain more than one cell. At least some of the cells are attached to each other, and some are specialized for different functions. In the simplest multicellular species, certain cells are specialized for producing or obtaining food while other cells are specialized for reproduction. The key point about **multicellularity** is that not all cells express the same genes.

A few species of bacteria are capable of aggregating and forming structures called fruiting bodies (see Chapter 28). Because cells in the fruiting bodies of these bacteria differentiate into specialized stalk cells and spore-forming cells, they are considered multicellular. But the vast majority of multicellular species are members of the Eukarya. Multicellularity arose independently in a wide array of eukaryotic lineages: the green plants, fungi, animals, brown algae, slime molds, and red algae.

To summarize, an array of novel morphological traits played a key role as protists diversified: the nucleus and endomembrane system, the mitochondrion, structures for protection and support, and multicellularity. Evolutionary innovations allowed protists to build and manage the eukaryotic cell in new ways.

Once a new type of eukaryotic cell or multicellular individual existed, subsequent diversification was often triggered by novel ways of finding food, moving, or reproducing. Let's consider each of these life processes in turn.

How Do Protists Obtain Food?

According to Chapter 28, bacteria and archaea can use a wide array of molecules as electron donors and electron acceptors during cellular respiration. They get these molecules by absorbing them directly from the environment. Other bacteria don't absorb their nutrition—instead, they make their own food via photosynthesis. Many groups of protists are similar to bacteria in the way they find food: They perform photosynthesis or absorb their food directly from the environment.

But one of the most important stories in the diversification of protists was the evolution of a novel method for finding food. Many protists ingest their food—they eat bacteria, archaea, or even other protists whole.

When ingestive feeding occurs, an individual takes in packets of food much larger than individual molecules. Thus, protists feed by (1) ingesting packets of food, (2) absorbing organic molecules directly from the environment, or (3) performing photosynthesis.

Some protists ingest food as well as performing photosynthesis—meaning that they use a combination of feeding strategies. In

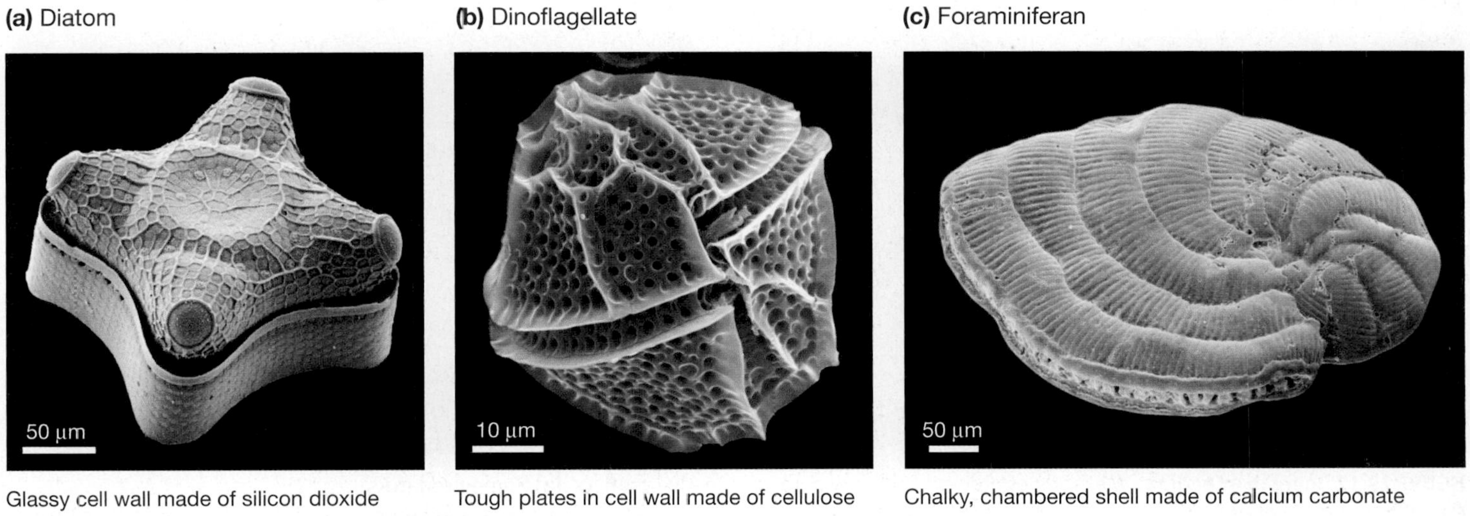

FIGURE 29.11 Hard Outer Coverings in Protists Vary in Composition.

addition, it's important to recognize that all three lifestyles—ingestive, absorptive, and photosynthetic—can occur within a single clade.

To drive this last point home, consider the monophyletic group called the alveolates. There are three major subgroups of this lineage, called dinoflagellates, apicomplexa, and ciliates. About half of the dinoflagellates are photosynthetic, while many others are parasitic. Apicomplexa are parasitic. Ciliates include many species that ingest prey, but some ciliates live in the guts of cattle or the gills of fish and absorb nutrients from their hosts. Other ciliate species make a living by holding algae or other types of photosynthetic symbionts inside their cells.

The punchline? Within each of the seven major lineages of eukaryotes, different methods for feeding helped trigger diversification.

INGESTIVE FEEDING Ingestive lifestyles are based on eating live or dead organisms or on scavenging loose bits of organic debris. Protists such as the cellular slime mold *Dictyostelium discoideum* are large enough to engulf bacteria and archaea; many protists are large enough to surround and ingest other protists or microscopic animals.

Feeding by engulfing is possible in protists that lack a cell wall. A flexible membrane and dynamic cytoskeleton give these species the ability to surround and "swallow" prey using long, fingerlike projections called **pseudopodia** ("false-feet"). The engulfing process is illustrated in **Figure 29.12a**.

Losing a cell wall and gaining the ability to move their plasma membrane around prey items or food particles was an important innovation during the evolution of protists:

- Bacteria and archaea are abundant in wet soils and aquatic habitats, where protists also live. Instead of competing with prokaryotes for sunlight or food molecules, protists could eat them.
- It was a prerequisite for the endosymbiosis events that led to the evolution of the mitochondrion and chloroplast.

Although many ingestive feeders actively hunt down prey and engulf them, others do not. Instead of taking themselves to food, these species attach themselves to a surface. Protists that feed in this way have cilia that surround the mouth and beat in a coordinated way. The motion creates water currents that sweep food particles into the cell (**Figure 29.12b**).

ABSORPTIVE FEEDING Absorptive feeding occurs when nutrients are taken up directly from the environment, across the plasma membrane. Absorptive feeding is common among protists.

Some protists that live by absorptive feeding are **decomposers**, meaning that they feed on dead organic matter, or **detritus**. But many of the protists that absorb their nutrition directly from the environment live inside other organisms. If they damage their host, the absorptive species is called a **parasite**.

PHOTOSYNTHESIS—ENDOSYMBIOSIS AND THE ORIGIN OF THE CHLOROPLAST You might recall from Chapter 10 and Chapter 28 that photosystems I and II evolved in bacteria, and that both photosystems occur in cyanobacteria. None of the basic machinery required for photosynthesis evolved in eukaryotes. Instead, they "stole" it—via endosymbiosis.

The endosymbiosis theory contends that the eukaryotic chloroplast originated when a protist engulfed a cyanobacterium. Once inside the protist, the photosynthetic bacterium provided its eukaryotic host with oxygen and glucose in exchange for protection and access to light. If the endosymbiosis theory is correct, today's chloroplasts trace their ancestry to cyanobacteria.

All of the photosynthetic eukaryotes have both a chloroplast and a mitochondrion. The evidence for an endosymbiotic origin for the chloroplast is even more persuasive than for mitochondria.

- Chloroplasts have the same list of bacteria-like characteristics presented earlier for mitochondria.
- There are many examples of endosymbiotic cyanobacteria living inside protists or animals today.
- The DNA sequences inside chloroplasts are extremely similar to cyanobacterial genes.

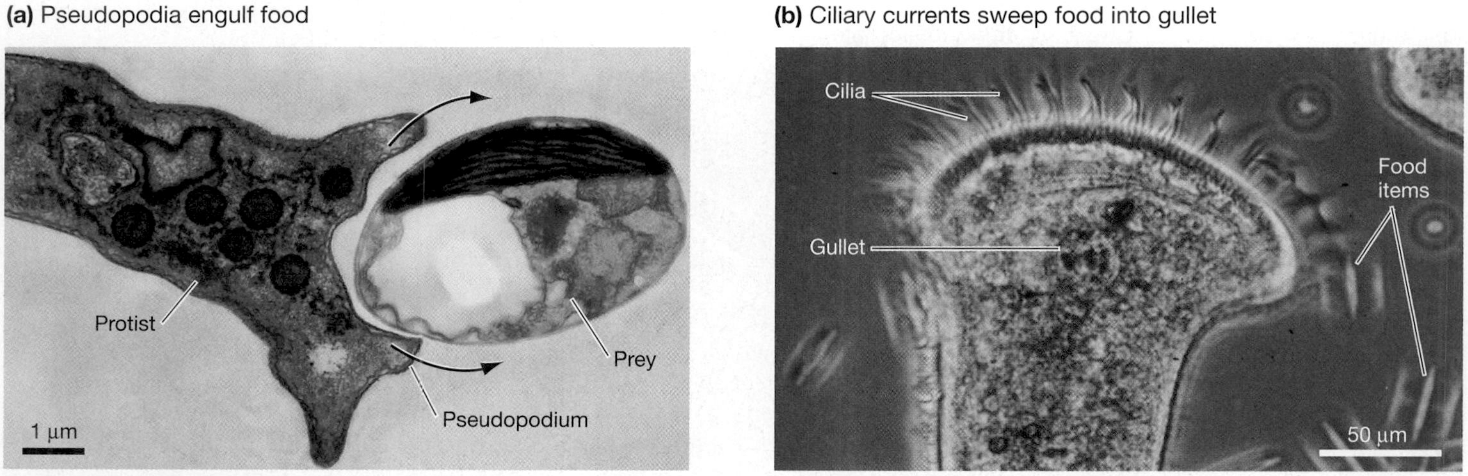

FIGURE 29.12 Ingestive Feeding. Methods of prey capture vary among ingestive protists. **(a)** Some predators engulf prey with pseudopodia; **(b)** other predators sweep them into their gullets with water currents set up by the beating of cilia. Note that the cells in part (a) have been colorized.

- The photosynthetic organelle of one group of protists, called the glaucophyte algae, has an outer layer containing the same constituent (peptidoglycan) found in the cell walls of cyanobacteria.
- Like chloroplasts, the cyanobacterium *Prochloron* contains chlorophylls *a* and *b* and has a system of internal membranes where the photosynthetic pigments and enzymes are located.

✔If you understand the evidence for the endosymbiotic origin of the chloroplast, you should be able to add a label indicating the location of chloroplast genes on the phylogenetic tree in Figure 29.7.

PHOTOSYNTHESIS—PRIMARY VERSUS SECONDARY ENDOSYMBIOSIS Which eukaryote originally obtained a photosynthetic organelle? Because all of the species in the Plantae have chloroplasts with two membranes, biologists infer that the original, or primary, endosymbiosis occurred in their common ancestor—a species that eventually gave rise to the glaucophyte algae, red algae, and green plants (green algae and land plants).

But chloroplasts also occur in three of the other major lineages of protists—the Excavata, Chromalveolates, and Rhizaria. In these species, the chloroplast is surrounded by more than two membranes—usually four.

To explain this observation, researchers hypothesize that the ancestors of these groups acquired their chloroplasts by ingesting photosynthetic protists that already had chloroplasts. This process, called secondary endosymbiosis, occurs when an organism engulfs a photosynthetic eukaryotic cell and retains its chloroplasts as intracellular symbionts. Secondary endosymbiosis is illustrated in **Figure 29.13**.

Figure 29.14 shows where primary and secondary endosymbiosis occurred on the phylogenetic tree of eukaryotes. Once

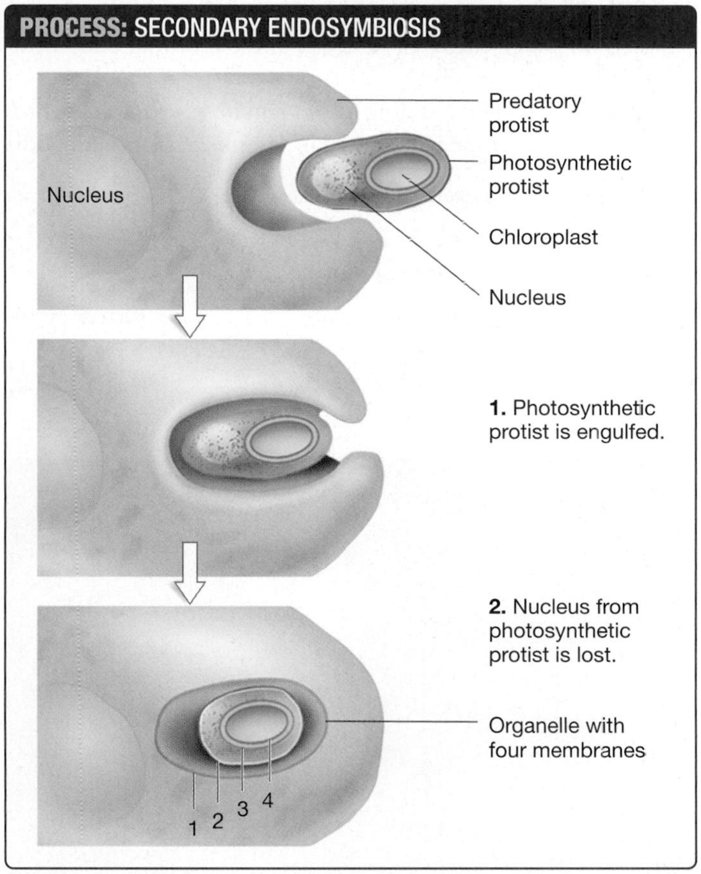

FIGURE 29.13 Secondary Endosymbiosis Leads to Organelles with Four Membranes. The chloroplasts found in some protists have four membranes and are hypothesized to be derived by secondary endosymbiosis. In species where chloroplasts have three membranes, biologists hypothesize that secondary endosymbiosis was followed by the loss of one membrane.

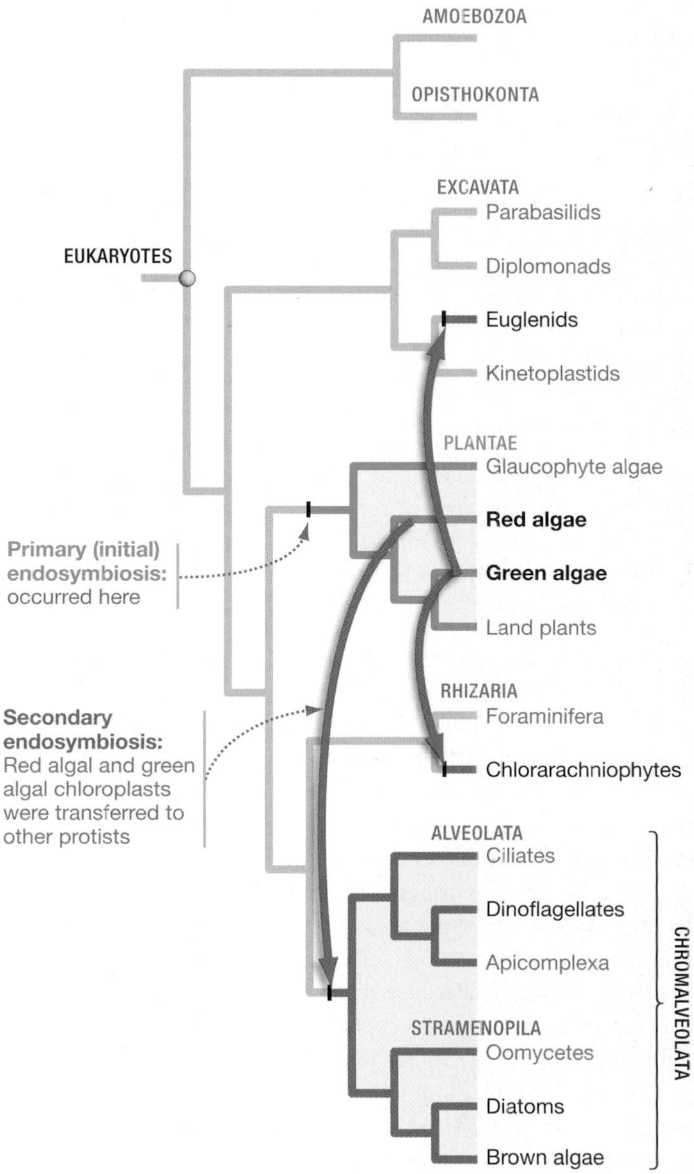

FIGURE 29.14 Photosynthesis Arose in Protists by Primary Endosymbiosis, Then Spread among Lineages via Secondary Endosymbiosis. Biochemical similarities link the chloroplasts found in chromalveolates with red algae and the chloroplasts found in euglenids and chlorarachniophytes with green algae. Note that among the chromalveolates, only the dinoflagellates, diatoms, and brown algae have photosynthetic species. In ciliates, apicomplexa, and oomycetes, the chloroplast has been lost or changed function.

TABLE 29.3 **Pigments in Photosynthetic Protists**

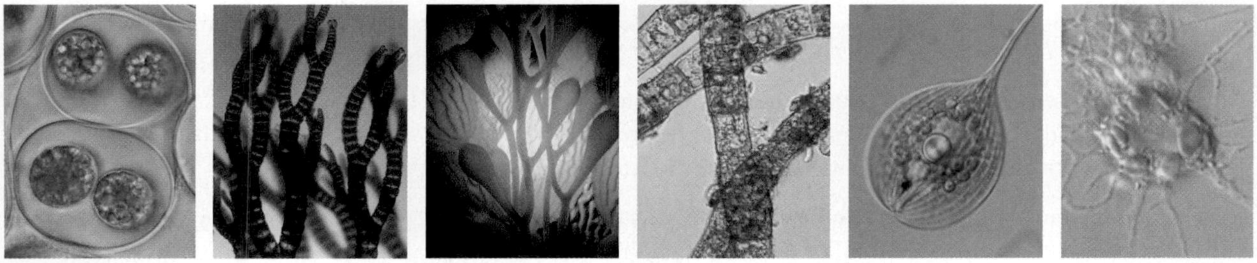

Lineage	Glaucophytes	Red algae	Brown algae, diatoms, dinoflagellates	Green algae	Euglenids	Chlorarachniophytes
Photosynthetic pigments	Chlorophyll *a*	Chlorophyll *a*	Chlorophyll *a* and *c*	Chlorophyll *a* and *b*	Chlorophyll *a* and *b*	Chlorophyll *a* and *b*
Accessory pigments	Phycocyanin, allophycocyanin, β-carotene	Primarily phycoerythrin	Fucoxanthin	Fucoxanthin, β-carotene	Xanthophylls, β-carotene	Xanthophylls

✔**QUESTION** Are the pigments listed in this table consistent with the primary endosymbiosis and secondary endosymbiosis hypotheses described in Figures 29.9 and 29.13? Explain your answer.

protists obtained the chloroplast, it was "swapped around" to new lineages via secondary endosymbiosis.

✔**If you understand endosymbiosis, you should be able to explain why the primary and secondary endosymbiosis events introduced in this chapter represent the most massive lateral gene transfers in the history of life, in terms of the number of genes moved at once.** (To review lateral gene transfer, see Chapter 20.)

PHOTOSYNTHESIS—DIVERSIFICATION IN PIGMENTS Many lineages of protists acquired the ability to photosynthesize independently of each other. In each case, the acquisition of a chloroplast triggered a radiation of photosynthetic species.

The key to understanding the diversification of photosynthetic eukaryotes is to realize that most of the photosynthetic lineages contain a unique collection of photosynthetic pigments. A particular photosynthetic pigment absorbs specific wavelengths of light. Thus, the presence of different pigments means that different species absorb different wavelengths in the electromagnetic spectrum. Because diverse combinations of pigments evolved, eukaryotic species could harvest unique wavelengths of light and avoid competition (**Table 29.3**).

Chapter 28 pointed out a similar phenomenon among photosynthetic bacteria. Like the nuclear envelope, mitochondrion, multicellularity, ingestive feeding, and other traits that originated with protists, the acquisition of the chloroplast was an innovation that triggered subsequent diversification.

How Do Protists Move?

Many protists actively move to find food. Predators such as the slime mold *Dictyostelium discoideum* crawl over a substrate in search of prey. Most of the unicellular, photosynthetic species are capable of swimming to sunny locations, though others drift passively in water currents. How are these crawling and swimming movements possible?

Amoeboid motion is a sliding movement observed in some protists. In the classic mode illustrated in **Figure 29.15**, pseudopodia

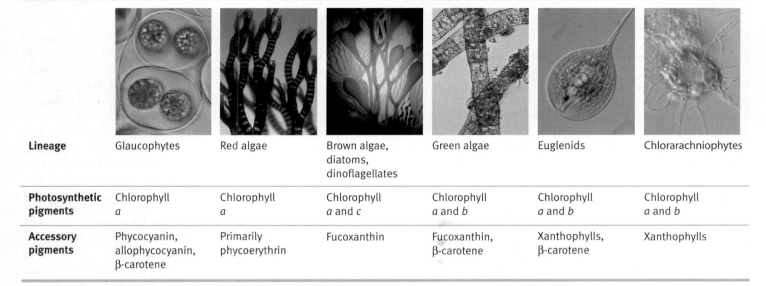

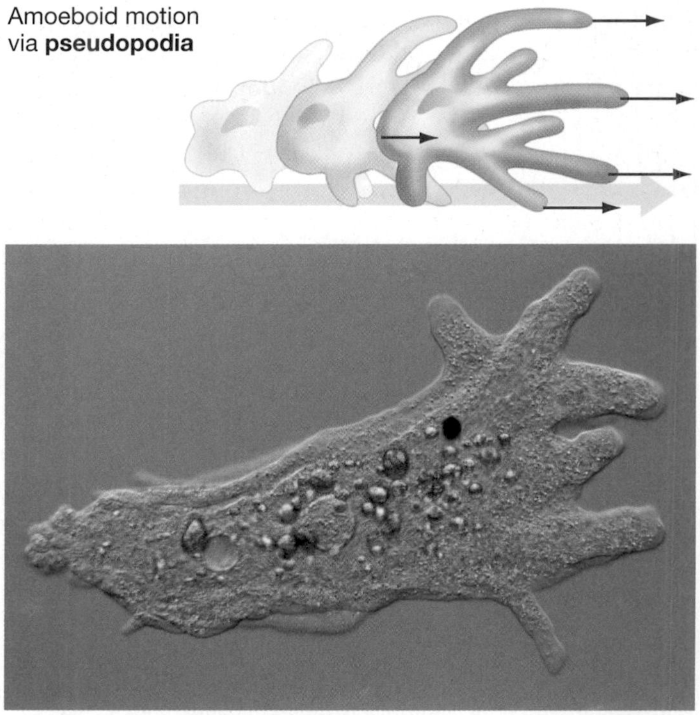

FIGURE 29.15 Amoeboid Motion Is Possible in Species That Lack Cell Walls. In amoeboid motion, long pseudopodia stream out from the cell. The rest of the cytoplasm, organelles, and external membrane follow.

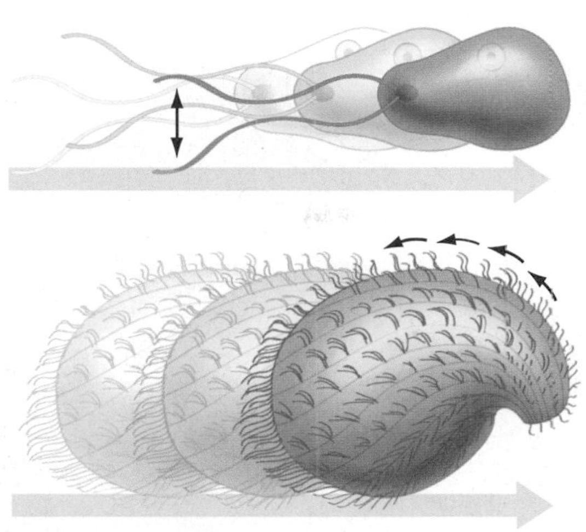

FIGURE 29.16 Many Protists Swim Using Flagella or Cilia.
(a) Flagella are long and few in number. **(b)** Cilia are short and numerous. In many cases they are used in swimming.

stream forward over a substrate, with the rest of the cytoplasm, organelles, and plasma membrane following. The motion requires ATP and involves interactions between proteins called actin and myosin inside the cytoplasm. The mechanism is related to muscle movement in animals, which is detailed in Chapter 46. But at the level of the whole cell, the precise sequence of events during amoeboid movement is still uncertain. The issue is attracting attention, because key immune system cells in humans use amoeboid motion as they hunt down and destroy disease-causing agents.

The other major mode of locomotion in protists involves flagella or cilia (**Figure 29.16a,** and **b**). Recall from Chapter 7 that flagella and cilia both consist of nine sets of doublet (paired) microtubules arranged around two central, single microtubules. Flagella are long and are usually found alone or in pairs; cilia are short and usually occur in large numbers on any one cell. Flagella and cilia can also be distinguished by the types of structures associated with the basal bodies where they originate.

Even closely related protists can use radically different forms of locomotion. For example, the ciliates swim by beating their cilia, the dinoflagellates swim by whipping their flagella, and mature cells in apicomplexa move by amoeboid motion (though their gametes swim via flagella). It is also common to find protists that do not exhibit active movement but instead float passively in water currents.

Movement is yet another example of the extensive diversification that occurred within each of the seven major monophyletic groups of eukaryotes.

How Do Protists Reproduce?

Sexual reproduction evolved in protists. As Chapter 12 pointed out, sexual reproduction can best be understood in contrast to asexual reproduction. The key issues are the type of nuclear division involved and the consequences for genetic diversity in the offspring produced.

- Asexual reproduction is based on mitosis in eukaryotic organisms and on fission in bacteria and archaea. It results in offspring that are genetically identical to the parent.
- Sexual reproduction is based on meiosis and fusion of gametes. It results in offspring that are genetically different from their parents and from each other.

Most protists undergo asexual reproduction routinely. But sexual reproduction occurs only intermittently in many protists—often at one particular time of year, or when individuals are crowded or food is scarce.

The evolution of sexual reproduction ranks among the most significant evolutionary innovations observed in eukaryotes.

SEXUAL VERSUS ASEXUAL REPRODUCTION The leading hypothesis to explain why meiosis evolved states that genetically variable offspring may be able to thrive if the environment changes. For example, offspring with genotypes different from those of their parents may be better able to withstand attacks by parasites that successfully attacked their parents (see Chapter 12). This is a key point because many types of parasites, including bacteria and viruses, have short generation times and evolve quickly.

Because the genotypes and phenotypes of parasites are constantly changing, natural selection is constantly favoring host individuals with new genotypes. The idea is that new offspring genotypes generated by meiosis may contain combinations of alleles that allow hosts to withstand attack by new strains of parasites. In short, many biologists view sexual reproduction as an adaptation to fight disease.

✔**If you understand the changing-environment hypothesis for the evolution of sex, you should be able to explain whether it is consistent with the observation that many protists undergo meiosis when food is scarce or the population density is high. In addition, if a damaging mutation occurs in a parent, it will be passed along to offspring that are produced asexually. But if sexual reproduction occurs, some offspring may be free of the mutation.**

Life Cycles—Haploid- versus Diploid-Dominated

A life cycle describes the sequence of events that occur as individuals grow, mature, and reproduce. The evolution of meiosis introduced a new event in the life cycle of protist species; what's more, it created a distinction between haploid and diploid phases in the life of an individual.

Recall from Chapter 12 that diploid individuals have two of each type of chromosome inside each cell, while haploid individuals have just one of each type of chromosome inside each cell. When meiosis occurs in diploid cells, it results in the production of haploid cells.

The life cycle of most bacteria and archaea is extremely simple: A cell divides, feeds, grows, and divides again. Bacteria and archaea are always haploid.

In contrast, virtually every aspect of a life cycle is variable among protists—whether meiosis occurs, whether asexual reproduction occurs, and whether the haploid or the diploid phase of the life cycle is the longer and more prominent phase.

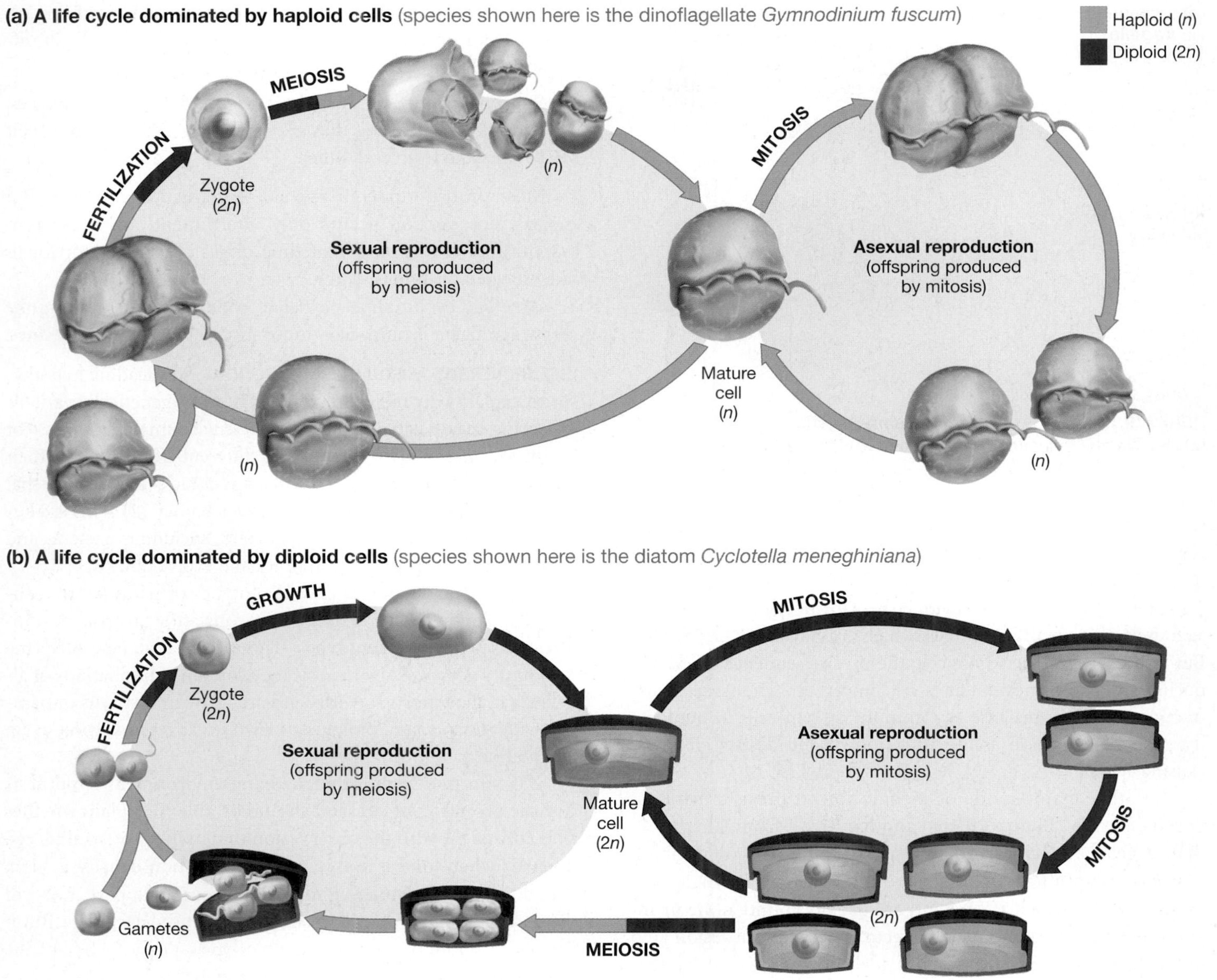

FIGURE 29.17 Life Cycles Vary Widely among Unicellular Protists. Many unicellular protists can reproduce by both sexual reproduction and asexual reproduction. The cell may be **(a)** haploid for most of its life or **(b)** diploid for most of its life.

Figure 29.17 illustrates some of this variation. Figure 29.17a depicts the haploid-dominated life cycle observed in many unicellular protists. The specific example given here is the dinoflagellate *Gymnodinium fuscum.*

To analyze a life cycle, start with **fertilization**—the fusion of two gametes to form a diploid zygote. Then trace what happens to the zygote. In this case, the diploid zygote undergoes meiosis. The haploid products of meiosis then grow into mature cells that eventually undergo asexual reproduction or produce gametes by mitosis.

Contrast the dinoflagellate cycle with the diploid-dominated life cycle of Figure 29.17b. The specific organism shown here is the diatom *Cyclotella meneghiniana.* Note that after fertilization, the diploid zygote develops into a sexually mature, diploid adult cell. Meiosis occurs in the adult and results in the formation of haploid gametes, which then fuse to form a diploid zygote. The important contrasts are (1) that meiosis occurs in the adult cell rather than in the zygote and (2) that gametes are the only haploid cells in the life cycle.

LIFE CYCLES—ALTERNATION OF GENERATIONS In contrast to the relatively simple life cycles of Figure 29.17, many multicellular protists have one phase in their life cycle that is based on a multicellular haploid form and another phase in their life cycle that is based on a multicellular diploid form. This alternation of multicellular haploid and diploid forms is known as **alternation of generations**.

- The multicellular haploid form is called a **gametophyte**, because specialized cells in this individual produce gametes by mitosis.

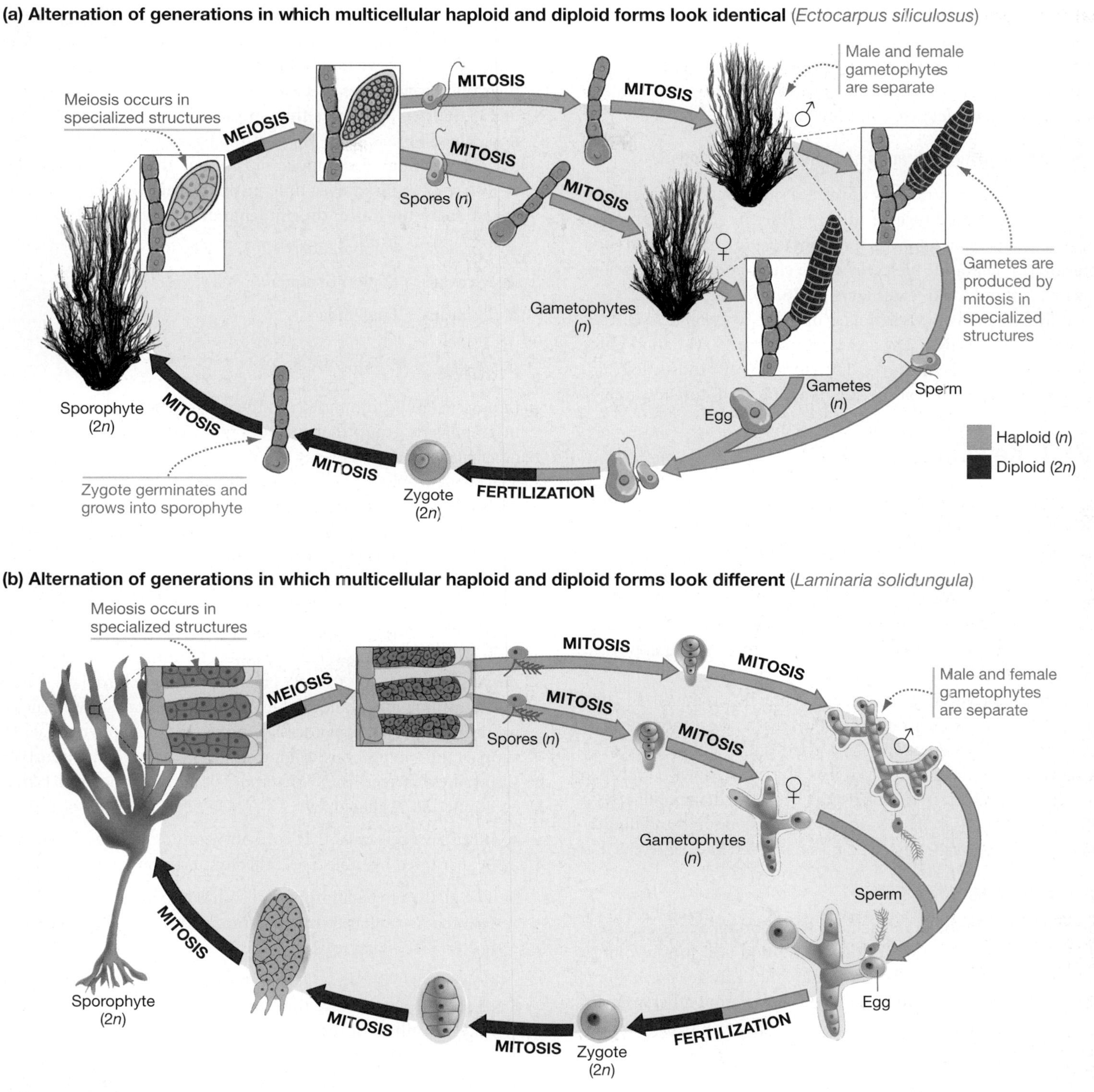

FIGURE 29.18 Alternation of Generations Occurs in Many Multicellular Protists.

- The multicellular diploid form is called a **sporophyte**, because it has specialized cells that undergo meiosis to produce haploid cells called spores.
- A spore is a single cell that develops into an adult organism but is not a product of fusion by gametes.

When alternation of generations occurs, a spore divides by mitosis to form a haploid, multicellular gametophyte. The haploid gametes produced by the gametophyte then fuse to form a diploid zygote, which grows into the diploid, multicellular sporophyte.

Note that gametophytes and sporophytes may be identical in appearance, as in the brown alga called *Ectocarpus siliculosus* (**Figure 29.18a**). In many cases, however, the gametophyte and sporophyte look different, as in the brown alga called *Laminaria solidungula* (**Figure 29.18b**). Among the protists, alternation of generations evolved independently in brown algae, red algae, and other groups.

To understand alternation of generations, some students find it helpful to memorize the following sentence: Sporophytes produce spores by meiosis; gametophytes produce gametes by mitosis. You will also find it helpful to go to the study area at *www.masteringbiology.com* and review.

MB **Web Activity** Alternation of Generations in a Protist

✔**If you understand how life cycles vary among multicellular protists, you should be able to (1) define the terms alternation of generations, gametophyte, sporophyte, spore, zygote, and gamete and (2) diagram a life cycle where alternation of generations occurs, without looking at Figure 29.18.**

Why does so much variation occur in the types of life cycles observed among protists? The answer is not known. Variation in life cycles is a major theme in the diversification of protists. Explaining why that variation exists remains a topic for future research.

29.4 Key Lineages of Protists

Each of the seven major Eukarya lineages has at least one distinctive morphological characteristic. But once an ancestor evolved a distinctive cell structure, its descendants diversified into a wide array of lifestyles. For example, parasitic species evolved independently in all seven major lineages. Photosynthetic species exist in most of the seven, and multicellularity evolved independently in four. Similar statements could be made about the evolution of life cycles and modes of locomotion.

In effect, each of the seven lineages represents a similar radiation of species into a wide array of lifestyles. Let's take a more detailed look at some representative taxa from six of the seven major groups, starting with an overview of the entire lineage and then looking in detail at selected subgroups. The seventh major lineage—the opisthokonts—is featured in the chapters on fungi and the animals. The lineage called green plants is analyzed in Chapter 30.

Amoebozoa

Species in the Amoebozoa lack cell walls and take in food by engulfing it. They move via amoeboid motion and produce large, lobe-like pseudopodia. They are abundant in freshwater habitats and in wet soils; some are parasites of humans and other animals.

Major subgroups in the lineage are lobose amoebae, cellular slime molds, and plasmodial slime molds. The cellular slime mold *Dictyostelium discoideum* is described in detail in **BioSkills 14** in Appendix A. ✔**You should be able to mark the origin of this lineage's amoeboid form on Figure 29.7 and explain whether it evolved independently of the amoeboid form in Rhizaria.**

■ Amoebozoa > Myxogastrida (Plasmodial Slime Molds)

Excavata

The unicellular species that form the Excavata are named for the morphological feature that distinguishes them—an "excavated" feeding groove found on one side of the cell. Because all excavates lack mitochondria, they were once thought to trace their ancestry to eukaryotes that existed prior to the origin of mitochondria. But researchers have found that excavates either have (1) genes in their nuclear genomes that are normally found in mitochondria, or (2) unusual organelles that appear to be vestigial mitochondria. These observations support the hypothesis that the ancestors of excavates had mitochondria, but that these organelles were lost or reduced over time in this lineage. ✔**You should be able to mark the loss of the mitochondrion on Figure 29.7.**

■ Excavata > Parabasalida

■ Excavata > Diplomonadida

■ Excavata > Euglenida

Plantae

Biologists are beginning to use the name **Plantae** to refer to the monophyletic group that includes glaucophyte algae, red algae, green algae, and land plants.

The glaucophyte algae are unicellular or colonial. They live in plankton or attached to substrates in freshwater environments—particularly in bogs or swamps. Some glaucophyte species have flagella or produce flagellated spores, but sexual reproduction has never been observed in the group. The chloroplasts of glaucophytes have a distinct bright blue-green color.

All of the lineages within Plantae are descended from a common ancestor that engulfed a cyanobacterium, beginning the endosymbiosis that led to the evolution of the chloroplast—their distinguishing morphological feature. This initial endosymbiosis probably occurred in an ancestor of today's glaucophyte algae. To support this hypothesis, biologists point to several important similarities between cyanobacterial cells and the chloroplasts that are found in the glaucophytes.

- Both have cell walls that contain peptidoglycan and that can be disrupted by molecules called lysozyme and penicillin.
- The glaucophyte chloroplast also has a membrane outside its wall that is similar to the membrane found in Gram-negative bacteria (see Chapter 28).

Consistent with these observations, phylogenetic analyses of DNA sequence data place the glaucophytes as the sister group to all other members of the Plantae. ✔**You should be able to mark the origin of the plant chloroplast on Figure 29.7.**

Chapter 30 introduces the green algae and land plants; here we consider just the red algae in detail.

■ Plantae > Rhodophyta (Red Algae)

Rhizaria

The rhizarians are single-celled amoebae that lack cell walls, though some species produce elaborate shell-like coverings. They move by amoeboid motion and produce long, slender pseudopodia. Over 11 major subgroups have been identified and named, including the planktonic organisms called actinopods, which synthesize glassy, silicon-rich skeletons, and the chlorarachniophytes, which obtained a chloroplast via secondary endosymbiosis and are photosynthetic. The best-studied and most abundant group is the Foraminifera. ✔You should be able to mark the origin of the rhizarian amoeboid form on Figure 29.7.

- Rhizaria > Foraminifera

Alveolata

Alveolates are distinguished by small sacs, called alveoli, that are located just under their plasma membranes. Although all members of this lineage are unicellular, the groups highlighted below—the Ciliata, the Dinoflagellata, and the Apicomplexa—are remarkably diverse in morphology and lifestyle. ✔You should be able to mark the origin of alveoli on Figure 29.7.

- Alveolata > Ciliata
- Alveolata > Dinoflagellata
- Alveolata > Apicomplexa

Stramenopila (Heterokonta)

Stramenopiles are sometimes called heterokonts, which translates as "different hairs." At some stage of their life cycle, all stramenopiles have flagella that are covered with distinctive hollow "hairs." The structure of these flagella is unique to the stramenopiles. ✔You should be able to mark the origin of the "hairy" flagellum on Figure 29.7. The lineage includes a large number of unicellular forms, although the brown algae are multicellular and include the world's tallest marine organisms, the kelp.

- Stramenopila > Oomycota (Water Molds)
- Stramenopila > Diatoms
- Stramenopila > Phaeophyta (Brown Algae)

Amoebozoa > Myxogastrida (Plasmodial Slime Molds)

The plasmodial slime molds got their name because individuals form a large, weblike structure (**Figure 29.19**) that consists of a single cell containing many diploid nuclei. Like oomycetes, the plasmodial slime molds were once considered fungi on the basis of their general morphological similarity (see Chapter 31).

Morphology The huge "supercell" form, with many nuclei in a single cell, occurs in few protists other than plasmodial slime molds. ✔You should be able to mark the origin of the supercell on Figure 29.7.

Feeding and locomotion Myxogastrida cells feed on decaying vegetation and move by amoeboid motion.

Reproduction When food becomes scarce, part of the amoeba forms a stalk topped by a ball-like structure in which nuclei undergo meiosis and form spores. The spores are then dispersed to new habitats by the wind or small animals. After spores germinate to form amoebae, two amoebae fuse to form a diploid cell that begins to feed and eventually grows into a supercell.

Human and ecological impacts Like cellular slime molds, plasmodial slime molds are important decomposers in forests. They help break down leaves, branches, and other dead plant material, releasing nutrients that they and other organisms can use.

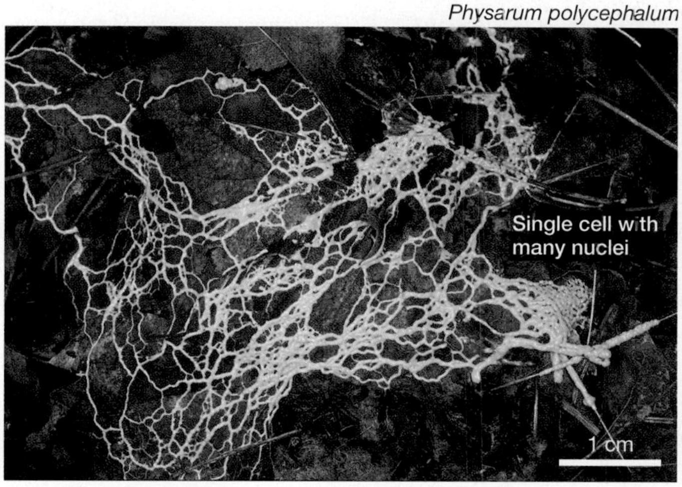

FIGURE 29.19 Plasmodial Slime Molds Are Important Decomposers in Forests.

Excavata > Parabasalida

No free-living parabasalids are known; all of the species described to date live inside animals. Of the 300 species of parabasalid, several live only in the guts of termites. Termites eat wood but cannot digest it themselves. Instead, the parabasalids produce enzymes that digest the cellulose in wood and release compounds that can be used by the termite host. The relationship between termites and parabasalids is considered mutualistic, because both parties benefit from the symbiosis.

Morphology Parabasalid cells lack a cell wall and mitochondria, and they have a single nucleus. A distinctive rod of cross-linked microtubules runs the length of the cell. The rod is attached to the basal bodies where a cluster of flagella arise (**Figure 29.20**). Although the number of flagella present varies widely, four or five is typical. ✔ **You should be able to mark the origin of the parabasalid structure for supporting flagella on Figure 29.7.**

Feeding and locomotion Parabasalids feed by engulfing bacteria, archaea, and organic matter. They swim using their flagella.

Reproduction All parabasalids reproduce asexually. Sexual reproduction also has been observed in a few species.

Human and ecological impacts *Trichomonas* infections can sometimes cause reproductive tract problems in humans, although members of this genus may also live in the gut or mouth of humans without causing harm. *Histomonas meleagridis* is an agricultural pest that causes disease outbreaks on chicken and turkey farms.

Trichomonas vaginalis

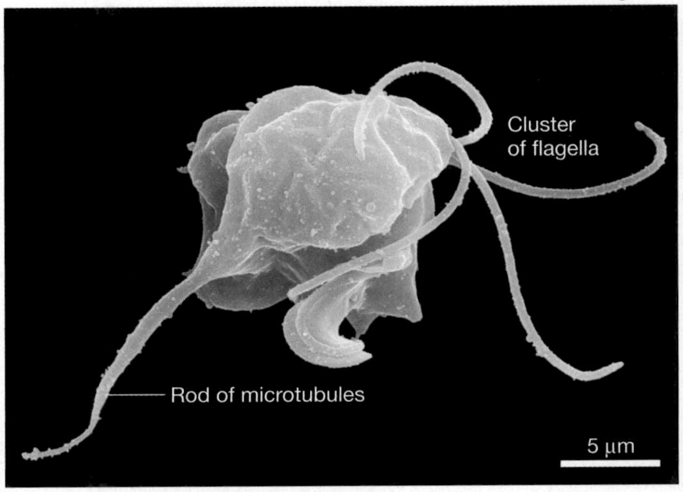

FIGURE 29.20 *Trichomonas* Causes the Sexually Transmitted Disease Trichomoniasis. The cell in this micrograph has been colorized.

Excavata > Diplomonadida

The first species of diplomonad known was described by the early microscopist Anton van Leeuwenhoek, who found *Giardia intestinalis* while examining samples of his own feces. About 100 species have been named to date. Many live in the guts of animal species without causing harm to their host; other species live in stagnant water habitats. In both types of environment, oxygen availability tends to be very low.

Morphology Each cell has two nuclei, which resemble eyes when viewed under the microscope (**Figure 29.21**). Each nucleus is associated with four flagella, for a total of eight flagella per cell. Some species lack the organelles called peroxisomes and lysosomes in addition to lacking functional mitochondria. All diplomonads lack a cell wall. ✔ **You should be able to mark the origin of the double nucleus on Figure 29.7.**

Feeding and locomotion Some are parasitic, but most ingest bacteria whole. They swim using their flagella.

Reproduction Only asexual reproduction occurs; meiosis has yet to be observed in members of this lineage.

Human and ecological impacts Both *Giardia intestinalis* and *G. lamblia* are common intestinal parasites in humans and cause giardiasis, or beaver fever. *Hexamida* is an agricultural pest that causes heavy losses in turkey farms each year.

Giardia lamblia

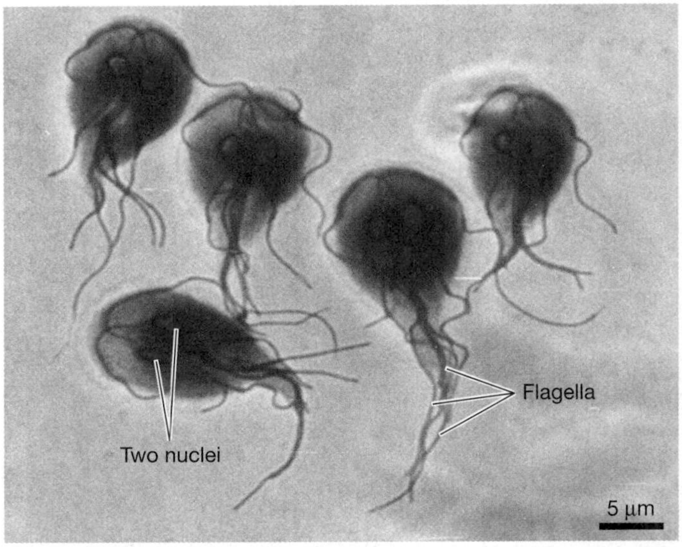

FIGURE 29.21 *Giardia* Causes Intestinal Infections in Humans.

Excavata > Euglenida

There are about 1000 known species of euglenid. Although most live in freshwater, a few are found in marine habitats. Fossil euglenids have been found in rocks over 410 million years old.

Morphology Most euglenids lack an external wall but have a unique system of interlocking protein molecules lying under the plasma membrane, which stiffen and support the cell.

Feeding and locomotion About one-third of the species have chloroplasts and perform photosynthesis, but most ingest bacteria and other small cells or particles. ✔ **You should be able to explain how the observation of ingestive feeding in euglenids relates to the hypothesis that this lineage gained chloroplasts via secondary endosymbiosis.** Instead of storing starch, the photosynthetic species synthesize a unique storage carbohydrate called paramylon. Some cells have a light-sensitive "eyespot" and use flagella to swim toward light (**Figure 29.22**). Other euglenids can inch along substrates using a unique form of movement based on extension of microtubules.

Reproduction Only asexual reproduction is known to occur among euglenids.

Human and ecological impacts Euglenids are important components of freshwater plankton and food chains.

Euglena velata

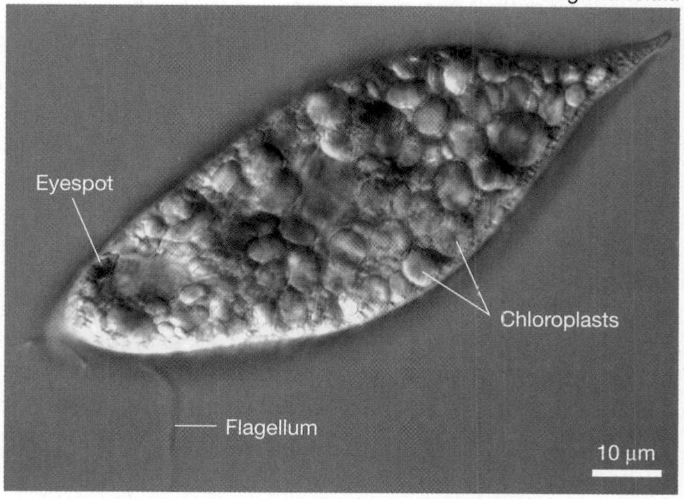

FIGURE 29.22 Euglenids Are Common in Ponds and Lakes.

Plantae > Rhodophyta (Red Algae)

The 6000 species of red algae live primarily in marine habitats. One species lives over 200 m below the surface; another is the only eukaryote capable of living in acidic hot springs. Although their color varies, many species are red because their chloroplasts contain large amounts of the accessory pigment phycoerythrin, which absorbs strongly in the blue and green portions of the visible spectrum. Because blue light penetrates water better than other wavelengths, red algae are able to live at considerable depth in the oceans. ✔ **You should be able to mark the origin of high phycoerythrin concentrations on Figure 29.7.**

Morphology Red algae cells have walls that are composed of cellulose and other polymers. A few species are unicellular, but most are multicellular. Many of the multicellular species are filamentous, but others grow as thin, hard crusts on rocks or coral. Some species have erect, leaf-like structures called thalli (**Figure 29.23**). Some species have cells with many nuclei.

Feeding and locomotion The vast majority of red algae are photosynthetic, though a few parasitic species have been identified. Red algae are the only type of algae that lack flagella.

Reproduction Asexual reproduction occurs through production of spores by mitosis. Alternation of generations is common, but the types of life cycles observed in red algae are extremely variable.

Human and ecological impacts On coral reefs, some red algae become encrusted with calcium carbonate. These species contribute to reef building and help stabilize the entire reef structure. Cultivation of *Porphyra* (or nori) for sushi and other foods is a billion-dollar-per-year industry in East Asia.

Mesophyllum lichenoides

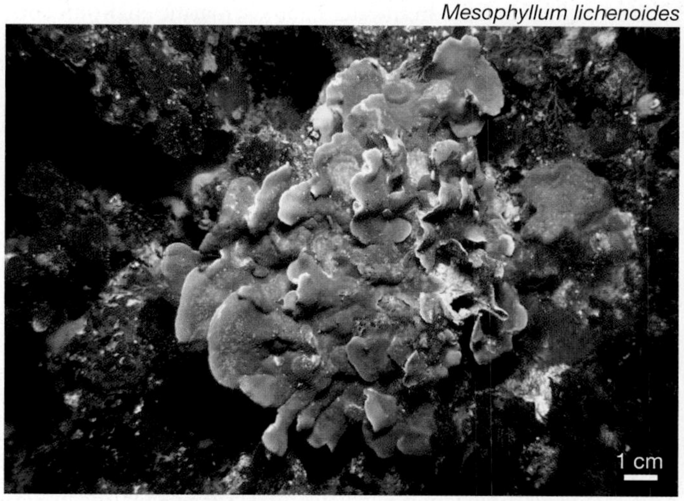

FIGURE 29.23 Red Algae Adopt an Array of Growth Forms.

Rhizaria > Foraminifera

Foraminifera, or forams, got their name from the Latin *foramen*, meaning "hole." Forams produce shells that have holes (see **Figure 29.24**) through which pseudopodia emerge. **✔ You should be able to mark the origin of the foram shell on Figure 29.7.** Fossil shells of foraminifera are abundant in marine sediments—there is a continuous record of fossilized forams that dates back 530 million years. A foram species was recently found living in sediments at a depth of 11,000 m below sea level. They are abundant in marine plankton as well as bottom habitats.

Morphology Foraminifera cells generally have multiple nuclei. The shells of forams are usually made of organic material stiffened with calcium carbonate ($CaCO_3$), and most species have several chambers. One species known from fossils was 12 cm long, but most species are much smaller. The size and shape of the shell are traits that distinguish foram species from each other.

Feeding and locomotion Like other rhizarians, forams feed by extending their pseudopodia and using them to capture and engulf bacterial and archaeal cells or bits of organic debris, which are digested in food vacuoles. Some species have symbiotic algae that perform photosynthesis and contribute sugars to their host. Forams simply float in the water.

Reproduction Asexual reproduction occurs by mitosis. When meiosis occurs, the resulting gametes are released into the open water, where pairs fuse to form a new individual.

Human and ecological impacts The shells of dead forams commonly form extensive sediment deposits when they settle out of the water, producing layers that eventually solidify into chalk, limestone, or marble. Geologists use the presence of certain foram species to date rocks—particularly during petroleum exploration.

Unidentified species

FIGURE 29.24 Forams Are Shelled Amoebae.

Alveolata > Ciliata

Ciliates were named for the cilia that cover them. Some 12,000 species are known from freshwater habitats, marine environments, and wet soils.

Morphology Ciliata cells have two distinctive nuclei: a large macronucleus and a small micronucleus (both are stained dark red in **Figure 29.25**). The macronucleus is polyploid and is actively transcribed. The micronucleus is diploid and is involved only in gene exchange between individuals. A few ciliates secrete an external skeleton. **✔ You should be able to mark the origin of the macronucleus/micronucleus structure on Figure 29.7.**

Feeding and locomotion Depending on the species, ciliates may be filter feeders, predators, or parasites. They use cilia to swim and have a mouth area where food is ingested.

Reproduction Ciliates divide to produce daughter cells asexually. They also undergo an unusual type of gene exchange called conjugation. During conjugation in ciliates, two cells line up side by side and physically connect. Micronuclei exchange between cells and fuse. The resulting nucleus eventually forms a new macronucleus and micronucleus.

Human and ecological impacts Ciliates are abundant in marine plankton and are important consumers. They are common in the digestive tracts of goats, sheep, cattle, and other grazers, where they feed on plant matter and help the host animal digest it.

Paramecium caudatum

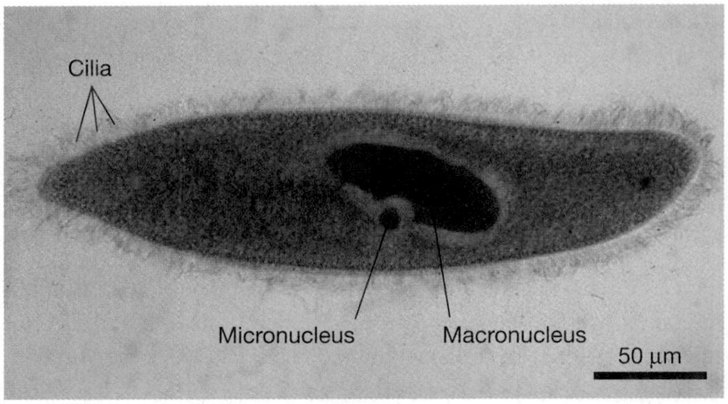

FIGURE 29.25 Ciliates Are Abundant in Freshwater Plankton.

Alveolata > Dinoflagellata

Most of the 4000 known species of dinoflagellates are ocean-dwelling plankton, although they are abundant in freshwater as well. Some species are capable of **bioluminescence**, meaning they emit light via an enzyme-catalyzed reaction (**Figure 29.26**).

Morphology Most dinoflagellates are unicellular, although some live in the aggregations of individual cells called colonies. Each species has a distinct shape, in some maintained by plates of cellulose near the surface of the cell. Unlike chromosomes of other eukaryotes, chromosomes in dinoflagellates are attached to the nuclear envelope at all times and do not contain histones.

Feeding and locomotion About half of the dinoflagellates are photosynthetic. Some of the photosynthetic species live in association with corals or sea anemones, but most are planktonic. The other species are predatory or parasitic. Dinoflagellates are distinguished by the arrangement of their two flagella: One flagellum projects out from the cell while the other runs around the cell in a groove. The two flagella are perpendicular to each other. Cells swim in a spinning motion using the two flagella. ✔ **You should be able to mark the origin of perpendicularly oriented flagella on Figure 29.7.**

Reproduction Both asexual and sexual reproduction occur. Cells that result from sexual reproduction may form tough cysts. A cyst is a resistant structure that can remain dormant until environmental conditions improve.

Human and ecological impacts Photosynthetic dinoflagellates are important primary producers in marine ecosystems—second only to diatoms in the amount of carbon they fix per year. A few species are responsible for harmful algal blooms (see Figure 29.4).

Noctiluca scintillans

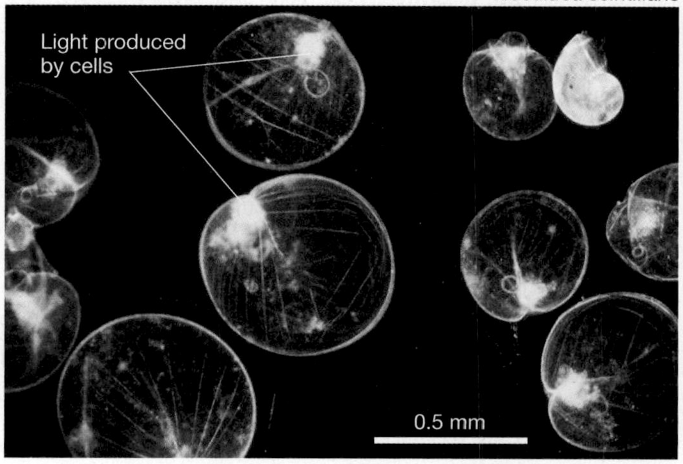

FIGURE 29.26 Some Dinoflagellate Species Are Bioluminescent.

Alveolata > Apicomplexa

All of the 5000 known species of Apicomplexa are parasitic. Species in the genus *Plasmodium* are well studied because they cause malaria in humans and other vertebrates.

Morphology Apicomplexa cells have a system of organelles at one end, called the apical complex, that is unique to the group. The apical complex allows the apicomplexan to penetrate the plasma membrane of its host. ✔ **You should be able to mark the origin of the apical complex on Figure 29.7.** The cells have a chloroplast-derived, nonphotosynthetic organelle with four membranes, indicating that apicomplexans descended from an ancestor that gained a chloroplast via secondary endosymbiosis.

Feeding and locomotion All absorb nutrition directly from their host. They lack cilia or flagella, but some species can move by amoeboid motion.

Toxoplasma gondii

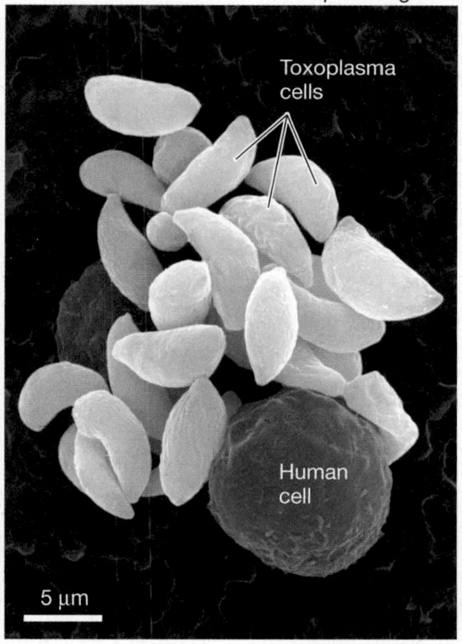

Reproduction Apicomplexans can reproduce sexually or asexually. In some species, the life cycle involves two distinct hosts, and cells must be transmitted from one host to the next.

Human and ecological impacts Various species in the genus *Plasmodium* infect birds, reptiles, or mammals and cause malaria. *Toxoplasma* is an important pathogen in people infected with HIV (**Figure 29.27**), and chicken farmers estimate that they lose $600 million per year controlling the apicomplexan *Eimeria*. Apicomplexans that parasitize insects have been used to control pests.

FIGURE 29.27 *Toxoplasma* Causes Infections in AIDS patients.

Stramenopila > Oomycota (Water Molds)

Based on morphology, oomycetes were thought to be fungi and were given fungus-like names, such as downy mildew and water molds. They are readily distinguished from fungi at the DNA level, are primarily aquatic, have cellulose instead of chitin as the primary carbohydrate in the cell walls, and have gametes with stramenopile-like flagella. The morphological similarities between oomycetes and fungi result from convergent evolution, because both groups make their living absorbing nutrition from living or dead hosts. ✔ You should be able to mark the origin of the absorptive lifestyle in oomycetes on Figure 29.7.

Morphology Some oomycetes are unicellular and some form long, branching filaments called **hyphae**. Species that form hyphae often have multinucleate cells. Cells have walls containing cellulose.

Feeding and locomotion Most species feed on decaying organic material in freshwater environments; a few are parasitic. Mature individuals are **sessile**—that is, permanently fixed to a substrate.

FIGURE 29.28 ***Phytophthora infestans* Infects Potatoes.** This is a colorized scanning electron micrograph.

Reproduction Most species are diploid throughout the majority of their life cycle. In aquatic species, spores that are produced via asexual or sexual reproduction have flagella and swim to find new food sources; in terrestrial species, spores swim in rainwater or are dispersed by the wind in dry conditions. Spores form in special structures, like those shown in **Figure 29.28**.

Human and ecological impacts Oomycetes are extremely important decomposers in aquatic ecosystems. Along with certain bacteria and archaea, they are responsible for breaking down dead organisms and releasing nutrients for use by other species. The parasitic species can be harmful to humans, however. The organism that caused the Irish potato famine was an oomycete. An oomycete parasite almost wiped out the French wine industry in the 1870s, and other species are responsible for epidemic diseases of trees, including the diebacks currently occurring in oaks found in Europe and the western United States, and in eucalyptus in Australia.

Stramenopila > Diatoms

Diatoms are unicellular or form chains of cells. About 10,000 species have been described to date, but some researchers claim that millions of species exist and have yet to be discovered.

Morphology Diatom cells are supported by silicon-rich, glassy cell walls (**Figure 29.29**) that sometimes form a box-and-lid arrangement (see Figure 29.11a). ✔ You should be able to mark the origin of the glassy cell wall on Figure 29.7.

Feeding and locomotion Diatoms are photosynthetic. Only their sperm cells have flagella and are capable of powered movement. In many species, the adult cells float in the water. But species living on a surface can glide via microtubules that project from their cell walls and that move in response to motor proteins.

Reproduction Diatoms divide by mitosis to reproduce asexually or by meiosis to form gametes that fuse to form a new individual. Many species can produce spores that are dormant during unfavorable growing conditions.

Human and ecological impacts In abundance, diatoms dominate the plankton of cold, nutrient-rich waters. They are found in virtually all aquatic habitats and are considered the most important producer of carbon compounds in fresh and salt water. Their cell walls settle into massive accumulations that are mined and sold commercially as diatomaceous earth, which is used in filtering applications and as an ingredient that adds bulk to polishes, paint, cosmetics, and other products.

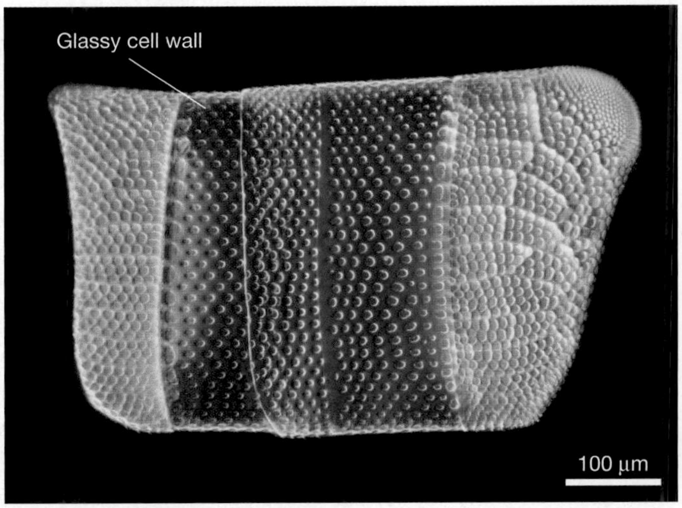

FIGURE 29.29 Diatoms Have Glass-Like Cell Walls.

Stramenopila > Phaeophyta (Brown Algae)

The color of brown algae is due to their unique suite of photosynthetic pigments. Over 1500 species have been described, most of them living in marine habitats.

Morphology The walls of Phaeophyta cells contain cellulose in addition to other complex polymers. All species are multicellular, with the body typically consisting of leaflike blades, a stalk known as a stipe, and a rootlike holdfast, which attaches the individual to a substrate (**Figure 29.30**). ✔ You should be able to mark the origin of multicellularity and brown algal photosynthetic pigments on Figure 29.7.

Feeding and locomotion Brown algae are photosynthetic and sessile, though reproductive cells may have flagella and be motile—capable of locomotion.

Reproduction Sexual reproduction occurs via the production of swimming gametes, which fuse to form a zygote. Most species exhibit alternation of generations. Depending on the species, the gametophyte and sporophyte stages may look similar or different.

Human and ecological impacts In many coastal areas with cool water, brown algae form forests or meadows that are important habitats for a wide variety of animals. In the Sargasso Sea, off Bermuda, floating brown algae form extensive rafts that harbor an abundance of animal species. The compound algin is purified from kelp and used in the manufacture of cosmetics and paint.

FIGURE 29.30 Many Brown Algae Have a Holdfast, Stalk, and Leaflike Blades.

CHAPTER 29 REVIEW

For media, go to the study area at www.masteringbiology.com

Summary of Key Concepts

Protists are a paraphyletic grouping that includes all eukaryotes except the land plants, fungi, and animals. Biologists study them because they are important medically, ecologically, and evolutionarily.

- Parasitic protists cause several important diseases in humans, including malaria.
- Toxin-producing protists that grow to high densities result in a harmful algal bloom.
- Protists are often tremendously abundant in marine and freshwater plankton and other habitats. Protists provide food for many organisms in aquatic ecosystems and fix so much carbon that they have a large impact on the global carbon budget.

✔ You should be able to explain what is "primary" about primary production by photosynthetic protists.

Key morphological innovations occurred as protists diversified: the nuclear envelope, multicellularity, and an array of structures that function in support and protection. In addition, the mitochondrion and chloroplast arose by endosymbiosis.

- Several morphological features are common to all or almost all eukaryotes, including the nucleus, endomembrane system, extensive cytoskeleton, and flagellum.
- Multicellularity evolved in several different protist groups independently.
- Structures that provide support or protection vary among lineages, and include cell walls, shells, internal support rods, and alveoli.
- Eukaryotes contain mitochondria or have genes indicating that their ancestors once contained mitochondria.
- Several types of data support the hypothesis that mitochondria originated as an endosymbiotic α-proteobacterium.
- The chloroplast's size, DNA structure, ribosomes, double membrane, and evolutionary relationships are consistent with the hypothesis that this organelle originated as an endosymbiotic cyanobacterium.
- After primary endosymbiosis occurred, chloroplasts were "passed around" to new lineages of protists via secondary endosymbiosis.

✔ You should be able to explain the original source of (**1**) the two membranes in a plant chloroplast, and (**2**) the four membranes in the chloroplast-derived organelle of *Plasmodium*.

Protists vary widely in the way they obtain food. Many species are photosynthetic; others obtain carbon compounds by ingesting food or parasitizing other organisms.

- Protists exhibit predatory, parasitic, or photosynthetic lifestyles, which evolved in many groups independently.
- The evolution of ingestive feeding was important for two reasons: (1) It allowed eukaryotes to obtain resources in a new way—by eating bacteria, archaea, and other eukaryotes; and (2) it made endosymbiosis and the evolution of mitochondria and chloroplasts possible.

✔ You should be able to propose a hypothesis to explain why the wide diversity of photosynthetic pigments observed among protists limits competition.

Protists vary widely in terms of how they reproduce. Sexual reproduction evolved in protists, and many protist species can reproduce both sexually and asexually.

- Protists undergo cell division based on mitosis and reproduce asexually.
- Many protists also undergo meiosis and sexual reproduction at some phase in their life cycle.
- Alternation of generations is common in multicellular species—meaning there are separate haploid and diploid forms of the same species. When alternation of generations occurs, haploid gametophytes produce gametes by mitosis; diploid sporophytes produce spores by meiosis.

✔ You should be able to explain how you can tell a gametophyte from a sporophyte, in terms of ploidy.

MB **Web Activity** Alternation of Generations in a Protist

Questions

✔ TEST YOUR KNOWLEDGE

Answers are available in Appendix B

1. Why are protists considered paraphyletic?
   a. They include many extinct forms, including lineages that no longer have any living representatives.
   b. They include some but not all descendants of their most recent common ancestor.
   c. They represent all of the descendants of a single common ancestor.
   d. Not all protists have all of the synapomorphies that define the Eukarya, such as a nucleus.
2. What material is *not* used by protists to manufacture hard outer coverings?
   a. cellulose
   b. lignin
   c. glass-like compounds that contain silicon
   d. mineral-like compounds such as calcium carbonate ($CaCO_3$)
3. What does amoeboid motion result from?
   a. interactions among actin, myosin, and ATP
   b. coordinated beats of cilia
   c. the whiplike action of flagella
   d. action by the mitotic spindle, similar to what happens during mitosis and meiosis
4. According to the endosymbiosis theory, what type of organism is the original ancestor of the chloroplast?
   a. a photosynthetic archaean
   b. a cyanobacterium
   c. an algal-like, primitive photosynthetic eukaryote
   d. a modified mitochondrion
5. Multicellularity is defined in part by the presence of distinctive cell types. At the cellular level, what does this criterion imply?
   a. Individual cells must be extremely large.
   b. The organism must be able to reproduce sexually.
   c. Cells must be able to move.
   d. Different cell types express different genes.
6. Why are protists an important part of the global carbon cycle and marine food chains?
   a. They have high species diversity.
   b. They are numerically abundant.
   c. They have the ability to parasitize humans.
   d. They have the ability to undergo meiosis.

✔ TEST YOUR UNDERSTANDING

Answers are available in Appendix B

1. Why is an extensive cytoskeleton and the lack of a cell wall required for ingestive feeding by engulfing?
2. Explain the logic behind the claim that the nuclear envelope is a synapomorphy that defines eukaryotes as a monophyletic group.
3. What is the relationship between meiosis and the alternation of generations?
4. Consider the endosymbiosis theory for the origin of the mitochondrion. What did each partner provide the other, and what did each receive in return?
5. Why was finding a close relationship between mitochondrial DNA and bacterial DNA considered particularly strong evidence in favor of the endosymbiosis theory?
6. The text claims that the evolutionary history of protists can be understood as a series of morphological innovations that established seven distinct lineages, each of which subsequently diversified based on innovative ways of feeding, moving, and reproducing. Explain how the Alveolata support this claim.

APPLYING CONCEPTS TO NEW SITUATIONS

Answers are available in Appendix B

1. Consider the following:
   - All living eukaryotes have mitochondria or have evidence in their genomes that they once had these organelles. Thus it appears that eukaryotes acquired mitochondria very early in their history.
   - The first eukaryotic cells in the fossil record correlate with the first appearance of rocks formed in an oxygen-rich ocean and atmosphere.

   How are these observations connected? (Hint: Before answering, glance at Chapter 9 and remind yourself what happens in a mitochondrion.)

2. Consider the following:
   - *Plasmodium* has an unusual organelle called an apicoplast. Recent research has shown that apicoplasts are derived from chloroplasts via secondary endosymbiosis and have a large number of genes encoded by chloroplast DNA.
   - Glyphosate is one of the most widely used herbicides. It works by poisoning an enzyme encoded by a gene in chloroplast DNA.
   - Biologists are testing the hypothesis that glyphosate could be used as an antimalarial drug in humans.

   How are these observations connected?

3. Suppose a friend says that we don't need to worry about global warming. Her claim is that increased temperatures will make planktonic algae grow faster and that carbon dioxide (CO_2) will be removed from the atmosphere faster. According to her, this carbon will be buried at the bottom of the ocean in calcium carbonate shells. As a result, the amount of carbon dioxide in the atmosphere will decrease and global warming will decline. Comment.

4. Biologists are beginning to draw a distinction between "species trees" and "gene trees." A species tree is a phylogeny that describes the actual evolutionary history of a lineage. A gene tree, in contrast, describes the evolutionary history of one particular gene, such as a gene required for the synthesis of chlorophyll *a*. In some cases, species trees and gene trees don't agree with each other. For example, the species tree for green algae indicates that their closest relatives are protists and plants. But the gene tree based on chlorophyll *a* from green algae suggests that this gene's closest relative is a bacterium, not a protist. What's going on? Why do these types of conflicts exist?

Earth has been called the Blue Planet, but it would be just as accurate to call it the Green Planet.

30 Green Algae and Land Plants

KEY CONCEPTS

- Green algae are an important source of oxygen and provide food for aquatic organisms; land plants release oxygen, hold soil and water in place, build soil, moderate extreme temperatures and winds, and provide food for terrestrial organisms.
- Land plants were the first multicellular organisms that could live with most of their tissues exposed to the air. A series of key adaptations allowed them to survive on land. In terms of total mass, plants dominate today's terrestrial environments.
- Once plants were able to grow on land, a sequence of important evolutionary changes made it possible for them to reproduce efficiently—even in extremely dry environments.

In terms of their total mass and their importance to other organisms, the green plants dominate terrestrial and freshwater habitats. When you walk through a forest or meadow, you are surrounded by green plants. If you look at pond or lake water under a microscope, green plants are everywhere.

The green plants comprise two major types of organisms: the green algae and the land plants. Green algae are important photosynthetic organisms in aquatic habitats—particularly lakes, ponds, and other freshwater settings. Land plants are the key photosynthesizers in terrestrial environments.

Although green algae have traditionally been considered protists, it is logical to study them along with land plants for two reasons: (1) They are the closest living relative to land plants and form a monophyletic group with them, and (2) the transition from aquatic to terrestrial life occurred when land plants evolved from green algae.

Land plants were the first organisms that could thrive with their tissues completely exposed to the air instead of being partially or completely submerged. Before they evolved, it is likely that the only life on the continents consisted of bacteria, archaea, and single-celled protists that thrive in wet soils. By colonizing the continents, plants transformed the nature of life on Earth. It was, in the words of Karl Niklas, "one of the greatest adaptive events in the history of life." Land plants made the Earth green.

30.1 Why Do Biologists Study the Green Algae and Land Plants?

Biologists study the green plants because they are fascinating, and because we couldn't live without them. Along with most other animals and fungi, humans are almost completely dependent on land plants for food. People rely on plants for other necessities of life as well—oxygen, fuel, building materials, and the fibers used in making clothing,

✔ When you see this checkmark, stop and test yourself. Answers are available in Appendix B.

paper, ropes, and baskets. But we also rely on land plants for important intangible value such as the aesthetic appeal of landscaping and bouquets. To drive this point home, consider that the sale of cut flowers generates over $1 billion each year in the United States alone.

Based on these observations, it is not surprising that agriculture, forestry, and horticulture are among the most important endeavors supported by biological science. Tens of thousands of biologists are employed in research designed to increase the productivity of plants and to create new ways of using them for the benefit of people. Research programs also focus on two types of land plants that cause problems for people: weeds that decrease the productivity of crop plants and newly introduced species that invade and then degrade natural areas.

Plants Provide Ecosystem Services

An **ecosystem** consists of all the organisms in a particular area, along with physical components of the environment such as the atmosphere, precipitation, surface water, sunlight, soil, and nutrients. Green algae and land plants provide **ecosystem services** because they enhance the life-supporting attributes of the atmosphere, surface water, soil, and other physical components of an ecosystem. Green plants alter the environment in ways that benefit many other organisms:

PLANTS PRODUCE OXYGEN Recall from Chapter 10 that plants perform oxygenic (literally, "oxygen-producing") photosynthesis. In this process, electrons that are removed from water molecules are used to reduce carbon dioxide (CO_2). In the process of stripping electrons from water, plants release oxygen molecules (O_2) as a by-product.

As Chapter 28 noted, oxygenic photosynthesis evolved in cyanobacteria and was responsible for the origin of an oxygen-rich atmosphere. The evolutionary success of plants continued this trend because plants add huge amounts of oxygen to the atmosphere.

Without the green plants, we and other aerobic organisms would be in danger of suffocating for lack of oxygen.

PLANTS BUILD AND HOLD SOIL Leaves and roots and stems that are not eaten when they are alive fall to the ground and provide food for worms, fungi, bacteria, archaea, protists, and other decomposers in the soil. These organisms add organic matter to the soil, which improves soil structure and the ability of soils to hold nutrients and water.

The extensive network of fine roots produced by trees, grasses, and other land plants helps hold soil particles in place. And by taking up nutrients in the soil, plants prevent the nutrients from being blown or washed away.

When areas are devegetated by grazing, farming, logging, or suburbanization, large quantities of soil and nutrients are lost to erosion by wind and water (**Figure 30.1**).

PLANTS HOLD WATER AND MODERATE CLIMATE Plant tissues take up and retain water. Intact forests, prairies, and wetlands also prevent rain from quickly running off a landscape: Plant leaves soften the physical impact of rainfall on soil, and plant organic matter builds the soil's water-holding capacity.

FIGURE 30.1 Devegetation Leads to Erosion.

When areas are devegetated, streams are more prone to flooding and groundwater is not replenished efficiently. It is common to observe streams alternately flooding and then drying up completely when the surrounding area is deforested.

By providing shade, plants reduce temperatures beneath them and increase relative humidity. They also reduce the impact of winds that dry out landscapes.

When plants are removed from landscapes to make way for farms or suburbs, habitats become much dryer and are subject to more extreme temperature swings.

PLANTS AS PRIMARY PRODUCERS Land plants are the dominant primary producers in terrestrial ecosystems. As Chapter 29 indicated, primary producers convert energy in sunlight into chemical energy. The sugars that land plants produce by photosynthesis support virtually all of the other organisms present in terrestrial habitats.

Plants are eaten by herbivores ("plant-eaters"), which range in size from insects to elephants. These consumers are eaten by carnivores ("meat-eaters"), ranging in size from the tiniest spiders to polar bears. Humans are an example of omnivores ("all-eaters"), organisms that eat both plants and animals. Omnivores feed at several different levels in the food chain. For example, people consume plants, herbivores such as chicken and cattle, and carnivores such as salmon and tuna.

Because of their role in primary production, land plants are the key to the carbon cycle on the continents. Plants take CO_2 from the atmosphere and reduce it to make sugars. Although green algae and land plants also produce a great deal of CO_2 as a result of cellular respiration, they fix much more CO_2 than they release.

The loss of plant-rich prairies and forests, due to suburbanization or conversion to agriculture has contributed to increased concentrations of CO_2 in the atmosphere. Higher carbon dioxide levels, in turn, are responsible for the rapid warming that is occurring worldwide (see Chapter 54).

Plants Provide Humans with Food, Fuel, Fiber, Building Materials, and Medicines

It is difficult to overstate the importance of plant research to the well-being of human societies. Plants provide our food supply as well as a significant percentage of the fuel, fibers, building materials, and medicines that we use.

FOOD Agricultural research began with the initial domestication of crop plants, which occurred independently at several locations around the world between 10,000 and 2000 years ago (**Figure 30.2a**). By actively selecting individuals with the largest and most nutritious seeds, leaves, or other plant parts year after year, our ancestors gradually changed the characteristics of certain wild species. This process is called **artificial selection**.

Figure 30.2b compares kernel size in modern maize and its wild ancestor—a grass called teosinte that is native to Mexico. Chapter 1 presented data indicating that artificial selection has been responsible for dramatic increases in the oil content of maize kernels over the past 100 years. Chapter 19 highlighted a current focus in agricultural research—the improvement of crop varieties through genetic engineering.

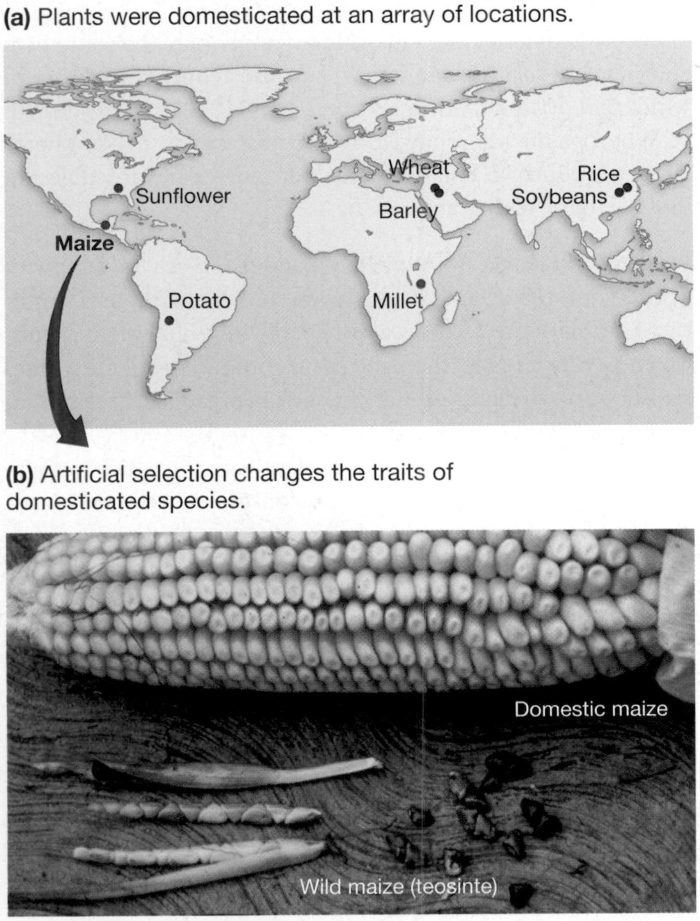

FIGURE 30.2 Crop Plants Are Derived from Wild Species via Artificial Selection. (a) Crop plants have originated on virtually every continent. **(b)** In some cases, artificial selection has changed domesticated species so radically that they bear little resemblance to their wild ancestors.

FUEL For perhaps 100,000 years, wood burning was the primary source of energy used by all humans. As **Figure 30.3** shows, however, wood has been replaced in the industrialized countries by other sources of energy. This graph shows the percentage of total energy needs in the United States fulfilled by wood, coal, and petroleum or natural gas, over time. Note that the first fuel to replace wood in the United States and many other industrialized countries has been coal, which forms from partially decayed plant material that is compacted over time by overlying sediments and hardened into rock.

Starting in the mid-1800s, people in England, Germany, and the United States began to mine coal deposits that originally formed during the Carboniferous period some 350–275 million years ago. The coal fueled blast furnaces that smelted vast quantities of iron ore into steel and powered the steam engines that sent trains streaking across Europe and North America. It is no exaggeration to claim that the organic compounds synthesized by plants more than 300 million years ago laid the groundwork for the Industrial Revolution.

Coal still supplies about 20 percent of the energy used in Japan and Western Europe and 80 percent of the energy used in China. Current research is focused on finding cleaner and more efficient ways of mining and burning coal.

FIBER AND BUILDING MATERIALS Although nylon and polyester derived from petroleum are increasingly important in manufacturing, cotton and other types of plant fibers are still important sources of raw material for clothing, rope, and household articles like towels and bedding.

Woody plants also provide lumber for houses and furniture. Relative to its density, wood is a stiffer and stronger building material than concrete, cast iron, aluminum alloys, or steel.

Woody plants also provide most of the fibers used in papermaking. The cellulose fibers refined from trees or bamboo and then used in paper manufacturing are stronger under tension (pulling) than nylon, silk, chitin, collagen, tendon, or bone—even though cellulose is 25 percent less dense. One line of research is focused on bioengineering reduced lignin content of

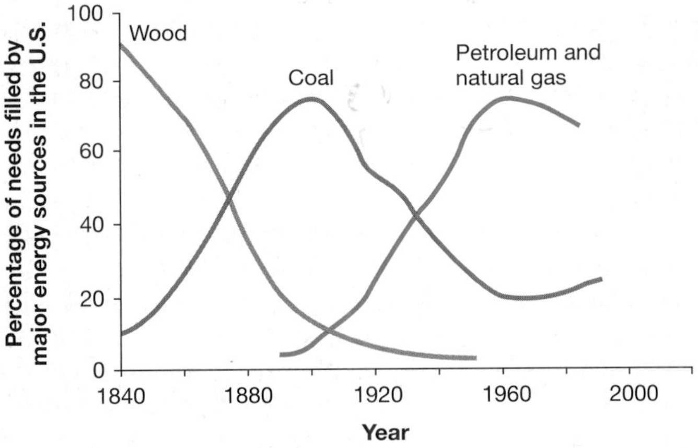

FIGURE 30.3 Changes over Time in Fuel Sources Used by an Industrialized Country. Although wood has declined in importance in the industrialized countries, it is still the primary cooking and heating fuel in many areas of the world.

TABLE 30.1 **Some Drugs Derived from Land Plants**

Compound	Source	Use
Atropine	Belladonna plant	Dilating pupils during eye exams
Codeine	Opium poppy	Pain relief, cough suppressant
Digitalin	Foxglove	Heart medication
Ipecac	Ipecac	Treating amoebic dysentery, poison control
Menthol	Peppermint	Cough suppressant, relief of stuffy nose
Morphine	Opium poppy	Pain relief
Papain	Papaya	Reduce inflammation, treat wounds
Quinine	Quinine tree	Malaria prevention
Quinidine	Quinine tree	Heart medication
Salicin	Aspen, willow trees	Pain relief (aspirin)
Steroids	Wild yams	Precursor compounds for manufacture of birth control pills and cortisone (to treat inflammation)
Taxol	Pacific yew	Ovarian cancer
Tubocurarine	Curare vine	Muscle relaxant used in surgery
Vinblastine, vincristine	Rosy periwinkle	Leukemia (cancer of blood)

certain woody species, so fewer pollutants need to be used in extracting lignin during the papermaking process.

PHARMACEUTICALS In both traditional and modern medicine, plants are a key source of drugs. **Table 30.1** lists some of the more familiar medicines derived from land plants; overall, it has been estimated that about 25 percent of the prescriptions written in the United States each year include at least one molecule derived from plants.

In most cases, plants synthesize these compounds in order to repel insects, deer, or other types of herbivores. For example, experiments have confirmed that morphine, cocaine, nicotine, caffeine, and other toxic compounds found in plants are effective deterrents to insect or mammalian consumers. Researchers continue to isolate and test new plant compounds for medicinal use in humans and domesticated animals.

30.2 How Do Biologists Study Green Algae and Land Plants?

Given the importance of plants to the planet in general and humans in particular, it is not surprising that knowing as much as possible about plants, including how they evolved, is a key component of biological science. To understand how green plants originated and diversified, biologists analyze (1) morphological traits, (2) the fossil record, and (3) phylogenetic trees estimated from similarities and differences in DNA sequences from homologous genes.

The three approaches are complementary and have produced a remarkably clear picture of how land plants evolved from green algae and then diversified. Let's consider each of these research strategies.

Analyzing Morphological Traits

The green algae include species that are unicellular, colonial, or multicellular and that live in marine, freshwater, or moist terrestrial habitats. The vast majority are aquatic. Although some land plants live in ponds or lakes or rivers, the vast majority live on land.

SIMILARITIES BETWEEN GREEN ALGAE AND LAND PLANTS The green algae have long been hypothesized to be closely related to land plants, because key morphological traits are similar in the two groups.

- Their chloroplasts contain the photosynthetic pigments chlorophyll *a* and *b* and the accessory pigment β-carotene.
- They have similar arrangements of the internal, membrane-bound sacs called thylakoids (see Chapter 10).
- Their cell walls, sperm, and peroxisomes are similar in structure and composition. (Recall from Chapter 7 that peroxisomes are organelles in which specialized oxidation reactions take place.)
- Their chloroplasts synthesize starch as a storage product.

Of all the green algal groups, the two most similar to land plants are the Coleochaetophyceae (coleochaetes) and Charophyceae (stoneworts; **Figure 30.4** on page 550). Because the species that make up these groups are multicellular and live in ponds and other types of freshwater environments, biologists hypothesize that land plants evolved from multicellular green algae that lived in freshwater habitats.

MAJOR MORPHOLOGICAL DIFFERENCES AMONG LAND PLANTS Based on morphology, the land plants were traditionally clustered into three broad categories:

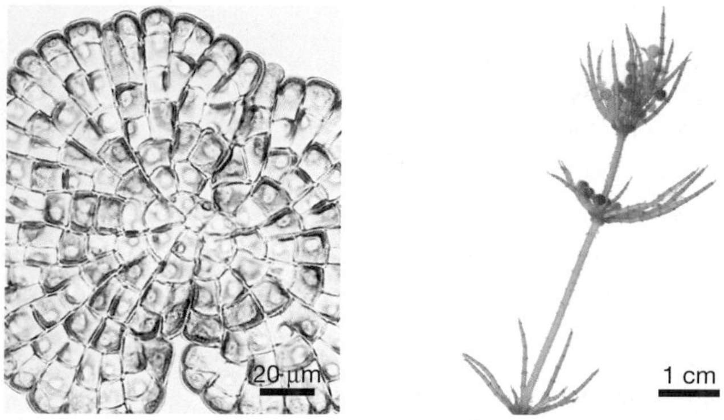

Coleochaetophyceae (coleochaetes) Charophyceae (stoneworts)

FIGURE 30.4 Most Green Algae Are Aquatic. Examples of species from the green algal lineages most closely related to the land plants.

1. Non-vascular plants (**bryophytes**) lack **vascular tissue**—specialized groups of cells that conduct water and nutrients from one part of the plant body to another. Mosses are bryophytes (**Figure 30.5a**).
2. Seedless vascular plants have vascular tissue but do not make seeds. A **seed** consists of an embryo and a store of nutritive tissue, surrounded by a tough protective layer. Ferns are seedless vascular plants (**Figure 30.5b**).
3. Seed plants have vascular tissue. The flowering plants, or **angiosperms** (literally, "encased seeds"), are seed plants (**Figure 30.5c**).

How are non-vascular plants, seedless vascular plants, and seed plants related to each other, and to green algae?

Using the Fossil Record

The first green plants in the fossil record are green algae in 700–725 million-year-old rocks. The first land plant fossils are found in rocks that are about 475 million years old. Because green algae appear long before land plants, the data support the hypothesis that land plants are derived from green algae.

At roughly the same time that green algae appeared and began to diversify, the oceans and atmosphere were starting to become oxygen-rich—as never before in Earth's history. Based on this time correlation, it is reasonable to hypothesize that the evolution of green algae contributed to the rise of oxygen levels on Earth. The origin of the oxygen atmosphere occurred not long before the appearance of animals in the fossil record and may have played a role in their origin and early diversification.

The fossil record of the land plants themselves is massive. In an attempt to organize and synthesize the database, **Figure 30.6** breaks it into five time intervals—each encompassing a major event in the diversification of land plants.

ORIGIN OF LAND PLANTS The oldest interval begins about 475 million years ago (mya), spans some 60 million years, and documents the origin of the group.

Most of the fossils dating from this period are microscopic. They consist of the reproductive cells called spores and sheets of a waxy coating called cuticle. Several observations support the hypothesis that these fossils came from green plants that were growing on land.

1. Cuticle is a watertight barrier that coats today's land plants and helps them resist drying.
2. The fossilized spores are surrounded by a sheetlike coating. Under the electron microscope, the coating material appears almost identical in structure to a watertight material called **sporopollenin**, which encases spores and pollen from modern land plants and helps them resist drying.
3. Fossilized spores that are 475 million years old have recently been found in association with spore-producing structures, called **sporangia** (singular: **sporangium**), that are similar to sporangia observed in some non-vascular plants.

SILURIAN-DEVONIAN EXPLOSION The second major interval in the fossil record of land plants is called the "Silurian-Devonian explosion." In rocks dated 416–359 mya, biologists find fossils from most of the major plant lineages. Virtually all of the adapta-

(a) Non-vascular plants **(b)** Seedless vascular plants **(c)** Seed plants

Do not have vascular tissue to conduct water and provide support (for example, mosses)

Have vascular tissue but do not make seeds (for example, ferns)

Have vascular tissue and make seeds (for example, flowering plants, or angiosperms)

FIGURE 30.5 Morphological Diversity in Land Plants.

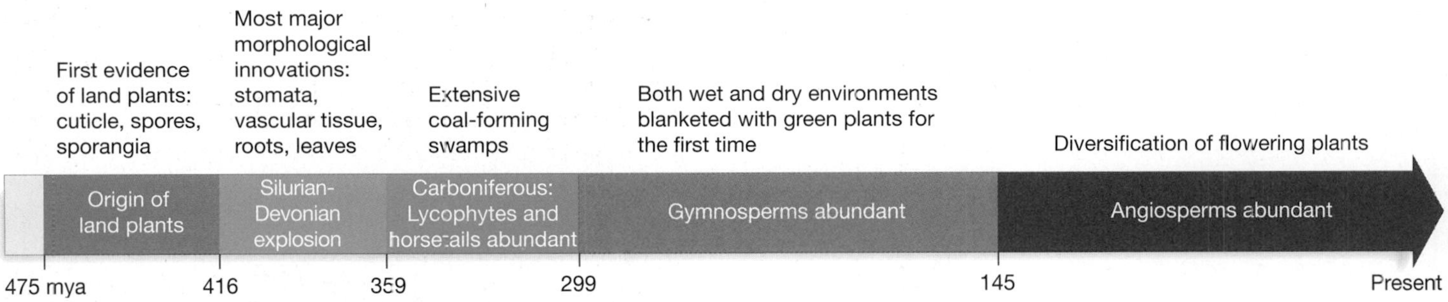

FIGURE 30.6 The Fossil Record of Land Plants Can Be Broken into Five Major Intervals. Note that the first insects are found in the fossil record at about 400 mya; the first terrestrial vertebrate animals at 365 mya, and the first mammals at 190 mya.

tions that allow plants to occupy dry, terrestrial habitats are present, including water-conducting tissue and roots.

According to the fossil record, plants colonized the land in conjunction with fungi that grew in a mutually beneficial association. The fungi grew belowground and helped provide land plants with nutrients from the soil; in return, the plants provided the fungi with sugars and other products of photosynthesis (see Chapter 31).

THE CARBONIFEROUS PERIOD The third interval in the fossil history of plants spans the aptly named Carboniferous period. In sediments dated from about 359 to 299 mya, biologists find extensive deposits of coal. Coal is a carbon-rich rock packed with fossil spores, branches, leaves, and tree trunks.

Most of these fossils are derived from the Lycophyta (the lycophytes or club mosses), Equisetophyta (horsetails), and ferns. Although the only living lycophytes and horsetails are small, during the Carboniferous these groups were species rich and included a wide array of tree-sized forms.

Because coal formation is thought to start only in the presence of water, the Carboniferous fossils indicate the presence of extensive forested swamps.

DIVERSIFICATION OF GYMNOSPERMS The fourth interval in land plant history is characterized by seed plants called **gymnosperms** ("naked-seeds"). Five major groups of gymnosperms are living today: Cycadophyta (cycads), Ginkgophyta (ginkgoes), Gnetophyta (gnetophytes), Pinophyta (pines, spruces, and firs), and other cone-bearing species (redwoods, junipers, yews).

Because gymnosperms grow readily in dry habitats, biologists infer that both wet and dry environments on the continents became blanketed with green plants for the first time during this interval. Gymnosperms are particularly prominent in the fossil record from 251 mya to 145 mya.

DIVERSIFICATION OF FLOWERING PLANTS The fifth interval in the history of land plants is still under way. This is the age of flowering plants—the angiosperms. The first flowering plants in the fossil record appear about 150 mya. The plants that produced the first flowers are the ancestors of today's grasses, orchids, daisies, oaks, maples, and roses.

According to the fossil record, then, the green algae appear first, followed by the non-vascular plants, seedless vascular plants, and seed plants. Organisms that appear late in the fossil record are often much less dependent on moist habitats than are groups that appear earlier. For example, the sperm cells of mosses and ferns swim to accomplish fertilization, while gymnosperms and angiosperms produce pollen grains that are transported via wind or insects and that then produce sperm.

These observations support the hypotheses that green plants evolved from green algae and that in terms of habitat use, the evolution of green plants occurred in a wet-to-dry trend.

To test the validity of these observations, biologists analyze data sets that are independent of the fossil record. Foremost among these are DNA sequences used to infer phylogenetic trees. Does the phylogeny of land plants confirm or contradict the patterns in the fossil record?

Evaluating Molecular Phylogenies

The phylogenetic tree in **Figure 30.7** on page 552 is a recent version of results emerging from laboratories around the world. The black bars across some branches show when key innovations occurred.

Each of the following bullets states a key observation about this tree, followed by a highlighted interpretation—often, a hypothesis that is supported by the observation.

- The green plants are monophyletic.
  Interpretation: A single common ancestor gave rise to all of the green algae and land plants.
- The initial splitting events on the tree, near the root, lead to lineages of green algae.
  Interpretation: Land plants evolved from green algae.
- Green algae are paraphyletic.
  Interpretation: The green algae include some but not all of the descendants of a single common ancestor.
- Charophyceae are the closest living relative to land plants.
  Interpretation: Land plants evolved from a multicellular green alga that lived in freshwater habitats.
- Land plants are monophyletic.
  Interpretation: There was only one transition from freshwater environments to land.
- The bryophytes or non-vascular plants are the earliest-branching groups among land plants.

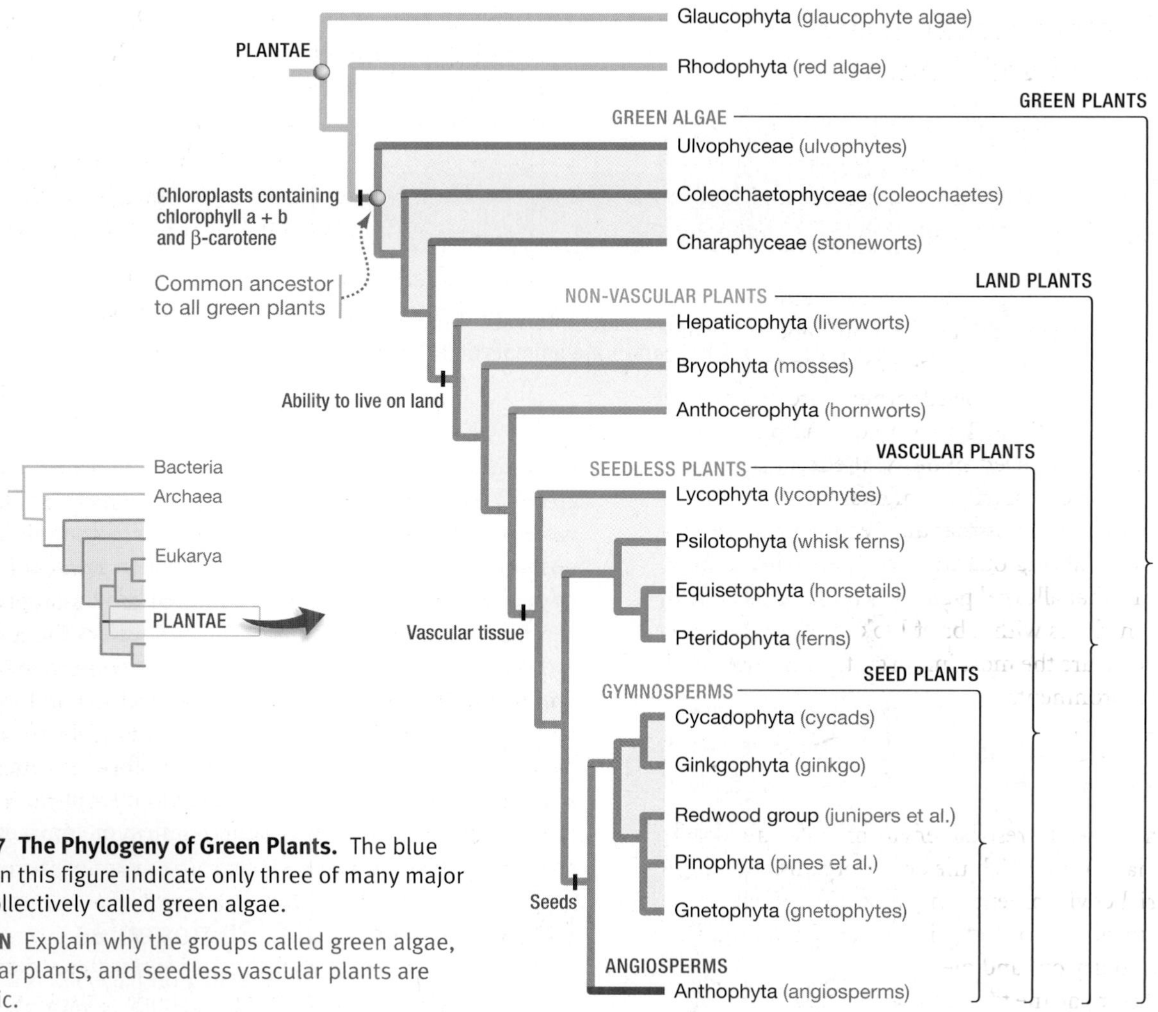

FIGURE 30.7 The Phylogeny of Green Plants. The blue branches on this figure indicate only three of many major lineages collectively called green algae.

✔**QUESTION** Explain why the groups called green algae, non-vascular plants, and seedless vascular plants are paraphyletic.

Interpretation: The non-vascular plants are the most ancient living group of plants.

- The non-vascular plants form a grade—meaning a sequence of lineages.
  Interpretation: The non-vascular plants are paraphyletic.
- The seedless vascular plants form a grade, but the vascular plants as a whole are monophyletic.
  Interpretation: Vascular tissue evolved once.
- The seed plants—the gymnosperms plus angiosperms—are monophyletic.
  Interpretation: The seed evolved once.
- The gymnosperms are a monophyletic group, as are the angiosperms.
  Interpretation: Among seed plants, there was a major divergence in how seeds develop—either "naked" (in gymnosperms) or protected inside a capsule (in angiosperms).

Although the tree in Figure 30.7 will undoubtedly change and improve as additional data accumulate, biologists are confident about its most fundamental message: The fossil record and the phylogenetic tree based on DNA sequence data agree on the order in which groups appeared. Land plant evolution began with non-vascular plants, proceeded to seedless vascular plants, and continued with the evolution of seed plants.

CHECK YOUR UNDERSTANDING

If you understand that . . .

- Biologists use the fossil record and phylogenetic analyses to study how green plants diversified.
- The data analyzed to date support the hypotheses that green plants are monophyletic and that land plants evolved from multicellular green algae that inhabited freshwater.
- The fossil record and molecular analyses agree that the non-vascular plants evolved first, followed by the seedless vascular plants and the seed plants.

✔ **You should be able to . . .**

Explain why (1) morphological data, (2) the fossil record, and (3) molecular phylogenies all support the hypothesis that land plants evolved from green algae.

Answers are available in Appendix B.

30.3 What Themes Occur in the Diversification of Land Plants?

Land plants have evolved from algae that grew on the muddy shores of ponds 475 million years ago to organisms that enrich the soil, produce much of the oxygen you breathe and most of the food you eat, and serve as symbols of health, love, and beauty. How did this happen?

Answering this question begins with recognizing the most striking trend in the phylogeny and fossil record of green plants: The most ancient groups in the lineage are dependent on wet habitats, while more recently evolved groups are tolerant of dry—or even desert—conditions. The story of land plants is the story of adaptations that allowed photosynthetic organisms to move from aquatic to terrestrial environments.

Let's first consider adaptations that allowed plants to grow in dry conditions without drying out and dying, and then analyze the evolution of traits that allowed plants to reproduce efficiently on land. This section closes with a brief look at the radiation of flowering plants, which are the most important plants in many of today's terrestrial environments.

The Transition to Land, I: How Did Plants Adapt to Dry Conditions?

For aquatic green algae, terrestrial environments are deadly. Compared with a habitat in which the entire organism is bathed in fluid, in terrestrial environments only a portion, if any, of the plant's tissues are in direct contact with water. Tissues that are exposed to air tend to dry out and die.

Once green plants made the transition to survive out of water, though, growth on land offered a bonanza of resources.

- *Light* The water in ponds, lakes, and oceans absorbs and reflects light. As a result, the amount of light available to drive photosynthesis is drastically reduced even a meter or two below the water surface.
- *Carbon dioxide* CO_2—the most important molecule required by photosynthetic organisms—is more abundant in the atmosphere and diffuses more readily there than it does in water

Natural selection favored early land plants with adaptations that solved the drying problem. These adaptations arose in two steps: (1) preventing water loss, which kept cells from drying out and dying; and (2) moving water from tissues with direct access to water to tissues without direct access. Let's examine each in turn.

PREVENTING WATER LOSS: CUTICLE AND STOMATA Section 30.2 pointed out that sheets of the waxy substance called cuticle are present early in the fossil record of land plants, along with encased spores. This observation is significant because the presence of cuticle in fossils is a diagnostic indicator of land plants.

Cuticle is a watertight sealant that covers the aboveground parts of plants and gives them the ability to survive in dry environments (**Figure 30.8a**). If biologists had to point to one innovation that made the transition to land possible, it would be the random mutations that led to the production of cuticle.

Covering surfaces with wax creates a problem, however, regarding the exchange of gases across those surfaces. Plants need to take in carbon dioxide (CO_2) from the atmosphere in order to perform photosynthesis. But cuticle is almost as impervious to CO_2 as it is to water.

Most modern plants solve this problem with a structure called a **stoma** ("mouth"; plural: **stomata**). A stoma consists of an opening surrounded by specialized **guard cells** (**Figure 30.8b**). The opening,

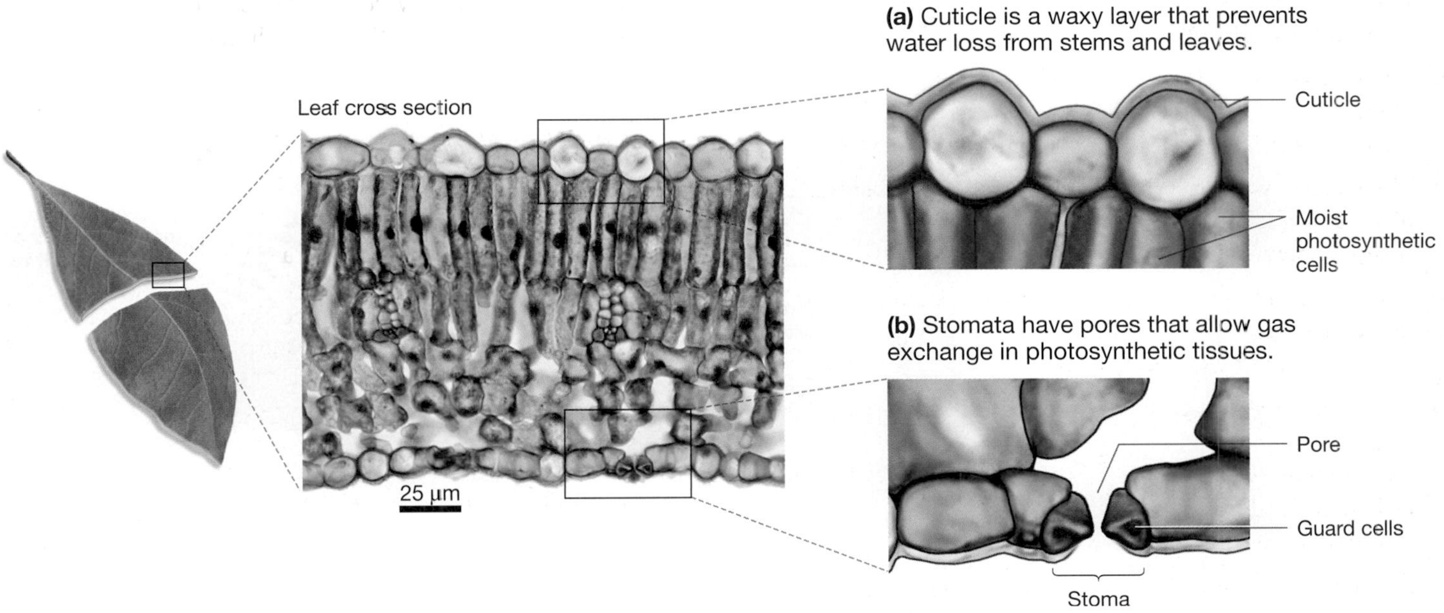

(a) Cuticle is a waxy layer that prevents water loss from stems and leaves.
Leaf cross section
Cuticle
Moist photosynthetic cells
(b) Stomata have pores that allow gas exchange in photosynthetic tissues.
Pore
Guard cells
25 μm
Stoma

FIGURE 30.8 Cuticle and Stomata Are the Most Fundamental Plant Adaptations to Life on Land. In these micrographs, leaf cells have been stained blue to make their structure more visible. **(a)** The interior of plant leaves and stems is extremely moist; cuticle prevents water from evaporating away. **(b)** Stomata create pores to allow CO_2 to diffuse into the interior of leaves and stems where cells are actively photosynthesizing, and to allow excess O_2 to diffuse out.

called a **pore**, opens or closes as the guard cells change shape. When guard cells become flaccid or "limp" due to an outflow of water, they close the stomata. Pores are closed in this way to limit water loss from the plant. But when guard cells become turgid or "taut" due to an inflow of water, they open the pore. Open stomata allow CO_2 to diffuse into the interior of leaves and stems where cells are actively photosynthesizing. They also allow excess O_2 to diffuse out. (The mechanism behind guard-cell movement is explored in Chapter 39.)

Stomata are present in all land plants except the liverworts, which have pores but no guard cells. These data suggest that the earliest land plants evolved pores that allowed gas exchange to occur at breaks in the cuticle-covered surface. Later, the evolution of guard cells gave land plants the ability to regulate gas exchange—and control water loss—by opening and closing their pores.

THE IMPORTANCE OF UPRIGHT GROWTH Once cuticle and stomata had evolved, plants could keep from drying out and thus keep photosynthesizing while exposed to air. Cuticle and stomata allowed plants to grow on the saturated soils of lake or pond edges. The next challenge? Defying gravity.

Multicellular green algae can grow erect because they float. They float because the density of their cells is similar to water's density. But outside of water, the body of a multicellular green alga collapses. The water that fills its cells is 1000 times denser than air. Although the cell walls of green algae are strengthened by the presence of cellulose, their bodies lack the structural support to withstand the force of gravity and to keep an individual erect in air.

Based on these observations, biologists hypothesize that the first land plants were small or had a low, sprawling growth habit. Besides lacking rigidity, the early land plants would have had to obtain water through pores or through a few cells that lacked cuticle—meaning they would have had to grow in a way that kept many or most of their tissues in direct contact with moist soil. If this hypothesis is correct, then competition for space and light would have become intense soon after the first plants began growing on land.

In a terrestrial environment, individuals that can grow erect have much better access to sunlight than individuals that are incapable of growing erect. But two problems have to be overcome for a plant to grow erect: (**1**) transporting water from tissues that are in contact with wet soil to tissues that are in contact with dry air, against the force of gravity; and (**2**) becoming rigid enough to avoid falling over in response to gravity and wind. As it turns out, vascular tissue solved both problems.

THE ORIGIN OF VASCULAR TISSUE Paul Kenrick and Peter Crane explored the origin of water-conducting cells and erect growth in plants by examining the extraordinary fossils found in a rock formation in Scotland called the Rhynie Chert. These rocks formed about 400 million years ago and contain some of the first large plant specimens in the fossil record—as opposed to the microscopic spores and cuticle found in older rocks.

The Rhynie Chert contains numerous plants that fossilized in an upright position. This indicates that many or most of the Rhynie plants grew erect. How did they stay vertical?

By examining fossils with the electron microscope, Kenrick and Crane established that species from the Rhynie Chert contained elongated cells that were organized into tissues along the length of the plant. Based on these data, the biologists hypothesized that the elongated cells were part of water-conducting tissue and that water could move from the base of the plants upward to erect portions through these specialized water-conducting cells.

- Some of the fossilized water-conducting cells had simple, cellulose-containing cell walls like the water-conducting cells found in today's mosses (**Figure 30.9a**).

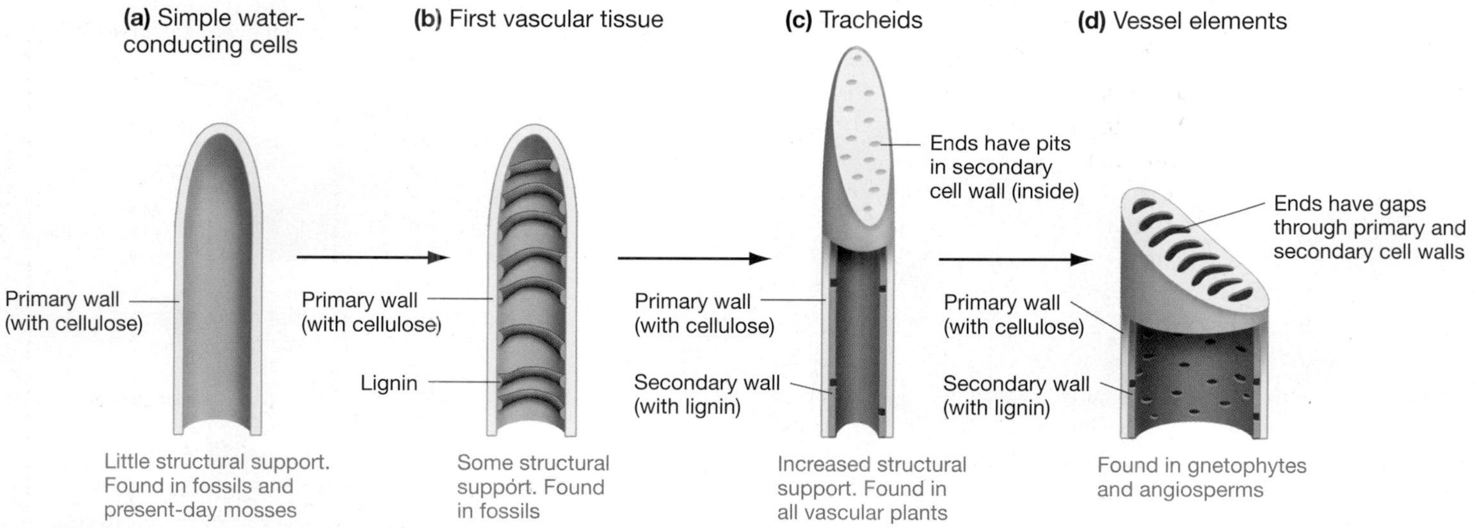

FIGURE 30.9 Evolutionary Sequence Observed in Water-Conducting Cells. According to the fossil record and the phylogeny of green plants, water-conducting cells became stronger over time due to the evolution of lignin and secondary cell walls. Efficient water transport was maintained through pits where the secondary cell wall is missing, or gaps where both the secondary and primary cell wall are absent.

✔**QUESTION** Biologists claim that vessels are more efficient than tracheids at transporting water, in part because vessels are shorter and wider than tracheids. Why does this claim make sense?

- Some of the water-conducting cells present in the early fossils had cell walls with thickened rings containing a molecule called lignin (**Figure 30.9b**).

Lignin is a complex polymer built from six-carbon rings. It is extraordinarily strong for its weight and is particularly effective in resisting compressing forces such as gravity.

These observations inspired the following hypothesis: The evolution of lignin rings gave stem tissues the strength to remain erect in the face of wind and gravity. Today, the presence of lignin in the cell walls of water-conducting cells is considered the defining feature of vascular tissue. The evolution of vascular tissue allowed early plants to support erect stems and transport water from roots to aboveground tissues.

ELABORATION OF VASCULAR TISSUE: TRACHEIDS AND VESSELS Once simple water-conducting tissues evolved, evolution by natural selection favored more complex types that were more efficient in providing support and transport.

In rocks that are about 380 million years old, biologists find the advanced water-conducting cells called tracheids. **Tracheids** are long, thin, tapering cells that have:

- a thickened, lignin-containing **secondary cell wall** in addition to a cellulose-based **primary cell wall**; and
- pits in the sides and ends of the cell where the secondary cell wall is absent, where water can flow efficiently from one tracheid to the next (**Figure 30.9c**).

The secondary cell wall gave tracheids the ability to provide better structural support, but water could still move through the cells easily because of the pits. Today, all vascular plants contain tracheids.

In fossils dated to 250–270 million years ago, biologists have documented the most advanced type of water-conducting cells observed in plants. **Vessel elements** are shorter and wider than tracheids, and their upper and lower ends have gaps where both the primary and secondary cell wall are missing. The width of vessels and the presence of open gaps reduces resistance and makes water movement extremely efficient (**Figure 30.9d**). In vascular tissue, vessel elements are lined up end to end to form a continuous pipelike structure.

In the stems and branches of some vascular plant species, tracheids or a combination of tracheids and vessels can form the extremely strong support material called **wood**. The anatomy of wood is explained in detail in Chapter 36.

All of the cell types shown in Figure 30.9 are dead when they mature, which means they lack cytoplasm. This feature allows water to move through the cells more efficiently (see Chapter 37). Taken together, the data summarized in the figure indicate that vascular tissue evolved in a series of gradual steps that provided increased structural support and increased efficiency in water transport.

Mapping Evolutionary Changes on the Phylogenetic Tree

Cuticle, stomata, and vascular tissue, were key adaptations that allowed early plants to colonize land. **Figure 30.10** summarizes how land plants adapted to dry conditions by mapping where major innovations occurred as t' versified. To review the concepts involved, go to at *www.masteringbiology.com*.

MB Web Activity Plant Evolution and the Phylogenetic Tree

As you study the tree in Figure 30.10, note that fundamentall, important adaptations to dry conditions—such as cuticle, pores, stomata, vascular tissue, and tracheids—evolved just once. But convergent evolution, which was introduced in Chapter 27, also occurred. When convergence occurs, similar traits evolve independently in two distinct lineages.

- Water-conducting cells evolved independently in mosses and in the vascular plants.
- Vessels evolved independently in gnetophytes and angiosperms.

As predicted by the convergence hypothesis, there are important differences in the development and anatomy in the water-conducting cells and vessels found in the different lineages.

The evolution of cuticle, stomata, and vascular tissue made it possible for plants to avoid drying out and to grow upright, while

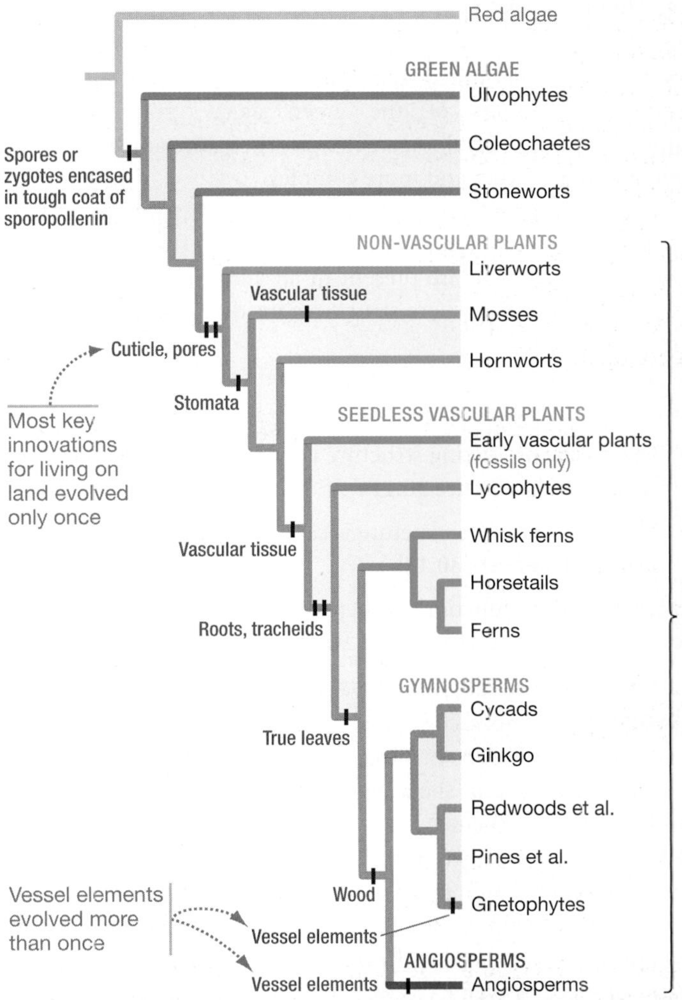

FIGURE 30.10 A Series of Evolutionary Innovations Allowed Plants to Adapt to Life on Land.

✔**QUESTION** Explain the logic that biologists used to map the location of the innovations on phylogenetic trees.

loving water from the base of the plant to its apex. Plants gained adaptations that allowed them not only to survive on land, but thrive. Now the question is, how did they reproduce?

The Transition to Land, II: How Do Plants Reproduce in Dry Conditions?

Section 30.2 introduced one of the key adaptations for reproducing on land: spores that resist drying because they are encased in a tough coat of sporopollenin. Sporopollenin-like compounds are found in the walls of some green algal zygotes; thick-walled, sporopollenin-rich spores appear early in the fossil record of land plants and occur in all land plants living today. Based on these observations, biologists infer that sporopollenin-encased spores were one of the innovations that made the initial colonization of land possible.

Two other innovations occurred early in land plant evolution and were instrumental for efficient reproduction in a dry environment: (1) Gametes were produced in complex, multicellular structures; and (2) the embryo was retained on the parent (mother) plant and was nourished by it.

PRODUCING GAMETES IN PROTECTED STRUCTURES The fossilized gametophytes of early land plants contain specialized reproductive organs called **gametangia** (singular: **gametangium**). Although members of the green algae group Charales (stoneworts) also develop gametangia, the gametangia found in land plants are larger and more complex.

The evolution of an elaborate gametangium was important because it protected gametes from drying and from mechanical damage. Gametangia are present in all land plants living today except angiosperms, where structures inside the flower perform the same functions.

In both the Charales and the land plants, individuals produce distinctive male and female gametangia.

- The sperm-producing structure is called an **antheridium** (plural: **antheridia**; **Figure 30.11a**).
- The egg-producing structure is called an **archegonium** (plural: **archegonia**; **Figure 30.11b**).

In terms of their function, antheridia and archegonia are analogous to the testes and ovaries of animals.

RETAINING AND NOURISHING OFFSPRING: LAND PLANTS AS EMBRYOPHYTES The second innovation that occurred early in land plant evolution involved the eggs that formed inside archegonia. Instead of shedding their eggs into the water or soil, land plants retain them.

Eggs are also retained in the green algal lineages that are most closely related to land plants: In Charales and other closely related groups, sperm swim to the egg, fertilization occurs, and the resulting zygote stays attached to the parent. Either before or after fertilization, the egg or zygote receives nutrients from the mother plant. But the parent plant dies each autumn as the temperature drops. The zygote remains on the dead parental tissue, settles to the bottom of the lake or pond, and overwinters. In spring, meiosis occurs, and the resulting spores develop into haploid adult plants.

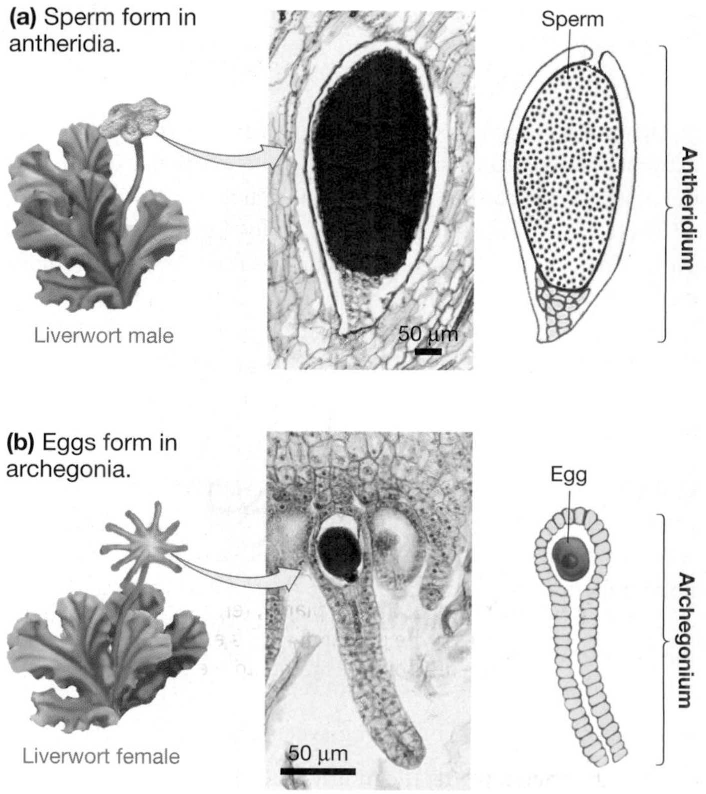

FIGURE 30.11 In All Land Plant Groups but Angiosperms, Gametes Are Produced in Gametangia. Gametangia are complex, multicellular structures that protect developing gametes from drying and mechanical damage.

In land plants, the zygote is also retained on the parent plant after fertilization. But in contrast to the zygotes of most green algae, the zygotes of all land plants begin to develop on the parent plant, forming a multicellular embryo that remains attached to the parent and can be nourished by it. This is important because land plant embryos do not have to manufacture their own food early in life. Instead, they receive most or all of their nutrients from the parent plant.

The retention of the embryo was such a key event in land plant evolution that the formal name of the group is **Embryophyta**—literally, the "embryo-plants." The retention of the fertilized egg in **embryophytes** is analogous to pregnancy in mammals, where offspring are retained by the mother and nourished through the initial stages of growth. Land plant embryos even have specialized **transfer cells**, which make physical contact with parental cells and facilitate the transfer of nutrients (**Figure 30.12**) much like the placenta that develops in a pregnant mammal.

Thick-walled spores, elaborate gametangia, and the embryophyte condition weren't the only key innovations associated with reproducing on land, though. In addition, all land plants undergo the phenomenon known as **alternation of generations**, introduced in Chapter 29.

ALTERNATION OF GENERATIONS When alternation of generations occurs, individuals represent a multicellular haploid phase or a multicellular diploid phase. The multicellular haploid stage

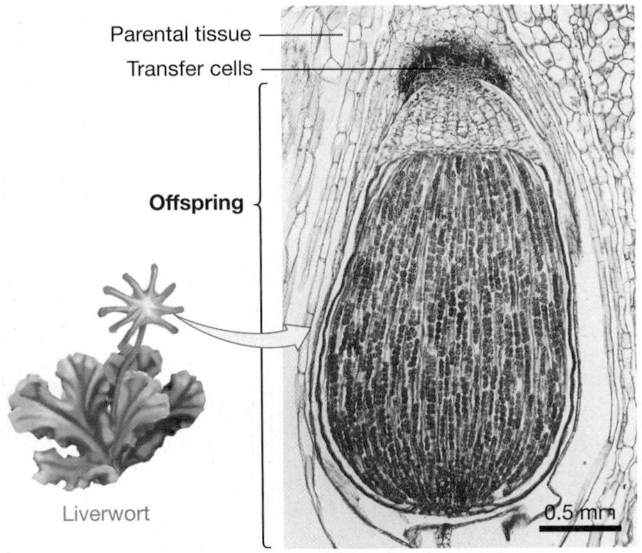

FIGURE 30.12 Land Plants Are Also Known as Embryophytes because Parents Nourish Their Young. In land plants, fertilization and early development take place on the parent plant. As embryos develop, transfer cells carry nutrients from the mother to the offspring.

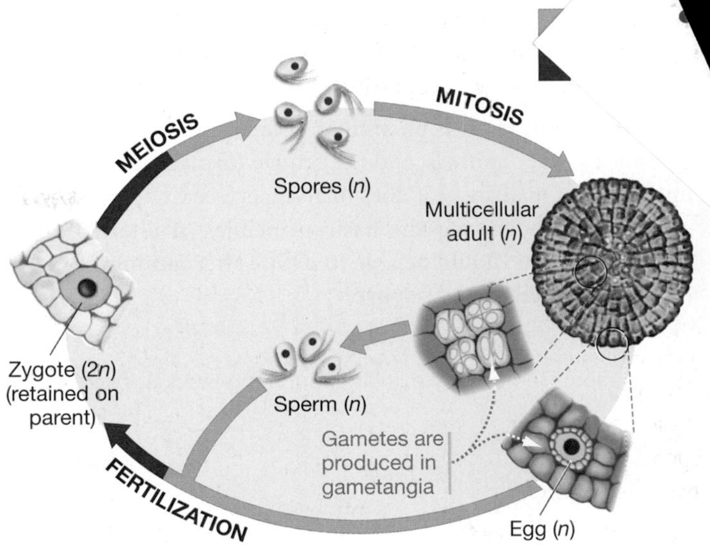

FIGURE 30.13 In Green Algae That Are Closely Related to Land Plants, Only the Zygote Is Diploid. The coleochaetes and stoneworts do not have alternation of generations. The multicellular stage is haploid.

is called the **gametophyte**; the multicellular diploid stage is called the **sporophyte**. The two phases of the life cycle are connected by distinct types of reproductive cells—gametes and spores.

Although alternation of generations is observed in a wide array of eukaryotic lineages and in some groups of green algae, it does not occur in the algal groups most closely related to land plants. In the coleochaetes and stoneworts, the multicellular form is haploid. Only the zygote is diploid. As **Figure 30.13** shows, the zygote undergoes meiosis to form haploid spores. After dispersing with the aid of flagella, the spores begin dividing by mitosis and eventually grow into an adult, haploid individual. You might recall that this haploid-dominant life cycle is common in protists and was diagrammed in Figure 29.17a.

These data suggest that alternation of generations originated in land plants independently of its evolution in other groups of eukaryotes, and that it originated early in their history—soon after they evolved from green algae. The adaptive significance of alternation of generations and its role in the successful colonization of land is still being debated, however. Keep this in mind as you study alternation of generations in land plants: You are analyzing one of the great unsolved problems in contemporary biology.

Alternation of generations always involves the same basic sequence of events, illustrated in **Figure 30.14**. To review how this type of life cycle works, put your finger on the sporophyte in the figure and trace the cycle clockwise to find the following five key events:

1. The sporophyte produces spores by meiosis. Spores are haploid.
2. Spores divide by mitosis and develop into a haploid gametophyte.
3. Gametophytes produce gametes by mitosis. Both the gametophyte and the gametes are haploid.
4. Two gametes unite during fertilization to form a diploid zygote.
5. The zygote divides by mitosis and develops into a multicellular, diploid sporophyte.

Once you've traced the cycle successfully, take a moment to compare and contrast zygotes and spores.

- Zygotes and spores are both single cells that divide by mitosis to form a multicellular individual.
- Zygotes result from the fusion of two cells, such as a sperm and an egg, but spores are not formed by the fusion of two cells.
- Zygotes produce sporophytes; spores produce gametophytes.

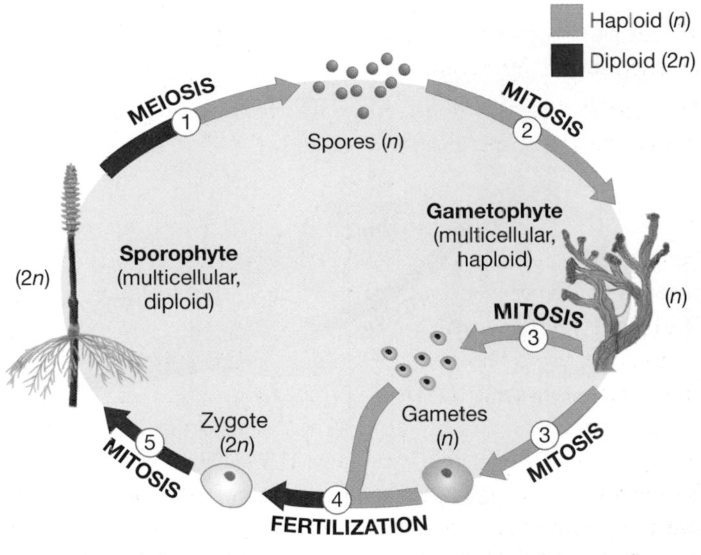

FIGURE 30.14 All Land Plants Undergo Alternation of Generations. Alternation of generations always involves the same sequence of five events.

- Spores are produced inside structures called sporangia; gametes are produced inside gametangia.

Alternation of generations can be a difficult topic to master, for two reasons: **(1)** It is unfamiliar because it does not occur in humans or other animals, and **(2)** gamete formation begins with mitosis—not meiosis, as it does in animals (see Chapter 12).

✔If you understand the basic principles of alternation of generations, you should be able to define alternation of generations, a sporophyte, and a spore.

THE GAMETOPHYTE-DOMINANT TO SPOROPHYTE-DOMINANT TREND IN LIFE CYCLES The five steps illustrated in Figure 30.14 occur in all species with alternation of generations. But in land plants, the relationship between the gametophyte and sporophyte is highly variable.

In non-vascular plants such as mosses, the sporophyte is small and short lived and is largely dependent on the gametophyte for nutrition (**Figure 30.15a**). When you see leafy-looking mosses growing on a tree trunk or on rocks, you are looking at

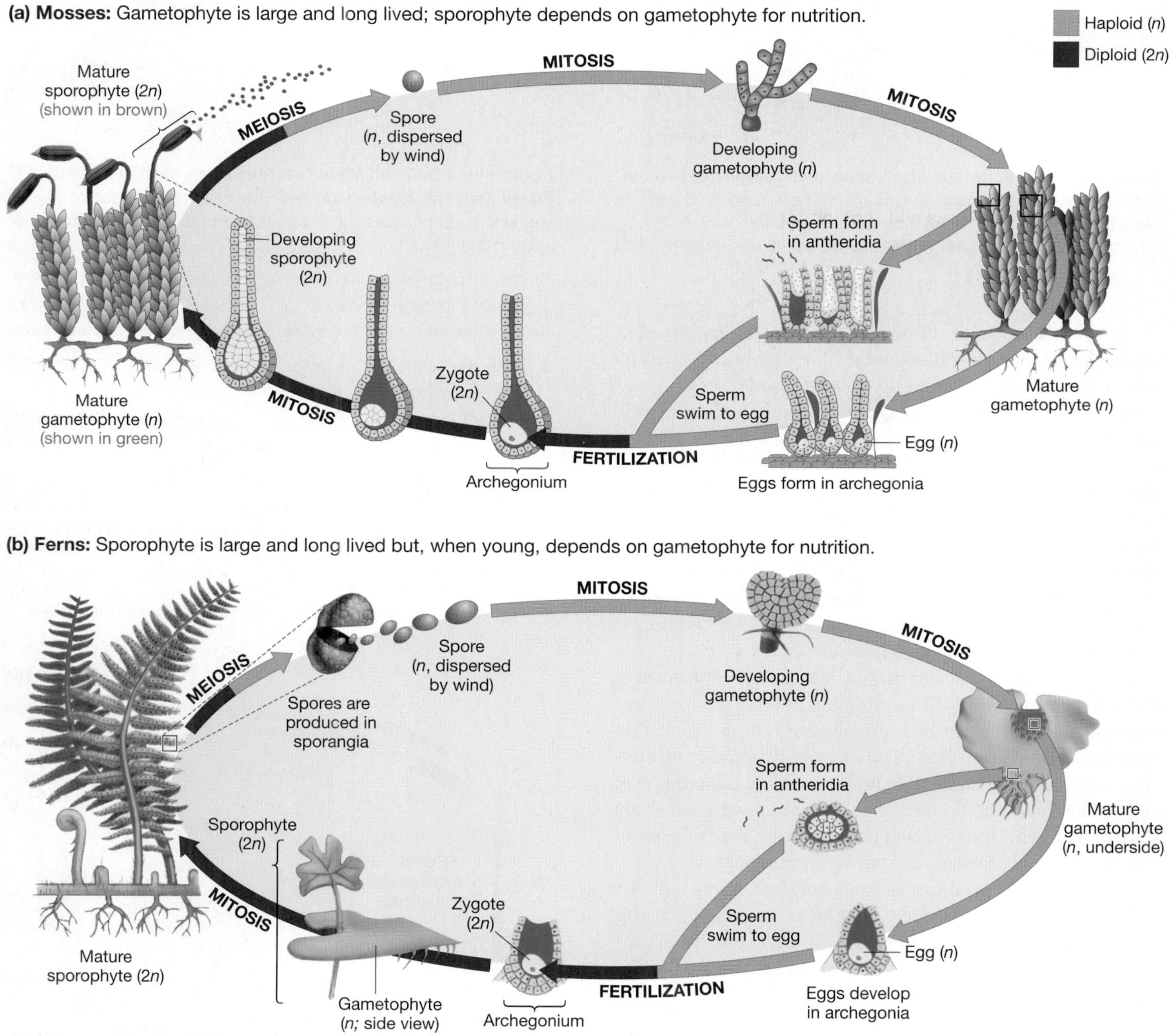

FIGURE 30.15 Gametophyte-Dominated Life Cycles Evolved Early; Sporophyte-Dominated Life Cycles Evolved Later. Like today's mosses, the earliest land plants in the fossil record have gametophytes that are much larger and longer lived than the sporophyte. In lineages that evolved later, such as ferns, the sporophyte is much larger and longer lived than the gametophyte.

✔QUESTION How can you tell that alternation of generations occurs in mosses and ferns?

gametophytes. Because the gametophyte is long lived and produces most of the food required by the individual, it is considered the dominant part of the life cycle.

In contrast, in ferns and other vascular plants the sporophyte is much larger and longer lived than the gametophyte (see **Figure 30.15b**). The ferns you see growing in gardens or forests are sporophytes. You'd have to hunt on your hands and knees to find their gametophytes, which are typically just a few millimeters in diameter. As you'll learn later in the chapter, the gametophytes of gymnosperms and angiosperms are even smaller—they are microscopic. Ferns and other vascular plants are said to have a sporophyte-dominant life cycle.

✔If you understand the difference between gametophyte-dominant and sporophyte-dominant life cycles, you should be able to examine the photos of hornworts (a non-vascular plant) and horsetails (a seedless vascular plant) in Figure 30.16, and identify which is the gametophyte and which is the sporophyte.

(a) Hornwort gametophytes and sporophytes

(b) Horsetail gametophyte and sporophyte

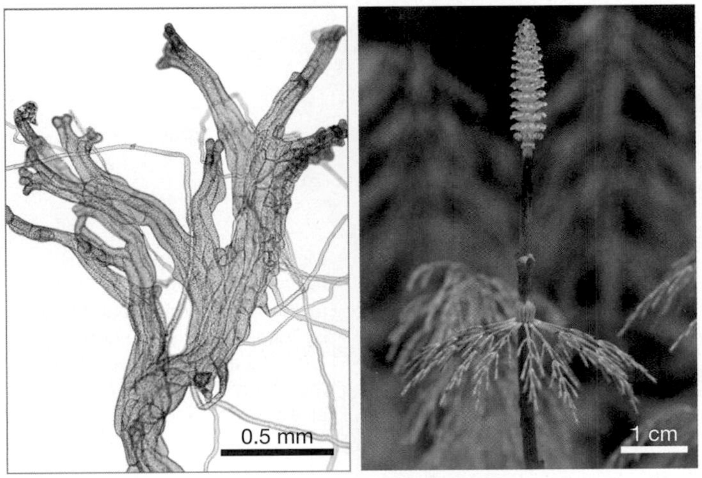

FIGURE 30.16 The Reduction of the Gametophyte Is One of the Strongest Trends in Land Plant Evolution. (a) A hornwort with spike-like sporophytes emerging from the gametophyte. **(b)** A horsetail species, both as a tiny, microscopic gametophyte and as a large, macroscopic sporophyte.

The transition from gametophyte-dominated ' sporophyte-dominated life cycles is one of the mc all trends in land plant evolution. To explain why it oc ologists hypothesize that sporophyte-dominated life cycl advantageous because diploid cells can respond to varying e ronmental conditions more efficiently than haploid cells can—particularly if the individual is heterozygous at many genes.

This idea has yet to be tested rigorously, however. If you came up with explanations for why alternation of generations evolved in land plants and why the gametophyte-dominant to sporophyte-dominant trend occurred, and if your ideas stood up to testing, the result would be celebrated worldwide.

HETEROSPORY In addition to sporophyte-dominated life cycles, another important innovation found in seed plants is called **heterospory**—the production of two distinct types of spore-producing structures and thus two distinct types of spores.

All of the non-vascular plants and most of the seedless vascular plants are **homosporous**—meaning that they produce a single type of spore. (Among the seedless vascular plants, some lycophytes and a few ferns are heterosporous.) Homosporous species produce spores that develop into bisexual gametophytes. Bisexual gametophytes produce both eggs and sperm (**Figure 30.17a**).

The two types of spore-producing structures found in heterosporous species are often found on the same individual (**Figure 30.17b**).

- **Microsporangia** are spore-producing structures that produce microspores. **Microspores** develop into male gametophytes, which produce the small gametes called sperm.
- **Megasporangia** are spore-producing structures that produce megaspores. **Megaspores** develop into female gametophytes, which produce the large gametes called eggs.

Thus, the gametophytes of seed plants are either male or female, but never both.

The evolution of heterospory was a key event in land plant evolution because it made possible one of the most important adaptations for life in dry environments—pollen.

POLLEN The non-vascular plants and the seedless vascular plants have male gametes that swim to the egg to perform fertilization. For a sperm cell to swim to the egg and fertilize it, there has to be a

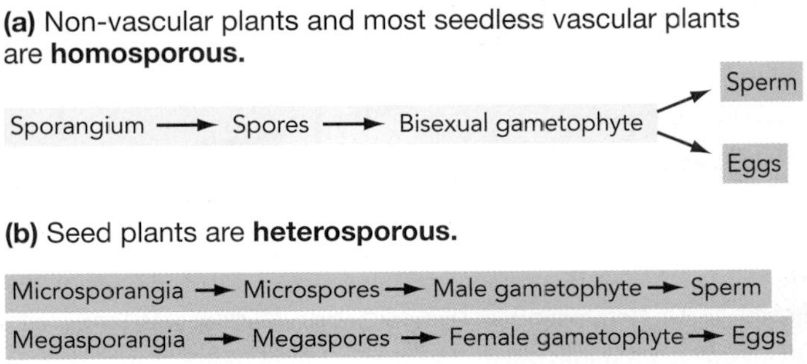

FIGURE 30.17 Heterosporous Plants Produce Male and Female Spores That Are Morphologically Distinct.

ontinuous sheet of water between the male and female gametophyte, a raindrop has to splash sperm onto a female gametophyte, or a wet insect has to carry sperm to a female gametophyte.

In species that live in dry environments, these conditions are rare. The land plants made their final break with their aquatic origins and were able to reproduce efficiently in dry habitats when a structure evolved that could move their gametes without the aid of water.

In heterosporous seed plants, the microspore germinates to form a tiny male gametophyte that is surrounded by a tough coat of sporopollenin, resulting in a **pollen grain**.

Pollen grains can be exposed to the air for long periods of time without dying from dehydration. They are also tiny enough to be carried to female gametophytes by wind or animals. Upon landing near the egg, the male gametophyte releases the sperm cells that accomplish fertilization.

When pollen evolved, then, heterosporous plants lost their dependence on water to accomplish fertilization. Instead of swimming to the egg as a naked sperm cell, their tiny gametophytes took to the skies.

SEEDS The evolution of large gametangia protected the eggs and sperm of land plants from drying. Embryo retention allowed offspring to be nourished directly by their parent, and pollen enabled fertilization to occur in the absence of water.

Retaining embryos has a downside, however: In ferns and horsetails, sporophytes have to live in the same place as their parent gametophyte. Seed plants overcome this limitation. Their embryos are portable and can disperse to new locations.

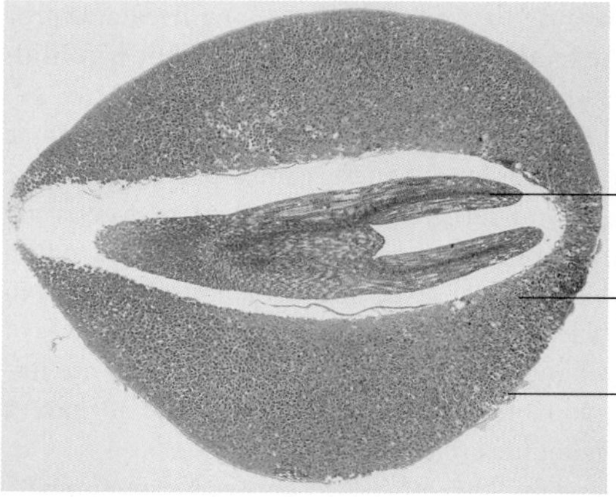

FIGURE 30.18 Seeds Contain an Embryo and a Food Supply and Can Be Dispersed. (Note that this specimen has been stained.)

A **seed** is a structure that includes an embryo and a food supply surrounded by a tough coat (**Figure 30.18**). Seeds allow embryos to be dispersed to a new habitat, away from the parent plant. Like a bird's egg, the seed provides a protective case for the embryo and a store of nutrients provided by the mother. Spores are an effective dispersal stage in non-vascular plants and seedless vascular plants, but they lack the stored nutrients found in seeds.

The evolution of heterospory, pollen, and seeds triggered a dramatic radiation of seed plants starting about 299 million years ago. To make sure that you understand these key processes and structures, study the life cycle of the pine tree pictured in **Figure 30.19**.

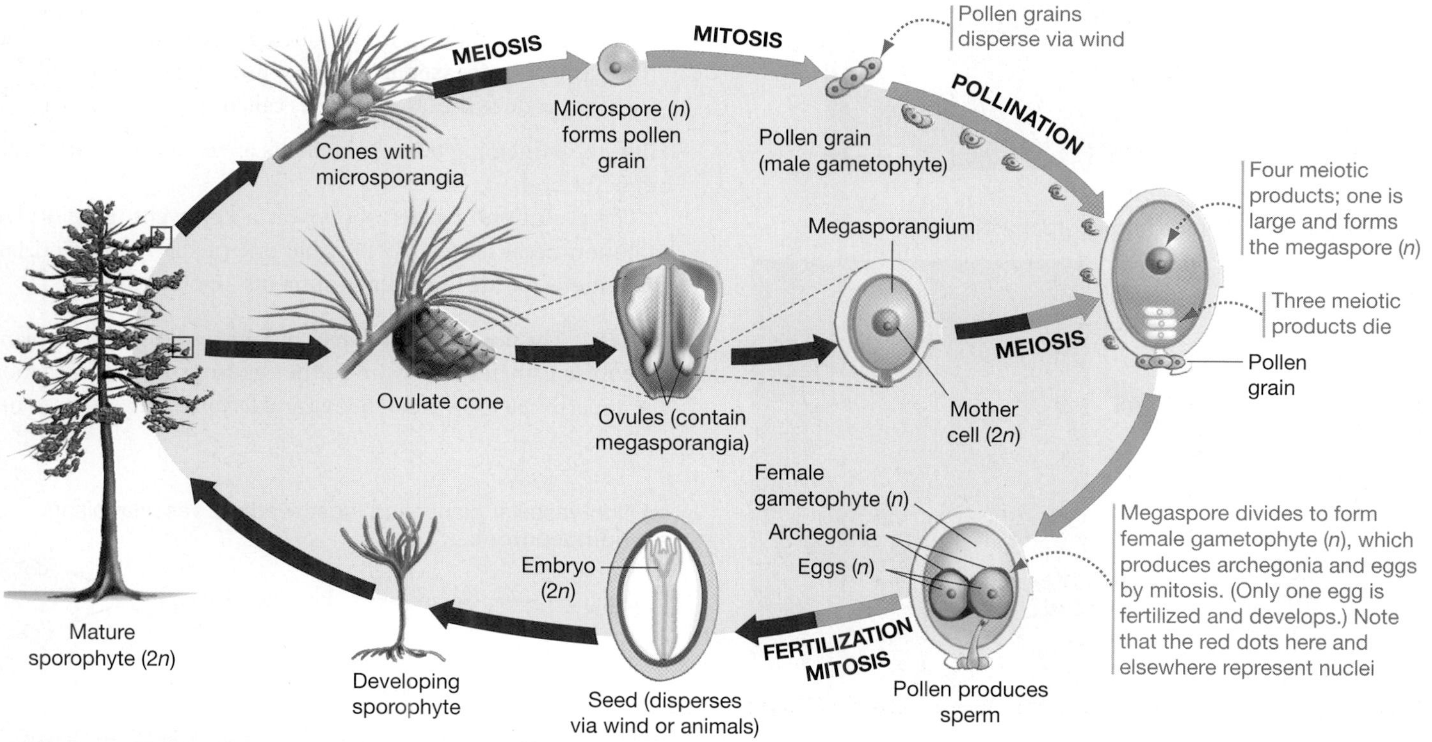

FIGURE 30.19 Heterospory in Gymnosperms: Microspores Produce Pollen Grains; Megaspores Produce Female Gametophytes.

✔**QUESTION** Compare the life cycle of the pine tree in Figure 30.19 with that of the fern pictured in Figure 30.15b. Is the gymnosperm gametophyte larger than, smaller than, or about the same size as a fern gametophyte? Compared with ferns, is the gymnosperm gametophyte more or less dependent on the sporophyte for nutrition?

- Starting with the sporophyte on the left, note that this and many other gymnosperm species have separate structures, called cones, where microsporangia and megasporangia develop. In this case, the two types of spores associated with heterospory develop in separate cones.
- The microsporangia contain a cell that divides by meiosis to form microspores, which then divide by mitosis to form pollen grains—tiny male gametophytes.
- Megasporangia are found inside protective structures called ovules. Megasporangia contain a mother cell that divides by meiosis to form a megaspore.
- The megaspore undergoes mitosis to form the female gametophyte, which contains egg cells.
- The female gametophyte stays attached to the sporophyte as pollen grains arrive and produce sperm that fertilize the eggs.
- Seeds mature as the embryo develops. Inside the seed, cells derived from the female gametophyte become packed with nutrients provided by the sporophyte.

When the seed disperses and germinates, the cycle of life begins anew.

FLOWERS Flowering plants, or angiosperms, are the most diverse land plants living today. About 250,000 species have been described, and more are discovered each year. Their success in terms of geographical distribution, number of individuals, and number of species revolves around a reproductive organ—the **flower**.

Flowers contain two key reproductive structures: the stamens and carpels illustrated on the left-hand side of **Figure 30.20**. Stamens and carpels are responsible for heterospory.

- A **stamen** includes a structure called an anther, where microsporangia develop. Meiosis occurs inside the microsporangia, forming microspores. Microspores then divide by mitosis to form pollen grains.
- A **carpel** contains a protective structure called an **ovary** where the ovules are found.

The presence of enclosed ovules inspired the name angiosperm ("encased-seed") as opposed to gymnosperm ("naked-seed").

As in gymnosperms, ovules contain the megasporangia. A cell inside the megasporangium divides by meiosis to form the megaspore, which then divides by mitosis to form the female gametophyte. When a pollen grain lands on a carpel and produces sperm, fertilization takes place, as shown on the right-hand side of Figure 30.20.

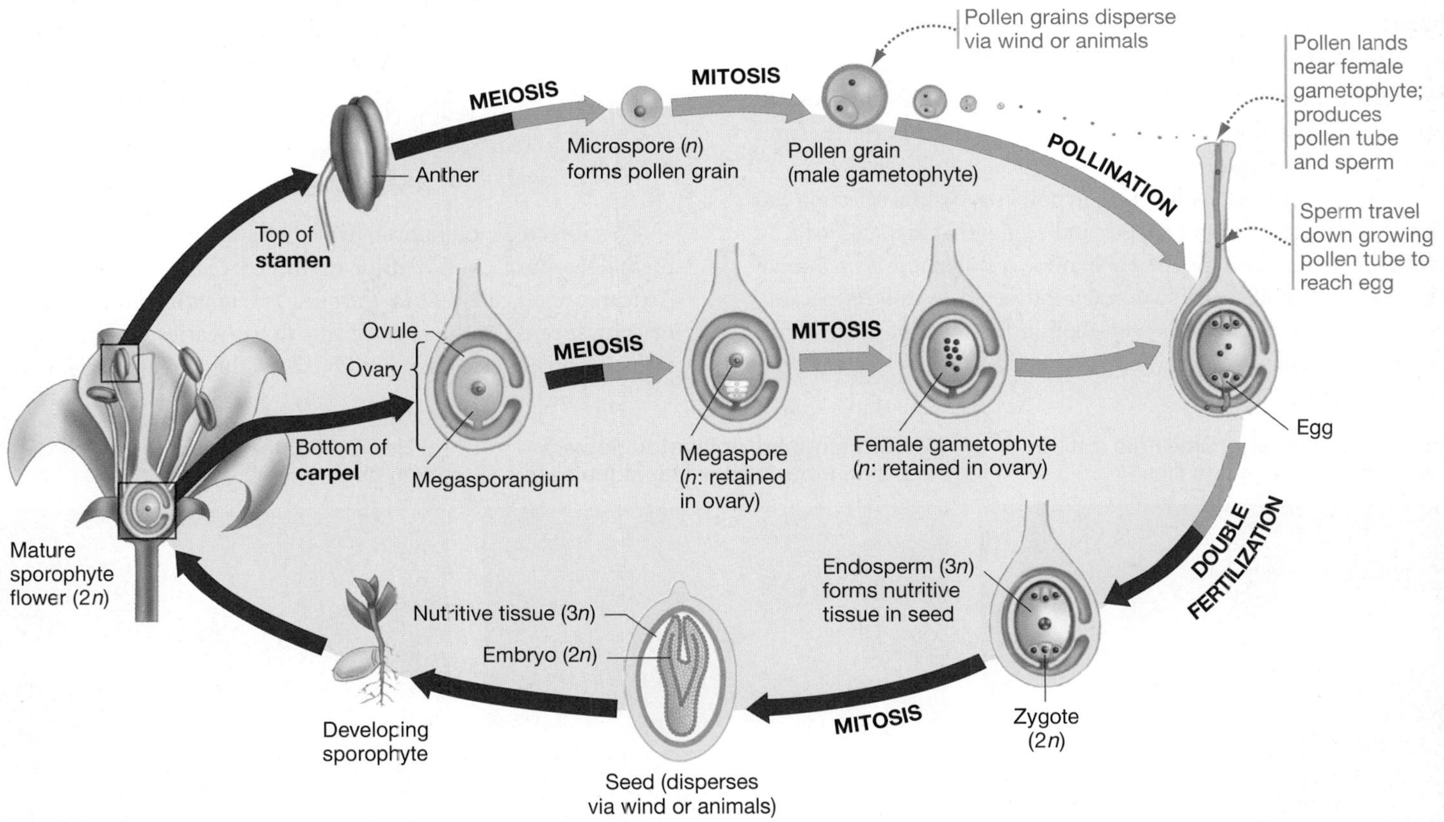

FIGURE 30.20 Heterospory in Angiosperms: Flowers Contain Microspores and Megaspores.

✔**QUESTION** Gymnosperm pollen grains typically contain from 4 to 40 cells; mature angiosperm pollen grains contain three cells. Gymnosperm female gametophytes typically contain hundreds of cells; angiosperm female gametophytes typically contain seven. In the fossil record, gymnosperms appear long before angiosperms. Do these observations conflict with the trend of reduced gametophytes during land plant evolution, or are they consistent with it? Explain your logic.

Angiosperm fertilization is unique, however, because it involves *two* sperm cells. One sperm fuses with the egg to form the zygote, while a second sperm fuses with two nuclei in the female gametophyte to form a triploid ($3n$) nutritive tissue called **endosperm**. The involvement of two sperm nuclei is called **double fertilization**.

The evolution of the flower, then, is an elaboration of heterospory. The key innovation was the evolution of the ovary, which helps protect female gametophytes from insects and other predators.

Double fertilization is another striking innovation associated with the flower, but its adaptive significance is still not well understood. Explaining the significance of double fertilization is another major challenge for biologists interested in understanding how land plants diversified.

POLLINATION BY INSECTS AND OTHER ANIMALS The story of the flower doesn't end with the ovary. Once stamens and carpels evolved, they became enclosed by modified leaves called **sepals** and **petals**. The four structures then diversified to produce a fantastic array of sizes, shapes, and colors—from red roses to blue violets. Specialized cells inside flowers also began producing a wide range of scents.

To explain these observations, biologists hypothesize that flowers are adaptations to increase the probability that an animal will perform **pollination**—the transfer of pollen from one individual's stamen to another individual's carpel. Instead of leaving pollination to an undirected agent such as wind, the hypothesis is that natural selection favored structures that reward an animal—usually an insect—for carrying pollen directly from one flower to another.

Under the directed-pollination hypothesis, natural selection has favored flower colors and shapes and scents that are successful in attracting particular types of pollinators. A pollinator is an animal that disperses pollen. Pollinators are attracted to flowers because flowers provide the animals with food in the form of protein-rich pollen or a sugar-rich fluid known as **nectar**. In this way, the relationship between flowering plants and their pollinators is mutually beneficial. The pollinator gets food; the plant gets sex (fertilization).

What evidence supports the hypothesis that flowers vary in size, structure, scent, and color in order to attract different pollinators?

The first type of evidence on this question is correlational in nature. In general, the characteristics of a flower correlate closely with the characteristics of its pollinator. A few examples will help drive this point home:

- *Scent* The carrion flower in **Figure 30.21a** produces molecules that smell like rotting flesh. The scent attracts carrion flies, which normally lay their eggs in animal carcasses. In effect, the plant tricks the flies. While looking for a place to lay their eggs on a flower, the flies become dusted with pollen. If the flies are already carrying pollen from a visit to a different carrion flower, they are likely to deposit pollen grains near the plant's female gametophyte. In this way, the carrion flies pollinate the plant.
- *Flower shape* Flowers that are pollinated by hummingbirds typically have petals that form a long, tubelike structure corresponding to the size and shape of a hummingbird's beak (**Figure 30.21b**). Nectar-producing cells are located at the base of the tube. When hummingbirds visit the flower, they insert their beaks and harvest the nectar. In the process, they transfer pollen grains attached to their throats or faces.
- *Flower color* Hummingbird-pollinated flowers tend to have red petals (Figure 30.21b), while bee-pollinated flowers tend to be purple or yellow (**Figure 30.21c**). Hummingbirds are attracted to red; bees have excellent vision in the purple and ultraviolet end of the spectrum.

The directed-pollination hypothesis also has strong experimental support. Consider, for example, recent work on a South African orchid called *Disa draconis*. The length of the long tube, or spur, located at the back of this flower varies among populations of this species. As predicted by the directed-pollination

(a) Carrion flowers: Smell like rotting flesh and attract carrion flies

(b) Hummingbird-pollinated flowers: Red, long tubes with nectar at the base

(c) Bee-pollinated flowers: Often bright purple

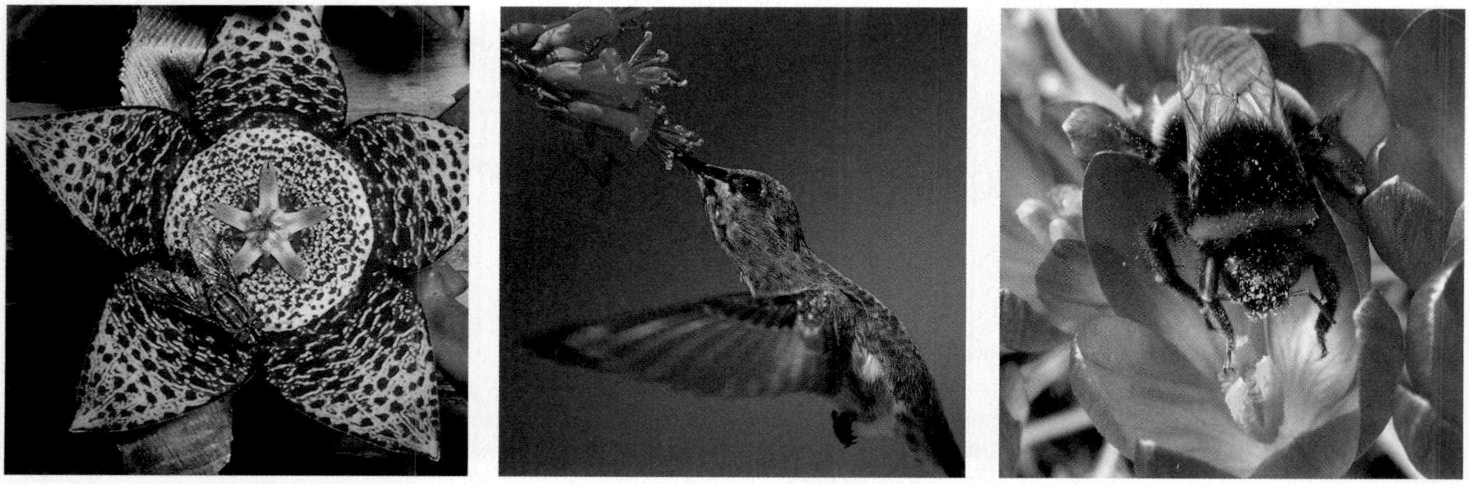

FIGURE 30.21 Flowers with Different Fragrances, Shapes, and Colors Attract Different Pollinators.

hypothesis, the length of the spur correlates with the length of the proboscis found in the insect pollinator. A proboscis is a specialized mouthpart found in some insects. When extended, it functions like a straw in sucking nectar or other fluids.

- Short-spurred *D. draconis* that grow in mountain habitats are pollinated by insect species that have a relatively short proboscis.
- Long-spurred members of this species that grow on low-lying sandplain habitats are pollinated by insects that have a particularly long proboscis.

To test the hypothesis that spur length affects pollination success, researchers artificially shortened the spurs of individuals in a long-spurred population (**Figure 30.22**). The biologists did this by tying off the spurs of randomly selected flowers with a piece of yarn, so that pollinating insects could not reach the end of the spur. The idea was that insects would not make contact with the flower's reproductive organs when they inserted their proboscis into the shortened tube. The biologists left nearby flowers alone but attached a piece of yarn near the spur, as a control treatment.

As the data in the "Results" box of Figure 30.22 indicate, flowers with short spurs received much less pollen and set much less seed than did flowers with normal-length spurs. These data strongly support the hypothesis that spur length is an adaptation that increases the frequency of pollination by particular insects.

Based on results like this, biologists contend that the spectacular diversity of angiosperms resulted, at least in part, from natural selection exerted by the equally spectacular diversity of insect, mammal, and bird pollinators. Mutations that change flower shape can change gene flow and start speciation (see Chapter 26).

FRUITS The evolution of the ovary was an important event in land plant diversification, but not only because it protected the female gametophytes of angiosperms. It also made the evolution of fruit possible.

A **fruit** is a structure that is derived from the **ovary** and encloses one or more seeds (**Figure 30.23a** on page 564). Tissues derived from the ovary are often nutritious and brightly colored (**Figure 30.23b**). Animals eat these types of fruits, digest

EXPERIMENT

QUESTION: Does flower shape affect reproductive success?

HYPOTHESIS: A correlation between flower spur length and pollinator proboscis length increases reproductive success in *Disa draconis*.

NULL HYPOTHESIS: Correlations between flower spur length and pollinator proboscis length have no effect on reproductive success in *Disa draconis*.

EXPERIMENTAL SETUP:

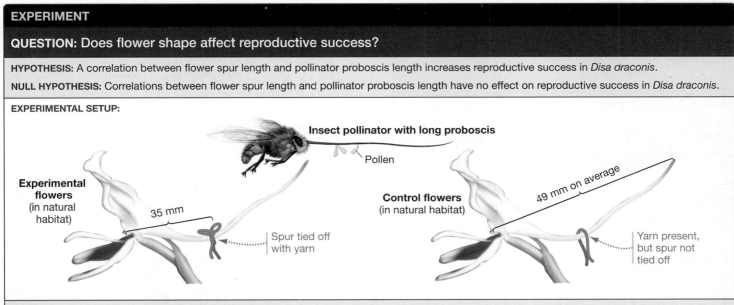

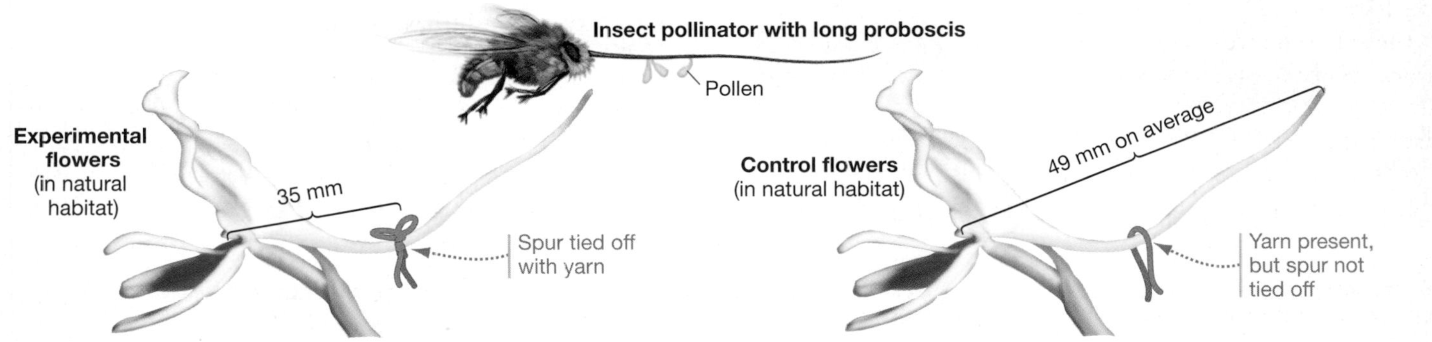

PREDICTION OF HYPOTHESIS TO BE TESTED: Altering spur length (and thus flower shape) will reduce reproductive success.

PREDICTION OF NULL HYPOTHESIS: Altering spur length (and thus flower shape) will have no impact on reproductive success.

RESULTS:

	Shortened spurs	Normal spurs
What fraction of flowers received pollen? ($n = 59$)	17%	46%
What fraction of flowers set seed? ($n = 56$)	18%	41%

Flowers with normal spurs have higher reproductive success

CONCLUSION: In this species, flower shape—specifically spur length—affects reproductive success.

FIGURE 30.22 The Adaptive Significance of Flower Shape: An Experimental Test.

SOURCE: Johnson, S. D. and K. E. Steiner. 1997. Long-tongued fly pollination and evolution of floral spur length in the *Disa draconis* complex (Orchidaceae). *Evolution* 51: 45–53.

✔**QUESTION** Why did the researchers bother to put yarn around the control flowers?

(a) Fruits are derived from ovaries and contain seeds.

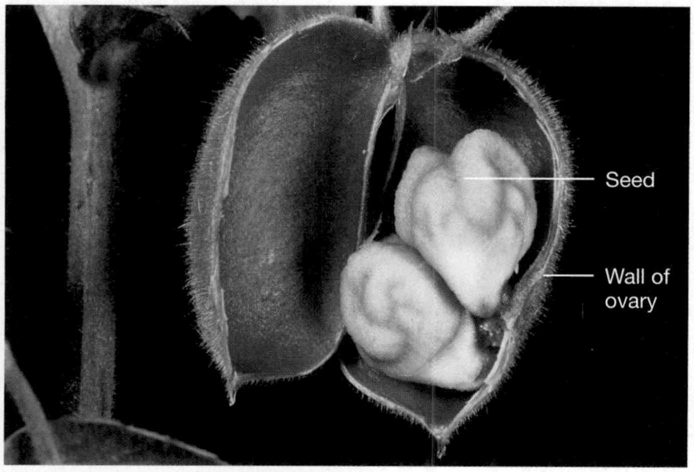

(b) Many fruits are dispersed by animals.

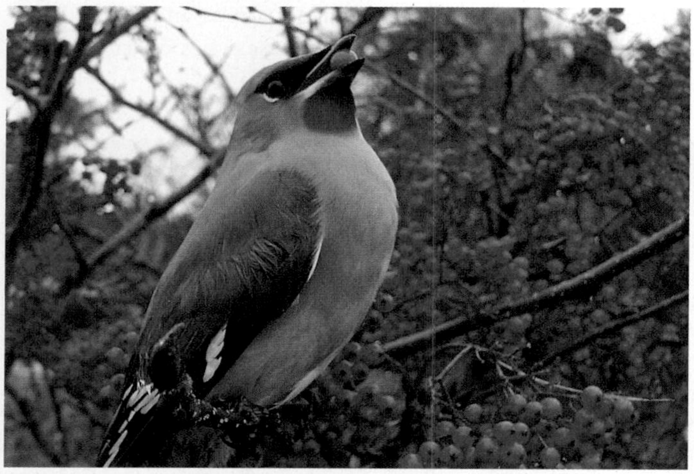

FIGURE 30.23 Fruits Are Derived from Ovaries Found in Angiosperms. **(a)** A pea pod is one of the simplest types of fruit. **(b)** The ovary wall often becomes thick, fleshy, and nutritious enough to attract animals that disperse the seeds inside.

the nutritious tissue around the seeds, and disperse seeds in their feces. In other cases, the tissues derived from the ovary help fruits disperse via wind or water. The evolution of flowers made efficient pollination possible; the evolution of fruits made efficient seed dispersal possible.

The list of adaptations that allow land plants to reproduce in dry environments is impressive; **Figure 30.24** summarizes them. Once land plants had vascular tissue and could grow efficiently in dry habitats, the story of their diversification revolved around traits that allowed sperm cells to reach eggs efficiently and helped seeds disperse to new locations.

The Angiosperm Radiation

For the past 125 million years, land plant diversification has really been about angiosperms. The Anthophyta or angiosperms represent one of the great adaptive radiations in the history of life. As Chapter 27 noted, an **adaptive radiation** occurs when a single lineage produces a large number of descendant species that are adapted to a wide variety of habitats.

The diversification of angiosperms is associated with three key adaptations: (1) vessels, (2) flowers, and (3) fruits. In combination, these traits allow angiosperms to transport water, pollen, and seeds efficiently. Based on these observations, it is not surprising that most land plants living today are angiosperms.

On the basis of morphological traits, the 250,000 species of angiosperms identified to date have traditionally been classified into two major groups: the monocotyledons, or **monocots**, and the dicotyledons, or **dicots**. Some familiar monocots include the grasses, orchids, palms, and lilies; familiar dicots include the roses, buttercups, daisies, oaks, and maples.

The names of the two groups were inspired by differences in a structure called the cotyledon. A **cotyledon** is the first leaf that is

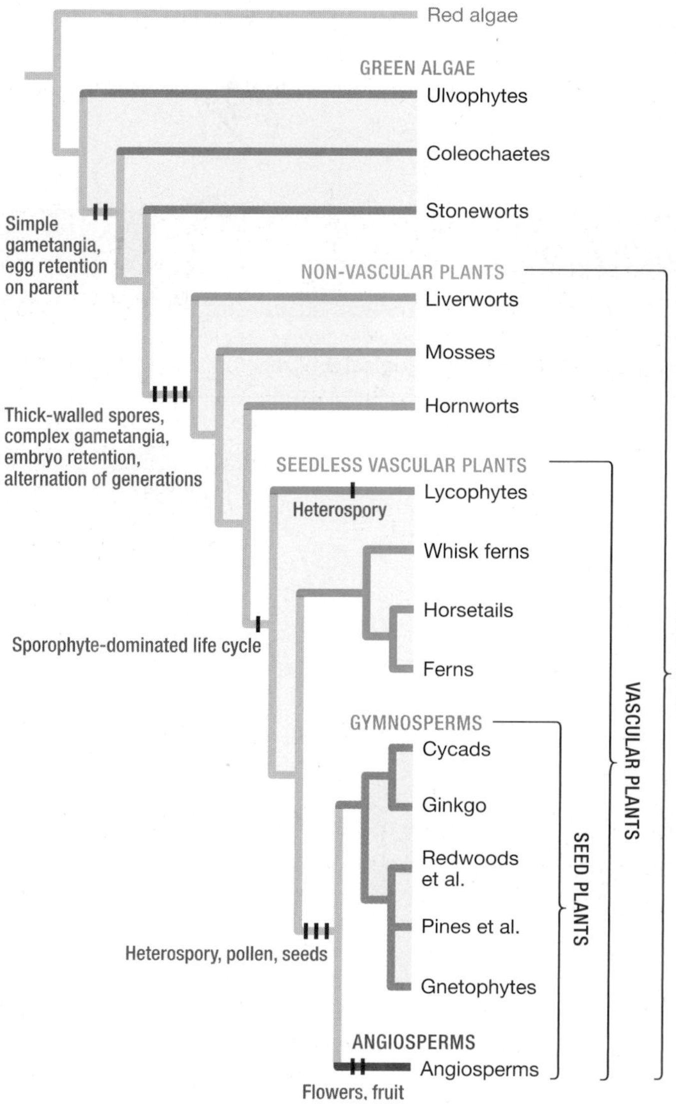

FIGURE 30.24 Evolutionary Innovations Allowed Plants to Reproduce Efficiently on Land.

✔**EXERCISE** Redwoods, pines, gnetophytes, and angiosperms are the only land plants that do not have flagellated sperm that swim to the egg (at least a short distance). Mark the loss of flagellated sperm on Figure 30.7.

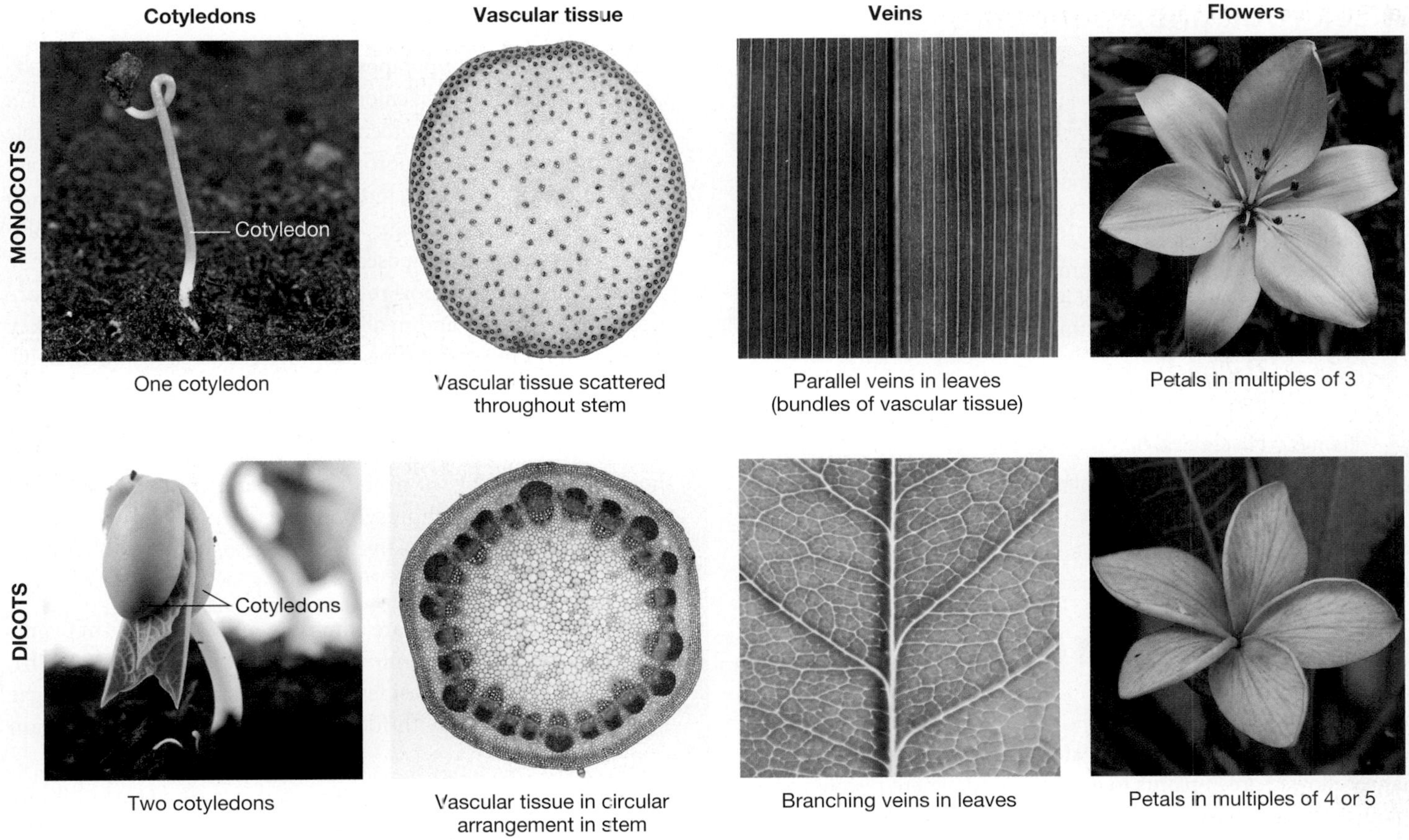

FIGURE 30.25 Four Morphological Differences between Monocots and Dicots. Note that the stem cross-sections, showing vascular tissue, have been stained.

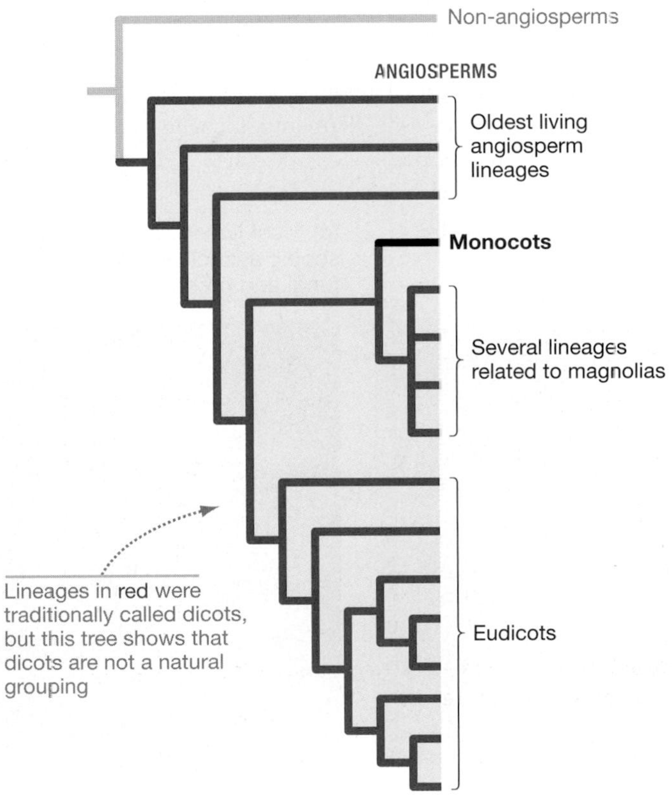

FIGURE 30.26 Monocots Are Monophyletic, Dicots Are Paraphyletic.

formed in an embryonic plant. As **Figure 30.25** shows, monocots have a single cotyledon (hence the "mono") while dicots have two cotyledons (hence the "di"). The figure also highlights other major morphological differences observed in monocots and dicots, including the arrangement of vascular tissue and leaf veins and the characteristics of flowers.

It would be misleading, however, to think that all species of flowering plants fall into one of these two groups—either monocots or dicots. Recent work has shown that dicots do not form a natural group consisting of a common ancestor and all of its descendants.

To drive this point home, consider the phylogeny illustrated in **Figure 30.26**. These relationships were estimated by comparing the sequences of several genes that are shared by all angiosperms. Notice that species with dicot-like characters are scattered around the angiosperm phylogenetic tree. Based on this analysis, biologists have concluded that although monocots are monophyletic, dicots are not. Dicots are paraphyletic.

Biologists have adjusted the names assigned to angiosperm lineages to reflect this new knowledge of phylogeny. The most important of these changes was identifying the **eudicots** ("true dicots") as a lineage that includes roses, daisies, and maples. Plant systematists continue to work toward understanding relationships throughout the angiosperm phylogenetic tree; there will undoubtedly be more name changes as knowledge grows.

CHECK YOUR UNDERSTANDING

If you understand that . . .

- Land plants were able to make the transition to growing in terrestrial environments, where sunlight and carbon dioxide are abundant, based on a series of evolutionary innovations.
- Adaptations for growing on land included cuticle, stomata, and vascular tissue.
- Adaptations for effective reproduction on land included gametangia, the retention of embryos on the parent, pollen, seeds, flowers, and fruits.

✓ You should be able to . . .

1. Explain why the evolution of cuticle and vascular tissue was important in survival or reproduction.
2. On Figure 30.7, map where the origin of pollen, flowers, and fruits occurred.

Answers are available in Appendix B.

30.4 Key Lineages of Green Algae and Land Plants

The evolution of cuticle, pores, stomata, and water-conducting tissues allowed green plants to grow on land, where resources for photosynthesis are abundant. Once the green plants were on land, the evolution of gametangia, retained embryos, pollen, seeds, and flowers enabled them to reproduce efficiently even in dry environments. The adaptations reviewed in Section 30.3 allowed the land plants to make the most important water-to-land transition in the history of life.

To explore green plant diversity in more detail, let's take a closer look at some major groups of green algae and land plants. The initial part of this section considers broad groupings of lineages; the "index cards" that follow focus on particular monophyletic groups.

Green Algae

The **green algae** are a paraphyletic group that totals about 7000 species. Their bright green chloroplasts are similar to those found in land plants. Specifically, green algal chloroplasts have a double membrane and contain chlorophylls *a* and *b* but relatively few accessory pigments. And like land plants, green algae synthesize starch in the chloroplast as a storage product of photosynthesis. They also have a cell wall that is composed primarily of cellulose.

Green algae are important primary producers in nearshore ocean environments and in all types of freshwater habitats. They are also found in several types of more exotic environments, including snowfields at high elevations, pack ice, and ice floes. These habitats are often splashed with bright colors due to large concentrations of unicellular green algae (**Figure 30.27a**). Although these cells live at near-freezing temperatures, they make all their own food via photosynthesis.

In addition, green algae live in close association with an array of other organisms.

- Unicellular green algae are common endosymbionts in planktonic protists that live in lakes and ponds (**Figure 30.27b**). The association is considered mutually beneficial: The algae supply the protists with food; the protists provide protection to the algae.
- **Lichens** are stable associations between green algae and fungi or between cyanobacteria and fungi, and are often found in terrestrial environments that lack soil, such as tree bark or bare rock (**Figure 30.27c**). The algae or cyanobacteria in a lichen are protected from drying by the fungus; in return they provide sugars produced by photosynthesis.

Of the 17,000 species of lichens described to date, about 85 percent involve green algae. The green algae that are involved are unicellular or grow in long filaments. Lichens are explored in more detail in Chapter 31.

(a) Green algae with red carotenoid pigments are responsible for pink snow.

(b) Many unicellular protists harbor green algae.

(c) Most lichens are a microscopic association between fungi and green algae.

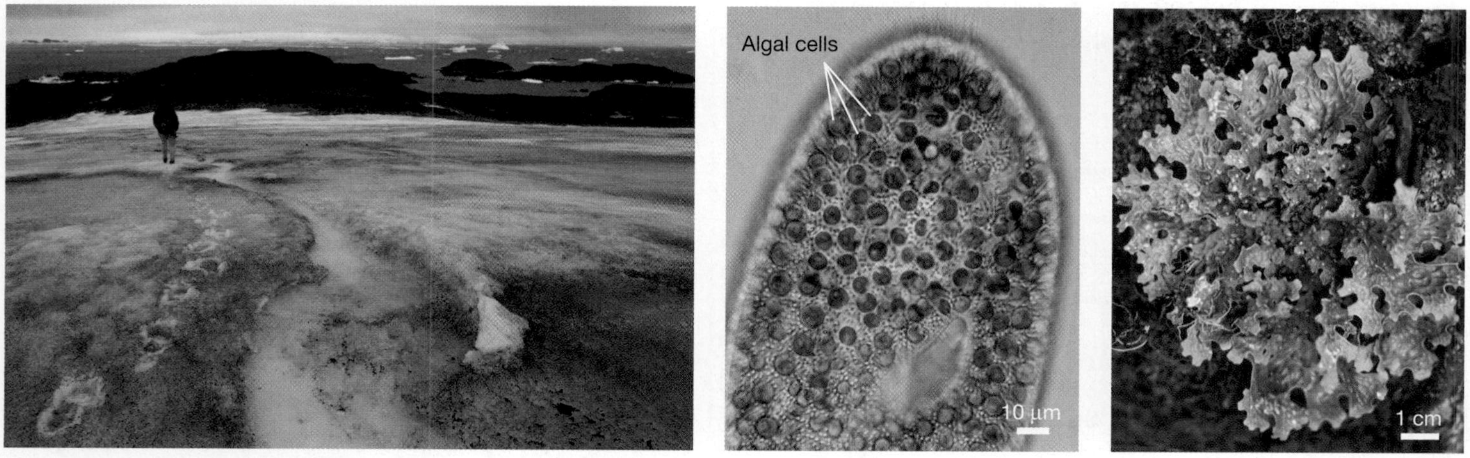

FIGURE 30.27 Some Green Algae Live in Unusual Environments.

Green algae are a large and fascinating group of organisms. In the section that closes the chapter, we'll take a closer look at just three of the many lineages.

- Green Algae > Ulvophyceae (Ulvophytes)
- Green Algae > Coleochaetophyceae (Coleochaetes)
- Green Algae > Charophyceae (Stoneworts)

Non-Vascular Plants ("Bryophytes")

The initial lineages to branch off the phylogeny of living land plants are sometimes known as the bryophytes or non-vascular plants.

All of the non-vascular plant species present today have a low, sprawling growth habit. In fact, it is unusual to find bryophytes more than 5 to 10 centimeters tall.

Individuals are anchored to soil, rocks, or tree bark by structures called **rhizoids**. Although simple water-conducting cells and tissues are found in some mosses, no bryophytes have vascular tissue with lignin-reinforced cell walls.

All bryophytes have flagellated sperm that swim to eggs through raindrops or small puddles on the plant surface. Spores are dispersed by wind.

- Non-Vascular Plants > Hepaticophyta (Liverworts)
- Non-Vascular Plants > Bryophyta (Mosses)
- Non-Vascular Plants > Anthocerophyta (Hornworts)

Seedless Vascular Plants

The seedless vascular plants are a paraphyletic group that forms a grade between the non-vascular plants and the seed plants. All species of seedless vascular plants have conducting tissues with cells that are reinforced with lignin, forming vascular tissue. Tree-sized lycophytes and horsetails are abundant in the fossil record, and tree ferns are still common inhabitants of certain habitats, such as mountain slopes in the tropics.

The sporophyte is the larger and longer-lived phase of the life cycle in all of the seedless vascular plants. The gametophyte is physically independent of the sporophyte, however. Eggs are retained on the gametophyte, and sperm swim to the egg with the aid of flagella. Thus, seedless vascular plants depend on the presence of water for reproduction—they need enough water to form a continuous layer that "connects" gametophytes and allows sperm to swim to eggs. Sporophytes develop on the gametophyte and are nourished by the gametophyte when they are small.

- Seedless Vascular Plants > Lycophyta (Lycophytes, or Club Mosses)
- Seedless Vascular Plants > Psilotophyta (Whisk Ferns)
- Seedless Vascular Plants > Equisetophyta (or Sphenotophyta) (Horsetails)
- Seedless Vascular Plants > Pteridophyta (Ferns)

Seed Plants

The seed plants are a monophyletic group consisting of the gymnosperms—cycads, ginkgo, redwoods, pines, and gnetophytes—and the angiosperms. The group is defined by two key synapomorphies: the production of seeds and the production of pollen grains.

- Seeds are a specialized structure for dispersing embryonic sporophytes to new locations. Seeds are the mature form of a fertilized ovule, the female reproductive structure that encloses the female gametophyte and egg cell.
- Pollen grains are tiny, sperm-producing gametophytes that are easily dispersed through air as opposed to water.

Seed plants are found in virtually every type of habitat, and they adopt every growth habit known in land plants. Their forms range from mosslike mats to shrubs and vines to 100-meter-tall trees. Seed plants can be **annual** (have a single growing season) or **perennial** (live for many years), with life spans ranging from a few weeks to almost five thousand years. The structure and function of seed plants is the focus of Chapters 36 through 40.

- Seed Plants > Gymnosperms > Cycadophyta (Cycads)
- Seed Plants > Gymnosperms > Ginkgophyta (Ginkgoes)
- Seed Plants > Gymnosperms > Redwood Group (Redwoods, Junipers, Yews)
- Seed Plants > Gymnosperms > Pinophyta (Pines, Spruces, Firs)
- Seed Plants > Gymnosperms > Gnetophyta (Gnetophytes)
- Seed Plants > Anthophyta (Angiosperms)

Green Algae > Ulvophyceae (Ulvophytes)

The Ulvophyceae are a monophyletic group composed of several diverse and important subgroups, with a total of about 4000 species. Members of this lineage range from unicellular to multicellular.

Many of the large green algae in habitats along ocean coastlines are members of the Ulvophyceae. *Ulva*, the sea lettuce (**Figure 30.28**), is a representative marine species. But there are also large numbers of unicellular or small multicellular species that inhabit the plankton of freshwater lakes and streams.

Reproduction Most ulvophytes reproduce both asexually and sexually. Asexual reproduction often involves production of spores that swim with the aid of flagella. Sexual reproduction usually results in production of a resting stage—a cell that is dormant in winter. In many species the gametes are not called eggs and sperm, because they are the same size and shape. In most species gametes are shed into the water, so fertilization takes place away from the parent plants.

Life cycle Many unicellular forms are diploid only as zygotes. Alternation of generations occurs in multicellular species. When alternation of generations occurs, gametophytes and sporophytes may look identical or different.

Human and ecological impacts Ulvophyceae are important primary producers in freshwater environments and in coastal areas of the oceans.

Ulva lactuca

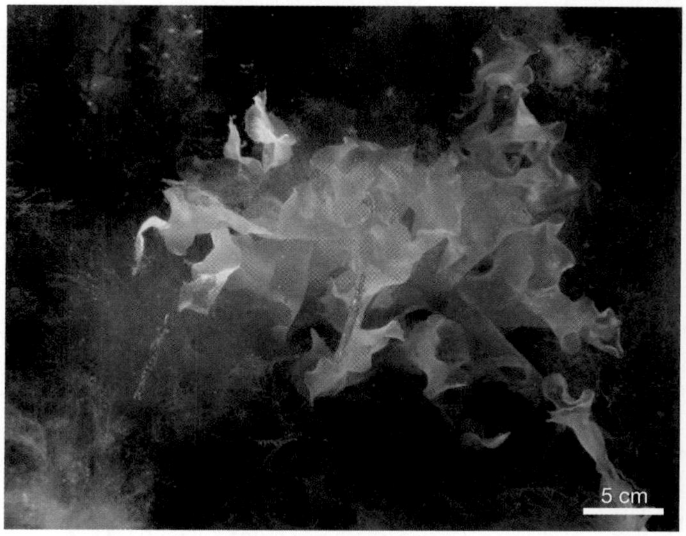

FIGURE 30.28 Green Algae Are Important Primary Producers in Aquatic Environments.

Green Algae > Coleochaetophyceae (Coleochaetes)

There are 19 species in this group. Most coleochaetes are barely visible to the unaided eye and grow as flat sheets of cells (**Figure 30.29**). They are considered multicellular because they have specialized photosynthetic and reproductive cells and because they contain **plasmodesmata**—structures introduced in Chapter 8 that connect adjacent cells.

The coleochaetes are strictly freshwater algae. They grow attached to aquatic plants such as water lilies and cattails or over submerged rocks in lakes and ponds. When they grow near beaches, they are often exposed to air when water levels drop in late summer.

Reproduction Asexual reproduction is common in coleochaetes and involves production of flagellated spores. During sexual reproduction, eggs are retained on the parent and are nourished after fertilization with the aid of transfer cells—a situation similar to that observed in land plants. In some species certain individuals are male and produce only sperm, while other individuals are female and produce only eggs.

Life cycle Alternation of generations does not occur. Multicellular individuals are haploid; the only diploid stage in the life cycle is the zygote.

Human and ecological impacts Because they are closely related to land plants, coleochaetes are studied intensively by researchers interested in how land plants made the water-to-land transition.

Coleochaete scutata

FIGURE 30.29 Coleochaetes Are Thin Sheets of Cells.

Green Algae > Charophyceae (Stoneworts)

There are several hundred species in this group. They are collectively known as stoneworts, because they commonly accumulate crusts of calcium carbonate ($CaCO_3$) over their surfaces. Like the coleochaetes, they have plasmodesmata and are multicellular. ✔**You should be able to mark the origin of plasmodesmata on Figure 30.7.** (They do not occur in ulvophytes.) Some species of stonewort can be a meter or more in length.

The stoneworts are freshwater algae. Certain species are specialized for growing in relatively deep waters, though most live in shallow water near lake beaches or pond edges.

Reproduction Sexual reproduction is common and involves production of prominent, multicellular gametangia similar to those observed in early land plants. In stoneworts, as in coleochaetes, the eggs are retained on the parent plant, which supplies eggs with nutrients prior to fertilization. ✔**You should be able to mark the origin of egg retention on Figure 30.7.** (Egg retention does not occur in ulvophytes.)

Life cycle Alternation of generations does not occur. Multicellular individuals are haploid; the only diploid stage in the life cycle is the zygote.

Human and ecological impacts Some species form extensive beds in lake bottoms or ponds and provide food for ducks and geese as well as food and shelter for fish (**Figure 30.30**). They are a good indicator that water is not polluted.

Chara species

FIGURE 30.30 Stoneworts Can Form Beds on Lake Bottoms.

Non-Vascular Plants > Hepaticophyta (Liverworts)

Liverworts got their name because some species native to Europe have liver-shaped leaves. According to the medieval *Doctrine of Signatures*, God indicated how certain plants should be used by giving them a distinctive appearance. Thus, liverwort teas were hypothesized to be beneficial for liver ailments. (They are not.) About 6500 species are known. They are commonly found growing on damp forest floors or riverbanks, often in dense mats (**Figure 30.31**), or on the trunks or branches of tropical trees.

Adaptations to land Liverworts are covered with cuticle. Some species have pores that allow gas exchange; in species that lack pores, the cuticle is thin.

Reproduction Asexual reproduction occurs when fragments of a plant are broken off and begin growing independently. Some species also produce small structures called **gemmae** asexually, during the gametophyte phase. Mature gemmae are knocked off the parent plant by rain and grow into independent gametophytes. During sexual reproduction, sperm and eggs are produced in gametangia.

Life cycle The gametophyte is the largest and longest-lived phase in the life cycle. Sporophytes are small, grow directly from the gametophyte, and depend on the gametophyte for nutrition. Spores are shed from the sporophyte and are carried away by wind or rain.

Human and ecological impacts When liverworts grow on bare rock or tree bark, their dead and decaying body parts contribute to the initial stages of soil formation. The liverwort *Marchantia* is an important model organism in plant biology.

Marchantia polymorpha

FIGURE 30.31 Liverworts Thrive in Moist Habitats.

Over 12,000 species of mosses have been named and described to date, and more are being discovered every year—particularly in the tropics. Mosses are informally grouped with other "bryophytes" (liverworts and hornworts) but are formally classified in their own monophyletic group: the phylum Bryophyta.

Although mosses are common in moist forests, they can also be abundant in more extreme environments, such as deserts and windy, treeless habitats in the Arctic, Antarctic, or mountaintops. In these severe conditions, mosses are able to thrive because their bodies can become extremely dry without dying. When the weather makes photosynthesis difficult, individuals dry out and become dormant, or inactive. Then when rains arrive or temperatures warm, the plants rehydrate and begin photosynthesis and reproduction.

Adaptations to land One subgroup of mosses contains simple conducting tissues consisting of cells that are specialized for the transport of water or food. But because these cells do not have walls that are reinforced by lignin, they are not considered true vascular tissue. ✔**You should be able to mark the origin of the simple water-conducting cells and tissues in this moss subgroup on Figure 30.7.** Because they lack true vascular tissue, most mosses are not able to grow much taller than a few centimeters.

Reproduction Asexual reproduction often occurs by fragmentation, meaning that pieces of gametophytes that are broken off by wind or a passing animal can begin growing independently. In many species, sexual reproduction cannot involve self-fertilization because the sexes are separate—meaning that an individual plant produces only eggs in archegonia or only sperm in antheridia. A typical sporophyte produces up to 50 million tiny spores. Spores are usually distributed by wind.

Life cycle The moss life cycle is similar to that of liverworts and hornworts: The sporophyte is retained on the much larger and longer-lived gametophyte and gets most of its nutrition from the gametophyte.

Human and ecological impacts Species in the genus *Sphagnum* are often the most abundant plant in wet habitats of northern environments (**Figure 30.32a**). Because *Sphagnum*-rich environments account for 1 percent of Earth's total land area, equivalent to half the area of the United States, *Sphagnum* species are among the most abundant plants in the world. *Sphagnum*-rich habitats are water-logged, nitrogen-poor, and often anaerobic, however, so the decomposition of dead mosses and other plants is slow. As a result, large deposits of semi-decayed organic matter, known as **peat**, accumulate. Researchers estimate that the world's peatlands store about 400 billion metric tons of carbon. If peatlands begin to burn or decay rapidly due to global warming, the CO_2 released will exacerbate the warming trend (see Chapter 54).

Peat is harvested as a traditional heating and cooking fuel in some countries (**Figure 30.32b**). It is also widely used as a soil additive in gardening, because *Sphagnum* can absorb up to 20 times its dry weight in water. This high water-holding capacity is due to the presence of large numbers of dead cells in the leaves of these mosses, which readily fill with water via pores in their walls.

(a) *Sphagnum* moss is abundant in northern wet habitats.

(b) Semi-decayed *Sphagnum* moss forms peat.

FIGURE 30.32 *Sphagnum* Mosses Are among the Most Abundant Plants in the World.

Non-Vascular Plants > Anthocerophyta (Hornworts)

Hornworts got their name because their sporophytes have a horn-like appearance (**Figure 30.33**) and because wort is the Anglo-Saxon word for plant. About 100 species have been described to date.

Adaptations to land Hornwort sporophytes have stomata. Research is under way to determine if they can open their pores to allow gas exchange or close their pores to avoid water loss during dry intervals.

Reproduction Depending on the species, gametophytes may contain only egg-producing archegonia or only sperm-producing antheridia, or both. Stated another way, individuals of some species are either female or male, while in other species each individual has both types of reproductive organs.

Life cycle The gametophyte is the longest-lived phase in the life cycle. Although sporophytes grow directly from the gametophyte, they are green because their cells contain chloroplasts. Sporophytes manufacture some of their own food but also get nutrition from the gametophyte. Spores disperse from the parent plant via wind or rain.

Human and ecological impacts Some species harbor symbiotic cyanobacteria that fix nitrogen.

Anthoceros laevis

FIGURE 30.33 Hornworts Have Horn-Shaped Sporophytes.

Seedless Vascular Plants > Lycophyta (Lycophytes, or Club Mosses)

Although the fossil record documents lycophytes that were 2 m wide and 40 m tall, the 1000 species of lycophytes living today are all small in stature (**Figure 30.34**). Most live on the forest floor or on the branches or trunks of tropical trees. Because of their appearance, they are often called ground pines or **club mosses**—even though they are neither pines nor mosses.

Adaptations to land Lycophytes are the most ancient land plant lineage with **roots**—a belowground system of tissues and organs that anchors the plant and is responsible for absorbing water and mineral nutrients. Roots differ from the rhizoids observed in bryophytes, because roots contain vascular tissue and thus are capable of conducting water and nutrients from belowground to the upper reaches of the plant. Unusual leaves called microphylls, which extend from the stems, are a synapomorphy found in lycophytes. ✔**You should be able to mark the origin of microphylls on Figure 30.7.**

Reproduction Asexual reproduction can occur by fragmentation or gemmae. During sexual reproduction, spores of some species give rise to bisexual gametophytes—meaning that each gametophyte produces both eggs and sperm. Self-fertilization is extremely rare in most of these species, however. In the genera *Selaginella* and *Isoetes*, in contrast, heterospory occurs and gametophytes are either male or female.

Life cycle The gametophytes of some species live entirely underground and get their nutrition from symbiotic fungi. In certain species, gametophytes live 6 to 15 years and give rise to a large number of sporophytes over time.

Human and ecological impacts Tree-sized lycophytes were abundant in the coal-forming forests of the Carboniferous period. In coal-fired power plants today, electricity is being generated by burning fossilized lycophyte and fern tissues.

Lycopodium species

FIGURE 30.34 Lycophytes Living Today Are Small in Stature.

Seedless Vascular Plants > Psilotophyta (Whisk Ferns)

Only two genera of whisk ferns are living today, and there are perhaps six distinct species. Whisk ferns are restricted to tropical regions and have no fossil record. They are extremely simple morphologically, with aboveground parts consisting of branching stems that have tiny, scale-like outgrowths instead of leaves (**Figure 30.35**).

Because they lack roots and leaves, the whisk ferns were considered an evolutionary "throwback"—a group that retained traits found in the earliest vascular plants. Molecular phylogenies challenge this hypothesis and support an alternative hypothesis: The morphological simplicity of whisk ferns is a derived trait—meaning that complex structures have been lost in this lineage.

Psilotum nudum

FIGURE 30.35 Psilotophytes Are Extremely Simple Morphologically.

Adaptations to land Whisk ferns lack roots. Some species gain most of their nutrition from fungi that grow in association with the whisk ferns' extensive underground stems called **rhizomes**. Other species grow in rock crevices or are **epiphytes** ("upon-plants"), meaning that they grow on the trunks or branches of other plants—in this case, in the branches of tree ferns. ✔**After reviewing where leaves and roots originated during land plant evolution (see Figure 30.10), you should be able to mark the loss of leaves and roots in whisk ferns on Figure 30.7.**

Reproduction Asexual reproduction occurs via the extension of rhizomes and the production of new aboveground stems. When spores mature, they are dispersed by wind and germinate into gametophytes that contain both archegonia and antheridia.

Life cycle Sporophytes may be up to 30 cm tall, but gametophytes are less than 2 mm long and live under the soil surface. Gametophytes absorb nutrients directly from the surrounding soil and from symbiotic fungi. Fertilization takes place inside the archegonium, and the sporophyte develops directly on the gametophyte.

Human and ecological impacts Some whisk fern species are popular landscaping plants, particularly in Japan. The same species can be a serious pest in greenhouses.

Seedless Vascular Plants > Equisetophyta (or Sphenophyta) (Horsetails)

Although horsetails are prominent in the fossil record of land plants, just 15 species are known today. All 15 are in the genus *Equisetum*. Translated literally, *Equisetum* means "horse-bristle." Both the scientific name and the common name, horsetail, come from the brushy appearance of the stems and branches in some species (**Figure 30.36**). Horsetails may be locally abundant in wet habitats such as stream banks or marsh edges.

Equisetum telmateia

FIGURE 30.36 Horsetails Have a Distinctive Brushy Appearance and Are Prominent in the Fossil Record.

Adaptations to land Horsetails have an adaptation that allows them to flourish in waterlogged, oxygen-poor soils: their stems are hollow, so oxygen readily diffuses down the stem to reach roots that cannot obtain oxygen from the surrounding soil. Horsetails are also distinguished by having whorled leaves and branches. ✔**You should be able to mark the origin of hollow stems, whorled branches, and whorled leaves on Figure 30.7.**

Reproduction Asexual reproduction is common in sporophytes and occurs via fragmentation or the extension of rhizomes. From these rhizomes, two types of erect, specialized stems may grow—stems that contain tiny leaves and chloroplast-rich branches and that are specialized for photosynthesis, or stems that bear clusters of sporangia and produce huge numbers of spores by meiosis.

Life cycle Gametophytes perform photosynthesis but are small and short lived. They normally produce both antheridia and archegonia, but in most cases the sperm-producing structure matures first. This pattern is thought to be an adaptation that minimizes self-fertilization and maximizes cross-fertilization.

Human and ecological impacts Horsetail stems are rich in silica granules. The glass-like deposits not only strengthen the stem but also made these plants useful for scouring pots and pans prior to the invention of other scrubbing tools—hence these plants are often called "scouring rushes."

With 12,000 species, ferns are by far the most species-rich group of seedless vascular plants. They are particularly abundant in the tropics. About a third of the tropical species are epiphytes, usually growing on the trunks or branches of trees. Species that can grow epiphytically live high above the forest floor, where competition for light is reduced, without making wood and growing tall themselves. The growth habits of ferns are highly variable among species, however, and ferns range in size from rosettes the size of your smallest fingernail to 20-meter-tall trees (**Figure 30.37**).

Adaptations to land Ferns are the only seedless vascular plants that have large, well-developed leaves—commonly called **fronds** because it is not yet clear whether they are homologous to the leaves found in seed plants. Leaves give the plant a large surface area, allowing it to capture sunlight for photosynthesis efficiently.

Reproduction In a few species, gametophytes reproduce asexually via production of gemmae. Typically, species that can reproduce via gemmae never produce gametes or sporophytes. In most species, however, sexual reproduction is the norm. Ferns are homosporous, but gametophytes develop as males or as females only, depending on light intensity, proximity to other gametophytes, and other environmental conditions.

Life cycle Although fern gametophytes contain chloroplasts and are photosynthetic, the sporophyte is typically the larger and longer-lived phase of the life cycle. In mature sporophytes, sporangia are usually found in clusters called **sori** on the undersides of leaves (**Figure 30.38**). The structure of the sporangia is a distinctive feature of ferns: It arises from a single cell and has a wall composed of a single cell layer. ✔You should be able to mark the evolution of the distinctive fern sporangium on Figure 30.7.

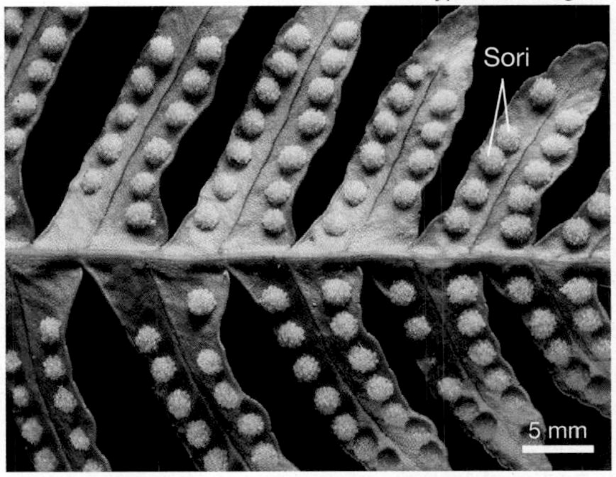

FIGURE 30.38 Fern Sporangia Are Often in Clusters (Sori) on the Underside of Leaves.

Human and ecological impacts In many parts of the world, people gather the young, unfolding fronds, or "fiddleheads," of ferns in spring as food. Ferns are also widely used as ornamental plants in landscaping.

FIGURE 30.37 Ferns Vary in Size and Are the Most Species-Rich Group of Seedless Vascular Plants.

Seed Plants > Gymnosperms > Cycadophyta (Cycads)

The cycads are so similar in overall appearance to palm trees, which are angiosperms, that cycads are sometimes called "sago palms." Although cycads were extremely abundant when dinosaurs were present on Earth 150–65 million years ago, only about 140 species are living today. Most are found in the tropics (**Figure 30.39**).

Adaptations to land Cycads do not make wood but are supported by stiff stems. They are unique among gymnosperms in having compound leaves—meaning that each leaf is divided into many smaller leaflets. **✓You should be able to mark the origin of the distinctive cycad leaf on Figure 30.7.**

Reproduction and life cycle Like other seed plants, cycads are heterosporous. Each sporophyte individual bears either microsporangia or megasporangia, but not both. Pollen is carried by insects (usually beetles or weevils) or, in some species, wind. Cycad seeds are large and often brightly colored. The colors attract birds and mammals, both of which eat and disperse the seeds.

Human and ecological impacts Cycads harbor large numbers of symbiotic cyanobacteria in specialized, aboveground root structures. The cyanobacteria are photosynthetic and fix nitrogen. The nitrogen acts as an important nutrient for nearby plants as well as the cycads themselves. Cycads are popular landscaping plants in some parts of the world.

Encephalartos transvenosus

FIGURE 30.39 Cycads Resemble Palms but Are Not Closely Related to Them.

Seed Plants > Gymnosperms > Ginkgophyta (Ginkgoes)

Although ginkgoes have an extensive fossil record, just one species is alive today. Leaves from the ginkgo, or maidenhair, tree are virtually identical in size and shape to those observed in fossil ginkgoes that are 150 million years old (**Figure 30.40**).

(a) Fossil ginkgo *Ginkgo huttoni*

(b) Living ginkgo *Ginkgo biloba*

FIGURE 30.40 The Ginkgo Tree Is a "Living Fossil."

Adaptations to land Unlike most gymnosperms, the ginkgo is **deciduous**—meaning that it loses its leaves each autumn. This adaptation allows plants to be dormant during the winter, when photosynthesis and growth are difficult.

Reproduction and life cycle Sexes are separate—individuals are either male or female. Pollen is transported by wind. Sperm have flagella, however. Once pollen grains land near the female gametophyte and mature, the sperm cells leave the pollen grain and swim to the egg cells.

Human and ecological impacts Although today's ginkgo trees are native to southeast China, they are planted widely as an ornamental. They are especially popular in urban areas, as they are tolerant of air pollution. In some countries, the inside of the seed is eaten as a delicacy.

Seed Plants > Gymnosperms > Redwood Group (Redwoods, Junipers, Yews)

The species in this lineage vary in growth form from sprawling juniper shrubs to the world's largest plants. Redwood trees growing along the Pacific Coast of North America can reach heights of over 115 m (375 ft) and trunk diameters of over 9.5 m (30 ft). These species were recently confirmed as a monophyletic group, independent of the Pinophyta, which also reproduce via cone-bearing structures, but the lineage still does not have a formal name.

Adaptations to land All of the species in this lineage are trees or large shrubs. Most have narrow leaves, which in many cases are arranged in overlapping scales (**Figure 30.41**). Narrow leaves have a small amount of surface area, which is not optimal for capturing sunlight and performing photosynthesis. But because the small surface area reduces water loss from leaves, many species in this lineage thrive in dry habitats or in cold environments where water is often frozen.

FIGURE 30.41 Some Species in the Redwood Group Have Scale-Like Leaves.

Reproduction and life cycle The species in this group are wind pollinated. As in all seed plants, the female gametophyte is retained on the parent. Thus, fertilization and seed development take place in the female cone. Like other gymnosperms, seeds do not form inside an encapsulated structure. Depending on the species, the seeds are dispersed by wind or by seed-eating birds or mammals.

Human and ecological impacts Redwoods, redcedar, whitecedar, and yellowcedar have wood that is highly rot-resistant and thus prized for making furniture, decks, house siding, or other applications where wood is exposed to the weather. Yew wood is often preferred for making traditional archery bows, and the berry-like cones of juniper are used to flavor gin. The anticancer drug taxol was originally found in Pacific yew trees.

Seed Plants > Gymnosperms > Pinophyta (Pines, Spruces, Firs)

The gymnosperms include two major lineages of cone-bearing species: the pines and allies discussed here, and the group that includes redwoods, junipers, yews, and cypresses. Species in both lineages have a reproductive structure called the cone, in which microsporangia and megasporangia are produced (**Figure 30.42**).

Pinophyta include the familiar pines, spruces, firs, Douglas fir, tamaracks, and true cedars. These are among the largest and most abundant trees on the planet, as well as some of the most long-lived. One of the bristlecone pines native to southwestern North America is at least 4750 years old.

Adaptations to land Pinophyta have needle-like leaves, with a small surface area that allows them to thrive in habitats where water is scarce. Pines are common on sandy soils that have poor water-holding capacity, and spruces and firs are common in cold environments where water is often frozen. All of the living species make wood as a support structure.

Reproduction and life cycle Sexes are separate, and pollen is transferred to female cones by the wind. Female cones take two years to mature, and are usually found high in trees. Male cones are usually found lower in the tree—possibly to reduce self-pollination from falling pollen.

Human and ecological impacts In terms of biomass, pines, spruces, firs, and other species in this group dominate forests that grow at high latitudes and high elevations—as well as sandy sites in warmer regions. Their seeds are key food sources for a variety of birds, squirrels, and mice, and their wood is the basis of the building products and paper industries in many parts of the world. The paper in this book was made from species in this group.

FIGURE 30.42 Pollen-Bearing Cones Produce Microsporangia; Ovulate Cones Produce Megasporangia.

Seed Plants > Gymnosperms > Gnetophyta (Gnetophytes)

The gnetophytes comprise about 70 species in three genera: *Gnetum* comprises vines and trees from the tropics; *Ephedra* is made up of desert-dwelling shrubs, including what may be the most familiar gnetophyte—the shrub called Mormon tea, which is common in the deserts of southwestern North America; *Welwitschia* contains a single species that probably qualifies as the world's most bizarre plant (**Figure 30.43**)—*W. mirabilis*, which is native to the deserts of southwest Africa. Although it has large belowground structures, the aboveground part consists of just two strap-like leaves, which grow continuously from the base and die at the tips. The leaves also split lengthwise as they grow and age.

Adaptations to land Gnetophytes have vessel elements in addition to tracheids. All of the living species make wood as a support structure.

Reproduction and life cycle The microsporangia and megasporangia are arranged in clusters at the end of stalks, similar to the way flowers are clustered in some angiosperms. Pollen is transferred by the wind or by insects. Double fertilization occurs in two of the three genera. As in other gymnosperms, seeds do not form inside an encapsulated structure.

Human and ecological impacts The drug ephedrine was originally isolated from species of *Ephedra* that are native to northern China and Mongolia. Ephedrine is used in the treatment of hay fever, colds, and asthma.

FIGURE 30.43 *Welwitschia* Is an Unusual Plant.

Seed Plants > Anthophyta (Angiosperms)

The flowering plants, or angiosperms, are far and away the most species-rich lineage of land plants. Over 250,000 species have already been described. They range in size from *Lemna gibba*—a floating, aquatic species that is less than half a millimeter wide—to massive oak trees. Angiosperms thrive in desert to freshwater to rain forest environments and are found in virtually every habitat except the deep oceans. They are the most common and abundant plants in most terrestrial environments.

The defining adaptation of angiosperms is the flower. **Flowers** are reproductive structures that hold either pollen-producing microsporangia or the megasporangia that produce megaspores and eggs, or both. Nectar-producing cells are often present at the base of the flower, and the color of petals helps to attract insects, birds, or bats that carry pollen from one flower to another (**Figure 30.44a**). Some angiosperms are pollinated by wind, however. Wind-pollinated flowers lack both colorful petals and nectar-producing cells (**Figure 30.44b**).

Adaptations to land In addition to flowers, angiosperms evolved vessels, the conducting cells that make water transport particularly efficient. Most angiosperms contain both tracheids and vessels.

Reproduction and life cycle Unlike gymnosperms, angiosperms have a carpel, a structure within the flower that contains an ovary. The ovary encloses one or more ovules, each of which encloses the female gametophyte. In most cases, male gametophytes are carried to female gametophytes by animal pollinators that are inadvertently dusted with pollen as they visit flowers to find food. Depending on the angiosperm species, self-fertilization may be common or absent. When the egg produced by the female gametophyte is fertilized, the ovule develops into a seed. When the ovary matures it forms a fruit, which contains the seed or seeds.

Human and ecological impacts It is almost impossible to overstate the importance of angiosperms to humans and other organisms. In most terrestrial habitats today, angiosperms supply the food that supports virtually every other species. For example, many insects eat flowering plants. Historically, the diversification of angiosperms correlated closely with the diversification of insects, which are by far the most species-rich lineage on the tree of life. It is not unusual for a single tropical tree to support dozens or even hundreds of insect species. Angiosperm seeds and fruits have also supplied the staple foods of virtually every human culture that has ever existed.

(Continued on next page)

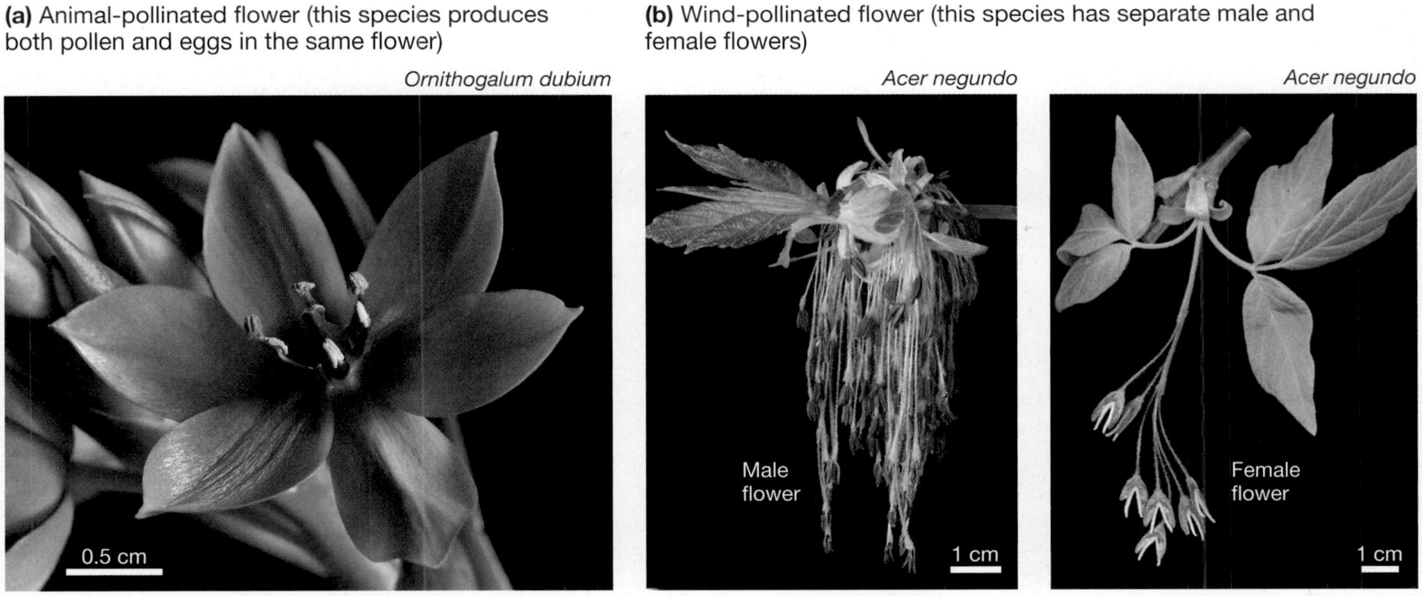

FIGURE 30.44 Wind-Pollinated Flowers Lack the Colorful Petals and Nectar Found in Most Animal-Pollinated Species.

CHAPTER 30 REVIEW

For media, go to the study area at www.masteringbiology.com

Summary of Key Concepts

Green algae are an important source of oxygen and provide food for aquatic organisms; land plants release oxygen, hold soil and water in place, build soil, moderate extreme temperatures and winds, and provide food for terrestrial organisms.

- Plants improve the quality of the environment for other organisms.
- Humans depend on plants for food, fiber, and fuel.

✓ You should be able to predict how the current and massive loss of plant species and plant communities will affect soils around the world.

Land plants were the first multicellular organisms that could live with most of their tissues exposed to the air. A series of key adaptations allowed them to survive on land. In terms of total mass, plants dominate today's terrestrial environments.

- The land plants evolved from green algae and colonized terrestrial environments in conjunction with fungi.
- The evolution of cuticle allowed plant tissues to be exposed to air without dying.
- The evolution of pores provided breaks in the cuticle and facilitated gas exchange, with CO_2 diffusing into leaves and O_2 diffusing out.
- The evolution of guard cells allowed plants to control the opening and closing of pores in a way that maximizes gas exchange and minimizes water loss.
- Vascular tissue conducts water and provides structural support that makes erect growth possible.
- True vascular tissue has cells that are dead at maturity and that have secondary cell walls reinforced with lignin.
- Erect growth is important because it reduces competition for light.
- Tracheids are water-conducting cells found in all vascular plants; in addition, angiosperms and gnetophytes have water-conducting cells called vessels.

✓ You should be able to explain why algae made the transition to terrestrial life just once, in light of the number of adaptations that were required.

(MB) **Web Activity** Plant Evolution and the Phylogenetic Tree

Once plants were able to grow on land, a sequence of important evolutionary changes made it possible for them to reproduce efficiently—even in extremely dry environments.

- All land plants are embryophytes, meaning that eggs and embryos are retained on the parent plant. Consequently, the developing embryo can be nourished by its mother.
- Seed plant embryos are dispersed from the parent plant to a new location, encased in a protective housing, and supplied with a store of nutrients.
- All land plants have alternation of generations.
- Over the course of land plant evolution, the gametophyte phase became reduced in terms of size and life span and the sporophyte phase became more prominent. In seed plants, male gametophytes are reduced to pollen grains and female gametophytes are reduced to tiny structures that produce an egg.

- The evolution of pollen was an important breakthrough in the history of life, because sperm no longer needed to swim to the egg—tiny male gametophytes could be transported through the air via wind or animals.

✔ **You should be able to discuss (compare and contrast) the advantages and disadvantages of having spores or seeds serve as the dispersal stage in a plant life cycle.**

Questions

✔ TEST YOUR KNOWLEDGE

Answers are available in Appendix B

1. Which of the following groups is definitely monophyletic?
   a. non-vascular plants
   b. green algae
   c. green plants
   d. seedless vascular plants
2. What is a difference between tracheids and vessels?
   a. Tracheids are dead at maturity; vessels are alive and are filled with cytoplasm.
   b. Vessels have openings (gaps) in the primary and secondary cell wall; tracheids have openings (pits) only in the secondary cell wall.
   c. Only tracheids have a thick secondary cell wall containing lignin.
   d. Only vessels have a thick secondary cell wall containing lignin.
3. Which of the following statements is *not* true?
   a. Green algae in the lineage called Charales are the closest living relatives of land plants.
   b. "Bryophytes" is a name given to the land plant lineages that do not have vascular tissue.
   c. The horsetails and the ferns form a distinct clade, or lineage. They have vascular tissue but reproduce via spores, not seeds.
   d. According to the fossil record and phylogenetic analyses, angiosperms evolved before the gymnosperms. Angiosperms are the only land plants with vessels.
4. The appearance of cuticle and stomata correlated with what event in the evolution of land plants?
   a. the first erect growth forms
   b. the first woody tissues
   c. growth on land
   d. the evolution of the first water-conducting tissues
5. What do seeds contain?
   a. male gametophyte and nutritive tissue
   b. female gametophyte and nutritive tissue
   c. embryo and nutritive tissue
   d. mature sporophyte and nutritive tissue
6. What is a pollen grain?
   a. male gametophyte
   b. female gametophyte
   c. male sporophyte
   d. sperm

✔ TEST YOUR UNDERSTANDING

Answers are available in Appendix B

1. Soils, water, and the atmosphere are major components of the abiotic (nonliving) environment. Describe how green plants affect the abiotic environment in ways that are advantageous to humans.
2. The evolution of cuticle presented land plants with a challenge that threatened their ability to live on land. Describe this challenge and explain why stomata represent a solution. Compare and contrast stomata with the pores found in liverworts. Explain why it is logical to observe that liverworts that lack pores have extremely thin cuticle.
3. Why was the evolution of lignin-reinforced cell walls significant?
4. Land plants may have reproductive structures that **(1)** protect gametes as they develop; **(2)** nourish developing embryos, **(3)** allow sperm to be transported in the absence of water, **(4)** provide stored nutrients and a protective coat so that offspring can be dispersed away from the parent plant, and **(5)** provide nutritious tissue around seeds that facilitates dispersal by animals. Name each of these five structures, and state which land plant group or groups have each structure.
5. What does it mean to say that a life cycle is gametophyte-dominant versus sporophyte-dominant?
6. Explain the difference between homosporous and heterosporous plants. Where are the microsporangium and megasporangium found in a tulip? What happens to the spores that are produced by these structures?

✔ APPLYING CONCEPTS TO NEW SITUATIONS

Answers are available in Appendix B

1. What is the significance of the observation that some members of the coleochaetes and stoneworts synthesize sporopollenin and/or lignin?
2. Vessel elements transport water much more efficiently than tracheids, but are much more susceptible than tracheids to being blocked by air bubbles. Suggest a hypothesis to explain why the vascular tissue of angiosperms consists of a combination of vessel elements and tracheids.
3. Angiosperms such as grasses, oaks, and maples are wind pollinated. The ancestors of these subgroups were probably pollinated by insects, however. As an adaptive advantage, why might a species "revert" to wind pollination? (Hint: Think about the costs and benefits of being pollinated by insects versus wind.) Why is it logical to observe that wind-pollinated species usually grow in dense stands containing many individuals of the same species? Why is it logical to observe that in wind-pollinated deciduous trees, flowers form very early in spring—before leaves form?
4. You have been hired as a field assistant for a researcher interested in the evolution of flower characteristics in orchids. Design an experiment to determine whether color, size, shape, scent, or amount of nectar is the most important factor in attracting pollinators to a particular species. Assume that you can change any flower's color with a dye and that you can remove petals or nectar stores, add particular scents, add nectar by injection, or switch parts among species by cutting and gluing.

The jack-o-lantern mushroom (*Omphalotus illudens*) is common on rotting logs and stumps in eastern North America. It is toxic to humans, but a derivative of one of the molecules responsible for its toxicity is being tested as an anticancer drug.

Fungi 31

Fungi are eukaryotes that grow as single cells or as large, branching networks of multicellular filaments. Familiar fungi include the mushrooms you've encountered in woods or lawns, the molds and mildews in your home, the organism that causes athlete's foot, and the yeasts used in baking and brewing.

Along with the land plants and animals, the **fungi** are one of three major lineages of large, multicellular eukaryotes that occupy terrestrial environments. When it comes to making a living, the species in these three groups use radically different strategies. Land plants make their own food through photosynthesis. Animals eat plants, protists, fungi, or each other. Fungi absorb their nutrition from other organisms—dead or alive.

Fungi that absorb nutrients from dead organisms are the world's most important decomposers. Although a few types of organisms are capable of digesting the cellulose in plant cell walls, fungi and a handful of bacterial species are the only organisms capable of completely digesting both the lignin and cellulose that make up wood. Without fungi, Earth's surface would be piled so high with dead tree trunks and branches that there would be almost no room for animals to move or plants to grow.

Other fungi specialize in absorbing nutrients from living organisms. When fungi absorb these nutrients without providing any benefit in return, they lower the fitness of their host organism and act as parasites. If you've ever had athlete's foot or a vaginal yeast infection, you've hosted a parasitic fungus.

The vast majority of fungi that live in association with other organisms benefit their hosts, however. In these cases, fungi are not parasites but **mutualists**. The roots of virtually every land plant in the world are colonized by an array of mutualistic fungi. In exchange for sugars that are synthesized by the host plant, the fungi provide the plant with water and key nutrients such as nitrogen and phosphorus. Without these nutrients, the host plants grow much more slowly or even starve.

KEY CONCEPTS

- Fungi are important in part because many species live in close association with land plants. They supply plants with key nutrients and decompose dead wood. They are the master recyclers of nutrients in terrestrial environments.
- All fungi make their living by absorbing nutrients from living or dead organisms. Fungi secrete enzymes so that digestion takes place outside their cells. Their morphology provides a large amount of surface area for efficient absorption.
- Many fungi have unusual life cycles. It is common for species to have a long-lived heterokaryotic stage, in which cells contain haploid nuclei from two different individuals. Although most species reproduce sexually, few species produce gametes.

✔ When you see this checkmark, stop and test yourself. Answers are available in Appendix B.

It is not possible to overstate the importance of these relationships between living land plants and the fungi that live in their roots. In the soils beneath every prairie, forest, and desert, an underground economy is flourishing. Plants are trading the sugar they manufacture for nitrogen or phosphorus atoms that are available from fungi. These plant-fungal associations are the world's most extensive bartering system. The soil around you is alive with an enormous network of fungi that are fertilizing the plants you see aboveground.

In short, fungi are the master traders and recyclers in terrestrial ecosystems. Some fungi release nutrients from dead plants and animals; others transfer nutrients they obtain to living plants. Because they recycle key elements such as carbon, nitrogen, and phosphorus and because they transfer key nutrients to plants, fungi have a profound influence on productivity and biodiversity. In terms of nutrient cycling on the continents, fungi make the world go around.

31.1 Why Do Biologists Study Fungi?

Given their importance to life on land and their intricate relationships to other organisms, it's no surprise that fungi are fascinating to biologists. But there are important practical reasons for humans to study fungi as well. They nourish the plants that nourish us. They affect global warming, because they are critical to the carbon cycle on land. Unfortunately, a handful of species can cause debilitating diseases in humans and crop plants. Let's take a closer look at some of the ways that fungi affect human health and welfare.

Fungi Provide Nutrients for Land Plants

Fungi that live in close association with plant roots are said to be **mycorrhizal** (literally, "fungal-root"; see **Figure 31.1a**). When biologists first discovered how extensive these fungal-plant associations are, they asked an obvious question: Does plant growth suffer if mycorrhizal fungi are absent?

Figure 31.1b shows a result typical of many experiments. In this case, seedlings were grown in the presence and absence of the mycorrhizal fungi normally found on their roots. The photographs document that this species grows three to four times faster in the presence of its normal fungal associates than it does without them.

For farmers, foresters, and ranchers, the presence of normal mycorrhizal fungi can mean the difference between profit and loss. Fungi are critical to the productivity of forests, croplands, and rangelands.

Fungi Speed the Carbon Cycle on Land

Fungi that make their living by digesting dead plant material are called **saprophytes** ("rotten-plants"). To understand why saprophytic fungi play a key role in today's terrestrial environments, recall from Chapter 30 that cells in the vascular tissues of land plants have secondary cell walls containing both lignin and cellulose. Wood forms when stems grow in girth by adding layers of lignin-rich vascular tissue.

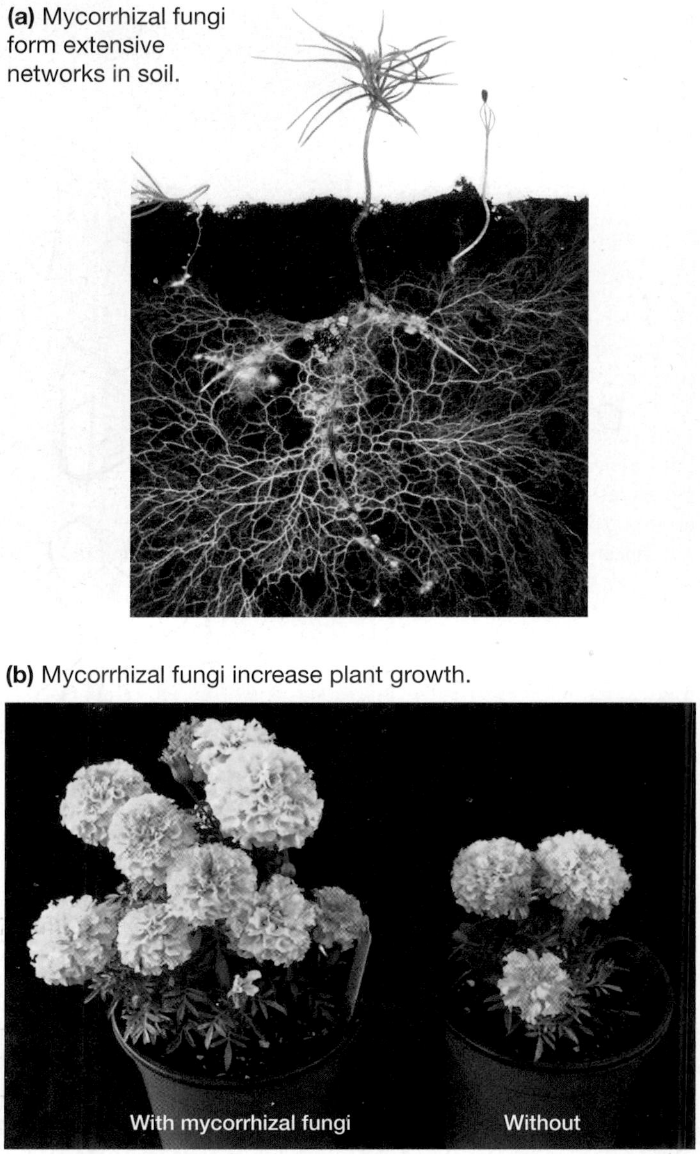

(a) Mycorrhizal fungi form extensive networks in soil.
(b) Mycorrhizal fungi increase plant growth.
With mycorrhizal fungi
Without

FIGURE 31.1 Plants Grow Better in the Presence of Mycorrhizal Fungi. (a) Root system of a larch tree seedling, with the mycelium from a mycorrhizal fungus visible. **(b)** Typical experimental results when plants are grown with and without their normal mycorrhizal fungi. (Fungi are not visible in the photo.)

When trees die, fungi are the organisms that break down wood into sugars and other small organic compounds. Fungi use these molecules as food. When fungi die or are eaten, the molecules are passed along to a wide array of other organisms.

Figure 31.2 highlights the role that fungi play as carbon atoms cycle through today's terrestrial environments. Note that there are two basic components of the **carbon cycle** on land:

1. the fixation of carbon by land plants—meaning that carbon in atmospheric CO_2 is reduced to cellulose, lignin, and other complex organic compounds in the bodies of plants; and
2. the release of CO_2 from plants, animals, and fungi as the result of cellular respiration—meaning the oxidation of glucose and production of the ATP that sustains life.

FIGURE 31.2 Fungi Speed Up the Cycling of Carbon Atoms through Terrestrial Ecosystems.

The fundamental point is that, for most carbon atoms, fungi connect the two parts of the cycle.

If fungi had not evolved the ability to digest lignin and cellulose soon after land plants evolved the ability to make these compounds, carbon atoms would have been sequestered in wood for millennia instead of being rapidly recycled into glucose molecules and CO_2. Terrestrial environments would be radically different than they are today and probably much less productive. On land, fungi make the carbon cycle turn much more rapidly than it would without fungi.

Fungi Have Important Economic Impacts

In humans, parasitic fungi cause athlete's foot, vaginitis, diaper rash, ringworm, pneumonia, and thrush, among other miseries. But even though these maladies can be serious, in reality only about 31 species of fungi—out of the hundreds of thousands of existing species—regularly cause illness in humans. Compared with the frequency of diseases caused by bacteria, viruses, and protists, the incidence of fungal infections in humans is low.

It would be easy to argue, in fact, that fungi have done more to promote human health than degrade it. The first antibiotic that was widely used, penicillin, was isolated from a fungus, and soil-dwelling fungi continue to be the source of many of the most important antibiotics prescribed against bacterial infections.

The major destructive impact that fungi have on people is through the food supply. Fungi known as rusts, smuts, mildews, wilts, and blights cause annual crop losses computed in the billions of dollars. These fungi are particularly troublesome in wheat, corn, barley, and other grain crops (**Figure 31.3a**). Saprophytic fungi are also responsible for enormous losses due to spoilage—particularly for fruit and vegetable growers (**Figure 31.3b**).

(a) Parasitic fungi infect corn and other crop plants.
(b) Saprophytic fungi rot fruits and vegetables.

FIGURE 31.3 Fungi Cause Problems with Crop Production and Storage. (a) A wide variety of grain crops are parasitized by fungi. Corn smut is a serious disease in sweet corn, although in Mexico the smut fungus is eaten as a delicacy. **(b)** Fungi decompose fruits and vegetables as well as leaves and tree trunks.

In nature, epidemics caused by fungi have killed 4 billion chestnut trees and tens of millions of American elm trees in North America. The fungal species responsible for these epidemics were accidentally imported on species of chestnut and elm native to other regions of the world. When the fungi arrived in North America and began growing in chestnuts and elms native to North America, the results were catastrophic. The local chestnut and elm populations had virtually no genetic resistance to the pathogens and quickly succumbed. The epidemics radically altered the composition of upland and floodplain forests in the eastern United States. Before these fungal epidemics occurred, chestnuts and elms dominated these habitats.

Fungi also have important positive impacts on the human food supply:

- Mushrooms are consumed in many cultures; in the industrialized nations they are used in sauces, salads, and pizza.
- The yeast *Saccharomyces cerevisiae* was domesticated thousands of years ago; today it and other fungi are essential to the manufacture of bread, soy sauce, tofu, cheese, beer, wine, whiskey, and other products. In most cases, domesticated fungi are used in conditions where the cells grow via fermentation, creating by-products like the CO_2 that causes bread to rise and beer and champagne to fizz.
- Enzymes derived from fungi are used to improve the characteristics of foods ranging from fruit juice and candy to meat.

To summarize, biologists study fungi because they affect a wide range of species in nature, including humans. What tools are helping researchers understand the diversity of fungi?

31.2 How Do Biologists Study Fungi?

About 80,000 species of fungi have been described and named to date, and about 1000 more are discovered each year. But the fungi are so poorly studied that the known species are widely regarded as a tiny fraction of the actual total.

To predict the actual number of fungal species alive today, David Hawksworth looked at the ratio of vascular plant species to fungal species in the British Isles—the area where the two groups are the most thoroughly studied. According to Hawksworth's analysis, there is an average of six species of fungus for every species of vascular plant on these islands. If this ratio holds worldwide, then the estimated total of 275,000 vascular plant species implies that there are 1.65 million species of fungi.

Although this estimate sounds large, recent data on fungal diversity suggest that it may be an underestimate. Consider what researchers found when they analyzed fungi growing on Barro Colorado Island, Panama: Living on the healthy leaves of just two tropical tree species were a total of 418 distinct morphospecies of fungi. (Recall from Chapter 26 that morphospecies are distinguished from each other by some aspect of morphology.) Because over 310 species of trees and shrubs grow on Barro Colorado, the data suggest that tens of thousands of fungi may be native to this island alone. If further work on fungal diversity in the tropics supports these conclusions, there may turn out to be many millions of fungal species.

This viewpoint of fungal diversity was reinforced by an analysis of the fungi living in conjunction with the roots of a single species of grass native to Eurasia. In this study, researchers used the direct sequencing approach, introduced in Chapter 28, to analyze the gene that codes for the RNA molecule in the small subunit of fungal ribosomes. The data showed that 49 phylogenetic species were living in conjunction with the grass roots. (The phylogenetic species concept was introduced in Chapter 26.) Most of the species had never before been described, and several represented completely new lineages of fungi. Biologists are only beginning to realize the extent of species diversity in fungi.

Let's consider how biologists are working to make sense of all this diversity, beginning with an overview of fungal morphology.

Analyzing Morphological Traits

Compared with animals and land plants, fungi have simple bodies. Only two growth forms occur among them: **(1)** single-celled forms called **yeasts** (**Figure 31.4a**), and **(2)** multicellular, filamentous structures called **mycelia** (singular: **mycelium**; **Figure 31.4b**). Many species of fungus grow either as a yeast or as a mycelium, but some regularly adopt both growth forms.

Because most fungi form mycelia and because this body type is so fundamental to the absorptive mode of life, most studies of fungal morphology have focused on them.

(a) Single-celled fungi are called yeasts.

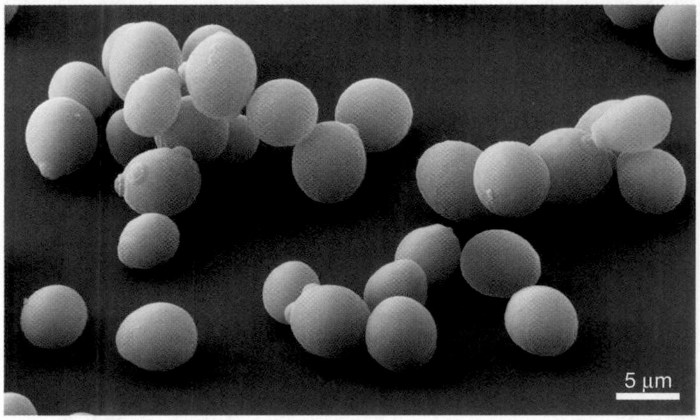

(b) Multicellular fungi have weblike bodies called mycelia.

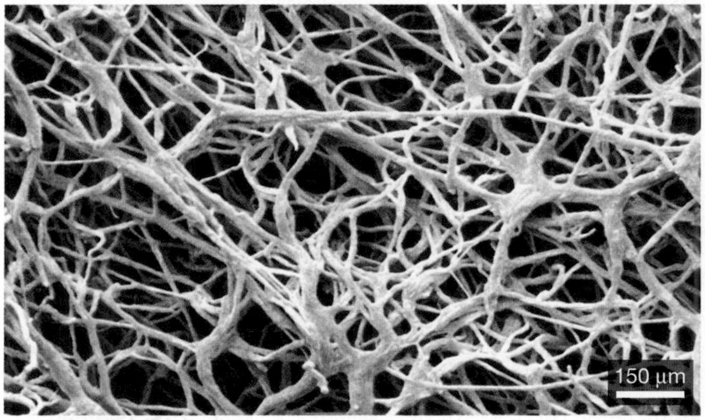

FIGURE 31.4 Fungi Have Just Two Growth Forms. Fungi grow **(a)** as single-celled yeasts and/or **(b)** as multicellular mycelia made up of long, thin, highly branched filaments. Both of these scanning electron micrographs are colorized.

THE NATURE OF THE FUNGAL MYCELIUM If food sources are plentiful, mycelia can be long lived and grow to be extremely large. Researchers discovered a mycelium growing across 1310 acres (6.5 km^2) in Oregon. This is an area substantially larger than most college campuses. The biologists estimated the individual's weight at hundreds of tons and its age at thousands of years, making it one of the largest and oldest organisms known.

Although most mycelia are much smaller and shorter lived than the individual in Oregon, all mycelia are dynamic. Mycelia constantly grow in the direction of food sources and die back in areas where food is running out. The body shape of a fungus can change almost continuously throughout its life. Recent data indicate that the individual filaments that make up a mycelium live, on average, only about five days.

THE NATURE OF HYPHAE The filaments within a mycelium are called **hyphae** (singular: **hypha**). Hyphae may be haploid or **heterokaryotic** ("different-kernel"), meaning that each cell contains several haploid nuclei from different parents. Most heterokaryotic hyphae are **dikaryotic** ("two-kernel"). In this case, two haploid nuclei, one from each parent, are present.

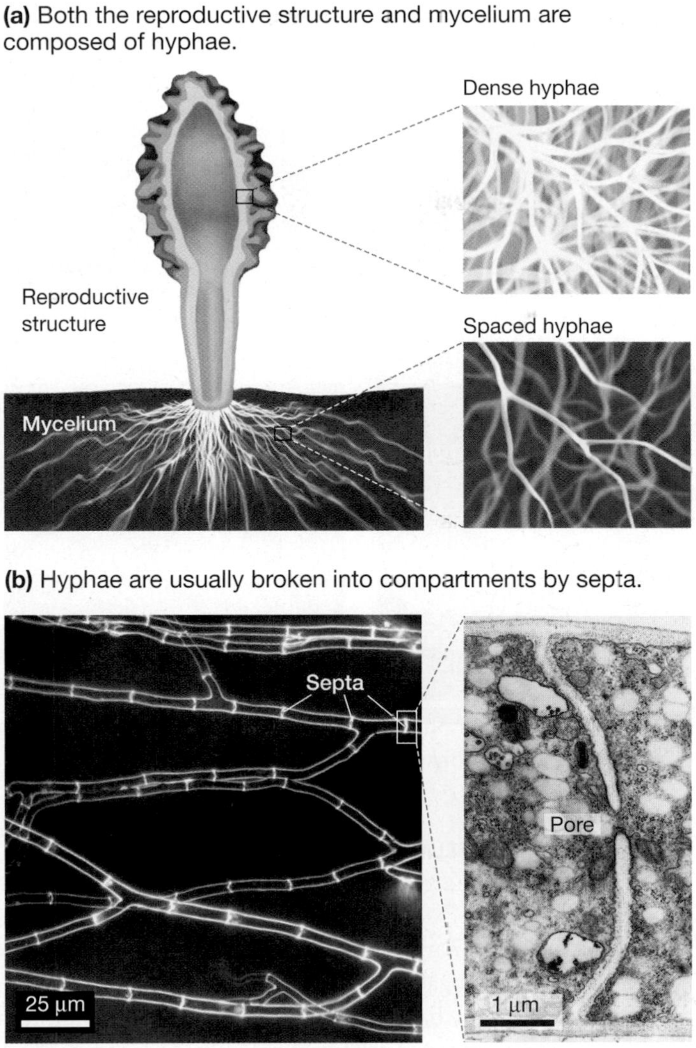
(a) Both the reproductive structure and mycelium are composed of hyphae.
Dense hyphae
Reproductive structure
Spaced hyphae
Mycelium
(b) Hyphae are usually broken into compartments by septa.
Septa
Pore
25 μm
1 μm

FIGURE 31.5 Multicellular Fungi Have Unusual Bodies. (a) The feeding structure of a fungus is a mycelium, which is made up of hyphae. In some species, hyphae come together to form multicellular structures such as mushrooms, brackets, or morels that emerge from the ground. **(b)** Hyphae are often divided into cell-like compartments by partitions called septa, which are broken by pores. As a result, the cytoplasm of different compartments is continuous.

As **Figure 31.5a** shows, hyphae are long, narrow filaments that branch frequently. In most fungi, each filament is broken into cell-like compartments by cross-walls called **septa** (singular: **septum; Figure 31.5b**).

Septa do not close off segments of hyphae completely. Instead, gaps called pores enable a wide variety of materials, even organelles and nuclei, to flow from one compartment to the next. Septa may have single large openings or a series of small gaps that give the septum a sieve-like appearance. Because nutrients, mitochondria, and even genes can flow though the entire mycelium—at least to a degree—the fungal mycelium is intermediate between a multicellular land plant or animal and an enormous single-celled organism.

Some fungal species are even **coenocytic** ("common-celled"; pronounced *see-no-SIT-ick*)—meaning that they lack septa entirely. Coenocytic fungi have many nuclei scattered throughout the mycelium. In effect, they are a single, gigantic cell.

It's also important to appreciate just how thin hyphae are. Plant root tips are typically about 1 mm in diameter, but fungal hyphae are typically less than 10 μm in diameter. Hyphae are 1/100th the size of a root tip. Fungal mycelia can penetrate tiny fissures in soil and absorb nutrients that are inaccessible to plant roots.

MYCELIA HAVE A LARGE SURFACE AREA Perhaps the most important aspect of mycelia and hyphae, however, is their shape. Because mycelia are composed of complex, branching networks of extremely thin hyphae, fungi have the highest surface-area-to-volume ratios observed in multicellular organisms.

To drive this point home, consider that the hyphae found in any fist-sized ball of rich soil typically have a surface area equivalent to half a page of this book. This surface area is important because fungi make their living via absorption, and a large surface area makes nutrient absorption extremely efficient.

The extraordinarily high surface area in a mycelium has a downside, however. The amount of water that evaporates from an organism is a function of its surface area—meaning that these organisms are prone to drying out. As a result, fungi are most abundant in moist habitats.

When conditions dry, the fungal mycelium may die back partially or completely. The reproductive cells called **spores** that are produced by sexual or asexual reproduction are resistant to drying, however. As a result, spores can endure dry periods, then germinate and resume growth when conditions improve. Mycelial growth is dynamic, and changes with moisture availability or food supply.

REPRODUCTIVE STRUCTURES Mycelia are an adaptation that supports the absorptive lifestyle of fungi. The only thick, fleshy structures that fungi produce are reproductive organs.

Mushrooms, puffballs, and other dense, multicellular structures that arise from mycelia do not absorb food. Instead, they function in reproduction. Typically they are the only part of a fungus that is exposed to the atmosphere, where drying is a problem. The mass of filaments on the inside of mushrooms is protected from drying by the densely packed hyphae forming the surface.

Relatively few species of fungi make the reproductive organs called mushrooms, however. Instead, most fungal species that undergo sexual reproduction produce one of four types of distinctive reproductive structures—only one of which is found inside mushrooms.

1. ***Swimming gametes and spores*** In certain species that live primarily in water or wet soils, the spores that are produced during asexual reproduction have flagella, as do the gametes produced during sexual reproduction (**Figure 31.6a** on page 584). These are the only motile cells known in fungi. Species with swimming gametes are traditionally known as chytrids (pronounced *KYE-trids*).
2. ***Zygosporangia*** In some species, haploid hyphae from two individuals meet and become yoked together as shown in **Figure 31.6b**. Cells from yoked hyphae fuse to form a distinctive

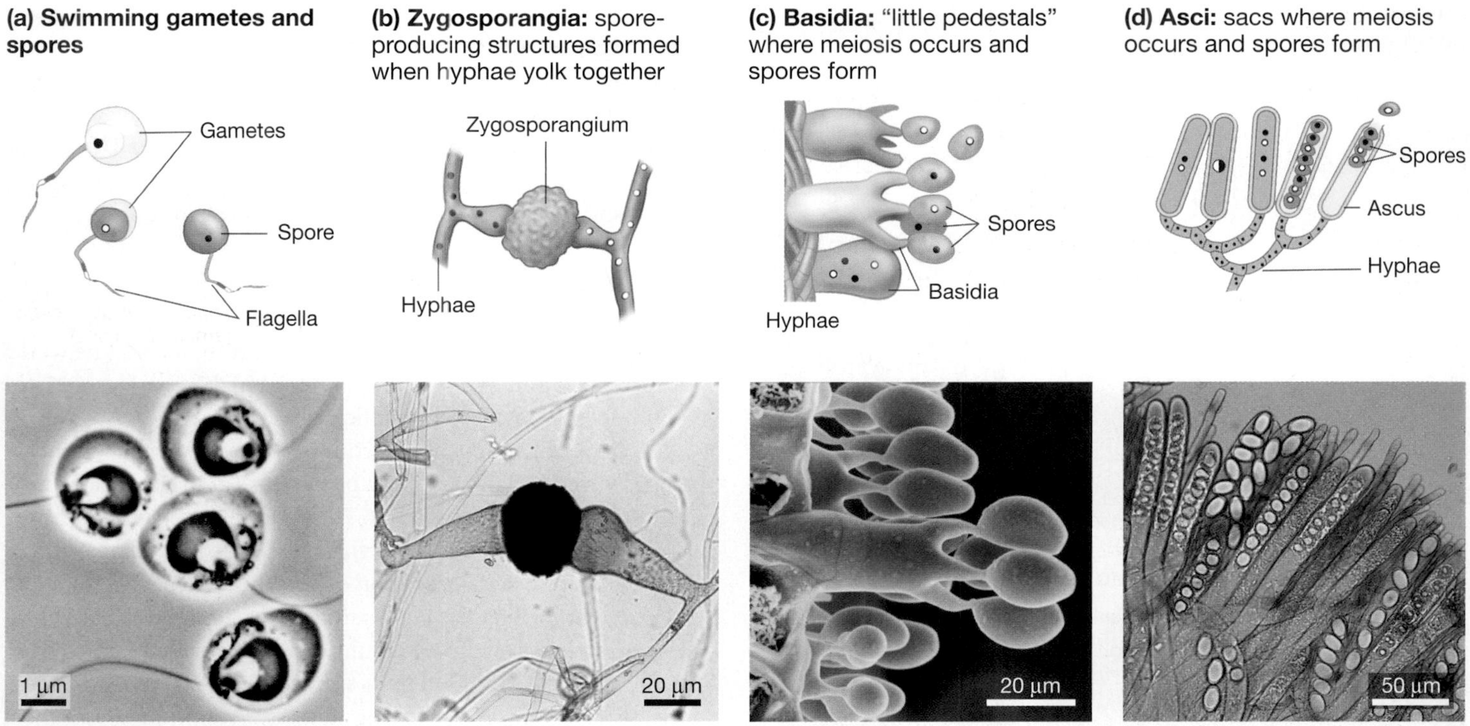

FIGURE 31.6 Four Distinct Reproductive Structures Are Observed in Fungi. The dots in the illustrations represent nuclei.

spore-producing structure called a **zygosporangium** (plural: **zygosporangia**; the Greek root *zygos* means to be yoked together like oxen). Species with a zygosporangium are traditionally known as zygomycetes.

3. ***Basidia*** Inside a mushroom, bracket, or puffball, specialized cells called **basidia** ("little-pedestals") form at the ends of hyphae and produce spores (**Figure 31.6c**). Species with basidia are traditionally called basidiomycetes.
4. ***Asci*** Inside cups, morels, and some other types of aboveground reproductive structures, specialized cells called **asci** (singular: **ascus**) form at the ends of hyphae and produce spores (**Figure 31.6d**). Species with asci are traditionally known as ascomycetes.

In sum, morphological studies allowed biologists to describe and interpret the mycelia growth habit as an adaptation that makes nutrient absorption extremely efficient. Careful analyses of morphological features also allowed researchers to identify four major types of reproductive structures.

Now the question is, which eukaryotes are most closely related to the fungi? And within the Fungi, do species that produce swimming gametes and spores, zygosporangia, basidia, and asci each form monophyletic groups—meaning that these distinctive reproductive structures evolved just once?

Evaluating Molecular Phylogenies

Researchers have sequenced and analyzed an array of genes to establish where fungi fit on the tree of life. The position of fungi as a whole is now well established; the position of lineages within fungi is still the subject of intense research.

FUNGI ARE CLOSELY RELATED TO ANIMALS **Figure 31.7** shows that fungi are much more closely related to animals than they are to land plants.

The close evolutionary relationship between fungi and animals explains why fungal infections in humans are much more difficult to treat than bacterial infections. Fungi and humans shared a common ancestor relatively recently. As a result, their enzymes and cell components are similar in structure and function. Drugs that disrupt fungal enzymes and cells are also likely to damage humans.

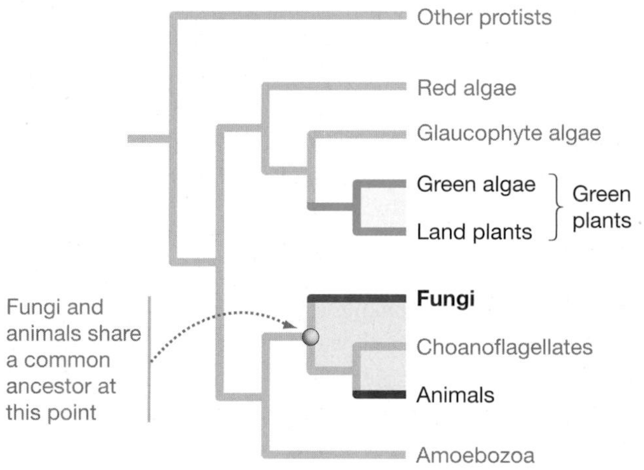

FIGURE 31.7 Fungi Are More Closely Related to Animals than to Land Plants. Phylogenetic tree showing the evolutionary relationships among the green plants, animals, fungi, and some groups of protists. (Choanoflagellates are solitary or colonial protists found in freshwater; they are introduced in Chapter 32.)

In addition to the DNA sequence data, three key morphological traits link animals and fungi:

1. Most animals and fungi synthesize the tough structural material called chitin (see Chapter 5). Chitin is a prominent component of the cell walls of fungi.
2. The flagella that develop in chytrid spores and in chytrid gametes are similar to those observed in animals: As in animals, the flagella in chytrids are single, are located at the back of reproductive cells, and move in a whiplash manner.
3. Both animals and fungi store food by synthesizing the polysaccharide glycogen. (Green plants, in contrast, synthesize starch as their storage product.)

WHAT IS THE RELATIONSHIP AMONG THE MAJOR FUNGAL GROUPS? To understand the relationships among species with swimming gametes, zygosporangia, basidia, and asci, biologists have sequenced a series of genes from an array of fungal species and used the data to estimate the phylogeny of the group. The results, shown in **Figure 31.8**, support several important conclusions:

- The single-celled eukaryotes called microsporidians are actually fungi.
  Interpretation: They are not distant relatives, as initially thought. This is important. Researchers are now testing the hypothesis that **fungicides**—molecules that are lethal to fungi—can cure microsporidian infections in bee colonies, silkworm colonies, and AIDS patients.
- The chytrids and zygomycetes form a paraphyletic group that is poorly resolved—meaning that the actual order of branching events among lineages with these reproductive structures is still not known (in Figure 31.8, they are collapsed into a polytomy—see **BioSkills 3** in Appendix A).
  Interpretation: Swimming gametes and the zygosporangium evolved more than once. Or, both structures were present in a common ancestor but then were lost in certain lineages.
- An important group called the Glomeromycota is monophyletic.
  Interpretation: The adaptations that allow these species to live in association with plant roots (discussed in Section 31.3) evolved once.
- The basidiomycetes are monophyletic—they form a lineage called Basidiomycota, or club fungi.
  Interpretation: The basidium evolved once.
- The ascomycetes are monophyletic—they form a lineage called Ascomycota, or sac fungi.
  Interpretation: The ascus evolved once.
- Together, the Basidiomycota and Ascomycota form a monophyletic group.
  Interpretation: Because basidiomycetes and ascomycetes are both dikaryotic, this growth habit evolved once.
- The sister group to fungi consists of protists called choanoflagellates and the animals.
  Interpretation: Because choanoflagellates and the most ancient groups of animals are aquatic, and because chytrids are aquatic, it is reasonable to conclude that the earliest fungi were aquatic, and that the switch to terrestrial life occurred early in the evolution of the Fungi.

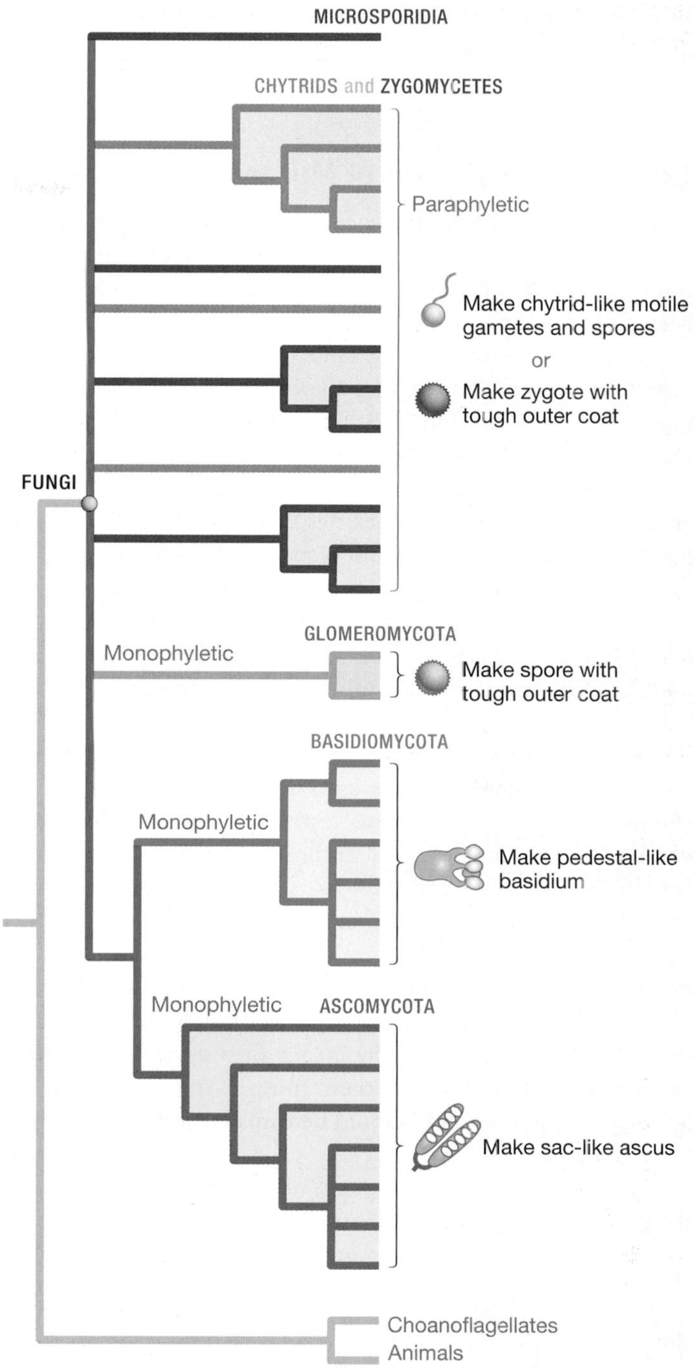

FIGURE 31.8 Phylogeny of the Fungi. A recent phylogenetic tree based on analyses of DNA sequence data. The icons represent the types of sexual reproductive structures observed in each major lineage.

Although progress on understanding the evolutionary history of fungi has been rapid, the phylogenetic tree in Figure 31.8 is still a work in progress. For example, it is not yet clear where microsporidians are placed relative to several lineages of chytrids and zygomycetes. (Microsporidians lack both swimming gametes and yoked hyphae.) Future work should clarify exactly how fungi diversified and how the diversification of fungi relates to the diversification of land plants.

Experimental Studies of Mutualism

Chapter 30 pointed out that the first plants in the fossil record are closely associated with fungal fossils. That chapter also claimed that the ability to absorb nutrients from fungi may have been crucial in the early evolution of land plants.

Close associations between land plants and fungi continue today. Researchers estimate that 90 percent of land plants live in physical contact with fungi. Stated another way, fungi and land plants often have a **symbiotic** ("together-living") relationship.

Although some species of fungi live in association with an array of different land plant species, some documented fungal-plant associations are specific. It is not uncommon for fungi to live in only a particular type of tissue, in one plant species.

Scientists categorize these symbiotic relationships as **mutualistic** if they benefit both species, **parasitic** if one species benefits at the expense of the other, or **commensal** if one species benefits while the other is unaffected.

To explore the nature of fungi-plant symbioses in detail, researchers have used isotopes as tracers for specific elements (**Figure 31.9**). For example, to test the hypothesis that fungi obtain food in the form of carbon-containing compounds from their plant associates, biologists have introduced radioactively labeled carbon dioxide into the air surrounding plants that do or do not contain symbiotic fungi. The labeled CO_2 molecules are incorporated into the sugars produced during photosynthesis, and the location of the radioactive atoms can then be followed over time by means of a device that detects radioactivity. If plants feed their fungal symbionts, then labeled carbon compounds should be transferred from the plant to the fungi.

To test the hypothesis that plants are receiving nutrients from their symbiotic fungi in return for sugars, researchers have added radioactive phosphorus atoms or the heavy isotope of nitrogen (^{15}N) to potted plants that do or do not contain symbiotic fungi. If fungi facilitate the transfer of nutrients from soil to plants, then plants grown in the presence of their symbiotic fungi should receive much more of the radioactive phosphorus or heavy nitrogen than do plants grown in the absence of fungi.

Experiments with isotopes used as tracers have shown that sugars and other carbon-containing compounds produced by plants via photosynthesis are transferred to their fungal symbionts. In some cases, as much as 20 percent of the sugars produced by a plant end up in their symbiotic fungi. In exchange, the symbiotic fungi facilitate the transfer of phosphorus or nitrogen—or both—from soil to the plant.

Because phosphorus and nitrogen are in extremely short supply in most environments, the nutrients supplied by symbiotic fungi are critical to the success of the plant. In this way, studies with isotopes have supported the hypothesis that most relationships between fungi and land plants are mutually beneficial.

CHECK YOUR UNDERSTANDING

If you understand that . . .

- The bodies of fungi are either single-celled yeasts or multicellular mycelia.
- During sexual reproduction, different groups of fungi produce distinct reproductive structures.

✔ You should be able to . . .

1. Explain why mycelia are interpreted as an adaptation to an absorptive lifestyle.
2. Identify the four types of reproductive structures observed in fungi.

Answers are available in Appendix B.

31.3 What Themes Occur in the Diversification of Fungi?

Why are there so many different species of fungi? This question is particularly puzzling given that fungi share a common attribute: They all make their living by absorbing food directly from their surroundings. In contrast to the diversity of food-getting strategies observed in bacteria, archaea, and protists, all fungi make their living in the same basic way. In this respect, fungi are like plants—virtually all of which make their own food via photosynthesis.

Chapter 30 showed that the diversification of land plants was driven not by novel ways of obtaining food, but by adaptations that allowed plants to grow and reproduce in a diverse array of terrestrial habitats. What drove the diversification of fungi? The answer is the evolution of novel methods for absorbing nutrients from a diverse array of food sources.

This section introduces a few of the ways that fungi go about absorbing nutrients from different food sources, as well as how they produce offspring. Let's explore the diversity of ways that fungi do what they do.

Fungi Participate in Several Types of Mutualisms

Not long after associations between fungi and the roots of land plants were discovered and shown to be mutualistic, researchers found that two types of plant-mycorrhizal interactions are particularly common, involving **ectomycorrhizal fungi (EMF)** and **arbuscular mycorrhizal fungi (AMF)**. The two major types of mycorrhizae have distinctive morphologies, geographic distributions, and functions.

- EMF are usually species from the Basidiomycota, though some ascomycetes participate.
- AMF include species from the Glomeromycota.

Mycorrhizae aren't the only type of symbiotic fungi found in plants, however. Researchers have also become interested in fungi that live in close association with the aboveground tissues of land plants—their leaves and stems. Fungi that live in the aboveground parts of plants are said to be **endophytic** ("inside-plants").

EXPERIMENT

QUESTION: Are mycorrhizal fungi mutualistic?

HYPOTHESIS: Host plants provide mycorrhizal fungi with sugars and other photosynthetic products. Mycorrhizal fungi provide host plants with phosphorus and/or nitrogen from the soil.

NULL HYPOTHESIS: No exchange of food or nutrients occurs between plants and mycorrhizal fungi. The relationship is not mutualistic.

EXPERIMENTAL SETUP:

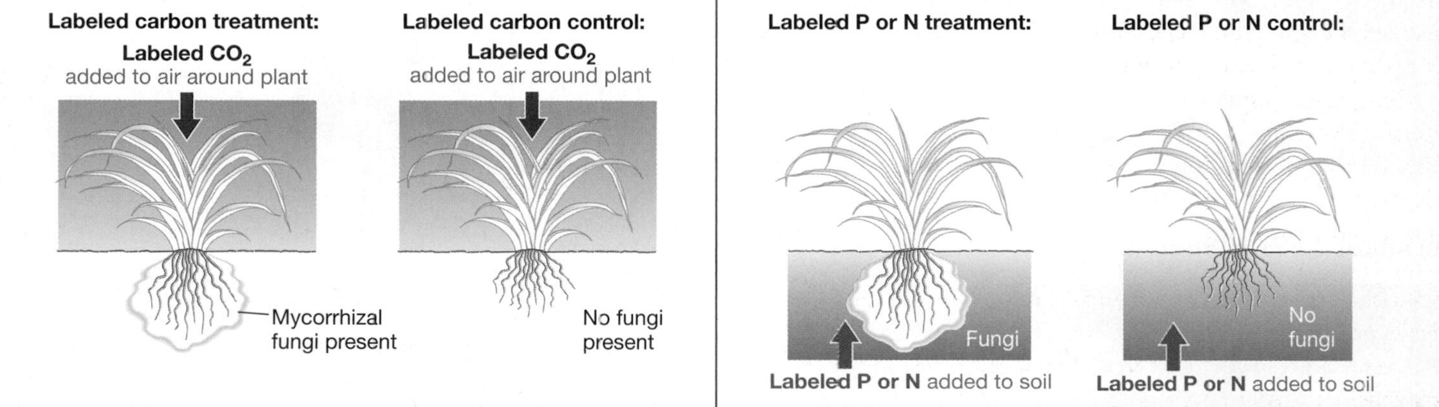

PREDICTION FOR LABELED CARBON: A large percentage of the labeled carbon taken up by the plant will be transferred to mycorrhizal fungi. In the control, little labeled carbon will be present in the soil surrounding the roots.

PREDICTION OF NULL HYPOTHESIS, LABELED CARBON: There will be no difference in the localization of carbon in the two treatments.

PREDICTION FOR LABELED P OR N: A large percentage of the labeled P or N taken up by the fungi will be transferred to the plant. In the control, little or no labeled P or N will be taken up by the plant.

PREDICTION OF NULL HYPOTHESIS, LABELED P OR N: There will be no difference between amounts of labeled P or N found in plant in presence or absence of fungi.

RESULTS:

CONCLUSION: The relationship between plants and mycorrhizal fungi is mutualistic. Plants provide mycorrhizal fungi with carbohydrates. Mycorrhizal fungi supply host plants with nutrients.

FIGURE 31.9 Experimental Evidence That Mycorrhizal Fungi and Plants Are Mutualistic. Sugars flow from plants to mycorrhizal fungi; key nutrients flow from mycorrhizal fungi to plants.

SOURCES: Bücking, H. and W. Heyser. 2001. Microautoradiographic localization of phosphate and carbohydrates in mycorrhizal roots of *Populus tremula* x *Populus alba* and the implications for transfer processes in ectomycorrhizal associations. *Tree Physiology* 21: 101–107.

✔**QUESTION** What do these "labeled nutrient" experiments tell you that you didn't already know from experiments like Figure 31.1, where plants are grown with and without mycorrhizae?

Recent research has shown that endophytic fungi are much more common and diverse than previously suspected. Further, data indicate that at least some species of endophytes are mutualistic.

These results support the general realization that most plants are covered with fungi—from the tips of their branches to the base of their roots. Throughout their lives, many or even most plants are involved in several distinct types of mutualistic relationships with fungi.

ECTOMYCORRHIZAL FUNGI (EMF) EMF like the one shown in **Figure 31.10a** on page 588 are found on many of the tree species in the temperate regions of the world (see Chapter 50), where warm summers alternate with cold winters. In this type of association, hyphae form a dense network that covers a plant's root tips. As the cross section in Figure 31.10a shows, individual hyphae penetrate between cells in the outer layer of the root, but hyphae do not enter the root cells.

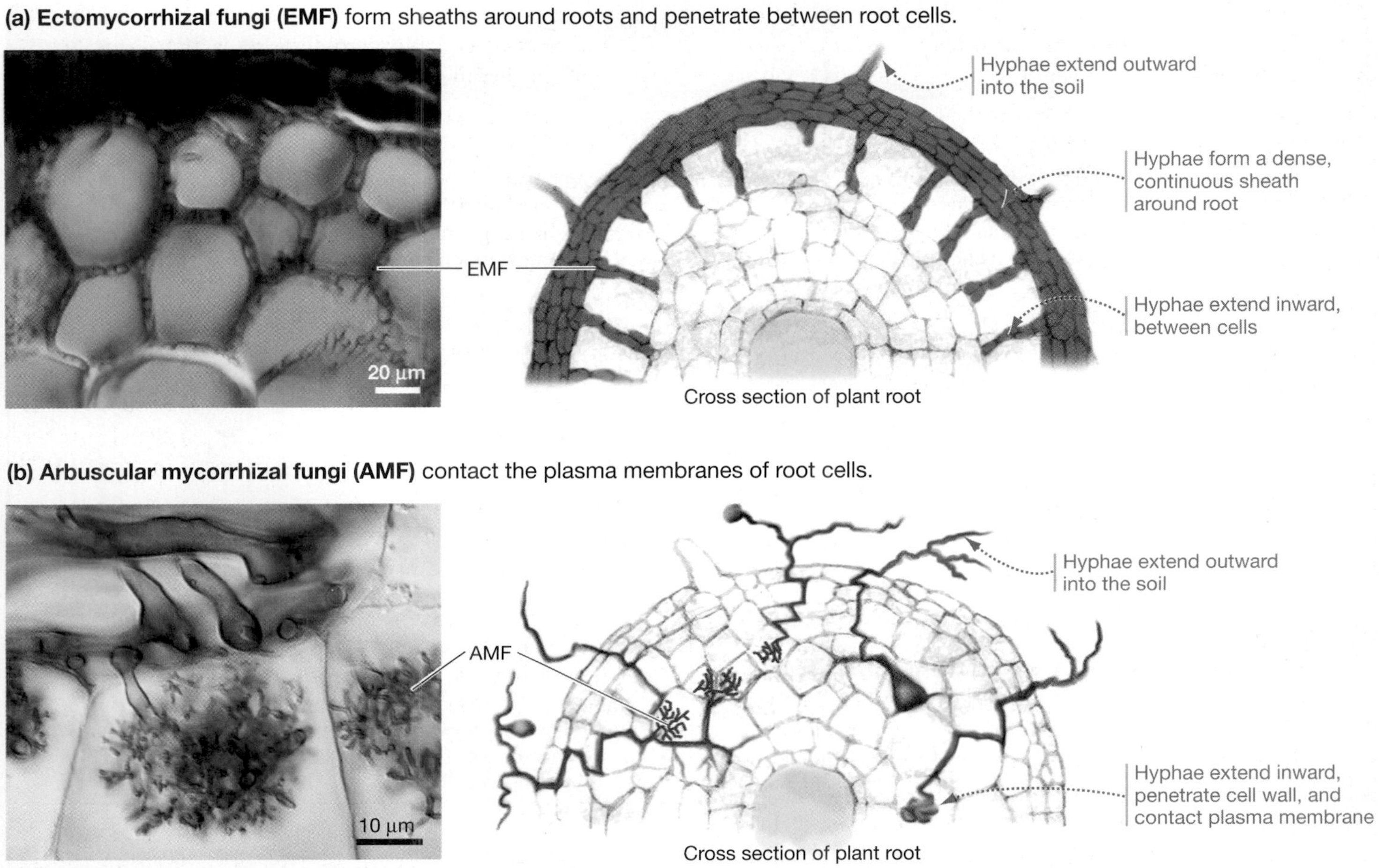

(a) Ectomycorrhizal fungi (EMF) form sheaths around roots and penetrate between root cells.
EMF
20 μm
Hyphae extend outward into the soil
Hyphae form a dense, continuous sheath around root
Hyphae extend inward, between cells
Cross section of plant root
(b) Arbuscular mycorrhizal fungi (AMF) contact the plasma membranes of root cells.
AMF
10 μm
Hyphae extend outward into the soil
Hyphae extend inward, penetrate cell wall, and contact plasma membrane
Cross section of plant root

FIGURE 31.10 Mutualistic Fungi Interact with the Roots of Plants in Two Distinct Ways. (a) Ectomycorrhizal fungi (EMF) form a dense network around the roots of plants. Their hyphae penetrate the intercellular spaces of the root but do not enter the root cells. **(b)** The hyphae of arbuscular mycorrhizal fungi (AMF) penetrate the walls of root cells, where they branch into bushy structures or balloon-like vesicles that are in close contact with the root cell's plasma membrane.

✔**EXERCISE** In the cross sections, add arrows and labels showing the direction of movement of N (nitrogen-containing compounds), P (phosphorus-containing compounds), and C (carbon-containing compounds).

The Greek root *ecto*, which refers to "outer," describes this association accurately: the fungi form an outer sheath on root tips that is often 0.1 mm thick. Hyphae also extend out from the sheath-like portion of the mycelium into the soil.

A recent genome sequencing project found that an EMF species contains genes for a large array of small proteins that are secreted from hyphae. This finding prompted the hypothesis that the secreted proteins act as signals to potential host plants. If so, establishing an association between an EMF and its host plant may involve a complex array of signals, which biologists are only beginning to understand.

How do these trees and fungi interact, once they start living together? In the habitats where EMF are abundant, nitrogen atoms tend to remain tied up in dead tissues, in amino acids and nucleic acids, instead of being available in the soil. The hyphae of EMF penetrate decaying material and release enzymes called peptidases that cleave the peptide bonds between amino acids in dead tissues. The amino acids released by this reaction are absorbed by the hyphae and transported to spaces between the root cells of trees, where they can be absorbed by the plant. EMF are also able to acquire phosphate ions that are bound to soil particles and transfer the ions to host plants. In return, the fungi receive sugars and other complex carbon compounds from the tree.

Researchers have found that when birch tree seedlings are grown with and without their normal EMF in pots filled with forest soil, only the seedlings with EMF are able to acquire significant quantities of nitrogen and phosphorus. Inspired by such data, a biologist has referred to EMF as the "dominant nutrient-gathering organs in most temperate forest ecosystems."

The hyphae of EMF are like an army of miners that discover, excavate, and deliver precious nuggets of nitrogen to trees. The productivity of the world's most important commercial forests depends on EMF.

ARBUSCULAR MYCORRHIZAL FUNGI (AMF) In contrast to the hyphae of EMF, the hyphae of arbuscular mycorrhizal fungi (AMF) grow *into* the cells of root tissue. The name *arbuscular* ("little-tree") was inspired by the bushy, highly branched hyphae, shown in **Figure 31.10b**, that form inside root cells.

AMF are also called **endomycorrhizal fungi**, because they penetrate the interior of root cell walls, or vesicular-arbuscular mycorrhizae (VAM), because the hyphae of some species form large, balloon-like vesicles inside root cells.

The key point is that the hyphae of AMF penetrate the cell wall and contact the plasma membrane of root cells directly. The highly branched hyphae inside the plant cell wall are thought to be an adaptation that increases the surface area available for exchange of molecules between the fungus and its host. However, AMF do not form a tight sheath around roots, as do EMF. Instead, they form a pipeline extending from inside plant cells in the root to the soil well beyond the root.

AMF are found in a whopping 80 percent of all land plant species. They are particularly common in grasslands and in the forests of tropical habitats. They are also widespread in temperate climates.

What do AMF do? Plant tissues decompose quickly in the grasslands and tropical forests where AMF flourish because the growing season is long and warm. As a result, nitrogen is often readily available to plants. Phosphorus is usually in short supply, though, because it tends to leach out of soils that experience high rainfall.

Based on these observations, biologists hypothesized that the most important function of AMF is to transfer phosphorus atoms from the soil to the host plant. Experiments with radioactive atoms confirmed the phosphate-transfer hypothesis by showing that AMF supply host plants with particularly large amounts of phosphorus, along with other nutrients. In return, host plants supply AMF with sugars and other forms of reduced carbon.

AMF are also extremely important in soil formation. The cell walls of their hyphae contain large quantities of a glycoprotein called **glomalin**. When the cells die, the glomalin enriches the organic matter in soil and helps bind organic compounds to sand or clay particles. Some recent estimates suggest that over 25 percent of all of the organic matter in soil consists of glomalin.

ARE ENDOPHYTES MUTUALISTS? Although endophytic fungi are relatively new to science, they are turning out to be both extremely common and highly diverse.

- Biologists in Brazil are examining tree leaves for the presence of fungi; each time they do, they are discovering several new species of endophytes.
- Recall from Section 31.2 that a study in Panama found hundreds of fungal species living in the leaves of just two tree species.

These newly discovered species are endophytes.

Recent research has shown that the endophytes found in some grasses produce compounds that benefit plants. The compounds deter or even kill herbivores. In exchange for the compounds, endophytes absorb sugars from the plant.

Based on these results, biologists have concluded that the relationship between endophytes and grasses is mutualistic. Similar types of anti-herbivore compounds have recently been documented in an endophyte that lives on morning glories.

In other types of plants, however, researchers have not been able to document benefits for the plant host. The current consensus is that at least some endophytic fungi may be commensals—meaning the fungi and the plants simply coexist with no observable effect, either deleterious or beneficial, on the host plant.

MUTUALISMS WITH OTHER SPECIES Do fungi take up residence with species other than land plants? The answer is yes.

- **Lichens** are a mutualistic partnership between a species of ascomycete and either a cyanobacterium or an alga. The nature of this relationship is explored in more detail in Section 31.4.
- Some ant species actively farm fungi inside their colonies. The ants fertilize and "weed" the fungal gardens, then harvest the fungi for food (see Chapter 32).

EMF mine nitrogen and some phosphorus for plants in temperate forests; AMF mine phosphorus for plants in grasslands and tropics as well as temperate regions. Other species of fungi engage in a wide array of symbiotic relationships.

What Adaptations Make Fungi Such Effective Decomposers?

The saprophytic fungi are master recyclers. Although bacteria and archaea are also important decomposers in terrestrial environments, fungi and a few bacterial species are the only organisms that can digest wood completely. Given enough time, fungi can turn even the hardest, most massive trees into soft soils.

How do fungi do it? You've already been introduced to two key adaptations:

- The large surface area of a mycelium makes nutrient absorption exceptionally efficient.
- Saprophytic fungi can grow toward the dead tissues that supply their food.

What other adaptations help fungi decompose plant tissues?

EXTRACELLULAR DIGESTION Large molecules such as starch, lignin, cellulose, proteins, and RNA cannot diffuse across the plasma membranes of hyphae. Only sugars, amino acids, nucleic acids, and other small molecules can enter the cytoplasm through hyphae. As a result, fungi have to digest their food before they can absorb it.

Instead of digesting food inside a stomach or food vacuole, as most animals and some protists do, respectively, fungi synthesize digestive enzymes and then secrete them outside their hyphae, into their food. Fungi perform **extracellular digestion**—digestion that takes place outside the organism. The simple compounds that result from enzymatic action are then absorbed by the hyphae.

As an example of how this process occurs, consider the enzymes responsible for digesting lignin and cellulose—the two most abundant organic molecules on Earth.

- Lignin comprises a family of extremely strong, complex polymers built from monomers that contain six-carbon rings. Recall from Chapter 30 that most lignin is found in the secondary cell walls of plant vascular tissues, where it furnishes structural support.
- Cellulose is a polymer of glucose and is found in the primary and secondary cell walls of plant cells.

Basidiomycetes can degrade lignin completely—to CO_2 and H_2O—as well as digest cellulose. Let's take a closer look at how they do it.

LIGNIN DEGRADATION Biologists have been keenly interested in understanding how basidiomycetes digest lignin. Paper manufacturers are also interested in this process because they need safe, efficient ways to degrade lignin in order to make soft, absorbent paper products.

To find out how lignin-digesting fungi do it, biologists began analyzing the proteins that these species secrete into extracellular space. After purifying these molecules, the investigators tested each protein for the ability to degrade lignin. Using this approach, investigators from two labs independently discovered an enzyme called lignin peroxidase.

Lignin peroxidase catalyzes the removal of a single electron from an atom in the ring structures of lignin. This oxidation step creates a free radical—an atom with an unpaired electron (see Chapter 2). This is an extremely unstable electron configuration, and leads to a series of uncontrolled and unpredictable reactions. These follow-on reactions split the polymer into smaller units.

Biologists have referred to this mechanism of lignin degradation as enzymatic combustion. The phrase is apt: The uncontrolled oxidation reactions triggered by lignin peroxidase are analogous to the uncontrolled oxidation reactions that occur when gasoline burns in a car engine.

The nonspecific nature of the reaction is remarkable, because virtually all of the other reactions catalyzed by enzymes are extremely specific. The lack of specificity makes sense, however. Unlike proteins, nucleic acids, and most other polymers with a regular and predictable structure, lignin is extremely heterogeneous. Over 10 types of covalent linkages are routinely found between the monomers that make up lignin. But once lignin peroxidase has created a free radical in the ring structure, any of these linkages can be broken.

The uncontrolled nature of the reactions has an important consequence. Instead of being stepped down an orderly electron transport chain as described in Chapter 9, the electrons involved in these reactions lose their potential energy in large, unpredictable jumps. As a result, the oxidation of lignin cannot be harnessed to drive the production of ATP and fuel the growth of hyphae. This conclusion is supported by an experimental observation: Fungi cannot grow with lignin as their sole source of food.

If wood-rotting fungi don't use lignin as food, why do they produce enzymes to digest it? The answer is simple. In wood, lignin forms a dense matrix around long strands of cellulose. Degrading the lignin matrix gives hyphae access to huge supplies of energy-rich cellulose. Saprophytic fungi are like miners. But instead of seeking out rare, gem-like nitrogen or phosphorus atoms as do EMF and AMF, the saprophytes use lignin peroxidase to blast away enormous lignin molecules, exposing rich veins of cellulose that can fuel growth and reproduction.

CELLULOSE DIGESTION Once lignin peroxidase has softened wood by stripping away its lignin matrix, the long strands of cellulose that remain can be attacked by enzymes called cellulases.

Like lignin peroxidase, cellulases are secreted into the extracellular environment by fungi. But unlike lignin peroxidase, cellulases are extremely specific in their action.

Biologists have purified seven different cellulases from the fungus *Trichoderma reesei*. Two of these enzymes catalyze a critical early step in digestion—they cleave long strands of cellulose into a disaccharide called cellobiose. The other cellulases are equally specific and also catalyze hydrolysis reactions. In combination, the suite of seven enzymes in *T. reesei* transforms long strands of cellulose into a simple monomer—glucose—that the fungus uses as a source of food.

Variation in Reproduction

Recall from Section 31.2 that fungi may produce swimming gametes and spores, yoked hyphae in which nuclei from different individuals fuse to form a zygote inside a protective structure, or specialized spore-producing cells called basidia and asci. As fungi diversified, they evolved an array of ways to reproduce as well as an array of sources for absorbing nutrition.

SPORES AS KEY REPRODUCTIVE CELLS The spore is the most fundamental reproductive cell in fungi. Spores are the dispersal stage in the fungal life cycle and are produced during both asexual and sexual reproduction. (Recall from Chapter 12 that asexual reproduction is based on mitosis, while sexual reproduction is based on meiosis.) Fungi produce spores in such prodigious quantities that it is not unusual for them to outnumber pollen grains in air samples.

If a spore falls on a food source and is able to germinate, a mycelium begins to form. As the fungus expands, hyphae grow in the direction in which food is most abundant. But if food begins to run out, mycelia respond by making spores, which are dispersed by wind or animals.

Why would mycelia reproduce when food is low? The leading hypothesis to answer this question is that spore production allows starving mycelia to disperse offspring to new habitats where more food might be available. Thus, spore production is favored by natural selection when individuals are under nutritional stress.

MULTIPLE MATING TYPES In many fungal lineages, hyphae come in several or many different mating types. Instead of having morphologically distinct males and females that produce sperm and eggs, hyphae of different mating types look identical.

Hyphae of the same mating type will not combine during sexual reproduction. When zygomycete hyphae grow close to each other, for example, they will not fuse unless the individuals have different alleles of one or more genes involved in mating. If chemical messengers released by two hyphae indicate that they are of different mating types, then fusion and zygosporangium formation follows.

In this way, mating types function as sexes. Instead of having just two sexes, a single fungal species may have tens of thousands. The basiomycete *Schizophyllum commune*, for example, is estimated to have 28,000 mating types.

Why so many? The leading hypothesis to explain the existence of mating types is that it helps generate genetic diversity in offspring. Genetic diversity, in turn, is known to be advantageous in fighting off infections and responding to changes in the environment (see Chapter 12).

HOW DOES FERTILIZATION OCCUR? Compared with green plants, protists, and animals, fertilization in fungi has important unique features:

- Only members of the Chytridiomycota produce gametes, and no fungus produces gametes that are different enough in size to be called sperm and egg.
- Fertilization occurs in two distinct steps in many fungi: **(1)** fusion of cells, and **(2)** fusion of nuclei from the fused cells. These two steps can be separated by long time spans and even long distances.

In many fungi, the process of sexual reproduction begins when hyphae from two individuals fuse to form a hybrid hypha. When the cytoplasm of two individuals fuses in this way, **plasmogamy** is said to occur (**Figure 31.11**).

In some cases, the nuclei from the two or more individuals stay independent after plasmogamy and grow into a heterokaryotic mycelium. The distinct nuclei function independently, even though gene expression must be coordinated in order for growth and development to occur. For example, the two nuclei divide as the hyphae expand, so each compartment that is divided by a septum contains one of each of the two nuclei. How the activities of the two nuclei are coordinated is a mystery.

In a heterokaryotic mycelium, one or more pairs of unlike nuclei may eventually fuse to form a diploid zygote. The fusion of nuclei is called **karyogamy**. The nuclei that are produced by karyogamy then divide by meiosis to form haploid spores.

As Chapter 29 indicated, most eukaryotes have life cycles dominated by haploid cells or diploid forms. Many fungi, in contrast, have a life cycle dominated by heterokaryotic cells.

✔If you understand the relationship between plasmogamy, heterokaryosis, and karyogamy, you should be able to compare and contrast these events with the life cycles of other eukaryotes. For example, which human cells undergo plasmogamy and karyogamy? Does heterokaryosis occur?

ASEXUAL REPRODUCTION As the left side of Figure 31.11 indicates, fungal species can reproduce asexually as well as sexually. There are, in fact, large numbers of ascomycetes that have never been observed to reproduce sexually.

During asexual reproduction, spore-forming structures are produced by a haploid mycelium, and spores are generated by mitosis. As a result, offspring are clones—meaning that they are genetically identical to their parent.

Four Major Types of Life Cycles

Among the sexually reproducing species of fungi, the presence of a heterokaryotic stage and the morphology of the spore-producing structure vary. Let's take a closer look at each of the four major types of life cycles that have been observed in fungi.

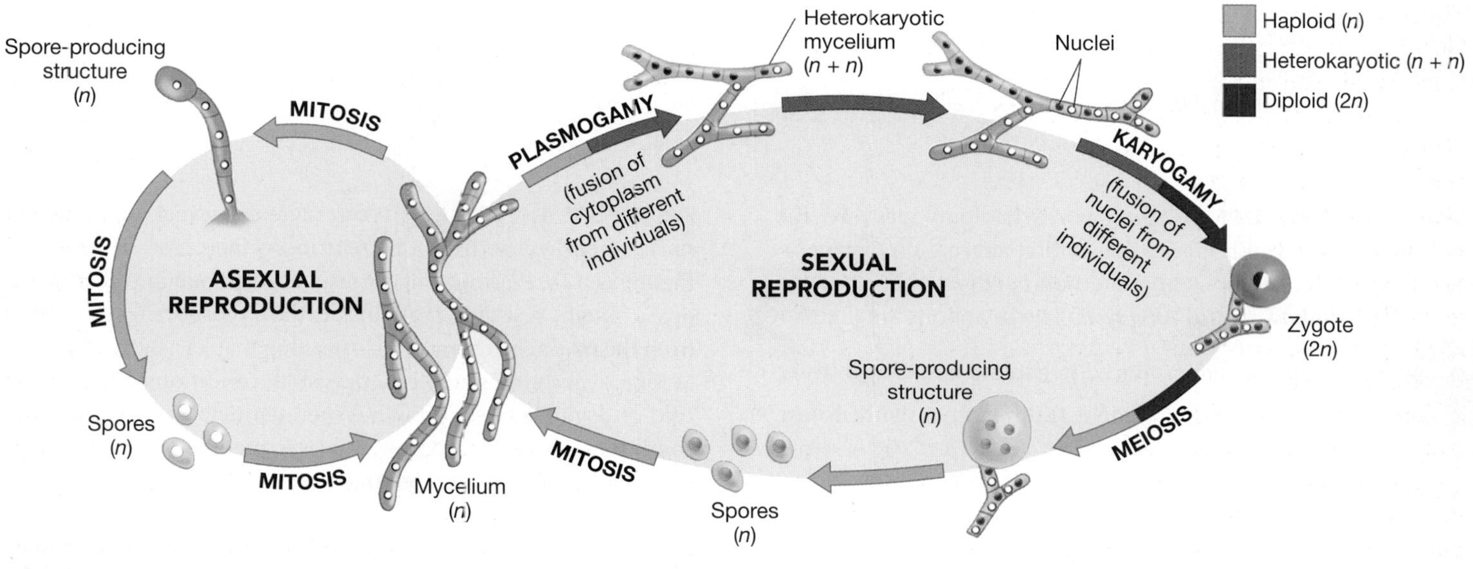

FIGURE 31.11 Fungi Have Unusual Life Cycles. A generalized fungal life cycle, showing both asexual and sexual reproduction.

✔**QUESTION** Fungi spend most of their lives feeding. Which is the longest-lived component of this life cycle?

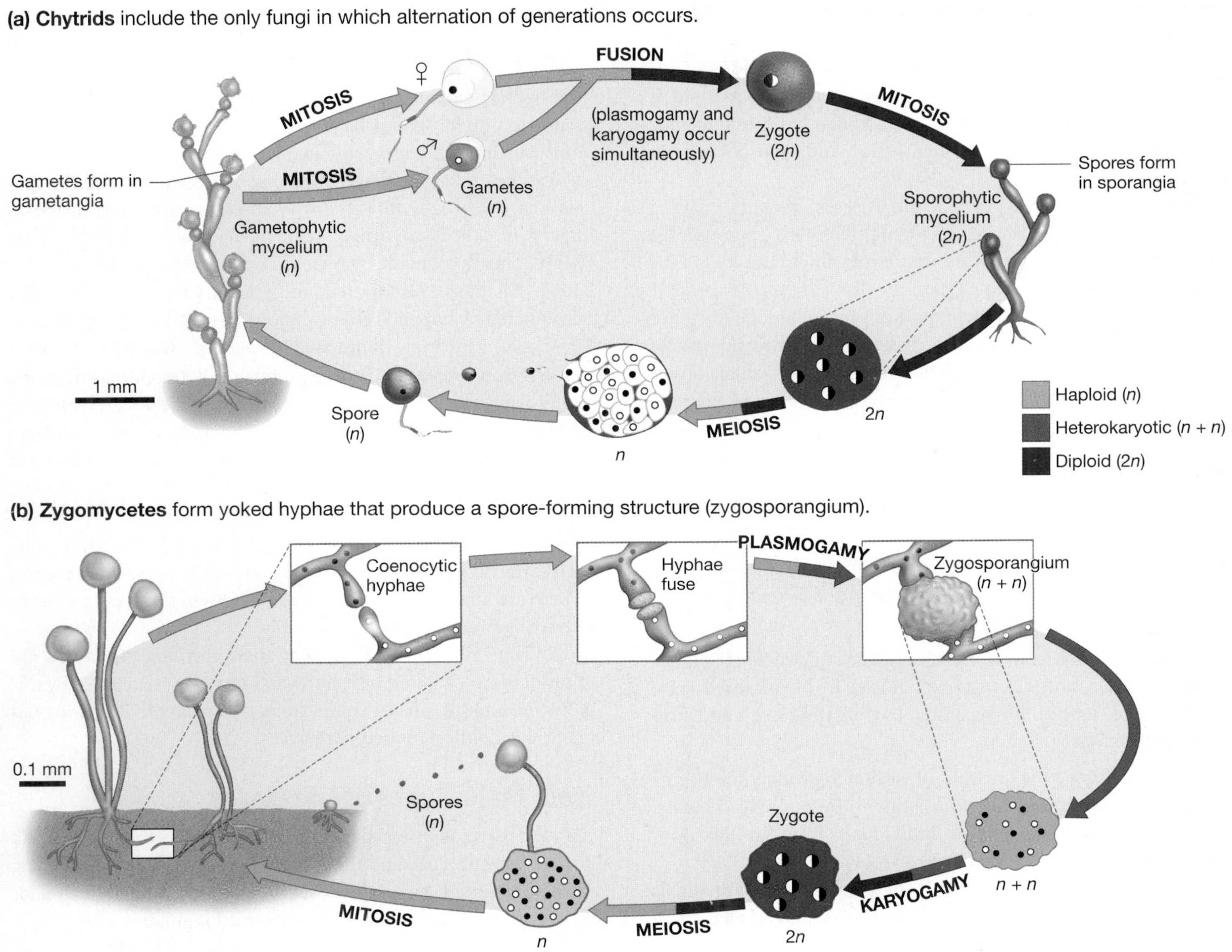

FIGURE 31.12 Variation in Sexual Reproduction in Fungi The sexual part of the life cycle in the four major groups of fungi: **(a)** chytrids, **(b)** zygomycetes, **(c)** Basidiomycota, and **(d)** Ascomycota.

✔**QUESTION** Asexual reproduction, as shown in Figure 31.11, is extremely common in zygomycetes and ascomycetes. What is the difference between mycelia produced via asexual versus sexual reproduction?

CHYTRIDIOMYCETE LIFE CYCLE The chytridiomycetes are the only type of fungi with species that exhibit alternation of generations. **Figure 31.12a** shows how alternation of generations occurs in the well-studied chytrid *Allomyces.* The key points are that:

- swimming gametes are produced in haploid adults by mitosis;
- gametes from the same individual or different individuals fuse to form a diploid zygote; and
- the zygote grows into a sporophyte.

The life cycle continues when meiosis occurs in the sporophyte, inside a structure called a **sporangium**. The haploid spores produced by meiosis disperse by swimming.

ZYGOMYCETE LIFE CYCLE In zygomycetes, sexual reproduction starts when hyphae from different individuals fuse, as shown in **Figure 31.12b**. Plasmogamy forms a zygosporangium that develops a tough, resistant coat. Inside the zygosporangium, nuclei from the mating partners fuse—meaning that karyogamy occurs.

The zygosporangium can persist if conditions become too cold or dry to support growth. When temperature and moisture conditions are again favorable, however, meiosis occurs. The meiotic products within the sporangium produce haploid spores.

When spores are released and germinate, they grow into new mycelia. These mycelia can reproduce asexually by making sporangia, with haploid spores being produced by mitosis and dispersed by wind.

(c) Basidiomycota have reproductive structures with many spore-producing basidia.

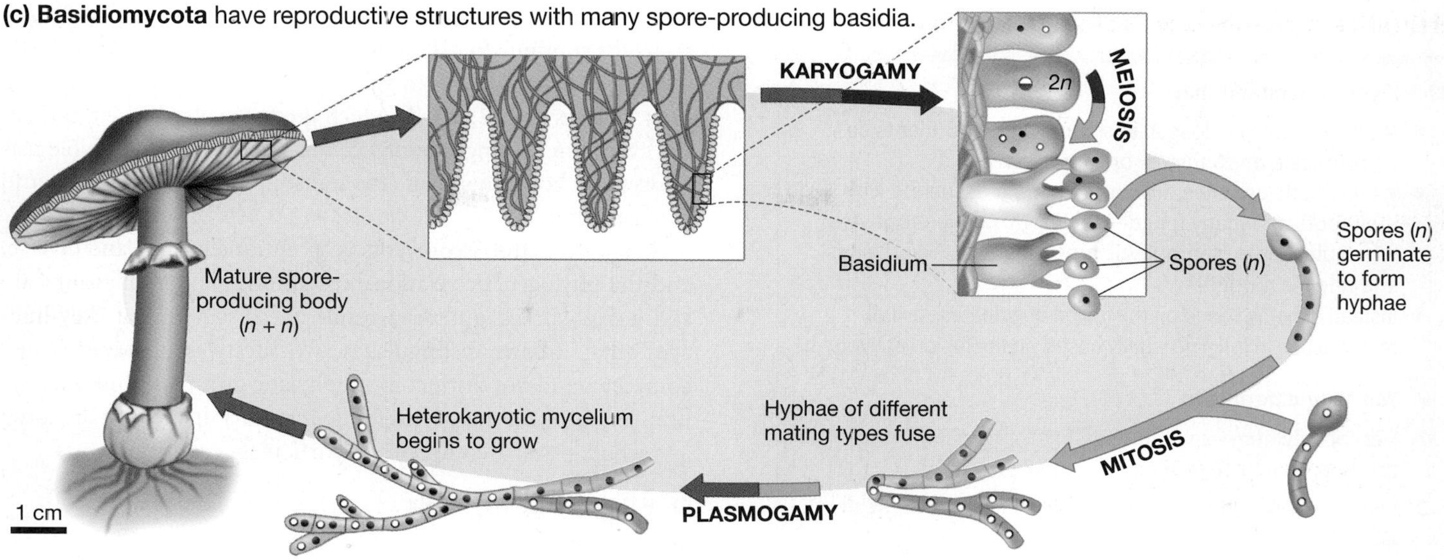

(d) Ascomycota have reproductive structures with many spore-producing asci.

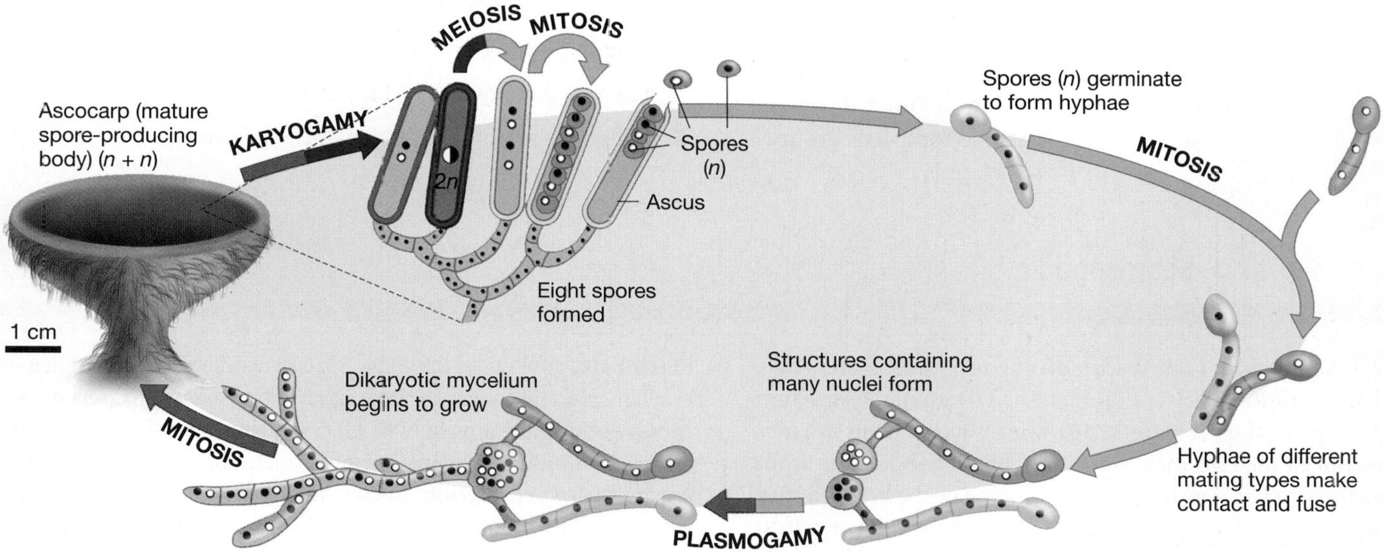

BASIDIOMYCOTA LIFE CYCLE Mushrooms, bracket fungi, and puffballs are reproductive structures produced by members of the Basidiomycota (**Figure 31.12c**). Even though their size, shape, and color vary enormously from species to species, all basidiomycete reproductive structures originate from the heterokaryotic hyphae of mated individuals.

Inside a mushroom or bracket or puffball, the pedestal-like, spore-producing cells called basidia form at the ends of heterokaryotic hyphae. Karyogamy occurs within the basidia. The diploid nucleus that results undergoes meiosis, and four haploid spores mature.

Spores are eventually ejected from the end of the basidia and are dispersed by the wind. It is not unusual for a single puffball or mushroom to produce a billion spores.

To review this sequence of life cycle events, go to the study area at *www.masteringbiology.com.*

(MB) **Web Activity** The Life Cycle of a Mushroom

ASCOMYCOTA LIFE CYCLE The reproductive structure in Ascomycota is produced by a dikaryotic hypha.

As **Figure 31.12d** illustrates, the process usually begins when hyphae or specialized structures from the same ascomycete species but from different mating types fuse, forming a cell containing many independent nuclei. A short dikaryotic hypha, containing one nucleus from each parent, emerges and eventually grows into a complex reproductive structure whose hyphae have the sac-like, spore-producing structures called asci at their tips.

After karyogamy occurs inside each ascus, meiosis takes place and haploid spores are produced. When the ascus matures, the spores inside are forcibly ejected. The spores are often picked up by the wind and dispersed.

CHECK YOUR UNDERSTANDING

If you understand that . . .

- Most fungi live in close association with land plants as mycorrhizae, endophytes, or saprophytes.
- When plants are alive, mycorrhizal fungi associate with their roots and many fungal species grow as endophytes.
- When plants die, saprophytic fungi degrade their tissues and release nutrients.
- Instead of being based on the fusion of gametes, sexual reproduction in fungi is usually based on the fusion of hyphae.

✔ You should be able to . . .

1. Describe evidence that mutualism occurs in EMF, AMF, and the endophytes of grasses.
2. Explain what happens inside a zygosporangium, basidium, and ascus.

Answers are available in Appendix B.

31.4 Key Lineages of Fungi

Based on the current estimate for the phylogeny of Fungi, it is clear that chytrids and zygomycetes do not form monophyletic groups and should not be named as single phyla (see Figure 31.8). Researchers continue to collect and analyze additional data—mostly DNA sequences—trying to find characters that **(1)** identify distinct lineages and **(2)** clarify which groups evolved when, early in the evolution of fungi. Resolving this issue would help biologists understand how fungi made the transition from living in water to living on land.

In addition, the Ascomycota is exceptionally large and diverse, and the phylogenetic relationships among major subgroups are still being worked out. As a result, the discussion of "key lineages" that follows distinguishes two lifestyles observed in ascomycetes and not distinct monophyletic groups within the phylum. Research on the phylogenetic relationships within the fungi as a whole—and ascomycetes in particular—continues.

- Fungi > Microsporidia
- Fungi > Chytrids
- Fungi > Zygomycetes
- Fungi > Glomeromycota
- Fungi > Basidiomycota (Club Fungi)
- Fungi > Ascomycota > Lichen-Formers
- Fungi > Ascomycota > Non-Lichen-Formers

Fungi > Microsporidia

All of the estimated 1200 species of microsporidians are single celled and parasitic. They are distinguished by a unique structure called the polar tube (**Figure 31.13**), which allows them to enter the interior of the cells they parasitize. The tube shoots out from the microsporidian, penetrates the membrane of the host cell, and acts as a conduit for the contents of the microsporidian cell to enter the host cell. Once inside, the microsporidian replicates and produces a generation of daughter cells, which go on to infect other host cells.

Microsporidians have a dramatically reduced genome. One species has the smallest genome known among eukaryotes—the total number of base pairs present is just half the number found in *E. coli*. Unlike other fungi, microsporidians lack functioning mitochondria. Like other fungi, the polysaccharide chitin is a prominent component of their cell wall. **✔ You should be able to mark the origin of the polar tube and the loss of mitochondria on Figure 31.8.**

Absorptive lifestyle Most microsporidians parasitize insects or fish. Because they enter the interior of host cells, they are called intracellular parasites.

Life cycle Variation in life cycles is extensive. Some species appear to reproduce only asexually. Others produce several different types of sexual or asexual spores, and some must successfully infect several different host species to complete their life cycle.

Human and ecological impacts Species from eight different genera can infect humans. In most cases, however, microsporidians cause serious infections only in AIDS patients and other individuals whose immune systems are not functioning well. Some microsporidians are serious pests in honeybee and silkworm colonies, while others cause severe infections in grasshoppers and are marketed as a biological control agent.

Nosema tractabile

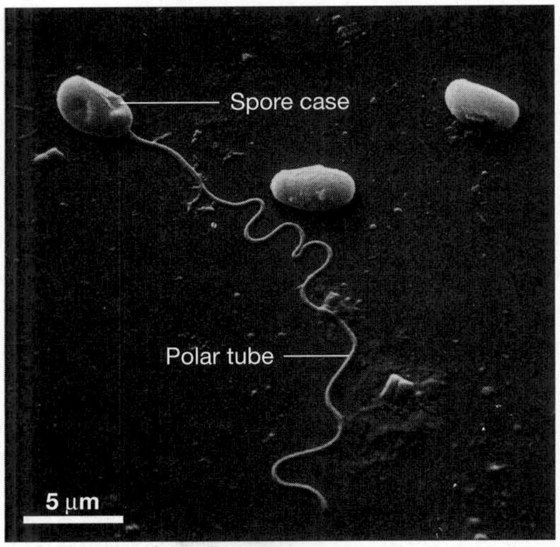

FIGURE 31.13 Microsporidia Infect Other Cells via a Polar Tube.

Fungi > Chytrids

Chytrids are largely aquatic and are common in freshwater environments. Some species live in wet soils, and a few have been found in desert soils that are wet only during a rainy season. Species found in dry soils have tough spores that endure harsh conditions. Spores from chytrids have been shown to germinate after a resting period of 31 years. Members of this group are the only fungi that produce motile cells. In each case, the motile cells use flagella to swim.

Absorptive lifestyle Many species of chytrids have enzymes that allow them to digest cellulose. As a result, these species are important decomposers of plant material in wet soils, ponds, and lakes. Many of the freshwater species are parasitic, and on occasion parasitic chytrids cause disease epidemics in algae or aquatic insects (including mosquitoes). Other species parasitize mosses, ferns, or flowering plants. Mutualistic chytrids are among the most important of the many organisms living in the guts of deer, cows, elk, and other mammalian herbivores, because the chytrids help these mammals digest their food (**Figure 31.14**). Chytrids produce cellulases that degrade the cell walls of grasses and other plants ingested by the animal, releasing sugars that are used by both the chytrids and their hosts.

Life cycle During asexual reproduction, most chytrid species produce spores that swim to new habitats via a flagellum. A few species reproduce sexually as well as asexually and exhibit alternation of generations. (Recall from Section 31.3 that these chytrids are the only fungal species to do so.) In species that reproduce sexually, plasmogamy and karyogamy may occur in a variety of ways: through fusion of hyphae, gamete-forming structures, or gametes.

Human and ecological impacts Chytrids are important decomposers in aquatic habitats and key mutualists in the guts of large, plant-eating mammals. Parasitic chytrids are also common. Biologists are investigating the possibility of using a chytrid that parasitizes mosquito larvae as a biological control agent; potato tubers are sometimes invaded and spoiled by a parasitic chytrid that causes black wart disease. Parasitic chytrids are largely responsible for catastrophic declines that have been occurring recently in amphibian populations all over the world.

Allomyces macrogynus

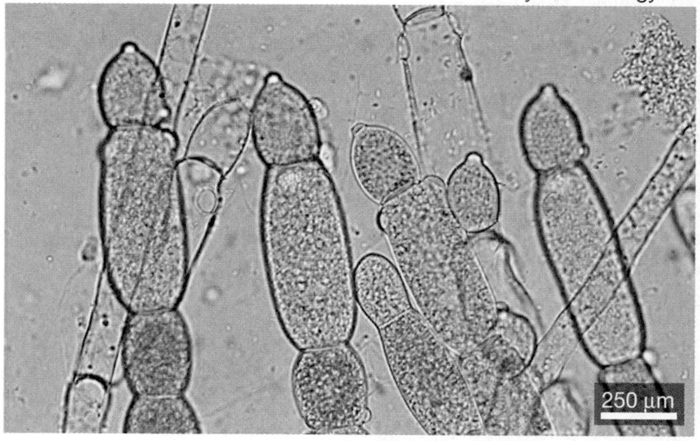

FIGURE 31.14 A Chytrid That is a Common Inhabitant of Soils and Ponds Worldwide.

Fungi > Zygomycetes

The zygomycetes ("yoked-fungi") are primarily soil-dwellers. Their hyphae yoke together and fuse during sexual reproduction and then form a durable, thick-walled zygosporangium.

Absorptive lifestyle Many members of zygomycete lineages are saprophytes and live in plant debris. Some parasitize other fungi, however, or are important parasites of insects and spiders.

Life cycle Asexual reproduction is extremely common. Ball-like sporangia are produced at the tips of hyphae that form stalks, and mitosis results in the production of spores which are dispersed by wind. During sexual reproduction, fusion of hyphae occurs only between individuals of different mating types. Fusion of hyphae from the same individual mycelium does not occur.

Human and ecological impacts The common bread mold *Rhizopus stolonifer* is a frequent household pest—probably the zygomycete that is most familiar to you. Saprophytic and parasitic members of these lineages are responsible for rotting strawberries and other fruits and vegetables and causing large losses in the fruit and vegetable industries. The black bread mold pictured in **Figure 31.15** is a familiar example of a saprophytic zygomycete. Some species of *Mucor* are used in the production of steroids for medical use. Species of *Rhizopus* and *Mucor* are used in the commercial production of organic acids, pigments, alcohols, and fermented foods.

(a) Black bread mold

Rhizopus nigricans

(b) A sporangium

FIGURE 31.15 Black bread molds are zygomycetes.

Fungi > Glomeromycota

Recent phylogenetic analyses have shown that the arbuscular mycorrhizal fungi (AMF) form a distinct phylum, indicating that they are a major monophyletic group. ✔ **You should be able to mark the origin of the AMF association on Figure 31.8.**

Absorptive lifestyle Recall from Section 31.3 that AMF absorb phosphorus-containing ions or molecules in the soil and transfer them, along with other nutrients and water, into the roots of trees, grasses, and shrubs in grassland and tropical habitats (**Figure 31.16**). In exchange, the host plant provides the symbiotic fungi with sugars and other organic compounds.

Life cycle Most species form spores underground. Glomeromycetes are difficult to grow in the laboratory, so their life cycle is not well known. No one has yet discovered a sexual phase in these species.

Human and ecological impacts Because grasslands and tropical forests are among the most productive habitats on Earth, the AMF are enormously important to both human and natural economies.

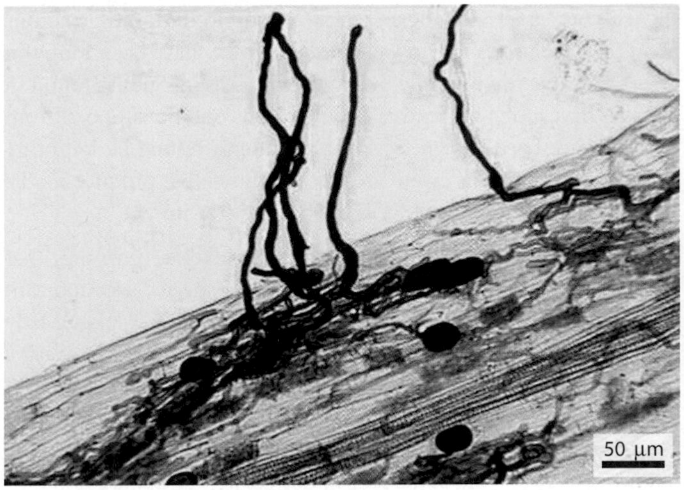

FIGURE 31.16 AMF Penetrate the Walls of Plant Root Cells.

Fungi > Basidiomycota (Club Fungi)

Although most basidiomycetes form mycelia and produce multicellular reproductive structures, some species have the unicellular growth form. The group is named for basidia, the club-like or pedestal-like cells where meiosis and spore formation occur. ✔ **You should be able to mark the origin of the basidium on Figure 31.8.** About 31,000 species of basidiomycetes have already been described, and more are being discovered each year.

Absorptive lifestyle Basidiomycetes are important saprophytes. Along with a few soil-dwelling bacteria, they are the only organisms capable of synthesizing lignin peroxidase and completely digesting wood. Some basidiomycetes are ectomycorrhizal fungi (EMF) that associate with trees in temperate forests. One subgroup consists entirely of parasitic forms called rusts, including species that cause serious infections in wheat and rye fields. The plant parasites called smut fungi are also basidiomycetes. Smuts specialize in infecting grasses; a few infect other fungi. Thus, the entire array of absorptive lifestyles found in fungi—saprophytic, mutualistic, and parasitic—is found within Basidiomycota.

Life cycle Asexual reproduction through production of spores is common in the Basidiomycota, although not as prevalent as in other groups of fungi. Asexual reproduction also occurs through growth and fragmentation of mycelia in the soil or in rotting wood, resulting in genetically identical individuals that are physically independent. During sexual reproduction, all basidiomycetes—even unicellular ones—produce basidia. In the largest subgroup in this lineage, basidia form in large, aboveground reproductive structures called mushrooms, brackets, earthstars, or puffballs (**Figure 31.17**). Basidiomycota and Ascomycota are the only fungi that have heterokaryotic (dikaryotic) mycelia. ✔ **You should be able to mark the origin of the dikaryotic condition on Figure 31.8.**

Human and ecological impacts EMF are enormously important in forestry. The temperate forests where these fungi are particularly abundant provide most of the hardwoods and softwoods used in building construction, furniture-making, and paper-making. Throughout the world, mushrooms are cultivated or collected from the wild as a source of food. The white button, crimini, and portabella mushrooms you may have seen in grocery stores are all varieties of the same species, *Agaricus bisporus*. Some of the toxins found in poisonous mushrooms are used in biological research; others have hallucinogenic effects on people and are used and traded illegally.

Lycoperdon species

FIGURE 31.17 Puffballs Can Produce Billions of Spores.

About half of the ascomycetes grow in symbiotic association with cyanobacteria and/or single-celled members of the green algae, forming the structures called lichens. Over 15,000 different lichens have been described to date; in most, the fungus involved is an ascomycete (although a few basidiomycetes participate as well). To name a lichen, biologists use the genus and species name assigned to the fungus that participates in the association. Most lichen-formers are found in a single monophyletic group within Ascomycota. ✔ **Based on this observation, you should be able to mark the origin of the lichen-forming habit on Figure 31.8.**

Absorptive lifestyle The fungus in lichens appears to protect the photosynthetic bacterial or algal cells. The fungal hyphae form a dense protective layer that shields the photosynthetic species and reduces water loss. In return, the cyanobacterium or alga provides carbohydrates that the fungus uses as a source of carbon and energy. However, the hyphae of some lichen-forming fungi have been observed to invade algal cells and kill them. This observation suggests a partially parasitic relationship in at least some lichens. The nature of lichen-forming associations is the subject of ongoing research.

Life cycle As **Figure 31.18a** shows, many lichens reproduce asexually via the production of small "mini-lichen" structures called soredia that contain both symbionts. Soredia disperse to a new location via wind or water and then develop into a new, mature individual via the growth of the algal or bacterial and fungal symbionts. In addition, the fungal partner may form asci. Spores that are shed from asci germinate to form a small mycelium. If the growing hyphae encounter enough of the appropriate algal or bacterial cells, a new lichen can form.

Human and ecological impacts In terms of their abundance and diversity, lichens dominate the Arctic and Antarctic tundras and are extremely common in boreal forests (see Chapter 50). They are the major food of caribou, as well as the most prevalent colonizers of bare rock surfaces throughout the world (**Figure 31.18b**). Rock-dwelling lichens are significant, because they break off mineral particles from the rock surface as they grow—launching the first step in soil formation. About 10,000 tons of lichens are processed annually and used in perfume production—either as a source of fragrant molecules or as a source of molecules that keep the fragrant components of perfumes from evaporating too rapidly.

(a) Cross section of a lichen, showing three layers

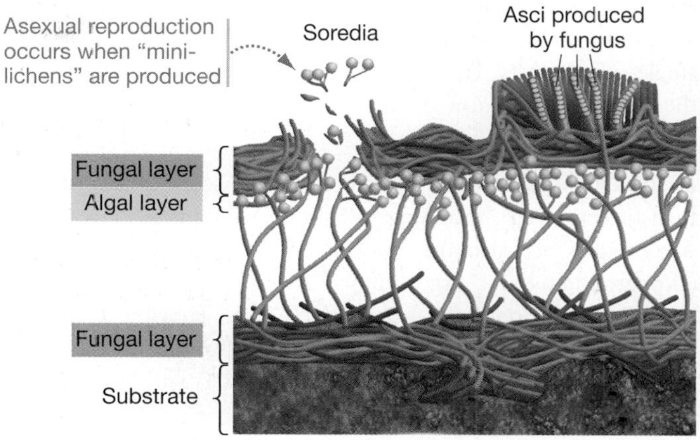

(b) Top view of lichen on a rock

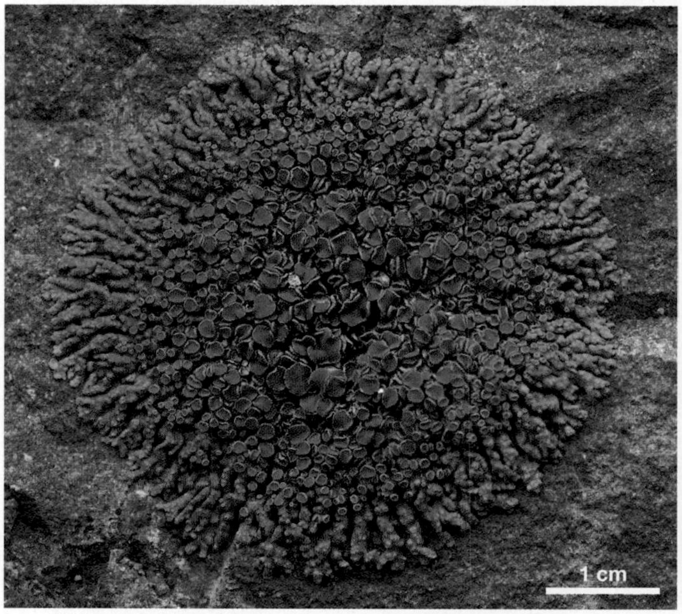

FIGURE 31.18 Lichens Are Associations between a Fungus and a Cyanobacterium or Green Alga. (a) In a lichen, cyanobacteria or green algae are enmeshed in a dense network of fungal hyphae. **(b)** Lichens often colonize surfaces, such as tree bark or bare rock, where other organisms are rare.

The ascomycetes that do not form lichens are found in virtually every terrestrial habitat, as well as some freshwater and marine environments. Although most ascomycetes form mycelia, many are single-celled yeasts. The ascus is a distinguishing characteristic of the Ascomycota. ✔ **You should be able to mark the origin of the ascus on Figure 31.8.**

Absorptive lifestyle A few members of the Ascomycota form mutualistic ectomycorrhizal fungi (EMF) associations with tree roots. Ascomycetes are also the most common endophytic fungi on aboveground tissues. Large numbers are saprophytic and are abundant in forest floors and in grassland soils. Parasitic forms are common as well. The entire array of absorptive lifestyles found in fungi has evolved within Ascomycota as well as within the Basidiomycota.

In addition, about 65 species in one ascomycete lineage are predatory—primarily on amoebae and other unicellular protists. Some are large enough to capture the microscopic animals called roundworms, however. The predatory ascomycetes trap their prey by means of sticky substances on their cell walls or catch prey in snares consisting of looped hyphae (**Figure 31.19**). Once a prey individual is captured, hyphae invade its body, digest it, and absorb the nutrients that are released. A predatory ascomycete has been found in fossilized material that is 100 million years old. ✔ **You should be able to mark the origin of predation on Figure 31.8.**

Life cycle The aboveground, ascus-bearing reproductive structures of these fungi may be a cup or saucer shape called an **ascocarp** (**Figure 31.20**). For that reason, these ascomycetes are known as **cup fungi**. It is also routine for spores called conidia to be produced asexually, at the ends of specialized hyphae called conidiophores.

Human and ecological impacts Some saprophytic ascomycetes can grow on jet fuel or paint and are used to help clean up contaminated sites. *Penicillium* is an important source of antibiotics, and *Aspergillus* produces citric acid used to flavor soft drinks and candy. Truffles and morels are so highly prized that they can fetch $1000 per pound, and the multibillion-dollar brewing, baking, and wine-making industries would collapse without the yeast *Saccharomyces cerevisiae*. A few parasitic ascomycetes cause infections in humans and other animals. In land plants, parasites from this group cause diseases including Dutch elm disease and chestnut blight.

Arthrobotrys anchonia

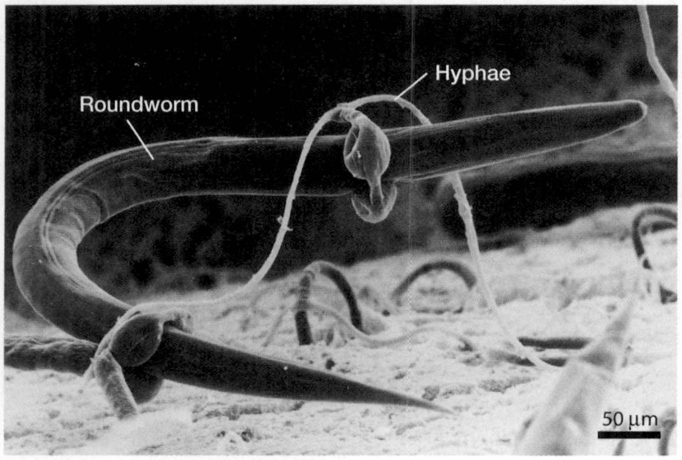

FIGURE 31.19 Some Ascomycetes Are Predatory.

Cookeina sulcipes

FIGURE 31.20 Ascomycetes Are Sometimes Called Cup Fungi.

CHAPTER 31 REVIEW

For media, go to the study area at www.masteringbiology.com

Summary of Key Concepts

Fungi are important in part because many species live in close association with land plants. They supply plants with key nutrients and decompose dead wood. They are the master recyclers of nutrients in terrestrial environments.

- Living plants are colonized by fungi.
- The roots of grasses and trees that grow in warm or tropical habitats are infiltrated by symbiotic glomeromycetes that supply the plant with phosphorus in exchange for sugars and other products of plant photosynthesis.
- The roots of trees in temperate habitats are colonized by basidiomycetes that exchange nitrogen for photosynthetic products.
- Many fungi live in leaves and stems; in some grasses and other species, these endophytic fungi secrete toxins that discourage herbivores.
- Parasitic fungi are responsible for devastating blights in crops and other plants.
- Once plants die, saprophytic fungi degrade the lignin and cellulose in wood and use nutrients from decaying plant material.
- Because they free up carbon atoms that would otherwise be locked up in wood, fungi speed up the carbon cycle in terrestrial habitats.

✔You should be able to predict the results of using fungicides to experimentally exclude fungi from 10 m × 10 m plots in a forest or grassland near your campus.

All fungi make their living by absorbing nutrients from living or dead organisms. Fungi secrete enzymes so that digestion takes place outside their cells. Their morphology provides a large amount of surface area for efficient absorption.

- Several adaptations make fungi exceptionally effective at absorbing nutrients from the environment.
- The two growth habits found among Fungi—single-celled "yeasts" or mycelia composed of long, filamentous hyphae—give fungal cells extremely high surface-area-to-volume ratios.
- Extracellular digestion, in which enzymes are secreted into food sources, enables fungi to break down extremely large molecules without ingesting them.
- Lignin decomposes through a series of uncontrolled oxidation reactions triggered by the enzyme lignin peroxidase.
- Cellulose digestion occurs in a carefully regulated series of steps, each catalyzed by a specific cellulase.

✔You should be able to explain why a large surface area increases the efficiency of absorption.

Many fungi have unusual life cycles. It is common for species to have a long-lived heterokaryotic stage, in which cells contain haploid nuclei from two different individuals. Although most species reproduce sexually, very few species produce gametes.

- Many species of fungi have never been observed to reproduce sexually. Most species can produce haploid spores either sexually or asexually, however.
- Sexual reproduction usually starts when hyphae from different individuals fuse—an event called plasmogamy. If the fusion of nuclei, or karyogamy, does not occur immediately, a heterokaryotic mycelium forms. Heterokaryotic cells may eventually produce spore-forming structures where karyogamy and meiosis take place.
- Four distinctive types of reproductive structures are known in fungi: (1) chytrids are aquatic fungi with motile gametes; (2) zygomycetes are soil-dwelling fungi with tough sporangia; (3) Basidiomycota have club-like, spore-forming structures; and (4) Ascomycota have sac-like, spore-forming structures.
- Recent analyses of DNA sequence data have revealed that the Basidiomycota, Ascomycota, and Glomeromycota each form monophyletic groups. Researchers are still trying to understand how the various groups of chytrids and zygomycetes are related to each other and to the microsporidians.

✔You should be able to explain why most fungi don't need to have gametes to accomplish sexual reproduction.

MB **Web Activity** The Life Cycle of a Mushroom

Questions

✔TEST YOUR KNOWLEDGE

Answers are available in Appendix B

1. The mycelial growth habit leads to a body with a high surface-area-to-volume ratio. Why is this important?
   a. Mycelia have a large surface area for absorption.
   b. The hyphae that make up mycelia are long, thin tubes.
   c. Most hyphae are broken up into compartments by walls called septa, although some exist as single, gigantic cells.
   d. Hyphae can infiltrate living or dead tissues.
2. What is plasmogamy?
   a. production of gametes, after fusion of hyphae from different individuals
   b. exchange of nutrients between symbiotic fungi and hosts
   c. fertilization—the fusion of cytoplasm and nuclei from different individuals
   d. fusion of the cytoplasm from different individuals, without nuclear fusion

3. The Greek root *ecto* means "outer." Why are ectomycorrhizal fungi, or EMF, aptly named?
   a. Their hyphae form tree-like branching structures inside plant cell walls.
   b. They are mutualistic.
   c. Their hyphae form dense mats that envelop roots but do not penetrate the walls of cells inside the root.
   d. They transfer nitrogen from outside their plant hosts to the interior.
4. The hyphae of AMF form bushy or balloon-like structures after making contact with the plasma membrane of a root cell. Why?
   a. They anchor the fungus inside the root, so the association is more permanent.
   b. They increase the surface area available for the transfer of nutrients.
   c. They produce toxins that protect the plant cells against herbivores.
   d. They break down cellulose and lignin in the plant cell wall.
5. What does it mean to say that a hypha is dikaryotic or heterokaryotic?
   a. Two nuclei fuse during sexual reproduction to form a zygote.
   b. Two or more independent nuclei, derived from different individuals, are present.
   c. The nucleus is diploid or polyploid—not haploid.
   d. It is extremely highly branched, which increases its surface area and thus absorptive capacity.
6. Very few organisms besides fungi have a heterokaryotic stage in their life cycle. Which of the following is another unusual aspect of the fungal life cycle?
   a. Some fungi exhibit alternation of generations—meaning that there is a multicellular diploid stage and a multicellular haploid stage.
   b. They produce eggs and sperm in approximately equal numbers, instead of many sperm and a few eggs.
   c. Spores have to fuse with each other before they develop into a new mycelium.
   d. Most varieties undergo sexual reproduction without producing eggs or sperm.

✔ TEST YOUR UNDERSTANDING

Answers are available in Appendix B

1. Explain why fungi that degrade dead plant materials are important to the global carbon cycle. Do you accept the text's statement that, without these fungi, "Terrestrial environments would be radically different than they are today and probably much less productive"? Why or why not?
2. Lignin and cellulose provide rigidity to the cell walls of plants. But in most fungi, chitin performs this role. Why is it logical that most fungi don't have lignin or cellulose in their cell walls?
3. Biologists claim that EMF and AMF species are better than plants at acquiring nutrients because they have a higher surface area and because they are more effective at acquiring phosphorus (P) and/or nitrogen (N). Compare and contrast the surface area of mutualistic fungi and plant roots. Explain why fungi are particularly efficient at acquiring P and N, compared to plants.
4. Using information from Chapters 3 through 6, list three key macromolecules found in plants that contain phosphorus. Explain why plant growth might be limited by access to P.
5. Compare and contrast the way that fungi degrade lignin with the way that they digest cellulose.
6. How is it possible for thousands of mating types to exist in a single species of fungus, instead of just two sexes like other eukaryotes?

✔ APPLYING CONCEPTS TO NEW SITUATIONS

Answers are available in Appendix B

1. The box on chytrids in Section 31.4 mentions that they may be responsible for massive die-offs currently occurring in amphibians. Review Koch's postulates in Chapter 28, then design a study showing how you would use Koch's postulates to test the hypothesis that chytrid infections are responsible for the frog deaths.
2. Some biologists contend that the ratio of plant species to fungus species worldwide is on the order of 1:6. Explain why you agree or disagree with this claim. In doing so, consider the analyses of endophytic, parasitic, lichen-forming, mycorrhizal, and saprophytic strategies presented in this chapter. Also, consider the diversity of tissues and organs available in plants.
3. Experiments indicate that cellulase genes are transcribed and translated together. If cells are selected to be extremely efficient at digesting cellulose, is this result logical? Would you predict that the gene that codes for lignin peroxidase is transcribed along with the cellulase genes? How would you test your prediction?
4. Many mushrooms are extremely colorful. Fungi do not see, so it is unlikely that colorful mushrooms are communicating with one another. One hypothesis is that the colors serve as a warning to animals that eat mushrooms, much like the bright yellow and black stripes on wasps. Design an experiment capable of testing this hypothesis.

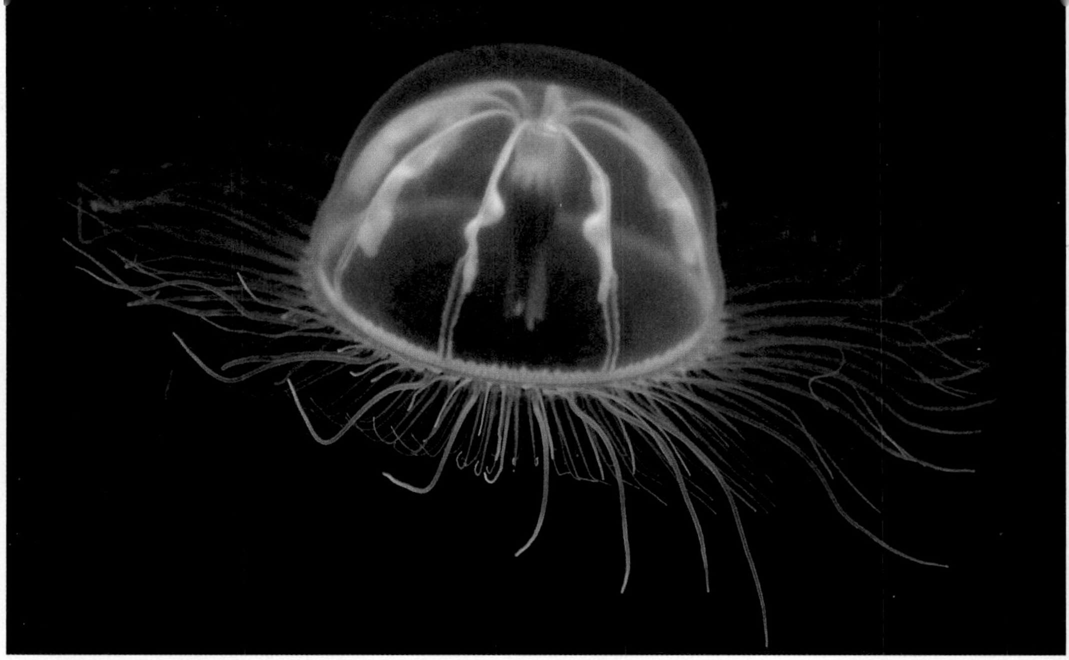

Jellyfish such as this hydromedusa are among the most ancient of all animals—they appear in the fossil record over 560 million years ago.

An Introduction to Animals 32

The **animals** are a monophyletic group of eukaryotes that share a series of traits:

- They are multicellular, with cells that lack cell walls but have an extensive extracellular matrix (ECM; see Chapter 8). The ECM includes proteins specialized for cell-cell adhesion and communication.
- They are heterotrophs, meaning that they obtain the carbon compounds they need from other organisms. Most ingest their food, rather than absorbing it across the body surface.

Animals are the only multicellular heterotrophs on the tree of life that ingest their food. Fungi are multicellular heterotrophs, but they absorb nutrients. Some slime molds could also be classified as multicellular heterotrophs that ingest food, but they are multicellular only during the reproductive phase of their life cycle—they are unicellular when they feed (see Chapter 29). As a result, animals are the largest predators, herbivores, and detritivores on Earth.

In addition, all animals move under their own power at some point in their life cycle, and all animals other than sponges have: (**1**) specialized cells called **neurons**—nerve cells—that transmit electrical signals to other cells; and (**2**) muscle cells that can change the shape of the body by contracting. In most animals, neurons connect to each other, forming a nervous system, and some neurons connect to muscle cells—which may contract in response to electrical signals from neurons. Muscles and neurons are adaptations that allow a large, multicellular body to move efficiently.

Over 1.2 million species of animals have been described and given scientific names to date, and biologists predict that tens of millions more have yet to be discovered. To analyze the almost overwhelming number and diversity of species in this lineage, this chapter presents a broad overview of how they diversified, along with more detailed information on the characteristics of the lineages that appear first in the fossil record. Chapters 33 and 34 follow up by focusing on the most species-rich groups.

KEY CONCEPTS

- Animals are multicellular, heterotrophic eukaryotes that lack cell walls and ingest their prey.
- Fundamental changes in morphology and development occurred as animals diversified.
- Recent phylogenetic analyses have shown that there are four major groups of animals: non-bilaterian lineages, two protostome groups (Lophotrochozoa and Ecdysozoa), and the deuterostomes.
- Within major groups of animals, evolutionary diversification was based on innovative ways of sensing the environment, feeding, and moving.
- Methods of sexual reproduction vary widely among animal groups, and many species can reproduce asexually. It is common for individuals to undergo metamorphosis during their life cycle.

✔ When you see this checkmark, stop and test yourself. Answers are available in Appendix B.

32.1 Why Do Biologists Study Animals?

If you ask biologists why they study animals, the first answer they'll give is, "Because they're fascinating." It's hard to argue with this statement. Consider ants.

- Ants live in colonies that may include millions of individuals. But colony-mates cooperate so closely in feeding, defense, and rearing young that each ant seems like a cell in a multicellular organism instead of an individual.
- Some ants live in trees, and protect their host trees by attacking giraffes and other grazing animals a million times their size. The trees provide oil-rich growths that the ants eat.
- Rancher ants protect the plant-sucking insects called aphids from spiders and other predators, and eat the sugar-rich honeydew that aphids secrete from their anus (**Figure 32.1a**).
- Farmer ants eat fungi that they plant, fertilize, and harvest in underground gardens (**Figure 32.1b**).
- Parasitic ants look and smell like a host ant species but enslave them, forcing the hosts to rear the young of the parasitic species instead of their own.

Based on observations like these, most people would agree that ants—and by extension, other animals—are indeed fascinating. But beyond pure intellectual interest, there are compelling practical reasons to study animals.

(a) "Rancher ants" tend aphids and eat their sugary secretions.

(b) "Farmer ants" cultivate fungi in gardens.

FIGURE 32.1 Examples of Sophisticated Behavior in Ants.

Biological Importance

Animals are key consumers in virtually every ecosystem—from the deep oceans to alpine ice fields and from tropical forests to arctic tundras. It is not possible to understand or preserve these ecosystems without understanding and preserving animals.

In addition to their ecological importance, animals are an extraordinarily diverse and species-rich lineage on the tree of life. Current estimates suggest that there are between 10 million and 50 million animal species living today. These species range in size and complexity from tiny sponges, which attach to a substrate and contain just a few cell types and simple tissues, to blue whales, which migrate tens of thousands of kilometers each year in search of food and contain trillions of cells, dozens of distinct tissues, an elaborate skeleton, and highly sophisticated sensory and nervous systems. A great deal of diversification has occurred in this lineage. To understand the history of life, biologists have to understand how animals came to be so diverse.

Role in Human Health and Welfare

The vast majority of research funded by government agencies, foundations, and corporations is motivated by a simple desire: improving people's standard of living. In some cases, understanding animals has a direct economic benefit; in other situations, the gains are through improving human health.

FOOD, MATERIALS, AND TRANSPORTATION Humans depend on wild and domesticated animals for food. In addition to domesticated mammals and birds, large industries are organized around the harvest of wild fish, mollusks (clams and mussels), and crustaceans (lobster, shrimp, crab).[1] In 2006, the global market value of fish and shellfish (mollusks and crustaceans) was $91.2 billion U.S. dollars. Most commercial fruit growers rely on bees and other animals to pollinate their crops.

Animals are also an important source of materials: fibers used to make clothing, blankets, and other articles, as well as the leather used in shoe-making and other industries.

In preindustrial societies, domesticated animals are the principal source of transportation and power. Horses, donkeys, oxen, and other domesticated animals transport people and cargo and do work such as plowing fields.

DISEASE TRANSMISSION Many animals play an important role in transmitting human diseases. Examples include the mosquito species that harbor the malaria parasite, and the rats that in earlier centuries carried the bacterium responsible for deadly epidemics of bubonic plague.

[1]In Chapters 32–34, common names are often given in parentheses after the formal name of a lineage. The common names indicate some of the more familiar members of the lineage in question.

When animals act as a transmission vector for a human disease, biologists work with physicians and public health officials on strategies to reduce animal–human transmission.

ANIMALS AS MODEL ORGANISMS Humans are animals, so efforts to understand human biology depend on advances in animal biology.

- Humans share a large portion of their genome with other animals. As a result, research on animals like fruit flies and zebrafish has revealed fundamental aspects of cellular and developmental biology that are shared by humans.
- Because genetic homologies between humans and other animals are so extensive, most drug testing and other types of biomedical research can be done on mice, rats, or primates instead of humans.

Tens of thousands of biologists are employed in studying or managing animals. How do researchers go about understanding the most fundamental aspects of animal diversity?

32.2 How Do Biologists Study Animals?

Biologists currently recognize about 30 **phyla**, or major lineages, of animals—the exact number is being debated and revised as additional information comes to light. **Table 32.1** lists 29 phyla that are included in most published analyses.

Each animal phylum has distinct morphological features—synapomorphies that identify it as a monophyletic group. At a more general level, however, groups of phyla are characterized by fundamental aspects of morphology and development that changed as animals diversified. What are these, and why are they important?

Analyzing Comparative Morphology

In essence, animals are moving and eating machines. To understand animal diversity, then, one of the basic tasks is to understand how these machines are put together. The origin and early evolution of animals was based on four aspects of the fundamental architecture, or **body plan**, of animals:

1. the origin and elaboration of tissues—especially the tissues found in embryos;
2. the origin and elaboration of the nervous system and the subsequent evolution of a cephalized body—one with a distinctive head region;
3. the evolution of a fluid-filled body cavity; and
4. variation in the events of early embryonic development.

Let's consider each in turn.

THE ORIGIN AND DIVERSIFICATION OF TISSUES All animals have groups of similar cells that are organized into the tightly integrated structural and functional units called **tissues**. Although sponges lack many types of tissues found in other animals, recent research has shown that one group of sponges have

TABLE 32.1 **An Overview of Major Animal Phyla**

Group and Phylum	Common Name or Example Taxa	Estimated Number of Species
Non-bilaterian groups		
Porifera	Sponges	7,000
Placozoa	Placozoans	1
Cnidaria	Jellyfish, corals, anemones, hydroids, sea fans	10,000
Ctenophora	Comb Jellies	100
Acoelomorpha	Acoelomate worms	10
Protostomes (lacks typical protostome development)		
Chaetognatha	Arrow Worms	100
Protostomes: Lophotrochozoa		
Rotifera	Rotifers	1,800
Platyhelminthes	Flatworms	20,000
Nemertea	Ribbon worms	900
Gastrotricha	Gastrotrichs	450
Acanthocephala	Acanthocephalans	1,100
Entoprocta	Entoprocts, kamptozoans	150
Gnathostomulida	Gnathostomulids	80
Annelida	Segmented worms	16,500
Molluska	Mollusks (clams, snails, octopuses)	94,000
Phoronida	Horseshoe worms	20
Bryozoa	Bryozoa, ectoprocts, moss animals	4,500
Brachiopoda	Brachiopods; lamp shells	335
Protostomes: Ecdysozoa		
Nematoda	Roundworms	25,000
Kinorhyncha	Kinorhynchs	150
Nematomorpha	Hair worms	320
Priapula	Priapulans	16
Onychophora	Velvet worms	110
Tardigrada	Water bears	800
Arthropoda	Arthropods (spiders, insects, crustaceans)	1,100,000
Deuterostomes		
Echinodermata	Echinoderms (sea stars, sea urchins, sea cucumbers)	7,000
Xenoturbellida	Xenoturbillidans	2
Hemichordata	Acorn worms	85
Chordata	Chordates (tunicates, lancelets, sharks, bony fish, amphibians, reptiles, mammals)	50,000

epithelium—a layer of tightly joined cells that covers the surface.[2] Other animals have an array of other tissue types as well as epithelium.

In animals other than sponges, the number of tissues that exist in an embryo varies. Animals whose embryos have two types of tissue are called **diploblasts** (literally, "two-buds"); animals whose embryos have three types are called **triploblasts** ("three-buds").

You might recall from Chapter 22 that these embryonic tissues are organized in layers, called **germ layers**. In diploblasts these germ layers are called **ectoderm** and **endoderm** (**Figure 32.2**). The Greek roots *ecto* and *endo* refer to outer and inner, respectively; the root *derm* means "skin." In most cases the outer and inner "skins" of diploblast embryos are connected by a gelatinous material that may contain some cells. In triploblasts, however, there is a germ layer called **mesoderm** between the ectoderm and endoderm. (The Greek root *meso* refers to middle.)

The embryonic tissues found in animals develop into distinct adult tissues, organs, and organ systems (see Chapter 22). In triploblasts, for example:

- *Ectoderm* gives rise to skin and the nervous system.
- *Endoderm* gives rise to the lining of the digestive tract.
- *Mesoderm* gives rise to the circulatory system, muscle, and internal structures such as bone and most organs.

In general, then, ectoderm produces the covering of the animal and endoderm generates the digestive tract. Mesoderm gives rise to the tissues in between. The same pattern holds in diploblasts, except that (**1**) muscle is simpler in organization and is derived from ectoderm, and (**2**) reproductive tissues are derived from endoderm.

Traditionally, two groups of animals have been recognized as diploblasts: the Cnidaria (pronounced *ni-DARE-ee-uh*)–which include the jellyfish, corals, sea pens, hydra, and anemones (**Figure 32.3a**)—and the Ctenophora (pronounced *ten-AH-for-ah*), or comb jellies (**Figure 32.3b**). Recent data, however, suggest that at least some cnidarians have mesoderm and are triploblastic. All other animals, from leeches to humans, are triploblastic.

The evolution of mesoderm was important because it gave rise to the first complex muscle tissue used in movement. Sponges lack muscle and are sessile, or nonmoving, as adults; their larvae (immature forms) move via cilia. Ctenophores have muscle cells that can change the body's shape, but both larvae and adults swim using cilia.

NERVOUS SYSTEMS, BODY SYMMETRY, AND CEPHALIZATION Sponges lack neurons, but cnidarians and ctenophores have nerve cells that are organized into a diffuse arrangement called a **nerve net** (**Figure 32.4a**). All other animals have a **central nervous system**, or CNS. In a CNS, some neurons are clustered into one or more large tracts or cords that project throughout the body; others are clustered into masses called **ganglia** (**Figure 32.4b**).

Nerve nets and central nervous systems are associated with two contrasting types of body symmetry: radial (literally, "spoke") symmetry or bilateral ("two-sides") symmetry. A body is symmetrical if it can be divided by a plane such that the resulting pieces are nearly identical.

Cnidarians and ctenophores, along with many sponges, have **radial symmetry**—meaning that they have at least two planes of symmetry. For example, almost any plane sectioned through the center of the hydra in Figure 32.4a produces two identical halves.

[2]For discussion and citations, see Nichols, S. and G. Wörheide. 2005. *Integrative and Comparative Biology* 45: 333–334; Elliott, G. R. D. and S. P. Leys. 2007. *Journal of Experimental Biology* 210: 3736–3748; Philippe, H., et al. 2009. *Current Biology* 19: 706–712.

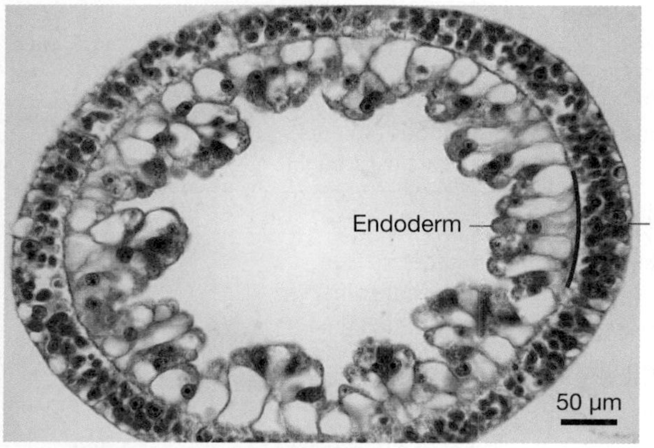

FIGURE 32.2 Diploblastic Animals Have Bodies Built from Ectoderm and Endoderm. This is a cross section through the tube-shaped portion of a hydra's body. The cells have been stained to make them more visible.

(a) Cnidaria include jellyfish, corals, and sea pens (shown).

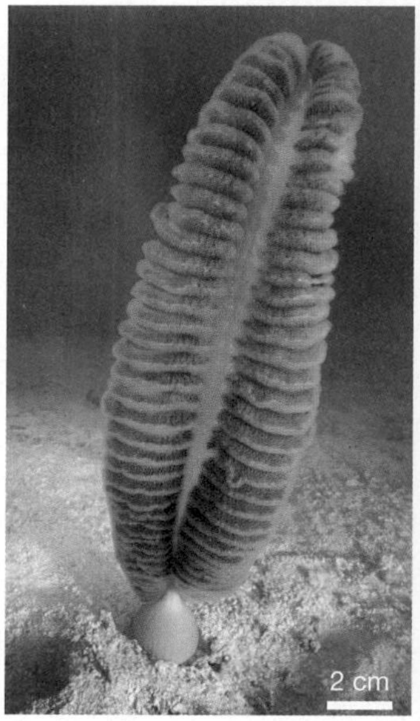

(b) Ctenophora are the comb jellies.

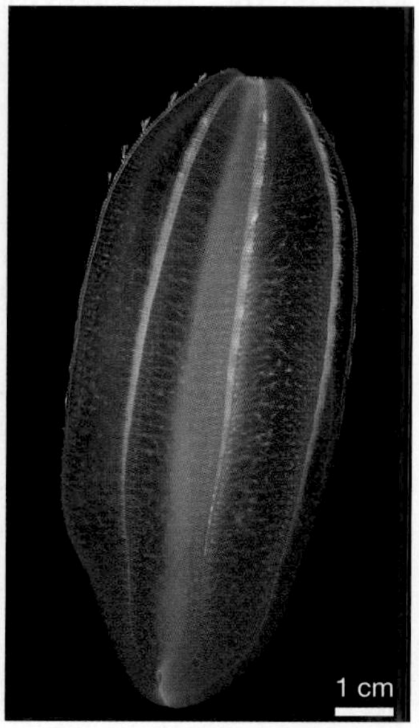

FIGURE 32.3 Cnidarians and Ctenophorans. (a) Like most members of the Cnidaria, this sea pen lives in marine environments. **(b)** Comb jellies are a major component of planktonic communities in the open ocean.

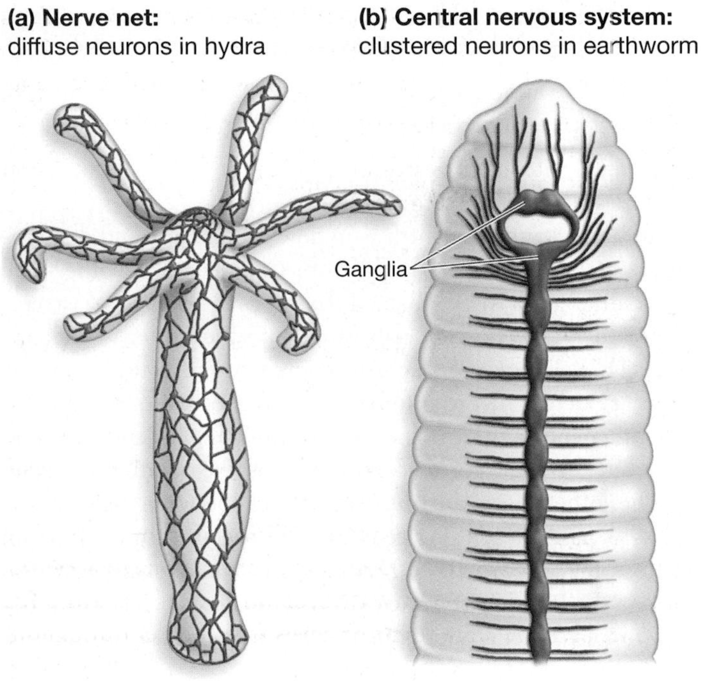

FIGURE 32.4 Associations between Body Symmetry and the Nervous System. **(a)** Radially symmetric animals, like this hydra, have a nerve net. **(b)** Bilaterally symmetric animals, like this earthworm, have a central nervous system.

Many of the radially symmetric animals living today either float in water or live attached to a substrate.

Organisms with **bilateral symmetry**, in contrast, have one plane of symmetry and tend to have a long, narrow body. The earthworm in Figure 32.4b, for example, only has one plane of symmetry—running lengthwise down its middle. Most of the bilaterally symmetric animals living today move through water or air or along or through a substrate.

How are symmetry and nervous systems related? The function of neurons and nervous systems is to transmit and process information in the form of electrical signals (see Chapter 45).

- *Radially symmetric organisms* are equally likely to encounter prey and other aspects of the environment in any direction. As a result, a diffuse nerve net can receive and send signals efficiently.
- *Bilaterally symmetric organisms* tend to encounter prey and other aspects of the environment at one end. As a result, it is advantageous to have many neurons concentrated at that end, with nerve tracts that carry information from there down the length of the body.

The evolution of bilateral symmetry occurred along with **cephalization**: the evolution of a head, or anterior region, where structures for feeding, sensing the environment, and processing information are concentrated. The large mass of neurons that is located in the head, and that is responsible for processing information from throughout the body, is called the cerebral ganglion or **brain**.

All triploblastic animals have bilateral symmetry except for species in the phylum Echinodermata (pronounced *ee-KINE-oh-der-ma-ta*)—a group where radial symmetry evolved independently of the sponges, cnidaria, and ctenophores. The echinoderms include species such as sea stars, sea urchins, feather stars, and brittle stars. Although their larvae are bilaterally symmetric, adult echinoderms are radially symmetric.

To explain the pervasiveness of bilateral symmetry, biologists point out that locating and capturing food is particularly efficient when movement is directed by a distinctive head region and powered by the rest of the body. In combination with the origin of mesoderm, a bilaterally symmetric body plan enabled rapid, directed movement and hunting. Lineages with a triploblastic, bilaterally symmetric, cephalized body had the potential to diversify into an array of formidable eating and moving machines.

WHY WAS THE EVOLUTION OF A BODY CAVITY IMPORTANT? A third architectural element that distinguishes animal phyla is the presence of an enclosed, fluid-filled cavity called a **coelom** (pronounced *SEE-lum*). The coelom creates a container for the circulation of oxygen and nutrients, along with space where internal organs can move independently of each other.

Although cnidarians and ctenophores have a central canal that functions in digestion and circulation, they do not have a coelom. The handful of triploblasts that do not have a coelom are called **acoelomates** ("no-cavity-form"; see **Figure 32.5a**); those that possess a coelom are known as **coelomates** (**Figure 32.5b**).

The coelom was a critically important innovation during animal evolution because an enclosed, fluid-filled chamber can act as an efficient **hydrostatic skeleton**. Soft-bodied animals with hydrostatic skeletons can move effectively even if they do not have fins or limbs.

Using a nematode (roundworm) as an example, movement is possible because the fluid inside the coelom stretches the body

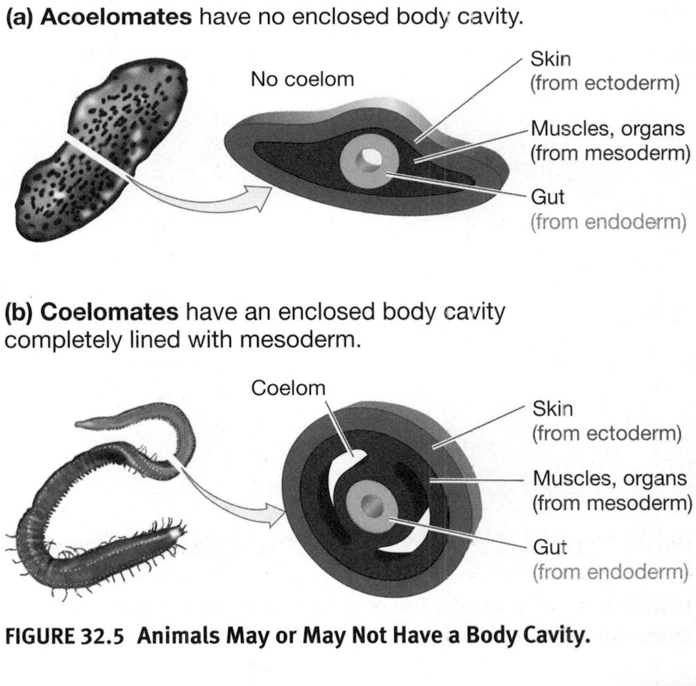

FIGURE 32.5 Animals May or May Not Have a Body Cavity.

wall—much like a water balloon—meaning that it is under tension (**Figure 32.6a**). This force exerts pressure on the fluid inside the coelom. When muscles in the body wall contract against the pressurized fluid, the force is transmitted through the fluid, changing the body's shape.

Figure 32.6b shows how coordinated muscle contractions and relaxations produce changes in the shape of a hydrostatic skeleton. The coordinated changes in body shape make writhing or swimming movements possible. ✔If you understand how a hydrostatic skeleton works, you should be able to demonstrate its function with a long, tube-shaped water balloon, using your fingers to pinch one side and then another to simulate muscle contractions.

By providing a hydrostatic skeleton, the evolution of the coelom gave bilaterally symmetric organisms the ability to move efficiently in search of food.

WHAT ARE THE PROTOSTOME AND DEUTEROSTOME PATTERNS OF DEVELOPMENT? Except for adult echinoderms, all of the coelomates—including juvenile forms of echinoderms—are bilaterally symmetric. This huge group of organisms is formally called the **Bilateria**. Based on distinctive events that occur early in the development of the embryo, the bilaterians can be split into two subgroups:

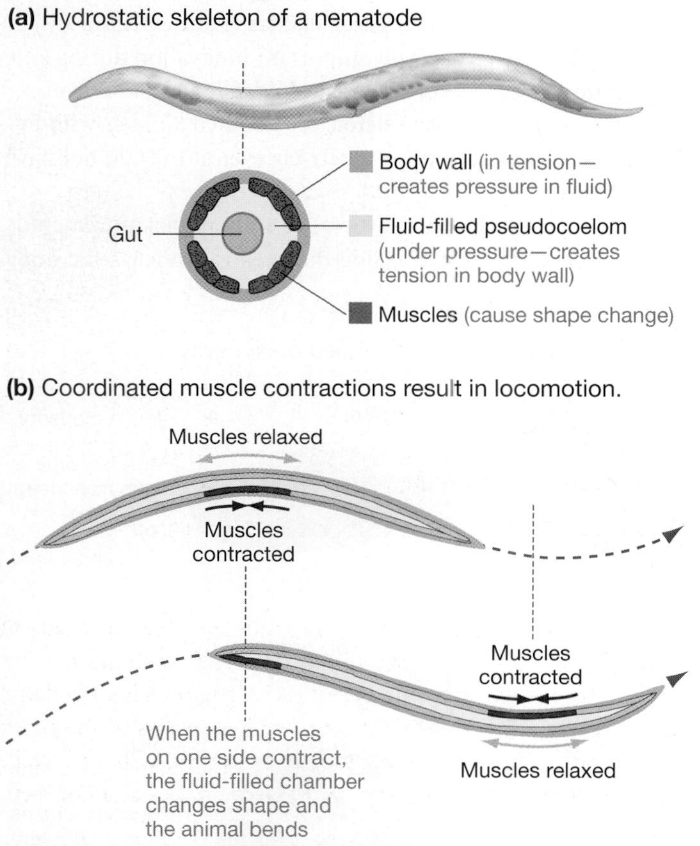

FIGURE 32.6 Hydrostatic Skeletons Allow Limbless Animals to Move.

✔**QUESTION** Suppose that all of the longitudinal muscles of this nematode contracted at the same time. What would happen?

1. **protostomes**, in which the mouth develops before the anus, and blocks of mesoderm hollow out to form the coelom; and
2. **deuterostomes**, in which the anus develops before the mouth, and pockets of mesoderm pinch off to form the coelom.

Translated literally, protostome means "first-mouth" and deuterostome means "second-mouth."

To understand the contrasts in how protostome and deuterostome embryos develop, recall from Chapter 22 that the three embryonic germ layers form during a process called gastrulation, which begins when cells move into the center of the embryo. The cell movements create a pore that opens to the outside (**Figure 32.7a**). In protostomes, this pore becomes the mouth. The other end of the gut, the anus, forms later. In deuterostomes, however, this initial pore becomes the anus; the mouth forms later.

The other developmental difference between the groups arises as gastrulation proceeds and the coelom begins to form. Protostome embryos have two blocks of mesoderm beside the gut. As the left side of **Figure 32.7b** indicates, their coelom begins to form when cavities open within each of the two blocks of

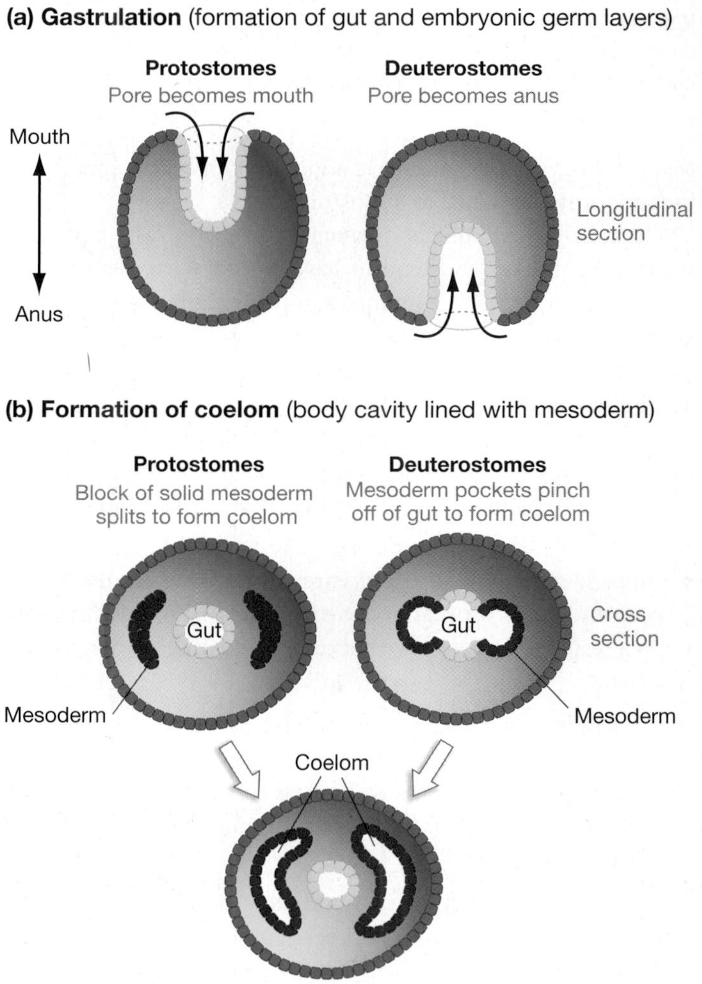

FIGURE 32.7 In Protostomes and Deuterostomes, Events in Early Development Differ. The differences between protostomes and deuterostomes show that there is more than one way to build a bilaterally symmetric, coelomate body plan.

mesoderm. In contrast, deuterostome embryos have layers of mesodermal cells located on either side of the gut. As the right side of Figure 32.7b shows, their coelom begins to form when these layers bulge out and pinch off to form fluid-filled pockets lined with mesoderm.

The vast majority of animal species, including the arthropods (insects, spiders, crustaceans), mollusks (clams, squid, octopuses), and annelids (segmented worms), are protostomes; humans and other chordates (fishes, amphibians, mammals) are deuterostomes.

In essence, the protostome and deuterostome patterns of development represent two distinct ways of achieving the same end—the construction of a bilaterally symmetric body that contains a cavity lined with mesoderm. The functional or adaptive significance of the differences—if any—is not known.

THE TUBE-WITHIN-A-TUBE DESIGN Over 99 percent of the animal species alive today are bilaterally symmetric triploblasts that have coeloms and follow either the protostome or deuterostome pattern of development. This combination of features turned out to be a spectacularly successful way to organize a moving and eating machine. To review basic features of the animal body plan, go to the study area at *www.masteringbiology.com.*

(MB) **Web Activity** The Architecture of Animals

Although it sounds complex to describe a certain animal as a "bilaterally symmetric, coelomic triploblast with protostome [or deuterostome] development," the bodies of most animals are actually extremely simple in form. The basic animal body is a tube within a tube. The inner tube is the individual's gut, and the outer tube forms the body wall (**Figure 32.8a**). The mesoderm in between forms muscles and organs.

In several animal phyla, individuals have long, thin, tubelike bodies that lack limbs. Animals with this body shape are commonly called **worms**. There are many wormlike phyla, including the nemerteans and chaetognaths shown in **Figure 32.8b**.

What about more complex-looking animals, such as grasshoppers and lobsters and horses? The body plan of these animals can also be thought of as a tube within a tube, except that the tube is mounted on legs. Consider that most animals with complex-looking bodies are relatively long and thin. They have an outer body wall that is more or less tubelike and an internal gut that runs from mouth to anus. The body cavity itself is filled with muscles and organs derived from mesoderm. Wings and legs are just efficient ways to move a tube-within-a-tube body around the environment.

Once evolution by natural selection produced the basic tube-within-a-tube design, most of the diversification of animals was triggered by the evolution of novel types of structures for moving and capturing food. Do data from phylogenetic analyses of DNA sequences support this view?

Evaluating Molecular Phylogenies

Perhaps the most influential paper ever published on the phylogeny of animals appeared in 1997. Using sequences from the gene that codes for the RNA molecule in the small subunit of the ribosome,

(a) The tube-within-a-tube body plan in an earthworm

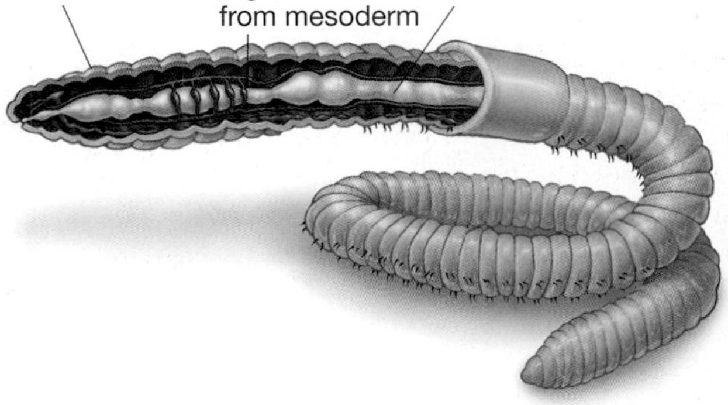

(b) Many animal phyla have wormlike bodies.

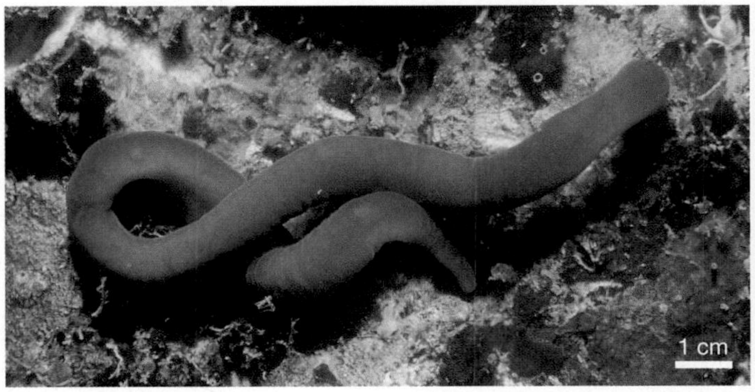

Ribbon worm (Nemertea)

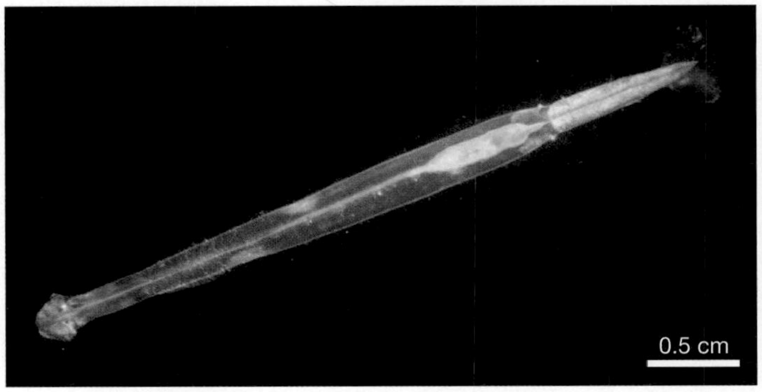

Arrow worm (Chaetognatha)

FIGURE 32.8 The Tube-within-a-Tube Body Plan Is Common in Animals.

Anna Marie Aguinaldo and colleagues estimated the phylogeny of species from 14 animal phyla. The results were revolutionary.

The phylogenetic tree in **Figure 32.9** on page 608 is an updated version of the 1997 result, based on further studies of the genes for ribosomal RNA and a large series of proteins. To analyze it, start from the root, work your way toward the tips, and note several key points:

- A group of protists called the choanoflagellates are the closest living relatives of animals. Choanoflagellates are not animals because they are not multicellular, but they share several key characteristics with sponges. Both are **sessile**, meaning that

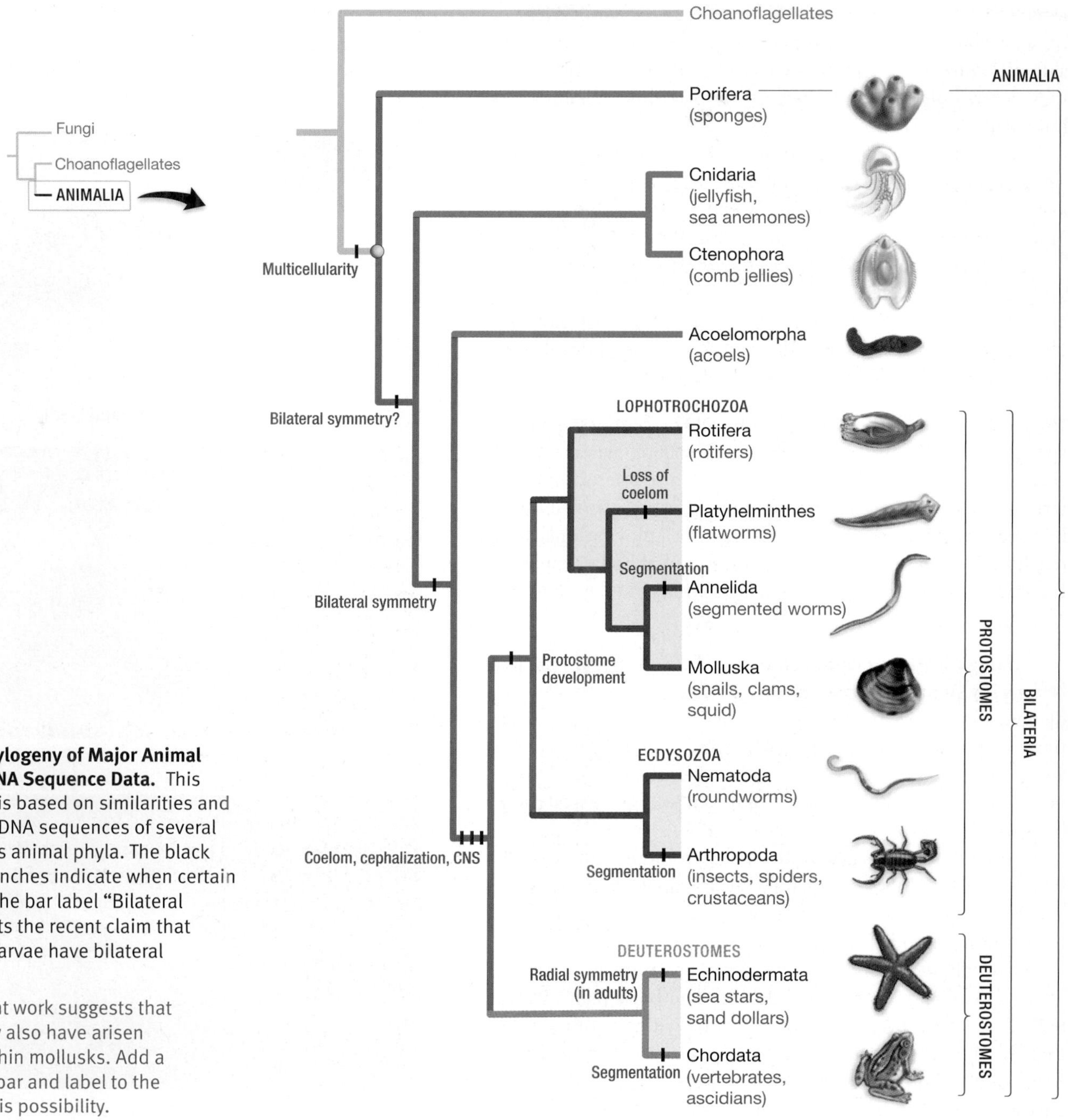

FIGURE 32.9 A Phylogeny of Major Animal Phyla Based on DNA Sequence Data. This phylogenetic tree is based on similarities and differences in the DNA sequences of several genes from various animal phyla. The black bars along the branches indicate when certain traits originated. The bar label "Bilateral symmetry?" relfects the recent claim that certain cnidarian larvae have bilateral symmetry.

✔**EXERCISE** Recent work suggests that segmentation may also have arisen independently within mollusks. Add a "Segmentation?" bar and label to the tree to indicate this possibility.

adults live permanently attached to a substrate. They also feed in a similar way, using cells with nearly identical morphology. As **Figure 32.10** shows, the beating of flagella creates water currents that bring organic debris toward the feeding cells of choanoflagellates and sponges. Sponge feeding cells are called **choanocytes**. In these feeding cells, food particles are trapped and ingested.

- Sponges are the sister group to all other animals. Stated another way, the initial split that occurred during animal diversification produced a lineage that evolved into today's sponges, and a lineage that evolved into all other animals living today.
- Ctenophora (jellyfish, hydra, anemones) and Cnidaria (comb jellies) are a monophyletic group.
- The placement of cnidarians and ctenophores as the sister group to acoelomorphs and bilaterians implies that endoderm and ectoderm were the first embryonic tissue types to evolve, and that radial symmetry evolved before bilateral symmetry.
- The closest living relative of bilaterally symmetric triploblasts, the Acoelomorpha, lack a coelom. This result indicates that the evolution of mesoderm preceded the evolution of the coelom.

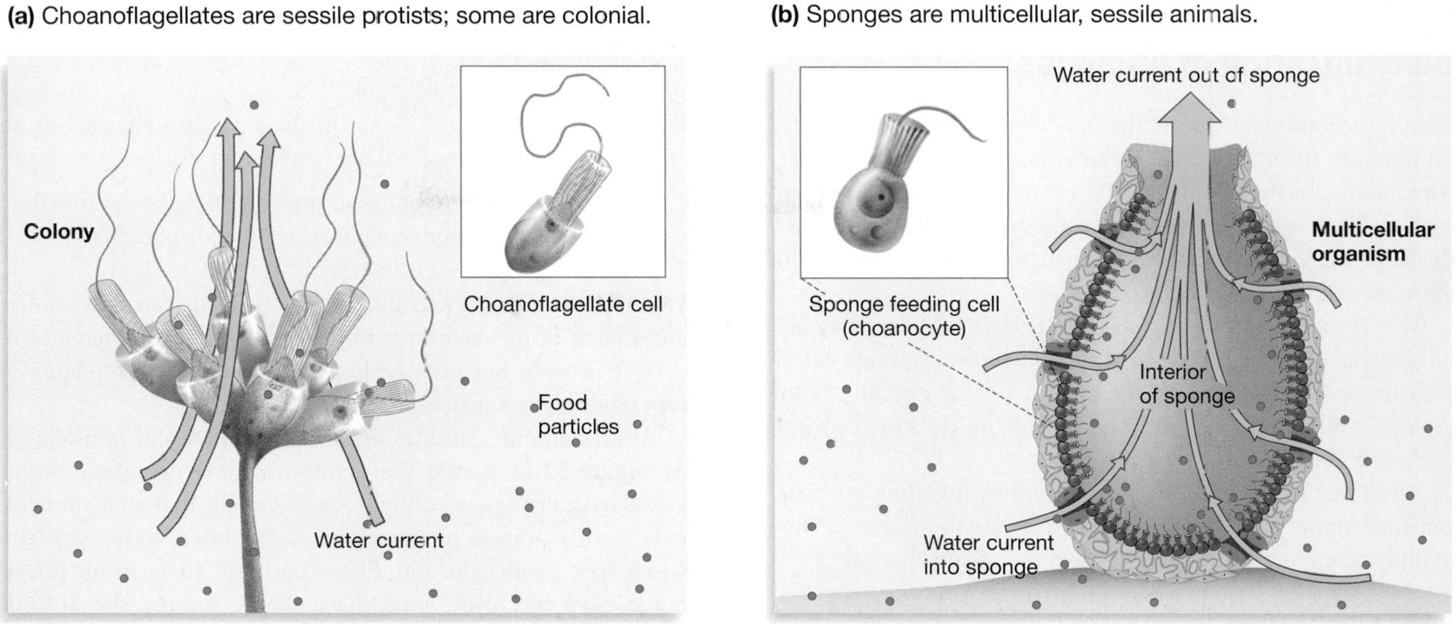

FIGURE 32.10 Choanoflagellates and Sponge Feeding Cells Are Almost Identical in Structure and Function.
(a) Choanoflagellates are suspension feeders. **(b)** A cross section of a simple sponge as it suspension feeds. The beating of flagella produces a water current that brings food into the body of the sponge, where it can be ingested by feeding cells.

- The split between the protostomes and the deuterostomes was a major event in the evolution of Bilateria.
- A fundamental split occurred within protostomes, forming two major subgroups with protostome development: (1) the Lophotrochozoa (pronounced *low-foe-tro-ko-ZOH-ah*) includes the mollusks and the annelids; and (2) the Ecdysozoa (pronounced *eck-die-so-ZOH-ah*) includes the arthropods and the nematodes. **Ecdysozoans** grow by shedding their external skeletons or outer coverings and expanding their bodies. **Lophotrochozoans** grow continuously when conditions are good.
- Species in the phylum Platyhelminthes (flatworms) are protostomes but lack a coelom—meaning that the coelom was lost during flatworm evolution. As Chapters 33 and 34 will detail, the coelom has also been reduced in an array of lineages—specifically, in groups that have limbs and thus do not rely on a hydrostatic skeleton for movement.
- During the evolution of bilaterians, segmentation arose several times independently. **Segmentation** is defined by the presence of repeated body structures, such as an earthworm's segments or a fish's vertebral column and ribs. Segmentation evolved independently in annelids (earthworms and other segmented worms) and arthropods (insects, spiders, and crustaceans), as well as in vertebrates. **Vertebrates** are a monophyletic lineage within the Chordata that is defined by the presence of a skull; many vertebrate species also have a backbone (see Chapter 34). Recent evidence suggests that segmentation also evolved independently in mollusks (snails, clams, squid). Segmentation is thought to be a particularly efficient way to organize a bilaterally symmetric body.
- **Invertebrates**, traditionally defined as all animals that are not vertebrates, are paraphyletic. The group includes some, but not all, of the descendants of a common ancestor.

Although biologists are increasingly confident that most or all of these conclusions are correct, the phylogeny of animals is still a work in progress. As data sets expand, it is likely that new analyses will not only confirm or challenge these results but also contribute other important insights into animal evolution. Stay tuned.

CHECK YOUR UNDERSTANDING

If you understand that . . .

- The origin and early diversification of animals was marked by changes in four fundamental features: the number of embryonic tissues present; the evolution of nervous systems and a bilaterally symmetric, cephalized body; the evolution of a body cavity; and protostome versus deuterostome patterns of development.

✓ **You should be able to . . .**

1. Explain why bilateral symmetry in combination with triploblasty and a coelom is responsible for the "tube-within-a-tube" design observed in most animals living today.
2. Explain why cephalization was important.

Answers are available in Appendix B.

32.3 What Themes Occur in the Diversification of Animals?

Within each animal phylum, the basic features of the body plan do not vary from species to species. For example, **mollusks** are triploblastic, bilaterally symmetric protostomes with a reduced coelom. Their body architecture features a muscular foot, a cavity called the visceral mass, and a structure called a mantle. But there are over 100,000 species of mollusk.

If the major animal lineages are defined by a particular body plan, what triggered the diversification of species within each lineage? In most cases, the answer to this question is the evolution of innovative methods for sensing the environment, feeding, and moving.

Recall that most animals get their food by ingesting other organisms. Animals are diverse because there are thousands of ways to find and eat the millions of different organisms that exist.

Sensory Organs

The evolution of a cephalized body was a major breakthrough in the evolution of animals. Along with a mouth and brain, a concentration of sensory organs in the head region—where the animal initially encounters the environment—is a key aspect of cephalization.

Chapter 46 explores the cells and organs responsible for sensing light (vision), specific molecules (smell and taste), and sound (hearing) in detail. Here the key point is to appreciate the diversity of sensory abilities and structures found in animals.

VARIATION IN SENSORY ABILITIES Certain senses are almost universal in animals. The common senses include touch and balance—meaning that animals can sense objects and gravity—along with a sense of smell, taste, and hearing. At least some ability to sense light and time is also common.

But as animals diversified, a wide array of more specialized sensory abilities evolved, including:

- *Magnetism* Many birds, sea turtles, sea slugs, and other animals can detect magnetic fields and use Earth's magnetic field as an aid in navigation.
- *Electric fields* Some aquatic predators, such as sharks, are so sensitive to electric fields that they can detect electrical activity in the muscles of passing prey.
- *Barometric pressure* Some birds can sense changes in air pressure, which may aid them in avoiding storms.

Variation in sensory abilities is important: it allows animals to collect information about a wide array of environments.

VARIATION IN SENSORY STRUCTURES Animals can sense different aspects of the environment because they have different sensory structures. But even within the same sensory modality or type, abilities and structures vary.

As an example, consider the array of eyes found in mollusks. As **Figure 32.11** shows, some mollusks have extremely simple eyes—basically a panel of light-sensitive cells that allow individuals to detect areas of light and dark. Others have eyes with lenses that bend light and allow the brain to form an image. Some have two eyes; some have many. Among the 100,000 species of mollusk, the structure of the eye is highly variable. In most cases, this variation in structure helps each species function efficiently in the environment it occupies.

Feeding

To organize the diversity of ways that animals find food, biologists distinguish *how* individuals eat from *what* they eat. In many cases, species that pursue different feeding strategies and food sources are found within a single lineage.

HOW ANIMALS FEED: FOUR GENERAL TACTICS Two simple ideas are key to understanding how animals from the same lineage can have the same basic body architecture but feed in radically different ways: (1) Their mouthparts vary, and (2) the structure of an animal's mouthparts correlates closely with its method of feeding. Keep these concepts in mind as you review the four general tactics that animals use to obtain food.

1. **Suspension feeders**, *also known as* **filter feeders**, *capture food by filtering out or concentrating particles floating in water or drifting through the air.*

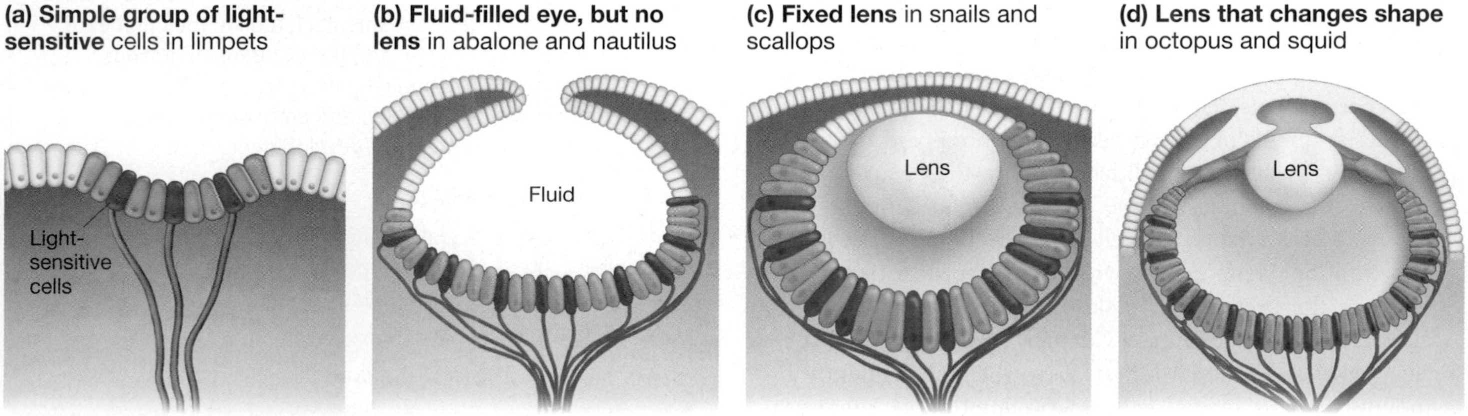

FIGURE 32.11 Variation in Eye Structure among Mollusk Eyes. Some mollusks, such as clams and mussels, lack eyes.

(a) Earthworms (Annelida) eat organic material within soil and detritus such as leaves on the surface of soil.

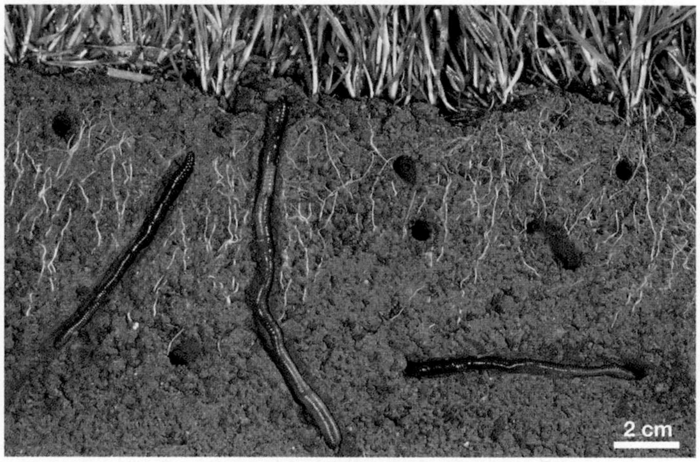

(b) Sea cucumbers (Echinodermata) use feeding tentacles to move detritus into their mouths.

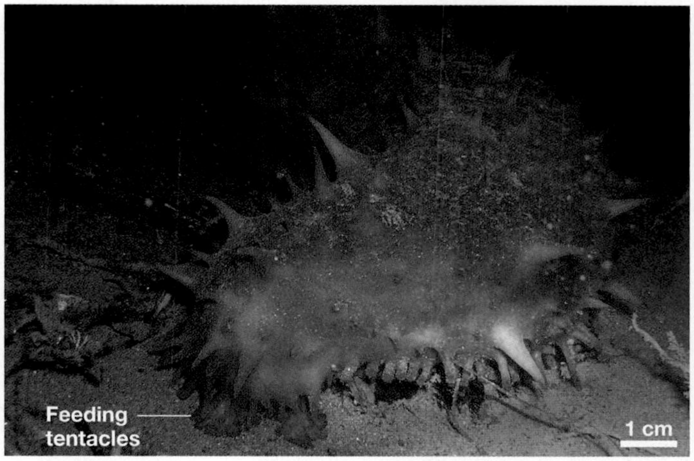

FIGURE 32.12 Deposit Feeders Eat Organic Material in Sediments and Other Deposits.

Suspension feeders employ a wide array of structures to trap suspended particles—usually small algae or animals or bits of detritus—and ingest them. Sponges, like the individual illustrated in Figure 32.10, are suspension feeders. So are clams and mussels, which pump water through their bodies and trap suspended food on their feathery gills—structures that also function in gas exchange. Baleen whales suspension feed by gulping water, squeezing it out between the horny baleen plates that line their mouths, and trapping shrimp-like organisms called krill inside.

Because particles float in water much more readily than in air, suspension feeding is particularly common in aquatic environments. Many suspension feeders are sessile.

2. **Deposit feeders** *ingest organic material that has been deposited within a substrate or on its surface.*

Many deposit feeders digest organic matter in the soil; their food consists of soil-dwelling bacteria, archaea, protists, and fungi, along with detritus that settles on the surface of the soil. Earthworms, for example, are annelids that swallow soil as well as leaves and other detritus on the surface of the soil (**Figure 32.12a**).

The seafloor is also rich in organic matter, which rains down from the surface and collects in food-rich deposits. These deposits are exploited by a wide array of segmented worms (annelids) as well as the echinoderms called sea cucumbers (**Figure 32.12b**).

Unlike suspension feeders, which are diverse in size and shape and use various trapping or filtering systems, deposit feeders are similar in appearance. They usually have simple mouthparts and their body shape is wormlike. Like suspension feeders, however, deposit feeders occur in a wide variety of lineages.

3. **Fluid feeders** *suck or mop up liquids like nectar, plant sap, blood, or fruit juice.*

Fluid feeders range from butterflies that feed on nectar with a straw-like proboscis (**Figure 32.13a**) to blowflies that feed on rotting fruit using a sponge-like mouthpart (**Figure 32.13b**). Fluid feeders are found in a wide array of lineages and often have mouthparts that allow them to pierce seeds, stems, skin, or other structures in order to withdraw the fluids inside.

4. **Mass feeders** *take chunks of food into their mouths.*

(a) Butterflies have an extensible, hollow proboscis.

(b) Blowflies have an extensible, sponge-like mouthpart.

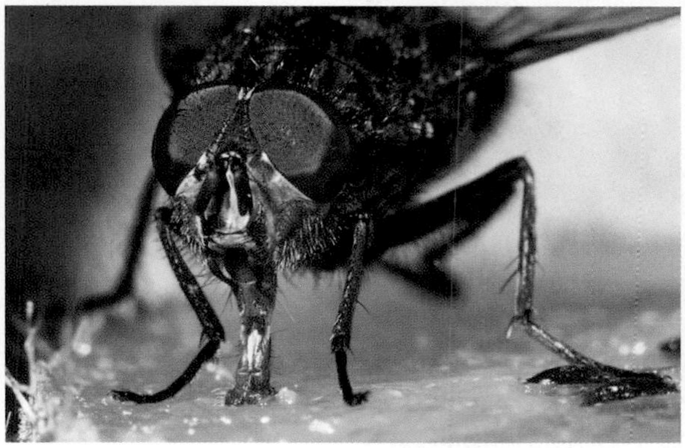

FIGURE 32.13 Fluid Feeders Drink Liquids.

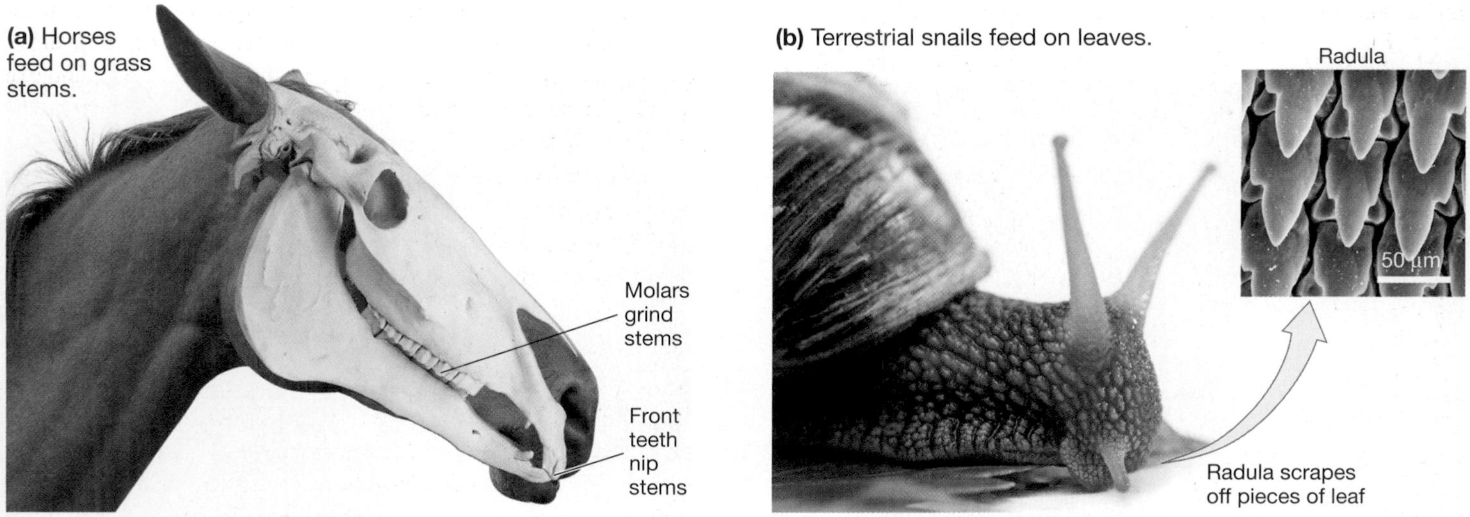

(a) Horses feed on grass stems.
Molars grind stems
Front teeth nip stems
(b) Terrestrial snails feed on leaves.

Radula scrapes off pieces of leaf

FIGURE 32.14 Food-Mass Feeders Ingest Bites or Lumps of Food.

In mass feeders, the structure of the mouthparts correlates with the type of food pieces that are harvested and ingested. Horses, for example, have sharp teeth in the front of the jaw for biting off grass stems and broad, flat molar teeth in the back of the jaw for mashing the coarse stems into a soft wad that can be swallowed (**Figure 32.14a**). In many snails, a feeding structure called a **radula** functions like a rasp or a file. The sharp plates on the radula move back and forth to scrape material away from a plant or alga so that it can be ingested (**Figure 32.14b**). Among vertebrates and snails, variation in tooth or radula structure correlates with the food source.

WHAT ANIMALS EAT: THREE GENERAL SOURCES Whether they feed by filtering, eating through deposits, or taking in fluids or masses, animals can be classified as (1) **herbivores** that feed on plants or algae, (2) **carnivores** that feed on animals, or (3) **detritivores** that feed on dead organic matter. Animals that eat both plants and animals are called **omnivores**.

In addition, herbivores and carnivores can be subclassified as (1) predators or (2) parasites.

1. **Predators** *are usually larger than their prey and kill them quickly.*

Predators use an array of mouthparts and hunting strategies. Many types of frogs, for example, are sit-and-wait predators. They sit still and wait for an insect or worm to move close, then capture it with a lightning-quick extension of their long, sticky tongue (**Figure 32.15a**). In contrast, wolves hunt by locating a prey organism and then running it down during an extended, long-distance chase (**Figure 32.15b**).

2. **Parasites** *are usually much smaller than their victims and often harvest nutrients without causing death.*

Endoparasites live inside their hosts and usually have simple, wormlike bodies. Tapeworms, of the phylum platyhelminthes

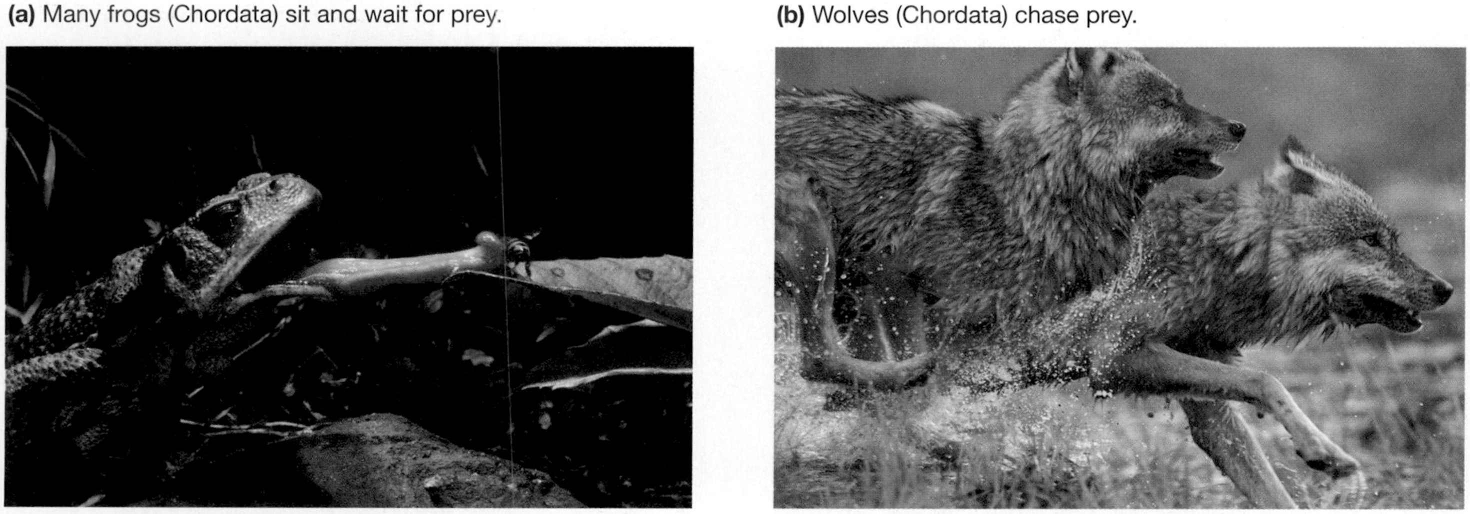

(a) Many frogs (Chordata) sit and wait for prey.
(b) Wolves (Chordata) chase prey.

FIGURE 32.15 Predators Kill and Eat Organisms.

(a) Tapeworms (Platyhelminthes) are endoparasites.

(b) Lice (Arthropoda) are ectoparasites.

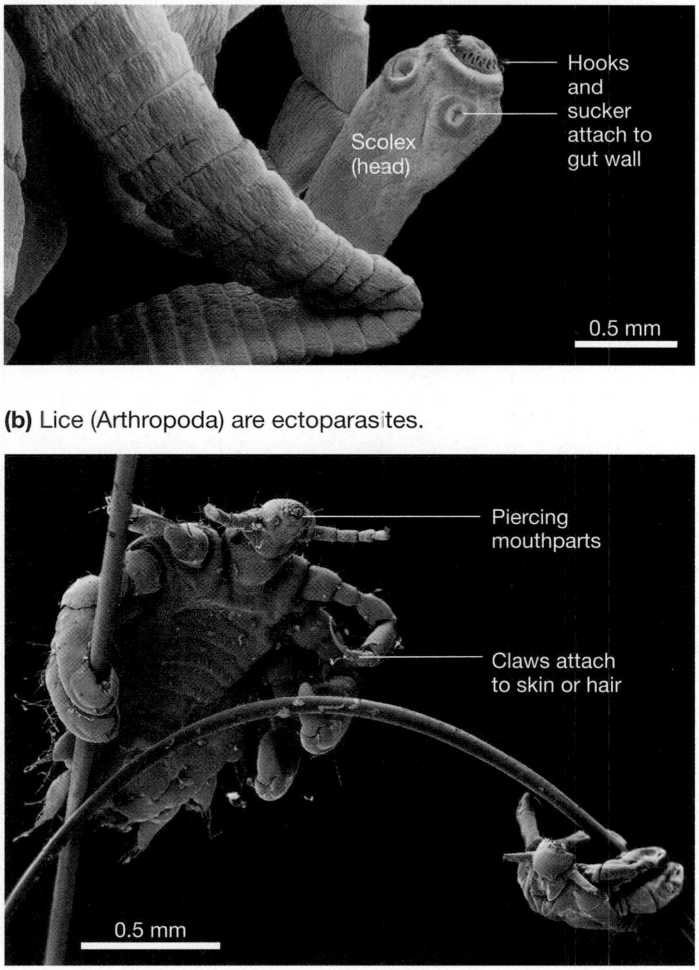

FIGURE 32.16 Parasites Take Nutrients from Living Animals.
(a) Tapeworms are common intestinal parasites of humans and other vertebrates. They absorb nutrition directly across their body wall. **(b)** The louse *Phthirus pubis* uses its clawlike legs to attach to the pubic region of humans, pierces the host's skin with its mouthparts, and feeds by sucking body fluids.

(flatworms), are endoparasites with no digestive system. Instead of a mouth, they have hooks or other structures on their head, called a scolex, that attach to their host's intestinal wall (**Figure 32.16a**). Instead of digesting food themselves, they absorb nutrients directly from their surroundings.

Ectoparasites live outside their hosts. Ectoparasites usually have limbs or mouthparts that allow them to grasp the host and mouthparts that allow them to pierce their host's skin and suck the nutrient-rich fluids inside (**Figure 32.16b**).

Movement

Animal locomotion has an array of important functions: finding food, finding mates, escaping from predators, and dispersing to new habitats. The ways that animals move in search of food and sex are highly variable; they burrow, slither, swim, fly, crawl, walk, or run. Movement is powered by cilia, flagella, or muscles that attach to a hard skeleton or compress a hydrostatic skeleton, enabling wriggling movements.

The hydrostatic skeleton is essential to locomotion in the many animal phyla with wormlike bodies. Another major innovation occurred as animals diversified, however. The limb made highly controlled, rapid movement possible.

TYPES OF LIMBS: UNJOINTED AND JOINTED Limbs are a prominent feature of species in many phyla. They are particularly important in two major lineages: the ecdysozoans and the vertebrates.

Some members of the Ecdysozoa, such as onychophorans (velvet worms), have unjointed, lobe-like limbs; others, such as arthropods, have more complex, jointed limbs (**Figure 32.17**). Jointed limbs make fast, precise movements possible and are a prominent limb type in vertebrates and arthropods.

ARE ALL ANIMAL APPENDAGES HOMOLOGOUS? Chapter 24 introduced the concept of homology, which is defined as similarity in traits due to inheritance from a common ancestor. Traditionally, biologists have hypothesized that the major types of jointed and unjointed animal limbs were not homologous.

To appreciate the logic behind this hypothesis, it's important to recognize just how diverse animal appendages are: They

(a) Velvet worms (Onychophora) have lobe-like limbs.

(b) Crabs (Arthropoda) have jointed limbs.

FIGURE 32.17 Unjointed and Jointed Limbs May Be Used in Locomotion.

(a) Polychaetes (Annelida) have parapodia.

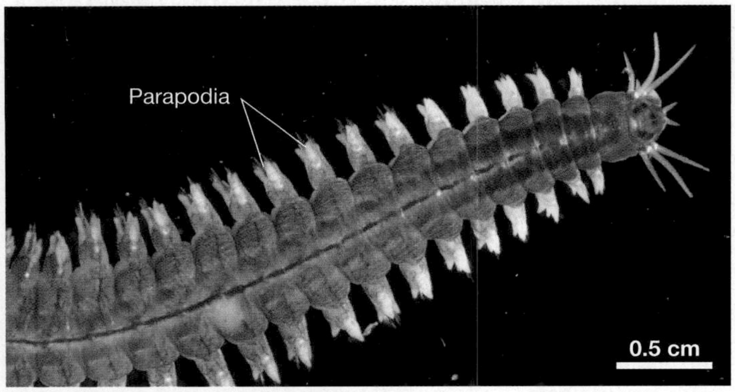

(b) Sea urchins (Echinodermata) have tube feet.

FIGURE 32.18 Some Species "Walk" on Parapodia or Tube Feet Instead of Legs.

range from the human arm to bristle-like structures called parapodia in segmented worms (**Figure 32.18a**) and the soft, extensible tube feet found in echinoderms (**Figure 32.18b**). Because the structure of animal appendages is so diverse, it was logical to maintain that at least some appendages evolved independently of each other. As a result, biologists predicted that completely different genes are responsible for each major type of appendage.

Recent results have challenged this view, however. The experiments in question involve a gene called *Distal-less,* which was originally discovered in fruit flies. (*Distal* means "away from the body.") *Distal-less,* or *Dll,* is aptly named. In fruit flies that lack this gene's normal protein product, only the most rudimentary limb buds form. The mutant limbs are "distal-less." The protein seems to deliver a simple message as a fruit-fly embryo develops: "Grow appendage out this way."

Biologists in Sean B. Carroll's lab set out to test the hypothesis that *Dll* might be involved in the initial phase of limb or appendage formation in other animals. As **Figure 32.19** shows, they used a fluorescent marker that sticks to the *Dll* gene product to locate tissues where the gene is expressed. When they introduced the fluorescent marker into embryos from annelids, arthropods, echinoderms, chordates, and other phyla, they found that it bound to *Dll* in all of them. More important, the highest concentrations of *Dll* gene products were found in cells that form appendages—even in phyla with

EXPERIMENT

QUESTION: Is the gene *Dll* involved in limb formation in species other than insects?

HYPOTHESIS: In all animals, *Dll* signals "grow appendage out here."

NULL HYPOTHESIS: *Dll* is not involved in the development of appendages in species other than insects.

EXPERIMENTAL SETUP:
Add a stain to developing embryos that will attach to *Dll* gene products (proteins), revealing their location. The stain can be fluorescent green or dark brown.

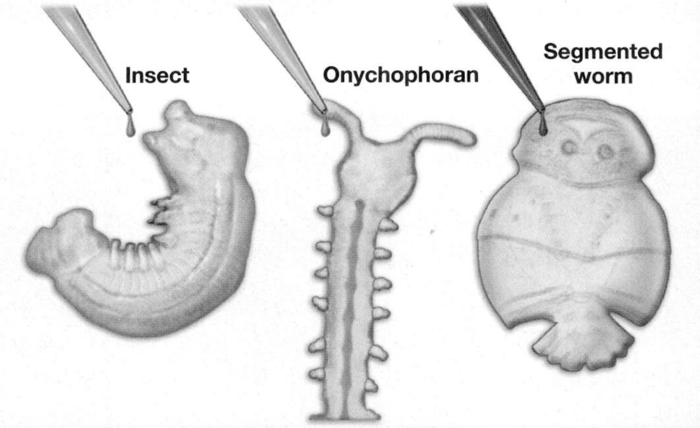

PREDICTION: In embryos from a wide array of species, stained *Dll* gene products will be localized to areas where appendages are forming.

PREDICTION OF NULL HYPOTHESIS: Stained *Dll* gene products will be localized to areas where appendages are forming only in insects.

RESULTS:

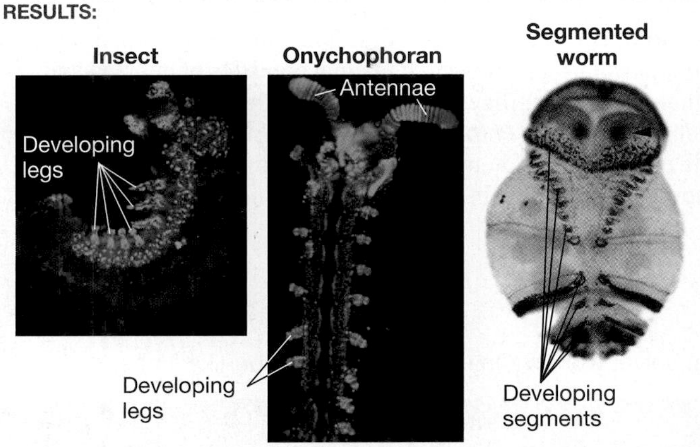

In species representing both Ecdysozoa (e.g., insect and onychophoran) and Lophotrochozoa (e.g., segmented worm), *Dll* is localized in areas of the embryo where appendages are forming.

CONCLUSION: The gene *Dll* is involved in limb formation in diverse species. The results suggest that all animal appendages may be homologous.

FIGURE 32.19 Experimental Evidence That All Animal Appendages Are Homologous.

SOURCE: Panganiban, G., et al. 1997. The origin and evolution of animal appendages. *Proceedings of the National Academy of Sciences USA* 94: 5162–5166.

✔**QUESTION** What results would have supported the null hypothesis?

(a) Internal fertilization in damselflies

(b) External fertilization in giant clams

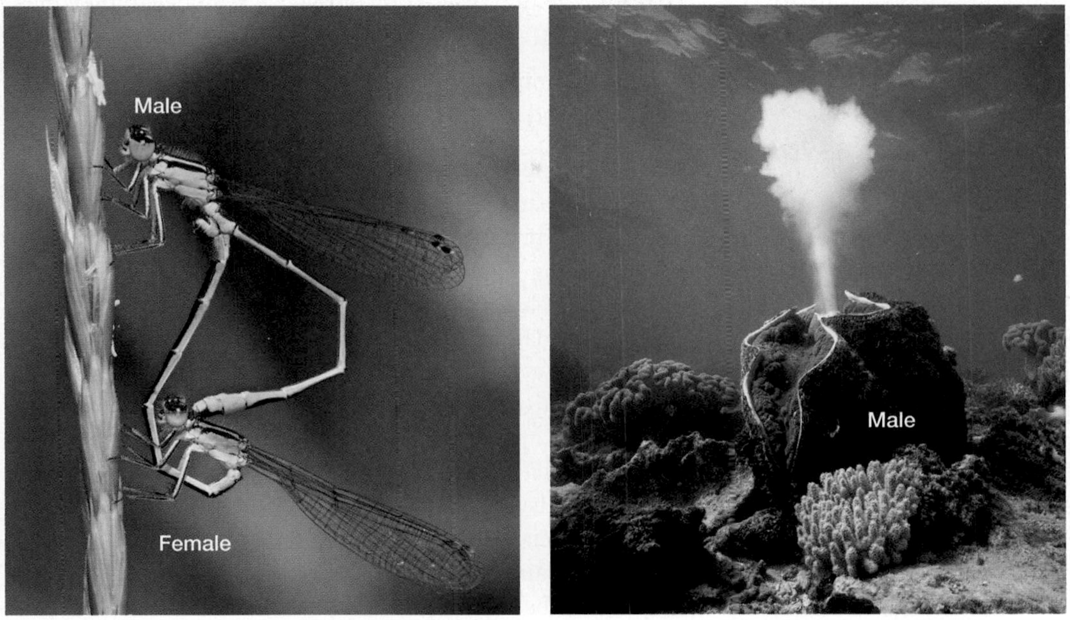

FIGURE 32.20 Fertilization Can Be Internal or External in Animals.
(a) When damselflies copulate, the male holds the female just behind her head with claspers and the female places the tip of her abdomen against the male's sperm transfer organ. The two can fly in this position. **(b)** A male giant clam releasing sperm into the ocean.

wormlike bodies that have extremely simple appendages. Other experiments have shown that *Dll* is also involved in limb formation in vertebrates.

Based on these findings, biologists are concluding that the protein product of the same gene marks the initial site of appendage growth in most if not all animals. The hypothesis is that all animal appendages have some degree of genetic homology and that they are all derived from appendages that were present in a common ancestor. The idea is that a simple appendage evolved early in the history of the Bilateria and that, subsequently, evolution by natural selection produced the diversity of limbs, antennae, and wings observed today.

Reproduction

An animal may be efficient at moving and eating, but if it does not reproduce, the alleles responsible for its effective locomotion and feeding will not increase in frequency in the population. As Chapter 24 emphasized, natural selection occurs when individuals with certain alleles produce more offspring than other individuals do. Organisms live to reproduce.

Given the array of habitats and lifestyles pursued by animals, it's not surprising that they exhibit a high degree of variation in how they reproduce. Chapter 48 details how gamete formation, fertilization, and early development occur in animals, but a few examples here will help drive home just how diverse animal reproduction is.

DOES FERTILIZATION OCCUR? At least some species in most animal phyla can reproduce asexually, through mitosis, as well as sexually (via meiosis and fusion of gametes).

In the lophotrochozoan phylum Rotifera, an entire lineage called the bdelloids (pronounced *DELL-oyds*) reproduces only asexually. Even certain fish, lizard, and snail species have never been observed to undergo sexual reproduction.

WHERE DOES FERTILIZATION OCCUR? When sexual reproduction does occur, fertilization may be internal or external.

When internal fertilization takes place, males typically insert a sperm-transfer organ into the body of a female (**Figure 32.20a**). In some cases, males produce sperm in packets, which females then pick up and insert into their own bodies. But in seahorses, females insert eggs into the male's body, where they are fertilized. (The male is pregnant for a time and then gives birth to live young.)

External fertilization is extremely common in aquatic species. Females lay eggs onto a substrate or into open water. Males shed sperm, which swim, on or near the eggs (**Figure 32.20b**).

WHERE DO EMBRYOS DEVELOP? Eggs or embryos may be retained in the female's body during development, or eggs may be laid outside to develop independently of the mother.

Mammals and other species that nourish embryos inside the body and give birth to live young are said to be **viviparous** ("live-bearing"); species that deposit fertilized eggs are **oviparous** ("egg-bearing"); and some species are **ovoviviparous** ("egg-live-bearing").

In ovoviviparous species, the females retain eggs inside their body during early development; but the growing embryos are nourished by yolk inside the egg and not by nutrients transferred directly from the mother, as in viviparous species. Ovoviviparous females then give birth to well-developed young.

Most mammals and a few species of sea stars, onychophorans, sharks, fish, amphibians, and lizards are viviparous; some snails, insects, reptiles, fishes, and sharks are ovoviviparous. But the vast majority of animals are oviparous.

Life Cycles

Besides reproducing in a variety of ways, animal life cycles are diverse. Perhaps the most spectacular innovation in animal life cycles involves the phenomenon known as **metamorphosis** ("change-form")—a change from an immature body type to an adult body type.

In discussing metamorphosis and other types of animal life cycles, biologists distinguish three types of life stages:

- **Larvae** (singular: **larva**) look radically different from adults, live in different habitats, and eat different foods. They are sexually immature—meaning that their reproductive organs are undeveloped.
- **Juveniles** look like adults and live in the same habitats and eat the same foods as adults, but are sexually immature.
- **Adults** are the reproductive stage in the life cycle.

In insects, either a larval or a juvenile stage occurs.

TWO TYPES OF INSECT METAMORPHOSIS In insects, the presence of a larval versus juvenile stage defines two distinct types of metamorphosis.

In **hemimetabolous** ("half-change") **metamorphosis**—also called **incomplete metamorphosis**—young are juveniles called nymphs that look like miniature versions of the adult. The aphid nymphs in **Figure 32.21a**, for example, shed their external skeletons several times and grow—gradually changing from wingless, sexually immature nymphs to sexually mature adults, some of which can fly. But throughout their life, aphids live in the same habitats and feed on the same food source in the same way: They suck sap.

In **holometabolous** ("whole change") **metamorphosis**—also called **complete metamorphosis**—young are larvae. As an example, consider the life cycle of the mosquito, illustrated in **Figure 32.21b**.

- Newly hatched mosquitoes live in quiet bodies of freshwater, where they suspension feed on bacteria, algae, and detritus.
- When a larva has grown sufficiently, the individual stops feeding and moving and secretes a protective case. The individual is now known as a **pupa** (plural: **pupae**). During pupation, the pupa's body is completely remodeled into a new, adult form.
- The adult mosquito flies and gets its nutrition as a parasite—taking blood meals from mammals and nectar from flowers.

WHAT IS THE ADAPTIVE SIGNIFICANCE OF METAMORPHOSIS? In insects, holometabolous metamorphosis is 10 times more common than hemimetabolous metamorphosis. Why?

The leading hypothesis is based on efficiency in feeding. Because juveniles and adults from holometabolous species feed on different materials in different ways and sometimes even in different habitats, they do not compete with each other.

An alternative hypothesis is based on the advantages of specialization. In many moths and butterflies, larvae are specialized for feeding, whereas adults are specialized for mating and feed rarely, if ever. Larvae are largely sessile, whereas adults are highly mobile. If specialization leads to higher efficiency in feeding and reproduction and thus higher fitness, then complete metamorphosis would be advantageous.

These hypotheses are not mutually exclusive, however—meaning that both could be correct, depending on the species being considered. Research continues.

A COMPLEX LIFE CYCLE IN CNIDARIA Metamorphosis is extremely common in marine animals, as well as in insects. Many marine fish, for example, have a larval stage—a period where young live near the ocean surface, far from the habitats where adults are found, and feed on microscopic prey.

In marine species that have limited or no movement as adults, larvae function as a dispersal stage. They are a little like the seeds of land plants—a life stage that allows individuals to move to new habitats, where they will not compete with their parents for space and other resources.

(a) Aphid:
Hemimetabolous metamorphosis

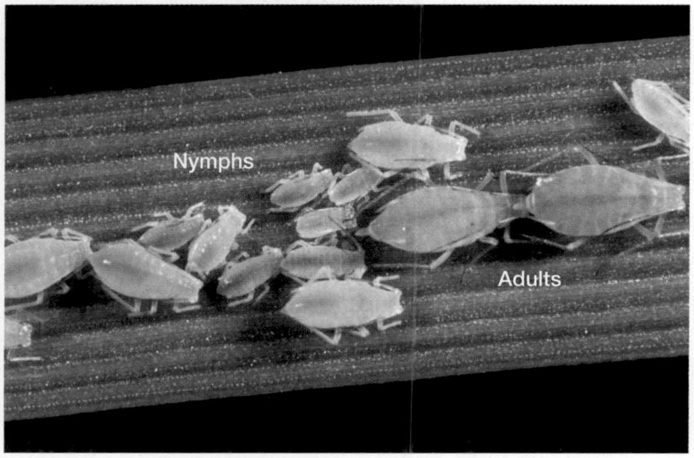

Nymphs look like miniature versions of adults and eat the same foods.

(b) Mosquito:
Holometabolous metamorphosis

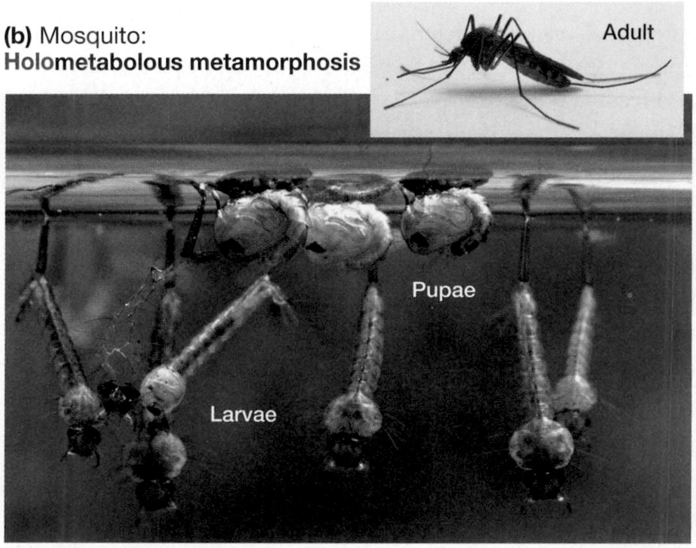

Larvae look substantially different from adults and eat different foods.

FIGURE 32.21 During Insect Metamorphosis, Individuals May or May Not Change Form Completely.

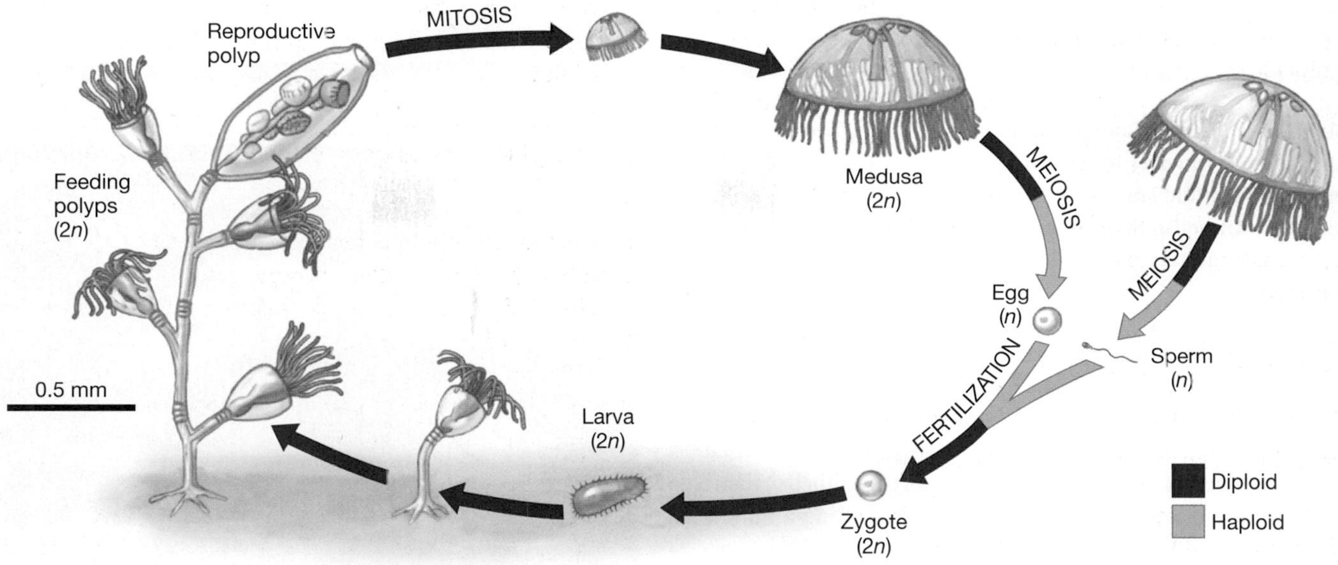

FIGURE 32.22 Cnidarian Life Cycles May Include a Polyp and Medusa Form. This is an example of a hydrozoan called *Obelia*. Colonies are often found attached to seaweed, shells, or rocks and contain hundreds of polyps.

✔**QUESTION** Does a protist- or plant-like alternation of generations occur in this species? Explain why or why not.

An even more complex life cycle is found in some cnidarians, which have three distinct body types during their life cycle: (**1**) A largely sessile form called a **polyp** that reproduces asexually, (**2**) a free-floating stage called a **medusa** (plural: **medusae**) that reproduces sexually, and (**3**) a larval form (**Figure 32.22**).

Polyps usually live attached to a substrate, suspension feed on detritus or small organisms, and frequently form large clusters of individuals called colonies. A **colony** is a group of identical individuals that are physically attached. Medusae, in contrast, float freely in the plankton and feed on crustaceans and other animals. Because polyps and medusae live in different habitats, the two stages of the life cycle exploit different food sources.

CHECK YOUR UNDERSTANDING

If you understand that . . .

- The story of animal evolution is based on two themes: (1) the evolution of a small suite of basic body plans, and (2) a diversification of species with the same basic body architecture, based on the evolution of innovative structures and methods for sensing the environment, capturing food, and moving.
- Methods for fertilization and parental care of embryos vary widely.
- Metamorphosis is common, with larvae often serving as a feeding or dispersal stage.

✔ **You should be able to . . .**

1. Explain how the mouthparts of deposit feeders and mass feeders are expected to differ.
2. Explain why external fertilization is particularly common in aquatic environments, and internal fertilization is particularly common in terrestrial environments.

Answers are available in Appendix B.

32.4 Key Lineages of Animals: Non-Bilaterian Groups

The goal of this chapter is to provide a broad overview of how animals diversified into such a morphologically diverse lineage. According to recent phylogenetic analyses, the closest living relatives to the animals are the choanoflagellates, a group of protists. The fossil record indicates that the phyla Porifera (sponges), Cnidaria (jellyfish and others), and Ctenophora are the most ancient of all animal lineages. The phylogeny of animals is consistent with the fossil record—the first sponges, cnidarians, and ctenophores appear at the base of the tree.

Let's explore the origins of animals by taking a more detailed look at lineages with similar-looking relatives that appear early in the fossil record, along with the Acoelomorpha—the lineages other than the Bilateria. The protostomes and deuterostomes are considered in Chapters 33 and 34, respectively.

- Porifera (Sponges)
- Cnidaria (Jellyfish, Corals, Anemones, Hydroids)
- Ctenophora (Comb Jellies)
- Acoelomorpha (Acoels)

Porifera (Sponges)

About 7000 species of sponges have been described to date. Although a few freshwater species are known, most are marine. All sponges are **benthic**, meaning that they live at the bottom of aquatic environments. Sponges are particularly common in rocky, shallow-water habitats of the world's oceans and in coastal areas of Antarctica.

The architecture of sponge bodies is built around a system of tubes and pores that create channels for water currents. Body symmetry varies among sponge species; most are asymmetrical—meaning that they have no plane of symmetry—but some species are radially symmetric (**Figure 32.23**). Sponges have specialized cell types and some species have well-organized epithelial tissue layers lining the inside and outside of the body, sealing a middle layer in between. ✔**You should be able to indicate the origin of epithelial tissue, which occurs in all animals, on Figure 32.9.** Sponges lack the other tissue types found in other animals, however. The "looseness" of sponge tissues is thought to be an adaptation that facilitates regeneration after wounding, the ability to remodel the body if water currents change, and the ability to move into and colonize new substrates as they become available. In many species, collagen fibers are augmented by **spicules**—stiff spikes of silica or calcium carbonate ($CaCO_3$)—to provide structural support for the body. One sponge species native to the Caribbean can grow to heights of 2 m.

Sponges have commercial and medical value to humans. The dried bodies of certain sponge species are able to hold large amounts of water and thus are prized for use in bathing and washing. In addition, researchers are increasingly interested in the array of toxins that sponges produce to defend themselves against predators and bacterial parasites—possibly for use in cancer chemotherapy.

Feeding Most sponges are suspension feeders. Their cells beat in a coordinated way to produce a water current that flows through small pores in the outer body wall, into chambers inside the body, and out through a single larger opening. As water passes by feeding cells, organic debris and bacteria, archaea, and small protists are filtered out of the current and then digested. Some deep-sea sponges are predators, however—they capture small crustaceans on hooks that project from the body.

Movement Most adult sponges are sessile, though a few species are reported to move at rates of up to 4 mm per day. Most species produce larvae that swim with the aid of cilia. Recent research has confirmed that at least one species can contract its body to expel waste products.

FIGURE 32.23 Some Sponges Form Radially Symmetric Tubes.

Reproduction Asexual reproduction occurs in a variety of ways, depending on the species. Some sponge cells are totipotent, meaning that small groups of adult cells have the capacity to develop into a complete adult organism. Thus, a fragment that breaks off an adult sponge has the potential to grow into a new individual. Although individuals of most species produce both eggs and sperm, self-fertilization is rare because individuals release their male and female gametes at different times. Fertilization usually takes place in the water, but some ovoviviparous species retain their eggs and then release mature, swimming larvae after fertilization and early development have occurred.

Cnidaria (Jellyfish, Corals, Anemones, Hydroids)

Although a few species of Cnidaria inhabit freshwater, the vast majority of the 11,000 species are marine. They are found in all of the world's oceans, occupying habitats from the surface to the substrate, and are important predators. The phylum comprises four main lineages: Hydrozoa (hydroids), Cubozoa (box jellyfish), Scyphozoa (jellyfish), and Anthozoa (anemones, corals, and sea pens).

Many cnidarians are radially symmetric diploblasts consisting of ectoderm and endoderm layers that sandwich gelatinous material known as **mesoglea**, which contains a few scattered ectodermal cells. Recent research indicates that at least some jellyfish are triploblastic and have bilaterally symmetric larvae. Cnidarians have a gastrovascular cavity instead of a flow-through gut—meaning there is only one opening to the environment for both ingestion and elimination of wastes.

Some cnidarians have a life cycle that includes both a sessile polyp form and a free-floating medusa (see Figure 32.22). Anemones and coral, however, exist only as polyps—never as medusae. Reef-building corals secrete outer skeletons of calcium carbonate that create the physical structure of a coral reef—one of the world's most productive habitats (see Chapter 54).

Feeding The morphological innovation that triggered the diversification of the cnidarians is a specialized cell, called a **cnidocyte**, which is used in prey capture. When cnidocytes sense a fish or other type of prey, the cells forcibly eject a barbed, spear-like structure, which may contain toxins. The barbs hold the prey, and the toxins subdue it until it can be brought to the mouth and ingested. Cnidocytes are commonly located near the mouths of cnidarians or on elongated structures called tentacles. Cnidarian toxins can be deadly to humans as well as to prey organisms; in Australia, twice as many people die each year from stings by box jellyfish as from shark attacks (**Figure 32.24**). ✔**You should be able to indicate the origin of cnidocytes on Figure 32.9.** Besides capturing prey actively, most species of coral and many anemones host photosynthetic dinoflagellates. The relationship is mutually beneficial, because the protists supply the cnidarian host with food in exchange for protection.

Movement Both polyps and medusae have simple, muscle-like tissue derived from ectoderm or endoderm, or in some cases mesoderm. In polyps, the gut cavity acts as a hydrostatic skeleton that works in conjunction with the muscle-like cells to contract or extend the body. Many polyps can also creep along a substrate, using muscle cells at their base. In medusae, the bottom of the bell structure is ringed with muscle-like cells. When these cells contract rhythmically, the bell pulses and the medusa moves by jet propulsion—meaning a forcible flow of water in the opposite direction of movement. Cnidarian larvae swim by means of cilia.

Chironex fleckeri

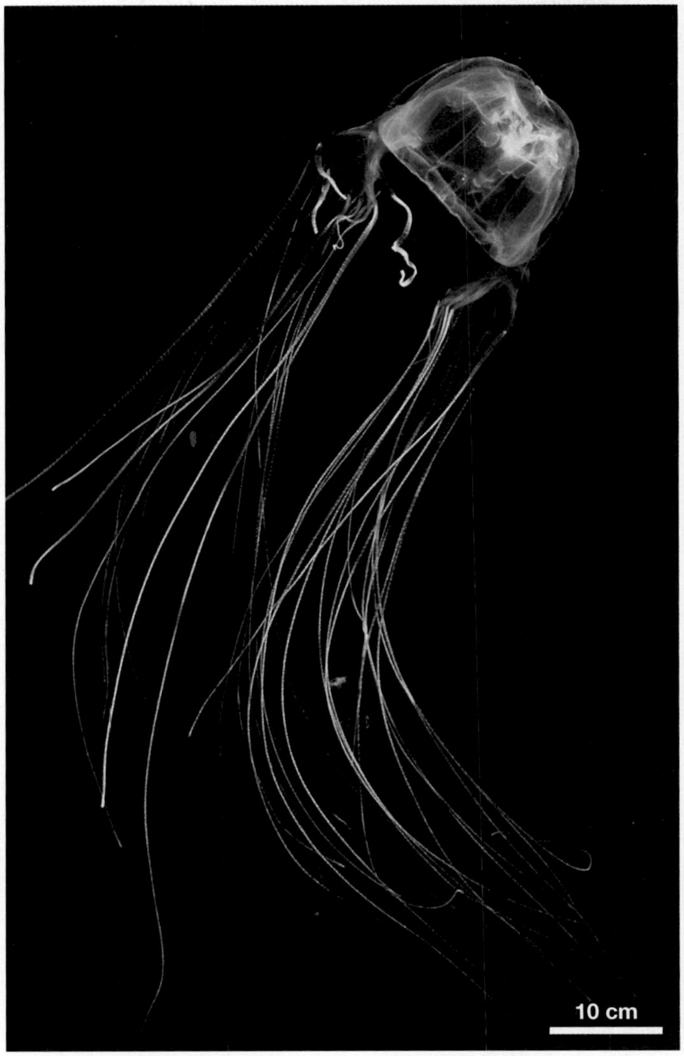

FIGURE 32.24 Some Jellyish Have Long Tentacles, Packed with Cnidocysts That Sting Prey.

Reproduction Polyps may produce new individuals asexually by (**1**) budding, in which a new organism grows out from the body wall of an existing individual; (**2**) fission, in which an existing adult splits lengthwise to form two individuals; or (**3**) fragmentation, in which parts of an adult regenerate missing pieces to form a complete individual. During sexual reproduction, gametes are usually released from the mouth of a polyp or medusa and fertilization takes place in the open water. Eggs hatch into larvae that become part of the plankton before settling and developing into a polyp.

Ctenophora (Comb Jellies)

Ctenophores are transparent, ciliated, gelatinous diploblasts that live in marine habitats (**Figure 32.25**). Although a few species live on the ocean floor, most are planktonic—meaning that they live near the surface. Only about 100 species have been described to date, but some are abundant enough to represent a significant fraction of the total planktonic biomass. Accidental introductions of the ctenophore *Mnemiopsis leidyi*, which preys on fish larvae, have devastated fish production in the Black Sea and Caspian Sea.

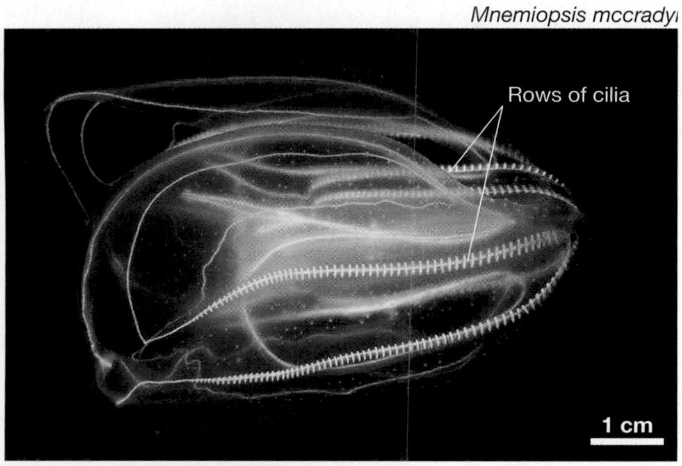

FIGURE 32.25 Ctenophores Are Planktonic Predators.

Feeding Ctenophores are predators. Feeding occurs in several ways, depending on the species. Some comb jellies have long tentacles covered with cells that release an adhesive when they contact prey. These tentacles are periodically wiped across the mouth so that captured prey can be ingested. In other species, prey can stick to mucus on the body and be swept toward the mouth by cilia. Still other species ingest large prey whole.

Movement Adults move via the beating of cilia, which occur in comblike plates. The plates form rows that run the length of the body. Ctenophores are the largest animals known to use cilia for locomotion. ✔You should be able to indicate the origin of swimming via rows of coordinated cilia on Figure 32.9.

Reproduction Most species have both male and female organs and routinely self-fertilize, though fertilization is external. Larvae are free swimming. The few species that live on the ocean floor undergo internal fertilization and brood their embryos until they hatch into larvae.

Acoelomorpha (Acoels)

As their name implies, the acoelomorphs lack a coelom. They are bilaterally symmetric worms that have distinct anterior and posterior ends and are triploblastic. Their nervous system is a nerve net, however, and they are not cephalized. They have a simple cavity where digestion occurs or no digestive system at all. In some species with a digestive cavity, the mouth is the only opening for the ingestion of food and elimination of waste products. Species that lack a digestive cavity absorb nutrients across their body wall. Most acoelomorphs are only a couple of millimeters long and live in mud or sand in marine environments (**Figure 32.26**).

Feeding Acoelomorphs feed on detritus or prey on small animals or protists that live in mud or sand.

Movement Acoelomorphs swim, glide along the surface, or burrow through substrates with the aid of cilia that cover either the entire body or the ventral surface.

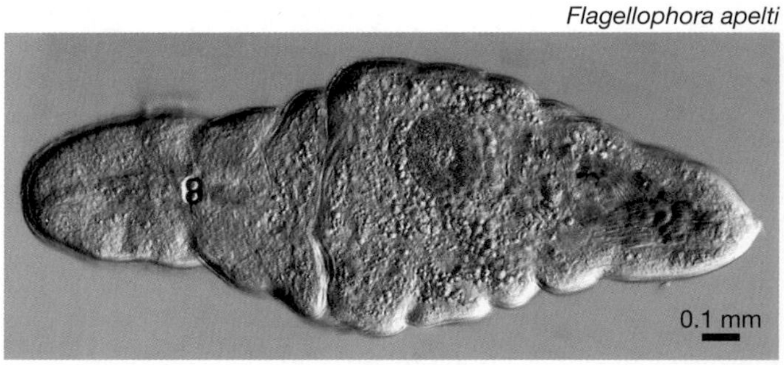

FIGURE 32.26 Acoelomorphs Are Small Worms That Live in Mud or Sand.

Reproduction Adults can reproduce asexually by fission (splitting in two) or by direct growth (budding) of a new individual from the parent's body. Individuals produce both sperm and eggs. Fertilization is internal, and fertilized eggs are laid outside the body.

CHAPTER 32 REVIEW

For media, go to the study area at www.masteringbiology.com

Summary of Key Concepts

Animals are multicellular, heterotrophic eukaryotes that lack cell walls and ingest their prey.

- The animals comprise 30–35 phyla and may number 10 million or more species.
- Biologists study animals because they are key consumers and because humans depend on them for food, materials, transportation, and power.

✔You should be able to predict how the total amount of plant material in a forest would change if all animals were excluded for one year.

Fundamental changes in morphology and development occurred as animals diversified.

- Sponges lack highly organized, complex tissues other than epithelium.
- Most cnidarians and ctenophores have radial symmetry, two embryonic germ layers, and neurons organized into a nerve net.
- Members of the Acoelomorpha have bilateral symmetry and three embryonic tissues but lack a body cavity, or coelom.
- Most animal species have bilateral symmetry, three embryonic tissues, and a coelom—features that gave rise to a "tube-within-a-tube" body plan—and a central nervous system.
- Bilaterians have cephalized bodies—meaning that a distinctive head region contains the mouth, brain, and sensory organs.
- Depending on the species involved, the tube-within-a-tube design is built in one of two fundamental ways—via the protostome or deuterostome patterns of development.

✔You should be able to draw a bilaterian that lacks limbs, and label the gut, outer body wall, muscle layers, head region, mouth, anus, brain, and major nerve tracts.

MB **Web Activity** The Architecture of Animals

Recent phylogenetic analyses have shown that there are four major groups of animals: non-bilaterian lineages, two protostome groups (Lophotrochozoa and Ecdysozoa), and the deuterostomes.

- Choanoflagellates—unicellular or colonial protists that filter feed—are the closest living relative to animals.
- Radial symmetry and diploblasty evolved before bilateral symmetry and mesoderm.
- Although the coelom was an important innovation in animal evolution, as it provided worm-like animals with a hydrostatic skeleton, it was lost or reduced in an array of important lineages.
- Segmented body plans arose independently as animals diversified.

✔You should be able to describe how protostomes and deuterostomes differ.

Within major groups of animals, evolutionary diversification was based on innovative ways of sensing the environment, feeding, and moving.

- Sensory abilities and sensory structures vary among species and correlate with their habitat and feeding method.
- Animals capture food in one of four ways: Suspension feeders filter organic material or small organisms from water; deposit feeders swallow sediments; fluid feeders lap or suck up liquids; food-mass feeders harvest packets of material.
- Even though the types of appendages used in animal locomotion range from simple lobe-like limbs to complex lobster legs, the genes that indicate where appendages develop may be homologous.

✔You should be able to state a hypothesis for why the structure of the eye varies so much among mollusks.

Methods of sexual reproduction vary widely among animal groups, and many species can reproduce asexually. It is common for individuals to undergo metamorphosis during their life cycle.

- Sexual reproduction may involve laying eggs or giving birth to well-developed young.
- In most cases, metamorphosis involves a dramatic change between the larval and adult body form, as well as stark contrasts between larvae and adults in the habitat used and feeding strategy.

✔You should be able to distinguish between a juvenile and larval form in insects and explain why the difference is important.

Questions

✔TEST YOUR KNOWLEDGE

Answers are available in Appendix B

1. What synapomorphy distinguishes animals as a monophyletic group, distinct from choanoflagellates?
   a. multicellularity
   b. movement via a hydrostatic skeleton
   c. growth by molting
   d. ingestive feeding
2. Which of the following patterns in animal evolution is correct?
   a. All triploblasts have a coelom.
   b. All triploblasts evolved from an ancestor that had a coelom.
   c. Sponges have epithelial tissues that line an enclosed, fluid-filled body cavity.
   d. Bilateral symmetry and cephalization evolved once.

3. Which of the following patterns in animal evolution is correct?
   - **a.** Segmentation evolved once.
   - **b.** The coelom was lost or reduced in many lineages.
   - **c.** Sponges lack tissues and are asymmetrical.
   - **d.** Radial symmetry evolved once.
4. Why do some researchers maintain that the limbs of all animals are homologous?
   - **a.** Homologous genes, such as *Dll*, are involved in their development.
   - **b.** Their structure—particularly the number and arrangement of elements inside the limb—is the same.
   - **c.** They all function in the same way—in locomotion.
   - **d.** Animal appendages are too complex to have evolved more than once.
5. In a "tube-within-a-tube" body plan, what is the interior tube?
   - **a.** ectoderm
   - **b.** mesoderm
   - **c.** the coelom
   - **d.** the gut
6. How are choanoflagellates and sponges similar?
   - **a.** Both are multicellular.
   - **b.** Both have a single tissue type: epithelium.
   - **c.** They share distinctive flagellated cells that function in suspension feeding.
   - **d.** Both groups are animals that appear early in the fossil record of the lineage.

✓ TEST YOUR UNDERSTANDING

Answers are available in Appendix B

1. Explain the difference between a diploblast and a triploblast. How was the evolution of mesoderm associated with the evolution of a coelom?
2. Explain how a hydrostatic skeleton works.
3. Why is it significant that animals are heterotrophic *and* multicellular?
4. Compare and contrast the types of mouthparts you would expect to find in herbivorous insect species that (1) suck plant fluids from stems, (2) eat their way through decaying plant material on the forest floor, (3) bite leaves, and (4) suck nectar from flowers.
5. Explain how an animal mother nourishes an embryo in oviparous species versus viviparous species.
6. Why are nerve nets associated with radial symmetry, while a CNS is associated with bilateral symmetry?

✓ APPLYING CONCEPTS TO NEW SITUATIONS

Answers are available in Appendix B

1. Suppose that a gene originally identified in nematodes (roundworms) is found to be homologous with a gene that can cause developmental abnormalities in humans. Would it be possible to study this same gene in fruit flies? Explain.
2. Ticks are arachnids (along with spiders and mites); mosquitoes are insects. Both ticks and mosquitoes are ectoparasites that make their living by extracting blood meals from mammals. Ticks undergo incomplete metamorphosis, while mosquitoes undergo complete metamorphosis. Based on these observations, would you predict ticks or mosquitoes to be the more successful group in terms of number of species, number of individuals, and geographic distribution? Explain why. How could you test your prediction?
3. Radial symmetry evolved in echinoderms (sea stars and sea urchins) from bilaterally symmetric ancestors. Predict whether the echinoderm nervous system is centralized or diffuse. Explain your logic.
4. Why are spiders unusual, in terms of the four general feeding tactics outlined in this chapter?

In numbers of individuals and species richness, protostomes are the most abundant and diverse of all animals—and the arthropods are by far the most abundant and diverse of the protostomes. The arthropods shown here come from all over the world.

Protostome Animals 33

Protostomes include some of the most familiar organisms on Earth. The phylum Arthropoda, for example, includes the insects, chelicerates (spiders and mites), crustaceans (shrimp, lobster, crabs, barnacles), and myriapods (millipedes, centipedes); the Mollusca comprises the snails, clams, chitons, and cephalopods (octopuses and squid).

Some protostome phyla are also particularly species-rich. Over 93,000 mollusks have been named thus far, and biologists estimate that there are over 10 million arthropod species living today—most of them unnamed and undescribed. Among insects alone, about 925,000 species have been formally identified to date.

Certain protostome lineages are also extremely abundant. A typical acre of pasture in England is home to almost 18 *million* individual beetles; the world population of ants is estimated to be 1 million billion individuals.

It is important to recognize, though, that not all protostome lineages have been as spectacularly successful as the Arthropoda and Mollusca. There are over 20 phyla of protostomes; some qualify as particularly obscure lineages on the tree of life. There are just 150 species in the phylum Kinorhyncha (mud dragons); the 80 species in the Gnathostomulida (pronounced *nath-oh-stoh-MEW-lida*) and the 450 species in the Gastrotricha (*GAS-troh-trika*) average less than 2 mm long.

Figure 33.1 on page 624 is a pie chart showing the relative numbers of species in various animal phyla. Note that the wedges labeled Chordata, Cnidaria, Echinodermata, Porifera, and "Other invertebrates" are small. Almost all animal species are protostomes.

As a group, protostomes live in just about every habitat that you might explore. They also include some of the most important model organisms in all of biological science: the fruit fly *Drosophila melanogaster* and the roundworm *Caenorhabditis elegans*. If you walk into a biology building on any university campus around the world, you are almost certain to find at least one lab where fruit flies or roundworms are being studied. If one of biology's most fundamental goals is to understand the diversity of life on Earth, then protostomes—particularly the arthropods and mollusks—demand our attention.

KEY CONCEPTS

- Molecular phylogenies support the hypothesis that protostomes are a monophyletic group divided into two major subgroups: the Lophotrochozoa and the Ecdysozoa.
- Although the members of many protostome phyla have limbless, wormlike bodies and live in marine sediments, the most diverse and species-rich lineages—Mollusca and Arthropoda—have bodies with distinctive, complex features. Mollusks and arthropods inhabit a wide range of environments.
- Key events triggered the diversification of protostomes, including several lineages making the water-to-land transition, a diversification in appendages and mouthparts, and the evolution of metamorphosis in both marine and terrestrial forms.

✔ When you see this checkmark, stop and test yourself. Answers are available in Appendix B.

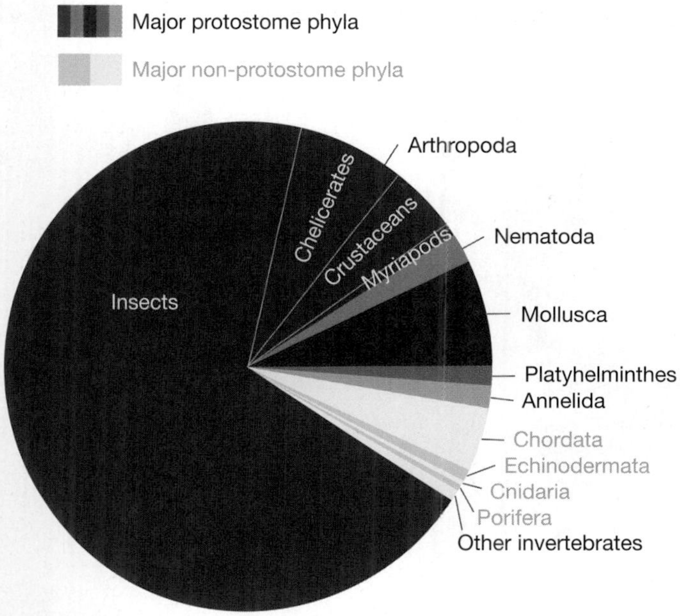

FIGURE 33.1 The Relative Abundances of Animal Lineages. Most animals are protostomes. About 70 percent of all known species of animals on Earth are insects, most of them beetles. (Humans and other vertebrates are deuterostomes, in the phylum Chordata.) This chapter focuses on the most species-rich lineages of protostomes.

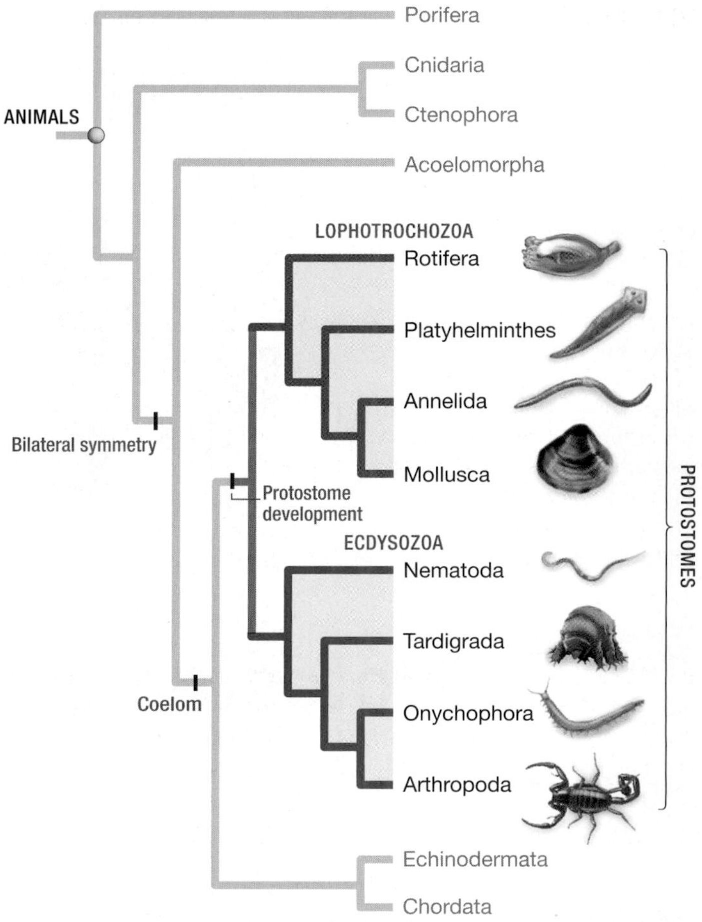

FIGURE 33.2 Protostomes Are a Monophyletic Group Comprising Two Major Lineages. There are 22 phyla of protostomes, but the eight major phyla shown account for about 99.5 percent of the known species.

33.1 An Overview of Protostome Evolution

There are two major groups of bilaterally symmetric, triploblastic, coelomate animals: the protostomes and deuterostomes. You might recall from Chapter 32 that protostome and deuterostome embryos develop in dramatically different ways. The protostome and deuterostome patterns of early development represent distinctive pathways for building a bilaterally symmetric body with a coelom.

When gastrulation occurs in protostomes, the initial pore that forms in the embryo becomes the mouth. If a **coelom** (a body cavity) forms later in development, it forms from openings that arise within blocks of mesodermal tissue. Phylogenetic studies have long supported the hypothesis that protostomes are a monophyletic group, meaning the protostome developmental sequence arose just once.

More recent analyses of DNA sequence data indicated that two major subgroups exist within the protostomes (**Figure 33.2**). The two monophyletic groups of protostomes are called the Lophotrochozoa and Ecdysozoa. (The lineages are pronounced *low-foe-tro-ko-ZOH-ah* and *eck-die-so-ZOH-ah*.)

What Is a Lophotrochozoan?

The 13 phyla of lophotrochozoans include the mollusks, annelids, and flatworms (Platyhelminthes; pronounced *plah-tee-hell-MIN-theez*). The group's name was inspired by morphological traits that are found in some, but not all, of the phyla in the lineage:

1. a feeding structure called a lophophore, which is found in three phyla; and
2. a type of larva called a trochophore, which is common to many of the phyla in the lineage.

As **Figure 33.3a** shows, a **lophophore** (literally, "tuft-bearer") is a specialized structure that rings the mouth and functions in suspension feeding. Lophophores are found in bryozoans (moss animals), brachiopods (lamp shells), and phoronids (horseshoe worms).

Trochophores are a type of larvae common to marine mollusks, annelids that live in the ocean, and several other phyla in the Lophotrochozoa. As **Figure 33.3b** shows, a **trochophore** (literally, "wheel-bearer") larva has a ring of cilia around its middle. These cilia allow swimming and in some species sweep food particles into the mouth. Recent analyses are suggesting that the trochophore originated early in the evolution of Lophotrochozoans, though different larval types evolved later in some groups.

Neither lophophores nor trochophores qualify as a synapomorphy in Lophotrochozoa, however. To date, biologists have yet

(a) **Lophophores** function in suspension feeding in adults.

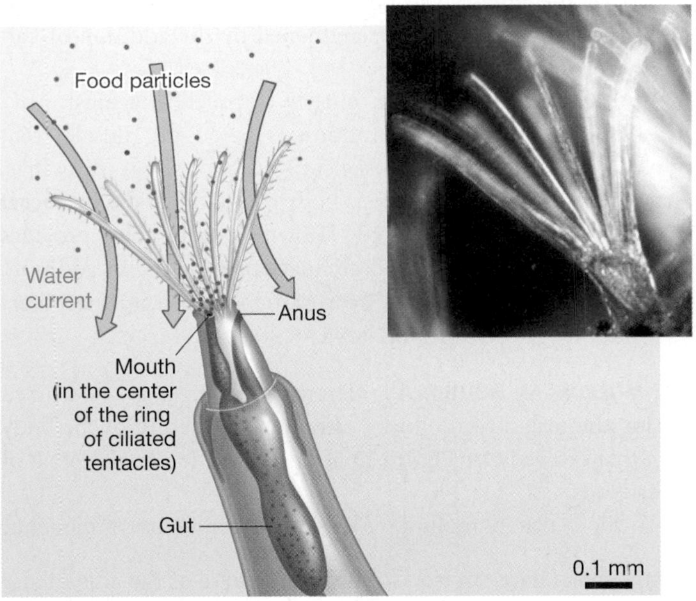

(b) **Trochophore** larvae swim and may feed.

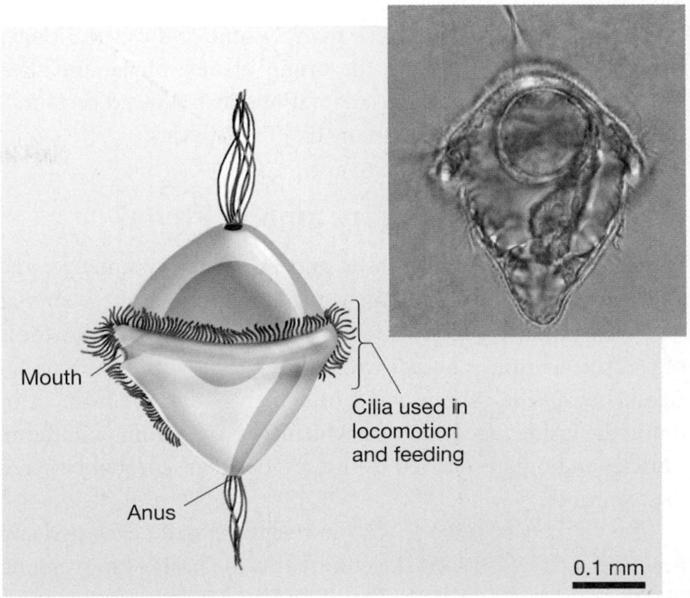

FIGURE 33.3 Lophotrochozoans Have Distinctive Traits. (a) Three phyla of lophotrochozoans have the feeding structure called a lophophore. **(b)** Many phyla of lophotrochozoans have the type of larva called a trochophore. In some cases the trochophore is a stage during larval development.

✔**EXERCISE** The word root *lopho*– means tuft; *trocho*– means wheel. Label the tuft and wheel in this figure.

to discover a morphological synapomorphy that distinguishes species in this lineage from all other protostomes.

In contrast, Ecdysozoa are defined by a clear synapomorphy: their method of growth. Like other animals, lophotrochozoans grow continuously and incrementally (**Figure 33.4a**). But ecdysozoans do not.

What Is an Ecdysozoan?

The most prominent of the seven ecdysozoan phyla are the roundworms (Nematoda) and arthropods (Arthropoda). All ecdysozoans grow by **molting**—that is, by shedding an exoskeleton or external covering. The Greek root *ecdysis,* which means "to slip out or escape," is appropriate because during a molt, an individual sheds its outer layer, or **cuticle**—called an **exoskeleton** if it is hard—and slips out of it (**Figure 33.4b**). Once the old covering is gone, the body expands and a larger cuticle or exoskeleton then forms. As ecdysozoans grow and mature, they undergo a succession of molts.

Growth by molting was required in ecdysozoans once they had evolved tough outer cuticles or thick exoskeletons. A stiff body covering was advantageous because it provides an effective structure for muscle attachment and affords protection. To support this claim, biologists point to the secretive behavior of ecdysozoans while individuals are molting. When crabs and other crustaceans have shed an old exoskeleton, their new exoskeleton takes several hours to harden. During this interval, individuals hide and do not feed or move about. In addition, experiments have shown that it is much easier for predators to attack and subdue individuals that are not protected by an exoskeleton during the intermolt period.

(a) Lophotrochozoans grow incrementally.

(b) Ecdysozoans grow by molting.

FIGURE 33.4 Lophotrochozoans and Ecdysozoans Differ in Their Mechanism of Growth. (a) Lophotrochozoans do not molt. The growth bands on this clam show periods of slow and rapid incremental growth. **(b)** Once an ecdysozoan such as this cicada has left its old exoskeleton, hours or days pass before the new exoskeleton is hardened. During this time, the individual is highly susceptible to predation or injury.

33.2 Themes in the Diversification of Protostomes

Protostomes have diverged into more than 22 phyla that are recognized by distinctive body plans or specialized mouthparts used in feeding. Some of these phyla then split into millions of different species. What drove all this diversification?

To answer this question, let's analyze some of the evolutionary innovations that resulted in the origin of new phyla, and then follow up by delving into the adaptations that allowed certain of these phyla to diversify into many different species.

How Do Body Plans Vary among Phyla?

All protostomes are triploblastic and bilaterally symmetric, and all protostomes undergo embryonic development in a similar way. You might recall from Chapter 32 that most protostome phyla have wormlike bodies with a basic tube-within-a-tube design. The outside tube is the skin, which is derived from ectoderm; the inside tube is the gut, which is derived from endoderm. Muscles and organs derived from mesoderm are located between the two tubes.

In the wormlike phyla, the coelom is well developed and functions as a hydrostatic skeleton that is the basis of movement. But the coelom is absent in flatworms, and in the most species-rich and morphologically complex protostome phyla—the Arthropoda and the Mollusca (snails, clams, squid)—it is drastically reduced.

A fully functioning coelom has two roles: providing space for fluids to circulate among organs, and providing a hydrostatic skeleton for movement. In arthropods and mollusks, other structures fulfill these functions. Species in these phyla don't have a fully functioning coelom for a simple reason: They don't need one. Let's take a closer look.

THE ARTHROPOD BODY PLAN Arthropods have segmented bodies that are organized into prominent regions called tagmata. In the grasshopper shown in **Figure 33.5a**, these regions are called the head, thorax, and abdomen. In addition, arthropods are distinguished by their jointed limbs and an exoskeleton made primarily of the polysaccharide chitin (see Chapter 5). In crustaceans, the exoskeleton is strengthened by the addition of calcium carbonate ($CaCO_3$).

Instead of being based on muscle contraction against a hydroskeleton, arthropod locomotion is based on muscles that apply force against the exoskeleton to move legs or wings. Arthropods also have a spacious body cavity called the **hemocoel** ("blood-hollow"; pronounced HEE-mah-seal) that provides space for internal organs and circulation of fluids. In addition, the hemocoel functions as a hydrostatic skeleton in caterpillars and other types of arthropod larva.

THE MOLLUSCAN BODY PLAN Dramatic changes also occurred during the origin of mollusks, although the nature of the body plan that evolved (see **Figure 33.5b**) is very different from that of arthropods.

The body plan of mollusks is based on three major components:

1. the **foot**, a large muscle located at the base of the animal and usually used in movement;
2. the **visceral mass**, the region containing most of the main internal organs; and
3. the **mantle**, an outgrowth of the body wall that covers the visceral mass, forming an enclosure called the mantle cavity.

Although the structure of the foot, visceral mass, and mantle vary widely among mollusk species, in all mollusks the visceral mass provides space for organs and the circulation of fluids. In some species the fluid enclosed by the visceral mass also functions in movement as a hydrostatic skeleton.

In many species the mantle secretes a shell made of calcium carbonate. Some mollusk species have a single shell; others have two, eight, or none at all.

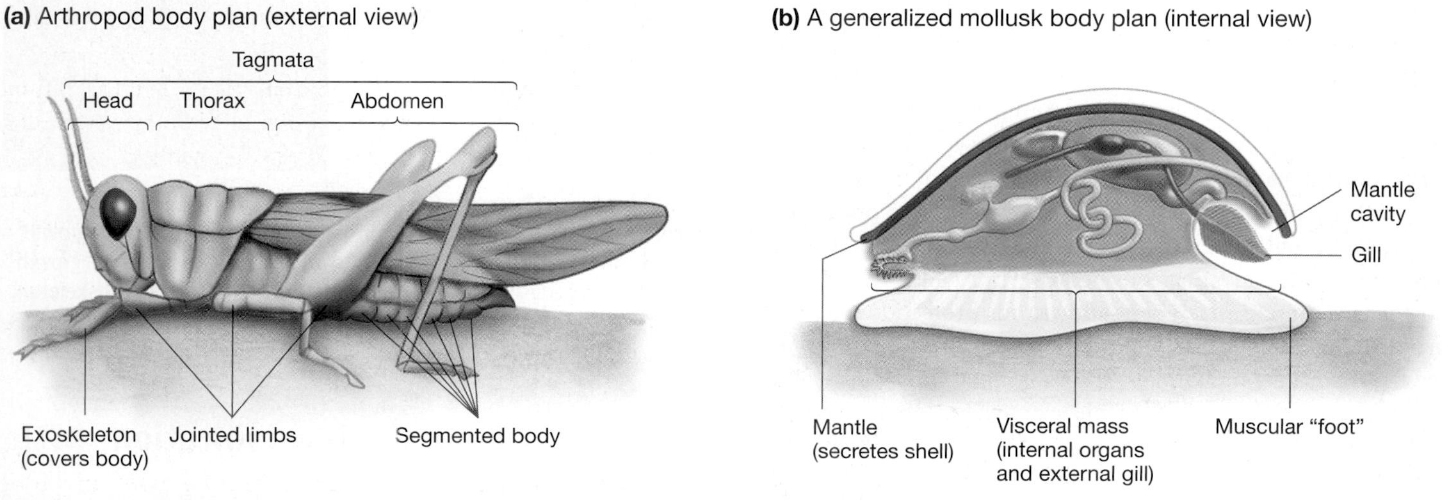

FIGURE 33.5 Arthropods and Mollusks Have Specialized Body Plans. (a) Arthropods have segmented bodies and jointed limbs, which enable these animals to move despite their hard outer covering, the exoskeleton. **(b)** The mollusk body plan is based on a foot, a visceral mass, and a mantle. Gills are located inside a cavity created by the mantle.

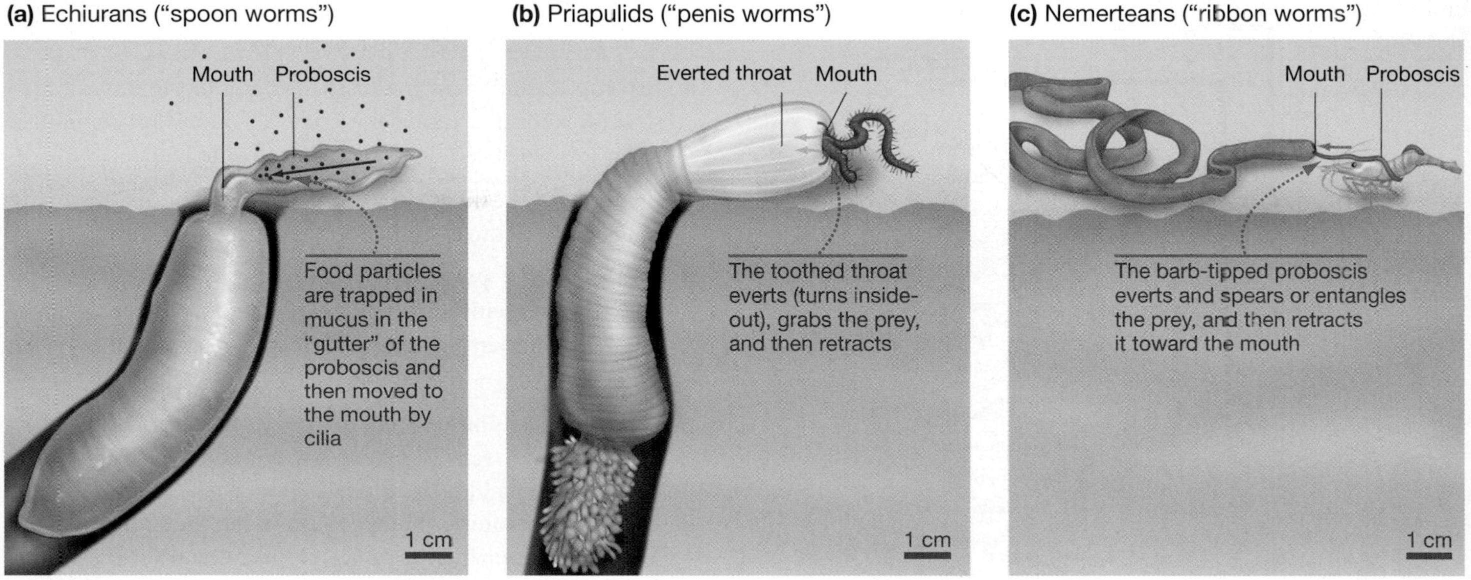
(a) Echiurans ("spoon worms")
Mouth
Proboscis
Food particles are trapped in mucus in the "gutter" of the proboscis and then moved to the mouth by cilia
1 cm
(b) Priapulids ("penis worms")
Everted throat
Mouth
The toothed throat everts (turns inside-out), grabs the prey, and then retracts
1 cm
(c) Nemerteans ("ribbon worms")
Mouth
Proboscis
The barb-tipped proboscis everts and spears or entangles the prey, and then retracts it toward the mouth
1 cm

FIGURE 33.6 Some Protostome Phyla with Wormlike Bodies Have Specialized Mouthparts.

✔**QUESTION** It is common to find species from an array of wormlike phyla in the same habitat. Suggest a hypothesis to explain why this is possible.

VARIATION AMONG BODY PLANS OF THE WORMLIKE PHYLA New types of body plans distinguish most of the phyla illustrated in Figure 33.2. In contrast, the general body plan is similar in most of the protostome lineages that have wormlike bodies. In many cases the wormlike lineages are distinguished by specialized mouthparts or feeding structures. To drive this point home, consider the three phyla illustrated in **Figure 33.6**.

- *Annelids (segmented worms)* Annelids include a group called the echiurans or spoon worms, which burrow into marine mud. Echiurans deposit feed using an extended structure called a **proboscis**, which forms a gutter leading to the mouth (**Figure 33.6a**). Cells in the gutter secrete mucus, which is sticky enough to capture pieces of detritus. The combination of mucus and detritus is then swept into the mouth by cilia on cells in the gutter.
- *Priapulids (penis worms)* Priapulids also burrow into the substrate, but act as sit-and-wait predators. When a polychaete (Annelida) or other prey item approaches, the priapulid everts its toothed, cuticle-lined throat—meaning that it turns its throat inside out—grabs the prey item, and retracts the structure to take in the food (**Figure 33.6b**).
- *Nemerteans (ribbon worms)* Nemerteans are active predators that move around the ocean floor in search of food. Instead of everting their throat to capture prey, they have a proboscis that can extend or retract (**Figure 33.6c**). Nemerteans spear small animals with their proboscis or wrap the extended proboscis around prey. The food is then pulled into the mouth.

Once a specific body plan had evolved, subsequent diversification was largely driven by adaptations that allowed protostomes to live on land or feed, move, or reproduce in novel ways. Recall that an **adaptation** is a trait that increases the fitness (reproductive success) of individuals relative to individuals without the trait.

Before going on to review the adaptations that drove diversification within certain lineages, go to the study area at *www.masteringbiology.com* to get an overview of variation among protostomes.

(MB) **Web Activity** Protostome Diversity

The Water-to-Land Transition

Protostomes are the most abundant animals in the surface waters and substrates of many marine and freshwater environments. But protostomes are common in virtually every terrestrial setting as well. Like the land plants and fungi, protostomes made the transition from aquatic to terrestrial environments.

To help put this achievement in perspective, recall from Chapter 30 that green plants made the move from freshwater to land just once. Data in Chapter 31 show that it is not yet clear whether fungi moved from aquatic habitats to land once or several times. Chapter 34 will show that only one lineage among deuterostomes moved onto land. But a water-to-land transition occurred multiple times as protostomes diversified.

EVIDENCE FOR MULTIPLE TRANSITIONS The evidence for multiple water-to-land transitions in protostomes is based on phylogenetic analyses, which support the hypothesis that the ancestors of the terrestrial lineages in each major subgroup of protostomes were aquatic.

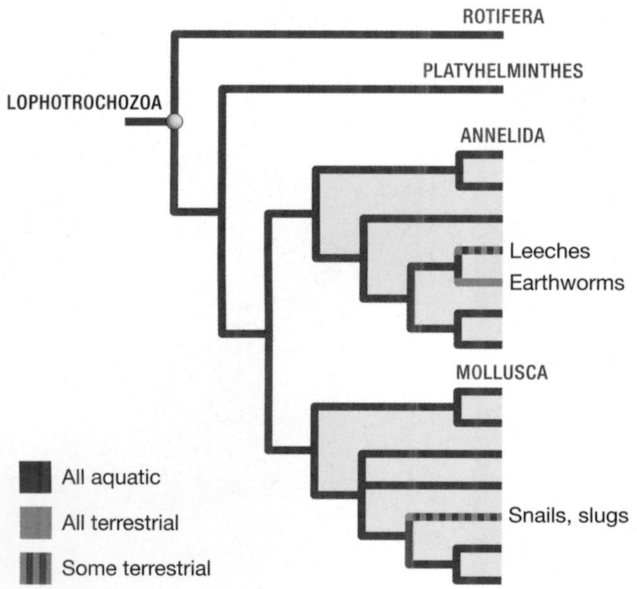

FIGURE 33.7 Evidence for a Water-to-Land Transition in the Lophotrochozoa. The appearance of isolated terrestrial groups among marine groups also occurs in the Ecdysozoa.

Figure 33.7, for example, shows a recent phylogeny of the annelids and mollusks within the Lophotrochozoa. Because almost all of the species in these two groups are aquatic, and because the outgroup species, the rotifers and platyhelminths are also aquatic, it is logical to infer that the ability to live on land evolved in the lineages leading to the snails/slugs and the earthworms. The alternative hypothesis—that terrestrial living was the ancestral state—would require that aquatic living arose fourteen times independently. Similar analyses show that terrestrial living evolved several times independently in the Ecdysozoa.

ADAPTATIONS TO TERRESTRIAL ENVIRONMENTS Why were so many different protostome groups able to move from water to land independently? The short answer is that it was easier for protostomes to accomplish this than for plants to do so. For example, land plants had to evolve roots and vascular tissue to transport water and support their stems. Similarly, the first terrestrial animals had to be able to support their body weight on land and move about. But the protostome groups that made the water-to-land transition already had hydrostatic skeletons, exoskeletons, appendages, or other adaptations for support and locomotion that happened to work on land as well as water.

To make the transition to land, new adaptations allowed protostomes to (1) exchange gases and (2) avoid drying out. As Chapter 44 will detail, land animals exchange gases with the atmosphere readily as long as a large, moist surface area is exposed to the air. The bigger challenge is to prevent the gas-exchange surface and other parts of the body from drying out. How do terrestrial protostomes solve this problem?

- Roundworms and earthworms live in humid soils or other moist environments and exchange gases across their body surface. They have a high surface-area-to-volume ratio, which increases the efficiency of gas exchange.
- Arthropods and many mollusks have gills or other respiratory structures located inside the body. This location minimized water loss when certain groups moved onto land.
- In mollusks, the mantle cavity that encloses the gills of aquatic snails evolved into the lung found in terrestrial snails.
- Insects evolved a waxy layer to minimize water loss from the body surface, with openings to respiratory passages that can be closed if the environment dries.

Unlike land plants and fungi, land animals can also move to moister habitats if the area they are in gets too dry.

Water-to-land transitions are important because they open up entirely new habitats and new types of resources to exploit. Based on this reasoning, biologists claim that the ability to live in terrestrial environments was a key event in the diversification of several protostome phyla.

Adaptations for Feeding

Protostomes include suspension, deposit, liquid, and food-mass feeders. Besides exploiting detritus, they prey on or parasitize plants, algae, or other animals. Exploiting a diversity of foods is possible because protostomes have such a wide variety of mouthparts for capturing and processing food.

Figure 33.6 highlighted mouthparts that distinguish some of the wormlike phyla. But within phyla, arthropods take the prize for mouthpart diversity. The mouthparts observed in this phylum vary in structure from tubes to pincers and allow the various species to pierce, suck, grind, bite, mop, chew, engulf, cut, or mash (**Figure 33.8**). All arthropods have the same basic body plan, but their mouthparts and food sources are highly diverse.

In many species of arthropods, the jointed limbs also play a key role in getting food. The crustaceans called krill and barnacles use their legs to sweep food toward their mouths as they suspension feed. Certain insects, crustaceans, spiders, and mollusks use their appendages to capture prey or hold food as it is being chewed or bitten by the mouthparts.

In most cases, larval and adult forms of the same species exploit different food sources. Metamorphosis is extremely common in protostomes and usually results in larvae and adults that live in different habitats and have different overall morphology and mouthparts.

Adaptations for Moving

Protostomes move in virtually every way known among animals. Aquatic larvae move with the aid of cilia, and wormlike protostomes that lack limbs move with the aid of a coelom that functions as a hydrostatic skeleton. Caterpillars, grubs, maggots, and other types of insect larvae with wormlike bodies have an enclosed, fluid-filled body cavity—the hemocoel—that also functions as a hydrostatic skeleton.

(a) Leaf-cutter ants cut leaves. (b) Sheep ticks pierce skin. (c) Nut weevils bore into hard fruit.

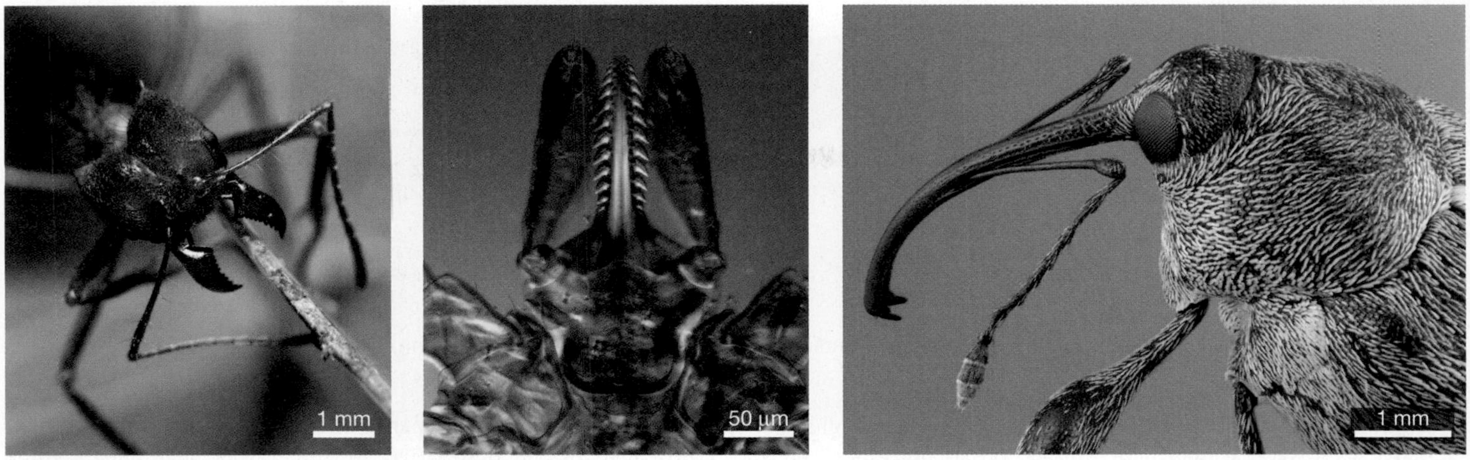

FIGURE 33.8 Arthropod Feeding Structures Are Diverse. Even though their mouthparts and food sources are diverse, all of these arthropods have segmented bodies that are organized into tagmata, with an exoskeleton and jointed appendages.

As arthropods diversified, however, a number of important evolutionary innovations allowed them to move in unique ways:

- ***Jointed limbs in arthropods*** Jointed limbs made rapid, precise walking, running, and jumping movements possible (**Figure 33.9a**). The jointed limb is a major reason arthropods have been so spectacularly successful in terms of species diversity, abundance of individuals, geographic range, and duration in the fossil record.
- ***The insect wing*** The insect wing is one of the most important adaptations in the history of life (**Figure 33.9b**). About two-thirds of the multicellular species living today are winged insects. In many cases, larvae lack wings and crawl or swim—only adults fly. According to data in the fossil record, insects were the first organisms that had wings and could fly.

  Like most insects today, the earliest insects had two pairs of wings. In most four-winged insects living today, however, the four wings function as two. Beetles fly with only their hindwings; butterflies, moths, bees, and wasps have hooked structures that make their two pairs of wings move together.

  Flies, in contrast, have a single pair of large wings that power flight. They also have a pair of small, winglike structures called halteres that provide stability during flight.
- ***The mollusk foot*** Snails and chitons are mollusks that have a muscular foot at the base of the body. Waves of muscle contractions sweep up or down the length of the large, muscular foot, allowing individuals to crawl along a surface (**Figure 33.9c**). Cells in the foot secrete a layer of mucus, which reduces friction and increases the efficiency of movement.

(a) Walking, running, and jumping (b) Flying (c) Gliding and crawling

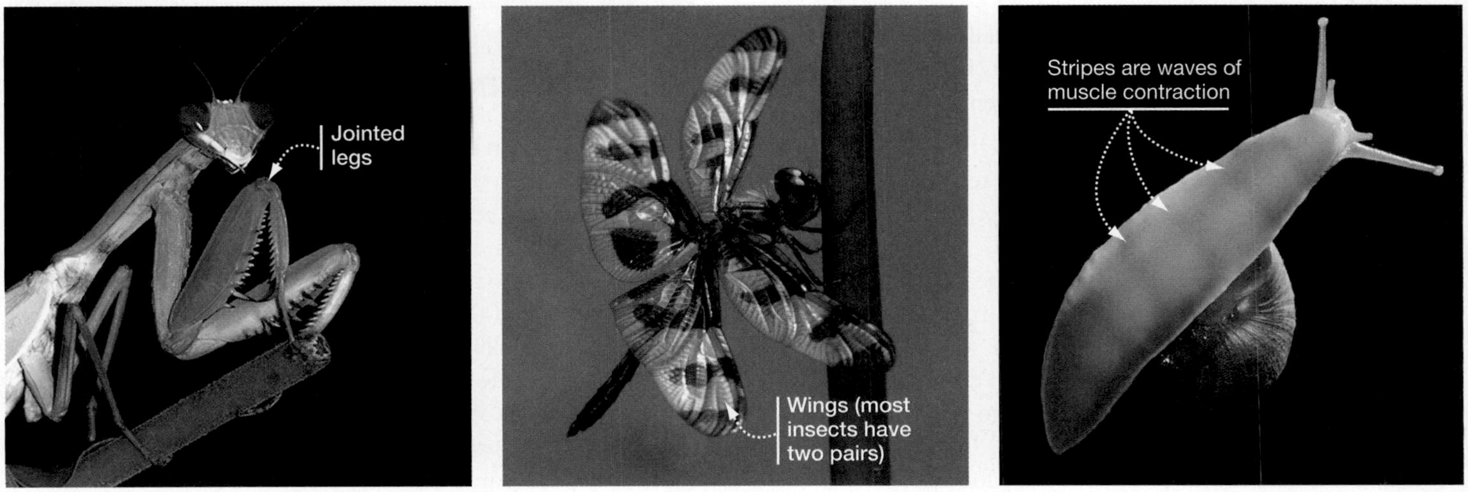

FIGURE 33.9 Protostome Locomotion Is Diverse. The evolution of **(a)** jointed limbs and **(b)** wings were key innovations in arthropod movement. **(c)** A wave of muscle contractions, up or down the length of the foot, allows mollusks like this snail to creep along a substrate.

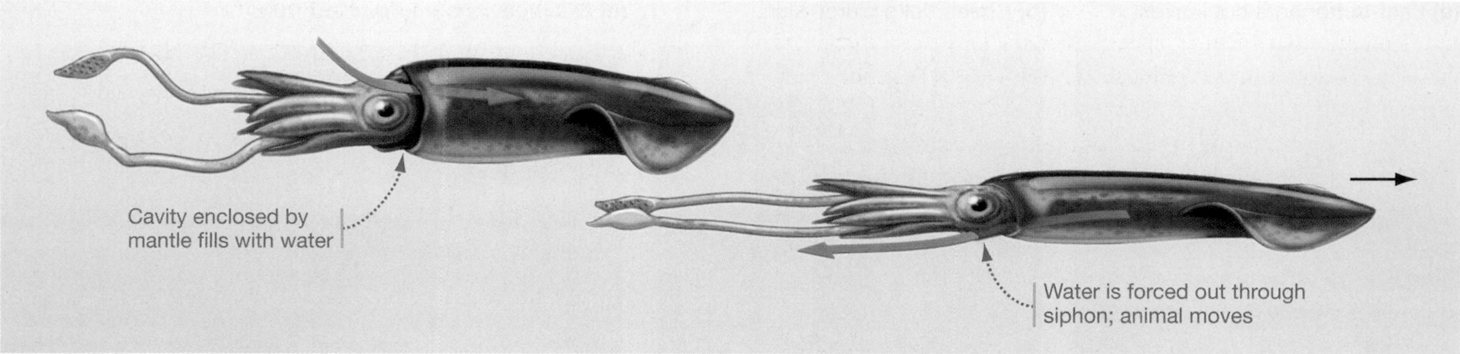
Cavity enclosed by mantle fills with water
Water is forced out through siphon; animal moves

FIGURE 33.10 Jet Propulsion in Mollusks. In jet propulsion, muscular contractions force the water out through a movable siphon.

- *Jet propulsion* Cephalopods (squid and octopus) are mollusks that have a mantle lined with muscle. When the cavity surrounded by the mantle fills with water and the mantle muscles contract, a stream of water is forced out of a tube called a **siphon**. The force of the water propels the squid (**Figure 33.10**). This mechanism, jet propulsion, evolved in squid long before human engineers thought of using the same principle to power aircraft.

Adaptations in Reproduction

When it comes to variation in reproduction and life cycles, protostomes do it all.

- Asexual reproduction by splitting the body lengthwise or by fragmenting the body is common in many of the wormlike phyla.
- Many crustacean and insect species reproduce asexually via **parthenogenesis** ("virgin-origin")—the production of unfertilized eggs, by mitosis, that develop into offspring.
- Sexual reproduction is often based on external fertilization in clams, bryozoans, brachiopods, and other groups.
- Sexual reproduction often begins with copulation and internal fertilization in groups that are capable of movement—such as crustaceans, snails, and insects—because males and females can meet.
- Females of ovoviviparous insects and snails bear live young. They do so by retaining fully formed eggs inside their bodies and then nourishing them via the nutrient-rich yolk that is inside the egg (see Chapter 32).

Although asexual reproduction is common, sexual reproduction is the predominant mode of producing offspring in most protostome groups.

Metamorphosis is another important aspect of reproduction in protostomes. In marine species, metamorphosis is hypothesized to be an adaptation that allows larvae to disperse to new habitats by floating or swimming in the plankton, where they can feed on food sources that are unavailable to adults. Metamorphosis is also common in insects, where it is hypothesized to be an adaptation that reduces competition for food between larvae and adults. The pupal stage that occurs in insects—when the larval body is remodeled into an adult (see Chapter 32)—does not occur in marine species, however.

In terrestrial environments, though, a more critical adaptation was an egg that resists drying. Insect eggs have a thick membrane that keeps moisture in, and the eggs of slugs and snails have a thin calcium carbonate shell that helps retain water. Desiccation-resistant eggs evolved repeatedly in populations that made the transition to life on land.

CHECK YOUR UNDERSTANDING

If you understand that . . .

- In protostomes with wormlike bodies, phyla vary in mouthpart structure and mode of feeding.
- Arthropods and mollusks have specialized body plans.
- A water-to-land transition occurred in several lineages independently.

You should be able to . . .

1. Describe the major features of the arthropod body and explain the advantage of a jointed limb versus an unjointed limb.
2. Name three problems that arise during a water-to-land transition in animals.

Answers are available in Appendix B.

33.3 Key Lineages: Lophotrochozoans

The lineages within Lophotrochozoa are united by shared mechanisms of early development and growth, but they are highly diverse in terms of their resulting morphology. To drive this point home, this section details key points about the biology of four key lophotrochozoan phyla: (1) Rotifera, (2) Platyhelminthes, (3) Annelida, and (4) Mollusca.

Because the Mollusca is so species-rich and diverse, the detailed descriptions provided later in this section consider the four most important lineages of mollusks separately: (1) **bivalves** (clams and mussels), (2) **gastropods** (slugs and snails), (3), **chitons**, and (4) **cephalopods** (squid and octopuses).

Unlike the lophotrochozoans as a whole, the mollusks are identified by distinctive characteristics. In addition to their lophotrochozoan mode of growth and development, mollusks have a specialized body plan based on a muscular foot, a visceral mass, and a mantle that in most species secretes a calcium carbonate shell. The coelom is much reduced in mollusks and functions only in reproduction and excretion of wastes. ✔**You should be able to indicate the origin of the molluscan body plan on Figure 33.2.**

A great deal of diversification occurred once the basic molluskan body plan had evolved. Although most live in marine environments, there are some terrestrial and freshwater forms. Most bivalves are suspension feeders; the other three groups of mollusks are herbivores or predators.

To review the diversity across and within lophotrochozoan phyla efficiently, the remainder of the section is organized as follows:

- Lophotrochozoans > Rotifera (Rotifers)
- Lophotrochozoans > Platyhelminthes (Flatworms)
- Lophotrochozoans > Annelida (Segmented Worms)
- Lophotrochozoans > Mollusca > Bivalvia (Clams, Mussels, Scallops, Oysters)
- Lophotrochozoans > Mollusca > Gastropoda (Snails, Slugs, Nudibranchs)
- Lophotrochozoans > Mollusca > Polyplacophora (Chitons)
- Lophotrochozoans > Mollusca > Cephalopoda (Nautilus, Cuttlefish, Squid, Octopuses)

Lophotrochozoans > Rotifera (Rotifers)

The 1800 rotifer species that have been identified thus far live in damp soils as well as marine and freshwater environments. They are important components of the plankton in freshwater and in brackish waters, where rivers flow into the ocean and water is slightly salty. Rotifers have a coelom, and most are less than 1 mm long. Although rotifers do not have a lophophore or a trochophore larval stage, their mode of growth and extensive similarities in DNA sequence identify them as a member of the lophotrochozoan lineage.

Feeding Rotifers have a cluster of cilia at their anterior end called a **corona** (**Figure 33.11**). In many species, the beating of the cilia in the corona makes suspension feeding possible by creating a current that sweeps microscopic food particles into the mouth. The corona is the signature morphological feature of this group. ✔**You should be able to indicate the origin of the corona on Figure 33.2.**

Philodina roseola

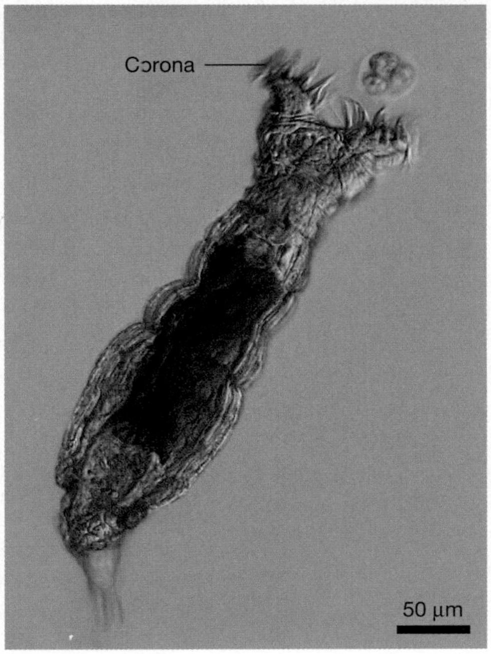

FIGURE 33.11 Rotifers Are Tiny, Aquatic Suspension Feeders.

Movement Although a few species of rotifers are sessile, most swim via the beating of cilia in the corona.

Reproduction Females produce unfertilized eggs by mitosis; the eggs then hatch into new, asexually produced individuals. Recall that the production of offspring via unfertilized eggs is termed parthenogenesis. In a group of rotifers called the bdelloids, only females have been found and all reproduction appears to be by parthenogenesis. Both sexual reproduction and asexual reproduction are observed in most rotifer species, however. Development is direct, meaning that fertilized eggs hatch and grow into adults without going through metamorphosis.

Lophotrochozoans > Platyhelminthes (Flatworms)

The flatworms are a large and diverse phylum. More than 400,000 species have been described in four major lineages within the phylum: (1) the free-living species called Turbellaria (**Figure 33.12a**, on page 632), (2) the endoparasitic tapeworms called Cestoda (**Figure 33.12b**), (3) the endoparasitic or ectoparasitic flukes, called Trematoda (**Figure 33.12c**), and (4) ectoparasites called Monogenea. Although a few turbellarian species are terrestrial, most live on the substrates of freshwater or marine environments. Tapeworms and other cestodes parasitize fish, mammals, or other vertebrates. Flukes parasitize vertebrates or mollusks; most monogeneans parasitize fish. In humans, a fluke is responsible for schistosomiasis—a serious public health issue in many developing nations.

(Continued on next page)

(a) Turbellarians are free living. *Pseudoceros ferrugineus*

(b) Cestodes are endoparasitic. *Taenia species*

(c) Trematodes are endoparasitic. *Dicrocoelium dendriticum*

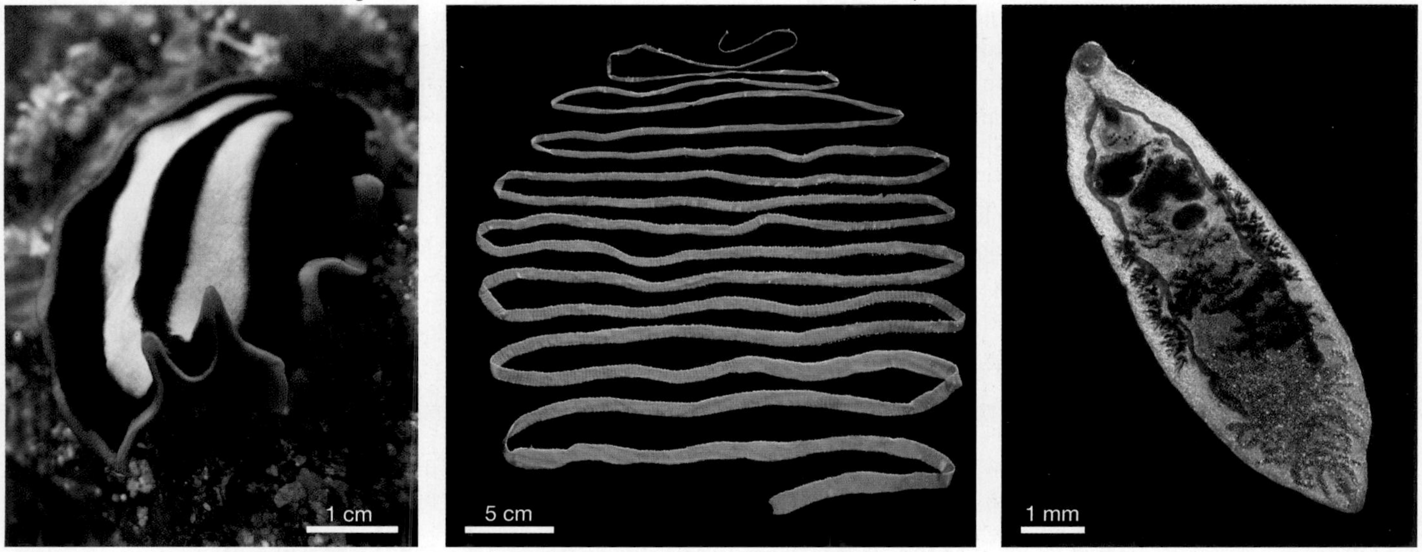

FIGURE 33.12 Flatworms Have Simple, Flattened Bodies.

Flatworms are named for the broad, flattened shape of their bodies. (The Greek roots *platy* and *helminth* mean "flat-worm.") Species in the Platyhelminthes are unsegmented and lack a coelom. They also lack structures that are specialized for gas exchange—taking in oxygen and ridding the body of carbon dioxide. Instead, gas exchange occurs across the body wall. Further, they do not have blood vessels for circulating oxygen and nutrients to their cells. Based on these observations, biologists interpret the flattened bodies of these animals as an adaptation that gives flatworms an extremely high surface-area-to-volume ratio. Because they have so much surface area, a large amount of gas exchange can occur directly across their body wall, with oxygen diffusing into the body from the surrounding water and carbon dioxide diffusing out. Because the volume of the body is so small relative to the surface area available, nutrients and gases can diffuse efficiently to all of the cells inside the animal. This body plan has a downside, however: Because the body surface has to be moist for gas exchange to take place, flatworms are restricted to environments where they are surrounded by fluid. ✔You should be able to indicate the origin of the flattened, acoelomate body plan on Figure 33.2.

Feeding Platyhelminthes lack a lophophore and have a digestive tract that is "blind"—meaning it has only one opening for ingestion of food and elimination of wastes.

Most turbellarians are hunters that prey on protists or small animals; others scavenge dead animals.

Tapeworms are strictly parasitic and feed on nutrients provided by hosts. They do not have a mouth or a digestive tract, and obtain nutrients by diffusion across their body wall.

Flukes are parasites and feed by gulping host tissues and fluids through a mouth. They have a blind digestive tract.

Movement In general, flatworms do not move much.

Some turbellarians can swim by undulating their bodies, and most can creep along substrates with the aid of cilia on their ventral surface.

Tapeworms move little; adult cestodes have hooked attachment structures at their anterior end that permanently attach them to the interior of their host.

Flukes move little.

Reproduction Turbellarians can reproduce asexually by splitting themselves in half. If they are fragmented as a result of a predator attack, the body parts can regenerate into new individuals. Most turbellarians contain both male and female organs and reproduce sexually by aligning with another individual and engaging in mutual and simultaneous fertilization.

In flukes and tapeworms, reproduction is similar: Individuals reproduce sexually and either cross-fertilize or self-fertilize. The reproductive systems and life cycles of flukes and tapeworms are extremely complex, however. In many cases the life cycle involves two or even three distinct host species, with sexual reproduction occurring in the **definitive host** and asexual reproduction occurring in one or more **intermediate hosts**. For example, humans are the definitive host of the blood fluke *Schistosoma mansoni*. Fertilized eggs are shed in a human host's feces and enter aquatic habitats if sanitation systems are poor. The eggs develop into larvae that infect snails. Inside the snail, asexual reproduction results in the production of a different type of larva, which emerges from the snail and burrows into the skin of humans who wade in infested water. Once inside a human, the parasite lives in blood and develops into a sexually mature adult.

Most **annelids** have a segmented body and a coelom that functions as a hydrostatic skeleton. ✔You should be able to indicate the origin of annelid segmentation on Figure 33.2. The 16,500 species that have been described thus far were traditionally divided into groups called Polychaeta (*pol-ee-KEE-ta*) and Clitellata. Recent analyses have shown the Polychaeta are paraphyletic, however. The current understanding of annelid phylogeny is consistent with the following points:

- The common ancestor of the annelids had a key synapomorphy in addition to segmentation: numerous, bristlelike extensions called **chaetae** that extend from appendages called **parapodia** (**Figure 33.13a**). (The name polychaetes means "many bristles".)
- Lineages that retained many chaetae are mostly marine. They range in size from species that are less than 1 mm long to species that grow to lengths of 3.5 m. They include a large number of burrowers in mud or sand, sedentary forms that secrete chitinous tubes, and active, mobile species.
- Chaetae were reduced or lost in a lineage called Clitellata.
- The Clitellata includes monophyletic groups called the oligochaetes (*oh-LIG-oh-keetes*) and leeches, or Oligochaeta and Hirudinea, respectively.
- Oligochaetes ("few-bristles") include the earthworms, which burrow in moist soils; an array of freshwater species; and a few marine forms. Oligochaetes lack parapodia and, as their name implies, have many fewer chaetae than do polychaetes (**Figure 33.13b**).
- The coelom of leeches is much smaller than that of other annelids and consists of a series of connected chambers. Leeches live in freshwater as well as marine and environments.
- Lineages called Sipunculida (peanut worms) and Echiura (spoon worms), which were traditionally thought to be independent phyla, are actually groups within Annelida. Segmentation and chaetae were lost in these species.

Feeding The polychaete lineages have a wide variety of methods for feeding: The burrowing forms deposit feed or suspension feed using mucous-lined nets; the sedentary forms suspension feed with the aid of a dense crown of tentacles; and the active forms graze on algae or hunt small animals, which they capture by everting their throats. Virtually all oligochaetes, in contrast, make their living by deposit feeding in soils. The tunnels that they make are critically important in aerating soil, and their feces contribute large amounts of organic matter. About half of the leeches are ectoparasites that attach themselves to fish or other hosts and suck blood and other body fluids (**Figure 33.13c**). Hosts are usually unaware of the attack, because leech saliva typically contains an anesthetic. The host's blood remains liquid as the leech feeds, because the parasite's saliva also contains an anticoagulant. Parasitic leeches are used by physicians to reduce swelling after surgery to reattach severed fingers or other body parts. The nonparasitic leech species are predators or scavengers.

Movement Polychaetes and oligochaetes crawl or burrow with the aid of their hydrostatic skeletons; the parapodia of polychaetes also act as paddles or tiny feet that aid in movement. Many polychaetes are excellent swimmers. Leeches can swim by using their hydrostatic skeletons to make undulating motions of the body.

Reproduction Asexual reproduction occurs in polychaetes and oligochaetes by transverse fission or fragmentation—meaning that body parts can regenerate a complete individual. Sexual reproduction in polychaetes may begin with internal or external fertilization, depending on the species. Most of the polychaete lineages have separate sexes and usually release their eggs directly into the water. Some species produce eggs that hatch into trochophore larvae, then go on to develop into more-complex larval forms. In oligochaetes and leeches, individuals produce both sperm and eggs and engage in mutual, internal cross-fertilization. Eggs are enclosed in a mucus-rich, cocoon-like structure; after fertilization, offspring develop directly into miniature versions of their parents.

(a) Polychaetes have many bristle-like chaetae. *Hermodice carunculata*

(b) Oligochaetes have few or no chaetae. *Chaetogaster* species

(c) Some leeches are ectoparasites. *Haemadipsa* species

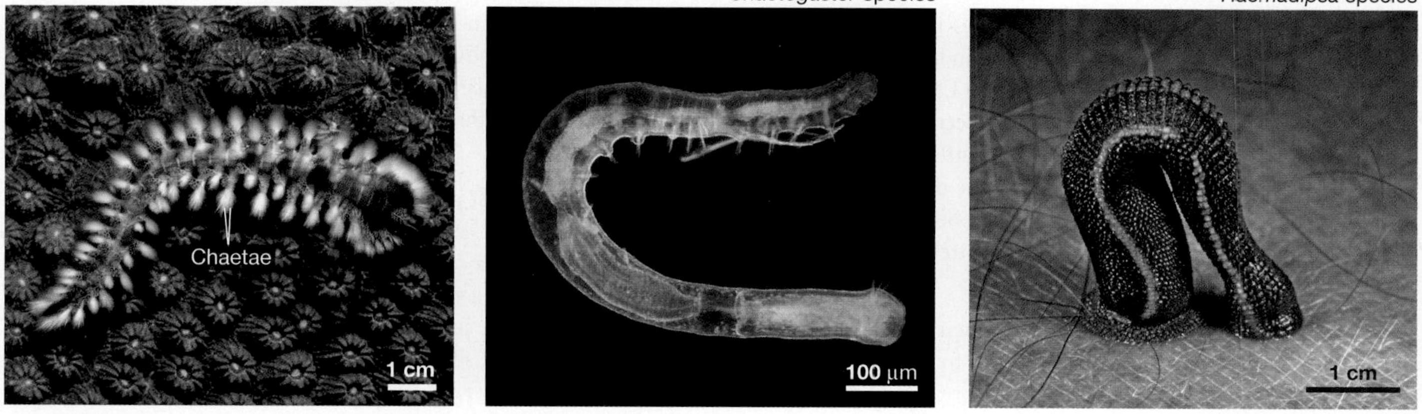

FIGURE 33.13 Annelids Are Segmented Worms.

The bivalves are so named because they have two separate shells made of calcium carbonate secreted by the mantle. The shells are hinged, and they open and close with the aid of muscles attached to them (**Figure 33.14a**). When the shell is closed, it protects the mantle, visceral mass, and foot. The bivalve shell is an adaptation that reduces predation.

Most bivalves live in the ocean, though there are many freshwater forms. Clams burrow into mud, sand, or other soft substrates and are sedentary as adults. Oysters and mussels are largely sessile as adults, but most of them live above the substrate, attached to rocks or other solid surfaces. Scallops are mobile and live on the surface of soft substrates. The smallest bivalves are freshwater clams that are less than 2 mm long; the largest bivalves are giant marine clams that may weigh more than 400 kg. All bivalves can sense gravity, touch, and certain chemicals; scallops even have eyes.

Because most bivalves live on or under the ocean floor and because they are covered by a hard shell, their bodies are often buried in sediment after death. Bivalves thus fossilize readily, and the Bivalvia lineage has the most extensive fossil record of any animal, plant, or fungal group. This large database has allowed biologists to conduct thorough studies of the evolutionary history of bivalves.

Bivalves are important commercially. Clams, mussels, scallops, and oysters are farmed or harvested from the wild in many parts of the world and used as food by humans. Pearl oysters that are cultivated or collected from the wild are the most common source of natural pearls used in jewelry.

Feeding Most bivalves are suspension feeders that take in any type of small animal or protist or detritus. Suspension feeding is based on a flow of water across gas-exchange structures called **gills**. The gills lie in the mantle cavity, between the mantle and the visceral mass, and consist of a series of thin membranes where particles are trapped. In many cases the water current flows through siphons, which are tubes formed by edges of the mantle—extending from the shell and forming a plumbing system. The siphons conduct water into the gills and then back out of the body, powered by the beating of cilia on the gills (**Figure 33.14b**). Bivalves are the only major group of mollusks that lack the feeding structure called a radula (see Chapter 32).

The efficiency of filter feeding in bivalves is illustrated by the impact that zebra mussels have had on certain lakes and streams in North America. After being introduced from Eurasia, zebra mussel populations have exploded in some aquatic ecosystems. The mussels are so efficient at removing organic matter that little food is left for native fish, whose populations decline dramatically as a result. Conversely, loss of native oysters from Chesapeake Bay has resulted in large accumulations of organic matter that have reduced water quality there.

Movement Clams burrow with the aid of their muscular foot, but are otherwise sedentary. Scallops are able to swim by clapping their shells together and forcing water to jet out, pushing them along. Scallop locomotion is similar to the swimming of cnidarian medusae, which move when muscular contractions force water out of their bell (see Chapter 32). Bivalves produce a swimming trochophore larva that is responsible for dispersing individuals to new locations.

Reproduction Most bivalves reproduce sexually. Eggs and sperm are shed into the water, and fertilized eggs develop into trochophore larvae. Trochophores then metamorphose into a distinct type of larva called a **veliger**, which continues to feed and swim before settling to the substrate and metamorphosing into an adult form that secretes a shell.

(a) Scallops live on the surface of the substrate and suspension feed.

Chlamys varia

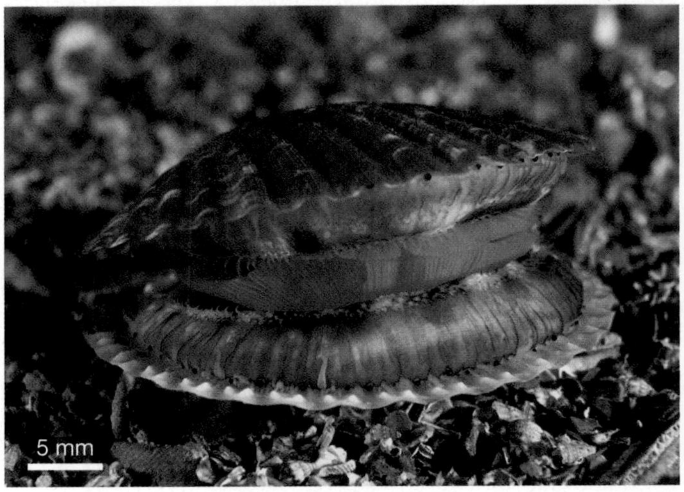
5 mm

(b) Most clams burrow in soft subtrates and suspension feed.

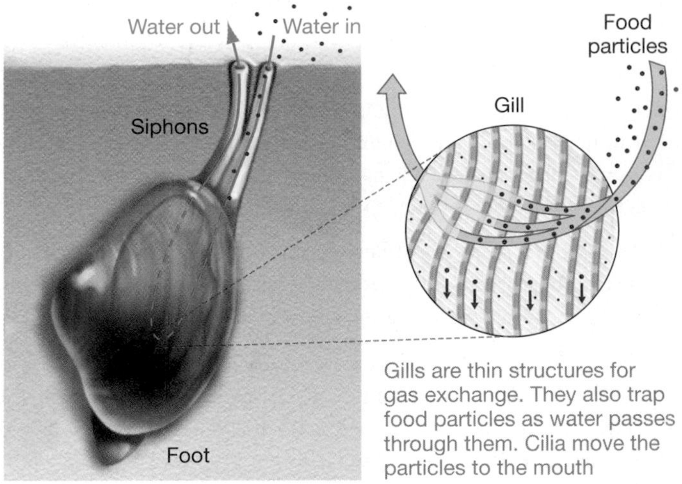
Water out
Water in
Food particles
Siphons
Gill
Foot
Gills are thin structures for gas exchange. They also trap food particles as water passes through them. Cilia move the particles to the mouth

FIGURE 33.14 Bivalves Have Two Shells.

The gastropods ("belly-feet") are named for the large, muscular foot on their ventral side. Most snails can retract their foot and body into a shell when they are attacked or when their tissues begin to dry out (**Figure 33.15a**). Land slugs and nudibranchs (pronounced *NEW-da-branks*)—also called sea slugs—lack shells, however. The shell was lost independently in the two groups; individuals often contain toxins or foul-tasting chemicals to protect them from being eaten. The bright colors of nudibranchs are thought to act as a warning to potential predators (**Figure 33.15b**). Gastropods are used as food in some cultures and are important in human disease because they serve as a host for flukes that also infect humans before completing their life cycle. About 70,000 species of gastropods are known.

Feeding Gastropods and other mollusks have a unique structure in their mouths called a radula. Recall from Chapter 32 that in many species the **radula** functions like a rasp to scrape away algae, plant cells, or other types of food. It is usually covered with teeth that are made of chitin and that vary in size and shape among species. Although most gastropods are herbivores or detritivores, specialized types of teeth allow some gastropods to act as predators. Species called drills, for example, use their radula to bore a hole in the shells of oysters or other mollusks and expose the visceral mass, which they then eat. Cone snails have highly modified, harpoon-like "teeth" mounted at the tip of an extensible proboscis and armed with poison. When a fish or worm passes by, the proboscis shoots out. The prey is speared by the tooth, subdued by the poison, and consumed.

Movement Waves of contractions down the length of the foot allow gastropods to move by creeping (see Figure 33.9). Sea butterflies are gastropods with a reduced or absent shell but a large, winglike foot that flaps and powers swimming movements.

Reproduction Females of some gastropod species can reproduce asexually by producing eggs parthenogenetically, but most reproduction is sexual. Sexual reproduction in some gastropods begins with internal fertilization. Most marine gastropods produce a veliger larva that may disperse up to several hundred kilometers from the parent. But in some marine species and all terrestrial forms, larvae are not free living. Instead, offspring remain in an egg case while passing through several larval stages and then hatch as miniature versions of the adults.

(a) Snails have a single shell, which they use for protection.
Maxacteon flammea

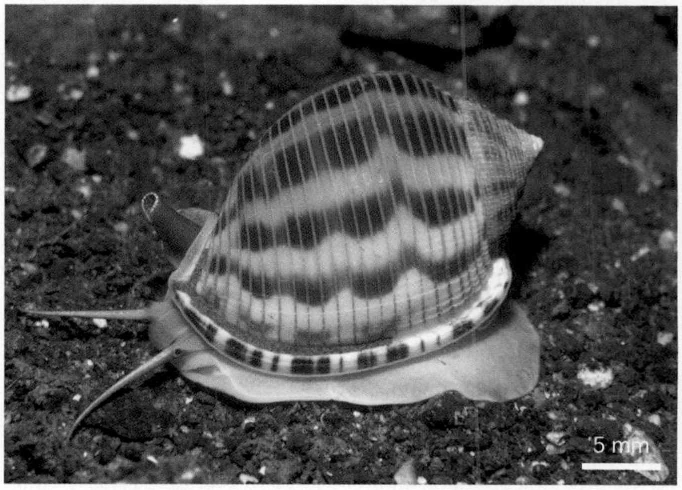
5 mm

(b) Land and sea slugs (nudibranchs) have lost their shells.
Chromodoris annae

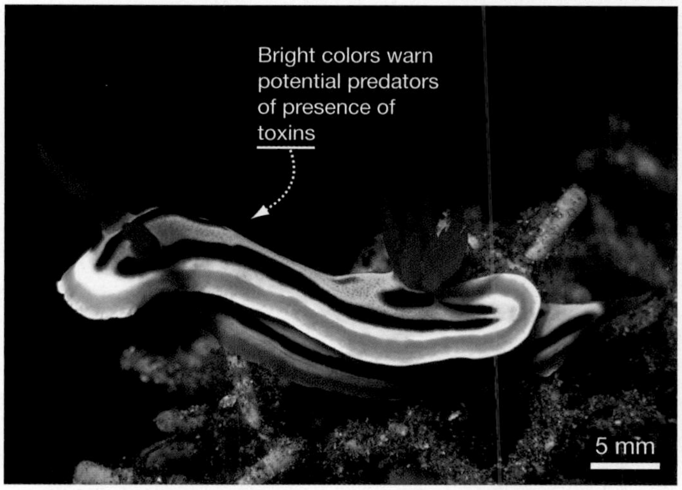
Bright colors warn potential predators of presence of toxins
5 mm

FIGURE 33.15 Gastropods Have a Single Shell or Lack Shells.

Lophotrochozoans > Mollusca > Polyplacophora (Chitons)

The Greek word roots that inspired the name *Polyplacophora* mean "many-plate-bearing." The name is apt because chitons (pronounced *KITE-uns*) have eight calcium carbonate plates along their dorsal side (**Figure 33.16**). The plates form a protective shell. The approximately 1000 species of chitons are marine. They are usually found on rocky surfaces in the intertidal zone, where rocks are periodically exposed to air at low tides.

Feeding Chitons have a radula and use it to scrape algae and other organic matter off rocks.

Movement Chitons move by creeping on their broad, muscular foot—as gastropods do.

Reproduction In most chiton species the sexes are separate and fertilization is external. In some species, however, sperm that are shed into the water enter the female's mantle cavity and fertilize eggs inside the body. Depending on the species involved, eggs may be enclosed in a membrane and released or retained until hatching and early development are complete. Most species have trochophore larvae.

Tonicella lineata

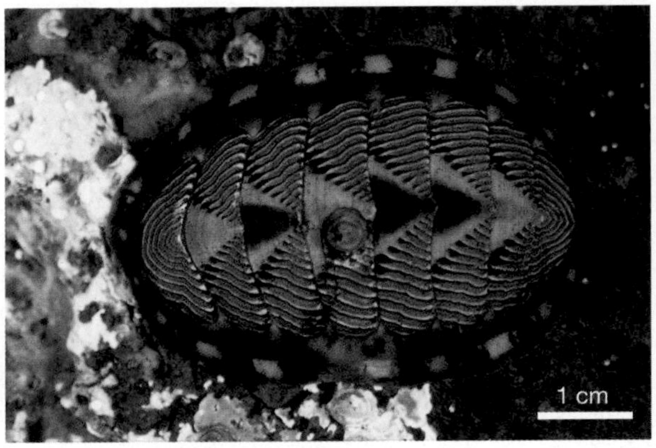

FIGURE 33.16 Chitons Have Eight Shell Plates.

Lophotrochozoans > Mollusca > Cephalopoda (Nautilus, Cuttlefish, Squid, Octopuses)

The cephalopods ("head-feet") have a well-developed "head" consisting of the visceral mass and a foot that is modified to form **tentacles:** long, thin, muscular extensions that aid in movement and prey capture (**Figure 33.17**). Except for the nautilus, the cephalopod species living today have either highly reduced shells or none at all. Most have large brains and image-forming eyes with sophisticated lenses.

Feeding Cephalopods are highly intelligent predators that hunt by sight and use their tentacles to capture prey—usually fish or crustaceans. They have a radula as well as a structure called a **beak**, which can exert powerful biting forces. Some cuttlefish and octopuses also inject poisons into their prey to subdue them.

Movement Cephalopods can swim by moving their fins to "fly" through the water, or by jet propulsion using the mantle cavity and siphon. They draw water into their mantle cavity and then force it out through a siphon (see Figure 33.10). Squid are built for speed and hunt small fish by chasing them down. Octopuses, in contrast, crawl along the substrate using their tentacles. They chase down crabs or other crustaceans, or they pry mussels or clams from the substrate and then use their beaks to crush the exoskeletons of their prey.

Reproduction Cephalopods have separate sexes, and some species have elaborate courtship rituals that involve color changes and interaction of tentacles. When a male is accepted by a female, he deposits sperm that are encased in a structure called a **spermatophore**. The spermatophore is transferred to the female, and fertilization is internal. Females lay eggs. When they hatch, juveniles develop directly into adults.

Octopus bimaculatus

FIGURE 33.17 Cephalopods Have Highly Modified Bodies.

33.4 Key Lineages: Ecdysozoans

The ecdysozoans were first recognized as monophyletic when investigators began using DNA sequence data to estimate the phylogeny of protostomes. Seven phyla are currently recognized in the lineage (see Table 32.1), including the Nematoda, Onychophora (*on-ee-KOFF-er-uh*) and the Tardigrada. Onychophora and Tardigrada are not species-rich phyla, but both are closely related to arthropods. They are similar to arthropods in having a segmented body and limbs. Unlike arthropods, their limbs are not jointed and they do not have an exoskeleton.

The onychophorans, or velvet worms, are small, caterpillar-like organisms that live in moist leaf litter and prey on small invertebrates. Onychophorans have lobe-shaped appendages and segmented bodies with a hemocoel (**Figure 33.18a**).

The tardigrades, or water bears, are microscopic animals that live in benthic (bottom) habitats of marine or freshwater environments. Large numbers of water bears can also be found in the film of water that covers moss or other land plants in moist habitats. Tardigrades have a reduced coelom but a prominent hemocoel, and they walk on their clawed, lobe-shaped legs (**Figure 33.18b**). Most feed by sucking fluids from plants or animals; others are **detritivores**. Tardigrades are notable for their ability to withstand extreme drying and pressure. In response to drying or other adverse conditions, individuals may reduce metabolism to near zero for a decade or more, then "reanimate" and resume normal activities when conditions improve.

(a) Onychophorans are called velvet worms.

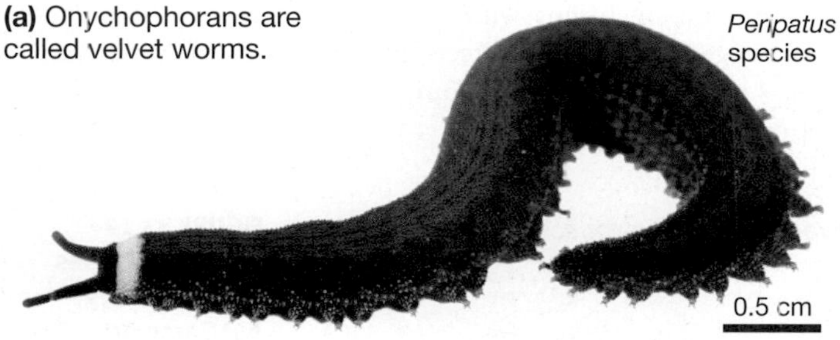

(b) Tardigrades are called water bears.

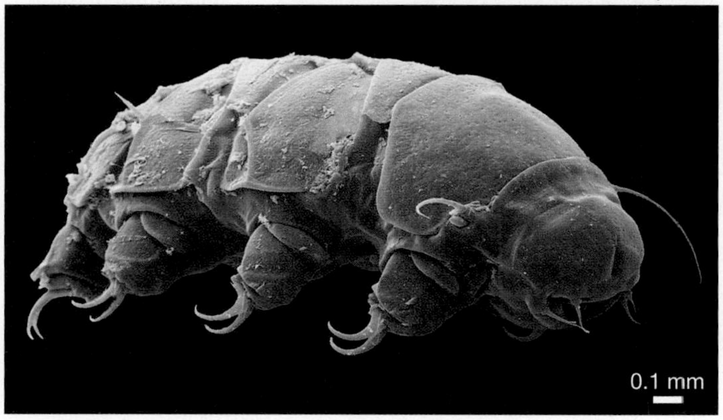

FIGURE 33.18 Onychophorans and Tardigrades Have Lobe-Shaped Limbs with Claws. The Onychophora and Tardigrada are terrestrial—though tardigrades live in wet habitats. The scanning electron micrograph in part (b) has been colorized.

In terms of duration in the fossil record, species diversity, and abundance of individuals, however, arthropods are the most important phylum within Ecdysozoa. They appear in the fossil record over 520 million years ago and have long been the most abundant animals observed in both marine and terrestrial environments. Well over a million living species have been described, and biologists estimate that millions or perhaps even tens of millions of arthropod species have yet to be discovered. In terms of species diversity, arthropods are easily the most successful lineage of eukaryotes.

Morphologically, **arthropods** are distinguished by segmented bodies and a complex, jointed exoskeleton. They have a highly reduced coelom but possess an extensive body cavity called a hemocoel. They also have an exoskeleton and paired, jointed appendages. The body is organized into distinct head and trunk tagmata in all arthropods; in many species there is an additional grouping of segments into two distinct trunk regions, usually called an abdomen and a thorax. ✔**You should be able to indicate the origin of lobe-shaped limbs and jointed limbs on Figure 33.2.**

Metamorphosis is common in arthropods. Their larvae have segmented bodies but may lack a hardened exoskeleton. As in other ecdysozoans, larval and adult forms grow by molting.

At least some segments along the arthropod body produce paired, jointed appendages. Arthropod appendages have an array of functions: sensing aspects of the environment, exchanging gases, feeding, or locomotion via swimming, walking, running, jumping, or flying. Arthropod appendages provide the ability to sense environmental stimuli and make sophisticated movements in response.

Most species also have sophisticated, image-forming compound eyes. A **compound eye** contains many lenses, each associated with a light-sensing, columnar structure. (Human and cephalopod eyes are **simple eyes**, meaning they have just one lens.) Most arthropods also have a pair of antennae on the head. **Antennae** are long, tentacle-like appendages that contain specialized receptor cells that function in the senses of touch and smell.

Although the phylogeny of arthropods is still being worked out, most data sets agree that the phylum as a whole is monophyletic; that the myriapods (millipedes and centipedes), chelicerates (spiders), insects, and crustaceans represent four major lineages within the phylum; and that crustaceans and insects are closely related.

Let's take a closer look at the Nematoda—another particularly diverse and species-rich phylum within Ecdysozoa—and the four major lineages within Arthropoda:

- Ecdysozoans > Nematoda (Roundworms)
- Ecdysozoans > Arthropoda > Myriapods (Millipedes, Centipedes)
- Ecdysozoans > Arthropoda > Insecta (Insects)
- Ecdysozoans > Arthropoda > Chelicerata (Spiders, Ticks, Mites, Horseshoe Crabs, Daddy Longlegs, Scorpions)
- Ecdysozoans > Arthropoda > Crustaceans (Shrimp, Lobster, Crabs, Barnacles, Isopods, Copepods)

Species in the phylum Nematoda are commonly called **nematodes** or **roundworms** (**Figure 33.19**). Roundworms are unsegmented worms with a coelom, a tube-within-a-tube body plan, and no appendages. They have no muscles that can change the diameter of the body—their body musculature consists solely of longitudinal muscles that shorten or lengthen the body upon contracting or relaxing, respectively. ✔You should be able to indicate the loss of circular muscles on Figure 33.2. Although some nematodes can grow to lengths of several meters, the vast majority of species are tiny—most are much less than 1 mm long. They lack specialized systems for exchanging gases and for circulating nutrients. Instead, gas exchange occurs across the body wall, and nutrients move by simple diffusion.

The nematode *Caenorhabditis elegans* is one of the most thoroughly studied model organisms in biology (see **BioSkills 14** in Appendix A). Species that parasitize humans have also been studied intensively. Pinworms, for example, infect about 40 million people in the United States alone, and *Onchocerca volvulus* causes an eye disease that infects around 20 million people in Africa and Latin America. Advanced infections by the roundworm *Wuchereria bancrofti* result in the blockage of lymphatic vessels, causing fluid accumulation and massive swelling—the condition known as elephantiasis. The disease trichinosis is caused by species in the genus *Trichinella* (**Figure 33.19a**). The parasites are passed from infected animals to humans by ingesting raw or undercooked meat. An array of symptoms results, including discomfort caused by *Trichinella* larvae encasing themselves in cysts within muscle or other types of tissue (**Figure 33.19b**).

Parasites represent a relatively small fraction of the 25,000 nematode species that have been described to date, however. The free-living nematodes are important ecologically because they are found in virtually every habitat known.

Nematodes are also fabulously abundant. Biologists have found 90,000 roundworms in a single rotting apple and have estimated that rich farm soils contain up to 9 billion roundworms per acre. Although they are not the most species-rich animal group, they are among the most abundant. The simple nematode body plan has been extraordinarily successful.

Feeding Roundworms feed on a wide variety of materials, including bacteria, fungi, plant roots, small protists or animals, or detritus. In most cases, the structure of their mouthparts is specialized in a way that increases the efficiency of feeding on a particular type of organism or material.

Movement Roundworms move with the aid of their hydrostatic skeleton. Most roundworms live in soil or inside a host, so when contractions of their longitudinal muscles cause them to wriggle, the movements are resisted by a stiff substrate. As a result, the worm pushes off the substrate, and the body moves.

Reproduction Sexes are separate in most nematode species. Sexual reproduction begins with internal fertilization and culminates in egg laying and the direct development of offspring. Individuals go through a series of four molts over the course of their lifetime.

(a) A nematode causes the parasitic disease trichinosis.

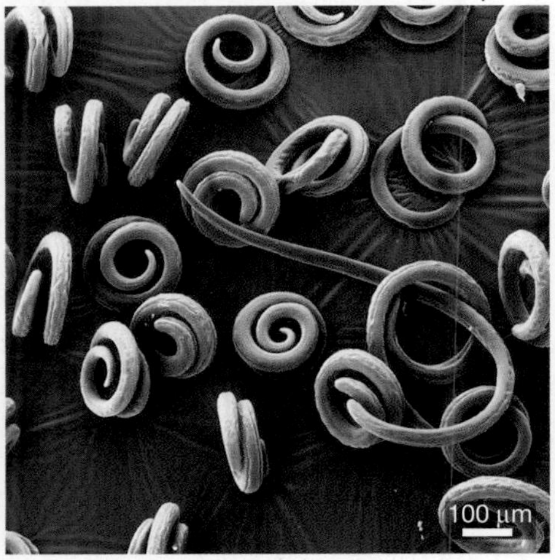

(b) *Trichinella spiralis* larvae encysted in muscle tissue

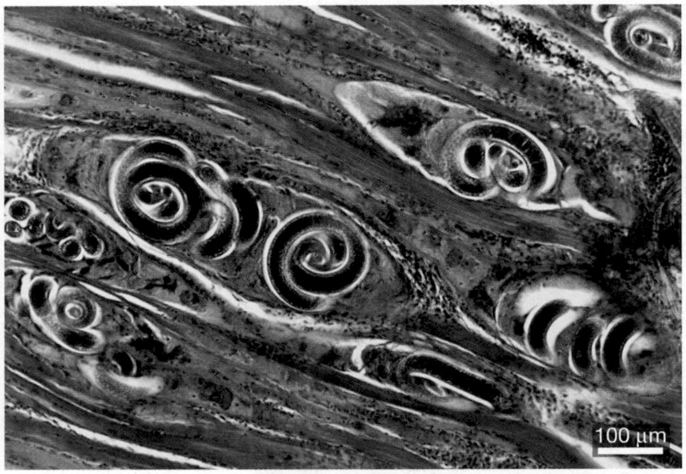

FIGURE 33.19 Most Nematodes Are Free Living, but Some Are Parasites. The scanning electron micrograph in part (a) has been colorized.

Ecdysozoans > Arthropoda > Myriapods (Millipedes, Centipedes)

The **myriapods** have relatively simple bodies, with a head region and a long trunk featuring a series of segments, each bearing one or two pairs of legs (**Figure 33.20**). If eyes are present, they consist of a few to many simple structures clustered on the sides of the head. The 11,600 species that have been described to date inhabit terrestrial environments all over the world.

Feeding Millipedes and centipedes have mouthparts that can bite and chew. These organisms live in rotting logs and other types of dead plant material that litters the ground in forests and grasslands. Millipedes are detritivores. Centipedes, in contrast, use a pair of poison-containing fangs just behind the mouth to hunt an array of insects. Large centipedes can inject enough poison to debilitate a human.

Movement Myriapods walk or run on their many legs; a few species burrow. Some millipedes have over 190 trunk sections, each with two pairs of legs, for a total of over 750 legs. Centipedes usually have fewer than 30 segments, with one pair of legs per segment.

Reproduction Myriapod sexes are separate, and fertilization is internal. Males deposit sperm in packets that are picked up by the female or transferred to her by the male. After females lay eggs in the environment, the eggs hatch into juveniles that develop into adults via a series of molts.

Scolopendra species

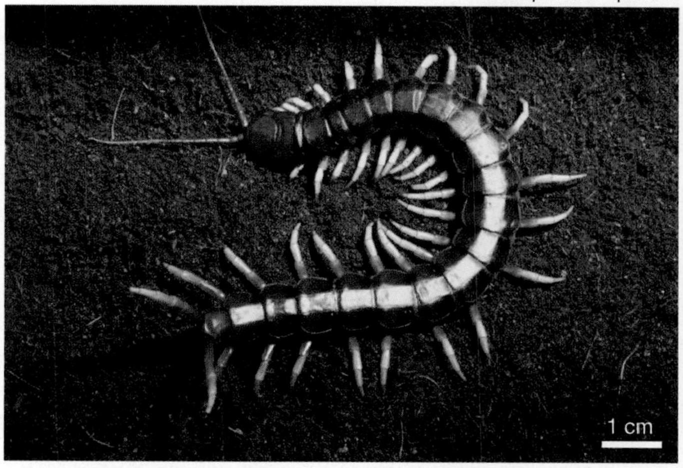

FIGURE 33.20 Myriapods Have One or Two Pairs of Legs on Each Body Segment.

Ecdysozoans > Arthropoda > Insecta (Insects)

About 925,000 species of insects have been named thus far, but it is certain that many more exist. In terms of species diversity and numbers of individuals, insects dominate terrestrial environments. In addition, the larvae of some species are common in freshwater streams, ponds, and lakes. **Table 33.1** on pages 640–641 provides detailed notes about eight of the most prominent lineages, called orders, of insects. The insect *Drosophila melanogaster* is one of the most thoroughly studied model organisms in biology (see **BioSkills 14** in Appendix A).

Insects are distinguished by having three tagmata—the head, **thorax**, and **abdomen** (see Figure 33.5). Three pairs of walking legs are located on the lateral surface of the thorax. Most species have one or two pairs of wings, mounted on the dorsal (back) side of the thorax. Typically the head contains four mouthpart structures, a pair of antennae that are used to touch and smell, and a pair of compound eyes.

Feeding Most insect species have four mouthpart structures (labrum, paired mandibles, paired maxillae, and labium) which have diversified in structure and function in response to natural selection. As a result, insects are able to feed in every conceivable manner and on almost every type of food source available on land. In species with holometabolous metamorphosis (see Chapter 32), larvae have wormlike bodies; most tunnel their way through food sources, though some feed by biting leaves. Adults, however, are predators or parasites on plant or animal tissues. Thus, diversification in food sources occurs even within species. Because so many insects make their living by feeding on plant tissues or fluids, the diversification of insects was closely correlated with the diversification of land plants. Insect predators usually eat other insects; insect parasites usually victimize other arthropods or mammals.

Movement Insects use their legs to walk, run, or swim, or they use their wings to fly. When insects walk or run, the sequence of movements usually results in three of their six legs maintaining contact with the ground at all times.

Reproduction Insect sexes are separate. Mating usually takes place through direct copulation, with the male inserting a sperm-transfer organ into the female. Most females lay eggs, but in a few species eggs are retained until hatching. Many species are also capable of reproducing asexually, through the production of unfertilized eggs via mitosis. In the vast majority of species, either incomplete or complete metamorphosis occurs, with complete metamorphosis being by far the most common.

TABLE 33.1 **Prominent Orders of Insects**

Order	Common Name	Number of Known Species	Description
Coleoptera ("sheath-winged") 	Beetles	350,000	**Key traits:** Hardened forewings, called elytra, protect the membranous hindwings that power flight. During flight, the elytra are held to the side and act as stabilizers. **Feeding:** Adults are important predators and scavengers. Larvae are called grubs and often have chewing mouth-parts. **Reproduction:** All have complete metamorphosis. **Notes:** The most species-rich lineage on the tree of life—about 25 percent of all described species are beetles. Range in length from 0.25 mm to 10 cm.
Lepidoptera ("scale-winged") 	Butterflies, moths	180,000	**Key traits:** Wings are covered with tiny, often colorful scales. The forewings and hindwings hook together and move as a unit in many groups. **Feeding:** Larvae usually have chewing mouthparts and either bore through food material or bite leaves; adults often feed on nectar. **Reproduction:** All have complete metamorphosis. **Notes:** Some species migrate long distances.
Diptera ("two-winged") 	Flies (including mosquitoes, gnats, midges)	120,000	**Key traits:** Reduced hindwings, called halteres, act as stabilizers during flight. **Feeding:** Adults are usually liquid feeders; larvae are called maggots and are often parasitic or feed on rotting material. **Reproduction:** All have complete metamorphosis. **Notes:** First flies in fossil record appear 240 million years ago.
Hymenoptera ("membrane-winged") 	Ants, bees, wasps	115,000	**Key traits:** Membranous forewings and hindwings lock together via tiny hooks, thus acting as a single wing during flight. **Feeding:** Most ants feed on plant material; most bees feed on nectar; wasps are predatory and often have parasitic larvae. **Reproduction:** All have complete metamorphosis. Males are haploid (they hatch from unfertilized eggs) and females are diploid (they hatch from fertilized eggs). Females deposit eggs via an ovipositor, which in some species is modified into a stinger used in defense. Larvae are often fed and protected by adults. **Notes:** Most species live in colonies, and many are "eusocial"—meaning that some individuals in the colony help raise the queen's offspring but never reproduce themselves.

TABLE 33.1 **Prominent Orders of Insects** *(continued)*

Order	Common Name	Number of Known Species	Description
Hemiptera ("half-winged") 	Bugs (including leaf hoppers, aphids, cicadas, scale insects)	85,000	**Key traits:** Some have a thickened forewing with a membranous tip; also mouthparts that are modified for piercing and sucking. **Feeding:** Most suck plant juices, but some are predatory. **Reproduction:** All have complete metamorphosis. **Notes:** Range in length from 1 mm to 11 cm.
Orthoptera ("straight-winged") 	Grasshoppers, crickets	20,000	**Key traits:** Large, muscular hind legs power movement by jumping. **Feeding:** Most have chewing mouthparts and are leaf eaters. **Reproduction:** All have incomplete metamorphosis. In many species, males give distinctive songs to attract mates. **Notes:** Range in length from 5 mm to 11.5 cm.
Trichoptera ("hairy-winged") 	Caddisflies	12,000	**Key traits:** Wings are covered with minute "hairs"—projections of the cuticle. **Feeding:** Adults have mouthparts adapted for sucking liquids, but have rarely been observed feeding; larvae are all aquatic and feed on detritus. Most larvae build a protective case of tiny stones and sticks. **Reproduction:** All have complete metamorphosis. **Notes:** Larvae and adults are important food sources for fish.
Odonata ("toothed") 	Dragonflies, damselflies	6,500	**Key traits:** Four membranous wings and long, slender abdomens. **Feeding:** Larvae and adults are predatory, often on flies. The "odon" in their name refers to strong, toothlike structures on their mandibles. **Reproduction:** Metamorphosis is hemimetabolous: nymphs are aquatic, adults are terrestrial. **Notes:** They hunt by sight and have reduced antennae (used for touch and smell) but enormous eyes with up to 28,000 lenses.

Most of the 70,000 species of chelicerates are terrestrial, although the horseshoe crabs and sea spiders are marine. The most prominent lineage within Chelicerata is the Arachnida (spiders, scorpions, mites, and ticks).

The chelicerate body consists of anterior and posterior regions (**Figure 33.21a**). The anterior region lacks antennae for sensing touch or odor but usually contains eyes. The group is named for a pair of appendages called **chelicerae**, found near the mouth. Depending on the species, the chelicerae are used in feeding, defense, copulation, movement, or sensory reception. Chelicerates have a total of 6 pairs of appendages.

Feeding Spiders, scorpions, and daddy longlegs capture insects or other prey. Although some of these species are active hunters, most spiders are sit-and-wait predators. They create sticky webs to capture prey, which fly or walk into the web and are subsequently trapped. The spider senses the vibrations of the struggling prey, pounces on it, and administers a toxic bite. Mites and ticks are ectoparasitic and use their piercing mouthparts to feed on host animals, or in the case of dust mites, on their dead skin (**Figure 33.21b**). Horseshoe crabs eat a variety of animals as well as algae and detritus. Most scorpions feed on insects; the largest scorpion species occasionally eat snakes and lizards.

Movement Like other arthropods, chelicerates move with the aid of muscles attached to an exoskeleton. They walk or crawl on their four pairs of jointed walking legs; some species are also capable of jumping. Horseshoe crabs and some other marine forms can swim slowly. Newly hatched spiders spin long, silken threads that serve as balloons and that can carry them on the wind more than 400 kilometers from the point of hatching.

Reproduction Sexual reproduction is the rule in chelicerates, and fertilization is internal in most groups. Courtship displays are extensive in many arthropod groups and may include both visual displays and the release of chemical odorants. In spiders, males use appendages called pedipalps to transfer sperm to females. These organs fit into the female reproductive tract in a "lock-and-key" fashion. Differences in the size and shape of male genitalia are often the only way to identify closely related species of spider.

Development is direct, meaning that metamorphosis does not occur. In spiders, males may present a dead insect as a gift that the female eats as they mate; in some species, the male himself is eaten as sperm is being transferred. In scorpions, females retain fertilized eggs. After the eggs hatch, the young climb on the mother's back. They remain there until they are old enough to hunt for themselves.

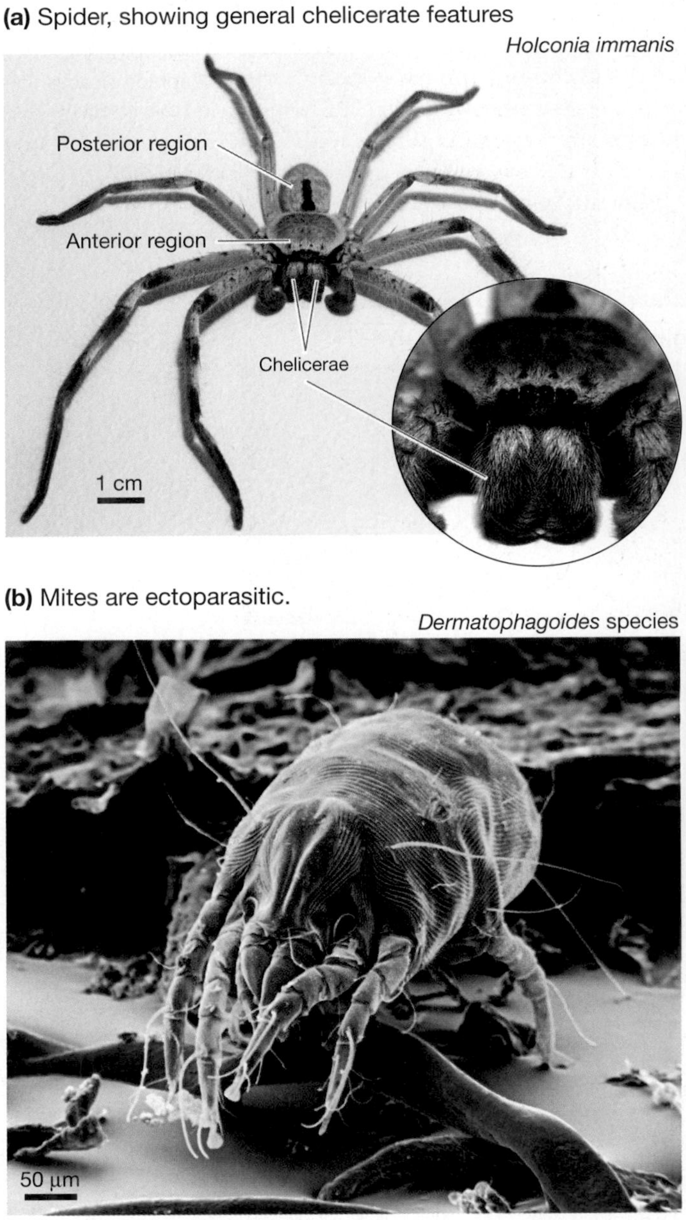

FIGURE 33.21 Chelicerates Have Two Body Regions. The scanning electron micrograph in part (b) has been colorized.

The 67,000 species of **crustaceans** that have been identified to date live primarily in marine and freshwater environments. A few species of crab and some isopods are terrestrial, however. (Terrestrial isopods are known as sowbugs, pillbugs, or roly-polies). Crustaceans are common in surface waters, where they are important consumers. They are also important grazers and predators in shallow-water benthic environments.

The segmented body of most crustaceans is divided into two distinct regions: (1) the cephalothorax, which combines the head and thorax, and (2) the abdomen. Many crustaceans have a **carapace**—a platelike section of their exoskeleton that covers and protects the cephalothorax (**Figure 33.22a**). They are the only type of arthropod with two pairs of antennae, and they have sophisticated, compound eyes—usually mounted on stalks (**Figure 33.22b**). Crustaceans and trilobites—a major lineage that did not survive the end-Permian mass extinction, 250 million years ago (see Chapter 27)—were the first arthropods to appear in the fossil record.

Feeding Most crustaceans have 4–6 pairs of mouthparts that are derived from jointed appendages, and as a group they use every type of feeding strategy known. Barnacles (**Figure 33.23**) and many shrimp are suspension feeders that use feathery structures located on head or body appendages to capture passing prey. Crabs and lobsters are active hunters, herbivores, and scavengers. Typically they have a pair of mouthparts called **mandibles** that bite or chew. Individuals capture and hold their food source with claws or other types of feeding appendages near their mouth and then use their mandibles to shred the food into small bits that can be ingested. As herbivores or detritivores, many species of crustaceans depend on algae for food.

Movement Crustacean limbs are highly diverse: species have many pairs of limbs and it is common for a species to have more than one type of limb. Limb structures in crustaceans include paddle-shaped forms used in swimming, feathery structures used in capturing food that is suspended in water, and slender legs that make sophisticated walking or running movements possible. Barnacles are one of the few types of sessile crustaceans. Adult barnacles cement their heads to a rock or other hard substrate, secrete a protective shell of calcium carbonate, and use their legs to capture food particles and transfer them to the mouth.

FIGURE 33.23 Barnacles Stand on Their Heads and Kick Food into Their Mouths.

Reproduction Most individual crustaceans are either male or female, and sexual reproduction is the norm. Fertilization is usually internal, and eggs are usually retained by the female until they hatch. Most crustaceans pass through several distinct larval stages; many species include a larval stage called a **nauplius**, which is usually planktonic. A nauplius has a single eye and appendages that develop into the two pairs of antennae, the mouthparts, and swimming legs of the adult.

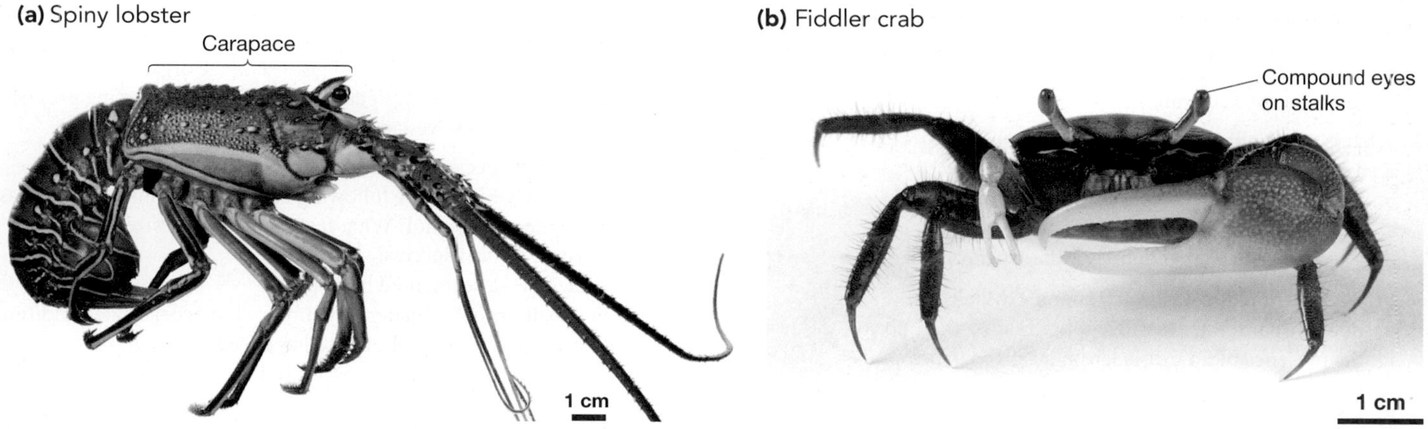

FIGURE 33.22 Crustaceans Are Named for the Hard, Calcified Exoskeltons of Lobsters and Crabs.

✔**EXERCISE** Circle the lobster's cephalothorax (combined head and thorax).

CHAPTER 33 REVIEW

For media, go to the study area at www.masteringbiology.com

Summary of Key Concepts

Molecular phylogenies support the hypothesis that protostomes are a monophyletic group divided into two major subgroups: the Lophotrochozoa and the Ecdysozoa.

- The protostomes comprise about 20 phyla and were originally identified because their embryos undergo early development in the same way.
- Lophotrochozoans grow incrementally. Some phyla in the lineage also have characteristic feeding structures called lophophores; many have a distinctive type of larvae called trochophores.
- Species in the Ecdysozoa grow by molting—meaning that they shed their old external covering and grow a new, larger one.

✔ You should be able to explain the logic behind the claim that the protostome pattern of development is a synapomorphy that identifies a monophyletic group.

Web Activity Protostome Diversity

Although the members of many protostome phyla have limbless, wormlike bodies and live in marine sediments, the most diverse and species-rich lineages—Mollusca and Arthropoda—have body plans with distinctive, complex features. Mollusks and arthropods inhabit a wide range of environments.

- All protostomes are bilaterally symmetric and triploblastic, but the nature of the body cavity is variable.
- The wormlike phyla of protostomes have well-developed coeloms that provide a hydrostatic skeleton used in locomotion. The coelom is absent in flatworms and reduced in Mollusca and Arthropoda.
- The mollusk's body includes a muscular foot, a visceral mass of organs, and a protective mantle.
- The arthropod body is segmented, divided into specialized tagmata such as the head, thorax, and abdomen of insects, and protected by an exoskeleton made primarily of chitin.

✔ You should be able to explain why flatworms lost a coelom, and why arthropods have a drastically reduced coelom.

Key events triggered the diversification of protostomes, including several lineages making the water-to-land transition, a diversification in appendages and mouthparts, and the evolution of metamorphosis in both marine and terrestrial forms.

- The transition to living on land occurred several times independently during the evolution of protostomes. It was facilitated by watertight shells and exoskeletons, the evolution of specialized structures to minimize water loss and manage gas exchange, and the ability to move to moist locations.
- Diversification in appendages and mouthparts gave protostomes the ability to move and find food in innovative ways.
- Complete metamorphosis allows offspring in sessile species to disperse to new habitats and enables larvae and adults to find food in different ways.

✔ You should be able to suggest two hypotheses for why complete metamorphosis is advantageous in clams.

Questions

✔ TEST YOUR KNOWLEDGE

Answers are available in Appendix B

1. Why is it logical to observe that mollusks have a drastically reduced coelom or no coelom at all?
   a. They evolved from flatworms, which also lack a coelom.
   b. They are the most advanced of all the protostomes.
   c. Their bodies are encased by a mantle, which may or may not secrete a shell.
   d. They have a muscular foot and visceral mass that fulfill the functions of a coelom.
2. What is a lophophore?
   a. a specialized filter-feeding structure
   b. the single opening in species with a blind gut
   c. a distinctive type of larva with a band of cilia
   d. a synapomorphy that defines lophotrochozoans
3. What is the function of the arthropod exoskeleton?
   a. Because hard parts fossilize more readily than do soft tissues, the presence of an exoskeleton has given arthropods a good fossil record.
   b. It has no well-established function. (Trilobites had an exoskeleton, and they went extinct.)
   c. It provides protection and functions in locomotion.
   d. It makes growth by molting possible.
4. Why is it logical that Platyhelminthes have flattened bodies?
   a. They have simple bodies and evolved early in the diversification of protostomes.
   b. They lack a coelom, so their body cannot form a rounded tube-within-a-tube design.
   c. A flat body provides a surface area for gas exchange, which compensates for their lack of gas-exchange organs.
   d. They are sit-and-wait predators that hide from passing prey by flattening themselves against the substrate.
5. In number of species, number of individuals, and duration in the fossil record, which of the following phyla of wormlike animals has been the most successful? What feature is hypothesized to be responsible for this success?
   a. Annelida—a segmented body plan
   b. Platyhelminthes—bilateral symmetry but acoelomate condition
   c. Phoronida—a lophophore used as a feeding apparatus
   d. Nemertea—a barbed proboscis used in prey capture
6. Which protostome phylum is distinguished by having body segments organized into tagmata?
   a. Mollusca
   b. Arthropoda
   c. Annelida
   d. Nematoda

✔TEST YOUR UNDERSTANDING

Answers are available in Appendix B

1. Describe the traits that distinguish the Lophotrochozoa and Ecdysozoa. Describe the traits that are common to both groups.
2. Did segmentation evolve once or multiple times during the evolution of protostomes? Explain the logic that justifies your answer.
3. Why were water-to-land transitions an important aspect of protostome diversification?
4. Explain why the evolution of the exoskeleton was such an important event in the evolution of arthropods.
5. Suggest a hypothesis to explain why the evolution of the wing was such an important event in the evolution of insects.
6. All of the wormlike protostome phyla have similar body plans. But there are many different worm-like phyla. What factors are responsible for this diversification?

✔APPLYING CONCEPTS TO NEW SITUATIONS

Answers are available in Appendix B

1. Recall that the phylum Platyhelminthes includes three major groups: the free-living turbellarians, the parasitic cestodes, and the parasitic trematodes. Because the parasitic forms are so simple morphologically, researchers suggest that they are derived from more complex, free-living forms. Draw a phylogeny of Platyhelminthes that would support the hypotheses that ancestral flatworms were morphologically complex, free-living organisms and that parasitism is a derived condition.
2. Brachiopoda is a phylum within the Lophotrochozoa. Even though they are not closely related to mollusks, brachiopods look and act like bivalve mollusks (clams or mussels). Specifically, brachiopods suspension feed, live inside two calcium carbonate shells that hinge together in some species, and attach to rocks or other hard surfaces on the ocean floor. How is it possible for brachiopods and bivalves to be so similar if they did not share a recent common ancestor?
3. The Mollusca includes a group of about 370 wormlike species called the Aplacophora. Although aplacophorans do not have a shell, their cuticle secretes calcium carbonate scales or spines. They lack a well-developed foot, and some species have a simple radula.
   - Predict where Aplacophora are found in the phylogenetic tree of Mollusca relative to bivalves, gastropods, chitons, and cephalopods. Explain your reasoning.
   - Predict where Aplacophora live and how they move. Explain your reasoning.
4. Consider the following: (a) The first arthropods to appear in the fossil record are marine crustaceans and trilobites, (b) terrestrial insects appeared relatively late in arthropod evolution, (c) although most crustaceans are aquatic, the crustacean groups called crabs and isopods each include both aquatic and terrestrial forms, and (d) analyses of DNA sequence data suggest that crustaceans and insects are closely related. Based on these observations, do you agree or disagree with the hypothesis that arthropods made the water-to-land transition more than once? Explain your reasoning.

In most habitats the "top predators"—animals that prey on other animals and aren't preyed upon themselves—are deuterostomes.

34 Deuterostome Animals

KEY CONCEPTS

- Echinoderms are radially symmetric as adults and have an endoskeleton and water vascular system. They are among the most important predators and herbivores in marine environments.
- All vertebrates have a skull and an extensive endoskeleton made of cartilage or bone; their diversification was driven in part by the evolution of the jaw and limbs. Vertebrates are the most important large-bodied predators and herbivores in marine and terrestrial environments.
- Humans are a tiny twig on the tree of life. Chimpanzees and humans diverged from a common ancestor that lived in Africa 6–7 million years ago. Since then, at least 14 humanlike species have existed.

The **deuterostomes** include the largest-bodied and some of the most morphologically complex of all animals. They range from the sea stars that cling to dock pilings, to the fish that dart in and out of coral reefs, to the wildebeests that migrate across the Serengeti Plains of East Africa.

The deuterostomes were initially grouped together because they all undergo early embryonic development in a similar way. When a humpback whale, sea urchin, or human is just beginning to grow, the gut starts developing from posterior to anterior—with the anus forming first and the mouth second. A coelom, if present, develops from outpocketings of mesoderm (see Chapter 32).

Today, most biologists recognize four phyla of deuterostomes: the Echinodermata, Hemichordata, Xenoturbellida, and Chordata (**Figure 34.1**).

- The echinoderms include the sea stars and sea urchins.
- The hemichordates, or "acorn worms," are probably unfamiliar to you—they burrow in marine sands or muds and make their living by deposit feeding or suspension feeding.
- A lone genus with two wormlike species, called *Xenoturbella,* was recognized as a distinct phylum in 2006.
- The chordates include the **vertebrates**, or animals with backbones. The vertebrates, in turn, comprise the hagfish, lampreys, sharks, bony fishes, amphibians, mammals, and reptiles (including birds).

Animals that are not vertebrates are collectively known as **invertebrates**.

Although deuterostomes share important features of embryonic development, their adult body plans and their feeding methods, modes of locomotion, and reproductive strategies are highly diverse. This chapter explores deuterostome diversity by introducing the echinoderms and chordates, and then taking a more in-depth look at the vertebrates and our own closest ancestors: the hominins.

✔ When you see this checkmark, stop and test yourself. Answers are available in Appendix B.

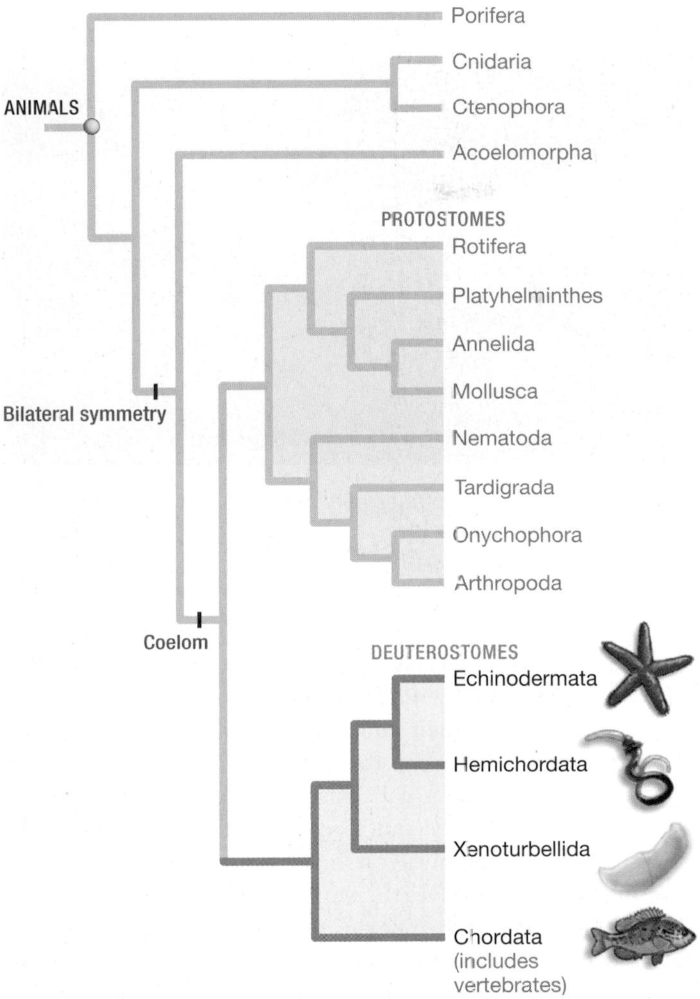

FIGURE 34.1 There Are Four Phyla of Deuterostomes. Vertebrates are in the phylum Chordata. Animals that are not vertebrates are collectively known as invertebrates. Over 95 percent of the known animal species are invertebrates.

✔**QUESTION** Are invertebrates monophyletic or paraphyletic?

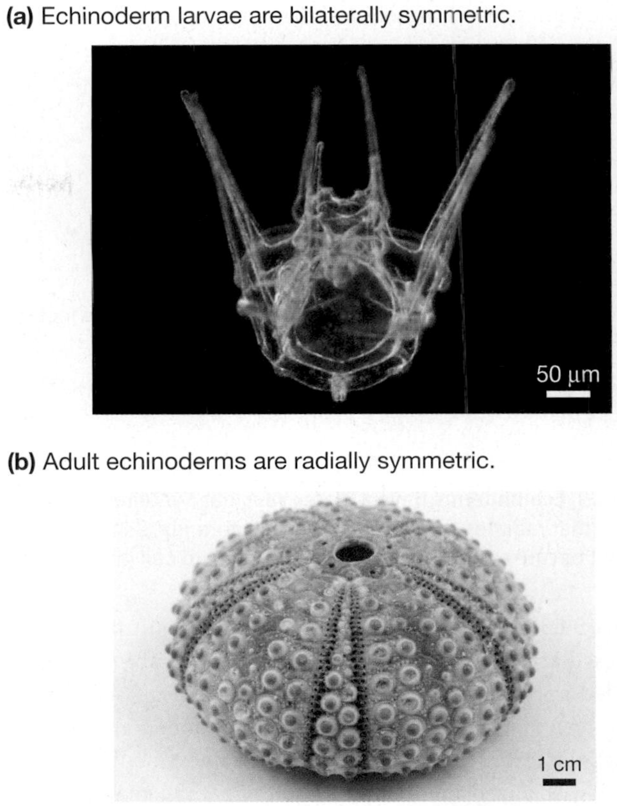

FIGURE 34.2 Body Symmetry Differs in Adult and Larval Echinoderms. Bilaterally symmetric sea urchin larvae **(a)** undergo metamorphosis and emerge as radially symmetric adults **(b)**.

34.1 What Is an Echinoderm?

The echinoderms (literally, "spiny-skins") were named for the spines or spikes observed in many species. Echinodermata is a large and diverse phylum but is exclusively marine.

In terms of numbers of species and range of habitats occupied, echinoderms are the second-most successful lineage of deuterostomes—next to the vertebrates. To date, biologists have cataloged about 7000 species of echinoderms.

Echinoderms are also abundant. In some deepwater environments, species in this phylum represent 95 percent of the total mass of organisms.

How are these animals put together?

The Echinoderm Body Plan

All deuterostomes are considered bilaterians, because they evolved from an ancestor that was bilaterally symmetric (see Chapter 32). Echinoderm larvae are also bilaterally symmetric (**Figure 34.2a**). But a remarkable event occurred early in the evolution of **echinoderms**: the origin of five-sided radial symmetry, called pentaradial symmetry, in adults.

Recall from Chapter 32 that radially symmetric animals are not cephalized. As a result, they tend to interact with the environment in all directions at once instead of facing the environment in one direction. If adult echinoderms are capable of movement, they tend to move equally well in all directions.

A second noteworthy feature of the echinoderm body is its **endoskeleton**: a hard structure, located just inside a thin layer of epidermal tissue, that provides protection and support (**Figure 34.2b**). As an individual is developing, cells secrete plates of calcium carbonate inside the skin. Depending on the species involved, the plates may remain independent and result in a flexible structure or fuse into a rigid case. If the plates do not fuse, they are connected by an unusual tissue that can be stiff or flexible—depending on conditions.

Another remarkable event in echinoderm evolution was the origin of a unique morphological feature: a series of branching, fluid-filled tubes and chambers called the **water vascular system**. In some groups, one of the tubes is open to the exterior where it meets the body wall, so seawater can flow into and out of the system. Inside, fluids move via the beating of cilia that line the interior of the tubes and chambers.

(a) Echinoderms have a water vascular system.

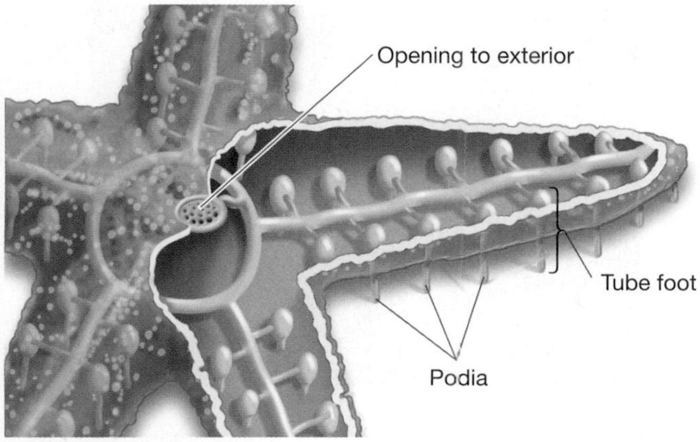

(b) Podia project from the underside of the body.

FIGURE 34.3 Echinoderms Have a Water Vascular System. **(a)** The water vascular system is a series of tubes and chambers that radiates throughout the body, forming a sophisticated hydrostatic skeleton. **(b)** Podia aid in movement because they extend from the body and can grab and release the substrate.

Figure 34.3a highlights a particularly important part of the water vascular system called tube feet. **Tube feet** are elongated, fluid-filled appendages. **Podia** (literally, "feet") are sections of the tube feet that project outside the body and make contact with the substrate (**Figure 34.3b**). The water vascular system forms a specialized hydrostatic skeleton that operates the podia. As podia extend and contract in a coordinated way along the base of an echinoderm, they alternately grab and release from the substrate. As a result, the individual moves.

Radial symmetry in adults, an endoskeleton of calcium carbonate, and the water vascular system are all synapomorphies—traits that identify echinoderms as a monophyletic group.

How Do Echinoderms Feed?

Depending on the lineage and species in question, echinoderms make their living by suspension feeding, deposit feeding, or harvesting algae or other animals. In most cases, an echinoderm's podia play a key role in obtaining food.

Many sea stars, for example, prey on bivalves. Clams and mussels respond to sea star attacks by contracting muscles that close their shells tight. But by clamping onto each shell with their podia, making their endoskeleton rigid, and pulling, sea stars are often able to pry the shells apart a few millimeters (**Figure 34.4a**). Once a gap exists, the sea star extrudes its stomach from its body and forces the stomach through the opening between the bivalve's shells. Upon contacting the visceral mass of the bivalve, the stomach of the sea star secretes digestive enzymes. It then begins to absorb the small molecules released by enzyme action. Eventually, only the shells of the prey remain.

Podia are also used in suspension feeding (**Figure 34.4b**). In most cases, podia are extended out into the water. When food particles contact them, the podia flick the food down to cilia, which sweep the particles toward the animal's mouth.

In deposit feeders, podia secrete mucus that is used to sop up food material on the substrate. The podia then roll the food-laden mucus into a ball and move it to the mouth.

(a) Sea star podia adhere to bivalve shells and pull them apart.

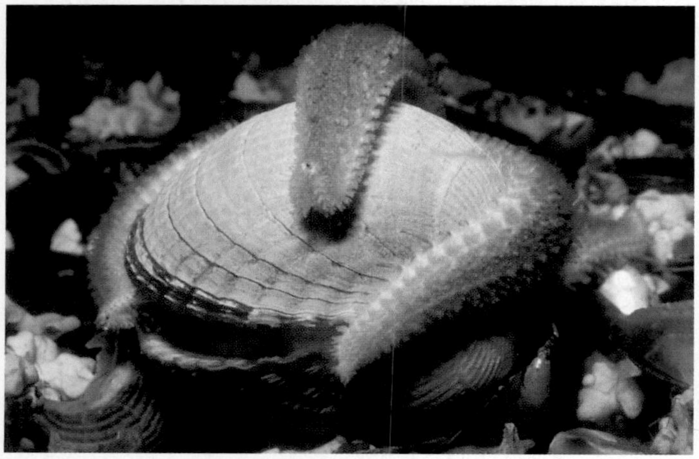

(b) Feather star podia trap particles during suspension feeding.

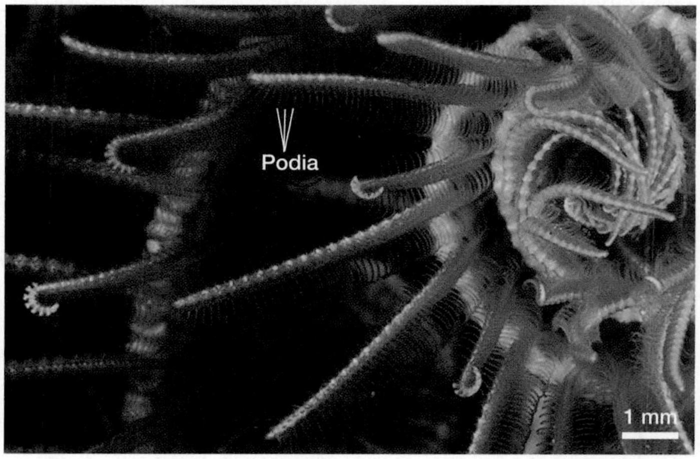

FIGURE 34.4 Echinoderms Use Their Podia in Feeding.

Key Lineages

The echinoderms living today make up five major lineages: (1) feather stars and sea lilies, (2) sea stars, (3) brittle stars and basket stars, (4) sea urchins and sand dollars, and (5) sea cucumbers (**Figure 34.5**). ✔**You should be able to indicate the origin of pentaradial symmetry in adults, the water vascular system, and the echinoderm endoskeleton on Figure 34.5.**

- Most feather stars and sea lilies are sessile suspension feeders.
- Brittle stars and basket stars have five or more long arms that radiate out from a small central disk. They use these arms to suspension feed, deposit feed by sopping up material with mucus, or capture small prey animals.
- Sea cucumbers are sausage-shaped animals that suspension feed or deposit feed with the aid of modified tube feet called tentacles that are arranged in a whorl around their mouths.

Sea stars and the sea urchins and sand dollars are described in detail in the boxes that follow.

- Echinodermata > Asteroidea (Sea Stars)
- Echinodermata > Echinoidea (Sea Urchins and Sand Dollars)

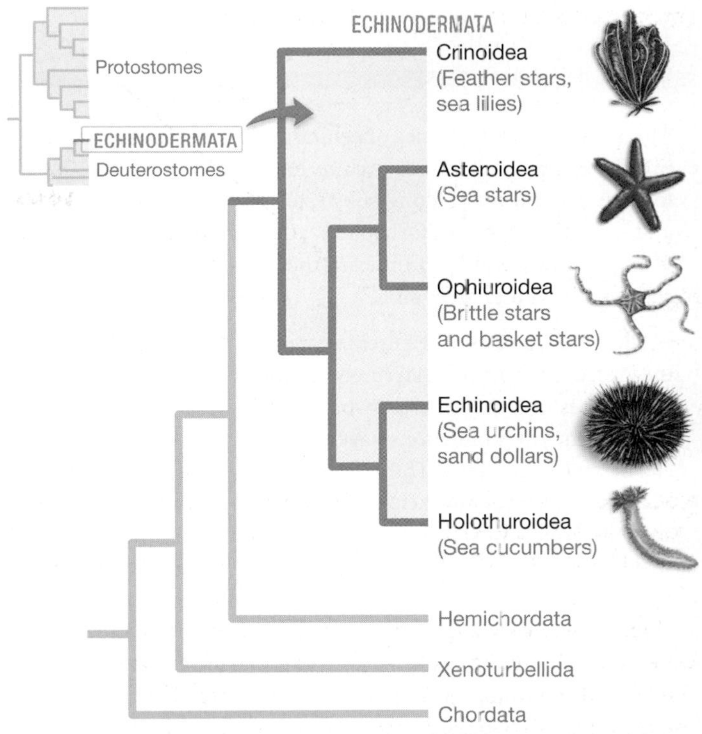

FIGURE 34.5 There Are Five Major Lineages of Echinoderms.

Echinodermata > Asteroidea (Sea Stars)

The 1700 known species of sea stars have bodies with five or more long arms—in some species up to 40—radiating from a central region that contains the mouth, stomach, and anus (**Figure 34.6**). Unlike brittle stars, though, the sea star's arms are not set off from the central region by clear, joint-like articulations. ✔**You should be able to indicate the origin of arms, and the origin of arms that are continuous with the central region as in sea stars, on Figure 34.5.**

When fully grown, sea stars can range in size from less than 1 cm in diameter to 1 m across. They live on hard or soft substrates along the coasts of all the world's oceans. Although the spines that are characteristic of some echinoderms are reduced to knobs on the surface of most sea stars, the crown-of-thorns star and a few other species have prominent, upright, movable spines.

Feeding Sea stars are predators or scavengers. Some species pull bivalves apart with their tube feet and feed by everting their stomach into the prey's visceral mass. In some cases, sea star predation on bivalves opens up space for other organisms—increasing biodiversity (see Chapter 53). Sponges, barnacles, and snails are also common prey. The crown-of-thorns sea star specializes in feeding on corals and is native to the Indian Ocean and western Pacific Ocean. Its population has skyrocketed recently—possibly because people are harvesting their major predator, a large snail called the triton, for its shell. Large crown-of-thorns star populations have led to the destruction of large areas of coral reef.

Movement Sea stars crawl with the aid of their tube feet.

Reproduction Sexual reproduction predominates in sea stars, and sexes are separate. At least one sea star arm is filled with reproductive organs that produce massive amounts of gametes—millions of eggs per female, in some species. Some species care for their offspring by holding fertilized eggs on their body until the eggs hatch. Larvae are long-lived and may spend weeks or months in the plankton, possibly dispersing thousands of kilometers. Most sea stars are capable of regenerating arms that are lost in predator attacks or storms. Some species can reproduce asexually by dividing the body in two, with each of the two individuals then regenerating the missing half.

Crossaster papposus

FIGURE 34.6 Sea Stars May Have Many Arms.

There are about 800 species of echinoids living today; most are sea urchins or sand dollars. Sea urchins have globe-shaped bodies and long spines and crawl along substrates. Sand dollars are flattened and disk-shaped, have short spines, and burrow in soft sediments. ✔You should be able to indicate the origin of globular or disc-shaped bodies on Figure 34.5.

Feeding Most types of sea urchins are herbivores. In some areas of the world, urchins are extremely important grazers on kelp and other types of algae. In cases when humans have removed predators and urchin populations have grown in response, their grazing can destroy kelp forests. The urchins chew away a structure called a holdfast, which anchors kelp to the substrate, causing them to float away and die. Most echinoids have a unique, jaw-like feeding structure in their mouths that is made up of five calcium carbonate teeth attached to muscles (**Figure 34.7a**). In many species, this apparatus can extend and retract as the animal feeds, allowing it to reach more food material. Sand dollars, in contrast, are suspension feeders. The beating of cilia on their spines creates tiny water currents which bring organic material toward the body. The food particles are then trapped in mucus and transported in long strings to the mouth. In this way, they harvest nutritious detritus found in sand or in other soft substrates (**Figure 34.7b**).

Movement Using their spines, sea urchins crawl and sand dollars burrow. Some species also use their podia to aid movement.

Reproduction Sexual reproduction predominates in sea urchins and sand dollars. Fertilization is external, and sexes are separate. As in sea stars, larvae are long-lived and may disperse long distances. The larvae of some species undergo asexual reproduction by dividing in two or budding off new individuals, before they undergo metamorphosis.

(a) Sea urchin *Lytechinus variegatus*

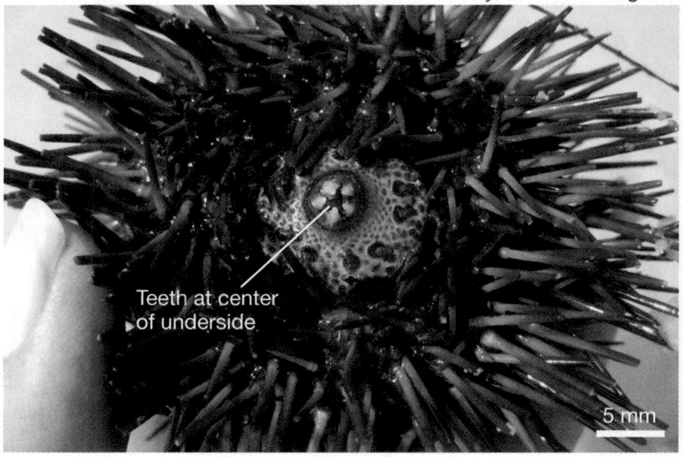

(b) Sand dollar *Echinarachnius parma*

FIGURE 34.7 Sea Urchins and Sand Dollars Are Closely Related.

CHECK YOUR UNDERSTANDING

If you understand that . . .

- Echinoderms have a distinctive body plan: pentaradial symmetry in adults, an endoskeleton, and a water vascular system.
- Most echinoderms use their podia to move, but they feed in a wide variety of ways—including using their podia to pry open bivalves, suspension feed, or deposit feed.

✔ You should be able to . . .

Suppose you were walking along the ocean and encountered an animal you had never seen before. Explain how you would figure out whether it was an echinoderm or not.

Answers are available in Appendix B.

34.2 What Is a Chordate?

The **chordates** are defined by the presence of four morphological features:

1. openings into the throat called **pharyngeal gill slits**;
2. a dorsal hollow **nerve cord** that runs the length of the body, comprised of projections from neurons;
3. a stiff and supportive but flexible rod called a **notochord**, which runs the length of the body; and
4. a muscular, post-anal tail—meaning a tail that contains muscle and extends past the anus.

Electrical signals carried by the dorsal hollow nerve cord coordinate muscle movement, and the notochord stiffens the muscular tail. Together, these traits create a "torpedo"—a long, streamlined animal that can swim forward rapidly. Because pharyngeal gill slits often function in suspension feeding, chordates are yet another illustration of a fundamental theme in animal evolution: diversification was driven by innovations that affect feeding and movement.

Three "Subphyla"

The phylum Chordata is made up of three major lineages: the (1) cephalochordates, (2) urochordates, and (3) vertebrates. All four defining characteristics of chordates—pharyngeal gill slits, dorsal hollow nerve cords, notochords, and muscular tails that extend past the anus—are found in these species at some stage in their life cycle.

THE CEPHALOCHORDATE BODY PLAN **Cephalochordates** are also called lancelets or amphioxus; they are small, mobile, torpedo-shaped animals with a "fish-like" appearance, and make their living by suspension feeding (**Figure 34.8a**). Adult cephalochordates live in ocean-bottom habitats, where they burrow in sand and suspension feed with the aid of their pharyngeal gill slits.

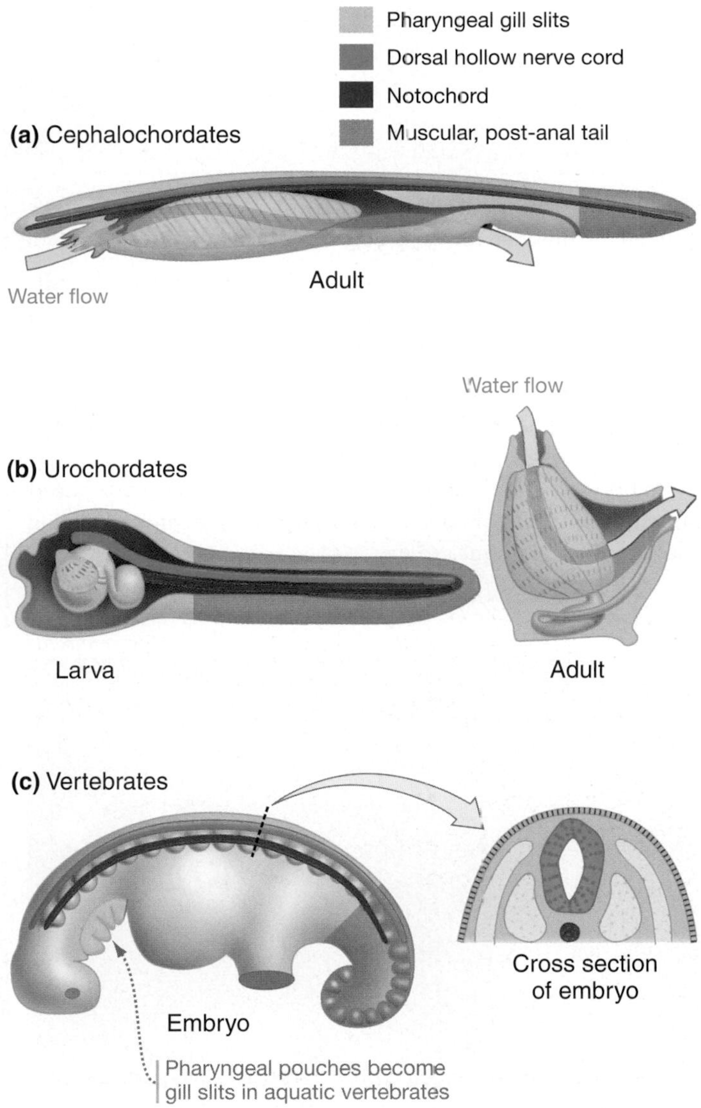

FIGURE 34.8 Four Traits Distinguish the Chordates.

The cephalochordates also have a notochord that functions as an endoskeleton. Because it stiffens their bodies, muscle contractions on either side result in fishlike movement when they swim during dispersal or mating.

THE UROCHORDATE BODY PLAN **Urochordates** are also called tunicates or sea squirts. As **Figure 34.8b** shows, pharyngeal gill slits are present in both larvae and adults and function in both feeding and gas exchange.

The dorsal hollow nerve cord, notochord, and tail are present in larvae and in the sexually mature forms of species that move as adults. Because the notochord stiffens the tail, muscular contractions on either side of a larva's tail wag it back and forth and result in swimming movements. As larvae swim or float in the upper water layers of the ocean, they drift to new habitats where food might be more abundant. As a result, larvae function as a dispersal stage in the life cycle.

THE VERTEBRATE BODY PLAN Vertebrates (**Figure 34.8c**) include the hagfish, lamprey, sharks, several lineages of fishes, amphibians, reptiles (including birds), and mammals. In vertebrates, the dorsal hollow nerve cord is the familiar spinal cord—a bundle of nerve cells that runs from the brain to the posterior of the body. Electrical signals from the spinal cord control movements and the function of organs (see Chapter 45).

Structures called pharyngeal pouches are homologous to pharyngeal gill slits and are present in all vertebrate embryos. In aquatic species, the creases between pouches open into gill slits and develop into part of the main gas-exchange organ—the **gills**. In terrestrial species, however, gill slits do not develop after the pharyngeal pouches form. In these species, the pharyngeal pouches are a vestigial trait (see Chapter 24).

A notochord appears in all vertebrate embryos. It no longer functions in body support and movement, however, as these roles are filled by the vertebral column. Instead, the notochard helps organize the body plan, early in development. You might recall from Chapter 22 that cells in the notochord secrete proteins that help induce the formation of somites—segmented blocks of tissue that form along the length of the body. Although the notochord itself disappears, cells in the somites later differentiate into the vertebrae, ribs, and skeletal muscles of the back, body wall, and limbs. In this way, the notochord is instrumental in the development of the trait that gave vertebrates their name.

Key Lineages: The Invertebrate Chordates

There are about 1600 species of tunicates, 24 species of lancelets, and over 50,000 vertebrates. Because they are so species-rich and diverse, vertebrates rate their own section in this chapter (Section 34.3). Here let's focus on the tunicates and lancelets. These lineages are sometimes referred to as the invertebrate chordates—meaning, members of the Chordata that lack vertebrae.

- Chordata > Cephalochordata (Lancelets)
- Chordata > Urochordata (Tunicates)

Chordata > Cephalochordata (Lancelets)

About two dozen species of cephalochordate have been described to date, all of them found in marine sands. Lancelets—also called amphioxus—have several characteristics that are intermediate between invertebrates and vertebrates. Chief among these is a notochord that is retained in adults, where it functions as an endoskeleton.

Feeding Adult cephalochordates feed by burrowing in sediment until only their head is sticking out into the water. They take water in through their mouth and trap food particles with the aid of mucus on their pharyngeal gill slits (**Figure 34.9**).

Movement Adults have large blocks of muscle arranged in a series along the length of the notochord. Lancelets are efficient swimmers because the flexible, rod-like notochord stiffens the body, making it wriggle when the blocks of muscle contract.

Reproduction Asexual reproduction is unknown, and individuals are either male or female. Gametes are released into the environment and fertilization is external.

Branchiostoma lanceolatum

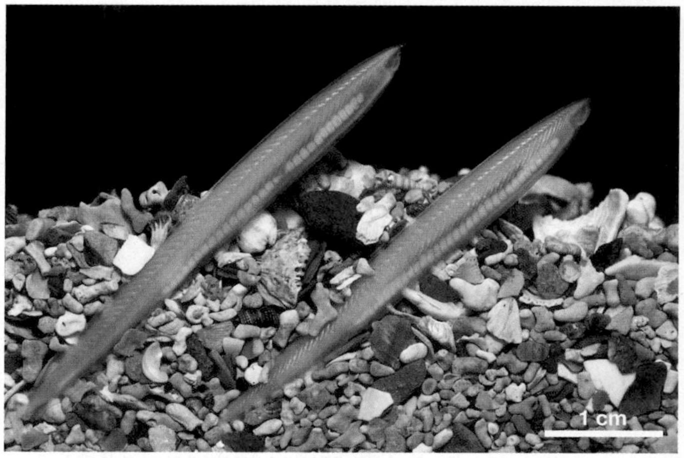

FIGURE 34.9 Lancelets Look Like Fish but Are Not Vertebrates.

Chordata > Urochordata (Tunicates)

The urochordates are also called tunicates and are comprised of two major sub-lineages: the sea squirts (or ascidians) and salps. All of the species described to date live in the ocean. Sea squirts live on the ocean floor (**Figure 34.10a**); salps live in open water (**Figure 34.10b**).

The synapomorphies that define the urochordates include an exoskeleton-like coat of polysaccharide, called a tunic, that covers and supports the body; a U-shaped gut; and two body openings, called siphons, where water enters and leaves during feeding.

Feeding Adult urochordates use their pharyngeal gill slits to suspension feed. A mucous sheet on the inside of the pharynx traps particles present in the water that enters one siphon, passes through the slits, and leaves through the other siphon.

Movement Larvae swim with the aid of the notochord, which stiffens the body and functions as a simple endoskeleton. Larvae are a dispersal stage and do not feed. Adults are sessile or float in currents.

Reproduction In most species, individuals produce both sperm and eggs. In some species, both sperm and eggs are shed into the water and fertilization is external; in other species, sperm are released into the water but eggs are retained, so that fertilization and early development are internal. Asexual reproduction by budding is also common in some groups.

(a) Sea squirt

Rhopalaea crassa

(b) Salp

Salpa maxima

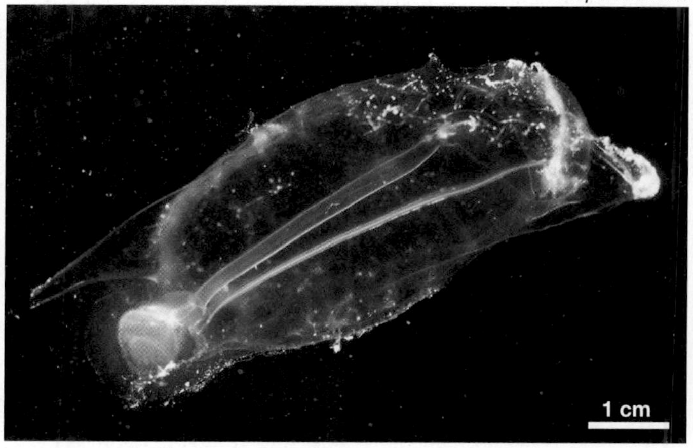

FIGURE 34.10 Sea Squirts and Salps Live in Different Habitats.

34.3 What Is a Vertebrate?

The vertebrates are a monophyletic group distinguished by two synapomorphies:

- a column of cartilaginous or bony structures called **vertebrae**, which form along the dorsal sides of the body; and
- a **cranium**, or skull—a bony, cartilaginous, or fibrous case that encloses the brain.

The vertebral column is important because it protects the spinal cord. The cranium is important because it protects the brain along with sensory organs such as eyes. Together, the vertebrae and cranium protect the central nervous system (CNS) and key sensory structures (see Chapter 32).

The vertebrate brain develops as an outgrowth of the most anterior end of the dorsal hollow nerve cord, and is important to the vertebrate lifestyle. Vertebrates are active predators and herbivores that can make rapid, directed movements with the aid of their endoskeleton. Coordinated movements are possible in part because vertebrates have large brains divided into three distinct regions:

1. a **forebrain**, housing the sense of smell;
2. a **midbrain**, associated with vision; and
3. a **hindbrain**, responsible for balance and, in some species, hearing.

Part of the forebrain is elaborated into a large structure called the **cerebrum**. In the jawed vertebrates, or **gnathostomes** (pronounced *NATH-oh-stomes*) the hindbrain consists of enlarged regions called the **cerebellum** and **medulla oblongata**. (The structure and functions of the vertebrate brain are analyzed in detail in Chapter 45.) The evolution of a large, three-part brain, protected by a hard cranium, was a key innovation in vertebrate evolution.

An Overview of Vertebrate Evolution

Vertebrates have been the focus of intense research for well over 300 years, in part because they are large and conspicuous and in part because they include the humans. All this effort has paid off in an increasingly thorough understanding of how the vertebrates diversified. Let's consider what data from the fossil record have to say about key events in vertebrate evolution and then examine the current best estimate of the phylogenetic relationships among vertebrates.

THE VERTEBRATE FOSSIL RECORD Both echinoderm and vertebrate fossils are present in rocks that formed during the Cambrian explosion (see Chapter 27). The vertebrate fossils in these rocks show that the earliest members of this lineage lived in the ocean about 540 million years ago, had streamlined, fishlike bodies, and appear to have had a skull made of cartilage. **Cartilage** is a strong but flexible tissue—found in your earlobes and the tip of your nose—that consists of scattered cells in a gel-like matrix of polysaccharides and protein fibers.

A cartilaginous endoskeleton is a basic vertebrate feature; only in the bony fishes and their descendants, the **tetrapods** (literally, "four-footed"), does the skeleton become composed of bone in the adult. **Bone** is a tissue consisting of cells and blood vessels encased in a matrix made primarily of a calcium- and phosphate-containing compound called hydroxyapatite, along with a small amount of protein fibers. In species with bony skeletons, growing bones are cartilaginous first and only become bony later in development.

Following the appearance of vertebrates, the fossil record documents a series of key innovations that occurred as this lineage diversified (**Figure 34.11**):

- *Bony exoskeleton* Fossil vertebrates from the early part of the Ordovician period, about 480 million years ago, are the first fossils to have bone. When bone first evolved, it did not occur in the endoskeleton. Instead, bone was deposited in scalelike plates that formed an exoskeleton—a hard structure that envelops the body. Most of your skull is made up of this type of dermal ("skin") bone—making your skull similar to the "head shields" in Ordovician fishes. Based on the fossils' overall morphology, biologists infer that these animals swam with the aid of a notochord and that they breathed and fed by gulping water and filtering it through their pharyngeal gill slits. Presumably, the bony plates helped provide protection from predators.

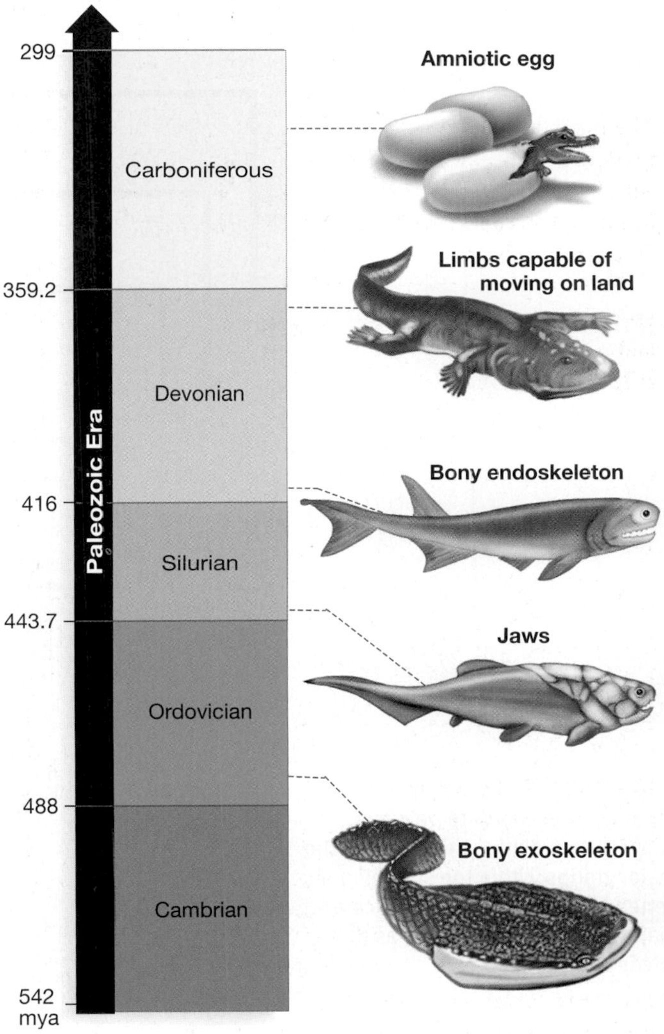

FIGURE 34.11 Timeline of the Early Vertebrate Fossil Record.

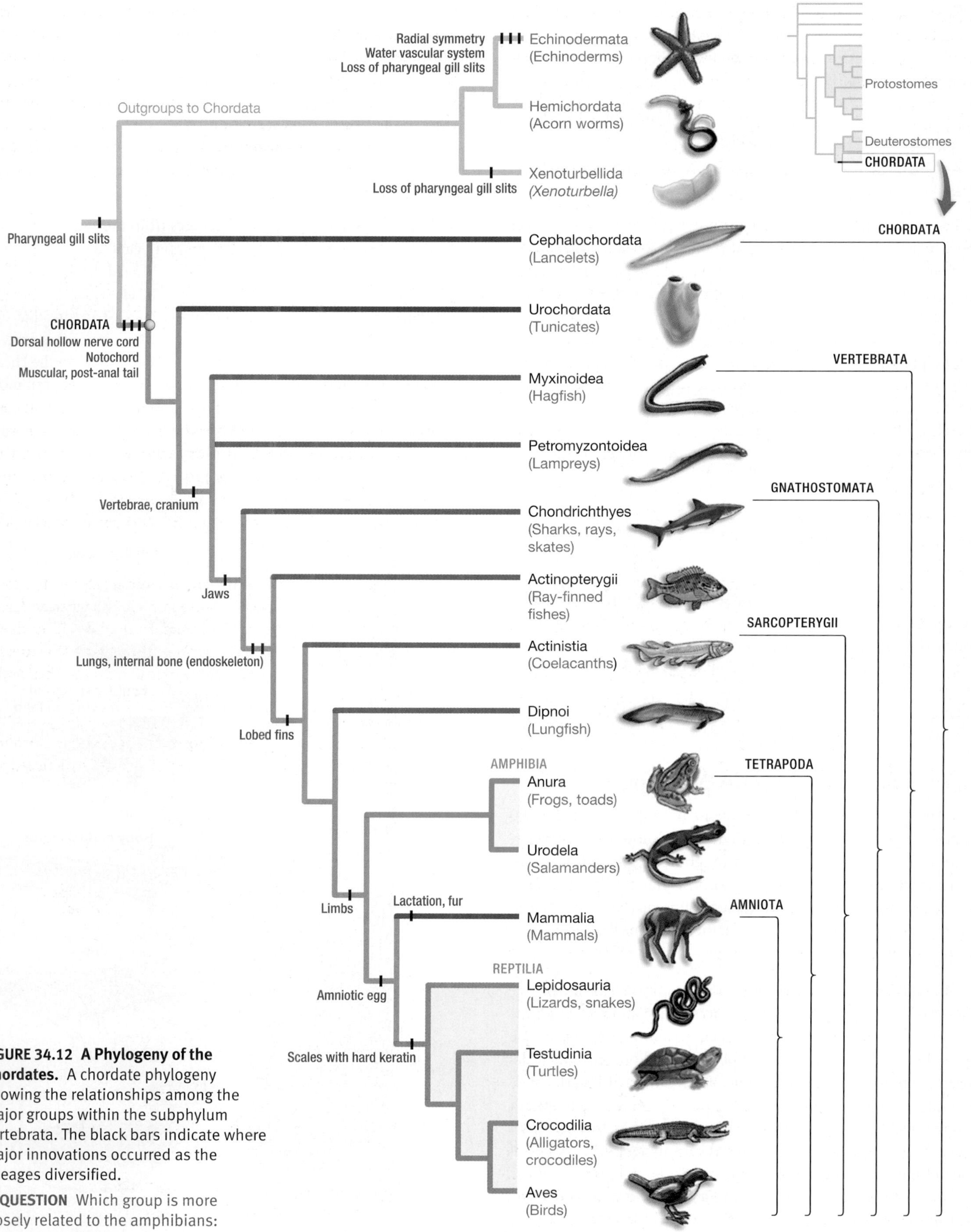

FIGURE 34.12 A Phylogeny of the Chordates. A chordate phylogeny showing the relationships among the major groups within the subphylum vertebrata. The black bars indicate where major innovations occurred as the lineages diversified.

✔**QUESTION** Which group is more closely related to the amphibians: mammals or birds?

- *Jaws* The first fish with jaws show up early in the Silurian, about 440 million years ago. The evolution of jaws was significant because it gave vertebrates the ability to bite, meaning that they were no longer limited to suspension feeding or deposit feeding. Instead, they could make a living as herbivores or predators. Soon after, jawbones with teeth appear in the fossil record. With jaws and teeth, vertebrates became armed and dangerous. The fossil record shows that a spectacular radiation of jawed fishes followed, filling marine and freshwater habitats.
- *Bony endoskeleton* In some lineages of fish found early in the Devonian period, the cartilaginous endoskeleton began to be stiffened by the deposition of bone. Unlike the dermal bone that had evolved earlier, the bony endoskeleton functioned to support movement—rapid swimming.
- *Limbs capable of moving on land* The next great event in the evolution of vertebrates was the transition to living on land. The first animals that had limbs and were capable of moving on land are dated to about 365 million years ago, late in the Devonian. These were the first of the tetrapods—animals with four limbs.
- *Amniotic egg* About 20 million years after the appearance of tetrapods in the fossil record, the first amniotes are present. The Amniota is a lineage of vertebrates that includes all tetrapods other than amphibians. The **amniotes** are named for a signature synapomorphy and adaptation: the amniotic egg. An **amniotic egg** is an egg that has membranes surrounding a food supply, a water supply, and a waste repository. These membranes provided support and extra surface area for gas exchange, allowing amniotes to produce larger, better-developed young.

To summarize, the fossil record indicates that vertebrates evolved through a series of major steps, beginning about 540 million years ago with vertebrates whose endoskeleton consisted of a notochord. The earliest vertebrates gave rise to cartilaginous fishes (sharks and rays) and several lineages of fish with bony skeletons. After the tetrapods emerged and amphibians resembling salamanders began to live on land, the evolution of the amniotic egg paved the way for the diversification of amniotes—specifically, reptiles and the animals that gave rise to mammals. Do phylogenetic trees estimated from analyses of DNA sequence data agree or conflict with these conclusions?

EVALUATING MOLECULAR PHYLOGENIES **Figure 34.12** provides a phylogenetic tree that summarizes the relationships among vertebrates, based on DNA sequence data. The labeled bars on the tree indicate where major innovations occurred. Although the phylogeny of deuterostomes continues to be a topic of intense research, researchers are increasingly confident that the relationships described in Figure 34.12 are accurate. Note that according to recent data, the tunicates are the closest living relatives of the vertebrates.

The overall conclusion from this tree is that the branching sequence inferred from morphological and molecular data correlates with the sequence of first appearances of groups in the fossil record. The sister groups to vertebrates lack the skull and vertebral column that define the vertebrates, and the sister groups to jawed vertebrates lack jaws and bony skeletons.

SEVERAL LINEAGES ARE FISHES The tree also indicates that there is no monophyletic group that includes all of the fishlike lineages. Instead, "fishy" organisms are a series of independent monophyletic groups. Taken together, they form what biologists call a **grade**—a sequence of lineages that are paraphyletic. You might recall from Chapter 30 that the non-vascular plants and seedless vascular plants also form grades and are paraphyletic.

✔If you understand the difference between a paraphyletic and monophyletic group, you should be able to draw what Figure 34.12 would look like if there actually were a monophyletic group called "The Fish."

To understand what happened during the subsequent diversification of vertebrates, let's explore some of the major innovations involved in feeding, movement, and reproduction in more detail.

Key Innovations

Among vertebrates, the most species-rich and ecologically diverse lineages are the ray-finned fishes and the tetrapods. Ray-finned fishes occupy habitats ranging from deepwater environments, which are perpetually dark, to shallow ponds that dry up each year. Tetrapods include the large herbivores and predators in terrestrial environments all over the world. About 27,000 species of ray-finned fish and about 27,000 species of tetrapods are living today (**Figure 34.13**).

THE VERTEBRATE JAW The most ancient vertebrates in the fossil record have relatively simple mouthparts. Today's hagfish and lampreys retain this trait: They lack jaws and cannot bite algae, plants, or animals efficiently. They have to make their living as ectoparasites or as deposit feeders that scavenge dead animals. Vertebrates were not able to harvest food by biting until jaws evolved.

The leading hypothesis for the origin of the jaw proposes that natural selection acted on mutations that affected the morphology of **gill arches**, which are curved regions of tissue between the

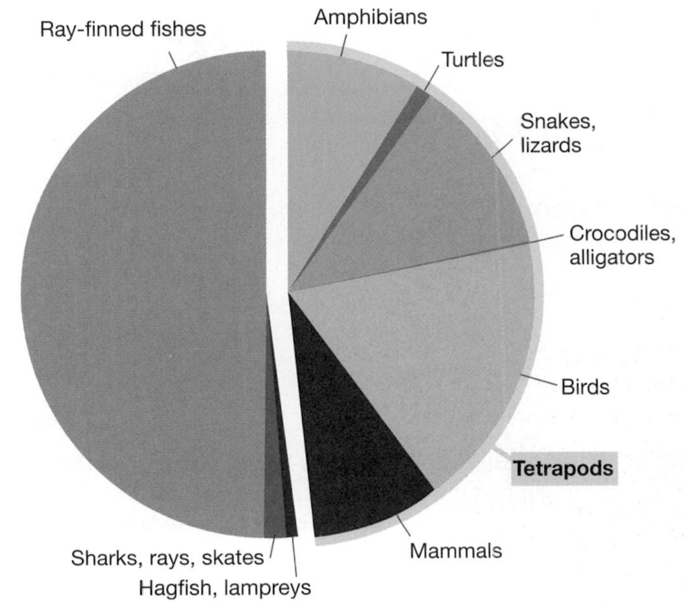

FIGURE 34.13 Relative Species Abundance among Vertebrates.

gills. The jawless vertebrates have bars of cartilage that stiffen these gill arches. The gill-arch hypothesis proposes that mutation and natural selection increased the size of the most anterior arch and modified its orientation slightly, producing the first working jaw (**Figure 34.14**). Three lines of evidence, drawn from comparative anatomy and embryology, support this hypothesis:

1. Both gill arches and jaws consist of flattened bars of bony or cartilaginous tissue that hinge and bend forward.
2. During development, the same population of cells gives rise to the muscles that move jaws and the muscles that move gill arches.
3. Unlike most other parts of the vertebrate skeleton, both jaws and gill arches are derived from specialized cells in the embryo called neural crest cells.

Taken together, these data support the hypothesis that jaws evolved from gill arches.

To explain why ray-finned fishes are so diverse in their feeding methods, biologists point to important modifications of the jaw:

- In most ray-finned fishes, the jaw is protrusible—meaning it can be extended to nip or bite out at food.

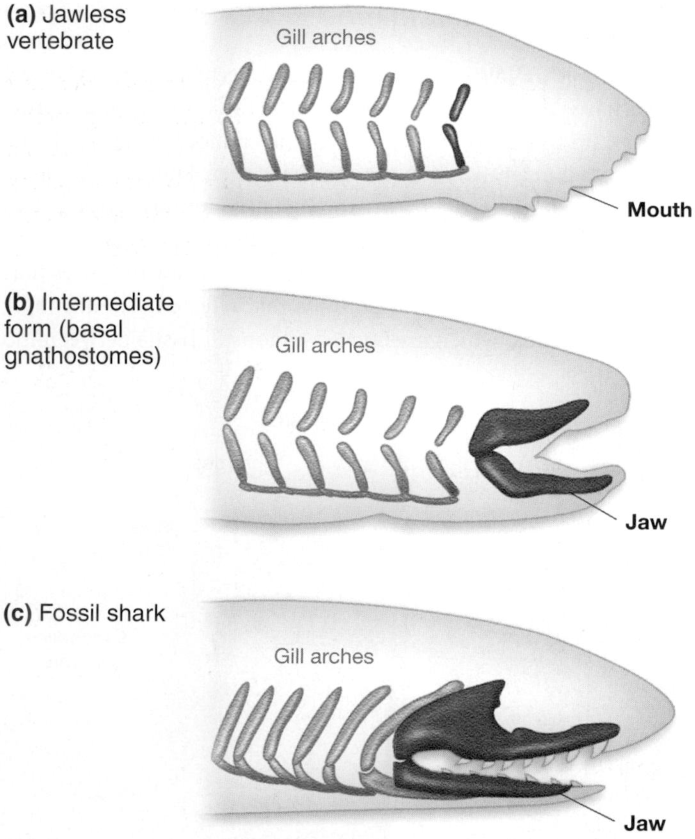

FIGURE 34.14 The Gill-Arch Hypothesis for the Evolution of the Jaw. **(a)** Gill arches support the gills in jawless vertebrates. **(b)** This intermediate form has yet to be found in the fossil record. **(c)** Fossil sharks that appeared later had more elaborate jaws.

✔**QUESTION** The transition from gill arches to jaws is complex. Would intermediate stages in the evolution of the jaw have any function?

- Several species-rich lineages of ray-finned fishes have a **pharyngeal** ("throat") **jaw**, consisting of modified gill arches, that functions as a second set of jaws. They are located in the back of the throat and make food processing particularly efficient.

(For more on the structure and function of pharyngeal jaws, see Chapter 43.)

To summarize, the radiation of ray-finned fishes was triggered in large part by the evolution of the jaw, by modifications that made it possible to protrude the jaw, and by the origin of the pharyngeal jaw. The story of tetrapods is different, however. Although jaw structure varies somewhat among tetrapod groups, the adaptation that triggered their initial diversification involved the ability to move and get to food, not to bite it and process it.

THE TETRAPOD LIMB To understand how tetrapods made the water-to-land transition, consider the morphology and behavior of their closest living relatives, the lungfish (**Figure 34.15**). Most living species of lungfish inhabit shallow, oxygen-poor water. To supplement the oxygen taken in by their gills, they have lungs and breathe air. (Note that many other fish have lungs, which function in gas exchange or maintaining buoyancy in water.) Some lungfish also have fleshy fins supported by bones and are capable of walking short distances along watery mudflats or the bottoms of ponds. In addition, some species can survive extended droughts by burrowing in mud.

Lungfish have a series of adaptations that allow them to survive on land for short periods. Could a lungfish-like organism have evolved into the first land-dwelling vertebrate with limbs?

Fossils provide strong links between the ancestors of today's lungfish and the earliest land-dwelling vertebrates. **Figure 34.16** shows a phylogeny of the species involved. Note that during the Devonian, the fossil record documents a series of species that indicate a gradual transition from a lobe-like fin to a limb that could support walking on land. The fossil record is actually more complete than shown here—the figure is just a sample of the tetrapod and tetrapod-like species known from this interval.

The figure highlights the number and arrangement of bones in the fossil fish fin and in early tetrapods. The color coding emphasizes that each fin or limb has a single bony element that is proximal (closest to the body; in blue) and then two bones that

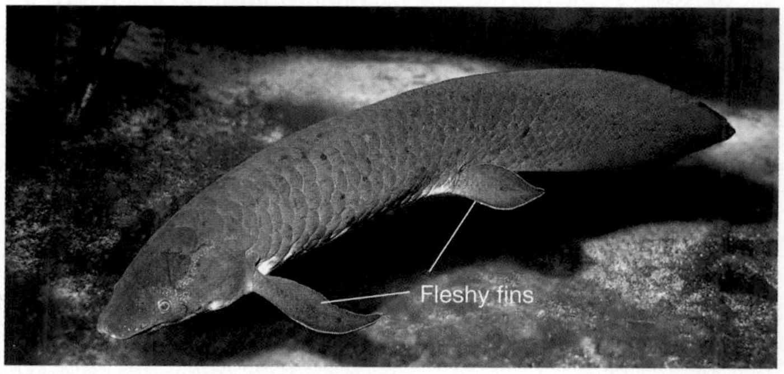

FIGURE 34.15 Lungfish Have Limb-like Fins. Some species of lungfish can walk or crawl short distances on their fleshy fins.

are distal (farther from the body) and arranged side by side (these are shown in red), followed by a series of distal elements (in orange). Because the structures are similar, and because no other animal groups have limb bones in this arrangement, the evidence for homology is strong.

To summarize, data from the fossil record support the hypothesis that mutation and natural selection gradually transformed fins into limbs as the first tetrapods became more and more dependent on terrestrial habitats.

The hypothesis that tetrapod limbs evolved from fish fins has also been supported by molecular genetic evidence. Recent work has shown that several regulatory proteins involved in the development of cartilaginous fish fins and ray-finned fish fins and the upper parts of mammal limbs are homologous. For example, the proteins produced by several different *Hox* genes (see Chapter 21) are found at the same times and in the same locations in fins and limbs.

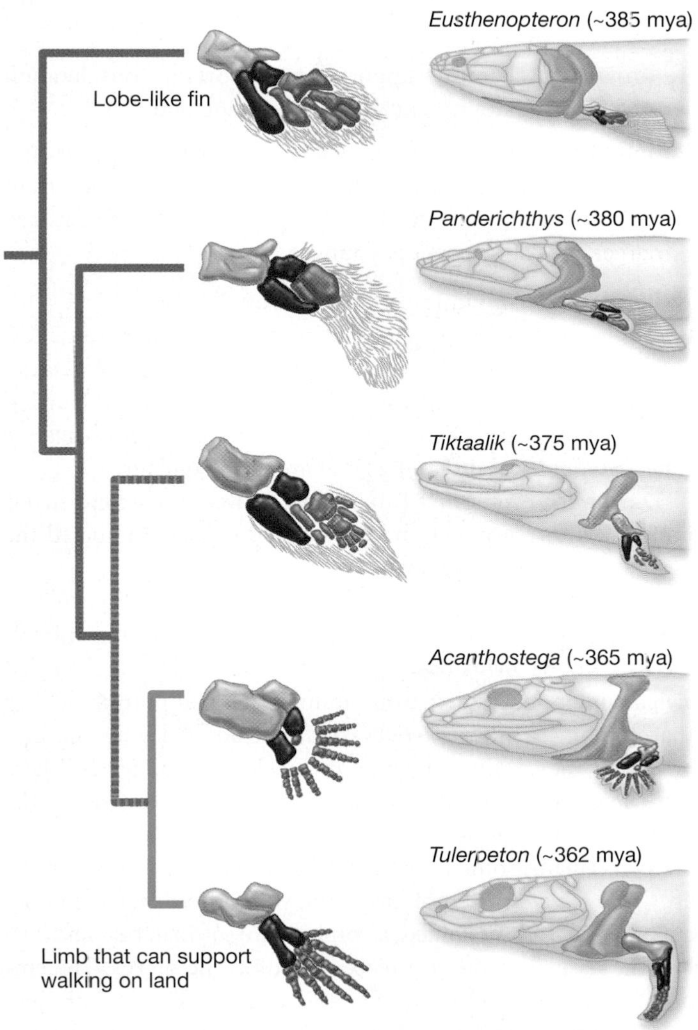

FIGURE 34.16 Evidence for a Fin-to-Limb Transition. The fossil record documents a series of species that indicate a gradual transition from a lobe-like fin to a limb that could support walking on land.

✔**QUESTION** This phylogeny was estimated from traits other than the morphology of the limb. Why is this important, in terms of addressing how limbs evolved?

These data suggest that these appendages are patterned by the same genes. As a result, the data support the hypothesis that tetrapod limbs evolved from fins.

FEATHERS AND FLIGHT Once the tetrapod limb evolved, natural selection elaborated it into structures that are used for running, gliding, crawling, burrowing, or swimming. In addition, wings and the ability to fly evolved independently in three lineages of tetrapods: the extinct flying reptiles called pterosaurs (pronounced *TARE-oh-sors*), the bats, and the birds.

How did flight evolve? The best data sets on this question involve feathered flight in birds. In the early 2000s, for example, Xing Xu and colleagues announced the discovery of a spectacular series of feathered dinosaur fossils (**Figure 34.17**). The newly discovered species address key questions about the evolution of birds, feathers, and flight:

- ***Did birds evolve from dinosaurs?*** On the basis of skeletal characteristics, all of these recently discovered fossil species clearly belong to a lineage of dinosaurs called the dromaeosaurs. Birds are part of the monophyletic group called dinosaurs.
- ***How did feathers evolve?*** The early fossils have an array of feather types that support Xu's model of feathers evolving in a series of steps, beginning with simple projections from the skin and culminating with the complex structures observed in today's birds.
- ***Did birds begin flying from the ground up or from the trees down?*** More specifically, did flight evolve with running species that began to jump and glide or make short flights, with the aid of feathers to provide lift? Or did flight evolve from tree-dwelling species that used feathers to glide from tree to tree, much as flying squirrels do today? This question is still unresolved.

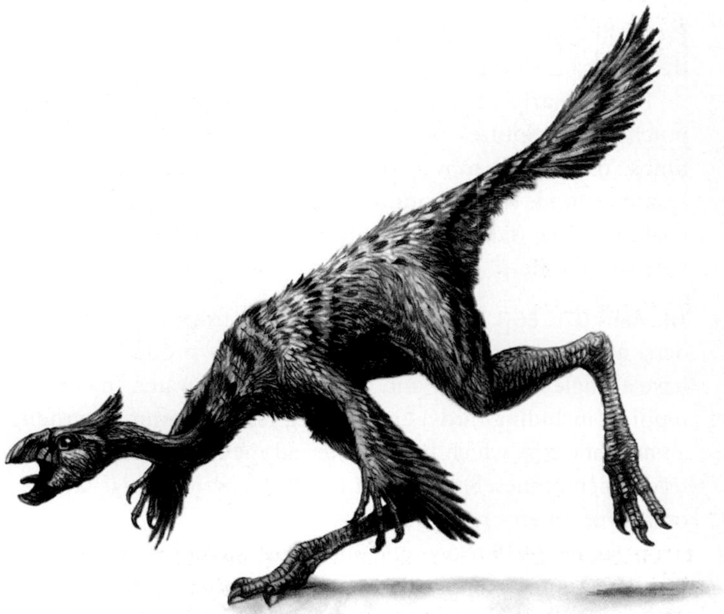

FIGURE 34.17 Feathers Evolved in Dinosaurs. An artist's depiction of what *Caudipteryx*, a feathered dinosaur, might have looked like in life.

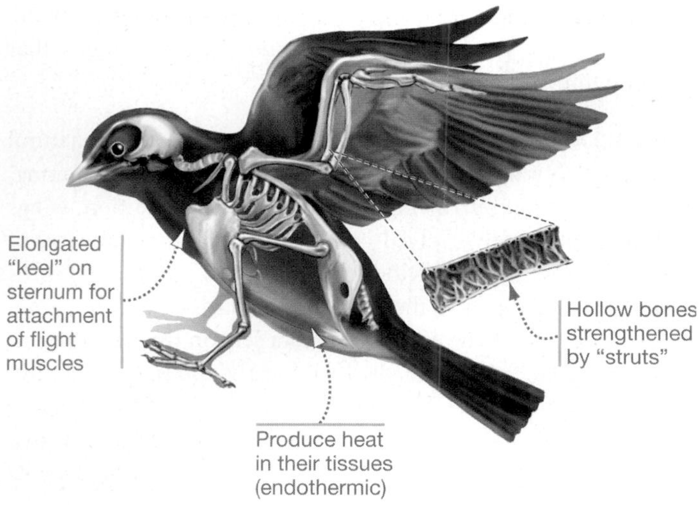
Elongated "keel" on sternum for attachment of flight muscles
Hollow bones strengthened by "struts"
Produce heat in their tissues (endothermic)

FIGURE 34.18 In Addition to Feathers, Birds Have Several Adaptations That Allow for Flight.

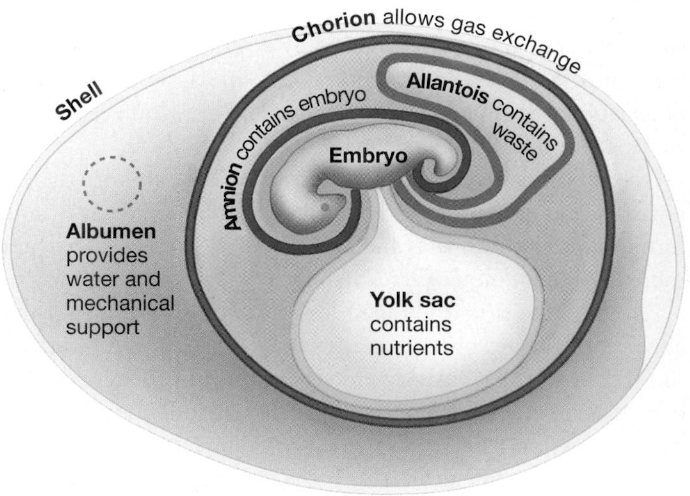
Chorion allows gas exchange
Shell
Amnion contains embryo
Allantois contains waste
Embryo
Albumen provides water and mechanical support
Yolk sac contains nutrients

FIGURE 34.19 An Amniotic Egg. Amniotic eggs have four membrane-bound sacs.

Once dinosaurs evolved feathers and took to the air, the fossil record shows that a series of adaptations made powered, flapping flight increasingly efficient (**Figure 34.18**).

- Most dinosaurs have a flat sternum (breastbone), but the bird sternum has a projection called the keel, which provides a large surface area to which flight muscles attach.
- Birds are extraordinarily light for their size, primarily because they have a drastically reduced number of bones and because their larger bones are thin-walled and hollow—though strengthened by bony "struts."
- Birds are capable of sustained activity year-round because they are endothermic—they maintain a high body temperature by producing heat in their tissues.

From dinosaurs that jumped and glided or floated from tree to tree, birds have evolved into extraordinary flying machines.

To summarize, the evolution of the jaw gave tetrapods the potential to capture and process a wide array of foods. With limbs, they could move efficiently on land—or even fly—in search of food. What about the other major challenge of terrestrial life? How did tetrapods produce offspring that could survive out of water?

THE AMNIOTIC EGG Fish and amphibians (frogs, toads, salamanders, and limbless species called caecilians) produce eggs that have a single membrane and lay their eggs in water. In contrast, reptiles (including birds) and the egg-laying mammals produce an **amniotic egg**, which has an external membrane and three internal membranes. Species that produce amniotic eggs lay them outside of water.

As **Figure 34.19** shows, the outermost membrane of an amniotic egg encloses a supply of water in a protein-rich solution called **albumen**. Albumen cushions the developing embryo and provides nutrients. The inner membranes surround the embryo itself, the yolk provided by the mother, and waste from the embryo. The additional membranes are thought to be advantageous because they:

1. provide mechanical support—an important consideration outside of a buoyant aquatic environment; and
2. increase the surface area available for exchange of gases and other materials. Efficient diffusion of molecules is important because it allows females to lay larger eggs that hatch into larger, more independent young.

In addition, amniotic eggs are surrounded by a shell. This layer is leathery in lizards and snakes. It is stiffened by some calcium carbonate deposits in turtles and crocodiles, and by extensive calcium carbonate deposits in birds. Lizards, snakes, turtles, and crocodiles bury their eggs in moist soils, but bird eggs are watertight and are laid in nests that are exposed to air.

During the evolution of mammals, however, a second major innovation in reproduction occurred—one that eliminated the need for any type of egg laying.

THE PLACENTA Recall from Chapter 32 that egg-laying animals are said to be **oviparous**; species that give birth are termed **viviparous**. In many viviparous animals, females produce an egg that contains a nutrient-rich yolk. Instead of laying the egg, however, the mother retains it inside her body. In these **ovoviviparous** species, the developing offspring depends on the resources in the egg yolk.

In caecilians, some lizards, and most mammals, however, the eggs that females produce have little yolk. After fertilization occurs and the egg is retained, a combination of maternal and embryonic tissues produces a placenta within the uterus. The embryo's contributions are the allantois and chorion—membranes that are involved in gas exchange in an amniotic egg. In the placenta, tissues derived from the allantois and chorion are also involved in the diffusion of gases, nutrients, and wastes.

The **placenta** is an organ that is rich in blood vessels and that facilitates a flow of oxygen and nutrients from the mother to the developing offspring (**Figure 34.20**). After a development period called **gestation**, the embryo emerges from the mother's body.

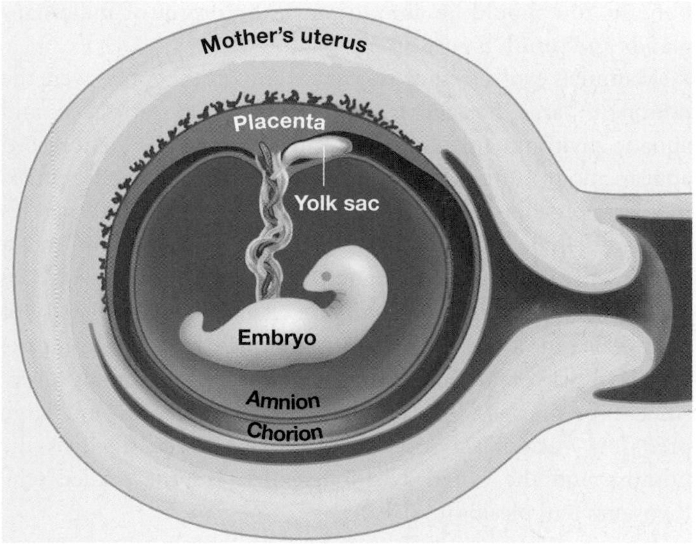

FIGURE 34.20 A Placenta Nourishes a Fetus Internally.

✔**QUESTION** Compare the relative size of the yolk sac here with that in the amniotic egg (Figure 34.19). Are they the same or different?

Why did viviparity and the placenta evolve? The leading hypotheses are that retaining the embryo inside the body has several advantages:

1. Offspring develop at a more constant, favorable temperature.
2. Offspring are protected.
3. Offspring are portable—mothers are not tied to a nest.

In effect, a placenta and viviparity are mechanisms for increasing the level of **parental care**: behavior by a parent that improves the ability of its offspring to survive. What other types of parental care evolved in vertebrates?

PARENTAL CARE In species that provide extensive parental care—even if only in the form of exceptionally large amounts of yolk in eggs—mothers participate in a fitness trade-off (see Chapter 24 and Chapter 41). They can produce fewer offspring, but the offspring that they do produce have a higher probability of survival.

In animals, mechanisms of parental care include incubating eggs to keep them warm during early development, keeping young warm and dry, supplying young with food, and protecting them from danger. In some insect and frog species, mothers carry eggs or newly hatched young on their bodies. In fishes, parents commonly guard eggs during development and fan them with oxygen-rich water.

Mammals and birds provide particularly extensive parental care. In both groups, the mother (and often the father, in birds) continues to feed and care for individuals after birth or hatching—sometimes for many years (**Figure 34.21**). Female mammals also **lactate**—meaning that they produce milk and use it to feed their offspring after birth.

Among large animals, the evolution of extensive parental care is hypothesized to be a major reason for the evolutionary success of mammals and birds. In both lineages, mothers produce relatively small numbers of large, high-quality offspring.

(a) Mammal mothers feed and protect newborn young.

(b) Many bird species have extensive parental care.

FIGURE 34.21 Parental Care in Mammals and Birds. (a) Female mammals feed and protect embryos inside their bodies until the young are well developed. Once the offspring is born, the mother feeds it milk until it is able to eat on its own. In some species, parents continue to feed and protect young for years. **(b)** In birds, one or both parents may incubate the eggs, protect the nest, and feed the young after hatching occurs.

CHECK YOUR UNDERSTANDING

If you understand that . . .

- Vertebrates have a distinctive body plan: bilateral symmetry with vertebrae that protect a spinal cord and a cranium that protects the brain.
- An array of key innovations occurred during the evolution of vertebrates: Jaws made it possible to bite and process food, limbs allowed tetrapods to move on land, and amniotic eggs were advantageous in terrestrial environments.

✔ **You should be able to . . .**

1. Explain the adaptive significance of the jaw.
2. Explain why the additional membranes in an amniotic egg allowed eggs to be larger.

Answers are available in Appendix B.

Key Lineages

The fossil record provides an increasingly clear picture of early vertebrate evolution; intensive and continuing research has explored the origins of key vertebrate innovations such as the jaw, the tetrapod limb, and the amniotic egg. Now that you've been introduced to key vertebrate features, let's take a detailed look at specific vertebrate lineages, beginning with an overview of the mammals and reptiles.

MAMMALIA **Mammals** are a monophyletic group, today comprising three major lineages: (1) the monotremes, (2) marsupials, and (3) eutherians (**Figure 34.22**). Monotremes lay eggs; marsupials have a poorly developed placenta but well-developed pouch for rearing offspring; eutherians have a well-developed placenta and extended pregnancy.

Mammals are easily recognized by the presence of hair or fur, which serves to insulate the body. Like birds, mammals are endotherms that maintain high body temperatures by oxidizing large amounts of food and generating large amounts of heat. Instead of insulating themselves with feathers, though, mammals retain heat because the body surface is covered with layers of hair or fur.

Endothermy evolved independently in birds and mammals. In both groups, endothermy is thought to be an adaptation that enables individuals to maintain high levels of activity—particularly at night or during cold weather.

In addition to being endothermic and having fur, mammals have **mammary glands**—a unique structure that makes lactation possible. The evolution of mammary glands gave mammals the ability to provide their young with particularly extensive parental care. Mammals are also the only vertebrates with facial muscles and lips—traits that make suckling milk possible—and the only vertebrates that have a lower jaw formed from a single bone. ✔You should be able to mark the origin of mammary glands and fur on Figure 34.22.

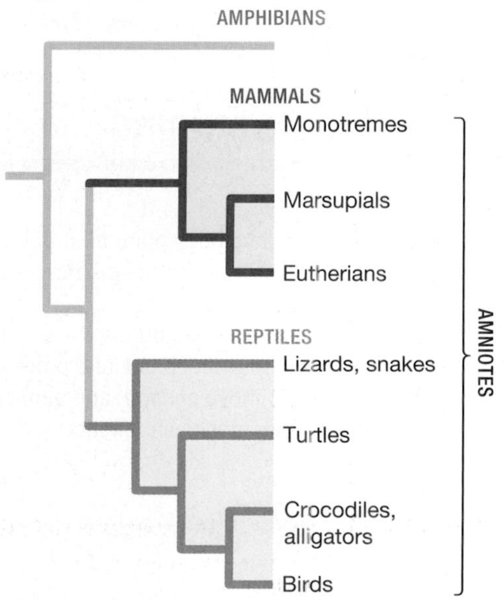

FIGURE 34.22 Mammals and Reptiles Are Monophyletic Groups.

Mammals evolved when dinosaurs and other reptiles were the dominant large herbivores and predators in terrestrial and aquatic environments. The earliest mammals in the fossil record appear about 195 million years ago; most were small animals that were probably active only at night. Many of the 4800 species of mammals living today have good nocturnal vision and a strong sense of smell, as their ancestors presumably did.

The adaptive radiation that gave rise to today's diversity of mammals did not take place until after the dinosaurs went extinct. Long after the dinosaurs were gone, the mammals diversified into lineages that included large herbivores and large predators—ecological roles that had once been filled by dinosaurs and the ocean-dwelling, extinct reptiles called ichthyosaurs and plesiosaurs.

REPTILIA The **reptiles** are a monophyletic group and represent one of the two major living lineages of amniotes—the other lineage consists of the extinct mammal-like reptiles and today's mammals. The major feature distinguishing the reptilian and mammalian lineages is the number and placement of openings in the side of the skull. These skull openings are important: Jaw muscles required for biting and chewing pass through them and attach to bones on the upper part of the skull.

Several features adapt reptiles for life on land. Their skin is made watertight by a layer of scales made of the protein keratin. Reptiles breathe air through well-developed lungs and lay shelled, amniotic eggs. In snakes and lizards, the egg has a leathery shell; in other reptiles, the shell includes some calcium carbonate.

In many reptiles, the sex of an individual is determined by the environment it experiences during early development. In certain species, for example, high temperatures produce mostly males while low temperatures produce mostly females.

The reptiles include the dinosaurs, pterosaurs (flying reptiles), and other extinct lineages that flourished from about 250 million years ago until the mass extinction at the end of the Cretaceous period, 65 million years ago. Today the Reptilia are represented by four major lineages: (1) lizards and snakes, (2) turtles, (3) crocodiles and alligators, and (4) birds.

Except for birds, almost all of the reptiles living today are **ectothermic** ("outside-heated")—meaning that individuals do not use internally generated heat to regulate their body temperature. It would be a mistake, however, to conclude that reptiles other than birds do not regulate their body temperature closely. Reptiles bask in sunlight, seek shade, and perform other behaviors to keep their body temperature at a preferred level.

To review the synapomorphies that biologists use to identify the major deuterostome and vertebrate groups, go to the study area at *www.masteringbiology.com*.

MB **Web Activity** Deuterostome Diversity

Now let's delve into a more detailed look at an array of fish lineages and major groups within Mammalia and Reptilia. All of the following groups are members of the Chordata and Vertebrata.

- Myxinoidea (Hagfish) and Petromyzontoidea (Lampreys)
- Chondrichthyes (Sharks, Rays, Skates)
- Actinopterygii (Ray-Finned Fishes)
- Actinistia (Coelacanths) and Dipnoi (Lungfish)
- Amphibia (Frogs, Salamanders, Caecilians)
- Mammalia > Monotremata (Platypuses, Echidnas)
- Mammalia > Marsupiala (Marsupials)
- Mammalia > Eutheria (Placental Mammals)
- Reptilia > Lepidosauria (Lizards, Snakes)
- Reptilia > Testudinia (Turtles)
- Reptilia > Crocodilia (Crocodiles, Alligators)
- Reptilia > Aves (Birds)

Chordata > Vertebrata > Myxinoidea (Hagfish) and Petromyzontoidea (Lampreys)

Although recent phylogenetic analyses indicate that hagfish and lampreys may belong to two independent lineages, some data suggest that they are a single group called the Cyclostomata ("round-mouthed"). The issue is not resolved, and will only be clarified with additional research. Because these animals are the only vertebrates that lack jaws, the 110 species in the two groups are still referred to as the jawless fishes.

The hagfish and lampreys are the only surviving members of the earliest branches at the base of the Vertebrata. Hagfish and lampreys have long, slender bodies and are aquatic. Hagfish are strictly marine; lampreys have both marine and freshwater forms. Most species are less than a meter long when fully grown. Adult hagfish lack any sort of vertebral column, but adult lampreys have small pieces of cartilage along the length of their dorsal hollow nerve cord. Both hagfish and lampreys have brains protected by a cranium, as do all vertebrates.

Feeding Hagfish are scavengers and predators (**Figure 34.23a**). They deposit feed on the carcasses of dead fish and whales, and some are thought to burrow through ooze at the bottom of the ocean, feeding on polychaete worms and other buried prey. Lampreys, in contrast, are ectoparasites. They attach to fish or other hosts by suction, and then use spines in their mouth and tongue to rasp a hole in the side of their victim (**Figure 34.23b**). Once the wound is open, they suck blood and other body fluids.

Movement Hagfish and lampreys have a well-developed notochord and swim by making undulating movements. Lampreys can also move themselves upstream, against the flow of water, by attaching their suckers to rocks and looping the rest of their body forward, like an inchworm. Both groups have tail fins and lamprey have a dorsal fin, but neither have the paired lateral appendages—meaning fins or limbs that emerge from each side—found in other vertebrates.

Reproduction Virtually nothing is known about hagfish mating or embryonic development. Some lampreys live in freshwater; others are **anadromous**—meaning they spend their adult life in the ocean, but swim up streams to breed. Fertilization is external, and adults die after breeding once. Lamprey eggs hatch into larvae that look and act like lancelets. The larvae burrow into sediments and suspension feed for several years before metamorphosing into free-swimming adults.

(a) Hagfish

Eptatretus stoutii

(b) Lampreys feeding on fish

Petromyzon marinus

FIGURE 34.23 Hagfish and Lampreys Are Jawless Vertebrates.

Chordata > Vertebrata > Chondrichthyes (Sharks, Rays, Skates)

The 970 species in this lineage are distinguished by a specialized type of reinforced cartilaginous skeleton (*chondrus* is the Greek word for cartilage). Along with jawed vertebrates, or gnathostomes (pronounced *NATH-oh-stomes*), sharks also have jaws. They, like the other fish-like gnathostomes, also have paired fins. Paired fins were an important evolutionary innovation because they stabilize the body during rapid swimming—keeping it from pitching up or down, yawing to one side or the other, or rolling. ✔You should be able to indicate the origin of paired fins and the chondrichthyan form of cartilage on Figure 34.12.

Most sharks, rays, and skates are marine, though a few species live in freshwater. Sharks have streamlined, torpedo-shaped bodies and an asymmetrical tail—the dorsal portion is longer than the ventral portion (**Figure 34.24a**). In contrast, the dorsal–ventral plane of the body in rays and skates is strongly flattened (**Figure 34.24b**).

Feeding A few species of ray and shark suspension feed on plankton, but most species in this lineage are predators. Skates and rays lie on the ocean floor and ambush passing animals; electric rays capture their prey by stunning them with electric discharges of up to 200 volts. Most sharks, in contrast, are active hunters that chase down prey in open water and bite them. The larger species of shark feed on large fish or marine mammals. Sharks are referred to as the "top predator" in many marine ecosystems, because they are at the top of the food chain—nothing eats them. Yet the largest of all sharks, the whale shark, is a suspension feeder. Whale sharks filter plankton out of water as it passes over their gills.

Movement Rays and skates swim by flapping their greatly enlarged pectoral fins. (Pectoral fins are located on the sides of an organism; dorsal fins are located on the dorsal surface.) Sharks swim by undulating their bodies from side to side and beating their large tails.

Reproduction Sharks use internal fertilization, and fertilized eggs may be shed into the water or retained until the young are hatched and well developed. In some viviparous species, embryos are attached to the mother by specialized tissues in a placenta, where the exchange of gases, nutrients, and wastes takes place. (A placenta evolved in certain shark lineages independently of its evolution elsewhere on the tree of life.) Skates are oviparous, but rays are viviparous.

(a) Sharks are torpedo shaped. *Prionace glauca*

(b) Skates and rays are flat. *Dasyatis americana*

FIGURE 34.24 Sharks and Rays Have Cartilaginous Skeletons.

Chordata > Vertebrata > Actinopterygii (Ray-Finned Fishes)

Actinopterygii (pronounced *ack-tin-op-teh-RIJ-ee-i*) means "ray-finned." Logically enough, these fish have fins that are supported by long, bony rods arranged in a ray pattern, in addition to a skeleton made of bone. Their bodies are covered with interlocking scales that provide a stiff but flexible covering, and many have a gas-filled **swim bladder**, which evolved from the lungs found in the earliest lineages of fish. The evolution of the swim bladder was an important innovation because it allowed ray-finned fishes to avoid sinking. Tissues are heavier than water, so the bodies of aquatic organisms tend to sink. Sharks and rays, for example, have to swim to avoid sinking. But ray-finned fishes have a bladder that changes in volume, depending on the individual's position. Gas is added to the bladder when a ray-finned fish swims down; gas is removed when the fish swims up. In this way, ray-finned fishes maintain neutral buoyancy in water of various depths and thus various pressures. ✔You should be able to indicate the origin of rayed fins and the swim bladder on Figure 34.12.

The actinopterygians are the most successful vertebrate lineage based on number of species, duration in the fossil record, and extent of habitats occupied. Almost 27,000 species of ray-finned

(Continued on next page)

fishes are known, including the smallest known vertebrate—a species where adult females average less than 8 mm in length.

The most important major lineage of ray-finned fishes is the Teleostei. About 96 percent of all living fish species, including familiar groups like the tuna, trout, cod, and goldfish, are teleosts (**Figure 34.25**). The teleosts underwent an adaptive radiation about the same time that mammals did.

Feeding Teleosts can suck food toward their mouths, grasp it with their protrusible jaws, and then process it with teeth on their jaws and with pharyngeal jaws in their throat. The size and shape of the mouth, the jaw teeth, and the pharyngeal jaw teeth all correlate with the type of food consumed. For example, most predatory teleosts have long, spear-shaped jaws armed with spiky teeth, as well as bladelike teeth on their pharyngeal jaws. Besides being major predators, ray-finned fishes are the most important large herbivores in both marine and freshwater environments.

Movement Like other fish, ray-finned fishes swim by alternately contracting muscles on the left and right sides of their bodies from head to tail, resulting in rapid, side-to-side undulations. Their bodies are streamlined to reduce drag in water. Teleosts have a flexible, symmetrical tail, which reduces the need to use their pectoral (side) fins as steering and stabilizing devices during rapid swimming.

Reproduction Most ray-finned fish species rely on external fertilization and are oviparous; some species have internal fertilization with external development; still others have internal fertilization and are viviparous. Although it is common for fish eggs to be released in the water and left to develop on their own, parental care occurs in some species. Parents may carry fertilized eggs on their fins, in their mouth, or in specialized pouches to guard them until the eggs hatch. In other cases the eggs are laid into a nest that is actively guarded and cared for. In freshwater teleosts, offspring develop directly; but marine species have larvae that are very different from adult forms. As they develop, marine fish larvae undergo a metamorphosis to the juvenile form, which then grows into an adult.

Holocentrus rufus

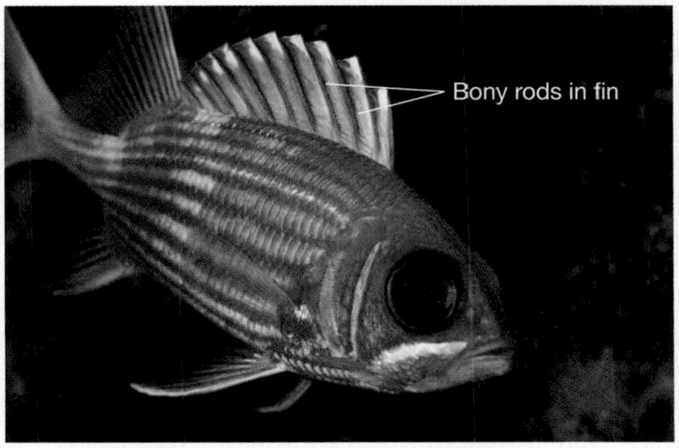

FIGURE 34.25 In Ray-Finned Fishes, Fins Are Supported by Long, Bony Rods.

Chordata > **Vertebrata** > Actinistia (Coelacanths) and Dipnoi (Lungfish)

Although coelacanths (pronounced *SEEL-uh-kanths*) and lungfish represent independent lineages, they are sometimes grouped together and called **lobe-finned fishes**. Lobe-finned fishes are common and diverse in the fossil record in the Devonian period, about 400 million years ago, but only eight species are living today. They are important, however, because they represent a crucial evolutionary link between the fishes and the tetrapods. Early fish in the fossil record have fins supported by stiff structures that extend from a base of bone. In ray-finned fish, the bony elements are reduced. But in lobefins, the bony elements extend down the fin, and branch—similar to the bony elements in the limbs of tetrapods (**Figure 34.26**). ✔You should be able to indicate the origin of extensive fin bones on Figure 34.12.

Coelacanths are marine and occupy habitats 150–700 m below the surface. In contrast, lungfish live in shallow, freshwater ponds and rivers (see Figure 34.15). As their name implies, lungfish have lungs and breathe air when oxygen levels in their habitats drop. Some species burrow in mud and enter a quiescent, sleeplike state when their habitat dries up during each year's dry season.

Feeding Coelacanths prey on fish. Lungfish are **omnivorous** ("all-eating"), meaning that they eat algae and plant material as well as animals.

Movement Coelacanths and lungfish swim by waving their bodies; some lungfish can also use their fins to walk along pond bottoms.

Reproduction Sexual reproduction is the rule, with fertilization internal in coelacanths and external in lungfish. Coelacanths are ovoviviparous; lungfish lay eggs. Lungfish eggs hatch into larvae that resemble juvenile salamanders.

Latimeria menadoensis

FIGURE 34.26 Coelacanths Are Lobe-Finned Fishes.

Amphibians are found throughout the world and occupy ponds, lakes, or moist terrestrial environments (**Figure 34.27a**). Translated literally, their name means "both-sides-living." The name is appropriate because adults of most species of amphibian feed on land but lay their eggs in water. In many species of amphibians, gas exchange occurs exclusively or in part across their moist, mucus-covered skin. ✔**You should be able to indicate the origin of "skin-breathing" on Figure 34.12.** The 5500 species of **amphibians** living today form three distinct clades: (**1**) frogs and toads, (**2**) salamanders, and (**3**) caecilians (pronounced *suh-SILL-ee-uns*).

Feeding Adult amphibians are carnivores.

- Most frogs are sit-and-wait predators that use their long, extendable tongues to capture passing prey.
- Salamanders also have an extensible tongue, which some species use in feeding.
- Terrestrial caecilians prey on earthworms and other soil-dwelling animals; aquatic forms eat vertebrates and small fish.

Movement Most amphibians have four well-developed limbs.

- In water, frogs and toads move by kicking their hind legs to swim; on land they kick their hind legs out to jump or hop.
- Salamanders walk on land; in water they undulate their bodies to swim.
- Caecilians lack limbs and eyes; terrestrial forms burrow in moist soils (**Figure 34.27b**).

Reproduction Reproduction is sexual, breeding occurs in water, and larvae undergo a dramatic metamorphosis into land-dwelling adults. For example, the fishlike tadpoles of frogs and toads develop limbs, and their gills are replaced with lungs.

- Frogs are oviparous and have external fertilization. In some species of frogs, parents may guard or even carry eggs. In many frogs, young develop in the water and suspension feed on plant or algal material.
- Salamanders have internal fertilization, and most species are oviparous. Salamander larvae are carnivorous.
- Caecilians have internal fertilization, and many species are viviparous.

(a) Frogs and other amphibians lay their eggs in water.

Bufo periglenes

(b) Caecilians are legless amphibians.

Gymnophis multiplicata

FIGURE 34.27 Amphibians Are the Most Ancient Tetrapods.

Chordata > Vertebrata > Mammalia > Monotremata (Platypuses, Echidnas)

The **monotremes** are the most basal lineage of mammals living today and are found only in Australia. They lay eggs and have metabolic rates—meaning a rate of using oxygen and oxidizing sugars for energy—that are lower than other mammals. Three species exist: one species of platypus and two species of echidna.

Feeding Monotremes have a leathery beak or bill. The platypus feeds on insect larvae, mollusks, and other small animals in streams (**Figure 34.28a**). Echidnas feed on ants, termites, and earthworms (**Figure 34.28b**).

Movement Platypuses swim with the aid of their webbed feet and walk when on land. Echidnas walk on their four legs.

Reproduction Platypuses lay their eggs in a burrow, while echidnas keep their eggs in a pouch on their belly. Young monotremes hatch quickly, and the mother must continue keeping them warm and dry for another four months. Like other mammals, monotremes produce milk and nurse their young. They lack well-defined nipples, however, and instead secrete milk from glands in the skin.

(a) Platypus

Ornithorhynchus anatinus

(b) Echidna

Tachyglossus aculeatus

FIGURE 34.28 Platypuses and Echidnas Are Egg-Laying Mammals.

Chordata > Vertebrata > Mammalia > Marsupiala (Marsupials)

The 275 known species of **marsupials** live in the Australian region and the Americas (**Figure 34.29**) and include the familiar opossums, kangaroos, wallabies, and koala. Although females have a placenta that nourishes embryos during development, the young are born after a short embryonic period and are poorly developed. They crawl from the opening of the female's reproductive tract to the female's nipples, where they suck milk. They stay attached to their mother until they grow large enough to move independently. ✔**You should be able to indicate the origin of the placenta and viviparity—traits that are also found in Eutherian mammals—on Figure 34.22.**

Didelphis virginiana

FIGURE 34.29 Marsupials Give Birth after a Short Embryonic Period. Opossums are the only marsupials in North America.

Feeding Marsupials are herbivores, omnivores, or carnivores. In many cases, convergent evolution has resulted in marsupials that are similar to placental species in morphology and way of life. For example, a recently extinct marsupial called the Tasmanian wolf was a long-legged hunter similar to the wolves of North America and northern Eurasia. A species of marsupial native to Australia specializes in eating ants and looks and acts much like the South American anteater, which is not a marsupial.

Movement Marsupials move by crawling, gliding, walking, running, or hopping.

Reproduction Marsupial young spend more time developing while attached to their mother's nipple than they do inside her body being fed via the placenta.

Chordata > Vertebrata > Mammalia > Eutheria (Placental Mammals)

The approximately 4300 species of **placental mammals**, or **eutherians**, are distributed worldwide. They are far and away the most species-rich and morphologically diverse group of mammals.

Biologists group placental mammals into 18 lineages called orders. The six most species-rich orders are the rodents (rats, mice, squirrels; 1814 species), bats (986 species), insectivores (hedgehogs, moles, shrews; 390 species), artiodactyls (pigs, hippos, whales, deer, sheep, cattle; 293 species), carnivores (dogs, bears, cats, weasels, seals; 274 species), and primates (lemurs, monkeys, apes, humans; 235 species).

Feeding The size and structure of the teeth correlate closely with the diet of placental mammals. Herbivores have large, flat teeth for crushing leaves and other coarse plant material; predators have sharp teeth that are efficient at biting and tearing flesh. The structure of the digestive tract also correlates with the placental mammals' diet. In some plant-eaters, for example, the stomach hosts unicellular organisms that digest cellulose and other complex polysaccharides.

Pongo abelii

FIGURE 34.30 Eutherians Are the Most Species-Rich and Diverse Group of Mammals.

Movement In placental mammals, the structure of the limb correlates closely with the type of movement performed. Eutherians fly, glide, run, walk, swim, burrow, or swing from trees (**Figure 34.30**). Hindlimbs are reduced or lost in aquatic groups such as whales and dolphins, which swim by undulating their bodies.

Reproduction Eutherians have internal fertilization and are viviparous. An extensive placenta develops from a combination of maternal and fetal tissues; and at birth, young are much better developed than in marsupials—some are able to walk or run minutes after emerging from the mother. ✔**You should be able to indicate the origin of delayed birth (extended development prior to birth) on Figure 34.22.** All eutherians feed their offspring milk until the young have grown large enough to process solid food. A prolonged period of parental care, extending beyond the nursing stage, is common as offspring learn how to escape predators and find food on their own.

Chordata > Vertebrata > Reptilia > Lepidosauria (Lizards, Snakes)

Most lizards and snakes are small reptiles with elongated bodies and scaly skin. ✔**You should be able to indicate the origin of scaly skin on Figure 34.22.** Most lizards have well-developed jointed legs, but snakes are limbless (**Figure 34.31**). The hypothesis that snakes evolved from limbed ancestors is partially supported by the presence of vestigial hip and leg bones in boas and pythons. There are about 7000 species of lizards and snakes alive now.

Feeding Small lizards prey on insects. Although most of the larger lizard species are herbivores, the 3-meter-long monitor lizard from the island of Komodo is a predator and scavenger that can eat deer. Snakes are carnivores; some subdue their prey by injecting poison through modified teeth called fangs. Snakes prey primarily on small mammals, amphibians, and invertebrates, which they swallow whole—usually headfirst.

Movement Lizards crawl or run on their four limbs. Snakes and lizards that are limbless burrow through soil, crawl over the ground, or climb trees by undulating their bodies.

Reproduction Although most lizards and snakes lay eggs, many are ovoviviparous; viviparity has also evolved numerous times in lizards. Most species reproduce sexually, but asexual reproduction, via the production of eggs by mitosis, is known to occur in six groups of lizards and one snake lineage.

Morelia viridis

FIGURE 34.31 Snakes Are Limbless Predators.

Chordata > Vertebrata > Reptilia > Testudinia (Turtles)

The 300 known species of turtles inhabit freshwater, marine, and terrestrial environments throughout the world. The testudines are distinguished by a shell composed of bony plates that fuse to the vertebrae and ribs (**Figure 34.32**). The shell functions in protection from predators—when threatened, turtles can withdraw their head and legs into it. ✔**You should be able to indicate the origin of the turtle shell on Figure 34.22.** The turtles' skulls are highly modified versions of the skulls of other reptiles. Turtles lack teeth, but their jawbone and lower skull form a bony beak. They range in size from the 2-m long, 900 kg leatherback sea turtle to the 8-cm long, 140 g speckled padloper tortoise.

Feeding Turtles are either carnivorous—feeding on whatever animals they can capture and swallow—or herbivorous. They may also scavenge dead material. Most marine turtles are carnivorous. Leatherback turtles, for example, feed primarily on jellyfish, and they are only mildly affected by the jellyfish's stinging cnidocytes (see Chapter 32). In contrast, species in the lineage of terrestrial turtles called the tortoises are plant-eaters.

Testudo hermanni

FIGURE 34.32 Turtles Have a Shell Consisting of Bony Plates.

Movement Turtles swim, walk, or burrow. Aquatic species usually have feet that are modified to function as flippers.

Reproduction All turtles are oviparous. Other than digging a nest prior to depositing eggs, parental care is lacking.

Chordata > Vertebrata > Reptilia > Crocodilia (Crocodiles, Alligators)

Only 24 species of crocodile and alligator are known. Most are tropical and live in freshwater or marine environments. They have eyes located on the top of their heads and nostrils located at the top of their long snouts—adaptations that allow them to sit semi-submerged in water for long periods of time, breathing air and monitoring activity around them by sight (**Figure 34.33**).

Feeding Crocodilians are predators. Like other toothed vertebrates, except mammals, their jaws are filled with conical teeth that are continually replaced as they fall out during feeding. Their usual method of killing small prey is by biting through the body wall. Large prey are usually subdued by drowning. Crocodilians eat amphibians, turtles, fish, birds, and mammals; one of their common hunting strategies is to leap out of the water to snatch unwary prey that have come to drink.

Movement Crocodiles and alligators walk on land. In water they swim with the aid of their large, muscular tails.

Reproduction Although crocodilians are oviparous, parental care is extensive. Eggs are laid in earth-covered nests that are guarded by the parents. When young inside the eggs begin to vocalize, parents dig them up and carry the newly hatched young inside their mouths to nearby water. Crocodilian young can hunt when newly hatched but stay near their mother for up to three years. ✔**You should be able to indicate the origin of extensive parental care—which also occurs in birds (and dinosaurs)—on Figure 34.22.**

Alligator mississippiensis

FIGURE 34.33 Alligators Are Adapted for Aquatic Life.

The fossil record provides conclusive evidence that birds descended from a lineage of dinosaurs that had a unique trait: **feathers**. In dinosaurs, feathers are hypothesized to have functioned as insulation and in courtship or aggressive displays. In birds, feathers insulate and are used for display but also furnish the lift, power, and steering required for flight. Birds have many other adaptations that make flight possible, including large breast muscles used to flap the wings. Bird bodies are lightweight because they have a reduced number of bones and organs and because their hollow bones are filled with air sacs linked to the lungs. Instead of teeth, birds have a horny beak. They are endotherms ("within-heating"), meaning that they have a high metabolic rate and use the heat produced, along with the insulation provided by feathers, to maintain a constant body temperature. The 9100 bird species alive today occupy virtually every habitat, including the open ocean (**Figure 34.34a**). ✔You should be able to indicate the origin of feathers, endothermy, and flight on Figure 34.22.

Feeding Plant-eating birds usually feed on nectar or seeds. Some birds are omnivores, although many are predators that capture insects, small mammals, fish, other birds, lizards, mollusks, or crustaceans. The size and shape of a bird's beak correlate closely with its diet. For example, predatory species such as falcons have sharp, hook-shaped beaks; finches and other seedeaters have short, stocky bills that can crack seeds and nuts; fish-eating species such as the great blue heron have spear-shaped beaks.

Movement Almost all species can fly, although flightlessness has evolved repeatedly in certain groups (**Figure 34.34b**). The size and shape of birds' wings correlate closely with the type of flying they do. Birds that glide or hover have long, thin wings; species that specialize in explosive takeoffs and short flights have short, stocky wings. Many seabirds are efficient swimmers, using their webbed feet to paddle or flapping their wings to "fly" underwater. Some ground-dwelling birds such as ostrich can run long distances at high speed.

Reproduction Birds are oviparous but provide extensive parental care. In most species, one or both parents build a nest and incubate the eggs. After the eggs hatch, parents feed offspring until they are large enough to fly and find food on their own.

(a) Albatross use their long, narrow wings to glide.

Diomedea salvini

(b) Flightless cormorants are native to the Galápagos Islands.

Phalacrocorax harrisi

FIGURE 34.34 Birds Are Feathered Descendants of Dinosaurs.

34.4 The Primates and Hominins

Although humans occupy a tiny twig on the tree of life, there has been a tremendous amount of research on human origins. This section introduces the lineage of mammals called the Primates, the fossil record of human ancestors, and data on the relationships among human populations living today.

The Primates

The lineage known as Primates consists of two main groups: prosimians and anthropoids.

- The **prosimians** ("before-monkeys") consist of the lemurs, found in Madagascar, and the tarsiers, pottos, and lorises of Africa and south Asia (**Figure 34.35a**). Most prosimians living today are relatively small in size, reside in trees, and are active at night.
- The Anthropoidea or **anthropoids** ("human-like") include the New World monkeys found in Central and South America, the Old World monkeys that live in Africa and tropical regions of Asia, the gibbons of the Asian tropics, and the Hominidae, or **great apes**—orangutans, gorillas, chimpanzees, and humans (**Figure 34.35b**).

(a) Prosimians live in Africa, Madagascar, and south Asia.

Lemur catta

(b) Anthropoids include New World monkeys, Old World monkeys, gibbons, and Great Apes.

Cacajao calvus

FIGURE 34.35 There Are Two Main Lineages of Primates.

The phylogenetic tree in **Figure 34.36** shows the evolutionary relationships among these groups.

WHAT MAKES A PRIMATE A PRIMATE? **Primates** tend to have hands and feet that are efficient at grasping, flattened nails instead of claws on the fingers and toes, brains that are large relative to overall body size, color vision, complex social behavior, and extensive parental care of offspring.

Along with other mammal groups that live in trees or make their living by hunting, primates have eyes located on the front of the face. Eyes that look forward provide better depth perception than do eyes on the sides of the face. The hypothesis here is that good depth perception is important in species that run or swing through trees and/or attack prey.

WHAT MAKES A GREAT APE A GREAT APE? The great apes are also called **hominids**. Compared with most types of primate, the hominids are relatively large bodied and have long arms, short legs, and no tail.

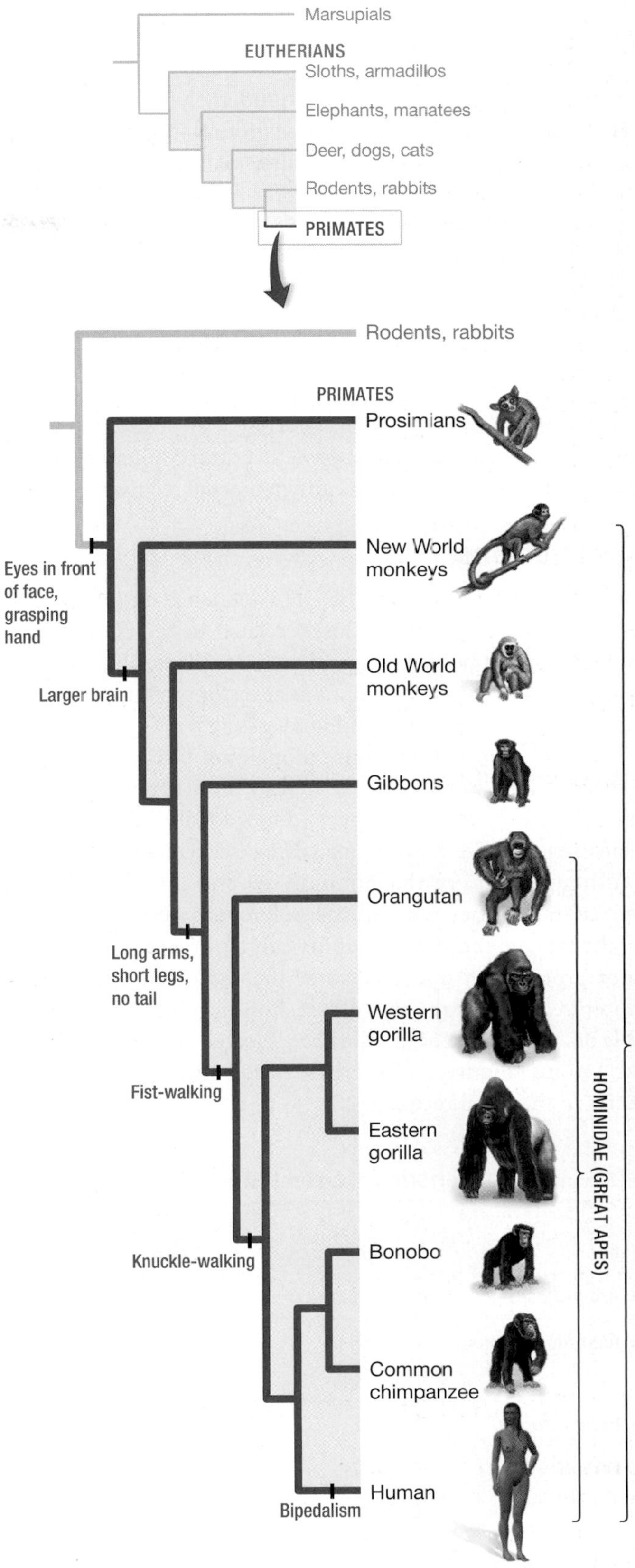

FIGURE 34.36 A Phylogeny of the Primates. Phylogenetic tree estimated from extensive DNA sequence data. According to the fossil record, humans and chimps shared a common ancestor 6 to 7 million years ago.

Although all of the great ape species except for the orangutans live primarily on the ground, they have distinct ways of walking.

- When orangutans come to the ground, they occasionally walk with their knuckles pressed to the ground. More commonly, though, they fist-walk—that is, they walk with the backs of their hands pressed to the ground.
- Gorillas and chimps only knuckle-walk. They also occasionally rise up on two legs—usually in the context of displaying aggression.
- Humans are the only great ape that is fully **bipedal** ("two-footed")—meaning they walk upright on two legs.

Bipedalism is the synapomorphy that defines the hominins. The **hominins** are a monophyletic group comprising *Homo sapiens* and more than a dozen extinct, bipedal relatives.

Fossil Humans

From extensive comparisons of DNA sequence data, it is now clear that humans are most closely related to the chimpanzees and that our next nearest living relatives are the gorillas. According to the fossil record, the common ancestor of chimps and humans lived in Africa about 7 million years ago.

The fossil record of hominins, though not nearly as complete as investigators would like, is rapidly improving. About 14 species have been found to date, and new fossils that inform the debate over the ancestry of humans are discovered every year.

Although naming the hominin species and interpreting their characteristics remain intensely controversial, most researchers agree that the hominins can be organized into four major groups that appeared after the recently characterized *Ardipithecus ramidus*—the oldest hominin known to date. **Table 34.1** summarizes key data for selected species within the four groups. **Figure 34.37** provides the time range of each species in the fossil record.

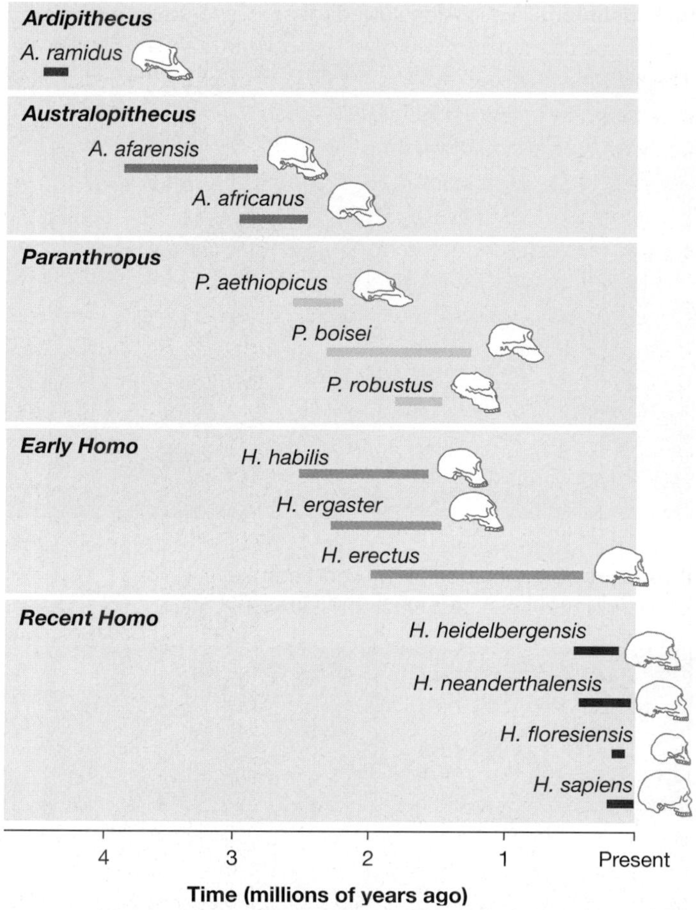

FIGURE 34.37 A Timeline of Human Evolution. The colored bars indicate the first appearance and last appearance in the fossil record of each species.

✔**QUESTION** How many species of hominin existed 2.2 million years ago (mya), 1.8 mya, and 100,000 years ago?

TABLE 34.1 **Characteristics of Selected Hominins**

Genus	Species	Location of Fossils	Estimated Average Braincase Volume (cm^3)	Estimated Average Body Size (kg)	Associated with Stone Tools?
■ ***Ardipithecus***	*A. ramidus*	Africa	325	50	no
■ ***Australopithecus***	*A. afarensis*	Africa	450	36	no
	A. africanus	Africa	450	36	no
■ ***Paranthropus***	*P. boisei*	Africa	510	44	no?
■ **Early *Homo***	*H. habilis*	Africa	550	34	yes
	H. ergaster	Africa	850	58	yes
	H. erectus	Africa, Asia	1000	57	yes
■ **Recent *Homo***	*H. heidelbergensis*	Africa, Europe	1200	62	yes
	H. neanderthalensis	Middle East, Europe, Asia	1500	76	yes
	H. floresiensis	Flores (Indonesia)	380	28	yes
	H. sapiens	Middle East, Europe, Asia	1350	53	yes

GRACILE AUSTRALOPITHECINES Four species of small hominins called gracile australopithecines have been identified thus far. The adjective *gracile*, or "slender," is appropriate because these organisms were slightly built. Adult males were about 1.5 meters tall, with an estimated weight of about 36 kg.

The gracile australopithecines are placed in the genus *Australopithecus* ("southern ape"). The name was inspired by the earliest specimens, which came from South Africa.

Several lines of evidence support the hypothesis that the gracile australopithecines were bipedal. The shape of the australopithecine knee and hip are consistent with bipedal locomotion, and the hole in the back of their skulls where the spinal cord connects to the brain is oriented downward (**Figure 34.38a**), just as it is in our species, *Homo sapiens*. In chimps, gorillas, and other vertebrates that walk on four feet, this hole is oriented backward.

ROBUST AUSTRALOPITHECINES Three species are grouped in the genus *Paranthropus* ("beside-human"). Like the gracile australopithecines, these robust australopithecines were bipedal. They were much stockier than the gracile forms, however—about the same height but an estimated 8–10 kilograms (20 pounds) heavier on average. In addition, their skulls were much broader and more robust.

All three species had massive cheek teeth and jaws, very large cheekbones, and a sagittal crest—a flange of bone at the top of the skull (**Figure 34.38b**). Because muscles that work the jaw attach to the sagittal crest and cheekbones, researchers conclude that these organisms had tremendous biting power and made their living by crushing large seeds or coarse plant materials. One species is nicknamed "nutcracker man."

The name *Paranthropus* was inspired by the hypothesis that the three known species are a monophyletic group that was a side branch during human evolution—an independent lineage that went extinct.

EARLY *HOMO* Species in the genus *Homo* are called **humans**. As **Figure 34.38c** shows, species in this genus have flatter and narrower faces, smaller jaws and teeth, and larger braincases than the earlier hominins do. (The **braincase** is the portion of the skull that encloses the brain.)

The appearance of early members of the genus *Homo* in the fossil record coincides closely with the appearance of tools made of worked stone—most of which are interpreted as handheld choppers or knives. Although the fossil record does not exclude the possibility that *Paranthropus* made tools, many researchers favor the hypothesis that extensive toolmaking was a diagnostic trait of early *Homo*.

RECENT *HOMO* The recent species of *Homo* date from 1.2 million years ago to the present. As **Figure 34.38d** shows, these species have even flatter faces, smaller teeth, and larger braincases than the early *Homo* species do. The 30,000-year-old fossil in the figure, for example, is from a population of *Homo sapiens* (our species) called the **Cro-Magnons**.

The Cro-Magnons were accomplished painters and sculptors who buried their dead in carefully prepared graves. There is also evidence that another species, the **Neanderthal** people (*Homo neanderthalensis*) made art and buried their dead in a ceremonial fashion.

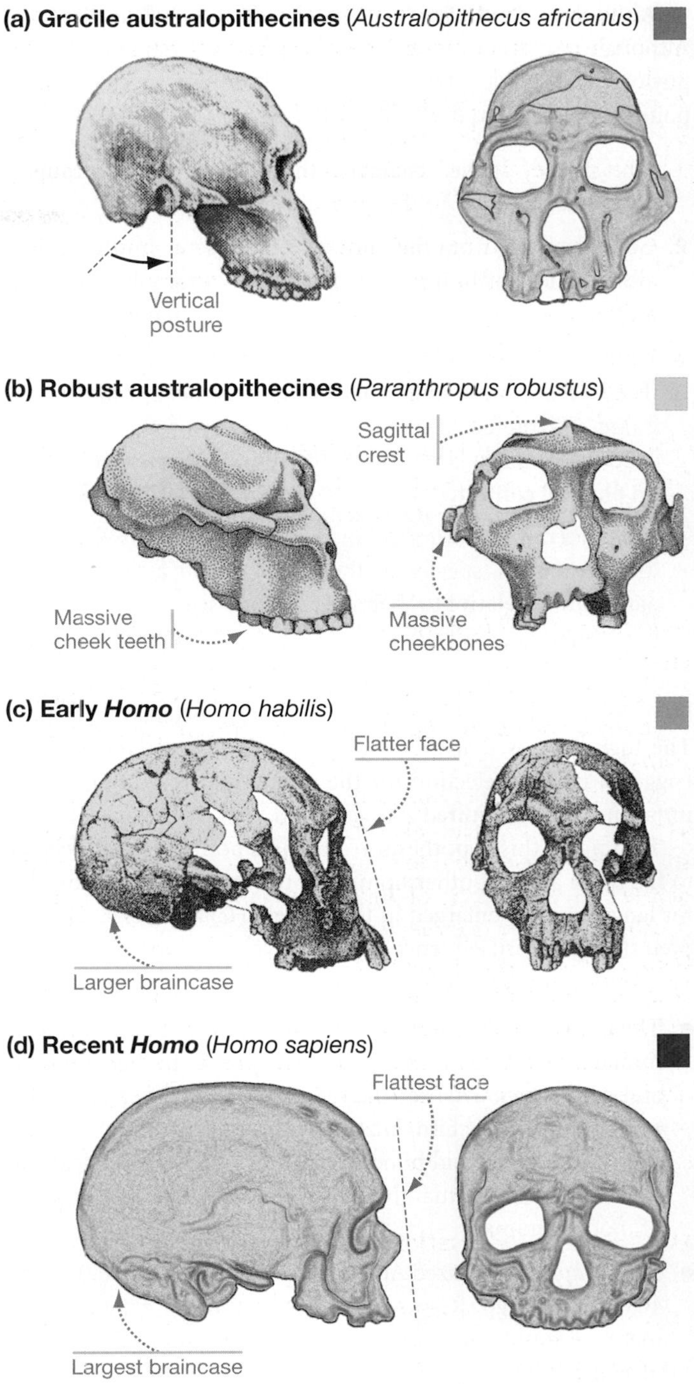

FIGURE 34.38 African Hominins Comprise Four Major Groups.

✔**QUESTION** The skulls are arranged as they appear in the fossil record, from most ancient to most recent—(a) to (d). How did the forehead and brow ridge of hominins change through time?

Perhaps the most striking recent *Homo*, though, is *H. floresiensis*. This species has been found only on the island of Flores in Indonesia, which was also home to a species of dwarfed elephants. *H. floresiensis* consisted of individuals that had braincases smaller than those of gracile australopithecines and were about a meter tall. Fossil finds suggest that the species inhabited the island from about 100,000 to 12,000 years before present and that dwarfed elephants were a major source of food.

WHAT CAN BE DEDUCED FROM THE HOMININ FOSSIL RECORD? Although researchers do not have a solid understanding of the phylogenetic relationships among the hominin species, several points are clear from the available data.

1. The shared, derived character that defines the hominins is bipedalism.
2. Several species from the lineage were present simultaneously during most of hominin evolution. For example, about 1.8 million years ago there may have been as many as five hominin species living in eastern and southern Africa.
3. Fossils from more than one species have been found in the same geographic location in rock strata of the same age. Thus, it is almost certain that different hominin species lived in physical contact.
4. Compared with the gracile and robust australopithecines and the great apes, species in the genus *Homo* have extremely large brains relative to their overall body size.

Why did humans evolve such gigantic brains? The leading hypothesis on this question is that early *Homo* began using symbolic spoken language along with initiating extensive tool use. The logic here is that increased toolmaking and language use triggered natural selection for the capacity to reason and communicate, which required a larger brain.

To support this hypothesis, researchers point out that, relative to the brain areas of other hominins, the brain areas responsible for language were enlarged in the earliest *Homo* species. There is even stronger fossil evidence for extensive use of speech in *Homo neanderthalensis* and early *Homo sapiens*:

- The hyoid bone is a slender bone in the voice box of modern humans that holds muscles used in speech. In Neanderthals and early *Homo sapiens*, the hyoid is vastly different in size and shape from a chimpanzee's hyoid bone. Researchers have found an intact hyoid bone associated with a 60,000-year-old Neanderthal individual that is virtually identical to the hyoid of modern humans.
- *Homo sapiens* colonized Australia by boat between 60,000 and 40,000 years ago. Researchers suggest that an expedition of that type could not be planned and carried out in the absence of symbolic speech.

To summarize, *Homo sapiens* is the sole survivor of an adaptive radiation that took place over the past 7 million years. Why all but one species went extinct is still a mystery, though competition for food and space may have played a part. Neanderthals, for example, flourished in Europe and central Asia prior to the arrival of *Homo sapiens* from Africa.

The Out-of-Africa Hypothesis

Our own species, *Homo sapiens*, is the only primate with a chin. The first *H. sapiens* fossils appear in African rocks that date to about 195,000 years ago. For some 130,000 years thereafter, the fossil record indicates that our species occupied Africa while *H. neanderthalensis* resided in Europe and the Middle East. Some evidence suggests that *H. erectus* may still have been present in Asia at that time.

In rocks dated between 60,000 and 30,000 years ago, however, *H. sapiens* fossils are found throughout Europe, Asia, Africa, and Australia. *H. erectus* had disappeared by this time, and *H. neanderthalensis* went extinct after coexisting with *H. sapiens* in Europe for thousands of years.

Phylogenetic trees that show the relationships among human populations living today agree with the pattern in the fossil record. As **Figure 34.39** shows, the first lineages to branch off lead to descendant populations that live in Africa today. Based on this observation, it is logical to infer that the ancestral population of modern humans also lived in Africa.

The tree shows that lineages subsequently branched off, leaving descendants that today live in Europe, central Asia, Polynesia, the Americas, and east Asia (**Figure 34.40**). The data suggest that modern humans originated in Africa, and that a population that left Africa split into a group that colonized Europe and Russia and a group that eventually spread throughout the rest of the world.

This scenario for the evolution of *H. sapiens* is called the **out-of-Africa hypothesis**. It contends that *H. sapiens* evolved independently of the earlier European and Asian species of *Homo*—meaning there was no interbreeding between *H. sapiens* and Neanderthals, *H. erectus*, or *H. floresiensis*. According to the out-of-Africa hypothesis, *H. sapiens* evolved its distinctive traits in Africa and then dispersed throughout the world.

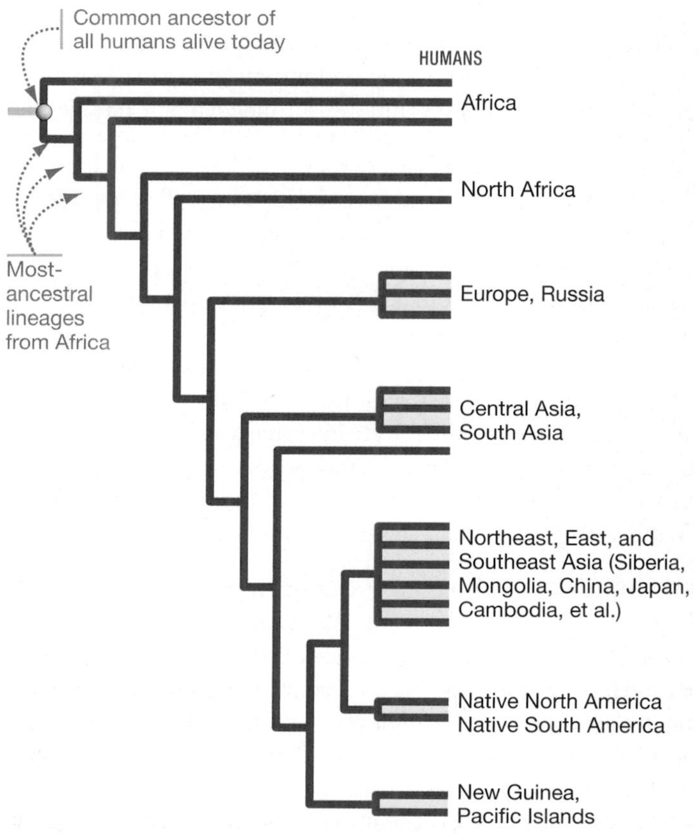

FIGURE 34.39 Phylogeny of Human Populations Living Today. The phylogeny was estimated from DNA sequence data.

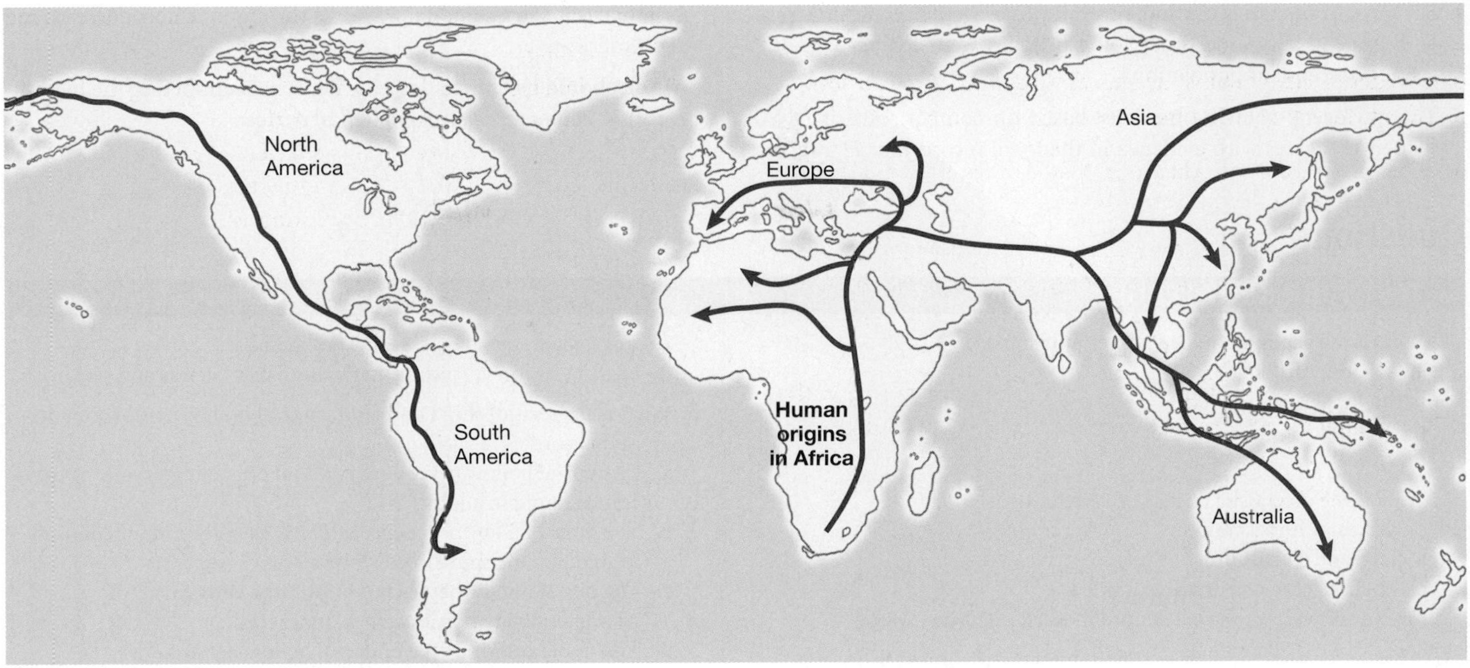

FIGURE 34.40 *Homo sapiens* Originated in Africa and Spread throughout the World. The phylogeny in Figure 34.39 supports the hypothesis that humans originated in Africa and spread out along the routes shown here.

CHAPTER 34 REVIEW

For media, go to the study area at www.masteringbiology.com

Summary of Key Concepts

Echinoderms are radially symmetric as adults and have an endoskeleton and water vascular system. They are among the most important predators and herbivores in marine environments.

- Echinoderm larvae are bilaterally symmetric but undergo a metamorphosis into radially symmetric adults.
- The echinoderm water vascular system is composed of fluid-filled tubes and chambers and extends from the body wall in projections called podia.
- Echinoderm podia can extend and retract in response to muscle contractions that move fluid inside the water vascular system. Many echinoderms move via their podia, and many species suspension feed, deposit feed, or act as predators with the aid of their podia.

✔You should be able to predict what happens when (1) sea stars that prey on mussels and clams are excluded from experimental plots, and (2) sea urchins that graze on kelp are excluded from experimental plots.

All vertebrates have a skull and an extensive endoskeleton made of cartilage or bone; their diversification was driven in part by the evolution of the jaw and limbs. Vertebrates are the most important large-bodied predators and herbivores in marine and terrestrial environments.

- Chordates are distinguished by the presence of pharyngeal gill slits, a dorsal hollow nerve cord, a notochord, and a muscular tail that extends past the anus.
- Vertebrates are distinguished by the presence of a cranium and vertebrae. In some groups of vertebrates, the body plan features an extensive endoskeleton composed of bone.
- Ray-finned fishes and tetrapods use their jaws to bite food and process it with teeth.
- Ray-finned fishes and tetrapods move when muscles attached to their endoskeletons contract or relax.
- Tetrapods can move on land because their limbs enable walking, running, or flying.
- The evolution of the amniotic egg allowed tetrapods to lay large eggs on land.
- Parental care was an important adaptation in some groups of ray-finned fishes and tetrapods—particularly mammals.

✔You should be able to explain why the tetrapod limb could have evolved in an animal that was aquatic.

MB **Web Activity** Deuterostome Diversity

Humans are a tiny twig on the tree of life. Chimpanzees and humans diverged from a common ancestor that lived in Africa 6–7 million years ago. Since then, at least 14 humanlike species have existed.

- The fossil record of the past 3.5 million years contains at least 14 distinct species of hominins. *Homo sapiens* is the sole surviving representative of an adaptive radiation.

- For most of human evolution, several species were present in Africa or Europe at the same time. Some lineages went extinct without leaving descendant populations.
- The phylogeny of living humans, based on comparisons of DNA sequences, agrees with evidence in the fossil record that *H. sapiens* originated in Africa and later spread throughout Europe, Asia, and the Americas.

✔**You should be able to describe evidence supporting the hypothesis that *Homo sapiens* originated in Africa.**

Questions

✔TEST YOUR KNOWLEDGE

Answers are available in Appendix B

1. What is the echinoderm endoskeleton made of?
a. calcium carbonate
b. bone
c. cartilage
d. chitin

2. What is the diagnostic trait(s) of vertebrates?
a. vertebrae and a cranium
b. jaws and spinal cord
c. endoskeleton constructed of bone
d. endoskeleton constructed of reinforced cartilage

3. Why are the pharyngeal jaws found in many ray-finned fishes important?
a. They allow the main jaw to be protrusible (extendable).
b. They make it possible for individuals to suck food toward their mouths.
c. They give rise to teeth that are found on the main jawbones.
d. They help process food.

4. Which of the following lineages make up the living Amniota?
a. reptiles and mammals
b. viviparous fishes
c. frogs, toads, salamanders, and caecilians
d. hagfish, lampreys, and cartilaginous fishes (sharks and rays)

5. What is the relationship between the dorsal hollow nerve cord and the spinal cord?
a. The dorsal hollow nerve cord is found only in embryos; the spinal cord is found in adults.
b. The dorsal hollow nerve cord stiffens the body and functions in movement; the spinal cord relays electrical signals.
c. The dorsal hollow nerve cord is found in early fossils of chordates; living species have a spinal cord.
d. The spinal cord is a type of dorsal hollow nerve cord.

6. Most species of hominins are known only from Africa. Which species have been found in other parts of the world as well?
a. early *Homo*—*H. habilis* and *H. ergaster*
b. *H. erectus*, *H. neanderthalensis*, and *H. floresiensis*
c. gracile australopithecines
d. robust australopithecines

✔TEST YOUR UNDERSTANDING

Answers are available in Appendix B

1. Explain how the water vascular system of echinoderms functions as a type of hydrostatic skeleton.

2. Explain why each of the four key synapomorphies that distinguish chordates is important in feeding or movement. Why is it possible to say that an animal is an "invertebrate chordate"?

3. The cells that make up jaws and gill arches are derived from the same population of embryonic cells. Why does this observation support the hypothesis that jaws evolved from gill arches in fish?

4. Describe genetic evidence that supports the hypothesis that the tetrapod limb evolved from the fins of lobe-finned fishes.

5. The text claims that "*Homo sapiens* is the sole survivor of an adaptive radiation that took place over the past 7 million years." Do you agree with this statement? Why or why not?

6. Explain how the evolution of the placenta and lactation in mammals improved the probability that their offspring would survive, compared to species without parental care.

✔APPLYING CONCEPTS TO NEW SITUATIONS

Answers are available in Appendix B

1. There is some evidence that pharyngeal gill slits occur in certain species of echinoderms that appear early in the fossil record. Explain the significance of this observation.

2. When feathers first evolved, did they function in flight? Explain your logic.

3. Xenoturbellidans have extremely simple, wormlike bodies. For example, they have a blind gut (only one opening) and no brain. Propose a hypothesis to explain how this simple body plan evolved in this lineage.

4. Mammals and birds are endothermic. Did they inherit this trait from a common ancestor, or did endothermy evolve independently in these two lineages? Provide evidence to support your answer.

Photomicrograph created by treating seawater with a fluorescing compound that binds to nucleic acids. The smallest, most abundant dots are viruses. The larger, numerous spots are bacteria and archaea. The largest splotches are protists.

Viruses 35

If you have ever been laid low by a high fever, cough, scratchy throat, body ache, and debilitating lack of energy, you may have wondered what hit you. What hit you was probably a **virus**: an obligate, intracellular parasite.

Why obligate? Viral replication is *completely* dependent on host cells. Why intracellular? Viruses must enter a host cell for replication to occur. Why parasite? Viruses reproduce at the expense of their host cells.

Viruses can also be defined by what they are not.

- They are not cells and are not made up of cells, so they are not considered organisms.
- They cannot manufacture their own ATP or amino acids or nucleic acids, and they cannot produce proteins on their own.

Viruses enter a **host cell**, take over its biosynthetic machinery, and use that machinery to manufacture a new generation of viruses. Outside of host cells, viruses simply exist.

When you have the flu, influenza viruses enter the cells that line your respiratory tract and use the machinery inside to make copies of themselves. Every time you cough or sneeze, you eject millions or billions of their offspring into the environment. If one of those viruses is lucky enough to be breathed in by another person, it may enter their respiratory tract cells and start a new infection.

Because they are not organisms, viruses are not given scientific (genus + species) names. Most biologists would argue that viruses are not alive, because they lack the five attributes of life introduced in Chapter 1. Yet viruses have a genome, they are superbly adapted to exploit the metabolic capabilities of their host cells, and they evolve. **Table 35.1** on page 676 summarizes some characteristics of viruses.

The diversity and abundance of viruses almost defy description. Each type of virus infects a specific unicellular species or cell type in a multicellular species, and nearly all organisms examined thus far are parasitized by at least one kind of virus. The bacterium

KEY CONCEPTS

- Viruses are tiny, noncellular parasites that infect virtually every type of cell known. They cannot perform metabolism on their own—meaning outside a parasitized cell—and are not considered to be alive.
- Although viruses are diverse morphologically, they can be classified as two general types: enveloped and nonenveloped.
- The viral replication cycle can be broken down into six steps: (1) attachment and entry into a host cell, (2) production of viral proteins, (3) replication of the viral genome, (4) assembly of a new generation of virions, (5) exit from the infected cell, and (6) transmission to a new host.
- In terms of diversity, the key feature of viruses is the nature of their genetic material.

✔ When you see this checkmark, stop and test yourself. Answers are available in Appendix B.

SUMMARY TABLE 35.1 **Viruses versus Organisms**

Characteristics	Viruses	Organisms
Hereditary material	DNA or RNA; can be single stranded or double stranded	DNA; always double stranded
Plasma membrane present?	No	Yes
Can carry out transcription independently?	No—even if a viral polymerase is present, transcription of viral genomes requires use of ATP and nucleotides provided by host cell	Yes
Can carry out translation independently?	No	Yes
Metabolic capabilities	Virtually none	Extensive—synthesis of ATP, reduced carbon compounds, vitamins, lipids, nucleic acids, etc.

Escherichia coli, which resides in the human intestine, is afflicted by several dozen types of viruses. The surface waters of the world's oceans teem with bacteria and archaea, yet viruses outnumber them in this habitat by a factor of 10 to 1. If you leaned over a boat and filled a wine bottle with seawater, it would contain about 10 billion virus particles, or **virions**—close to one and a half times the world's population of humans.

35.1 Why Do Biologists Study Viruses?

Any study of life's diversity would be incomplete unless it included a look at the acellular parasites that exploit that diversity. But viruses are also important from a practical standpoint. To health-care workers, agronomists, and foresters, these parasites are a persistent—and sometimes catastrophic—source of misery and economic loss.

The nature of viruses has been understood only since the 1940s, but they have been the focus of intense research ever since. Biologists study viruses because they cause illness and death. In the human body, virtually every system, tissue, and cell can be infected by one or more kinds of virus (**Figure 35.1**). Much research on viruses is motivated by the desire to minimize the damage they can cause.

Recent Viral Epidemics in Humans

Physicians and researchers use the term **epidemic** (literally, "upon-people") to describe a disease that rapidly affects a large number of individuals over a widening area. Viruses have caused the most devastating epidemics in recent human history. During the eighteenth and nineteenth centuries, it was not unusual for

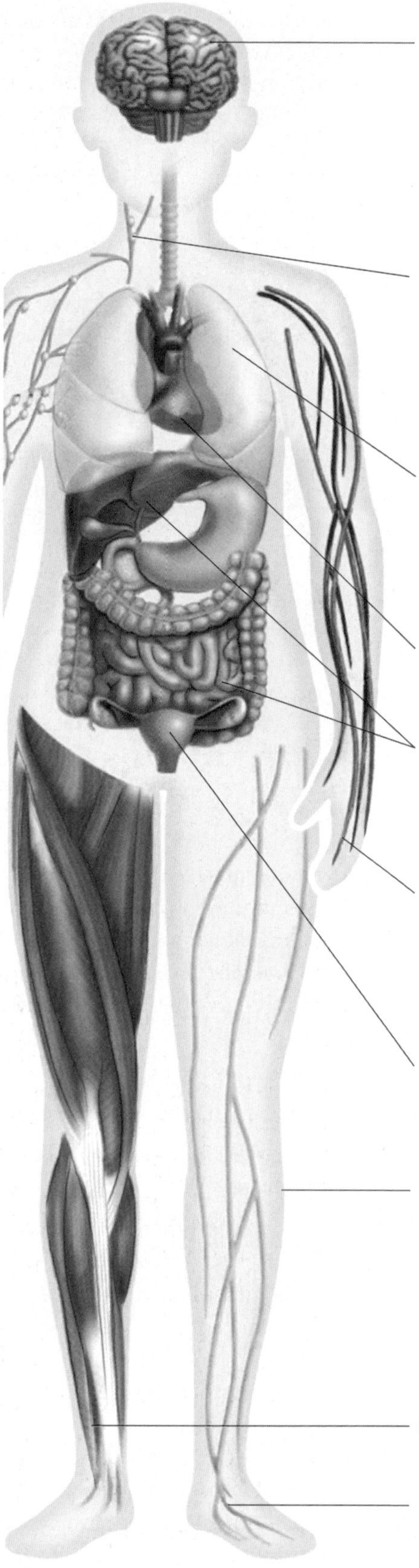

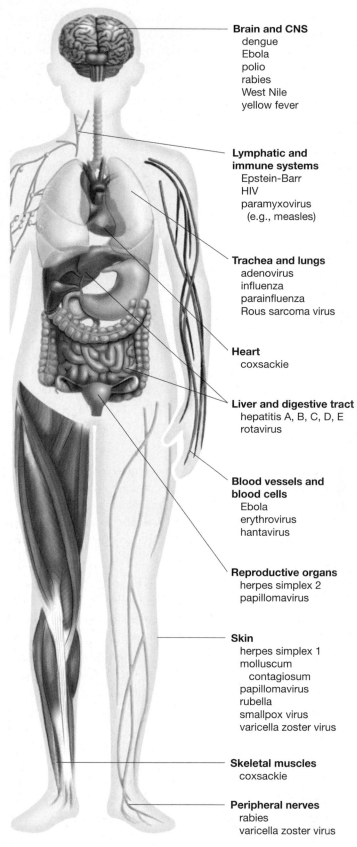

FIGURE 35.1 Human Organs and Systems That Are Parasitized by Viruses.

✔**EXERCISE** Choose two viruses that you are familiar with. Name the location of the infection, and describe how the virus is transmitted.

Native American tribes to lose 90 percent of their members over the course of a few years to measles, smallpox, and other viral diseases spread by contact with European settlers. To appreciate the impact of these epidemics, think of 10 close friends and relatives—then remove nine.

An epidemic that is worldwide in scope is called a **pandemic**. The influenza outbreak of 1918–1919, called "Spanish flu," qualifies as the most devastating pandemic recorded to date. The strain of influenza virus that emerged in 1918 infected people worldwide and was particularly **virulent**—meaning it tended to cause severe disease. Within hours of showing symptoms, the lungs of previously healthy people often became so heavily infected that affected individuals suffocated to death. Most victims were between the ages of 20 and 40. The viral outbreak occurred just as World War I was drawing to a close and killed far more people than did the conflict itself. For example, ten times as many Americans died of influenza than were killed in combat in the war. Worldwide, the Spanish flu is thought to have killed 20–50 million people.

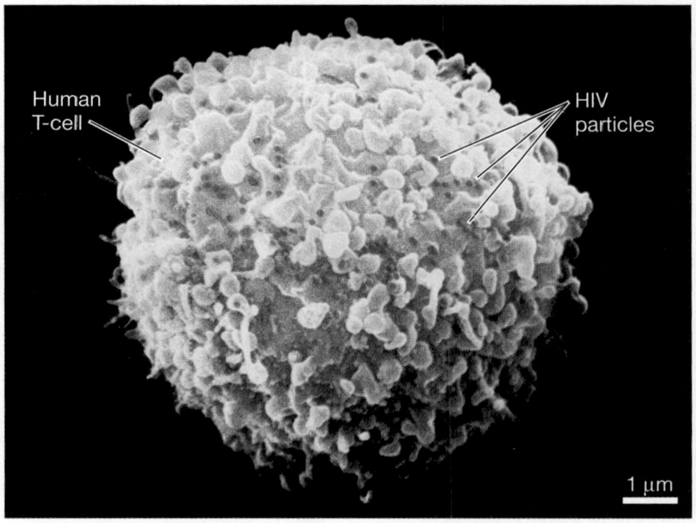

FIGURE 35.2 HIV-Infected T cell. Scanning electron micrograph showing HIV particles (false-colored green) emerging from an infected human T lymphocyte (a type of immune system cell).

Current Viral Pandemics in Humans: HIV

In terms of the total number of people affected, the measles and smallpox epidemics among native peoples of the Americas and the 1918 influenza outbreak are almost certain to be surpassed by the incidence of AIDS. **Acquired immune deficiency syndrome (AIDS)** is an affliction caused by the **human immunodeficiency virus (HIV)**.

HIV is now the most intensively studied of all viruses. Since the early 1980s, governments and private corporations from around the world have spent hundreds of millions of dollars on HIV research. Given this virus's current and projected impact on human populations around the globe, the investment is justified.

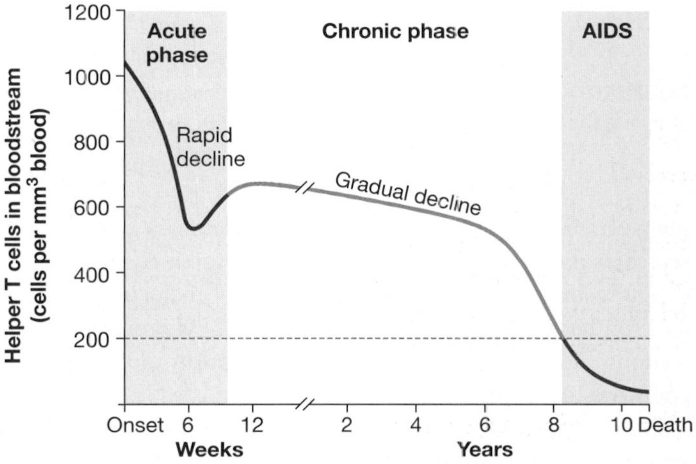

FIGURE 35.3 T-Cell Counts Decline during an HIV Infection. Graph of changes in the number of T cells that are present in the bloodstream over time, based on data from a typical patient infected with HIV. The acute phase may be associated with symptoms such as fever. Few or no disease symptoms occur in the chronic phase. AIDS typically occurs when T-cell counts dip below $200/mm^3$ of blood.

HOW DOES HIV CAUSE DISEASE? Like other viruses, HIV parasitizes specific types of cells. The cells most affected by HIV are called helper T lymphocytes and macrophages (**Figure 35.2**). These cells are components of the **immune system**, which is the body's defense system against disease. Chapter 49 explains just how crucial helper T lymphocytes and macrophages are to the immune system's response to invading bacteria and viruses.

If an HIV virion succeeds in infecting a helper T lymphocyte or macrophage and reproduces inside, the cell dies as hundreds of new virions are released and infect more cells. Although the body continually replaces helper T lymphocytes and macrophages, the number produced does not keep pace with the number being destroyed by HIV. As a result, the total number of helper T lymphocytes in the bloodstream gradually declines as an HIV infection proceeds (**Figure 35.3**).

When the T-cell count drops, the immune system's responses to invading bacteria and viruses become less and less effective. Eventually, too few helper T lymphocytes are left to fight off pathogens efficiently, and bacteria and viruses begin to multiply unchecked. In almost all cases, one or more of these infections proves fatal. HIV kills people indirectly—by making them susceptible to pneumonia, an array of eukaryotic parasites, and unusual types of cancer.

WHAT IS THE SCOPE OF THE AIDS PANDEMIC? Researchers with the United Nations AIDS program estimate that AIDS has already killed 28 million people worldwide. HIV infection rates have been highest in east and central Africa, where one of the greatest public health crises in history is now occurring. In Botswana, for example, blood-testing programs have confirmed that over 20 percent of individuals carry HIV. Although there may be a lag of as much as 8–12 years between the initial infection and the onset of illness, virtually all people who become infected with the virus will die of AIDS.

Currently, the UN estimates the total number of HIV-infected people worldwide at about 33 million. An additional 2.7 million

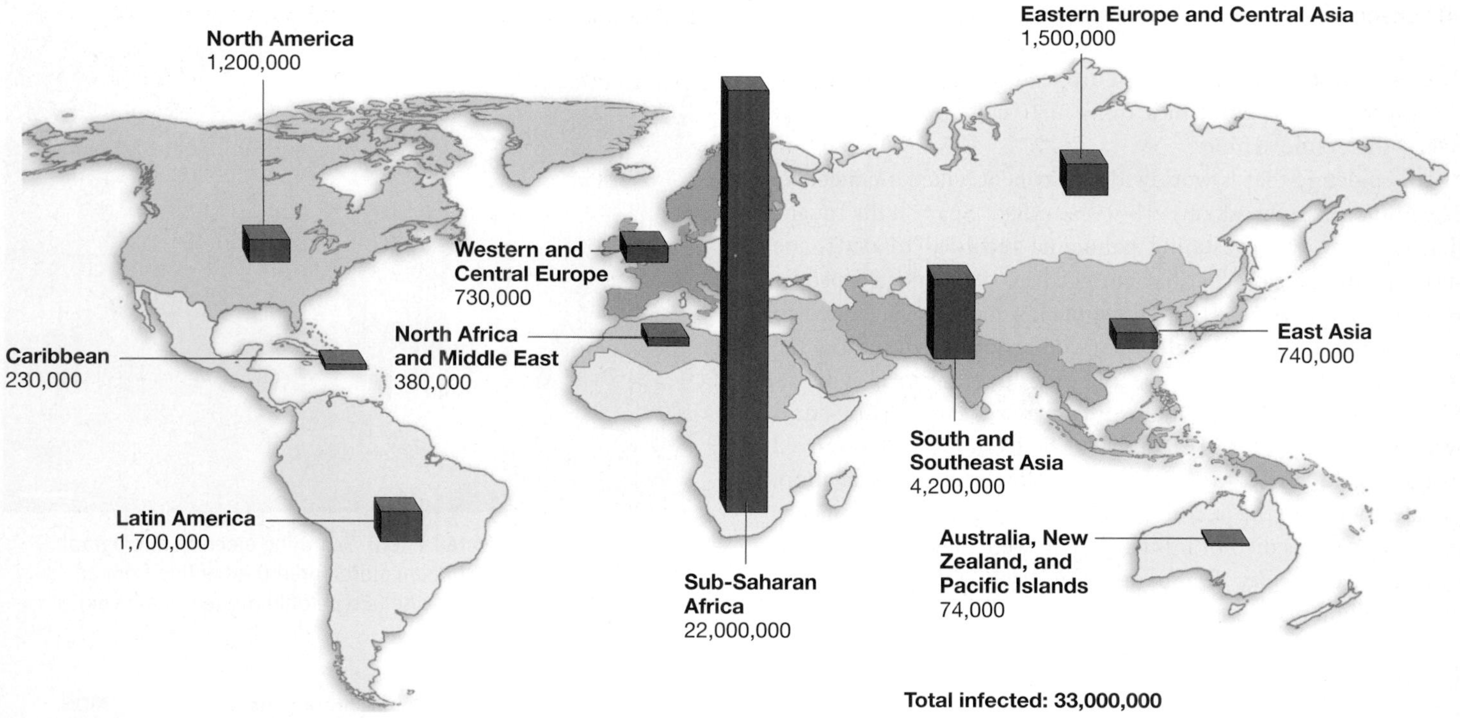

FIGURE 35.4 Geographic Distribution of HIV Infections. Data, compiled by the United Nations AIDS program, showing the numbers of people living with HIV in December 2008, by geographic area.

people are infected each year, and the pandemic is growing. Researchers are particularly alarmed because the focus of the epidemic is shifting from its historical center of incidence—central and southern Africa—to south and east Asia (**Figure 35.4**). Infection rates are growing in some of the world's most populous countries—particularly India and China.

Most viral and bacterial diseases afflict the very young and the very old. But because HIV is primarily a sexually transmitted disease, young adults are most likely to contract the virus and die. People who become infected in their late teens or twenties die of AIDS in their twenties or thirties. Tens of millions of people are being lost in the prime of their lives. Physicians, politicians, educators, and aid workers all use the same word to describe the epidemic's impact: staggering.

35.2 How Do Biologists Study Viruses?

Many researchers who study viruses focus on two goals: (**1**) developing vaccines that help hosts fight off disease if they become infected and (**2**) developing antiviral drugs that prevent a virus from replicating efficiently inside the host. Both types of research begin with attempts to isolate the virus in question.

Isolating viruses takes researchers into the realm of nanobiology, in which structures are measured in billionths of a meter. (One nanometer, abbreviated nm, is 10^{-9} meter.) Viruses range from about 20 to 300 nm in diameter. They are dwarfed by eukaryotic cells and even by bacterial cells (**Figure 35.5**). Millions of viruses can fit on the period at the end of this sentence.

If virus-infected cells can be grown in culture or harvested from a host individual, researchers can usually isolate the virus by passing the cells through a filter. The filters used to study viruses have pores that are large enough for viruses to pass through but are too small to admit cells.

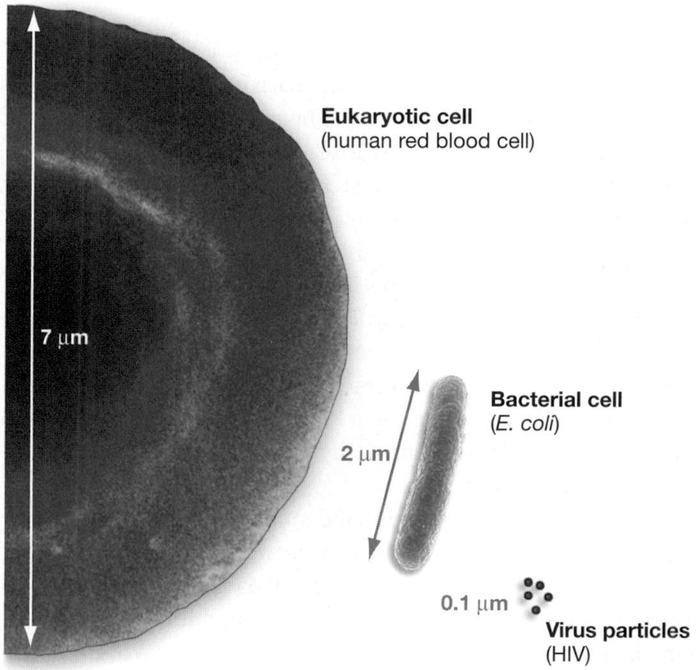

FIGURE 35.5 Viruses Are Tiny.

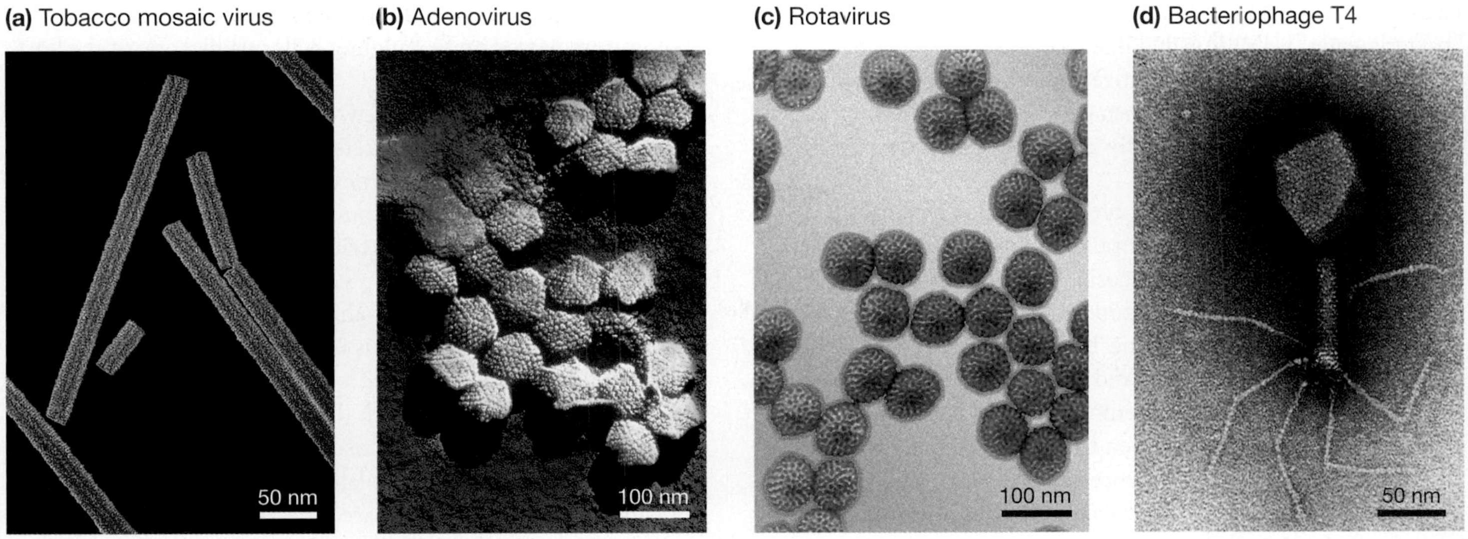

FIGURE 35.6 Viruses Vary in Size and Shape. Virus shapes include **(a)** rods, **(b)** polyhedrons, **(c)** spheres, and **(d)** complex shapes with "heads" and "tails." The viruses in these micrographs have been colorized.

To test the hypothesis that the solution passing through the filter contains viruses that cause a specific disease, researchers expose uninfected host cells to this filtrate. If exposing host cells to the filtrate results in infection, then the hypothesis that the virus caused the disease is supported.

In this way, researchers can isolate a virus and confirm that it is the causative agent of infection. Recall from Chapter 28 that these steps are inspired by Koch's postulates, which established the criteria for linking a specific infectious agent with a specific disease.

Once biologists have isolated a virus, how do they study and characterize it? Let's begin with morphological traits, then consider how viral replication cycles vary.

Analyzing Morphological Traits

To see a virus, researchers usually rely on transmission electron microscopy (see **BioSkills 10** in Appendix A). Only the very largest viruses, such as the smallpox virus, are visible with a light microscope. Electron microscopy has revealed that viruses come in a wide variety of shapes, and many viruses can be identified by shape alone (**Figure 35.6**).

In terms of overall structure, most viruses fall into just two general categories. Most viruses are (1) enclosed by just a shell of protein called a **capsid** or (2) enclosed by both a capsid and a membrane-like **envelope**. Regarding their morphology, then, the important distinction among viruses is whether they are nonenveloped or enveloped.

Nonenveloped viruses consist of genetic material and possibly one or more enzymes inside a capsid—a protein coat. The nonenveloped virus illustrated in **Figure 35.7a** is an adenovirus. You undoubtedly have adenoviruses on your tonsils or in other parts of your upper respiratory passages right now. As the micrograph in Figure 35.6d shows, the morphology of nonenveloped viruses may be complex.

Enveloped viruses also have genetic material inside a capsid or bound to capsid proteins, but the capsid is surrounded by an envelope. The envelope consists of viral proteins embedded in a phospholipid bilayer derived from a membrane found in a host cell—specifically, the host cell in which the virion was manufactured (**Figure 35.7b**). Later sections in the chapter will detail how most of these viruses obtain their envelope from an infected host cell.

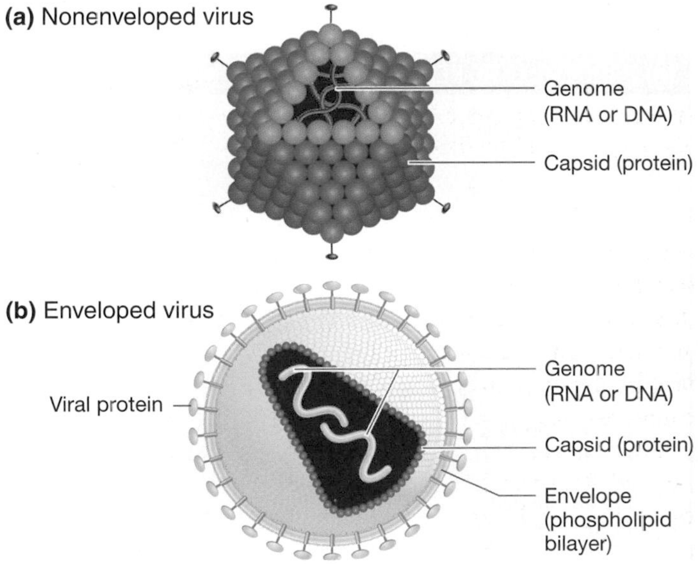

FIGURE 35.7 Viruses Are Nonenveloped or Enveloped. (a) A protein coat forms the exterior of nonenveloped viruses. **(b)** In enveloped viruses, the exterior is composed of a membranous sphere. Inside this envelope, the hereditary material is enclosed by a protein coat or bound to proteins.

Analyzing Variation in Growth Cycles: Replicative and Latent Growth

Although it is likely that millions of types of virus exist, they all infect their host cells in one of two general ways: via replicative growth or in a dormant form referred to as latency in animal viruses or lysogeny in bacteriophages. A **bacteriophage** (literally, "bacteria-eater") is a virus that infects bacteria.

All viruses undergo replicative growth; some can halt the replication cycle and enter a dormant state as well. Both types of viral infection begin when part or all of a virion enters the interior of a host cell.

Figure 35.8a shows the **lytic cycle**, or replicative growth, of a bacteriophage. Note that the viral genome enters the host cell and it is transcribed, using nucleoside triphosphates provided by the host. The host cell manufactures viral proteins and begins to replicate the viral genome using host or viral enzymes. When synthesis of the viral genome and viral proteins is complete, a new generation of virions assembles inside the host cell. The replicative cycle is complete when the virions exit the cell—usually killing the host cell in the process.

Figure 35.8b diagrams the **lysogenic cycle**, or lysogeny, of a bacteriophage. Note that only certain types of viruses are capable of this type of latent infection. During lysogenic growth, viral DNA becomes incorporated into the host's chromosome. Once the viral genome is in place, it is replicated by the host's DNA polymerase each time the cell divides. Copies of the viral genome are passed on to daughter cells just like one of the host's own genes.

In the lysogenic state, no new virions are being produced and no unrelated cells are being infected. The virus is simply transmitted from one generation to the next along with the host's genes.

✔**If you understand this concept, you should be able to explain the contrast in how bacteriophage genes are replicated via lytic versus lysogenic growth.**

In some viruses that infect animal cells, a state called **latency** can exist. When an animal virus is latent, the genome just resides in the cell without replicating or producing new virions. Depending on the virus involved, the genome may or may not be integrated into the host chromosome.

It is not possible to treat a latent infection with drugs called **antivirals**, because the viruses are quiescent. But even lytic infections are notoriously difficult to treat because viruses use so many of the host cell's enzymes during the lytic replication cycle. Drugs that disrupt these enzymes usually harm the host much more than they harm the virus.

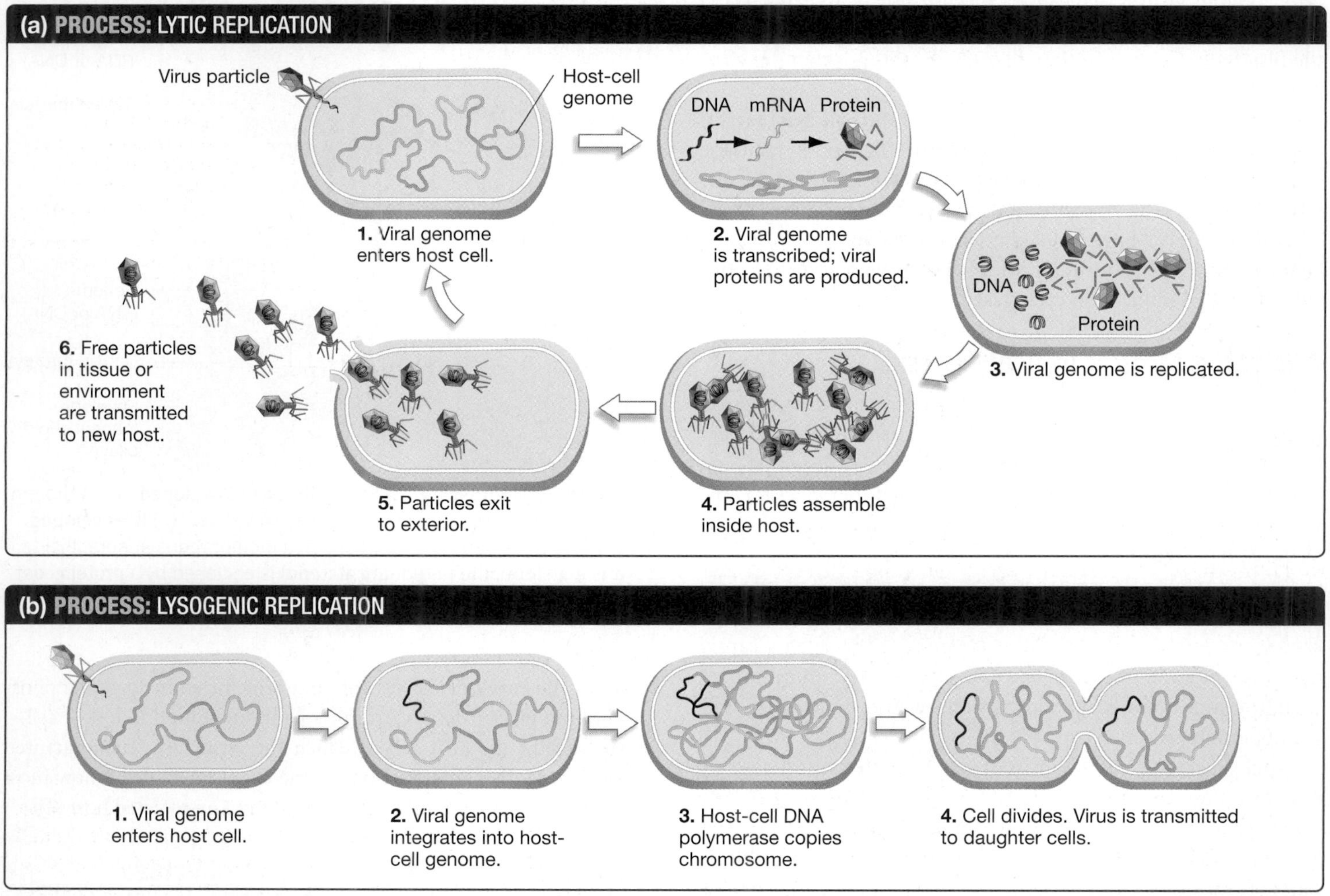
(a) PROCESS: LYTIC REPLICATION
Virus particle
Host-cell genome
DNA mRNA Protein
1. Viral genome enters host cell.
2. Viral genome is transcribed; viral proteins are produced.
DNA
Protein
3. Viral genome is replicated.
4. Particles assemble inside host.
5. Particles exit to exterior.
6. Free particles in tissue or environment are transmitted to new host.
(b) PROCESS: LYSOGENIC REPLICATION
1. Viral genome enters host cell.
2. Viral genome integrates into host-cell genome.
3. Host-cell DNA polymerase copies chromosome.
4. Cell divides. Virus is transmitted to daughter cells.

FIGURE 35.8 Bacteriophages May Replicate via Lytic or Lysogenic Cycles, or Both. (a) All viruses follow the same general lytic replication cycle, which results in a new generation of virions and host cell death. **(b)** Some bacteriophages are also capable of lysogeny, meaning their genome can become integrated into the host-cell chromosome.

✔**EXERCISE** Compare and contrast a lysogenic bacteriophage to the transposable elements introduced in Chapter 20.

To understand viral diversity and how antiviral drugs are developed, let's consider each phase of the lytic cycle in more detail.

Analyzing the Phases of the Replicative Cycle

Six phases are common to replicative growth in virtually all viruses: (1) attachment onto a host cell and entry into the interior, (2) transcription of the viral genome and production and processing of viral proteins, (3) replication of the viral genome, (4) assembly of a new generation of virions, (5) exit from the infected cell, and (6) transmission to a new host. The corresponding steps are depicted in Figure 35.8a.

Each virus has a particular way of entering a host cell and completing the subsequent phases of the cycle. Let's take a closer look.

HOW DO VIRUSES ENTER A CELL? The replication cycle of a virus begins when a free virion enters a target cell. This is no simple task. All cells are protected by a plasma membrane, and many cells also have a cell wall. How do viruses breach these defenses, insert themselves into the cytoplasm inside, and begin an infection?

Most plant viruses enter host cells after a sucking or biting insect has disrupted the host cell wall with its mouthparts. In contrast, viruses that parasitize bacterial cells or that attack animal cells gain entry by binding to a specific molecule on the cell wall or plasma membrane.

After binding to a bacterial cell wall, bacteriophages use an enzyme called lysozyme to degrade the wall. (Lysozyme is also found in human tears, where it acts as an antibiotic in the eye.) The genome of the nonenveloped virus enters the host cell through the hole created by lysozyme. A portion of the capsid seals the hole, and the remainder of the capsid remains on the cell wall or membrane.

Enveloped viruses and noneveloped viruses that attack animal cells bind to one or more specific proteins in the host cell's plasma membrane. When this happens, the capsid of a nonenveloped virus is taken up or the viral envelope and the host's plasma membrane fuse, and the capsid enters the cell.

To appreciate how investigators identify the proteins that viruses use to enter host cells, consider research on HIV. In 1981—right at the start of the AIDS epidemic—biomedical researchers realized that people with AIDS had few or no T lymphocytes possessing **CD4**, a particular membrane protein. These cells, called helper T lymphocytes, are symbolized $CD4^{+}$. This discovery led to the hypothesis that CD4 functions as the receptor or "doorknob" that HIV uses to enter host cells. The doorknob hypothesis predicts that if CD4 is blocked, then HIV will not be able to enter host cells.

Two teams tested this hypothesis (**Figure 35.9**):

- They grew large populations of T lymphocytes in culture and created 160 separate samples.
- They added an antibody to one of the cell-surface proteins found on T lymphocytes, with each of the 160 samples receiving a different antibody. Recall from earlier chapters that **antibodies** are proteins that bind to specific parts of specific proteins (see **BioSkills 9** in Appendix A).
- They added HIV virions to each sample.

EXPERIMENT

QUESTION: Does the CD4 protein function as the "doorknob" that HIV uses to enter host cells?

HYPOTHESIS: CD4 is the membrane protein that HIV uses to enter cells.

NULL HYPOTHESIS: CD4 is not the membrane protein that HIV uses to enter cells.

EXPERIMENTAL SETUP:

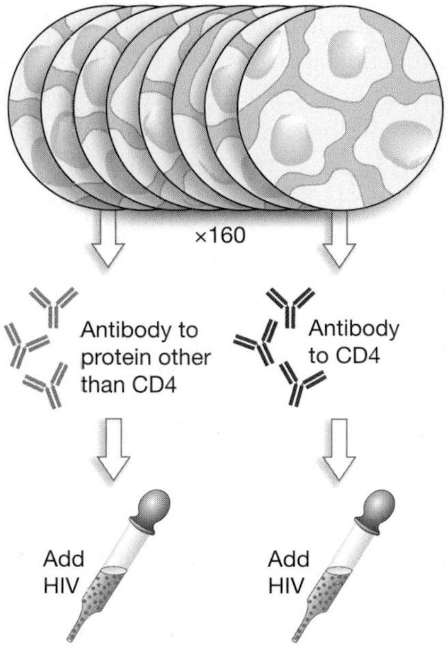

1. Take 160 identical samples of helper T cells from a large population of T cells growing in culture.

2. Add a different antibody to each sample of cells — each antibody will "block" a specific membrane protein.

3. Add a constant number of HIV virions to all samples. Incubate cultures under conditions optimal for virus entry.

PREDICTION: HIV will not infect cells with antibody to CD4 but will infect other cells.

PREDICTION OF NULL HYPOTHESIS: HIV will infect cells with antibody to CD4.

RESULTS:

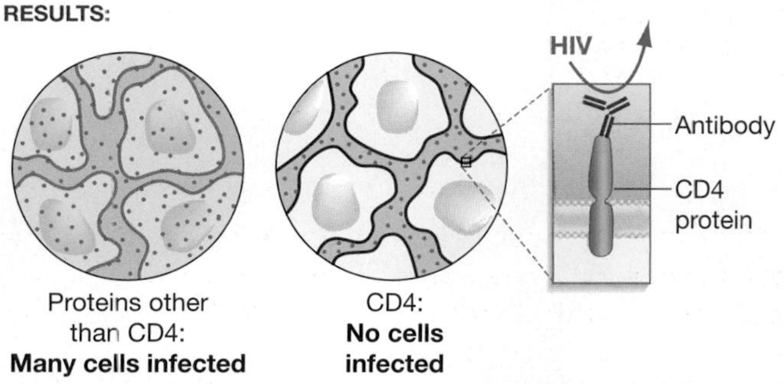

CONCLUSION: HIV uses CD4 proteins as the "doorknob" to enter helper T cells. Thus, only cells with CD4 on their surface can be infected by HIV.

FIGURE 35.9 Experiments Confirmed That CD4 Is the Receptor Used by HIV to Enter Host Cells. In this experiment, the antibodies added to each culture bound to a specific protein found on the surface of T lymphocytes. Antibody binding blocked the membrane protein, so the protein could not be used by HIV to gain entry to the cells.

SOURCE: Maddon, P. J., et al. 1986. The T4 gene encodes the AIDS virus receptor and is expressed in the immune system and the brain. *Cell* 47: 333–348.

✔**QUESTION** Does this experiment show that CD4 is the only membrane protein required for HIV entry? Explain why or why not.

The key point here is that each of the 160 antibodies blocked a different cell-surface protein. If one of the antibodies used in the experiment happened to bind to the receptor used by HIV, that antibody would cover up the receptor. In this way, the antibody would prevent HIV from entering the cell and protect that cell from infection.

This approach led both research teams to reach exactly the same result: Only antibodies to CD4 protected the cells from viral entry.

Later work confirmed that HIV virions can enter cells only if the virions also bind to a second membrane protein, called a **co-receptor**, in addition to CD4. In most individuals, proteins called CXCR4 and CCR5 function as co-receptors. If the proteins in a virion's envelope bind to both CD4 and a co-receptor, the lipid bilayers of the virion's envelope and the plasma membrane of the T lymphocyte fuse (**Figure 35.10**). When fusion occurs, HIV has breached the cell boundary. The viral capsid enters the cytoplasm and infection proceeds.

The discovery of the proteins required for viral entry inspired a search for compounds that would block them and prevent HIV from entering cells. Drugs that act in this way are called fusion inhibitors. Molecules that block the HIV envelope protein or CCR5 are currently in use, and molecules that block CD4 are now being tested.

PRODUCING VIRAL PROTEINS Viruses lack the ribosomes and tRNAs necessary for translating their own mRNAs into proteins. For a virus to make the proteins required to produce a new generation of virions, it must exploit the host cell's biosynthetic machinery.

Production of viral proteins begins soon after a virus enters a cell and continues after the viral genome is replicated. Viral mRNAs and proteins are produced and processed in one of two ways, depending on whether the proteins end up in the outer envelope of a virion or in the capsid.

RNAs that code for a virus's envelope proteins are translated into proteins as if they were the RNAs of the cell's own membrane proteins (see Chapter 7). As **Figure 35.11a** shows, these viral mRNAs are translated by ribosomes attached to the endoplasmic reticulum (ER). Afterward the resulting proteins are transported to the Golgi apparatus, where carbohydrate groups are attached, producing glycoproteins. In some viruses, such as HIV, the finished glycoproteins are then transported to the plasma membrane, where they are ready to be assembled into new virions.

In contrast, a different route is taken by RNAs that code for proteins that make up the capsid or inner core of a virion (see **Figure 35.11b**). These RNAs are translated by ribosomes in the cytoplasm, just as non–membrane-bound cellular mRNAs are. In some viruses, long polypeptide sequences called polyproteins are later cut into functional proteins by a viral enzyme called **protease**. This enzyme cleaves viral polyproteins at specific locations—a critical step in the production of finished viral proteins. In HIV, the resulting protein fragments are assembled into new viral capsids near the host cell's plasma membrane.

The discovery that HIV produces a protease triggered a search for drugs that would block the enzyme. This search got a huge boost when researchers succeeded in visualizing the three-dimensional structure of HIV's protease, using the X-ray crystallographic techniques introduced in **BioSkills 10** in Appendix A. The enzyme has an opening in its interior adjacent to the active site (**Figure 35.12a**). Viral polyproteins fit into the opening and are cleaved at the active site.

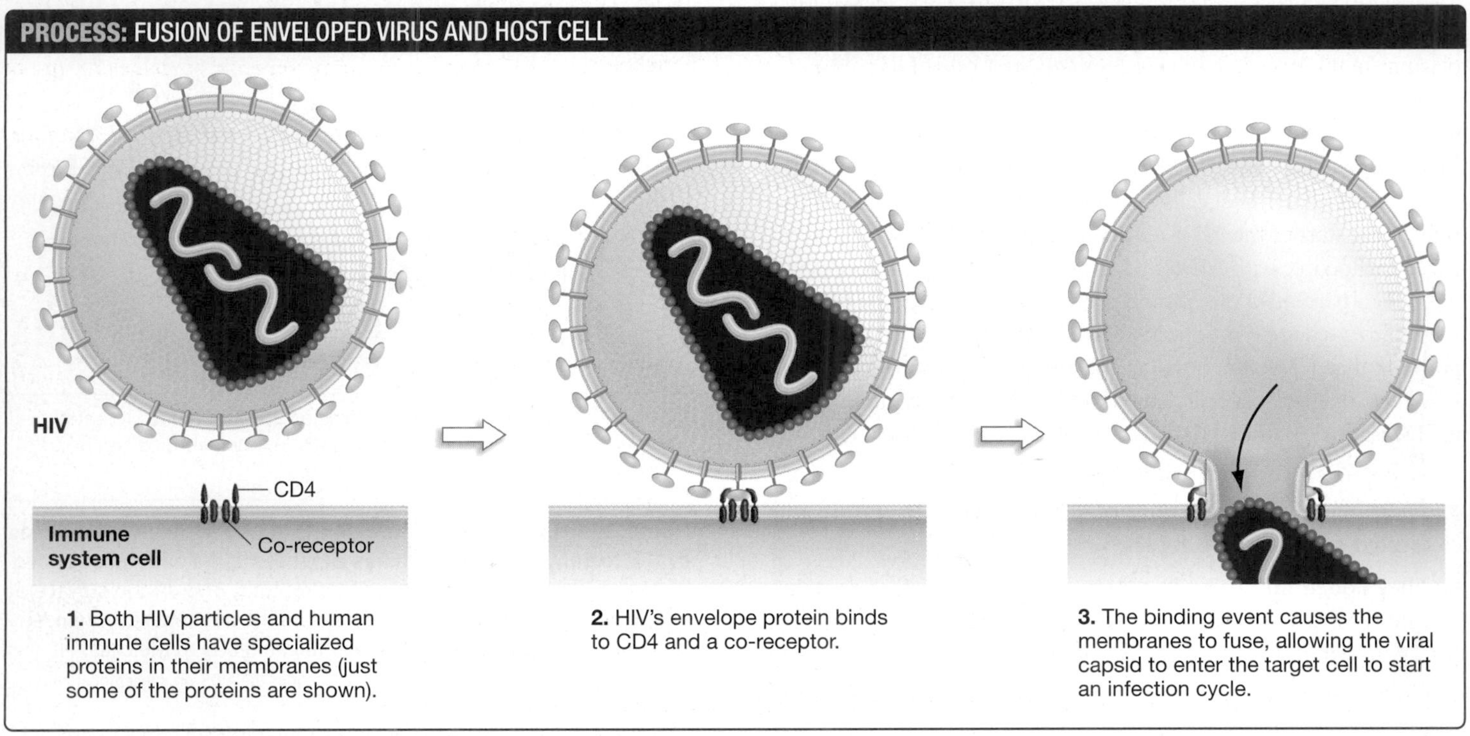

PROCESS: FUSION OF ENVELOPED VIRUS AND HOST CELL
HIV
CD4
Immune system cell
Co-receptor
1. Both HIV particles and human immune cells have specialized proteins in their membranes (just some of the proteins are shown).
2. HIV's envelope protein binds to CD4 and a co-receptor.
3. The binding event causes the membranes to fuse, allowing the viral capsid to enter the target cell to start an infection cycle.

FIGURE 35.10 Enveloped Viruses Bind to Membrane Proteins of Target Cells.

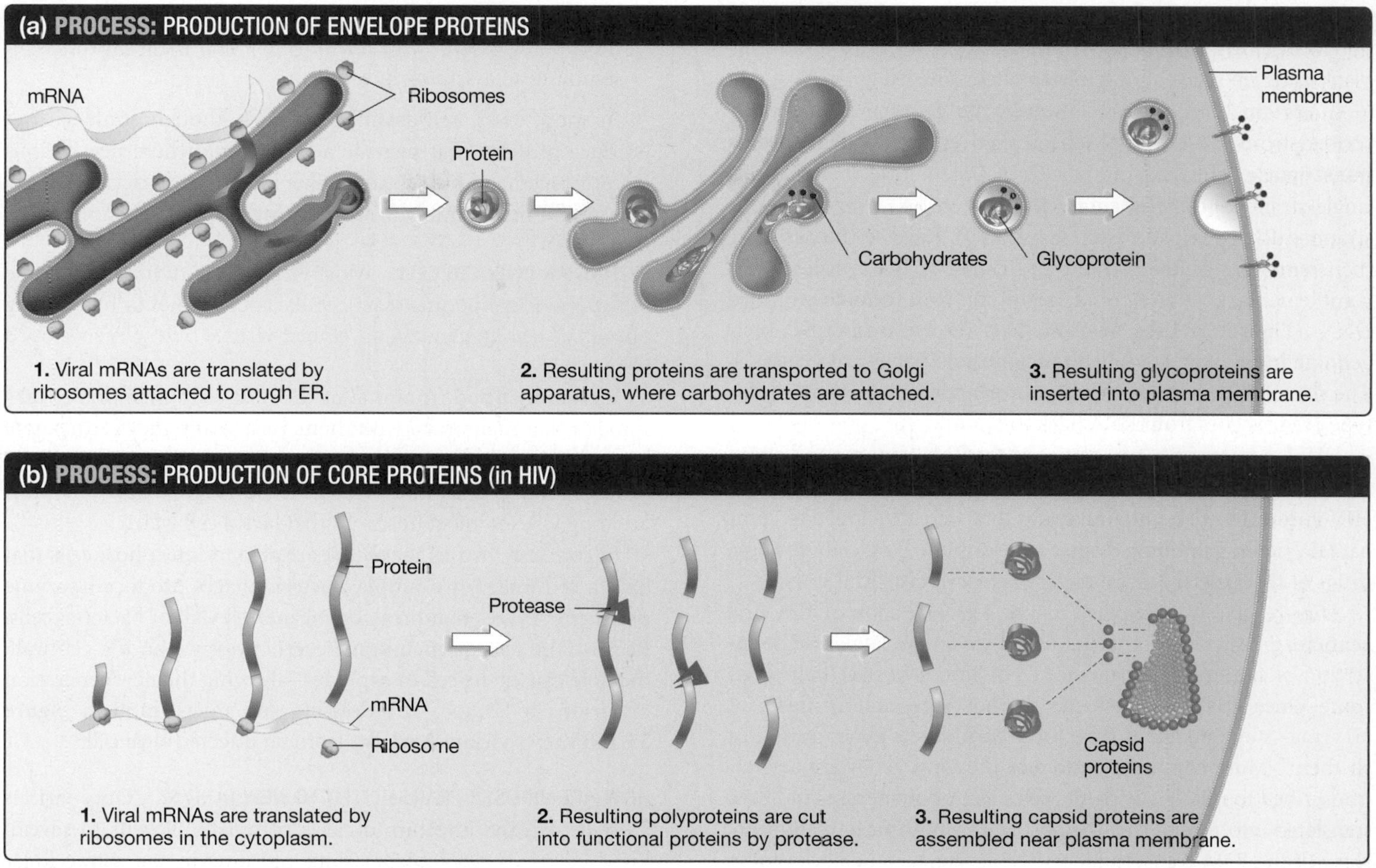
(a) PROCESS: PRODUCTION OF ENVELOPE PROTEINS
mRNA
Ribosomes
Protein
Plasma membrane
Carbohydrates
Glycoprotein
1. Viral mRNAs are translated by ribosomes attached to rough ER.
2. Resulting proteins are transported to Golgi apparatus, where carbohydrates are attached.
3. Resulting glycoproteins are inserted into plasma membrane.
(b) PROCESS: PRODUCTION OF CORE PROTEINS (in HIV)
Protein
Protease
mRNA
Ribosome
Capsid proteins
1. Viral mRNAs are translated by ribosomes in the cytoplasm.
2. Resulting polyproteins are cut into functional proteins by protease.
3. Resulting capsid proteins are assembled near plasma membrane.

FIGURE 35.11 Production of Viral Proteins.

Based on these data, researchers immediately began searching for molecules that could fit into the opening and prevent protease from functioning by blocking the active site (**Figure 35.12b**). Several HIV protease inhibitors are being used to reduce viral replication.

HOW DO VIRUSES COPY THEIR GENOMES? Viruses must copy their genes to make a new generation of virions and continue an infection. Many viruses contain a polymerase that copies the viral genome inside the infected host cell, using nucleoside triphosphates provided by the host cell. Some DNA viruses, for example, contain a viral DNA polymerase that makes a copy of the viral genome.

In other viruses, however, the genome consists of RNA. In most viruses that have an RNA genome, copies of the genome are

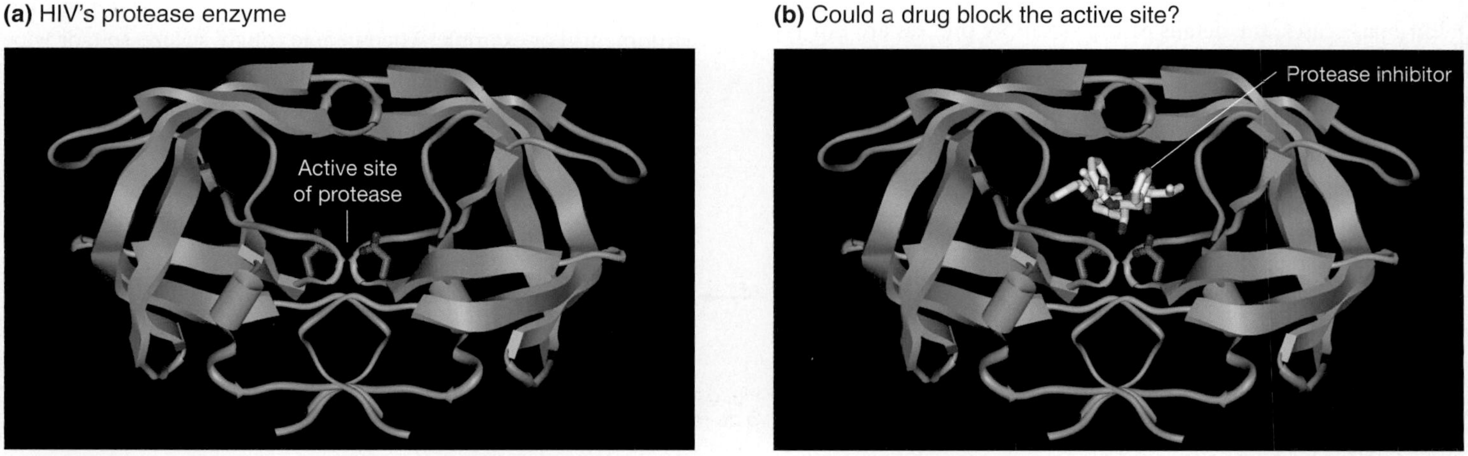
(a) HIV's protease enzyme
Active site of protease
(b) Could a drug block the active site?
Protease inhibitor

FIGURE 35.12 The Three-Dimensional Structure of HIV's Protease. **(a)** Ribbon diagram depicting the three-dimensional shape of HIV's protease enzyme. **(b)** Once protease's structure was solved, researchers began synthesizing compounds that were predicted to fit into the active site and prevent the enzyme from working.

synthesized by the viral enzyme **RNA replicase**, which is an RNA polymerase. RNA replicase synthesizes RNA from an RNA template, using ribonucleotide triphosphates provided by the host cell.

In certain RNA viruses, however, the genome is first transcribed from RNA to DNA by a viral enzyme called **reverse transcriptase**. This enzyme is a DNA polymerase that makes a single-stranded **complementary DNA**, or **cDNA**, from a single-stranded RNA template (see Chapter 19). Reverse transcriptase then removes the RNA strand and catalyzes the synthesis of a complementary DNA strand, resulting in a double-stranded DNA (**Figure 35.13**). Viruses that reverse-transcribe their genome in this way are called **retroviruses** ("backward viruses"). The name is apt because the flow of genetic information in this type of virus goes from RNA back to DNA.

HIV is a retrovirus. Two copies of the RNA genome and about 50 molecules of reverse transcriptase lie inside the capsid of each HIV virion. The first antiviral drugs that were developed to combat HIV act by inhibiting reverse transcriptase. Logically enough, drugs of this type are called reverse transcriptase inhibitors.

After reverse transcriptase makes a cDNA copy of the viral genome, another viral enzyme called integrase catalyzes the insertion of the double-stranded cDNA into a host-cell chromosome. Once it is integrated into the host's genome in this way, HIV may stay in a latent state for a period—its genes may "just sit there." More commonly, though, the viral genes are actively transcribed to mRNA, using the cell's RNA polymerase, and then translated into proteins by the host cell's ribosomes. In addition, some of the RNA transcripts serve as genomes to be packaged in the next generation of HIV virions.

ASSEMBLY OF NEW VIRIONS Once the viral genome has been replicated and viral proteins are produced, they are transported to locations where a new generation of virions assembles inside the infected host cell.

During assembly, the capsid forms around the viral genome. In many cases, copies of non-capsid proteins like viral DNA or RNA polymerases or reverse transcriptases are also packaged inside the capsid. And in enveloped viruses, envelope proteins become inserted into the host cell's plasma membrane.

In many cases, the details of the assembly process are not yet well understood.

- HIV and other enveloped viruses appear to assemble while attached to the inside surface of the host's plasma membrane.
- Bacteriophages and other nonenveloped viruses with complex morphologies appear to assemble in a step-by-step process resembling an assembly line.

In most cases, self-assembly occurs—though some viruses produce proteins that provide a scaffolding where new virions are put together. To date, researchers have yet to develop drugs that inhibit the assembly process during an HIV infection.

EXITING AN INFECTED CELL Most viruses leave a host cell in one of two ways: by budding from cellular membranes or by bursting out of the cell. In general, enveloped viruses bud; nonenveloped viruses burst.

Viruses that bud from one of the host cell's membranes take some of that membrane with them. As a result, they incorporate host-cell phospholipids into their envelope, along with envelope proteins encoded by the viral genome (**Figure 35.14a**). Most budding viruses infect host cells that lack a cell wall.

In contrast, viruses that burst are able to infect host cells that have a cell wall. For example, bacteriophages produce lysozyme or other enzymes that break down the cell wall of bacterial cells. Because the plasma membrane exerts pressure on the cell wall, the hole causes the cell to explode—allowing the new generation of virions to escape. The drawing and micrograph in **Figure 35.14b** shows virions bursting from an infected algal cell.

HOW ARE VIRUSES TRANSMITTED TO NEW HOSTS? Once virions are released, the infection cycle is complete. Dozens to several hundred newly assembled virions are now in the extracellular space. What happens next?

If the host cell is part of a multicellular organism, the new generation of virions begins traveling through the body—often via the bloodstream or lymphatic system. There, they may be bound by antibodies produced by the immune system. In vertebrates, this binding marks the virions for destruction. But if a virion contacts an appropriate host cell before it encounters antibodies, then the virion will infect that cell. This starts the replication cycle anew.

What if the virus has infected a unicellular organism, or if the virus leaves a multicellular host entirely and enters the external environment? For example, when people cough, sneeze, spit, or wipe a runny nose, they help rid their body of viruses and bacteria. But they also project the pathogens into the environment, sometimes directly onto an uninfected host.

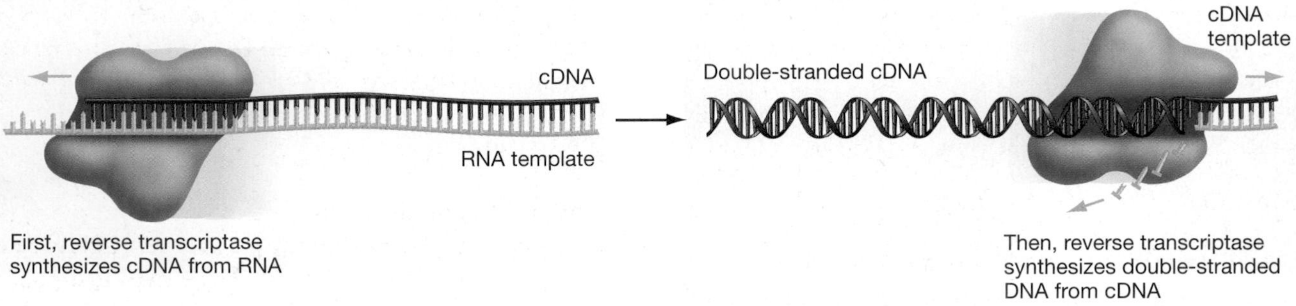
cDNA
RNA template
First, reverse transcriptase synthesizes cDNA from RNA
Double-stranded cDNA
cDNA template
Then, reverse transcriptase synthesizes double-stranded DNA from cDNA

FIGURE 35.13 Reverse Transcriptase Catalyzes Synthesis of a Double-Stranded DNA from an RNA Template. The DNA produced by reverse transcriptase is called a cDNA because its base sequence is complementary to the RNA template.

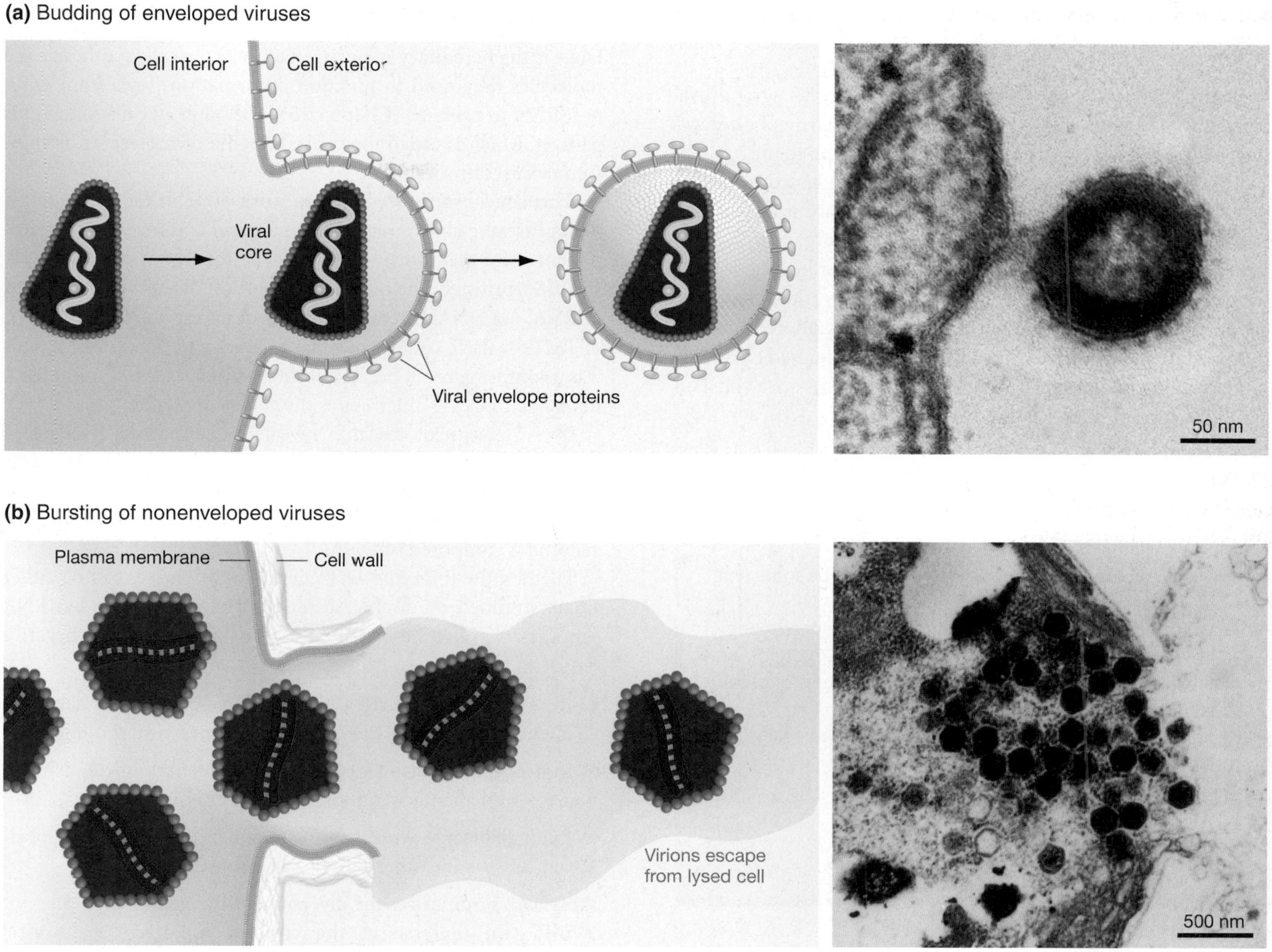

FIGURE 35.14 Viruses Leave Infected Cells by Budding or Bursting. Bursting kills the host cell; budding kills the host cell if it is extensive.

✔**QUESTION** Propose a hypothesis to explain why budding can be fatal to the host cell.

From the virus's point of view, a new host represents an unexploited habitat brimming with resources in the form of target cells. The situation is analogous to that of a multicellular animal dispersing to a new habitat and colonizing it. Viruses that successfully colonize a new host replicate and increase in number. The alleles carried by these successful colonists increase in frequency in the total population. In this way, natural selection favors alleles that allow viruses to do two things: (1) replicate within a host and (2) be transmitted to new hosts.

To review the phases of the HIV replicative cycle, including transmission, go to the study area at *www.masteringbiology.com*.

(MB) **Web Activity** The HIV Replicative Cycle

For agricultural scientists, physicians, and public health officials, reducing the likelihood of **transmission**—the spread of pathogens from one individual to another—is often an effective way to combat a viral disease. For example, HIV is transmitted from person to person via body fluids such as blood, semen, or vaginal secretions. Faced with decades of disappointing results in drug and vaccine development, public health officials are aggressively promoting preventive medicine, through:

- condom use;
- aggressive treatment of venereal diseases, because the lesions caused by chlamydia, genital warts, and gonorrhea encourage the transmission of HIV-contaminated blood during sexual intercourse; and
- promoting sexual abstinence or monogamy.

In the case of viruses that are transmitted through more casual person-to-person contact, the most effective means of reducing transmission involve the simple steps that public health officials recommend during flu and cold outbreaks.

- frequent and thorough hand washings,

- covering the mouth and nose when coughing or sneezing, and
- staying home during illness.

Halting transmission is also important in the agricultural context. When virus outbreaks threaten crops, farmers often respond by removing infected plants and destroying them.

The effectiveness of preventive medicine underscores one of this chapter's fundamental messages: Viruses are a fact of life. Every organism is victimized by viruses; every organism has defenses against them. But the tree of life will never be free of these parasites. Mutation and natural selection guarantee that viral genomes will continually adapt to the defenses offered by their hosts, regardless of whether those defenses are devised by an immune system or by biomedical researchers. Viruses are a constant threat for every organism alive.

CHECK YOUR UNDERSTANDING

If you understand that . . .

- All organisms and cell types are parasitized by some type of virus.
- Viruses are specialized for infecting certain species and cell types, because for a virus to enter a cell, proteins on the exterior of the virus have to bind to proteins or other components on the surface of the host cell.
- After infecting a cell, viruses may replicate or become latent.

✔ You should be able to . . .

Explain how HIV enters a host cell.

Answers are available in Appendix B.

35.3 What Themes Occur in the Diversification of Viruses?

If viruses can infect virtually every type of organism and cell known, how can biologists possibly identify themes that help organize viral diversity? The answer is that, in addition to being identified as enveloped or nonenveloped, viruses can be categorized by the nature of their hereditary material—in essence, the type of molecule that serves as the genome. The single most important aspect of viral diversity is the variation that exists in their genetic material.

Two other points are critical to recognize about viral diversity: (1) Biologists do not have a solid understanding of how viruses originate, but (2) it is certain that viruses will continue to diversify. Because viral polymerases have high error rates and because viruses lack error repair enzymes, mutation rates are extremely high. Viruses change constantly—giving them the potential to evolve rapidly.

Let's analyze the diversity of molecules that make up viral genes, then consider hypotheses to explain where viruses come from and recent data on how viruses are evolving before our eyes.

The Nature of the Viral Genetic Material

DNA is the hereditary material in all cells. As cells synthesize the molecules they need to function, information flows from DNA to mRNA to proteins (Chapter 15). Although all cells follow this pattern, which is called the central dogma of molecular biology, some viruses break it.

This conclusion traces back to work done in the 1950s, when biologists were able to separate the protein and nucleic acid components of a plant virus known as the tobacco mosaic virus, or TMV. Surprisingly, the nucleic acid portion of this virus consisted of RNA, not DNA. Later experiments demonstrated that the RNA of TMV, by itself, could infect plant tissues and cause disease. This was a confusing result because it showed that, in this virus at least, RNA—not DNA—functions as the genetic material.

Subsequent research revealed an amazing diversity of viral genome types. In some groups of viruses, such as the agents that cause measles and flu, the genome consists of RNA. In others, such as the viruses that cause herpes and smallpox, the genome is composed of DNA.

Further, the RNA and DNA genomes of viruses can be either single stranded or double stranded. The single-stranded RNA genomes can also be classified as "positive sense" or "negative sense" or "ambisense."

- In a **positive-sense virus**, the genome contains the same sequences as the mRNA required to produce viral proteins.
- In a **negative-sense virus**, the base sequences in the genome are complementary to those in viral mRNAs.
- In an **ambisense virus**, some sections of the genome are positive sense while others are negative sense.

Table 35.2 summarizes the diversity of viral genome types.

✔If you understand the concept of diversity in viral genomes, you should be able to name two types of viral genomes that are not based on double-stranded DNA. For each of these genomes, you should be able to explain the steps in information flow from gene to protein.

Finally, the number of genes found in viruses varies widely. The tymoviruses that infect plants contain as few as three genes, but the genome of smallpox can code for up to 353 proteins.

Where Did Viruses Come From?

No one knows how viruses originated. To address this question, biologists are currently considering three hypotheses to explain where viruses came from.

ORIGIN IN PLASMIDS AND TRANSPOSABLE ELEMENTS? Like viruses, the plasmids introduced in Chapter 19 and the transposable elements introduced in Chapter 20 are acellular, mobile genetic elements that replicate with the aid of a host cell. Certain viruses are actually indistinguishable from plasmids except for one feature: They encode proteins that form a capsid and allow the genes to exist outside of a cell.

Some biologists hypothesize that simple viruses, plasmids, and transposable elements represent "escaped gene sets." This hy-

SUMMARY TABLE 35.2 The Diversity of Viral Genomes

Key: ss = single stranded; ds = double stranded; (+) = positive sense (genome sequence is the same as viral mRNA); (−) = negative sense (genome sequence is complementary to viral mRNA)

Genome	Strand	Example(s)	Host	Result of Infection	Notes
(+)ssRNA	+	TMV	Tobacco plants	Tobacco mosaic disease (leaf wilting)	TMV was the first RNA virus to be discovered.
(−)ssRNA	−	Influenza	Many mammal and bird species	Influenza	The negative-sense ssRNA viruses transcribe their genomes to mRNA via RNA replicase.
dsRNA	+ / −	Phytovirus	Rice, corn, and other crop species	Dwarfing	Double-stranded RNA viruses are transmitted from plant to plant by insects. Many can also replicate in their insect hosts.
ssRNA that requires reverse transcription for replication	+	Rous sarcoma virus	Chickens	Sarcoma (cancer of connective tissue)	These are called retroviruses. Rous sarcoma virus was identified as a cancer-causing agent in 1911—decades before any virus was seen.
ssDNA—can be (+), (−), or (+) and (−)	+ or −	φX174	Bacteria	Death of host cell	The genome for φX174 is circular and was the first complete genome ever sequenced.
dsDNA that is replicated through an RNA intermediate	+ / −	Hepatitis B virus	Humans	Hepatitis	These are called "reversiviruses."
dsDNA that is replicated by DNA polymerase	+ / −	Baculovirus Smallpox Bacteriophage	Insects Humans Bacteria	Death Smallpox Death	These include the largest viruses in terms of genome size and overall size.

pothesis states that mobile genetic elements are descended from clusters of genes that physically escaped from bacterial or eukaryotic chromosomes long ago.

According to this hypothesis, the escaped gene sets took on a mobile, parasitic existence because they happened to encode the information needed to replicate themselves at the expense of the genomes that once held them. In the case of viruses, the hypothesis is that the escaped genes included the instructions for making a protein capsid and possibly envelope proteins. According to the escaped-gene hypothesis, it is likely that each of the distinct types of RNA viruses and some of the DNA viruses represent distinct "escape events."

To support the escaped-genes hypothesis, researchers would probably have to discover a brand-new virus that originated in this way, or viruses that had so recently derived from intact bacterial or eukaryotic genes that the viral DNA sequence still strongly resembled the DNA sequence of those genes.

ORIGIN IN SYMBIOTIC BACTERIA? Some researchers contend that DNA viruses with large genomes trace their ancestry back to free-living bacteria that once took up residence inside eukaryotic cells. The idea is that these organisms degenerated into viruses by gradually losing the genes required to synthesize ribosomes, ATP, nucleotides, amino acids, and other compounds.

Although this idea sounds speculative, it cannot be dismissed lightly. Chapter 28 introduced species in the genus *Chlamydia*, which are bacteria that live as parasites inside animal cells. And Chapter 29 provided evidence that the organelles called mitochondria and chloroplasts, which reside inside eukaryotic cells, originated as intracellular symbionts. Investigators contend that, instead of evolving into intracellular symbionts that aid their host cell, DNA viruses became parasites capable of replication and transmission from one host to another.

To support the degeneration hypothesis, researchers have pointed to mimivirus, which contains some genes involved in protein synthesis. It is still not clear, though, whether these genes are remnants from an ancestral cell or were acquired from a host cell that was infected relatively recently. If the degeneration hypothesis is correct, then mimivirus should eventually lose these genes, and become completely dependent on host cell machinery to make its proteins.

ORIGIN AT THE ORIGIN OF LIFE? Recently, some researchers have started to discuss a third alternative to explain the origin of viruses: that they trace their ancestry back to the first, RNA-based forms of life on Earth. If this hypothesis is correct, then the RNA genomes of some viruses are descended from genes found in early inhabitants of the RNA world (see Chapter 4).

To support this hypothesis, advocates point to the ubiquity of viruses—which suggests that they have been evolving along with organisms since life began. But currently, there is no widely accepted view of where viruses came from. Because viruses are so diverse, all three hypotheses may be valid.

Emerging Viruses, Emerging Diseases

Although it is not known how the various types of virus originated, it is certain that viruses will continue to diversify. With alarming regularity, the front pages of newspapers carry accounts of deadly viruses that are infecting humans for the first time.

- In 1993 a hantavirus that normally infects mice suddenly afflicted dozens of people in the southwestern United States. Nearly half of the people who developed hantavirus pulmonary syndrome died.
- In 1995 the Ebola virus, a variant of a monkey virus, caused a wave of infections in the Democratic Republic of Congo. By the time the outbreak subsided, over 200 cases had been reported; 80 percent were fatal.
- Numerous reports of avian flu infecting and killing humans caused worldwide alarm in 2005–2006.

HIV, hantavirus pulmonary syndrome, Ebola, and avian flu are examples of **emerging diseases**: new illnesses that suddenly affect significant numbers of individuals in a host population. In these cases, the causative agents were considered emerging viruses because they had switched from their traditional host species to a new host—humans.

USING PHYLOGENETIC TREES TO UNDERSTAND EMERGING VIRUSES How do researchers know that an emerging virus has "jumped" to a new host? The answer is to analyze the evolutionary history of the virus in question, and estimate a phylogenetic tree that includes its close relatives.

HIV, for example, belongs to a group of viruses called the lentiviruses, which infect a wide range of mammals, including house cats, horses, goats, and primates. (*Lenti* is a Latin root that means "slow"; here it refers to the long period observed between the start of an infection by these viruses and the onset of the diseases they cause.) Consider the conclusions that can be drawn from the phylogenetic tree of HIV, shown in **Figure 35.15**.

1. ***There are immunodeficiency viruses.*** Many of HIV's closest relatives parasitize cells that are part of the immune system. Several of them cause diseases with symptoms reminiscent of AIDS. For unknown reasons, though, HIV's closest relatives don't appear to cause disease in their hosts. These viruses infect monkeys and chimpanzees and are called simian immunodeficiency viruses (SIVs).
2. ***There are two HIVs.*** There are two distinct types of human immunodeficiency viruses, called HIV-1 and HIV-2. Although both can cause AIDS, HIV-1 is far more virulent and is the better studied of the two.

   HIV-1's closest known relatives are immunodeficiency viruses isolated from chimpanzees that live in central Africa.

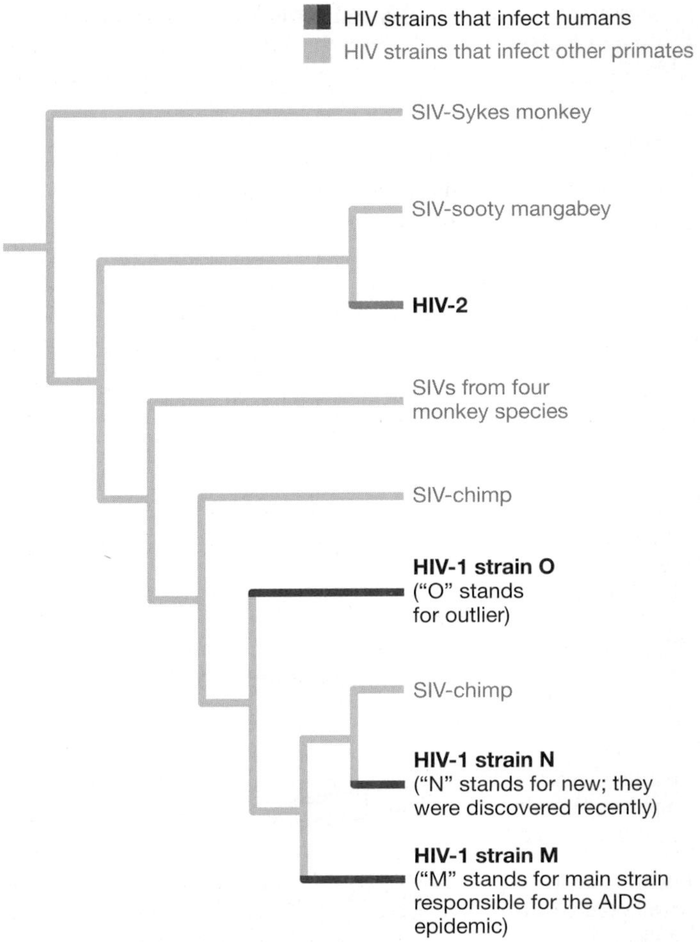

FIGURE 35.15 Phylogeny of HIV Strains and Types. Phylogenetic tree showing the evolutionary relationships among some of the immunodeficiency viruses that infect primates—including chimpanzees, humans, and several species of monkeys.

✔**EXERCISE** On the appropriate branches, indicate where an SIV jumped to humans (draw and label bars across the appropriate branches).

   In contrast, HIV-2's closest relatives are immunodeficiency viruses that parasitize monkeys called sooty mangabeys.

   In central Africa, where HIV-1 infection rates first reached epidemic proportions, contact between chimpanzees and humans is extensive. Chimps are hunted for food and kept as pets. Similarly, sooty mangabeys are hunted and kept as pets in west Africa, where HIV-2 infection rates are highest.

   To make sense of these observations, biologists suggest that HIV-1 is a descendant of viruses that infect chimps, and HIV-2 is a descendant of viruses that infect sooty mangabeys. The "jumps" between host species probably occurred when a human cut up a monkey or chimpanzee for use as food.

3. ***Multiple "jumps" have occurred.*** Several strains of HIV-1 exist. A virus **strain** consists of populations that have similar characteristics. The most important HIV-1 strains are called O for outlier (meaning, the most distant group relative to other

strains), N for new, and M for main. Because few mechanisms exist for the virus to jump from humans back to chimps or monkeys, it is likely that each of these strains represents an independent origin of HIV from a chimp SIV strain.

The last point is particularly important. The existence of distinct strains suggests that HIV-1 has jumped from chimps to humans several times. It may do so again in the future.

RESPONDING TO AN OUTBREAK Physicians become alarmed when they see a large number of patients with identical and unusual disease symptoms in the same geographic area over a short period of time. The physicians report these cases to public health officials, who take on two urgent tasks: (1) identifying the agent that is causing the new illness and (2) determining how the disease is being transmitted.

Several strategies can be used to identify the pathogen responsible for an emerging disease. In the case of the outbreak of hantavirus pulmonary syndrome, officials recognized strong similarities between symptoms in the U.S. cases and symptoms caused by a hantavirus called Hantaan virus native to northeast Asia. The Hantaan virus rarely causes disease in North America; its normal host is rodents.

To determine whether a Hantaan-like virus was responsible for the U.S. outbreak, researchers began capturing mice in the homes and workplaces of afflicted people. About a third of the captured rodents tested positive for the presence of a Hantaan-like virus. Genome sequencing studies confirmed that the virus was a previously undescribed type of hantavirus. Further, the sequences found in the mice matched those found in infected patients. Based on these results, officials were confident that a rodent-borne hantavirus was causing the wave of infections.

The next step in the research program, identifying how the agent is being transmitted, is equally critical. If a virus that normally parasitizes a different species suddenly begins infecting humans, if it can be transmitted efficiently from person to person, and if it causes serious illness, then the outbreak has the potential to become an epidemic. But if transmission takes place only between the normal host and humans, as is the case with rabies, then the number of cases will probably remain low. To date, for example, avian flu has not been transmitted efficiently from person to person—only inefficiently from birds to people.

Determining how a virus is transmitted takes old-fashioned detective work. By interviewing patients about their activities, researchers called epidemiologists decide whether each patient could have acquired the virus independently. Were the individuals infected with hantavirus in contact with mice? Was the illness showing up in health-care workers who were in contact with infected individuals, implying that it was being transmitted from human to human?

In the hantavirus outbreak, public health officials concluded that no human-to-human transmission was taking place. Health-care workers did not become ill, and because patients had not had extensive contact with one another, it was likely that each had acquired the virus independently. The outbreak also coincided with a short-term, weather-related explosion in the local mouse population.

The most likely scenario was that people had acquired the pathogen by inhaling dust or handling food that contained remnants of mouse feces or urine. In short, hantavirus did not have the potential to cause an epidemic. The best medicine was preventive: Homeowners were advised to trap mice in living areas, wear dust masks while cleaning any area where mice might have lived, and store food in covered jars.

The Ebola virus, in contrast, was clearly transmitted from person to person. Many doctors and nurses were stricken after tending to patients with the virus. Infections that originate in hospitals usually spread when carried from patient to patient on the hands of caregivers. But by carefully observing the procedures that were being followed by hospital staff, researchers concluded that the Ebola virus was being transmitted only through direct contact with body fluids (blood, urine, feces, or sputum).

The Ebola outbreak was brought under control when hospital workers raised their standards of hygiene and insisted on the immediate disposal or disinfection of all contaminated bedding, utensils, and equipment. The situation with the avian flu outbreak of 2005–2006 was similar—person-to-person transmission was extremely rare and based only on extensive, direct contact with body fluids.

Had the Ebola virus or avian flu virus been transmitted by casual contact, such as touch or inhalation—as is the common cold virus—a massive epidemic could have ensued.

CHECK YOUR UNDERSTANDING

If you understand that . . .

- Among viruses, several different types of molecules serve as the genetic material.
- Emerging viruses have switched host species. They become dangerous if they are rapidly transmitted among host individuals.

✓ You should be able to . . .

1. Explain how the negative-sense, single-stranded RNA genome of influenza viruses, which is complementary in sequence to viral mRNA, is copied during the replicative cycle.
2. State whether a mutation that allowed avian flu virus to spread via airborne virions coughed out by infected individuals would make the virus more or less dangerous to humans. Explain your logic.

Answers are available in Appendix B.

35.4 Key Lineages of Viruses

Because scientists are almost certain that viruses originated multiple times throughout the history of life, there is no such thing as the phylogeny of all viruses. Stated another way, there is no single phylogenetic tree that represents the evolutionary history of viruses as there is for the organisms discussed in previous chapters. Instead, researchers focus on comparing base sequences in the genetic material of small, closely related groups of viruses and using these data to reconstruct the phylogenies of particular lineages, exemplified by the tree in Figure 35.15.

To organize the diversity of viruses on a larger scale, researchers group them into seven general categories based on the nature of their genetic material. This approach to organizing viruses is called the Baltimore classification, because it was proposed by David Baltimore in 1971. Within these seven general categories, biologists also identify a total of about 70 virus families that are distinguished by (1) the structure of the virion (often whether it is enveloped or nonenveloped), and (2) the nature of the host species.

Although they do not have formal scientific names, viruses within families are grouped into distinct genera for convenience. Within genera, biologists identify and name types of virus, such as HIV, the measles virus, and smallpox. Within each of these viral types, populations with distinct characteristics may be identified and named as strains. The strain is the lowest, or most specific, level of taxonomy for viruses.

To get a sense of viral diversity, let's survey some of the most important groups that can be identified by the nature of their genetic material.

- Double-Stranded DNA (dsDNA) Viruses
- RNA Reverse-Transcribing Viruses (Retroviruses)
- Double-Stranded RNA (dsRNA) Viruses
- Negative-Sense Single-Stranded RNA ([−]ssRNA) Viruses
- Positive-Sense Single-Stranded RNA ([+]ssRNA) Viruses

Double-Stranded DNA (dsDNA) Viruses

The double-stranded DNA viruses are a large group, composed of some 21 families and 65 genera. Smallpox (**Figure 35.16**) is perhaps the most familiar of these viruses. Although smallpox had been responsible for millions of deaths throughout human history, it was eradicated by vaccination programs. Smallpox is currently extinct in the wild; the only remaining samples of the virus are stored in research labs.

Genetic material As their name implies, the genes of these viruses consist of a single molecule of double-stranded DNA. The molecule may be linear or circular.

Host species These viruses parasitize hosts from throughout the tree of life, with the notable exception of land plants. They include virus families called the T-even and λ bacteriophages, some of which infect *E. coli*. In addition, the pox viruses, herpesviruses, and adenoviruses—which include types that parasitize humans—have genomes consisting of double-stranded DNA.

Replication cycle In most double-stranded DNA viruses that infect eukaryotes, viral genes have to enter the nucleus to be replicated. In many DNA viruses, the viral genes are replicated only during S phase, when the host cell's chromosomes are being replicated. As a result, these types of viruses can sustain an infection only in cells that are actively dividing, such as the cells lining the respiratory tract or urogenital canal. Some of these viruses are capable of inducing replication in infected cells.

FIGURE 35.16 Smallpox Is a Double-Stranded DNA Virus.

RNA Reverse-Transcribing Viruses (Retroviruses)

The genomes of the RNA reverse-transcribing viruses are composed of single-stranded RNA. There is only one family, called the retroviruses.

Genetic material Virions have two copies of their single-stranded RNA genome.

Host species Species in this group are known to parasitize only vertebrates—specifically birds, fish, or mammals. HIV is the most familiar virus in this group. The Rous sarcoma virus, the mouse mammary tumor virus, and the murine (mouse) leukemia virus are other retroviruses that have also been studied intensively. Rous sarcoma virus was the first virus shown to be associated with the development of cancer (in chickens); the mouse viruses were the first viruses known to increase the risk of cancer in mammals (**Figure 35.17**). In most cases, viruses that are associated with cancer development carry genes that contribute to uncontrolled growth of the cells they infect.

Replication cycle Retroviruses contain the enzyme reverse transcriptase inside their capsid. When the virus's RNA genome and reverse transcriptase enter a host cell's cytoplasm, the enzyme catalyzes the synthesis of a single-stranded cDNA from the original RNA. Reverse transcriptase then makes this cDNA double stranded. The double-stranded DNA version of the genome enters the nucleus with a viral protein called integrase. Integrase catalyzes the integration of the viral genes into a host chromosome. The virus may remain quiescent for a period, but eventually the genes are transcribed to RNA to begin lytic growth and the production of a new generation of virions.

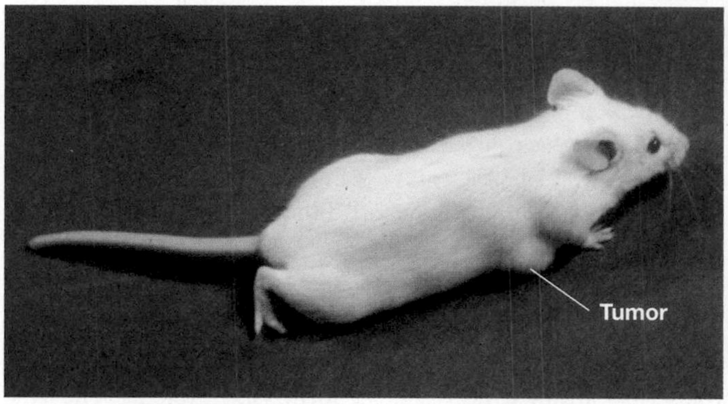

FIGURE 35.17 Some Retroviruses, Such as the Mouse Mammary Tumor Virus, Are Associated with Cancer.

Double-Stranded RNA (dsRNA) Viruses

There are 7 families of double-stranded RNA viruses and a total of 22 genera. Most of the viruses in this group are nonenveloped.

Genetic material In some families, virions have a genome consisting of 10–12 double-stranded RNA molecules; in other families, the genome is composed of just 1–3 RNA molecules.

Host species A wide variety of hosts, including fungi, land plants, insects, vertebrates, and bacteria, are victimized by viruses with double-stranded RNA genomes. Particularly prominent are viruses that cause disease in rice, corn, sugarcane, and other crops. **Figure 35.18** shows rice plants that are being attacked by a double-stranded RNA virus. Infections are also common in *Penicillium*, a filamentous fungus that produces the antibiotic penicillin. Reovirus and rotavirus infections, the leading cause of infant diarrhea in humans, are responsible for over 110 million cases and 440,000 deaths each year.

Replication cycle Once inside the cytoplasm of a host cell, the double-stranded genome of these viruses serves as a template for the synthesis of single-stranded RNAs, which are then translated into viral proteins. Some of the proteins form the capsids for a new generation of virions. Copies of the genome are created when a viral enzyme converts the single-stranded transcripts into double-stranded RNA.

FIGURE 35.18 dsRNA Viruses, Such as the Ragged Stunt Virus, Parasitize a Wide Array of Organisms—Here, Rice.

Negative-Sense Single-Stranded RNA ([−]ssRNA) Viruses

There are 7 families and 30 genera in this group. Most members of this group are enveloped, but some negative-sense single-stranded RNA viruses lack an envelope.

Genetic material The sequence of bases in a negative-sense RNA virus is opposite in polarity to the sequence in a viral mRNA. Stated another way, the single-stranded virus genome is complementary to the viral mRNA. Depending on the family, the genome may consist of a single RNA molecule or up to eight separate RNA molecules.

Host species A wide variety of plants and animals are parasitized by viruses that have negative-sense single-stranded RNA genomes. If you have ever suffered from the flu, the mumps, or the measles, then you are painfully familiar with these viruses (**Figure 35.19**). The Ebola, Hantaan, and rabies viruses also belong to this group.

Replication cycle When the genome of a negative-sense single-stranded RNA virus enters a host cell, a viral RNA polymerase uses that genome as a template to make new viral mRNA. Certain viral transcripts are then translated to form viral proteins. The viral mRNAs also serve as a template for the synthesis of new copies of the negative-sense single-stranded RNA genome.

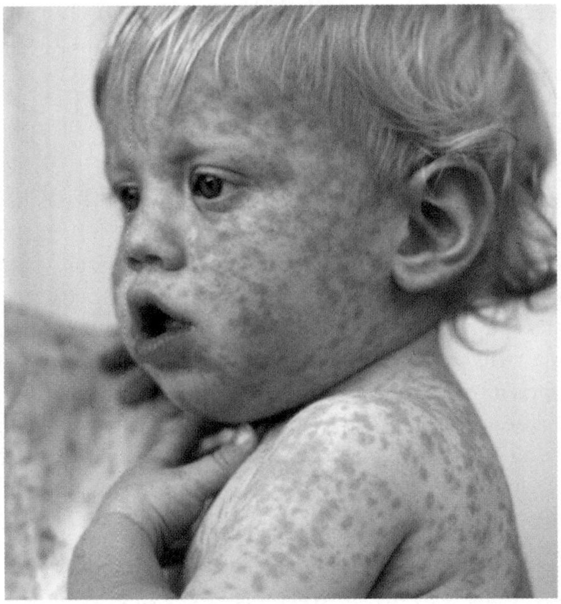

FIGURE 35.19 (−)ssRNA Viruses, Such as the Measles Virus, Cause Some Common Childhood Diseases.

Positive-Sense Single-Stranded RNA ([+]ssRNA) Viruses

This is the largest group of viruses known, with 21 families and 81 genera.

Genetic material The sequence of bases in a positive-sense RNA virus is the same as that of a viral mRNA. Stated another way, the genome does not need to be transcribed in order for proteins to be produced. Depending on the species, the genome consists of one to three RNA molecules.

Host species Most of the commercially important plant viruses belong to this group. Because they kill groups of cells in the host plant and turn patches of leaf or stem white, they are often named mottle viruses, spotted viruses, chlorotic (meaning, lacking chlorophyll) viruses, necrotic (meaning, killed cells surrounded by intact tissue) viruses, or mosaic viruses. **Figure 35.20** shows a healthy cowpea leaf and a cowpea leaf that has been attacked by a positive-sense single-stranded RNA virus. Some species in this group of viruses specialize in parasitizing bacteria, fungi, or animals, however. A variety of human maladies, including the common cold, polio, and hepatitis A, C, and E, are caused by positive-sense RNA viruses.

Replication cycle When the genome of these viruses enters a host cell, the single-stranded RNA is immediately translated into a protein, often a polyprotein, which is cleaved to generate other functional distinct, individual proteins. These proteins include enzymes that make copies of the genome. The new generation of virions is assembled within the host cell.

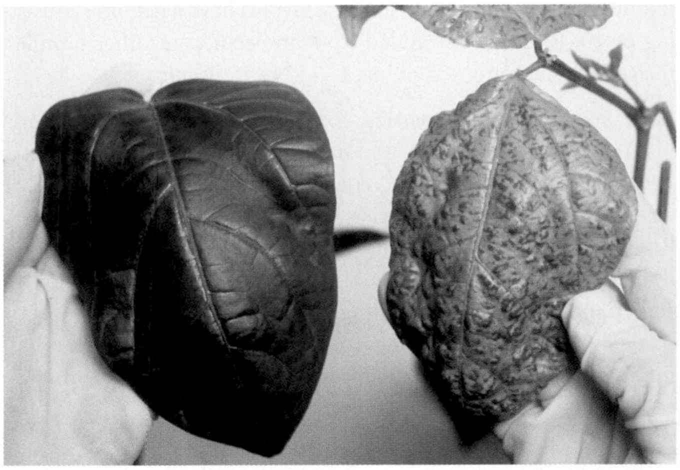

FIGURE 35.20 (+)ssRNA Viruses, Such as the Cowpea Mosaic Virus, Cause Important Plant Diseases.

CHAPTER 35 REVIEW

For media, go to the study area at www.masteringbiology.com

Summary of Key Concepts

Viruses are tiny, noncellular parasites that infect virtually every type of cell known. They cannot perform metabolism on their own—meaning outside a parasitized cell—and are not considered to be alive.

- Viruses cause illness and death in plants, fungi, bacteria, and archaea, as well as in humans and other animals.
- Viruses are specialists—different types of viruses infect particular species and types of cells.

✓You should be able to explain how it is possible for viruses to evolve by natural selection if they are not alive. (Hint: review Darwin's four postulates in Chapter 24.)

Although viruses are diverse morphologically, they can be classified as two general types: enveloped and nonenveloped.

- Both nonenveloped and enveloped viruses have a capsid made of protein.
- A capsid usually encloses viral enzymes as well as the viral genome.
- Nonenveloped viruses consist of a naked capsid, but in enveloped viruses, the capsid is surrounded by a membranous envelope.

✓You should be able to explain how enveloped viruses obtain their envelope.

The viral replication cycle can be broken down into six steps: (1) attachment and entry into a host cell, (2) production of viral proteins, (3) replication of the viral genome, (4) assembly of a new generation of virions, (5) exit from the infected cell, and (6) transmission to a new host.

- Viral entry into a host depends on specific interactions between viral proteins and compounds on the surface of the host cell.
- The protein-production and genome-replication phases of the replication cycle depend on enzymes, chemical energy, and molecules provided by the host cell.
- The assembly phase of the cycle is not well understood.
- Nonenveloped viruses exit a cell by lysis; enveloped viruses exit a cell by budding. Host cells are usually killed in the process.
- Once they are released, virions can infect new cells in the same multicellular organism or be transmitted to a new host.
- Some viruses may enter a lysogenic or latent phase, when they do not replicate by making more virions, but may transmit genetic material to daughter cells when the host divides.

✓You should be able to explain how at least three of the six phases of a viral replicative cycle can be stopped, using HIV as an example.

MB **Web Activity** The HIV Replicative Cycle

In terms of diversity, the key feature of viruses is the nature of their genetic material.

- The genomes of viruses may consist of double-stranded DNA, single-stranded DNA, double-stranded RNA, or one of several types of single-stranded RNA.
- Viral genomes are small. They do not code for ribosomes, and most viral genomes do not code for the enzymes needed to translate their own proteins or to perform other types of biosynthesis.

✓You should be able to explain how a virus that has a single-stranded DNA genome is able to replicate its genes.

Questions

✓TEST YOUR KNOWLEDGE

Answers are available in Appendix B

1. How do viruses that infect animals enter an animal's cells?
   - **a.** The viruses pass through a wound.
   - **b.** The viruses bind to a membrane component.
   - **c.** The viruses puncture the cell wall.
   - **d.** The viruses lyse the cell.
2. What does reverse transcriptase do?
   - **a.** It synthesizes proteins from mRNA.
   - **b.** It synthesizes tRNAs from DNA.
   - **c.** It synthesizes DNA from RNA.
   - **d.** It synthesizes RNA from DNA.
3. What do host cells provide for viruses?
   - **a.** nucleoside triphosphates and amino acids
   - **b.** ribosomes
   - **c.** ATP
   - **d.** all of the above
4. When do most enveloped virions acquire their envelope?
   - **a.** during entry into the host cell
   - **b.** during budding from the host cell
   - **c.** as they burst from the host cell
   - **d.** as they integrate into the host cell's chromosome
5. What reaction does protease catalyze?
   - **a.** polymerization of amino acids into peptides
   - **b.** cutting of polyprotein chains into functional proteins
   - **c.** folding of long peptide chains into functional proteins
   - **d.** assembly of virions
6. What feature distinguishes the seven major categories of viruses?
   - **a.** the method of exiting a cell
   - **b.** the method of entering a cell
   - **c.** the nature of the host
   - **d.** the nature of the genome

✓TEST YOUR UNDERSTANDING

Answers are available in Appendix B

1. The outer surface of a virus consists of either a membrane-like envelope or a protein capsid. Which type of outer surface does HIV have? Which type does adenovirus have? How does the outer surface of a virus correlate with its mode of exiting a host cell, and why?
2. Compare the morphological complexity of HIV with that of bacteriophage T4. Which virus would you predict has the larger genome? Explain the logic behind your hypothesis.
3. Compare and contrast bacteriophage growth rates during lytic versus lysogenic growth. Is it possible for viral populations to increase if viral DNA remains in the lysogenic state?
4. Explain why viral diseases are more difficult to treat than diseases caused by bacteria.
5. Why does the diversity of viral genome types support the hypothesis that viruses originated several times independently?
6. What types of data convinced researchers that HIV originated when a simian immunodeficiency virus "jumped" to humans?

✓APPLYING CONCEPTS TO NEW SITUATIONS

Answers are available in Appendix B

1. Suppose you could isolate a virus that parasitizes the pathogen *Staphylococcus aureus*—a bacterium that causes acne, boils, and a variety of other afflictions in humans. How could you test whether this virus might serve as a safe and effective treatment?
2. If you were in charge of the government's budget devoted to stemming the AIDS epidemic, would you devote most of the resources to drug development, vaccine development, or preventive medicine? Defend your answer.
3. Bacteria fight viral infections with restriction endonucleases (bacterial enzymes; introduced in Chapter 19). Restriction endonucleases cut up, or break, viral DNA at specific sequences. The enzymes do not cut a bacterium's own DNA, because the bases in the bacterial genome are protected from the enzyme by methylation (the addition of a CH_3 group). Generate a hypothesis to explain why members of the Eukarya do not have restriction endonucleases.
4. Consider these two contrasting definitions of life:
   a. An entity is alive if it is capable of replicating itself via the directed chemical transformation of its environment.
   b. An entity is alive if it is an integrated system for the storage, maintenance, replication, and use of genetic information.

   According to these definitions, are viruses alive? Explain.

This young hawksbill sea turtle is tangled in a fishing net—now a common cause of mortality. Increasingly, ecology is the study of how humans are altering the environment in ways that make life difficult for other organisms.

An Introduction to Ecology 50

Ecology is the study of how organisms interact with their environment. Except for the episode 65 million years ago, when a mountain-sized asteroid struck Earth, environments are changing more rapidly right now than they have at any time in the past 3.5 billion years. Efforts to maintain human health and welfare depend on our ability to cope with these environmental changes. Ecology has become one of the most important and dynamic fields in biological science.

Ecology is also among the most synthetic fields in biological science. In this unit, you'll be applying what you've learned about biochemistry, physiology, evolution, and the diversity of life.

Ecology's primary goal is to understand the distribution and abundance of organisms. How do interactions with other organisms and the physical environment dictate why certain species live where they do, and how many individuals live there? Some biologists ask why orangutans are restricted to forests in Borneo and why their numbers are declining so rapidly. Other biologists create mathematical models to predict how quickly the human immunodeficiency virus will increase in India, or how long it will take for the current human population of 6.8 billion to double.

Distribution and abundance are core questions in today's world. Global climate change is having dramatic effects on species' distributions, and expanding human populations are causing the abundance of many species to crash. Let's delve in.

KEY CONCEPTS

- Ecology's goal is to explain the distribution and abundance of organisms. It is the branch of biology that provides a scientific foundation for conservation efforts.
- Physical structure—particularly water depth—is the primary factor that limits the distribution and abundance of aquatic species. Climate—specifically, both the average value and annual variation in temperature and in moisture—is the primary factor that limits the distribution and abundance of terrestrial species.
- Climate varies with latitude, elevation, and other factors—such as proximity to oceans and mountains. Climate is changing rapidly around the globe.
- A species' distribution is constrained by historical and biotic factors, as well as by abiotic factors such as physical structure and climate.

50.1 Areas of Ecological Study

To understand why organisms live where they do and in what numbers, biologists break ecology into several levels of analysis. This is a common strategy in biological science. Cell biologists study how cells work at different levels of organization—from individual molecules to multimolecular machines and membranes to multicellular organisms.

Physiologists analyze processes at the level of ions and molecules as well as whole cells, tissues and organs, and complete systems.

In ecology, researchers work at four main levels: (**1**) organisms, (**2**) populations, (**3**) communities, and (**4**) ecosystems. The levels aren't rigid or exclusive; researchers routinely draw on all four to explore a particular question. Let's examine each.

Organismal Ecology

At the organismal level, researchers explore the morphological, physiological, and behavioral adaptations that allow individuals to live in a particular area. Consider sockeye salmon, for example. After spending four or five years feeding and growing in the ocean, these salmon travel hundreds or thousands of kilometers to return to the stream where they hatched (**Figure 50.1a**). Females create nests in the gravel stream bottom and lay eggs. Nearby males compete for the chance to fertilize eggs as they are laid. When breeding is finished, all the adults die.

Biologists want to know how these individuals interact with their physical surroundings and with other organisms in and around the stream. How do individuals cope with the transition from living in salt water to living in freshwater? Which females get the best nesting sites and lay the most eggs, and which males are most successful in fertilizing eggs?

Population Ecology

A **population** is a group of individuals of the same species that lives in the same area at the same time. When biologists study population ecology, they focus on how the numbers of individuals in a population change over time.

For example, researchers use mathematical models to predict the future of salmon populations (**Figure 50.1b**). Many of these populations have declined as their habitats have become dammed or polluted. If the factors that affect population size can be described accurately enough, mathematical models can assess the impact of proposed dams, changes in weather patterns, altered harvest levels, or specific types of protection efforts.

Community Ecology

A biological **community** consists of the species that interact with each other within a particular area. Community ecologists ask questions about the nature of the interactions between species and the consequences of those interactions. The work might concentrate on predation, parasitism, and competition, or explore how groups of species respond to fires, floods, and other disturbances.

As an example, consider the interactions among salmon and other species in the marine and stream communities where they live. When they are at sea, salmon eat smaller fish and are themselves hunted and eaten by orcas, sea lions, humans, and other mammals; when they return to freshwater to breed, they are preyed on by bears and bald eagles (**Figure 50.1c**). In both marine and freshwater habitats, salmon are subject to parasitism and disease. They are also heavily affected by disturbances—particularly changes in their food supply due to overfishing and the damming, logging, or suburbanization of breeding streams.

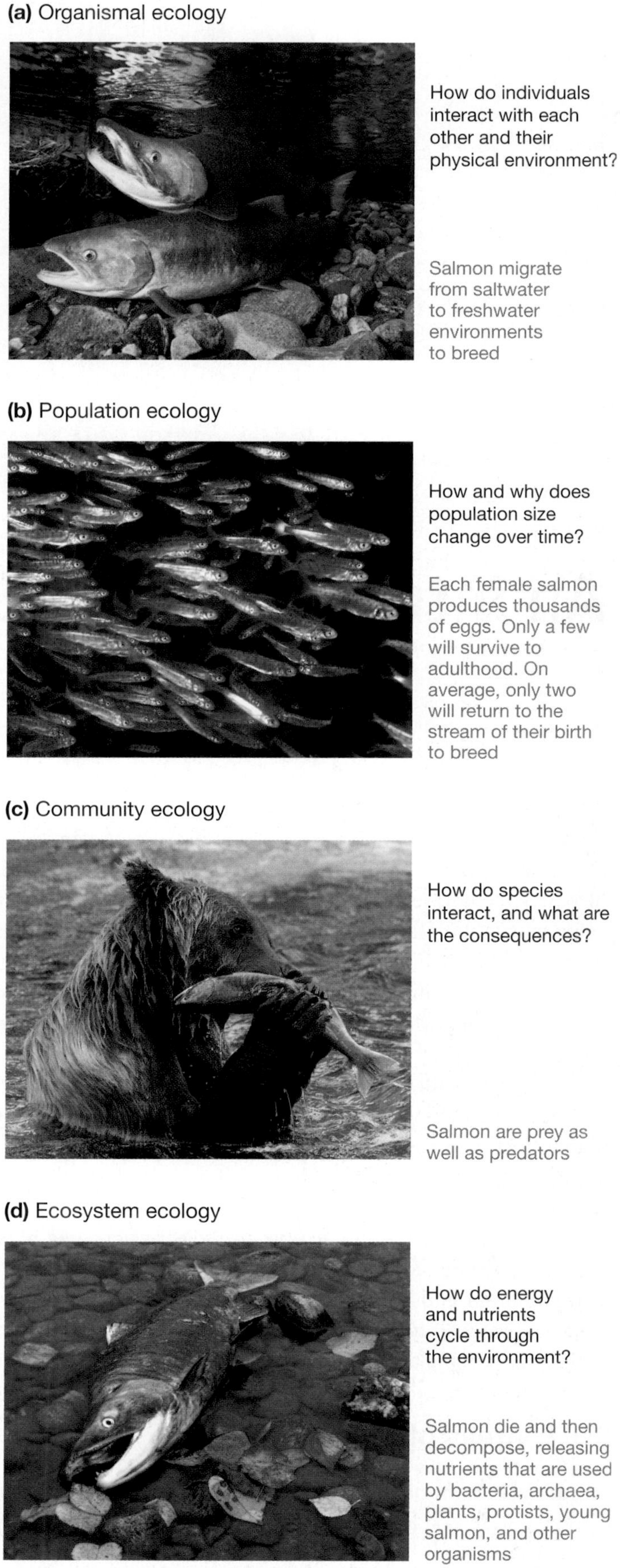

(a) Organismal ecology
How do individuals interact with each other and their physical environment?
Salmon migrate from saltwater to freshwater environments to breed
(b) Population ecology
How and why does population size change over time?
Each female salmon produces thousands of eggs. Only a few will survive to adulthood. On average, only two will return to the stream of their birth to breed
(c) Community ecology
How do species interact, and what are the consequences?
Salmon are prey as well as predators
(d) Ecosystem ecology
How do energy and nutrients cycle through the environment?
Salmon die and then decompose, releasing nutrients that are used by bacteria, archaea, plants, protists, young salmon, and other organisms

FIGURE 50.1 Biologists Study Ecology at Four Main Levels.

Ecosystem Ecology

Ecosystem ecology is an extension of organismal, population, and community ecology. An **ecosystem** consists of all the organisms in a particular region along with nonliving components. These physical, or **abiotic** (literally, "not-living"), components include air, water, and soil. At the ecosystem level, biologists study how nutrients and energy move among organisms and through the surrounding atmosphere and soil or water.

Because humans are adding massive amounts of energy and nutrients to ecosystems all over the world, this work has direct public policy implications. Ecosystem ecologists are responsible for assessing the impact of pollution and increased temperature on the distribution and abundance of species—in fact, on Earth's ability to support life.

Salmon are interesting to study at the ecosystem level because they link marine and freshwater ecosystems. They harvest nutrients in the ocean and then, when they migrate, die, and decompose, transport those molecules to streams and streamside forests (**Figure 50.1d**). In this way, salmon transport chemical energy and nutrients from one habitat to another. Because salmon are sensitive to pollution and to changes in water temperature, human-induced changes in marine and freshwater ecosystems have a large impact on their populations.

How Do Ecology and Conservation Efforts Interact?

The four levels of ecological study are synthesized and applied in conservation biology. **Conservation biology** is the effort to study, preserve, and restore threatened populations, communities, and ecosystems (see Chapter 55).

Ecologists study how interactions between organisms and their environments result in a particular species being found in a particular area at a particular population size. Conservation biologists apply these data to preserve species and restore environments.

Conservation biologists are like physicians. But instead of drawing on results from research in molecular biology and physiology, they draw on research in ecology and evolution. Instead of prescribing drugs for sick people, they prescribe remedies for environments, so they produce a diversity of species, clean air, pure water, and productive soils.

50.2 Types of Aquatic Ecosystems

If ecology is the study of how species interact with their environment, what makes up the environment? Every environment has both physical and biological components. The abiotic or physical aspects include temperature, precipitation, sunlight, and wind. The **biotic** ("living") components are members of the same or different species.

Chapters 51–53 focus on biological interactions—how individuals interact with offspring, mates, competitors, predators, prey, and parasites. This chapter focuses on how the physical environment affects organisms. Organisms have a restricted set of physical conditions in which they can survive and thrive. Why?

Let's answer this question by analyzing the physical attributes of aquatic ecosystems in this section, moving on to terrestrial ecosystems in Section 50.3, and considering how global warming is affecting both types of environments in Section 50.4. In freshwater and salt water, three key physical factors affect the distribution and abundance of organisms: nutrient availability, water depth, and water movement.

Nutrient Availability

In many aquatic ecosystems, nutrients such as nitrogen and phosphorus are in short supply. If water is moving, nutrients tend to be washed away. If water is still, nutrients tend to fall to the bottom and collect in the form of debris. Nutrient levels are important: They limit growth rates in the photosynthetic organisms that provide food for other species.

Let's consider two important events that affect nutrient availability in lakes and oceans. In both cases, they are mechanisms that bring nutrients from the bottom up to the water surface, where they can nourish the growth of photosynthetic species.

OCEAN UPWELLING In the oceans, nutrients in the sunlit surface waters are constantly lost in the form of dead organisms that rain down into the depths. In certain coastal regions of the world's oceans, however, nutrients are brought up to the surface by currents that cause upwellings.

Figure 50.2 shows how this happens off the coast of Peru. Here the prevailing winds blow along the coastline, pushing surface water to the north. Because the Earth rotates constantly, this wind-driven water current is slowly moved offshore. As the surface water moves away from the coast, it is steadily replaced by water moving up from the ocean bottom.

The upwelling water is nutrient rich. In effect, it recycles nutrients that earlier had fallen to the ocean floor. The nutrients

PROCESS: OCEAN UPWELLING

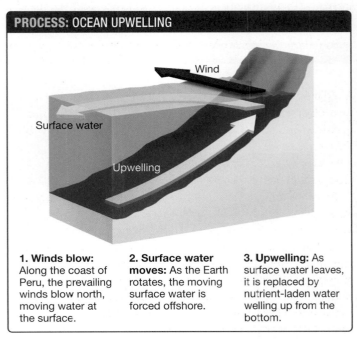

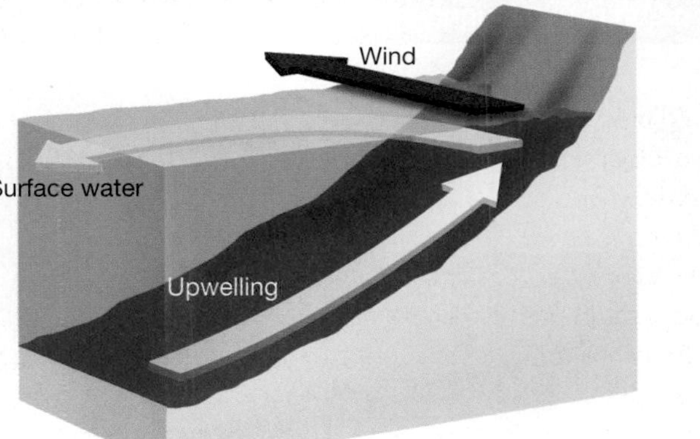

1. Winds blow: Along the coast of Peru, the prevailing winds blow north, moving water at the surface.

2. Surface water moves: As the Earth rotates, the moving surface water is forced offshore.

3. Upwelling: As surface water leaves, it is replaced by nutrient-laden water welling up from the bottom.

FIGURE 50.2 Ocean Upwellings Bring Nutrients to the Surface.

support luxuriant growths of photosynthetic cells, which feed an array of small grazing animals, which in turn feed small fish called anchoveta. When currents are favorable and upwelling is steady, the anchoveta fishery is the most productive in the world.

LAKE TURNOVER Because they were scoured by glaciers as recently as 10,000 years ago, the higher latitudes of the world host many lakes. Each year, these bodies of water undergo remarkable changes known as the spring and fall **turnovers**.

The spring and fall lake turnovers occur in response to changes in air temperature. To see how this happens, examine the temperature profiles in **Figure 50.3**.

Note that in winter, water at the surface of a northern lake is locked up in ice at a temperature of 0°C. The water just under the ice is slightly warmer and is relatively oxygen rich, because it was exposed to the atmosphere as winter set in. Because water is most dense at 4°C, the water at the bottom of the lake is at 4–5°C. The bottom water is also oxygen poor, because organisms that decompose falling organic material use up available oxygen in the process.

A gradient in temperature such as this is called a **thermocline** ("heat-slope"). Thermal stratification is said to occur.

The ice begins to melt, however, when spring arrives. The temperature of the water on the surface rises until it reaches 4°C. The water at the surface of the lake is now heavier than the water below it. As a result, the water on the surface sinks. The water at the bottom of the lake is displaced and comes to the surface, completing the spring turnover.

During the spring turnover, water at the bottom of the lake carries sediments and nutrients from the bottom up to the surface. This flush of nutrients triggers a rapid increase in the growth of photosynthetic algae and bacteria that biologists call the spring bloom.

When temperatures cool in the fall, the water at the surface reaches a temperature of 4°C and sinks, displacing water at the bottom and creating the fall turnover. The fall turnover is important because it brings oxygen-rich water from the surface down to the bottom, and because it again brings nutrients from the bottom up to the sunlit surface waters.

Without the spring and fall turnovers, most freshwater nutrients would remain on the bottom of lakes. These aquatic ecosystems would be much less productive, as a result.

Water Flow

Water movement is a critical factor in aquatic ecosystems because it presents a physical challenge. It can literally sweep organisms off their feet.

Organisms that live in fast-flowing streams have to cope with the physical force of the water, which constantly threatens to move them downstream. Marine organisms that live in intertidal regions are exposed to the air periodically each day, as well as to violent wave action during storms.

Nutrient and light availability influence productivity in aquatic ecosystems; water movement has an effect on productivity and is a physical force that organisms have to content with. The boxes that follow summarize the types of freshwater and marine environments that organisms occupy, with an emphasis on how they are affected by variation in water depth and movement.

Water Depth

Water absorbs and scatters light, so the amount and types of wavelengths available to organisms change dramatically as water depth increases. Light has a major influence on **productivity**—the total amount of carbon fixed by photosynthesis per unit area per year.

At the water surface, all wavelengths of light are equally available. But as **Figure 50.4a** shows, ocean water specifically removes light in the blue and red regions of the visible spectrum. This is important because wavelengths in the blue and red regions are required for photosynthesis in many species (see Chapter 10).

PROCESS: LAKE TURNOVER

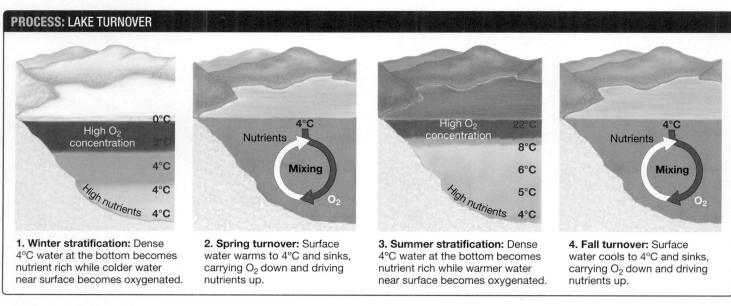

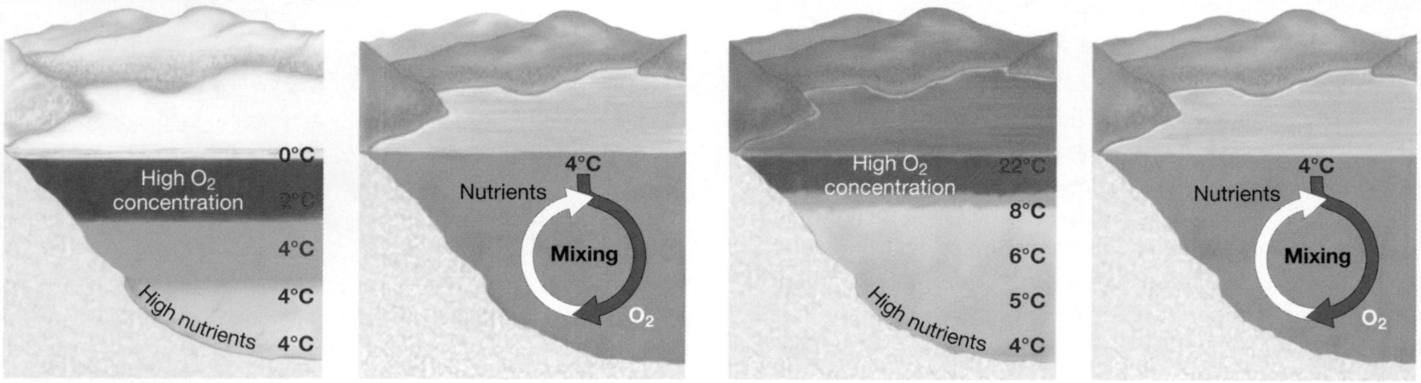

1. Winter stratification: Dense 4°C water at the bottom becomes nutrient rich while colder water near surface becomes oxygenated.

2. Spring turnover: Surface water warms to 4°C and sinks, carrying O_2 down and driving nutrients up.

3. Summer stratification: Dense 4°C water at the bottom becomes nutrient rich while warmer water near surface becomes oxygenated.

4. Fall turnover: Surface water cools to 4°C and sinks, carrying O_2 down and driving nutrients up.

FIGURE 50.3 In Temperate Regions, Lakes Turn Over Each Spring and Fall.

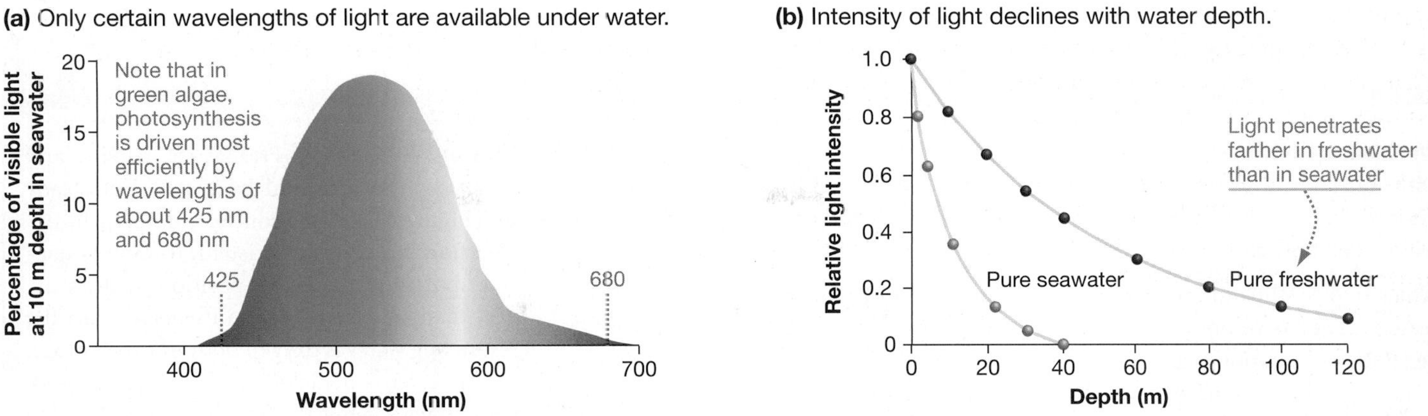

FIGURE 50.4 Availability of Light Changes Dramatically with Increasing Water Depth.

✔**QUESTION** Would you expect to find photosynthetic organisms at greater depths in freshwater or seawater? Explain your answer.

The quantity of light available to organisms also diminishes rapidly with increasing depth. The graph in **Figure 50.4b** plots water depth along the *x*-axis and light intensity—relative to the surface—on the *y*-axis. Each data point is a reading of light intensity at a particular depth.

Note that in pure seawater, the total amount of light available at a depth of 10 m is less than 40 percent of what it is at the surface; virtually no light reaches depths greater than 40 m (131 ft). In seawater that contains organisms or debris, light penetration is dramatically less than in pure seawater.

Freshwater Environments > Lakes and Ponds

Lakes and ponds are distinguished from each other by size. Ponds are small; lakes are large enough that the water in them can be mixed by wind and wave action. Most lakes and ponds occur in high latitudes—they formed in depressions that were created by the scouring action of glaciers thousands of years ago. In the tropics, most lakes consist of old river channels. Elsewhere, many of the lakes and ponds that exist now were dug recently by people.

Water Depth Biologists describe the structure of lakes and ponds by naming five zones (**Figure 50.5**).

- The **littoral** ("seashore") **zone** consists of the shallow waters along the shore, where flowering plants are rooted.
- The **limnetic** ("lake") **zone** is offshore and comprises water that receives enough light to support photosynthesis.
- The **benthic** ("depths") **zone** is made up of the substrate.
- Regions of the littoral, limnetic, and benthic zones that receive sunlight are part of the **photic zone**.
- Portions of a lake or pond that do not receive sunlight make up the **aphotic zone**.

Water Flow and Nutrient Availability Water movement in lakes and ponds is driven by wind and temperature. The littoral and limnetic zones are typically much warmer and better oxygenated than the benthic zone, simply because they receive so much more solar radiation and are in contact with oxygen in the atmosphere. The benthic zone, in contrast, is relatively nutrient rich because dead and decomposing bodies sink and accumulate there. In many cases, movement of nutrients from the benthic to littoral zones is dependent on seasonal turnovers.

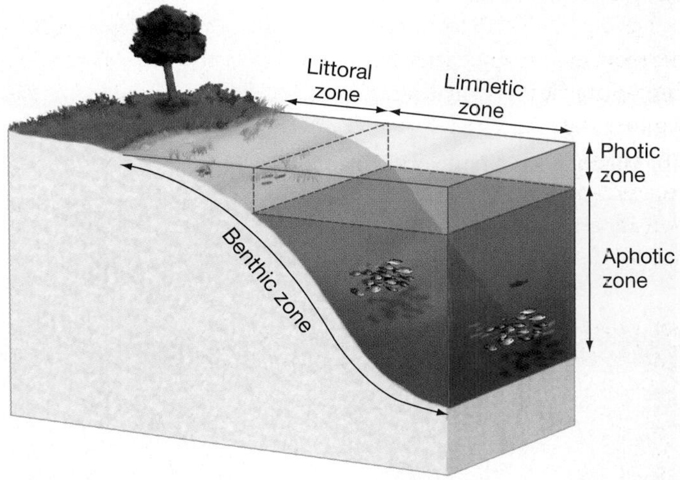

FIGURE 50.5 Lakes Have Distinctive Zones Defined by Water Depth and Distance from Shore.

Organisms Cyanobacteria, algae, and other microscopic organisms, collectively called **plankton**, live in the photic zone, as do the fish and small crustaceans—organisms related to shrimp and crabs—that eat them. In shallow parts of the photic zone, rooted plants are common. Animals (invertebrates and fish) that consume dead organic matter, or **detritus**, are particularly abundant in the benthic zone.

✔**You should be able to predict which zone of lakes and ponds produces the most grams of organic material per square meter per year, and explain your logic.**

Wetlands are shallow-water habitats where the soil is saturated with water for at least part of the year. They are usually distinguished from terrestrial habitats by the presence of "indicator plants" that only grow in saturate soils.

Water Depth Wetlands are distinct from lakes and ponds for two reasons: They have only shallow water, and they have **emergent vegetation**—meaning plants that grow above the surface of the water. All or most of the water in wetlands receives sunlight; emergent plants capture sunlight before it strikes the water.

(a) Bogs are stagnant and acidic.

Water Flow and Nutrient Availability Freshwater marshes and swamps are wetland types characterized by a slow but steady flow of water. **Bogs** (**Figure 50.6a**), in contrast, develop in depressions where water flow is low or nonexistent. If water is stagnant, oxygen is used up during the decomposition of dead organic matter faster than it enters via diffusion from the atmosphere. As a result, bog water is oxygen poor or even anoxic. Once the oxygen in the water is depleted, decomposition slows. Organic acids and other acids build up, lowering the pH of the water. At low pH, nitrogen becomes unavailable to plants. As a result, bogs are nutrient poor. Marshes and swamps, in contrast, are relatively nutrient rich.

Organisms The combination of acidity, lack of available nitrogen, and anoxic conditions makes bogs extremely unproductive habitats. Marshes and swamps, in contrast, offer ample supplies of oxygenated water and sunlight, along with nutrients available from rapid decomposition. As a result, they are extraordinarily productive. **Marshes** lack trees and typically feature grasses, reeds, or other nonwoody plants (**Figure 50.6b**); **swamps** are dominated by trees and shrubs (**Figure 50.6c**). Because their physical environments are so different, there is little overlap in the types of species found in bogs, marshes, and swamps.

✔You should be able to explain why carnivorous plants, which capture and digest insects, are relatively common in bogs but rare in marshes and swamps.

(b) Marshes have nonwoody plants.

(c) Swamps have trees and shrubs.

FIGURE 50.6 Wetland Types Are Distinguished by Water Flow and Vegetation.

Freshwater Environments > Streams

Streams are bodies of water that move constantly in one direction. Creeks are small streams; rivers are large. In terms of the environment that is available to organisms, the major physical variables in streams are the speed of the current and the availability of oxygen and nutrients.

Water Depth Most streams are shallow enough that sunlight reaches the bottom. Availability of sunlight is usually not a limiting factor for organisms.

Water Flow and Nutrient Availability The structure of a typical stream varies along its length (**Figure 50.7**). Where it originates at a mountain glacier, lake, or spring, a stream tends to be cold, narrow, and fast. As it descends toward a lake, ocean, or larger river, a stream accepts water from tributaries and becomes larger, warmer, and slower.

Oxygen levels tend to be high in fast-moving streams because water droplets are exposed to the atmosphere when moving water splashes over rocks or other obstacles. Oxygen from the atmosphere diffuses into the droplets. In contrast, slow-moving streams that lack rapids tend to become relatively oxygen poor. Also, cold water holds more oxygen than warm water does (see Chapter 44).

Slow-moving streams tend to be more nutrient rich than fast-moving streams, because decaying matter does not flush away as quickly. In many cases, streams are dependent on nutrients that arrive via falling leaves or other organic matter, or on soil particles that wash in from surrounding terrestrial communities.

Organisms It is rare to find photosynthetic organisms in small, fast-moving streams; nutrient levels tend to be low and most of the organic matter present consists of leaves and other materials that fall into the water from outside the stream. Fish, insect larvae, mollusks, and other animals have adaptations that allow them to maintain their positions in the fast-moving portions of streams. As streams widen and slow down, conditions become more favorable for the growth of algae and plants, and the amount of organic matter and nutrients increases. As a result, the same stream often contains completely different types of organisms near its source and near its end, or mouth.

✔You should be able to explain why, in terms of their ability to perform cellular respiration, fish species found in cold, fast-moving streams tend to be much more active than fish species found in warm, slow-moving streams.

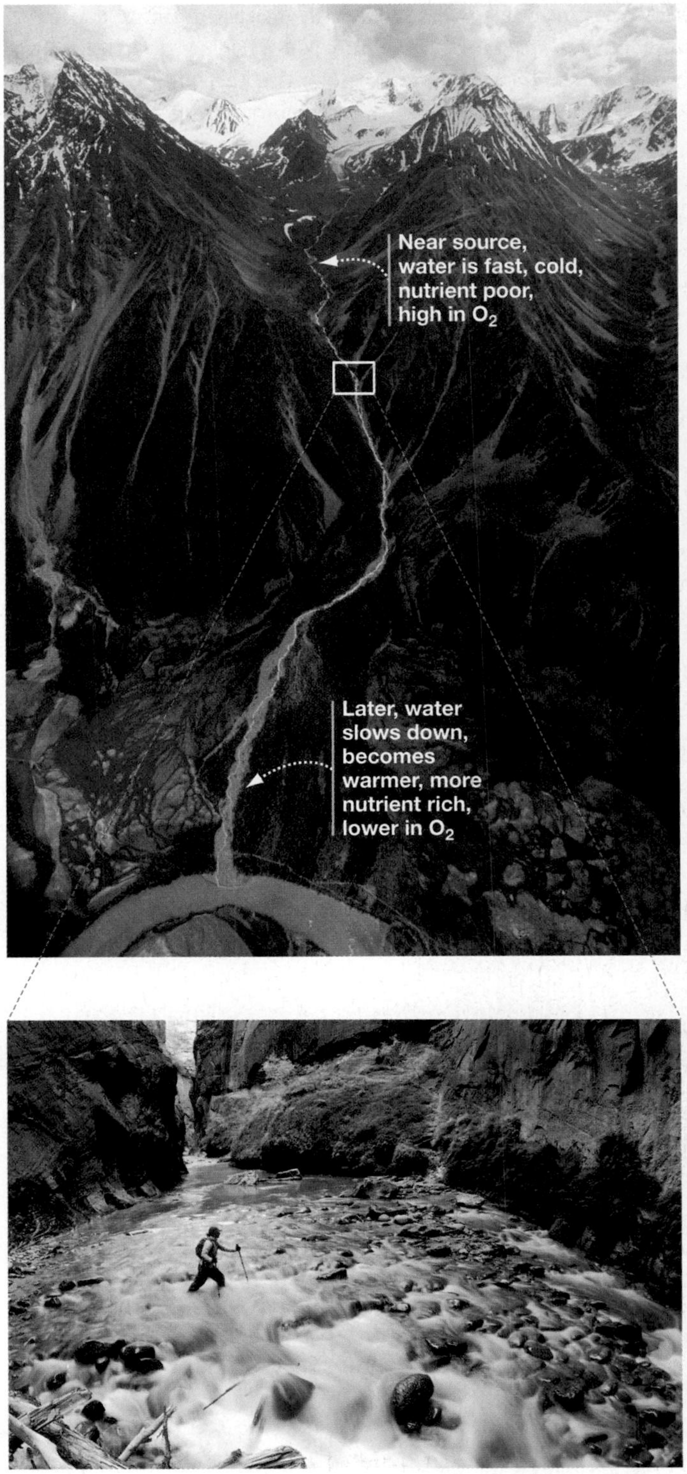
Near source, water is fast, cold, nutrient poor, high in O_2
Later, water slows down, becomes warmer, more nutrient rich, lower in O_2

FIGURE 50.7 Distinctive Environments Appear along a Stream's Length.

Freshwater/Marine Environments > Estuaries

Estuaries form where rivers meet the ocean—meaning that freshwater mixes with salt water (**Figure 50.8**). In essence, an estuary includes slightly saline marshes as well as the body of water that moves in and out of these environments. Salinity varies with **(1)** changes in river flows—it declines when the river floods and increases when the river ebbs—and **(2)** with proximity to the ocean. Salinity has dramatic effects on osmosis and water balance (see Chapters 37 and 42); species that live in estuaries have adaptations that allow them to cope with variations in salinity.

Water Depth Most estuaries are shallow enough that sunlight reaches the substrate. Water depth may fluctuate dramatically, however, in response to tides, storms, and floods.

Water Flow and Nutrient Availability Water flow in estuaries fluctuates daily and seasonally due to tides, storms, and floods. The fluctuation is important because it alters salinity, which in turn affects which types of organisms are present. Estuaries are nutrient rich, because nutrient-laden sediments are deposited when flowing river water slows as it enters the ocean.

Organisms Because the water is shallow and sunlit, and because nutrients are constantly replenished by incoming river water, estuaries are among the most productive environments on Earth. They are often packed with young fish, which feed on abundant vegetation and plankton while hiding from predators.

✔You should be able to explain why estuaries and freshwater marshes contain few of the same species, even though they are both shallow-water habitats filled with rooted plants.

FIGURE 50.8 Estuaries Are Highly Productive Environments.

Marine Environments > The Ocean

The world's oceans form a continuous body of salt water and are remarkably uniform in chemical composition. Regions within an ocean vary markedly in their physical characteristics, however, with profound effects on the organisms found there.

Water Depth Biologists describe the structure of an ocean by naming six regions (**Figure 50.9**).

- The **intertidal** ("between tides") **zone** consists of a rocky, sandy, or muddy beach that is exposed to the air at low tide but submerged at high tide.
- The **neritic zone** extends from the intertidal zone to depths of about 200 m. Its outermost edge is defined by the end of the **continental shelf**—the gently sloping, submerged portion of a continental plate.
- The **oceanic zone** is the "open ocean"—the deepwater region beyond the continental shelf.
- The bottom of the ocean is the **benthic zone**.
- The intertidal and sunlit regions of the neritic, oceanic, and benthic zones make up a **photic zone**.
- Areas that do not receive sunlight are in an **aphotic zone**.

Water Flow and Nutrient Availability Water movement in the ocean is dominated by different processes at different depths. In the intertidal zone, tides and wave action are the major influences. In the neritic zone, currents that bring nutrient-rich water from the benthic zone of the deep ocean toward shore have a heavy impact. Throughout the ocean, large-scale currents circulate water in the oceanic zone in response to prevailing winds and the Earth's rotation.

In general, nutrient availability is dictated by water movement. The neritic and intertidal zones are relatively nutrient rich, because they receive nutrients from rivers and upwellings. The oceanic zone, in contrast, constantly loses nutrients due to a steady rain of dead organisms drifting to the benthic zone.

(Continued on next page)

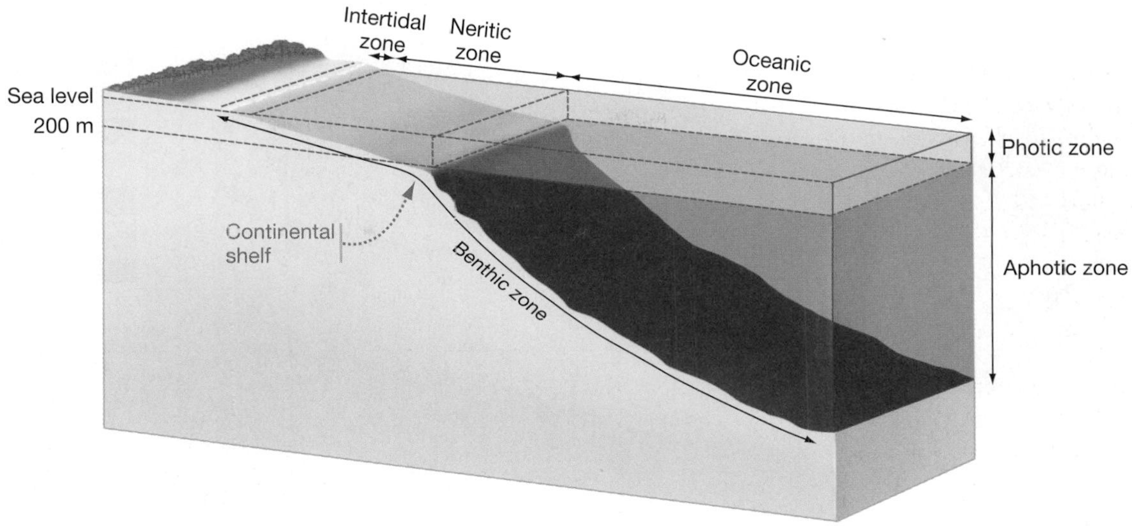

FIGURE 50.9 Oceans Have Distinctive Zones Defined by Water Depth and Distance from Shore.

Organisms Each zone in the ocean is populated by distinct species that are adapted to the physical conditions present. Organisms that live in the intertidal zone must be able to withstand physical pounding from waves and desiccation and high temperatures at low tide. Productivity is high, however, due to the availability of sunlight and nutrients contributed by estuaries as well as by currents that sweep in nutrient-laden sediments from offshore areas.

Productivity is also high on the outer edge of the neritic zone, due to nutrients contributed by upwellings at the edge of the continental plate. Almost all of the world's major marine fisheries exploit organisms that live in the neritic zone. In the tropics, shallow portions of the neritic zone may host **coral reefs**. Because the water is warm and sunlight penetrates to the ocean floor in these habitats, coral reefs are among the most productive environments in the world (see Chapter 54).

If coral reefs are the rain forests of the ocean, then the oceanic zone is the desert. Sunlight is abundant in the photic zone of the open ocean, but nutrients are extremely scarce. When the photosynthetic organisms and the animals that feed on them die, their bodies drift downward out of the photic zone and are lost. In the open ocean, the benthic zone can be several kilometers below the surface, and there is no mechanism for bringing nutrients back up from the bottom. The aphotic zone of the open ocean is also extremely unproductive because light is absent and photosynthesis is impossible. Most organisms present in the aphotic zone—including the sea cucumbers highlighted in Chapter 34—survive on the rain of dead bodies from the photic zone.

✓You should be able to predict whether animals that live in the aphotic zone have functioning eyes.

CHECK YOUR UNDERSTANDING

If you understand that . . .

- Because nutrients tend to sink to the bottom of aquatic ecosystems, productivity depends on mechanisms to bring nutrients back up to the surface—as well as on the availability of light.
- Aquatic environments are distinguished by aspects of physical structure—primarily the depth of water and the rate at which it flows.

✓ You should be able to . . .

1. Explain why coastal environments are the most productive areas in the ocean.
2. Explain what happens when water at the surface of a lake cools to 4°C and the water below it is slightly warmer.

Answers are available in Appendix B.

50.3 Types of Terrestrial Ecosystems

If you could walk from the equator in South America to the North Pole, you would notice startling changes in the organisms around you. Lush tropical forests with broad-leaved evergreen trees would give way to seasonally dry forests and then to deserts. The deserts would yield to the vast grasslands of central North America, which terminate at the boreal forests of the subarctic. If you pressed on, you would reach the end of the trees and the beginning of the most northerly community—the arctic tundra.

If you walked from sea level to the top of the Andes mountains of Peru, you would experience similar types of changes. Wet forests along the coast would give way to drier forests at higher elevations, where trees might drop their leaves for part of the year. Near the peaks, trees would thin and eventually yield to alpine tundras filled with hardy, low-growing plants.

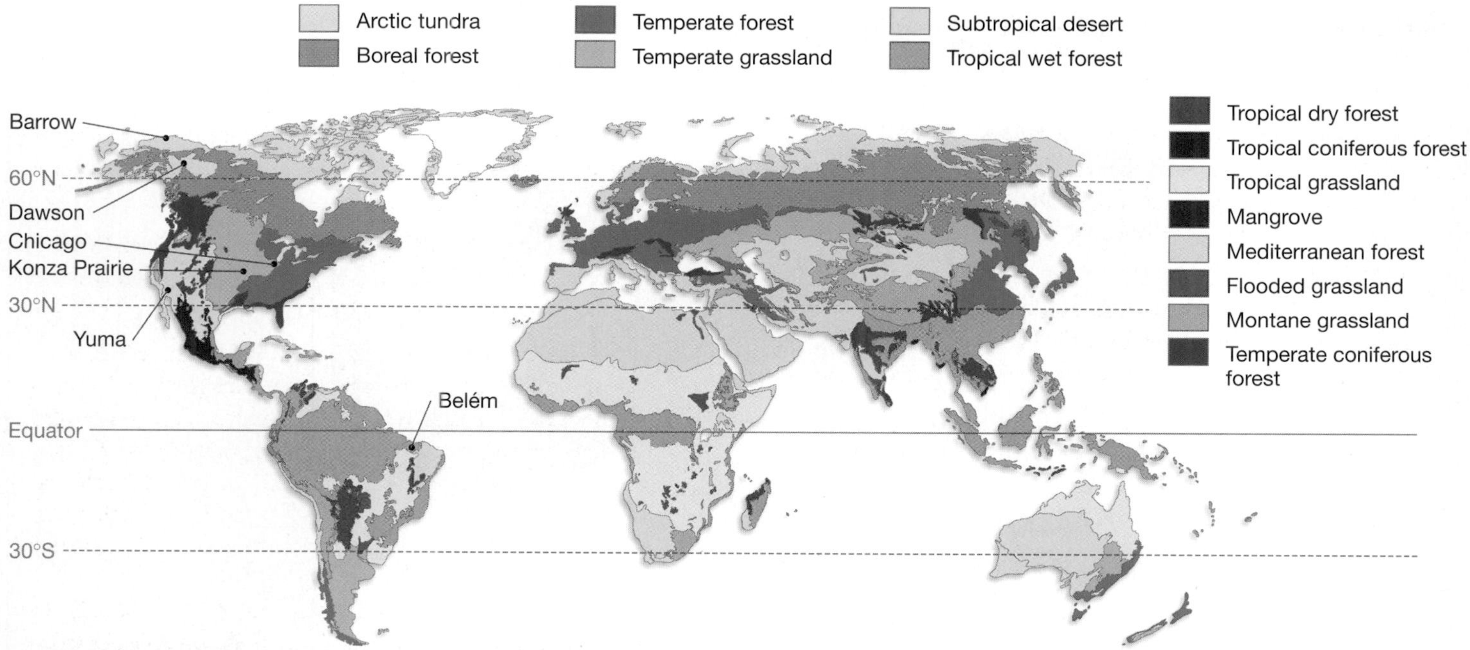

FIGURE 50.10 Distinct Biomes Are Found throughout the World. A recent classification identified the 14 major biomes shown here. The locations indicated with black dots are highlighted in this section.

Broad-leaved evergreen forests, deserts, and grasslands are **biomes**: major groupings of plant and animal communities defined by a dominant vegetation type. **Figure 50.10** shows the global distribution of the most common types of biomes.

Each of the biomes found around the world is associated with a distinctive set of abiotic conditions. Just as water depth and movement have an overriding influence on aquatic ecosystems, the type of biome present in a terrestrial region depends on **climate**—the prevailing, long-term weather conditions found in an area. **Weather** consists of the specific short-term atmospheric conditions of temperature, precipitation, sunlight, and wind.

- Temperature is critical because the enzymes that make life possible work at optimal efficiency only in a narrow range of temperatures (see Chapter 3). Temperature also affects the availability of moisture: Water freezes at low temperatures and evaporates rapidly at high temperatures.
- Moisture is significant because it is required for life, and because terrestrial organisms constantly lose water to the environment through evaporation or transpiration. To stay alive, they must reduce water loss and replace lost water.
- Sunlight is essential because it is required for photosynthesis.
- Wind is important because it exacerbates the effects of temperature and moisture. Wind increases heat loss due to evaporation and convection, and it increases water loss due to evaporation and transpiration. It also has a direct physical impact on organisms such as birds, flying insects, and plants—it pushes them around.

Of the four components of climate, variation in temperature and moisture are far and away the most important in determining plant distribution and abundance. More specifically, the nature of the biome that develops in a particular region is governed by (1) average annual temperature and precipitation, and (2) annual variation in temperature and precipitation. Each biome contains species that are adapted to a particular temperature and moisture regime. The amount and variability of heat and precipitation structure terrestrial environments, much as water depth and flow rate structure aquatic environments.

Biologists are particularly concerned with how temperature and water influence **net primary productivity (NPP)**. NPP is defined as the total amount of carbon that is fixed per year minus the amount that is oxidized during cellular respiration. Fixed carbon that is consumed in cellular respiration provides energy for the organism but is not used for growth—that is, production of **biomass**.

NPP is key because it represents the organic matter that is available as food for other organisms. In terrestrial environments NPP is often estimated by measuring **aboveground biomass**—the total mass of living plants, excluding roots.

Photosynthesis, plant growth, and NPP are maximized on land when temperatures are warm and conditions are wet. Conversely, photosynthesis cannot occur efficiently at low temperatures or under drought stress. Photosynthetic rates are maximized in warm temperatures, when enzymes work efficiently, and in humid weather, when stomata can remain open and CO_2 is readily available.

To help you understand how temperature and precipitation influence biomes, let's take a detailed look at the six biomes that represent Earth's most extensive vegetation types, ranging from the wet tropics to the arctic. In each case, you'll be analyzing temperature and precipitation data from a specific location, typical of the biome in question. Data plotted in red indicate the daily average temperature in each biome, graphed throughout the year; bar charts plotted in purple indicate the average monthly precipitation.

Tropical wet forests—also called tropical rain forests—are found in equatorial regions around the world. Plants in this biome have broad leaves as opposed to narrow, needle-like leaves, and are evergreen. Older leaves are shed throughout the year, but there is no complete, seasonal loss of leaves.

Temperature The data in **Figure 50.11** are from Belém, Brazil, where mean monthly temperatures never drop below 25°C and never exceed 30°C. Compared to other biomes, tropical wet forests show almost no seasonal variation in temperature. This is important because temperatures are high enough to support growth throughout the year.

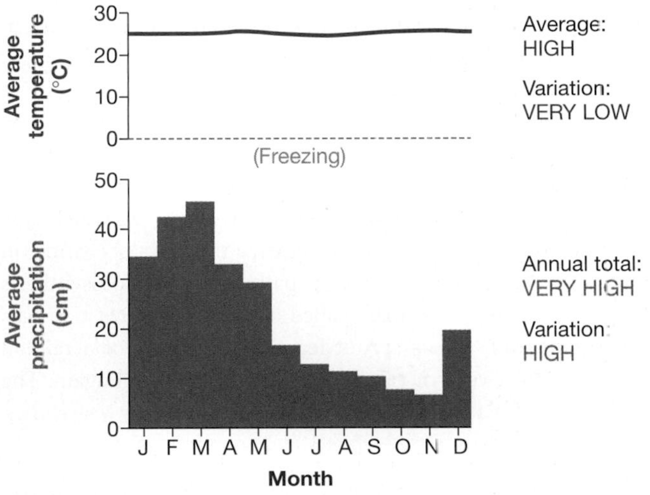

FIGURE 50.11 Climate Data from Belém, Brazil.

Precipitation Even in the driest month of the year, November, this region receives over 5 cm (2 in.) of rainfall—considerably more than the *annual* rainfall of many deserts.

Vegetation Favorable year-round growing conditions produce riotous growth, leading to extremely high productivity and aboveground biomass (**Figure 50.12**). Tropical wet forests are also renowned for their species diversity. It is not unusual to find over 200 tree species in a single 100 m × 100 m study plot. And based on counts of the insects and spiders collected from single trees, some biologists contend that the world's tropical wet forests may hold up to 30 million species of arthropods alone.

The diversity of plant sizes and growth forms in wet forest communities produces extraordinary structural diversity. In tropical wet forests, a few extremely large trees tower above a layer of large trees that form a distinctive and continuous **canopy** (the uppermost layers of branches). From the canopy to the ground, there is a complex assortment of vines, **epiphytes** (plants that grow entirely on other plants), small trees, shrubs, and herbs. This diversity of growth forms presents a wide array of habitat types for animals.

✔You should be able to explain why the presence of vines and epiphytes increases the productivity of tropical wet forests.

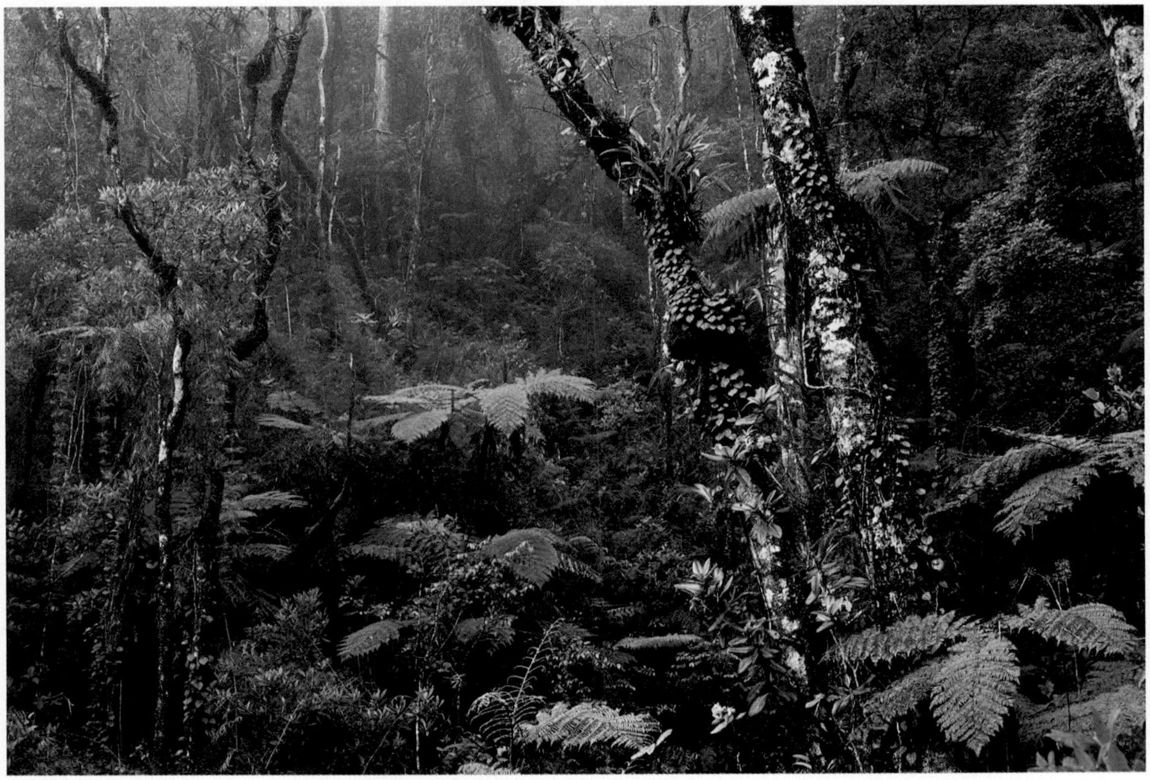

FIGURE 50.12 Tropical Wet Forests Are Extremely Rich in Species.

Subtropical deserts are found throughout the world in two distinctive locations: 30 degrees latitude, or distance from the equator, both north and south. Most of the world's great deserts—including the Sahara, Gobi, Sonoran, and Australian outback—lie on or about 30° N or 30° S latitude—a pattern that is explained in Section 50.4.

Temperature The data in **Figure 50.13** are from Yuma, Arizona, in the Sonoran Desert of southwestern North America. Mean monthly temperatures in subtropical deserts vary more than in tropical wet climates, but in Yuma at least, temperatures still never fall below freezing. (Freezing nighttime temperatures are common elsewhere in the Sonora and in many other subtropical deserts.)

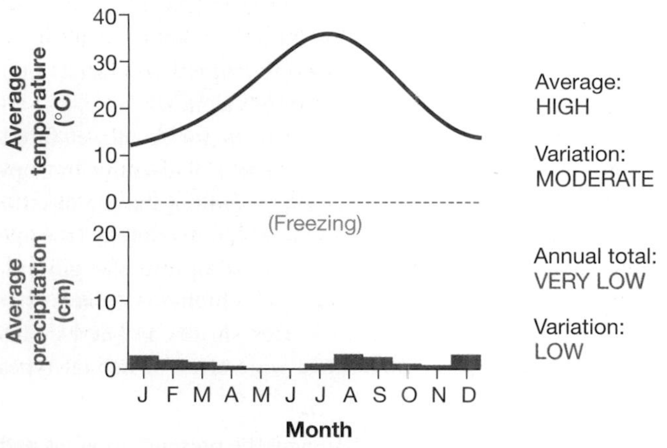

FIGURE 50.13 Climate Data from the Sonoran Desert.

Precipitation The most striking feature of the subtropical desert climate is low precipitation. The average annual precipitation in Yuma is just 7.5 cm (3 in.).

Vegetation The scarcity of water in deserts has profound implications. Conditions are often too dry to support photosynthesis, so the productivity of desert communities is a tiny fraction of average values for tropical forest communities. Further, as the photo in **Figure 50.14** shows, individual plants are widely spaced—a pattern that may reflect intense competition for water.

Desert species adapt to the extreme temperatures and aridity in one of two ways: Growing at a low rate year-round, or breaking dormancy and growing rapidly in response to any rainfall. Cacti can grow year-round because they possess adaptations to cope with hot, dry conditions: small leaves or no leaves (in which case photosynthetic cells are located in stems); a thick, waxy coating on leaves and stems; and the CAM pathway for photosynthesis (Chapter 10). The desert shrub called ocotillo, in contrast, is usually dormant. Individuals sprout leaves within days of a rainfall and drop them a week or two later when soils dry out again. The seeds of annual plants can also lie dormant for many years, then germinate after a rain.

✔You should be able to explain why most desert plant species have no leaves or leaves with a small surface area.

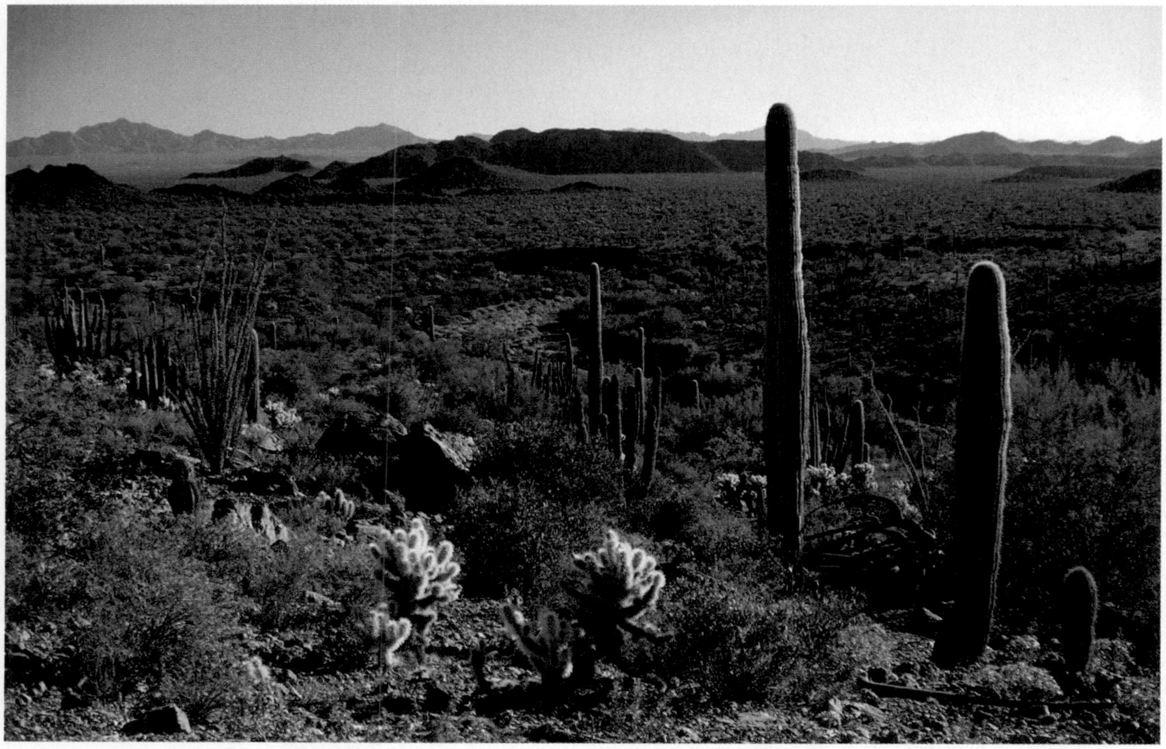

FIGURE 50.14 Subtropical Deserts Are Characterized by Widely Spaced Plants.

Temperate grassland communities are found throughout central North America and the heartland of Eurasia (see Figure 50.10). In North America they are commonly called prairies; in central Eurasia they are known as steppes.

Temperature A region of the world is called **temperate** if it has pronounced annual fluctuations in temperature—typically hot summers and cold winters. Temperature variation is important because it dictates a well-defined growing season. In the temperate zone, plant growth is possible only in spring, summer, and fall months when moisture and warmth are adequate. **Figure 50.15** presents climate data for a typical grassland community—in this case, the Konza Prairie near Manhattan, Kansas.

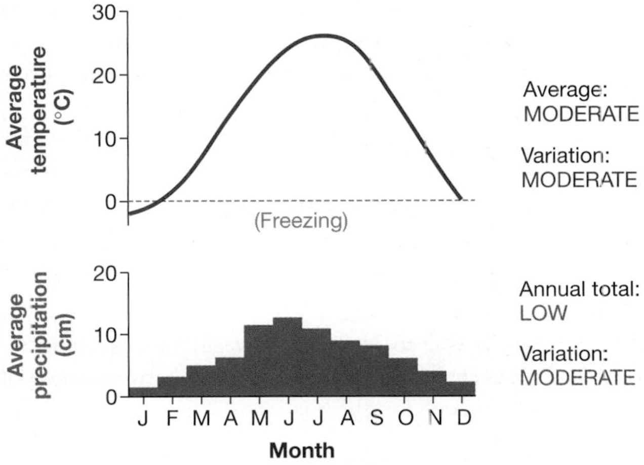

FIGURE 50.15 Climate Data from Konza Prairie.

Precipitation Precipitation at Konza is over four times greater than that in Yuma, Arizona. Nonetheless, conditions are still slightly too hot and dry to support forests; average annual precipitation is 83.4 cm (32.5 in.).

Vegetation Grasses are the dominant life-form in temperate grasslands (**Figure 50.16**) for one of two reasons: (**1**) Conditions are too dry to enable tree growth, or (**2**) encroaching trees are burned out by fires. Prairie fires are ignited by lightning strikes or by native people managing land for game animals that graze on grasses and herbaceous (nonwoody) plants. In contrast to deserts, plant life is extremely dense—virtually every square centimeter is filled.

Although the productivity of temperate grasslands is generally lower than that of forest communities, grassland soils are often highly fertile. The subsurface is packed with roots and rhizomes, which add organic material to the soil as they die and decay. Further, grassland soils retain nutrients, because rainfall is low enough to keep key ions from dissolving and leaching out of the soil. It is no accident, then, that the grasslands of North America and Eurasia are the breadbaskets of those continents. The conditions that give rise to natural grasslands are ideal for growing wheat, corn, and other cultivated grasses.

✔You should be able to explain why fires are much more common in grasslands than they are in deserts, even though deserts are drier.

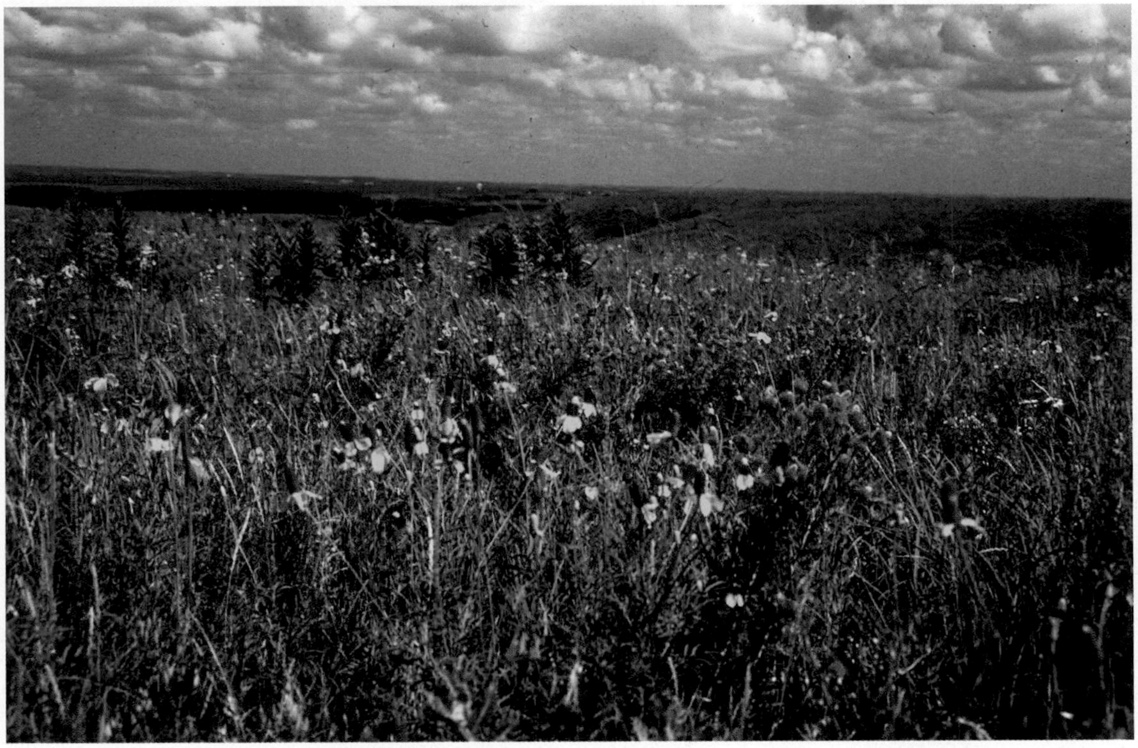

FIGURE 50.16 Temperate Grasslands Have More Biomass Belowground than Aboveground. Although most of the plants in this photo are less than half a meter high, their roots may reach depths of 6 meters or more.

In temperate areas with relatively high precipitation, grasslands give way to forests. Temperate forests are the most common biome found in eastern North America, western Europe, east Asia, Chile, and New Zealand.

Temperature As **Figure 50.17** shows, temperate forests experience a period in which mean monthly temperatures fall below freezing and plant growth stops. The data in this graph are from Chicago, Illinois, which was forested before being settled.

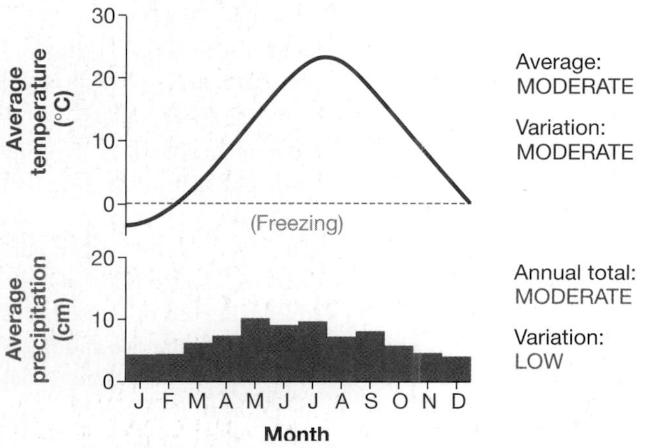

FIGURE 50.17 Climate Data from Chicago Temperate Forest.

Precipitation Compared to grassland climates, precipitation in temperate forests is moderately high and relatively constant throughout the year. Chicago, for example, has an annual precipitation of 85 cm (34 in.); during most months, precipitation exceeds 5 cm (2 in.).

Vegetation In North America and Europe, temperate forests are dominated by deciduous species, which are leafless in winter and grow new leaves each spring and summer (**Figure 50.18**). Needle-leaved evergreens are also common. But in the temperate forests of New Zealand and Chile, broad-leaved evergreens predominate.

Most temperate forests have productivity levels that are lower than those of tropical forests yet higher than those of deserts or grasslands. The level of diversity is also moderate. A temperate forest in southeastern North America may have more than 20 tree species; similar forests to the north may have fewer than 10.

The contrast in productivity between tropical and temperate forests is due to the fluctuation in temperature in temperate regions. Temperate forests are less productive simply because temperatures do not support photosynthesis year-round. The contrast in productivity may also be responsible for the contrast in species richness (see Chapter 53).

✔You should be able to describe the vegetation found in regions with climate characteristics intermediate between those of grasslands and temperate forests.

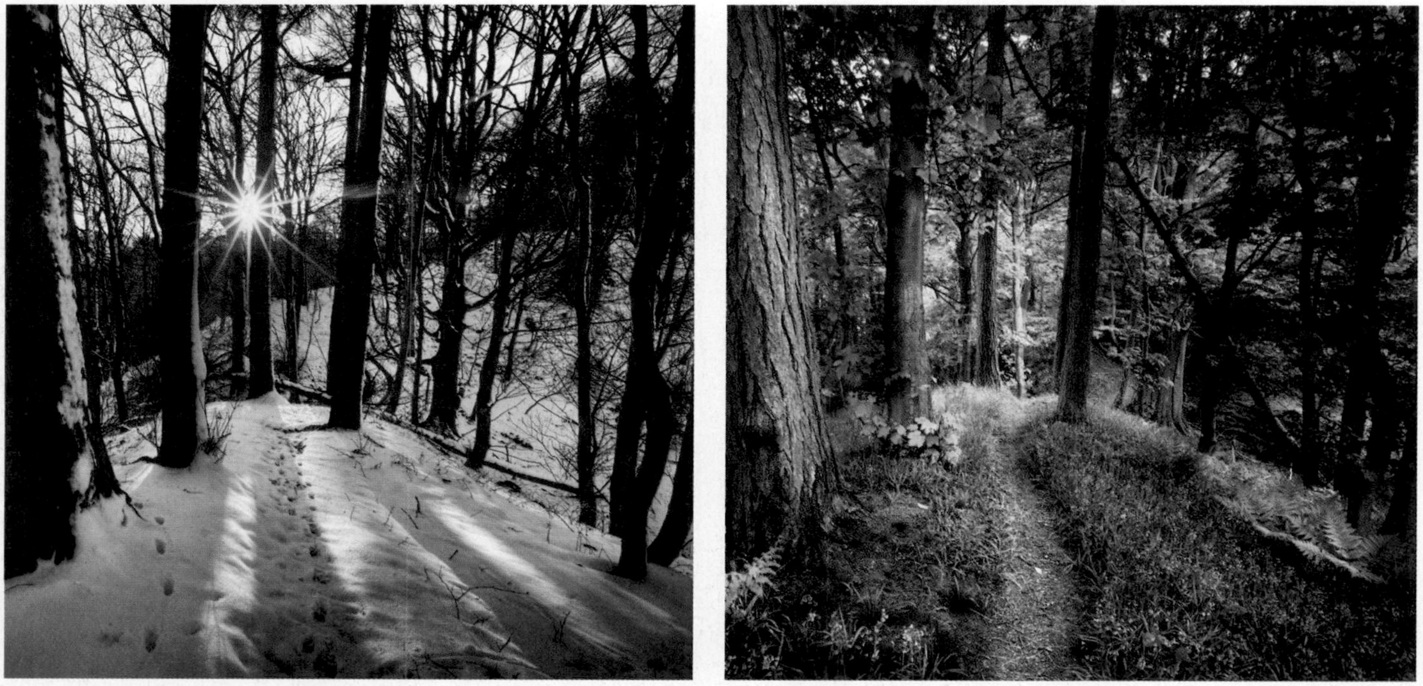

FIGURE 50.18 Many Temperate Forests Are Dominated by Broad-leaved Deciduous Trees.

The boreal forest, or **taiga**, stretches across most of Canada, Alaska, Russia, and northern Europe. Because these regions are just south of the Arctic Circle, they are referred to as subarctic.

Temperature The data in **Figure 50.19** are from Dawson, in the Yukon Territory of Canada. The region is characterized by very cold winters and cool, short summers. Temperature variation is extreme; in the course of a year, subarctic areas may be subject to temperature ranges of more than 70°C.

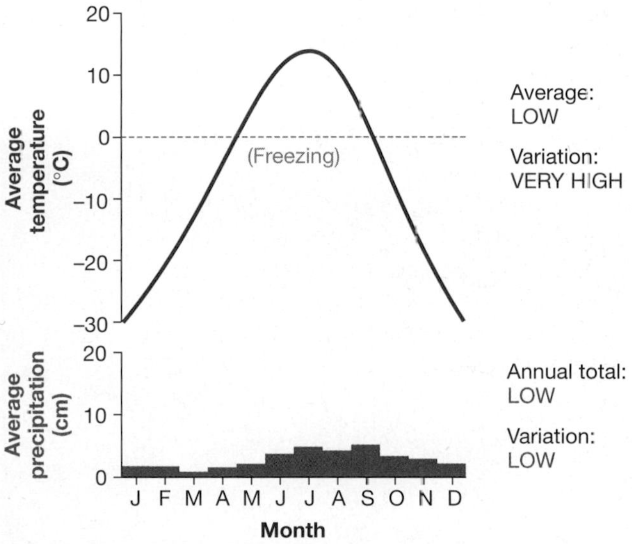

FIGURE 50.19 Climate Data from Yukon Territory.

Precipitation Annual precipitation in boreal forests is low, but temperatures are so cold that evaporation is minimal; as a result, moisture is usually abundant enough to support tree growth.

Vegetation Boreal forests are dominated by highly cold-tolerant conifers, including pines, spruce, fir, and larch trees. Except for larches, these species are evergreen. Two hypotheses have been offered to explain why evergreens predominate in cold environments, even though they do not photosynthesize in winter. The first is that evergreens can begin photosynthesizing early in the spring, even before the snow melts, when sunshine is intense enough to warm their needles. The second hypothesis is based on the observation that boreal forest soils tend to be acidic and contain little available nitrogen. Because leaves are nitrogen rich, species that must produce an entirely new set of leaves each year might be at a disadvantage. To date, however, these hypotheses have not been tested rigorously.

Based on these observations, it is not surprising that the productivity of boreal forests is low. Aboveground biomass is high, however, because slow-growing tree species may be long lived and gradually accumulate large standing biomass. Boreal forests also have exceptionally low species diversity. The boreal forests of Alaska, for example, typically contain seven or fewer tree species.

The other salient characteristic of boreal forests is that they lack most elements of structure seen in tropical or even temperate forests. They often have just a tree layer and a ground layer (**Figure 50.20**).

✔You should be able to predict how the global distribution of boreal forests will change in response to global warming.

FIGURE 50.20 Boreal Forests Are Dominated by Needle-Leaved Evergreen Trees.

The arctic **tundra** biome, which lies poleward from the subarctic, is found throughout the arctic regions of the Northern Hemisphere and in regions of Antarctica that are not covered in ice.

Temperature Tundra develops in climatic conditions like those in Barrow, on the northern coast of Alaska (**Figure 50.21**). The growing season is 10–12 weeks long at most; for the remainder of the year, temperatures are below freezing.

Precipitation Precipitation on the arctic tundra is extremely low. The annual precipitation in Barrow is actually less than that in the Sonoran Desert of southwestern North America. Because of the extremely low evaporation rates, however, many arctic soils are saturated year-round.

Vegetation The arctic tundra is treeless. The leading explanation is that the short growing season prevents the production of large amounts of non-photosynthetic tissue. Also, tall plants that poke above the snow in winter experience substantial damage from wind-driven snow and ice crystals. Woody shrubs such as willows, birch, and blueberries are common, but they rarely exceed the height of a child. Most tundra species hug the ground.

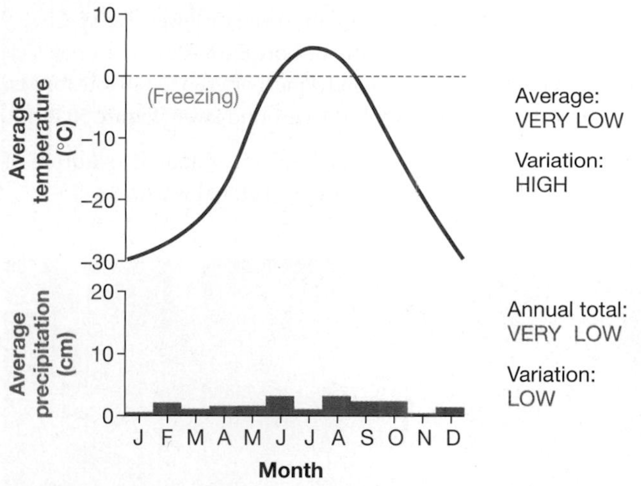

FIGURE 50.21 Climate Data from Barrow, Alaska.

Arctic tundra has low species diversity, low productivity, and low aboveground biomass. Most tundra soils are in the perennially frozen state known as **permafrost**. The low temperatures inhibit both the release of nutrients from decaying organic matter and the uptake of nutrients into live roots. Unlike desert biomes, however, the ground surface in tundra communities is completely covered with plants or lichens (**Figure 50.22**). Animal diversity also tends to be low in the arctic tundra, although insect abundance—particularly of biting flies—can be staggeringly high.

✔**You should be able to explain why the types of vegetation in arctic tundras are similar to the types found in alpine tundras, which occur at high elevation.**

FIGURE 50.22 Arctic Tundra Features Low-Growing Plants.

CHECK YOUR UNDERSTANDING

If you understand that . . .

- In terrestrial environments, distinctive biomes develop based on two aspects of temperature and moisture: average value and variability over the year.

✔ **You should be able to . . .**

Predict whether tropical dry forests, which have high year-round temperatures but a distinct dry season, are more or less productive than tropical wet forests.

Answers are available in Appendix B.

50.4 The Role of Climate and the Consequences of Climate Change

Each type of aquatic environment and terrestrial biome hosts species that are adapted to the abiotic conditions present at that location. Global warming—due to rising concentrations of atmospheric CO_2 (see Chapter 54)—is having a profound impact on these abiotic factors. Runoff from melting glaciers and pole ice is changing water depths along coasts, and warming ocean waters are disturbing long-established patterns in the direction and intensity of ocean currents. On land, higher air temperatures are altering growing seasons and water availability in biomes around the globe.

To understand how global climate change is affecting the distribution and abundance of organisms, biologists begin by studying why climate varies in predictable ways around the planet. Then they ask how changes in these climate patterns are affecting the organisms present, with the aim of predicting the consequences of continued change. Let's do the same.

Global Patterns in Climate

Simple questions often have fascinating answers. This turns out to be true for questions about why climate varies around the globe. For example, why are some parts of the world warmer and wetter than others? Why do seasons exist?

WHY ARE THE TROPICS WET? One of the most striking weather patterns on Earth involves precipitation. When average annual rainfall is mapped for regions around the globe, it is clear that areas along the equator receive the most moisture. In contrast, locations about 30 degrees latitude north and south of the equator are among the driest on the planet. Why do these patterns occur?

A major cycle in global air circulation, called a **Hadley cell**, is responsible for making the Amazon River basin wet and the Sahara Desert dry. Hadley cells are named after George Hadley, who in 1735 conceived of the idea of enormous air circulation patterns.

To understand how Hadley cells work, put your finger on the equator in **Figure 50.23a** and follow the arrows.

- Air that is heated by the strong sunlight along the equator expands, lowering its pressure and causing it to rise. Note that this warm air holds a great deal of moisture, because warm water molecules tend to stay in vapor form instead of condensing into droplets.
- As air rises above the equator, it radiates heat to space. It also expands into the larger volume of the upper atmosphere, which lowers its density and temperature—a phenomenon known as adiabatic cooling.
- As rising air cools, its ability to hold water declines. When water vapor cools, water condenses.

The result? High levels of precipitation occur along the equator.

The story is just beginning, however.

- As more air is heated along the equator, the cooler, "older" air above Earth's surface is pushed poleward.
- When the air mass has cooled enough, its density increases and it begins to sink.
- As it sinks, it absorbs more and more solar radiation reflected from Earth's surface and begins to warm while being pushed to the surface by higher-density air above it.
- As the air warms, it also gains water-holding capacity. Thus, the air approaching the Earth "holds on" to its water and little rain occurs where it returns to the surface.

The result? A band of deserts in the vicinity of 30° latitude north and south. These areas are bathed in warm, dry air.

Similar air circulation cells occur between 30° latitude and 60° latitude, and between 60° latitude and the poles (**Figure 50.23b**).

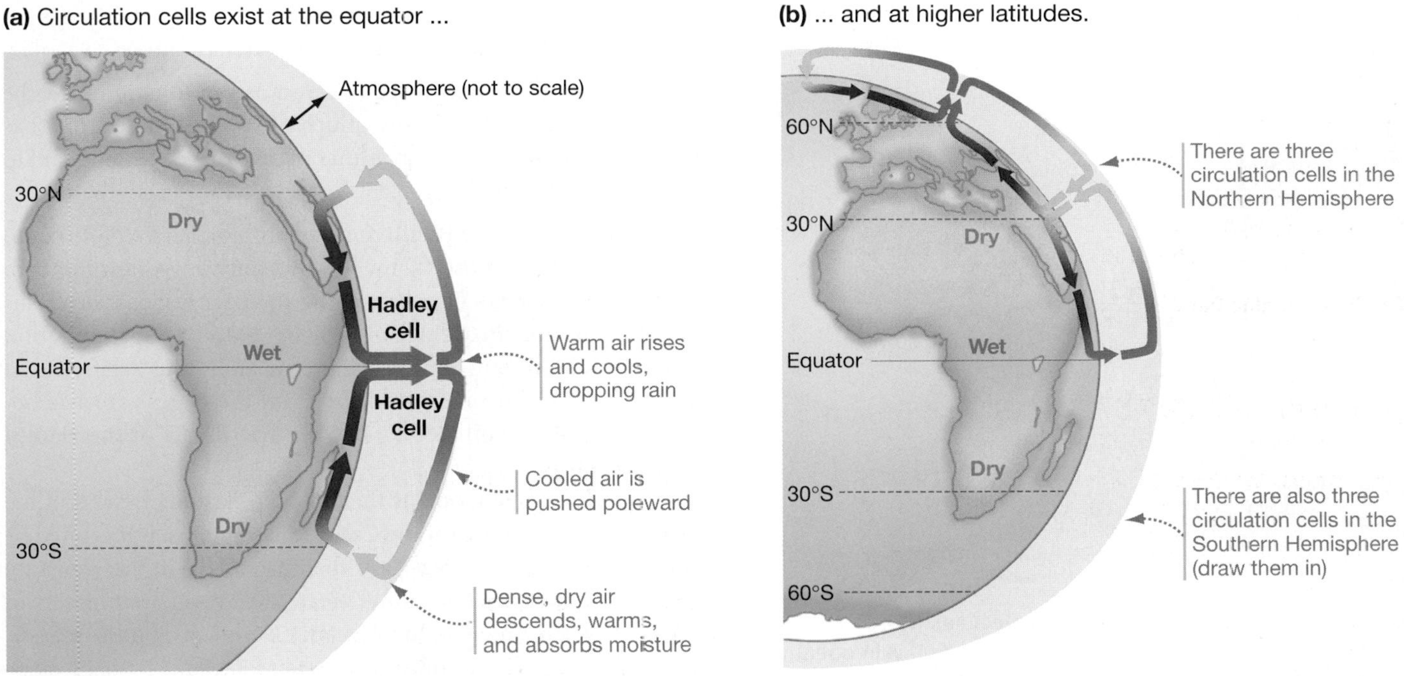

FIGURE 50.23 Global Air Circulation Patterns Affect Rainfall. Hadley cells explain why the tropics are much wetter than regions near 30° latitude north and south. Regions with rising air, such as those at the equator, tend to be wetter than regions with descending air.

✔**QUESTION** Is air at the North Pole wet or dry? Is your answer consistent with the data in Figure 50.21?

✔**If you understand how Hadley cells work, you should be able to add diagrams to Figure 50.24b showing the three cells that occur in the Southern Hemisphere.**

For more review on tropical atmospheric circulation, go to the study area at *www.masteringbiology.com.*

(MB) **Web Activity** Tropical Atmospheric Circulation

WHY ARE THE TROPICS WARM AND THE POLES COLD? In general, areas of the world are warm if they receive a large amount of sunlight per unit area; they are cold if they receive a small amount of sunlight per unit area. Over the course of a year, regions at or near the equator receive much more sunlight per unit area—and thus much more energy in the form of heat—than regions that are closer to the poles.

- At the equator, the Sun is often directly overhead. As a result, sunlight strikes the surface at or close to a 90° angle. At these angles, Earth receives a maximum amount of solar radiation per unit area (**Figure 50.24**).
- Because Earth's surface slopes away from the equator, the Sun strikes the surface at lower and lower angles moving toward the poles. When sunlight arrives at a low angle, much less energy is received per unit area.

The pattern of decreasing average temperature with increasing latitude is a result of Earth's spherical shape.

WHAT CAUSES SEASONALITY IN WEATHER? The striking global patterns in temperature and moisture that we've just reviewed are complicated by the phenomenon of seasons. Seasons are defined as regular, annual fluctuations in temperature, precipitation, or both.

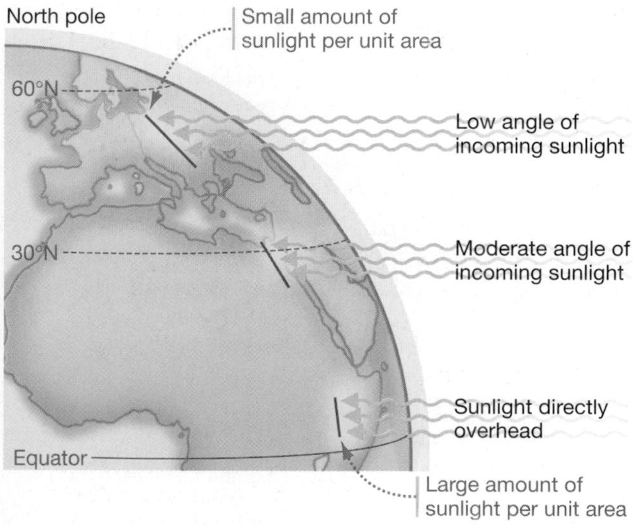

FIGURE 50.24 Solar Radiation per Unit Area Declines with Increasing Latitude. Over the course of a year, the Sun is frequently almost directly overhead at the equator. As a result, equatorial regions receive a large amount of solar radiation per unit area. At latitudes greater than 23.5°, the Sun is never directly overhead. As a result, high-latitude regions receive less solar radiation per unit area than do low-latitude regions.

As **Figure 50.25** shows, seasonality occurs because Earth is tilted on its axis by 23.5 degrees. As a result of this incline, the Northern Hemisphere is tilted toward the Sun in June. In this position, it faces the Sun most directly. In June, the Northern Hemisphere presents its least-acute angle to the Sun and receives the largest amount of solar radiation per unit area.

The Southern Hemisphere, in contrast, is tilted away from the Sun in June. As a result, it presents a steep angle for incoming sunlight and receives its smallest quantity of solar radiation per unit area.

The upshot? In June it is summer in the Northern Hemisphere and winter in the Southern Hemisphere. Conversely, in December it is summer in the Southern Hemisphere, and winter in the Northern Hemisphere.

In March and September, the equator faces the Sun most directly. During these months, the tropics receive the most solar radiation. If Earth did not tilt on its axis, there would be no seasons.

MOUNTAINS AND OCEANS: PHYSICAL FEATURES HAVE REGIONAL EFFECTS ON CLIMATE The broad patterns of climate that are dictated by global heating patterns, Hadley cells, and seasonality are overlain by regional effects. The most important of these are due to the presence of mountain ranges and proximity to an ocean.

Figure 50.26 shows how mountain ranges affect regional climate, using the Cascade Mountain range along the Pacific Coast of North America as an example.

- In this area of the world, the prevailing winds are from the west. These winds bring moisture-laden air from the Pacific Ocean onto the continent.
- As the air masses begin to rise over the mountains, the air cools and releases large volumes of water as rain.
- Once cooled air has passed the crest of a mountain range, the air is relatively dry because much of its moisture content has already been released. Areas that receive this dry air are said to be in a **rain shadow**.

The area along the Pacific Coast from northern California to southeast Alaska hosts some of the only high-latitude rain forests in the world. One area along the Pacific Coast of Washington State gets an average of 340 cm (134 in.) of precipitation each year. Mountain ranges also occur along the Pacific Coast of Southern California and Mexico; but these areas are arid because they are dominated by dry air that is part of the Hadley circulation.

Logically enough, one of the few high-latitude deserts in the world exists to the east of the Cascade Mountains. It is created by a rain shadow and averages less than 25 cm (10 in.) of moisture annually. Similar rain shadows create the Great Basin desert of western North America, the Atacama Desert of South America, and the dry conditions of Asia's Tibetan Plateau.

While the presence of mountain ranges tends to produce extremes in precipitation, the presence of an ocean has a moderating influence on temperature. To understand why this is so, recall from Chapter 2 that water has an extremely high **specific heat**,

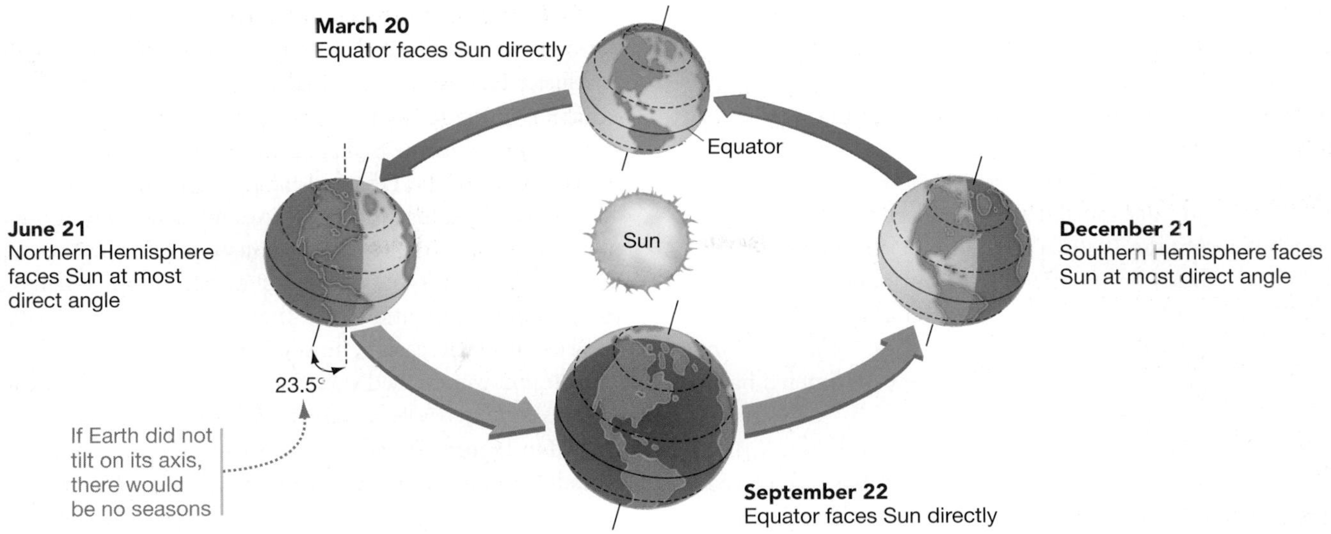
March 20
Equator faces Sun directly
Equator
June 21
Northern Hemisphere faces Sun at most direct angle
Sun
December 21
Southern Hemisphere faces Sun at most direct angle
23.5°
If Earth did not tilt on its axis, there would be no seasons
September 22
Equator faces Sun directly

FIGURE 50.25 Earth's Orbit and Tilt Create Seasons at High Latitudes. Seasons occur due to annual variation in the amount of solar radiation that different parts of the Earth receive.

meaning that it has a large capacity for storing heat energy. Because water molecules form hydrogen bonds with each other, it takes a great deal of heat energy to boil water or melt ice—in fact, to change the temperature of water at all. The hydrogen bonds have to be broken before the water molecules themselves can begin to absorb heat and move faster, which we measure as increased temperature.

Because water has a high specific heat, it can absorb a great deal of heat from the atmosphere in summer, when the water temperature is cooler than the air temperature. As a result, the ocean moderates summer temperatures on nearby landmasses. Similarly, the ocean releases heat to the atmosphere in winter, when the water temperature is warmer than the air temperature. As a result, islands and coastal areas have much more moderate climates than do nearby areas inland.

How Will Global Climate Change Affect Ecosystems?

Climate—particularly the amount and variability of heat and precipitation—has a dramatic effect on terrestrial ecosystems. Climate also has a significant impact on aquatic ecosystems. Heat and precipitation influence ocean currents and lake turnover, as well as water levels in lakes, streams, and oceans.

It is now well established that CO_2 pollution is causing climate change (see Chapter 54). Climate has changed frequently and radically throughout the history of life, but recent increases in average temperatures around the globe, combined with a projected increase of up to 5.8°C over the next 100 years, represents one of the most rapid periods of climate change since life began.

Biologists use three tools to predict how global warming will affect aquatic and terrestrial ecosystems.

1. *Simulation studies* are based on computer models of weather patterns in local regions. By increasing average temperatures in these models, researchers can predict how wind and rainfall patterns, storm frequency and intensity, and other aspects of weather and climate will change. In effect, simulation studies predict how temperature and precipitation profiles, like those in this chapter's "Terrestrial biomes boxes," will change.
2. *Observational studies* are based on long-term monitoring at fixed sites around the globe. Researchers are documenting changes in key physical variables—water depth, water flow, temperature, and precipitation—as well as changes in the

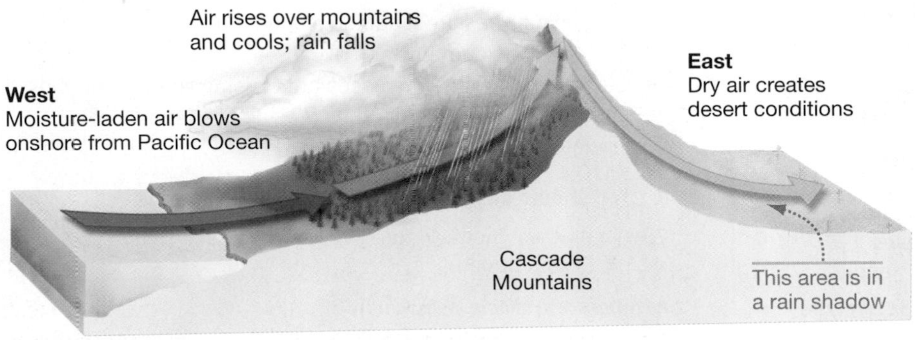
Air rises over mountains and cools; rain falls
West
Moisture-laden air blows onshore from Pacific Ocean
East
Dry air creates desert conditions
Cascade Mountains
This area is in a rain shadow

FIGURE 50.26 Mountain Ranges Create Rain Shadows. When moisture-laden air rises over a mountain range, the air cools enough to condense water vapor. As a result, moisture falls to the ground as precipitation. On the other side of the mountain, little water is left to provide precipitation.

distribution and abundance of organisms. By extrapolation, they can predict how change may continue at these sites.

3. *Experiments* are designed to simulate changed climate conditions and to record responses by the organisms present.

EXPERIMENTS THAT MANIPULATE TEMPERATURE Even though the scientific community did not become alert to the reality of global warming until the 1980s, a large number of multiyear experiments have already been done to test how increased temperature affects organisms in nature. Because arctic regions are projected to experience particularly dramatic changes in temperature and precipitation, many of these experiments have been done in the tundra biome.

Figure 50.27 shows how the most successful of these experiments are done. Transparent, open-topped chambers are placed at random locations around a study site, and the characteristics of the vegetation inside the chambers are recorded annually. The same measurements are made in randomly assigned study plots that lack chambers.

Across experiments, the average annual temperature inside the open-topped chambers increases by 1–3°C. This increase is consistent with conservative projections for how temperatures will change over the next century.

In an analysis of data from 11 such study sites scattered throughout the arctic in North America and Eurasia, several strong patterns emerged:

- Overall, species diversity decreases.
- Compared with control plots, grasses and shrubs increase inside the chambers, and mosses and lichens decrease.

These results support simulation and observational studies predicting that arctic tundra environments are giving way to boreal forest. Biomes are changing right before our eyes.

FIGURE 50.27 Experimental Increases in Temperature Change Species Composition in Arctic Tundra. The transparent, open-topped chambers shown here are an effective way to increase average annual temperature. Inside the chambers, species composition changes relative to otherwise similar sites nearby.

✔**QUESTION** Why does a clear chamber like this increase average temperature inside?

EXPERIMENTS THAT MANIPULATE PRECIPITATION Recent simulation and observational studies have converged on a remarkable conclusion: Increases in average global temperature are increasing variability in temperature and precipitation. Stated another way, global climate change is making climates more extreme. Average temperature and average annual precipitation in many regions may not be changing substantially, even though variation is.

Thus far, the frequency of monsoon rains, cyclones, hurricanes, and other types of storm events is not changing. The amount of precipitation and the wind speeds associated with these events is increasing, however.

To test how increased variation in climate will affect a temperate grassland ecosystem, researchers set up the experiment described in **Figure 50.28**. The goal was to keep the average annual rainfall unchanged during the growing season, but increase variability.

The biologists increased the severity of storms by capturing rain that fell over study plots during some storms and then applying that water to the same plots during later storms. The bar charts in the "Experimental setup" section of the figure show how much water entered each type of plot as the growing season progressed.

The results from the first four years of the experiment are plotted in the bar charts in the "Results" section. Note that:

- Average soil moisture and average overall NPP declined in the experimental plots, compared with similar plots where rainfall was not manipulated.
- On average, overall species diversity is increasing in the high-variation plots because the grass that produces most of the biomass at this site is doing poorly in the experimental plots—opening up room for other species to grow.

The experiment is ongoing, however, because it is possible that long-term changes will differ from short-term changes, due to continued alterations in the species present.

Experiments like this are documenting that climate change causes dramatic alterations in the composition and distribution of biomes. Your world is changing.

CHECK YOUR UNDERSTANDING

If you understand that . . .

- Global climate patterns are dictated by differential heating of Earth's surface, Hadley cells, and seasonality. These general patterns are modified at a regional scale by the presence of mountains and oceans.
- Global temperature increases are causing changes in the distribution of biomes and increasing the severity of storm events.

✔ **You should be able to . . .**

Predict how the biome where you live will change over the course of your lifetime.

Answers are available in Appendix B.

EXPERIMENT

QUESTION: How does increasing variation in rainfall—without changing total amount—affect grassland biomes?

HYPOTHESIS: Increasing variability in rainfall affects soil moisture, net primary productivity (NPP), and species diversity.

NULL HYPOTHESIS: Increasing variability in rainfall has no effect on grassland biomes.

EXPERIMENTAL SETUP:

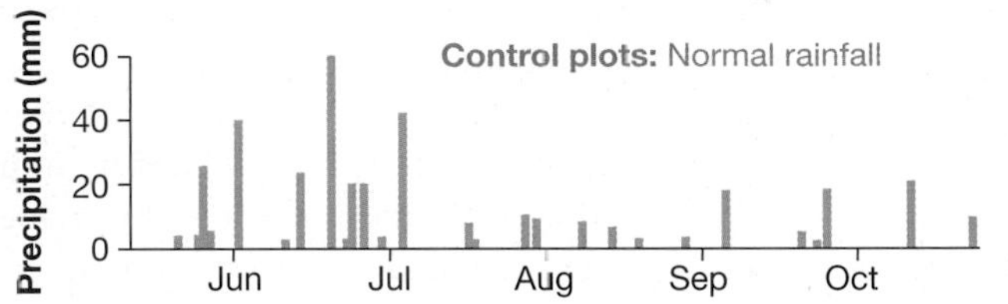

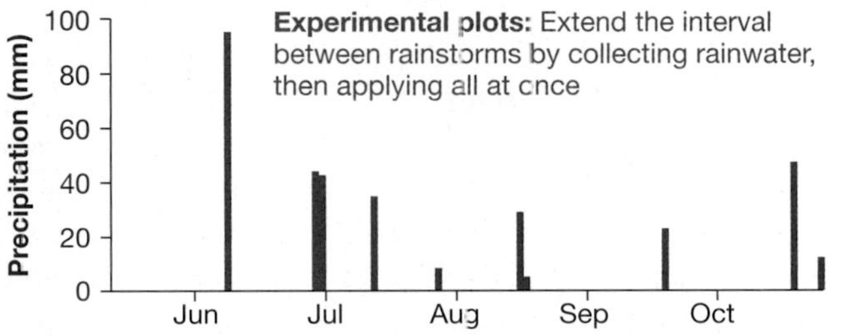

Record soil moisture, aboveground biomass (as an index of NPP), and species diversity, over a 4-year period.

PREDICTION: Soil moisture, NPP, and/or species diversity will differ between experimental plots and control plots.

PREDICTION OF NULL HYPOTHESIS: Soil moisture, NPP, and species diversity will not differ between experimental plots and control plots.

RESULTS:

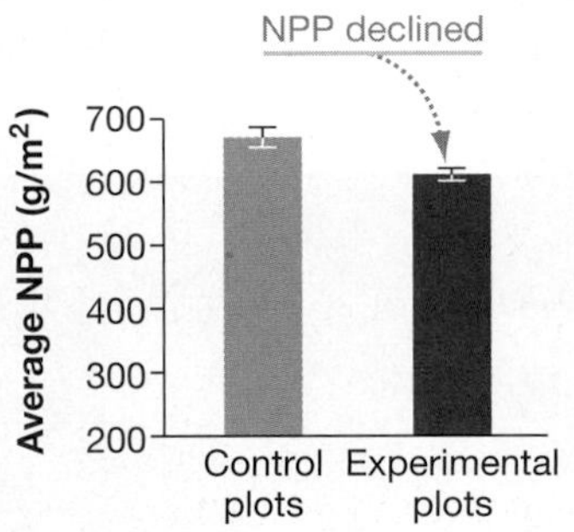

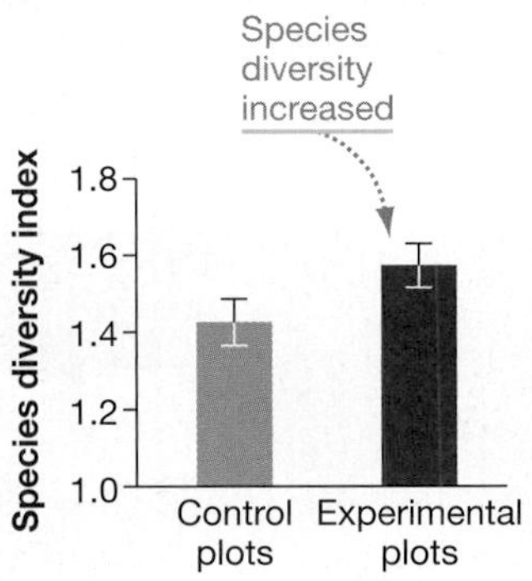

CONCLUSION: Altering variability in rainfall has dramatic effects on soil moisture (data not shown), NPP, and species diversity.

FIGURE 50.28 Experimental Increases in Rainfall Variability Change Species Composition in Temperate Grasslands.

SOURCE: Knapp, A. K. et al. 2002. Rainfall variability, carbon cycling, and plant species diversity in a mesic grassland. *Science* 298: 2202–2205.

✔**QUESTION** Of all the possible data to collect, why did the researchers decide to measure NPP and species diversity?

50.5 Biogeography: Why Are Organisms Found Where They Are?

Understanding why organisms are found where they are is one of ecology's most basic tasks. It is a fundamental aspect of predicting the consequences of climate change, and is required for designing nature preserves in a rational and effective way.

The study of how organisms are distributed geographically is called **biogeography**. Chapter 53 analyzes general patterns in how species numbers vary around the world, Chapter 54 presents data showing that the geographic ranges of many species are changing rapidly in response to global climate change, and Chapter 55 analyzes "hotspots" where species diversity is particularly high. Here, let's focus on how interactions with the abiotic and biotic environments affect where a particular species lives.

Abiotic Factors

The most fundamental observation about the **range**, or geographic distribution, of organisms is simple: No one species can survive the full range of environmental conditions present on Earth. *Thermus aquaticus* can thrive in hot springs with temperatures above 70°C because its enzymes do not denature at near-boiling temperatures. But the same cells would die instantly if they were transplanted to the frigid waters near polar ice—even though vast numbers of other archaea are present. The cold-water species have enzymes that work efficiently at near-freezing temperatures.

No organism thrives in both hot springs and cold ocean water. The reason is fitness trade-offs—a concept introduced in Chapters 24 and 41. No enzyme can function well in both extremely high temperatures and extremely low temperatures. Because of fitness trade-offs, organisms tend to be adapted to a limited set of physical conditions—to a particular temperature and moisture regime on land, or a particular water depth and movement regime in water.

But adaptation to a proscribed set of abiotic conditions is just the start of the story of explaining why a particular species lives where it does. For example, the temperature and moisture regimes of southern California, in North America, are almost identical to those found in the Mediterranean region of Europe and north Africa. The two regions are part of the same biome. But the species present at the sites are almost completely different.

Why? To understand a species' distribution thoroughly, biologists examine historical and biotic factors in addition to the physical conditions present.

The Role of History

In studying a species' range, one of the first factors to consider is the history of dispersal. **Dispersal** is the movement of an individual from its place of origin to the location where it lives and breeds as an adult.

If a particular species is missing from an area, it may be due to a physical barrier to dispersal—a river, a glacier, a mountain range, or an ocean. The plant species found in California are different from those found in southern Spain because the Pacific and Atlantic Oceans prevent the dispersal of seeds.

Alfred Russel Wallace was among the first to document this phenomenon. In addition to codiscovering evolution by natural selection (see Chapter 24), Wallace founded the study of biogeography.

THE WALLACE LINE: BARRIERS TO DISPERSAL While working in the Malay archipelago, Wallace realized that the plants and animals native to the more northern and western islands were radically different from the species found on the more southern and eastern islands. Sumatra, to the northwest, has tigers, rhinos, and other species with close relatives in southern Asia. New Guinea, to the southeast, has tree kangaroos and other species with close relatives on Australia. But both islands are part of the tropical wet forest biome.

This biogeographical demarcation, now known as the **Wallace line**, separates species with Asian and Australian affinities (**Figure 50.29**). It exists because a deep trench in the ocean maintained a water barrier to dispersal, even when ocean levels dropped during the most recent glaciations. As a result, landforms on either side of the line remained isolated at a time when most of the other islands became connected, encouraging dispersal.

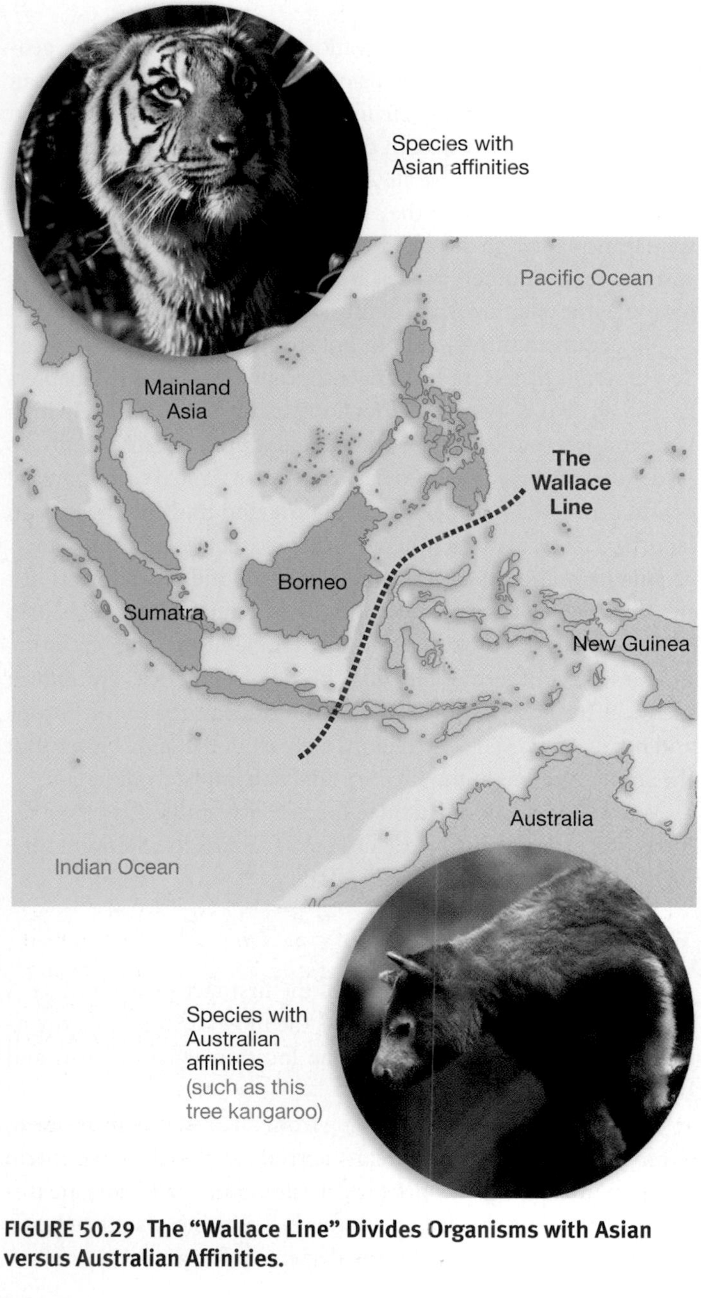

FIGURE 50.29 The "Wallace Line" Divides Organisms with Asian versus Australian Affinities.

HUMANS AS DISPERSAL AGENTS Unfortunately, humans are now breaking down many natural barriers to dispersal. This is why public health officials closely monitor annual flu outbreaks in south China—where most new strains originate. New flu strains disperse all over the planet in a matter of weeks or days, transported in the respiratory passages of infected airplane passengers. Prior to the invention of air travel, disease epidemics in humans were much more likely to be confined to relatively small geographic areas.

Similarly, humans have transported thousands of plants, birds, insects, and other species across physical barriers to new locations—sometimes purposefully and sometimes by accident. One accidental introduction has had disastrous consequences for the arid shrublands and grasslands of the western United States.

EXOTIC AND INVASIVE SPECIES In 1889 a native plant of Eurasia called cheatgrass was accidentally introduced to western North America in a shipment of crop seeds. Cheatgrass is an annual plant with seeds that germinate over winter, so that plants grow and set seed in early spring. As a result, cheatgrass seedlings get a "jump" on many North American plant species, which initiate growth later. Early growth allows cheatgrass to use more of the available water and nutrients. Within 30 years, cheatgrass had taken over tens of thousands of square miles of habitat.

If an **exotic species**—one that is not native—is introduced into a new area, spreads rapidly, and eliminates native species, it is said to be an **invasive species**. In North America alone, dozens of invasive species have had devastating effects on native plants and animals. Examples include kudzu (**Figure 50.30**), purple loosestrife, reed canary grass, garlic mustard, Russian thistle (also known as

FIGURE 50.30 Introduced Species Can Cause Problems. Kudzu, imported from Asia, has taken over parts of the southeastern U.S.

tumbleweed), European starlings, African honeybees, and zebra mussels. Similar lists could be compiled for other regions of the world.

Regarding the distribution of organisms, the story of invasive species is straightforward: They and other introduced species did not exist in certain parts of the world until recently, simply because they had never been dispersed there.

Only about 10 percent of the species introduced to an area actually become common, however, and a smaller percentage than that increases in population enough to be considered invasive. Why is it that some species introduced into an area do not thrive? Conversely, why are invasives so successful?

Biotic Factors

The distribution of a species is often limited by biotic factors—meaning interactions with other organisms. As an example, recall the data presented in Chapter 26 on the current and former distributions of hermit warblers and Townsend's warblers along the Pacific coast of North America.

- Both hermit warblers and Townsend's warblers live in evergreen forests.
- Experiments have shown that male Townsend's warblers directly attack male hermit warblers and evict them from breeding territories.
- Historical data indicate that the geographic range of Townsend's warblers has been expanding steadily at the expense of hermit warblers.

These observations support the hypothesis that a biotic factor—competition with another species—is limiting the range of hermit warblers. Competition is not the only biotic factor to consider, however.

- The moth pictured in **Figure 50.31** lays its eggs only in the flowers of yucca plants. As a result, this species does not exist outside the range of yucca plants.

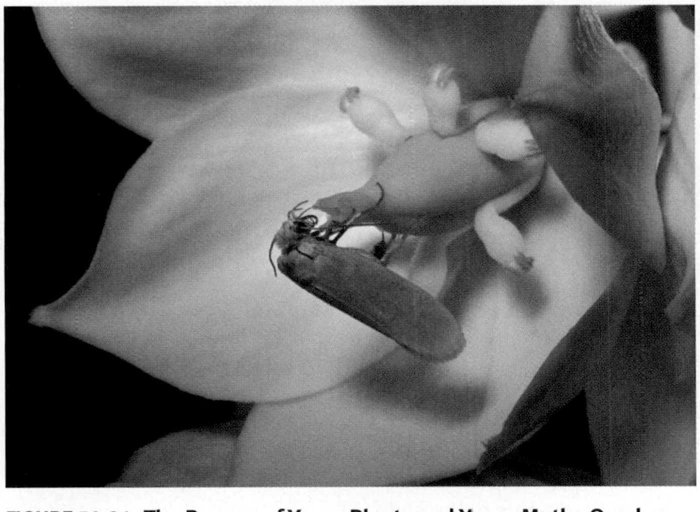

FIGURE 50.31 The Ranges of Yucca Plants and Yucca Moths Overlap. Yucca moths lay their eggs only in the flowers of yucca plants, so the moths cannot live where yucca plants are absent.

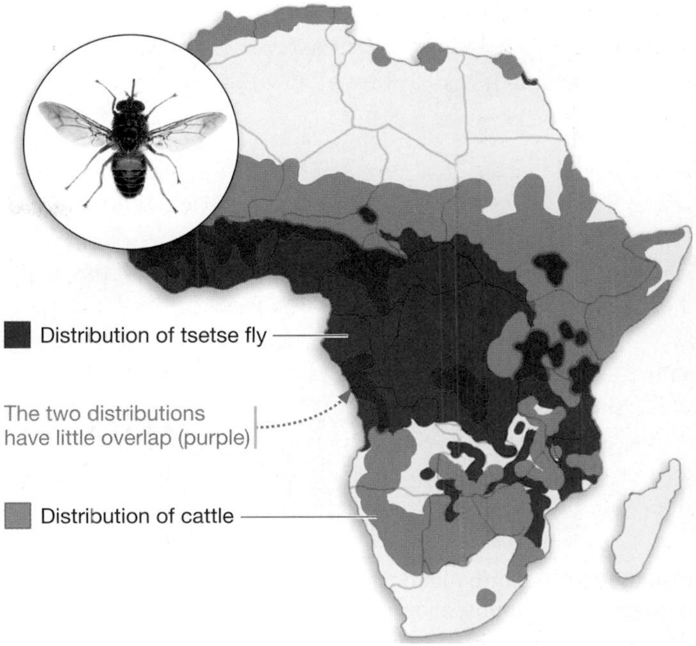

FIGURE 50.32 In Africa, Tsetse Flies Limit the Distribution of Cattle. Tsetse flies carry a disease that is fatal to cattle.

- In Africa, the range of domestic cattle is limited by the distribution of tsetse (pronounced *TSEE-tsee*) flies. Tsetse flies transmit a parasite that causes the disease trypanosomiasis, which is fatal in cattle (**Figure 50.32**).
- Most of the songbirds native to Hawaii are limited to alpine habitats—above the range of the mosquitoes that transmit avian malaria.

Biotic and Abiotic Factors Interact

In many cases, introduced species may not become established in a region because of abiotic factors—particularly temperature and moisture. An area may simply be too cold or too dry for an organism to survive. For example, even though the invasive plant kudzu grows prolifically throughout the southeastern United States, it is killed when the soil freezes to a depth of a foot or more. As a result, it has not expanded its range farther north and west.

It is often difficult to separate the effects of biotic and abiotic factors on a species' range, however. To drive this point home, consider how cheatgrass became established in North America.

Cheatgrass does not grow in wet grasslands because it does not compete well against the tall species that thrive there. But it has been able to invade two important types of biomes in North America: the arid, shrub-dominated habitats known as sage-steppe, and dry temperate grasslands. In each case, the reasons for the success of cheatgrass were different.

THE ROLE OF FIRE IN CHEATGRASS'S SPREAD Cheatgrass has been able to invade sage-steppe habitats with the aid of fire. Fire is an abiotic factor that can kill the shrubs that dominate sage-steppe.

Fires tend to be rare in sage-steppe biomes because plants are widely spaced. This distribution usually prevents fires from

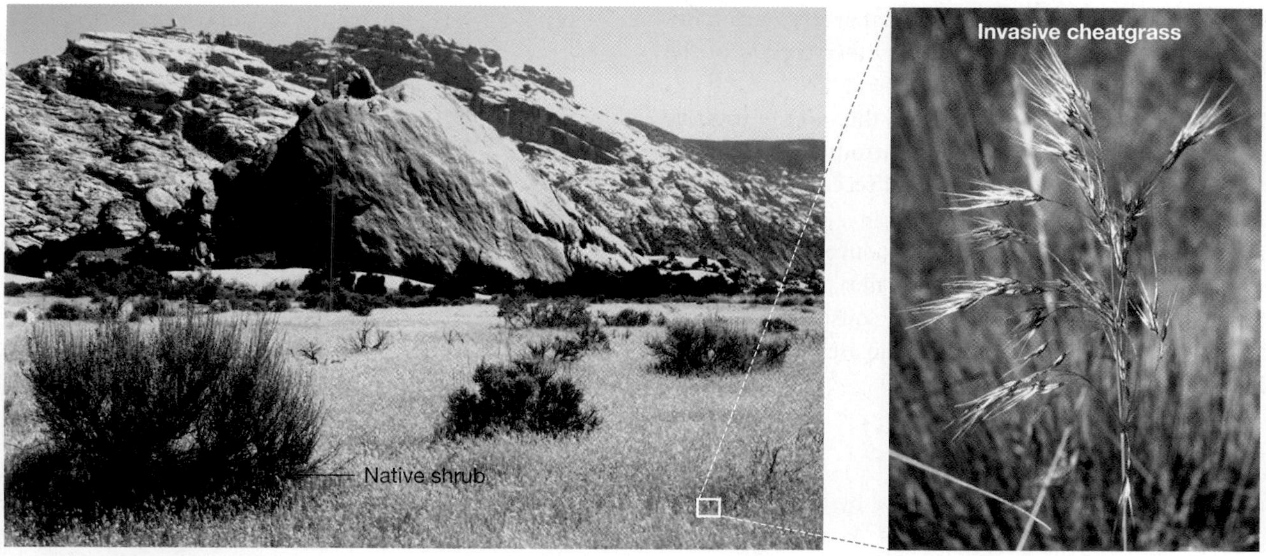

FIGURE 50.33 In North America, Sage-Steppe Has Been Invaded by Cheatgrass. Cheatgrass fills in spaces between native shrubs, allowing fires to burn through the habitat.

spreading and affecting a large area. But when this biome has been invaded by cheatgrass and begins to dry in late summer, beds of dead cheatgrass offer a continuous, extremely flammable source of fuel (**Figure 50.33**).

If a fire gets started, it does not affect the cheatgrass for two reasons:

1. Cheatgrass is an annual, so there is no living tissue exposed once the growing season is over.
2. Cheatgrass seeds sprout readily in soils that have been depleted of organic matter by fire.

Because sagebrush and other shrubs are perennials, their tissues are heavily damaged or killed by hot fires. By promoting the frequency and intensity of fire in sage-steppe biomes, the presence of cheatgrass creates an environment that favors the spread of cheatgrass.

How has this species been able to invade dry grassland biomes?

THE ROLE OF OVERGRAZING IN CHEATGRASS'S SPREAD Prior to the arrival of cheatgrass, grasslands in arid areas of the American West were dominated by the types of bunchgrasses shown in **Figure 50.34**. Bunchgrasses do not form a continuous sod; instead, they grow in compact bunches.

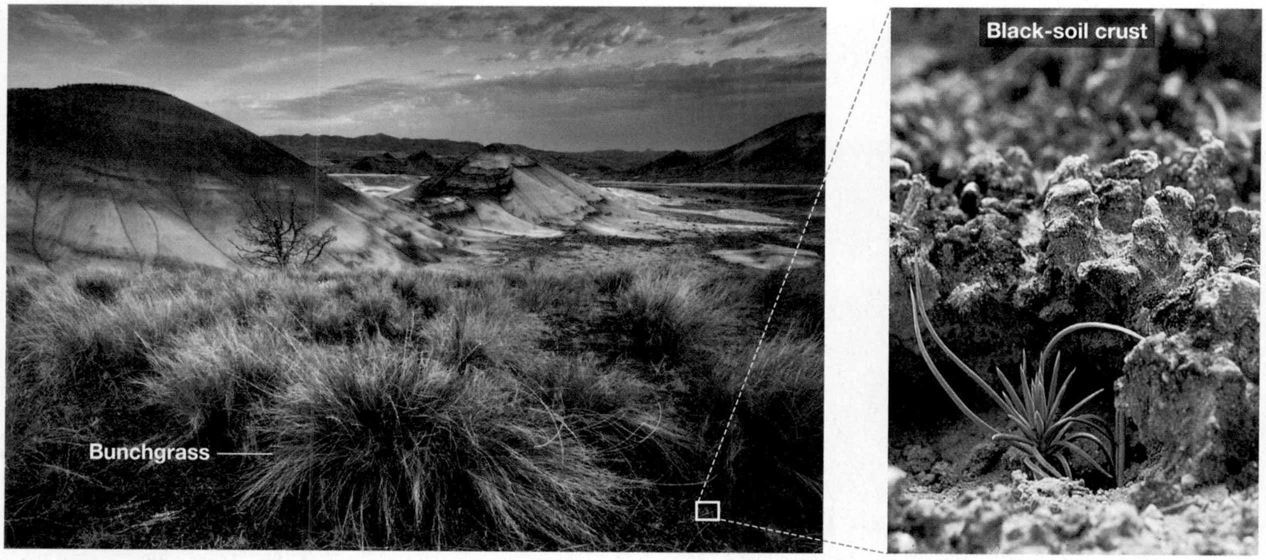

FIGURE 50.34 Arid Grasslands Feature Bunchgrasses and Black-Soil Crust. Bunchgrasses grow in compact bunches instead of forming a sod. The black-soil crust that forms between bunchgrasses is composed of cyanobacteria, archaea, and other microscopic organisms that add organic matter and nutrients to the soil.

✔**QUESTION** Suggest a hypothesis to explain why grasses with the bunching growth habit tend to be found in dry habitats, as compared to grasses that form sods.

The spaces between bunchgrasses are occupied by what biologists call "black-soil crust." This material is a community of soil-dwelling cyanobacteria, archaea, and other microorganisms that grow when moisture is available. Many of these species also fix nitrogen (see Chapter 38). The nitrogen becomes available to the surrounding plants when the microbial cells die and decompose.

When European settlers arrived and began grazing cattle in this biome, however, both the black soils and bunchgrasses were affected. Cattle grazed on the bunchgrasses until the grasses died, and their hooves compacted and disrupted the black soils.

Neither component of the ecosystem was allowed to recover, in part because intensive grazing continued and in part because cheatgrass arrived:

- Cheatgrass shades out black-soil crust organisms. When black-soil crust is gone, the amount of nitrogen available to bunchgrasses is reduced.
- Cheatgrass competes successfully with bunchgrasses for water and nutrients, because it completes its growth early in the spring—before the native bunchgrasses have broken dormancy and begun to grow.
- As in sage-steppe habitats, the expansion of cheatgrass increased the frequency and severity of fires, which further degraded the integrity of black-soil crust communities.

To summarize, changes in abiotic and biotic factors allowed cheatgrass to invade native biomes in North America and thrive as an invasive species. Currently, managing cheatgrass is the major problem facing conservation biologists and rangeland managers in the western United States.

The story of cheatgrass is particularly dramatic. But it is important to remember that the range of every species on Earth is limited by a combination of historical, abiotic, and biotic factors. The range of a particular species depends on dispersal ability, the capacity to survive climatic conditions, and the ability to find food and avoid being eaten.

Climate and other abiotic factors were the focus of this chapter; the next several chapters focus on the biotic aspects of ecology. How do interactions among organisms of the same species and interactions among different species affect geographic range and population size?

CHAPTER 50 REVIEW

For media, go to the study area at www.masteringbiology.com

Summary of Key Concepts

Ecology's goal is to explain the distribution and abundance of organisms. It is the branch of biology that provides a scientific foundation for conservation efforts.

- Biologists study ecology at four main levels: (1) organisms, (2) populations, (3) communities, and (4) ecosystems.
- Organismal ecology's goal is to understand how individuals interact with their physical surroundings and with other organisms.
- Population ecology focuses on how and why populations grow or decline.
- Community ecology is the study of how different species interact.
- Ecosystem ecology analyzes interactions between organisms and their abiotic environment—particularly the flow of energy and nutrients.
- To preserve endangered species and communities, biologists apply analytical methods and results from all four levels.

✓You should be able to explain why these four levels are hierarchical.

Physical structure—particularly water depth—is the primary factor that limits the distribution and abundance of aquatic species. Climate—specifically, both the average value and annual variation in temperature and in moisture—is the primary factor that limits the distribution and abundance of terrestrial species.

- Water depth is a crucial factor in aquatic ecosystems because the wavelengths of sunlight that are required for photosynthesis do not penetrate water efficiently. Thus, only certain zones in a lake or ocean can support photosynthesis, and organisms in other aquatic zones must depend on food that rains out of the sunlit areas near the surface.
- Distinct climatic regimes are associated with distinct types of terrestrial vegetation, or biomes. Biomes represent the major types of environments available to terrestrial organisms.
- Because photosynthesis is most efficient when temperatures are warm and water supplies are ample, the productivity and degree of seasonality in biomes varies with temperature and moisture.

✓You should be able to explain how global warming may affect terrestrial biomes due to (1) increased evaporation from the oceans, and (2) increased transpiration from plants.

Climate varies with latitude, elevation, and other factors—such as proximity to oceans and mountains. Climate is changing rapidly around the globe.

- An array of abiotic factors affect organisms, including the chemistry of water in aquatic habitats; the nature of soils in terrestrial habitats; and climate, which is a combination of temperature, moisture, sunlight, and wind.
- Climate varies around the globe because sunlight is distributed asymmetrically, with equatorial regions receiving on average more solar radiation than do regions toward the poles.
- Hadley cells create bands of wet and dry habitats, and the tilt of Earth's axis causes seasonality in the amount of sunlight that non-equatorial regions receive.
- The presence of mountains can create local areas of wet or dry habitats, and proximity to an ocean moderates temperatures in nearby terrestrial habitats.

- Global climate change is causing temperature profiles and precipitation patterns to change in ecosystems throughout the planet.

✔You should be able to explain how climate would be expected to vary around the Earth if the planet were cylindrical and spun on an axis that was at a right angle to the Sun.

(MB) **Web Activity** Tropical Atmospheric Circulation

A species' distribution is constrained by historical and biotic factors, as well as by abiotic factors such as physical structure and climate.

- The range, or geographic distribution, of a particular species depends on dispersal ability, the capacity to survive climatic conditions, and the ability to find food, combat diseases, and avoid being eaten.
- For each species, a unique combination of historical factors, abiotic factors, and biotic factors dictates where individuals live and the size of the population.

✔You should be able to predict what would happen to the range of cattle in Africa if the tsetse fly were eradicated.

Questions

✔TEST YOUR KNOWLEDGE

Answers are available in Appendix B

1. Where do rain shadows exist?
   a. the part of a mountain that receives prevailing winds and heavy rain
   b. the region beyond a mountain range that receives dry air masses
   c. the region along the equator where precipitation is abundant
   d. the region near 30° N and 30° S latitude that receives hot, dry air masses
2. A region receives less than 5 cm (2 in.) of precipitation annually and has temperatures that never drop below freezing. Which type of biome is present?
   a. subtropical desert
   b. temperate grassland
   c. boreal forest
   d. arctic tundra
3. What is the predominant type of vegetation in a tropical wet forest?
   a. shrubs and bunchgrasses
   b. herbs, grasses, and vines
   c. broad-leaved deciduous trees
   d. broad-leaved evergreen trees
4. The littoral zone of a lake is most similar to which of the following environments?
   a. oceanic zone
   b. intertidal zone
   c. neritic zone
   d. marsh
5. Typically, where are oxygen levels highest and nutrient levels lowest in a stream?
   a. near its source
   b. near its mouth, or end
   c. where it flows through a swamp or marsh
   d. where it forms an estuary
6. Why are certain exotic species considered "invasive"?
   a. They are found in areas where they are not native.
   b. They were introduced by humans—often accidentally.
   c. They spread aggressively and displace native species.
   d. They benefit from increased grazing and wildfires.

✔TEST YOUR UNDERSTANDING

Answers are available in Appendix B

1. Name the four main levels of study in ecology. Write a question about the ecology of humans that is relevant to each level.
2. In June, does the Northern Hemisphere receive more or less solar radiation than the equator does? Explain your answer.
3. Explain why the Australian Outback receives so little rainfall.
4. Contrast the productivities of the intertidal, neritic, and oceanic zones of marine environments. Explain why large differences in productivity exist.
5. Compare and contrast the biomes found at increasing elevation on a mountain with the biomes found at increasing latitude in Figure 50.10.
6. Explain why productivity is much lower in bogs than it is in marshes or lakes that receive the same amount of solar radiation.

✔APPLYING CONCEPTS TO NEW SITUATIONS

Answers are available in Appendix B

1. Mars has an even more pronounced tilt than Earth does. Does Mars experience seasons? Why or why not?
2. Mountaintops are closer to the Sun than the surrounding area is. Why are mountaintops cold?
3. The southwest corners of some Caribbean islands are extremely dry, even though other areas of the islands receive substantial rainfall. Explain this observation. Base your answer on a hypothesis (one that you propose) about the direction of the prevailing winds in this region of the world.
4. Edinburgh, Scotland, and Moscow, Russia, are at the same latitude, yet these two cities have very different climates—Moscow has much colder winters. Explain why.

This chapter is part of The Big Picture. See how on pages 1126–1127.

This male satin bowerbird (on the left) is decorating his display area with objects of his preferred color: blue. A female, on the right, is inspecting the bower. She may mate with the male if she finds him, and the display, attractive enough.

Behavioral Ecology 51

To biologists, **behavior** is action—specifically, the response to a stimulus. Pond-dwelling bacteria swim toward drops of blood that fall into the water. Time-lapse movies of growing sunflowers document the steady movement of their flowering heads throughout the day, as they turn to face the shifting Sun. Filaments of the fungus *Arthrobotrys* form loops, then release a molecule that attracts roundworms. When a roundworm touches a loop, the fungal filaments swell—ensnaring the worm. The fungus then grows into the worm's body and digests it.

Among animals, action may be frequent and even spectacular. Peregrine falcons reach speeds of up to 320 km/hr (200 mph) when they dive in pursuit of ducks or other flying prey. Some Arctic terns migrate over 32,000 km (20,000 miles) each year. Deep inside a hive, in complete darkness, honeybees communicate the position and nature of food sources a mile away—by "dancing."

Ecology is the study of how organisms interact with their physical and biological environments; behavioral biology is the study of how organisms respond to particular stimuli from those environments. The action in behavior is this response. What does an *Anolis* lizard do when it gets too hot? How does that same individual respond when a member of the same species approaches it? Or when a house cat approaches?

Although all organisms respond in some way to signals from their environment, most behavioral research is performed on vertebrates, arthropods, or mollusks. With their sophisticated nervous systems and skeletal-muscular systems, these animals can sense, process, and respond rapidly to a wide array of environmental stimuli.

51.1 An Introduction to Behavioral Biology

Behavior is a particularly synthetic field in biological science. To understand why animals and other organisms do what they do, researchers have to ask questions about

KEY CONCEPTS

- Biologists analyze behavior at the proximate and ultimate levels—the genetic and physiological mechanisms and how they affect fitness.
- Individuals can behave in a wide range of ways; which behavior occurs depends on current conditions.
- Foraging patterns may vary with genotype; foraging decisions maximize energy gain and minimize costs.
- Sexual behavior depends on surges of sex hormones; females choose mates that provide good alleles and needed resources.
- Animals navigate using an array of cues; the costs of migration can be offset by benefits in food availability.
- Animals communicate with movements, odors, or other stimuli; communication can be honest or deceitful.
- Individuals that behave altruistically are usually helping relatives or individuals that help them in return.

✔ When you see this checkmark, stop and test yourself. Answers are available in Appendix B.

genetics, hormonal signals, neural signaling, natural selection, evolutionary history, and ecological interactions.

To make sense of this diversity, it's helpful to recognize that behavioral ecologists ask questions and test hypotheses at two fundamental levels—proximate and ultimate. Let's explore what these two levels are about, then consider the question of whether animals make decisions about how they should respond to a particular situation.

Proximate and Ultimate Causation

Most behavioral studies start by carefully observing what animals do in response to specific problems or situations. As research progresses, investigators use experimental approaches to probe the proximate and ultimate causes of behavior.

- **Proximate** (or mechanistic) **causation** explains *how* actions occur in terms of the neurological, hormonal, and skeletal-muscular mechanisms involved.
- **Ultimate** (or evolutionary) **causation** explains *why* actions occur—based on their evolutionary consequences and history. Does a certain species perform its courtship display because its ancestors happened to behave that way? Is a particular behavior currently adaptive, meaning that it increases an individual's fitness—its ability to produce offspring? If so, how does it help individuals produce offspring in a particular environment?

To illustrate the proximate and ultimate levels of causation, consider the spiny lobster in **Figure 51.1**. These animals spend the day hiding in cracks or holes in coral reefs. At night they emerge and wander away from the reef in search of clams, mussels, crabs, or other sources of food. Before dawn, they return to one of several dens they use on a regular basis. How do they find their way back?

FIGURE 51.1 Spiny Lobsters Can Find Their Way Back to Their Home Burrow. Spiny lobsters wander widely in search of food at night, then return to their hiding places in coral reefs before dawn.

✔**QUESTION** Compared to other lobsters, spiny lobsters have extremely long antennae for their body size. Antennae are involved in sensing chemicals, water currents, and objects in the environment. Generate a hypothesis to explain how long antennae function at the proximate level, and why long antennae are advantageous at the ultimate level.

This question is proximate in nature—it asks how an individual does what it does. Although research is continuing, recent experiments support the hypothesis that spiny lobsters have receptors in their brains that detect changes in Earth's magnetic field, and that they navigate by using information on how the orientation and strength of the field changes as they move about.

At the ultimate level of causation, biologists want to know why the behavior occurs. The leading hypothesis to explain homing behavior is that the ability to navigate allows spiny lobsters to search for food over a wide area under cover of darkness, then return to a safe refuge before sharks and other large predators that hunt by sight can find them.

It's important to recognize that efforts to explain behavior at the proximate and ultimate levels are complementary. To understand what an organism is doing, biologists want to know how the behavior happens and why.

Behavioral biologists work at the interface of many fields and at both proximate and ultimate levels. And in many cases, their work is ecological in nature. Questions about behavior are almost always aimed at understanding how individuals cope with changing physical conditions or how they interact with individuals of their own or other species—in short, how they interact with their environment.

Conditional Strategies and Decision Making

In some cases, animals respond to a change in their environment in a simple, highly predictable way. When a kangaroo rat is feeding at night and hears the sound of a rattlesnake's rattle, it jumps back—away from the direction of the stimulus. The jump-back behavior is performed the same way every time.

INNATE, INFLEXIBLE BEHAVIOR IS RARE Highly inflexible, stereotyped behavior patterns like a kangaroo rat's jump-back are called **fixed action patterns**, or **FAPs**. FAPs are examples of what biologists call **innate behavior**. Innate behavior is inherited—meaning that it is passed from parents to offspring—and shows little variation based on learning or the individual's condition.

When researchers first began to explore animal behavior experimentally, they focused on these types of simple actions. This research strategy is common in biological science. If you are addressing a new question—whether it's how DNA is transcribed or how ions cross plasma membranes—it's helpful to start by studying the simplest possible system. Once simple systems are understood, investigators can delve into more complex situations and questions.

Behavioral biology started the same way. Early work focused on inflexible behavioral responses.

MOST BEHAVIOR IS FLEXIBLE AND CONDITION-DEPENDENT Even though virtually all species studied to date show some degree of innate behavior, it is much more common for an individual's behavior to (1) change in response to learning, and (2) show flexibility in response to changing environmental conditions. Organisms are not stimulus-response machines.

Most animals have a range of actions that they can perform in response to a situation. Animals take in information from the environment and, based on that information, make decisions about what to do. Animals make choices.

If sharks and other predators are particularly abundant, spiny lobsters are likely to spend less time looking for food and more time hiding in a protected refuge. If predators are rare or absent, spiny lobsters might even begin to emerge and look for food during the day. Biologists characterize behavior like this as condition-dependent.

To link condition-dependent behavior to fitness—and thus the ultimate level of explanation—biologists use a framework called **cost-benefit analysis**. Animals appear to weigh the costs and benefits of responding to a particular situation in various ways. Costs and benefits are measured in terms of their impact on fitness—the ability to produce offspring.

Note the phrase "appear to weigh" in the previous paragraph. It's important to recognize that the decisions made by non-human organisms are not—as far as is known—conscious. A spiny lobster does not sit in its hole, counting predators and calculating the likelihood of being eaten. Instead, evolution has shaped its genome to direct a nervous and endocrine system that is capable of (1) taking in information and (2) directing behavior that is likely to pass that genome on to the next generation.

Five Questions in Behavioral Ecology

The following five sections of this chapter introduce some of the most prominent questions in behavioral ecology—the study of how individuals interact with their environment. In each case, the question is posed in terms of a decision that an animal has to make. The ensuing discussion is divided into research on how the question has been answered at the proximate level and the ultimate level.

The goals of this chapter are to explore some key topics in behavioral ecology and illustrate the spirit and synthetic nature of the field. Let's begin with an introduction to research on a resource that every animal has to find in its environment: food.

51.2 What Should I Eat?

When animals seek food, they are **foraging**. In some cases, the food source that a species exploits is extremely limited. Giant pandas, for example, subsist almost entirely on the shoots and leaves of bamboo. But individuals must still make decisions about which bamboo plants to harvest and which parts of a particular plant to eat and which to ignore.

In most cases, animals have a relatively wide range of foods that they exploit over the course of their lifetime. To understand why individuals eat the way they do, let's consider (1) research on genetic variation in foraging behavior and (2) the costs and benefits of feeding at various distances from home.

Foraging Alleles in *Drosophila melanogaster*

When Marla Sokolowski was working as an undergraduate research assistant, she noticed that some of the fruit-fly larvae she was studying tended to move after feeding at a particular location, while others tended to stay put. She reared these "rovers" and "sitters" to adulthood, bred them, and found that the offspring of rovers also tended to be rovers, while offspring of sitters tended to be sitters (**Figure 51.2**). Thus, this aspect of behavior is inherited. In fruit flies, variation in larval foraging behavior is at least partly due to genetic variation among individuals.

Using information from the *Drosophila melanogaster* genome sequence (see Chapter 20) and genetic mapping techniques introduced in Chapter 19, Sokolowski found and cloned a gene associated with the rover-sitter difference. She named it *foraging* (*for*).

Follow-up work showed that the protein product of *for* is involved in a signal transduction cascade (see Chapters 8 and 47)—meaning it is involved in the response to a cell-cell signal. Rovers and sitters tend to behave differently when they are foraging because they have different alleles of the *for* gene.

Genes that are homologous to *for* have now been found in honeybees and the roundworm *Caenorhabditis elegans*. Currently, researchers are focused on finding out which signaling pathway uses the *for* gene product and why different alleles change feeding behavior. Presumably, the signaling pathway is associated with feeding or nervous system development. When this aspect of fly feeding is thoroughly understood at the proximate level, biologists will know why variants of the same protein make larvae more likely to move or stay in place after feeding.

By altering population density experimentally and documenting the reproductive success of rovers and sitters, Sokolowski has shown that the rover allele is favored when population density is high and food is in short supply. In contrast, the sitter allele reaches high frequency in low-density populations, where food is abundant. Rovers do better at high density because they are more likely to find unused patches of food; sitters do better at low density because they don't waste energy moving around.

These results are exciting because they link a proximate mechanism with an ultimate outcome. In this case, the presence of certain alleles is responsible for a difference in fitness in specific types of habitats.

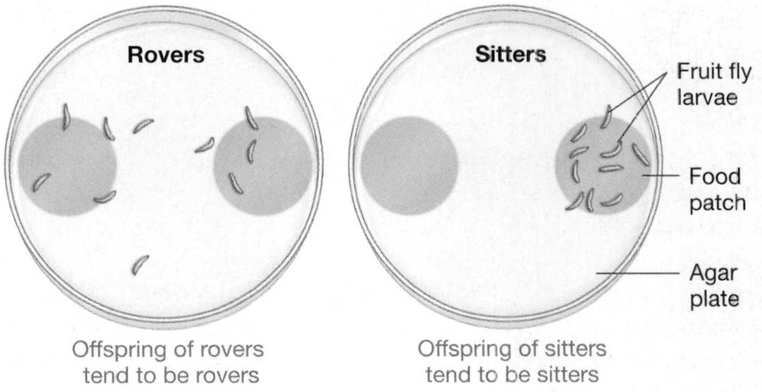

Rovers
Sitters
Fruit fly larvae
Food patch
Agar plate
Offspring of rovers tend to be rovers
Offspring of sitters tend to be sitters

FIGURE 51.2 Foraging Behavior in Fruit-Fly Larvae Is Heritable.

Optimal foraging in White-Fronted Bee-Eaters

In *Drosophila* larvae, foraging is a relatively inflexible behavior—individuals don't "decide" to be rovers or sitters. But in *Drosophila* adults and in most animal species, foraging behavior is much more sophisticated.

When biologists set out to study why animals forage in a particular way, they usually start by assuming that individuals make decisions that maximize the amount of usable energy they take in, given the costs of finding and ingesting their food and the risk of being eaten while they're at it. This claim—that animals maximize their feeding efficiency—is called **optimal foraging**.

To test whether optimal foraging actually occurs, a researcher began studying feeding behavior in a bird called the white-fronted bee-eater (**Figure 51.3a**). Mated pairs of this species dig tunnels in riverbanks and raise their young inside. Appropriate riverbanks are rare, so many pairs build tunnels in the same location, forming a colony. To feed their young, pairs have to leave the colony, capture insects, and bring the prey back to the nest. Each pair defends a specific area where only they find food.

The key observation here is that some individuals forage a few meters from the colony, while some have to look for food hundreds of meters away. It takes only a few seconds to make a round trip to the closest feeding territories, versus several minutes to the farthest territories. Thus, the fitness cost of each feeding trip varies widely among individuals.

If optimal foraging occurs, how do individuals maximize their benefits, given the cost they have to pay in time and energy? Individuals that have to find food far from the colony stay away longer on each trip. And as the data in **Figure 51.3b** show, individuals foraging far from the colony bring back a much larger mass of insects on each trip, on average, than do individuals that fly a short distance each time.

White-fronted bee-eaters appear to make decisions that maximize the energy they deliver to their offspring, given their costs in finding food. This behavior is highly flexible and condition dependent. If the same individual had a territory close to the colony instead of far away, it would make different decisions. Its behavior would change, but in each case it would be acting in a way that maximizes its fitness.

CHECK YOUR UNDERSTANDING

If you understand that . . .

- In *Drosophila* larvae, foraging behavior depends in large part on variation at a single gene. The rover allele is favored at high population density; the sitter allele is favored at low density. In this case, biologists understand both the proximate and ultimate mechanisms responsible for variation in foraging behavior.
- In white-fronted bee-eaters, foraging behavior appears to minimize fitness costs and maximize fitness benefits.

✔ **You should be able to . . .**

Suppose that a *for*-like gene with rover and sitter alleles existed in white-fronted bee-eaters (with rover alleles making individuals more likely to wander far in search of food), and explain which allele you would expect to be common in small colonies versus large colonies.

Answers are available in Appendix B.

51.3 Who Should I Mate With?

Biologists sometimes claim that almost all animal behavior revolves around food or sex. Section 51.2 focused on food; here the subject is sex.

Let's start by considering the proximate mechanisms responsible for triggering sexual activity in lizards, and then go on to explore research on how a female barn swallow chooses the indi-

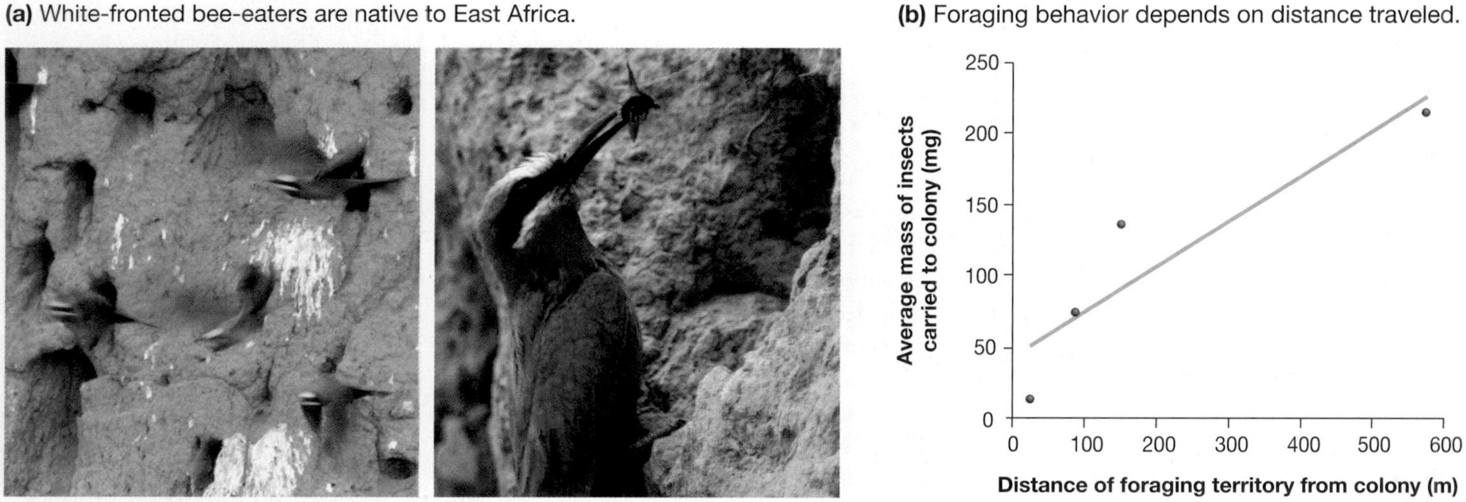

FIGURE 51.3 White-Fronted Bee-Eaters Vary Their Foraging Behavior Depending on Conditions.

vidual she is going to mate with. Biologists want to know how sexual activity occurs, in terms of underlying hormonal mechanisms, and why variation in mate choice affects fitness.

Sexual Activity in *Anolis* Lizards

Anolis carolinensis lives in the woodlands of the southeastern United States. After spending the winter under a log or rock, males emerge in January and establish breeding territories (**Figure 51.4a**). Females become active a month later, and the breeding season begins in April. By May, females are laying an egg every 10–14 days. By the time the breeding season is complete three months later, the eggs produced by a female will total twice her body mass.

Figure 51.4b graphs how sexual condition changes throughout the year, by plotting the size of the testes in males and ovaries of females—relative to an individual's overall body size. Individuals come into breeding condition in spring, with males experiencing growth of gonads long before females do. In both sexes, the gonads shrink in size during the fall and winter.

TESTOSTERONE AND ESTRADIOL What causes these dramatic seasonal changes in behavior? The proximate answer is sex hormones—testosterone in males and estradiol in females. Testosterone is produced in the testes of males, and estradiol in the ovaries of females (see Chapter 48).

The evidence for effects of sex hormones is direct. Testosterone injections induce courtship activity in castrated males; estradiol injections induce sexual activity in females whose ovaries have been removed.

This result leaves some key questions unanswered, however. *Anolis* lizards are most successful if they reproduce early in the spring. In springtime their food supplies are increasing, and snakes and other predators are not yet hunting lizards to feed their own young. If testosterone and estradiol levels induced sexual activity in *Anolis* lizards at the wrong time of year, it would be a disaster for the individuals involved.

In addition, it is critical for all *Anolis* individuals in a population to start their sexual activity at the same time, so that females can find males that are ready to breed, and vice versa. What environmental cues trigger the production of sex hormones in early spring? How do male and female *Anolis* synchronize their sexual readiness?

TESTING THE EFFECTS OF LIGHT AND SOCIAL STIMULATION To answer these questions, David Crews brought a large group of sexually inactive adult lizards into the laboratory during the winter and divided them into five treatment groups. The physical environment was exactly the same in all treatments. Each individual received identical food, and in all treatments high "daytime" temperatures were followed by slightly lower "nighttime" settings. Crews also continued to monitor the condition of lizards that remained in natural habitats nearby.

To test the hypothesis that changes in day length signal the arrival of spring and trigger initial changes in sex hormones, Crews exposed the five treatment groups in the laboratory to artificial lighting that simulated the long days and short nights of spring.

To test the hypothesis that social interactions among individuals are responsible for synchronizing sexual behavior, he varied the social setting among the five treatment groups:

1. a single isolated female;
2. a group of females;
3. a single female with a single male;
4. a single female with a group of castrated (nonbreeding) males; and
5. a single female with a group of uncastrated (breeding) males.

Each week, Crews examined the ovaries of females in each treatment. He also monitored the ovaries of females in nearby natural habitats. These females were not exposed to springlike conditions.

As **Figure 51.5** on page 1024 shows, the differences in the animals' reproductive systems were dramatic. The data points on this graph represent the average percentage of females with mature egg-producing structures, in each treatment; the points are connected by lines to help you keep track of the different treatments and see the overall patterns. Note two key points:

(a) Courtship display of a male *Anolis* lizard

(b) Changes in sexual organs through the year

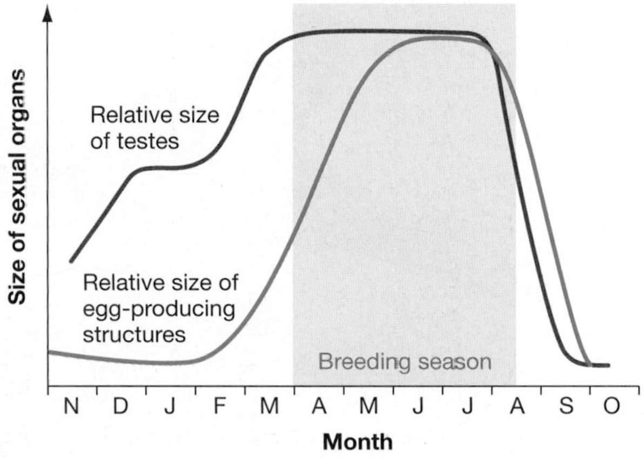

FIGURE 51.4 Sexual Behavior in *Anolis* Lizards Is Seasonal.

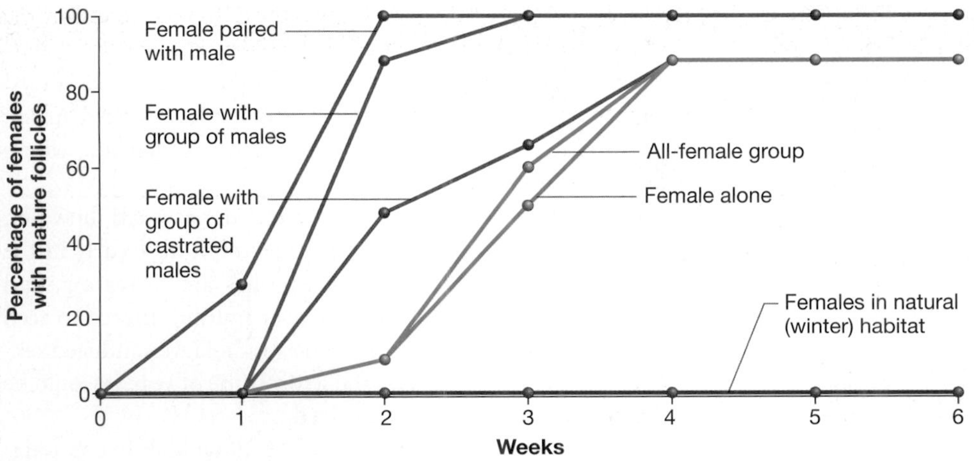

FIGURE 51.5 Exposure to Springlike Conditions and Breeding Males Stimulates Female Lizards to Produce Eggs. The percentage of female lizards with mature follicles, plotted over time. Five treatment groups were exposed to springlike light and temperature in the lab; data are also plotted for females left in a natural (winter) habitat. Each data point represents the average value for a group of 6–10 females.

✔**QUESTION** Some critics contend that the females that remained outside do not represent a legitimate control in this experiment. What conditions would represent a better control?

1. Females that were exposed to springlike conditions began producing eggs; females in the field that were not exposed to springlike conditions did not.
2. Females that were exposed to breeding males began producing eggs much earlier than did the females placed in the other treatment groups.

These results support the hypothesis that two types of stimulation are necessary to produce the hormonal changes that lead to sexual behavior. Females need to experience springlike light and temperatures *and* exposure to breeding males.

VISUAL CUES FROM MALES TRIGGER FEMALE READINESS What is it about breeding males that triggers earlier estradiol release in females? What male signal causes the difference in female egg production?

When males court females, they bob up and down and extend a brightly colored patch of skin called a dewlap (Figure 51.4a). To test the hypothesis that this visual stimulation triggers changes in estradiol production, Crews repeated the previous experiment but added a twist: He placed some females with males that had intact dewlaps, and other females with males whose dewlaps had been surgically removed.

The result? Females who were living with dewlap-less males were slow to produce eggs. They were just as slow, in fact, as the females in the first experiment that had been grouped with castrated males. These latter females had not been courted at all.

These data suggest that the dewlap is a key visual signal. The experiments succeeded in identifying the environmental cues—long day length and bobbing dewlaps—that trigger hormone production and the onset of sexual behavior.

How Do Female Barn Swallows Choose Mates?

Given a choice, who should an individual choose for a mate? According to theory and data presented in Chapter 25:

- Females are usually the gender that is pickiest about mate choice.
- Females choose males that contribute good alleles and/or resources to their offspring.

To see this type of decision making in action, consider an experiment on mate choice by female barn swallows native to Denmark. Although both male and female barn swallows help to build the nest and feed the young, there is a significant amount of sexual dimorphism—males are slightly larger and more brightly colored, and their outer tail feathers are about 15 percent longer than the same feathers in females (**Figure 51.6**).

As Chapter 25 indicated, research on other bird species has shown that individuals with particularly long tails, bright colors, and energetic courtship displays are usually the healthiest and best fed in the population. Because it takes large amounts of energy and resources to produce traits like these, they serve as an honest signal of male quality.

Recent work has shown that barn swallows with particularly long tails are more efficient in flight and thus more successful in finding food. Tail length is a heritable trait—meaning that certain alleles affect tail length. If a female barn swallow mates with a long-tailed male, then, she should be picking an individual that will pass high-fitness alleles on to her offspring *and* be able to help her rear those offspring efficiently.

A researcher tested this prediction by altering tail length experimentally—in this case, by cutting off tail feathers and re-gluing

Male barn swallows have longer tail feathers and slightly brighter colors than females do

FIGURE 51.6 Male Barn Swallows Have Elongated Tail Feathers.

EXPERIMENT

QUESTION: Does tail length in barn swallows affect female choice of mates?

HYPOTHESIS: Females prefer to mate with the longest-tailed males.

NULL HYPOTHESIS: Tail length in males has no influence on female choice of mates.

EXPERIMENTAL SETUP:

Capture many males at the same site and randomly assign them to 1 of 4 groups:

Short tail: Outermost tail feathers cut

Control 1: Outermost tail feathers cut and glued back in place

Control 2: Males caught and handled, but no change in tail feathers

Elongated tail: Outermost feathers cut and elongated with feathers from other males, glued in place

Release males and document which males are first to mate, have a second nest that season, and raise the most offspring.

RESULTS:

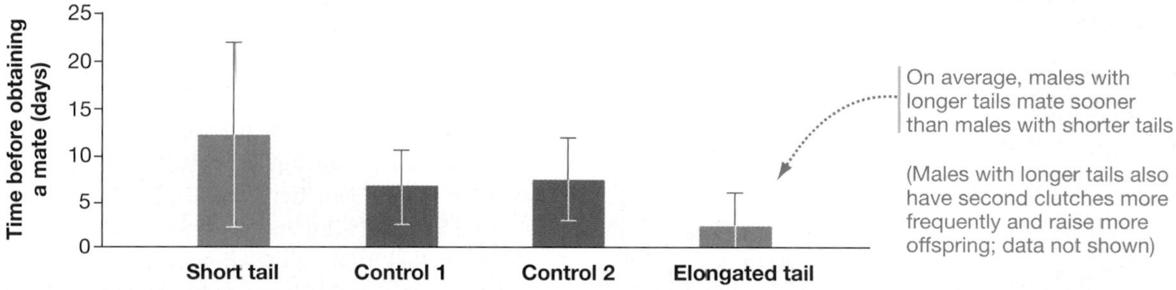
Time before obtaining a mate (days)
25
20
15
10
5
0
Short tail
Control 1
Control 2
Elongated tail
On average, males with longer tails mate sooner than males with shorter tails
(Males with longer tails also have second clutches more frequently and raise more offspring; data not shown)

CONCLUSION: Females prefer to mate with long-tailed males.

FIGURE 51.7 Barn Swallow Females Prefer to Mate with Long-Tailed Males.

SOURCE: Møller, A. P. 1988. Female choice selects for male sexual tail ornaments in the monogamous swallow. *Nature* 332: 640–642.

✔**QUESTION** Why did the investigator bother to artificially lengthen and shorten tails of males? Why didn't he just record natural variation in tail length and observe mating success?

them. The operation is painless and does not appear to affect flying ability.

As the "Experimental setup" section of **Figure 51.7** shows, the biologist caught a large series of males and randomly assigned them to one of four groups: either shortened tails, "sham-operated" unchanged tails, untouched tails, or elongated tails. When he documented how long it took males in each group to find a mate, a strong trend emerged: As the bar graphs in the "Results" section of Figure 51.7 indicate, males with elongated tails mated earliest on average—meaning that females preferred them.

This finding is important. Pairs that nest early have time to complete a second nest later in the season and thus produce the most offspring. Longer-tailed males have higher fitness than shorter-tailed males.

Data on mate choice in barn swallows reinforce a central theme of this chapter: Animals usually make decisions in a way that maximizes their fitness.

CHECK YOUR UNDERSTANDING

If you understand that . . .

- In *Anolis* lizards, springlike light conditions and male dewlap displays cause estradiol release and sexual readiness in females.
- In barn swallows, females that prefer to mate with long-tailed males have highest fitness.

✔ **You should be able to . . .**

1. Explain why it is valid to claim that *both* springlike light conditions and male dewlap displays are required to bring female *Anolis* into optimal breeding condition.
2. Generate a hypothesis to explain the proximate mechanism responsible for bringing female barn swallows into breeding condition.

Answers are available in Appendix B.

51.4 Where Should I Live?

Most animals select the food they eat and the individuals they mate with. In most cases, they also select the habitat where they live.

Biologists have explored an array of questions related to habitat selection:

- whether juveniles should disperse from the area where they were raised or whether they should stay;
- how large an area should be defended against competitors in species that maintain a **territory** for breeding or feeding; and
- whether individuals have higher fitness when they occupy crowded, high-quality habitats or uncrowded, poor-quality habitats.

Here there is space to address just one question related to these types of "lifestyle" decisions: the proximate and ultimate mechanisms responsible for migration. In ecology, **migration** is defined as the long-distance movement of a population associated with a change of seasons. At the proximate level, how do animals know where to go? At the ultimate level, why is it worth the cost in time, energy, and risk of predation or accident?

How Do Animals Find Their Way on Migration?

To organize research into how animals find their way during migratory movements, biologists distinguish three categories of navigation:

1. **Piloting** is the use of familiar landmarks.
2. **Compass orientation** is movement that is oriented in a specific direction.
3. **True navigation** is the ability to locate a specific place on Earth's surface.

A thought experiment will help you understand these categories.

Suppose you grew up on the shores of Lake Superior but had never seen a map of the region. One day, you are transported to the middle of this enormous lake. Because you were blindfolded and made to spin in place en route, you are disoriented and have no idea where you are. But you are given a magnetic compass.

The compass tells you where north, south, east, and west are. Unfortunately this information will not help you find home, because you have no idea where you are in relation to home. You are stuck because you have no mechanism of true navigation.

Now suppose that a helicopter flying overhead drops you a map. The map has an X marking your current position and an H marking your home. The map also has a scale and a compass symbol. From this information, you determine that home is 100 miles to the west.

The map has solved your navigational problem. The compass has solved your orientation problem. Now that you know you live to the west, you can use the compass to find west and paddle home. As you approach the shore, you recognize familiar landmarks. Using them to pilot gets you the rest of the way home.

Organisms other than humans do not have manufactured compasses or printed maps. How do they navigate? Although biologists know little about the nature of a map sense (true navigation) in animals, piloting and compass orientation are increasingly well understood.

PILOTING A substantial amount of evidence suggests that at least some migratory animals use piloting to find their way.

For example, offspring follow their parents—south in the fall and north in the spring—in some species of migratory birds and mammals. Young appear to memorize the route.

Piloting even plays a role in species with more sophisticated navigational abilities. Homing pigeons, for example, can find their way home even when they are released in strange terrain a long distance away. But if these birds are equipped with frosted spectacles so that they see the world as a foggy haze, they only return to within a mile or so of their home—they do not actually find it. Based on these experiments, researchers have concluded that homing pigeons use piloting to navigate the final stages of a journey home.

COMPASS ORIENTATION How do animals perform compass orientation? To date, most research on compass orientation has been done on migratory birds, including whooping cranes (**Figure 51.8**). To determine where north is, these animals appear to use the Sun during the day and the stars at night. Let's consider each type of compass orientation separately, because they work differently.

The Sun is difficult to use as a compass reference because its position changes during the day. It rises in the east, is due south at noon (in the Northern Hemisphere), and sets in the west. To use the Sun as a compass reference, then, an animal must have an internal clock that defines morning, noon, and evening. Fortunately, most animals have such a clock. The **circadian clock** that exists in organisms maintains a 24-hour rhythm of chemical activity. The clock is set by the light–dark transitions of day and night. It tells individuals enough about the time of day that they can use the Sun's position to find magnetic north.

FIGURE 51.8 An Ultralight Guides Young Whooping Cranes to Winter Nesting Grounds. These human-reared birds will use compass orientation to find their own way back north again. Successful migration is key to increasing the number of these endangered birds in the wild.

The situation is actually simpler on clear nights, because migratory birds in the Northern Hemisphere can use the North Star to find magnetic north and select a direction for migration. But what if the weather is cloudy and neither the Sun nor stars are visible?

Under these conditions, migratory birds appear to orient using Earth's magnetic field. Exactly how they detect magnetism is under debate. One hypothesis contends that the birds can detect magnetism by their visual system, through an unknown molecular mechanism. An alternative hypothesis maintains that individuals have small particles of magnetic iron—the mineral called magnetite—in their bodies. Changes in the positions of magnetic particles, in response to Earth's magnetic field, could then be detected and provide reliable information for compass orientation.

Although research on mechanisms of compass orientation continues, one important point is clear: Birds and perhaps other organisms have multiple mechanisms of finding a compass direction. At least some species can use a Sun compass, a star compass, and a magnetic compass. Which system an individual uses depends on the weather and other circumstances.

To analyze another set of experiments on how animals find their way, go to the study area at *www.masteringbiology.com.*

Web Activity Homing Behavior in Digger Wasps

Why Do Animals Move with a Change of Seasons?

In their extent and the navigation challenge they pose, migratory movements can be spectacular:

- Arctic terns nest in tundras in the far north. One population flies south along the coast of Africa to wintering grounds off Antarctica, and then flies back north along the eastern coast of South America. Chicks follow their parents south but make the return trip on their own. The journey totals over 32,000 km (20,000 miles).
- Many of the monarch butterflies native to North America spend the winter in the mountains of central Mexico or southwest California. Tagged individuals are known to have flown over 3000 km (1870 miles) to get to a wintering area. In spring, monarchs begin the return trip north. They do not live to complete the trip, however. Instead, mating takes place along the way, and females lay eggs en route. Although adults die on the journey, offspring continue to head north. After several generational cycles of reproduction and death in the original habitats in northern North America have passed, individuals again migrate south to overwinter. Instead of a single generation making the entire migratory cycle, then, the round trip takes several generations.
- Salmon that hatch in rivers along the Pacific Coast of North America and northern Asia migrate to the ocean when they are a few months to several years old, depending on the species. After spending several years feeding and growing in the North Pacific Ocean, they return to the stream where they hatched. There they mate and die.

In most cases, it is relatively straightforward to generate hypotheses for why migration exists at the ultimate level.

- Arctic terns feed on fish that are available in different parts of the world at different seasons. Thus, individuals that do not migrate might starve; individuals that do migrate should achieve higher reproductive success.
- Monarch butterflies may achieve higher reproductive success by migrating to wintering areas and new breeding areas than by trying to overwinter in northern North America.
- Salmon eggs and young are safer and thrive better in freshwater habitats than they do in the ocean, but adult salmon can find much more food in saltwater habitats than in freshwater and thus grow to bigger size, which confers a benefit in competition for nest sites and mates.

Testing these hypotheses rigorously is more difficult, as researchers have to be able to compare the fitness of individuals that do and do not migrate. Recently, however, researchers have documented the evolution of sockeye salmon populations that do not migrate to the ocean—in watersheds where migratory sockeye were introduced by humans. The nonmigratory populations, called kokanee, are much smaller-bodied than the migratory populations—presumably because there is less food available in the freshwater habitats.

The fitness of migrant and nonmigrant sockeye salmon appears be similar in this case, due to fitness trade-offs (see Chapter 24). In salmon, there is a large benefit of large body size in competing for nesting sites and mates. Increased access to food is a benefit of migration. There is a high cost, however, in time, energy, and predation risk.

At the ultimate level, researchers are also struggling to understand how a nonmigratory population can evolve into a migratory one that travels to a completely different part of the world. It's conceivable that habitat changes occurring in response to global climate change (see Chapter 54) may eventually furnish an example. Changes in the range and arrival times of migratory species have already been documented; it may only be a matter of time before a nonmigratory population begins to migrate, to cope with habitat changes caused by global warming.

51.5 How Should I Communicate?

The song of a bird, the bobbing dewlap of a male *Anolis carolinensis*, and the sentences in this text all have the same overall goal: communication. In biology, **communication** is defined as any process in which a signal from one individual modifies the behavior of a recipient individual. A **signal** is any information-containing behavior or characteristic. Communication is a crucial component of animal behavior. It creates a stimulus that elicits a response.

By definition, communication is a social process. For communication to occur, it is not enough that a signal is sent; the signal must be received and acted on. A lizard's bobbing dewlap does not qualify as communication unless another individual sees it and responds to the message.

Honeybee Language

Honeybees are highly social animals that live in hives. Inside the hive, a queen bee lays eggs that are cared for by individuals called workers. Besides caring for young and building and maintaining the hive, workers obtain food for themselves and other members of the colony by gathering nectar and pollen from flowering plants.

Biologists noticed that bees appear to recruit to food sources, meaning that if a new source is discovered by one or a few individuals, many more bees begin showing up over time. This observation inspired the hypothesis that successful food-finders have some way of communicating the location of food to other individuals.

At the ultimate level, it's straightforward to come up with a hypothesis for why bees communicate about food sources. If more workers find a rich food source quickly, the total amount of resources harvested, and the number of offspring produced by the hive, should increase. But at the proximate level, the task is more difficult. How do bees "talk" to each other?

THE DANCE HYPOTHESIS In the 1930s Karl von Frisch began studying bee communication by observing bees that built hives inside the glass-walled chambers he had constructed. He found that if he placed a feeder containing sugar water near one of these observation hives, a few of the workers began moving in a circular pattern on the vertical, interior walls of the hive.

Von Frisch called these movements the "round dance" (**Figure 51.9a**). Other bee workers appeared to follow the progress of the dance by touching the displaying individual as it danced, and to respond by flying away from the hive in search of the food source.

To investigate the function of these movements further, von Frisch placed feeders containing sugar water at progressively greater distances from the hive. Using this technique, he was able to get bees to visit feeders at a distance of several kilometers from the hive. By catching bees at the feeders and dabbing them with paint, he could individually mark successful food-finders. Follow-up observations at the hive confirmed that marked foragers (1) danced when they returned to the hive, and (2) returned to the food source with greater and greater numbers of unmarked bees.

To explain these data, von Frisch proposed that the round dance contained information about the location of food. Because bee hives are completely dark, he hypothesized that workers got information from the dance by touching the dancer and following the dancer's movements.

COMMUNICATING DIRECTIONS AND DISTANCES When von Frisch began placing feeders at longer distances from the hive, he found that successful food-finders used a new type of display. He named these movements the "waggle dance," because they combined circular movements like those of the round dance with short, straight runs (**Figure 51.9b**). During these runs, the dancer vigorously moved her abdomen from side to side.

These observations supported the hypothesis that both the round dance and waggle dance communicate information about food sources. Recent work has shown that the round and waggle dances are actually the same type of behavior—round dances just have a short waggle phase.

Three key observations allowed von Frisch to push our understanding of bee language result further. He noticed that:

- The orientation of the waggle part of the dance varied.
- The direction of the waggle run correlated with the direction of the food source from the hive.
- The length of the straight, "waggling" run was proportional to the distance the foragers had to fly to reach the feeder.

By varying the location of the food source and observing the orientation of the waggle dance given by marked workers, von Frisch was able to confirm that dancing bees were communicating the position of the food relative to the current position of the Sun. For example:

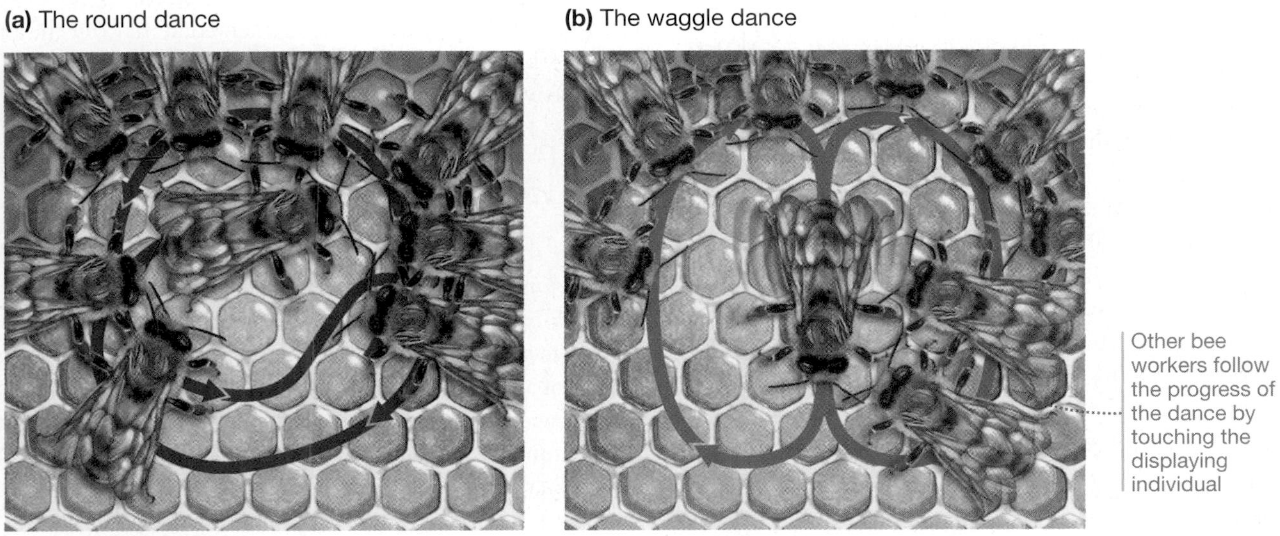
(a) The round dance
(b) The waggle dance
Other bee workers follow the progress of the dance by touching the displaying individual

FIGURE 51.9 Honeybees Perform Two Types of Dances. (a) During the round dance, successful food-finders move in a circle. **(b)** During the waggle dance, successful foragers move in a circle but then make straight runs through the circle. During the straight part of the dance, the dancer waggles her abdomen.

- If food is directly away from the Sun's current position, marked bees give the waggle portion of their dance directly downward (**Figure 51.10a**).
- If the food is 90 degrees to the right of the Sun, marked bees waggle 90 degrees to the right of vertical (**Figure 51.10b**).

These results are nothing short of astonishing. Honeybees do not have large brains, yet they are capable of symbolic language. What's more, they are able to interpret the angle of the waggle dance performed on a vertical surface and to respond by flying horizontally along the corresponding angle.

Further work has confirmed that the dance language of bees includes several modes of communication. In addition to the tactile information in the movements themselves, bees make sounds during the dance and give off scents that indicate the nature of the food source.

Modes of Communication

As the honeybee example shows, animals can use several modes of communication at once. And in addition to relying on tactile and olfactory signals, communication can be

- acoustic, as in the song of crickets or birds; or
- visual, as in the color patterns of birds or fish.

At the ultimate level, which mode of communication confers highest fitness?

One of the most general observations about communication is that the type of signal used by an organism correlates with its habitat. For example, sound travels much farther than light in aquatic habitats. Based on this observation, it is logical to observe that humpback whales rely on songs for long-distance communication. In some cases, humpback whale songs can travel hundreds of kilometers. Groups of whales use acoustic communication to keep together as they move from summer feeding areas to winter breeding grounds, and individual males sing to attract mates and warn rivals.

Correlations between habitat and mode of communication are also observed in visual, olfactory, and tactile communication. Animals that are active during the day and that live in open or treeless habitats tend to rely on visual communication during courtship and territorial displays. Bats, wolves, and other animals that are active at night communicate via sound or scent. Ants and termites that live underground rely on olfactory and tactile communication.

Each mode of communication has advantages and disadvantages. Songs and calls can carry information over long distances, but are short lived. Thus, they have to be repeated to be effective, and frequent repetition requires a large expenditure of time and energy. In addition, acoustic communication attracts predators. It is no surprise that when a hawk or falcon approaches a marsh inhabited by courting red-winged blackbirds, things get very quiet. Communication systems have been honed by natural selection to maximize their benefits and minimize their costs.

When Is Communication Honest or Deceitful?

At the ultimate level, one of the questions that biologists ask about communication concerns the quality of the information. Is the signal reliable?

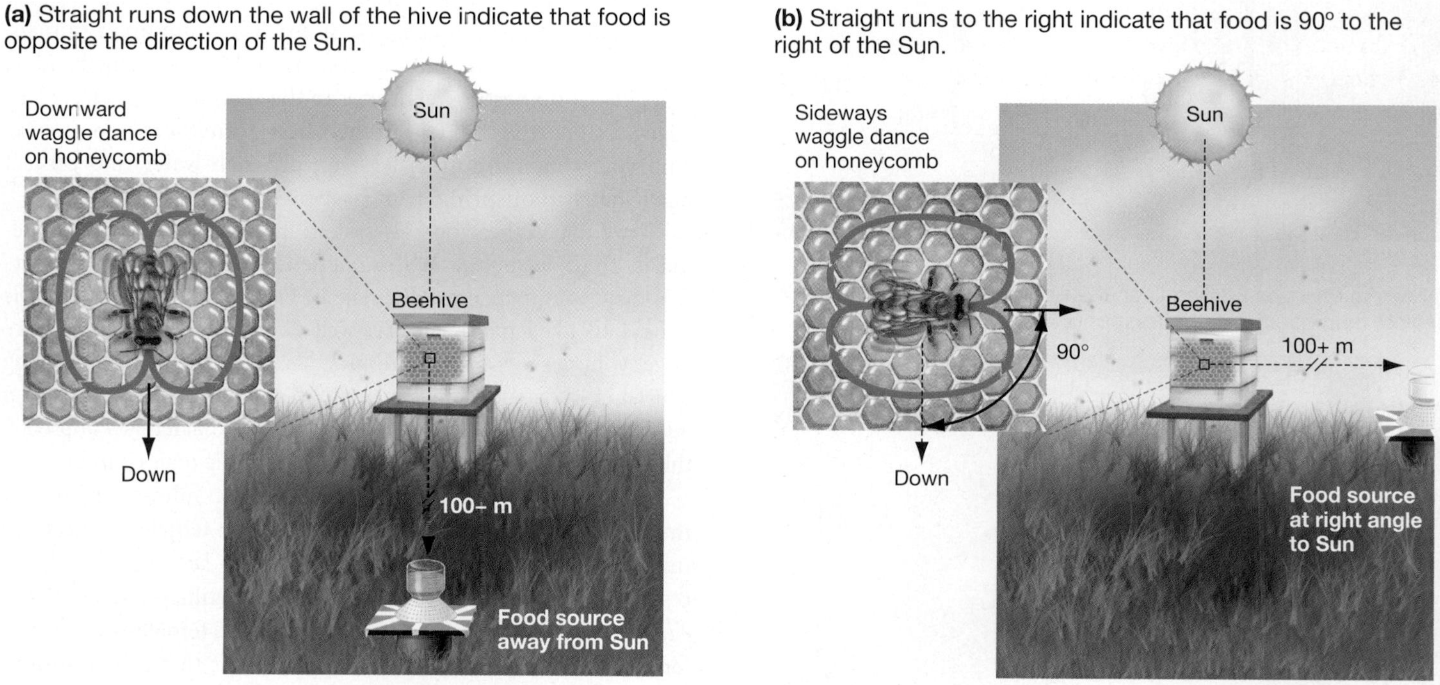

(a) Straight runs down the wall of the hive indicate that food is opposite the direction of the Sun.
Downward waggle dance on honeycomb
Sun
Beehive
Down
100+ m
Food source away from Sun
(b) Straight runs to the right indicate that food is 90° to the right of the Sun.
Sideways waggle dance on honeycomb
Sun
Beehive
90°
100+ m
Down
Food source at right angle to Sun

FIGURE 51.10 The Direction of the Waggle Dance Indicates the Location of Food Relative to the Sun. Waggle dances are done only when the food is at least 100 m from the hive.

✔**EXERCISE** The length of the straight run in the waggle dance indicates the relative distance of the food. Diagram a waggle dance that indicates food in the direction of the Sun and twice as far away as the food sources indicated by the dances drawn here.

As far as is known, honeybees are always honest. Stated another way, honeybee dances consistently "tell the truth" about the location of food sources. This observation is logical, because the honeybees that occupy a hive are closely related and cooperate closely in the rearing of offspring. As a result, it is advantageous for an individual to convey information accurately. If a food-finder provided inaccurate or misleading information to its hivemates, fewer offspring would be reared and the food-finder's fitness would be reduced.

In many cases, however, natural selection has favored the evolution of deceitful communication. Recent research on deceitful communication has highlighted just how complex interactions between signalers and receivers can be. A few examples will help drive this point home.

DECEIVING INDIVIDUALS OF ANOTHER SPECIES **Figure 51.11** illustrates two of the hundreds of examples of deceitful communication that have been documented among members of different species. Many of the best-studied instances involve predation:

- The anglerfish in Figure 51.11a has an appendage that dangles near its mouth and looks remarkably like a minnow. If another fish approaches this "lure" and attempts to eat it, the anglerfish attacks.
- Male and female fireflies flash a species-specific signal to each other during courtship. Predatory *Photuris* fireflies can mimic the pattern of flashes given by females of several other species. A *Photuris* female attracts a male of different species with the appropriate set of flashes, and then attacks and eats him (Figure 51.11b).

(a) Anglerfish use a "lure" to attract prey.

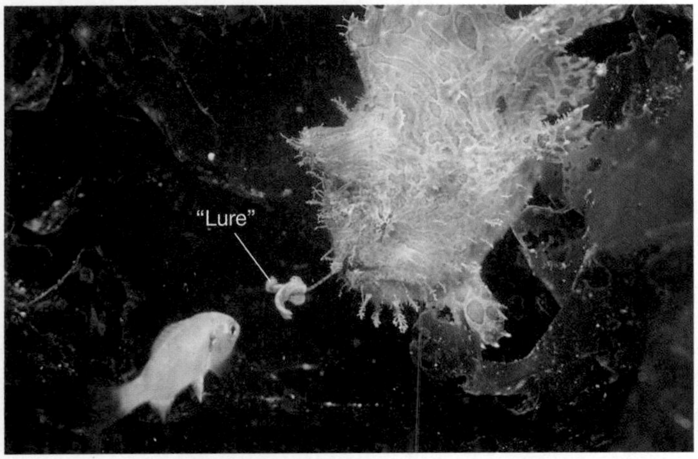

(b) Female *Photuris* fireflies flash the courtship signal of another species, then eat males that respond.

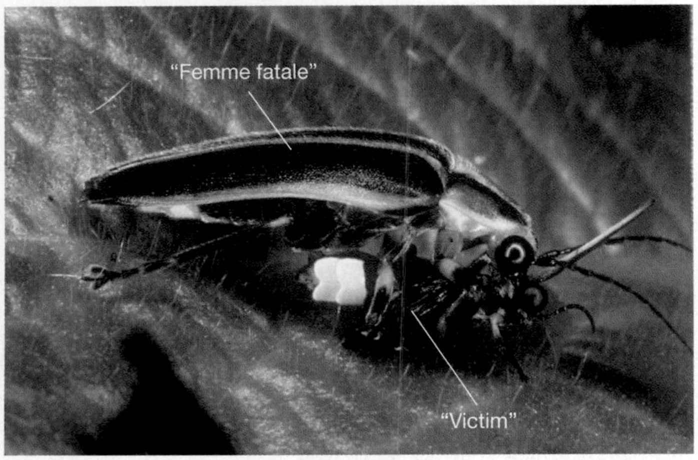

FIGURE 51.11 Deceitful Communication Is Common in Nature.

In each of these examples, individuals increase their fitness by providing inaccurate or misleading information to members of a different species. Deceitful communication is also known in plants, where it functions in pollination. For example, several orchid species have flowers that look and smell like female wasps and that accomplish pollination by "fooling" male wasps into attempting copulation.

In all of these cases, though, it's important to realize that the signaler is not consciously "lying." Instead, natural selection has simply favored certain behavioral or morphological traits that effectively communicate deceitful information.

DECEIVING INDIVIDUALS OF THE SAME SPECIES In some cases, natural selection has favored the evolution of traits or actions that deceive members of an organism's own species. When mantis shrimps molt, for example, they lack any external covering and cannot use their large claws. Thus, they are unable to defend the cavities where they live. But if another mantis shrimp approaches with the intent of evicting the individual and taking over the cavity, the molting individual bluffs. It raises its claws in the normal aggressive display and may even lunge at the intruder.

Perhaps the best-studied type of deceit in nonhuman animals, however, involves the mating system of bluegill sunfish. Male bluegills set up nesting territories in the shallow water along lake edges, fertilize the eggs laid in their nests, care for the developing embryos by fanning them with oxygen-rich water, and protect newly hatched offspring from large predators.

Some males cheat on this system, however, by mimicking females. To understand how this happens, examine the female, normal male, and female-mimic male in **Figure 51.12**. Female-mimic males look like females but have well-developed testes and produce large volumes of sperm. Mimics also act like females during courtship movements with territory-owning males. They even adopt the usual egg-laying posture. When normal females approach the nest and begin courtship, the female-mimic males join in.

The territory-owning male tolerates the mimic, apparently thinking that he is successfully courting two females at the same time. But when the actual female begins to lay eggs, the mimic responds by releasing sperm—and thus fertilizing some of the eggs—and then darting away. In this way, the female-mimic male fathers offspring but does not help care for them. In addition, female-mimics do not have to expend time and energy growing to large size and making and defending a nest.

WHEN DOES DECEPTION WORK? In analyzing deceitful communication in bluegill sunfish and other species, researchers point

FIGURE 51.12 In Bluegill Sunfish, Some Males Look and Act Like Females.

out that in most cases, deceit works only when it is relatively rare. The logic behind this hypothesis runs as follows: If deceit becomes extremely common, then natural selection will strongly favor individuals that can detect and avoid or punish liars. But if liars are rare, then natural selection will favor individuals that are occasionally fooled but are more commonly rewarded by responding to signals in a normal way.

In mantis shrimps, for example, molting occurs only once or twice a year. Thus, the vast majority of contests over hiding places involve individuals that have an intact exoskeleton and claws that qualify as a deadly weapon. As a result, intruders usually benefit by interpreting threat displays as honest and responding to the stimulus by retreating. If molting were frequent and bluffing behavior common, then intruders would benefit by challenging territory owners, which would likely be defenseless.

CHECK YOUR UNDERSTANDING

If you understand that . . .

- Communication is an exchange of information between individuals.
- In most instances, the mode of communication that animals use maximizes the probability that the information will be transferred efficiently.
- Communication can be honest or deceitful, depending on the nature of the information received.

✔ You should be able to . . .

1. Explain the costs and benefits of auditory, olfactory, and visual communication.
2. Explain why natural selection favors individuals that can detect deceitful communication and either avoid "liars" or punish them.

Answers are available in Appendix B.

51.6 When Should I Cooperate?

The types of behavior reviewed thus far all have a key common element: They help individuals respond to environmental stimuli in a way that increases their fitness. There is a type of behavior that appears to contradict this pattern, however: altruism.

Altruism is behavior that has a fitness cost to the individual exhibiting the behavior and a fitness benefit to the recipient of the behavior. It is the formal term for self-sacrificing behavior. Altruism decreases an individual's ability to produce offspring but helps others produce more offspring.

The existence of altruistic behavior appears to be paradoxical. If certain alleles make an individual more likely to be altruistic, those alleles should be selected against. Does altruism actually occur?

Kin Selection

There is no question that self-sacrificing behavior occurs in nature. For example, black-tailed prairie dogs perform a behavior called alarm calling (**Figure 51.13**). These burrowing mammals live in large communities, called towns, throughout the Great Plains region of North America. When a badger, coyote, hawk, or other predator approaches a town, some prairie dogs give alarm calls that alert other prairie dogs to run to mounds and scan for the threat.

Giving these calls is risky. In several species of ground squirrels and prairie dogs, researchers have shown that alarm-callers draw attention to themselves by calling and are in much greater danger of being attacked than non-callers are.

HAMILTON'S RULE How can natural selection favor the evolution of self-sacrificing behavior? William D. Hamilton answered this question by creating a mathematical model to assess how an allele that contributes to altruistic behavior could increase in frequency in a population.

To model the fate of altruistic alleles, Hamilton represented the fitness cost of the altruistic act to the actor as C and the fitness benefit to the recipient as B. Both C and B are measured in

FIGURE 51.13 Black-Tailed Prairie Dogs Are Highly Social. Prairie dogs live with their immediate and extended family within large groups called towns. The individual with the upright posture has spotted an intruder and may give an alarm call.

units of offspring produced. His model showed that the allele could spread if

$$Br > C$$

where r is the **coefficient of relatedness**. The coefficient of relatedness is a measure of how closely the actor and beneficiary are related. Specifically, r measures the fraction of alleles in the actor and beneficiary that are identical by descent—that is, inherited from the same ancestor (see **Box 51.1**).

The formulation $Br > C$ is called **Hamilton's rule.** It states that altruistic behavior is most likely when three conditions are met:

- The fitness benefits of altruistic behavior are high for the recipient.
- The altruist and recipient are close relatives.
- The fitness costs to the altruist are low.

When Hamilton's rule holds, alleles associated with altruistic behavior will be favored by natural selection—because close relatives are very likely to have copies of the altruistic alleles. They will increase in frequency in the population.

INCLUSIVE FITNESS Hamilton's rule is important because it shows that individuals can pass their alleles on to the next generation not only by having their own offspring, but also by helping close relatives produce more offspring. To capture this point, biologists refer to direct fitness and indirect fitness.

- Direct fitness is derived from an individual's own offspring.
- Indirect fitness is derived from helping relatives produce more offspring than they could produce on their own.

The combination of direct and indirect fitness components is called **inclusive fitness.**

BOX 51.1 QUANTITATIVE METHODS: Calculating the Coefficient of Relatedness

The coefficient of relatedness, r, varies between 0.0 and 1.0. If two individuals have no identical alleles that were inherited from the same ancestor, then their r value is 0.0. Because every allele in pairs of identical twins is identical, their coefficient of relatedness is 1.0.

What about other relationships? **Figure 51.14a** shows how r is calculated between half-siblings. (To review what the boxes, circles, and lines in a pedigree mean, see Figure 13.22.) Half-siblings share one parent. Thus, r represents the probability that half-siblings share alleles as a result of inheriting alleles from their common parent. It is critical to realize that in each parent-to-offspring link of descent, the probability of any particular allele being transmitted is 1/2. This is so because meiosis distributes alleles from the parent's diploid genome to their haploid gametes randomly. Thus, half the gametes produced by a parent get one of the alleles present at each gene, and half the gametes produced get the other allele. Half-siblings are connected by two such parent-to-offspring links. The overall probability of two half-siblings sharing the same allele by descent is $1/2 \times 1/2 = 1/4$ (To review rules for combining probabilities, see **BioSkills 13** in Appendix A.)

To think about this calculation in another way, focus on the red arrows in Figure 51.14a. The left arrow represents the probability that the mother transmits a particular allele to her son. The right arrow represents the probability that the mother transmits the same allele to her daughter. Both probabilities are 1/2. Thus, the probability that the mother transmitted the same allele to both her son and daughter is $1/2 \times 1/2 = 1/4$.

Figure 51.14b shows how r is calculated between full siblings. The challenge here is to calculate the probability of two individuals sharing the same allele as a result of inheriting it through their mother or through their father. The probability that full siblings share alleles as a result of inheriting them from one parent is 1/4. Thus, the probability that full siblings share alleles inherited from either their mother or their father is $1/4 + 1/4 = 1/2$.

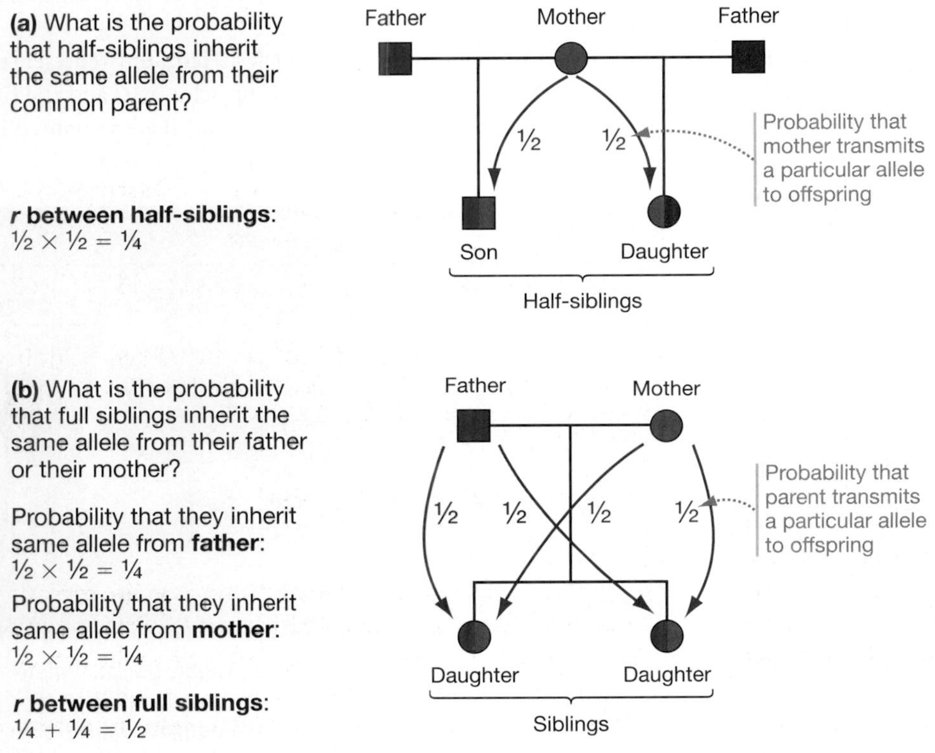

(a) What is the probability that half-siblings inherit the same allele from their common parent?
r between half-siblings:
½ × ½ = ¼
Father
Mother
Father
½
½
Probability that mother transmits a particular allele to offspring
Son
Daughter
Half-siblings
(b) What is the probability that full siblings inherit the same allele from their father or their mother?
Probability that they inherit same allele from father:
½ × ½ = ¼
Probability that they inherit same allele from mother:
½ × ½ = ¼
r between full siblings:
¼ + ¼ = ½
Father
Mother
½
½
½
½
Probability that parent transmits a particular allele to offspring
Daughter
Daughter
Siblings

FIGURE 51.14 The Coefficient of Relatedness (r) Is Calculated from Information in Pedigrees.

✔**QUESTION** What is the r between first cousins?

Biologists use the term **kin selection** to refer to natural selection that acts through benefits to relatives. Kin selection results in increased indirect fitness.

✔**If you understand kin selection, you should be able to explain why it is likely to be common in humans. You should also be able to explain whether kin selection occurs in plants.**

TESTING HAMILTON'S RULE Does Hamilton's rule work? Do animals really favor relatives when they act altruistically? To test the kin-selection hypothesis, a researcher studied which of the inhabitants of a black-tailed prairie dog town were most likely to give alarm calls.

Within a large prairie dog town, individuals live in small groups called coteries that share the same underground burrow. Members of each coterie defend a territory inside the town.

By tagging offspring that were born over several generations, the researcher identified the genetic relationships among individuals in the town. More specifically, the researcher determined, within each coterie, whether each individual had:

1. no close genetic relatives in its coterie;
2. no offspring in the coterie but at least one sibling, cousin, uncle, aunt, niece, or nephew; or
3. at least one offspring or grandoffspring in the coterie.

The kin-selection hypothesis predicts that individuals who do not have close genetic relatives nearby will rarely give an alarm call. To evaluate this prediction, the biologist recorded the identity of callers during 698 experiments. In these studies, a stuffed badger was dragged through the colony on a sled.

Were prairie dogs with close relatives nearby more likely to call, or did kinship have nothing to do with the probability of alarm calling? The bar charts in **Figure 51.15** show the average proportion of times that individuals in each of the three categories called during an experiment. The data indicate that black-tailed prairie dogs are much more likely to call if they live in a coterie that includes close relatives.

This same pattern—of preferentially dispensing help to kin—has been observed in many other species of social mammals and birds. Most cases of self-sacrificing behavior that have been analyzed to date are consistent with Hamilton's rule and are hypothesized to be the result of kin selection.

EXPERIMENT

QUESTION: Do black-tailed prairie dogs prefer to help relatives when they give an alarm call?

HYPOTHESIS: Individuals give an alarm call only when close relatives are near.

NULL HYPOTHESIS: The presence of relatives has no influence on the probability of alarm calling.

EXPERIMENTAL SETUP:

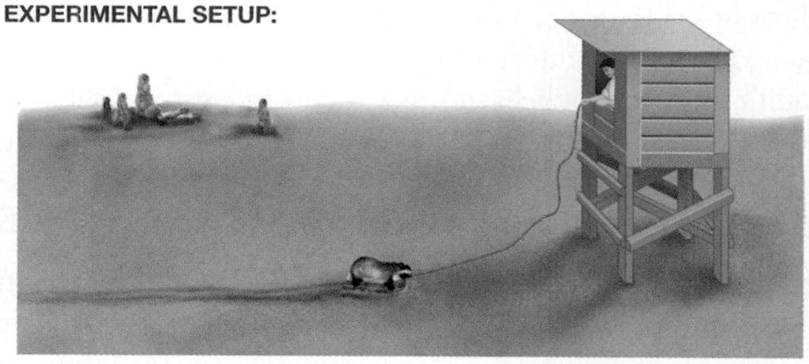

1. Determine relationships among individuals in prairie dog coterie.

2. Drag stuffed badger across territory of coterie.

3. From observation tower, record which members of coterie give an alarm call.

4. Repeat experiment 698 times. Each individual prairie dog coterie is tested 6–9 times over 3-year period.

PREDICTION OF KIN-SELECTION HYPOTHESIS: Individuals in coteries that contain a close genetic relative are more likely to give an alarm call than are individuals in coteries that do not contain a close genetic relative.

PREDICTION OF NULL HYPOTHESIS: The presence of relatives in coteries will not influence the probability of alarm calling.

RESULTS:

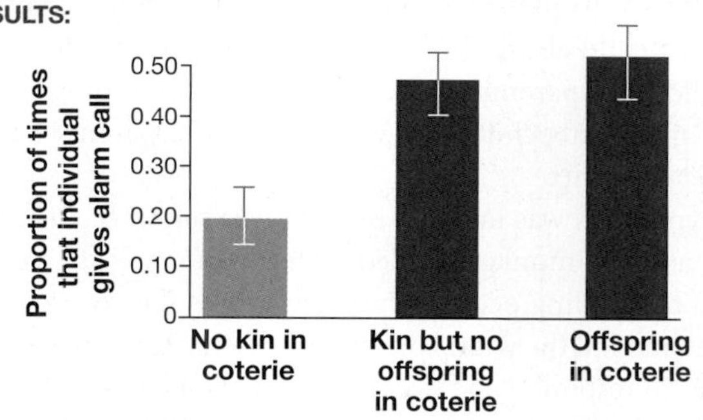

CONCLUSION: Alarm calling usually benefits relatives.

FIGURE 51.15 Experimental Evidence That Black-Tailed Prairie Dogs Are More Likely to Give Alarm Calls if Relatives Are Nearby.

SOURCE: Hoogland, J. L. 1995. *The Black-tailed Prairie Dog*. Chicago, IL: University of Chicago Press.

✔**QUESTION** What would be an appropriate control in this experiment, and what hypothesis would it test?

Reciprocal Altruism

During long-term studies of highly social animals, such as lions, chimpanzees, and vampire bats, biologists have observed nonrelatives helping each other. Chimps and other primates, for example, may spend considerable time grooming unrelated members of their social group—cleaning their fur and removing ticks and other parasites from their skin. In vampire bats, individuals that have been successful in finding food are known to regurgitate blood meals to non-kin that have not been successful and that are in danger of starving.

How can self-sacrificing behavior like this evolve if kin selection is not acting? The leading hypothesis to explain altruism among nonrelatives is called **reciprocal altruism**: an exchange of fitness benefits that are separated in time. Reciprocal altruists help individuals who have either helped them in the past or are likely to help them in the future.

Data that have been collected so far support the reciprocal-altruism hypothesis in at least some instances.

- Among vervet monkeys, individuals are most likely to groom unrelated individuals that have groomed or helped them in the past.
- Vampire bats are most likely to donate blood meals to non-kin that have previously shared food with them.

Reciprocal altruism is also widely invoked as an explanation for the helpful and cooperative behavior commonly observed among unrelated humans.

To summarize, altruistic behavior increases the fitness of individuals by (1) favoring kin, or (2) increasing the likelihood of receiving help in the future from non-kin. Altruism is a flexible, condition-dependent behavior that occurs in a surprisingly wide array of species that live in social groups.

An Extreme Case: Abuse of Non-Kin in Humans

Hamilton's rule specifies when individuals should be altruistic to relatives; reciprocal altruism explains why it can be adaptive, in some circumstances, for nonrelatives to cooperate. What about the opposite of cooperation—abuse or other types of violence?

Martin Daly and Margo Wilson helped pioneer the use of "selection thinking" in studies of human behavior. They were among the first to ask how understanding kin selection and other aspects of natural selection can inform research on humans.

Daly and Wilson have been particularly interested in studying the "dark side" of human behavior—homicide, domestic violence, and adultery—from a biological perspective. Under certain conditions, can these types of events be consistent with actions favored by natural selection?

To answer this question, consider data that these researchers gathered on child abuse. To begin, they hypothesized that selection should favor parents who invest resources in biological children but not in stepchildren—with whom the parents have no genetic relationship.

The hypothesis was inspired by the observation of infanticide in lions and in monkeys called Hanuman langurs. In these species, infant-killing occurs when new males take over a group of females and kill the young offspring present. The females come into estrus in response, and bear young by the new males.

The Daly-Wilson hypothesis predicts that child abuse should be much more common in households containing a stepparent than it is in homes containing only biological parents. Further, Daly and Wilson predicted that, if stepparents abuse their stepchildren, infants should be most at risk. The logic behind this prediction is that very young children demand the most time and resources and are least able to defend themselves.

Are the data consistent with these predictions? Daly and Wilson analyzed the most extreme form of child abuse—the killing of children by parents. Using a database on all homicides reported in Canada between 1974 and 1983, they found 341 cases in which a child was killed by a biological parent and 67 in which the perpetrator was a stepparent. But because households containing biological parents are much more common, they realized that they needed to compare the *rate* of violence in the two types of households, rather than the absolute number.

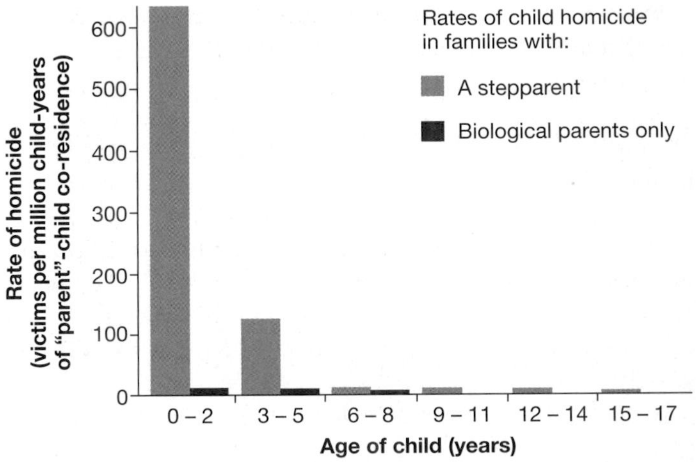

FIGURE 51.16 Rates of Child Homicide Are Higher in Families Containing a Stepparent. Rates of child homicide expressed as victims per "million child-years" in Canadian households.

Daly and Wilson expressed the homicide rate as the number of children killed per million "child-years." One child-year is a year that a child of a particular age lived in a particular family situation. A 5-year-old living with the same parents its entire life would represent five child-years.

The results are graphed in **Figure 51.16**. Note that children who live with a stepparent are at much higher risk of abuse than are children who live with biological parents. More specifically, kids who are less than 2 years old are 70 times more likely to be killed in a household with a stepparent than in a household with only biological parents. To put this number in perspective, consider that smokers are 11 times more likely to develop lung cancer than are nonsmokers.

The data have been helpful in raising awareness among social workers and clinicians—to be especially alert to indications of violence in households with stepparents where very young children are present.

CHECK YOUR UNDERSTANDING

If you understand that . . .

- Hamilton's rule states that alleles for altruistic behavior increase in frequency if the fitness cost of the behavior for the actor is low, the fitness benefit to the recipient is high, and the actor and recipient are closely related.
- Reciprocal altruism is based on an exchange of fitness benefits separated in time.

✔ **You should be able to . . .**

1. Use Hamilton's rule to describe situations in which a black-tailed prairie dog is likely and unlikely to give an alarm call.
2. Explain why reciprocal altruism has been observed only in species where individuals have good memories and live in small, long-lived social groups.

Answers are available in Appendix B.

In interpreting these results, though, it is *extremely* important to keep four points in mind:

1. Violence occurs among biological kin as well as non-kin.
2. The vast majority of stepparents are solicitous and generous with their stepchildren.
3. In today's culture, the behavior is undoubtedly pathological rather than adaptive, because child abusers are jailed and treated as abhorrent.
4. Violence is a condition-dependent behavior. Our genetic heritage and neural and hormonal systems make an enormous range of behavior possible. Ask yourself: What kinds of life conditions would drive a person to commit such a heinous act? What can friends, family members, and society in general do to help individuals keep from getting into circumstances that lead to violence?

CHAPTER 51 REVIEW

For media, go to the study area at www.masteringbiology.com

Summary of Key Concepts

Biologists analyze behavior at the proximate and ultimate levels—the genetic and physiological mechanisms and how they affect fitness.

- At the proximate level, experiments and observations focus on understanding how specific gene products, neuron activity, and hormonal signals cause behavior.
- At the ultimate level, researchers seek to understand the adaptive significance of behavior, or how it enables individuals to survive and reproduce.
- By combining proximate and ultimate viewpoints, biologists can seek a comprehensive understanding of how and why animals do what they do.

✔ You should be able to explain why fruit-fly larvae vary in foraging behavior, at both the proximate and ultimate levels.

Individuals can behave in a wide range of ways; which behavior occurs depends on current conditions.

- Foraging in white-fronted bee-eaters, mate choice in barn swallows, and the onset of sexual behavior in *Anolis* lizards are examples of flexible, condition-dependent behaviors triggered by changing environmental stimuli.
- In most cases, animals behave in a way that maximizes their fitness under current conditions.

✔ You should be able to relate Hamilton's rule to the concept of condition-dependence in behavior.

Foraging patterns may vary with genotype; foraging decisions maximize energy gain and minimize costs.

- At the proximate level, the rover and sitter phenotypes in fruit-fly larvae are one of the best-understood examples of variation in foraging behavior. Different alleles at a single gene increase the probability that an individual will move or stay after feeding.
- The rover allele has highest fitness at high population density; the sitter allele has highest fitness at low population density.
- In white-fronted bee-eaters, individuals that live far from the colony minimize the costs of foraging and maximize the benefits by increasing the length of time they look for food and bringing back larger amounts of food with each trip.

✔ You should be able to discuss the costs and benefits of roving versus sitting in fruit-fly larvae.

Sexual behavior depends on surges of sex hormones; females choose mates that provide good alleles and needed resources.

- In *Anolis* lizards, males and females come into breeding condition in spring, in response to surges of the sex hormones testosterone and estradiol.
- Female *Anolis* come into breeding condition fastest if they are exposed to springlike temperatures *and* courtship displays from males.
- Female barn swallows prefer to mate with long-tailed males, because the longest-tailed males are the best-fed and most disease-free.

✔ You should be able to generate a hypothesis for why vigorous bobbing displays of a brightly colored dewlap might be a reliable signal of male quality in *Anolis* lizards.

Animals navigate using an array of cues; the costs of migration can be offset by benefits in food availability.

- Many animal species can find their way by following familiar landmarks.
- When navigating, many animal species get compass information from the Sun, the stars, and Earth's magnetic field.
- Animals usually migrate to find seasonally available food. The benefits of the food sources have to outweigh the energetic costs and danger of the migratory movement.

✔ You should be able to explain why migration in some birds is considered condition dependent, based on the observation that some species do not migrate if people supply food in feeders.

Animals communicate with movements, odors, or other stimuli; communication can be honest or deceitful.

- Honeybees have a symbolic language—their dancing movements communicate accurate information about the direction and distance of food sources.
- Deceitful communication can be adaptive—as when anglerfish fool prey into attacking a "lure" attached to the anglerfish's face or when molting mantis shrimps try to bluff their way into retaining ownership of a hiding place.

✔ You should be able to predict the consequences for deceitful communication in bluegill sunfish if most large, territory-owning males were fished out of a lake.

Web Activity Homing Behavior in Digger Wasps

Individuals that behave altruistically are usually helping relatives or individuals that help them in return.

- When altruistic behavior is directed toward close relatives, alleles that lead to self-sacrificing may increase in frequency due to kin selection.
- Some animals that live in close-knit social groups engage in reciprocal altruism—meaning they exchange help over time.

✔ **You should be able to describe the characteristics of a species for which altruistic behavior is expected to be extremely common.**

Questions

✔ TEST YOUR KNOWLEDGE

Answers are available in Appendix B

1. What do proximate explanations of behavior focus on?
   a. how displays and other types of behavior have changed through time, or evolved
   b. the functional aspect of a behavior, or its "adaptive significance"
   c. genetic, neurological, and hormonal mechanisms of behavior
   d. appropriate experimental methods when studying behavior
2. What unit(s) do biologists use when analyzing the costs and benefits of behavior?
   a. proximate versus ultimate
   b. body size (e.g., large versus small)
   c. time and energy
   d. fitness—the ability to survive and produce offspring
3. What is "reciprocal" about reciprocal altruism?
   a. It is based on an exchange of fitness benefits.
   b. The events involved are separated in time.
   c. The individuals involved usually know each other well.
   d. The individuals involved have to have good memories, to keep track of which individuals they have helped in the past.
4. Which of the following statements about the waggle dance of the honeybee is not correct?
   a. The length of a waggling run is proportional to the distance from the hive to a food source.
   b. Sounds and scents produced by the dancer provide information about the nature of the food source.
   c. The dancer uses no elements of the round dance.
   d. The orientation of the waggling run provides information about the direction of the food from the hive, relative to the Sun's position.
5. Why are biologists convinced that the sex hormone testosterone is required for normal sexual activity in male *Anolis* lizards?
   a. Male *Anolis* lizards with larger testes court females more vigorously than do males with smaller testes.
   b. The testosterone molecule is not found in female *Anolis* lizards.
   c. Male *Anolis* lizards whose gonads had been removed did not develop dewlaps.
   d. Male *Anolis* lizards whose gonads had been removed did not court females.
6. What does Hamilton's rule specify?
   a. why animals do things "for the good of the species"
   b. why reciprocal altruism can lead to fitness gains for unrelated individuals
   c. how alleles that favor self-sacrificing acts increase in frequency via kin selection
   d. the conditions under which more complex behaviors evolve from simpler behaviors

✔ TEST YOUR UNDERSTANDING

Answers are available in Appendix B

1. Discuss the proximate and ultimate causes of homing behavior by spiny lobsters.
2. What does it mean to say that an animal is foraging in an optimal way?
3. Why do traits like long tails in male barn swallows serve as honest signals of male quality?
4. What environmental stimuli cause changes in hormone levels that lead to egg laying in *Anolis* lizards? What biologically relevant information do these stimuli provide?
5. For an animal to navigate, it must have a "map" and a "compass." Explain why both types of information are needed. Describe three types of compasses that have been identified in migratory or homing species.
6. How are kin selection and reciprocal altruism different? Which should be more common, and why?

✔ APPLYING CONCEPTS TO NEW SITUATIONS

Answers are available in Appendix B

1. To date, the rover and sitter alleles have only been shown to affect foraging behavior in fruit-fly larvae. Design an experiment to test the hypothesis that adults with the rover allele tend to fly farther in search of food sources than adults with the sitter allele.
2. Most tropical habitats are highly seasonal. But instead of alternating warm and cold seasons, there are alternating wet and dry seasons. Most animal species breed during the wet months. If you were studying a species of *Anolis* native to the tropics, what environmental cue would you simulate in the lab to bring them into breeding condition? How would you simulate this cue?
3. A biologist once remarked that he'd be willing to lay down his life to save two brothers or eight cousins. Explain what he meant.
4. Based on the theory of reciprocal altruism, predict the conditions under which people are expected to donate blood.

This chapter is part of The Big Picture. See how on pages 1126–1127.

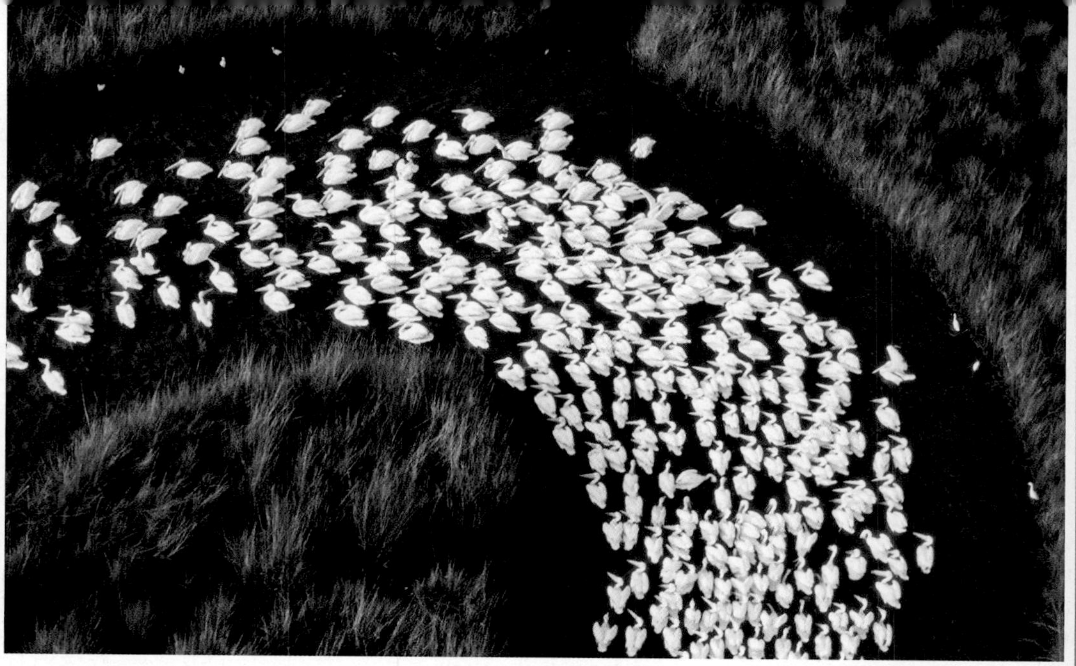

A flock of white pelicans swimming in a channel of the Mississippi River. This chapter explores how and why growth rates in populations change through time.

Population Ecology 52

If you asked a biologist to name two of today's most pressing global issues, she might say global warming and extinction of species. If you were asked to name two of the most pressing issues facing your region, you might say traffic and the price of housing. All four problems have a common cause: recent and dramatic increases in the size of the human population.

In 1940, for example, Mexico City had a population of about 1.6 million. The city grew to 5.4 million in 1960, 13.9 million in 1980, and over 19 million in 2007. How much bigger will it get over the next 50 years? And how large will the entire human population be by the time your kids are in college?

A **population** is a group of individuals of the same species that live in the same area at the same time; **population ecology** is the study of how and why the number of individuals in a population changes over time.

With the explosion of human populations across the globe, the massive destruction of natural habitats, and the resulting threats to species throughout the tree of life, population ecology has become a vital field in biological science. The mathematical and analytical tools introduced in this chapter help biologists predict changes in population size and design management strategies to save threatened species.

Let's start by considering some of the basic tools that biologists use to study populations, then follow up with examples of how biologists study changes in the population size of humans and other species over time. The chapter concludes by asking how all of these elements fit together in efforts to limit human population growth and save endangered species.

KEY CONCEPTS

- Life tables summarize how likely it is that individuals of each age class in a population will survive and reproduce.
- The growth rate of a population can be calculated from life-table data or from the direct observation of changes in population size over time.
- Researchers observe a wide variety of patterns when they track changes in population size over time, ranging from growth rates that slow when populations are at high density, to regular cycles, to continued growth independent of population size.
- Data from population ecology studies help biologists evaluate prospects for endangered species and design effective management strategies, as well as to predict changes in human populations.

52.1 Demography

The number of individuals present in a population depends on four processes: birth, death, immigration, and emigration.

✔ When you see this checkmark, stop and test yourself. Answers are available in Appendix B.

- Populations grow due to births—here, meaning any form of reproduction—and **immigration**, which occurs when individuals enter a population by moving from another population.
- Populations decline due to deaths and **emigration**, which occurs when individuals leave a population to join another population.

Analyzing birthrates, death rates, immigration rates, and emigration rates is fundamental to **demography**: the study of factors that determine the size and structure of populations through time.

To predict the future of a population, biologists have to know something about its makeup: how many individuals of each age are alive, how likely individuals of different ages are to survive to the following year, how many offspring are produced by females of different ages, and how many individuals of different ages immigrate and emigrate each **generation**—the average time between a mother's first offspring and her daughter's first offspring.

If a population consists primarily of young individuals with a high survival rate and reproductive rate, the population size should increase over time. But if a population comprises chiefly old individuals with low reproductive rates and low survival rates, then it is almost certain to decline over time. To understand a population's dynamics, biologists turn to the data contained in a life table.

Life Tables

A **life table** summarizes the probability that an individual will survive and reproduce in any given time interval over the course of its lifetime. Life tables were invented almost 2000 years ago; in ancient Rome they were used to predict food needs. In modern times, life tables have been the domain of life-insurance companies, which have a strong financial interest in predicting the likelihood of a person dying at a given age. More recently, biologists are using life tables to study the demographics of endangered species.

***LACERTA VIVIPARA*: A CASE STUDY** To understand how researchers use life tables, consider the lizard *Lacerta vivipara* (**Figure 52.1**). *L. vivipara* is a common resident of open, grassy habitats in western Europe. As their name suggests—vivipara means "live birth"—most populations are ovoviviparous and give birth to live young (see Chapter 32).

FIGURE 52.1 *Lacerta vivipara* Are Native to Europe. Females in most populations bear live young, though females in *L. vivipara* populations from northern Spain and southwest France lay eggs.

Researchers set out to estimate the life table of a low-elevation population in the Netherlands, with the goal of comparing the results to data that other researchers had collected from *L. vivipara* populations in the mountains of Austria and France and in lowland habitats in Britain and Belgium. They wanted to know whether populations that live in different environments vary in basic demographic features.

To begin the study, the researchers visited their study site daily during the seven months that these lizards are active during the year. Each day the researchers captured and marked as many individuals as possible.

Because this program of daily monitoring continued for seven years, the biologists were able to document the number of young produced by each female in each year of her life. If a marked individual was not recaptured in a subsequent year, they assumed that it had died sometime during the previous year.

The data allowed researchers to calculate the number of individuals that survived each year in each particular age group as well as how many offspring each female produced. What did the numbers reveal?

SURVIVORSHIP **Survivorship**—a key component of a life table—is defined as the proportion of offspring produced that survive, on average, to a particular age. For example, suppose 1000 *L. vivipara* are born in a particular year. These individuals represent a **cohort**—a group of the same age that can be followed through time. How many individuals would survive to age 1, age 2, age 3, and so on?

As **Table 52.1** shows, survivorship from birth to age 1 was 0.424 in the Netherlands population. If 1000 females were born in a particular year in this population, on average 424 would still be alive one year later. Survivorship from birth to age 2 was 0.308, meaning an average of 308 female lizards would survive for two years.

To analyze general patterns in survivorship, biologists plot the logarithm of the number of survivors versus age. Using a logarithmic scale on the *y*-axis is a matter of convenience—it makes patterns easier to see (see **BioSkills 7** in Appendix A). The resulting graph is called a **survivorship curve**.

Studies on a wide variety of species indicate that three general types of survivorship curves exist (**Figure 52.2a**).

1. Humans have what biologists call a type I survivorship curve. In this pattern, survivorship throughout life is high—most individuals approach the species' maximum life span.
2. Type II survivorship curves occur in species where individuals have about the same probability of dying in each year of life. Blackbirds and other songbirds have this type of curve.
3. Many plants have Type III curves—a pattern defined by extremely high death rates for seeds and seedlings but high survival rates later in life.

TABLE 52.1 **Life Table for *Lacerta vivipara* in the Netherlands**

Age	Number Alive	Survivorship	Fecundity	Survivorship × Fecundity = Average Number of Offspring Produced per Female of Age *x*
0	1000	1.000	0.00	0.00
1	424	0.424	0.08	0.03
2	308	0.308	2.94	0.91
3	158	0.158	4.13	0.65
4	57	0.057	4.88	0.28
5	10	0.010	6.50	0.07
6	7	0.007	6.50	0.05
7	2	0.002	6.50	0.01

SOURCE: Data are from H. Strijbosch and R. C. M. Creemers, 1988. Comparative demography of sympatric populations of *Lacerta vivipara* and *Lacerta agilis*. *Oecologia* 76: 20–26. With kind permission of Springer Science and Business Media.

Figure 52.2b provides a graph for you to plot survivorship of *L. vivipara*, using the data reported in Table 52.1. Note the log-log scale on the graph.

FECUNDITY The number of female offspring produced by each female in a population is termed **fecundity**. In most cases, biologists only keep track of females when calculating life-table data because the number of males present rarely affects population dynamics. There are almost always enough males present to fertilize all of the females in breeding condition, so growth rates depend entirely on females.

Because researchers documented the reproductive output of the same *L. vivipara* lizard females year after year, they were able to calculate a quantity called **age-specific fecundity**: the average number of female offspring produced by a female in age class *x*. An **age class** is a group of individuals of a specific age—for example, all female lizards between 4 and 5 years old.

As **Box 52.1** on page 1040 shows, data on survivorship and fecundity allow researchers to calculate the growth rate of a population. How do data on the Netherlands populations compare with populations in different types of habitats?

The Role of Life History

The life-table data in Table 52.1 are interesting because they contrast with results from other populations of *L. vivipara*. Consider data on fecundity:

- In the Netherlands, almost no 1-year-old female *L. vivipara* reproduce.
- In Brittany, France, 50 percent of 1-year-old female lizards reproduce.
- In the mountains of Austria, females don't begin breeding until they are 4 years old.

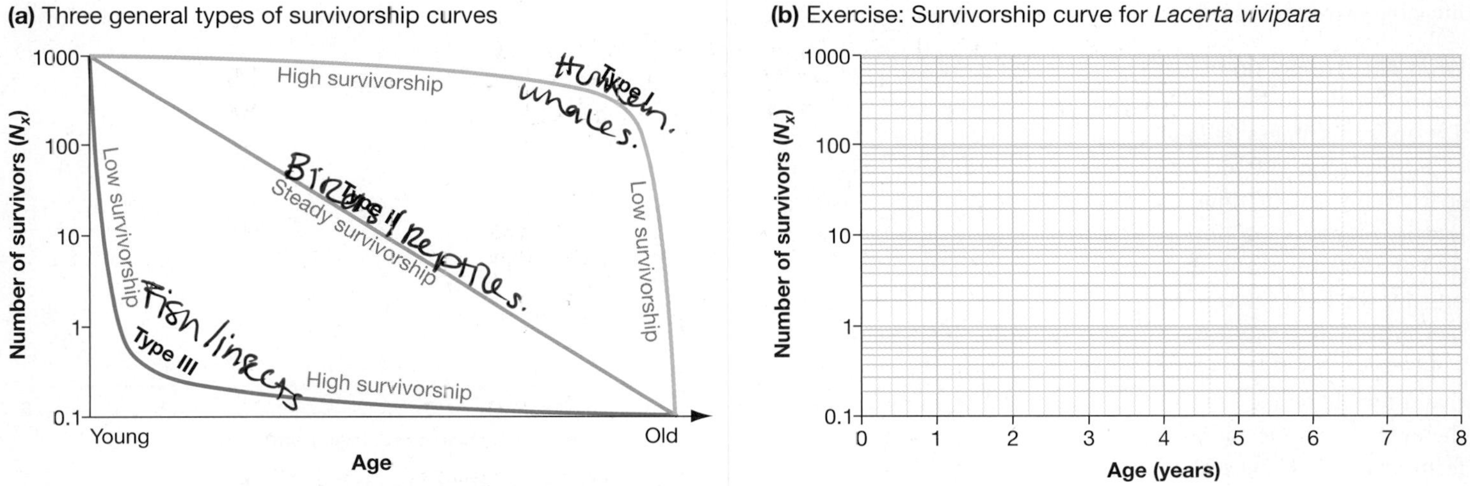

FIGURE 52.2 Survivorship Curves Identify When Mortality Rates Are Low, Steady, or High.

✔**EXERCISE** Fill in the graph in part (b), with data on survivorship of *Lacerta vivipara* given in Table 52.1. Compare the shape of the curve to the generalized graphs in part (a).

BOX 52.1 QUANTITATIVE METHODS: Using Life Tables to Calculate Population Growth Rates

If immigration and emigration are not occurring, the data in a life table can be used to calculate a population's growth rate. This is logical, because survivorship and fecundity are ways to express death rates and birthrates—the other two factors influencing population size. To see how a population's growth rate can be estimated from life-table data, let's look at each component in a life table more carefully. Recall that life tables usually focus exclusively on females.

Survivorship is symbolized as l_x, where x represents the age class being considered. Survivorship for age class x is calculated by dividing the number of females in that age class (N_x) by the number of females that existed as offspring (N_0):

$$l_x = \frac{N_x}{N_0} \qquad \textbf{(Eq. 52.1)}$$

Age-specific fecundity is symbolized m_x, where x again represents the age class being considered. Age-specific fecundity is calculated as the total number of female offspring produced by females of a particular age, divided by the total number of females of that age class present. It represents the average number of female offspring produced by a female of age x.

Documenting age-specific survivorship and fecundity allows researchers to calculate the average number of female offspring produced by females during each age of life, as $l_x m_x$. For example, if 30 percent of the females in a population live to be 2 years old, and if 2-year-old females have an average of 2 offspring, then the average number of births at age 2, per female born into the population, is $l_x m_x = 0.30 \times 2 = 0.6$.

Summing $l_x m_x$ values over the entire life span of a female gives the net reproductive rate, R_0, of a population:

$$R_0 = \sum_{i=0}^{x} l_x m_x \qquad \textbf{(Eq. 52.2)}$$

Here the Greek letter sigma (Σ) stands for "sum." The expression on the right-hand side of the equals sign is read, "sum the $l_x m_x$ values from each year of life, starting from birth or hatching ($i = 0$), through age x.

The **net reproductive rate** represents the growth rate of a population per generation. The logic behind the equation for R_0 is that the growth rate of a population per generation equals the average number of female offspring that each female produces over the course of her lifetime. A female's average lifetime reproduction, in turn, is a function of survivorship and fecundity at each age class. In Table 52.1, R_0 is the sum of the survivorship $\times$ fecundity values in the right column.

If R_0 is greater than 1, then the population is increasing in size. If R_0 is less than 1, then the population is declining. ✔If you understand these concepts, you should be able to use the data in Table 52.1 to calculate R_0 in the Netherlands population of *L. vivipara*,[1] state how many female offspring an average *L. vivipara* female produces over the course of her lifetime,[2] and describe whether the population is growing, stable, or declining.[3]

Answers are given below.

[1]$R_0 = 2.0$; [2]2.00; [3]The population is growing rapidly.

The contrasts are equally stark in terms of survivorship. In the Austrian population, most females live much longer than do individuals in either lowland population—in the Netherlands or France.

In Brittany, fecundity is high but survivorship is low; in Austria, fecundity is low but survivorship is high. The population in the Netherlands is intermediate. In this species, key aspects of the life table vary dramatically among populations.

WHAT ARE FITNESS TRADE-OFFS? Why isn't it possible for *L. vivipara* females to have both high fecundity and high survival? The answer is **fitness trade-offs** (see Chapters 24 and 41).

Fitness trade-offs occur because every individual has a restricted amount of time and energy at its disposal—meaning that its resources are limited. If a female lizard devotes a great deal of energy to producing a large number of offspring, it is not possible for her to devote that same energy to her immune system, growth, nutrient stores, or other traits that increase survival.

To drive this point home, **Figure 52.3** graphs the probability that a female of a species survives to the following year versus the average number of eggs laid—or clutch size, a measure of fecundity—in 10 species of birds.

- There are no points in the upper right corner of the graph because it is not possible to have high fecundity and high survivorship.
- There are no points in the lower left corner of the graph because species with low survivorship and low fecundity have low population growth and go extinct.

A female can maximize fecundity, maximize survival, or strike a balance between the two.

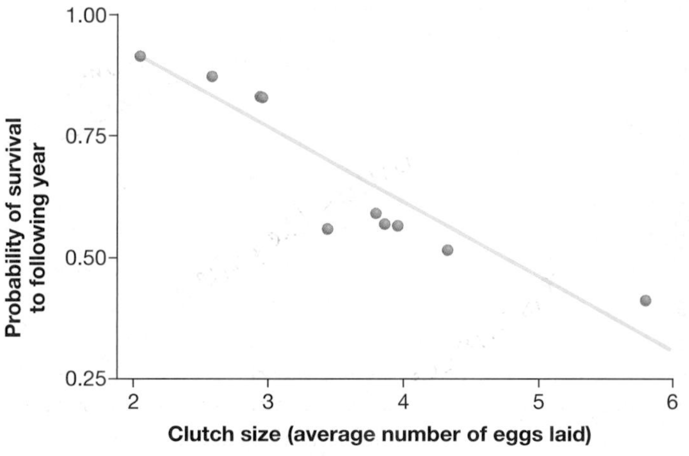

FIGURE 52.3 There Is a Trade-Off between Survival and Reproduction. Each data point on this graph represents a different species of bird. A statistical analysis called regression was used to create a "line of best fit" to these data points.

LIFE HISTORY IS BASED ON RESOURCE ALLOCATION An organism's **life history** describes how an individual allocates resources to growth, reproduction, and activities or structures that are related to survival. Traits such as survivorship, age-specific fecundity, age at first reproduction, and growth rate are all aspects of an organism's life history. Understanding variation in life history is all about understanding fitness trade-offs.

In almost all cases, biologists find that life history is shaped by natural selection in a way that maximizes an individual's fitness in its environment. For example, Chapter 41 provided experimental evidence that there is a fitness trade-off between egg size and egg number in side-blotched lizards. Females in that study population strike a balance between egg size and egg number that is optimal for their habitat.

In *L. vivipara* populations, biologists contend that females who live a long time but mature late and have few offspring each year have high fitness in cold, high-elevation habitats, such as Austria. In these habitats, females have to reduce their reproductive output and put more energy into traits that increase survival in a harsh environment. In contrast, females who have short lives but mature early and have large numbers of offspring each year do better in warm, low-elevation habitats, such as Brittany.

✔If you understand the role that fitness trade-offs play in determining life history patterns, you should be able to: (1) predict how survivorship and fecundity should compare in *L. vivipara* in the warmest part of their range (southwest France and northern Spain), and (2) comment on why females in these populations lay many eggs instead of giving birth to a relatively small number of live young.

PATTERNS ACROSS SPECIES The data you've reviewed thus far all involve comparisons within species. But the same patterns exist when many different species are compared.

In general, individuals from species with high fecundity tend to grow quickly, reach sexual maturity at a young age, and produce many small eggs or seeds. The mustard plant *Arabidopsis thaliana*, for example, germinates and grows to sexual maturity in just 4 to 6 weeks. In this species, individuals usually live only a few months but may produce as many as 10,000 tiny seeds. They live fast and die young.

In contrast, individuals from species with high survivorship tend to grow slowly and invest resources in traits that reduce damage from enemies and increase their own ability to compete for water, sunlight, or food. A coconut palm, for example, may take a decade to mature but live 60–70 years and produce offspring each year. Coconut palms invest resources in making stout stems that allow them to grow tall, a relatively extensive root system, large fruits (coconuts), molecules that make herbivores sick, and enzymes that reduce infections from disease-causing fungi. These traits increase survivorship but decrease fecundity.

An *Arabidopsis thaliana* plant and a coconut palm represent two ends of a broad continuum of life-history characteristics (**Figure 52.4**). How does a species' location on this continuum affect its growth rate?

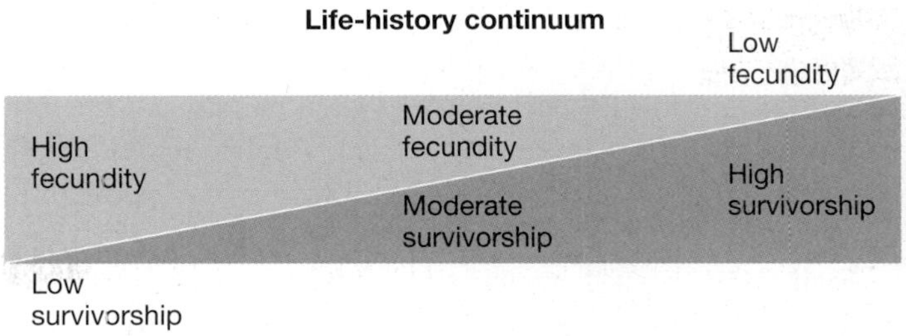

FIGURE 52.4 Life-History Traits Form a Continuum. Every organism can be placed somewhere on this life-history continuum. The placement of a species is most meaningful when it is considered relative to closely related species.

✔QUESTION In plants, what types of traits would you expect to see on the left, middle, and right of the continuum? Consider growth habit (herbaceous, shrub, or tree) as well as relative disease- and predator-fighting ability, seed size, seed number, and body size.

52.2 Population Growth

The most fundamental questions that biologists ask about populations involve growth or decline in numbers of individuals. For conservationists, analyzing and predicting changes in population size is fundamental to managing threatened species.

Recall that four processes affect a population's size: Births and immigration add individuals to the population; deaths and emigration remove them. It follows that a population's overall growth rate is a function of birthrates, death rates, immigration rates, and emigration rates.

Quantifying the Growth Rate

A population's growth rate is the change in the number of individuals in the population (ΔN) per unit time (Δt). Here the Greek symbol Δ (delta) means change.

If no immigration or emigration is occurring, then a population's growth rate is equal to the number of individuals (N) in the population times the difference between the birthrate per individual (b) and death rate per individual (d). The difference between the birthrate and death rate per individual is called the per-capita rate of increase and is symbolized r. (Per capita means "for each individual.")

If the per-capita birthrate is greater than the per-capita death rate, then r is positive and the population is growing. But if the per-capita death rate begins to exceed the per-capita birthrate, then r becomes negative and the population declines. Within populations, r varies through time. Its value can be positive, negative, or 0.

When conditions are optimal for a particular species—meaning birthrates per individual are as high as possible and death rates per individual are as low as possible—then r reaches a maximal value called the **intrinsic rate of increase,** r_{max}. When this occurs, a population's growth rate is expressed as

$$\Delta N/\Delta t = r_{max}N$$

In species such as *Arabidopsis* and fruit flies, which breed at a young age and produce many offspring each year, r_{max} is high. In contrast, r_{max} is low in species such as giant pandas and coconut palms, which take years to mature and produce few offspring each year. Stated another way, r_{max} is a function of a species' life-history traits.

Each species has a characteristic r_{max} that does not change. But at any specific time, a population has an instantaneous growth rate, or per-capita rate of increase, symbolized by r. r_{max} tells you what the maximum growth rate is; r tells you what it is at a particular time. "Little r" is less than or equal to r_{max}.

The instantaneous growth rate of a population at a particular time is actually likely to be much lower than r_{max}. A population's r is also likely to be different from r values of other populations of the same species, and to change over time. The instantaneous growth rate is dynamic.

Exponential Growth

The graph in **Figure 52.5** plots changes in population size, for various values of r, under the condition known as exponential growth. **Exponential population growth** occurs when r does not change over time.

The key point about exponential growth is that the growth rate does not depend on the number of individuals in the population. Biologists say that this type of population growth is **density independent**.

It's important to emphasize that exponential growth adds an increasing number of individuals as the total number of individuals, N, gets larger. As an extreme example, an r of 0.02 per year in a population of 1 billion adds over 20 million individuals per year. The same growth rate in a population of 100 adds just over 2 individuals per year. Even if r is constant, the number of individuals added to a population is a function of N. The rate of increase is the same, but the number of individuals added is not.

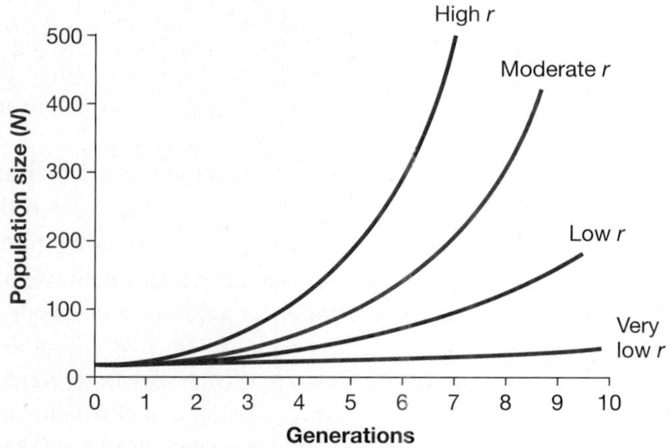

FIGURE 52.5 Exponential Growth Is Independent of Population Size. When the per-capita growth rate r does not change over time, exponential growth occurs. Population size may increase slowly or rapidly, depending on the size of r.

In nature, exponential growth is observed in two circumstances: **(1)** a few individuals found a new population in a new habitat, or **(2)** a population has been devastated by a storm or some other type of catastrophe and then begins to recover, starting with a few surviving individuals. But it is not possible for exponential growth to continue indefinitely.

If *Arabidopsis* populations grew exponentially for a long period of time, they would eventually fill all available habitat. When **population density**—the number of individuals per unit area—gets very high, the population's per-capita birthrate will decrease and the per-capita death rate will increase, causing r to decline. Stated another way, growth is often **density dependent**.

Logistic Growth

To analyze what happens when growth is density dependent, biologists use a parameter called carrying capacity. **Carrying capacity, *K*,** is defined as the maximum number of individuals in a population that can be supported in a particular habitat over a sustained period of time.

The carrying capacity of a habitat depends on a large number of factors: food, space, water, soil quality, and resting or nesting sites. Carrying capacity can change from year to year, depending on conditions.

A LOGISTIC GROWTH EQUATION If a population of size N is below the carrying capacity K, then the population should continue to grow. More specifically, a population's growth rate is proportional to $(K - N)/K$:

$$\frac{\Delta N}{\Delta t} = r_{max} N\left(\frac{K - N}{K}\right)$$

This expression is called a logistic growth equation.

In the expression $(K - N)/K$, the numerator defines the number of additional individuals that can be accommodated in a habitat with carrying capacity K; dividing $K - N$ by K turns this number of individuals into a proportion. Thus $(K - N)/K$ describes the proportion of "unused resources and space" in the habitat. It can be thought of as the environment's resistance to growth.

- When N is small, then $(K - N)/K$ is close to 1 and the growth rate should be high.
- As N gets larger, $(K - N)/K$ gets smaller.
- When N is at carrying capacity—meaning that $K = N$—then $(K - N)/K$ is equal to 0 and growth stops.

Thus, as a population approaches a habitat's carrying capacity, its growth rate should slow.

The logistic growth equation describes **logistic population growth**, or changes in growth rate that occur as a function of population size. Just as exponential growth is density independent, logistic growth is density dependent. **Box 52.2** explores population growth models and how they are applied in more detail.

BOX 52.2 QUANTITATIVE METHODS: Developing and Applying Population Growth Equations

To explore how biologists model changes in population size in more detail, let's consider data on whooping cranes, which are large, wetland-breeding birds native to North America (**Figure 52.6**). Whooping cranes may live more than 20 years in the wild, but it is common for females to be six or seven years old before they breed for the first time. Female cranes lay just two eggs per year and usually rear just one chick. Based on these life-history characteristics, we would expect the growth rate of whooping crane populations to be extremely low.

Conservationists monitor this species closely, because hunting and habitat destruction reduced the total number of individuals in the world to about 20 in the mid-1940s. Since then, intensive conservation efforts have resulted in a current total population of about 462. That number includes a group of 253 individuals that breeds in Wood Buffalo National Park in the Northwest Territories of Canada.

In addition to the Wood Buffalo population, two new populations of cranes have been established recently by releasing offspring raised in captivity. One of these groups lives near the Atlantic coast of Florida year-round; the other group migrates from breeding areas in northern Wisconsin to wintering areas along Florida's Gulf coast.

According to the biologists who are managing whooping crane recovery, the species will no longer be endangered when the two newly established populations have at least 25 breeding pairs—meaning there would be about 125 individuals in each—and when neither population needs to be supplemented with captive-bred young in order to be self-sustaining. How long will this take?

FIGURE 52.6 Whooping Cranes Have Been the Focus of Intensive Conservation Efforts.

Discrete Growth

Whooping cranes breed once per year, so the simplest way to express a crane population's growth rate is to compare the number of individuals at the start of one breeding season to the number at the start of the following year's breeding season.

When populations breed during discrete seasons, their growth rate can be calculated as for whooping cranes. To create a general expression for how populations grow over a discrete time interval, biologists use N to symbolize population size. N_0 is the population size at time zero (the starting point), and N_1 the population size one breeding interval later. In equation form, the growth rate is given as

$$N_1/N_0 = \lambda \qquad \textbf{(Eq. 52.3)}$$

The parameter λ (lambda) is called the **finite rate of increase**. (In mathematics, a *finite rate* refers to an observed rate over a given period of time. A *parameter* is a variable or constant term that affects the shape of a function but does not affect its general nature.) The current population size at Wood Buffalo is 253. Suppose that biologists count 265 cranes on the breeding grounds next year. The population growth rate could be calculated as $265/253 = 1.047 = \lambda$. Stated another way, the population will have grown at the rate of 4.7 percent per year. Rearranging the expression in Equation 52.3 gives

$$N_1 = N_0\lambda \qquad \textbf{(Eq. 52.4)}$$

Stated more generally, the size of the population at the end of year t will be given by

$$N_t = N_0\lambda^t \qquad \textbf{(Eq. 52.5)}$$

This equation summarizes how populations grow when breeding takes place seasonally. The size of the population at time t is equal to the starting size, times the finite rate of increase multiplied by itself t times.

In a sense, λ works like the interest rate at a bank. For species that breed once per year, the "interest" on the population is compounded annually. A savings account with a 5 percent annual interest rate increases by a factor of 1.05 per year.

If a population is growing, then its λ is greater than one. The population is stable when λ is 1.0 and declining when λ is less than 1.0.

If a population's age structure is stable, meaning that the proportion of females in each age class is not changing over time, then its finite rate of increase also has a simple relationship to its net reproductive rate: $\lambda = R_0/g$, where g is the generation time. In essence, dividing the net reproductive rate by generation time transforms it into a discrete rate.

Continuous Growth

The parameters λ and r—the finite rate of increase and the per-capita growth rate—have a simple mathematical relationship. The best way to understand their relationship is to recall that λ expresses a population's growth rate over a discrete interval of time. In contrast, r gives the population's per-capita growth rate at any particular instant. This is why r is also called the instantaneous rate of increase. The relationship between the two parameters is given by

$$\lambda = e^r \qquad \textbf{(Eq. 52.6)}$$

where e is the base of the natural logarithm, or about 2.72. (For help with using logarithms, see **BioSkills 7** in Appendix A. Also, note that the relationship between any finite rate and any instantaneous rate is given by finite rate = $e^{\text{instantaneous rate}}$.)

Substituting Equation 52.6 into Equation 52.5 gives

$$N_t = N_0 e^{rt} \qquad \textbf{(Eq. 52.7)}$$

This expression summarizes how populations grow when they breed continuously, as do humans and bacteria, instead of at defined intervals. For species that breed continuously, the "interest" on the population is compounded continuously. When the growth rates λ and r are equivalent,

(continued)

(continued)

however, the differences between discrete and continuous growth are negligible.

Because r represents the growth rate at any given time, and because r and λ are so closely related, biologists routinely calculate r for species that breed seasonally. In the whooping crane example, $\lambda = 1.047$ per year $= e^r$. To solve for r, take the natural logarithm of both sides. (**BioSkills 7** in Appendix A explains how to calculate a natural logarithm using your calculator.) In this case, $r = 0.046$ per year.

The instantaneous rate of increase, r, is also directly related to the net reproductive rate, R_0, introduced in Box 52.1. In most cases, r is calculated as $\ln R_0/g$, where g is the generation time. Thus, r can be calculated from life-table data. It is a more useful measure of growth rate than R_0, because r is independent of generation time.

To summarize, biologists have developed several ways of calculating and expressing a population's growth rate. Growth rate expressed as λ has the advantage of being easy to understand, and R_0 has the advantage of being calculated directly from life-table data. Although r is slightly more difficult conceptually, it is the most useful expression for growth rate, because it is independent of generation time and is relevant for species that breed either seasonally or continuously.

Applying the Models

To get a better feel for r and for Equation 52.7, consider the following series of questions about whooping cranes. The key to answering these questions is to realize that Equation 52.7 has just four parameters. Given three of these parameters, you can calculate the fourth. ✔If you understand this concept, you should be able to solve the following four problems:

1. If 20 individuals were alive in 1941 and 462 existed in 2009, what is r? Here $N_t = 462$, $N_0 = 20$, and $t = 68$ years. Substitute these values into Equation 52.7 and solve for r. Then check your answer at the end of this box.[1]
2. In the most recent report issued by the biologists working on the Wood Buffalo crane recovery program, it is estimated that the flock should be able to sustain an r of 0.046 for the foreseeable future. If the flock currently contains 253 individuals, how long will it take that population to double? Here $N_0 = 253$ and $N_t = 2 \times 253 = 506$. In this case, you solve Equation 52.7 for t. Then check your answer.[2]
3. In 2002 a pair of birds in the flock that lives in Florida year-round successfully raised offspring (nine years after the first cranes were introduced there). In 2003 another pair of cranes in this population bred successfully. If the number of breeding pairs continues to double each year, how long will it take to reach the goal of 25 breeding pairs? (Note that $\lambda = 2.0$ if a population is doubling each year.) Given that $N_0 = 2$ and $N_t = 25$, solve for t again, and check your answer.[3]
4. The whooping crane flock that migrates between Wisconsin and Florida was founded in 2001. If its development is like that of the resident flock in Florida, the first successful breeding attempt will occur in 2010—nine years after the initial introduction. Suppose that the instantaneous growth rate for the number of breeding pairs in this population will be 0.05. In what year will the breeding population reach 25 pairs? This is the year that whooping cranes should come off the endangered species list. Here $N_0 = 1$, $N_t = 25$, and $r = 0.05$. Solve for t, and add this number of years to 2010; then check your answer.[4]

[1] $r = 0.046$; [2] $t = 15$ years; [3] $t = 3.64$ years; [4] 2074

GRAPHING LOGISTIC GROWTH **Figure 52.7a** illustrates density-dependent growth in a hypothetical population. The graph plots changes in population size over time, and has three sections.

1. Initially, growth is exponential—meaning that r is constant.
2. With time, N increases to the point where competition for resources or other density-dependent factors begins to occur. As a result, the growth rate begins to decline.
3. When the population is at the habitat's carrying capacity, the growth rate is 0—the graph of population size versus time is flat.

This is exactly what happened in an experiment on laboratory populations of ciliates (see Chapter 29): *Paramecium aurelia* and *P. caudatum*. An investigator placed 20 individuals from one of the *Paramecium* species into 5 mL of a solution. He created many replicates of these 5-mL environments for each species separately. He kept conditions as constant as possible by adding the same number of bacterial cells every day for food, washing the solution every second day to remove wastes, and maintaining the pH at 8.0.

In the graphs in **Figure 52.7b**, each data point represents population size for one of the species—the average of the replicates in the experiment. Note that both species exhibited logistic growth in this environment. The carrying capacity differed in the two species, however. The maximum density of *P. aurelia* averaged 448 individuals per mL, but that for *P. caudatum* averaged just 128 individuals per mL. When the two species are grown together—one species gets eliminated (see Chapter 53).

In both of the *Paramecium* species in this experiment, exponential growth could be sustained for only about five days. What factors cause growth rates to change?

What Limits Growth Rates and Population Sizes?

Population sizes change as a result of two general types of factors:

- *Density-independent factors* alter birthrates and death rates irrespective of the number of individuals in the population, and usually involve changes in the abiotic environment—variation in weather patterns, or catastrophic events such as cold snaps, hurricanes, volcanic eruptions, or drought.

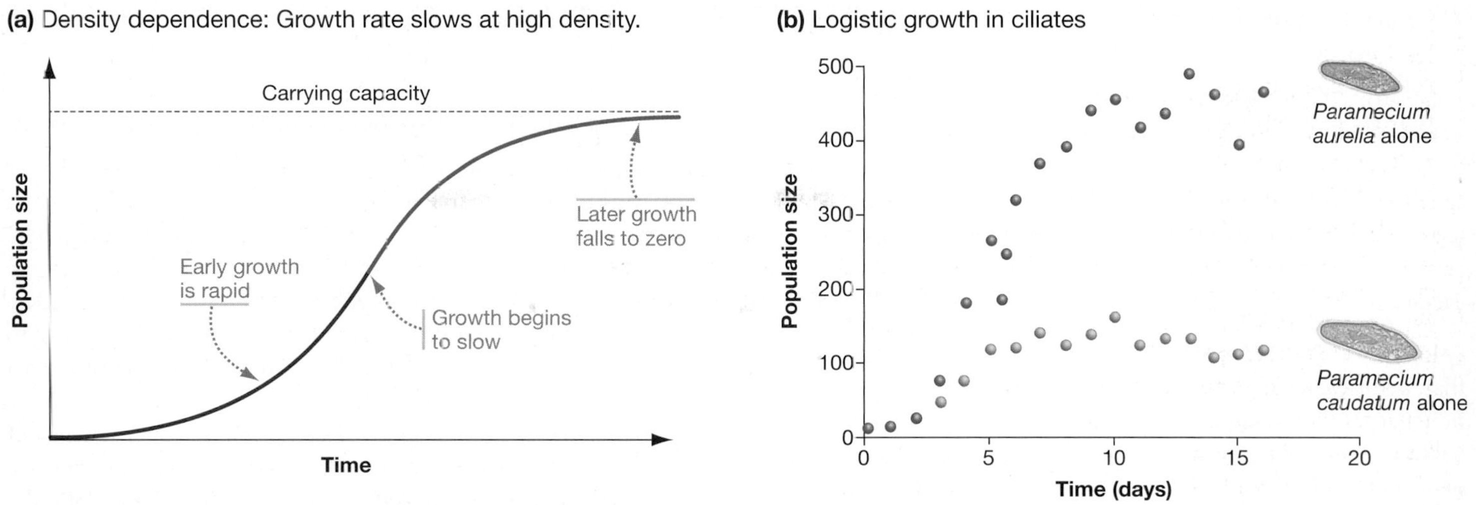

FIGURE 52.7 Logistic Growth Is Dependent on Population Size. (a) A curve illustrating the pattern predicted by the logistic growth equation. This pattern often occurs when a small number of individuals colonizes an unoccupied habitat. Initially, r is high because competition for resources is low to nonexistent. Carrying capacity depends on the quality of the habitat and can vary over time. **(b)** Data from laboratory experiments with two species of *Paramecium*.

- *Density-dependent factors* change in intensity as a function of population size, and are usually biotic in nature. When trees crowd each other, they have less water, nutrients, and sunlight at their disposal and make fewer seeds.

A CLOSER LOOK AT DENSITY DEPENDENCE To get a better understanding of how biologists study the density-dependent factors that affect population size, consider the data in **Figure 52.8**:

- The graph in Figure 52.8a presents results of an experimental study of a coral-reef fish called the bridled goby. Each data point represents an identical artificial reef, constructed by a researcher from the rubble of real coral reefs. The initial density, plotted along the x-axis, represents the marked bridled gobies that were released on each artificial reef at the start of the experiment. The proportion surviving, plotted on the vertical axis, represents the introduced individuals that were still living at that reef 2.5 months later. This graph shows a strong density-dependent relationship in survivorship.
- The graph in Figure 52.8b is from a long-term study of song sparrows on Mandarte Island, British Columbia. Each data point represents a different year. The density of females,

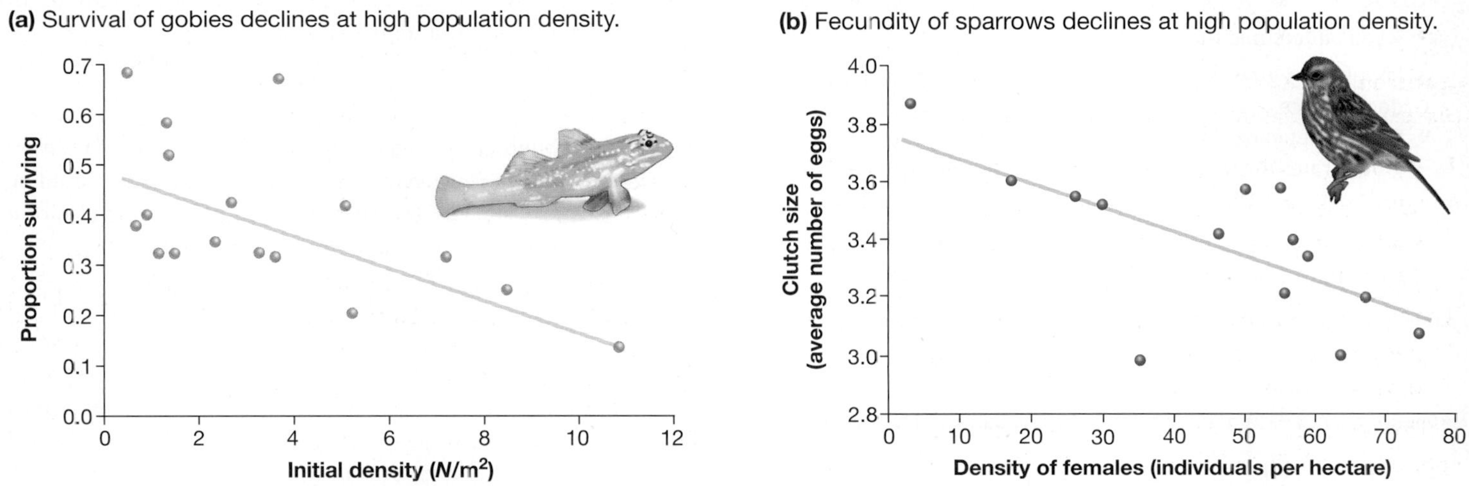

FIGURE 52.8 Density-Dependent Growth Results from Changes in Survivorship and Fecundity. (a) When bridled gobies are introduced at different population densities to artificial reefs, survivorship is highest at low density. **(b)** In the song sparrow population on Mandarte Island, there is a strong negative correlation between population density in a particular year and the average number of eggs (clutch size) produced by females.

✔**QUESTION** When song sparrow populations are at high density and extra food is provided to females experimentally, average clutch size is much higher than expected from the data in part (b). Based on these data, state a hypothesis to explain why average clutch size declines at high density.

plotted along the horizontal axis, is the number of females that bred on the island; the clutch size, plotted on the vertical axis, is the average number of eggs laid by each female. This graph indicates a strong density-dependent relationship in fecundity.

Density-dependent changes in survivorship and fecundity cause logistic population growth. In this way, density-dependent factors define a particular habitat's carrying capacity. If gobies get crowded, they die or emigrate. If song sparrows are crowded, the average number of eggs and offspring that they produce declines.

CARRYING CAPACITY IS NOT FIXED It's important to recognize that *K* varies among species and populations. *K* varies because for any particular species, some habitats are better than other habitats due to differences in food availability, space, and other density-dependent factors. Stated another way, *K* varies in space. It also varies with time, as conditions in some years are better than in others.

In addition, the same habitat may have a very different carrying capacity for different species—as the experiment with growing paramecia in 5-mL vials showed. The same area will tend to support many more individuals of a small-bodied species than of a large-bodied species, for example, simply because large individuals demand more space and resources.

These simple observations help explain the variation in total population size that exists among species and among populations of the same species. To review the basic principles of exponential growth, logistic growth, and factors that limit population size, go to the study area at *www.masteringbiology.com.*

(MB) **BioFlix™** Population Ecology, **Web Activity** Modeling Population Growth

CHECK YOUR UNDERSTANDING

If you understand that . . .

- Populations grow exponentially unless slowed by density dependence.
- Density-dependent factors that influence population growth rates include competition for resources such as food and sunlight.

✔ You should be able to . . .

Propose a density-dependent factor that limits growth in *Paramecium aurelia* and design an experiment to test this hypothesis.

Answers are available in Appendix B.

52.3 Population Dynamics

The tools introduced in the previous two sections provide a foundation for exploring how biologists study **population dynamics**—changes in populations through time. Research on population dynamics has uncovered a wide array of patterns in natural populations in addition to exponential and logistic growth. Let's start by considering one of the most important patterns observed: the "blinking on-and-off" of fragmented populations.

How Do Metapopulations Change through Time?

If you browse through an identification guide for trees or birds or butterflies, you'll likely find range maps indicating that most species occupy a broad area. In reality, however, the habitat preferences of many species are restricted, and individuals occupy only isolated patches within that broad area. The meadows occupied by Glanville fritillaries—an endangered species of butterfly native to the Åland islands off the coast of Finland (**Figure 52.9**)—are a well-studied example of this pattern.

If individuals from a species occupy many small patches of habitat, so that they form many independent populations, they are said to represent a **metapopulation**—a population of populations.

Glanville fritillaries exist naturally as metapopulations. But because humans are reducing large, contiguous areas of forest and grasslands to isolated patches or reserves, more and more species are being forced into a metapopulation structure. Recent research on Glanville fritillaries by Ilkka Hanski and colleagues illustrates the consequences of metapopulation structure for endangered species.

METAPOPULATIONS SHOULD BE DYNAMIC Research on Glanville fritillaries began with a survey of meadow habitats on the islands. Because fritillary caterpillars feed on just two types of host plant, *Plantago lanceolata* and *Veronica spicata*, the team was able to pinpoint potential butterfly habitats.

Hanski's group estimated the fritillary population size within each patch of habitat by counting the number of larval webs in each. Of the 1502 meadows that contained the host plants, 536 had Glanville fritillaries. The patches ranged from 6 m^2 to 3 ha in area. Most had only a single breeding pair of adults and one larval group, but the largest population contained hundreds of pairs of breeding adults and 3450 larvae.

Figure 52.10 illustrates how a metapopulation like this is expected to change over the years.

- Given enough time, each population within the larger metapopulation is expected to go extinct. The cause could be a catastrophe, such as a storm or an oil spill; it could also be a disease outbreak or a sudden influx of predators.
- Migration from nearby populations can reestablish populations in empty habitat fragments.

In this way, there is a balance between extinction and recolonization within a metapopulation. Even though subpopulations blink on and off over time, the overall population is maintained at a stable number of individuals.

AN EXPERIMENTAL TEST Does this metapopulation model correctly describe the population dynamics of fritillaries? To answer this question, Hanski's group conducted a mark-recapture study. They caught and marked many individuals, released them, and revisited fritillary habitats, later, to recapture as many individuals as possible (see Box 52.3 on page 1048).

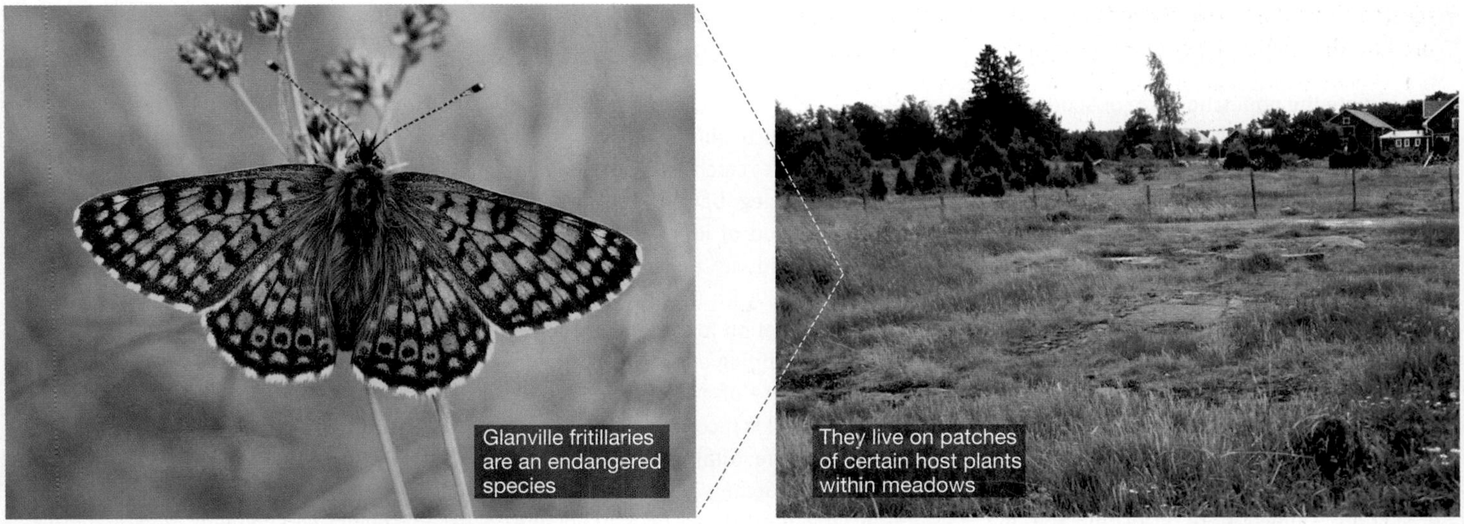

Glanville fritillaries are an endangered species
They live on patches of certain host plants within meadows

FIGURE 52.9 Glanville Fritillaries Live in Patches of Meadow. The Glanville fritillary is declining rapidly in many parts of Europe. In Finland, the only remaining individuals occupy the Åland islands.

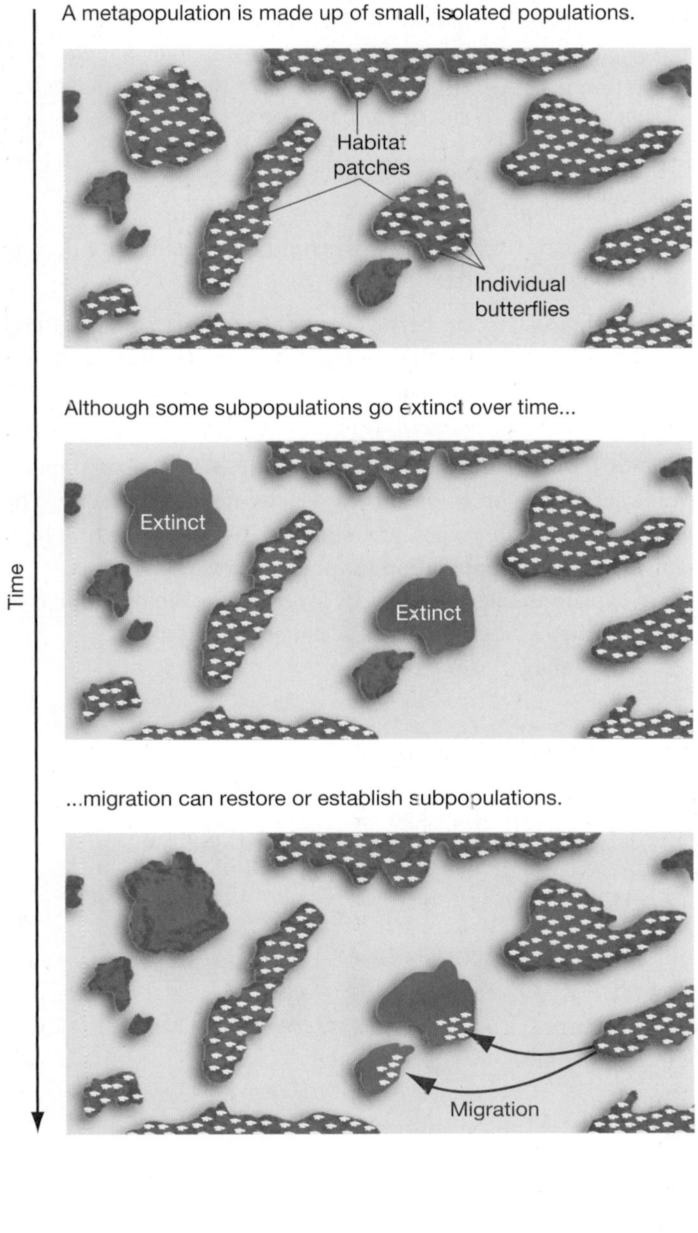

FIGURE 52.10 Metapopulation Dynamics Depend on Extinction and Recolonization. The overall population size of a metapopulation stays relatively stable even if subpopulations go extinct. These populations may be restored by migration, or unoccupied habitats might be colonized.

Of the 1731 butterflies that the biologists marked and released, 741 were recaptured over the course of the summer study period. Of the recaptured individuals, 9 percent were found in a previously unoccupied patch. This migration rate is high enough to suggest that patches where a population has gone extinct will eventually be recolonized.

Hanski and co-workers repeated the survey two years after their initial census. Just as the metapopulation model predicted, some populations had gone extinct and others had been created. On average, butterflies were lost from 200 patches each year; 114 unoccupied patches were colonized and newly occupied each year. The overall population size was relatively stable even though constituent populations came and went.

To summarize, the history and future of a metapopulation is driven by the birth and death of populations, just as the dynamics of a single population are driven by the birth and death of individuals. As the final section in this chapter will show, these dynamics have important implications for saving endangered species.

Why Do Some Populations Cycle?

Analyzing population cycles has been a particularly productive way to understand how intraspecific ("within-species") factors—such as competition for nutrients and space—interact with interspecific interactions, such as predation and disease.

BOX 52.3 QUANTITATIVE METHODS: Mark-Recapture Studies

To estimate the population size of sedentary (largely immobile) organisms, researchers sample appropriate habitats by counting the individuals that occur along transects or inside rectangular plots—called quadrats—set up at random locations in the habitat. These counts can be extrapolated to the entire habitat area to estimate the total population size. In addition, they can be compared to later censuses to document trends over time.

In contrast, estimating the total population size of organisms that are mobile and that do not congregate into herds or flocks or schools is much more of a challenge. In species such as the Glanville fritillary, the population inside sample quadrats or along transects changes constantly as individuals move in and out. Further, it can be difficult to track whether a particular individual has already been counted.

If individuals can be captured and then tagged in some way, however, the total population size of a mobile species can be estimated by using a research approach called mark-recapture. To begin a mark-recapture study, researchers catch individuals and mark them with leg bands, ear tags, or some other method of identification. After the marked individuals are released, they are allowed to mix with the unmarked animals in the population for a period of time. Then a second trapping effort is conducted, and the percentage of marked individuals that were captured is recorded.

To estimate the total population from these data, researchers make a key assumption: The percentage of marked and recaptured individuals is equal to the percentage of marked individuals in the entire population. This assumption should be valid when no bias exists regarding which individuals are caught in each sample attempt. It is important that individuals do not learn to avoid traps after being caught once and that they do not emigrate or die as a result of being trapped.

The relationship between marked and unmarked individuals can be concisely expressed algebraically:

$$\frac{m_2}{n_2} = \frac{m_1}{N}$$

In this equation, m_2 is the number of marked animals in the second sample (the recapture), n_2 is the total number of animals (marked and unmarked) in the second sample, m_1 is the number of marked animals in the first sample, and N is the total population size. Having measured m_2, n_2, and m_1, the researcher can estimate N. ✔If you understand how to estimate population size in a mark-recapture study, you should be able to do so in this example: Suppose researchers marked 255 animals and later were able to trap a total of 162 individuals in the population, of which 78 were marked. What is the estimate for total population size? Solve for N (give the nearest integer) and check your answer.[1]

[1] $N = 530$

Figure 52.11 shows a classic case of population cycling that involves two species: snowshoe hare and lynx in northern Canada.

The *x*-axis on this graph plots a 50-year interval; the *y*-axes show changes in hare population density (on the left, with a logarithmic scale) and changes in lynx density (on the right). Data for hares are plotted in tan; data for lynx are plotted in purple. Note several key points:

1. The scales on the *y*-axes are different—there are many more hares than lynx. (The *y*-axis for hares is logarithmic, to compress the values so they can be compared more easily to changes in the lynx population.)
2. The hare and lynx populations cycle every 11 years on average.
3. Changes in lynx density lag behind changes in hare density by about two years.

Snowshoe hares are herbivores—they subsist on leaves and stems that grow close to the ground. Lynx are predators, and subsist mainly on snowshoe hares.

IS IT FOOD OR PREDATION? Most hypotheses to explain population cycles hinge on some sort of density-dependent factor. The idea is that food shortages, predation, or disease intensify at high density and cause population numbers to crash.

To explain the hare-lynx cycle, for example, biologists pushed two hypotheses based on density dependent factors:

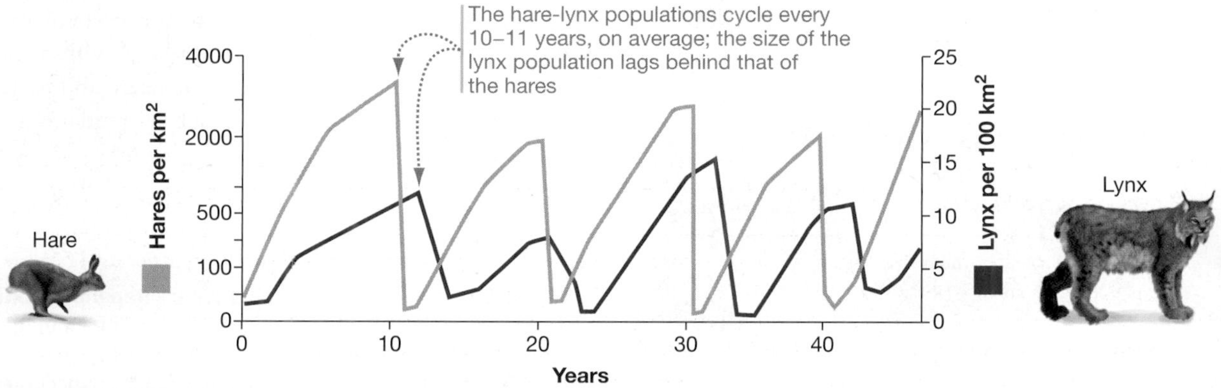

Figure 52.11 The Hare and Lynx Population Cycles Are Synchronous, but Lagged.

1. Hares use up all their food when their populations reach high density and starve; in response, lynx also starve.
2. Lynx populations reach high density in response to increases in hare density. At high density, lynx eat so many hares that the prey population crashes.

Stated another way, hares control lynx population size, or lynx control hare population size. The cycle could be controlled by intraspecific competition or interspecific interactions.

A FIELD EXPERIMENT To test these hypotheses, researchers set up a series of 1-km^2 study plots in boreal forest habitats that were as identical as possible (**Figure 52.12**).

- Three plots were left as unmanipulated controls.
- One plot was ringed with an electrified fence with a mesh that excluded lynx but allowed hares to pass freely.
- Two plots received additional food for hares year-round.
- One plot had a predator-exclusion fence and was also supplemented with food for hares year-round.

The biologists then monitored the size of the hare and lynx populations over an 11-year period, or enough time for a complete cycle in the two populations.

The graphs in Figure 52.12 plot average hare density in the lynx-exclusion, food-addition, and combination plots relative to controls. A value of 1 on the *y*-axis means that hare density in the experimental and control plots were the same. Each bar represents the average population density over a six-month interval; data are plotted for eight years of the study.

The graphs should convince you that plots with predators excluded showed higher hare populations than did control plots during the "decline" phase of the cycle. This result supports the hypothesis that predation by lynx reduces hare populations. Plots with

EXPERIMENT

QUESTION: What factors control the hare-lynx population cycle?

HYPOTHESIS: Predation, food availability, or a combination of those two factors controls the hare-lynx cycle.

NULL HYPOTHESIS: The hare-lynx cycle isn't driven by predation, food availability, or a combination of those two factors.

EXPERIMENTAL SETUP:

Document hare population in seven study plots (similar boreal forest habitats, each 1 km^2) for duration ofone cycle (1987–1994).

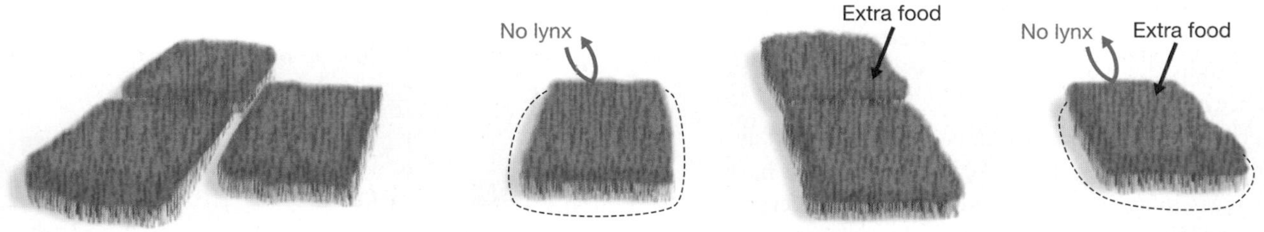

3 plots: Unmanipulated controls

1 plot: Electrified fence excludes lynx but allows free access by hares.

2 plots: Supply extra food for hares.

1 plot: Electrified fence excludes lynx but allows free access by hares; supply extra food for hares.

PREDICTION: Hare populations in at least one type of manipulated plot will be higher than the average population in control plots.

PREDICTION OF NULL HYPOTHESIS: Hare populations in all of the plots will be the same.

RESULTS:

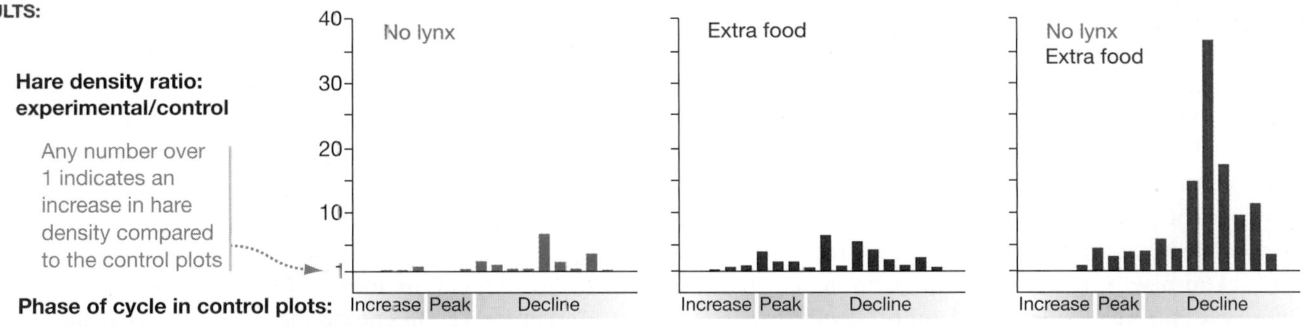

CONCLUSION: Hare populations are limited by both predation and food availability. When predation and food limitation occur together, they have a greater effect than either factor does independently.

FIGURE 52.12 Experimental Evidence That Predation and Food Availability Drive the Hare-Lynx Cycle.

SOURCE: Krebs, C. J. et al. 1995. Impact of food and predation on the snowshoe hare cycle. *Science* 269: 1112–1115.

✔**QUESTION** Why didn't the research group do more replicates of each treatment?

supplemental food also had much higher populations than controls during the peak and decline phases. But the plot with supplemental food *and* predators excluded had a hare population that got as much as 35 times more dense than the population in the control plots.

Note that most of the effects occurred during the decline phase of the cycle. As control plots were crashing, the experimental plots retained some or many hares.

These data support the hypothesis that hare populations are limited by availability of food as well as by predation, and that food availability and predation intensity interact—meaning the combined effect of food and predation is much larger than their impact in isolation. The leading hypothesis to explain this combined effect is that when hares are at high density, individuals are weakened by nutritional stress and are more susceptible to predation.

Research on population cycles has increased our understanding of density dependence factors in population dynamics. Now let's consider how the arrangement of populations in space affects overall population dynamics.

How Does Age Structure Affect Population Growth?

As the life-table data presented in Section 52.1 showed, age has a dramatic effect on the probability that an individual will reproduce and survive to the following year. But the previous analysis of life-table data left out a critical point about the ages of individuals in a population: A population's **age structure**—meaning the proportion of individuals that are at each possible age—has a dramatic influence on the population's growth over time.

To see how changes in age structure can affect population dynamics, let's consider two case histories for which biologists have documented changes in age structure. One example concerns a flowering plant, and the other focuses on our own species.

AGE STRUCTURE IN A WOODLAND HERB The common primrose, pictured in **Figure 52.13a**, grows in the woodlands of Western Europe. Primroses grow on the forest floor and can germinate, mature, and produce offspring only in the relatively high-light environment created when a tall tree falls and opens a gap in the forest canopy.

The sunny spaces where primroses thrive are short lived, because mature trees and saplings in and around the gap grow and fill the space. As they do, light levels in the gap decline and the environment becomes increasingly unsuitable for herbs such as primrose.

To understand how primrose populations respond to this ephemeral woodland environment, researchers hypothesized that populations in new gaps and populations in large gaps—meaning those with the highest light availability—would experience high rates of reproduction and an age structure characterized by large numbers of juveniles.

The researchers also hypothesized that as light levels declined in a particular gap due to the growth of surrounding trees, primrose reproduction would decline. As a result, the age structure of populations in older gaps should be characterized by a large proportion of individuals in adult stages, with fewer juveniles.

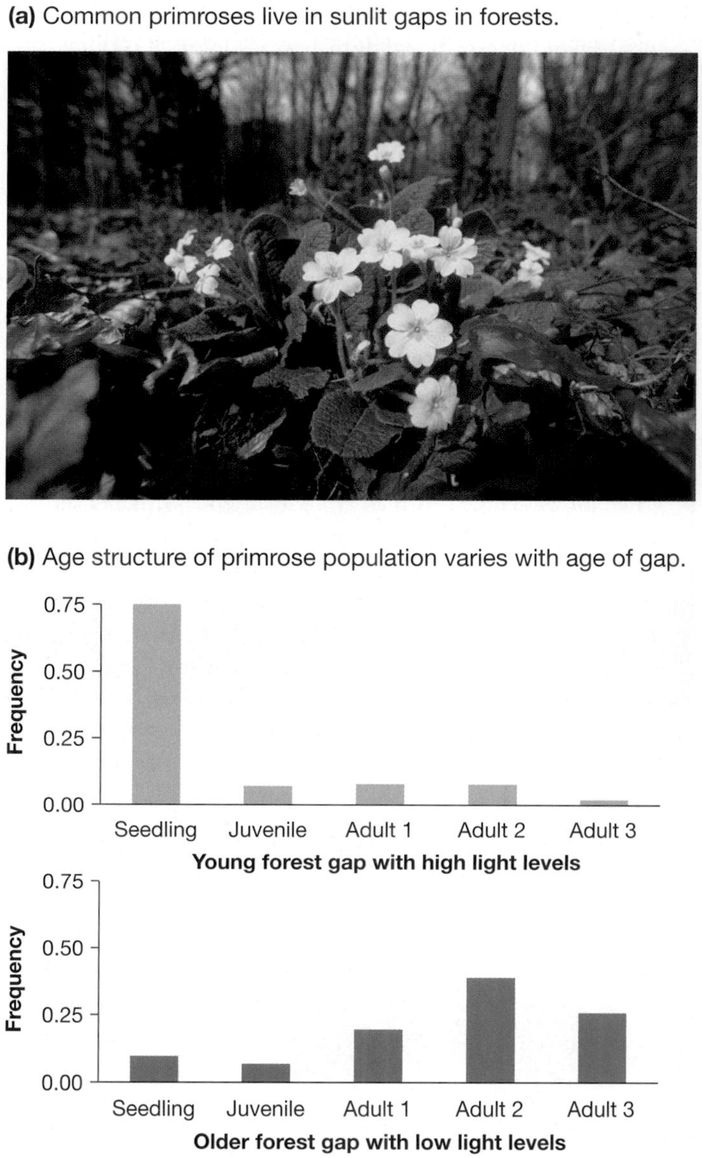

FIGURE 52.13 The Age Structure of Common Primrose Populations Changes over Time. In the graphs, "Adult 1," "Adult 2," and "Adult 3" individuals are of increasingly large size and thus age.

To test these hypotheses, the investigators selected eight common primrose populations growing in a range of light levels, from high to low. The team established 1-m^2 study plots inside each gap and studied approximately 350 individuals per population. Each individual was assigned to one of five ages: seedlings, juveniles, or adults in one of three size classes.

As **Figure 52.13b** shows, the data supported the hypothesis that in gaps with high light levels, populations have a larger proportion of juveniles than do gaps with low light levels. In addition, the proportion of juvenile individuals declined over time in all gaps as light levels declined. These data have three important messages about the dynamics of primrose populations:

1. Populations that are dominated by juveniles should experience rapid growth. Population size should then decline with time due to a density-independent factor: shading by trees.
2. The long-term trajectory of the overall primrose population in an area may depend primarily on the frequency and severity of windstorms that knock down trees and create sunlit gaps. If so, it means that the dynamics of primrose populations are governed by an abiotic, density-independent factor.
3. In a large tract of forest, the primrose population will consist of a large number of subpopulations, each found in canopy gaps. Some of these subpopulations are likely to be tiny, and each subpopulation will arise, grow, and then go extinct over time. Primroses have metapopulation structure.

AGE STRUCTURE IN HUMAN POPULATIONS As another example of how age structure affects population dynamics, consider humans. To study age structure in our species, researchers use stacks of horizontal bars to plot the number of males and the number of females in each age cohort—often in 5-year intervals between birth (0) and 100. The resulting graph is called a population pyramid or age pyramid.

In countries where industrial and technological development took place generations ago, like Sweden, survivorship has been high and fecundity low for several decades. An age pyramid like that in **Figure 52.14a** results. The most striking pattern in these data is that there are similar numbers of people in most age classes. The evenness occurs because about the same number of infants are being born each year, and because most survive to old age.

In contrast, the age distribution is bottom-heavy in less-developed nations like Honduras (**Figure 52.14b**). These populations are dominated by the very young because they are undergoing rapid growth—with more children being born recently than exist in older age classes.

Analyzing an age pyramid can give biologists important information about a population's history. But studying age distributions can also help researchers predict a population's future.

Look again at Figure 52.14a, and note that the white lines and lighter bars show what the age structure of Sweden is projected to be in 2050. Modest changes in the age distribution will occur because survivorship has increased while fecundity remains the same. But the population is not expected to grow quickly because only modest numbers of individuals reach reproductive age. The projections highlight a major public policy concern in the industrialized countries: how to care for an increasingly aged population.

Now consider Figure 52.14b again. Due to dramatic improvements in health care, most young Hondurans survive to reproductive age. The projected age distribution in 2050 does not continue to flare out, however, because fecundity is declining. The projections illustrate a major public policy concern in less-developed countries: providing education and jobs for an enormous group of young people who will be reaching adulthood during your lifetime.

POPULATION "MOMENTUM" The data in Figure 52.14b make another important point: Because of recent and rapid population growth in developing countries, overall population size will increase dramatically in these nations over the course of your lifetime. Honduras will have many more people in 2050 than it does now.

A large part of this increase will be due to increased survivorship. But the number of offspring being born each year is also expected to stay high, even though fecundity is predicted to *decline*.

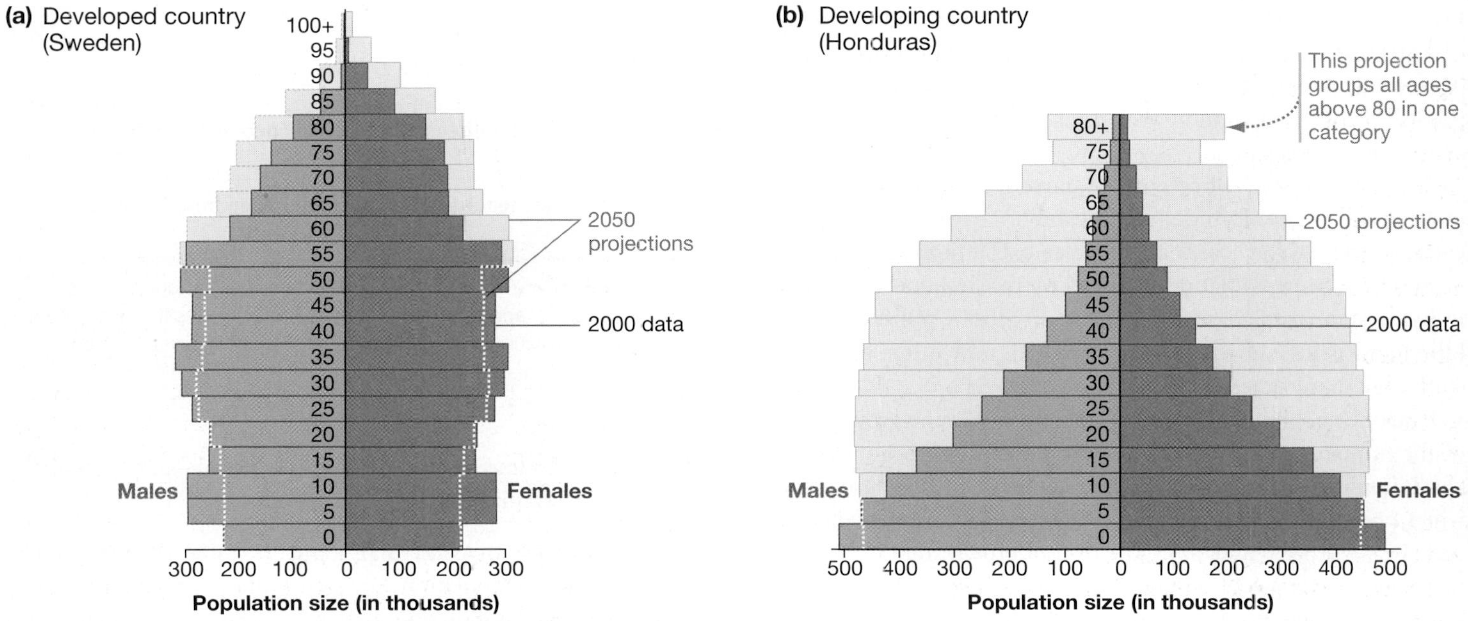

FIGURE 52.14 The Age Structure of Human Populations Varies Dramatically.

This result seems paradoxical, but is based on a key observation: There are now so many young women in these populations that the overall number of births will stay high, even though the average number of children *per female* is much less than it was a generation ago.

To capture this point, biologists say that these populations have momentum or inertia. Combined with high survivorship, age structures like these make continued increases in the total human population almost inevitable.

Analyzing Change in the Growth Rate of Human Populations

Figure 52.15 plots changes in the human population over the past 250 years. Although the curve's shape looks superficially similar to exponential growth, the growth rate for humans has—until recently—actually *increased* over time since 1750, leading to a steeply rising curve.

It is almost impossible to overemphasize just how dramatically the human population has grown recently. **Table 52.2** presents data that will allow you to calculate (**1**) how long it has taken this population to add each billion people, (**2**) determine how many times the population has doubled—from a starting point in the year A.D. 1420—and (**3**) figure out how many years the population took to double each time.

To put your calculations on doubling time in perspective, consider that until your grandparents' generation, no individual had lived long enough to see the human population double. But many members of the generation born in the early 1920s have lived to see the population *triple*. And to drive home the impact of continued population doubling, consider the following: If you were given a penny on January 1, then \$0.02 on January 2, \$0.04 on January 3, and so on, you would be handed \$10,737,418 on January 31.

HOW LARGE IS THE CURRENT HUMAN POPULATION? As this book goes to press, the world population is estimated at over 6.8 billion. About 77 million additional people—equivalent to the current population of Egypt, or more than double the state of California—are being added each year.

The consequences of recent and current increases in human population are profound. In addition to being the primary cause of the habitat losses and species extinctions analyzed in Chapter 55, overpopulation is linked by researchers to declines in living standards, mass movements of people, political instability, and acute shortages of water, fuel, and other basic resources in many parts of the world.

The one encouraging trend in the data is that the growth rate of the human population has already peaked and begun to decline. The highest growth rates occurred between 1965 and 1970, when populations increased at an average of 2.04 percent per year. Between 1990 and 1995, the overall growth rate in human populations averaged 1.46 percent per year; the current growth rate is 1.2 percent annually.

In humans, *r* may be undergoing the first long-term decline in history. The question is: Will the human population stabilize or begin to decline in time to prevent global—and potentially irreversible—damage?

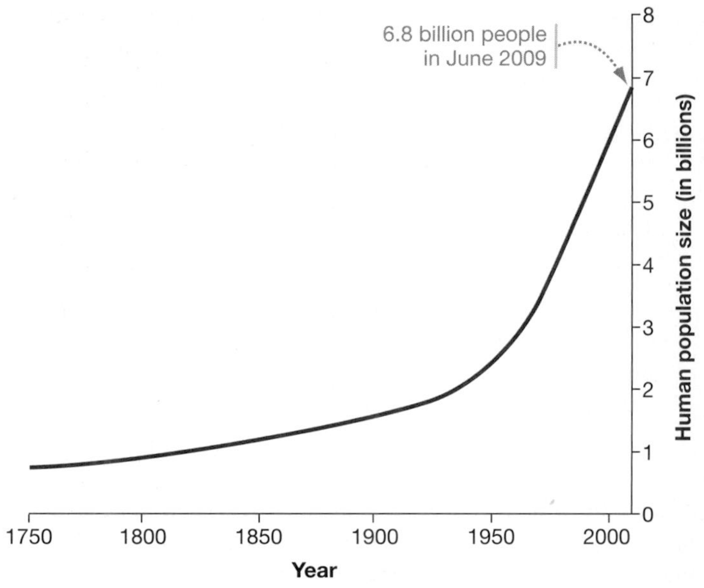

FIGURE 52.15 The Human Population Has Been Growing Rapidly. Graph based on estimates from historical and census data.

TABLE 52.2 **Milestones in Human Population Growth**

Date	Human Population (in millions)
1420	375
1720	750
1800	1000
1875	1500
1927	2000
1961	3000
1974	4000
1987	5000
1999	6000
2009	6800
Number of years required to add 1 billion people to 2009 level:	
Number of years required to double population from 2009 level:	

✔**EXERCISE** Using the data provided here and Equations 52.5 and 52.7 in Box 52.2, and assuming that growth continues at the same rate as from 1999 to 2009, fill in the correct number of years in the blank cells at the bottom of the table.

WILL HUMAN POPULATION SIZE PEAK IN YOUR LIFETIME? The United Nations Population Division makes regular projections for how human population size will change between now and 2050. For most readers of this book, these projections describe what the world will look like as you reach your early 60s.

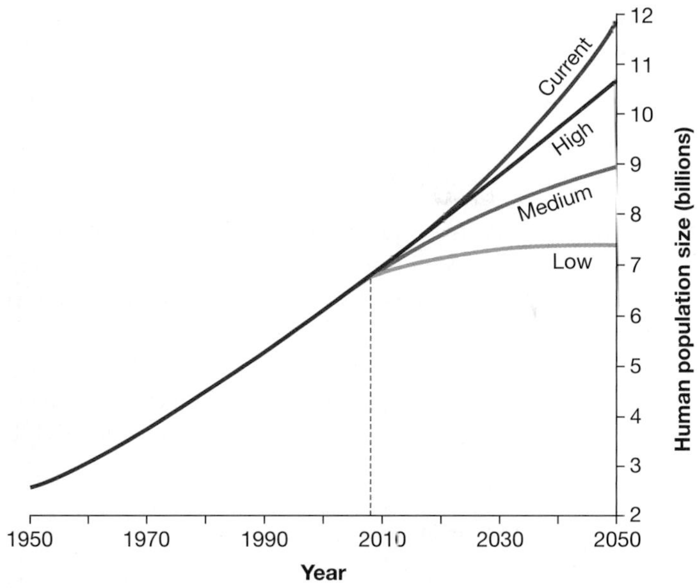

FIGURE 52.16 Projections for Human Population Growth to 2050. The "Current," "High," "Medium," and "Low" labels refer to fertility rates.

The UN projections are based on four scenarios, which hinge on different values for fertility rates—the average number of children that each woman has during her lifetime. Currently, the worldwide average is 2.7. This represents an enormous reduction from the fertility rates during the 1950s, which averaged 5.0 children per woman. **Figure 52.16** shows how total population size is expected to change by 2050 if average fertility continues to decline and averages 2.5 (high), 2.1 (medium), or 1.7 (low) children per woman.

The middle number in the projections, 2.1, is the replacement rate in developed countries, where survivorship is high. The **replacement rate** is the average fertility required for each woman to produce exactly enough offspring to replace herself and her offspring's father. When this fertility rate is sustained for a generation, $r = 0$ and there is **zero population growth (ZPG)**.

A glance at Figure 52.16 should convince you that the four scenarios are starkly different. If average fertility around the world stays at its present level, world population will be closing in on 12 billion about the time you might consider retiring. This is close to double the current population. To realize what this might mean, imagine what traffic or home prices in your city would be like if double the number of people were using the roads and looking for housing.

The "high-fertility" projection assumes that fertility rates continue to drop but average 2.5 children per woman over this interval. This model predicts that world population in 2050 will reach nearly 11 billion—a 75 percent increase over today's population—with no signs of peaking. The low-fertility projection, in contrast, predicts that the total human population in 2050 will be about 7.4 billion and will already have peaked.

THE ROLE OF FERTILITY RATES The UN's population projections make an important point: The future of the human population hinges on fertility rates—on how many children each of the women living today decides to have. Those decisions, in turn, depend on a wide array of factors, including how free women are to choose their family size and how much access women have to education.

Why education? When women are allowed to become educated, they tend to delay when they start having children and have a smaller overall family size. Access to education and reliable birth control methods (see Chapter 48), in addition to overall economic development and access to quality health care, will play a large role in determining how world population changes over your lifetime.

To summarize, humans may be approaching the end of a period of rapid growth that lasted well over 500 years. How quickly growth rates decline and how large the population eventually becomes will be decided primarily by changes in fertility rates.

To review concepts on human population growth, go to the study area at *www.masteringbiology.com.*

MB **Web Activity** Human Population Growth and Regulation

CHECK YOUR UNDERSTANDING

If you understand that . . .

- Density-dependent factors such as competition for food and predation intensity can drive population cycles in some species.
- In a metapopulation, subpopulations "blink on and off" over time as some go extinct and others reestablish via colonization.
- A population's age structure reflects changes in survivorship and fertility rates over time.
- Age structure impacts population dynamics because it dictates how many females of reproductive age are in the population.

✔ **You should be able to . . .**

1. Explain why the age structure of human populations differs between developed and developing nations.
2. Explain why changes in fertility rates have such a dramatic impact on projections for the total human population in 2050.

Answers are available in Appendix B.

52.4 How Can Population Ecology Help Endangered Species?

When designing programs to save species threatened with extinction, conservationists draw heavily on concepts and techniques from population ecology. As habitat destruction pushes species throughout the world into decline, the study of population growth rates and population dynamics has taken on an increasingly applied tone.

To illustrate one of the ways that biologists apply the theory and results introduced in this chapter, let's analyze how an understanding of life-table data and geographic structure can help direct conservation action.

Using Life-Table Data

Collecting and analyzing demographic data such as age-specific survivorship and fecundity are important for saving endangered species and for other applied problems. To understand why, suppose that you were in charge of reintroducing a population of lizards to a nature reserve, and that research conducted when this species occupied the site previously documented survivorship and fecundity.

Your initial plan is to take 1000 newly hatched females from a captive breeding center and release them into the habitat, along with enough males for breeding to occur. Is this population likely to become established and grow, or will you need to keep introducing offspring that have been raised in captivity?

MAKING POPULATION PROJECTIONS To answer this question, you need to use the life-table data in **Figure 52.17a** and calculate:

1. How many adults survive to each age class each year?
2. How many offspring are produced by each adult age class—each year over the course of several years?

Figure 52.17b starts these calculations for you. The fate of the original 1000 females is indicated in red. Note that in the second year, just 330 of these individuals have survived; 40 are left as 3-year-olds.

How many offspring did this generation produce? Remember that just 330 of the original 1000 females are expected to remain after one year. As the first purple number in Figure 52.17b

(a) Life table

Age (x)	Survivorship (l_x)	Fecundity (m_x)
0 (birth)	----	0.0
1	0.33	3.0
2	0.2	4.0
3	0.04	5.0
4		

(b) Fate of first-generation females

Year	0 (newborns)	1-year-olds	2-year-olds	3-year-olds	Total population size (N)
1st	1000 (just introduced)				1000
2nd	990 (= 330 × 3.0)	330 (= 1000 × 0.33)			1320 (= 990 + 330)
3rd	800 (= 200 × 4.0)		200 (= 1000 × 0.20)		
4th	200 (= 40 × 5.0)			40 (= 1000 × 0.04)	
5th					

(c) Fate of first- and second-generation females

Year	0 (newborns)	1-year-olds	2-year-olds	3-year-olds	Total population size (sum across all rows)
1st	1000				1000
2nd	990	330			1320
3rd	800 + 981 (981 = 327 × 3.0)	327 (= 990 × 0.33)	200		2308 (= 800 + 981 + 327 + 200)
4th	200 + 792 (792 = 198 × 4.0)		198 (= 990 × 0.20)	40	
5th	195 (195 = 39 × 5.0)			39 (= 990 × 0.04)	

FIGURE 52.17 Life-Table Data Can Be Used to Project the Future of a Population. (a) Life table providing age-specific survivorship and fecundity for a hypothetical population of lizards. **(b)** Predicted fate of 1000 one-year-old females introduced into a habitat just before the breeding season. The number of individuals in this cohort is shown in red; the number of offspring they produce each year is indicated in purple. **(c)** Extension of the data in part (b), indicating how many of the offspring produced by the original females in their first year survived in subsequent years, shown in purple, and how many offspring they produced in each subsequent year, shown in green.

✔**EXERCISE** Assume that all 4-year-old females die after producing three young. Fill in the 4th year in the figure.

indicates, these survivors have an average of 3.0 female offspring apiece. Thus, they contribute 990 new female offspring to the population. In the second year, the 200 females that are left from the original cohort have an average of 4.0 female offspring each and contribute 800 juveniles. In their third year, the 40 surviving females average 5.0 offspring and contribute 200 juveniles.

Figure 52.17c extends the calculations by showing what happens as the offspring of the original females begin to breed. Their offspring are shown in green. ✔**By adding subsequent generations and continuing the analysis, you should be able to predict whether the population will stay the same, decline, or increase over time.** In this way, life-table data can be used to predict the future of populations.

ALTERING VALUES FOR SURVIVORSHIP AND FECUNDITY Part of the value of a population projection based on life-table data—like the one begun in Figure 52.17—is that it allows biologists to alter values for survivorship and fecundity at particular ages and assess the consequences. For example, suppose that a predatory snake began preying on juvenile lizards. According to the model in Figure 52.17, what would be the impact of a change in juvenile mortality rate?

Analyses like this allow biologists to determine which aspects of survivorship and fecundity are especially sensitive for particular species. The studies done to date support some general conclusions:

- Whooping cranes, sea turtles, spotted owls, and many other endangered species have high juvenile mortality, low adult mortality, and low fecundity. In these species, the fate of a population is extremely sensitive to increases in adult mortality. Based on this insight, conservationists have recently begun an intensive campaign to reduce the loss of adult female sea turtles in fishing nets. Previously, most conservation action had focused on protecting eggs and nesting sites.
- In humans and other species with high survivorship in most age classes, rates of population growth are extremely sensitive to changes in age-specific fecundity. Because of this, programs to control human population growth focus on two issues: lowering fertility rates through the use of birth control, and delaying the age of first reproduction by improving women's access to education.

In some or even most cases, however, the population projections made from life-table data may be too simplistic to be useful. For example, conservationists may need to expand the basic demographic models to account for occasional disturbances such as fires or storms or disease outbreaks. And what about species where overall population dynamics are dictated by a metapopulation structure?

Preserving Metapopulations

Habitat destruction caused by suburbanization and other human activities leaves small populations isolated in pockets of intact habitat. Work on Glanville fritillaries and other species has shown that a small, isolated population—even one within a nature preserve—is unlikely to survive over the long term. What can be done about this?

In Glanville fritillaries, data collected by Hanski's group have identified the attributes of subpopulations that are most likely to persist: They are (1) large, (2) occupy larger geographical areas, and (3) are closer to neighboring populations (and hence more likely to be colonized). In addition, fritillary populations with low genetic diversity—probably due to inbreeding—are more likely to go extinct than are populations of the same size that have higher genetic diversity.

Results like these have important messages for conservation biologists:

- Areas that are being protected for threatened species should be substantial enough in area to maintain large populations that are unlikely to go extinct in the near future. Large populations are more likely than small populations to endure disasters such as storms, and they are much less susceptible to the damaging effects of inbreeding.
- When it is not possible to preserve large tracts of land, an alternative is to establish systems of smaller tracts that are connected by corridors of habitat, so that migration between patches is possible. Constructing "wildlife corridors" is particularly important for species that do not fly, and when patches of appropriate habitat are separated by highways, subdivisions, rivers, or other obstacles that are dangerous for dispersing animals to cross. Corridors are also essential for plants with limited seed dispersal. In this case, it may take years for a plant species to "grow across" a corridor and disperse to new habitats.
- If the species that is threatened exists as a metapopulation, it is crucial to preserve at least some patches of unoccupied habitat to provide future homes for immigrants. If a population is lost from a preserve, the habitat should continue to be protected so it can be colonized in the future. This can be difficult to do—nonscientists are often skeptical about preserving scarce land that does not (yet) contain the species of interest.

These results also have a more general message: Although traditional population growth models such as the expressions for exponential and logistic growth are simple and elegant, the factors they ignore—immigration and emigration—are crucial to understanding the dynamics of most populations. Because metapopulation structure is common, biologists must use more sophisticated models to predict the fates of populations.

CHAPTER 52 REVIEW

For media, go to the study area at www.masteringbiology.com

Summary of Key Concepts

Life tables summarize how likely it is that individuals of each age class in a population will survive and reproduce.

- Three general types of survivorship curves have been observed: (1) high survivorship throughout life, (2) a constant probability of dying at each year of life, and (3) high mortality in juveniles followed by high survivorship in adults.
- The resources available to an individual are always limited, so any increase in allocation of resources to survival and competitive ability necessitates a decrease in the resources allocated to reproduction. This trade-off between survival and reproduction is the most fundamental aspect of a species' life history.
- In analyzing life history, there is a continuum between populations or species with high survivorship and low fecundity and populations or species with low survivorship and high fecundity

✔You should be able to explain why a life table is relevant only to a particular population at a particular time, and predict how the first life tables—constructed in ancient Rome—differ from the current life table for the same population.

The growth rate of a population can be calculated from life-table data or from the direct observation of changes in population size over time.

- Exponential growth occurs when the per-capita growth rate, *r*, does not change over time.
- During logistic growth, growing populations approach the carrying capacity of their environment and *r* declines to 0.
- Density dependence in population growth is due to competition for resources, disease, predation, or other factors that increase in intensity when population size is high. At high density, survivorship and fecundity decrease.

✔You should be able to draw a logistic growth curve, describe how *r* changes along the length of the curve, and suggest factors responsible for each change in *r*.

(MB) **BioFlix™** Population Ecology, **Web Activity** Modeling Population Growth

Researchers observe a wide variety of patterns when they track changes in population size over time, ranging from growth rates that slow when populations are at high density, to regular cycles, to continued growth independent of population size.

- Predator-prey interactions and food availability are density-dependent factors that drive the population cycle of snowshoe hares.
- Age structure has a profound impact on population dynamics. A population with few juveniles and many adults past reproductive age, like the human populations of developed nations, may be declining or stable in size. In contrast, a population with a large proportion of juveniles is likely to increase rapidly in size.
- The total human population has been increasing rapidly since about 1750 and is currently over 6.8 billion. In countries where survivorship has increased due to medical advances, fecundity has decreased.
- Based on various scenarios for average female fertility, the total human population in 2050 is expected to be between 7 and 11 billion.

✔You should be able to explain why human populations are projected to continue rapid growth, even though fertility rates are declining.

(MB) **Web Activity** Human Population Growth and Regulation

Data from population ecology studies help biologists evaluate prospects for endangered species and design effective management strategies, as well as to predict changes in human populations.

- Human activities are isolating populations into metapopulations occupying small, fragmented habitats.
- The dynamics of a metapopulation are driven by the birth and death of populations, just as the dynamics of a population are driven by the birth and death of individuals.
- Migration among habitat patches is essential for the stability of a metapopulation, so conservationists are trying to preserve unoccupied patches of habitat and establish corridors that link habitat fragments.

✔You should be able to explain why a small, isolated population is virtually doomed to extinction, but why population size may be stable in a species consisting of a large number of small, isolated populations.

Questions

✔TEST YOUR KNOWLEDGE

Answers are available in Appendix B

1. What is the defining feature of exponential growth?
   a. The population is growing very quickly.
   b. The growth rate is constant.
   c. The growth rate increases rapidly over time.
   d. The growth rate is very high.
2. What four factors define population growth?
   a. age-specific birthrates, age-specific death rates, age structure, and metapopulation structure
   b. survivorship, age-specific mortality, fecundity, death rate
   c. mark-recapture, census, quadrat sampling, transects
   d. births, deaths, immigration, emigration
3. In what populations does exponential growth tend to occur?
   a. in populations that colonize new habitats
   b. in populations that experience intense competition
   c. in populations that experience high rates of predation
   d. in metapopulations

4. Which of the following is *not* a reason that population growth declines as population size approaches the carrying capacity?
   a. Climate becomes unfavorable.
   b. Competition for resources increases.
   c. Predation rates increase.
   d. Disease rates increase.
5. If most individuals in a population are young, why is the population likely to grow rapidly in the future?
   a. Death rates will be low.
   b. The population has a skewed age distribution.
   c. Immigration and emigration can be ignored.
   d. Many individuals will begin to reproduce soon.
6. Why have population biologists become particularly interested in the dynamics of metapopulations?
   a. because humans exist as a metapopulation
   b. because whooping cranes exist as a metapopulation
   c. because many populations are becoming restricted to small islands of habitat
   d. because metapopulations explain why populations occupying large, contiguous areas are vulnerable to extinction

TEST YOUR UNDERSTANDING

Answers are available in Appendix B

1. Explain Equations 52.4 and 52.5 in words.
2. Explain why the growth rate of species with a type I survivorship curve depends primarily on fertility rates. Explain why the growth rate of species with a type III survivorship curve is extremely sensitive to changes in adult survivorship.
3. Offer a hypothesis to explain why humans have undergone near-exponential growth for over 500 years. Why can't exponential growth continue indefinitely? Describe two documented examples of density-dependent factors that influence population growth in natural populations.
4. Explain why biologists want to maintain (a) "habitat corridors" that connect populations in a metapopulation, and (b) unoccupied habitat that is appropriate for the species in question.
5. Make a rough sketch of the age distribution in developing versus developed countries, and explain why the shapes are different. How is AIDS, which is a sexually transmitted disease, affecting the age distribution in countries hard hit by the epidemic?
6. Compare the pros and cons of using R_0, λ, and r to express growth rates. What is the difference between r (the per-capita rate of increase) and r_{max} (the maximum or intrinsic growth rate)?

APPLYING CONCEPTS TO NEW SITUATIONS

Answers are available in Appendix B

1. When wild plant and animal populations are logged, fished, or hunted, only the oldest or largest individuals tend to be taken. Many of the commercially important species are long-lived and are slow to begin reproduction. If harvesting is not regulated carefully and exploitation is intense, what impact does harvesting have on a population's age structure? How might harvesting affect the population's life-table and growth rate?
2. Design a system of nature preserves for an endangered species of beetle whose larvae feed on only one species of sunflower. The sunflowers tend to be found in small patches that are scattered throughout dry grassland habitats.
3. The population of the United States is projected to increase dramatically over the next 50 years, even though fertility rates are only slightly above replacement level and the age distribution is largely stable. How is this possible?
4. In most species the sex ratio is at or near 1.00, meaning that there is an approximately equal number of males and females. In China, however, there is a strong preference for male children. According to the 2000 census there, the sex ratio of newborns is almost 1.17, meaning that close to 117 boys are born for every 100 girls. Based on these data, researchers project that there will soon be about 50 million more men than women of marriageable age in China. Discuss how this skewed sex ratio might affect the population growth rate in China.

This chapter explores how species that live in coral reefs and other communities interact with each other, and how communities change over time.

53 Community Ecology

KEY CONCEPTS

- Interactions among species, such as competition, consumption, and mutualism have two main outcomes: (1) They affect the distribution and abundance of the interacting species, and (2) they are agents of natural selection and thus affect the evolution of the interacting species. The nature of interactions between species frequently changes over time.
- The assemblage of species found in a biological community changes over time and is primarily a function of climate and chance historical events.
- Species richness is higher in large islands near continents than in small, isolated islands, due to differences in immigration and extinction. Species richness is also higher in the tropics and lower toward the poles, but the mechanism responsible for this pattern is still controversial.

Chapter 52 focused on the dynamics of populations—how and why they grow or decline, and how they change over time and through space. That chapter considered each population as an isolated entity.

In reality, though, populations of different species form complex assemblages called communities. A biological **community** consists of interacting species, usually living within a defined area.

This chapter's task is to analyze the dynamics of biological communities—how they develop and change over time. Biologists want to know how communities work, and how to manage them in a way that will preserve species and create an environment that people want to live in.

How do species inside communities interact with each other, and what are the long-term consequences? What happens to communities when they are disturbed by fires or flood, and why is the number of species higher in some areas than others? Let's delve in.

53.1 Species Interactions

The species in a community interact almost constantly. Organisms eat each other, pollinate one another, exchange nutrients, compete for resources, and provide habitats for each other. In many cases, a population's fate is tightly linked to the other species that share its habitat.

To study species interactions, biologists focus on analyzing the effects on the fitness of the individuals involved. Recall from Chapter 24 that **fitness** is defined as the ability to survive and produce offspring.

Does the relationship between two species provide a fitness benefit to members of one species (a "+" interaction) but hurt members of the other species (a "−" interaction)? Or does the association have no effect on the fitness of a participant (a "0" interaction)?

✔ When you see this checkmark, stop and test yourself. Answers are available in Appendix B.

- **Competition** occurs when individuals use the same resources—resulting in lower fitness for both (−/−).
- **Consumption** occurs when one organism eats or absorbs nutrients from another. The interaction increases the consumer's fitness but decreases the victim's fitness (+/−).
- **Mutualism** occurs when two species interact in a way that confers fitness benefits to both (+/+).
- **Commensalism** occurs when one species benefits but the other species is unaffected (+/0).

Of the four general types of interactions, commensalism is the least-studied. An example is the birds that follow moving army ants in the tropics. As the ants march along the forest floor, they hunt insects and small vertebrates. As they do, birds follow and pick off prey species that fly or jump up out of the way of the ants (**Figure 53.1**).

Three Themes

The rest of this section focuses on the most common types of species interactions: competition, consumption, and mutualism. As you analyze each type of species interaction, watch for three key themes:

1. ***Species interactions can affect the distribution and abundance of a particular species.*** Chapter 50 explored how species interactions can affect geographic ranges, using the examples of the yucca plant/yucca moth mutualism and the impact of a disease carried by tsetse flies on the distribution of cattle in Africa. Chapter 52 introduced impacts on population size, with data on how predation by lynx affects hare populations in northern Canada.
2. ***Species act as agents of natural selection when they interact.*** Deer are fast and agile in response to natural selection exerted by their major predators, wolves and cougars. The speed and agility of deer, in turn, promote natural selection that favors wolves and cougars that are fast and that have superior eyesight, senses of smell, and hunting strategies. To capture this point, biologists say that **coevolution** is occurring.

   As species interact over time, it is common to observe a **coevolutionary arms race**—a repeating cycle of reciprocal adaptation. In humans, an arms race occurs when one nation develops a new weapon, which prompts a rival country to develop a defensive weapon, which pushes the original country to manufacture an even more powerful weapon, and so on. In biology, coevolutionary arms races occur between predators and prey, parasites and hosts, and other types of interacting species.
3. ***The outcome of interactions among species is dynamic and conditional.*** Consider the relationship between army ants and birds that follow them, which is usually commensal. If bird attacks start to force other insects into the path of the ants, then both ants and birds benefit and the relationship becomes mutualistic. But if birds begin to steal prey that would otherwise be taken by ants, then the relationship becomes parasitic. The outcome of the interaction may depend on the number and types of prey, birds, and ants present and may change over time.

Competition

Competition lowers the fitness of the individuals involved for a simple reason: When competitors use resources, those resources are not available to help individuals survive better and produce more offspring.

Competition that occurs between members of the same species is called **intraspecific** (literally, "within species") **competition**. Intraspecific competition for space, sunlight, food, and other resources intensifies as a population's density increases. As a result, intraspecific competition is a major cause of density-dependent growth (see Chapter 52). **Interspecific** ("between species") **competition** occurs when individuals from different species use the same limiting resources.

USING THE NICHE CONCEPT TO ANALYZE INTERSPECIFIC COMPETITION Early work on interspecific competition focused on the concept of the niche. A **niche** can be thought of as the range of resources that the species is able to use or the range of conditions it can tolerate.

(a) Ants stir up insects while hunting ...
(b) ... antbirds tag along and benefit.

FIGURE 53.1 Commensals Gain a Fitness Advantage but Don't Affect the Species They Depend On. Birds that associate with army ants are commensals. They have no measurable fitness effect on the ants, but gain from the association by capturing insects that try to fly or climb out of the way of the ants.

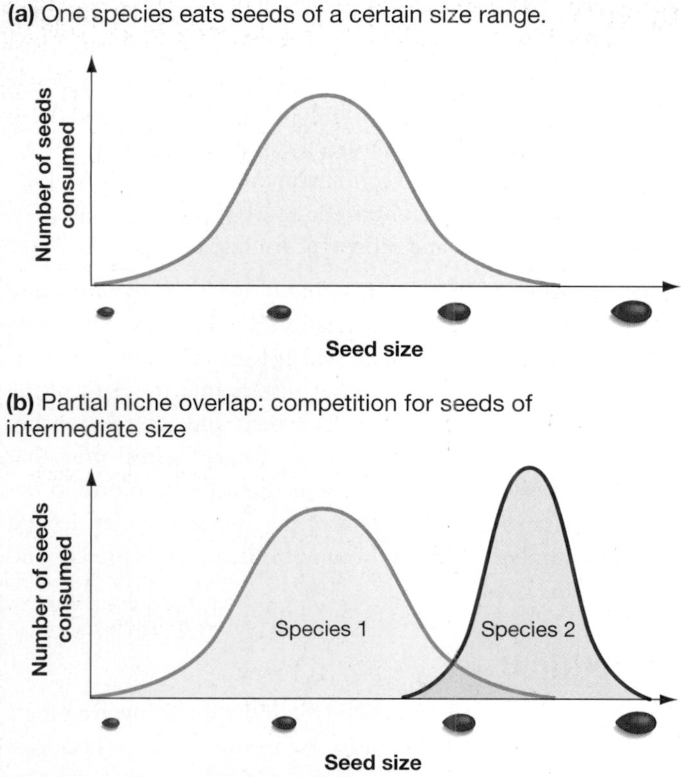

FIGURE 53.2 Niche Overlap Leads to Competition. (a) A graph describing one aspect of a species' fundamental niche, meaning the range of resources that it can use or range of conditions it can tolerate. **(b)** Competition occurs when the niches of different species overlap.

G. Evelyn Hutchinson proposed that a species' niche could be envisioned by plotting its habitat requirements along a series of axes. **Figure 53.2a**, for example, represents one niche axis for a hypothetical species. In this case, the habitat requirement plotted is the size of seeds eaten by members of this population, which might be a function of mouth or tooth size. Other niche axes could represent other types of foods used or the temperatures, humidity, and other environmental conditions tolerated by the species. In these types of graphs, the *y*-axis represents the number or proportion of resources used, averaged over the population.

Interspecific competition occurs when the niches of two species overlap. The two species plotted in **Figure 53.2b**, for example, compete for seeds of intermediate size. When competition occurs, each individual will get fewer seeds on average, and average fitness in each population will decline.

WHAT HAPPENS WHEN ONE SPECIES IS A BETTER COMPETITOR? G. F. Gause claimed it is not possible for species with the same niche to coexist. This hypothesis, called the **competitive exclusion principle**, was inspired by a series of experiments Gause did with similar species of the unicellular pond-dweller *Paramecium*.

- When *P. caudatum* and *P. aurelia* grew in separate laboratory cultures, both species exhibited logistic growth (see Chapter 52).
- When the two species grew in the same culture together, only the *P. aurelia* population exhibited a logistic growth pattern. *P. caudatum*, in contrast, was driven to extinction (**Figure 53.3**).

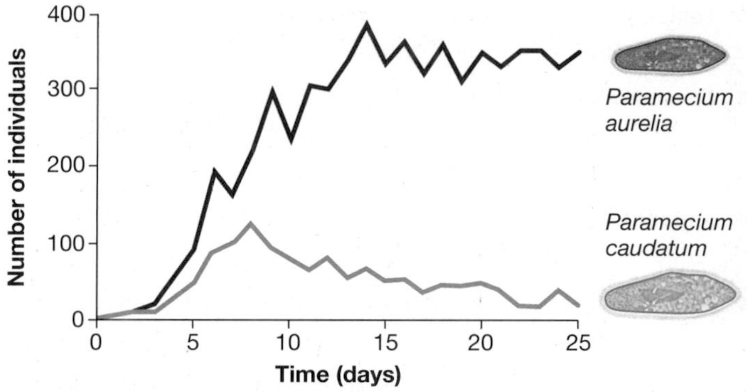

FIGURE 53.3 Competitive Exclusion in Two Species of *Paramecium*.

Gause's result is a product of asymmetric competition. When **asymmetric competition** occurs, one species suffers a much greater fitness decline than the other species does. Under **symmetric competition**, each of the interacting species experiences a roughly equal decrease in fitness. The distinction is graphed in **Figure 53.4a**, where the fitness of two hypothetical species, called A and B, is plotted for habitats where A and B occur separately and together.

If asymmetric competition occurs and the two species have completely overlapping niches (**Figure 53.4b**), then the stronger competitor is likely to drive the weaker competitor to extinction. But if the niches do not overlap completely, then the species that is the weaker competitor should be able to retreat to an area of non-overlap. In cases like this, an important distinction arises.

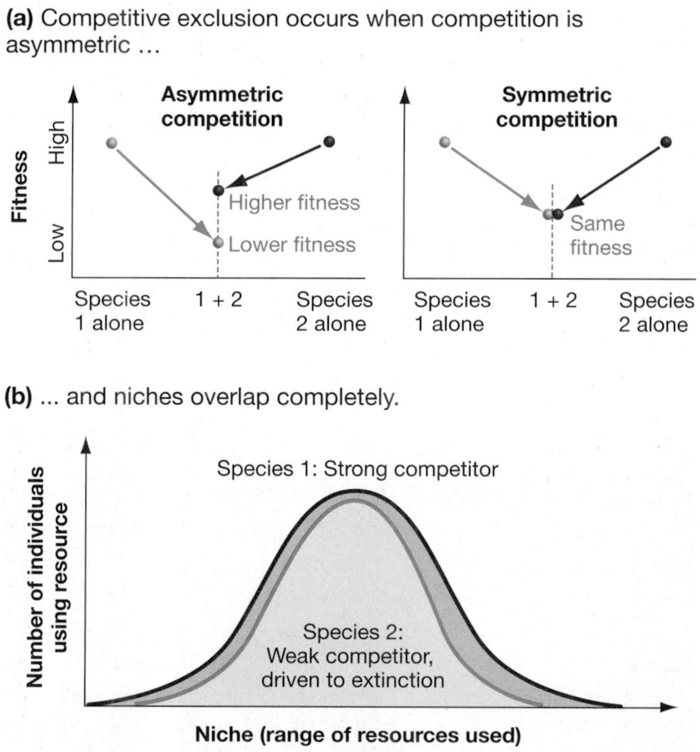

FIGURE 53.4 Asymmetric Competition and Niche Overlap Explain Competitive Exclusion. If two species have completely overlapping niches, there is no refuge for the weaker competitor. It may be driven to extinction.

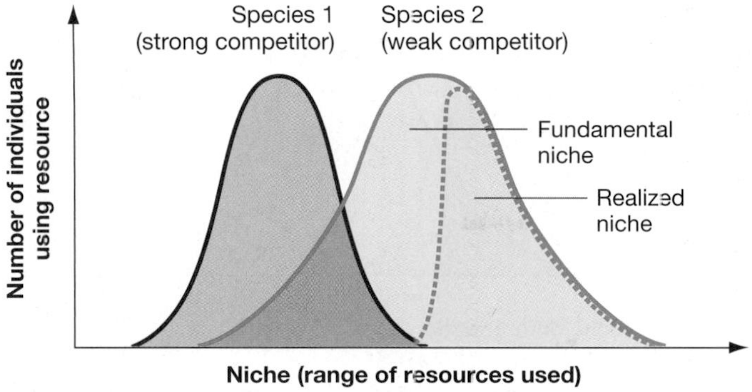

FIGURE 53.5 Fundamental Niches Are Broader than Realized Niches.

- A species' **fundamental niche** is the combination of resources or areas used or conditions tolerated in the absence of competitors.
- A species' **realized niche** is the portion of resources or areas used or conditions tolerated when competition occurs (**Figure 53.5**).

In Gause's laboratory cultures, competition between *P. aurelia* and *P. caudatum* was asymmetric and competitive exclusion occurred. How can researchers study competition in the field, under natural conditions?

EXPERIMENTAL STUDIES OF COMPETITION Joseph Connell began a classic study of competition in the late 1950s, after observing an interesting pattern in an intertidal rocky shore in Scotland. He noticed that there were two species of barnacles with distinctive distributions. The adults of one species, *Chthamalus stellatus*, occurred in an upper intertidal zone. The adults of the other species, *Semibalanus balanoides*, were restricted to a lower intertidal zone. The upper intertidal zone is a more severe environment for barnacles, because it is exposed to the air for longer periods at low tide each day.

Although adult barnacles live attached to rocks, larvae are mobile. The young of both species were found together in the lower intertidal zone, however. Why aren't *Chthamalus* adults found there as well?

To answer this question, Connell considered two hypotheses:

1. Adult *Chthamalus* are competitively excluded from the lower intertidal zone.
2. Adult *Chthamalus* are absent from the lower intertidal zone because they do not thrive in the physical conditions there.

Connell tested these hypotheses by performing the experiment shown in **Figure 53.6.** He began by removing a number of rocks that had been colonized by *Chthamalus* from the upper intertidal zone and transplanting them into the lower intertidal zone. He screwed the rocks into place and allowed *Semibalanus* larvae to colonize them. Once the spring colonization period was over, Connell divided each rock into two groups. In one half, he removed all *Semibalanus* that were in contact with or next to a *Chthamalus*.

This experimental design allowed Connell to document *Chthamalus* survival in the absence of competition with *Semibalanus*, and compare it with survival during competition.

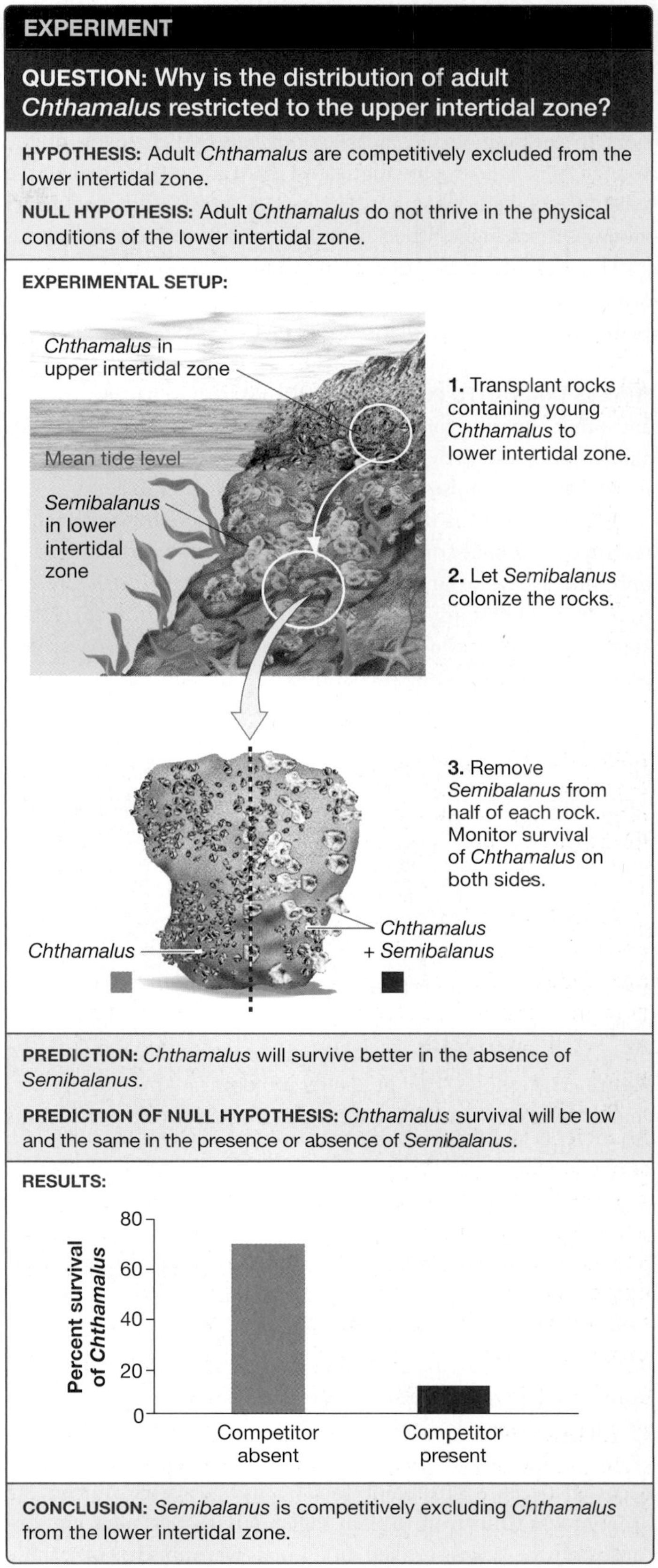
EXPERIMENT
QUESTION: Why is the distribution of adult Chthamalus restricted to the upper intertidal zone?
HYPOTHESIS: Adult Chthamalus are competitively excluded from the lower intertidal zone.
NULL HYPOTHESIS: Adult Chthamalus do not thrive in the physical conditions of the lower intertidal zone.
EXPERIMENTAL SETUP:
Chthamalus in upper intertidal zone
Mean tide level
Semibalanus in lower intertidal zone
1. Transplant rocks containing young Chthamalus to lower intertidal zone.
2. Let Semibalanus colonize the rocks.
3. Remove Semibalanus from half of each rock. Monitor survival of Chthamalus on both sides.
Chthamalus
Chthamalus + Semibalanus
PREDICTION: Chthamalus will survive better in the absence of Semibalanus.
PREDICTION OF NULL HYPOTHESIS: Chthamalus survival will be low and the same in the presence or absence of Semibalanus.
RESULTS:
Percent survival of Chthamalus
80
60
40
20
0
Competitor absent
Competitor present
CONCLUSION: Semibalanus is competitively excluding Chthamalus from the lower intertidal zone.

FIGURE 53.6 Experimental Evidence for Competitive Exclusion.

SOURCE: Connell, J. H. 1961. The influence of interspecific competition and other factors on the distribution of the barnacle *Chthamalus stellatus*. *Ecology* 42: 710–723.

✔**QUESTION** Why was it important to carry out both treatments on the same rock? Why not use separate rocks?

This is a common experimental strategy in competition studies: One of the competitors is removed, and the response by the remaining species is observed.

Connell's results support the hypothesis of competitive exclusion. In the unmodified areas, *Semibalanus* killed many of the young *Chthamalus* by growing against them and lifting them off the substrate. As the bar graph in Figure 53.6 shows, *Chthamalus* survival was much higher when all of the *Semibalanus* were removed.

The fundamental niches of these two species overlap. *Semibalanus* is a superior competitor for space, but *Chthamalus* is able to thrive in a smaller realized niche.

FITNESS TRADE-OFFS IN COMPETITION Why haven't *Semibalanus* and other superior competitors taken over the world? Biologists answer this question by invoking the concept of **fitness trade-offs**—inevitable compromises in adaptation (see Chapter 24).

The key insight is that the ability to compete for a particular resource—like space on rocks or edible seeds of a certain size—is only one aspect of an organism's niche. If individuals are extremely good at competing for a particular resource, then they are probably less good at enduring drought conditions, warding off disease, or preventing predation.

In the case of *Semibalanus* and *Chthamalus* growing in the intertidal, the fitness trade-off is rapid growth and success in competing for space versus the ability to endure the harsh physical conditions of the upper intertidal. *Semibalanus* are fast-growing and large; *Chthamalus* grow slowly but can survive long exposures to the air and to intense sun and heat. Neither species can do both things well. Fitness trade-offs limit the ability of superior competitors to spread.

MECHANISMS OF COEXISTENCE: NICHE DIFFERENTIATION It's important to realize that because competition is a −/− interaction, there is strong natural selection on both species to avoid it. **Figure 53.7** shows the predicted outcome: An evolutionary change in traits reduces the amount of niche overlap, and thus the amount of competition.

An evolutionary change in resource use, caused by competition, is called **niche differentiation** or resource partitioning. The change that occurs in species' traits, and that allows individuals to exploit different resources, is called **character displacement**. Character displacement makes niche differentiation possible.

Peter and Rosemary Grant recently documented character displacement occurring in Galápagos finches. You might recall, from Chapter 24, that the Grants observed dramatic increases in average beak size and body size of a *Geospiza fortis* (medium ground finch) population during a drought in 1977. The changes occurred because the major food source available during the drought was fruits from a plant called *Tribulus cistoides,* and because only the largest-beaked individuals were able to crack these fruits and eat the seeds inside.

In 1982, however, individuals from a species called *Geospiza magnirostris* (large ground finch) arrived on the island and began breeding. *G. magnirostris* are about twice the size of *G. fortis* and use *T. cistoides* fruits as their primary food source.

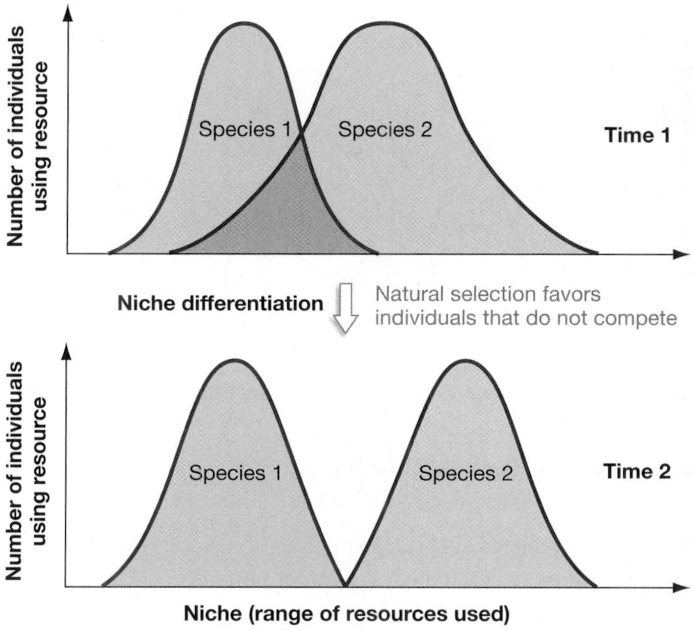

FIGURE 53.7 Over Time, Competition Can Lead to Niche Differentiation.

✔**QUESTION** Why is "resource partitioning" a good name for the outcome of this process?

A severe drought recurred in 2003 and 2004; most finches died of starvation. When the Grants measured the surviving *G. fortis* they made a remarkable observation: In contrast to the 1977 drought, only the *smallest*-beaked individuals survived (**Figure 53.8**).

Data on feeding behavior indicated that *G. magnirostris* were outcompeting *G. fortis* for *T. cistoides* fruits. Only *G. fortis* that could eat extremely small seeds efficiently could survive.

Evolution had occurred: Average beak size in the *G. fortis* population decreased. This is a dramatic example of character displacement. Natural selection favored medium ground finches that did not compete with *G. magnirostris*; the outcome was character displacement and niche differentiation. Theory developed decades earlier had made a successful prediction about a natural experiment.

COMPETITION AND CONSERVATION One of the goals of conservation biology is to keep biological communities intact. One of the major threats to communities is invasive species—a topic that was introduced in Chapter 50 and is revisited in Chapter 55.

Recent experiments have shown that communities that contain a large number of different species are more resistant to invasion than communities with a smaller number of species. Presumably, the presence of additional species means that a higher proportion of the resources available in a particular habitat are being used—making it more difficult for a new species to invade and take over.

This result—that competition can help communities resist invasion—is important for biologists who are trying to restore natural communities. The message is that as many native species as possible should be reintroduced, to create restored communities that resist invasion more effectively.

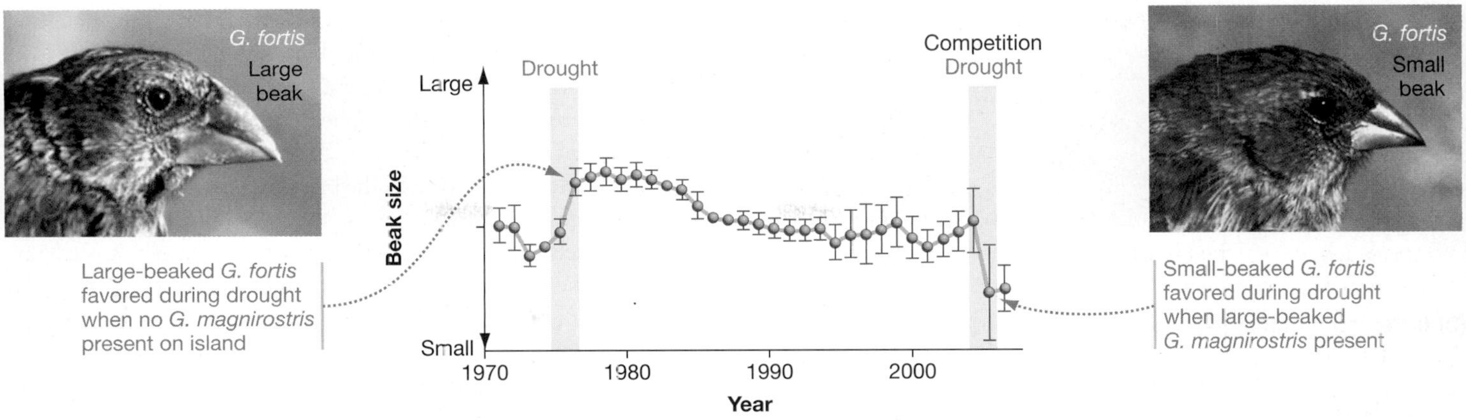

FIGURE 53.8 A Natural Experiment in Niche Differentiation. In *Geospiza fortis,* the data shown here would produce a shift in niche like that shown as a hypothetical case in Figure 53.7.

Consumption

Consumption occurs when one organism eats another. There are three major types of consumption:

1. **Herbivory** takes place when **herbivores** ("plant-eaters") consume plant tissues. Bark beetles mine cambium; cicadas suck sap; caterpillars chew leaves.
2. **Parasitism** occurs when a **parasite** consumes relatively small amounts of tissue or nutrients from another individual, called the **host**. Parasitism often occurs over a long period of time. It is not necessarily fatal, and parasites are usually small relative to their host. An array of worms and unicellular protists parasitizes humans; ticks parasitize cattle and other large mammals. In some cases, though, the term parasitism is used more broadly, to denote use of other resources. For example, birds and insects that lay their eggs in other species' nests and induce the other species to raise the young are called social parasites.
3. **Predation** occurs when a predator kills and consumes all or most of another individual. The consumed individual is called the prey. Woodpeckers eat bark beetles; finches eat seeds; ladybird beetles devour aphids; wasps kill caterpillars.

To illustrate how biologists analyze the impact of consumption, let's consider a series of questions about predators, herbivores, parasites, and their victims.

CONSTITUTIVE DEFENSES In fitness terms, consumption is costly for prey and beneficial for consumers. Prey individuals do not passively give up their lives to increase the fitness of their predators, however. Natural selection strongly favors traits that allow individuals to avoid being eaten.

- Prey may hide or run, fly, or swim away.
- Many plants lace their tissues with cocaine, nicotine, caffeine, or other compounds that are toxic to consumers.
- There is safety in numbers—schooling and flocking behavior is an effective way to reduce the risk of predation, in part because predators become confused when they dive or swim or run into a mass of prey.
- Some prey fight back: Selection has favored individuals with the ability to sequester or spray toxins, or employ weaponry such as sharp spines or kicking hooves.

Traits like these are called **standing** or **constitutive defenses**, because they are always present (**Figure 53.9**).

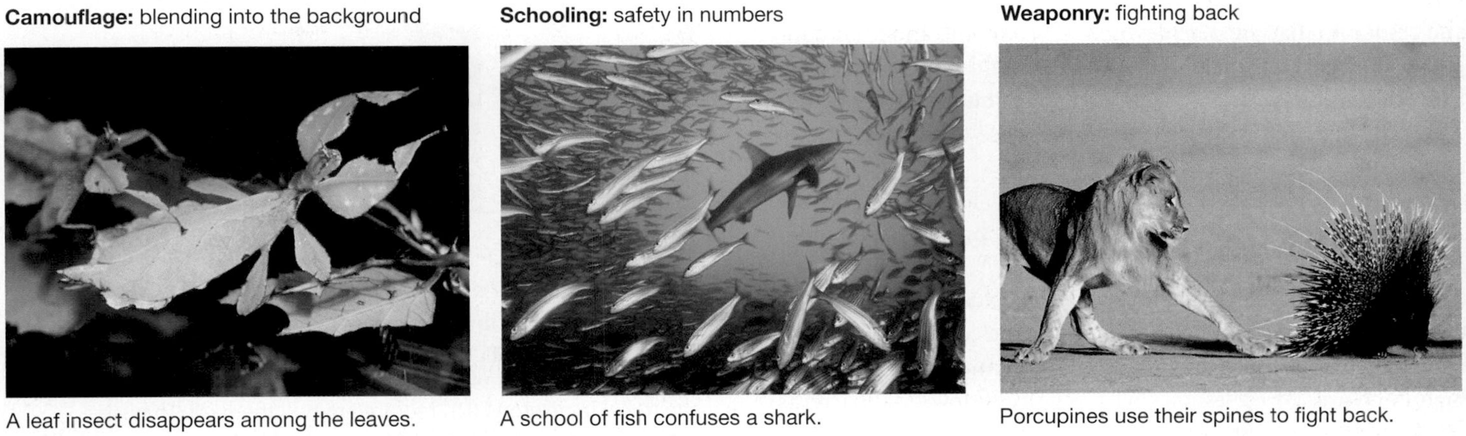
Camouflage: blending into the background
Schooling: safety in numbers
Weaponry: fighting back
A leaf insect disappears among the leaves.
A school of fish confuses a shark.
Porcupines use their spines to fight back.

FIGURE 53.9 Constitutive Defenses Are Always Present.

FIGURE 53.10 There Are Two Basic Forms of Mimicry in Prey Species.

Some of the best-studied constitutive defenses involve the phenomenon called mimicry. **Mimicry** occurs when one species closely resembles another species. The paper wasp on the far left of **Figure 53.10a**, for example, is dangerous because of its stinger. It also looks like other dangerous insects—particularly other wasps and bees. When harmful prey species resemble each other, **Müllerian mimicry** is said to occur.

To explain the existence of Müllerian mimics, biologists propose that the existence of similar-looking dangerous prey in the same habitat increases the likelihood that predators will learn to avoid them. In this way, Müllerian mimicry should reduce the likelihood of dangerous individuals being attacked.

Wasps also act as a model for harmless species, however—such as the moth, beetle, and fly shown in **Figure 53.10b**. To explain these mimics, biologists propose that predators avoid the harmless mimics because they mistake them for a dangerous wasp. This resemblance is known as **Batesian mimicry**.

INDUCIBLE DEFENSES Although constitutive defenses can be extremely effective, they are expensive in terms of the energy and resources that must be devoted to producing and maintaining them. Based on this observation, it should not be surprising to learn that many prey species have **inducible defenses**: defensive traits that are produced only in response to the presence of a predator.

Induced defenses decline if predators leave the habitat. Inducible defenses are efficient energetically, but they are slow—it takes time to produce them.

To see how inducible defenses work, consider research on the blue mussels that live in an estuary along the coast of Maine. Biologists had documented that predation on mussels by crabs was high in an area of the estuary with relatively slow tidal currents (a "low-flow" area) but low in an area of the estuary with relatively rapid tidal currents (a "high-flow" area). The researchers hypothesized that if blue mussels possess inducible defenses, then heavily defended prey individuals should occur in the low-flow area, where predation pressure is higher, but not in the high-flow area, where water movement reduces the number of crabs present.

To evaluate this hypothesis, the biologists measured mussel shell characteristics in the two areas. As the bar graphs in **Figure 53.11** show, mussels in the high-predation area were more strongly attached to their base or substrate and had thicker shells than did mussels in the low-predation area. These traits make the mussels more difficult to remove from the substrate and harder to crush. Thick shells and tight attachment are antipredator defenses.

The data in Figure 53.11 are correlational in nature, however, so they are open to interpretation. A critic of the inducible defense explanation could offer a reasonable alternative hypothesis—for example, that only constitutive defenses exist, and that crabs have eliminated weakly attached mussels with thin shells from the low-flow areas. The observed differences could also be due to variation in light, temperature, or other abiotic factors that might affect mussel traits but have nothing to do with predation.

To test the inducible-defense hypothesis more rigorously, biologists carried out the experiment diagrammed in **Figure 53.12**. The tank on the right allowed the researchers to measure shell growth in mussels that were "downstream" from crabs. As predicted by the induced-defense hypothesis, the mussels exposed to a crab in this way developed significantly thicker shells than did mussels that were not exposed to a crab. These results suggest that, even without direct contact, mussels can sense the presence of crabs and increase their investment in defenses.

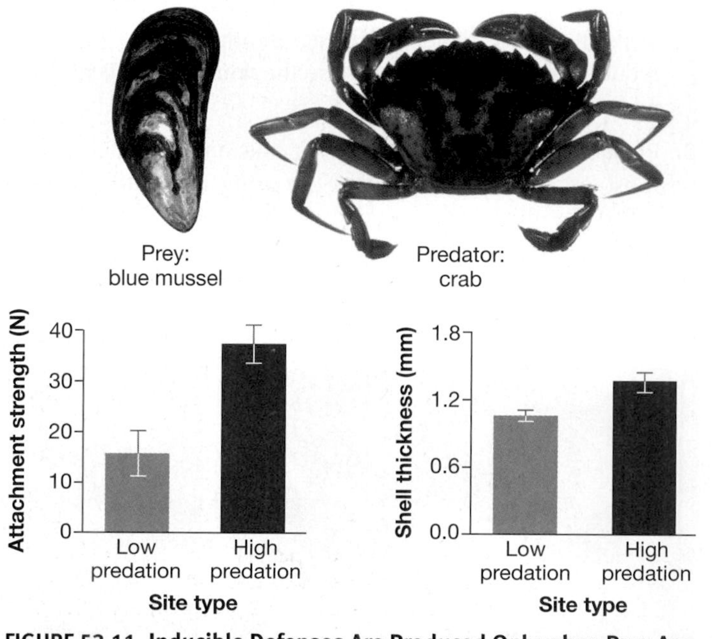

FIGURE 53.11 Inducible Defenses Are Produced Only when Prey Are Threatened. To measure how strongly mussels are attached to the substrate, researchers drilled a hole in their shells and measured the force, in newtons (N), required to pull them off. They also measured how large mussel shells were relative to an individual's size by dividing shell weight by soft-tissue weight.

EXPERIMENT

QUESTION: Are mussel defenses induced by the presence of crabs?

HYPOTHESIS: Mussels increase investment in defense in the presence of crabs.

NULL HYPOTHESIS: Mussels do not increase investment in defense in the presence of crabs.

EXPERIMENTAL SETUP:

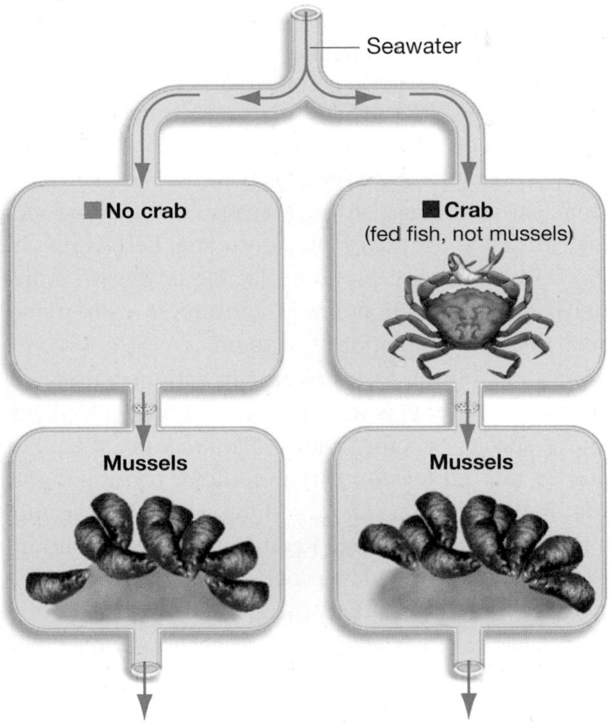

PREDICTION: Mussels downstream of the crab tank will have thicker shells than mussels downstream of the empty tank.

PREDICTION OF NULL HYPOTHESIS: Mussels in the two tanks will have shells of equal thickness.

RESULTS:

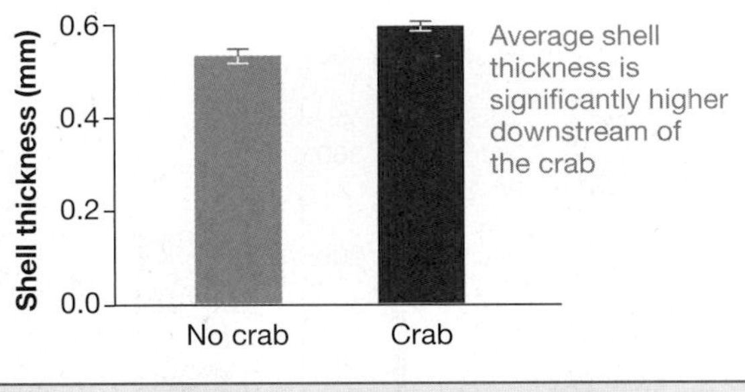

CONCLUSION: Mussels increase investment in defense when they detect crabs. Shell thickness is an inducible defense.

FIGURE 53.12 Experimental Evidence for Inducible Defenses.

SOURCE: Leonard, G. H., M. D. Bertness, and P. O. Yund. 1999. Crab predation, waterborne clues, and inducible defenses in the blue mussel, *Mytolus edulis*. *Ecology* 80: 1–14.

✔**QUESTION** Why did the researchers feed the crabs fish instead of mussels?

ARE ANIMAL PREDATORS EFFICIENT ENOUGH TO REDUCE PREY POPULATIONS? Prey are typically smaller than predators, have larger litter or clutch sizes, and tend to begin reproduction at a younger age. As a result, they have a much larger intrinsic growth rate. You might recall from Chapter 52 that this quantity, symbolized r_{max}, is the maximum growth rate that a population can achieve under ideal conditions.

If prey reproduce rapidly and are also well defended, can predators kill enough of them to reduce the prey population significantly?

Data from predator-removal programs—in which wolves, cougars, coyotes, or other predators are actively killed by human hunters—suggest that the answer is yes. For example, when a wolf control program in Alaska during the 1970s decreased predator abundance, the population of moose, on which wolves prey, tripled. This observation suggests that wolves had reduced this moose population far below the carrying capacity—the number that could be supported by the available space and food (see Chapter 52).

Other types of experiments are consistent with this result. Recall from Chapter 52 that the regular population cycles of snowshoe hares appear to be driven, at least in part, by density-dependent increases in predation. Taken together, the data available to date indicate that in many instances, predators are efficient enough to reduce prey populations below carrying capacity.

WHY DON'T HERBIVORES EAT EVERYTHING—WHY IS THE WORLD GREEN? If predators affect the size of populations in prey that can run or fly or swim away, then consumers should have a devastating impact on plants and on mussels, anemones, sponges, and other sessile (nonmoving) animals.

In some cases, this prediction turns out to be correct. For example, consider the results of a recent **meta-analysis**—a study of studies, meaning an analysis of a large number of data sets on a particular question. Biologists who compiled the results of more than 100 studies on herbivory found that the median percentage of mass removed from aquatic algae by herbivores was 79 percent. Herbivores eat the vast majority of algal food available in aquatic biomes. The figure dropped to just 30 percent for aquatic plants, however, and only 18 percent for terrestrial plants.

Why don't herbivores eat more of the food available on land? Stated another way, why is the world green? Biologists routinely consider two hypotheses:

1. ***Top-down control hypothesis: Herbivore populations are limited by predation and disease.***

The "top-down" name is inspired by the food chain concept introduced in Chapter 29 and explored in detail in Chapter 54. In a food chain, herbivores are "on top of" plants.

The logic of top-down control is simple: Predators and parasites remove herbivores that eat plants. As a result, a great deal of plant material remains uneaten.

In a recent test of this idea, researchers monitored herbivory on islands created by a 1986 dam project in Venezuela. On some small islands in the new lake, predators disappeared. On the predator-free islands, there are now many more herbivores than on similar sites nearby where predators are present. As predicted by the top-down control hypothesis, a much higher percentage of the total plant material is being eaten on the predator-free islands. For example, predator-free islands had just 25 percent of the small trees found on islands that contained predators.

2. ***Bottom-up limitation hypothesis: Plant tissues are well-defended or offer poor nutrition.***

The "bottom-up" name reflects the position of plants as the base or bottom of a food chain. The idea here is that plant tissues aren't eaten because they are toxic, or that herbivore numbers are limited because plant tissues offer poor nutrition.

The poor-nutrition aspect of this hypothesis was inspired by the observation that plant tissues have less than 10 percent of the nitrogen found in animal tissues, by weight. If the growth and reproduction of herbivores are limited by the availability of nitrogen, then their populations will be low and the impact of herbivory on plants relatively slight. Herbivores could eat more plant material to gain nitrogen, but at a cost—they would be exposed to predation and expend energy processing the food. When nitrogen concentration in plants is increased experimentally by fertilization, it is not unusual to see significant increases in herbivore growth rates.

In addition, most plant tissues are defended by weapons such as thorns, prickles, or hairs, or by potent poisons such as nicotine, caffeine, and cocaine. Manufacturing defensive compounds seems like the perfect solution to the problem of herbivory. In practice, however, plants face a complex challenge in defending themselves.

To drive this point home, consider data on interactions between cottonwood trees and two of their herbivores: beavers and leaf beetles. Cottonwoods resprout after they are cut down by a beaver (**Figure 53.13a**), and resprouted trees contain high concentrations of a defensive compound that deters further attack by beavers (**Figure 53.13b**). This is an induced defense.

Unfortunately for cottonwoods, leaf beetle larvae eat this defensive compound readily. In fact, the beetle larvae appear to use the anti-beaver compound to defend themselves against their major predator: ants. The bar graph in **Figure 53.13c** is from an experiment where researchers placed beetle larvae on an ant mound, and recorded how long they stayed alive. If larvae have fed on leaves of cut and resprouted cottonwoods, they survive ant attacks much longer than if they have fed on normal cottonwood leaves.

The punchline? Top-down control, nitrogen limitation, and effective defense are all important factors in limiting the impact of herbivory. The mix of factors that keeps the world green varies from species to species and habitat to habitat.

ADAPTATION AND ARMS RACES Over the long term, how do species that interact via consumption affect each other's evolution? When predators and prey or herbivores and plants interact over time, coevolutionary arms races result. Traits that increase feeding efficiency evolve in predators and herbivores. In response, traits that make prey unpalatable or elusive evolve. In counter-response, selection favors consumers with traits that overcome the prey adaptation. And so on.

To see a coevolutionary arms race in action, consider interactions between humans and the most serious of all human parasites—species in the genus *Plasmodium*. Recall from Chapter 29 that *Plasmodium* are unicellular protists that cause malaria. Malaria kills at least a million people a year, most of them pre–school-age children. Recent data suggest that humans and *Plasmodium* are locked in a coevolutionary arms race.

In West Africa, for example, there is a strong association between an allele called *HLA-B53* and protection against malaria. In liver cells that are infected by *Plasmodium*, HLA-B53 proteins on the surface of the liver cell display a parasite protein called cp26 (**Figure 53.14**). The display is a signal that immune system cells can read. It means "I'm an infected cell. Kill me before they

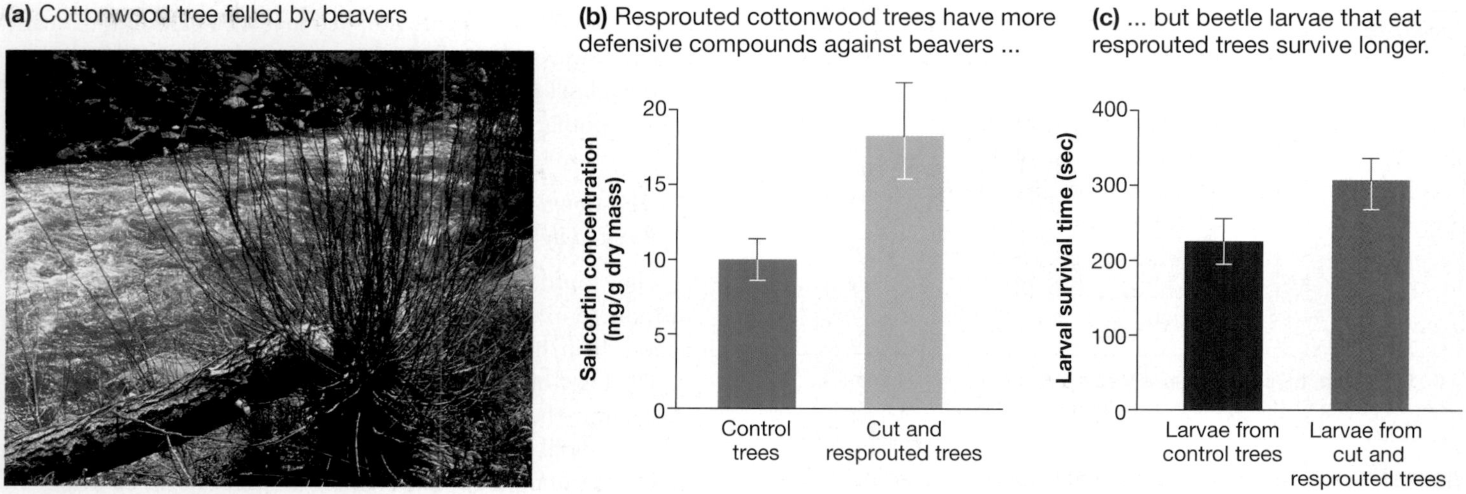

FIGURE 53.13 Defensive Compounds in Cottonwood Trees Discourage Beavers but Favor Beetles.

✔**QUESTION** Beetle larvae survive better on resprouted versus control trees, so they should damage resprouted trees more. If natural selection has maximized cottonwood fitness, which presents the greater overall threat to the trees, beavers or beetle larvae?

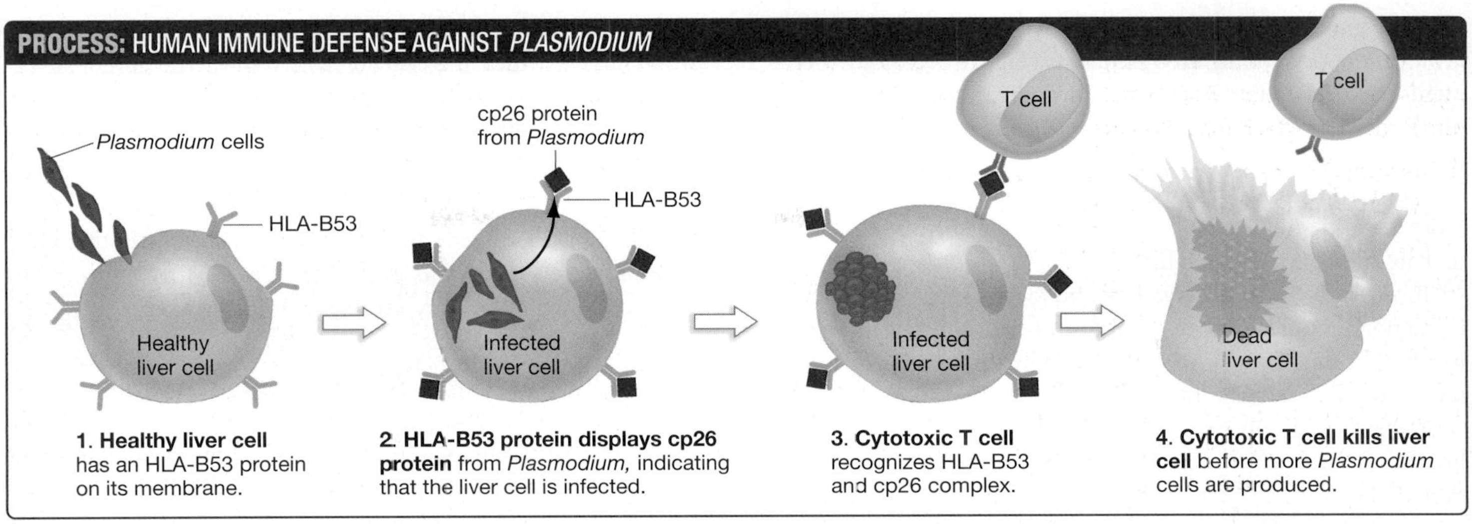
PROCESS: HUMAN IMMUNE DEFENSE AGAINST PLASMODIUM
Plasmodium cells
HLA-B53
Healthy liver cell
cp26 protein from Plasmodium
HLA-B53
Infected liver cell
T cell
Infected liver cell
T cell
Dead liver cell
1. Healthy liver cell has an HLA-B53 protein on its membrane.
2. HLA-B53 protein displays cp26 protein from Plasmodium, indicating that the liver cell is infected.
3. Cytotoxic T cell recognizes HLA-B53 and cp26 complex.
4. Cytotoxic T cell kills liver cell before more Plasmodium cells are produced.

FIGURE 53.14 Interactions between the Human Immune System and *Plasmodium*. If HLA-B53 binds to a particular *Plasmodium* protein, then infected cells are recognized and destroyed.

TABLE 53.1 **Some *Plasmodium* Strains Are Particularly Effective at Infecting Humans**

Plasmodium Strain	Infection Rate	Interpretation
cp26	Low	HLA-B53 binds to these proteins. Immune response is effective.
cp29	Low	
cp27	High	HLA-B53 does not bind to these proteins. Immune response is not as effective.
cp28	Average	

kill all of us." Immune system cells destroy the liver cell before the parasite cells inside can multiply.

In this way, people who have at least one copy of the *HLA-B53* allele are better able to beat back malarial infections. People who have the *HLA-B53* allele appear to be winning the arms race against malaria.

Follow-up research has shown that the arms race is far from won, however. *Plasmodium* populations in West Africa now have a variety of alleles for the protein recognized by HLA-B53. Some of these variants bind to HLA-B53 and trigger an immune response in the host, but others escape detection (**Table 53.1**). To make sense of these observations, researchers suggest that natural selection has favored the evolution of *Plasmodium* strains with weapons that counter HLA-B53.

The arms race continues. To review the *Plasmodium*-human interaction, go to the study area at *www.masteringbiology.com*.

(MB) **Web Activity** Life Cycle of a Malaria Parasite

CAN PARASITES MANIPULATE THEIR HOSTS? To thrive, parasites do not just have to invade tissues and grow while evading defensive responses by their host. They also have to be transmitted to new hosts. To a parasite, an uninfected host represents uncolonized habitat, teeming with resources. What have biologists learned about how parasites are transmitted to new hosts?

To answer this question, consider species of tree-dwelling ants (**Figure 53.15a**) that are parasitized by nematodes (roundworms). When nematodes have matured inside a host individual, they lay eggs in the ant's posterior-most body region. The exoskeleton in the area becomes translucent, and the yellow eggs inside make the structure look reddish instead of the normal black color (**Figure 53.15b**). Infected ants also hold the region up

(a) Uninfected ant **(b)** Infected ant

(c) Infected ants resemble berries that are eaten by birds.

FIGURE 53.15 A Parasite That Manipulates Host Behavior. When the tropical ant *Cephalotes atratus* becomes infected with parasitic nematodes, its appearance and behavior changes.

✔**EXERCISE** Design an experiment to test the hypothesis that infected ants are more likely than uninfected ants to be eaten by birds.

in a "flagging" posture, making them look like the berries that grow on the trees occupied by the ants (**Figure 53.15c**). Experiments have shown that fruit-eating birds are much more likely to pluck the posterior-most body region from infected ants than uninfected ants.

Why is this interesting? To complete their life cycle, the nematodes have to grow inside birds, before being shed in bird feces that are subsequently eaten by ants.

To interpret these observations, biologists suggest that these nematodes not only change the appearance of the ants, they also manipulate their behavior. The changes make the parasite much more likely to be transmitted to a new host.

Biologists have now cataloged a large number of case studies like this, from an array of parasites and hosts. Parasite manipulation of hosts is another example of extensive coevolution in species that interact.

USING CONSUMERS AS BIOCONTROL AGENTS Research on the dynamics of predator-prey interactions and other aspects of consumption has paid off in practical benefits: In some cases, biologists have been able to control pests by introducing predators or parasites.

The invasive plant purple loosestrife, for example, had succeeded in taking over large expanses of wetlands in North America. Through careful experimentation, however, biologists were able to find two insect species that feed exclusively on purple loosestrife. By introducing the herbivores into wetlands, researchers have been able to reduce purple loosestrife infestations dramatically.

In agriculture and forestry, the use of predators and parasites as biocontrol agents is a key part of **integrated pest management**: strategies to maximize crop and forest productivity while using a minimum of insecticides or other types of potentially harmful compounds.

Mutualism

Mutualisms are +/+ interactions that involve a wide variety of organisms and rewards.

- Bees visit flowers to harvest nectar and pollen. The bees benefit by using nectar and pollen as a food source; flowering plants benefit because in the process of visiting flowers, foraging bees carry pollen from one plant to another and accomplish pollination. Chapters 30 and 40 detailed some of the adaptations found in flowering plants that increase the efficiency of pollination.
- One of the most important of all mutualisms occurs between mycorrhizal fungi and plant roots. Chapter 31 reviewed experimental evidence indicating that mycorrhizal fungi receive sugars and other carbon-containing compounds in exchange for nitrogen or phosphorus acquired by the plant partner. The plant feeds the fungus; the fungus fertilizes the plant.
- Chapters 28 and 38 introduced the mutualism between nitrogen-fixing bacteria and the plant species that house them in their tissues. The host plants provide sugars and protection to the bacteria; the bacteria supply ammonia or nitrate in return.
- Chapter 32 cited mutualisms between rancher ants and aphids, in which the rancher ants protect the aphids in exchange for sugar-rich honeydew, and between farmer ants and fungi, in which the farmer ants cultivate fungi for food.
- **Figure 53.16a** shows ants in the genus *Crematogaster*, which live in acacia trees native to Africa. The ants live in bulbs at the base of acacia thorns and feed on small structures that grow from leaf tips. The ants protect the tree by attacking and biting herbivores, and by cutting vegetation from the ground below the host tree.
- **Figure 53.16b** illustrates cleaner shrimp in action. These shrimp pick external parasites from the jaws and gills of fish. In this mutualism, one species receives dinner while the other obtains medical attention.

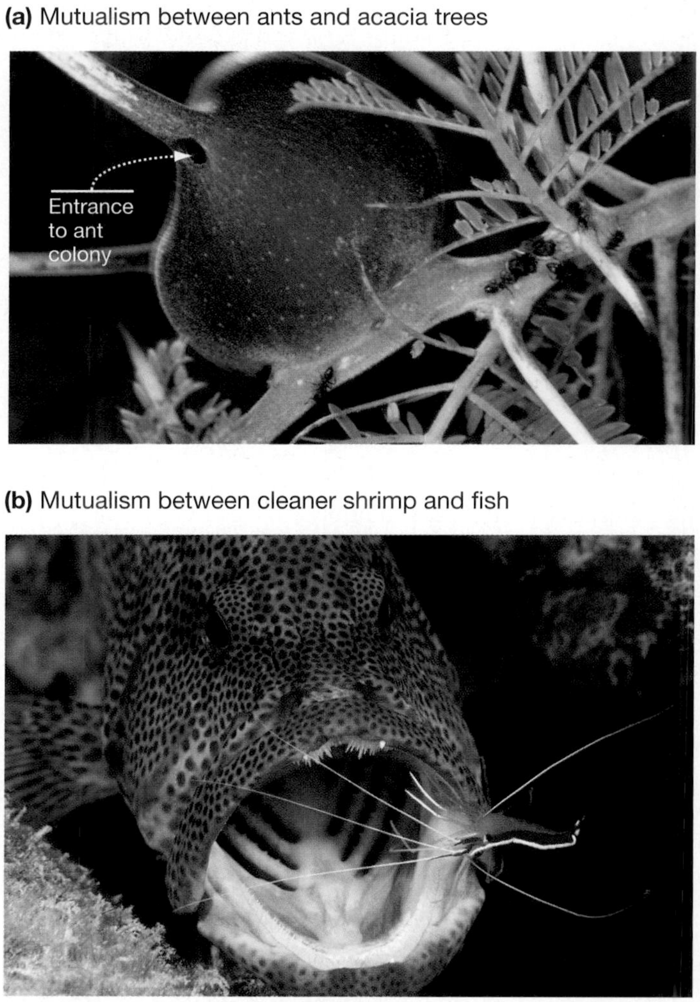
(a) Mutualism between ants and acacia trees
Entrance to ant colony
(b) Mutualism between cleaner shrimp and fish

FIGURE 53.16 Mutualisms Take Many Forms. (a) In certain species of acacia tree, ants in the genus *Crematogaster* live in large bulbs at the base of spines and attack herbivores that threaten the tree. The ants eat nutrient-rich tissue produced at the tips of leaves. **(b)** Cleaner shrimp remove and eat parasites that take up residence on the gills of fish.

✔**EXERCISE** Name a possible cost of these mutualistic associations to the acacia tree, the *Crematogaster*, the cleaner shrimp, and the host fish.

THE ROLE OF NATURAL SELECTION IN MUTUALISM As the examples you've just reviewed show, the rewards from mutualistic interactions range from the transportation of gametes to food, housing, medical help, and protection. It is important to note, however, that even though mutualisms benefit both species, the interaction does not involve individuals from different species being altruistic or "nice" to each other.

Expanding on a point that Charles Darwin introduced in 1862, Judith Bronstein described mutualisms as "a kind of reciprocal parasitism; that is, each partner is out to do the best it can by obtaining what it needs from its mutualist at the lowest possible cost to itself." Her point is that the benefits received in a mutualism are a by-product of each individual pursuing its own self-interest by maximizing its fitness—its ability to survive and reproduce.

In this light, it is not surprising that some species "cheat" on mutualistic systems. For example, deceit pollination occurs when certain species of plants produce a showy flower but no nectar reward. Pollinators receive no reward for making a visit and carrying out pollination. Evolutionary studies show that deceit pollinators evolved from ancestral species that did provide a reward. Over time, a +/+ interaction evolved into a +/− interaction.

MUTUALISMS ARE DYNAMIC A recent experimental study of mutualism provides another good example of the dynamic nature of these interactions. This study focused on ants and treehoppers. Ants are insects that live in colonies; treehoppers are small, herbivorous insects that feed by sucking sugar out of the phloem of plants. Treehoppers excrete the sugary solution honeydew from their posteriors. The honeydew, in turn, is harvested for food by ants.

It is clear that ants benefit from this association. But do the treehoppers? Biologists hypothesized that the ants might protect the treehoppers from their major predator, jumping spiders. These spiders feed heavily on juvenile treehoppers.

To test the hypothesis that ants protect treehoppers, the researchers studied ant-treehopper interactions over a three-year period. As **Figure 53.17** shows, the researchers marked out a 1000-m^2 study plot. Each year they removed the ants from one group of the treehopper host plants inside the plot but left the others alone to serve as a control. Then they compared the growth and survival of treehoppers on plants with and without ants.

Recall that this is a common research strategy for studying species interactions. To assess the fitness costs or benefits of the interaction, researchers remove one of the participants experimentally and document the effect on the other participant's survival and reproduction, compared with the survival and reproduction of control individuals that experience a normal interaction.

In both the first and third years of the study, the number of treehopper young on host plants increased in the treatment with ants but showed a significant decline in the treatment with no ants. This result supports the hypothesis that treehoppers benefit from the interaction with ants because the ants protect the treehoppers from predation by jumping spiders.

In the second year of the study, however, the researchers found a very different pattern. There was no difference in offspring survival, adult survival, or overall population size between treehopper populations with ants and those without ants.

Why? The researchers were able to answer this question because they also measured the abundance of spiders that prey on treehoppers in each of the three years. Their census data showed that in the second year of the study, spider populations were very low.

EXPERIMENT

QUESTION: Is the relationship between ants and treehoppers mutualistic?

HYPOTHESIS: Ants harvest food from treehoppers. In return, they protect treehoppers from jumping spiders.

NULL HYPOTHESIS: Ants harvest food from treehoppers but are not beneficial to treehopper survival.

EXPERIMENTAL SETUP:

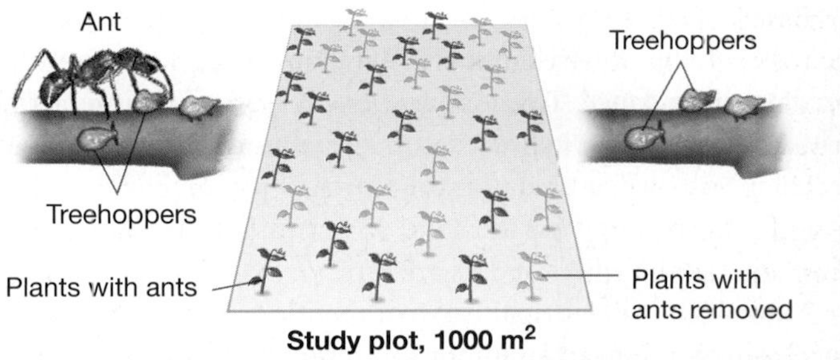

PREDICTION: Treehopper reproduction will be higher when ants are present than when ants are absent.

PREDICTION OF NULL HYPOTHESIS: There will be no difference in the number of young treehoppers on the plants.

RESULTS:

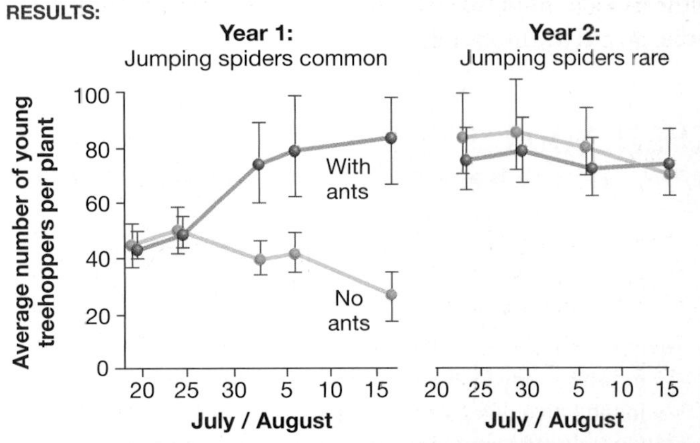

CONCLUSION: Treehoppers benefit from the interaction with ants only when jumping spiders are common.

FIGURE 53.17 Experimental Evidence That the Treehopper-Ant Interaction Is Mutualistic. The data points on the graphs represent average values from many study plots.

SOURCE: Cushman, J. H. and T. G. Whitham. 1989. Conditional mutualism in a membracid-ant association: temporal, age-specific, and density-dependant effects. *Ecology* 70: 1040–1047.

✔**QUESTION** The researchers assigned the treatments to different plants at random, and did many replicates of each treatment. Why was this important?

SUMMARY TABLE 53.2 **Species Interactions**

Type of Interaction	Fitness Effects	Short-Term Impact: Distribution and Abundance	Long-Term Impact: Coevolution
Competition	−/−	Reduces population size of both species; if competition is asymmetric, competitive exclusion reduces range of one species and may even lead to extinction.	Niche differentiation via selection to reduce competition
Consumption	+/−	Impact on prey population depends on prey density and effectiveness of defenses.	Strong selection on prey for effective defense; strong selection on consumer for traits that overcome defenses
Mutualism	+/+	Population size and range of both species are dependent on each other.	Strong selection on both species to maximize fitness benefits and minimize fitness costs of relationship
Commensalism	+/0	Population size and range of commensal may depend on size and distribution of host.	Strong selection on commensal to increase fitness benefits in relationship; no selection on host

Based on these results, the investigators concluded that the benefits of the ant-treehopper interaction depend entirely on predator abundance. Treehoppers benefit from their interaction with ants in years when predators are abundant but are unaffected in years when predators are scarce. If producing honeydew is costly to treehoppers, then the +/+ mutualism changes to a +/− interaction when spiders are rare.

Mutualism is like parasitism, competition, and other types of species interactions in an important respect: The outcome of the interaction depends on current conditions. Because the costs and benefits of species interactions are fluid, an interaction between the same two species may range from parasitism to mutualism to competition. **Table 53.2** summarizes the fitness effects, short-term impacts on population size, and long-term evolutionary aspects of species interactions.

CHECK YOUR UNDERSTANDING

If you understand that . . .

- Natural selection favors the evolution of traits that decrease competition, resulting in changes in a species' niche over time.
- Consumers may reduce the population size of the species on which they feed and often exert strong natural selection for effective defense mechanisms.
- Mutualisms benefit the species involved and can lead to highly coevolved associations, such as mycorrhizal fungi and symbiotic nitrogen-fixing bacteria.

✔ **You should be able to . . .**

1. Explain why resource partitioning (niche differentiation) does not involve a conscious choice on the part of the individuals involved.
2. Explain what a coevolutionary arms race is and give an example.

Answers are available in Appendix B.

53.2 Community Structure

In a community, each species interacts with dozens or hundreds or even thousands of other species. You, for example, are a walking, talking, biological community that includes trillions of bacterial and archaeal cells. Your skin is covered with them; they pack your gut lining and thrive in your mouth and teeth and respiratory tract. Some of the species present parasitize you, some are mutualistic, many are commensal.

Unfortunately, we know relatively little about the structure and dynamics of your personal biological community—although promising research has begun on questions about changes in species composition over time, competition for resources, and responses to disturbance.

Fortunately, biologists have been studying forest, grassland, and marine communities for over 80 years. Time will tell whether the general principles that have emerged from this research also apply to the community that you carry around with you.

How Predictable Are Communities?

The first question that biologists asked about woodlands and deserts and shorelines concerned structure: Do biological communities have a tightly prescribed organization and composition, or are they merely loose assemblages of species?

THE CLEMENTS-GLEASON DICHOTOMY Beginning with a paper published in 1936, Frederick Clements hypothesized that biological communities are stable, integrated, and orderly entities with a highly predictable composition. His reasoning was that species interactions are so extensive—and coevolution so important—that communities have become highly integrated and interdependent units in nature. Stated another way, the species within a community cannot live without each other.

In developing this hypothesis, Clements likened the development of a plant community to the development of an individual organism. He argued that communities develop over time by passing through a series of predictable stages dictated by extensive interactions among species. Eventually, this developmental progression culminates in a final stage known as a **climax community**. The climax community is stable—it does not change over time.

According to Clements, the nature of a climax community is determined by the area's climate. He predicted that if a fire or other disturbance destroys the climax community, it will reconstitute itself by repeating its predictable developmental stages.

Henry Gleason, in contrast, contended that the community found in a particular area is neither stable nor predictable. He claimed that plant and animal communities are ephemeral associations of species that just happen to share similar climatic requirements.

According to Gleason, it is largely a matter of chance whether a similar community develops in the same area after a disturbance occurs. Gleason downplayed the role of biotic factors, such as species interactions, in structuring communities. Instead, chance historical events—for example, which seeds and juvenile animals happened to arrive after a disturbance—were key elements in determining which species are found at a particular location.

Which viewpoint is more accurate? Let's consider experimental and observational data.

AN EXPERIMENTAL TEST To explore the predictability of community structure experimentally, biologists constructed 12 identical ponds (**Figure 53.18**). They filled the ponds at the same time with water that contained enough chlorine to kill any preexisting organisms. The Clements and Gleason hypotheses make clear and contrasting predictions:

- If community structure is predictable, then each pond should develop the same community of species once the chlorine vaporized and made the water habitable.
- If community structure is unpredictable, then each pond should develop a different community.

To test these predictions, the researchers sampled water from the ponds repeatedly for one year. They measured temperature, chemical makeup, and other physical characteristics of the water and recorded the diversity and abundance of each planktonic species by examining the samples under the microscope. Species that make up **plankton** live near the surface of water and swim little, if at all.

Their results? The red dots on the graph in Figure 53.18 indicate whether a particular species was found in each of the 12 ponds. Note that they found a total of 61 planktonic species in all of the ponds. Individual ponds had only 31 to 39 species apiece, however. This observation is important. Each pond contained just half to two-thirds of the total number of species that lived in the experimental area and that were available for colonization. A number of species occurred in most or all of the 12 ponds, but each pond had a unique species assemblage. Why?

To explain their results, the researchers contended that some species are particularly good at dispersing and are likely to colonize all or most of the available habitats. Other species disperse more slowly and tend to reach only one or a few of the available habitats. Further, the investigators proposed that the arrival of certain competitors or predators early in the colonization process

EXPERIMENT

QUESTION: Are communities predictable or unpredictable?

COMMUNITIES-ARE-PREDICTABLE HYPOTHESIS: The group of species present at a particular site is highly predictable.

COMMUNITIES-ARE-UNPREDICTABLE HYPOTHESIS: The group of species present at a particular site is highly unpredictable.

EXPERIMENTAL SETUP:

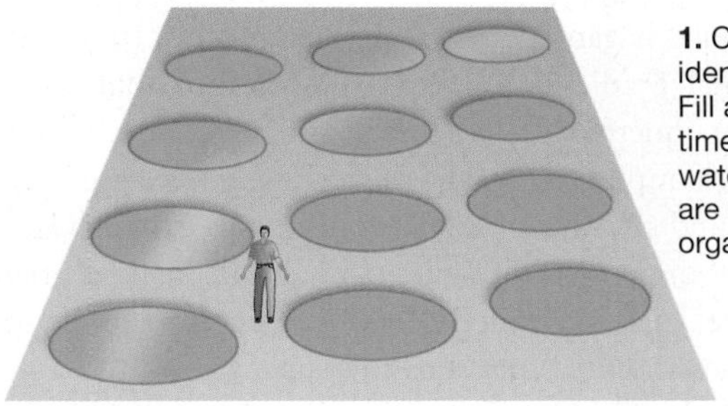

1. Construct 12 identical ponds. Fill at the same time and sterilize water so that there are no preexisting organisms.

2. Examine water samples from each pond. Identify each plankton species present in each sample.

COMMUNITIES-ARE-PREDICTABLE PREDICTION: Identical plankton communities will develop in all 12 ponds.

COMMUNITIES-ARE-UNPREDICTABLE PREDICTION: Different plankton communities will develop in different ponds.

RESULTS (AFTER 1 YEAR):

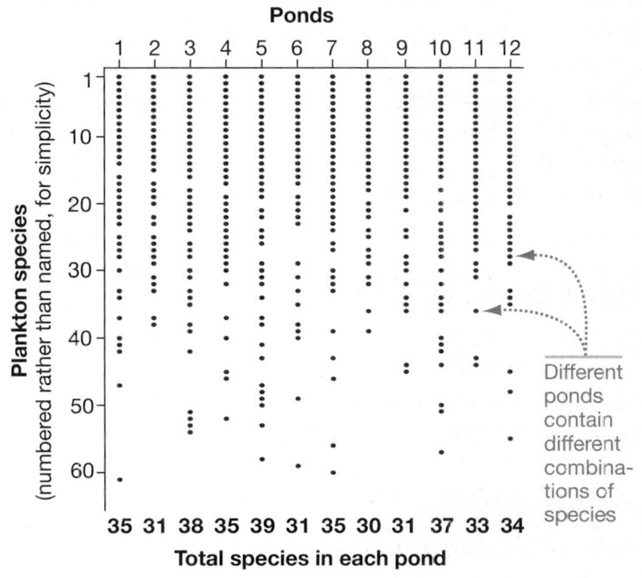

CONCLUSION: Although about half of the species present appear in all or most ponds, each pond has a unique composition. Both hypotheses are partially correct.

FIGURE 53.18 Experimental Evidence That Identical Communities Do Not Develop in Identical Habitats.

SOURCE: Jenkins, D. G. and A. L. Buikema. 1998. Do similar communities develop in similar sites? A test with zooplankton structure and function. *Ecological Monographs* 68: 421–443.

✔**QUESTION** At least 20 species are found in most or all of the ponds. Does this observation suggest that community composition is predictable or unpredictable?

greatly affects which species are able to invade successfully later. As a result, the specific details of community assembly and composition are somewhat contingent and difficult to predict. At least to some degree, communities are a product of chance and history.

MAPPING CURRENT AND PAST SPECIES' DISTRIBUTIONS If communities are predictable assemblages, then the ranges of species that make up a particular community should be congruent—meaning that the same group of species should almost always be found growing together.

When biologists began documenting the ranges of tree species along elevational gradients, however, they found that species came and went independently of each other. As you go up a mountain, for example, you might find white oak trees growing with hickories and chestnuts. As you continue to gain elevation, chestnuts might disappear but white oaks and hickories remain. Farther upslope, hickories might drop out of the species mix while white pines and red pines start to appear.

Data on the historical composition of plant communities supported these observations. Studies of fossil pollen documented that the distribution of plant species and communities at specific locations throughout North America has changed radically since the end of the last ice age about 11,000 years ago.

An important pattern emerged from these data: Species do not come and go in the fossil record in tightly integrated units. Instead, the ranges of individual species tended to change independently of one another.

For example, at a study site in Mirror Lake, New Hampshire, the percentage of pollen present from pines and oaks rose and fell in tandem over the past 14,000 years—just as predicted for a tightly integrated, pine-oak forest community. In contrast, the percentage of pollen from spruce, fir, cedar, beech, and most other tree species changed independently of each other. In general, studies of fossil pollen suggest that the composition of most plant communities has been dynamic and contingent on historical events, rather than static.

The overall message of research on community structure is that Clements's position was too extreme and that Gleason's view may be closer to being correct. Although both biotic interactions and climate are important in determining which species exist at a certain site, chance and history also play a large role.

How Do Keystone Species Structure Communities?

Even though communities are not predictable assemblages dictated by obligatory species interactions, biotic interactions are still important. The presence of certain consumers can have an enormous impact on the species present. In some cases, the structure of an entire community can change dramatically if a single species of predator or herbivore is removed from a community or added to it.

As an example of this research, consider an experiment that Robert Paine conducted in intertidal habitats of the Pacific Northwest of North America. In this environment, the sea star *Pisaster ochraceous* is an important predator (**Figure 53.19a**). Its preferred food is the California mussel *Mytilus californianus*.

When Paine removed *Pisaster* from experimental areas, what had been diverse communities of algae and invertebrates became overgrown with solid stands of mussels (**Figure 53.19b**). *M. californianus* is a dominant competitor, but its populations had been held in check by sea star predation.

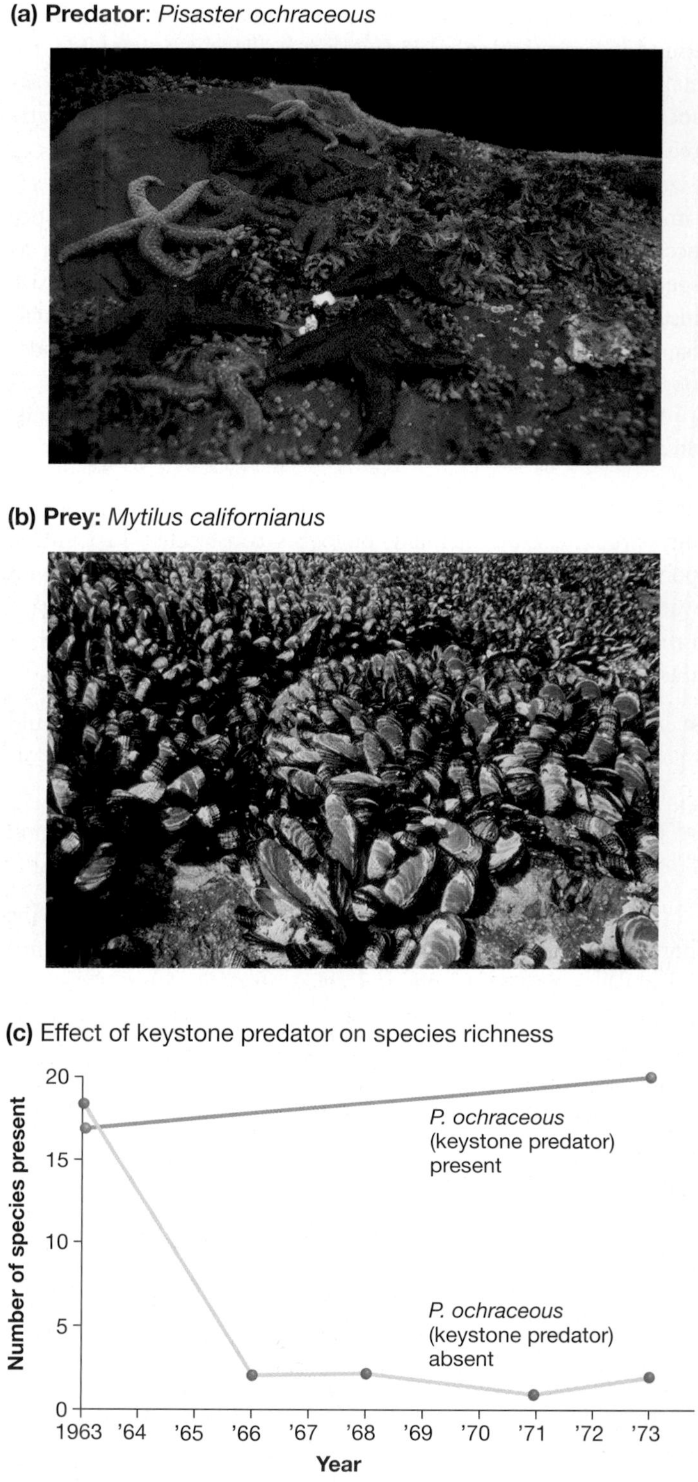

FIGURE 53.19 Keystone Predation Alters Community Structure in a Rocky Intertidal Habitat. Each data point in part (c) represents the average number of species present in several study plots.

The graph in **Figure 53.19c** plots the number of species present in experimental plots where *P. ochraceous* was present or excluded. When the predator was absent, the species richness of the habitat changed radically.

To capture the effect that a predator such as *Pisaster* can have on a community, Paine coined the term "keystone species." A **keystone species** has a much greater impact on the distribution and abundance of the surrounding species than its abundance and total biomass would suggest.

Sea stars and other keystone species vary in abundance in space or over time. As a result, the composition and structure of their communities vary. Communities are ever-changing entities whose composition is difficult to predict.

53.3 Community Dynamics

Once biologists had a basic understanding of how species interact and how communities are structured, they turned to questions about how communities change through time. Like cells, individuals, and species, communities can be described in one word: dynamic.

Disturbance and Change in Ecological Communities

Community composition and structure may change radically in response to changes in abiotic and biotic conditions. Biologists have become particularly interested in how communities respond to disturbance. A **disturbance** is any event that removes biomass from a community. (Recall from Chapter 50 that biomass is the mass of living organisms.)

Forest fires, windstorms, floods, the fall of a large canopy tree, disease epidemics, and short-term explosions in herbivore numbers all qualify as disturbances. These events are important because they alter light levels, nutrients, unoccupied space, or some other aspect of resource availability.

Biologists have come to realize that the impact of disturbance is a function of three factors: (1) the type of disturbance, (2) its frequency, and (3) its severity—for example, the speed and duration of a flood, or the intensity of heat during a fire. In addition, most communities experience a characteristic type of disturbance. In most cases, disturbances occur with a predictable frequency and severity.

To capture this point, biologists refer to a community's **disturbance regime**. For example, fires kill all or most of the existing trees in a boreal forest every 100 to 300 years, on average. In contrast, numerous small-scale tree falls, usually caused by windstorms, occur in temperate and tropical forests every few years.

HOW DO RESEARCHERS DETERMINE A COMMUNITY'S DISTURBANCE REGIME? Ecologists use two strategies to determine a community's natural pattern of disturbance. The first approach is based on obtaining data in a short-term analysis, and extrapolating to predict the longer-term pattern.

How is this done? As an example, an observational study might document that 1 percent of all boreal forest on Earth burns in a given year. Assuming that fires occur randomly, researchers project that any particular piece of boreal forest has a 1 in 100 chance of burning each year. According to this reasoning, fires will recur in that particular area every 100 years, on average.

This extrapolation approach is straightforward to implement, but it has important drawbacks. In boreal forests, for example, fires do not occur randomly in either space or time. They are more likely in some areas than in others, and they tend to occur in particularly dry years. Unless sampling is extensive, it is difficult to avoid errors caused by extrapolating from particularly disturbance-prone or disturbance-free areas or years.

The second approach is based on reconstructing the history of a particular site. For example, researchers can estimate the frequency and impact of storms by finding wind-killed trees and determining their date of death. (Investigators do so by comparing patterns in the growth rings of the dead trees with those of living individuals nearby; see Chapter 36.) The disturbances that have been most extensively studied using historical techniques, however, are forest fires.

Forest fires often leave a layer of burned organic matter and charcoal on the surface of the ground. As a result, researchers can dig a soil pit, find charcoal layers, and use radioisotope dating to establish when the fires occurred. It's also possible to date the death of trees killed by fire by comparing their growth rings with those of living trees. Further, trees that are not killed by fire are often scarred. When a fire burns close enough to kill a patch of cambium tissue, a scar forms. These fire scars occur most often at the tree's base, where dead leaves and twigs accumulate and furnish fuel. Fire frequency can be estimated by counting the number of growth rings that occur between fire scars on the same tree.

WHY IS IT IMPORTANT TO UNDERSTAND DISTURBANCE REGIMES? To appreciate why biologists are so interested in understanding disturbance regimes, consider a recent study on the fire history of giant sequoia groves in California (**Figure 53.20**). Giant sequoias

FIGURE 53.20 Giant Sequoia after a Fire.

grow in small, isolated groves on the western side of the Sierra Nevada range. Individuals live more than a thousand years, and many have been scarred repeatedly by fires.

A biologist obtained samples of cross sections through the bases of 90 giant sequoias in five different groves. As **Figure 53.21a** shows, the cross sections contained numerous rings that had been scarred by fire. To determine the date of each disturbance, the researcher counted tree rings back from the present.

The graph in **Figure 53.21b** plots the average number of fires that occurred each century in the groves, from the year 0 A.D. to 1900. He found that in most of the groves, 20–40 fires had occurred each century. The data indicated that each tree had been burned an average of 64 times.

This study established that fires are extremely frequent in these forests. Because not enough time would pass for large amounts of fuel to accumulate between fires, they were probably of low severity.

Partly because of this work, the biologists responsible for managing sequoia groves now set controlled fires or let low-intensity natural fires burn instead of suppressing them immediately. To maintain communities in good condition, biologists have to ensure that the normal disturbance regime occurs. Otherwise, community composition changes dramatically.

Similarly, studies of disturbance regimes along the Colorado River in southwestern North America inspired land managers to begin releasing a huge pulse of water from the reservoirs behind dams, at intervals. The floods that have resulted were designed to mimic the natural disturbance regime. According to follow-up studies, the artificial floods appear to have benefited the plant and animal communities downstream by depositing nutrients and creating additional sandbar habitat.

(a) Fire scars in the growth rings

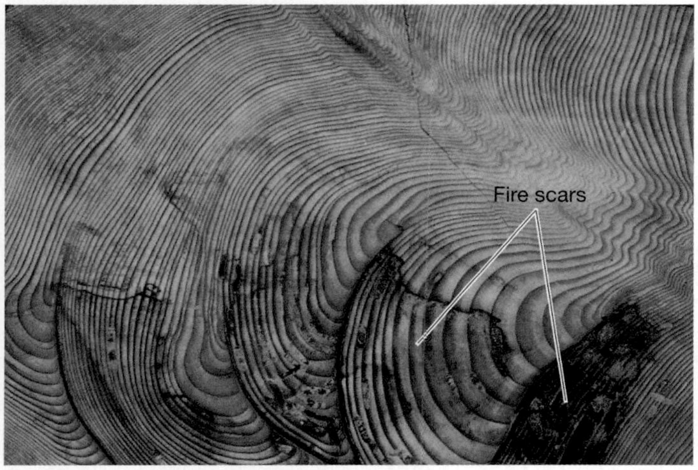

(b) Reconstructing history from fire scars

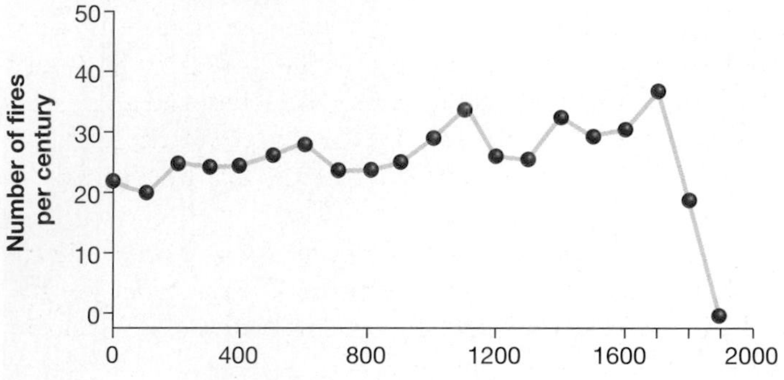

FIGURE 53.21 History of Disturbance in a Fire-Prone Community. Because trees form one ring (light band/dark band) every year, researchers can count the rings to determine how often fires have occurred during the last 2000 years in giant sequoia groves.

Succession: The Development of Communities after Disturbance

Severe disturbances remove all or most of the organisms from an area. The recovery that follows is called **succession**. The name was inspired by the observation that certain species succeed others over time.

- **Primary succession** occurs when a disturbance removes the soil and its organisms as well as organisms that live above the surface. Glaciers, floods, volcanic eruptions, and landslides often initiate primary succession.
- **Secondary succession** occurs when a disturbance removes some or all of the organisms from an area but leaves the soil intact. Fire and logging are examples of disturbances that initiate secondary succession.

As **Figure 53.22** shows, a sequence of plant communities develops as succession proceeds. Early successional communities are dominated by species that are short lived and small in stature, and that disperse their seeds over long distances. Late successional communities are dominated by species that tend to be long lived, large, and good competitors for resources such as light and nutrients.

The specific sequence of species that appears over time is called a successional pathway. What determines the pattern and rate of species replacement during succession at a particular time and place? To answer this question, biologists focus on three factors:

1. the particular traits of the species involved;
2. how the species interact; and
3. historical and environmental circumstances, such as the size of the area involved and weather conditions.

Let's consider each in turn.

THE ROLE OF SPECIES TRAITS Species traits, such as dispersal capability and the ability to withstand extreme dryness, are particularly important early in succession. As common sense would predict, recently disturbed sites tend to be colonized by plants and animals with good dispersal ability. When these organisms arrive, however, they often have to endure harsh environmental conditions.

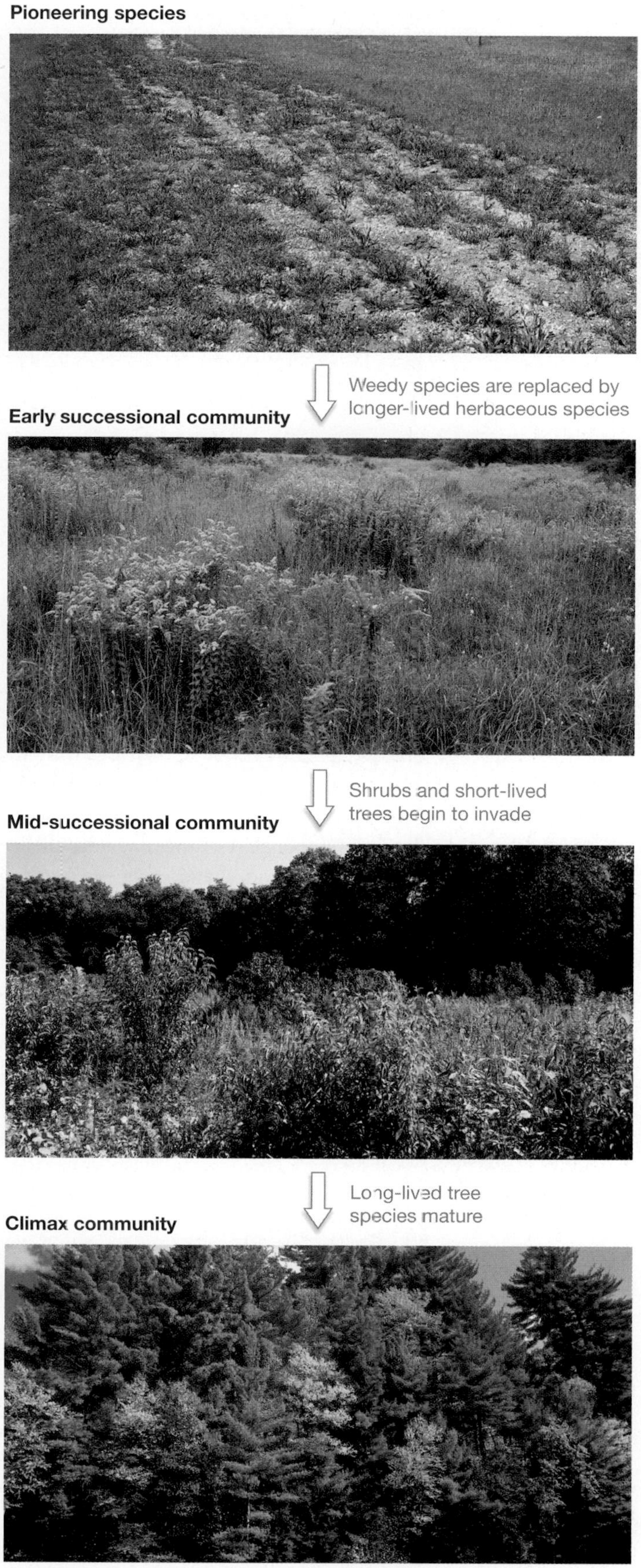
Pioneering species
Weedy species are replaced by longer-lived herbaceous species
Early successional community
Shrubs and short-lived trees begin to invade
Mid-successional community
Long-lived tree species mature
Climax community

FIGURE 53.22 Succession in Midlatitude Temperate Forests. Succession leads to the development of a temperate forest from a disturbed state (in this case, an abandoned agricultural field).

Pioneering species tend to have "weedy" life histories. A **weed** is a plant that is adapted for growth in disturbed soils. Because soil is often disturbed when biomass is removed, weeds tend to thrive at the start of succession.

Early successional species devote most of their energy to reproduction and little to competitive ability. They are at the "high fecundity, low survivorship" end of the life history continuum introduced in Chapter 52.

More specifically, pioneering species have small seeds, rapid growth, and a short life span, and they begin reproducing at an early age. As a result, they have a high reproductive rate (high r_{max}—see Chapter 52). But in addition, they can tolerate severe abiotic conditions such as high light levels, poor nutrient availability, and desiccation.

THE ROLE OF SPECIES INTERACTIONS Once colonization is under way, the course of succession tends to depend less on how species cope with aspects of the abiotic environment and more on how they interact with other species. This change occurs because plants that grow early in succession change abiotic conditions in a way that makes the conditions less severe.

For example, because plants provide shade, they reduce temperatures and increase humidity. Their dead bodies also add organic material and nutrients to the soil. As abiotic conditions improve, biotic interactions become more important.

During succession, existing species can have one of three effects on subsequent species:

- **Facilitation** takes place when the presence of an early arriving species makes conditions more favorable for the arrival of certain later species, by providing shade or nutrients.
- **Tolerance** means that existing species do not affect the probability that subsequent species will become established.
- **Inhibition** occurs when the presence of one species inhibits the establishment of another. For example, a plant species that requires high light levels to germinate may be inhibited late in succession by the presence of mature trees that prevent sunlight from reaching the forest floor.

THE ROLE OF CHANCE AND HISTORY In addition to species traits and species interactions, the pattern and rate of succession depend on the historical and environmental context in which they occur. For example, the communities that developed after forest fires disturbed Yellowstone National Park in 1988 depended on the size of the burned patch and how hot the fire had been at that location.

Succession is also affected by the particular weather or climate conditions that occur during the process. Variation in weather and climate causes different successional pathways to occur in the same place at different times.

Analyzing species traits, species interactions, and the historical/environmental context provides a useful structure for understanding why particular successional pathways occur. To see this theoretical framework in action, let's examine data on the course of primary succession that has occurred in Glacier Bay, Alaska.

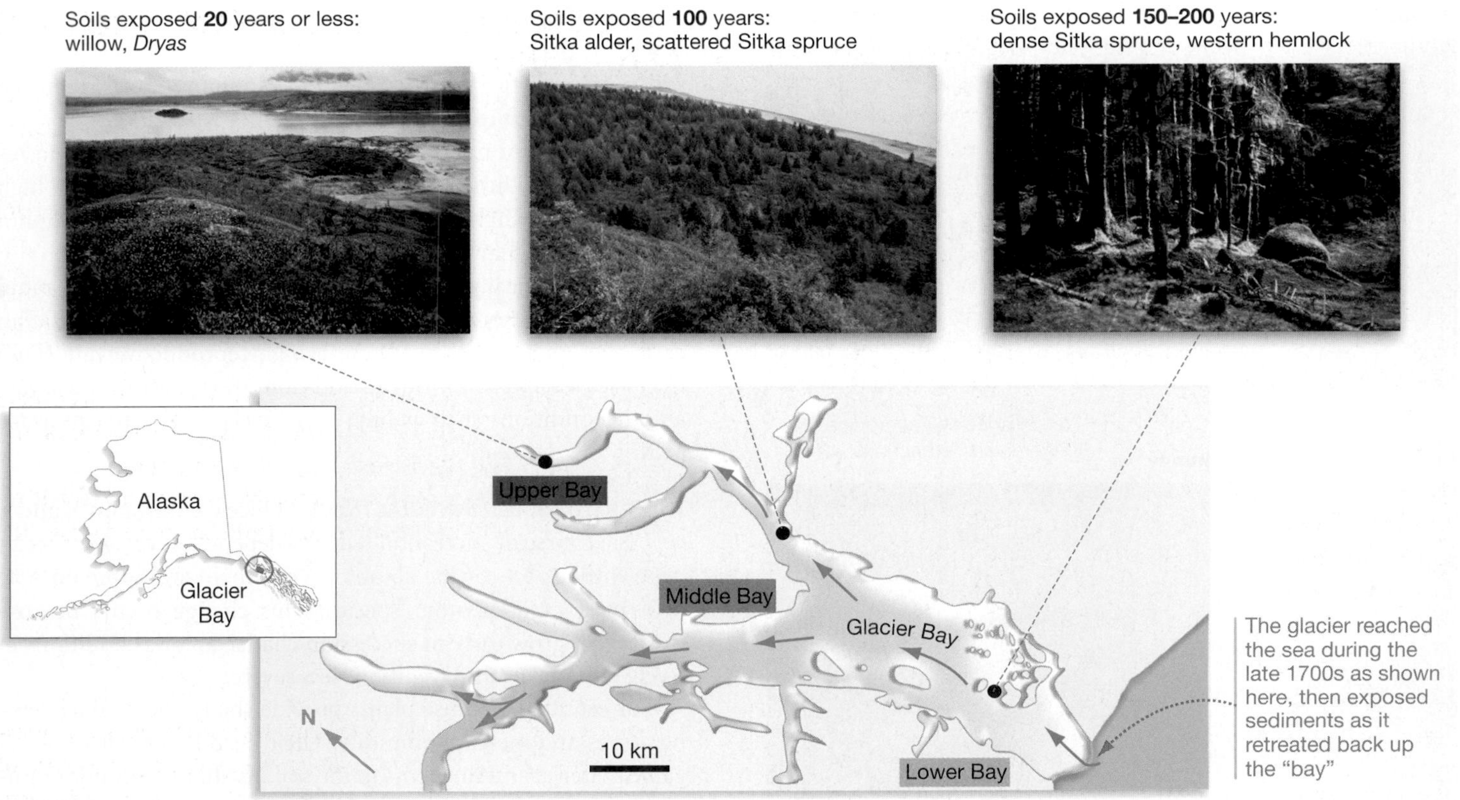

FIGURE 53.23 Successional Stages in Glacier Bay.

A CASE HISTORY: GLACIER BAY, ALASKA An extraordinarily rapid and extensive glacial recession is occurring at Glacier Bay (**Figure 53.23**). In just 200 years, glaciers that once filled the bay have retreated approximately 100 km, exposing tracts of barren glacial sediments to colonization. Because of this event, Glacier Bay has become an important site for studying succession.

Figure 53.23 shows the plant communities found in the area.

- Locations that have been ice free for 20 years or less do not have a continuous plant cover. Instead, they host scattered individuals of willow and a small shrub called *Dryas*.
- Areas that have been deglaciated for about 100 years are inhabited by dense thickets of a shrub called Sitka alder and scattered Sitka spruce trees.
- The oldest sites, ranging from 150 to 200 years of soil exposure, are dense forests of Sitka spruce and western hemlock.

These observations inspired a hypothesis for the pattern of succession in Glacier Bay: With time, the youngest communities of *Dryas* and willow succeed to alder thickets, which subsequently become dense spruce-hemlock forests. The key claim is that there is a single successional pathway throughout the bay.

Research has challenged this hypothesis, however. Biologists who reconstructed the history of each community by studying tree rings found that at least two distinct successional pathways have occurred:

1. In the lower part of the bay, soon after the ice retreated, Sitka spruce began growing and quickly formed dense forests. Western hemlock arrived after spruce and is now common.
2. At middle-aged sites in the upper part of the bay, alder thickets were dominant for several decades, and spruce is just beginning to become common. These forests will probably never be as dense as the ones in the lower bay, however, and there is no sign that western hemlock has begun to establish itself.

In contrast, the youngest sites in the uppermost part of the bay may be following a third pathway. Alder thickets became dominant fairly early, but spruce trees are scarce. Instead, cottonwood trees are abundant.

To review these successional patterns, go to the study area at *www.masteringbiology.com.*

Web Activity Primary Succession

Then consider a question that these data raise: How do species traits, species interactions, and dispersal patterns interact to generate the observed successional pathways?

- *Species traits* Western hemlock is abundant at older sites but largely absent from young ones. This is logical, because its seeds germinate and grow only in soils containing a substantial amount of organic matter. In addition, young hemlock trees can tolerate deep shade but not bright sunlight. Western hemlock does not tolerate early successional conditions, but thrives later in succession.
- *Species interactions* Sitka alder facilitates the growth of Sitka spruce. The facilitating effect occurs because symbiotic bacteria that live inside nodules on the roots of alder convert

atmospheric nitrogen (N_2) to nitrogen-containing molecules that alder use to build proteins and nucleic acids. When alder leaves fall and decay or roots die, the nitrogen becomes available to spruce. Although spruce trees are capable of invading and growing without the presence of alder, they grow faster when alder stands have added nitrogen to the soil. Note that spruce gains from the association. Alder is either unaffected, or negatively affected if it becomes shaded by spruce.

Competition is another important species interaction. For example, shading by alder reduces the growth of spruce until spruce trees are tall enough to protrude above the alder thicket. Once the spruce trees breach the alder canopy, however, alder dies out, because it is unable to compete with spruce trees for light.

- *Historical and environmental context* According to geological evidence, glacial ice was more than 1100 m thick in the upper part of Glacier Bay during the mid-1700s. Because forests grow to an elevation of only 700 m or 800 m in this part of Alaska, the glacier eliminated all of the existing forests. But in the lower part of the bay, the ice was substantially thinner. As a result, some forests remained on the mountain slopes beyond the ice, even at the height of glaciation.

  The difference is crucial. In the lower part of the bay, nearby forests provided a ready source of Sitka spruce and western hemlock seed. But in the upper part of the bay, no forests remained to contribute seed. As a result, two dramatically different successional pathways were set in motion.

To summarize, successional pathways are determined by the adaptations that certain species have to their abiotic environment, interactions among species, and the history of the site. Species traits and species interactions tend to make succession predictable; history and chance events contribute a degree of unpredictability to succession.

CHECK YOUR UNDERSTANDING

If you understand that . . .

- Disturbance is a normal part of communities.
- The impact of a disturbance depends on its type, frequency, and severity.
- After a disturbance occurs, a succession of species and communities replaces the individuals that were lost.
- The exact sequence of species observed is a function of their traits, their interactions, and the history of the site.

✓ You should be able to . . .

1. Explain why early successional species alter the environment in ways that make growing conditions more difficult for themselves.
2. Explain why the presence or absence of a species like alder, where nitrogen-fixation occurs, might alter the course of a succession.

Answers are available in Appendix B.

53.4 Species Richness in Ecological Communities

The diversity of species present is a key feature of biological communities, and it can be quantified in two ways.

- **Species richness** is a simple count of how many species are present in a given community.
- **Species diversity** is a weighted measure that incorporates a species' relative abundance as well as its presence or absence (see **Box 53.1** on page 1079).

At scales above relatively small study plots, it is rare to have data on relative abundance. As a result, biologists sometimes use species richness and species diversity interchangeably.

To introduce research on richness and diversity, let's focus on a simple question: Why are some communities more species rich than others?

Predicting Species Richness: The Theory of Island Biogeography

When researchers first began counting how many species are present in various areas, a strong pattern emerged: Larger patches of habitat contain more species than do smaller patches of habitat. The observation is logical because large areas should contain more types of niches and thus support higher numbers of species.

Early work on species richness highlighted another pattern, however—one that was harder to explain. Islands in the ocean have smaller numbers of species than do areas of the same size on continents. Why?

As it turned out, answering this question has had far-reaching implications in ecology and conservation biology. This often happens in biological science: Research on a seemingly esoteric question—often done out of the sheer joy of discovery—turns out to have important implications for completely unrelated issues, or even applications to practical problems that had previously been intractable.

THE ROLE OF IMMIGRATION AND EXTINCTION Robert MacArthur and Edward O. Wilson tackled this question by assuming that speciation occurs so slowly that the number of species present on an island is a product of just two events: immigration and extinction. The rates of both of these processes, they contended, should vary with the number of species present on an island.

- Immigration rates should decline as the number of species on the island increases, because individuals that arrive are more likely to represent a species that is already present. In addition, competition should prevent new species from becoming established when many species are already present on an island—meaning that immigration should seldom result in establishment of a new population.
- Extinction rates should increase as species richness increases, because niche overlap and competition for resources will be more intense.

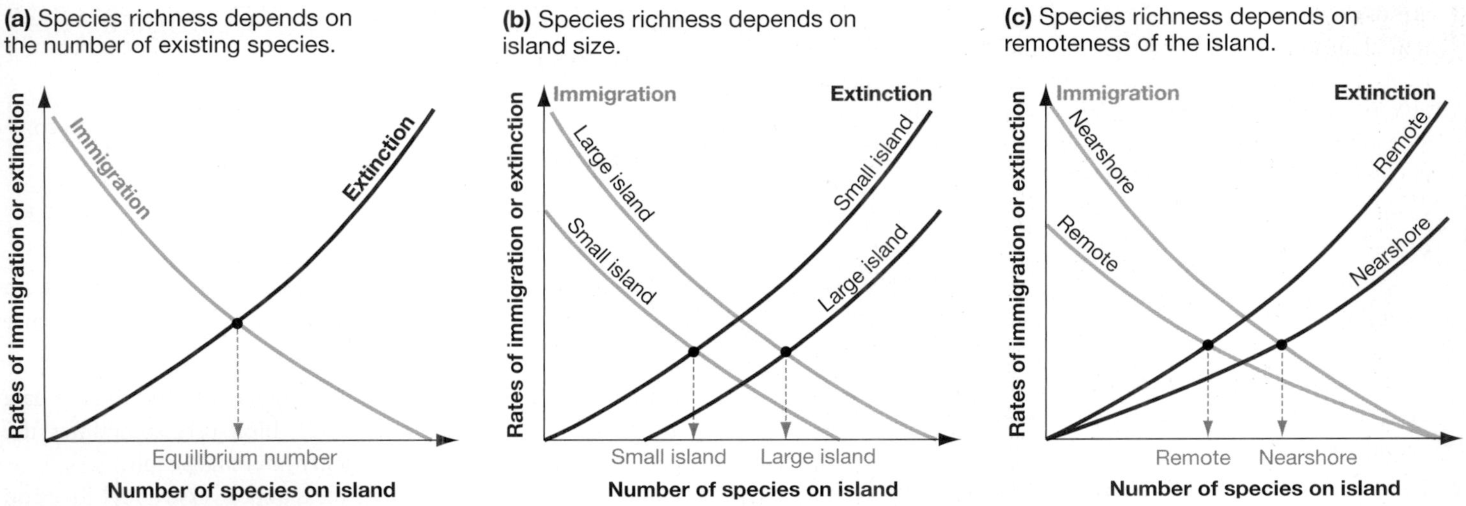

FIGURE 53.24 Species Richness Varies as a Function of Island Characteristics.

✔**QUESTION** Suppose that the mainland habitats closest to an island were wiped out by suburbanization. How would this affect the curves in part (c)?

The graphs in **Figure 53.24a** illustrate these two predictions. Note that when the two processes interact, the result is an equilibrium—a balance between the arrival of new species and the extinction of existing ones. If species richness changes due to a hurricane or fire or some other disturbance, then continued immigration and extinction should restore the equilibrium value.

THE ROLE OF ISLAND SIZE AND ISOLATION MacArthur and Wilson also realized that immigration and extinction rates should vary as a function of island size and how far the island was to a continent or other source of immigrants.

- Immigration rates should be higher on large islands that are close to mainlands, because immigrants are more likely to find large islands that are close to shore than small ones that are far from shore.
- Extinction rates should be highest on small islands that are far from shore, because fewer resources are available to support large populations, and because fewer individuals arrive to keep the population going.

Their theory makes two predictions: Species richness should be higher on (**1**) larger islands than smaller islands (**Figure 53.24b**), and (**2**) nearshore islands versus remote islands (**Figure 53.24c**).

APPLYING THE THEORY The MacArthur-Wilson model, called the theory of island biogeography, is important for several reasons:

- It is relevant to a wide variety of island-like habitats such as alpine meadows, lakes and ponds, and caves. It is also relevant to national parks and other reserves that are surrounded by human development, forming islands of habitat.
- It is relevant to species that have the metapopulation structure introduced in Chapter 52. Immigration and extinction rates in subpopulations should follow the dynamics predicted by the theory.
- It made specific predictions that could be tested. For example, researchers have measured species richness on tiny islands, removed all of the species present, and then measured whether the same number of species recolonized the island and reached an equilibrium number. Predictions about immigration and extinction rates have also been measured by observing the same islands over time.
- It can help inform decisions about the design of natural preserves. In general, the most-species-rich reserves should be ones that are (**1**) relatively large, and (**2**) located close to other relatively large habitat areas.

Global Patterns in Species Richness

Biologists have long understood that large habitat areas tend to be species rich, and the theory of island biogeography has been successful in framing thinking about how species richness should vary among island-like habitats. But researchers have had a much more difficult time explaining what may be the most striking pattern in species richness.

THE LATITUDINAL GRADIENT Biologists who began cataloging the flora and fauna of the tropics in the mid-1800s quickly recognized that these habitats contain many more species than do temperate or subarctic environments. Data compiled in the intervening years have confirmed the existence of a strong latitudinal gradient in species diversity—for communities as a whole as well as for many taxonomic groups.

In birds, mammals, fish, reptiles, many aquatic and terrestrial invertebrates, orchids, and trees, species diversity declines as latitude increases. **Figure 53.26** graphs data on the number of vascular plant species present from the equator toward the poles. Although this pattern is not universal—a number of marine groups, as well as shorebirds, show a *positive* relationship between latitude and diversity—it is widespread. Why does it occur?

BOX 53.1 QUANTITATIVE METHODS: Measuring Species Diversity

To measure species diversity, a biologist could simply count the number of species present in a community. The problem is that simple counts provide an incomplete picture of diversity. The relative abundance of species is also an important component of diversity. To grasp this point, consider the composition of the three hypothetical communities shown in **Figure 53.25**. These communities are nearly identical in species richness but differ greatly in the relative abundance of each species.

These data can be used to compare two measures of species diversity. Species richness is simply the number of species found in a community. In this case, communities 1 and 2 have equal species richness and community 3 is lower in richness by one species. It is important to note, however, that communities 2 and 3 have similar relative abundances of each species, or what biologists call high *evenness*. Community 1, in contrast, is highly uneven. Fifty-five percent of the individuals in community 1 belong to species A, and other species are relatively rare. An uneven community has lower effective diversity than its species richness would indicate.

To take evenness into account, other diversity indices have been developed. A simple example is the Shannon index, given by the following equation:

$$H' = -\sum_{i=1}^{n} p_i \ln p_i$$

In this equation, p_i is the proportion of individuals in the community that belong to species i. The index is summed over all the species in the study.

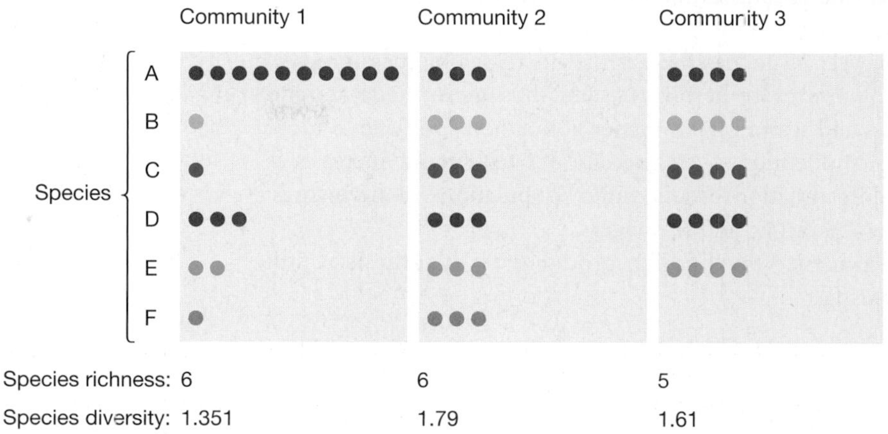

FIGURE 53.25 Species Diversity Can Be Quantified.

The Shannon index of species diversity for the three hypothetical communities shown in Figure 53.25 was calculated by **(1)** computing the proportion of individuals in each community that belong to each species, **(2)** taking the natural logarithm of each of these proportions (see **BioSkills 7** in Appendix A for an introduction to logarithms), **(3)** multiplying each natural logarithm times the proportion for each species, and **(4)** summing the total across the six species in the community. **✔If you understand the equation, you should be able to do these calculations and get the same result given in Figure 53.25.** Notice that while communities 1 and 2 have the same species richness, community 2 has higher diversity because of its greater evenness. Community 3 has lower species richness than community 1 but higher diversity.

✔If you understand how to use and interpret the Shannon index, you should be able to calculate species richness and the Shannon diversity index for 3 communities that "double" Communities 1–3 in Figure 53.25. For example, one of the three new communities should be identical to Community 1, except that there are 2 species with 10 individuals present like species "A", 2 species with 1 individual present like species "B", 2 species with 1 individual present like species "C", 2 species with 3 individuals present like species "D", and so on. There are a total of 12 species in two of the communities, and 10 in the third. **✔You should also be able to compare and contrast your results with the richness and diversity values given in Figure 53.25.**

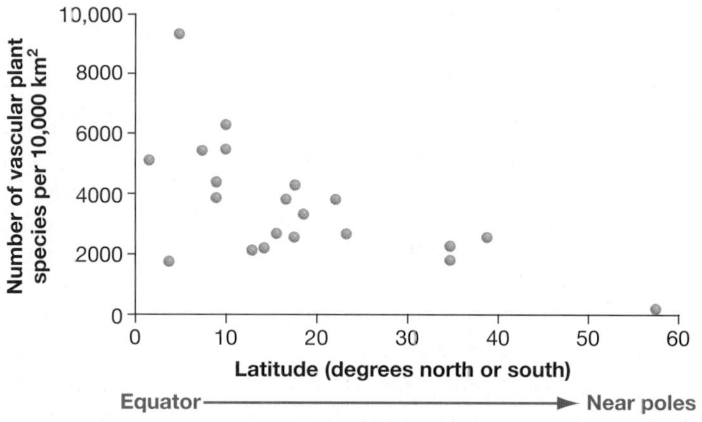

FIGURE 53.26 There Is a Strong Latitudinal Gradient in Species Richness of Vascular Plants.

To explain why species diversity might decline with increasing latitude, biologists have to consider two fundamental principles:

1. The causal mechanism must be abiotic, because latitude is a physical phenomenon produced by Earth's shape. The explanation must be a physical factor that varies predictably with latitude and that could produce changes in species diversity.
2. The species diversity of a particular area is the sum of four processes: speciation, extinction, immigration (colonization), and emigration (dispersal). Thus, the latitudinal gradient must be caused by an abiotic factor that affects the rate of speciation, extinction, immigration, or emigration in a way that would lead to more species in the tropics and fewer near the poles.

Over 30 hypotheses have been proposed to explain the latitudinal gradient. Let's consider four of the most prominent ideas.

THE HIGH-PRODUCTIVITY HYPOTHESIS High productivity in the tropics could promote high diversity by increasing speciation rates and decreasing extinction rates. (Recall from Chapter 50 that productivity is the total amount of photosynthesis per unit area per year.) The logic here is that increased biomass production supports more herbivores, and thus more predators and parasites and scavengers. If higher population sizes lead to more intense competition, then speciation rates should increase as niche differentiation occurs within populations of herbivores, predators, parasites, and scavengers.

The evidence on the high-productivity hypothesis is still inconclusive.

- It is supported by global patterns in productivity that will be introduced in Chapter 54. For example, productivity is extremely high in biomes such as tropical rain forests and coral reefs, which have high species diversity.
- It is not supported by experimental addition of fertilizer to aquatic or terrestrial communities. In most of these studies, productivity increases but species diversity decreases. Critics contend that the time and spatial scale of these experiments is too small to be relevant to global patterns, however.
- It is not supported in some highly productive habitats, such as estuaries (see Chapter 50), where species diversity is often extremely low.

THE ENERGY HYPOTHESIS The energy hypothesis is an extension of the high-productivity hypothesis. Recognizing that some highly productive habitats are species-poor, investigators proposed that the key factor can't be productivity alone.

The energy hypothesis claims that high temperatures increase species diversity by increasing productivity *and* the likelihood that organisms can tolerate the physical conditions in a region. The idea is that high energy inputs—rather than high nutrient inputs—are the key to high speciation rates and low extinction rates.

Although the energy hypothesis is relatively new, some data support it. For example, biologists who analyzed data on gastropods and other marine invertebrates have documented a strong correlation between the temperature of marine waters and species diversity.

THE AREA AND AGE HYPOTHESIS Temperate and arctic latitudes were repeatedly scoured by ice sheets over the last 2 million years, but tropical regions were not. In addition, the land area of the tropics has historically been larger than the land area at high latitudes. Thus, tropical regions have had more time and more space for speciation to occur than other regions have.

Recent data suggest, however, that tropical forests were dramatically reduced in size by widespread drying trends during the ice ages. Existing forests may be much younger than originally thought. If so, then the contrast in the age of northern and southern habitats may not be enough to explain the dramatic difference in species diversity.

THE INTERMEDIATE DISTURBANCE HYPOTHESIS Across habitats and biomes, species diversity is much higher in mid-successional communities than in pioneer or mature communities. The **intermediate disturbance hypothesis** was inspired by this observation. It holds that regions with a moderate type, frequency, and severity of disturbance should have high species richness and diversity.

The logic here is that, with intermediate levels of disturbance, communities will contain pioneering species as well as species better adapted to late-successional conditions. For example, recent studies have confirmed that tree falls and canopy gaps occur regularly in tropical forests, and that fires occur in these biomes occasionally. As yet, however, there are no convincing data showing that intermediate levels of disturbance are more likely to occur in the tropics than they are at higher latitudes.

What can you conclude from all this? The current consensus is that each of the factors discussed here may influence diversity. It is tempting to look for single causes to explain important patterns and events. But in this case, multiple causation is probably more realistic.

CHAPTER 53 REVIEW

For media, go to the study area at www.masteringbiology.com

Summary of Key Concepts

Interactions among species, such as competition, consumption, and mutualism have two main outcomes: (1) They affect the distribution and abundance of the interacting species, and (2) they are agents of natural selection and thus affect the evolution of the interacting species. The nature of interactions between species frequently changes over time.

- To categorize the different types of interactions that occur among species, biologists consider whether each participant experiences a net fitness cost or benefit from the interaction. These costs and benefits depend on the conditions that prevail at a particular time and place and may change through time.
- Competition occurs when the niches of two species overlap—meaning they use the same resources. Competition may result in the complete exclusion of one species (the weaker competitor). It may also result in niche differentiation, in which competing species evolve traits that allow them to exploit different resources or live in different areas.
- Consumption occurs when consumers eat other individuals, which resist through standing defenses or inducible defenses. Predators are efficient enough to reduce the size of many prey populations. Levels of herbivory are relatively low in terrestrial ecosystems, however, because predation and disease limit herbivore populations,

because plants provide little nitrogen, and because many plants contain toxic compounds or other types of defenses.

- Parasites are consumers that generally spend all or part of their life cycle in or on their host (or hosts). They usually have traits that allow them to escape host defenses and even manipulate host behavior to increase their likelihood of transmission to a new host. In turn, hosts have evolved counter-adaptations that help fight off parasites.
- Mutualism provides participating individuals with food, shelter, transport of gametes, or defense against predators. For each species involved, the costs and benefits of mutualism may vary over time and from place to place.

✔You should be able to give an example of how competition can evolve into commensalism, and how a mutualistic relationship can evolve into a parasitic one.

(MB) **Web Activity** Life Cycle of a Malaria Parasite

The assemblage of species found in a biological community changes over time and is primarily a function of climate and chance historical events.

- Historical and experimental evidence support the view that communities are dynamic rather than static, and that their composition is neither entirely predictable nor stable over time.
- Each community has a characteristic disturbance regime—meaning a type, severity, and frequency of disturbance that it experiences.
- Three types of factors influence the pattern of succession that occurs after a disturbance: (1) A species' physiological traits influence when it can successfully join a community; (2) interactions among species influence if and when a species can become established; and (3) the historical and environmental context of the site affects which species are present.

✔You should be able to explain why climate makes the three successional pathways documented in Glacier Bay similar, and why chance historical events make them different.

(MB) **Web Activity** Primary Succession

Species richness is higher in large islands near continents than in small, isolated islands, due to differences in immigration and extinction. Species richness is also higher in the tropics and lower toward the poles, but the mechanism responsible for this pattern is still controversial.

- Large islands that are close to continents have high immigration rates due to proximity and space available for new species, and low extinction rates. Isolated islands have low immigration rates, and small islands have high extinction rates due to small amounts of available habitat and low population sizes.
- The high species richness observed in the tropics results from a combination of factors—the most important of which may be high temperatures that increase productivity and provide relatively benign abiotic conditions.

✔You should be able to suggest a hypothesis to explain an exception to the latitudinal gradient rule—specifically, that most species of shorebirds breed at high latitudes, not the tropics. (Note that many shorebird species nest near the abundant lakes found in arctic tundra.)

Questions

✔TEST YOUR KNOWLEDGE

Answers are available in Appendix B

1. What is competitive exclusion?
   a. the evolution of traits that reduce niche overlap and competition
   b. interactions that allow species to occupy their fundamental niche
   c. the degree to which the niches of two species overlap
   d. the claim that species with the same niche cannot coexist

2. What is niche differentiation?
   a. the evolution of traits that reduce niche overlap and competition
   b. interactions that allow species to occupy their fundamental niche
   c. the degree to which the niches of two species overlap
   d. the claim that species with the same niche cannot coexist

3. Why is the phrase "coevolutionary arms race" an appropriate way to characterize the long-term effects of species interactions?
   a. Both plants and animals have evolved weapons for defense that are so effective that many plants are not eaten and predators cannot reduce prey populations to extinction.
   b. Adaptations that give one species a fitness advantage in an interaction are likely to be countered by adaptations in the other species that eliminate this advantage.
   c. In all species interactions except for mutualism, at least one species loses (suffers decreased fitness).
   d. Even mutualistic interactions can become parasitic if conditions change. As a result, interacting species are always "at war."

4. Why are inducible defenses advantageous?
   a. They are always present; thus, an individual is always able to defend itself.
   b. They make it impossible for a consumer to launch surprise attacks.
   c. They result from a coevolutionary arms race.
   d. They make efficient use of resources, because they are produced only when needed.

5. Which of the following is *not* correlated with species diversity?
   a. latitude
   b. productivity
   c. longitude
   d. island size

6. What is net primary productivity?
   a. an individual's lifetime reproductive success (lifetime fitness)
   b. an individual's average annual reproductive success
   c. the total amount of photosynthesis that occurs in an area of a given size per year
   d. the amount of energy that is stored in standing biomass per year

✓ TEST YOUR UNDERSTANDING

Answers are available in Appendix B

1. The text claims that species interactions are conditional and dynamic. Do you agree with this statement? Why or why not? Cite specific examples to support your answer.
2. State two hypotheses that have been proposed to explain the low level of herbivory in terrestrial plant communities. Are these hypotheses mutually exclusive? (In other words, can both be correct?) Explain why or why not.
3. Biologists have tested the hypotheses that communities are highly predictable versus highly unpredictable. State the predictions that these hypotheses make with respect to (a) changes in the distribution of the species in a particular community over time, and (b) the communities that should develop at sites where abiotic conditions are identical. Which hypothesis appears to be more accurate?
4. What is a disturbance? Consider the role of fire in a forest. Compare and contrast the consequences of high-frequency versus low-frequency fire, and high versus low severity of fire.
5. Summarize the life-history attributes of early successional species. Why are these attributes considered adaptations?
6. Explain why high productivity should lead to increased species richness in habitats such as tropical rain forests and coral reefs.

✓ APPLYING CONCEPTS TO NEW SITUATIONS

Answers are available in Appendix B

1. Some insects harvest nectar by chewing through the wall of the structure that holds the nectar. As a result, they obtain a nectar reward, but pollination does not occur. Suppose that you observed a certain bee species obtaining nectar in this way from a particular orchid species. Over time, how might you expect the characteristics of the orchid population to change in response to this bee behavior?
2. Using this chapter's information on fire regimes in giant sequoia groves, propose a management plan for Sequoia National Park. Explain the logic behind your plan.
3. Suppose that a two-acre lawn on your college's campus is allowed to undergo succession. Describe how species traits, species interactions, and the site's history might affect the community that develops.
4. Design an experiment to test the hypothesis that increasing species richness increases a community's productivity.

 This chapter is part of The Big Picture. See how on pages 1126–1127.

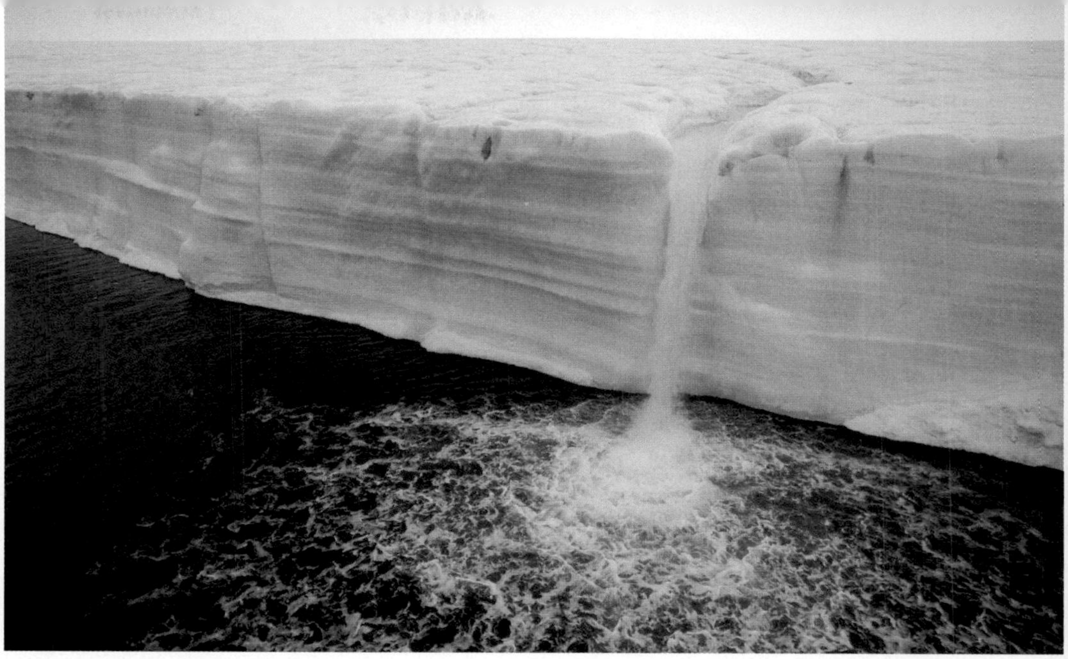

Glaciers all over the world are melting at unprecedented rates, raising sea levels. This chapter explores how humans are altering ecosystems all over the planet.

Ecosystems 54

The data in this chapter carry a simple, but important, message: Some of the most urgent problems facing your generation are rooted in ecosystem ecology. Global warming, acid rain, a hole in the atmosphere's ozone layer, estrogen-mimicking pollutants, and nitrate pollution (see Chapter 28) are just a few of the issues that ecosystem ecologists are documenting.

The chapter has a simple theme, as well: In nature, everything is connected. This is because the components of an **ecosystem**—the species present, along with abiotic components such as the soil, climate, water, and atmosphere—are linked by flows of energy and nutrients. When you change the species that are present or the abiotic environment, you change the ecosystem.

Humans are removing species and adding massive amounts of energy and nutrients to ecosystems all over the planet. Chapter 55 considers the role of biodiversity and the consequences of species loss; the focus here is on how people are changing climate, water, and other abiotic aspects of ecosystems.

To understand the consequences of altering the abiotic environment, let's consider (**1**) how energy moves among the components of an ecosystem, (**2**) how nutrients move among the components of an ecosystem, and (**3**) why global warming is occurring.

KEY CONCEPTS

- An ecosystem has four components: (1) the abiotic environment, (2) primary producers, (3) consumers, and (4) decomposers. These components are linked by the movement of energy and nutrients.
- As energy flows from producers to consumers and decomposers in a food web, much of it is lost. The productivity of terrestrial ecosystems is limited by warmth and moisture; nutrient availability is the key constraint in aquatic ecosystems.
- To analyze nutrient cycles, biologists focus on the nature of the reservoirs where elements reside and the processes that move elements between reservoirs. Nutrient addition by humans is increasing productivity and causing pollution.
- The burning of fossil fuels has led to rapid global warming; rapid ecological and evolutionary changes are being observed in response.

54.1 How Does Energy Flow through Ecosystems?

A **primary producer** is an **autotroph** (literally, "self-feeder")—an organism that can synthesize its own food from inorganic sources. Primary producers are the conduit where energy enters ecosystems.

In most ecosystems, primary producers use solar energy to manufacture food via photosynthesis. But in deep-sea hydrothermal vents and iron-rich rocks deep below Earth's surface, primary producers use the chemical energy contained in inorganic compounds

✔ When you see this checkmark, stop and test yourself. Answers are available in Appendix B.

such as hydrogen (H_2), methane (CH_4), or hydrogen sulfide (H_2S) to make food (see Chapter 28).

Primary producers form the basis of ecosystems because they transform the energy in sunlight or inorganic compounds into the chemical energy stored in sugars. The primary producers use this chemical energy in two ways:

- *Maintenance* Organisms have to use chemical energy to replace worn-out cell components, transport material inside cells and between cells, synthesize hormones and other chemical signals, and produce molecules and structures that function in defense against disease and attack. It takes energy just to stay alive.
- *Growth* Chemical energy that isn't required for maintenance can be used for growth and reproduction. Energy that is invested in new tissue or offspring is called **net primary productivity (NPP)**.

In addition, the use of chemical energy is never 100 percent efficient—some energy is inevitably lost as heat.

Why Is NPP So Important?

NPP results in **biomass**—organic material that non-photosynthetic organisms can eat. NPP is critically important because it represents the amount of energy that is available to the other living components of an ecosystem: consumers and decomposers.

- **Consumers** eat living organisms. **Primary consumers** eat primary producers; **secondary consumers** eat primary consumers; **tertiary consumers** eat secondary consumers, and so on. You are a primary consumer when you eat a salad, a secondary consumer when you eat beef, and a tertiary consumer when you eat tuna.
- **Decomposers**, or **detritivores**, obtain energy by feeding on the remains of other organisms or waste products. Dead animals and dead plant tissues ("plant litter") are collectively known as **detritus**. Many fungi and bacteria are decomposers, but humans rarely act as detritivores.

Figure 54.1 illustrates how energy moves around the four components of an ecosystem. You could draw a parallel diagram showing how nitrogen, phosphorus, and other nutrients move among ecosystem components, with soil or water acting as the source of the key atoms. The four components of an ecosystem are linked by the movement of energy and nutrients.

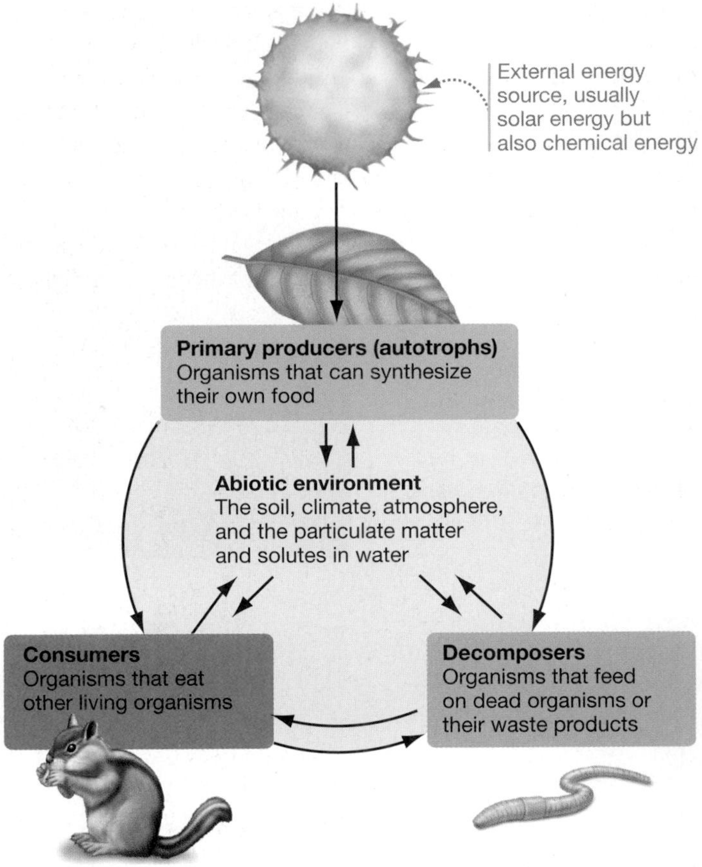

FIGURE 54.1 The Four Components of an Ecosystem Interact. The arrows represent energy. A similar diagram could be drawn to represent the flow of nutrients.

Solar Power: Transforming Incoming Energy to Biomass

Figure 54.1 has an arrow from the Sun to a leaf. How big is this arrow, that is, how much of the energy that enters an ecosystem gets converted into biomass? To answer this question, let's look at data from the world's most intensively studied ecosystem.

Since the early 1960s, a team of researchers has been studying how energy and nutrients flow through a temperate forest ecosystem in the northeastern United States: the Hubbard Brook Experimental Forest in New Hampshire.

From June 1, 1969, to May 31, 1970, 1,254,000 kcal of solar radiation per square meter (m^2) entered this ecosystem in the form of sunlight. If this amount of energy were available in the form of electricity, it would easily power two 150-watt lightbulbs that burned continuously for one year.

By documenting rates of photosynthesis in a large sample of forest plants, the biologists calculated that the plants used 10,400 kcal/m^2 of this energy in photosynthesis. This value represents **gross primary productivity**: the total amount of photosynthesis in a given area and time period.

The team also calculated **gross photosynthetic efficiency**—the ratio of gross primary productivity to the total incoming solar radiation. At Hubbard Brook, efficiency was 10,400 kcal/m^2 $\div$ 1,254,000 kcal/m^2 $=$ 0.008 or 0.8 percent. This value is typical of other ecosystems as well.

Why is efficiency so low? The answer has several components.

- Plants in temperate biomes have drastically reduced photosynthetic rates in winter, although sunlight is available.
- If conditions get dry during the summer, stomata close to conserve water. Photosynthesis slows dramatically due to a lack of CO_2.
- Even when conditions are ideal, the pigments that drive photosynthesis can absorb only a fraction of the wavelengths available and thus a fraction of the total energy received.

Although overall efficiency is low, a great deal of photosynthesis is still occurring. What happened to all of the products? After calculating photosynthetic rates in an array of plants, the researchers also measured growth—the production of new biomass. The data allowed them to estimate that of gross primary production, only about 45 percent went to NPP. The other 55 percent was used for maintenance activities or was lost as heat.

Biologists have done similar studies in other ecosystems. The general pattern is that only a tiny fraction of incoming sunlight is converted to chemical energy, and only a fraction of this gross primary productivity is used to build biomass. Now let's consider the next question: What happens to NPP?

Trophic Structure

Biomass represents chemical energy. Every time a monarch caterpillar eats a leaf or a fungus absorbs molecules from decaying wood, chemical energy flows from a primary producer to a primary consumer or decomposer. To describe these energy flows, biologists identify distinct feeding levels in an ecosystem. Organisms that obtain their energy from the same type of source occupy the same **trophic** ("feeding") **level**.

FOOD CHAINS AND FOOD WEBS A **food chain** connects the trophic levels in a particular ecosystem. In doing so, it describes how energy moves from one trophic level to another. **Figure 54.2** shows five trophic levels in a **decomposer food chain**—with **primary decomposers**, which feed on plant detritus, at the second trophic level—and four trophic levels in a **grazing food chain**, with primary consumers at the second trophic level.

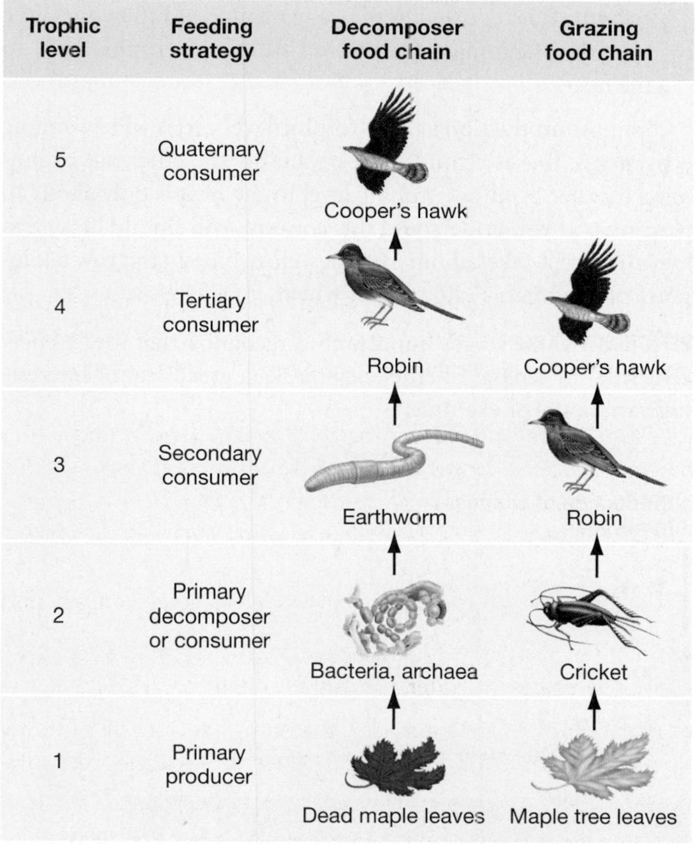

Trophic level
Feeding strategy
Decomposer food chain
Grazing food chain
5
Quaternary consumer
Cooper's hawk
4
Tertiary consumer
Robin
Cooper's hawk
3
Secondary consumer
Earthworm
Robin
2
Primary decomposer or consumer
Bacteria, archaea
Cricket
1
Primary producer
Dead maple leaves
Maple tree leaves

FIGURE 54.2 Trophic Levels Identify Steps in Energy Transfer. Many other species exist at each level in this temperate-forest ecosystem.

Note several key points:

- At higher trophic levels, the grazing and decomposer food chains often merge. For example, the robin in the figure functions as a secondary consumer in the grazing food chain and a tertiary consumer in the decomposer food chain.
- It is common for consumers—such as humans and robins—to feed at multiple trophic levels.

Because most consumers feed on a wide array of organisms and at multiple trophic levels, food chains are embedded in more complex **food webs**. Food webs attempt to include many or most of the species eaten by a sample of organisms in an ecosystem, joined by arrows indicating a flow of energy. For example:

- The Cooper's hawk in Figure 54.2 feeds on a wide array of small birds and small mammals, in addition to robins.
- The robin in the figure eats a wide array of insects, wormlike species, and spiders, as well as fruits and seeds (primary producers).

A food web would include these organisms, along with estimates of the amount of energy transferred at each link. Food webs are a compact way of summarizing energy flows and documenting the complex trophic interactions that occur in ecosystems.

ENERGY FLOW TO GRAZERS VERSUS DECOMPOSERS How much biomass gets eaten dead versus alive? More formally, how much energy is consumed by primary consumers versus primary decomposers?

The answer to this question varies enormously among habitats. At Hubbard Brook, for example, only about 24 percent of NPP is eaten by consumers—76 percent is uneaten until the primary producer dies. These values are actually unusual for forest ecosystems, where the percentage of NPP consumed can be as low as 5–7 percent. In a forest, much of the biomass is tied up in indigestible wood and isn't transferred to other organisms until it decays. The percentage of NPP that gets consumed in marine habitats, where most of the primary production is done by algae, is much higher—often 35–40 percent.

This is an important point, for two reasons:

1. If you want to understand energy flow in a forest, you have to understand decomposers. Decomposition is less of a factor in marine ecosystems, where primary consumers are paramount.
2. The decomposer food chain is "leaky." At Hubbard Brook, for example, large amounts of energy leave the forest ecosystem in the form of detritus that washes into streams. In these streams, photosynthesis by aquatic algae and plants introduced only about 10 kcal/m^2/yr, versus 6039 kcal/m^2/yr that washed in from the surrounding forest. In marine ecosystems, energy leaves in the form of dead bodies that rain down to the ocean floor.

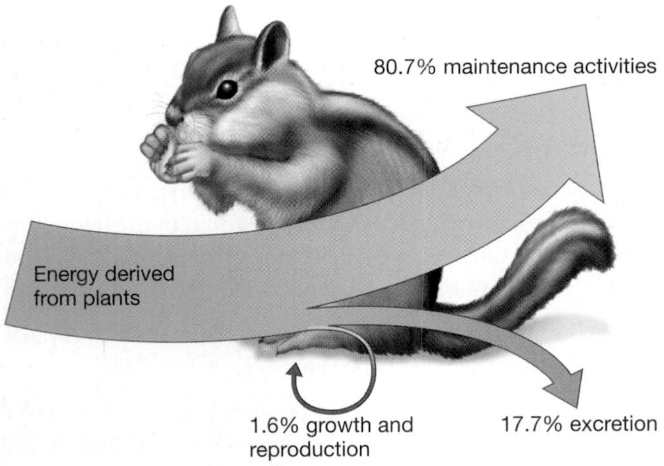

FIGURE 54.3 How Do Consumers Use Primary Production?

✔**QUESTION** Aphids, small plant-eating insects, are ectothermic (acquire heat from outside sources), largely sedentary, and feed on high-quality foods containing almost no indigestible molecules. Describe how the percentages given for chipmunks here would compare in aphids.

The "dead versus alive" percentage can even vary with conditions. In a recent experiment done in small ponds, for example, researchers found that much more NPP got consumed when light levels were low and nutrient levels were high. Under these conditions, algae flourished and were readily eaten by grazing animals. But when light levels were high and nutrient levels were low, most primary production was done by cyanobacteria, which many grazers avoided. Most of the cyanobacteria died without being eaten; their biomass entered the decomposer food chain. In an ecosystem like this, the dynamics of energy flow could change season to season or even week to week.

Energy Transfer between Trophic Levels

All ecosystems share a characteristic pattern: The total biomass produced each year declines from lower trophic levels up to the higher levels. To understand why, let's go back to the Hubbard Brook experimental forest.

Figure 54.3 illustrates what happened to the energy obtained by consumers, using chipmunks as an example. Chipmunks are small rodents that are seed predators. On average, these primary consumers harvest 31 kcal/m^2 of energy each year at Hubbard Brook. Of that total, 17.7 percent is unused—meaning that it is not digested and absorbed. Instead, it is excreted. In addition, 80.7 percent of total energy consumed is used for maintenance. Just 1.6 percent of the yearly 31 kcal/m^2 goes into the production of new chipmunk tissue by growth and reproduction.

The same pattern would be seen in a secondary consumer—a Cooper's hawk, for example, that eats chipmunks and other primary consumers. Consider the energy that a hawk has to expend to catch its prey. This energy cannot be used to make more hawk biomass.

The general point is simple: Because only a fraction of the total energy consumed is used for growth and reproduction, the amount of biomass produced at the second trophic level must be less than the amount at the first trophic level; productivity at the third trophic level must be less than that at the second.

THE PYRAMID OF PRODUCTIVITY The data in **Figure 54.4** graphs the biomass produced at trophic levels 1–4 in a temperate deciduous forest, with a separate horizontal bar representing each trophic level. Note that the graph reports two numbers:

1. Productivity is a rate, measured here in units of grams of biomass produced per square meter of area each year (g/m^2/year).
2. Efficiency is a ratio and thus dimensionless—the units cancel out. (It is reported here as a percentage.) Efficiency is the fraction of biomass transferred from one trophic level to the next.

Biomass production at each trophic level varies widely among ecosystems. But as a rough rule of thumb, the efficiency of biomass transfer from one trophic level to the next is only about 10 percent. ✔**If you understand this concept, you should be able to explain why it takes about 10 times more energy to grow a kilogram of beef than a kilogram of wheat.**

EFFICIENCY VARIES It's important to recognize that the "10 percent rule" in ecological efficiency masks a great deal of interesting variation. For example:

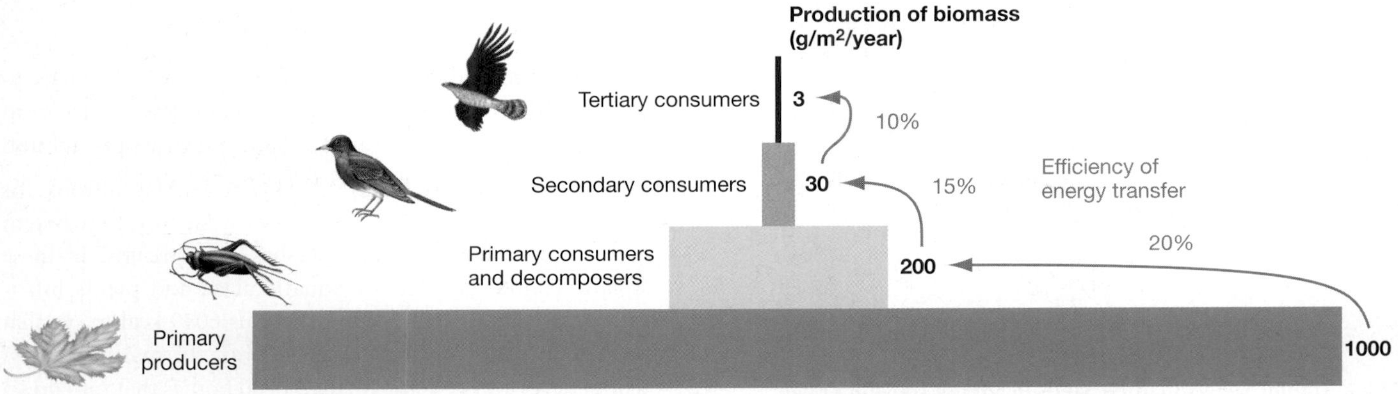

FIGURE 54.4 Productivity Declines at Higher Trophic Levels. This pattern is called the pyramid of productivity.

- Large mammals are more efficient at producing biomass than small mammals, because they have a smaller surface-area-to-volume ratio and lose less heat (see Chapter 41).
- Compared to endotherms, biomass production is much more efficient in ectotherms. You might recall from Chapter 41 that ectotherms rely principally on heat gained from the environment and do not oxidize sugars to keep warm. As a result, they devote much less energy to maintenance than endotherms do. Chipmunks devote just 1.5 percent of their energy intake to growth; caterpillars and other ectotherms may devote 20–40 percent.

✓If you understand this concept, you should be able to explain why it may be more efficient to eat crustaceans and fish than beef, if the animals you are consuming feed at the same trophic level.

Trophic Cascades and Top-Down Control

What happens when a component of a food chain or food web changes? You were introduced to this question briefly in Chapter 53, when you reviewed the keystone species concept. You might recall that the sea star *Pisaster ochraceous* primarily eats mussels. If *Pisaster* is excluded from a habitat, mussels take over—dramatically changing which species are present.

When a consumer (*Pisaster*) limits a prey population (mussels) in this way, biologists say that **top-down control** is occurring (see Chapter 53). The name was inspired by the relationships diagrammed in a food chain. As a secondary consumer, *Pisaster* is "on top" of mussels, which are primary consumers.

When changes in top-down control cause conspicuous effects two or three links away in a food web, biologists say that a **trophic cascade** has occurred. Wolves, for example, were reintroduced to the Greater Yellowstone Ecosystem of western North America in 1995. Wolves are a "top predator"—meaning that nothing eats them.

Although the Yellowstone wolf population has grown to only about 150 individuals, their presence has led to far-reaching changes in the food web (**Figure 54.5**).

- In Yellowstone, wolves primarily feed on elk. As elk numbers declined and individuals changed their feeding habits due to fear of being in the open, populations of favorite elk foods like aspen, cottonwood, and willow have increased.
- The changes in willow and other species triggered an increase in beavers, which compete with elk for these plants.
- Beavers dam streams, forming ponds that provide habitat for frogs, turtles, and an array of other species, whose numbers have increased.
- Wolves do not tolerate coyotes. As coyotes declined, mouse populations increased. The increase in mouse populations led to larger populations of hawks that prey on mice.

✓If you understand the concept of a trophic cascade, you should be able to change the relative size of the arrows in Figure 54.5 to indicate whether the energy transferred in each link increased or decreased as the wolf population grew to its present size.

Analogous changes have been observed in ecosystems where humans have *removed* species. Atlantic cod and other large predators, for example, have virtually disappeared from large parts of the Atlantic Ocean due to overfishing. Shrimp and crab—species that cod used to eat—have increased in response; the primary producers that shrimp and crab eat have decreased.

The message from these studies is clear: Changing the composition of food webs can have profound effects. The underlying

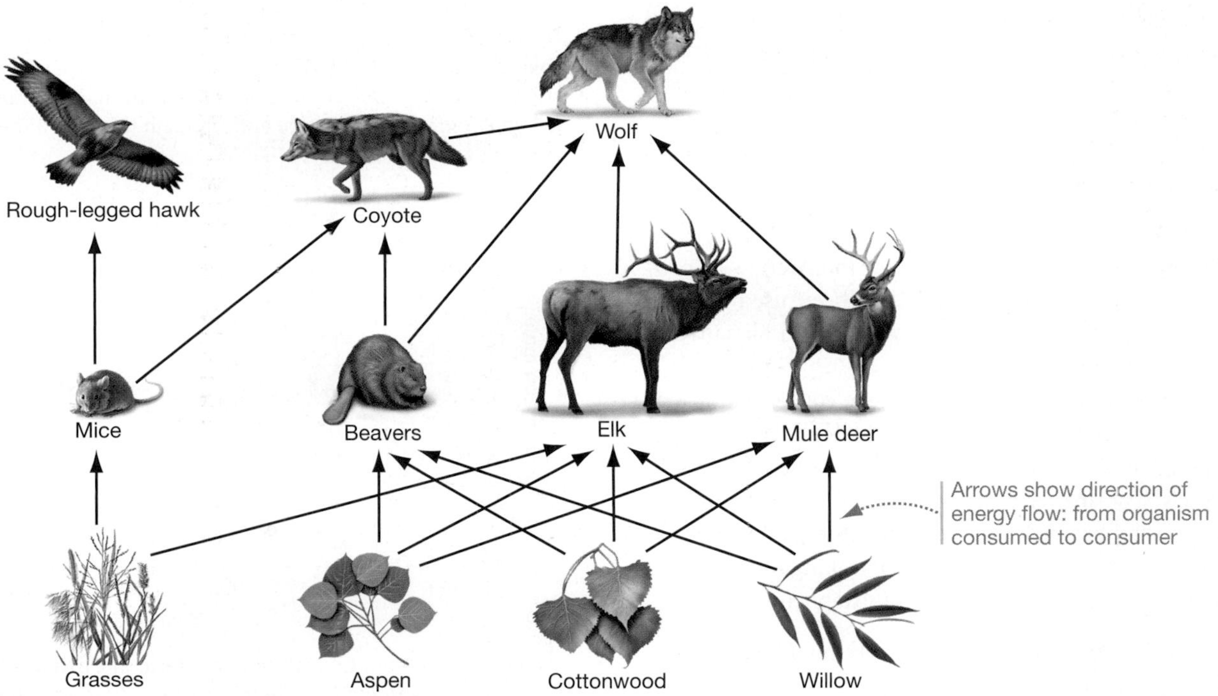

FIGURE 54.5 Reintroduction of Wolves Has Had a Dramatic Effect on the Food Web in Yellowstone.
This food web shows only a few of the hundreds of species affected by the reintroduction of wolves.

cause is energy flow—one of the factors that links components in an ecosystem. It's an important issue because humans are removing large predators from a wide array of ecosystems, through hunting or habitat destruction.

Biomagnification

Before leaving the topic of trophic interactions, let's take a brief look at the fate of certain molecules that are transferred through a food web along with chemical energy. As an example, consider the insecticide toxaphene.

Toxaphene is a cocktail of structurally similar organic compounds that include chlorine atoms. For decades toxaphene was widely used as an insecticide in cotton and soybean fields, until data on its toxicity to humans and wildlife led to it being banned in the U.S. in 1986. Although toxaphene is now illegal in most countries, it continues to be used in some parts of the world.

WHAT IS A POP? Toxaphene tends to persist in the environment without breaking down into harmless substances. It is a POP—a persistent organic pollutant. Toxaphene and other POPs undergo **biomagnification**—meaning that they increase in concentration at higher levels in a food chain.

Biomagnification begins when persistent atoms or molecules are taken up from air or water by primary producers and passed on to primary, secondary, and tertiary consumers. In many cases, POPs or other substances are sequestered in certain parts of a consumer's body instead of being excreted. Toxaphene, for example, tends to become localized in fat cells because it is lipid-soluble. In vertebrates it is also sequestered in the liver—the organ responsible for processing toxins.

Remember that organisms at each trophic level eat about 10 g of tissue to make 1 g of their own tissue. Even if toxaphene is present at extremely low concentration in primary producers, it is sequestered at much higher concentration in primary consumers—simply because they take in so much primary producer tissue and retain the toxaphene. The effect is magnified at each succeeding trophic level. In many cases, concentrations get high enough in secondary and tertiary consumers to poison them.

TOXAPHENE IN THE ARCTIC The most dramatic example of toxaphene biomagnification is occurring in the Arctic—a "pristine" ecosystem where toxaphene has never been used directly. Instead, the molecules blow into the ecosystem on air currents introduced in Chapter 50.

As the data in **Figure 54.6** show, toxaphene concentrations increase dramatically at higher levels in the Arctic food chain. Are the levels high enough to poison humans and other species at the highest trophic levels? The answer to this question is highly controversial. The answer is important, however, because the Inuit people native to the Arctic depend on fish such as burbot, along with seals and other mammals that that show high levels of toxaphene, for food.

ESTROGEN-MIMICS Researchers who are working on biomagnification in the Arctic and elsewhere are particularly concerned about organic compounds that bind to receptors for the vertebrate sex hormone estradiol, which is an estrogen (see Chapter 48). In laboratory experiments, high concentrations of these estrogen-mimicking substances "feminize" individuals—meaning that male gonads begin producing female-specific compounds or cell types.

There is now good evidence that estrogen-mimicking POPs are impacting fish populations in the Arctic and elsewhere. Problems are also showing up in some human populations. As the graph in **Figure 54.7** shows, certain communities in the Arctic—in this

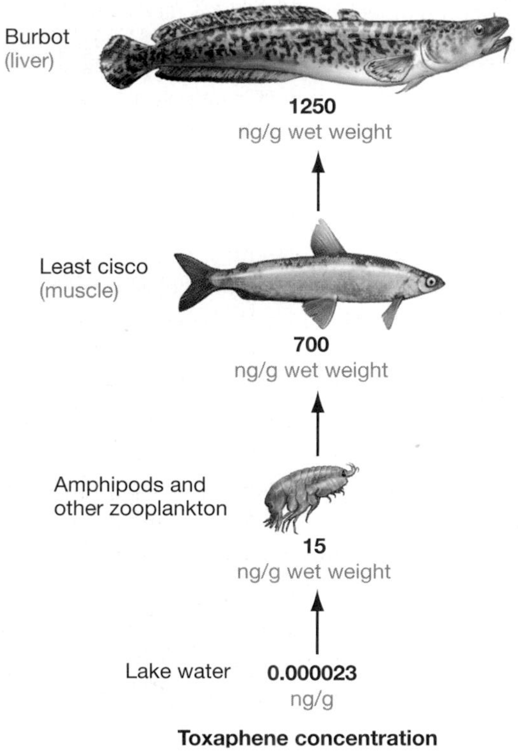

FIGURE 54.6 Biomagnification Can Lead to High Concentrations of Toxic Substances at High Trophic Levels. Note that the toxaphene concentrations reported here are usually from lipid or blubber. Researchers report POP concentrations in this way if the molecule in question is lipid-soluble.

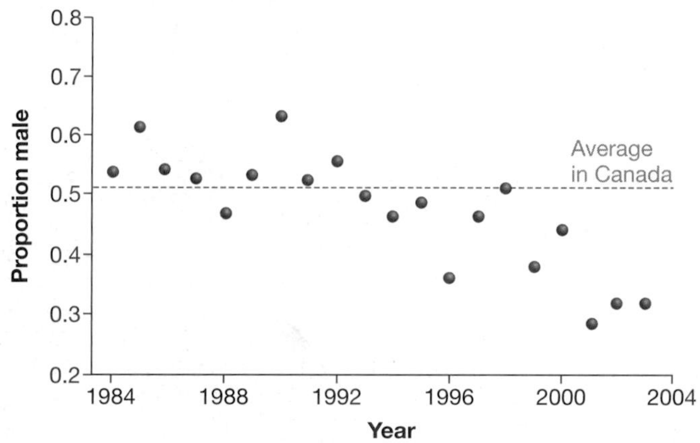

FIGURE 54.7 Recent Changes in Sex-Ratio at Birth in an Arctic First Nations Community. The data shown here are from the Aamjiwnaang First Nation (native) community near the village of Sarnia, in northern Ontario.

case, one located next to a petrochemical plant—have recently experienced aberrant sex ratios. The dotted line on the graph shows the average proportion of males among newborns throughout Canada; the data points indicate that in this Arctic community, many more girls are now being born than boys.

Is the problem in this community due to estrogen-mimics? If so, which molecule is to blame? The data compiled to date indicate that toxaphene has relatively mild estrogenic activity. Although other compounds have been shown to have strong estrogenic activity in fish and some reptiles, to date there is no well-documented tie between particular compounds and sex ratio or other "feminization" problems in humans.

In the meantime, it is clear that biomagnification is a serious side effect of basic patterns in energy flow through ecosystems. Toxins are passed through food webs along with energy and nutrients. Research continues.

Global Patterns in Productivity

The data presented thus far should convince you that NPP is a fundamental attribute of an ecosystem, that energy flows through food webs in complex and dynamic ways, and that understanding trophic interactions has important practical consequences for humans. Now let's step back and consider two "big picture" questions about NPP: Where is most of it produced, and what limits it?

WHICH GEOGRAPHICAL AREAS ARE MOST PRODUCTIVE? **Figure 54.8** summarizes data on NPP from around the globe. Note that the highest NPP on land is indicated in red, yellow, and green; the highest NPP in the oceans is indicated in blue. In general, NPP on land is much higher than it is in the oceans.

The leading hypothesis to explain this pattern is simply that much more light is available to drive photosynthesis on land than in marine environments. As data presented in Chapter 50 indicated, water quickly absorbs many of the light wavelengths that are used in photosynthesis. This simple observation also underlines the importance of the evolution of land plants, highlighted in Chapter 30. Before green plants invaded the land, NPP in terrestrial environments may have been as low as it is in today's oceans.

The terrestrial ecosystems with highest productivity are located in the wet tropics. With the exception of the world's major deserts, NPP on land declines from the equator toward the poles. Chapter 50 pointed out that deserts occur at 30 degrees of latitude north and south of the equator, due to global patterns in air circulation that affect precipitation and thus plant growth. Otherwise, NPP declines with latitude.

You might recall from Chapter 53 that the equator-to-poles gradient in NPP is thought to be important in explaining the equator-to-poles gradient in species richness that occurs in most types of plants and animals. If so, then areas of high productivity should also be "hotspots" for high species richness (see Chapter 55).

Productivity patterns in marine ecosystems contrast strongly with those of terrestrial environments, however. Marine productivity is highest along coastlines, and it can be as high near the poles as it is in tropics. The oceanic areas colored deep blue in Figure 54.8 include most of the world's most important fisheries. The oceanic zones, introduced in Chapter 50, have extremely low NPP. Typically, a square meter (m^2) of open ocean produces a maximum of 35 g (1.2 oz) of organic matter each year. In terms of productivity per m^2, the open ocean is a desert.

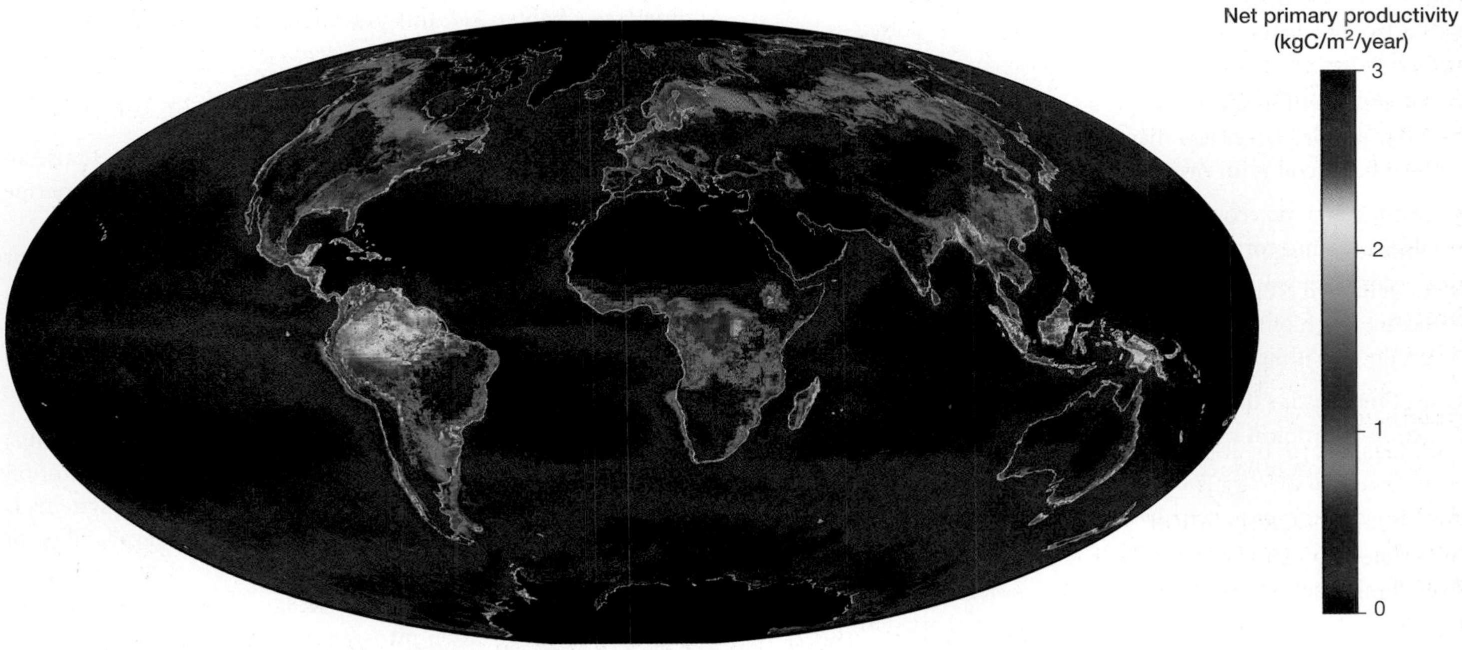

FIGURE 54.8 Net Primary Productivity Varies among Geographic Regions. The terrestrial ecosystems with the highest primary productivity are found in the tropics; tundras and deserts have the lowest. The highest productivity in the oceans occurs in nutrient-rich coastal areas.

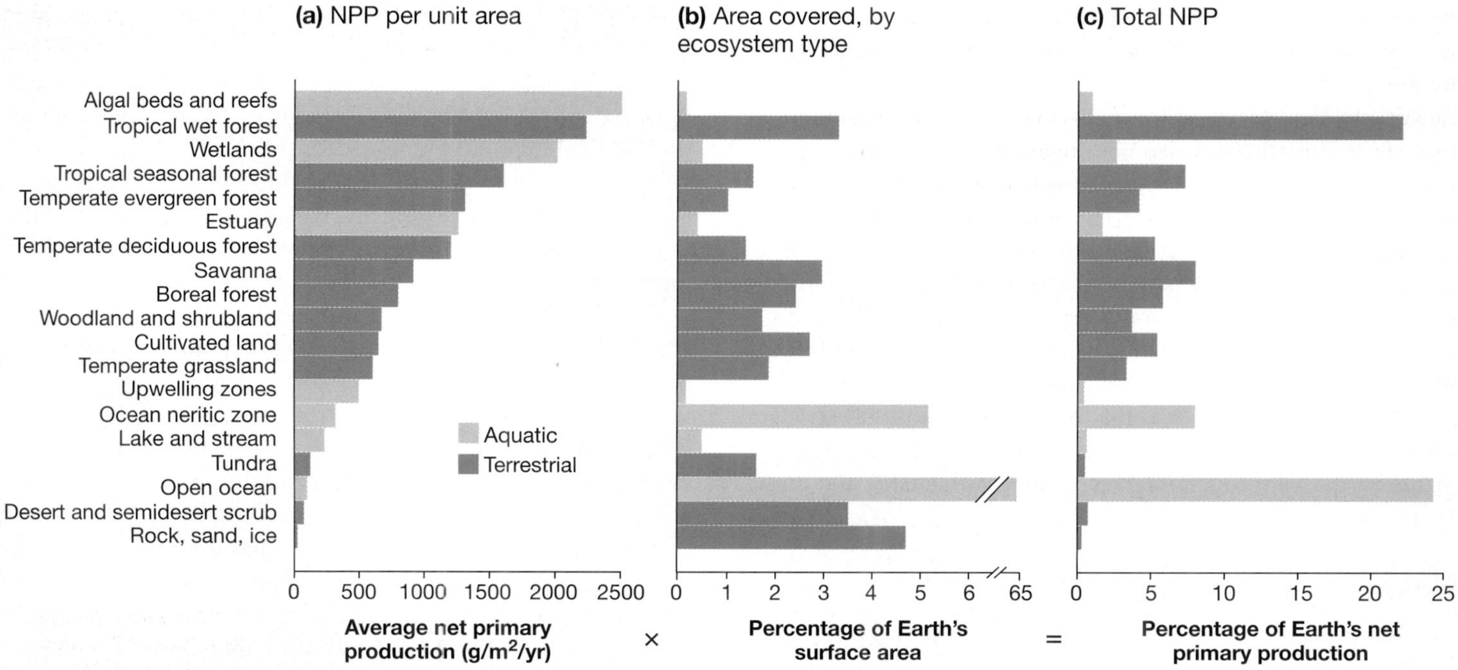

FIGURE 54.9 Net Primary Productivity Varies among Ecosystems. **(a)** Among biomes, average annual NPP per square meter varies by over three orders of magnitude. **(b)** Most of Earth's surface is covered by open ocean. The most common terrestrial habitat consists of unvegetated rock, sand, or ice. **(c)** Even though it has a low NPP per square meter, the open ocean is so vast that it is responsible for over 25 percent of Earth's total NPP.

WHICH BIOMES ARE MOST PRODUCTIVE? **Figure 54.9** presents NPP data a different way—organized by biome instead of by geography. Figure 54.9a provides data on average NPP per square meter per year for each biome; Figure 54.9b documents the total area that is covered by each type of ecosystem; and Figure 54.9c presents the percentage of the world's total productivity—the result of multiplying the data in part (a) by the data in part (b). In each case, data from different types of biomes are plotted as horizontal bars. Note that:

- Tropical wet forests and tropical seasonal forests (which have a dry season) cover less than 5 percent of Earth's surface but together account for over 30 percent of total NPP.
- Among aquatic ecosystems, the most productive habitats are algal beds and coral reefs, wetlands, and estuaries.
- Even though NPP per square meter is extremely low in the open ocean, this biome is so extensive in terms of area that its total production is high.

One of the reasons that biologists are so concerned about the destruction of tropical rain forests and coral reefs is that they are the world's most productive habitats.

THE "HUMAN APPROPRIATION" Of the total NPP available on our planet, how much are humans using? A research group answered this question recently in terms of potential NPP—meaning, NPP that would exist without human activity. Using data from a wide array of sources, they estimated that humans are currently appropriating 23.8 percent of potential NPP. Of this:

- 53 percent is directly harvested and used,
- 40 percent is lost due to suburbanization and other land-use changes, and
- 7 percent is lost due to human-induced fires.

This analysis is astonishing. One species—our own—is appropriating almost a quarter of the planet's biomass. Given increases in human population projected to occur over your lifetime (see Chapter 52), this percentage is almost guaranteed to grow—probably substantially.

What Limits Productivity?

The patterns highlighted in Figure 54.8 and Figure 54.9 raise an interesting question: What limits NPP in terrestrial and marine ecosystems?

The short answer is that productivity is limited by any factor that limits the rate of photosynthesis. Based on information in Chapter 10 and Unit 7, the candidates are temperature and the availabilities of water, sunlight, and nutrients. Different limiting factors prevail in different environments, however.

A TERRESTRIAL-MARINE CONTRAST In terrestrial environments, NPP is lowest in deserts and arctic regions. This observation suggests that the overall productivity of terrestrial ecosystems is limited by a combination of temperature and availability of water and sunlight.

At a local level, however, NPP on land is also limited by nutrient availability—often nitrogen or phosphorus. When biologists fertilize a terrestrial ecosystem with nitrogen or phosphorus, NPP almost always increases. Recent research has also shown that NPP in temperate and boreal forests is increasing, due to nitrogen that is blowing in from agricultural fields hundreds or thousands of miles away, where it was originally applied as a fertilizer.

To explain why the productivity of marine habitats is so much higher along coastlines than in deepwater regions, biologists focus on nutrient limitation. As Chapter 50 pointed out, the shallow water along coasts receives nutrients from two major sources:

1. rivers that carry and deposit nutrients from terrestrial ecosystems; and
2. nearshore ocean currents that bring nutrients from the cold, deep water of the oceanic zone back up to the surface.

Both of these sources are absent in the surface waters of the open ocean. In addition, nutrients found in organisms near the surface of the open ocean—where sunlight is abundant—constantly fall to dark, deeper waters in the form of dead cells, and are lost.

IRON-FERTILIZATION EXPERIMENTS Trace elements such as zinc, iron, and magnesium are particularly rare in the open ocean. These atoms are important because they are required as enzyme cofactors (see Chapter 3). Iron, for example, is essential to the proteins that are involved in electron transport chains (see Chapters 9 and 10). On the basis of these observations, biologists have proposed that the productivity of open-ocean ecosystems could be increased dramatically by fertilizing them with iron or other types of trace elements.

The results of one iron-fertilization experiment are shown in **Figure 54.10**. In this graph, each data point represents the average of a large number of sample sites from surface waters inside the treatment area, surface waters outside the treatment area, or water 30 m below the surface and inside the treatment area.

The graph indicates that about a week following the experimental addition of iron to the surface of an experimental area, there were large increases in the concentration of chlorophyll *a*—an indicator of the presence of photosynthetic cells. The increase continued for at least five days in surface waters compared to both water sampled from 30 m below the surface and water outside the experimental area.

Consistent with this result, recent research on ocean waters that are naturally enriched by nearby iron-containing rocks indicates that these regions have exceptionally high NPP.

Results like these provide strong support for the hypotheses that:

1. NPP in marine ecosystems is limited primarily by the availability of nutrients, and
2. iron is particularly important in the open ocean.

These data have also inspired a novel proposal to fight global warming by reducing CO_2 concentrations in the atmosphere (see Section 54.3). The photosynthetic cells that grow in response to iron addition use large amounts of atmospheric CO_2 that dissolves in seawater. Some researchers suggest that fertilizing the open oceans could increase NPP in these deserts enough to reduce atmospheric CO_2 and slow global warming.

EXPERIMENT

QUESTION: Is net primary productivity (NPP) in the open ocean limited by nutrients?

HYPOTHESIS: NPP in the open ocean is limited by availability of iron.

NULL HYPOTHESIS: NPP in the open ocean is not limited by availability of iron.

EXPERIMENTAL SETUP:

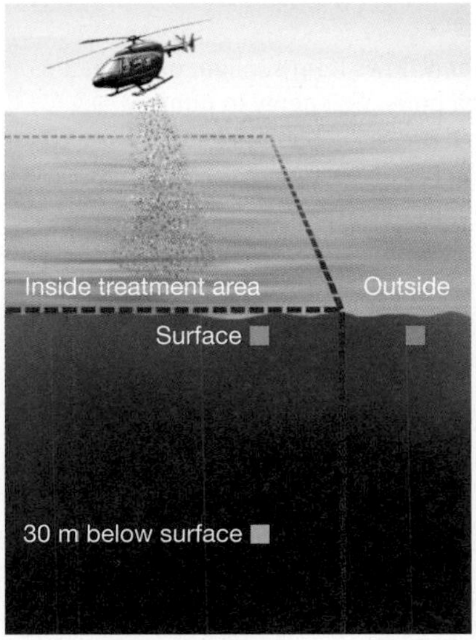

1. Add 350 kg iron (as $FeSO_4$) to a treatment area—a patch of ocean 8 km × 10 km.

2. Take water samples for a two-week period outside and inside the treatment area, at surface and at a depth of 30 m, and record amount of chlorophyll *a* present (as indicator of NPP).

PREDICTION: Amount of chlorophyll *a* near the surface inside the treatment area will increase relative to amounts outside the treatment area or at 30 m below the surface.

PREDICTION OF NULL HYPOTHESIS: Amount of chlorophyll *a* will be the same in all measurements.

RESULTS:

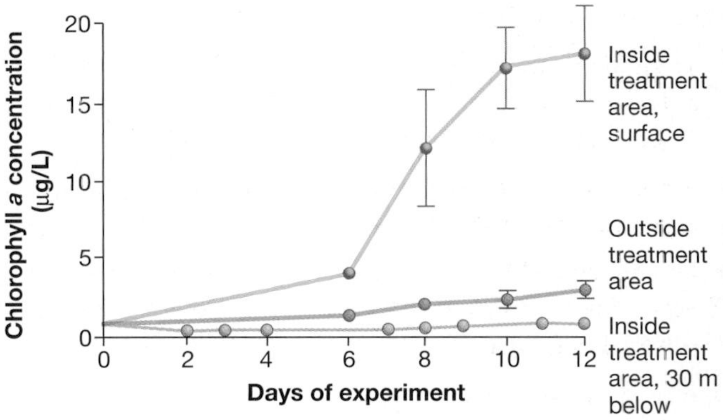

CONCLUSION: NPP in the open ocean is limited by the scarcity of nutrients—specifically, iron.

FIGURE 54.10 Fertilization with Iron Increases NPP in the Open Ocean.

SOURCE: Tsuda, A., et al. 2003. A mesoscale iron enrichment in the western subarctic Pacific induces a large centric diatom bloom. *Science* 300: 958–961.

✔**QUESTION** Most of the increased chlorophyll *a* was present in a single species of diatom (see Chapter 29). What happens to the biomass present in diatoms after they die?

This proposal is being hotly debated right now. Could iron fertilization be done on a large enough scale to have an impact? If so, what would it cost and who would pay? When the photosynthetic cells die and fall to the ocean bottom, would they have negative effects on other parts of the marine ecosystem? At present, we simply lack the data to answer these questions conclusively.

CHECK YOUR UNDERSTANDING

If you understand that . . .

- In an ecosystem, energy flows from sunlight or inorganic compounds with high potential energy to primary producers, and from there to consumers and decomposers.
- Productivity diminishes at each subsequent trophic level in the ecosystem.
- In most ecosystems, NPP is limited by conditions that limit the rate of photosynthesis: temperature and/or the availability of sunlight, water, and nutrients.

✔ You should be able to . . .

1. Explain why productivity diminishes with increasing trophic levels.
2. Predict what will happen to the food web in an ecosystem where global warming leads to an increase in NPP.

Answers are available in Appendix B.

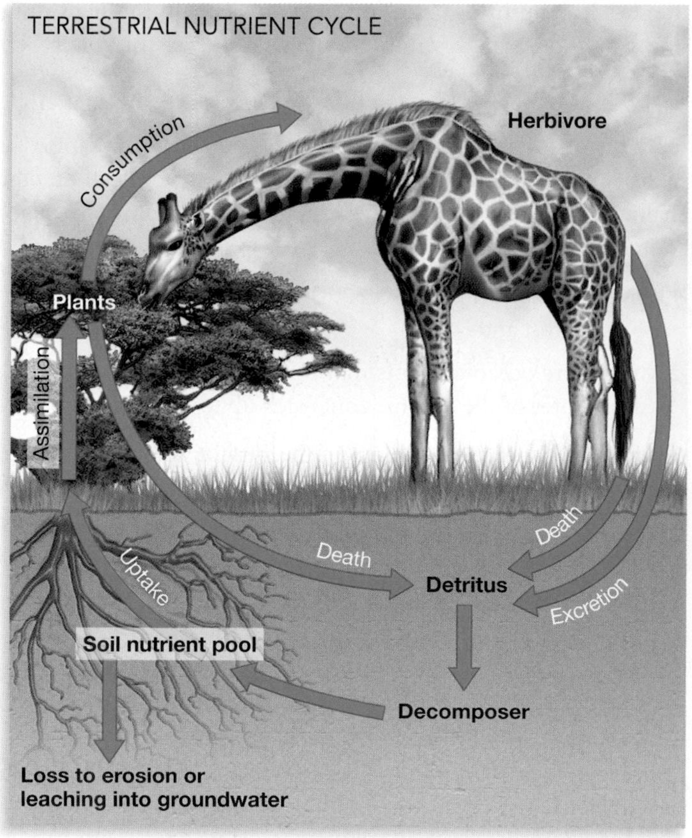

FIGURE 54.11 Generalized Terrestrial Nutrient Cycle. Nutrients cycle from organism to organism in an ecosystem via assimilation, consumption, and decomposition. Nutrients are exported from ecosystems when water or organisms leave the area.

54.2 How Do Nutrients Cycle through Ecosystems?

Energy and POPs are not the only entities that get transferred when one organism eats another. Tissues also contain carbon (C), nitrogen (N), phosphorus (P), calcium (Ca), and other elements that act as nutrients.

Atoms are constantly reused as they move through trophic levels, but they also spend time suspended in air, dissolved in water, or held in soil. The path that an element takes as it moves from abiotic systems through producers, consumers, and decomposers and back again is referred to as its **biogeochemical** ("life-Earth-chemical") **cycle**. Because humans are disturbing biogeochemical cycles on a massive scale, research on the topics is exploding.

Nutrient Cycling within Ecosystems

To understand the basic features of a biogeochemical cycle, study the generalized terrestrial nutrient cycle in **Figure 54.11**. Start with the "Uptake" arrow, which indicates nutrients that are taken up from the soil by plants and assimilated into plant tissue; carbon enters primary producers as carbon dioxide from the atmosphere. Note that if the plant tissue is eaten, the nutrients pass to the consumer food web; if the plant tissue dies, the nutrients enter the compartment labeled "Detritus."

What happens as detritus decomposes? Nutrients that reside in plant litter, animal excretions, and dead animal bodies are used by bacteria, archaea, roundworms, fungi, and other primary decomposers. These microscopic decomposers and the carbon-containing compounds that they release combine to form what biologists call soil organic matter.

Soil organic matter is a complex mixture of partially and completely decomposed detritus. Completely decayed organic material is called **humus** because it is rich in a family of carbon-containing molecules called humic acids. Chapter 38 described other components of the soil.

Eventually, decomposition converts the nutrients in soil organic matter to an inorganic form. For example, cellular respiration by soil-dwelling bacteria and archaea converts the nitrogen present in amino acids to ammonium (NH_4^+) or nitrate (NO_3^-) ions. Once this step is accomplished, the nutrients are available for uptake by plants. Uptake by plants, following decomposition, highlights the cyclical nature of nutrient flow through ecosystems.

WHAT FACTORS CONTROL THE RATE OF NUTRIENT CYCLING? Of the many links in a nutrient cycle, the decomposition of detritus most often limits the overall rate at which nutrients move through an ecosystem. Until decomposition occurs, nutrients stay tied up in intact tissues.

The decomposition rate, in turn, is influenced by two types of factors:

1. abiotic conditions such as oxygen availability, temperature, and precipitation; and
2. the quality of the detritus as a nutrient source for the fungi, bacteria, and archaea that accomplish decomposition.

To appreciate the importance of abiotic conditions on the decomposition rate, consider the difference in detritus accumulation between boreal forests and tropical wet forests. Chapter 50 indicated that boreal forests are found in areas where temperature is low. As a result, soils in these ecosystems are cold and wet. But tropical wet forests occur in areas where temperatures and rainfall are high. Soils tend to remain moist and warm all year long.

Figure 54.12 illustrates typical soils from boreal forests and tropical wet forests. Notice that the uppermost part of the soil in a boreal forest consists of partially decomposed detritus and organic matter. In a tropical forest, this layer is virtually absent.

The contrast occurs because the cold and wet conditions in boreal forests limit the metabolic rates of decomposers. As a result, decomposition fails to keep up with the input of detritus, and organic matter accumulates there. In the tropics, conditions are so favorable for fungi, bacteria, and archaea that decomposition keeps pace with detrital inputs. Nutrients cycle slowly through boreal forests but rapidly through wet tropical forests.

The quality of detritus also exerts a powerful influence on the growth of decomposers, and thus the decomposition rate. Decomposition is inhibited if detritus is (1) low in nitrogen, or (2) high in lignin.

When nitrogen is scarce, cells have a difficult time making additional proteins and nucleic acids. Chapter 31 noted that lignin—an important constituent of wood—is extremely difficult to digest. The presence of lignin is a major reason why wood takes much longer to decompose than leaves do. Only basidiomycete fungi and a few bacteria have the enzymes required to completely degrade lignin.

The presence of oxygen is also critically important. If soils or water become depleted of oxygen, decomposition slows dramatically. Fungi can only perform cellular respiration using oxygen as the final electron acceptor (see Chapter 9). And even though many bacteria and archaea can thrive in the absence of oxygen and continue decomposition, the growth rates of anaerobic species are a tiny fraction of rates in aerobic bacteria and archaea (to understand why, see Chapter 28).

FIGURE 54.12 Temperature and Moisture Affect Decomposition Rates.

SOURCES OF NUTRIENT LOSS AND GAIN Although nutrients are constantly reused in ecosystems, they can also be lost—much as energy can. Nutrients leave an ecosystem whenever biomass leaves.

- If an herbivore eats a plant and moves out of the ecosystem before excreting the nutrients or dying, the nutrients are lost.
- Nutrients leave ecosystems when flowing water or wind removes particles or inorganic ions and deposits them somewhere else.

Several of the major impacts that humans have on ecosystems—such as farming, logging, burning, and soil erosion—accelerate nutrient loss. Farming and logging remove biomass in the form of plants; burning releases nutrients to the atmosphere; soil erosion moves nutrients that are bound to soil particles or dissolved in water.

For an ecosystem to function normally, nutrients that are lost must be replaced. There are three major mechanisms to replace lost nutrients:

1. Atoms that act as nutrients are released as rocks weather (see Chapter 38).
2. Nutrients can also blow in on soil particles or arrive as solutes in streams.
3. Nitrogen—a particularly important nutrient in maintaining NPP—is added when the nitrogen-fixing bacteria introduced in Chapter 38 convert molecular nitrogen (N_2) in the atmosphere to usable nitrogen in ammonium or nitrate ions.

In ecosystems that humans manage, nutrients usually have to be actively replaced to maintain productivity. People accomplish this by adding fertilizers or planting nitrogen-fixing crops.

AN EXPERIMENTAL STUDY The ideas presented thus far are logical and provide a strong conceptual foundation for thinking about nutrient cycles. But to explore nutrient movement in more depth, researchers have turned to experiments.

One of the first large-scale experiments on nutrient cycling took place at the Hubbard Brook Experimental Forest. A group of researchers there began by choosing two similar **watersheds**—areas drained by a single stream—for study. Before starting the experiment, they documented that 90 percent of the nutrients in the watersheds was in soil organic matter; an additional 9.5 percent was in plant biomass. The nutrients in the soil were either dissolved in water or attached to small particles of sand or clay (see Chapter 38).

EXPERIMENT

QUESTION: How does the presence of vegetation affect the rate of nutrient export in a temperate-forest ecosystem?

HYPOTHESIS: Presence of vegetation lowers the rate of nutrient export because it increases soil stability and recycling of nutrients.

NULL HYPOTHESIS: Presence of vegetation has no effect on the rate of nutrient export.

EXPERIMENTAL SETUP:

1. Choose two similar watersheds. Document nutrient levels in soil organic matter, plants, and streams.

2. Devegetate one watershed, and leave the other intact.

3. Monitor the amount of dissolved substances in streams.

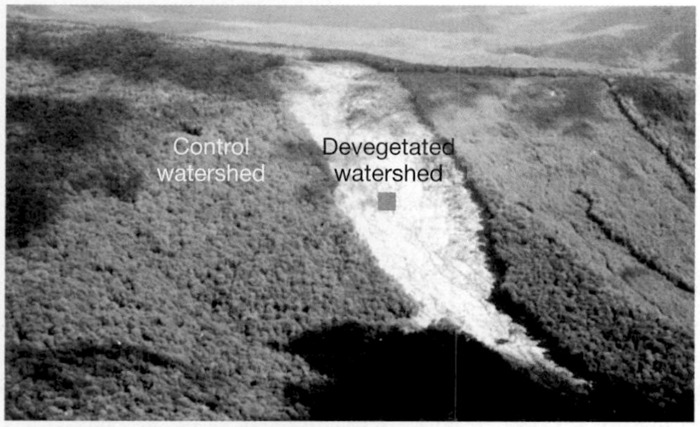

PREDICTION: Amount of dissolved substances in stream in devegetated watershed will be much higher than amount of dissolved substances in stream in control watershed.

PREDICTION OF NULL HYPOTHESIS: No difference will be observed in amount of dissolved substances in the two streams..

RESULTS:

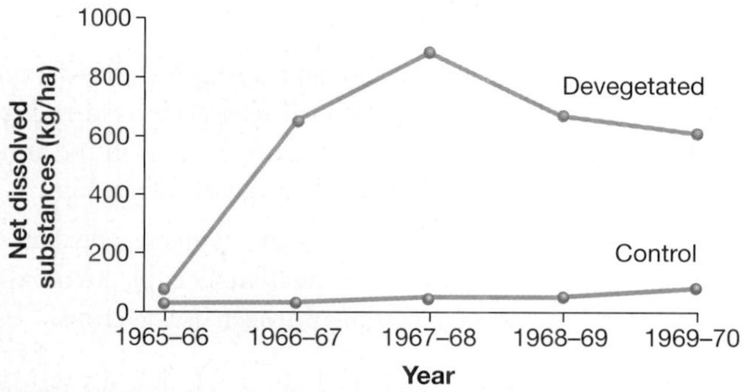

CONCLUSION: Presence of vegetation limits nutrient loss. Removing vegetation leads to large increases in nutrient export.

FIGURE 54.13 Experimental Evidence That Deforestation Increases the Rate of Nutrient Loss from Ecosystems.

SOURCE: Likens, G. E., F. H. Bormann, N. M. Johnson, D. W. Risher, and R. S. Pierce. 1970. Effects of cutting and herbicide treatment on nutrient budgets in the Hubbard Brook watershed-ecosystem. *Ecological Monographs* 40: 23–47.

✔**QUESTION** Propose a hypothesis to explain why the level of nutrient export from the experimental site appears to peak and then decline.

To begin the experimental treatment, they cut all vegetation, including the trees, in one of the two watersheds (**Figure 54.13**). In each of the following three years, the clear-cut watershed was treated with an herbicide to prevent vegetation from regrowing. As a result, the experimental watershed was devegetated. Researchers left the other watershed untreated to serve as a control.

✔**If you understand the general principles of nutrient cycling, you should be able to name which arrow in Figure 54.11 was removed by the devegetation study.** The purpose of the experiment was to remove this link and document the consequences.

The graph in Figure 54.13 documents the amount of total dissolved substances that washed out of the streams in each watershed over the course of the four years of devegetation. Losses from the devegetated site were typically over 10 times as high as they were from the control site.

Why? Plant roots were no longer able to hold soil particles in place, and plant roots no longer took up and recycled nutrients that were dissolved in soil water. Instead, massive quantities of nitrate and other nutrients washed out of the soil and were lost to the ecosystem.

Long-term devegetation of this type has occurred in formerly forested areas of the Middle East, North Africa, and elsewhere, due to intensive farming and grazing. Nutrients that used to occur in those soils were lost. Today, the productivity of these regions is a tiny fraction of what it once was.

Global Biogeochemical Cycles

When nutrients leave one ecosystem, they enter another. In this way, the movement of ions and molecules among ecosystems links local biogeochemical cycles into one massive global system. Local and global cycles interact when water, organisms, or wind move nutrients.

As an introduction to how these global biogeochemical cycles work, let's consider the global water, carbon, and nitrogen cycles. These cycles have recently been heavily modified by human activities—with serious ecological consequences.

To understand how biogeochemical cycling works on a global scale, researchers focus on three fundamental questions:

1. What are the nature and size of the reservoirs—areas or "compartments" where elements are stored for a period of time? In the case of carbon, the biomass of living organisms is an important reservoir, as are sediments and soils. Another significant carbon reservoir is buried in the form of coal and oil.
2. How fast does the element move between reservoirs, and what processes are responsible for moving elements from one compartment to another? The global photosynthetic rate, for example, measures the rate of carbon flow from carbon dioxide (CO_2) in the atmosphere to living biomass. When fossil fuels burn, carbon that was buried in coal or petroleum moves into the atmosphere as CO_2.
3. How does one biogeochemical cycle interact with another biogeochemical cycle? For example, researchers are trying to understand how changes in the nitrogen cycle affect the carbon cycle.

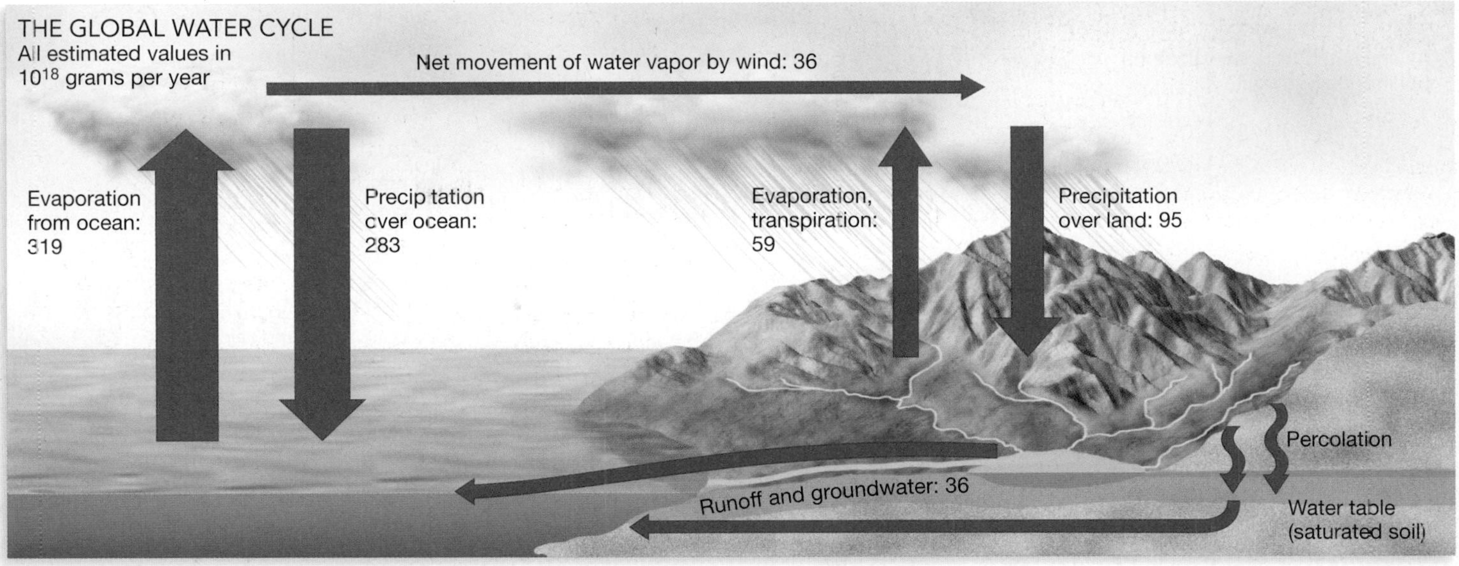

FIGURE 54.14 The Global Water Cycle.

✔**QUESTION** Predict how the amount of water evaporated from the ocean is changing in response to global warming. Discuss one possible consequence of this change.

THE GLOBAL WATER CYCLE A simplified version of the **global water cycle** appears in **Figure 54.14**. The diagram shows the estimated amount of water that moves between major components of the cycle over the course of a year.

To analyze this cycle, begin with evaporation of water out of the ocean, and the subsequent precipitation of water back into the ocean. For the marine component of the cycle, evaporation exceeds precipitation—meaning that, over the oceans, there is a net gain of water to the atmosphere.

When this water vapor moves over the continents, it is joined by a small amount of water that evaporates from lakes and streams and a large volume of water that is transpired by plants. The total volume of water in the atmosphere over land is balanced by the amount of rain and other forms of precipitation that falls on the continents.

The cycle is completed by the water that moves from the land back to the oceans via streams and **groundwater**—water that is found in soil. Overall, there is no net gain or loss of water.

What impact are humans having? Perhaps the simplest and most direct impacts concern rates of groundwater replenishment.

- Asphalt and concrete surfaces reduce the amount of water that percolates from the surface to enter deep soil layers.
- When grasslands and forests are converted to agricultural fields, root systems that hold water are lost. More water runs off into streams and less percolates into groundwater.
- Irrigated agriculture, along with industrial and household use, is removing massive amounts of water from groundwater storage and bringing it to the surface.

The **water table**—the level where soil is saturated with stored water—is dropping on every continent. Between 1986 and 2006, the water table north of Beijing, China, experienced drops averaging almost 61 m (200 ft). Throughout India, water tables are falling at a rate of between 1 m and 3 m per year. Similar rates are being documented in Yemen, parts of Mexico, and states in the southern Great Plains region of the United States. The city of Bangkok, which is located just above sea level, is sinking up to 7.5 cm each year because so much water is being pumped out of the ground below it (**Figure 54.15**).

Humans are mining water in many parts of the world—meaning that populations have already grown beyond carrying capacity for water. Lack of water is exacerbating political tensions in several areas of the world, including the Middle East.

FIGURE 54.15 Flooding Becomes More Frequent as Bangkok Sinks. This man is walking through an antique market during a recent flood. Some studies suggest that much of Bangkok may be permanently underwater by 2100.

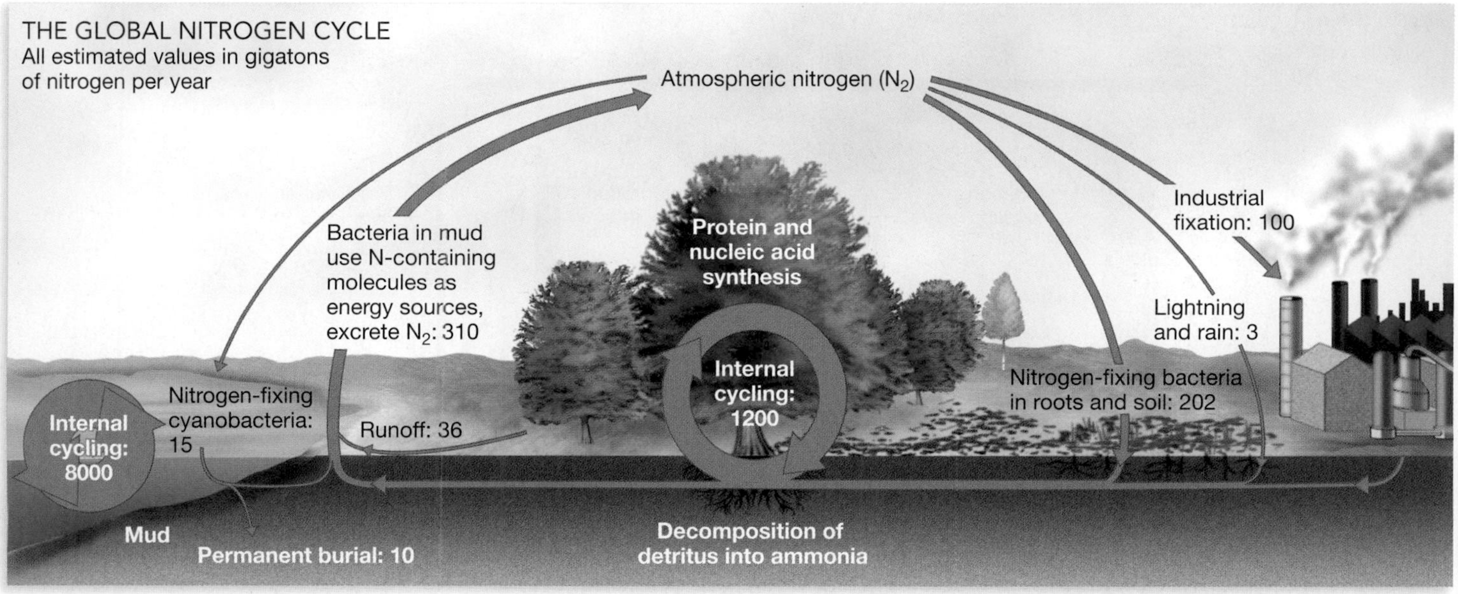

FIGURE 54.16 The Global Nitrogen Cycle. Nitrogen enters ecosystems as ammonia or nitrate via fixation from atmospheric nitrogen. It is exported in runoff and as nitrogen gas given off by bacteria that use nitrogen-containing compounds as an electron acceptor.

THE GLOBAL NITROGEN CYCLE **Figure 54.16** illustrates the **global nitrogen cycle**. As Chapter 38 noted, nitrogen is added to ecosystems in a usable form only when it is reduced, or "fixed," meaning that it is converted from N_2 to ammonium or nitrate ions. Nitrogen fixation results from lightning-driven reactions in the atmosphere and from enzyme-catalyzed reactions in bacteria that live in the soil and oceans.

The nitrogen cycle has been profoundly altered by human activities, primarily through the addition of massive amounts of reduced nitrogen. The graphs in **Figure 54.17** show current estimates for the amount of nitrogen fixed by natural sources versus human sources. The amount of nitrogen fixation from human sources is now approximately equal to the amount of nitrogen fixation from natural sources. As the labels on the right side of the graph indicate, there are three major sources of this human-fixed nitrogen:

1. industrially produced fertilizers;
2. the cultivation of crops, such as soybeans and peas that harbor nitrogen-fixing bacteria; and
3. the production of nitric oxide during the combustion of fossil fuels.

Adding nitrogen to terrestrial ecosystems usually increases productivity. But overfertilization with nitrogen has important downsides.

- Chapter 28 detailed how nitrogen-laced runoff from agricultural fields has led to the formation of oxygen-free "dead zones" in aquatic ecosystems.
- When nitrogen is added to experimental plots in grasslands in midwestern North America, a few competitively dominant species grow rapidly and displace other species that do not respond to nitrogen inputs as strongly. Productivity increases, but species diversity decreases.

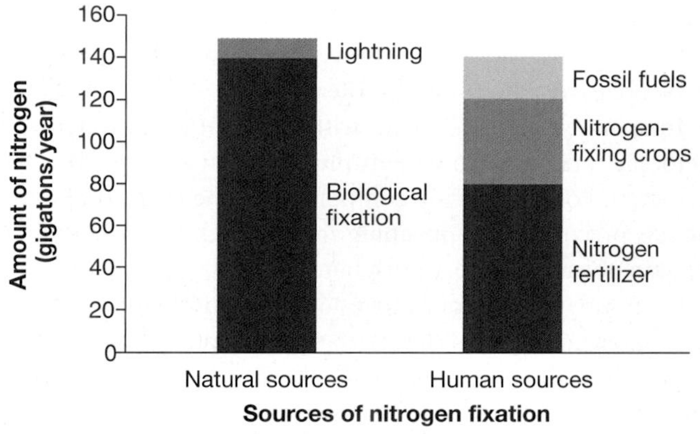

FIGURE 54.17 Humans Are Adding Large Amounts of Nitrogen to Ecosystems. Human activities now fix almost as much nitrogen each year as natural sources do. Thus, human activities have almost doubled the total amount of nitrogen available to organisms.

THE GLOBAL CARBON CYCLE The **global carbon cycle** documents the movement of carbon among terrestrial ecosystems, the oceans, and the atmosphere (**Figure 54.18**). Of these three reservoirs, the ocean is by far the largest. The atmospheric reservoir is also important despite its relatively small size, because carbon moves into and out of it rapidly—through organisms.

In both terrestrial and aquatic ecosystems, photosynthesis is responsible for taking carbon out of the atmosphere and incorporating it into tissue. Cellular respiration, in contrast, releases carbon that has been incorporated into living organisms to the atmosphere, in the form of carbon dioxide.

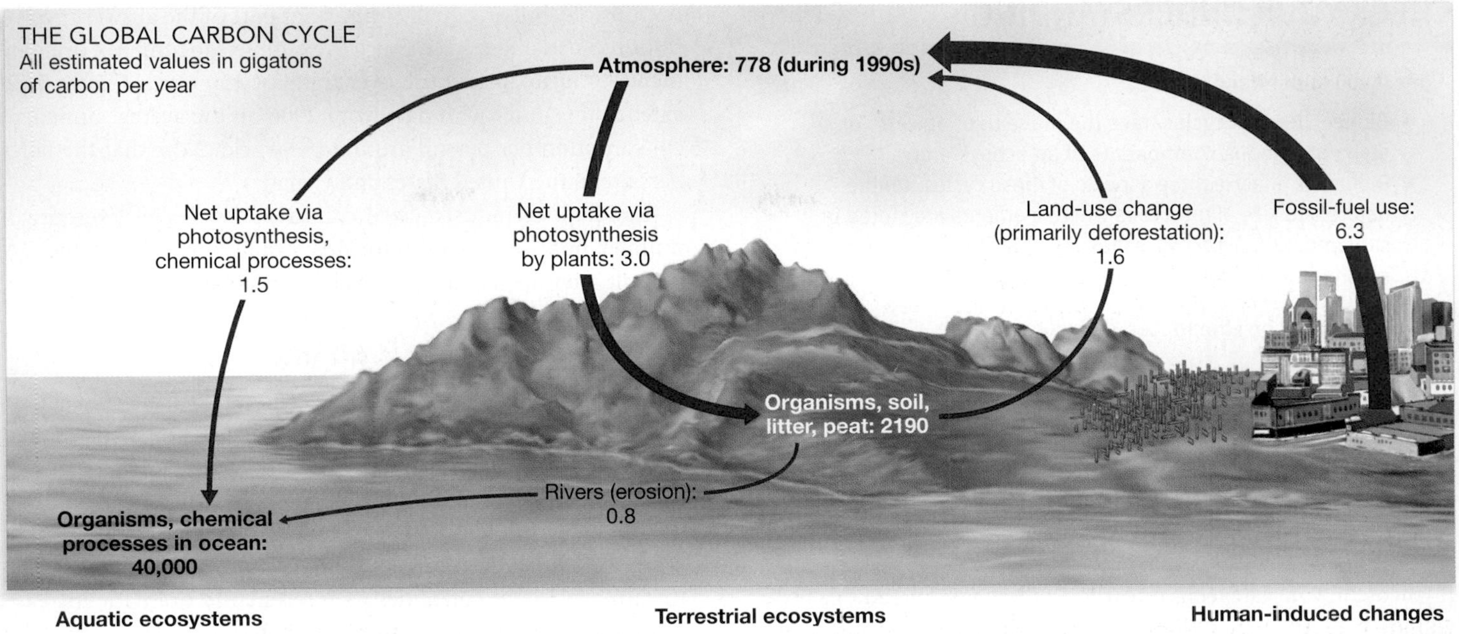

FIGURE 54.18 The Global Carbon Cycle. The arrows indicate how carbon moves into and out of ecosystems. Deforestation and the use of fossil fuels are adding 7.9 gigatons of carbon to the atmosphere each year. Of that 7.9 gigatons, 2 gigatons are fixed by photosynthesis in terrestrial ecosystems and 2 gigatons are fixed by physical and chemical processes in the oceans. The remainder—3.9 gigatons—is added to the atmosphere.

How have humans changed the carbon cycle? Burning fossil fuels moves carbon from an inactive geological reservoir, in the form of petroleum or coal, to an active reservoir—the atmosphere. When you burn gasoline, you are releasing carbon atoms that have been locked up in petroleum reservoirs for hundreds of millions of years.

Land-use changes have also altered the global carbon cycle. Deforestation and suburbanization: **(1)** reduce an area's net primary productivity—reducing its ability to sequester CO_2, and (2) release CO_2 directly when fire is used for clearing or when dead limbs, twigs, and stumps are left to decompose.

Figure 54.19a highlights the dramatic increase in carbon released from fossil-fuel burning over the past century; **Figure 54.19b** shows corresponding changes in CO_2 concentrations. Note that CO_2 levels in the atmosphere have now risen to levels far above the 280 ppm that was typical prior to 1860.

These changes in the global carbon cycle are important because carbon dioxide functions as a **greenhouse gas**: It traps heat that has been radiated from Earth and keeps it from being lost to space, similar to the way the glass of a greenhouse traps heat.

More specifically, carbon dioxide is one of several gases in the atmosphere that absorb and reflect the infrared wavelengths radiating from Earth's surface. Increases in amounts of greenhouse gases are warming Earth's climate by increasing the atmosphere's heat-trapping potential.

To review these carbon cycle concepts, go to the study area at *www.masteringbiology.com.* To review their consequences, go on to the chapter's next section.

(MB) **BioFlix™** The Carbon Cycle, **Web Activity** The Global Carbon Cycle

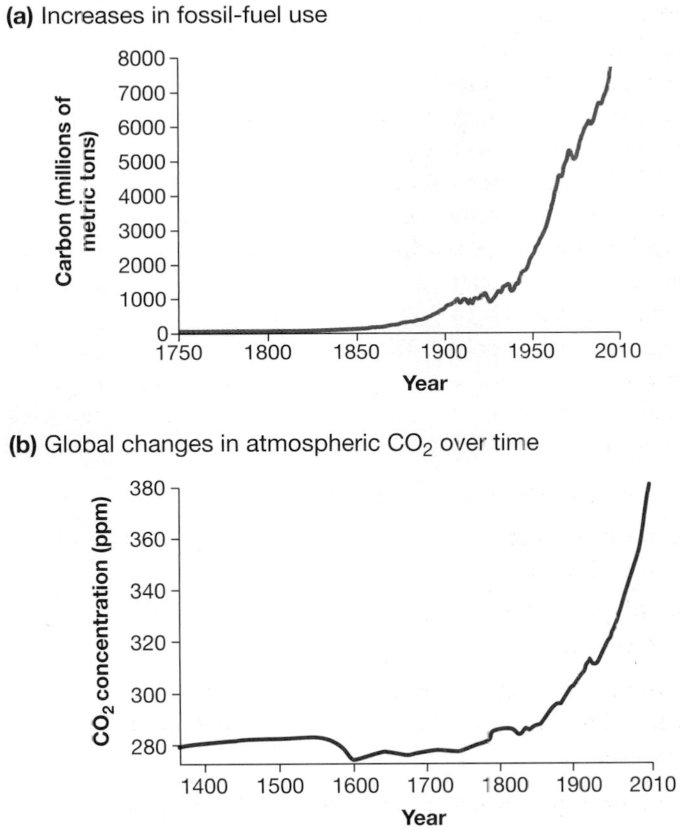

FIGURE 54.19 Humans Are Causing Increases in Atmospheric Carbon Dioxide. (a) Rates of carbon flow from fossil-fuel burning have increased as human populations have increased. **(b)** For centuries, average CO_2 concentrations in the atmosphere were about 280 parts per million (ppm).

CHECK YOUR UNDERSTANDING

If you understand that . . .

- Biogeochemical cycles trace the movement of water or elements among components of an ecosystem.
- Biologists analyze three aspects of these cycles: (1) the nature and size of reservoirs where elements are stored or captured, (2) the processes responsible for moving elements among reservoirs, and (3) interactions among cycles.

✓ You should be able to . . .

Explain how human activities are altering the global water cycle.

Answers are available in Appendix B.

54.3 Global Warming

If you are in your late teens or early twenties as you read this text, you are part of a generation that will experience the most traumatic episode of environmental change in human history. The trauma has two sources: the massive loss of species, documented in Chapter 55, and **global warming**.

Why is the environment deteriorating so rapidly? The explosion in human population size analyzed in Chapter 52 is partly responsible. But an equally important part of the answer involves resource use. Figure 54.19a, for example, documents dramatic increases in fossil-fuel use over the past 150 years. **Figure 54.20** extends this point with data from 2006 on the average annual oil consumption per person around the world. Note that the tallest bars are in the United States and Canada.

Residents of industrialized countries may be relatively few in number, but they burn extraordinary quantities of fossil fuels. As a result, they are largely responsible for global warming.

Understanding the Problem

What evidence backs the claim that humans are altering climate on a global scale? In explaining climate change, the chain of causation begins with the direct link between the CO_2 released by human fossil fuel use and increases in atmospheric CO_2. It continues with a link between high atmospheric CO_2 and increased trapping of infrared radiation that would otherwise leave the atmosphere; it concludes with recent and dramatic increases in average temperatures around the globe.

In 1988 an international group of scientists, called the Intergovernmental Panel on Climate Change (IPCC), was formed to evaluate the consequences of well-documented increases in CO_2 and other heat-trapping gases. The group has since produced a series of reports summarizing the state of scientific knowledge on the issue.

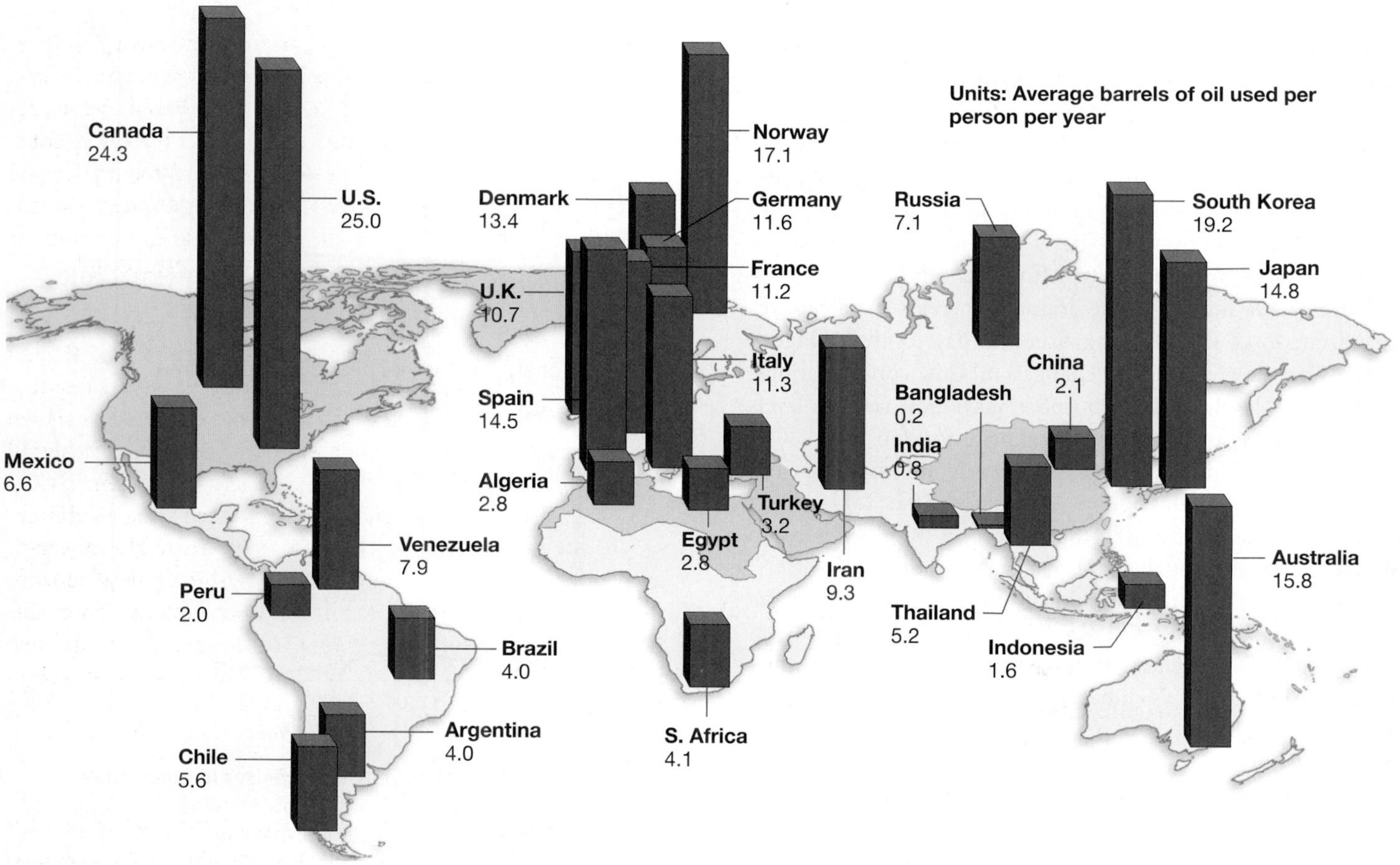

FIGURE 54.20 Per-Capita Oil Consumption Varies among Countries. These data are from 2006.

- The 1998 IPCC report concluded that current evidence suggests a "discernible human influence on climate." This was the IPCC's first statement supporting the hypothesis that rising CO_2 concentrations due to human activities are having a measurable impact on climate.
- The IPCC's next major report, released in 2001, stated that "There is new and stronger evidence that most of the warming observed over the last 50 years is attributable to human activities."
- In their most recent report, issued in 2007, the IPCC declared that evidence for global warming is unequivocal and that it is "very likely" due to human-induced changes in greenhouse gases.

Two points are important to note:

1. Climate change has occurred throughout Earth's history. What is unusual is the extremely rapid *rate* at which temperatures are changing currently.
2. The temperature increases being observed are global averages, reported on a per-year basis. Some regions may actually get cooler, and extensive month-to-month or year-to-year variation will occur. But averaged over the entire planet and over time, Earth is already much warmer than it was just a few decades ago, and is projected to get much more so.

How much will average temperatures rise in our lifetimes? The answer depends in part on whether CO_2 and other greenhouse gases continue to increase. **Figure 54.21** shows data on recent changes in atmospheric carbon dioxide, recorded at a research station in Mauna Loa, Hawaii. To date, there is no indication that the rate of increase is slowing. Although an array of regulations or incentives to reduce CO_2 emissions has been proposed by international bodies, no agreement has been reached yet.

Predicting the future state of a system as complex and variable as Earth's climate is extremely difficult, however. Global climate models are the primary tools that scientists use to make these projections. A global climate model is based on a large series of equations that describe how the concentrations of various gases in the atmosphere, solar radiation, transpiration rates, and other parameters interact to affect climate.

The models currently being used suggest that average global temperature will undergo additional increases of 1.1−6.4°C (2.0−11.5°F) by the year 2100. The low number is based on models that assume no further increase in greenhouse gases over present levels; the high number is based on models that assume continued intensive use of fossil fuels and increases in greenhouse gases.

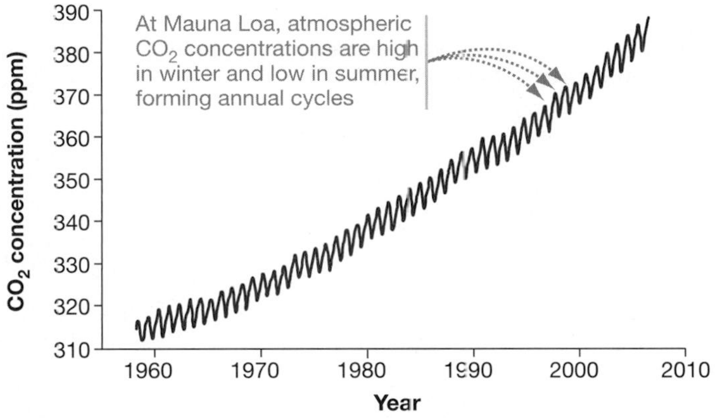

FIGURE 54.21 Recent Changes in Atmospheric CO_2. These data are from the Mauna Loa Observatory in Hawaii. Because this research station is far from large-scale human influences, it should accurately represent the average condition of the atmosphere in the Northern Hemisphere.

✔**QUESTION** Why are atmospheric CO_2 concentrations low in the Northern Hemisphere in summer and high in winter?

Positive and Negative Feedback

How will ecosystems respond to global warming? Answering this question is difficult, because ecosystems respond to warming in ways that increase or decrease CO_2 concentrations, and thus exacerbate or mitigate warming.

DOCUMENTING POSITIVE FEEDBACK Positive feedback occurs when changes due to global warming result in a release of additional greenhouse gases—meaning that global warming should get worse. Two types of positive feedback have been documented thus far.

- Warmer and drier climate conditions have increased the frequency of forest fires in the Rocky Mountains of North America as well as in the boreal forests of Canada. Forest fires release CO_2, which leads to even more warming.
- Traditionally, tundras sequester carbon in the form of soil organic matter, because decomposition rates are extremely low. During a series of warm summers in the 1980s, however, researchers found that decomposition rates increased sufficiently to release carbon from stored soil organic matter and to produce a net flow of carbon to the atmosphere. Greenhouse gases that are trapped in "permanently" frozen soils, or permafrost, could also be released when tundras warm.

DOCUMENTING NEGATIVE FEEDBACK Negative feedback occurs when changes due to global warming result in increased uptake and sequestration of CO_2 and other greenhouse gases—meaning that global warming should be reduced. Recent experiments have shown that the growth rates of several tree species and some agricultural crops increase in direct response to increasing atmospheric CO_2. Because CO_2 is required for photosynthesis, it can act as a fertilizer.

Will ecosystem responses increase or decrease global warming? Currently, it is not clear whether positive or negative feedbacks will predominate. Researchers are working to answer this question by using computer models, experiments like the two studies highlighted in Chapter 50, and analyses of how ecosystems responded to past climate changes.

Impact on Organisms

Even though global temperatures have risen only slightly in comparison with projections for the next 50–100 years, biologists have already documented dramatic impacts on organisms.

GEOGRAPHIC RANGE CHANGES The geographic ranges of many organisms—including the malaria parasite—are changing. Malaria is considered a tropical disease, but the range of its causative agent, the protist *Plasmodium falciparum*, is moving north.

As another example, consider recent data on the ranges of small crustaceans, called copepods, in the North Atlantic. Copepods are key primary consumers in marine plankton. Copepods have been studied intensively because they are important commercially: They are eaten by fish that are key food sources for tertiary-consumer species of fish that are harvested and eaten by humans.

The maps in **Figure 54.22a** show changes that occurred in a 40-year period, beginning in 1960, in the number of copepod species typical of warm marine waters off southern Europe versus the number of copepod species typical of cold northern European waters. The key observation is that southern species have moved north, while the range of cold-water species has declined.

PHENOLOGY The timing of seasonal events, or **phenology**, is changing in many biomes. Recent studies have confirmed that migratory bird species are arriving on their breeding grounds in Europe earlier than in previous decades, and that fungi are fruiting earlier in some seasonal environments.

The graph in **Figure 54.22b** provides another example. Note that the *x*-axis plots time, starting in the mid-1930s and continuing through the late 1990s. The *y*-axis records the date of first flowering for a plant native to the Midwestern United States. Over approximately 60 years, the first day of the year when flowers were produced by *Baptisia leucantha* plants moved up by about 10 days. To interpret this observation, biologists hypothesize that the climate warmed enough to promote growth and flowering earlier in the year.

CORAL BLEACHING Coral reefs are among the most productive and species-rich ecosystems in the world. Reefs are large, undersea structures built by colonies of corals—cnidarians (see Chapter 32) that secrete calcium carbonate skeletons. Coral reefs are extremely productive in part because corals house symbiotic algae that perform photosynthesis.

In addition to overfishing and pollution, the integrity of coral reefs is threatened by two issues associated with global warming:

1. When atmospheric CO_2 dissolves, it reacts with water to form bicarbonate (H_2CO_3), which then drops a proton to form carbonic acid (HCO_3^-). Increased CO_2 levels in the atmosphere are increasing the rate of this reaction, and the protons that are released are lowering the pH of the oceans. Laboratory experiments have shown that corals form calcium carbonate skeletons much more slowly when pH drops. Recent data indicate that reef formation has slowed in Australia's Great Barrier Reef—perhaps in part because of acidification.
2. For reasons that are not yet clear, increases in water temperature cause reef-building corals to expel their photosynthetic algae. When this "coral bleaching" continues due to sustained exposure to warm water, corals begin to die of starvation.

EXTINCTIONS Biologists are concerned that rapid changes in climate may lead to extinction events.

- As the climate warms, alpine tundra habitats are disappearing. Plants that have restricted geographic ranges and grow on mountaintops may "run out of habitat" and go extinct.
- Researchers are concerned that habitats may change so rapidly that long-lived trees with limited dispersal will not be able to keep up—their ranges will not be able to change fast enough.
- Polar bears hunt seals that come up for air at holes in pack ice. As pack ice retreats throughout the Arctic, polar bears are dying of starvation. If present trends continue, polar bears may be extinct in the wild by 2100.

To date, though, the strongest evidence that global warming has caused species extinctions concerns amphibians. There is

(a) Cold-water copepods are declining in the North Atlantic.

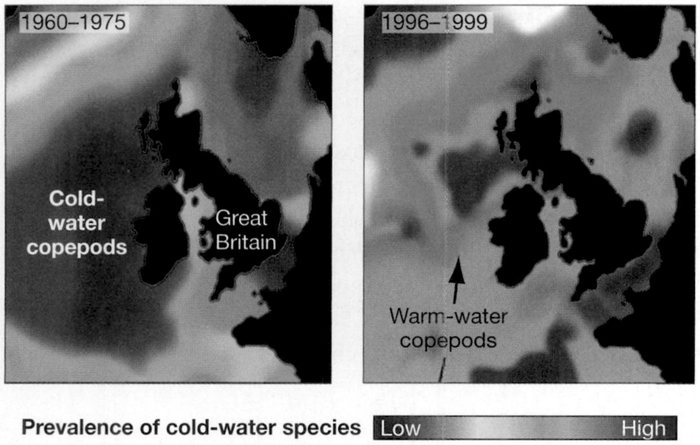

(b) Flowering times for some species in midwestern North America are earlier in the year.

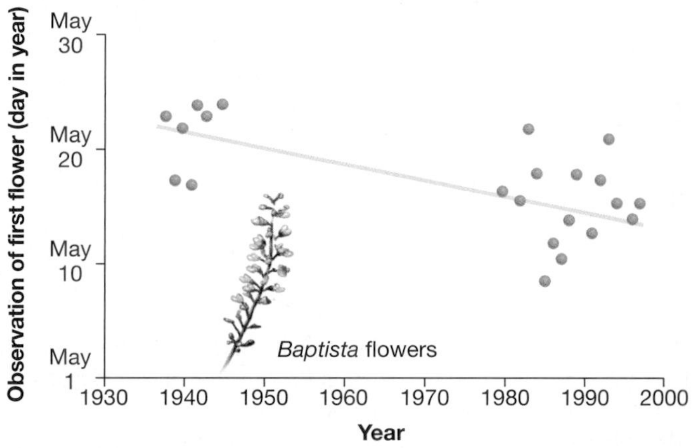

FIGURE 54.22 Global Warming Is Causing Changes in Geographic Ranges and Behavior. (a) Most species of copepod are restricted to waters of a certain temperature. **(b)** In a data set from Wisconsin, 19 of the 55 events tracked over a 61-year period occurred significantly earlier. Only one event occurred later.

convincing data supporting the hypothesis that most of the 122 frog species that have gone extinct recently succumbed to a parasitic fungus. Data published in 2006 suggest that global warming has increased the frequency and severity of these fungal infections—either because higher temperatures have stressed frogs and made them less resistant to infection, because higher temperatures create better growing conditions for the fungus, or both.

EVOLUTIONARY CHANGE WITHIN POPULATIONS In some populations, changing temperatures are already causing allele frequencies to change—meaning that some species are evolving in response to global warming.

The best data to date come from the fruit fly *Drosophila subobscura*. Recent work has shown that alleles that increase fitness in hot habitats have increased in frequency in Europe, South America, and North America independently.

To summarize, global warming has already caused significant changes in geographic ranges, phenology, and the health of coral reefs, as well as evolution in the form of extinctions and allele frequency changes. If the climate models published to date are correct, it will cause many more changes over the course of your lifetime.

Productivity Changes

Global warming—sometimes in combination with human-induced changes in the nitrogen cycle—is having another effect: altering NPP. For example, experiments have documented that warming temperatures, the addition of nitrogen and other nutrients, and rising CO_2 levels can all increase NPP in some terrestrial ecosystems.

Biologists are monitoring the issue closely because NPP is so central to the functioning of ecosystems. As the first two sections of this chapter emphasized, NPP is the basis of both energy flows and nutrient flows in ecosystems.

HOW IS NPP CHANGING? **Figure 54.23** shows how average annual NPP has changed globally, in (1) terrestrial environments between 1982 and 1999, and (2) marine environments from

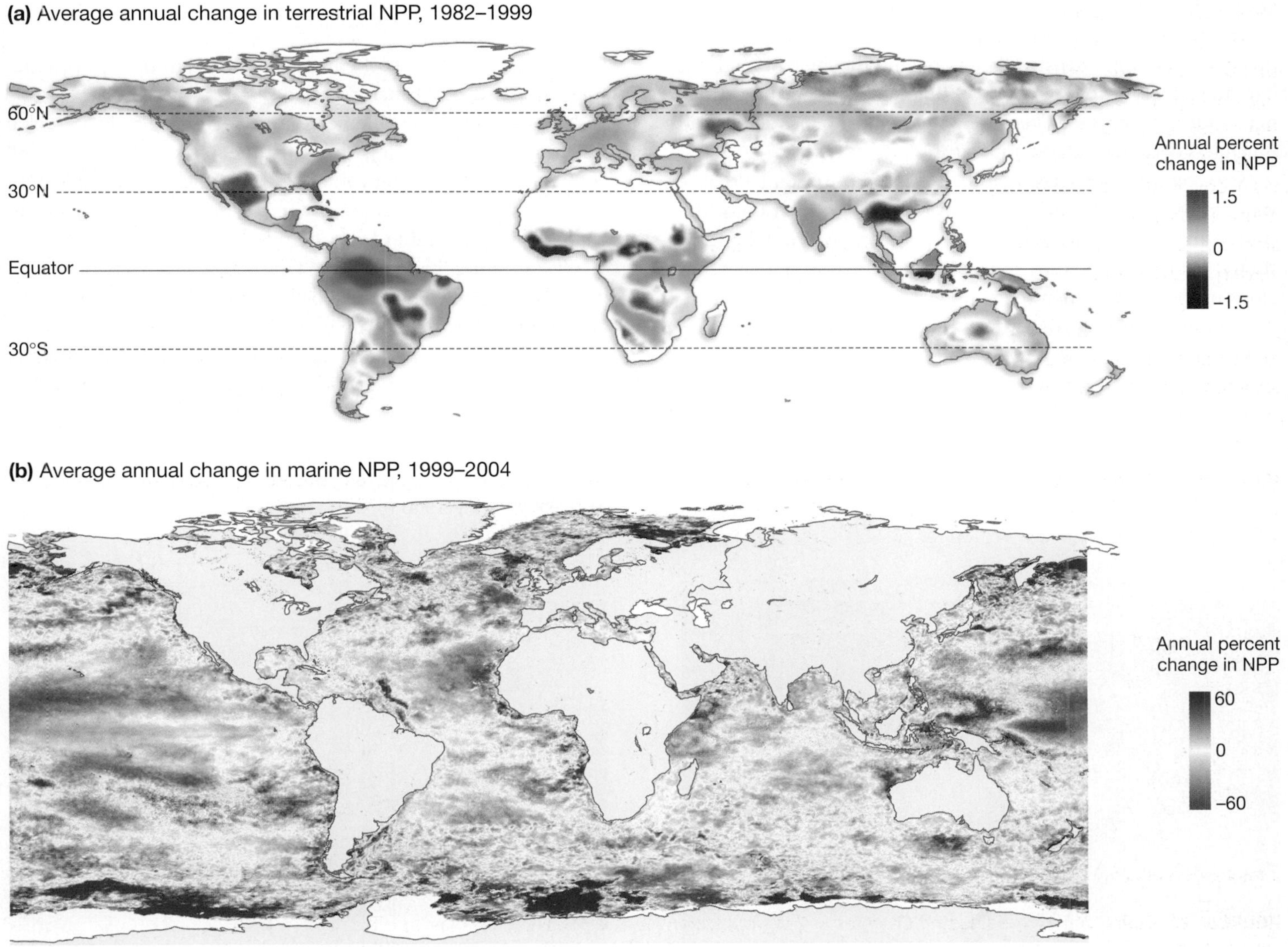

FIGURE 54.23 Consequences of Global Warming: Recent Changes in Terrestrial and Marine NPP.

1999 to 2004. These are the most recent intervals for which data are available.

On land, global NPP increased by over 6 percent during the 17-year study interval. In the oceans, though, the 1999–2004 data show a convincing correlation between increased surface water temperatures and decreased productivity. The data highlight a stark contrast: NPP is increasing on land but decreasing in the oceans.

The overall increase in terrestrial productivity is thought to be due to rising temperatures, increased rainfall in the tropics, and CO_2 fertilization—all factors that increase the rate of photosynthesis in plants. On land, increased NPP should exert a powerful negative feedback on global warming.

It's important to note, though, that global changes in NPP are underlain by considerable local variation. On land, productivity during the 1982–1999 interval tended to increase near the equator and at 60°N latitude, but to decrease in the Arctic and an array of more localized areas. In the ocean, productivity during the 1999–2004 interval dropped dramatically in large areas of the pelagic zone, but increased in a number of other regions.

WHY IS OVERALL PRODUCTIVITY DECLINING IN THE OCEANS? The leading explanation for the drop in marine productivity hinges on changes in water density. To understand the logic behind this hypothesis, recall from Chapter 50 that lakes can become stratified. Stratification occurs because water is most dense at 4°C and much less dense at higher temperatures.

Seawater also becomes stratified. This is important because water in the benthic zone is nutrient rich, due to the rain of decomposing organic material from the surface. As **Figure 54.24a** shows, water in the benthic zone is at 4°C year-round in large regions of the ocean. When the temperature of surface water rises due to global warming, the water at the surface becomes even less dense than benthic water.

Here's the key: When surface water becomes lighter, water currents are much less likely to be strong enough to overcome the density difference and bring nutrient-rich, 4°C water all the way up to the surface, where nutrients can spur the growth of photosynthetic bacteria and algae (**Figure 54.24b**). In this way, global warming is causing surface waters to become more nutrient poor.

LOCAL AND GLOBAL CONSEQUENCES What are the consequences of these local and global changes in NPP? The short answer is, "It depends."

In local areas of the ocean, the consequences of dramatically increased productivity have usually been negative.

1. Chapter 28 detailed how nitrate ions—derived from fertilizers applied to cornfields in midwestern North America—are washing into the Gulf of Mexico. Increased nitrate concentrations have stimulated the growth of planktonic organisms, whose subsequent decomposition has used up available oxygen and triggered the formation of an anoxic "dead zone."
2. Warmer water, in combination with increases in nitrate and other fertilizers of human origin, has increased the frequency or intensity of harmful algal blooms in the ocean—large populations of dinoflagellates that release toxins into the surrounding water (see Chapter 29).

On a global level, lower NPP in the ocean means that the productivity of the world's fisheries may decline.

It is also not clear whether the negative feedback on global warming that is occurring on land—due to increased NPP—will compensate for the decline in NPP occurring in the oceans. It remains to be seen whether atmospheric CO_2 will rise or fall due to changes in the amount of CO_2 being fixed by photosynthesis.

Overall, then, biologists don't yet know whether changes in productivity will be beneficial or detrimental to ecosystems. The answer should be forthcoming, however. Because global temperatures continue to rise and large-scale additions of nitrogen, phosphorus, and carbon dioxide are ongoing, humans are implementing a global-scale experiment on the effects of altering NPP.

(a) Much of the ocean is stratified by density and temperature.

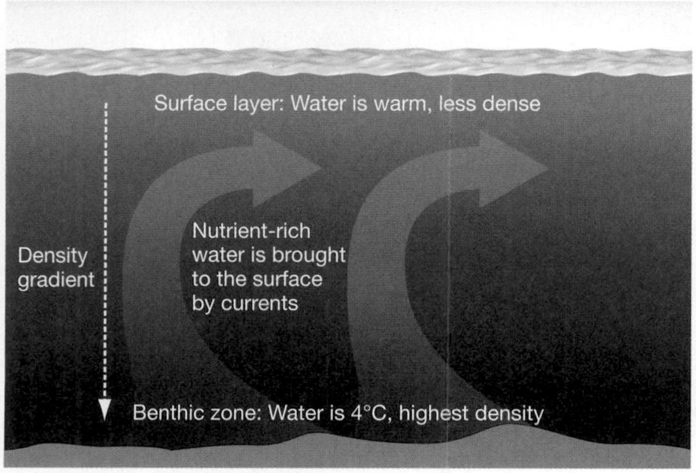

(b) Global warming increases the density gradient, making it less likely for layers to mix.

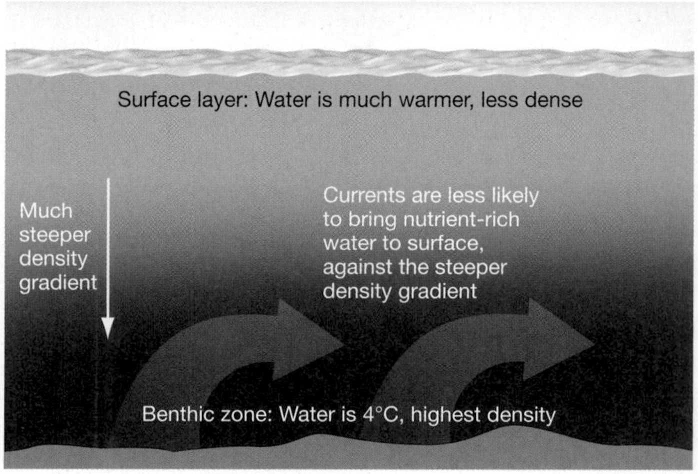

FIGURE 54.24 Temperature Stratification in the Ocean May Affect Productivity.

CHAPTER 54 REVIEW

For media, go to the study area at www.masteringbiology.com

Summary of Key Concepts

An ecosystem has four components: (1) the abiotic environment, (2) primary producers, (3) consumers, and (4) decomposers. These components are linked by the movement of energy and nutrients.

- An ecosystem consists of species that interact in a food web, along with their abiotic environment.
- As energy flows through ecosystems and as nutrients cycle through them, energy and nutrients are exchanged between biotic and abiotic components of the ecosystem.
- Net primary productivity (NPP) is the foundation of ecosystems, because it represents the energy available for consumption by other organisms. Humans are appropriating a large percentage of NPP.

✔You should be able to draw the relationships among the four components of an ecosystem where no photosynthesis occurs—the outside source of energy is iron- and sulfur-containing compounds released from magma (liquid rock).

As energy flows from producers to consumers and decomposers in a food web, much of it is lost. The productivity of terrestrial ecosystems is limited by warmth and moisture; nutrient availability is the key constraint in aquatic ecosystems.

- Organisms that acquire energy from the same type of source are said to occupy the same trophic level. Because energy transfer from one trophic level to the next is inefficient, ecosystems have a pyramid of productivity: Biomass production is highest at the lowest trophic level and lower at each higher trophic level.
- The feeding relationships among species in a particular ecosystem are described by a food chain or food web. Changes in the species composition of a food web can cause trophic cascades; persistent toxins can be magnified in concentration as they are passed up trophic levels in a food web.
- Among terrestrial ecosystems, NPP is highest in tropical wet forests and tropical dry forests; among aquatic ecosystems, productivity is highest in coral reefs, wetlands, and estuaries.
- Although productivity is extremely low in the oceanic zone, the area covered by this ecosystem is so extensive that it accounts for the highest percentage of Earth's overall productivity.

✔You should be able to describe the nature of a food chain where a minimum of energy would be lost.

To analyze nutrient cycles, biologists focus on the nature of the reservoirs where elements reside and the processes that move elements between reservoirs. Nutrient addition by humans is increasing productivity and causing pollution.

- Nutrients move through ecosystems in biogeochemical cycles.
- The rate of nutrient cycling is strongly affected by the rate of decomposition of detritus. The decomposition rate, in turn, is affected by abiotic environmental conditions such as temperature and by the quality of the detritus.
- Nutrients "leak" from ecosystems when biomass leaves. Experiments have shown that the loss of vegetation greatly increases the rate of nutrient loss.
- Human activities have doubled the amount of usable nitrogen that enters ecosystems each year. These increases have led to increased productivity, but also to pollution and to loss of biodiversity.

✔You should be able to describe conditions under which decomposition rates in terrestrial and marine environments are extremely low—creating a carbon sink due to a buildup of organic matter.

MB **BioFlix™** The Carbon Cycle, **Web Activity** The Global Carbon Cycle

The burning of fossil fuels has led to rapid global warming; rapid ecological and evolutionary changes are being observed in response.

- Average global temperatures are increasing rapidly because land-use changes and burning of fossil fuels have increased the flow of carbon in the form of CO_2 into the atmosphere, and because carbon dioxide acts as a greenhouse gas.
- Both positive and negative feedbacks are occurring in response to global warming.
- Biologists have documented an array of consequences, including extinctions and changes in productivity, allele frequencies, and geographic ranges.

✔You should be able to propose two mechanisms that humans could adopt to reduce global warming and explain the logic behind each.

Questions

✔TEST YOUR KNOWLEDGE

Answers are available in Appendix B

1. What is the difference between a community and an ecosystem?
   - **a.** An ecosystem comprises a community and the abiotic environment.
   - **b.** An ecosystem is a type of community.
   - **c.** An ecosystem includes only the plant community or communities present in an environment.
   - **d.** An ecosystem includes only the abiotic aspects of a particular environment.
2. Which of the following ecosystems would you expect to have the highest primary production?
   - **a.** subtropical desert
   - **b.** temperate grassland
   - **c.** boreal forest
   - **d.** tropical dry forest

3. Most of the net primary productivity that is consumed is used for what purpose?
   a. respiration by primary consumers
   b. respiration by secondary consumers
   c. growth by primary consumers
   d. growth by secondary consumers
4. What is biomagnification?
   a. biological imaging that provides images or data at fine spatial scales
   b. conversion of an energy input (e.g., sunlight) to chemical energy in the form of sugars, by primary producers
   c. accumulation of certain molecules at high concentration at upper trophic levels
   d. "overfertilization"—increases in nutrients that lead to pollution (detrimental effects)
5. Which of the following is normally the longest-lived reservoir for carbon?
   a. atmosphere CO_2
   b. marine plankton (primary producers *and* consumers)
   c. petroleum
   d. wood
6. Devegetation has what effect on ecosystem dynamics?
   a. It increases belowground biomass.
   b. It increases nutrient export.
   c. It increases rates of groundwater recharge (penetration of precipitation to the water table).
   d. It increases the pool of soil organic matter.

✔ TEST YOUR UNDERSTANDING

Answers are available in Appendix B

1. Explain the difference between a food chain and a food web. What do the arrows in each type of diagram represent?
2. Suppose that cougars were reintroduced to a region where deer populations are currently so high that most tree saplings and other low-growing plant species are consumed. Using the wolf reintroduction in Yellowstone as a model, predict the trophic cascade that might occur. (Cougars eat deer.)
3. Explain why decomposition rates are higher in some ecosystems than in others, and give examples. How do decomposers regulate nutrient availability in an ecosystem?
4. Define "bleaching" and ocean acidification, and explain how each is affecting coral reefs.
5. Name one positive feedback on global warming and one negative feedback. Explain why each is occurring.
6. Why are the open oceans nutrient poor? Why are coastal areas and intertidal habitats relatively nutrient rich?

✔ APPLYING CONCEPTS TO NEW SITUATIONS

Answers are available in Appendix B

1. Suppose you had a small set of experimental ponds at your disposal and an array of pond-dwelling algae, plants, and animals. How could you use a radioactive isotope of phosphorus to study nutrient cycling in these experimental ecosystems?
2. Fertilizing the open oceans with iron is controversial, because some researchers predict that it could produce dead zones in what are now valuable fishing areas near coastlines. Explain how this could happen, assuming that ocean currents move water under fertilized areas in the open ocean to coastal regions. (Hint: review the discussion of dead zones in Chapter 28.)
3. Suppose that herbivores were removed from a temperate deciduous forest ecosystem (e.g., by fencing to exclude herbivorous mammals, and spraying insecticides to kill herbivorous insects). Predict what would happen to the rate of nitrogen cycling. Explain the logic behind your prediction.
4. During the Carboniferous Period, rates of decomposition slowed even though plant growth was extensive (probably due to the formation of vast, oxygen-poor swamp habitats). As a result, large amounts of biomass accumulated in terrestrial environments (much of this biomass is now coal). The fossil record indicates that atmospheric oxygen increased, atmospheric carbon dioxide increased, and global temperatures dropped. Explain why.

This chapter is part of The Big Picture. See how on pages 1126–1127.

An enormous carved stone statue—a portion of which has fallen on its side—on Easter Island. Archaeological data indicate that Easter Island was once covered by extensive forests that supported a thriving human population.

Biodiversity and Conservation Biology 55

When Europeans arrived on Easter Island—also known as Rapa Nui—about 3500 km (2180 miles) west of Chile, they found about 1000 people and a desolate, treeless landscape dotted with gigantic stone statues. Things were not always this way. Archaeological digs and pollen samples indicate that the island was once covered with lush forest dominated by the world's biggest palm tree—now extinct—along with what may have been the largest colonies of breeding seabirds in the Pacific Ocean.

About 800 years ago, Easter Island was a lush tropical paradise that supported a thriving population estimated at 10,000 people. What happened?

According to Jared Diamond, the archaeological and fossil data indicate that an ecological disaster happened. Diamond hypothesizes that the human population grew to massive size as people used canoes carved from palm logs to harvest the abundant seabirds, fish, and dolphins. The islanders had the leisure time to carve gigantic statues and roll them into place on beds of palm logs.

But once all the palms were cut, the system collapsed. Without palm trunks to make canoes, dolphin hunting and fishing were much less efficient. Soil erosion from the deforested slopes may have cut into the productivity of banana and sweet potato plantations. The population crashed to a tenth of its former size, and the great statues fell. On Easter, biodiversity losses had direct impacts on human health and welfare.

Easter is part of a larger island: Earth. The species and resources on our planet are limited, but the human population is growing unchecked. Are we headed for an Easter Island–like collapse?

Many biologists are afraid that the answer might be yes. This chapter's goal is to analyze the data responsible for this fear, and introduce the positive steps that are being taken to avert an Easter Island–like scenario. Let's begin by defining **biodiversity**—the thing that biologists want to protect.

KEY CONCEPTS

- Biodiversity is quantified at the level of allelic diversity, species diversity, and ecosystem diversity.
- According to recent analyses, the sixth mass extinction in the history of life is occurring due to habitat loss, overexploitation, and global climate change.
- Humans depend on biodiversity for the products that wild species provide and for ecosystem services that protect the quality of the abiotic environment.
- Solutions to the biodiversity crisis include protecting key habitats, lowering human population growth and resource use, restoring ecosystems, mitigating climate change, and supporting sustainable development.

✓ When you see this checkmark, stop and test yourself. Answers are available in Appendix B.

55.1 What Is Biodiversity?

Perhaps the simplest way to think about biodiversity is in terms of the tree of life—the phylogenetic tree of all organisms, introduced in Chapter 1 and explored in detail in Chapters 28 through 35. If biologists are eventually able to estimate the complete tree of life, using the techniques reviewed in Chapter 27, the branches would represent all of the lineages of organisms living today; the tips would represent all of the species.

When biodiversity increases, branches and tips are added and the tree of life gets fuller and bushier. When extinctions occur, tips and perhaps branches are removed; the tree of life becomes thinner and sparser.

Coordinated, multinational efforts are under way to estimate the tree of life—particularly major lineages such as the fungi, brown algae, and land plants. These and other efforts to estimate phylogenies are producing a better picture of the relationships among lineages and species, and a clearer understanding of how extinction is affecting biodiversity. But even though the tree of life is a compelling way to document biodiversity, it does not tell the entire story.

Biodiversity Can Be Measured and Analyzed at Several Levels

To get a complete understanding of the diversity of life, biologists recognize and analyze biodiversity at the genetic, species, and ecosystem level. Let's consider each in turn.

GENETIC DIVERSITY **Genetic diversity** is the total genetic information contained within all individuals of a species. It is measured as the number and relative frequency of all alleles present in a species. At the genetic level, biodiversity is everywhere around you, from the variety of apples at the grocery store to the genetically based variation in coloration and singing ability of sparrows in your backyard.

Because no two members of the same species are genetically identical, most species are the repository of an immense array of alleles. For example, distinct populations of the same species often contain unique alleles—alleles that are maintained by genetic drift or have been favored by natural selection, as adaptations to local conditions.

Recently, efforts to catalog genetic diversity have been boosted by two technical breakthroughs:

1. the ability to sequence the entire genome of multiple members of the same species, and
2. the research protocol called environmental sequencing.

Both approaches were introduced in Chapter 20. You might recall that in environmental sequencing, researchers take a sample from the soil or water present in a habitat and sequence all or most of the genes present. It's a way to document genetic diversity even if the species present are unknown.

SPECIES DIVERSITY As its name implies, **species diversity** is based on the variety of species on Earth. Chapter 53 pointed out that species diversity is measured by quantifying the number and relative frequency of species in a particular region.

Recent efforts to document species diversity have used a new technique called **bar coding**: the use of a well-characterized gene sequence to identify distinct species, using the phylogenetic species concept introduced in Chapter 26. The idea is that variation in one gene—for example, a portion of the DNA found in the mitochondria of mammals, specifically the gene for cytochrome *c* oxidase subunit 1—allows biologists to recognize different species, much as a slight variation in a bar code allows a grocery store scanner to recognize different products. The specific gene used as a bar code varies among lineages. Researchers who are contributing to this effort submit sequence data from "bar coding genes" to centralized databases.

There is an additional aspect of species diversity, however, that biologists call taxonomic diversity. Taxonomic diversity is important because some lineages on the tree of life are extremely species rich. Prominent examples include the African cichlid fish, introduced in Chapter 27, and the 35,000 species of orchids. Other lineages, in contrast, are extremely species poor. Some, in fact, are represented by a single species (**Figure 55.1**).

Some biologists argue that it is particularly important to preserve populations from species-poor lineages, because they are the last living representatives of their lineages. If those populations are wiped out, an entire lineage is lost forever. Such is the case with the Yangtze River dolphin in Figure 55.1b. Most biologists concede that the species is now extinct.

ECOSYSTEM DIVERSITY **Ecosystem diversity** is the array of biotic communities in a region along with abiotic components, such as soil, water, and nutrients. Ecosystem diversity is more difficult to define and measure than genetic diversity or species diversity, however, because ecosystems do not have sharp boundaries. Recall from Chapter 54 that ecosystems are complex and dynamic assemblages of organisms that interact with each other and their nonliving environment.

Attempts to measure ecosystem diversity focus on capturing the array of biotic communities in a region, along with variation in the physical conditions present. Areas around estuaries, for example, tend to have high ecosystem diversity due to the combination of stream, wetland, intertidal, coastal, and upland habitats (see Chapter 50).

Note that the three levels of biodiversity are not independent of each other. When an estuary is dredged or filled, for example, biodiversity is affected at all three levels: genetic and species diversity change due to the different numbers and types of individuals present, and ecosystem diversity is altered due to the change in abiotic conditions.

CHANGE THROUGH TIME Biodiversity can be recognized and quantified on several distinct levels, but it is also dynamic.

- Mutations that create new alleles increase genetic diversity; natural selection, genetic drift, and gene flow may eliminate certain alleles or change their frequency, leading to an increase or decrease in overall genetic diversity.

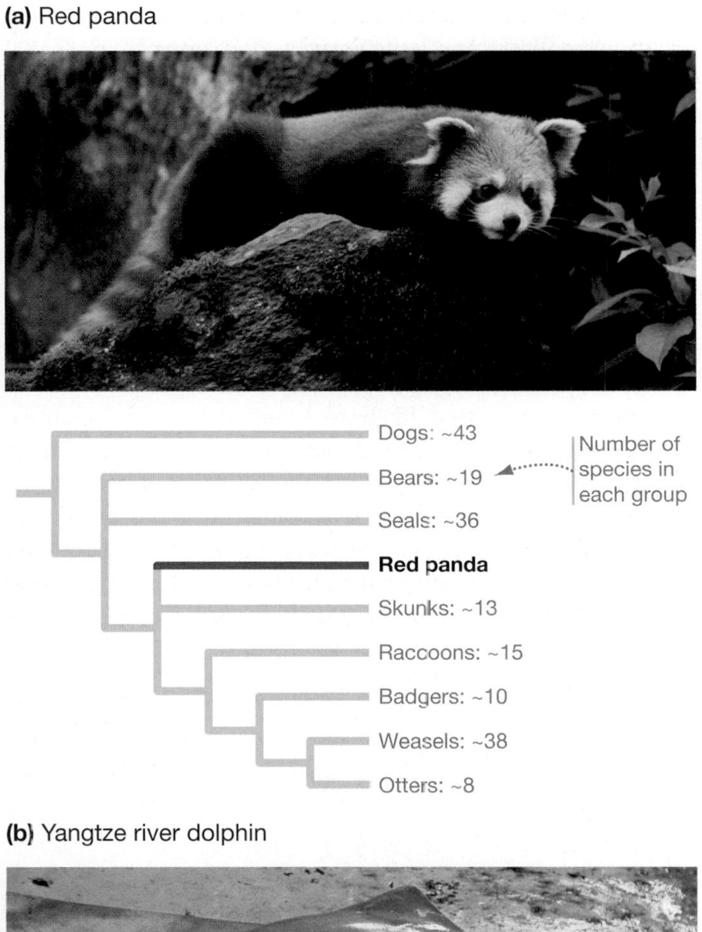

FIGURE 55.1 Phylogenetically Distinct Species May Be High-Priority Targets for Conservation. The red panda and the Yangtze river dolphin have few close relatives and represent distinct branches on the tree of life. If conservation measures attempt to preserve taxonomic diversity, these species would be a high priority for preservation. Unfortunately, it is too late for the Yangtze river dolphin—the species is now considered extinct.

- Speciation increases species diversity; extinction decreases it.
- Changes in climate or other physical conditions can result in the formation of new ecosystems, as can the evolution of new species that interact in novel ways. Disturbances such as volcanic eruptions, human activities, and glaciation can destroy ecosystems.

Biodiversity is not static. It has been changing since life on Earth began.

How Many Species Are Living Today?

One of the simplest questions about biodiversity is also one of the most difficult to answer: How many species are there on Earth? The answer is not known. Given the massive effort that it would take to document every form of life on Earth, the answer will probably never be known.

Biologists are well aware, though, that the approximately 1.5 million species cataloged to date represent a tiny fraction of the number actually present. Chapter 31, for example, noted that in Britain, an average of six fungal species live on each well-studied plant species. Because over 275,000 plant species have been described, a conservative estimate based on the 6:1 fungi:plant ratio—assuming that the ratio is universal—suggests that 1.65 million species of fungi exist, not counting species that don't live on plants. But only 80,000 have been described thus far, in total.

The issue is even more acute in poorly studied lineages like bacteria and archaea, where researchers routinely discover dozens of new species in each environmental sequencing or direct sequencing study (see Chapter 20 and Chapter 28). For example, a recent direct sequencing study estimated that over 500 species of bacteria live in the human mouth, but only about half had been described and named already. Even among well-studied groups like birds and mammals, new species are discovered almost every year.

Given that only a fraction of the organisms alive have been discovered to date, how can biologists go about estimating the total number of species on Earth? One approach is based on intensive surveys of species-rich groups at small sites; a second strategy is based on attempts to identify all of the species present in a particular region.

TAXON-SPECIFIC SURVEYS Terry Erwin and J. C. Scott published a classic study of species diversity in a single taxon. They began by estimating the number of insect species that live in the canopy of a single tropical tree (**Figure 55.2** on page 1108). To do this, they used an insecticide to knock down insects from the top of a *Luehea seemannii* tree and identified over 900 beetle species among the individuals that fell. Most of the species were new to science.

To use these data as an indicator of global arthropod species diversity, Erwin and Scott used the logic outlined in **Box 55.1** on page 1108 to extrapolate from the diversity found on a single tree to the diversity on all known tropical trees. Their analysis came up with an estimate of 30 million insect species.

More recent single-taxon surveys have sampled larger geographic areas. For example, Philippe Bouchet and colleagues conducted a massive survey of marine mollusks in coral-reef habitats along the west coast of New Caledonia, a tropical island

FIGURE 55.2 Estimating Species Richness via Intensive Local Sampling. A small sample of the insects collected from a single tree in the Amazonian rain forest.

in the southwest Pacific Ocean. The team spent more than a year collecting mollusks at 42 sites over a total area of almost 300 square kilometers. The survey represented the most thorough sampling effort ever made to determine the species diversity of mollusks, and produced more than 127,000 individuals representing 2738 species. These numbers far exceed the mollusk diversity recorded for any comparable-sized area. The species total was 2 to 3 times the total number of mollusk species that had been reported for similar habitats in the region.

In reporting their findings, the biologists emphasized that 20 percent of the species found were represented by a single specimen. This observation suggests that many species are exceedingly rare and thus likely to be missed by less-intensive sampling efforts. And when the investigators compared their data with a survey in progress at a second site, different in reef structure but only 200 kilometers away from the original site, they found that only 36 percent of species were shared between the two sites. Taken together, the results support the hypothesis that the 93,000 known mollusk species may represent just a third to a half of the actual total.

ALL-TAXA SURVEYS The first effort to find and catalog all of the species present in a large area is now under way. The location is the 2200 km^2 Great Smoky Mountains National Park in the southeastern United States. A consortium of biologists, volunteers, and research organizations initiated this All Taxa Biodiversity Inventory in 1998.

To date, the survey has discovered over 890 species that are new to science and over 6300 species that had never before been found in the park. When the inventory is complete, in 2015, biologists will have the best database ever assembled on the biodiversity present in a region.

BOX 55.1 QUANTITATIVE METHODS: Extrapolation Techniques

To estimate the world total of insect species from the specimens found on a single tree, Erwin and Scott performed the following extrapolation:

1. Estimate the number of beetle species that live only on *L. seemannii*. Based on earlier work with insects on this tree species, they estimated the number of unique beetle species at 160.
2. Estimate the number of beetle species found on all tropical trees, assuming that there are 50,000 species and assuming that each tree species hosts 160 unique species of beetle.

   $160 \times 50{,}000 = 8$ million beetle species in tropical trees worldwide

3. Estimate the total number of insect species, based on the estimate that 40 percent (0.40) of all known insect species are beetles.

   8 million/0.40 = 20 million insect species in tropical trees worldwide

4. Estimate the total number of insect species worldwide, assuming that about 33 percent of insects live on the ground instead of in trees. If there are 20 million species in trees, then there would be an additional 10 million on the ground.

   20 million + 10 million = 30 million insect species in all habitats worldwide

The total number is shockingly large—only about a million species are known to date.

It is important to be cautious—even skeptical—about extrapolation-based estimates like this, however, as they rely on a large number of assumptions and estimates. Changes in these assumptions and estimates, when multiplied through the steps in the calculations, can lead to large differences in the final outcome.

For example, suppose additional research showed that *L. seemannii* is unusual, and that there are actually only an average of 100 unique species of beetle that live on each species of tropical tree. Now do the same calculation to estimate the total of insect species worldwide. What do you get?[1]

Erwin and Scott were aware of this point; the purpose of their study was to stimulate discussion and more research.

[1]18.7 million

CHECK YOUR UNDERSTANDING

If you understand that . . .

- Biologists document biodiversity at the genetic, species, and ecosystem levels.

✔ You should be able to . . .

Outline a study that would quantify species diversity on your campus.

Answers are available in Appendix B.

55.2 Where Is Biodiversity Highest?

If documenting the extent of genetic, species, and ecosystem diversity is the most fundamental goal of research on biodiversity, then the second-most important aim is to understand its geographic distribution. Chapter 53 introduced the most prominent geographic pattern in the distribution of biodiversity: In most taxonomic groups, species richness is highest in the tropics and declines toward the poles.

Tropical rain forests are particularly species rich. Even though they represent just 7 percent of Earth's land area, they are thought to contain at least 50 percent of all species present. As Chapter 53 noted, understanding why tropical forests are so species rich and why a latitudinal gradient occurs is an active research area.

Hotspots of Biodiversity and Endemism

Biologists have also been working to understand the distribution of species richness at finer spatial scales than global gradients. The map in **Figure 55.3a**, for example, was constructed by dividing the world's landmasses into a grid of cells 1° latitude by 1° longitude. In the tropics, this translates to rectangles approximately 111 km by 111 km (69 mi by 69 mi). Because lines of longitude converge at the poles, the rectangles are about 111 km by 55 km at 60° latitude. Using published data, researchers then plotted how many species of birds breed in each cell in the grid. These data are consistent with the latitudinal gradient in species richness, but indicate that some areas of the tropics are much more species rich than others.

An organization called Conservation International pioneered these types of analyses and promoted the term **biodiversity hot spot** to characterize regions that are extraordinarily rich in biodiversity. In terms of bird species richness, the Andes mountains, along the west coast of South America, the Amazon River basin in the central part of South America, portions of East Africa, and southwest China are important hotspots.

Researchers have also been keenly interested in understanding which regions of the world have a high proportion of **endemic species**—meaning, species that are found in an area and nowhere else. **Figure 55.3b** maps the location of endemic species of breeding birds, based on the same cells and data as the species richness map in Figure 55.3a.

(a) Biodiversity hotspots in terms of **species richness of birds**

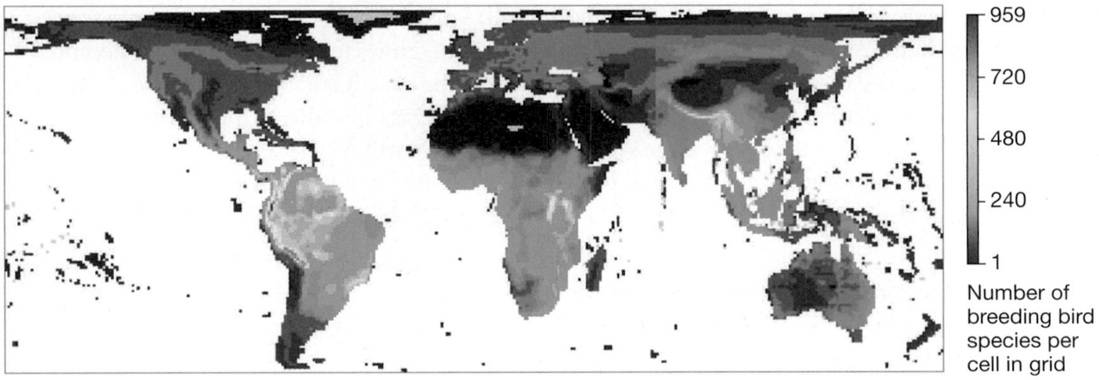

(b) Biodiversity hotspots in terms of **endemic species of birds**

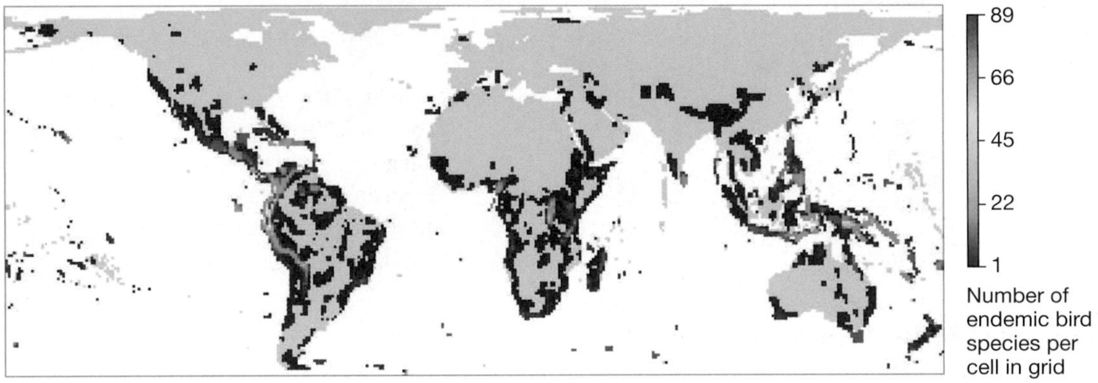

FIGURE 55.3 Biodiversity Hotspots Have Extraordinarily High Species Richness. Both maps are based on a grid of cells that is 1° longitude by 1° latitude.

Conservation Hotspots

Mapping species richness hotspots and centers of endemism can inspire research on *why* certain regions contain many species or a high proportion of endemic taxa. But in addition, biologists are studying the geographic distribution of biodiversity as a way of focusing conservation efforts.

Starting in the year 1999, for example, a multinational team coordinated by Conservation International set out to identify regions of the world that meet two criteria:

1. They contain at least 1500 endemic vascular plant species (>0.5 percent of the world's total).
2. They have lost at least 70 percent of their traditional or primary vegetation.

The goal was to identify regions of the world that are in most urgent need of conservation action—areas where efforts to preserve habitat would have the highest return on investment. The idea was that efforts to protect a diversity of primary producers would guarantee the protection of a wide array of primary, secondary, and tertiary consumers.

The 34 regions that currently meet the two "conservation hotspot" criteria are shown in **Figure 55.4**. Although these areas represent just 2.3 percent of the Earth's land area, they contain over 50 percent of all known plant species and 42 percent of all known terrestrial vertebrate species, as endemics.

The 34 hotspots were once much more extensive than they are now. Overall, 86 percent of their habitat area has already been destroyed. In effect, much of the world's biodiversity is being squeezed into a few shrinking spaces.

It's also important to note that key regions with high plant species diversity—specifically the rain forests found in the Amazon, the Congo River basin, and New Guinea—are not included as conservation hotspots, because they have not yet lost 70 percent of their primary vegetation. Effectively protecting these areas along with the hotspots defined on the map would provide protection for over 65 percent of all land plants in just 5 percent of the Earth's land surface.

55.3 Threats to Biodiversity

No species lasts forever. Climate change, disease, competition from newly arrived species, and habitat alteration are all natural processes. They have been happening since life began. Extinction, like death, is a fact of life.

If extinction is natural, why are biologists so concerned about habitat and species conservation? The answer is rate.

Today, species are vanishing faster than at virtually any other time in Earth's history. Modern rates of extinction are 100 to 1000 times greater than the average, or "background," rate recorded in the fossil record over the past 550 million years.

Either directly or indirectly, current extinctions are being caused by the demands of a rapidly growing human population—currently at over 6.8 billion as this book goes to press and increasing by about 77 million people per year. The island that is planet Earth is much more crowded than it used to be.

Based on the data in hand, the vast majority of biologists agree that the sixth mass extinction in the history of multicellular life is occurring. Between the time you are reading this and the time your great-grandchildren are grown, human impacts on the planet promise to equal or exceed those of the gigantic asteroid that smashed into Earth 65 million years ago.

Changes in the Nature of the Problem

Most extinctions that have occurred over the past 1000 years took place on islands and were either caused by overexploitation—meaning, overhunting or overfishing—or the introduction of **exotic species**—nonnative competitors, diseases, or predators.

You might recall, from Chapter 50, that if an exotic species is introduced to a new area, grows to large population size, and disrupts species native to the area, it is called an invasive. That chapter discussed how cheatgrass has altered plant communities in large sections of western North America.

The brown tree snake pictured in **Figure 55.5** is a dramatic example of an invasive species that has caused extinctions. After being

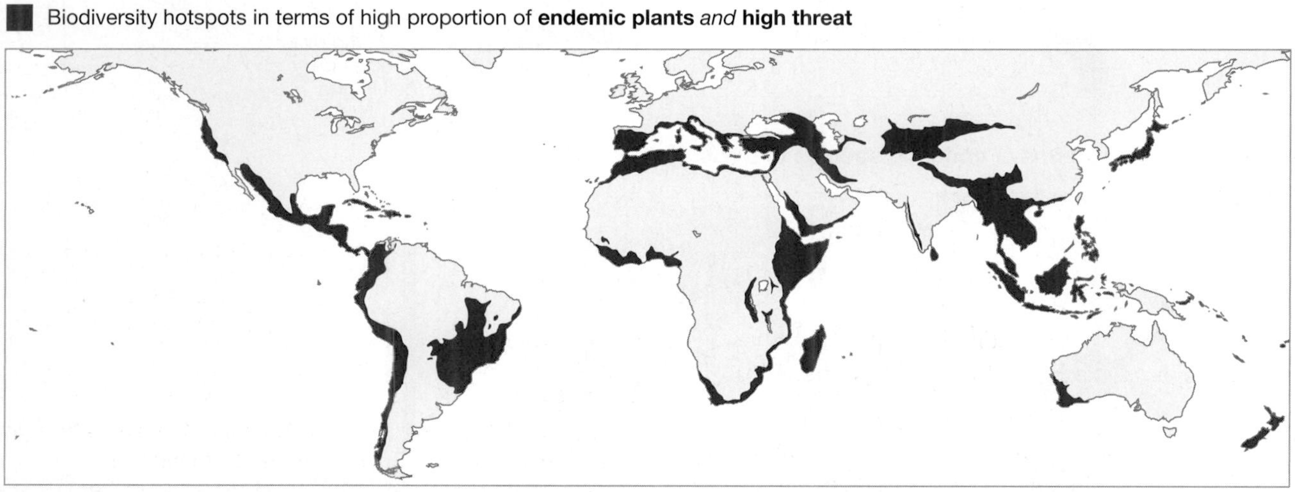

FIGURE 55.4 Conservation Hotspots Identify Priorities for Immediate Action. The areas shown in red contain a high diversity of vascular plant species and have lost at least 70 percent of their original vegetation.

FIGURE 55.5 Invasive Species Are Destructive. The brown tree snake has extinguished dozens of species of birds on Guam.

introduced to the island of Guam, predation by the expanding snake population caused the extinction of dozens of bird species.

The most dramatic example of extinctions due to overexploitation and exotic species involves birds native to islands in the South Pacific. Recent research on fossils found on islands throughout the region suggests that about 2000 bird species were wiped out as people colonized this area about A.D. 1200. Many of these extinctions occurred due to predation by humans or by rats, pigs, and other animals introduced by humans. A loss of 2000 species is significant—about 10,000 bird species are left today.

Starting in the twentieth century, though, two patterns in species loss began to change:

1. **Endangered species**, which are almost certain to go extinct unless effective conservation programs are put in place, are now more likely to live on continents than islands.
2. Although overexploitation and invasives continue to be serious problems, habitat destruction and global climate change are now the primary threats to species.

The bar graphs in **Figure 55.6** show the results of a recent analysis on causes of endangerment for 488 endangered species native to Canada. Several patterns jump out of the data:

- Habitat loss is the single most important factor in the decline of these species. It is a significant issue for over 90 percent of endangered species in terrestrial environments.
- Virtually all of the endangered species are affected by more than one factor. This is important, because it means that conservation biologists may have to solve more than one problem for any given species to recover.
- Overexploitation is the dominant problem for marine species, while pollution plays a large role for freshwater species.
- Factors beyond human control can be important. These include predation or competition with native species, natural disturbances such as fires or droughts, or the fact that some species have narrow niches and have historically been rare. Stated another way, background extinctions will continue to occur.

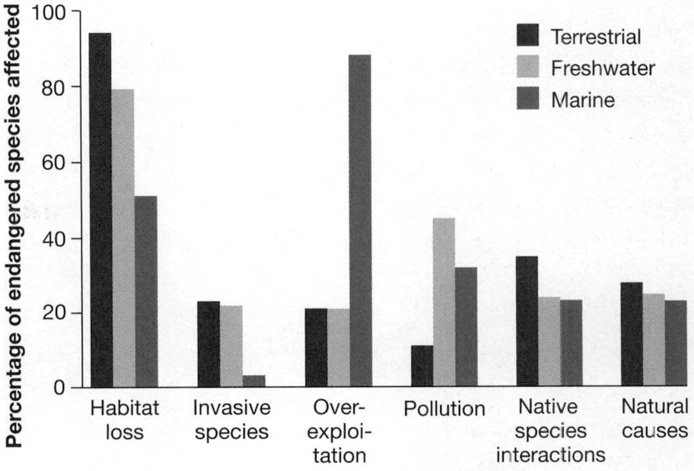

FIGURE 55.6 Endangerment Is Caused by Multiple Factors. These data are from Canada.

✔QUESTION Why do the data for terrestrial, freshwater, and marine species each add up to more than 100 percent?

Do these conclusions hold up for other parts of the world? Although invasive species are a major problem in some regions, most analyses completed to date are broadly consistent with the Canadian data. For example, overexploitation is a grave concern for marine fisheries worldwide. Two-thirds of harvestable species are now considered fully exploited or depleted, and a recent study of the global fishing industry concluded that 90 percent of individuals in large-bodied fish species from bottom or open-water habitats have been already removed from the world's oceans. The species affected include tuna, swordfish, marlin, cod, halibut, and flounder. Overhunting has also emerged recently as a dire threat to many mammal populations in Africa and southeast Asia (**Figure 55.7**).

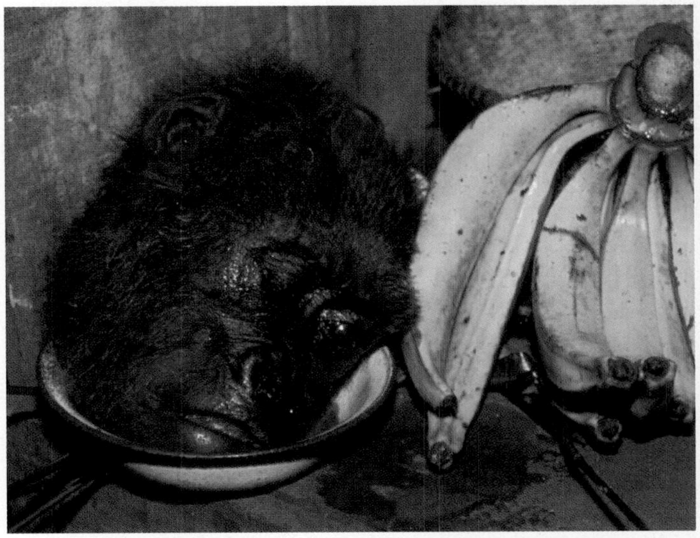

FIGURE 55.7 Overexploitation Continues to Be a Problem in Some Regions. The recent growth of the "bushmeat" trade has devastated many populations of primates and other mammals in Africa. Overexploitation is also a major cause of endangerment in marine species worldwide (see Figure 55.6).

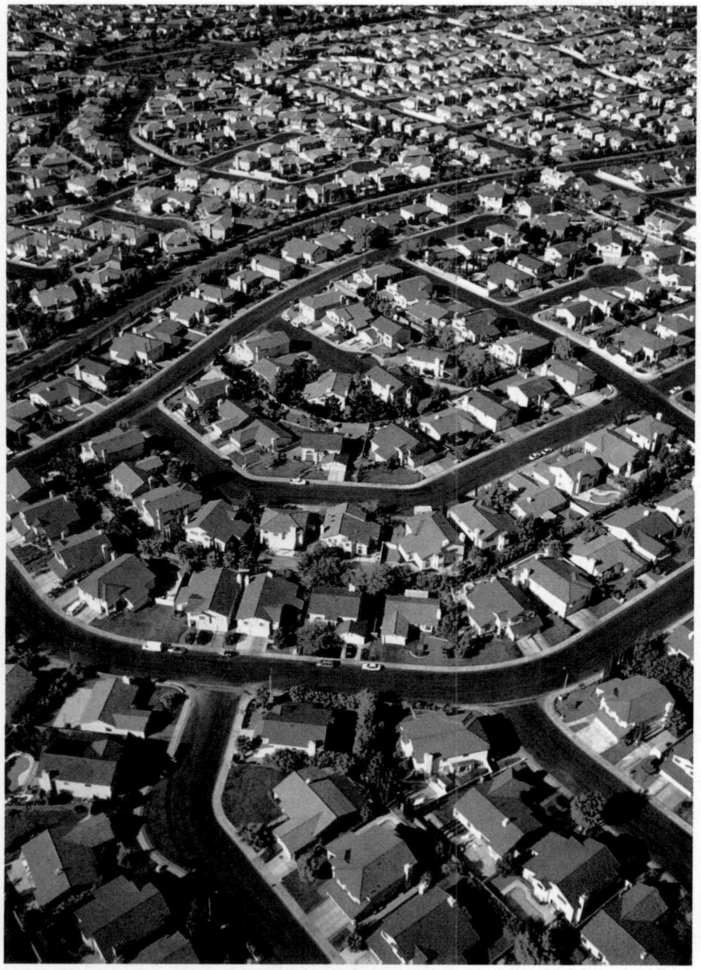

FIGURE 55.8 Suburbanization Is a Major Cause of Habitat Loss in North America.

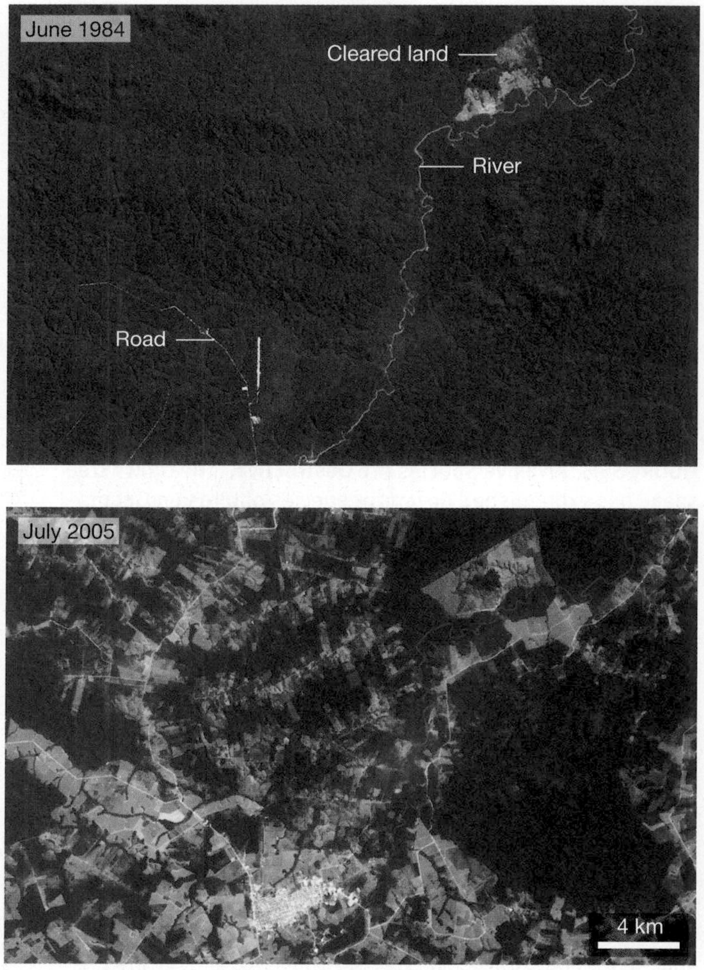

FIGURE 55.9 Rates of Deforestation in Tropical Wet Forests Are High. In the satellite image from Rondônia, Brazil, at top, the dark green areas are intact forest. In the lower image of the same location 21 years later, the light areas have been burned or logged and converted to agricultural fields and pastures.

HABITAT DESTRUCTION Humans cause **habitat destruction** by logging and burning forests, damming rivers, dredging or filling estuaries and wetlands, plowing prairies, grazing livestock, excavating minerals, and building housing developments, golf courses, shopping centers, office complexes, airports, and roads (**Figure 55.8**).

On a global scale, one of the most important types of habitat destruction is deforestation—especially the conversion of primary forests to agricultural fields and human settlements. To appreciate the extent of deforestation in some areas, consider satellite images of wet tropical rain forest in the Brazilian state of Rondônia that were made in 1984 and in 2005 (**Figure 55.9**).

Analyses of satellite photos such as these have shown that as many as 3 million hectares, an area about 10 percent larger than the state of Maryland, was deforested in the Amazon each year during the 1990s. (A hectare, abbreviated ha, is 100 meters by 100 meters, approximately the size of two football fields.)

More recent analyses suggest that the global rate of deforestation has now slowed compared to peak rates in the 1990s. The latest report from the United Nations Food and Agriculture Organization, released in 2005, states that the world's forests experienced a net loss of about 7.3 million ha/year between 2000 and 2005. This is down from the average net loss of 8.9 million ha/year during the previous decade.

Conservationists greeted this report with caution. Net losses dropped because of extensive forest regrowth in China, North America, and Europe, even though over 8 million ha/year of primary—meaning, previously uncut—forest were lost in South America, southeast Asia, and Africa each year from 2000 to 2005.

More specifically, the total area of wet tropical forest dropped by 2.36 percent during 2000–2005. If this rate of tropical deforestation continues, over 28 percent of the wet tropical forest that exists today will be gone in your lifetime, including almost half of the Brazilian Amazon. These projections are extremely conservative, as the human population in Brazil, Indonesia, and other key countries is still increasing rapidly.

In essence, the forests of the Amazon and southeast Asia are experiencing the same type of deforestation that occurred in European and North American forests from 1700 to 1900. Forest loss in South America and Africa is particularly important, however, because it is occurring in biodiversity hotspots.

HABITAT FRAGMENTATION In addition to destroying natural areas outright, human activities fragment large, contiguous areas of natural habitats into small, isolated fragments. **Habitat fragmentation** concerns biologists for several reasons.

1. It can reduce habitats to a size that is too small to support some species. This is especially true for keystone predators (see Chapters 53 and 54) such as wolves, mountain lions, grizzly bears, and bluefin tuna, which need vast natural spaces in which to feed, find mates, and reproduce successfully. Large animal species demand large spaces.
2. By creating islands of habitat in a sea of human-dominated landscapes, fragmentation reduces the ability of individuals to disperse from one habitat to another. Isolated populations are vulnerable to the types of chance catastrophes and genetic problems detailed later in the chapter.
3. Fragmentation creates large amounts of "edge" habitat. The edges of intact habitat are subject to invasion by weedy species and are exposed to more intense sunlight and wind, creating difficult conditions for plants.

The decline in habitat quality caused by fragmentation is being documented in a long-term experiment in a tropical wet forest. In an area near Manaus, Brazil, that was slated for clear-cutting, a research group set up 66 square, 1-hectare experimental plots that remained uncut. Thirty-nine of these study plots were located in fragments designed to contain 1, 10, or 100 hectares of intact forest. Twenty-seven of the plots were set up nearby, in continuous wet forest. As the "Experimental setup" section of **Figure 55.10** shows, the distribution of the study plots allowed the research team to monitor changes inside forest fragments of different sizes and to compare these changes with conditions in unfragmented forest. Note that many of the study plots were located on forest edges.

When the research group surveyed the plots at least 10 years after the initial cut, they recorded two predominant effects:

1. a rapid loss of species diversity, especially from the smaller fragments; and
2. a startling drop in **biomass**, or the total amount of fixed carbon, in the study plots located near the edges of logged fragments. As the graph in the "Results" section of Figure 55.9 shows, most of the loss occurred within 2–4 years of the fragmentation event.

In tropical wet forests, most of the biomass is concentrated in large trees. Due to exposure to high winds and dry conditions, the edges of the experimental plots contained many downed and dying trees. Follow-up work showed that in edge habitats and in the fragmented patches, large-seeded, slow-growing trees typical of undisturbed forests are being replaced by early successional, "weedy" species (see Chapter 53). A large number of bird and understory plant species have disappeared as well.

To review the effects of habitat fragmentation, go to the study area at *www.masteringbiology.com.*

(MB) **Web Activity** Habitat Fragmentation

EXPERIMENT

QUESTION: How does fragmentation affect the quality of tropical wet forest habitats?

HYPOTHESIS: Fragmentation reduces the quality of wet forest habitats.

NULL HYPOTHESIS: Fragmentation has no effect on the quality of wet forest habitats.

EXPERIMENTAL SETUP:

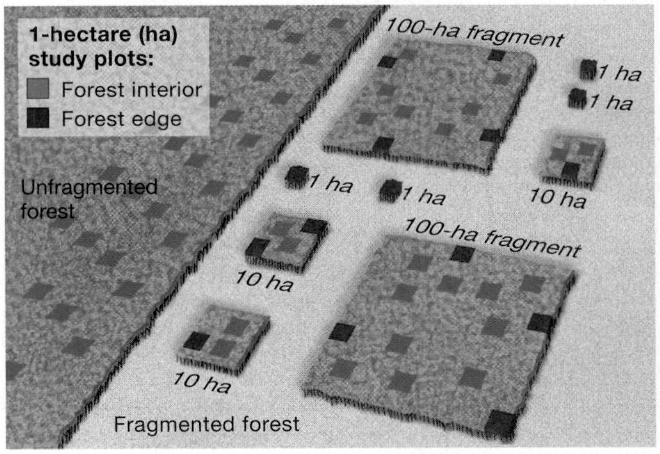

PREDICTION: Species diversity and biomass will decline in forest fragments compared with those of the forest interior, particularly along edges of fragments.

PREDICTION OF NULL HYPOTHESIS: Species diversity and biomass will be the same inside forest fragments and along edges as in the forest interior.

RESULTS:

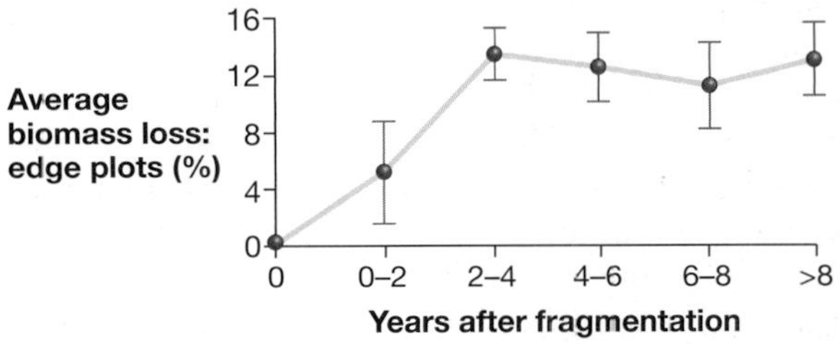

CONCLUSION: Biomass declines sharply along edges of forest fragments.

FIGURE 55.10 Experimental Evidence for Edge Effects in Fragmented Forests. Researchers tracked 66 study plots among four 1-hectare fragments, three 10-hectare fragments, and two 100-hectare fragments.

SOURCE: Laurance, W. F., P. Delamônica, S. G. Laurance, H. L. Vasconcelos, and T. E. Lovejoy. 2000. Rainforest fragmentation kills big trees. *Nature* 404: 836.

✔**QUESTION** The locations of the fragments and study plots were assigned at random. Why was this important?

The take-home message of this experiment and other research is clear: When habitats are fragmented, the quality *and* quantity of habitat decline drastically. In addition to losing over 8 million hectares of primary forest in South America, southeast Asia, and Africa each year to clear-cutting, we are losing large amounts of high-quality habitat throughout the world to fragmentation.

STOCHASTIC AND GENETIC PROBLEMS IN SMALL POPULATIONS In effect, habitat loss and habitat fragmentation are forcing many species into the metapopulation structure introduced in Chapter 52. The small, isolated populations that make up a metapopulation are much more likely than large populations to be wiped out for two reasons:

1. Catastrophic events such as storms, disease outbreaks, or fires exterminate small populations more readily than large populations. Biologists refer to episodes like this as stochastic—meaning, chance—events.
2. Small populations suffer from inbreeding depression and random loss of alleles due to genetic drift.

Why do genetic problems occur? Small populations become inbred simply because so few potential mates are available. Over time, virtually all individuals become related. In most populations, inbreeding leads to lowered fitness, due to deleterious recessive alleles being homozygous more frequently. Chapter 25 introduced this phenomenon, known as inbreeding depression.

In addition, genetic drift is much more pronounced in small populations than large populations (see Chapter 25). Because it leads to the random loss or fixation of alleles, drift reduces genetic variation in populations. This is a concern, because if the environment changes due to the evolution of a new disease or global warming, the population may lack alleles that confer high fitness in the new environment.

When biologists document fitness declines in small, isolated populations, one option is to increase gene flow experimentally by importing individuals of the same species from a different population. For example, **Figure 55.11** shows data on annual population growth rate (λ; see Chapter 52) in a small population of bighorn sheep isolated on a refuge in northwest Montana.

As you study the graph, note that from the time the population was founded in 1922 until 1985, average growth rate declined steadily. During this time, the average population size was about 42 individuals. After introducing 5 new individuals in 1985 and 10 additional sheep in 1990–1994, though, λ began increasing and population size has increased. In some cases at least, introducing new alleles may counteract the effects of genetic drift and inbreeding.

CLIMATE CHANGE As evidence that global warming is affecting the distribution and abundance of species, biologists have begun to ask how climate change might change extinction rates. Chapter 54 highlighted several issues:

- loss of coral reefs due to high-temperature-induced "bleaching";
- loss of habitat for species native to arctic and alpine tundras and pack-ice habitats;
- problems with trees and other slowly dispersing species being unable to track rapid changes in climate; and
- ocean acidification—due to increased carbon dioxide concentrations and formation of carbonic acid—inhibiting the ability of corals, crustaceans, and other animals to make calcium carbonate skeletons.

In addition, the melting of polar ice caps and glaciers, due to increased global temperature, is causing sea levels to rise. The most recent analyses project that sea levels will rise an additional 0.8 to 2.0 m by the year 2100. The consequences for species that rely on beach and nearshore habitats are unclear.

Taken together, overexploitation, habitat loss and fragmentation, and climate change pose daunting threats to biodiversity. Just how many extinctions are all of these problems projected to cause?

How Can Biologists Predict Future Extinction Rates?

If the most fundamental question about biodiversity is how many species are alive, then the most basic question in conservation biology is how rapidly species are going extinct. Both questions have been difficult to answer.

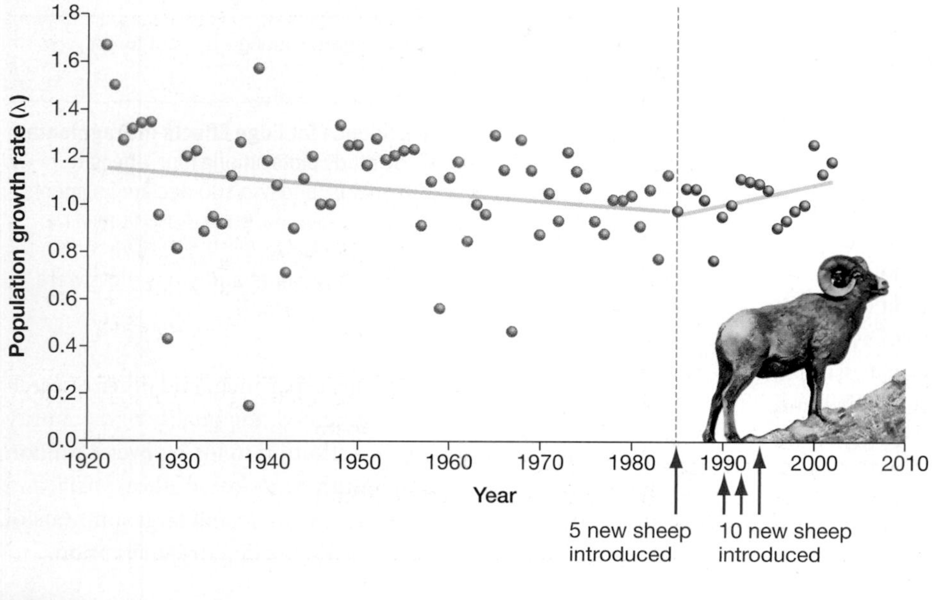

FIGURE 55.11 Experimental Evidence that Gene Flow Can Alleviate Genetic Problems in Small Populations. The *y*-axis plots the finite rate of increase (λ) in an isolated population of bighorn sheep, before and after unrelated individuals were introduced.

✔**EXERCISE** Draw a horizontal line indicating the value of λ where population size is constant (no growth).

Box 55.2 introduces an approach called Population Viability Analysis, or PVA, that researchers use to estimate the probability that a particular species will go extinct. To predict extinction rates for groups of species or over large areas, biologists use two different approaches. The first is based on direct counts of species that are known to have gone extinct recently or are in imminent danger of going extinct. The second is based on predicting the consequences of habitat loss and climate change, using existing data on the relationship between the size of habitats and the number of species present.

BOX 55.2 QUANTITATIVE METHODS: Population Viability Analysis

A population viability analysis, or PVA, is a model that estimates the likelihood that a population will avoid extinction for a given time period. In most cases, a PVA attempts to combine data on:

- age-specific survivorship and fecundity (see Chapter 52),
- geographic structure, and
- rate and severity of habitat disturbance.

Typically, a population is considered viable if the analysis predicts that it has a 95 percent probability of surviving for at least 100 years.

To get a feel for how a PVA works, consider an endangered marsupial called Leadbeater's possum (**Figure 55.12a**), which inhabits old-growth forests in southeastern Australia and relies on dead trees for nest sites. The goal of the PVA was to assess the effects that logging these habitats might have on the viability of the species.

To get started, researchers used life-table data—estimates of age-specific survival and average fecundity per female per year—to project how populations would grow if they were confined to scattered patches of habitat. The analysis allowed them to assess the impact of migration between patches—which can found new populations and introduce genetic diversity—and to simulate the effects of logging, fires, storms, and other disturbances.

Figure 55.12b illustrates the results. Each data point on the graphs represents the size of the overall population in a particular year. The four curves describe the changes in population size predicted to occur over time, based on different assumptions about the migration rates between groups of possums in patches of old-growth forest.

Note that when migration is high, the overall population size is predicted to stabilize at approximately 65 individuals. When no migration among fragments occurs, the final population size is predicted to be fewer than 20 individuals. The PVA showed that reducing the size and number of remaining old-growth fragments would reduce the possibility for migration and lead to rapid population declines.

Like all analyses based on simulations, though, a PVA makes many assumptions about future events and is only as accurate as the data entered into it. In particular, the usefulness of PVAs has been challenged because data on age-specific fecundity and survivorship and other basic demographic information are lacking or poorly documented in many endangered species.

A group of biologists defended the approach, however, by analyzing 21 long-term studies that have been completed on threatened populations of animals. To do their analysis, the researchers split each of the 21 data sets in two. Data from the first half of each study were used to run a PVA on each of the 21 species. After comparing the predictions of the PVAs to the actual data from the second half of each study, the researchers found that the correspondence was extremely close. This result strengthens confidence in PVA as a good predictive method for land managers.

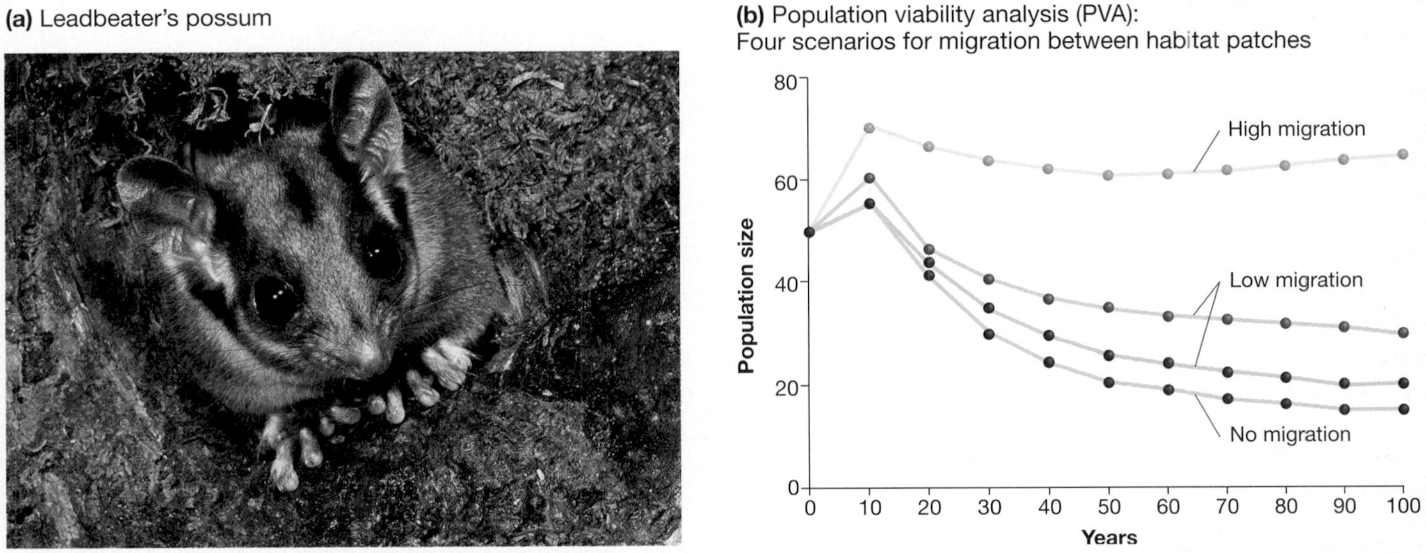

FIGURE 55.12 A PVA Can Predict the Consequences of Habitat Loss and Fragmentation. (a) Leadbeater's possum, a marsupial that breeds in old-growth forests of southeast Australia. **(b)** The migration rates modeled here range from no migration to 5 percent of the individuals migrating each year.

ESTIMATES BASED ON DIRECT COUNTS The best information on current extinction rates comes from studies on the best-studied of all major lineages on the tree of life: birds. Based on data in the fossil record, background extinctions in lineages like birds are estimated to occur at the rate of 1 extinction per million species alive per year. Of the approximately 10,000 bird species known, then, one species should go extinct every 100 years due to normal or background events.

Recent analyses have focused on species that were wiped out over the past 1000 years, coincident with the expansion of Polynesian people throughout the Pacific and Europeans throughout the world. These data suggest that birds have been going extinct at a rate 100 times background levels—meaning that a bird species has gone extinct every year, on average.

Although conservation programs targeted at birds have lowered this rate in the last few decades, estimates based on the numbers of species that are currently endangered suggest that the rate could reach 1000–1500 extinctions per million species alive per year by the end of this century. If this prediction holds, you may live long enough to see 10 or more species of bird go extinct each year.

Similar trends are occurring in listings of other well-studied groups.

- A recently published analysis of all 5487 known species of mammal living today—an effort that took five years and involved over 1700 experts from 130 countries—concluded that almost 25 percent are currently threatened with extinction.
- A comparison of comprehensive surveys of butterfly populations on the British Isles, conducted from 1970 to 1982 and repeated from 1995 to 1999, showed that 71 percent of the species are declining.

The message from these analyses is that the current rate of extinction is already many times higher than background or normal levels. Because human populations are growing so rapidly, it is poised to increase dramatically in the near future.

SPECIES–AREA RELATIONSHIPS A second strategy for predicting the future of Earth's biodiversity focuses on the consequences of the most urgent problem: habitat loss. Given reasonable projections of how much habitat will be lost over a given time period, biologists can estimate rates of extinction based on well-documented **species–area relationships**.

Figure 55.13 gives an example of a species–area curve—a graph that plots habitat area on the *x*-axis and the number of species present on the *y*-axis. In this case, the habitat areas are islands in the Bismarck Archipelago near New Guinea, in the South Pacific, and the species documented are birds. Each data point represents the number of species found on a different island. Note that the relationship between habitat area and the number of species present is linear when both axes are logarithmic. (For help interpreting logarithms, see **BioSkills 7** in Appendix A.)

The "log-linear" relationship turns out to be typical for other habitats and taxonomic groups as well. When biologists have plotted species–area relationships for plants, butterflies, mammals, or birds from islands or continental habitats around the globe, the relationship is consistently described by a function of the form $S = cA^z$, where:

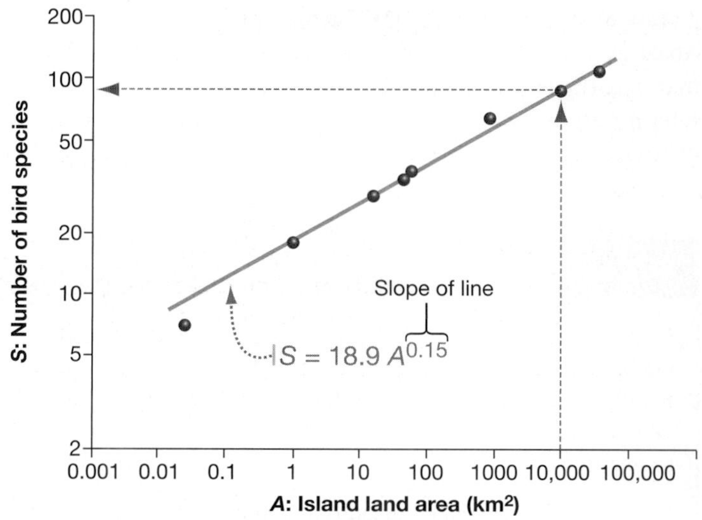

FIGURE 55.13 Species–Area Plots Quantify the Relationship between Species Richness and Habitat Area. The dashed arrows on this graph show the number of bird species that are expected to live on an island with an area of 10,000 km^2.

- S is the number of species.
- A represents the habitat area.
- The c term is a constant that scales the data. Its value is high in species-rich areas, such as coral reefs or tropical wet forests, and low in species-poor areas, such as arctic tundras.
- The exponent z represents the slope of the line on a log-log plot. Thus, z describes how rapidly species numbers change with area. Typically, z is about 0.25.

For example, the solid line drawn through the points in Figure 55.13, from the Bismark Archipelago, is described by the function $S = (18.9)A^{0.15}$.

To understand how biologists use species–area curves to project extinction rates, ask yourself: If sea level rise and deforestation wiped out 90 percent of the habitats in part of the Bismark Archipelago—for example, if A were reduced from 10,000 km^2 to 1000 km^2—what percentage of species would disappear?

According to the graph, the number of bird species present would drop from about 75 to about 53. Thus, the answer is roughly 30 percent. If z were higher, as it is in most regions of the world, the species loss would be much greater.

An international consortium of researchers recently studied a sample of habitats from around the globe and projected how the ranges of 1103 of the plants and animals present would change as a result of global warming. Then they used species–area curves to project how many species would be lost, as habitat is altered due to global warming. According to the "medium severity" projection for rates of global warming, about 25 percent of the species in their sample will be extinct or nearly extinct by the year 2050.

CHECK YOUR UNDERSTANDING

If you understand that . . .

- Until recently, most human-induced extinctions have occurred on islands and were due to overhunting or the impact of introduced species.
- Although invasive species and overexploitation remain a major threat to biodiversity, the majority of current problems are on continents and are due to habitat destruction and fragmentation—due to direct human activities or the effects of global warming.
- Current and projected extinction rates can be estimated from direct counts and from an analysis of species–area relationships.

✔ You should be able to . . .

1. Explain why fragmentation reduces habitat quality and leads to genetic problems.
2. Use the species–area relationship to compare the proportion of species lost when 90 percent of a 100,000 km^2 habitat is destroyed in a region where $c = 19$ and $z = 0.20$ versus the same-sized region where $c = 19$ and $z = 0.25$.

Answers are available in Appendix B.

55.4 Why Is Biodiversity Important?

One of the questions that biologists have to address about biodiversity is one that you may have heard from friends or relatives: Who cares? People are concerned about their grades or job or health or car or love life. Why should they worry about something as abstract as biodiversity?

The answer has two parts: one economic and one biological. Let's consider each in turn.

Economic Benefits of Biodiversity

Wild species have provided the raw material to fuel the development of human societies ever since the first large-scale, highly organized cultures began to develop about 10,000 years ago. People felled vast tracts of forest for building materials and fuel; selectively bred wild plants to yield food and fibers; domesticated animals to provide food, labor, and material goods; harvested the oceans for protein; and processed plants, animals, and fungi as sources of medicines.

The direct use of biodiversity continues today:

- *Seed banks* Agricultural scientists are preserving diverse strains of crop plants in **seed banks**—long-term storage facilities—and continue to use wild relatives of domesticated species in breeding programs aimed at improving crop traits. In addition, genetic engineering techniques are being used to transfer alleles from a diverse array of species into crop plants. In some cases these efforts have reduced dependence on pesticides, increased resistance to diseases and drought, and improved nutritional value and overall crop yields (see Chapter 19).
- *Pollination of crops* The production of almonds, apples, cherries, chocolate, alfalfa, and an array of other crops depends on the presence of wild pollinators. In the United States alone, insect-pollinated crops produce $40 billion worth of products annually.
- *Bioprospecting* Research programs collectively known as **bioprospecting** focus on assessing bacteria, archaea, plants, fungi, and frogs as novel sources of drugs or ingredients in consumer products. Bioprospecting has benefited from the recent explosion of genetic and phylogenetic information, because biologists can now search genomes from a wide array of species to find alleles with desired functions. For example, recent advances in biotechnology facilitated the development of a new painkiller from the paralyzing sting of tropical cone snails, and a blood anticoagulant from the saliva of vampire bats.
- *Bioremediation* Strategies for cleaning up oil spills, abandoned mines, and contaminated industrial sites are incorporating **bioremediation**—the use of bacteria, archaea, and plants to metabolize pollutants and render them harmless (see Chapter 28).
- *Ecotourism* Recreation based on visiting wild places, or **ecotourism**, is a major industry internationally and is growing rapidly. In South Africa, for example, the number of tourists visiting wildlife preserves increased from 454,428 in 1986 to over 6 million in 1999. In 2004, ecotourism grew three times faster than the tourism industry as a whole.
- *Flood control* High ecosystem diversity—particularly the presence of forests or grasslands on steep slopes and the presence of wetlands in low-lying areas—dramatically reduces flood damage and the danger posed by mudslides.

In addition, the benefits of biodiversity extend beyond the direct use of diverse genes and species by humans to include **ecosystem services**—processes that increase the quality of the abiotic environment.

Recall from Chapter 10 that green plants and other photosynthetic organisms produce the oxygen we breathe, and from Chapter 30 that the presence of land plants builds soil, reduces soil erosion, moderates local temperature and wind conditions, and increases the volume of water retained in lakes, streams, and soils. In 1997 a team of economists led by Robert Costanza estimated the value of these services at US$33 trillion—almost twice the gross national product of all nations combined.

The air you breathe, the water you drink, and the soils that grow food you eat are services that ecosystems provide. The quality of your life is tied to the quality of those ecosystems.

Biological Benefits of Biodiversity

The direct and indirect economic benefits of biodiversity are impressive. But what about the biological value? Do biological communities and ecosystems function better when species diversity is high?

The effort to answer this question began in the mid-1980s. Since that time, it has blossomed into one of the most active research frontiers in community ecology and conservation biology.

BIODIVERSITY INCREASES PRODUCTIVITY **Figure 55.14** summarizes a classic experiment on how species richness affects one of the most basic aspects of an ecosystem function: the production of biomass. As Chapter 54 emphasized, net primary productivity is important because it is the source of chemical energy used by species throughout a food web.

As the "Experimental setup" portion of the figure shows, David Tilman and colleagues classified 32 grassland plant species into five functional categories. The functional groups differ by the timing of their growing season and by whether they allocate most of their resources to manufacturing woody stems or seeds. The researchers planted plots with a mixture of between 0 and 32 randomly chosen species, representing from 0 to 5 functional categories, and measured the amount of aboveground biomass produced in each.

The results, graphed in the "Results" section of Figure 55.14, indicate that both the number and type of species present had impor-

EXPERIMENT

QUESTION: Do high species richness and high functional diversity of species increase aspects of ecosystem function such as net primary productivity (NPP)?

HYPOTHESIS: NPP increases with increasing species richness and with increasing functional diversity of species.

NULL HYPOTHESIS:

EXPERIMENTAL SETUP:

Plant a total of 289 experimental plots, each with up to 32 species and up to 5 functional groups:

Cool-season grasses: Grow in spring | **Warm-season grasses:** Grow in summer | **Legumes:** Fix nitrogen | **Woody plants:** Trees, shrubs | **Forbs:** Lots of seeds

Examples of experimental plots:

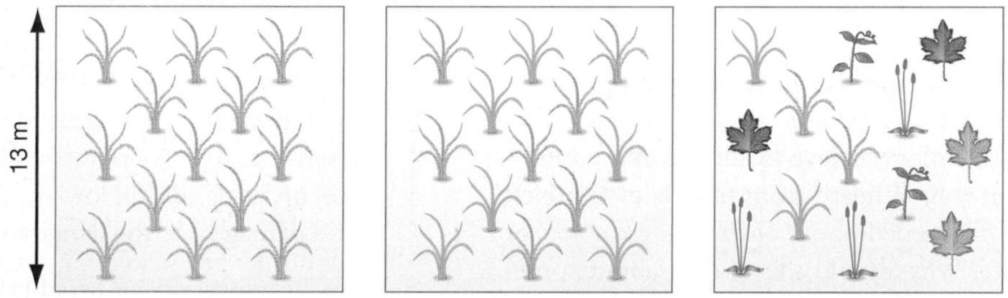

1 species, 1 functional group | 2 species, 1 functional group | 6 species, 4 functional groups

PREDICTION:

PREDICTION OF NULL HYPOTHESIS:

RESULTS:

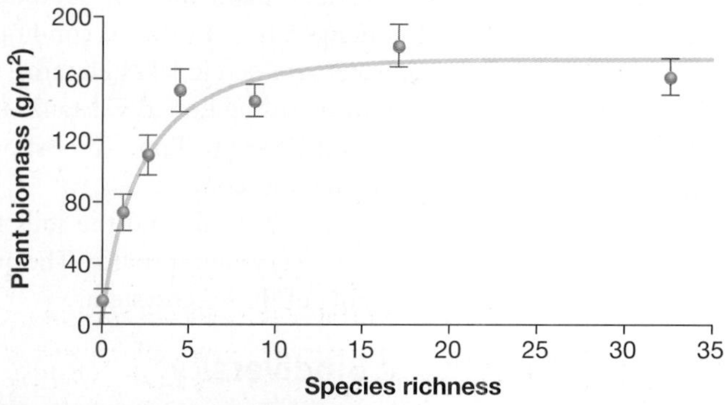

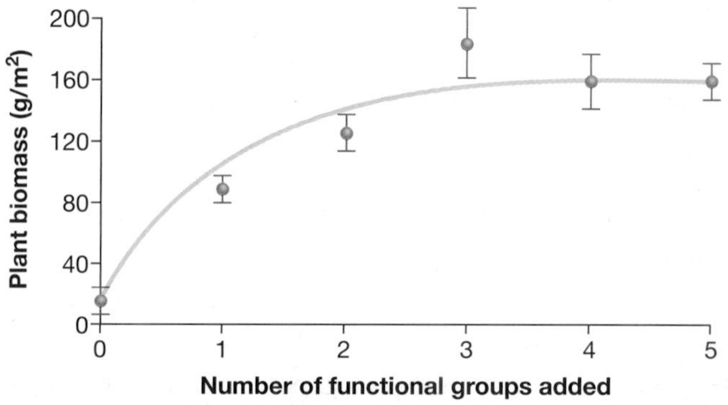

CONCLUSION: In this plant community, NPP increases with increasing species richness and increasing functional diversity of plants, at least up to a point.

FIGURE 55.14 Evidence that the Productivity of Ecosystems Depends on the Number and Types of Species Present.

SOURCE: Tilman, D., J. Knops, D. Wedin, P. Reich, M. Ritchie, and E. Siemann. 1997. The influence of functional diversity and composition on ecosystem processes. *Science* 277: 1300–1302.

✔**EXERCISE** Write in the null hypothesis and both sets of predictions.

tant effects on biomass production. Experimental plots with more species and with a wider diversity of functional groups were more productive.

Note that total biomass leveled off as species richness and functional diversity increased, however. This observation suggests that increasing species diversity improves ecosystem function only up to a point. This pattern suggests that at least some species in ecosystems are redundant in terms of their ability to contribute to productivity.

Follow-up experiments in an array of ecosystems not only supported the conclusion that species richness has a positive impact on NPP, but showed that several causal mechanisms may be at work.

- *Resource use efficiency* Diverse assemblages of plant species make more efficient use of the sunlight, water, and nutrients available and thus lead to greater overall productivity. For example, some prairie plants extract water near the soil surface, while others use water available a meter or more below the surface. When species diversity is high, more overall water is used and more photosynthesis can occur.
- *Facilitation* Certain species or functional groups facilitate the growth of other species by providing them with nutrients, partial shade, or other benefits. In prairies, the presence of onion-family plants may discourage herbivores, or decaying roots from nitrogen-fixing species may fertilize other species.
- *Sampling effects* In many habitats it is common to observe that one or two species are extremely productive. If the number of species in a study plot is low, it is likely that the "big producer" will be missing and NPP will be low. But if the number of species in a study plot is high, then it is likely the big producer is present and NPP will be high. Simply due to sampling, then, high-species plots will tend to outproduce low-species plots.

The three mechanisms listed are not mutually exclusive—several can operate at the same time. Further, long-term experiments have shown that the reason for the pattern can change over time. As plants mature, resource use efficiency and facilitation may become more important than sampling effects.

DOES BIODIVERSITY LEAD TO STABILITY? When biologists refer to the stability of a community, they mean its ability to:

1. withstand a disturbance without changing,
2. recover to former levels of productivity or species richness after a disturbance, and
3. maintain productivity and other aspects of ecosystem function as conditions change over time.

Resistance is a measure of how much a community is affected by a disturbance. **Resilience** is a measure of how quickly a community recovers following a disturbance.

To test the hypothesis that diversity increases resistance and resilience, the team that did the work highlighted in Figure 55.14 followed up on a natural experiment. A **natural experiment** occurs when comparison groups are created by an unplanned, unmanipulated change in conditions.

In 1987–1988 a severe drought hit their study site. When the drought ended, the team asked whether species richness affected the response to the disturbance. The results are shown in **Figure 55.15**. Note that the graph's *y*-axis documents the change in total biomass that occurred from the year prior to the drought until the height of the drought. This quantity reflects how resistant the community is to disturbance. A completely resistant community would show no change. As predicted, drought resistance appeared to be higher in more diverse communities than in less diverse ones.

EXPERIMENT

QUESTION: Are more species-rich communities more stable than less-diverse communities?

HYPOTHESIS: Resistance to disturbance should increase with increasing species richness.

NULL HYPOTHESIS: There is no relationship between species richness and resistance.

EXPERIMENTAL SETUP:

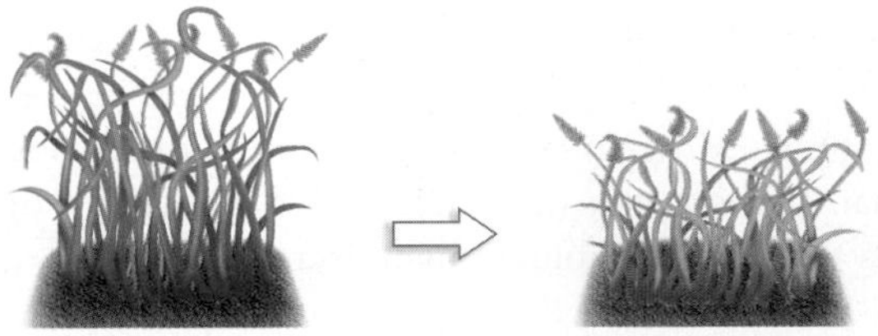

Compare biomass of experimental plots before drought and at peak of drought. (This was a natural experiment—severe drought just happened to occur during study.)

PREDICTION: Plots that were more species rich before the drought will be more resistant to change.

PREDICTION OF NULL HYPOTHESIS: All the plots will have similar resistance regardless of species richness before the drought.

RESULTS:

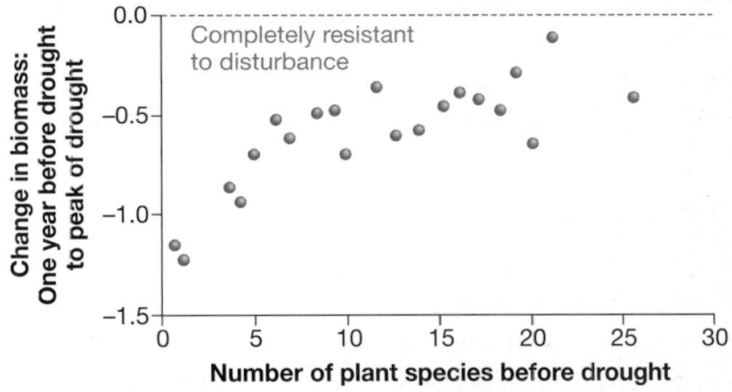

CONCLUSION: Resistance to disturbance increases with increasing species richness.

FIGURE 55.15 Evidence that Species-Rich Communities Resist Disturbance.

SOURCE: Tilman, D. and J. A. Downing. 1994. Biodiversity and stability in grasslands. *Nature* 367: 363–365.

✔**QUESTION** The 0.0 value on the graph's vertical axis is labeled "Completely resistant to disturbance." Why?

In addition, the group analyzed the change in biomass in each plot four years after the drought versus biomass prior to the drought. This analysis focused on how resilient the community is. A completely resilient community would recover quickly from the disturbance and have the same biomass at both times. The data indicated that most plots containing five or fewer species showed a significant lowering of biomass after the disturbance, indicating that they had not recovered. But in all of the plots that contained more than five species, biomass after the drought was the same as biomass prior to the disturbance. This is strong evidence that species-rich areas recover from disturbance faster than species-poor areas.

More recent experiments on California grasslands suggest that species-rich ecosystems may also be less prone to invasion by exotic species. The hypothesis here is that species-rich communities use resources more efficiently—leaving less room for invasives. Taken together, these results lead biologists to be increasingly confident that species richness has a strongly positive effect on how ecosystems function.

In North American grasslands, at least, communities that are more diverse appear to be more productive, more resistant to disturbance and invasion, and more resilient than communities that are less diverse. Increased species richness increases the services provided by ecosystems. By implication, biologists can infer that if ecosystems are simplified by extinctions, then productivity and other attributes might decrease.

An Ethical Dimension?

The data reviewed in this section suggest that there are sound economic and biological reasons to preserve species. People live better and ecosystems work better when biodiversity is high.

But in addition, an array of biologists, philosophers, and religious leaders argue that humans have an ethical obligation to preserve species and ecosystems. Their position is that organisms have intrinsic worth, and that humans diminish the world by extinguishing species and destroying ecosystems.

If this is so, then extinction is a moral issue as well as a biological and economic problem. One of the most important reasons to preserve biodiversity may simply be that it is the right thing to do.

CHECK YOUR UNDERSTANDING

If you understand that . . .

- Experimental plots with higher species richness tend to have higher productivity and show increased stability in response to disturbance or changed abiotic conditions.

✓ You should be able to . . .

Design an experiment to test the hypothesis that species-rich areas are more effective at building soil, retaining soil nutrients, and minimizing soil erosion than are species-poor areas.

Answers are available in Appendix B.

55.5 Preserving Biodiversity

Extinction is irreversible. The only solution to the biodiversity crisis is to prevent the loss of alleles, species, and ecosystems.

Solving any global problem requires a common goal to be defined. In the case of preserving biodiversity, the objective is to sustain diverse communities in natural landscapes while supporting the extraction of resources required to maintain the health and well-being of the human population.

What's working? And what can you do to help?

Designing Effective Protected Areas

When biologists recognized the threat to biodiversity, they joined with government agencies, economists, community leaders, private landowners, and others to set aside protected areas. To assess the progress of this effort, the World Conservation Union sponsors the World Parks Congress every decade.

- In 1992 the Congress met in Caracas, Venezuela, and established a goal of setting aside 10 percent of Earth's land surface in protected areas.
- In 2003 the Congress met in Durban, South Africa, and announced that this goal had been surpassed: Protected areas covered 11.5 percent of Earth's terrestrial surface.

Will these protected areas be enough to avert a mass extinction?

GAP ANALYSIS PROGRAM Researchers are using a geographic approach called the Gap Analysis Program (GAP) to assess the effectiveness of the current system of protected areas. A GAP analysis tries to identify gaps between geographic areas that are particularly rich in biodiversity and areas that are actually managed for the preservation of biodiversity.

One GAP analysis combined data sets on the distribution of mammals, birds, amphibians, and freshwater turtles with a map of world protected areas. The analysis revealed that many species' ranges occur completely outside any protected areas. It also pinpointed regions in Mexico, Madagascar, and elsewhere where the gap between species' ranges and protected areas is particularly high.

To date, most GAP analyses suggest that, because relatively few species richness hotspots are included in existing protected areas, the 11.5 percent of Earth's surface area that is now being managed for biodiversity will not be enough to conserve many species. Efforts to fill these gaps are now focused on preserving species richness hotspots and centers of endemism like those illustrated in Figure 55.3.

In addition to working on where reserves are set up, biologists are focusing more attention on how reserves are designed. For example, it is often impossible to preserve large areas of high-quality habitat. In most cases, biologists have to accept the fact that habitats are fragmented and species exist as metapopulations. Given this reality, what is the best way to prevent small, isolated populations from going extinct?

WILDLIFE CORRIDORS The leading hypothesis in reserve design is that strips of undeveloped habitat, called **wildlife corridors**, should connect preserves that would otherwise be isolated. In some cases wildlife corridors are as simple as a walkway under a major highway, so that animals could move from one side to the other without being killed.

By facilitating the movement of individuals, the goals of corridors are to:

1. allow areas to be recolonized if a species was lost, and
2. encourage gene flow that introduces new alleles—counteracting the deleterious effects of genetic drift and inbreeding.

Do corridors actually work? To answer this question, a research team established a series of restored natural areas in the middle of a large, species-poor pine plantation. This was the equivalent of restoring patches of grassland habitat in the middle of an enormous cornfield. Some of the restoration sites were connected by corridors, while others were isolated. By monitoring the composition of species inside each habitat patch over time, the group was able to show that connected patches are steadily gaining more species over time compared to unconnected patches (**Figure 55.16**). This is strong evidence that wildlife corridors can increase overall species richness in a metapopulation.

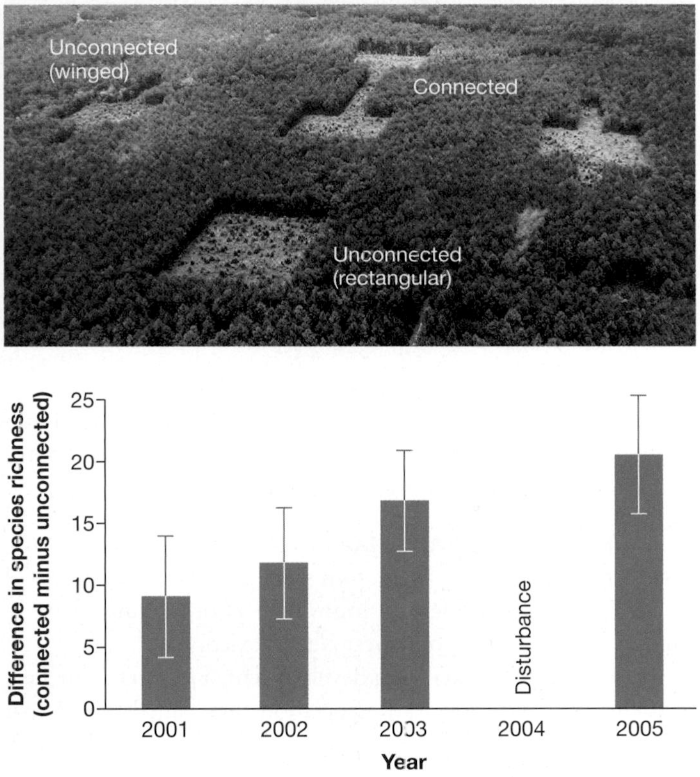

FIGURE 55.16 Experimental Evidence that Wildlife Corridors Work. Species richness could not be assessed in 2004 because the study plots were burned as part of a program to restore natural habitats.

✔**EXERCISE** Draw and label bars showing what this graph would look like if corridors had no effect on species richness.

Beyond Protected Areas: A Comprehensive Approach

Given the magnitude of the current extinction rate, the continued growth in human population size and resource use, and the difficulty of setting up enough preserved areas in the world's conservation hotspots, conservation biologists are advocating a multipronged strategy. Preserving land will not be enough. At least four other strategies need to be an important part of a lasting solution to the biodiversity crisis.

SUSTAINABLE DEVELOPMENT In almost every case, the underlying causes of the biodiversity crisis are political and economic pressures that encourage short-term overexploitation of land and other resources and discourage **sustainability**—the managed use of resources at a rate only as fast as the rate at which they are replaced.

You may be familiar with efforts to develop renewable sources of energy, promote agricultural and forestry practices that maintain productivity without the addition of synthetic fertilizers or groundwater, and recycle metals and other raw materials. In essence, the challenge is to create ways for humans to live off the resources that are being produced continuously on Earth, rather than mining resources that have been stored for centuries or millennia.

The simple fact is that many of these stored resources—including petroleum, groundwater, and topsoils—are becoming exhausted.

The need to promote sustainable development has inspired a new field of endeavor, called sustainability science, that blends economics, political and social science, engineering, and applied biological sciences like forestry and agriculture.

STABILIZING HUMAN POPULATION SIZE AND RESOURCE USE Most biologists recognize that it may be impossible to preserve high species diversity and high-functioning ecosystems if two things happen:

- the human population grows to 13 billion or more—from its current 6.8 billion—over the next century, as projected by the "high-fertility" scenario introduced in Chapter 52; and
- individuals in the industrialized nations continue to use fossil fuels and other resources at the current rate (see Chapter 54).

It is not clear, however, that any human population in history has ever voluntarily and consciously limited its growth and resource use. As a species, we are breaking new ground.

Recent drops in fertility—documented in Chapter 52—and recent efforts to promote sustainable development are grounds for optimism. Efforts to limit family size and resource use are critical to solving the biodiversity crisis.

EX SITU CONSERVATION It may take decades for human population size and resource use to stabilize and for sustainable development practices to become widespread. In the meantime, thousands of species may be lost.

One approach to coping with this situation is referred to as **ex situ conservation**—the preservation of species in zoos, aquaria, wildlife ranches, seed banks, or other artificial settings. Translated literally, ex situ means "out of place." The name was inspired by a contrast with in situ or "in place" conservation—the establishment of protected areas where populations can be maintained in their natural habitat.

Some species—including Pere David's deer, Przewalski's horse, and the Arabian oryx—are extinct in the wild and exist only in captivity. In other cases, captive breeding and reintroduction programs have succeeded in creating wild populations of endangered species. Chapter 52 highlighted the successful effort to establish new flocks of whooping cranes, using eggs and young produced in zoos and other breeding centers. The peregrine falcon, California condor, and several other species have been rescued from the brink of extinction by ex situ conservation and reintroduction to the wild.

ECOSYSTEM RESTORATION In many areas of the world, ecosystems are already heavily degraded or lost. The state of Illinois, for example, was estimated to contain 22 million acres of prairie when European settlers arrived; today only about 2000 acres of this biome remain. When natural areas are badly degraded or gone, biologists turn to restoration.

Thousands of large-scale and small-scale ecosystem restoration and reforestation projects are now occurring around the globe. In 2004, Wangari Maathai won the Nobel Peace Prize to honor her work founding the Green Belt Movement—a reforestation organization that has now sponsored the planting of over 1 billion trees, beginning in Kenya and expanding internationally (**Figure 55.17**).

In the tropics, one of the most successful ecological restoration efforts focused on the seasonally dry forest at the Area de Conservación Guanacaste in northwestern Costa Rica. The primary task facing conservation area staff there was to stop human-caused fires. Their efforts since the mid-1980s have succeeded in transforming a vast swath of marginally productive ranching land into an increasingly popular ecotourist destination and a water source for neighboring farms and ranches (**Figure 55.18**).

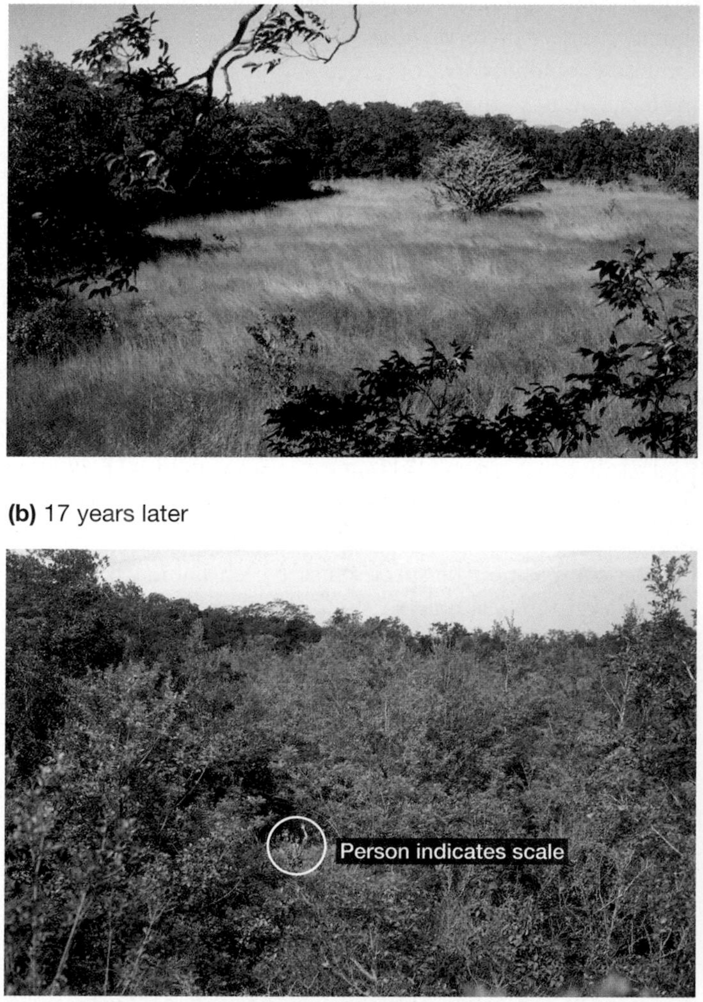

FIGURE 55.18 Using Fire Prevention to Restore Forest Ecosystems in the Guanacaste Reserve. Before and after views of the same region, 17 years apart, in the Area de Conservación Guanacaste, Costa Rica. To appreciate how quickly the trees grew, notice the person standing near the middle of the photo in (b).

FIGURE 55.17 The Greenbelt Movement Is a Global Reforestation Effort. These women are tending a tree nursery in Kenya.

A CASE HISTORY: THE MALPAI BORDERLANDS What does a successful, multipronged conservation strategy look like in practice? As an example, consider an innovative effort taking place in southeast Arizona and southwest New Mexico.

In response to threats from development, a group of ranchers banded together to form an association they called the Malpai Borderlands group. To preserve biodiversity and their livelihood, the ranchers:

- protected 42,000 acres of their land with conservation easements, which are legal agreements that are placed on the titles to the property and prevent the land from being developed for vacation homes or other uses incompatible with ranching and wildlife preservation;

FIGURE 55.19 Using Fire to Restore Grassland Ecosystems in the Malpai Borderlands. The same region, in the Malpai Borderlands of the Southwestern United States, before and after a prescribed burn. The burn occurred in a wet year and resulted in a dramatic increase in native grasses.

- set up cooperative "grassbanks," or pastureland that is made available to ranchers whose own grazing areas are suffering from short-term drought and need relief from continued grazing; and
- reintroduced fire to the area via prescribed (intentional) burns—a restoration tool that removes encroaching woody shrubs and encourages the growth of native grasses (**Figure 55.19**).

The ranchers have also been cooperating with each other and with biologists to monitor or reintroduce native vegetation and animals such as the thick-billed parrot, bighorn sheep, Mexican jaguar, Yaqui chub, and Chiricahua leopard frog to the region.

Today, the Malpai Borderlands group is considered a model of innovative action by private individuals, in cooperation with governmental agencies and nongovernmental organizations (NGOs). The project benefits local people as well as biodiversity.

It can be done. Your generation is facing the most serious global environmental crisis in the history of our species. The decisions you make, ranging from how many children you have to how much you drive a car, will have far-reaching consequences. Change happens one person at a time.

As someone with a background in biology, you have the intellectual tools to make an important, positive impact. In addition to studying life, we have an obligation to help save it.

CHAPTER 55 REVIEW

For media, go to the study area at www.masteringbiology.com

Summary of Key Concepts

Biodiversity is quantified at the level of allelic diversity, species diversity, and ecosystem diversity.

- Genetic diversity is well characterized within some species, but biologists are only beginning to use genome sequencing techniques to explore the extent of genetic diversity in ecosystems.
- Research on the total number of species alive today is also at a preliminary stage, with estimates ranging from 10 to 100 million.
- To date, the message from efforts to characterize biodiversity is that it is much more extensive than expected and that a great deal remains to be learned.

✔You should be able to explain why environmental sequencing is an effective way to catalog biodiversity at the genetic level.

According to recent analyses, the sixth mass extinction in the history of life is occurring due to habitat loss, overexploitation, and global climate change.

- Historically, most human-caused extinctions have occurred on islands because of direct exploitation or the introduction of exotic herbivores and predators.
- Habitat destruction is currently the leading cause of extinctions, with climate change projected to be an important new mechanism. Most extinctions are now occurring on continents.
- Experiments in the Brazilian Amazon have shown that habitat loss leads not only to a rapid decline in biodiversity but also to a decline in the quality of the remaining habitats due to fragmentation.
- To estimate how many species will go extinct in the near future, biologists combine data on current rates of habitat loss—usually estimated from satellite images taken over time—with data on the average number of species found in habitats of a given size.
- According to data on the rates of extinction in birds and other well-studied groups, it is likely that 60 percent of all species will be wiped out within 500 years.

✔You should be able to explain why small, geographically isolated populations are at high risk of extinction.

MB Web Activity Habitat Fragmentation

Humans depend on biodiversity for the products that wild species provide and for ecosystem services that protect the quality of the abiotic environment.

- Species diversity is important for maintaining the productivity of natural ecosystems and their ability to build and hold soil, moderate local climates, retain and cycle nutrients, retain surface water and recharge groundwater, prevent flooding, and produce oxygen.
- Humans gain direct economic benefits from fishing, forestry, agriculture, tourism, and other activities that depend on biodiversity.
- At the ecosystem level, experiments have shown that high species richness increases aspects of ecosystem function such as productivity, resistance to disturbance, and ability to recover from disturbance.

✔You should be able to explain why high species richness can lead to high productivity, in terms of the efficiency of resource use.

Solutions to the biodiversity crisis include protecting key habitats, lowering human population growth and resource use, restoring ecosystems, mitigating climate change, and supporting sustainable development.

- One conservation priority is to preserve habitats in biodiversity and conservation hotspots, and link preserves with habitat corridors that facilitate gene flow.
- The major threats to biodiversity—habitat destruction and climate change—will not lessen until the human population stabilizes and resource use becomes sustainable.
- Ecosystem restoration efforts are having a positive impact on biodiversity around the world.

✔You should be able to explain why the 11 percent of land area protected thus far will not be enough to avert a mass extinction.

Questions

✔TEST YOUR KNOWLEDGE

Answers are available in Appendix B

1. What does a species–area plot show?
   a. the overall distribution, or area, occupied by a species
   b. the relationship between the body size of a species and the amount of territory or home range it requires
   c. the number of species found, on average, in tropical versus northern areas
   d. the number of species found, on average, in a habitat of a given size
2. What does a GAP analysis do?
   a. It compares the current distributions of species with the locations of preserved habitats.
   b. It quantifies *g*ross *a*boveground *p*roductivity.
   c. It uses data on the rates at which lists of threatened species are growing to project the rate of future extinctions.
   d. It uses genome sequencing techniques to quantify genetic (allelic) diversity in an ecosystem.
3. What is resilience?
   a. resistance to change during a disturbance
   b. the ability to recover from a disturbance
   c. production of aboveground biomass
   d. total biomass production (above- and belowground)
4. What is a biodiversity "hotspot"?
   a. an area where an all-taxon survey is under way
   b. an area where an environmental sequencing study has been completed
   c. a habitat with high NPP
   d. an area with high species richness
5. Why do small populations become inbred?
   a. They are often part of a metapopulation structure.
   b. Genetic drift becomes a prominent evolutionary force.
   c. Over time, all individuals become increasingly related.
   d. Natural selection does not operate efficiently in small populations.
6. What is the primary cause of endangerment in marine environments?
   a. overexploitation
   b. pollution
   c. global warming
   d. invasive species

✔TEST YOUR UNDERSTANDING

Answers are available in Appendix B

1. How has the nature of the threats to biodiversity changed over the past 400 years?
2. What is "sustainable" about sustainable development?
3. Biologists claim that the all-taxa survey now under way at the Great Smoky Mountains National Park in the United States will improve their ability to estimate the total number of species living today. Discuss the benefits and limitations that this data set will provide in understanding the extent of global biodiversity.
4. How are species–area curves used to relate rates of habitat destruction to projected extinction rates?
5. What evidence supports the hypotheses that species richness increases ecosystem functions, such as productivity, resistance to disturbance, and resilience?
6. Explain why the construction of wildlife corridors can help maintain biodiversity in a fragmented landscape.

✓ APPLYING CONCEPTS TO NEW SITUATIONS

Answers are available in Appendix B

1. What are ecosystem services? Does anyone own them or pay for them—that is, can you make money off them? How does this affect efforts to protect ecosystems?
2. You are helping design a series of reserves in a tropical country. List the steps you would recommend for gathering data and creating a plan that would protect a large amount of biodiversity in a small amount of land.
3. The maps to the right chronicle the loss of old-growth forest (>200 years old) that occurred in Warwickshire, England, and in the United States. In your opinion, under what conditions is it ethical for conservationists who live in these countries to lobby government officials in Brazil, Indonesia, and other tropical countries to slow the rate of loss of old-growth forest?
4. Make a list of characteristics that would render a species particularly vulnerable to extinction by humans. Make a list of characteristics that would render a species particularly resistant to pressure from humans. Try to think of an example of each type of species.

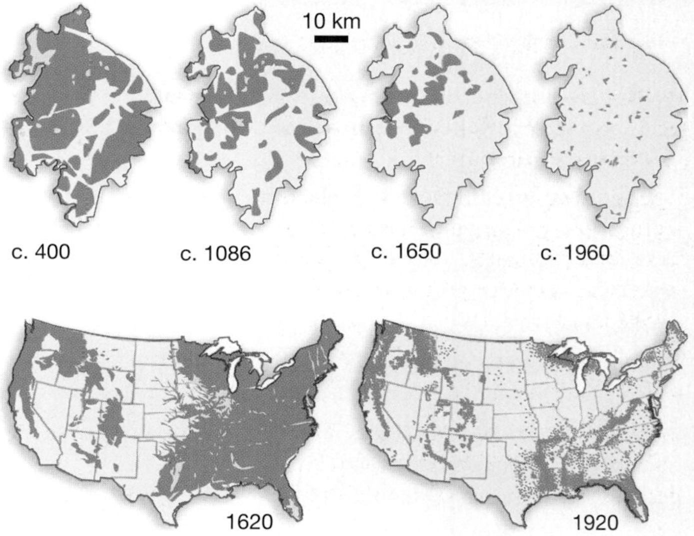

This chapter is part of The Big Picture. See how on pages 1126–1127.

The Big Picture

Ecology is the study of (1) how organisms interact with the biotic and abiotic components of their environment, and (2) the distribution and abundance of organisms that result from that interaction.

Human activities are causing radical changes in population sizes, climate, soils, water, and atmosphere. As a result, virtually every box and arrow in this concept map is being affected.

In addition to documenting the changes that are occurring, biologists are educating the public and policy makers about the value of biodiversity and healthy ecosystems, and researching the most effective ways to preserve species and keep ecosystems functioning normally.

Note that each box in the concept map indicates the chapter or section where you can go for review. Also, be sure to do the blue exercises in the Check Your Understanding box below.

CHECK YOUR UNDERSTANDING

If you understand the big picture . . .

✔ You should be able to . . .

1. Add a "Habitat loss" box that is connected to the "Species" box. Explain how this connection affects community structure, species richness, and primary productivity.
2. Add a "Global warming" box that is connected to the "CO_2" box. Explain how this connection affects community structure, species richness, and primary productivity.
3. Fill in the blue ovals with appropriate linking verbs or phrases.

Answers are available in Appendix B.

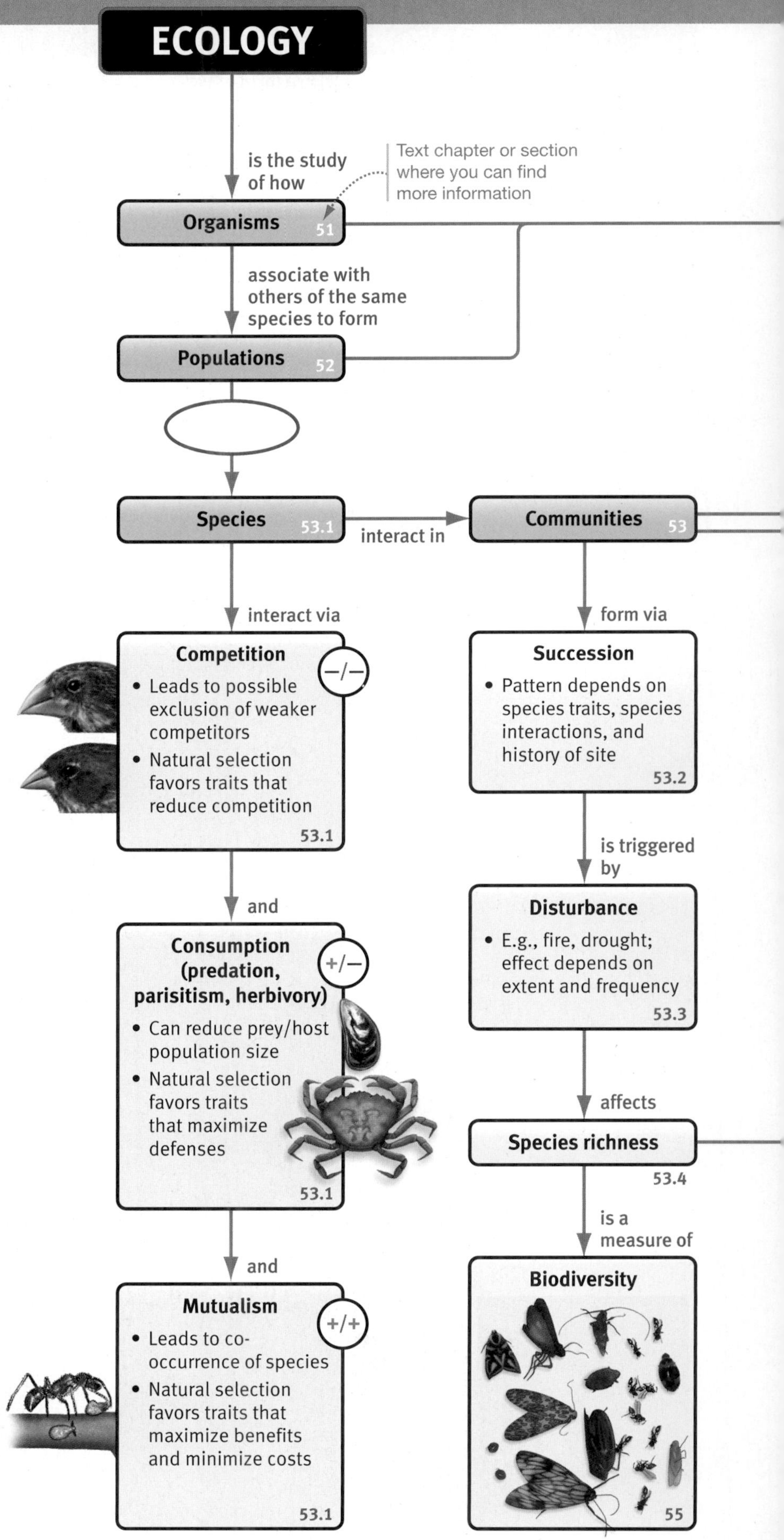

ECOLOGY
is the study of how
Text chapter or section where you can find more information
Organisms 51
associate with others of the same species to form
Populations 52
Species 53.1
interact in
Communities 53
interact via
Competition
–/–
• Leads to possible exclusion of weaker competitors
• Natural selection favors traits that reduce competition
53.1
and
Consumption (predation, parisitism, herbivory)
+/–
• Can reduce prey/host population size
• Natural selection favors traits that maximize defenses
53.1
and
Mutualism
+/+
• Leads to co-occurrence of species
• Natural selection favors traits that maximize benefits and minimize costs
53.1
form via
Succession
• Pattern depends on species traits, species interactions, and history of site
53.2
is triggered by
Disturbance
• E.g., fire, drought; effect depends on extent and frequency
53.3
affects
Species richness
53.4
is a measure of
Biodiversity
55

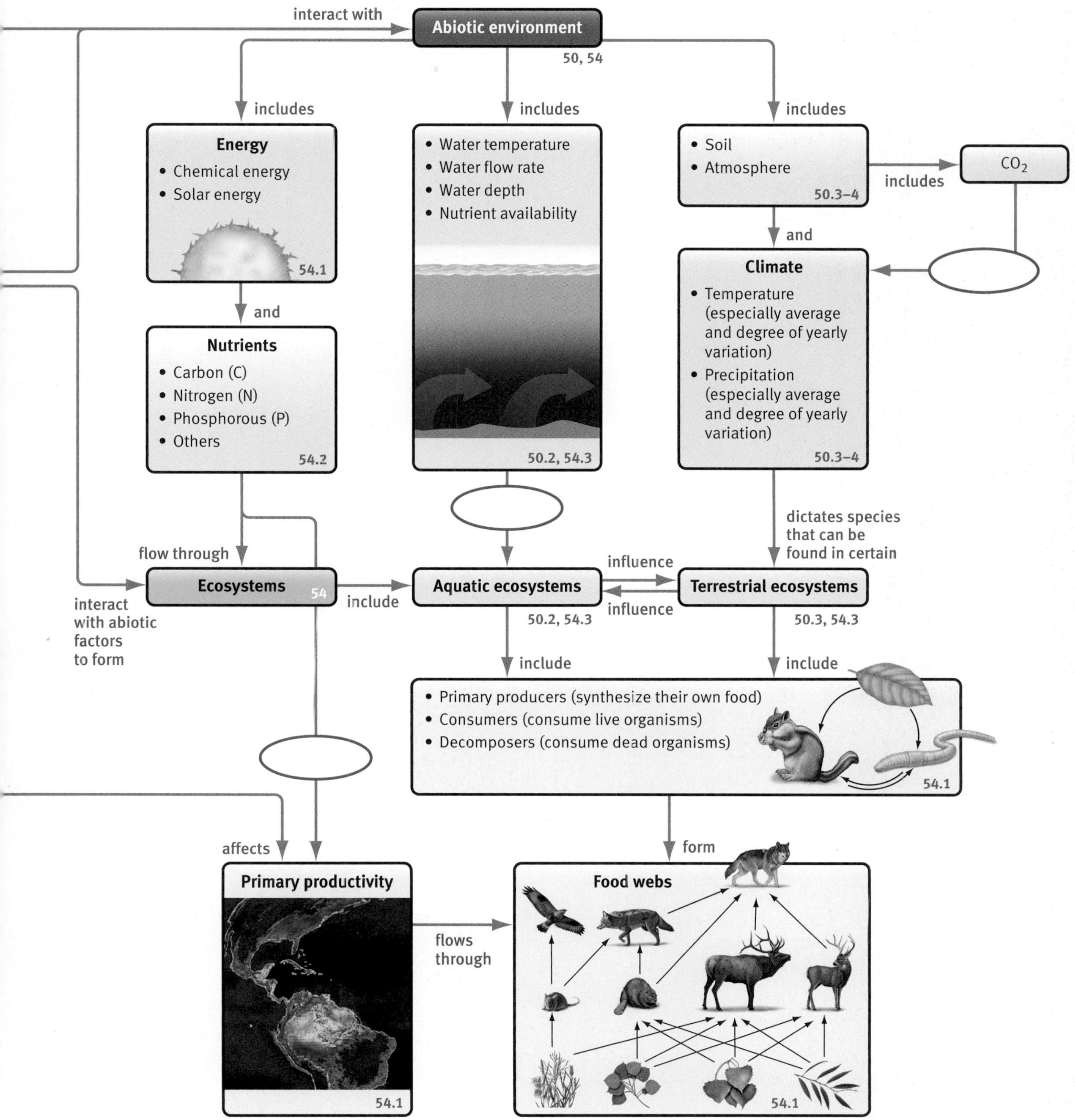
interact with
Abiotic environment
50, 54
includes
includes
includes
Energy
• Chemical energy
• Solar energy
54.1
and
Nutrients
• Carbon (C)
• Nitrogen (N)
• Phosphorous (P)
• Others
54.2
• Water temperature
• Water flow rate
• Water depth
• Nutrient availability
50.2, 54.3
• Soil
• Atmosphere
50.3–4
includes
CO_2
and
Climate
• Temperature (especially average and degree of yearly variation)
• Precipitation (especially average and degree of yearly variation)
50.3–4
dictates species that can be found in certain
flow through
Ecosystems
54
interact with abiotic factors to form
include
Aquatic ecosystems
50.2, 54.3
influence
influence
Terrestrial ecosystems
50.3, 54.3
include
include
• Primary producers (synthesize their own food)
• Consumers (consume live organisms)
• Decomposers (consume dead organisms)
54.1
affects
form
Primary productivity
54.1
flows through
Food webs
54.1

APPENDIX A BioSkills

BIOSKILLS 1

THE METRIC SYSTEM

The metric system is the system of units of measure used in every country of the world but three (Liberia, Myanmar, and the United States). It is also the basis of the SI system used in scientific publications.

The popularity of the metric system is based on its consistency and ease of use. These attributes, in turn, arise from the system's use of the base 10. For example, each unit of length in the system is related to all other measures of length in the system by a multiple of 10. There are 10 millimeters in a centimeter; 100 centimeters in a meter; 1000 meters in a kilometer.

Measures of length in the English system, in contrast, do not relate to each other in a regular way. Inches are routinely divided into 16ths; there are 12 inches in a foot; 3 feet in a yard; 5280 feet (or 1760 yards) in a mile.

TABLE B1.2 **Prefixes Used in the Metric System**

Prefix	Abbreviation	Definition
nano-	n	$0.000\,000\,001 = 10^{-9}$
micro-	μ	$0.000\,001 = 10^{-6}$
milli-	m	$0.001 = 10^{-3}$
centi-	c	$0.01 = 10^{-2}$
deci-	d	$0.1 = 10^{-1}$
—	—	$1 = 10^{0}$
kilo-	k	$1000 = 10^{3}$
mega-	M	$1\,000\,000 = 10^{6}$
giga-	G	$1\,000\,000\,000 = 10^{9}$

TABLE B1.1 **Metric System Units and Conversions**

Measurement	Unit of Measurement and Abbreviation	Metric System Equivalent	Converting Metric Units to English Units
Length	kilometer (km)	$1 \text{ km} = 1000 \text{ m} = 10^{3} \text{ m}$	1 km = 0.62 mile
	meter (m)	1 m = 100 cm	1 m = 1.09 yards = 3.28 feet = 39.37 inches
	centimeter (cm)	$1 \text{ cm} = 0.01 \text{ m} = 10^{-2} \text{ m}$	1 cm = 0.3937 inch
	millimeter (mm)	$1 \text{ mm} = 0.001 \text{ m} = 10^{-3} \text{ m}$	1 mm = 0.039 inch
	micrometer (μm)	$1\ \mu\text{m} = 10^{-6} \text{ m} = 10^{-3} \text{ mm}$	
	nanometer (nm)	$1 \text{ nm} = 10^{-9} \text{ m} = 10^{-3}\ \mu\text{m}$	
Area	hectare (ha)	$1 \text{ ha} = 10{,}000 \text{ m}^2$	1 ha = 2.47 acres
	square meter (m^2)	$1 \text{ m}^2 = 10{,}000 \text{ cm}^2$	$1 \text{ m}^2 = 1.196$ square yards
	square centimeter (cm^2)	$1 \text{ cm}^2 = 100 \text{ mm}^2 = 10^{-4} \text{ m}^2$	$1 \text{ cm}^2 = 0.155$ square inch
Volume	liter (L)	1 L = 1000 mL	1 L = 1.06 quarts
	milliliter (mL)	$1 \text{ mL} = 1000\ \mu\text{L} = 10^{-3} \text{ L}$	1 mL = 0.034 fluid ounce
	microliter (μL)	$1\ \mu\text{L} = 10^{-6} \text{ L}$	
Mass	kilogram (kg)	1 kg = 1000 g	1 kg = 2.20 pounds
	gram (g)	1 g = 1000 mg	1 g = 0.035 ounce
	milligram (mg)	$1 \text{ mg} = 1000\ \mu\text{g} = 10^{-3} \text{ g}$	
	microgram (μg)	$1\ \mu\text{g} = 10^{-6} \text{ g}$	
Temperature	Kelvin (K)*		$K = °C + 273.15$
	degrees Celsius (°C)		$°C = \frac{5}{9}(°F - 32)$
	degrees Fahrenheit (°F)		$°F = \frac{9}{5}°C + 32$

*Absolute zero is −273.15°C = 0 K

If you have grown up in the United States and are accustomed to using the English system, it is extremely important to begin developing a working familiarity with metric units and values. The tables and questions below should help you get started with this process.

✔Questions

1. Some friends of yours just competed in a 5-kilometer run. How many miles did they run?
2. An American football field is 120 yards long, while rugby fields are 144 meters long. In yards, how much longer is a rugby field than an American football field?
3. What is your normal body temperature in degrees Celsius? (Normal body temperature is 98.6°F.)
4. What is your current weight in kilograms?
5. A friend asks you to buy a gallon of milk. How many liters would you buy to get approximately the same volume?

BIOSKILLS 2

READING GRAPHS

Graphs are the most common way to report data, for a simple reason. Compared to reading raw numerical values in a table or list, a graph makes it much easier to understand what the data mean.

Learning how to read and interpret graphs is one of the most basic skills you'll need to acquire as a biology student. As when learning piano or soccer or anything else, you need to understand a few key ideas to get started and then have a chance to practice—a *lot*—with some guidance and feedback.

Getting Started

To start reading a graph, you need to do three things: read the axes, figure out what the data points represent—that is, where they came from—and think about the overall message of the data. Let's consider each in turn.

What Do the Axes Represent?

Graphs have two axes: one horizontal and one vertical. The horizontal axis of a graph is also called the *x*-axis or the abscissa. The vertical axis of a graph is also called the *y*-axis or the ordinate. Each axis represents a variable that takes on a range of values. These values are indicated by the ticks and labels on the axis. Note that each axis should *always* be clearly labeled with the unit or treatment it represents.

Figure B2.1 shows a scatterplot—a type of graph where continuous data are graphed on each axis. Continuous data can take an array of values over a range. In contrast, discrete data can take only a restricted set of values. If you were graphing the average height of men and women in your class, height is a continuous variable but gender is a discrete variable.

In the example in the figure, the *x*-axis represents time in units of generations of maize; the *y*-axis represents the average percentage of the dry weight of a maize kernel that is protein.

To create a graph, researchers plot the independent variable on the *x*-axis and the dependent variable on the *y*-axis (Figure B2.1a). The terms independent and dependent are used because the values on the *y*-axis depend on the *x*-axis values. In our example, the researchers wanted to show how the protein content of maize kernels in a study population changed over time. Thus, the protein concentration plotted on the *y*-axis depended on the year (generation) plotted on the *x*-axis. The value on the *y*-axis always depends on the value on the *x*-axis, but not vice versa.

In many graphs in biology, the independent variable is either time or the various treatments used in an experiment. In these cases, the *y*-axis records how some quantity changes as a function of time or as the outcome of the treatments applied to the experimental cells or organisms.

What Do the Data Points Represent?

Once you've read the axes, you need to figure out what each data point is. In our maize kernel example, the data point in Figure B2.1b represents the average percentage of protein found in a sample of kernels from a study population in a particular generation.

If it's difficult to figure out what the data points are, ask yourself where they came from—meaning, how the researchers got them. You can do this by understanding how the study was done and by understanding what is being plotted on each axis. The *y*-axis will tell you what they measured; the *x*-axis will usually tell you when they measured it or what group was measured. In some cases—for example, in a plot of average body size versus average brain size in primates—the *x*-axis will report a second variable that was measured.

What Is the Overall Trend or Message?

Look at the data as a whole, and figure out what they mean. Figure B2.1c suggests an interpretation of the maize kernel example. If the graph shows how some quantity changes over time, ask yourself if that quantity is increasing, decreasing, fluctuating up and down, or staying the same. Then ask whether the pattern is the same over time or whether it changes over time.

When you're interpreting a graph, it's extremely important to limit your conclusions to the data presented. Don't extrapolate beyond the data, unless you are explicitly making a prediction based on the assumption that present trends will continue. For example, you can't say that average % protein content was increasing in the population before the experiment started, or that it will continue to increase in the future. You can only say what the data tell you.

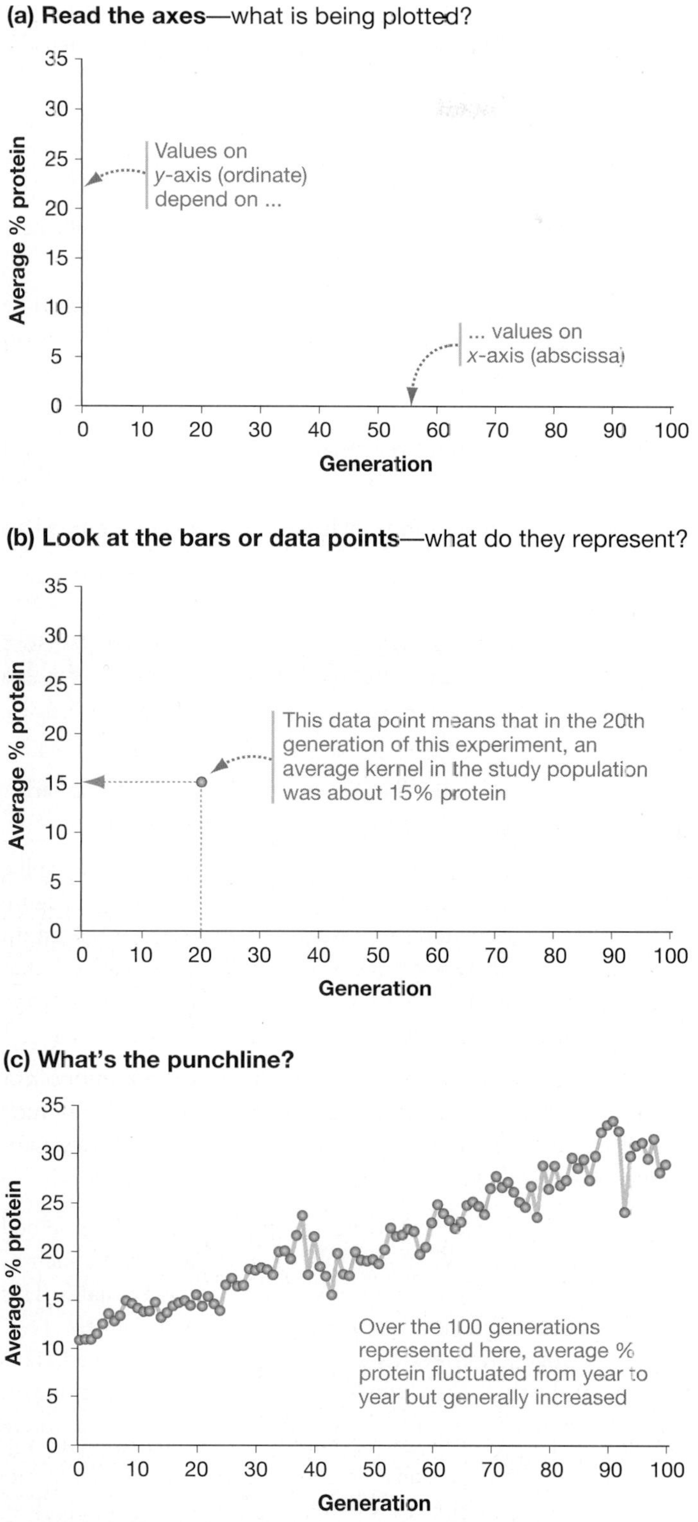

FIGURE B2.1 Scatterplots Are Used to Graph Continuous Data.

Types of Graphs

Most of the graphs in this text are scatterplots like the one shown here, where individual data points are plotted.

Sometimes the data points in a scatterplot will be by themselves, sometimes they will be connected by dot-to-dot lines to help make the overall trend clearer, as in this figure, and sometimes they will have a smooth line through them. A smooth line through data points—sometimes straight, sometimes curved—is a mathematical "line of best fit." A line of best fit represents a mathematical function that summarizes the relationship between the *x* and *y* variables. It is "best" in the sense of fitting the data points most precisely.

Scatterplots are the most appropriate type of graph when the data have a continuous range of values and you want to show individual data points. But you will also come across two other major types of graphs in this text:

- *Bar charts* plot data that have discrete or categorical values instead of a continuous range of values. In many cases the bars might represent different treatment groups in an experiment, as in **Figure B2.2a**. In this graph, the height of the bar indicates the average value.
- *Histograms* illustrate frequency data and can be plotted as numbers or percentages. **Figure B2.2b** shows an example where height is plotted on the *x*-axis and the number of students in a population is plotted on the *y*-axis. Each rectangle indicates the number of individuals in each interval of height, which reflects the relative frequency, in this population, of people whose heights are in that interval. The measurements could also be recalculated so that the *y*-axis would report the proportion of people in each interval. Then the sum of all the bars would equal 100 percent.

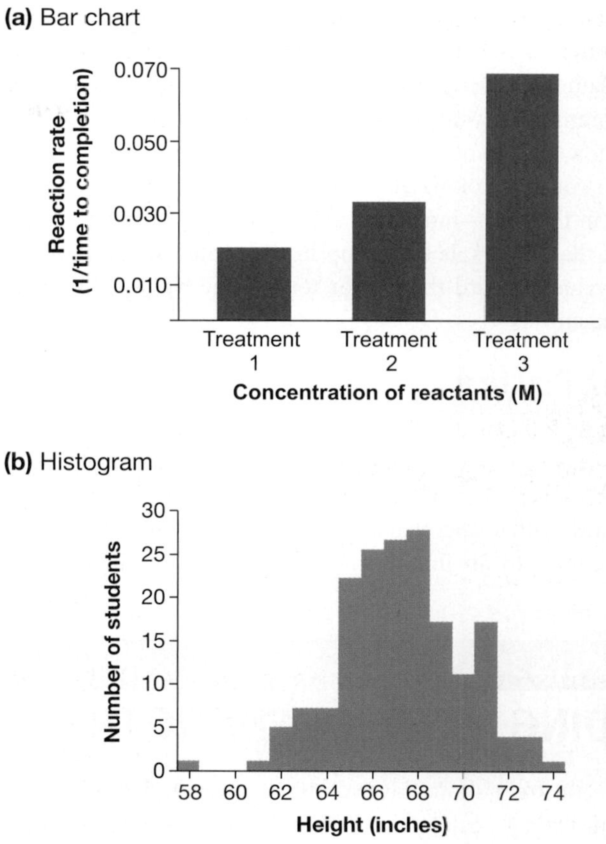

FIGURE B2.2 Bar Charts and Histograms. (a) Bar charts are used to graph data that are discontinuous or categorical. **(b)** Histograms show the distribution of frequencies or values in a population.

When you are looking at a bar chart that plots values from different treatments in an experiment, ask yourself if these values are the same or different. If the bar chart reports averages over discrete ranges of values, ask what trend is implied—as you would for a scatterplot.

When you are looking at a histogram, ask whether there is a "hump" in the data—indicating a group of values that are more frequent than others. Is the hump in the center of the distribution of values, toward the left, or toward the right? If so, what does it mean?

Getting Practice

Working with this text will give you lots of practice with reading graphs—they appear in almost every chapter. In many cases we've inserted an arrow to represent your instructor's hand at the whiteboard, with a label that suggests an interpretation or draws your attention to an important point on the graph. In other cases, you should be able to figure out what the data mean on your own or with the help of other students or your instructor.

✔Questions

1. What is the total change in average percent protein in maize kernels, from the start of the experiment until the end?
2. What was the trend in average percent protein in maize kernels between generation 37 and generation 42?
3. Would the conclusions from the bar chart in Figure B2.2a be different if the data and label for Treatment 3 were put on the far left and the data and label for Treatment 1 on the far right?
4. In Figure B2.2b, about how many students in this class are 70 inches tall?
5. What is the most common height in the class graphed in Figure B2.2b?

BIOSKILLS 3

READING A PHYLOGENETIC TREE

Phylogenetic trees show the evolutionary relationships among species, just as a genealogy shows the relationships among people in your family. They are unusual diagrams, however, and it can take practice to interpret them correctly.

To understand how evolutionary trees work, consider **Figure B3.1**. Notice that a phylogenetic tree consists of branches, nodes, and tips.

- Branches represent populations through time. In this text, branches are drawn as horizontal lines. In most cases the length of the branch is arbitrary and has no meaning, but in some cases branch lengths are proportional to time (if so, there will be a scale at the bottom of the tree). The vertical lines on the tree represent splitting events, where one group broke into two independent groups. Their length is arbitrary—chosen simply to make the tree more readable.
- Nodes (also called forks) occur where an ancestral group splits into two or more descendant groups (see point B in Figure B3.1). Thus, each node represents the most recent common ancestor of the two or more descendant populations that emerge from it. If more than two descendant groups emerge from a node, the node is called a polytomy (see node C).
- Tips (also called terminal nodes) are the tree's endpoints, which represent groups living today or a dead end—a branch ending in extinction. The names at the tips can represent species or larger groups such as mammals or conifers.

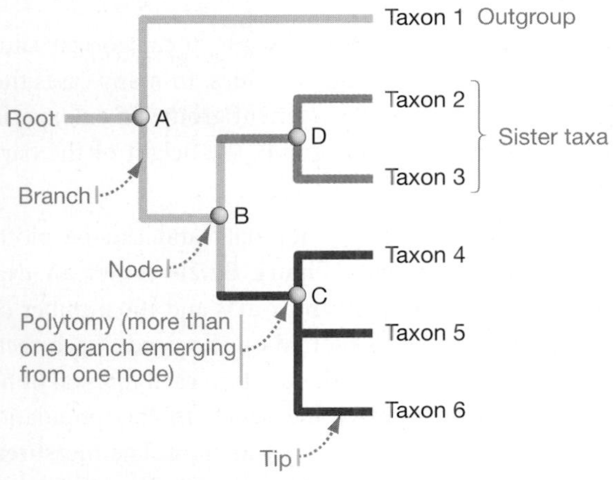

FIGURE B3.1 Phylogenetic Trees Have Roots, Branches, Nodes, and Tips.

✔**EXERCISE** Circle all four monophyletic groups present.

Recall from Chapter 1 that a taxon (plural: taxa) is any named group of organisms. A taxon could be a single species, such as *Homo sapiens*, or a large group of species, such as Primates. Tips connected by a single node on a tree are called sister taxa.

The phylogenetic trees used in this text are all rooted. This means that the first, or most basal, node on the tree—the one on the far left in this book—is the most ancient. To determine where the root on a tree occurs, biologists include one or more outgroup species when they are collecting data to estimate a particular phylogeny. An outgroup is a taxonomic group that is known to have diverged prior to the rest of the taxa in the study.

In Figure B3.1, "Taxon 1" is an outgroup to the monophyletic group consisting of taxa 2–6. A monophyletic group consists of an ancestral species and all of its descendants. The root of a tree is placed between the outgroup and the monophyletic group being studied. This position in Figure B3.1 is node A.

Understanding monophyletic groups is fundamental to reading and estimating phylogenetic trees. Monophyletic groups may also be called lineages or clades and can be identified using the "one-snip test": If you cut any branch on a phylogenetic tree, all of the branches and tips that fall off represent a monophyletic group. Using the one-snip test, you should be able to convince yourself that the monophyletic groups on a tree are nested. In Figure B3.1, for example, the monophyletic group comprising node A and taxa 1–6 contains a monophyletic group consisting of node B and taxa 2–6, which includes the monophyletic group represented by node C and taxa 4–6.

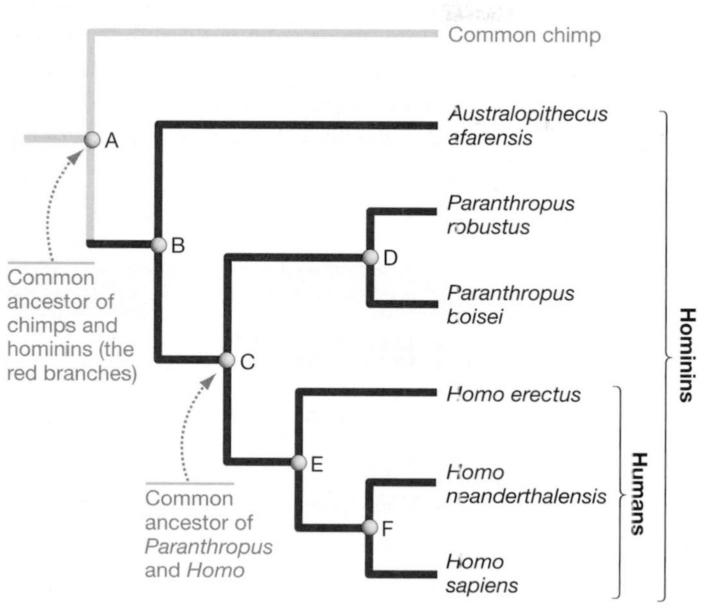

FIGURE B3.2 An Example of a Phylogenetic Tree. A phylogenetic tree showing the relationships of species in the monophyletic group called hominins.

✔**EXERCISE** All of the hominins walked on two legs—unlike chimps and all of the other primates. Add a mark on the phylogeny to show where upright posture evolved, and label it "origin of walking on two legs." Circle and label a pair of sister species. Label an outgroup to the monophyletic group called humans (species in the genus *Homo*).

To put all these new terms and concepts to work, consider the phylogenetic tree in **Figure B3.2**, which shows the relationships between common chimpanzees and six human and humanlike species that lived over the past 5–6 million years. Chimps functioned as an outgroup in the analysis that led to this tree, so the root was placed at node A. The branches marked in red identify a monophyletic group called the hominins.

To practice how to read a tree, put your finger at the tree's root, at the far left, and work your way to the right. At node A, the ancestral population split into two descendant populations. One of these populations eventually evolved into today's chimps; the other gave rise to the six species of hominins pictured. Now continue moving your finger toward the tips of the tree until you hit node C. It should make sense to you that at this splitting event, one descendant population eventually gave rise to two *Paranthropus* species, while the other became the ancestor of humans—species in the genus *Homo*. If multiple branches emerge from a node, creating a polytomy, it means that the populations split from one another so quickly that it is not possible to tell which split off earlier or later.

As you study Figure B3.2, you should consider a couple of important points.

1. There are many equivalent ways of drawing this tree. For example, this version shows *Homo sapiens* on the bottom. But the tree would be identical if the two branches emerging from node E were rotated 180°, so that the species appeared in the

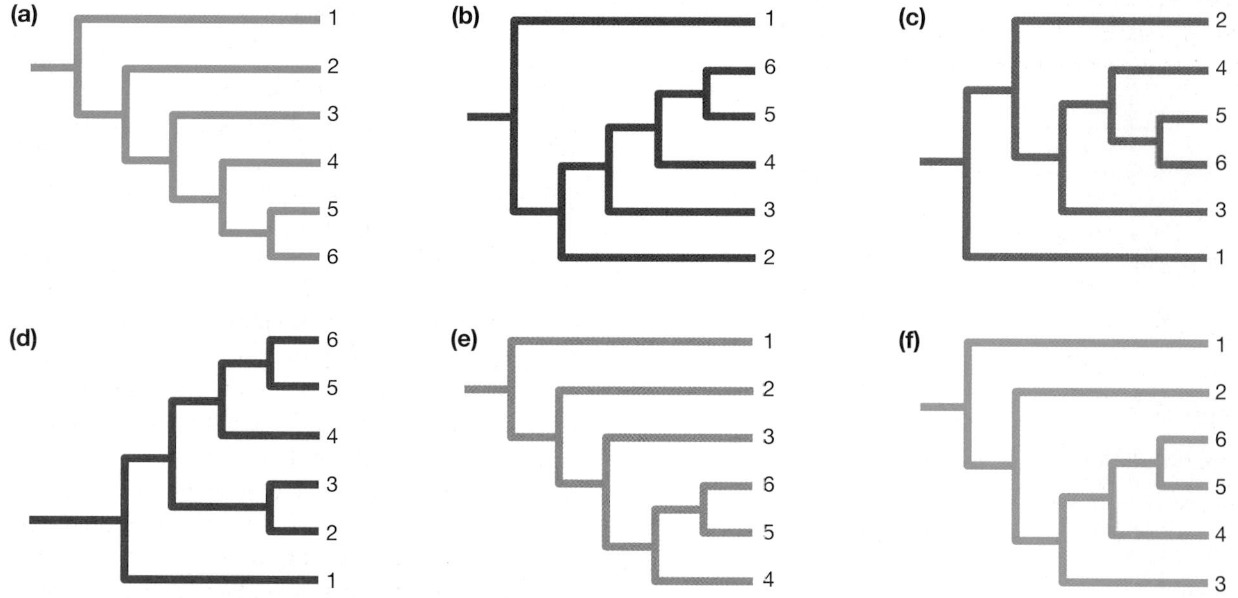

FIGURE B3.3 Alternative Ways of Drawing the Same Tree.

✔**QUESTION** Five of these six trees describe exactly the same relationships among taxa 1 through 6. Identify the tree that is different from the other five.

order *Homo sapiens, Homo neanderthalensis, Homo erectus.* Trees are read from root to tips, not from top to bottom or bottom to top.

2. No species on any tree is any higher or lower than any other. Chimps and *Homo sapiens* have been evolving exactly the same amount of time since their divergence from a common ancestor—neither species is higher or lower than the other. It is legitimate to say that more-ancient groups like *Australopithecus afarensis* have traits that are ancestral or more basal—meaning, that appeared earlier in evolution—compared to traits that appear in *Homo sapiens*, which are referred to as more derived.

Figure B3.3 on page B:5 presents a chance to test your tree-reading ability. Five of the six trees shown in this diagram are identical in terms of the evolutionary relationships they represent. One differs. The key to understanding the difference is to recognize that the ordering of tips does not matter in a tree—only the ordering of nodes (branch points) matters. You can think of a tree like a mobile: the tips can rotate without changing the underlying relationships.

BIOSKILLS 4

SOME COMMON LATIN AND GREEK ROOTS USED IN BIOLOGY

Greek or Latin Root	English Translation	Example Term
a, an	not	anaerobic
aero	air	aerobic
allo	other	allopatric
amphi	on both sides	amphipathic
anti	against	antibody
auto	self	autotroph
bi	two	bilateral symmetry
bio	life, living	bioinformatics
blast	bud, sprout	blastula
co	with	cofactor
cyto	cell	cytoplasm
di	two	diploid
ecto	outer	ectoparasite
endo	inner, within	endoparasite
epi	outer, upon	epidermis
exo	outside	exothermic
glyco	sugary	glycolysis
hetero	different	heterozygous
homo	alike	homozygous
hydro	water	hydrolysis

Greek or Latin Root	English Translation	Example Term
hyper	over, more than	hypertonic
hypo	under, less than	hypotonic
inter	between	interspecific
intra	within	intraspecific
iso	same	isotonic
logo, logy	study of	morphology
lyse, lysis	loosen, burst	glycolysis
macro	large	macromolecule
meta	change, turning point	metamorphosis
micro	small	microfilament
morph	form	morphology
oligo	few	oligopeptide
para	beside	parathyroid gland
photo	light	photosynthesis
poly	many	polymer
soma	body	somatic cells
sym, syn	together	symbioticm, synapsis
trans	across	translation
tri	three	trisomy
zygo	yoked together	zygote

✔Questions

Provide literal translations of the following terms:

1. heterozygote
2. glycolysis
3. morphology
4. trisomy

BIOSKILLS 5

USING STATISTICAL TESTS AND INTERPRETING STANDARD ERROR BARS

When biologists do an experiment, they collect data on individuals in a treatment group and a control group, or several such comparison groups. Then they want to know whether the individuals in the two (or more) groups are different. For example, Chapter 2 introduces an experiment in which student researchers measured how fast a product formed when they set up a reaction with three different concentrations of reactants. Each treatment—meaning, each combination of reactant concentrations—was replicated many times.

Figure B5.1 graphs the average reaction rate for each of the three treatments in the experiment. Note that Treatments 1, 2, and 3 represent increasing concentrations of reactants. The thin "I-beams" on each bar indicate the standard error of each average. The standard error is a quantity that indicates the uncertainty in the calculation of an average.

For example, if two trials with the same concentration of reactants had a reaction rate of 0.075 and two trials had a reaction rate of 0.025, then the average reaction rate would be 0.50. In this case, the standard error would be large. But if two trials had a reaction rate of 0.051 and two had a reaction rate of 0.049, the average would still be 0.050, but the standard error would be small.

In effect, the standard error quantifies how confident you are that the average you've calculated is the average you'd observe if you did the experiment under the same conditions an extremely large number of times. It is a measure of precision.

Once they had calculated these averages and standard errors, the students wanted to answer a question: Does reaction rate increase when reactant concentration increases?

After looking at the data, you might conclude that the answer is yes. But how could you come to a conclusion like this objectively, instead of subjectively?

The answer is to use a statistical test. This can be thought of as a three-step process.

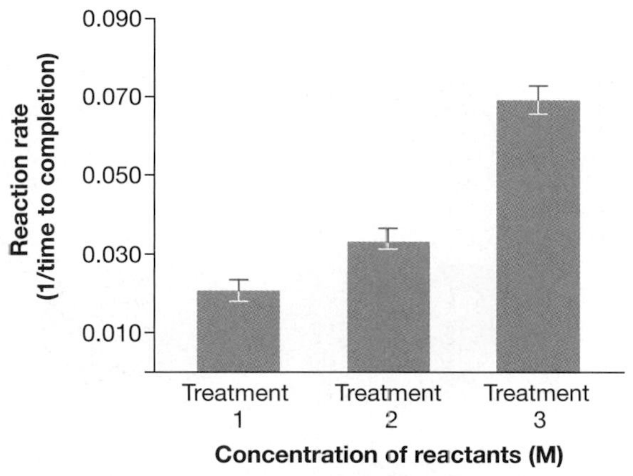

FIGURE B5.1 Standard Error Bars Indicate the Uncertainty in an Average.

1. Specify the null hypothesis, which is that reactant concentration has no effect on reaction rate.
2. Calculate a test statistic, which is a number that characterizes the size of the difference among the treatments. In this case, the test statistic compares the actual differences in reaction rates among treatments to the difference predicted by the null hypothesis. The null hypothesis predicts that there should be no difference.
3. The third step is to determine the probability of getting a test statistic as large as the one calculated just by chance. The answer comes from a reference distribution—a mathematical function that specifies the probability of getting various values of the test statistic if the null hypothesis is correct. (If you take a statistics course, you'll learn which test statistics and reference distributions are relevant to different types of data.)

You are very likely to see small differences among treatment groups just by chance—even if no differences actually exist. If you flipped a coin 10 times, for example, you are unlikely to get exactly five heads and five tails, even if the coin is fair. A reference distribution tells you how likely you are to get each of the possible outcomes of the 10 flips if the coin is fair, just by chance.

In this case, the reference distribution indicated that if the null hypothesis of no actual difference in reaction rates is correct, you would see differences as large as those observed only 0.01 percent of the time just by chance. By convention, biologists consider a difference among treatment groups to be statistically significant if you have less than a 5 percent probability of observing it just by chance. Based on this convention, the student researchers were able to claim that the null hypothesis is not correct for reactant concentration. According to their data, the reaction they studied really does happen faster when reactant concentration increases.

It is likely that you'll be doing actual statistical tests early in your undergraduate career. To use this text, though, you only need to be aware of what statistical testing does. And you should take care to inspect the standard error bars on graphs in this book. As a *very* rough rule of thumb, averages often turn out to be significantly different, according to an appropriate statistical test, if there is no overlap between two times the standard errors.

✔Question

Suppose you estimated the average height of students in your class by sampling two individuals at random. Then suppose you estimated the same quantity by sampling every individual that showed up for class on a particular day. Which estimate of the average is likely to have the smallest standard error, and why?

READING CHEMICAL STRUCTURES

If you haven't had much chemistry yet, learning basic biological chemistry can be a challenge. One of the stumbling blocks is simply being able to read chemical structures efficiently and understand what they mean. This skill will come much easier once you have a little notation under your belt and you understand some basic symbols.

Atoms are the basic building blocks of everything in the universe, just as cells are the basic building blocks of your body. Every atom has a 1- or 2-letter symbol. **Table B6.1** shows the symbols for most of the atoms you'll encounter in this book. You should memorize these. The table also offers details on how the atoms form bonds as well as how they are represented in visual models.

When atoms attach to each other by covalent bonding, a molecule forms. Biologists have a couple of different ways of representing molecules—you'll see each of these in the book and in class.

- A molecular formula like those in **Figure B6.1a** simply lists the atoms present in a molecule, with subscripts indicating how many of each atom are present. If the formula has no subscript, only one of that type of atom is present. A methane (natural gas) molecule, for example, can be written as CH_4. It consists of one carbon atom and four hydrogen atoms.
- Structural formulas like those in **Figure B6.1b** show which atoms in the molecule are bonded to each other, with each bond indicated by a dash. The structural formula for methane indicates that each of the four hydrogen atoms forms one covalent bond with carbon, and that carbon makes a total of four covalent bonds. Single covalent bonds are symbolized by a single dash; double bonds are indicated by two dashes.

Even simple molecules have distinctive shapes, because different atoms make covalent bonds at different angles. Ball-and-stick and space-filling models show the geometry of the bonds accurately.

TABLE B6.1 **Some Attributes of Atoms Found in Organisms**

Atom	Symbol	Number of Bonds It Can Form	Standard Color Code*
Hydrogen	H	1	white
Carbon	C	4	black
Nitrogen	N	3	blue
Oxygen	O	2	red
Sodium	Na	1	—
Magnesium	Mg	2	—
Phosphorus	P	5	orange or purple
Sulfur	S	2	yellow
Chlorine	Cl	1	—
Potassium	K	1	—
Calcium	Ca	2	—

*In ball-and-stick or space-filling models.

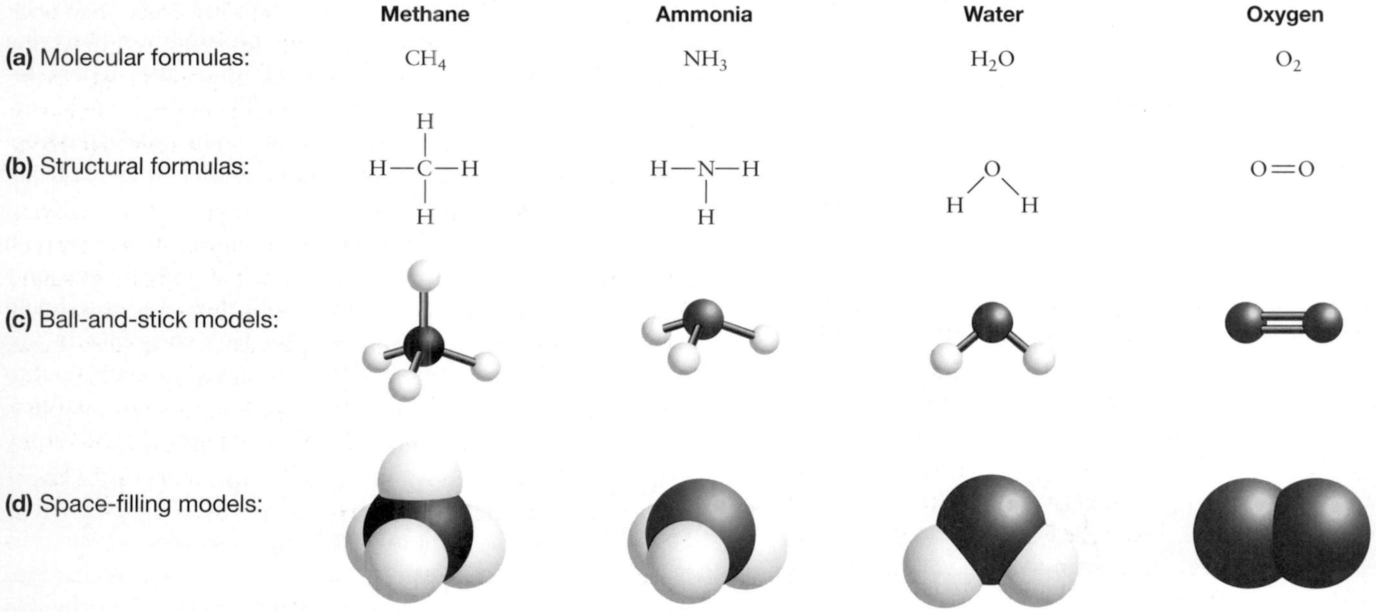

FIGURE B6.1 Molecules Can Be Represented in Several Different Ways.

✔**EXERCISE** Carbon dioxide consists of a carbon atom that forms a double bond with each of two oxygen atoms, for a total of four bonds. It is a linear molecule. Write carbon dioxide's molecular formula, then draw its structural formula, a ball-and-stick model, and a space-filling model.

- In a ball-and-stick model, a stick is used to represent each covalent bond (see **Figure B6.1c**).
- In space-filling models, the atoms are simply stuck onto each other in their proper places (see **Figure B6.1d**).

To learn more about a molecule when you look at a chemical structure, ask yourself three questions:

1. ***Is the molecule polar—meaning that some parts are more negatively or positively charged than others?*** Molecules that contain nitrogen or oxygen atoms are often polar, because these atoms have such high electronegativity (see Chapter 2). This trait is important because polar molecules dissolve in water.
2. ***Does the structural formula show atoms that might participate in chemical reactions?*** For example, are there charged atoms or amino or carboxyl (−COOH) groups that might act as a base or an acid?
3. ***In ball-and-stick and especially space-filling models of large molecules, are there interesting aspects of overall shape?*** For example, is there a groove where a protein might bind to DNA, or a cleft where a substrate might undergo a reaction in an enzyme?

BIOSKILLS 7

USING LOGARITHMS

You have probably been introduced to logarithms and logarithmic notation in algebra courses, and you will encounter logarithms at several points in this course. Logarithms are a way of working with powers—meaning, numbers that are multiplied by themselves one or more times.

Scientists use exponential notation to represent powers. For example,

$$a^x = y$$

means that if you multiply a by itself x times, you get y. In exponential notation, a is called the base and x is called the exponent. The entire expression is called an exponential function.

What if you know y and a, and you want to know x? This is where logarithms come in. You can solve for exponents using logarithms. For example,

$$x = \log_a y$$

This equation reads, x is equal to the logarithm of y to the base a. Logarithms are a way of working with exponential functions. They are important because so many processes in biology (and chemistry and physics, for that matter) are exponential in nature. To understand what's going on, you have to describe the process with an exponential function and then use logarithms to work with that function.

Although a base can be any number, most scientists use just two bases when they employ logarithmic notation: 10 and e. Logarithms to the base 10 are so common that they are usually symbolized in the form $\log y$ instead of $\log_{10} y$. A logarithm to the base e is called a natural logarithm and is symbolized ln (pronounced *EL-EN*) instead of log. You write "the natural logarithm of y" as $\ln y$. The base e is an irrational number (like π) that is approximately equal to 2.718. Like 10, e is just a number. But both 10 and e have qualities that make them convenient to use in biology (and chemistry, and physics).

Most scientific calculators have keys that allow you to solve problems involving base 10 and base e. For example, if you know y, they'll tell you what $\log y$ or $\ln y$ are—meaning that they'll solve for x in our example above. They'll also allow you to find a number when you know its logarithm to base 10 or base e. Stated another way, they'll tell you what y is if you know x, and y is equal to e^x or 10^x. This is called taking an antilog. In most cases, you'll use the inverse or second function button on your calculator to find an antilog (above the log or ln key).

To get some practice with your calculator, consider the equation

$$10^2 = 100$$

If you enter 100 in your calculator and then press the log key, the screen should say 2. The logarithm tells you what the exponent is. Now press the antilog key while 2 is on the screen. The calculator screen should return to 100. The antilog solves the exponential function, given the base and the exponent.

If your background in algebra isn't strong, you'll want to get more practice working with logarithms—you'll see them frequently during your undergraduate career. Remember that once you understand the basic notation, there's nothing mysterious about logarithms. They are simply a way of working with exponential functions, which describe what happens when something is multiplied by itself a number of times—like cells that divide and then divide again and then again.

Using logarithms will also come up when you are studying something that can have a large range of values, like the concentration of hydrogen ions in a solution or the intensity of sound that the human ear can detect. In cases like this, it's convenient to express the numbers involved as exponents. Using exponents makes a large range of numbers smaller and more tractable. For example, instead of saying that hydrogen ion concentration in a solution can range from 1 to 10^{-14}, the pH scale allows you to simply say that it ranges from 1 to 14. Instead of giving the actual value, you're expressing it as an exponent. It just simplifies things.

✓Questions

In Chapter 52, you'll use the equation $N_t = N_0 e^{rt}$.

1. What type of function does this equation describe?
2. After taking the natural logarithm of both sides, how would you write the equation?

BIOSKILLS 8

MAKING CONCEPT MAPS

A concept map is a graphical device for organizing and expressing what you know about a topic. It has two main elements: **(1)** concepts that are identified by words or short phrases and placed in a box or circle, and **(2)** labeled arrows that physically link two concepts and explain the relationship between them. The concepts are arranged hierarchically on a page, with the most general concepts at the top and the most specific ideas at the bottom.

The combination of a concept, a linking word, and a second concept is called a proposition. Good concept maps also have cross-links—meaning, labeled arrows that connect different elements in the hierarchy, as you read down the page.

Concept maps were initially developed by Joseph Novak in the early 1970s and have proven to be an effective studying and learning tool. They can be particularly valuable if constructed by a group, or when different individuals exchange and critique concept maps they have created independently. Although concept maps vary widely in quality and can be graded using objective criteria, there are many equally valid ways of making a high-quality concept map on a particular topic.

When you are asked to make a concept map in this text, you will usually be given at least a partial list of concepts to use. As an example, suppose you were asked to create a concept map on experimental design and were given the following concepts: results, predictions, control treatment, experimental treatment, controlled (identical) conditions, conclusions, experiment, hypothesis to be tested, null hypothesis. One possible concept map is shown in **Figure B8.1**.

Good concept maps have four qualities:

- They exhibit an organized hierarchy, indicating how each concept on the map relates to larger and smaller concepts.
- The concept words are specific—not vague.
- The propositions are accurate.
- There is cross-linking between different elements in the hierarchy of concepts.

As you practice making concept maps, go through these criteria and use them to evaluate your own work, as well as the work of fellow students.

✔Exercises

1. In many cases, investigators contrast a hypothesis being tested with an alternative hypothesis that does not qualify as a null hypothesis. Add an "Alternative hypothesis" concept to the map in Figure B8.1, along with other concepts and labeled linking arrows needed to indicate its relationship to other information on the map.
2. Add a box for the concept "Statistical testing" (see **BioSkills 5**) along with appropriately labeled linking arrows.

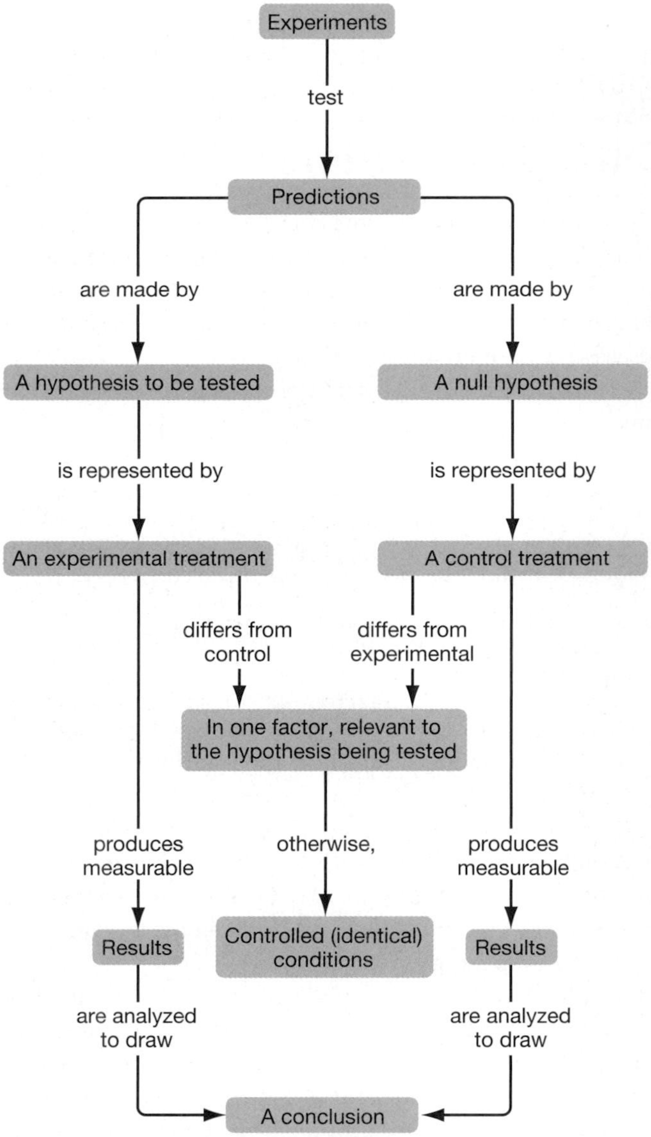

FIGURE B8.1 A Concept Map on Principles of Experimental Design.

BIOSKILLS 9

SEPARATING AND VISUALIZING MOLECULES

To study a molecule, you have to be able to isolate it. This is a two-step process: the molecule has to be separated from other molecules in a mixture, and then physically picked out or located in a purified form. This BioSkill focuses on the techniques that biologists use to separate nucleic acids and proteins and then find the particular one they are interested in.

Using Electrophoresis to Separate Molecules

In molecular biology, the standard technique for separating proteins and nucleic acids is called gel electrophoresis or, simply, electrophoresis (literally "electricity-moving"). You may be using electrophoresis in a lab for this course, and you will certainly be analyzing data derived from electrophoresis in this text.

The principle behind electrophoresis is simple. Both proteins and nucleic acids carry a charge. As a result, these molecules move when placed in an electric field. Negatively charged molecules move toward the positive electrode; positively charged molecules move toward the negative electrode.

To separate a mixture of macromolecules so that each can be isolated and analyzed, researchers place the sample in a gelatinous substance. More specifically, the sample is placed in a "well"—a slot in a sheet or slab of the gelatinous substance. The "gel" itself consists of long molecules that form a matrix of fibers. The presence of the fibers keeps molecules in the sample from moving around randomly, but the gelatinous matrix also has pores through which the molecules can pass.

When an electrical field is applied across the gel, the molecules in the well move through the gel toward an electrode. Molecules that are smaller or more highly charged for their size move faster than do larger or less highly charged molecules. As they move, then, the molecules separate by size and by charge. Small and highly charged molecules end up at the bottom of the gel; large, less-charged molecules remain near the top.

An Example "Run"

Figure B9.1 shows the electrophoresis setup used in an experiment investigating how RNA molecules polymerize In this case, the investigators wanted to document how long RNA molecules became over time, when ribonucleotide triphosphates were present in a particular type of solution.

Step 1 shows how they loaded samples of macromolecules, taken on different days during the experiment, into wells at the top of the gel slab. This is a general observation: Each well holds a different sample. In this and many other cases, the researchers also filled a well with a sample containing fragments of known size, called a size standard or "ladder."

In step 2 the researchers immersed the gel in a solution that conducts electricity and applied a voltage across the gel. The molecules in each well started to run down the gel, forming a lane. After several hours of allowing the molecules to move, they removed the electric field (step 3). By then, molecules of different size and charge had separated from one another. In this case,

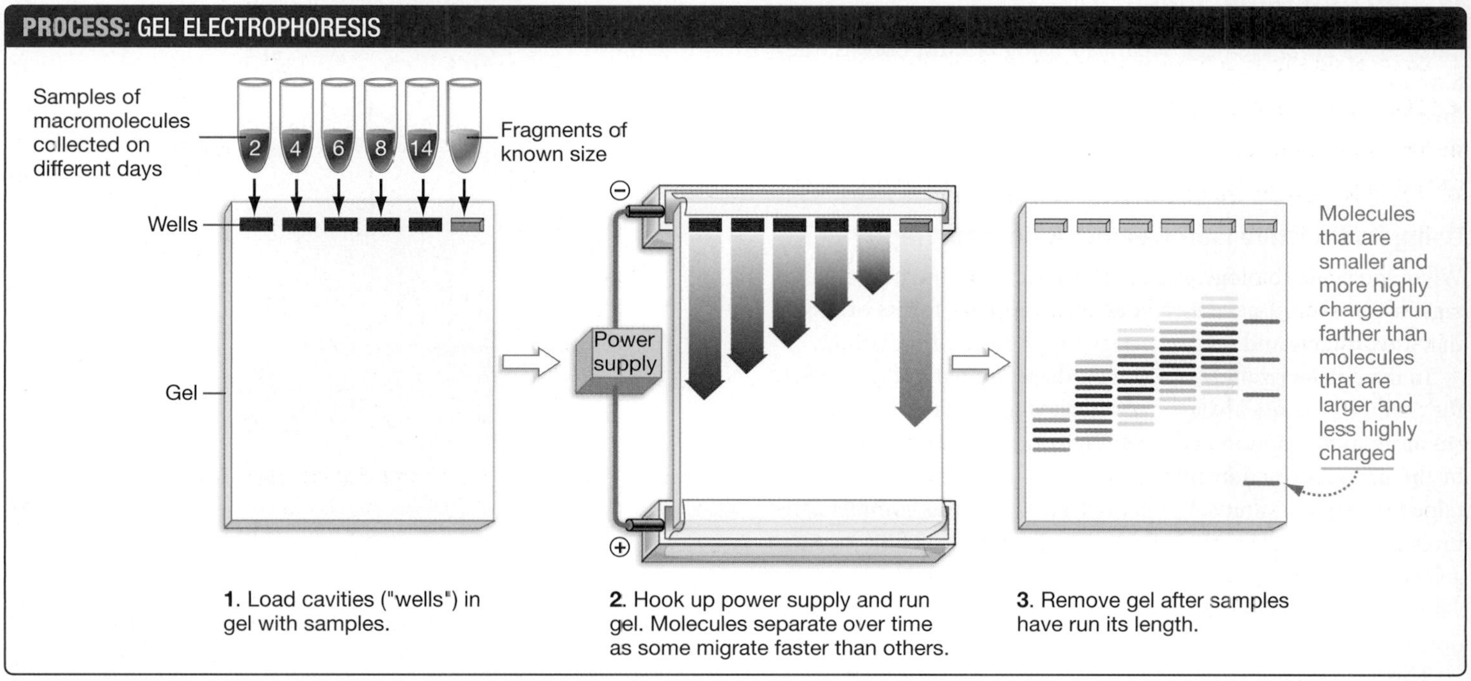

PROCESS: GEL ELECTROPHORESIS
Samples of macromolecules collected on different days
2
4
6
8
14
Fragments of known size
Wells
Gel
Power supply
Molecules that are smaller and more highly charged run farther than molecules that are larger and less highly charged
1. Load cavities ("wells") in gel with samples.
2. Hook up power supply and run gel. Molecules separate over time as some migrate faster than others.
3. Remove gel after samples have run its length.

FIGURE B9.1 Macromolecules Can Be Separated via Gel Electrophoresis.

✔**QUESTION** DNA and RNA run toward the positive electrode. Why are these molecules negatively charged?

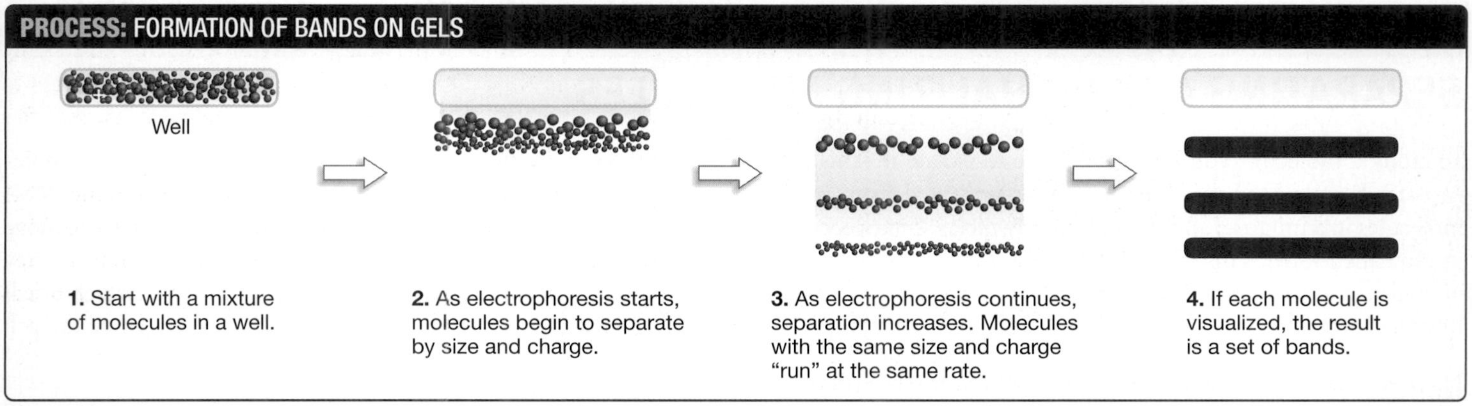

PROCESS: FORMATION OF BANDS ON GELS
Well
1. Start with a mixture of molecules in a well.
2. As electrophoresis starts, molecules begin to separate by size and charge.
3. As electrophoresis continues, separation increases. Molecules with the same size and charge "run" at the same rate.
4. If each molecule is visualized, the result is a set of bands.

FIGURE B9.2 On a Gel, Alike Molecules Form Bands.

small RNA molecules had reached the bottom of the gel. Above them were larger RNA molecules, which had run more slowly.

Why Do Separated Molecules Form Bands?

When researchers visualize a particular molecule on a gel, using techniques described below, the image that results consists of bands: shallow lines that are as wide as a lane in the gel. Why?

To understand the answer, study **Figure B9.2**. The top panel shows the original mixture of molecules. In this cartoon, the size of each dot represents the size of each molecule. The key is to realize that the original sample contains many copies of each specific molecule, and that these copies run down the length of the gel together—meaning, at the same rate—because they have the same size and charge.

It's that simple: Alike molecules form a band because they stay together.

Visualizing Molecules

Once molecules have been separated using electrophoresis, they have to be detected. Unfortunately, proteins and nucleic acids are invisible unless they are tagged in some way. Let's consider two of the most common tagging systems, then consider how researchers can tag and visualize specific molecules of interest and not others.

Using Radioactive Isotopes and Autoradiography

When molecular biology was getting under way, the first types of tags in common use were radioactive isotopes—forms of atoms that are unstable and release energy in the form of radiation.

In the polymerization experiment diagrammed in Figure B9.1, for example, the researchers had attached a radioactive phosphorus atom to the monomers—ribonucleotide triphosphates—used in the original reaction mix. Once polymers formed, they contained radioactive atoms. When electrophoresis was complete, the investigators visualized the polymers by laying X-ray film over the gel. Because radioactive emissions expose film, a black dot appears wherever a radioactive atom is located in the gel. So many black dots occur so close together that the collection forms a dark band.

This technique for visualizing macromolecules is called autoradiography. The autoradiograph that resulted from the polymerization experiment is shown in **Figure B9.3**. The samples, taken on days 2, 4, 6, 8, and 14 of the experiment, are labeled along the bottom. The far right lane contains macromolecules of known size; this lane is used to estimate the size of the molecules in the experimental samples. The bands that appear in each sample lane represent the different polymers that had formed.

Reading a Gel

One of the keys to interpreting or "reading" a gel is to realize that darker bands contain more radioactive markers, indicating the presence of many radioactive molecules. Lighter bands contain fewer molecules.

To read a gel, then, you look for **(1)** the presence or absence of bands in some lanes—meaning, some experimental samples—versus others, and **(2)** contrasts in the darkness of the bands—meaning, differences in the number of molecules present.

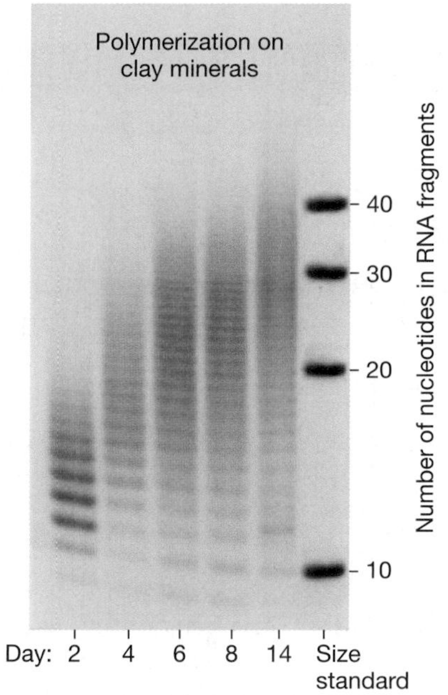

FIGURE B9.3 Autoradiography Is a Technique for Visualizing Macromolecules. The molecules in a gel can be visualized in a number of ways. In this case, the RNA molecules in the gel exposed an X-ray film because they had radioactive atoms attached. When developed, the film is called an autoradiograph.

For example, several conclusions can be drawn from the data in Figure B9.3. First, a variety of polymers formed at each stage. After the second day, for example, polymers from 12 to 18 monomers long had formed on the clay particles used in this experiment. Second, the overall length of polymers produced increased with time. At the end of the fourteenth day, most of the RNA molecules were between 20 and 40 monomers long.

Starting in the late 1990s and early 2000s, it became much more common to tag nucleic acids with fluorescent tags. Once electrophoresis is complete, the presence of fluorescence can be detected by exposing the gel to an appropriate wavelength of light; the fluorescent tag fluoresces or glows in response (fluorescence is explained in Chapter 10).

Fluorescent tags have important advantages over radioactive isotopes: (**1**) They are safer to handle. (**2**) They are faster—you don't have to wait hours or days for the radioactive isotope to expose a film. (**3**) They come in multiple colors, so you can tag several different molecules in the same experiment and detect them independently.

Using Nucleic Acid Probes

In many cases, researchers want to find one specific molecule—a certain DNA sequence, for example—in the collection of molecules on a gel. How is this possible? The answer hinges on using a particular molecule as a probe.

Chapter 19 explains how probes work in detail. Here it's enough to get the general idea: A probe is a marked molecule that binds specifically to your molecule of interest. The "mark" is often a radioactive atom or a fluorescent tag.

If you are looking for a particular DNA or RNA sequence on a gel, for example, you can expose the gel to a single-stranded probe that binds to the target sequence by complementary base pairing. Once it has bound, you can detect the band through autoradiography or fluorescence.

- Southern blotting is a technique for making DNA fragments that have been run out on a gel single-stranded, transferring them from the gel to a nylon filter, and then probing them to identify segments of interest. The technique was named after its inventor, Edwin Southern.
- Northern blotting is a technique for transferring RNA fragments from a gel to a nylon filter, and then probing them to detect target segments. The name is a lighthearted play on Southern blotting—the protocol from which it was derived.

Using Antibody Probes

How can researchers find a particular protein out of a large collection of different proteins? The answer is to use an antibody. An antibody is a protein that binds specifically to a section of a different protein. The structure of antibodies and their function in the immune system are explored in detail in Chapter 49.

To use an antibody as a probe, investigators attach a tag molecule—often an enzyme that catalyzes a color-forming reaction—to the antibody and allow it to react with proteins in a mixture. The antibody will stick to the specific protein that it binds to, and then can be visualized thanks to the tag it carries.

If the proteins in question have been separated by gel electrophoresis, the result is called a Western blot. The name Western is an extension of the Southern and Northern pattern.

✔Question

Suppose you've been given a gel that has been stained for "RNA X." One lane contains no bands. Two lanes have a band in the same location, even though one of the bands is barely visible and the other is extremely dark. The fourth lane has a light band located above the bands in the other lanes. Interpret the gel.

BIOSKILLS 10

BIOLOGICAL IMAGING: MICROSCOPY AND X-RAY CRYSTALLOGRAPHY

A lot of biology happens at levels that can't be detected with the naked eye. Biologists use an array of microscopes to study small multicellular organisms, individual cells, and the contents of cells. And to understand what individual macromolecules or multimolecular machines like ribosomes look like, researchers use data from a technique called X-ray crystallography.

You'll probably use dissecting microscopes and compound light microscopes to view specimens during your labs for this course, and throughout this text you'll be seeing images generated from other types of microscopy and from X-ray crystallographic data. One of the fundamental skills you'll be acquiring as an introductory student, then, is a basic understanding of how these techniques work. The key is to recognize that each approach for visualizing microscopic structures has strengths and weaknesses. As a result, each technique is appropriate for studying certain types or aspects of cells or molecules.

Light Microscopy

If you use a dissecting microscope during labs, you'll recognize that it works by magnifying light that bounces off a whole specimen—often a live organism. You'll be able to view the specimen in three dimensions, which is why these instruments are sometimes called stereomicroscopes, but the maximum magnification possible is only about 20 to 40 times normal size (20× to 40×).

To view smaller objects, you'll probably use a compound microscope. Compound microscopes magnify light that is passed *through* a specimen. The instruments used in introductory labs are usually capable of 400× magnifications; the most sophisticated compound microscopes available can achieve magnifications of about 2000×. This is enough to view individual bacterial or eukaryotic cells and see large structures inside cells, like condensed chromosomes (see Chapter 11). To prepare a specimen for viewing under a compound light microscope, the tissues or cells are usually sliced to create a section thin enough for light to pass through efficiently. The section is then dyed to increase contrast and make structures visible. In many cases, different types of dyes are used to highlight different types of structures.

Electron Microscopy

Until the 1950s, the compound microscope was the biologist's only tool for viewing cells directly. But the invention of the electron microscope provided a new way to view specimens. Two basic types of electron microscopy are now available: one that allows researchers to examine cross sections of cells at extremely high magnification, and one that offers a view of surfaces at somewhat lower magnification.

Transmission Electron Microscopy (TEM)

The transmission electron microscope is an extraordinarily effective tool for viewing cell structure at high magnification. TEM forms an image from electrons that pass through a specimen, just as a light microscope forms an image from light rays that pass through a specimen.

Biologists who want to view a cell under a transmission electron microscope begin by "fixing" the cell, meaning that they treat it with a chemical agent that stabilizes the cell's structure and contents while disturbing them as little as possible. Then the researcher permeates the cell with an epoxy plastic that stiffens the structure. Once this epoxy hardens, the cell can be cut into extremely thin sections with a glass or diamond knife. Finally, the sectioned specimens are impregnated with a metal—often lead. (The reason for this last step is explained shortly.)

Figure B10.1a outlines how the transmission electron microscope works. A beam of electrons is produced by a tungsten filament at the top of a column and directed downward. (All of the air is pumped out of the column, so that the electron beam isn't scattered by collisions with air molecules.) The electron beam passes through a series of lenses and through the specimen. The lenses are actually electromagnets, which alter the path of the beam much like a glass lens in a dissecting or compound microscope bends light. The lenses magnify and focus the image on a screen at the bottom of the column. There the electrons strike a coating of fluorescent crystals, which emit visible light in response—just like a television screen. When the microscopist moves the screen out of the way and allows the electrons to expose a sheet of black-and-white film, the result is a micrograph—a photograph of an image produced by microscopy.

The image itself is created by electrons that pass through the specimen. If no specimen were in place, all the electrons would pass through and the screen (and micrograph) would be uniformly bright. Unfortunately, cell materials by themselves would also appear fairly uniform and bright. This is because an atom's ability to deflect an electron depends on its mass. In turn, an atom's mass is a function of its atomic number. The hydrogen, carbon, oxygen, and nitrogen atoms that dominate biological molecules have low atomic numbers. This is why cell biologists must saturate cell sections with lead solutions. Lead has a high atomic number and scatters electrons effectively. Different macromolecules take up lead atoms in different amounts, so the

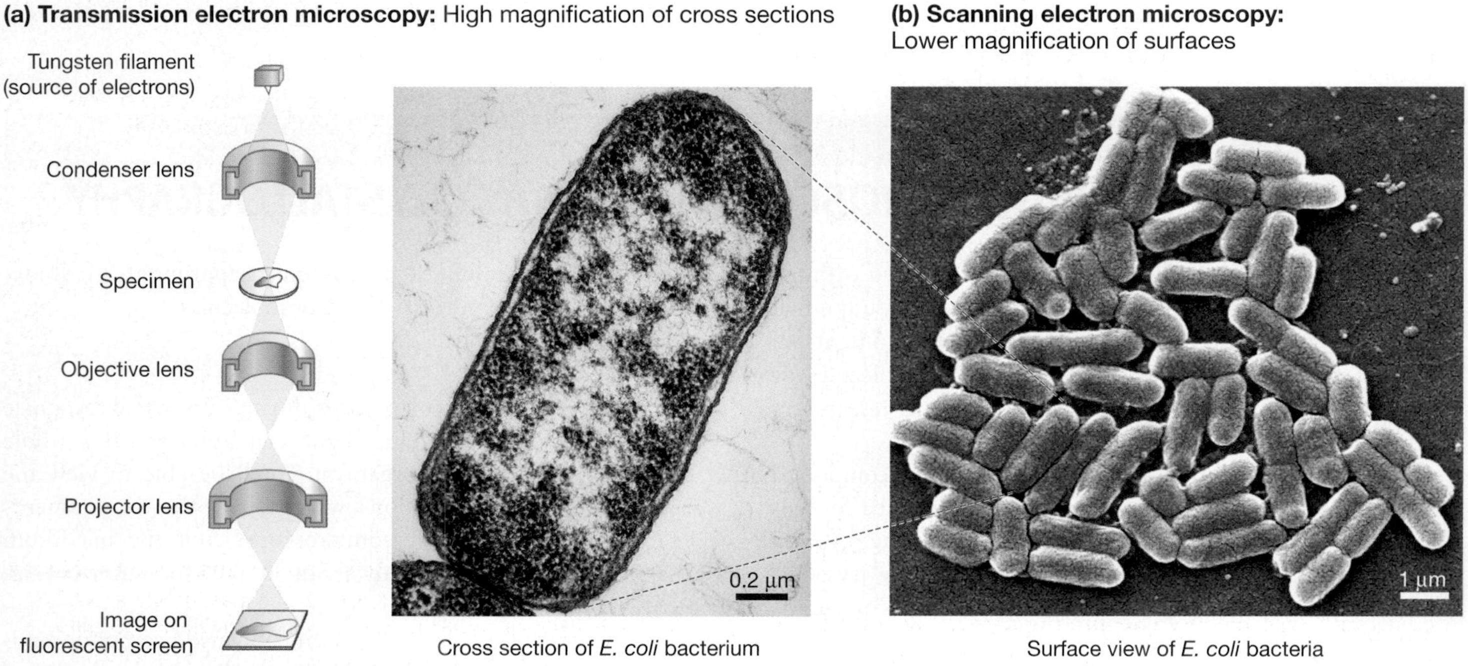

FIGURE B10.1 There Are Two Basic Types of Electron Microscopy.

metal acts as a "stain" that produces contrast. With TEM, areas of dense metal scatter the electron beam most, producing dark areas in micrographs.

The advantage of TEM is that it can magnify objects up to 250,000×—meaning that intracellular structures are clearly visible. The downsides are that researchers are restricted to observing dead, sectioned material, and they must take care that the preparation process does not distort the specimen.

Scanning Electron Microscopy (SEM)

The scanning electron microscope is the most useful tool biologists have for looking at the surfaces of structures. Materials are prepared for scanning electron microscopy by coating their surfaces with a layer of metal atoms. To create an image of this surface, the microscope scans the surface with a narrow beam of electrons. Electrons that are reflected back from the surface or that are emitted by the metal atoms in response to the beam then strike a detector. The signal from the detector controls a second electron beam, which scans a TV-like screen and forms an image magnified up to 50,000 times the object's size.

Because SEM records shadows and highlights, it provides images with a three-dimensional appearance (**Figure B10.1b**). It cannot magnify objects nearly as much as TEM can, however.

Studying Live Cells and Real-Time Processes

Until the 1960s, it was not possible for biologists to get clear, high-magnification images of living cells. But a series of innovations over the past 50 years has made it possible to observe organelles and subcellular structures in action.

The development of video microscopy, where the image from a light microscope is captured by a video camera instead of by an eye or a film camera, proved revolutionary. It allowed specimens to be viewed at higher magnification, because video cameras are more sensitive to small differences in contrast than are the human eye or still cameras. It also made it easier to keep live specimens functioning normally, because the increased light sensitivity of video cameras allows them to be used with low illumination, so specimens don't overheat. And when it became possible to digitize video images, researchers began using computers to remove out-of-focus background material and increase image clarity.

A more recent innovation was the use of a fluorescent molecule called green fluorescent protein, or GFP, which allows researchers to tag specific molecules or structures and follow their movement over time. GFP is naturally synthesized in jellyfish that fluoresce, or emit light. By affixing GFP molecules to another protein and then inserting it into a cell, investigators can follow the protein's fate over time and even videotape its movement. For example, researchers have videotaped GFP-tagged proteins being transported from the rough ER through the Golgi apparatus and out to the plasma membrane. This is cell biology: the movie.

GFP's influence has been so profound that the researchers who developed its use in microscopy were awarded the 2008 Nobel prize in Chemistry.

Visualizing Structures in 3-D

The world is three-dimensional. To understand how microscopic structures and macromolecules work, it is essential to understand their shape and spatial relationships. Consider three techniques currently being used to reconstruct the 3-D structure of cells, organelles, and macromolecules.

- *Confocal microscopy* is carried out by mounting cells that have been treated with one or more fluorescing tags on a microscope slide and then focusing a beam of ultraviolet light at a specific depth within the specimen. The fluorescing tag emits visible light in response. A detector for this light is then set up at exactly the position where the emitted light comes into focus. The result is a sharp image of a precise plane in the cell being studied (**Figure B10.2**). By altering the focal plane, a researcher can record images from an array of depths in the specimen; a computer can then be used to generate a 3-D image of the cell.
- *Electron tomography* uses a transmission electron microscope to generate a 3-D image of an organelle or other subcellular structure. The specimen is rotated around a single axis, with the researcher taking many "snapshots." The individual images are then pieced together with a computer. This technique has provided a much more accurate view of mitochondrial structure than was possible using traditional TEM (see Chapter 7).
- *X-ray crystallography, or X-ray diffraction analysis,* is the most widely used technique for reconstructing the 3-D structure of molecules. As its name implies, the procedure is based on bombarding crystals of a molecule with X-rays. X-rays are scattered in precise ways when they interact with the electrons surrounding the atoms in a crystal, producing a diffraction pattern that can be recorded on X-ray film or other types of

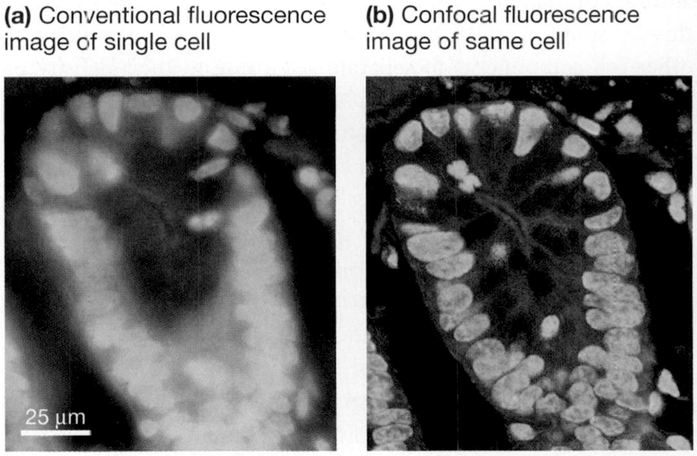

(a) Conventional fluorescence image of single cell
(b) Confocal fluorescence image of same cell
25 μm

FIGURE B10.2 Confocal Microscopy Provides Sharp Images of Living Cells. **(a)** The conventional image of this mouse intestinal cell is blurred, because it results from light emitted by the entire cell. **(b)** The confocal image is sharp, because it results from light emitted at a single plane inside the cell.

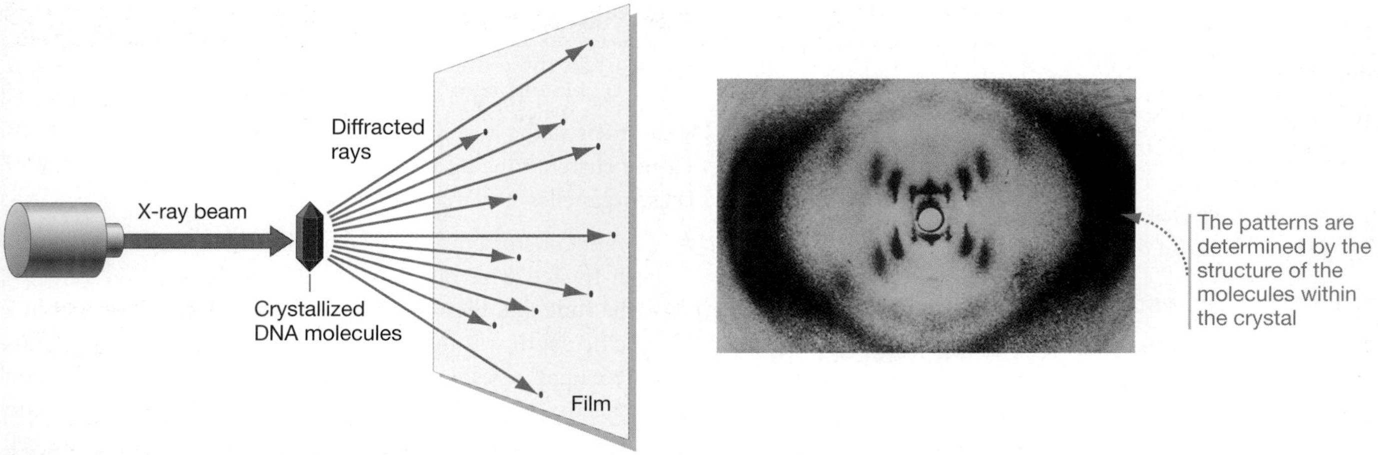

FIGURE B10.3 X-Ray Crystallography. When crystallized molecules are bombarded with X-rays, the radiation is scattered in distinctive patterns. The photograph at the right shows an X-ray film that recorded the pattern of scattered radiation from DNA molecules.

detectors (**Figure B10.3**). By varying the orientation of the X-ray beam as it strikes a crystal and documenting the diffraction patterns that result, researchers can construct a map representing the density of electrons in the crystal. By relating these electron-density maps to information about the primary structure of the nucleic acid or protein, a 3-D model of the molecule can be built. Virtually all of the molecular models used in this book were built from X-ray crystallographic data.

✔Questions

1. Suppose you are looking at a transmission electron micrograph of a cancerous cell from the human liver. No mitochondria are present. Does this mean that the cell lacks mitochondria?
2. X-ray crystallography is time consuming and technically difficult. Why is the effort to understand the structure of biological molecules worthwhile? What's the payoff?

BIOSKILLS 11

SEPARATING CELL COMPONENTS BY CENTRIFUGATION

Biologists use a technique called differential centrifugation to isolate specific cell components. Differential centrifugation is based on breaking cells apart to create a complex mixture and then separating components in a centrifuge. A centrifuge accomplishes this by spinning cells in a solution that allows molecules and other cell components to separate according to their density or size and shape. The individual parts of the cell can then be purified and studied in detail, in isolation from other parts of the cell.

The first step in preparing a cell sample for centrifugation is to release the organelles and cell components by breaking the cells apart. This can be done by putting them in a hypotonic solution, by exposing them to high-frequency vibration, by treating cells with a detergent, or by grinding them up. Each of these methods breaks apart plasma membranes and releases the contents of the cells.

The resulting pieces of plasma membrane quickly reseal to form small vesicles, often trapping cell components inside. The solution that results from the homogenization step is a mixture of these vesicles, free-floating macromolecules released from the cells, and organelles. A solution such as this is called a cell extract or cell homogenate.

When a cell homogenate is placed in a centrifuge tube and spun at high speed, the components that are in solution tend to move outward, along the red arrow in **Figure B11.1a**. The effect is similar to a merry-go-round, which seems to push you outward in a straight line away from the spinning platform. In response to this outward-directed force, the solution containing the cell homogenate exerts a centripetal (literally, "center-seeking") force that pushes the homogenate away from the bottom of the tube. Larger, denser molecules or particles resist this inward force more readily than do smaller, less dense ones and so reach the bottom of the centrifuge tube faster.

To separate the components of a cell extract, researchers often perform a series of centrifuge runs. Steps 1 and 2 of **Figure B11.1b** illustrate how an initial treatment at low speed causes larger, heavier parts of the homogenate to move below smaller, lighter parts. The material that collects at the bottom of the tube is called the pellet, and the solution and solutes left behind form the supernatant ("above swimming"). The supernatant is placed in a fresh tube and centrifuged at increasingly higher speeds and longer durations. Each centrifuge run continues to separate cell components based on their size and density.

To accomplish even finer separation of macromolecules or organelles, researchers frequently follow up with centrifugation at extremely high speeds. One strategy is based on filling the centrifuge tube with a series of sucrose solutions of increasing

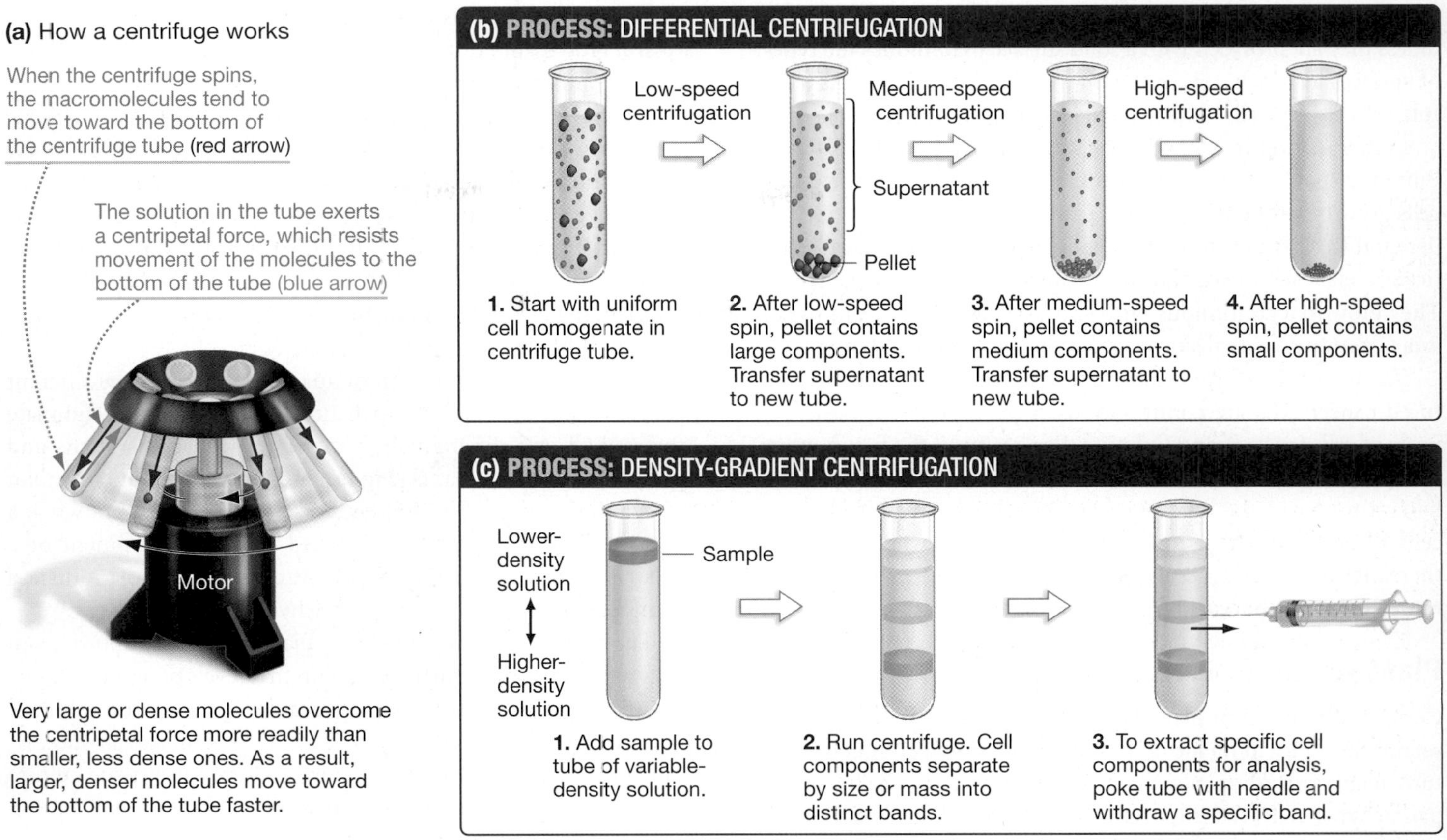
(a) How a centrifuge works
When the centrifuge spins, the macromolecules tend to move toward the bottom of the centrifuge tube (red arrow)
The solution in the tube exerts a centripetal force, which resists movement of the molecules to the bottom of the tube (blue arrow)
Motor
Very large or dense molecules overcome the centripetal force more readily than smaller, less dense ones. As a result, larger, denser molecules move toward the bottom of the tube faster.
(b) PROCESS: DIFFERENTIAL CENTRIFUGATION
Low-speed centrifugation
Medium-speed centrifugation
High-speed centrifugation
Supernatant
Pellet
1. Start with uniform cell homogenate in centrifuge tube.
2. After low-speed spin, pellet contains large components. Transfer supernatant to new tube.
3. After medium-speed spin, pellet contains medium components. Transfer supernatant to new tube.
4. After high-speed spin, pellet contains small components.
(c) PROCESS: DENSITY-GRADIENT CENTRIFUGATION
Lower-density solution
Higher-density solution
Sample
1. Add sample to tube of variable-density solution.
2. Run centrifuge. Cell components separate by size or mass into distinct bands.
3. To extract specific cell components for analysis, poke tube with needle and withdraw a specific band.

FIGURE B11.1 Cell Components Can Be Separated by Centrifugation. (a) The forces inside a centrifuge tube allow cell components to be separated. **(b)** Through a series of centrifuge runs made at increasingly higher speeds, an investigator can separate fractions of a cell homogenate by size via differential centrifugation. **(c)** A high-speed centrifuge run can achieve extremely fine separation among cell components by density-gradient centrifugation.

density (**Figure B11.1c**). The density gradient allows cell components to separate on the basis of small differences in size, shape, and density. When the centrifuge run is complete, each cell component occupies a distinct band of material in the tube, based on how quickly each moves through the increasingly dense gradient of sucrose solution during the centrifuge run. A researcher can then collect the material in each band for further study.

✔Questions

1. Electrophoresis separates molecules by charge and size. What is the physical basis for separating molecules or cell components via centrifugation?
2. Compared to other types of centrifugation, why would extremely high-speed centrifugation allow similar molecules to separate?

BIOSKILLS 12

CELL AND TISSUE CULTURE METHODS

For researchers, there are important advantages to growing plant and animal cells and tissues outside the organism itself. Cell and tissue cultures provide large populations of a single type of cell or tissue and the opportunity to control experimental conditions precisely.

Animal Cell Culture

The first successful attempt to culture animal cells occurred in 1907, when a researcher cultivated amphibian nerve cells in a drop of fluid from the spinal cord. But it was not until the 1950s and 1960s that biologists could routinely culture plant and animal cells in the laboratory. The long lag time was due to the difficulty of re-creating conditions that exist in the intact organism precisely enough for cells to grow normally.

To grow in culture, animal cells must be provided with a liquid mixture containing the nutrients, vitamins, and hormones that stimulate growth. Initially, this mixture was serum, the liquid portion of blood; now serum-free media are available for certain cell types. Serum-free media are preferred because they are much more precisely defined chemically than serum.

In addition, many types of animal cells will not grow in culture unless they are provided with a solid surface that mimics the types of surfaces to which cells in the intact organisms adhere. As a result, cells are typically cultured in flasks (**Figure B12.1a**, left).

Even under optimal conditions, though, normal cells display a finite life span in culture. In contrast, many cultured cancerous cells grow indefinitely. In culture, cancerous cells also do not adhere tightly to the surface of the culture flask. These characteristics correlate with two features of cancerous cells in organisms: They grow in a continuous, uncontrolled fashion and can break away from the original site to infiltrate new tissues and organs.

Because of their immortality and relative ease of growth, cultured cancer cells are commonly used in research on basic aspects of cell structure and function. For example, the first human cell type to be grown in culture was isolated in 1951 from a malignant tumor of the uterine cervix. These cells are called HeLa cells in honor of their donor, Henrietta Lacks, who died soon thereafter from her cancer. HeLa cells continue to grow in laboratories around the world (Figure B12.1a, right).

Plant Tissue Culture

Certain cells found in plants are totipotent—meaning that they retain the ability to divide and differentiate into a complete, mature plant, including new types of tissue. These cells, called parenchyma cells, are important in wound healing and asexual reproduction. But they also allow researchers to grow complete adult plants in the laboratory, starting with a small number of parenchyma cells.

Biologists who grow plants in tissue culture begin by placing parenchyma cells in a liquid or solid medium containing all the nutrients required for cell maintenance and growth. In the early days of plant tissue culture, investigators found not only that specific growth signals called hormones were required for successful growth and differentiation but also that the relative abundance of hormones present was critical to success.

The earliest experiments on hormone interactions in tissue cultures were done with tobacco cells in the 1950s by Folke Skoog and co-workers. These researchers found that when the hormone called auxin was added to the culture by itself, the cells enlarged but did not divide. But if the team added roughly equal amounts of auxin and another growth signal called cytokinin to the cells, the cells began to divide and eventually formed a callus, or an undifferentiated mass of parenchyma cells.

By varying the proportion of auxin to cytokinins in different parts of the callus and through time, the team could stimulate the growth and differentiation of root and shoot systems and produce whole new plants (**Figure B12.1b**). A high ratio of auxin to cytokinin led to the differentiation of a root system, while a high ratio of cytokinin to auxin led to the development of a shoot system. Eventually Skoog's team was able to produce a complete plant from just one parenchyma cell.

The ability to grow whole new plants in tissue culture from just one cell has been instrumental in the development of genetic engineering (see Chapter 19). Researchers insert recombinant genes into target cells, test the cells to identify those that successfully express the recombinant genes, and then use tissue culture techniques to grow those cells into adult individuals with a novel genotype and phenotype.

✔Questions

1. What is a limitation of how experiments on HeLa cells are interpreted?
2. State a disadvantage of doing experiments on plants that have been propagated from single cells growing in tissue culture.

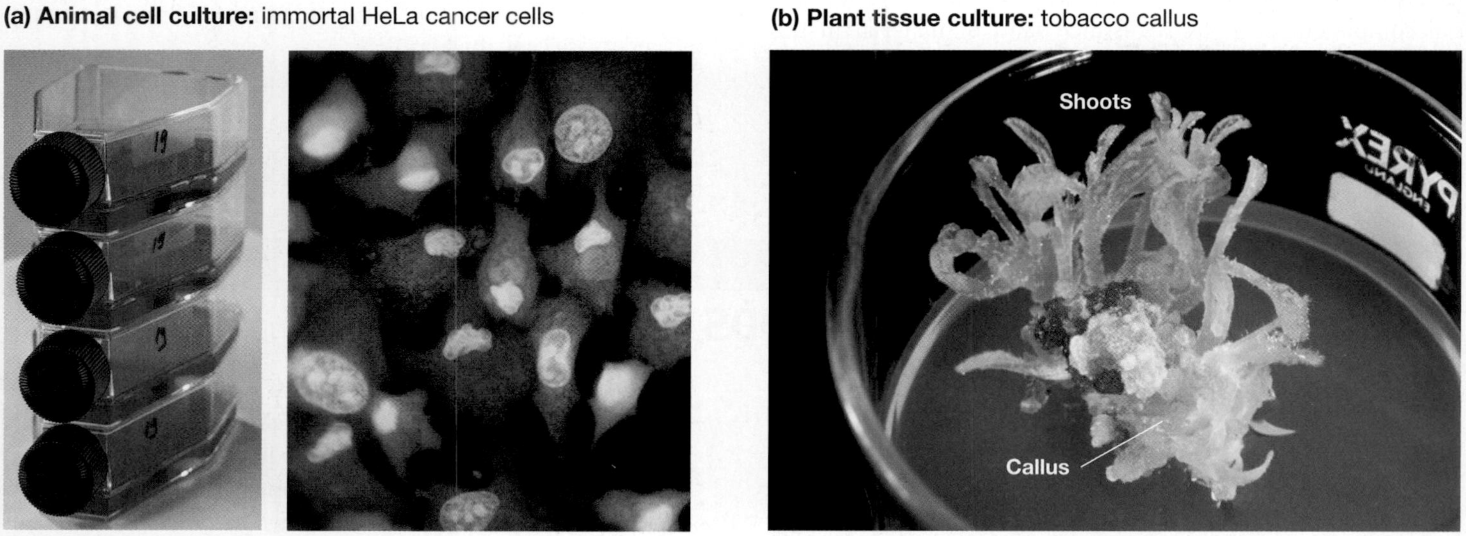
(a) Animal cell culture: immortal HeLa cancer cells
(b) Plant tissue culture: tobacco callus
Shoots
Callus
PYREX

FIGURE B12.1 Animal and Plant Cells Can Be Grown in the Lab.

BIOSKILLS 13

COMBINING PROBABILITIES

In several cases in this text, you'll need to combine probabilities from different events in order to solve a problem. One of the most common applications is in genetics problems. For example, Punnett squares work because they are based on two fundamental rules of probability. Each rule pertains to a distinct situation.

The Both-And Rule

The both-and rule—also known as the product rule or multiplication rule—applies when you want to know the probability that two or more independent events occur together. Let's use the rolling of two dice as an example. What is the probability of rolling two sixes? These two events are independent, because the probability of rolling a six on one die has no effect on the probability of rolling a six on the other die. (In the same way, the probability of getting a gamete with allele *R* from one parent has no effect on the probability of getting a gamete with allele *R* from the other parent. Gametes fuse randomly.)

The probability of rolling a six on the first die is 1/6. The probability of rolling a six on the second die is also 1/6. The probability of rolling a six on *both* dice, then, is $1/6 \times 1/6 = 1/36$. In other words, if you rolled two dice 36 times, on average you would expect to roll two sixes once.

In the case of a cross between two parents heterozygous at the *R* gene, the probability of getting allele *R* from the father is 1/2 and the probability of getting *R* from the mother is 1/2. Thus, the probability of getting both alleles and creating an offspring with genotype *RR* is $1/2 \times 1/2 = 1/4$.

The Either-Or Rule

The either-or rule—also known as the sum rule or addition rule—applies when you want to know the probability of an event happening when there are several different ways for the same event or outcome to occur. In this case, the probability that the event will occur is the sum of the probabilities of each way that it can occur.

For example, suppose you wanted to know the probability of rolling either a one or a six when you toss a die. The probability of drawing each is 1/6, so the probability of getting one or the other is $1/6 + 1/6 = 1/3$. If you rolled a die three times, on average you'd expect to get a one or a six once.

In the case of a cross between two parents heterozygous at the *R* gene, the probability of getting an *R* allele from the father and an *r* allele from the mother is $1/2 \times 1/2 = 1/4$. Similarly, the probability of getting an *r* allele from the father and an *R* allele from the mother is $1/2 \times 1/2 = 1/4$. Thus, the combined probability of getting the *Rr* genotype in either of the two ways is $1/4 + 1/4 = 1/2$.

✔Questions

1. Suppose that four students each toss a coin. What is the probability of four "tails"?
2. After a single roll of a die, what is the probability of getting either a two, a three, or a six?

BIOSKILLS 14

MODEL ORGANISMS

Research in biological science starts with a question. In most cases, the question is inspired by an observation about a cell or an organism. To answer it, biologists have to study a particular species. Study organisms are often called model organisms, because investigators hope that they serve as a model for what is going on in a wide array of species.

Model organisms are chosen because they are convenient to study and because they have attributes that make them appropriate for the particular research proposed. They tend to have some common characteristics:

- ***Short generation time and rapid reproduction*** This is important because it makes it possible to produce offspring quickly and perform many experiments in a short amount of time—you don't have to wait long for individuals to grow.
- ***Large numbers of offspring*** This is particularly important in genetics, where many offspring phenotypes and genotypes need to be assessed to get a large sample size.
- ***Small size, simple feeding and habitat requirements*** These attributes make it relatively cheap and easy to maintain individuals in the lab.

The following notes highlight just a few model organisms supporting current work in biological science.

Escherichia coli

Of all model organisms in biology, perhaps none has been more important than the bacterium *Escherichia coli*—a common inhabitant of the human gut. The strain that is most commonly worked on today, called K-12 (**Figure B14.1a** on page B:20), was originally isolated from a hospital patient in 1922.

During the last half of the twentieth century, key results in molecular biology originated in studies of *E. coli*. These results include the discovery of enzymes such as DNA polymerase, RNA polymerase, DNA repair enzymes, and restriction endonucleases; the elucidation of ribosome structure and function; and the initial characterization of promoters, regulatory transcription factors, regulatory sites in DNA, and operons. In many cases, initial discoveries made in *E. coli* allowed researchers to confirm that homologous enzymes and processes existed in an array of organisms, often ranging from other bacteria to yeast, mice, and humans.

The success of *E. coli* as a model for other species inspired Jacques Monod's claim that "Once we understand the biology of *Escherichia coli*, we will understand the biology of an elephant." The genome of *E. coli* K-12 was sequenced in 1997, and the strain continues to be a workhorse in studies of gene function, biochemistry, and particularly biotechnology. Much remains to be learned, however. Despite over 60 years of intensive study, the function of about a third of the *E. coli* genome is still unknown.

In the lab, *E. coli* is usually grown in suspension culture, where cells are introduced to a liquid nutrient medium, or on plates containing agar—a gelatinous mix of polysaccharides. Under optimal growing conditions—meaning before cells begin to get crowded and compete for space and nutrients—a cell takes just 30 minutes on average to grow and divide. At this rate, a single cell can produce a population of over a million descendants in just 10 hours. Except for new mutations, all of the descendant cells are genetically identical.

Dictyostelium discoideum

The cellular slime mold *Dictyostelium discoideum* is not always slimy, and it is not a mold—meaning a type of fungus. Instead, it is an amoeba. Amoeba is a general term that biologists use to characterize a unicellular eukaryote that lacks a cell wall and is extremely flexible in shape. *Dictyostelium* has long fascinated biologists because it is a social organism. Independent cells sometimes aggregate to form a multicellular structure.

Under most conditions, *Dictyostelium* cells are haploid (n) and move about in decaying vegetation on forest floors or other habitats. They feed on bacteria by engulfing them whole. When these cells reproduce, they can do so sexually by fusing with another cell then undergoing meiosis, or asexually by mitosis, which is more common. If food begins to run out, the cells begin to aggregate. In many cases, tens of thousands of cells cohere to form a 2-mm-long mass called a slug (**Figure B14.1b**). (This is not the slug that is related to snails.)

After migrating to a sunlit location, the slug stops and individual cells differentiate according to their position in the slug. Some form a stalk; others form a mass of spores at the tip of the stalk. (A spore is a single cell that develops into an adult organism, but it is not formed from gamete fusion like a zygote.) The entire structure, stalk plus mass of spores, is called a fruiting body. Cells that form spores secrete a tough coat and represent a durable resting stage. The fruiting body eventually dries out, and the wind disperses the spores to new locations, where more food might be available.

Dictyostelium has been an important model organism for investigating questions about eukaryotes:

- Cells in a slug are initially identical in morphology but then differentiate into distinctive stalk cells and spores. Studying this process helped biologists better understand how cells in plant and animal embryos differentiate into distinct cell types.
- The process of slug formation has helped biologists study how animal cells move and how they aggregate as they form specific types of tissues.
- When *Dictyostelium* cells aggregate to form a slug, they stick to each other. The discovery of membrane proteins responsible for cell-cell adhesion helped biologists understand some of the general principles of multicellular life highlighted in Chapter 8.

Arabidopsis thaliana

In the early days of biology, the best-studied plants were agricultural varieties such as maize (corn), rice, and garden peas. When biologists began to unravel the mechanisms responsible for oxygenic photosynthesis in the early to mid-1900s, they relied on green algae that were relatively easy to grow and manipulate in the lab—often the unicellular species *Chlamydomonas reinhardii*—as an experimental subject.

Although crop plants and green algae continue to be the subject of considerable research, a new model organism emerged in the 1980s and now serves as the preeminent experimental subject in plant biology. That organism is *Arabidopsis thaliana*, commonly known as thale cress or wall cress (**Figure B14.1c**).

Arabidopsis is a member of the mustard family, or Brassicaceae, so it is closely related to radishes and broccoli. In nature it is a weed—meaning a species that is adapted to thrive in habitats where soils have been disturbed.

One of the most attractive aspects of working with *Arabidopsis* is that individuals can grow from a seed into a mature, seed-producing plant in just four to six weeks. Several other attributes make it an effective subject for study: It has just five chromosomes, has a relatively small genome with limited numbers of repetitive sequences, can self-fertilize as well as undergo cross-fertilization, can be grown in a relatively small amount of space and with a minimum of care in the greenhouse, and produces up to 10,000 seeds per individual per generation.

Arabidopsis has been instrumental in a variety of studies in plant molecular genetics and development, and it is increasingly popular in ecological and evolutionary studies. In addition, the entire genome of the species has now been sequenced, and studies have benefited from the development of an international "*Arabidopsis* community"—a combination of informal and formal associations of investigators who work on *Arabidopsis* and use regular meetings, e-mail, and the Internet to share data, techniques, and seed stocks.

Saccharomyces cerevisiae

When biologists want to answer basic questions about how eukaryotic cells work, they often turn to the yeast *Saccharomyces cerevisiae.*

S. cerevisiae is unicellular and relatively easy to culture and manipulate in the lab (**Figure B14.1d**). In good conditions, yeast cells grow and divide almost as rapidly as bacteria. As a result, the

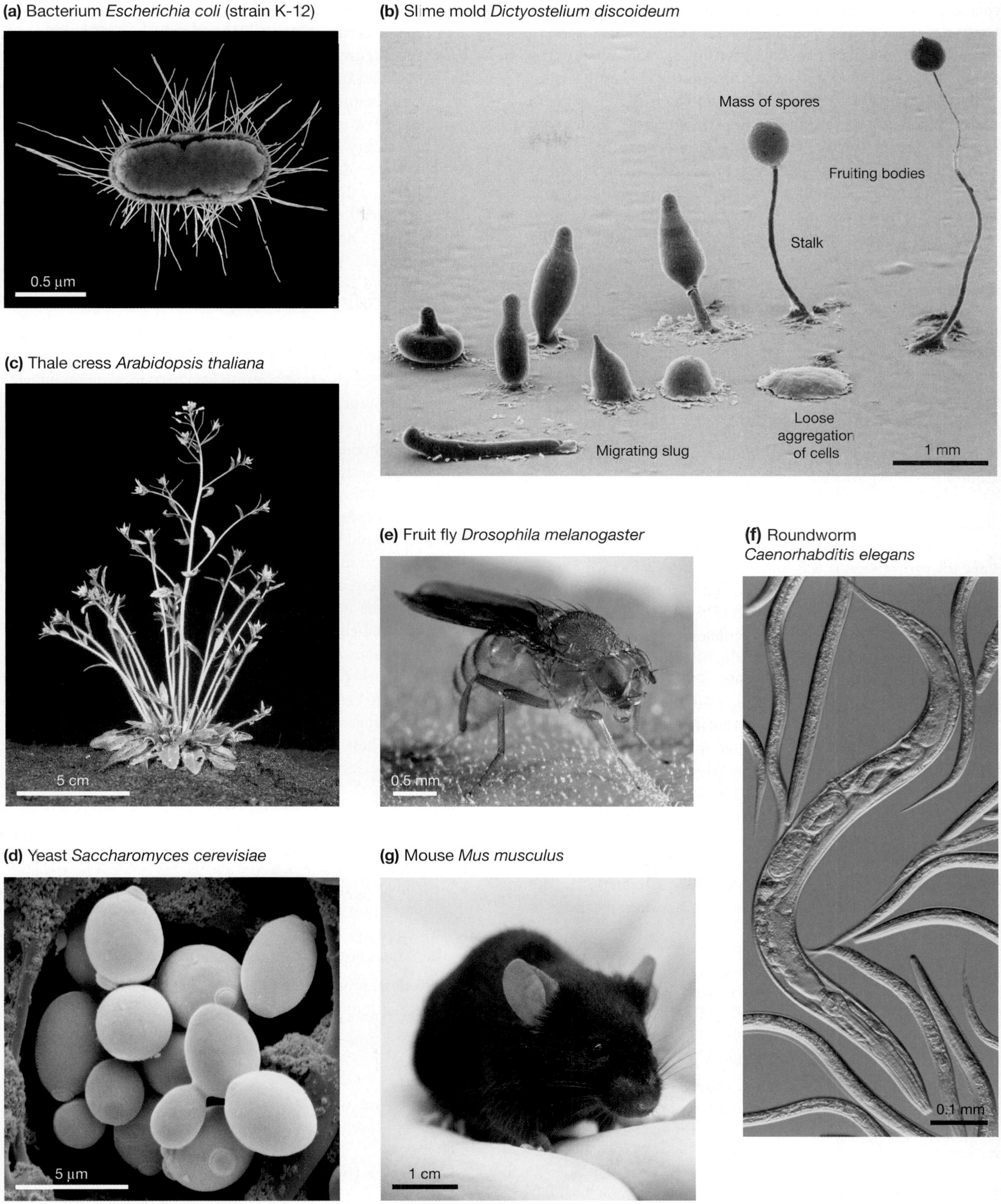

FIGURE B14.1 Model Organisms.

✔**QUESTION** *E. coli* is grown at a temperature of 37°C. Why?

species has become the organism of choice for experiments on control of the cell cycle and regulation of gene expression in eukaryotes. For example, research has confirmed that several of the genes controlling cell division and DNA repair in yeast have homologs in humans; and when mutated, these genes contribute to cancer. Strains of yeast that carry these mutations are now being used to test drugs that might be effective against cancer.

S. cerevisiae has become even more important in efforts to interpret the genomes of organisms like rice, mice, zebrafish, and humans. It is much easier to investigate the function of particular genes in *S. cerevisiae* by creating mutants or transferring specific alleles among individuals than it is to do the same experiments in mice or zebrafish. Once the function of a gene has been established in yeast, biologists can look for the homologous gene in other eukaryotes. If such a gene exists, they can usually infer that it has a function similar to its role in *S. cerevisiae.* It was also the first eukaryote with a completely sequenced genome.

Drosophila melanogaster

If you walk into a biology building on any university campus around the world, you are almost certain to find at least one lab where the fruit fly *Drosophila melanogaster* is being studied (**Figure B14.1e**).

Drosophila has been a key experimental subject in genetics since the early 1900s. It was initially chosen as a focus for study by T. H. Morgan because it can be reared in the laboratory easily and inexpensively, matings can be arranged, the life cycle is completed in less than two weeks, and females lay a large number of eggs. These traits made fruit flies valuable subjects for breeding experiments designed to test hypotheses about how traits are transmitted from parents to offspring (see Chapter 13).

More recently, *Drosophila* has also become a key model organism in the field of developmental biology. The use of flies in developmental studies was inspired in large part by the work of Christianne Nüsslein-Volhard and Eric Wieschaus, who in the 1980s isolated flies with genetic defects in early embryonic development. By investigating the nature of these defects, researchers have gained valuable insights into how various gene products influence the development of eukaryotes (see Chapter 21). The complete genome sequence of *Drosophila* has been available to investigators since the year 2000.

Caenorhabditis elegans

The roundworm *Caenorhabditis elegans* emerged as a model organism in developmental biology in the 1970s, due largely to work by Sydney Brenner and colleagues. (*Caenorhabditis* is pronounced *see-no-rab-DIE-tiss.*)

C. elegans was chosen for three reasons: **(1)** Its cuticle (soft outer layer) is transparent, making individual cells relatively easy to observe (**Figure B14.1f**); **(2)** adults have exactly 959 nonreproductive cells; and, most important, **(3)** the fate of each cell in an embryo can be predicted because cell fates are invariant among individuals. For example, when researchers examine a 33-cell *C. elegans* embryo, they know exactly which of the 959 cells in the adult will be derived from each of those 33 embryonic cells.

In addition, *C. elegans* are small (less than 1 mm long), are able to self-fertilize or cross-fertilize, and undergo early development in just 16 hours. The entire genome of *C. elegans* has now been sequenced.

Mus musculus

Because the house mouse *Mus musculus* is the most important model organism among mammals, it is especially prominent in biomedical research—where researchers need to work on individuals with strong genetic and developmental similarities to humans.

The house mouse was an intelligent choice of model organism in mammals because it is small and thus relatively inexpensive to maintain in captivity, and because it breeds rapidly. A litter can contain 10 offspring and generation time is only 12 weeks—meaning that several generations can be produced in a year. Descendants of wild house mice have been selected for docility and other traits that make them easy to handle and rear; these populations are referred to as laboratory mice (**Figure B14.1g**).

Some of the most valuable laboratory mice are strains with distinctive, well-characterized genotypes. Inbred strains are virtually homogenous genetically (see Chapter 25) and are useful in experiments where gene-by-gene or gene-by-environment interactions have to be controlled. Other populations carry mutations that knock out genes and cause diseases similar to those observed in humans. These individuals are useful for identifying the cause of genetic diseases and testing drugs or other types of therapies.

✔Questions

Which model organisms described here would be the best choice for the following studies? In each case, explain your reasoning.

1. A study of why specific cells in an embryo die at certain points in normal development. One goal is to understand the consequences for the individual when programmed cell death does not occur when it is supposed to.
2. A study of proteins that are required for cell-cell adhesion.
3. Research on a gene suspected to be involved in the formation of breast cancer in humans.

Answers

CHAPTER 1

Check Your Understanding (CYU)

CYU p. 5 The data points would all be about 11 percent, indicating no change in average kernel protein content over time. **CYU p. 8** From the sequence data provided, Species A and B differ only in one nucleotide of the rRNA sequence (position 10 from left). Species C differs from Species A and B in four nucleotides (positions 1, 2, 9, and 10). A correctly drawn phylogenetic tree would indicate that Species A and B appear to be closely related, with Species C being more distantly related [**see Figure A1.1**]. **CYU p. 12** The key here is to test predation rates during the hottest part of the day (when desert ants actually feed) versus other parts of the day. The experiment would best be done in the field, where natural predators are present. One approach would be to capture a large number of ants, divide the group in two, and measure predation rates (number of ants killed per hour) when they are placed in normal habitat during the hottest part of the day versus an hour before (or after). (1) The control group here is the normal condition—ants out during the hottest part of the day. If you didn't include a control, a critic could argue that predation did or did not occur because of your experimental setup or manipulation, not because of differences in temperature. (2) You would need to make sure that there is no difference in body size, walking speed, how they were captured and maintained, or other traits that might make the ants in the two groups more or less susceptible to predators. They should also be put out in the same habitat, so the presence of predators is the same in the two treatments.

You Should Be Able To (YSBAT)

YSBAT p. 5 Average kernel protein content declined, from 11 percent to a much lower value. (The experiment is still continuing; to date the low-selection line is about 5 percent.) **YSBAT p. 12** (1) A critic could argue that the ants weren't navigating normally, because they had been caught and released and transferred to a new channel. (2) A critic could argue that the ants can't navigate normally on their manipulated legs.

Caption Questions and Exercises

Figure 1.2 If Pasteur had done any of the things listed, he would have had more than one variable in his experiment. This would allow critics to claim that he got different results because of the differences in broth types, heating, or flask types—not the difference in exposure to preexisting cells. The results would not be definitive. **Figure 1.4** Molds and other fungi are more closely related to green algae because they differ from plants at two positions (5 and 8 from left), but differ from green algae at only one position (8). **Figure 1.6** The eukaryotic cell is roughly 10 times (more exactly, 8.5 times) the size of the prokaryotic cell. **Figure 1.8** You should be skeptical, because the results could be due to the single individual being sick or injured or unusual in some other way.

Summary of Key Concepts

KC 1.1 Dead cells cannot maintain a difference in conditions inside the cell versus outside, replicate, use energy, or process information. **KC 1.2** Observations on thousands of diverse species supported the claim that all organisms consist of cells. The hypothesis that all cells come from preexisting cells was supported when Pasteur showed that new cells do not arise and grow in a boiled liquid unless they are introduced from the air. **KC 1.3** If seeds with higher protein content leave the most offspring, then individuals with low protein in their seeds will become rare over time. **KC 1.4** A newly discovered species can be classified as a member of the Bacteria if the sequence of its rRNA contains some features found only in Bacteria. The same logic applies to classifying a new species in the Archaea or Eukarya. **KC 1.5** (1) A hypothesis is an explanation of how the world works; a prediction is an outcome you should observe if the hypothesis is correct. (2) Experiments are convincing because they measure predictions from two opposing hypotheses. Both predicted actions cannot occur, so one hypothesis will be supported while the other will not.

Test Your Knowledge

1. d; **2.** d; **3.** c; **4.** b; **5.** d; **6.** c

Test Your Understanding

1. That the entity they discovered replicates, processes information, acquires and uses energy, is cellular, and that its populations evolve. **2.** [**See Figure A1.2**] **3.** The rule ensured that two different organisms would never end up with the same name. **4.** Over time, traits that increased the fitness of individuals in this habitat became increasingly more frequent in the population. **5.** Individuals with certain traits selected, in the sense that they produce the most offspring. **6.** Yes. If evolution is defined as "change in the characteristics of a population over time," then those organisms that are most closely related should have experienced less change over time. On a phylogenetic tree, species with substantially similar rRNA sequences would be diagrammed with a closer common ancestor—one that had the sequences they inherited—than the ancestors shared between species with dissimilar rRNA sequences.

Applying Concepts to New Situations

1. A scientific theory is not a guess—it is an idea whose validity can be tested with data. Both the cell theory and the theory of evolution have been validated by large bodies of observational and experimental data. **2.** If all eukaryotes living today have a nucleus, then it is logical to conclude that the nucleus arose in a common ancestor of all eukaryotes, indicated by the arrow you should have added to the figure on page 14 [**see Figure A1.3**]. If it had arisen in a common ancestor of Bacteria or Archaea, then species in those groups would have had to lose the trait—an unlikely event. **3.** The data set was so large and diverse that it was no longer reasonable to argue that noncellular life-forms would be discovered. **4.** Yes. The heritable traits that confer resistance to HIV should increase over time.

CHAPTER 2

Check Your Understanding (CYU)

CYU p. 21 [**See Figure A2.1**] **CYU p. 33** (1) Gibbs equation: $\Delta G = \Delta H - T\Delta S$. ΔG symbolizes the change in the Gibbs free energy. ΔH represents the difference in potential energy between the products and the reactants. T represents the temperature (in degrees Kelvin) at which the reaction is taking place. ΔS symbolizes the change in entropy (disorder). (2) When ΔH is negative—meaning that the reactants have lower potential energy than the products—and when ΔS is positive, meaning that the products have higher entropy (are more disordered) than the reactants.

You Should Be Able To (YSBAT)

YSBAT p. 22 (1) $^{\delta+}H{-}O^{\delta-}{-}H^{\delta+}$. (2) If water were linear, the partial negative charge on oxygen would have partial positive charges on either side. Compared to the actual, bent molecule, the partial negative charge would be much less exposed and less able to participate in hydrogen bonding. **YSBAT p. 26** The pH 8 solution has 100 times fewer protons than the pH 6 solution. **YSBAT p. 30** (1) As T increases, ΔG is more likely to be negative, making the reaction spontaneous. (2) Exothermic reactions may be nonspontaneous if they result in a decrease in entropy—meaning that the products are more ordered than the reactants (ΔS is negative).

FIGURE A1.1

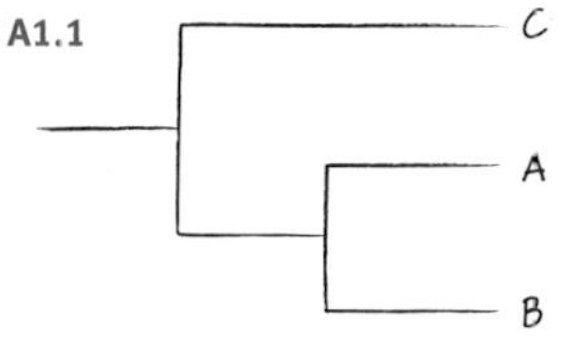

FIGURE A1.2

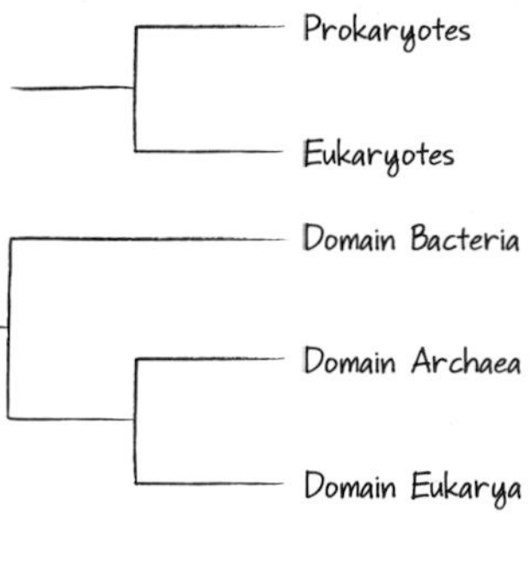

FIGURE A1.3

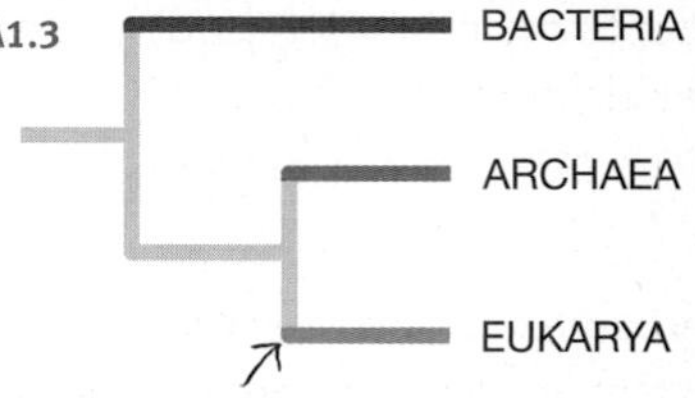

FIGURE A2.1

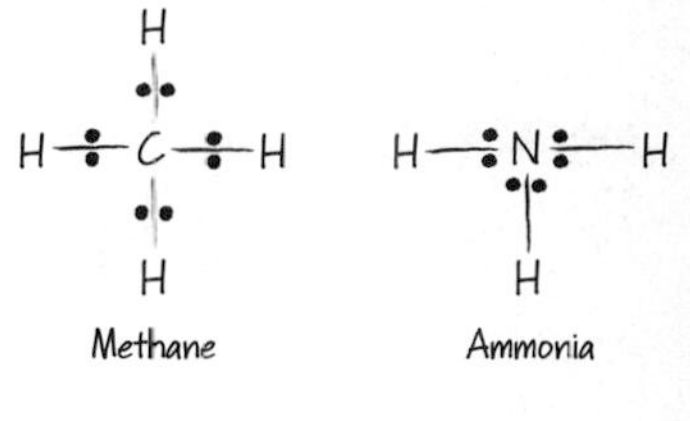

Caption Questions and Exercises

Figure 2.7 Oxygen and nitrogen have high electronegativity. They hold shared electrons more tightly than C, H, and many other atoms, resulting in polar bonds. **Figure 2.13** Oils are nonpolar. They have long chains of carbon atoms bonded to hydrogen atoms, which share electrons evenly because their electronegativities are similar. When an oil and water are mixed, the polar water molecules interact with each other via hydrogen bonding much more strongly than they interact with the nonpolar oil molecules, which interact with themselves instead. **Figure 2.16** The coffee becomes less acidic because milk is more alkaline (pH 6.5) than black coffee (pH 5). **Table 2.2** (Row 1) "Cause": Electrostatic attraction between partial charges on water molecules and opposite charges on ions; hydrogen bonds between water and other polar molecules. (Row 2) "Biological Consequences": Ice floats on denser liquid water. If it were denser and sank, oceans and lakes would fill with ice. (Row 4) "Cause": Liquid water must absorb lots of heat energy to break hydrogen bonds and change to a gas. One possible concept map relating the structure of water to its properties is shown below [see Figure A2.2]. **Figure 2.19** [See Figure A2.3] **Figure 2.22** The reaction rate at a specific set of reactant concentrations, averaged over many replicates. **Table 2.3** All the functional groups in Table 2.3, except the sulfhydryl group (−SH), are highly polar. The sulfhydryl group is only very slightly polar.

Summary of Key Concepts

KC 2.1 [See Figure A2.4] **KC 2.2** Because these functional groups are polar, water forms hydrogen bonds with them. **KC 2.3** Chemical energy—meaning, potential energy stored in chemical bonds—is transformed to heat. **KC 2.4** Spontaneous chemical reactions lead toward disorder and a release of energy. Cells are full of highly ordered molecules that contain high-energy bonds. Energy must be added to make these molecules and maintain life. **KC 2.5** Electrons in carbon-carbon bonds are held loosely and equally between the carbon atoms, whereas electrons in a carbon-oxygen bond are held tightly by the oxygen. Because of this difference, molecules with carbon-carbon bonds have more potential energy than carbon dioxide with its two carbon-oxygen bonds. Carbon-carbon bonds are often present in large, highly ordered molecules. The entropy of a group of such molecules, a measure of their disorder, is much less than the entropy of a group of simple, less-ordered molecules such as carbon dioxide.

Test Your Knowledge

1. b; **2.** a; **3.** a; **4.** d; **5.** d; **6.** d

Test Your Understanding

1. The reaction lowers the pH of the solution by releasing extra H^+ into the solution. If additional CO_2 is added, the sequence of reactions would be driven to the right, which would make the ocean more acidic. **2.** No. Shells that are farther from the protons (positive charges) in the nucleus house electrons that have greater potential energy than shells closer to the nucleus. **3.** [See Figure A2.5] The electron orbitals surrounding oxygen are oriented in the shape of a tetrahedron. Two of these orbitals are occupied by unshared electrons and the other two are shared with hydrogen atoms, resulting in the bent shape of the water molecule. Due to its greater electronegativity, oxygen attracts the shared electrons more strongly than does hydrogen, so it has a partial negative charge while the hydrogen atoms have a partial positive charge. **4.** In a covalent bond, the shared electrons are extremely close to two nuclei. In a hydrogen bond, the (+) and (−) charges are only partial, and the attraction is at a much greater distance. **5.** Water forms large numbers of hydrogen bonds, which must be broken before the molecules can start moving faster—meaning that their temperature has increased. **6.** The overall shape of an organic molecule is determined by its carbon framework. The functional groups attached to the carbons determine the molecule's chemical behavior, because these groups are likely to interact with other molecules.

Applying Concepts to New Situations

1. In CO_2, the double bonds between the carbon and each of the oxygen atoms lock the molecule into a linear shape that is nonpolar. In water, the partial charges on

FIGURE A2.2

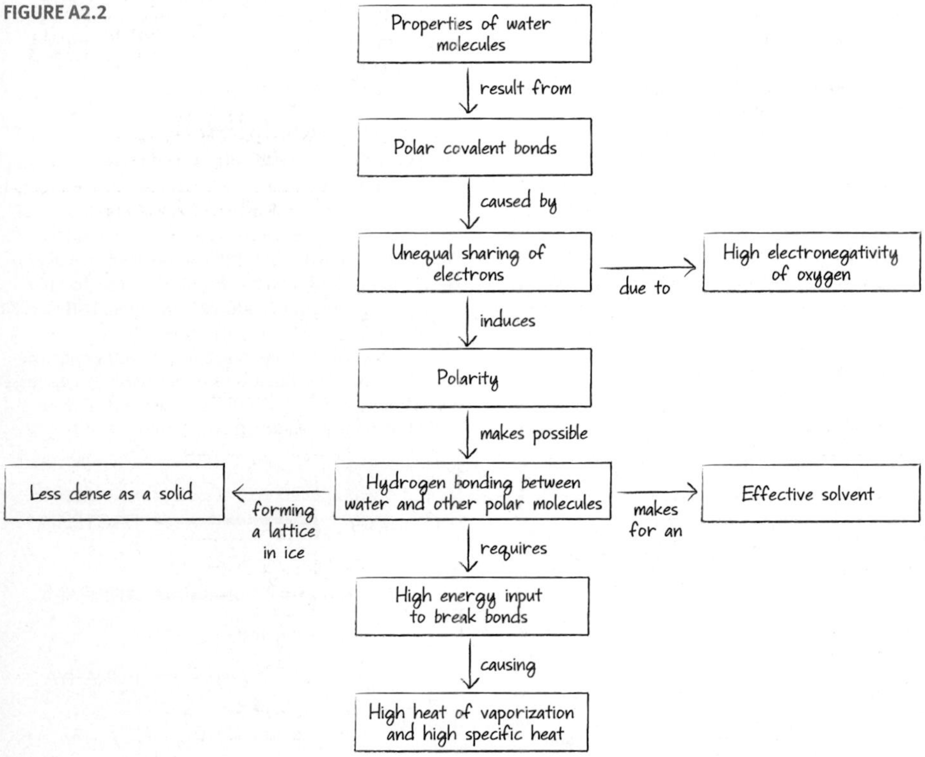

FIGURE A2.4

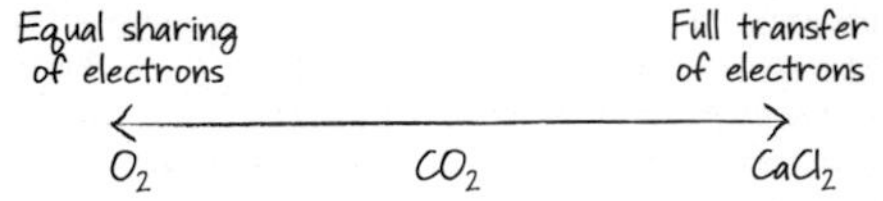

FIGURE A2.5

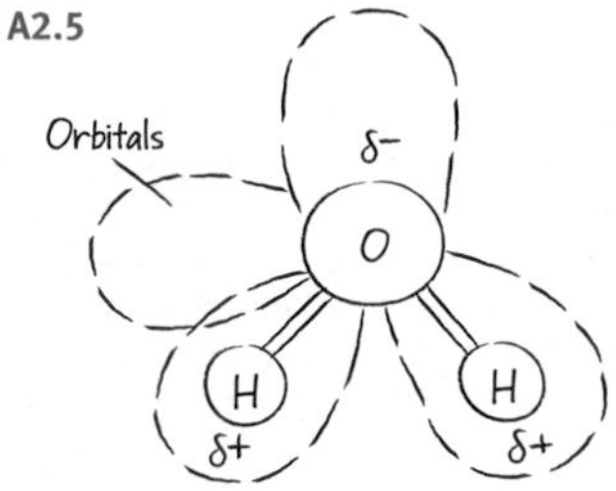

FIGURE A2.3

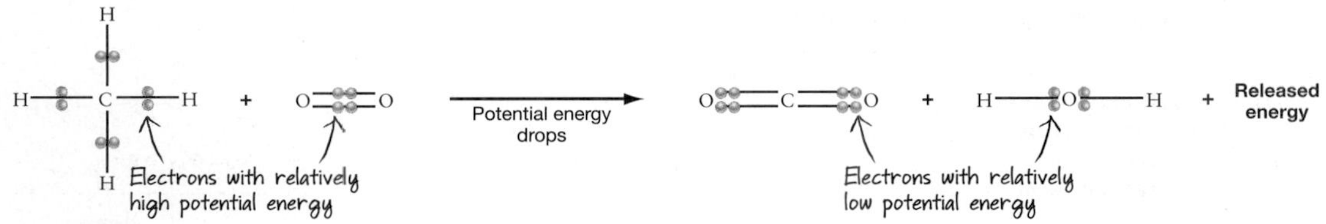

oxygen and hydrogen allow water to form hydrogen bonds with other polar or charged substances. In addition, the single covalent bonds in water are much easier to break than the double covalent bonds in carbon dioxide, making it more reactive. **2.** When oxygen is covalently bonded to carbon or hydrogen, the bond electrons spend more time near the oxygen atom. In contrast, carbon and hydrogen have roughly similar electronegativities, and they tend to share the electrons of a covalent bond more equally. **3.** The Sun will burn out when the mass of all of its component atoms is finally converted to energy. **4.** In hot weather, water absorbs large amounts of heat due to its high specific heat and high heat of vaporization. In cold weather, water releases the large amount of heat that it has absorbed.

CHAPTER 3

Check Your Understanding (CYU)

CYU p. 44 [See Figure A3.1] **CYU p. 51** Secondary, tertiary, and quaternary structure all depend on bonds and other interactions between amino acids that are linked in a chain (primary structure). **CYU p. 56** (1) Substrates can react only if they meet in a precise orientation. (2) The shape change alters the shape of the active site so that substrates no longer fit or can't be oriented correctly for a reaction to occur.

You Should Be Able To (YSBAT)

YSBAT p. 50 The sequence of amino acids in the two polypeptide chains constitutes the primary structure. Each chain contains α-helices and a β-pleated sheet (secondary structures), and is folded into a specific three-dimensional shape (tertiary structure). The overall three-dimensional shape of the dimeric Cro protein, formed by association of two identical polypeptides, constitutes its quaternary structure. **YSBAT p. 53** More—because catalysts lower the activation energy required for the reaction to proceed, more molecules at any given temperature have enough kinetic energy to supply the activation energy. **YSBAT p. 54** (1) orienting substrates, (2) transition state, (3) R-groups, (4) structure.

Caption Questions and Exercises

Figure 3.1 The water-filled flask is the ocean; the gas-filled flask is the atmosphere; the condensed water droplets are rain; the electrical sparks are lightning. **Figure 3.3** The green R-groups contain mostly C and H, which have roughly equal electronegativities. Electrons are evenly shared in C—H bonds and C—S bonds, so the groups are nonpolar. Most of the pink R-groups have a highly electronegative oxygen atom with a partial negative charge, making them polar. Cysteine has a sulfur that is slightly more electronegative than hydrogen, so it will be less polar than the other pink groups. **Figure 3.6** A polar covalent bond. The electrons are shared in the C—N bond, but because N is more electronegative than C, the bond is polar. **Figure 3.16** In order for a chemical reaction to proceed, the kinetic energy of the reactant molecules and atoms must be greater than the activation energy. Only the molecules to the right of the line you drew will be able to react. **Figure 3.18** No—a catalyst only changes the activation energy, not the overall free energy change. **Figure 3.22** [See Figure A3.2]

Summary of Key Concepts

KC 3.1 Many possible correct answers, including: the presence of an active site in an enzyme that is precisely shaped to fit a substrate or substrates in the correct orientation for a reaction to occur; the "donut" shape of porin allowing certain substances to pass through it; the butterfly shape of TATA-box binding protein being precisely the right size for a DNA molecule to fit. **KC 3.2** R-groups containing partial charges or full charges can form hydrogen bonds with water, make the amino acid soluble. Nonpolar R-groups do not interact with water and make the amino acid insoluble. **KC 3.3** [See Figure A3.3] **KC 3.4** When a catalyst is present, the reactants and products don't change—only the transition state changes. Thus, the overall free energy change in the reaction is the same—only the activation energy changes.

Test Your Knowledge

1. d; **2.** c; **3.** a; **4.** b; **5.** a; **6.** b

Test Your Understanding

1. The shape of reactant molecules (the key) fits into the active site of an enzyme (the lock). The model assumed that the enzyme is rigid; in fact it is flexible and dynamic. **2.** It will fold in on itself, away from water, when placed in an aqueous solution. **3.** Both are mechanisms that regulate enzymes; the difference is whether the regulatory molecule binds at the active site (competitive inhibition) or away from the active site (allosteric regulation). **4.** Energy is required. Polymerization is a nonspontaneous reaction because the product molecules have lower entropy and greater potential energy than the reactants. **5.** Proteins are highly variable in overall shape and chemical properties due to variation in the composition of R-groups and the array of secondary through quaternary structures that are possible. This variation allows them to fulfill many different roles in the cell. Diversity in the shape and reactivity of active sites makes them effective catalysts. **6.** The function of an enzyme depends on the shape and chemical reactivity of its active site. Temperature and pH affect enzyme function because they can change the shape and chemical reactivity of the active site—either by disrupting bonds or interfering with acid-base reactions.

Applying Concepts to New Situations

1. Yes—within limits. As Figure 3.23 shows, some bacteria thrive in very hot or highly acidic environments on Earth. But extremely low pHs or high temperatures are likely to denature proteins. **2.** The data suggest that the enzyme and substrate form a transition state that requires a change in the shape of the active site, with each movement corresponding to one reaction. **3.** Without the coenzyme, the free radical–containing transition state would not be stabilized and the reaction rate would drop dramatically. **4.** Residues in the active site changed in a way that made the drugs less likely to bind and prevent the normal substrate from binding.

FIGURE A3.1

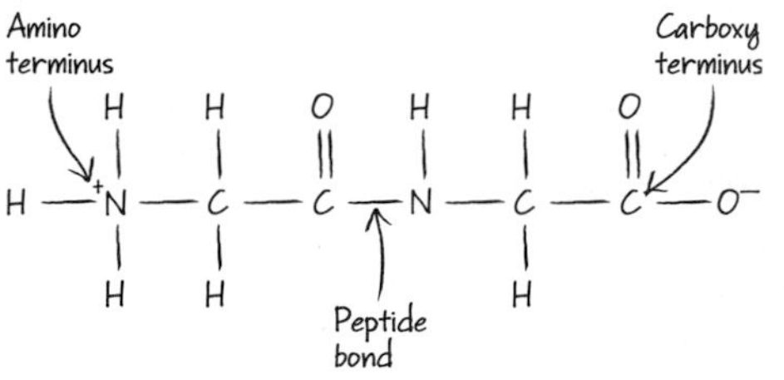

FIGURE A3.2

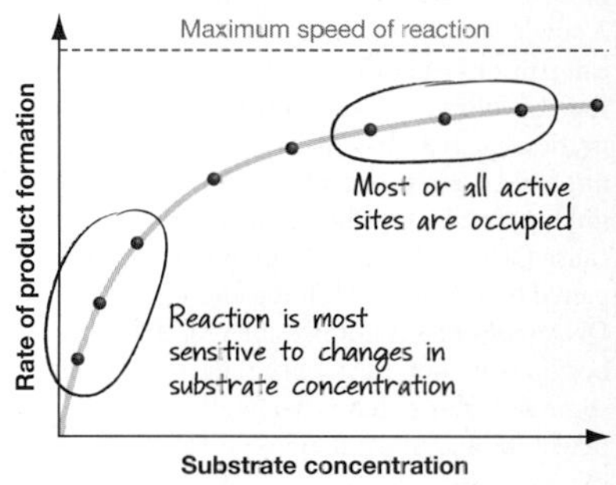

FIGURE A3.3

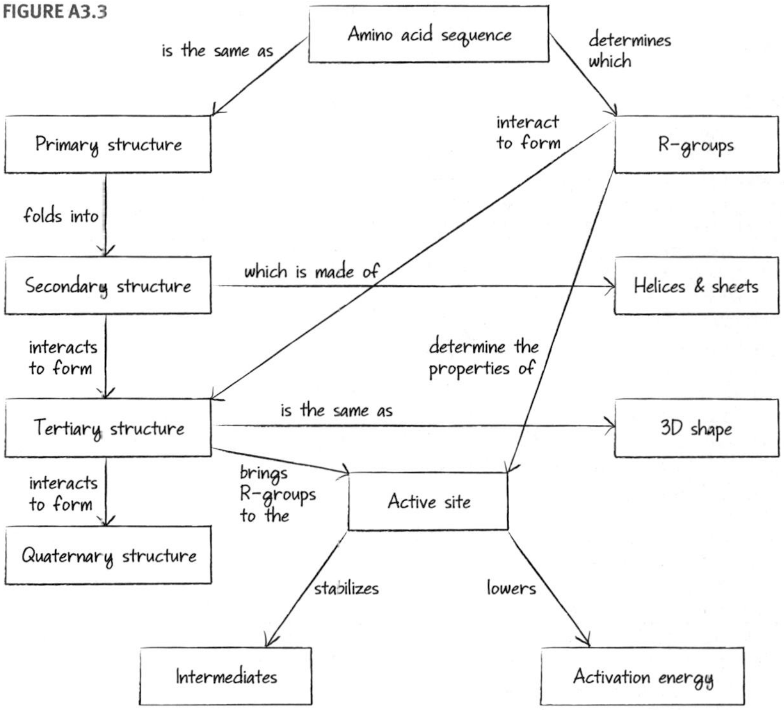

CHAPTER 4

Check Your Understanding (CYU)

CYU p. 62 [See Figure A4.1] **CYU p. 66** [See Figure A4.2]

You Should Be Able To (YSBAT)

YSBAT p. 60 [See Figure A4.3] **YSBAT p. 64** If the two strands are parallel, the nitrogenous bases are not aligned in a way that allows hydrogen bond formation. No helix would form.

Caption Questions and Exercises

Figure 4.3 5′ UAGC 3′. **Figure 4.9** Endergonic—energy must be added (as heat) for the reaction to occur.

Summary of Key Concepts

KC 4.1 (1) Both polymerize via energy-demanding condensation reactions that add a monomer to a growing chain. But proteins polymerize via formation of peptide bonds while nucleic acids polymerize via formation of phosphodiester linkages. (2) Proteins have an amino terminus at one end and a carboxyl terminus at the other end, while nucleic acids have a 5′ end (with an unlinked phosphate group) and a 3′ end (with an unlinked hydroxyl group). (3) Both have a "backbone"; in proteins it consists of amino and carbonyl groups, while in nucleic acids it consists of phosphate groups and sugars. **KC 4.2** C-G pairs involve three hydrogen bonds, so they are more stable than A-T pairs with just two bonds. **KC 4.3** A single-stranded RNA molecule has unpaired bases that can pair up with other bases on the same RNA strand, thereby folding the molecule into stem-and-loop configurations with a particular three-dimensional shape. The unpaired bases also can bind to another RNA molecule, linking the two together in a quaternary structure. Because DNA molecules are double stranded, with no unpaired bases, internal folding and interaction with other DNA molecules is not possible. **KC 4.4** If the copied ribozymes are not exactly like the template molecule, then some are likely to have changes that make them more efficient as catalysts. If there were no variation—meaning, if copying were exact—there could be no improvement in efficiency or other change over time.

Test Your Knowledge

1. c; **2.** d; **3.** a; **4.** b; **5.** a; **6.** c

Test Your Understanding

1. [See Figure A4.4] **2.** The addition of the phosphate groups raises the potential energy of the resulting monomers (nucleoside triphosphates) enough that the polymerization reaction is exergonic. **3.** Nucleic acids are directional because the two ends of the polymer are different. One end has a free phosphate group on a 5′ carbon; the other end has a free hydroxyl group bonded to a 3′ carbon. **4.** DNA is a more stable molecule than RNA because it lacks a hydroxyl group on the 2′ carbon and is therefore resistant to degradation, and because the two sugar-phosphate backbones are linked by many hydrogen bonds between nitrogenous bases. DNA is more stable than proteins because it is symmetrical and has few exposed chemical groups that could participate in chemical reactions. **5.** Hairpin structures form when folding of the sugar-phosphate backbone allows the bases in one part of an RNA strand to align with bases in another segment of the same RNA strand in an antiparallel fashion, so they can hydrogen-bond and form a double helix. **6.** DNA has limited catalytic ability because it (1) lacks functional groups that can stabilize transition states and (2) has a regular structure that is not conducive to forming active sites. RNA molecules can catalyze some reactions because they (1) have exposed hydroxyl functional groups that can stabilize transition states and (2) can fold into shapes that create active sites. Proteins are the most effective catalysts because (1) amino acids have a wide range of chemical reactivity, and (2) they fold into an enormous diversity of shapes that can create active sites.

Applying Concepts to New Situations

1. Yes—if the complementary bases lined up over the entire length of the two strands, they would twist into a double helix analogous to a DNA molecule. The same types of hydrogen bonds and hydrophobic interactions would occur as observed in the "stem" portions of hairpins in single-stranded RNA. **2.** An RNA replicase would undergo replication and be able to evolve. It would process information in the sense of copying itself, and would use energy to drive endergonic polymerization reactions. It would not be bound by a membrane, however, and would not be able to acquire energy. It would best be considered as an intermediate step between nonlife and "true life," as outlined in Chapter 1. **3.** It is unlikely that bases would align properly for hydrogen bonding to occur, so hydrophobic interactions would probably be more important. **4.** No single right answer (these are opinion questions). Note that when biologists design studies about the origin of life, they base their experimental conditions on the best available current knowledge about early Earth conditions. New information about these conditions might require a revision of the experiment that leads to different results and conclusions.

FIGURE A4.1

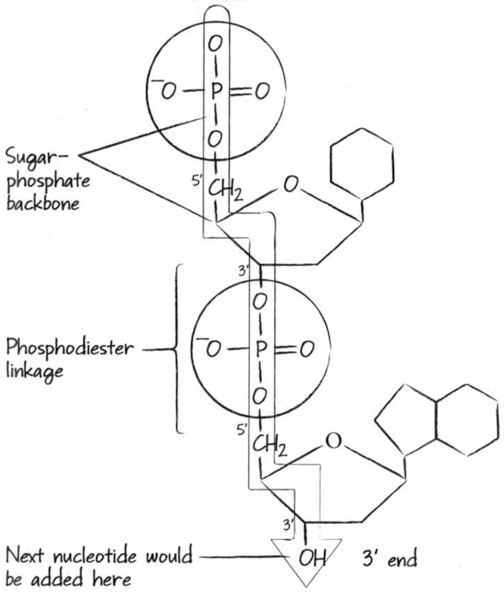

FIGURE A4.2

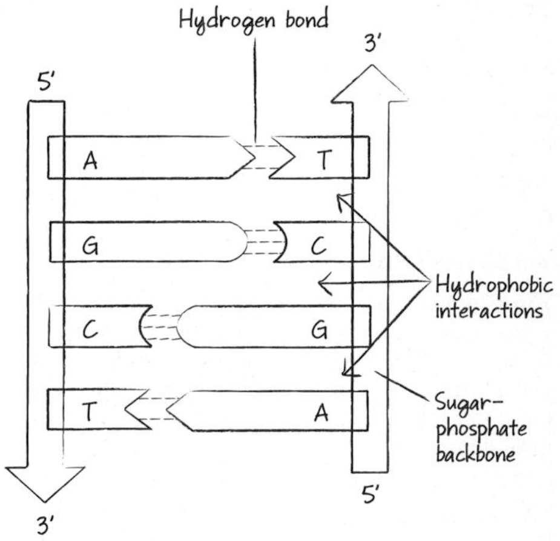

FIGURE A4.3

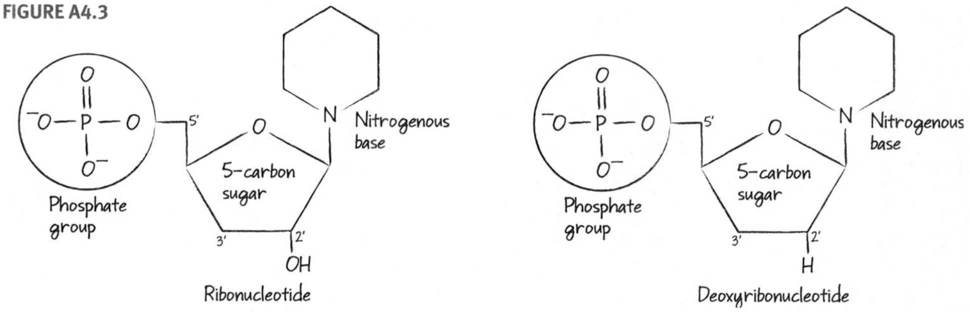

FIGURE A4.4

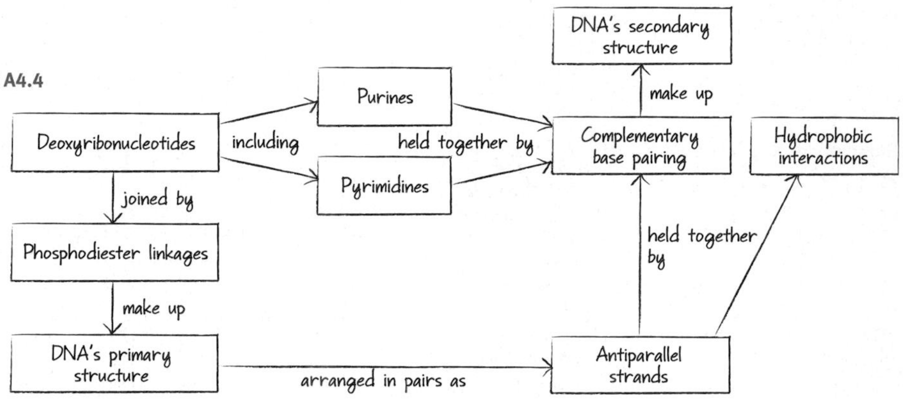

CHAPTER 5

Check Your Understanding (CYU)

CYU p. 73 [See Figure A5.1] **CYU p. 76** They could differ in (1) location of linkages (e.g., 1,4 or 1,6); (2) types of linkages (e.g., α or β); (3) the sequence of the monomers (e.g., two galactose then two glucose, versus alternating galactose and glucose); and/or (4) whether the four monomers are linked in a line or whether they branch. **CYU p. 79** (1) Aspect 1: The β-1,4-glycosidic linkages in these molecules are difficult to degrade. Aspect 2: When individual molecules of these carbohydrates align, bonds form between them and produce tough fibers or sheets. (2) Most are probably being broken down into glucose, which in turn is being broken down in reactions that lead to the synthesis of ATP. In short, the carbohydrates are providing you with chemical energy.

Caption Questions and Exercises

Figure 5.2 [See Figure A5.2] **Figure 5.7** All of the C–C and C–H bonds should be circled.

Summary of Key Concepts

KC 5.1 Molecules have to interact in an extremely specific orientation in order for a reaction to occur. Changing the location of a functional group by even one carbon can mean that the molecule will undergo completely different types of reactions. **KC 5.2** The orientations of the α- versus β-glycosidic linkages are different, so the molecules in a chain end up in a spiral versus linear arrangement. **KC 5.3** (1) Polysaccharides used for energy storage are formed entirely from glucose monomers joined by α-glycosidic linkages; structural polysaccharides are comprised of glucose or other sugars joined by β-glycosidic linkages. (2) The monomers in energy-storage polysaccharides are linked in a spiral arrangement; the monomers in structural polysaccharides are linked in a linear arrangement. (3) Energy-storage polysaccharides may branch; structural polysaccharides do not. (4) Individual chains of energy-storage polysaccharides do not associate with each other; adjacent chains of structural polysaccharides are linked by hydrogen bonds or covalent bonds.

Test Your Knowledge

1. d; **2.** a; **3.** c; **4.** a; **5.** d; **6.** b

Test Your Understanding

1. Carbohydrates are ideal for signaling the identity of the cell because they are so diverse structurally. This diversity enables them to serve as very specific identity tags for cells. **2.** When you compare the glucose monomers in an α-1,4-glycosidic linkage versus a β-1,4-glycosidic linkage, the linkages are located on opposite sides of the plane of the glucose rings (e.g., "above" versus "below" the plane), and the glucose monomers are linked in the same orientation versus having every other glucose flipped in orientation. β-1,4-glycosidic linkages are much more difficult for enzymes to break, so they resist degradation. **3.** Starch and glucose both consist of glucose monomers joined by α-1,4-glycosidic linkages, and both function as storage carbohydrates. Starch is composed of an unbranched amylose and branched amylopectin, while glycogen is even more highly branched. **4.** The electrons in the C=O bonds of carbon dioxide molecules are held tightly by the highly electronegative oxygen atoms, so have low potential energy. The electrons in the C–C and C–H bonds of carbohydrates are shared equally and have much higher potential energy. **5.** They have β-1,4-glycosidic linkages, which are resistant to degradation and put glucose monomers in positions where hydrogen bonds can form between adjacent strands, forming strong fibers. **6.** Glycogen has α-1,4-glycosidic linkages with α-1,6-glycosidic linkages at branch points; cellulose has β-1,4-glycosidic linkages and is not branched—instead, it forms hydrogen bonds with adjacent cellulose molecules. Glycogen is an energy-storage molecule; cellulose is a structural polysaccharide.

Applying Concepts to New Situations

1. Carbohydrates are energy-storage molecules, so minimizing their consumption may reduce total energy intake. Lack of available carbohydrate also forces the body to use fats for energy, reducing the amount of fat that is stored. **2.** Lactose is made up of glucose and galactose. **3.** Amylase breaks down the starch in the cracker into glucose monomers, which stimulate the sweet receptors in your tongue. **4.** When bacteria contact lysozyme, their cell walls, which contain peptidoglycan, begin to degrade, leading to the death of the bacteria.

CHAPTER 6

Check Your Understanding (CYU)

CYU p. 85 (1) Fats consist of three fatty acids linked to glycerol; steroids have a distinctive four-ring structure with a side group attached; phospholipids have a hydrophilic, phosphate-containing "head" region and a hydrocarbon tail. (2) In cholesterol, the hydrocarbon steroid rings are hydrophobic; the hydroxyl group is hydrophilic. In phospholipids, the phosphate-containing head group is hydrophilic; the fatty acid chains are hydrophobic. **CYU p. 89** [See Table A6.1] **CYU p. 92** [See Figure A6.1] **CYU p. 99** Passive transport does not require an

FIGURE A5.1

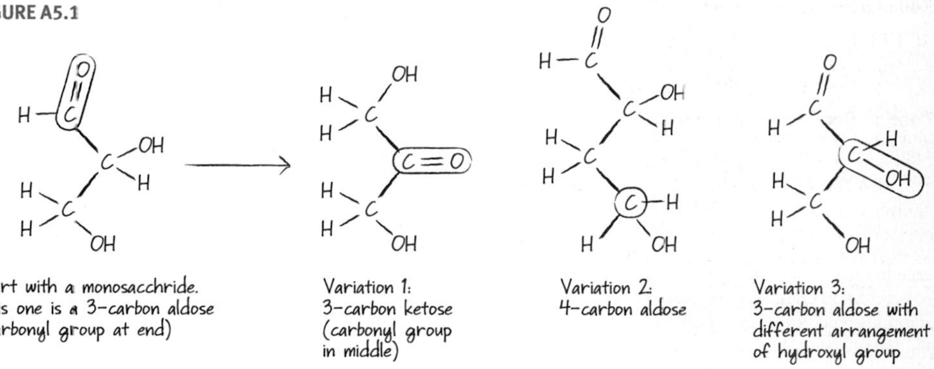

FIGURE A5.2

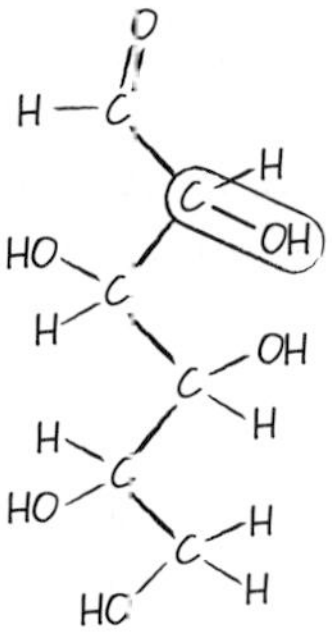

FIGURE A6.1

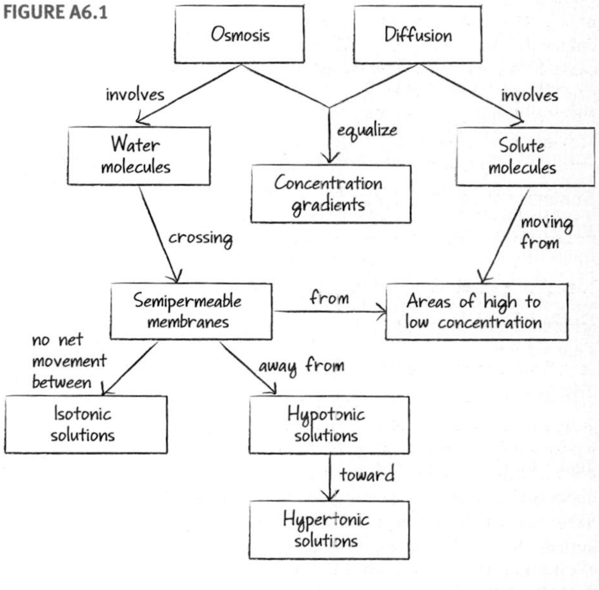

TABLE A6.1

Factor	Effect on permeability	Reason
Temperature	Decreases as temperature decreases.	Hydrophobic tails move more slowly, increasing the strength of hydrophobic interactions (membrane is more dense).
Cholesterol	Decreases as cholesterol content increases.	Cholesterol molecules fill in the spaces between the fatty-acid tails, making the membrane more tightly packed.
Length of hydrocarbon tails	Decreases as length of hydrocarbon tails increases.	Longer hydrocarbon tails have stronger hydrophobic interactions, making the tails pack together more tightly.
Saturation of hydrocarbon tails	Decreases as degree of saturation increases.	Saturated fatty acids have straight hydrocarbon tails that pack together and form many hydrophobic interactions, leaving few gaps.

expenditure of energy—it happens as a result of existing concentration or electrical gradients. Active transport is active in the sense of requiring energy. In cotransport, a second ion or molecule is transported against its electrochemical gradient with ("co") an ion that is transported along its electrochemical gradient.

You Should Be Able To (YSBAT)

YSBAT p. 84 Fats have many C—C and C—H bonds, which have high free energy. They are hydrophobic because they contain long hydrocarbon chains and lack any highly polar groups. **YSBAT p. 86** No, because both are relatively large molecules that are polar or carry a charge. **YSBAT p. 91** (1) Add a large number of green dots to the inside of the liposome. (2) Add a large number of green dots to the solution outside the liposome. **YSBAT p. 94** Your arrow should point out of the cell. There is no concentration gradient, but the outside has a net positive charge, which favors outward movement of negative ions. **YSBAT p. 96** [See Figure A6.2]

Caption Questions and Exercises

Figure 6.3 At the polar hydroxyl group in cholesterol and the polar head group in phospholipids. **Figure 6.14** Higher, because less water would have to move to the right side to achieve equilibrium. **Figure 6.21** The electrical gradient would reverse, because the inside of the membrane would be more positive than the outside. The concentration gradient for sodium would also reverse; sodium would diffuse to the exterior. **Figure 6.22** No—the 10 replicates where no current was recorded probably represent instances where the CFTR protein was damaged and not functioning properly. (In general, no experimental method works "perfectly.") **Figure 6.27** "Diffusion": description as given; no proteins involved. "Facilitated diffusion": Passive movement of ions or molecules that cannot cross a phospholipid bilayer readily, along a concentration gradient. Facilitated by channels or transporters. "Active transport": Active (energy-demanding) movement of ions or molecules against an electrochemical gradient.

Summary of Key Concepts

KC 6.1 Highly permeable bilayers consist of phospholipids with short, unsaturated hydrocarbon tails. **KC 6.2** (1) The solute will diffuse until both sides are at equal concentrations. (2) Water will diffuse toward the side with the higher solute concentration. **KC 6.3** [See Figure A6.3]

Test Your Knowledge

1. b; **2.** a; **3.** b; **4.** d; **5.** c; **6.** d

Test Your Understanding

1. No, because they have no polar end to interact with water. Instead, these lipids would collect and float on the surface of water, or collect in droplets suspended in water, reducing their interaction with water to a minimum. **2.** Hydrophilic, phosphate-containing head groups interact with water; hydrophobic, fatty-acid tails associate with each other. A bilayer has a lower potential energy and thus is more stable than are independent phospholipids in solution. **3.** Ethanol's polarity reduces the speed at which it can cross a membrane, but its small size and lack of charge increase the speed at which it crosses membranes. Ethanol probably crosses a membrane more slowly than water (which is also polar, but smaller) but faster than larger polar molecules, and much faster than charged molecules. **4.** If no membrane separates the two solutions, then the constant, random motion of solute and water molecules causes even mixing of the two solutions. No net diffusion of water molecules will occur. **5.** Only hydrophobic amino acids interact with the nonpolar lipid tails in the interior of a phospholipid bilayer. Figure 3.3 lists the following nonpolar, hydrophobic amino acids: glycine, alanine, valine, leucine, isoleucine, methionine, phenylalanine, tryptophan, and proline. According to Table 3.1, tyrosine and cysteine are also relatively hydrophobic, suggesting that they might also be found in the interior of transmembrane proteins. **6.** CO_2 will diffuse into the cell; water will move out via osmosis. Na^+ and K^+ will enter the cell through gramicidin by facilitated diffusion. Cl^- will not move.

Applying Concepts to New Situations

1. Flip-flops should be rare, because they require polar head group to pass through the hydrophobic portion of the lipid bilayer. To test this prediction, you could monitor the number of dyed phospholipids that transfer from one side of the membrane to the other in a given period of time. **2.** The kinks in unsaturated hydrocarbon tails keep membranes fluid, even at low temperature, because they prevent extensive hydrophobic interactions. In hot environments, it would be advantageous for phospholipids to have saturated tails, to prevent membranes from being too fluid. **3.** Adding a methyl group makes a drug more hydrophobic and thus more likely to pass through a lipid bilayer. Adding a charged group make it hydrophilic and reduces its ability to pass through the lipid bilayer. **4.** Detergents are amphipathic. Their hydrophobic ends interact with grease while their hydrophilic ends interact with water. This forms a bridge between the grease and the water, effectively making the grease dissolve in water and allowing it to be washed away.

CHAPTER 7

Check Your Understanding (CYU)

CYU p. 105 (1) Photosynthetic membranes increase food production by providing a large surface area to hold the pigments and enzymes required for photosynthesis. (2) The presence of a magnetic mineral, which changes position as the cell moves through a magnetic field, allows the cell to move in a directed way. (3) The layer of thick, strong material stiffens the cell and provides protection from mechanical damage. **CYU p. 115** (1) Both organelles contain specific sets of enzymes, and the interior of lysosomes is acidic. Peroxisomes contain catalase and other enzymes that process fatty acids and toxins via oxidation reactions. Lysosomal enzymes digest macromolecules, releasing monomers that can be recycled into new macromolecules. (2) From top to bottom, the cells in the last column should read as follows: administrative/information hub, protein factory, large molecule manufacturing and shipping (protein synthesis and folding center, protein finishing and shipping line, fat factory, waste processing and recycling center), fatty-acid processing and detox center, warehouse, power station, food-manufacturing facility, support beams, perimeter fencing with secured gates, and leave blank. **CYU p. 122** (1) Proteins that lack a signal sequence will not be delivered to the ER. They are released into the cytosol. (2) The proteins will not be secreted—they will be found in lysosomes. **CYU p. 128** The products will not be able to move through the system normally, because their microtubule "roads" are absent.

You Should Be Able To (YSBAT)

YSBAT p. 128 (1) The microtubules at the bottom of the axoneme would slide to the right, but the axoneme would not bend. (2) Nothing would happen (dynein wouldn't move).

Caption Questions and Exercises

Figure 7.1 [See Figure A7.1] **Figure 7.15** The vesicles deliver digestive enzymes that are required for the mature lysosome to function. **Figure 7.16** Storing the toxins in vacuoles prevents the toxins from damaging the plant's own organelles and cells. **Figure 7.19** The cell wall is outside the plasma membrane. **Figure 7.21** [See Figure A7.2] **Figure 7.23** "Prediction": The labeled tail region fragments or the labeled core region fragments of the nucleoplasmin protein will be found in the cell nucleus. "Prediction of null hypothesis": No labeled fragments of the nucleoplasmin protein will be found in the nucleus of the cell. "Conclusion": The "Send to nucleus" zip code is in the tail region of the nucleoplasmin protein.

FIGURE A6.2

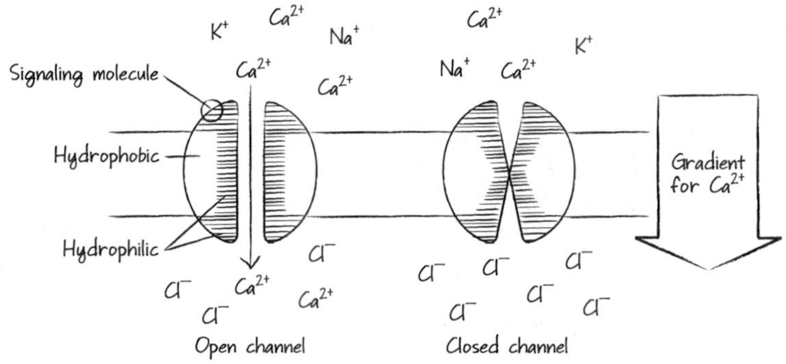

FIGURE A6.3

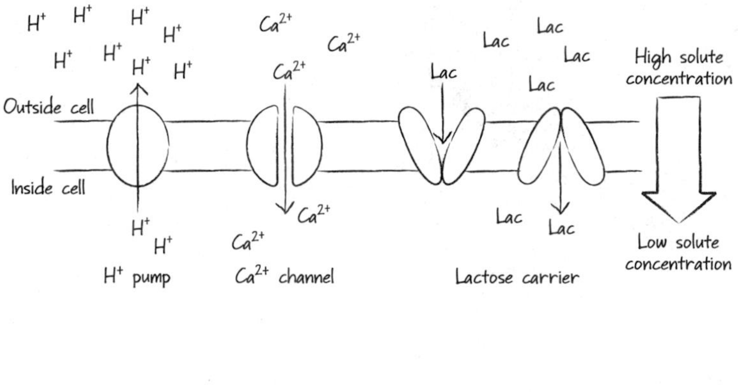

Summary of Key Concepts

KC 7.1 (1) They will be unable to make ATP and will die. (2) They will be limp and less able to resist attack by predators. In many environments they will be unable to resist the osmotic pressure of water entering the cytoplasm and will burst. (3) They will be unable to synthesize new proteins and will die. **KC 7.2** As a group, proteins have complex and highly diverse shapes and chemical properties. Individual proteins, then, can recognize and bind to molecular zip codes in a very specific way (like a key fitting into a lock). **KC 7.3** Actin filaments are made up of two strands of actin monomers, microtubules are made up of tubulin protein dimers that form a tube, intermediate filaments are made up of a variety of different protein monomers. Actin filaments and microtubules exhibit polarity (or directionality), with new subunits constantly being added or subtracted at either end (but added faster to the plus end). All three elements provide structural support, but only actin filaments and microtubules are involved in movement and cell division. **KC 7.4** Micrographs of cells at increasing time intervals after the pulse treatment represent time-lapse photography—a set of still images that show change through time.

Test Your Knowledge

1. a; **2.** c; **3.** c; **4.** b; **5.** a; **6.** d

Test Your Understanding

1. All cells are bound by a plasma membrane, are filled with cytoplasm, carry their genetic information (DNA) in chromosomes, and contain ribosomes (the sites of protein synthesis). Some prokaryotes have organelles not found in plants or animals, such as a magnetite-containing structure. Plant cells have chloroplasts, vacuoles, and a cell wall. Animal cells contain lysosomes and lack a cell wall. **2.** Ribosome in cytoplasm (protein is synthesized) → Rough ER (protein is folded and initially modified) → Transport vesicle → Golgi apparatus (protein is glycosylated; has molecular tag indicating destination) → Transport vesicle → Plasma membrane → Extracellular space. **3.** When a globular "head" section of kinesin binds and releases ATP, it undergoes a conformational change that swings it forward and binds it to the microtubule. The two globular head regions alternate between swinging and binding movements, causing the protein to "walk" down the microtubule. **4.** Changes in the length of microfilaments and microtubules contribute to changes in the size and shape of the cytoskeleton. **5.** Plasma membrane proteins are produced by the endomembrane system. Cells that function in membrane transport processes should have an extensive endomembrane system, including RER. **6.** All microtubules have the same basic structure, consisting of tubulin dimers that form a tube with plus and minus ends. The microtubules involved in structural support have no associated motor proteins; those involved in vesicle movement and flagella do. The microtubules that make up flagella are grouped in a 9 doublets + 2 single-microtubules arrangement. Structural and vesicle transport microtubules are single structures.

Applying Concepts to New Situations

1. Since mannose-6-phosphate serves as a lysosome tag in the endomembrane system, material that is ingested by phagocytic cells of the immune system might be targeted to lysosomes in the same way. To test this hypothesis, you could expose immune system cells to a population of bacteria that have labeled proteins, isolate the proteins after ingestion, and test to see whether they now have a mannose-6-phosphate tag attached. **2.** The proteins must receive a molecular zip code that binds to a receptor on the surface of peroxisomes. They could diffuse randomly to peroxisomes, or be transported in a directed way by vesicles or specialized motor proteins. **3.** (a) "Housekeeping"—destruction of macromolecules

FIGURE A7.1

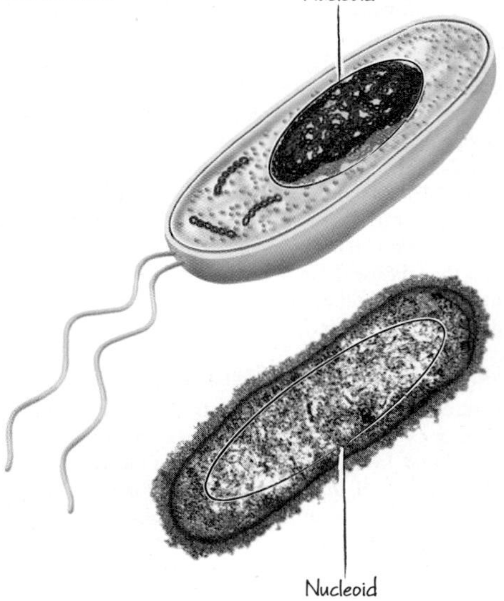

FIGURE A7.2

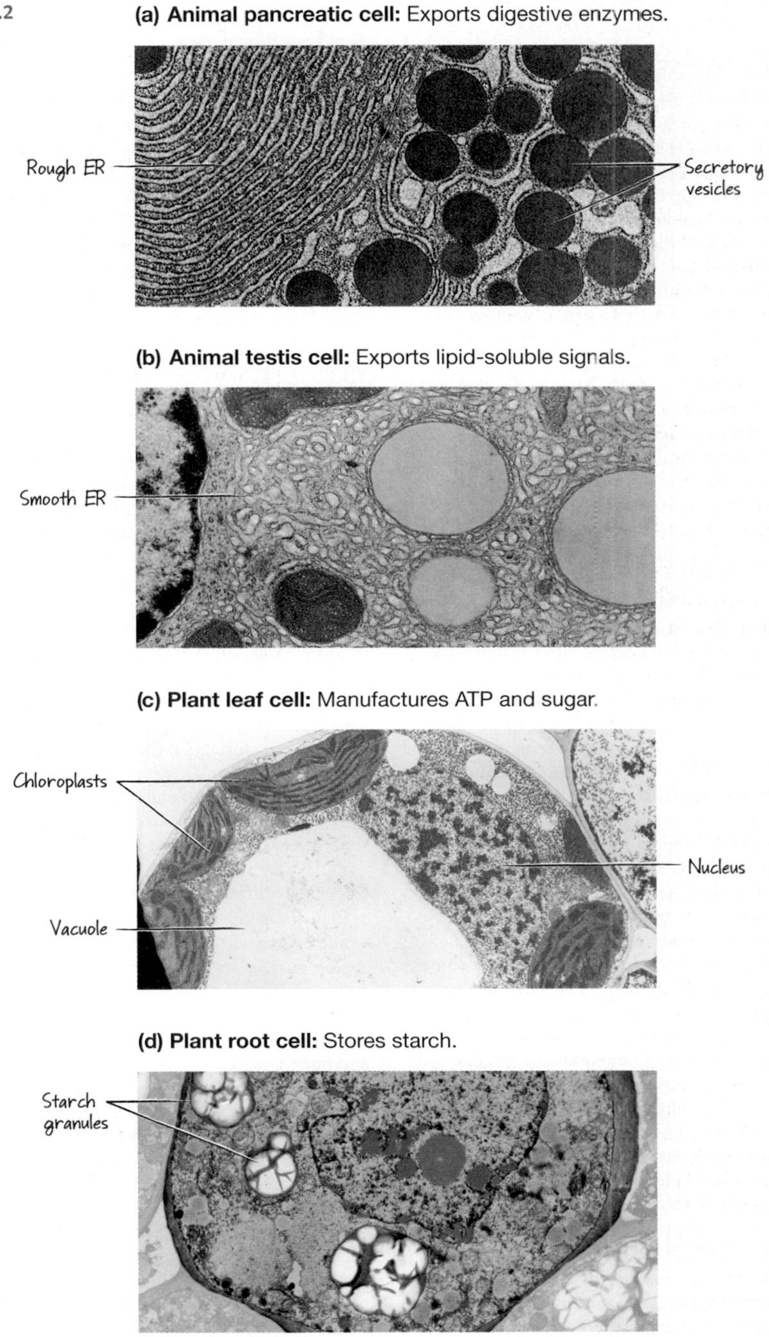

that are dangerous or no longer needed. (b) Structural support (e.g., wood is made up of cells with extensive cell walls). (c) Detoxification of dangerous compounds via oxidation reactions, or processing of stored fatty acids. **4.** The cell wall prevents the cells from bursting in hypotonic (aqueous) environments, or protects the cell from damage. Expose mutant and normal individuals to (1) a hypotonic environment, and (2) a eukaryote or virus that routinely attacks and kills this archaeal species. If the first hypothesis is correct, the mutant cells will burst but the normal cells will live. If the second hypothesis is correct, more mutant cells will die than normal cells.

CHAPTER 8

Check Your Understanding (CYU)

CYU p. 134 Plant cell walls and animal ECMs are both fiber composites. In plant cell walls the fiber component consists of cross-linked cellulose fibers and the ground substance is pectin. In animal ECMs the fiber component consists of collagen fibrils and the ground substance is gel-forming polysaccharides. **CYU p. 139** (1) The three structures differ in composition, but their function is similar. The middle lamella in plants is composed of pectins that glue adjacent cells together. Tight junctions are made up of membrane proteins that line up and "stitch" adjacent cells together. Desmosomes are "rivet-like" structures composed of proteins that link the cytoskeletons of adjacent cells. (2) The plasma membranes of adjacent plant cells are continuous at plasmodesmata, so cytoplasm, smooth endoplasmic reticulum, and other cell components are shared. Gap junctions connect adjacent animal cells by forming protein-lined pores. The openings allow ions and small molecules to be shared between cells. **CYU p. 146** (1) Each cell-cell signal binds to a specific receptor protein. A cell can respond to a signal only if it has the appropriate receptor. Different types of cells have different types of receptors. (2) Signals are amplified if one or more steps in a signal transduction pathway, involving either second messengers or a phosphorylation cascade, results in the activation of multiple downstream molecules.

You Should Be Able To (YSBAT)

YSBAT p. 138 (1) Yes—molecules can pass through a middle lamella because it is made up of gelatinous material that is not watertight. (2) Developing muscle cells could not adhere normally and muscle tissue would not form properly. The embryo would die. **YSBAT p. 143** The spy is the signal that arrives at the receptor (the castle gate). The guard is the G protein–linked receptor in the plasma membrane, and the queen is the G protein. The commander of the guard is the enzyme that catalyzes production of a second messenger (the soldiers). **YSBAT p. 144** (1) The red dominos (RTK components) would be the first dominos in the chain and the first to be tipped to initiate the domino cascade. The black domino (Ras) would be the second domino in line. Then there would be green, blue, pink, and yellow dominos (other kinases) set up in branching patterns to represent how Ras phosphorylates and activates a number of different kinases. Each of those colored dominos would be followed by a line of dominos of the same color representing the target proteins of those kinases. (2) Tipping the red domino would knock down the black domino (Ras) and would initiate a branching domino cascade that is dependent on the black domino's falling. The red domino represents how RTK activates Ras and how Ras initiates a cascade of phosphorylation.

Caption Questions and Exercises

Figure 8.9 "Prediction of null hypothesis": Cells will not adhere, or will adhere randomly with respect to cell type and species. **Figure 8.10** "Prediction": Cells treated with an antibody that blocks membrane proteins involved in adhesion will not adhere. "Prediction of null hypothesis": All cells will adhere normally. **Figure 8.12** Top: Tight junctions; Middle: Desmosome; Bottom: Gap junctions.

Summary of Key Concepts

KC 8.1 (1) Adjacent cells fall apart from each other. (2) Nearby cells fall apart from each other; both cells and tissues are weaker and more susceptible to damage. **KC 8.2** Epithelium separates compartments, creating an inside and outside. If tight junctions degrade, materials from the outside will diffuse in and materials from the inside will diffuse out. **KC 8.3** Adrenalin binds to both heart and liver cells, but the activated receptors trigger different signal transduction pathways and lead to different cell responses. **KC 8.4** If only one step existed in a signal transduction pathway, it would be difficult for the process to be regulated by many different pathways. Multiple steps provide points for different pathways to regulate the response.

Test Your Knowledge

1. b; **2.** b; **3.** a; **4.** b; **5.** d; **6.** c

Test Your Understanding

1. The cross-linked fiber components withstand tension; the ground substance withstands compression. **2.** If each enzyme in the cascade phosphorylates many enzymes in the next step of the cascade, the initial signal will be amplified many times over. **3.** Although both are made up of membrane-spanning proteins, tight junctions seal adjacent animal cells together while gap junctions allow a flow of material between them. Middle lamellae are pectin-rich layers that glue adjacent plant cells together; plasmodesmata are gaps in the cell walls of adjacent cells where the plasma membrane is continuous, allowing exchange of materials between them. **4.** When dissociated cells from two sponge species were mixed, the cells sorted themselves into distinct aggregates that contained only cells of the same species. By blocking membrane proteins with antibodies and isolating cells that would *not* adhere, researchers found that specialized groups of proteins, including cadherins, are responsible for selective adhesion. **5.** Signal crosses plasma membrane and binds to intracellular receptor (reception) → receptor changes conformation, and signal-receptor complex moves to target site (processing) → signal-receptor complex binds to a target molecule (e.g., a gene or membrane pump), which changes its activity (response) → signal falls off receptor or is destroyed; receptor changes to inactive conformation (deactivation). **6.** Information from different signals may conflict or be reinforcing. "Cross-talk" between signaling pathways allows cells to integrate information from many signals at the same time, instead of responding to each signal in isolation.

Applying Concepts to New Situations

1. If an animal lacked an extracellular matrix, its cells would be structurally weak. Aggregations of similar cell types would be unable to form tissues, so individuals would probably develop as a mass of poorly connected or unconnected cells. A plant species with a similar disability would experience similar problems, but would also burst if placed in a hypotonic environment. **2.** (a) The response would have to be extremely local—the activated signal-receptor complex would have to affect nearby proteins. (b) Little or no amplification could occur, because there could never be more than one molecule that triggers the response. (c) The only way to regulate the response would be to block the receptor or make it more responsive to the signal. **3.** Chitin forms chains that can cross-link with one another. Therefore, the fungal cell wall likely has a structure similar to that of the plant cell wall (see Figure 8.2, left) except that chitin, rather than cellulose, is the major fiber component. **4.** Antibody binding would prevent activation of the associated G protein. Even if the appropriate signal were present, no second messenger could be produced and no cell response could occur. Permanent binding by a drug would either activate the G protein constantly or block activation completely.

CHAPTER 9

Check Your Understanding (CYU)

CYU p. 153 (1) In part, because its three phosphate groups have four negative charges in close proximity. The electrons repulse each other, raising their potential energy. (2) Electrons in C—O bonds are held more tightly than electrons in C—H bonds, so they have lower potential energy. **CYU p. 161** (1) and (2) are combined with the answer to CYU p. 166 below. (3) Start with 12 dimes on glucose. (These dimes represent the 12 electrons that will be moved to electron carriers during redox reactions throughout glycolysis and the citric acid cycle.) Move two dimes to the NADH box generated by glycolysis and the other 10 dimes to the pyruvate box. Then move these 10 dimes through the pyruvate dehydrogenase box, placing two of them in the NADH box next to pyruvate dehydrogenase and the remaining eight dimes in the acetyl CoA box. Next move the eight dimes in the acetyl CoA box through the citric acid cycle, placing six of them in the NADH box and two in the $FADH_2$ box generated during the citric acid cycle. (4) These boxes are marked with stars in the diagram. **CYU p. 166** [See Figure A9.1] To illustrate the chemiosmotic mechanism, take the dimes (electrons) piled on the NADH and $FADH_2$ boxes and move them through the ETC. While moving these dimes, also move pennies from the mitochondrial matrix to the intermembrane space. As the dimes exit the ETC, add them to oxygen to generate water. Once all of the pennies have been pumped by the ETC into the intermembrane space, move them through ATP synthase back into the mitochondrial matrix to fuel the formation of ATP. **CYU p. 168** Electron acceptors such as oxygen have a much lower electronegativity than pyruvate. Donating an electron to O_2 causes a greater drop in potential energy, making it possible to generate much more ATP per molecule of glucose.

You Should Be Able To (YSBAT)

YSBAT p. 150 (1) Ribulose and carbon dioxide will not react because the reaction is endergonic and cannot proceed without the input of energy. (2) It is "activated" because it now has high potential energy. (3) The endergonic reaction (ribulose + CO_2) is coupled to the exergonic reaction (ATP hydrolysis) via the phosphorylation of ribulose. **YSBAT p. 152** (1) They are oxidized. (2) They are reduced. (3) Oxygen. (4) The reactants have higher energy. "Energy" should be added to the product side of the equation with "Release of energy" written below it. **YSBAT p. 163** "Indirect" is accurate because the energy released during glucose oxidation is stored in reduced electron carriers (instead of being used to produce ATP directly). The energy released as electrons from these carriers flow through the ETC is used to generate a proton gradient across a membrane. These protons then diffuse down their concentration gradient through ATP synthase located in the membrane, which drives ATP synthesis.

Caption Questions and Exercises

Figure 9.6 In glucose, put two dots in the middle of each C—C or C—H bond, and two dots next to the O in the C—O bond. In O_2, put four electrons in the middle of the double bond. In CO_2, put four electrons next to each O. In H_2O, put two electrons near O in each O—H bond. The electrons are held more tightly in the products than in the reactants, so potential energy has decreased. The difference in potential energy is released, so the reaction is exergonic. **Figure 9.8** *Glycolysis:* "What goes in" = glu-

FIGURE A9.1

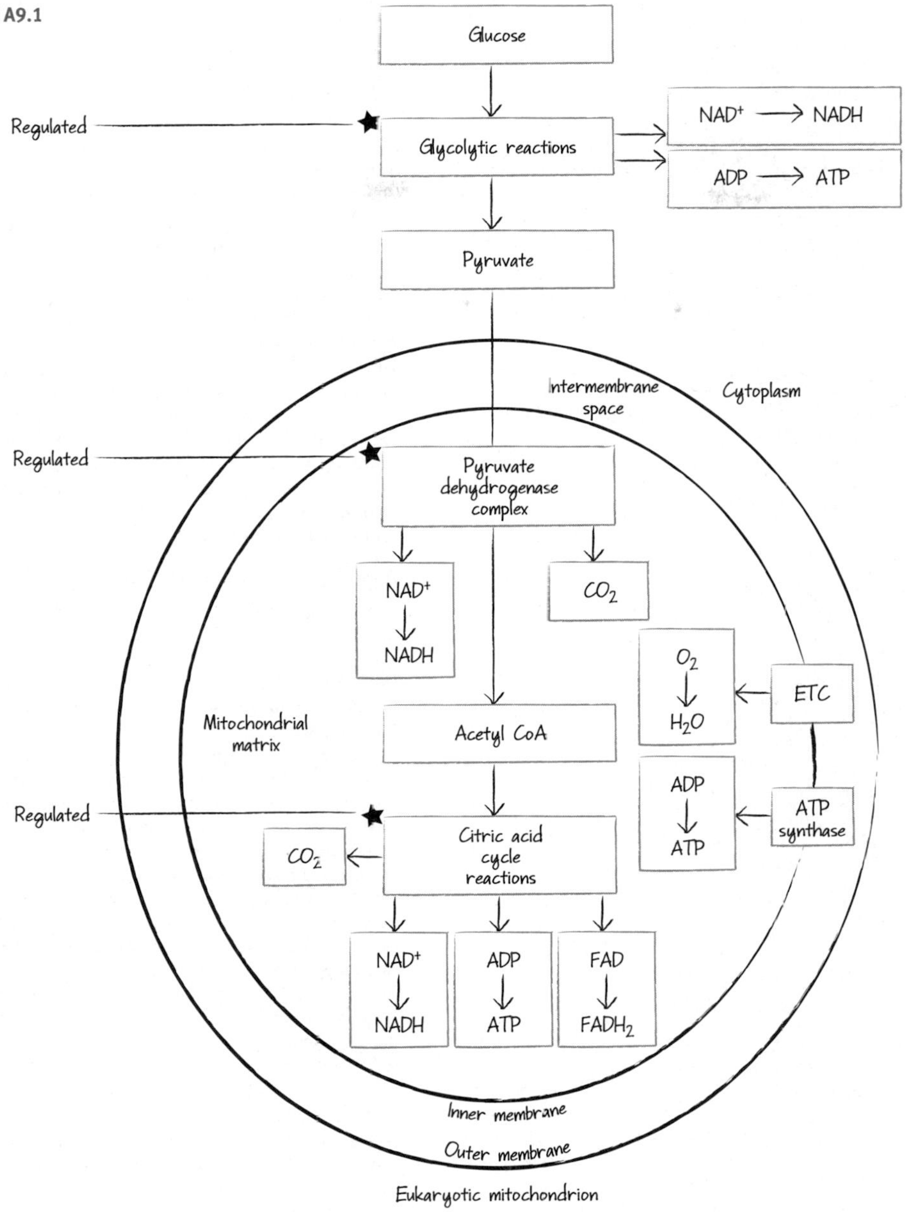

cose, NAD^+, ADP, inorganic phosphate; "What comes out" = pyruvate, NADH, ATP. *Pyruvate processing:* "What goes in" = pyruvate, NAD^+; "What comes out" = NADH, CO_2, acetyl CoA. *Citric acid cycle:* "What goes in" = acetyl CoA, NAD^+, FADH, GDP or ADP, inorganic phosphate; "What comes out" = NADH, $FADH_2$, ATP or GTP, CO_2. *Electron transport and chemiosmosis:* "What goes in" = NADH, $FADH_2$, O_2; "What comes out" = ATP, H_2O, NAD^+, FAD. **Figure 9.12** If the regulatory site had a higher affinity for ATP than the active site, then ATP would always be bound at the regulatory site, and glycolysis would always proceed at a very slow rate. **Figure 9.14** "Positive control": AMP, NAD^+, CoA (reaction substrates). "Negative control by feedback inhibition": acetyl CoA, NADH, ATP (reaction products). **Figure 9.16** An allosteric regulator binds somewhere other than the active site and causes shape changes that slow the reaction or speed it up. A competitive inhibitor binds at the active site and keeps the substrate from binding. **Figure 9.18** Production of NADH and $FADH_2$. (In the line representing free energy, there are greater drops at each occurrence of NADH and $FADH_2$ versus ATP formation.) **Figure 9.20** Yes, because the artificial vesicles include only an ATP-synthesizing enzyme and a proton pump that creates a proton-motive force. The only energy source in this experiment is the proton gradient, so the result strongly supports the chemiosmotic hypothesis. (Note: This experiment does not test whether all ATP production in the ETC is through chemiosmosis. It does not preclude the possibility that the ETC also performs some substrate-level phosphorylation.) **Figure 9.21** The proton gradient arrow should start above in the inner membrane space and point down across the membrane into the mitochondrial matrix. *Complex I:* "What goes in" = NADH and H^+; "What comes out" = NAD^+, e^-, H^+. *Complex II:* "What goes in" = $FADH_2$; "What comes out" = FAD, e^-. *Complex III:* "What goes in" = e^-; "What comes out" = e^-. *Complex IV:* "What goes in" = e^-, H^+, O_2; "What comes out" = H_2O, H^+. **Figure 9.26** [See Figure A9.2]

Summary of Key Concepts

KC 9.1 Phosphorylation adds two negative charges in a small area. The electrical repulsion that results raises the protein's potential energy and its tertiary structure. **KC 9.2** The free-energy drop from glucose to oxygen (or another electron acceptor, during cellular respiration) is much greater than the free energy drop from glucose to pyruvate (during fermentation). Thus, there is more free energy available to use in synthesizing ATP. **KC 9.3** NADH would decrease if a drug poisoned the acetyl CoA-to-citrate enzyme, but increase if a drug poisoned ATP synthase. **KC 9.4** [See Figure A9.3] **KC 9.5** Organisms that produce ATP by fermentation grow more slowly than those that produce ATP via cellular respiration

FIGURE A9.2

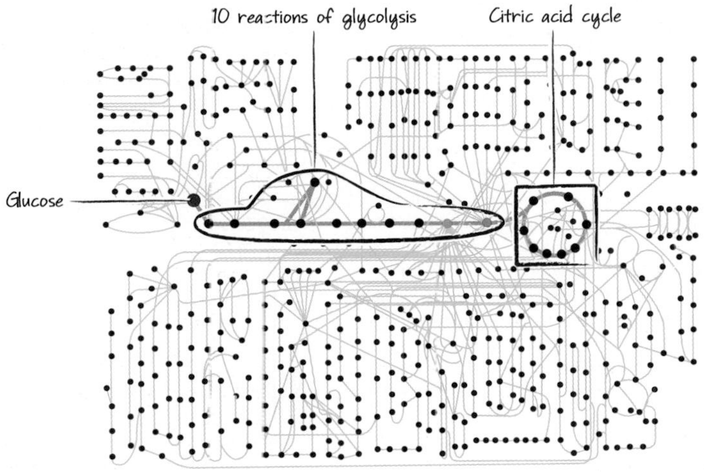

FIGURE A9.3

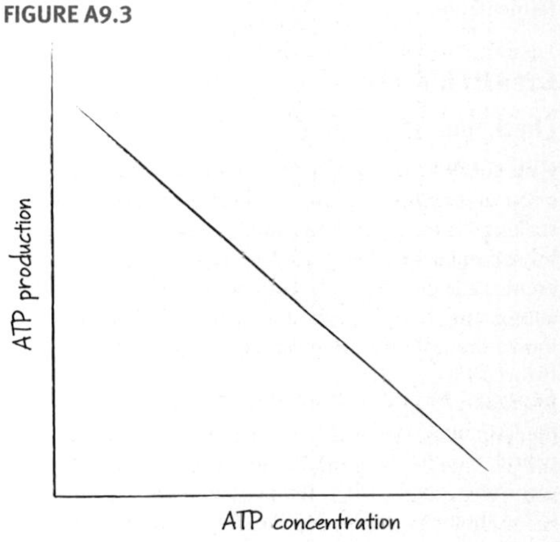

simply because fermentation produces fewer ATP molecules per glucose molecule than cellular respiration does.

Test Your Knowledge

1. b; **2.** b; **3.** d; **4.** c; **5.** a; **6.** b

Test Your Understanding

1. NADH and $FADH_2$ get their electrons from the intermediates of glycolysis, pyruvate processing, and the TCA cycle and deliver them to the ETC, where they reduce O_2. **2.** Both processes produce ATP from ADP and P_i, but substrate-level phosphorylation occurs when enzymes remove a "high-energy" phosphate from a substrate and directly transfer it to ADP, while oxidative phosphorylation is based on electrons moving through an ETC and production of a proton-motive force that drives ATP synthase. **3.** Aerobic respiration is much more productive because oxygen has extremely high electronegativity compared to other electron acceptors, resulting in a greater release of energy during electron transport and more proton pumping. **4.** Glycolysis → Pyruvate processing → TCA cycle → ETC and chemiosmosis. The first three steps are responsible for glucose oxidation; the final step produces the most ATP. **5.** Electron transport makes oxidative phosphorylation possible (oxidative phosphorylation occurs when electrons have been transported through the ETC and a proton gradient has been established). ATP synthase consists of an F_o unit and F_1 unit joined by a stalk. When protons flow through the F_o unit, the stalk and F_1 unit spin. The motion drives the synthesis of ATP from ADP and P_i. **6.** Stored carbohydrates can be broken down into glucose that enters the glycolytic pathway. If carbohydrates are absent, products from fat and protein catabolism can be used to fuel cellular respiration or fermentation. If ATP is plentiful, anabolic reactions use intermediates of the glycolytic pathway and the TCA cycle to synthesize carbohydrates, fats, and proteins.

Applying Concepts to New Situations

1. When complex IV is blocked, electrons can no longer be transferred to oxygen, the final acceptor, and cellular respiration stops. Fermentation could keep glycolysis going, but it is inefficient and unlikely to fuel a cell's energy needs over the long term. Cells that lack the enzymes required for fermentation would die first. **2.** Because mitochondria with few cristae would have fewer electron transport chains and ATP synthase molecules, they would produce much less ATP than mitochondria with numerous cristae. **3.** When oxygen is unavailable for cellular respiration, yeast cells switch to fermentation, which occurs in the cytosol. They are unlikely to expend large amounts of energy and materials to maintain mitochondria. **4.** The potential energy drop between glucose and pyruvate is relatively small. When pyruvate acts as the electron acceptor during fermentation, a great deal of potential energy remains in the alcohol or other fermentation products.

CHAPTER 10

Check Your Understanding (CYU)

CYU p. 179 When a pigment absorbs energy and an electron is excited in response, the event is analogous to striking the tuning fork. When an excited electron transmits resonance energy to another pigment molecule, the event is analogous to touching a vibrating tuning fork to another and making it vibrate. **CYU p. 184** (1) Your model should conform to the relationships shown in Figure 10.13. (2) Put four dimes on the water-splitting reaction ($2\ H_2O \rightarrow 4\ H^+ + O_2$). When four photons hit the PSII antenna complex, move the dimes to pheophytin, then to the PSII ETC (plastoquinone and the cytochrome complex). The dimes go to plastocyanin, then to the antenna complex of PSI. When four photons strike, move the dimes to the PSI ETC and ferrodoxin, then combine them with $NADP^+ + 2\ H^+$ to form NADPH. **CYU p.190** (1) (a) C_4 plants use PEP carboxylase to fix CO_2 into organic acids in mesophyll cells. These organic acids are then pumped into bundle-sheath cells, where they release carbon dioxide to rubisco. (b) CAM plants take in CO_2 at night, and have enzymes that fix it into organic acids stored in the central vacuoles of photosynthesizing cells. During the day, the organic acids are processed to release CO_2 to rubisco. (c) By diffusion through a plant's stomata when they are open. (2) Sucrose can be transported to other parts of the plant because it is water soluble. Starch is not water soluble, so it cannot be transported. It acts as a storage product, instead.

You Should Be Able To (YSBAT)

YSBAT p. 181 Photosystem II is very similar to the ETC in mitochondria. Both use the energy released from electron transfer to pump protons, generating a proton gradient that is used to generate ATP through ATP synthase. The high-energy electrons for PSII come from chlorophyll via water, however, while they come from NADH and $FADH_2$ for the ETC. The eventual products of PSII are ATP and oxygen; the eventual products of the mitochondrial ETC are ATP and water. **YSBAT p. 182** Light → Antenna complex → Reaction center → Pheophytin → ETC → Proton gradient → ATP synthase. Electrons from water are donated to the reaction center. When they absorb resonance energy and reduce pheophytin, they convert electromagnetic energy to chemical energy. **YSBAT p. 183** (1) Plastocyanin transfers electrons that move through PSII to the reaction center of PSI. (2) After they are excited by a photon and donated to the initial electron acceptor. (3) The electrons that originate from PSII's reaction center end up in NADPH—they are not cycled back to form water. **YSBAT p. 186** Rubisco joins CO_2 with RuBP to form a six-carbon intermediate that is immediately hydrolyzed to yield two molecules of 3-phosphoglycerate, which participate in reactions that yield G3P.

Caption Questions and Exercises

Figure 10.5 [See Figure A10.1] **Figure 10.7** The energy state corresponding to a photon of green light would be located between the energy states corresponding to red and blue photons. **Figure 10.10** Yes—otherwise, changes in the rate of photosynthesis could be due to changes in the density of photosynthetic cells, not differences in wavelengths received. **Figure 10.17** The researchers didn't have any basis on which to predict these intermediates. They needed to perform the experiment to identify them. **Figure 10.24** One arrow should point from sucrose to starch, another from starch to sucrose. Note that the interconversion requires transport between the cytosol and chloroplast.

Summary of Key Concepts

KC 10.1 The Calvin cycle depends on the ATP and NADPH produced by PSI and PSII, so it is not independent of light. **KC 10.2** Oxygen is produced by a critical step in photosynthesis: splitting water to provide electrons to PSII. If oxygen production increases, it means that more electrons are moving through the photosystems. **KC 10.3** The "final" product of the reaction sequence reacts to form the molecule that is the substrate for the initial reaction in the sequence. **KC 10.4** Both forms of rubisco would increase carbohydrate production: Oxygen would not compete with CO_2 for binding to the active site of rubisco and the competing reactions of photorespiration would not occur, or rubisco would fix 10 times as much CO_2 in a given period of time.

Test Your Knowledge

1. d; **2.** c; **3.** a; **4.** c; **5.** b; **6.** d

Test Your Understanding

1. In PSII, it occurs when excited electrons from the reaction center are accepted by pheophytin, reducing it. In PSI, it occurs when excited electrons from the reaction center are accepted by ferredoxin. **2.** CO_2 is fixed when rubisco catalyzes the reaction between carbon dioxide and ribulose bisphosphate to form 3-phosphoglycerate. Subsequent reactions that produce sugar require phosphorylation by ATP and high-energy electrons from NADPH. **3.** PSII and PSI both have reaction centers, an energy transformation step, and an ETC. PSII produces ATP while PSI produces NADPH, and only PSII splits water to obtain electrons. Plastocyanin transfers electrons between the two photosystems, connecting them. **4.** Photorespiration occurs when levels of CO_2 are low and O_2 are high. Less sugar is produced because (1) CO_2 doesn't participate in the initial reaction catalyzed by rubisco, and (2) when rubisco catalyzes the reaction with O_2 instead, one of the products is eventually broken down to CO_2 in a process that uses ATP. **5.** See Figure 10.22b and Figure 10.23. **6.** Photosynthesis in chloroplasts produces glucose, which is processed in mitochondria to produce ATP.

Applying Concepts to New Situations

1. Both organelles have a double membrane, extensive internal membranes, ETCs that include cytochromes and quinones, and ATP synthase. Only chloroplasts have chlorophyll and other pigment molecules that absorb light. Chloroplasts use NADPH as an electron carrier; mitochondria use NADH and $FADH_2$. The Calvin cycle occurs only in chloroplasts; the TCA cycle occurs only in mitochondria. **2.** Because CO_2 and O_2 levels minimized the impact of photorespiration when rubisco evolved, the hypothesis is credible. But once O_2 levels increased, any change in rubisco that minimized photorespiration would give individuals a huge advantage over organisms with "old" forms of rubisco. There has been plenty of time for such changes to occur, making the "holdover" hypothesis less credible. **3.** Carotenoids should be located close to the chlorophyll molecules in the reaction center, so they can pass energy along and neutralize free radicals that could damage chlorophyll. One way to test this hypothesis is to isolate thylakoid membranes and test for the presence of carotenoid and chlorophyll molecules. **4.** No—they are unlikely to have the same complement of photosynthetic pigments. Different wavelengths of light are available in various layers of a forest and water depths. It is logical to predict that plants and algae have pigments that absorb the available wavelengths efficiently. One way to test this hypothesis would be to isolate pigments from species in different locations, and test the absorbance spectra of each.

THE BIG PICTURE: ENERGY

Check Your Understanding (CYU), p. 192

1. Photosynthesis uses H_2O as a substrate and releases O_2 as a by-product; cellular respiration uses O_2 as a sub-

FIGURE A10.1

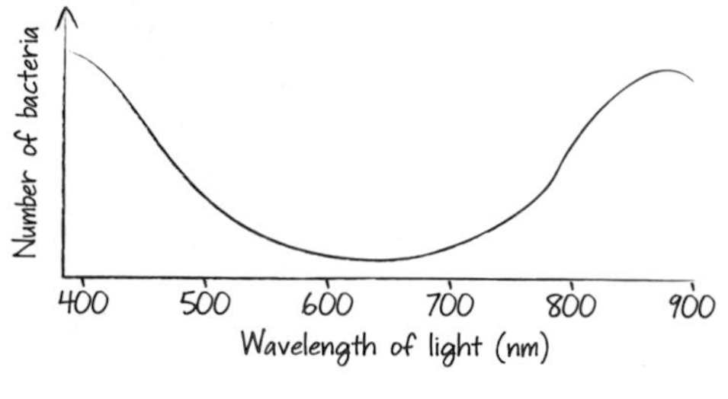

strate and releases H_2O as a by-product. **2.** Photosynthesis uses CO_2 as a substrate; cellular respiration releases CO_2 as a by-product. **3.** CO_2 fixation would essentially stop; CO_2 would continue to be released by cellular respiration. CO_2 levels in the atmosphere would increase rapidly, and production of new plant tissue would cease—meaning that most animals would quickly starve to death. **4.** ATP "is used by" the Calvin cycle; photosystem I "yields" NADPH.

CHAPTER 11

Check Your Understanding (CYU)

CYU p. 202 (1) [See Figure A11.1] (2) Interphase, prophase, prometaphase, metaphase, anaphase, telophase. Sister chromatids condense in prophase, move to the middle of the cell in metaphase, break apart in anaphase. **CYU p. 206** (1) $G_1 \rightarrow S \rightarrow G_2 \rightarrow M \rightarrow G_1$, etc. Checkpoints occur at the end of G_1 and G_2 and during M phase. (2) MPF Cdk levels are fairly constant throughout the cycle, but MPF cyclin increases. MPF activity increases at the end of G_2, initiating M phase, and declines at the end of M phase.

You Should Be Able To (YSBAT)

YSBAT p. 197 (1) Genes are segments of chromosomes that code for proteins. (2) Chromosomes are made of chromatin. (3) Sister chromatids are copies of the same chromosome, joined together. **YSBAT p. 200** [See Table A11.1] **YSBAT p. 204** MPF activates proteins that get mitosis under way. MPF consists of cyclin and a Cdk, and is inactivated by phosphorylation (at two sites) and activated by dephosphorylation (at one site). Enzymes that degrade cyclin reduce MPF levels. **YSBAT p. 208** [See Figure A11.2]

Caption Questions and Exercises

Figure 11.6 Unreplicated. **Figure 11.10** Fuse two interphase cells. If the regulatory molecule hypothesis is correct, neither cell should start M phase. But if the fusion event itself is the trigger, then at least one cell should start M phase. **Figure 11.13** In effect, MPF turns itself off after it is activated. If this didn't happen, the cell would undergo mitosis again right away. **Figure 11.15** After 1990, the death rate from lung cancer decreased in men but not in women. One hypothesis to explain the data is that rates of cigarette smoking declined in men but increased in women.

Summary of Key Concepts

KC 11.1 The gap phases give the cell time to replicate organelles and grow prior to division, as well as perform the normal functions required to stay alive. **KC 11.2** Kinetochore microtubules originate in the cytoplasm but chromosomes are in the nucleus. If the nuclear envelope did not disintegrate, the microtubules could not reach the chromosomes. **KC 11.3** The cell cycle will move forward, from G_1 to S phase, triggering cell division. **KC 11.4** A wide variety of defects, and multiple defects, cause a cell to become cancerous. No one drug or therapy can cure all of the defects involved.

Test Your Knowledge

1. b; **2.** d; **3.** d; **4.** d; **5.** c; **6.** c

Test Your Understanding

1. [See Figure A11.3] For daughter cells to have identical complements of chromosomes, all the chromosomes must be replicated during the S phase, the spindle apparatus must connect with the kinetochores of each replicated chromosome in prometaphase, and the sister chromatids of each replicated chromosome must separate in anaphase. **2.** One possible concept map is shown

FIGURE A11.1

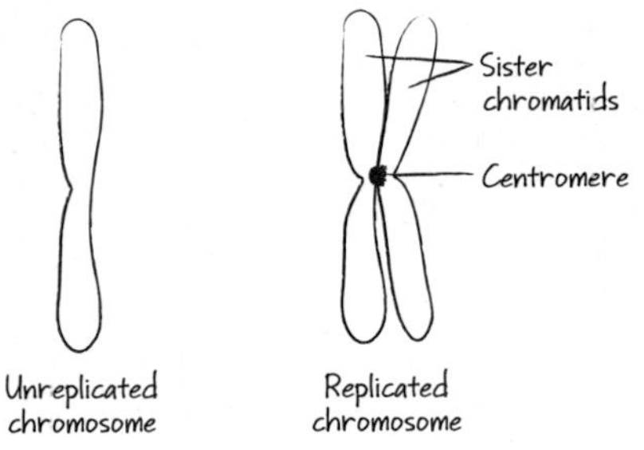

FIGURE A11.2

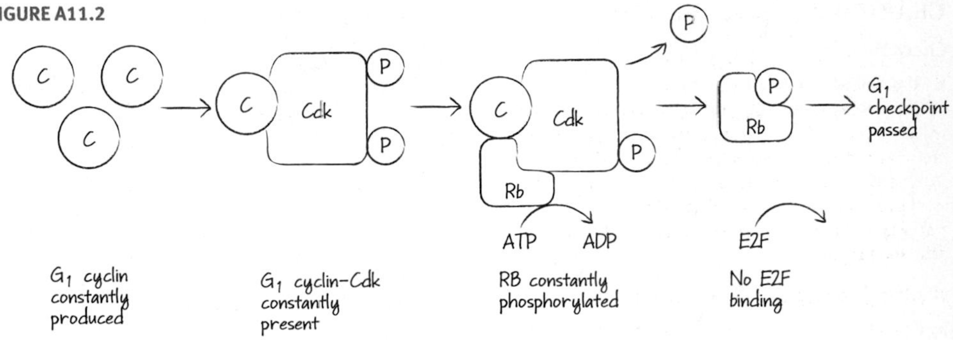

FIGURE A11.3

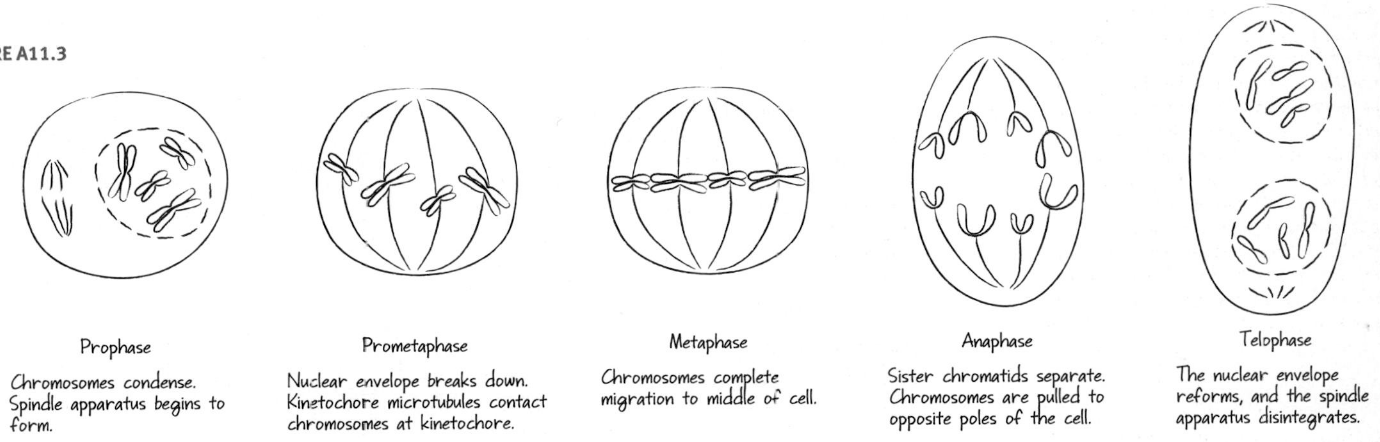

Prophase
Chromosomes condense. Spindle apparatus begins to form.
Prometaphase
Nuclear envelope breaks down. Kinetochore microtubules contact chromosomes at kinetochore.
Metaphase
Chromosomes complete migration to middle of cell.
Anaphase
Sister chromatids separate. Chromosomes are pulled to opposite poles of the cell.
Telophase
The nuclear envelope reforms, and the spindle apparatus disintegrates.

TABLE A11.1

	Prophase	Prometaphase	Metaphase	Anaphase	Telophase
Mitotic spindles	Grow	Contact chromosomes	Move chromosomes	Shorten	Break down
Nuclear envelope	Disintegrates	Nonexistent	Nonexistent	Nonexistent	Re-forms
Chromosomes	Condense	Attach to kinetochore microtubules	Move to metaphase plate	Sister chromatids separate	Collect at opposite poles

below [see Figure A11.4]. **3.** Cell fusion experiments suggested that something in mitotic cells initiated mitosis in other cells. Microinjection experiments suggested that this "something" is in the cytoplasm of M-phase cells. **4.** Protein kinases phosphorylate other proteins. Phosphorylation changes a protein's shape, altering its function (activating or inactivating it). As a result, protein kinases regulate the function of other proteins. **5.** Cyclin concentrations cycle during the cell cycle. At high concentration, cyclins bind to a specific cyclin-dependent kinase (or Cdk), forming an active protein kinase, such as MPF. **6.** If a cancer has not yet metastasized, the tumor can be completely removed by surgery.

Applying Concepts to New Situations

1. It is efficient: If the cell is not going to divide, there is no reason to invest energy in replication. **2.** A single cell with many identical nuclei in it. **3.** Labeled cells are in S phase, so 4 hours passed between the end of S phase and the start of M phase. **4.** Cancer requires many defects. Older cells have had more time to accumulate defects. Individuals with a genetic predisposition to cancer start out with some cancer-related defects, but this does not mean that the additional defects required for cancer to occur will develop.

CHAPTER 12

Check Your Understanding (CYU)

CYU p. 220 (1) Use four long and four short pipe cleaners (or pieces of cooked spaghetti) to represent the chromatids of two replicated homologous chromosomes (four total chromosomes). Mark two long and two short ones with a colored marker pen to distinguish maternal and paternal copies of these chromosomes. Twist identical pipe cleaners (e.g., the two long colored ones) together to simulate replicated chromosomes. Arrange the pipe cleaners to depict the different phases of meiosis I as follows: *Early prophase I:* Align sister chromatids of each homologous pair to form two tetrads. *Late prophase I:* Form one or more crossovers between non-sister chromatids in each tetrad. (This is hard to simulate with pipe cleaners—you'll have to imagine that each chromatid now contains both maternal and paternal segments.) *Metaphase I:* Line up homologous pairs (the two pairs of short pipe cleaners and the two pairs of long pipe cleaners) at the metaphase plate. *Anaphase I:* Separate homologs. Each homolog still consists of sister chromatids joined at the centromere. *Telophase I and cytokinesis:* Move homologs apart to depict formation of two haploid cells, each containing a single replicated copy of two different chromosomes. (2) During anaphase I, homologs (not sister chromatids, as in mitosis) are separated, making the cell products of meiosis I haploid. **CYU p. 223** (1) [See Figure A12.1] Maternal chromosomes are white and paternal chromosomes are black. Daughter cells with many other possible combinations of chromosomes than shown could result from meiosis of this parent cell. (2) Asexual reproduction generates no appreciable genetic diversity. Self-fertilization is preceded by meiosis so it generates gametes, through crossing over and independent assortment, that have combinations of alleles not present in the parent. Outcrossing generates the most genetic diversity among offspring because it produces new combinations of alleles from two different individuals.

You Should Be Able To (YSBAT)

YSBAT p. 213 $n = 12$; the organism is diploid with $2n = 24$ (in females; 23 in males). **YSBAT p. 214** [See Figure A12.2] Because the two sister chromatids are identical and attached, it is sensible to consider them as parts of a single chromosome. **YSBAT p. 216** [See Figure A12.3] **YSBAT p. 218** Crossing over would not occur and the daughter cells produced by meiosis would be diploid, not haploid. There would be no reduction division. **YSBAT p. 221** Each gamete would inherit either all maternal or all paternal chromosomes. This would limit genetic variation in the offspring by precluding the many possible gametes containing various combinations of maternal and paternal chromosomes.

Caption Questions and Exercises

Figure 12.11 [See Figure A12.4] **Figure 12.12** Asexually: 64 (16 individuals from generation three produce 4 offspring per individual). Sexually: 16 (8 individuals from generation three form 4 couples; each couple produces 4 offspring). **Figure 12.13** Sampling widely reduces the possibility that the results are due to an environmental factor—other than the presence of parasites—that differs between sexually and asexually reproducing populations.

Summary of Key Concepts

KC 12.1 Haploid cells cannot undergo meiosis, because there is no homolog to synapse with—haploid cells can-

FIGURE A11.4

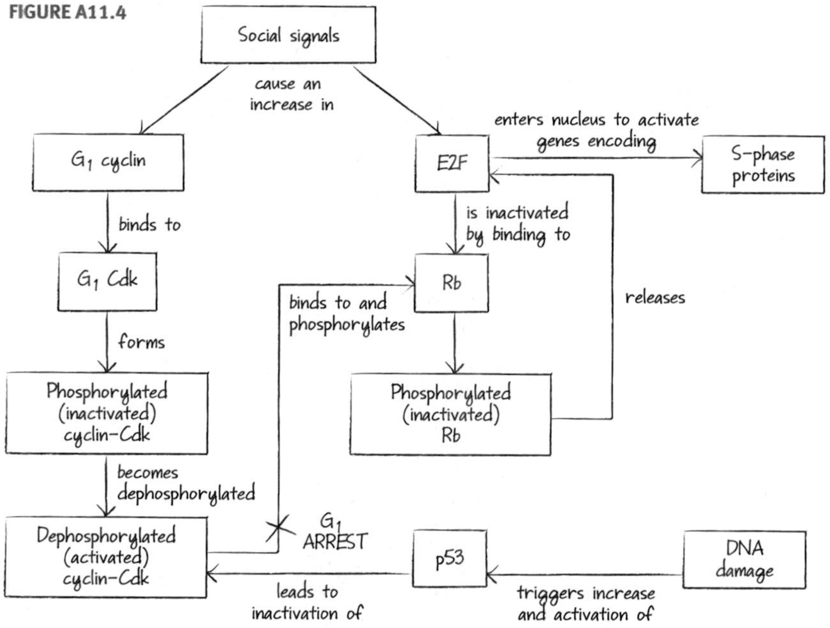

FIGURE A12.1

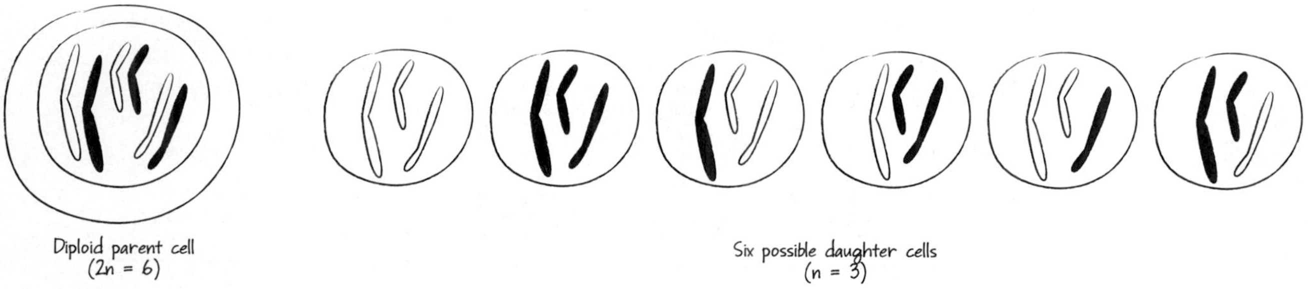

not undergo a reduction in ploidy. **KC 12.2** Monozygotic twins are genetically identical—the cells that originally gave rise to the two individuals were the product of mitosis. Dizygotic twins each arise from gametes that are the product of meiosis. Genetically, they are no more alike than any pair of siblings. **KC 12.3** Sexual reproduction will likely occur during seasons when conditions are changing rapidly, because genetically diverse offspring may have an advantage in the new conditions. **KC 12.4** The resulting gametes will be diploid instead of haploid.

Test Your Knowledge

1. b; **2.** a; **3.** b; **4.** b; **5.** d; **6.** a

Test Your Understanding

1. Homologous chromosomes are similar in size, shape, and gene content, and originate from different parents. Sister chromatids are exact copies of a chromosome that are generated when chromosomes are replicated (S phase of the cell cycle). **2.** Refer to Figure 12.6 as a guide for this exercise. The four pens represent the chromatids in one replicated homologous pair; the four pencils, the chromatids in a different homologous pair. To simulate meiosis II, make two "haploid cells"—each with a pair of pens and a pair of pencils representing two replicated chromosomes (one of each type in this species). Line them up in the middle of the cell, then separate the two pens and the two pencils in each cell such that one pen and one pencil goes to each of four daughter cells. **3.** Meiosis I is a reduction division because homologs separate—daughter cells have just one of each type of chromosome instead of two. Meiosis II is not a reduction division because sister chromatids separate—daughter cells have unreplicated chromosomes instead of replicated chromosomes, but still just one of each type. **4.** Tetraploids produce diploid gametes, which combine with a haploid gamete from a diploid individual to form a triploid offspring. Mitosis proceeds normally in triploid cells because replicated chromosomes align independently before sister chromatids separate. But during meiosis in a triploid, homologous chromosomes can't pair up correctly. The third set of chromosomes does not have a homologous partner to pair with. **5.** Asexually produced individuals are genetically identical, so if one is susceptible to a new disease, all are. Sexually produced individuals are genetically unique, so if a new disease strain evolves, at least some plants are likely to be resistant to it. **6.** If homologs or sister chromatids do not separate normally, the daughter cells will receive one too many chromosomes or one too few. The resulting chromosome sets are considered unbalanced because they do not have the normal number of copies of each gene.

Applying Concepts to New Situations

1. The gibbon would have 22 chromosomes in each gamete, and the siamang would have 25. Each somatic cell of the offspring would have 47 chromosomes. The offspring should be sterile because it has an unbalanced set of chromosomes, so not all of the chromosomes would have a homolog to pair with in meiosis prophase I. **2.** One in eight. **3.** Aneuploidy is the major cause of spontaneous abortion. If spontaneous abortion is rare in older women, it would result in a higher incidence of aneuploid conditions such as Down syndrome in older women, as recorded in the figure. **4.** (a) Such a study might be done in the laboratory, controlling conditions in identical tanks. A population of snails can be established in each tank. A parasite could be added to one tank, and then changes in the frequency of sexual reproduction in the snail populations can be monitored and observed over time. (b) In nature, all of the forces that can act to produce change in a population are present. A laboratory test is only an approximation of what occurs in the wild. For example, sexually reproduced offspring might only have an advantage when the parasite-ridden population is also stressed by food shortages or high or low temperatures. If so, an experiment in the lab with ample food and preferred temperatures would give an incorrect result.

CHAPTER 13

Check Your Understanding (CYU)

CYU p. 239 See answers to Genetics Problems on p. A:14. **CYU p. 243** (1) [**See Figure A13.1**] Segregation of alleles occurs when homologs that carry those alleles are separated during anaphase I. One allele ends up in each

FIGURE A12.2

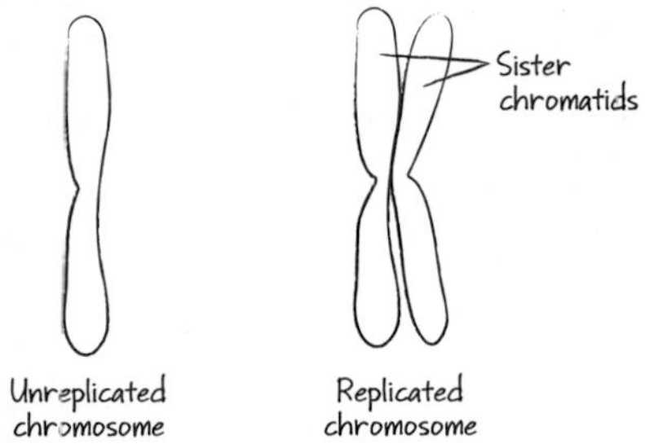

FIGURE A12.3

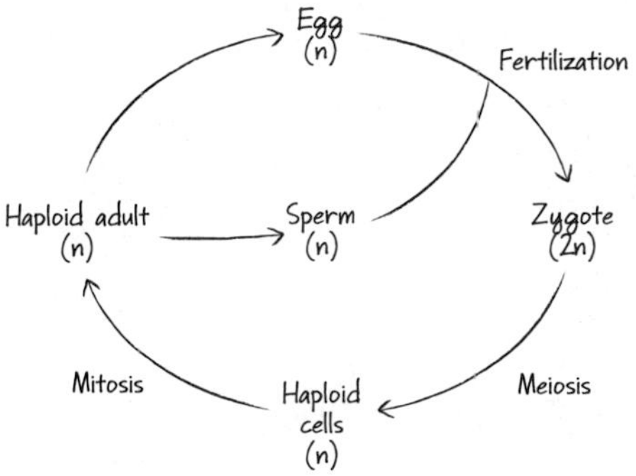

FIGURE A12.4

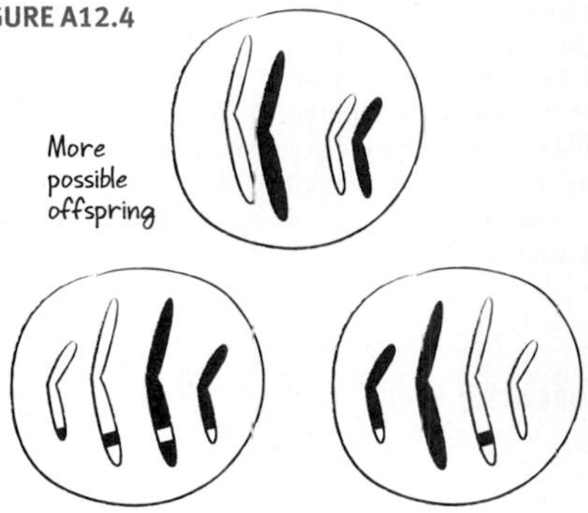

FIGURE A13.1

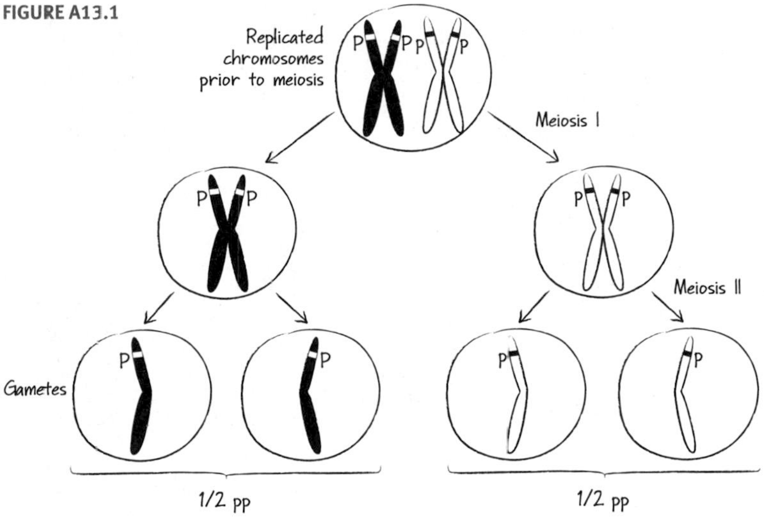

daughter cell. (2) [**See Figure A13.2**] Independent assortment occurs because homologous pairs line up randomly at the metaphase plate during metaphase I. The figure shows two alternative arrangements of homologs in metaphase I. As a result, it is equally possible for a gamete to receive the following four combinations of alleles: *YR, Yr, yR, yr*. **CYU p. 250** (1) The comb phenotype results from interactions between alleles at two different genes, not a single gene. Matings between rose and pea comb chickens produce F_2 offspring that may have a new combination of alleles, and thus new phenotypes.
(2) Kernel color in wheat is influenced by alleles at many different genes, not a single gene. F_2 offspring have a normal distribution of phenotypes, not a 3:1 ratio.

You Should Be Able To (YSBAT)

YSBAT p. 236 Punnett squares are an efficient way to predict offspring genotypes from parents of known genotype. The phenotype ratios are 1:1 round:wrinkled; the genotype ratios are 1:1 *Yy:yy*. **YSBAT p. 238** *AABb* → *AB* and *Ab*. *PpRr* → *PR, Pr, pR,* and *pr*. *AaPpRr* → *APR, APr, ApR, Apr, aPR, apR, apr,* and *aPr*.

Caption Questions and Exercises

Figure 13.3 An experiment is a failure if you didn't learn anything from it. That is not the case here. **Figure 13.4** No—the outcome (the expected offspring genotypes that the Punnett square generates) will be the same.
Figure 13.12 X^{WY}, X^{Wy}, X^{wY}, X^{wy}. **Figure 13.13** Random chance (or perhaps red-eyed, gray-bodied males don't survive well). **Figure 13.17** The gene colored orange is *ruby*; the gene colored blue is *miniature wings*.
Figure 13.22 If both parents are heterozygous (carriers), on average 1/4 of their children will be affected. If the allele is rare, then it is unlikely for the spouse of an affected (homozygous) parent to be a carrier—hence, children will not be affected—they will only get one allele from their affected parent. **Figure 13.24** All affected females would have all affected sons and daughters, no matter what the father's genotype. Affected fathers would transmit the allele to half of their daughters, so on average half would be affected if the mother was unaffected.

Summary of Key Concepts

KC 13.1 *B* and *b*. **KC 13.2** *BR, Br, bR,* and *br*, in equal proportions. **KC 13.3** The *B* and *b* alleles are located on different but homologous chromosomes, which separate into different daughter cells during meiosis I. The *BbRr* notation indicates that the *B* and *R* genes are on different chromosomes. As a result, the chromosomes line up independently of each other in metaphase of meiosis I. The *B* allele is equally likely to go to a daughter cell with *R* as with *r*; likewise the *b* allele is equally likely to go to a daughter cell with *R* as with *r*. **KC 13.4** Instead of two alleles on the same chromosome being transmitted together, crossing over separates them so they are transmitted independently of each other. They are no longer physically linked.

Genetics Questions

1. d; **2.** b; **3.** a; **4.** d; **5.** a; **6.** a; **7.** d; **8.** b; **9.** b; **10.** d

Genetics Problems

1. 3/4; 1/256 (see **BioSkills 13** in Appendix A); 1/2 (the probabilities of transmitting the alleles or having sons does not change over time). **2.** Your answer to the first three parts should conform to the F_1 and F_2 crosses diagrammed in Figure 13.5b, except that different alleles and traits are being analyzed. The recombinant gametes would be *Yi* and *yI*. Yes—there would be some individuals with round, green seeds and with wrinkled, yellow seeds. **3.** Cross 1: non-crested (*Cc*) × non-crested (*Cc*) = 14 non-crested (*C_*); 7 crested (*cc*). Cross 2: crested (*cc*) × crested (*cc*) = 22 crested (*cc*). Cross 3: non-crested (*Cc*) × crested (*cc*) = 7 non-crested (*Cc*); 6 crested (*cc*). Non-crested (*C*) is the dominant allele. **4.** This is a dihybrid cross that yields progeny phenotypes in a 9:3:3:1 ratio. Let *O* stand for the allele for orange petals and *o* the allele for yellow petals; let *S* stand for the allele for spotted petals and *s* the allele for unspotted petals. Start with the hypothesis that *O* is dominant to *o*, that *S* is dominant to *s*, that the two genes are found on different chromosomes so they assort independently, and that the parent individual's genotype is *OoSs*. If you do a Punnett square for the *OoSs* × *OoSs* mating, you'll find that progeny phenotypes should be in the observed 9:3:3:1 proportions. **5.** Let *D* stand for the normal allele and *d* for the allele responsible for Duchenne-type muscular dystrophy. The woman's family has no history of the disease, so her genotype is almost certainly *DD*. The man is not afflicted, so he must be *DY*. (The trait is X-linked, so he has only one allele; the "*Y*" stands for the Y chromosome.) Their children are not at risk. The man's sister could be a carrier, however—meaning she has the genotype *Dd*. If so, then half of the second couple's male children are likely to be affected. **6.** Your stages of meiosis should look like Figure 12.6, except with $2n = 4$ instead of $2n = 6$. The *A* and *a* alleles could be on the red and blue versions of the longest chromosome, and the *B* and *b* alleles could be on the red and blue versions of the smallest chromosomes. The places you draw them are the locations of the *A* and *B* genes, but each chromosome has only one allele. Each pair of red and blue chromosomes is a homologous pair. Sister chromatids bear the same allele (e.g., both sister chromatids of the long blue chromosomes might bear the *a* allele). Chromatids from the longest and shortest chromosomes are not homologous. To identify the events that result in the principles of segregation and independent assortment, see Figures 13.7 and 13.8 and substitute *A, a,* and *B, b* for *R, r* and *Y, y*. **7.** Half of their offspring should have the genotype iI^A and the type A blood phenotype. The other half of their offspring should have the genotype iI^B and the type B blood phenotype. Second case: the genotype and phenotype ratios would be 1:1:1:1 I^AI^B (type AB) : I^Ai (type A) : I^Bi (type B) : *ii* (type O). **8.** Because the children of Tukan and Valco had no eyes and smooth skin, you can conclude that the allele for eyelessness is dominant to eyes and the allele for smooth skin is dominant to hooked skin. *E* = eyeless, *e* = two eye sets, *S* = smooth skin, *s* = hooked skin. Tukan is *eeSS*; Valco is *EEss*. The children are all *EeSs*. Grandchildren with eyes and smooth skin are *eeS_*. Assuming that the genes are on different chromosomes, one-fourth of the children's gametes are *ee* and three-fourths are *S_*. So $\frac{1}{4}$ *ee* × $\frac{3}{4}$ *S_* × 32 = 6 children would be expected to have two sets of eyes and smooth skin. **9.** Although the mothers were treated as children by a reduction of dietary phenylalanine, they would have accumulated phenylalanine and its derivatives once they

FIGURE A13.2

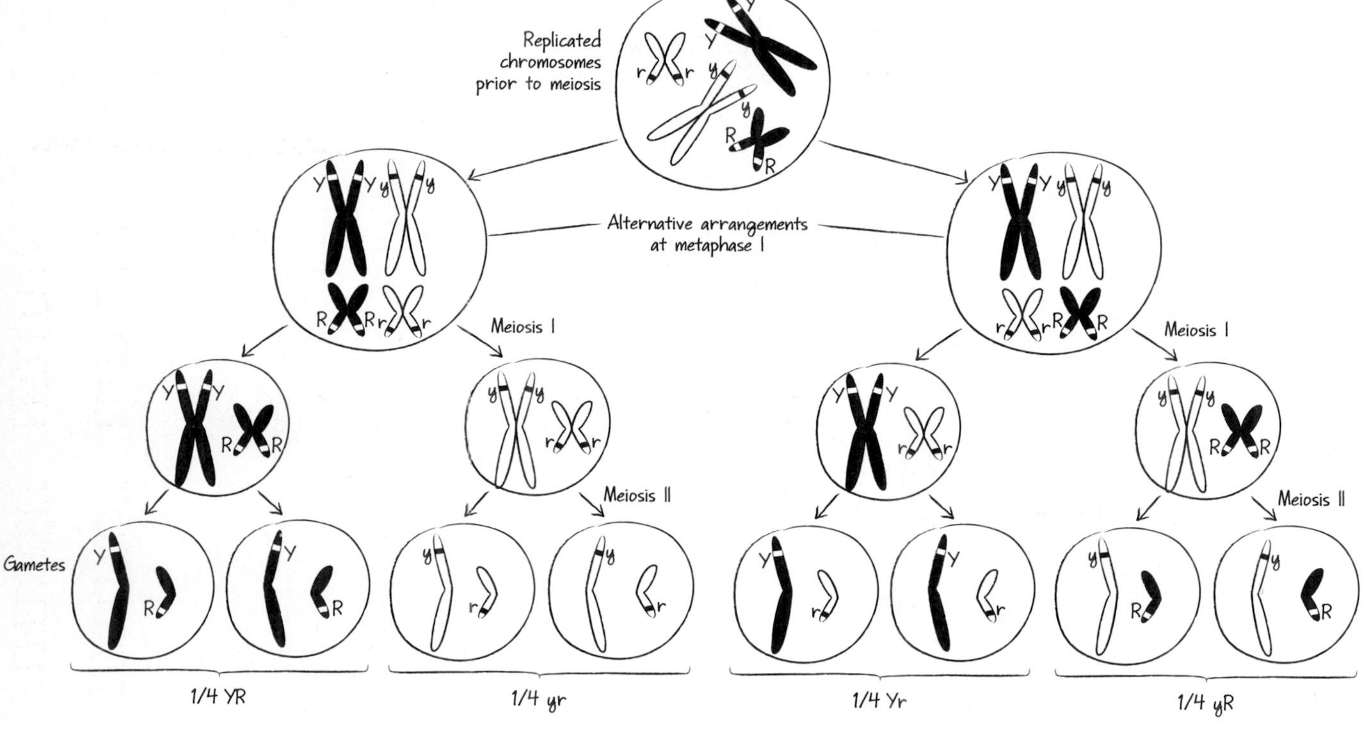

went off the low-phenylalanine diet as young adults. Children born of such mothers were therefore exposed to high levels of phenylalanine during pregnancy. For this reason, a low-phenylalanine diet is recommended for such mothers throughout the pregnancy. **10.** According to Mendel's model, palomino individuals should be heterozygous at the locus for coat color. If you mated palomino individuals, you would expect to see a combination of chestnut, palomino, and cremello offspring. If blending inheritance occurred, however, all of the offspring should be palomino. **11.** Because this is an X-linked trait, the father who has hemophilia could not have passed the trait on to his son. Thus, the mother in couple 1 must be a carrier and have passed the recessive allele on to her son, who is XY and affected. To educate a jury about the situation, you should draw what happens to the X and Y during meiosis, then make a drawing showing the chromosomes in couple 1 and couple 2, with a Punnett square showing how these chromosomes get passed to the affected and unaffected children. **12.** The curved-wing allele is autosomal recessive; the lozenge-eye allele is sex-linked (specifically, X-linked) recessive. Let *L* be the allele for long wings and *l* be the allele for curved wings; let X^R be the allele for red eyes and X^r the allele for lozenge eyes. The female parent is LlX^RX^r; the male parent is LlX^RY. **13.** Albinism indicates the absence of pigment, so let *b* stand for an allele that gives the absence of blue and *y* for an allele that gives the absence of yellow pigment. If blue and yellow pigment blend to give green, then both green parents are *BbYy*. The green phenotype is found in *BBYY, BBYy, BbYY,* and *BbYy* offspring. The blue phenotype is found in *BByy* or *Bbyy* offspring. The yellow phenotype is observed in *bbYY* or *bbYy* offspring. Albino offspring are *bbyy*. The phenotypes of the offspring should be in the ratio 9:3:3:1 as green:blue:yellow:albino. Two types of crosses yield *BbYy* F_1 offspring: *BByy* × *bbYY* (blue × yellow) and *BBYY* × *bbyy* (green × albino). **14.** Autosomal dominant.

CHAPTER 14

Check Your Understanding (CYU)

CYU p. 268 (1) DNA polymerase adds only nucleotides to the free 3′-OH on a strand. Primase synthesizes a short RNA sequence that provides the free 3′ end necessary for DNA polymerase to start working. (2) The helix is already opened and the downstream portions relaxed. **CYU p. 271** (1) Telomerase is not needed—bacterial chromosomes don't shorten after replication because they are circular. DNA polymerase can synthesize a complementary strand all the way around. (2) Otherwise there would be no template to synthesize the extension of a single DNA strand.

You Should Be Able To (YSBAT)

YSBAT p. 264 [See Figure A14.1] The new strands grow in opposite directions, each in 5′ → 3′ direction. **YSBAT p. 266** Helicase, topoisomerase, single-strand DNA-binding proteins, primase, and DNA polymerase are all required for leading-strand synthesis. If any one of these proteins is nonfunctional, then DNA replication will not occur. **YSBAT p. 267** [See Figure A14.2] If DNA ligase was defective, then the leading strand would be continuous, and the lagging strand would have gaps in it where the Okazaki fragments had not been joined.

Caption Questions and Exercises

Figure 14.2 The lack of radioactive protein in the pellet (after centrifugation) is strong evidence; they could also make micrographs of infected bacterial cells before and after agitation. **Figure 14.3** 5′ TAG 3′. **Figure 14.5** The same two bands should appear, but the upper band (DNA containing only ^{14}N) should get bigger and darker and the lower band (hybrid DNA) should get smaller and lighter in color, as there is less and less heavy DNA in each succeeding generation. **Figure 14.13** As long as the RNA template could bind to the "overhanging" section of single-stranded DNA, any sequence could produce a longer strand. For example, 5′ CCCAUUCCC 3′ would work just as well. **Figure 14.15** They are much lower in energy. **Figure 14.17** Exposure to UV radiation can cause formation of thymine dimers. If thymine dimers are not repaired, they represent mutations. If such mutations occur in genes controlling the cell cycle, cells can grow abnormally, resulting in cancers.

Summary of Key Concepts

KC 14.1 Complementary bases hydrogen bond, and hydrophobic interactions occur between bases stacked inside the double helix. **KC 14.2** The bases added during DNA replication are shown in color type.

Original DNA:	CAATTACGGA
	GTTAATGCCT
Replicated DNA:	CAATTACGGA
	GTTAATGCCT
	CAATTACGGA
	GTTAATGCCT

KC 14.3 DNA polymerase synthesizes DNA by adding deoxyribonucleotides to the free 3′-OH group of an existing nucleotide sequence (primer), using single-stranded DNA as a template. Primase synthesizes short RNA sequences, using single-stranded DNA as a template. In contrast to DNA polymerase, primase can begin synthesis de novo, that is, without a primer. Telomerase adds deoxyribonucleotides to the unreplicated 3′ end of the lagging strand, using as the template a short RNA molecule that is associated with the enzyme. **KC 14.4** If errors in DNA aren't corrected, they represent mutations. When DNA repair systems fail, the mutation rate increases. As the mutation rate increases, the chance that one or more cell cycle genes will be mutated increases. Mutations in these genes often result in uncontrolled cell division, ultimately leading to cancer.

Test Your Knowledge

1. d; **2.** c; **3.** d; **4.** b; **5.** c; **6.** d

Test Your Understanding

1. Whether DNA or proteins were labeled. **2.** There are two replication forks at each point of origin—replication proceeds in both directions at the same time. **3.** On the lagging strand, DNA polymerase moves away from the replication fork. When helicase unwinds a new section of DNA, primase must build a new primer on the lagging strand (closer to the fork) and another polymerase molecule must begin synthesis at this point. On the leading strand, DNA polymerase moves in the same direction as helicase, so synthesis can continue without interruption from one primer (at the origin of replication). **4.** Telomerase binds to the 3′ overhang at the end of a chromosome. Once bound, it begins catalyzing the addition of deoxyribonucleotides to the overhang in the 5′ → 3′ direction, lengthening the overhang. This allows primase, DNA polymerase, and ligase to catalyze the addition of deoxyribonucleotides to the lagging strand in the 5′ → 3′ direction, restoring the lagging strand to its original length. **5.** First, DNA polymerases are highly selective in matching complementary bases. Second, DNA polymerase III proofreads by recognizing mismatched bases, cutting out the incorrect base, and replacing it with the correct base. Third, proteins of the mismatch repair system excise incorrect bases and insert the proper ones. **6.** DNA is a symmetrical double helix. A mismatch (e.g., A-G or C-T) or damage to a base pair (e.g., thymine dimer) will cause a distortion or kink in the molecule that is recognizable by repair enzymes. Because the bases on the two strands are complementary, repair enzymes are able to replace the damaged bases easily and accurately.

Applying Concepts to New Situations

1. [See Figure A14.3] Primase would not have to build primers on the leading and lagging strands, because DNA polymerase III would be able to use a "raw" single-stranded template. It is likely that telomerase would still

FIGURE A14.1

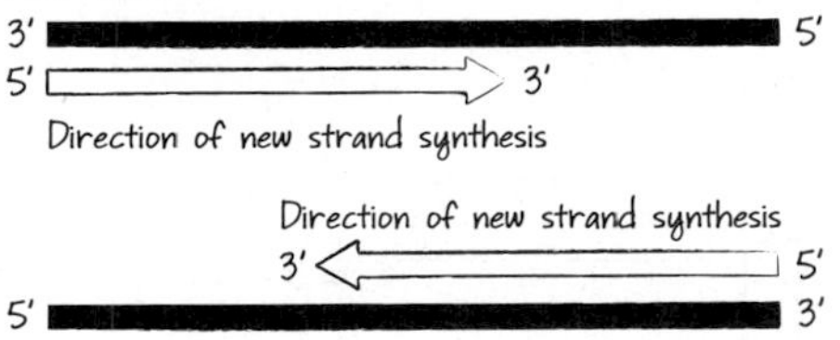

FIGURE A14.2

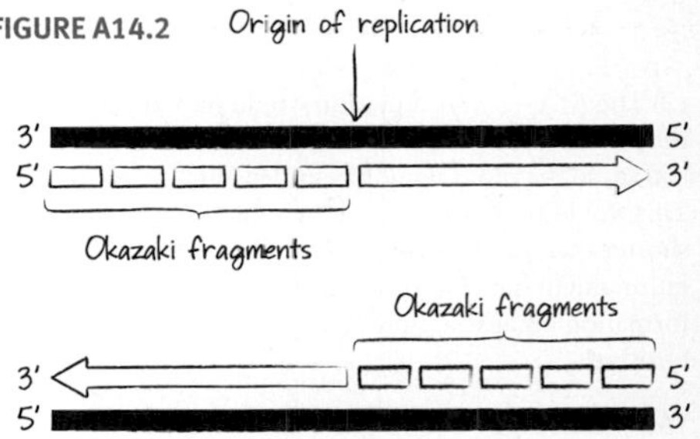

FIGURE A14.3

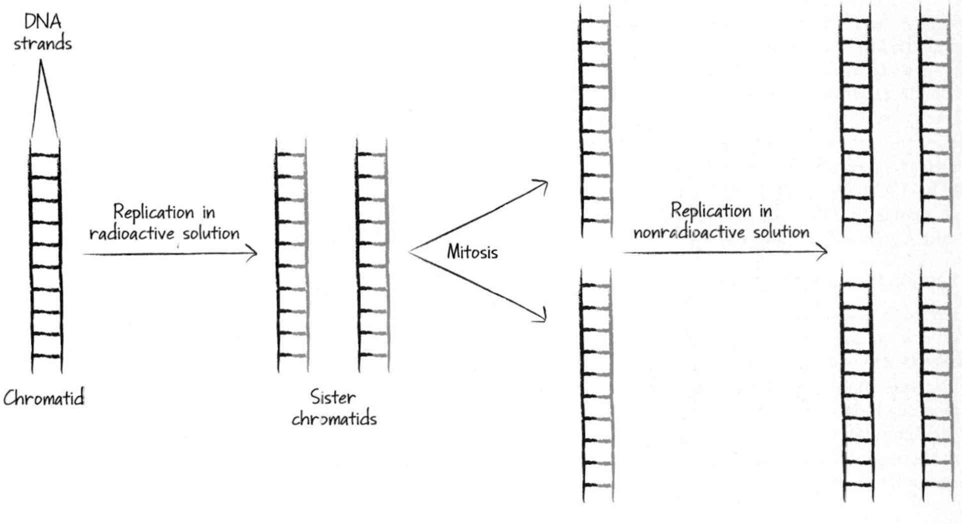

be required because, like primase, DNA polymerase would probably not have room to attach to the last end of the lagging strand, leaving a single-stranded overhang. **2.** (a) In **Figure A.14** below, the gray lines represent DNA strands containing radioactivity. Following replication in a radioactive solution, one DNA strand of each sister chromatid is radioactive. (b) [**See Figure A14.4**] **3.** (a) The double mutant of both *uvrA* and *recA* is most sensitive to UV light; the single mutants are in between; and the wild type is least sensitive. (b) The *recA* gene contributes more to UV repair through most of the UV dose levels. But at very high UV doses, the *uvrA* gene is somewhat more important than the *recA* gene. **4.** Many types of cancer arise because of mutations, especially those in genes involved in controlling cell growth. If chemicals that increase mutation rates in bacteria also increase mutation rates in humans, then the probability of mutations in cell control genes, and hence of cancer formation, also increases.

CHAPTER 15

Check Your Understanding (CYU)

CYU p. 282 If an error in transcription changes the sequence of bases in an mRNA, it may change the protein that is subsequently produced—and thus change the cell's phenotype. Changes in the sequence of other types of RNAs might also affect the phenotype. **CYU p. 285** (1) Remember that DNA and mRNA sequences are written in the $5' \rightarrow 3'$ direction unless indicated otherwise. Thus, to convert DNA triplets (codons) to mRNA codons, the DNA sequence is read "backwards" ($3' \rightarrow 5'$) and U (rather than T) is the base transcribed from A.

DNA codon	mRNA codon	Amino acid
ATA	UAU	Tyrosine
GTA	UAC	Tyrosine
TTA	UAA	Stop
GCA	CGU	Arginine

(2) The ATA → GTA mutation would have no effect on the protein. The ATA → TTA mutation introduces a stop codon, so the resulting polypeptide would be shortened. This would result in synthesis of a mutant protein much shorter than the original protein. The ATA → GCA mutation might have a profound effect on the protein's conformation because arginine's structure is different from tyrosine's.

You Should Be Able To (YSBAT)

YSBAT p. 285 (1) The codons in Figure 15.5 are translated correctly. (2) [**See Figure A15.1**] (3) There are many possibilities (just pick alternate codons for one or more of the amino acids); one is an mRNA sequence of (running $5' \rightarrow 3'$) GCG-AAC-GAU-UUC-CAG. To get the corresponding DNA sequence, write this sequence but substitute Ts for Us: GCG-AAC-GAT-TTC-CAG. This strand also runs in the $5' \rightarrow 3'$ direction. Now write the complementary bases, which will be in the $3' \rightarrow 5'$ direction: CGC-TTC-CTA-AAG-GTC. When this second strand is transcribed by RNA polymerase, it will produce the mRNA given with the proper $5' \rightarrow 3'$ orientation.

Caption Questions and Exercises

Figure 15.1 No, it could not make citrulline from ornithine without enzyme 2. Yes, it would no longer need enzyme 2 to make citrulline. **Figure 15.2** Many possibilities: strain of fungi used, exact method for creating mutants and harvesting spores to plant, exact growing conditions (temperature, light, recipe for growth medium—including concentration of supplemented amino acids), objective criteria for determining growth or no-growth. **Figure 15.5** $4 \times 4 \times 4 \times 4 = 256$ amino acids. A 4-base code is unlikely to evolve because only 20 amino acids are commonly used in the synthesis of proteins—there is no selective advantage for a code to be more complicated than necessary. **Figure 15.7** [**See Figure A15.1**] **Figure 15.8** The DNA sequence changes only at a single location, or point.

Summary of Key Concepts

KC 15.1 Not all genes code for polypeptides or proteins. A better statement might be something like, "one-gene, one RNA or polypeptide product." **KC 15.2** Transcription means to copy, and mRNA is a short-lived copy of the information in DNA. Translation means to change languages, and proteins are a different "chemical language" than nucleic acids (DNA and RNA). **KC 15.3** If a change in a DNA sequence doesn't change the amino acid specified by the corresponding mRNA codon, then the change in genotype doesn't change the phenotype. Many changes in the third positions of codons do not change the phenotype, because the genetic code is redundant. **KC 15.4** The new mutation produced light coat colors, which allowed beach-dwelling mice to be camouflaged and survive better. If the same mutation occurred in a woodland-dwelling population, the light-colored individuals would probably be spotted by predators and eaten.

Test Your Knowledge

1. d; **2.** a; **3.** d; **4.** a; **5.** d; **6.** b

Test Your Understanding

1. In Morse code, different combinations of dots and dashes code for the complexity of the English language. In the genetic code, different combinations of bases code for the complexity of proteins in the cell. **2.**

Substrate 1 $\xrightarrow{A}$ Substrate 2 $\xrightarrow{B}$ Substrate 3 $\xrightarrow{C}$

Substrate 4 $\xrightarrow{D}$ Substrate 5 $\xrightarrow{E}$ BiologicalSciazine

Substrate 3 would accumulate. Hypothesis: The individuals have a mutation in the gene for enzyme D. **3.** They supported an important prediction of the hypothesis: Losing a gene (via mutation) resulted in loss of an enzyme. **4.** It puts the three-nucleotide codons out of register. If the first base is deleted from the sequence ATG-CGA-GAC-TTA, then the reading frame becomes TGC-GAG-ACT-TA. **5.** In a triplet code, addition or deletion of 1–2 bases disrupts the reading frame "downstream" of the mutation site(s), resulting in a dysfunctional protein. But addition or deletion of 3 bases restores the reading frame—the normal sequence is disrupted only between the first and third mutation. The resulting protein is altered, but may still be able to function normally. Only a triplet code would show these patterns. **6.** A point mutation changes the nucleotide sequence of an existing allele, creating a new one, so it always changes the genotype. But because the genetic code is redundant, some point mutations do not change the resulting mRNA codons and thus do not change the protein product.

Applying Concepts to New Situations

1. [**See Figure A15.2**] **2.** Every copying error would result in a mutation that would change the amino acid sequence of the protein and would likely affect its function. **3.** Before the central dogma was understood, DNA was known to be the hereditary material, but no one knew how particular sequences of bases resulted in the production of RNA and protein products. The central dogma clarified

FIGURE A14.4

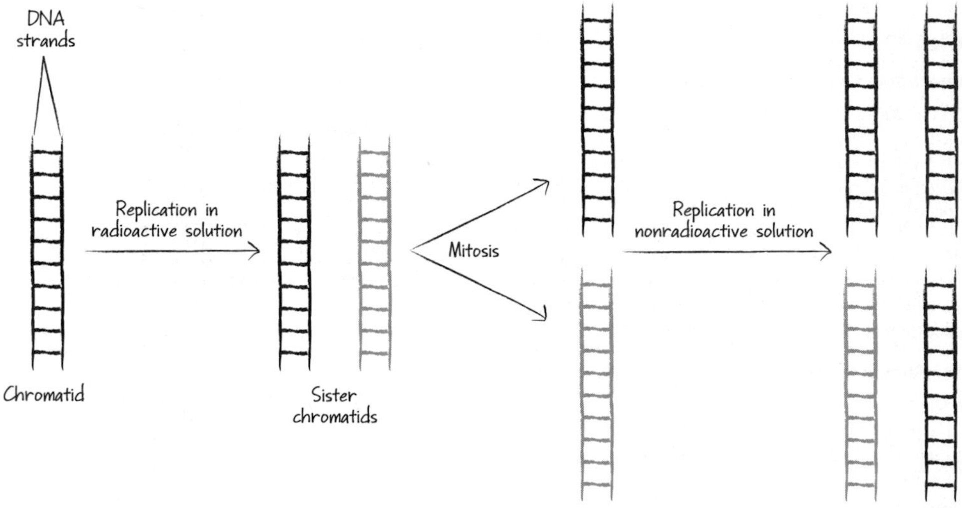

FIGURE A15.1

mRNA sequence:
5′ AUG-CCC-CUG-GAG-GGG-GUU-AGA-CAU 3′

Amino acid sequence:
Met-Pro-Leu-Glu-Gly-Val-Arg-His

FIGURE A15.2

Bottom DNA strand:
5′ AACTT-TAC(start)-GGG-CAA-ACC-ACT-AGC-CCA-ATG-TCG-ATC(stop)-AGTTTC 3′

mRNA sequence:
5′ AUG-CCC-GUU-UGG-AGA-UCG-GGU-UAC-AGC-UAG 3′

Amino acid sequence:
Met-Pro-Val-Trp-Arg-Ser-Gly-Tyr-Ser

how genotypes produce phenotypes. **4.** A triplet code that overlapped by one base might work as follows:

```
5′ AUGUUACGGAAUUGA 3′
   GUU
     AUC
       CGG
         GAA
           AUU
             UGA
```

CHAPTER 16

Check Your Understanding (CYU)

CYU p. 293 (1) Different types of sigma proteins have different amino acid sequences and overall structures, so they are able to bind to stretches of DNA that differ in their base sequence. (2) NTPs are required because the three phosphate groups raise the monomer's potential energy enough to make the polymerization reaction exergonic. **CYU p. 295** (1) The subunits contain both RNA (the "ribonucleo" in the name) and proteins. (2) The cap and tail protect mRNAs from degradation and facilitate translation. **CYU p. 304** E is for Exit—the site where empty tRNAs are ejected; P is for Peptidyl (or peptide bond)—the site where peptide bond formation takes place; A is for Aminoacyl—the site where aminoacyl tRNAs enter.

You Should Be Able To (YSBAT)

YSBAT p. 299 (1) The amino acid attaches on the top left of the L-shaped structure. (2) The anticodon is antiparallel in orientation to the mRNA codon, and contains the complementary bases.

Caption Questions and Exercises

Figure 16.1 RNA is synthesized in the 5′ → 3′ direction; the DNA template is "read" 3′ → 5′. **Figure 16.5** There would be no loops—the molecules would match up exactly. **Figure 16.12** If the amino acids stayed attached to the tRNAs, the gray line in the graph would stay high and the green line low. If the amino acids were transferred to some other cell component, the gray line would decline but the green line would be low.

Summary of Key Concepts

KC 16.1 The new sequence would probably not bind sigma or basal transcription factors as effectively as the original sequence. If so, transcription rate at that gene would probably decrease. **KC 16.2** Bacterial genes lack introns, so their RNA products do not require splicing. And because bacterial mRNAs are translated immediately, they do not need a cap and tail. **KC 16.3** These RNAs transfer amino acids to growing polypeptides; they transfer information in nucleic acids (an mRNA codon) to a protein product. **KC 16.4** An aminoacyl tRNA synthetase catalyzes the addition of an amino acid to a tRNA, forming an aminoacyl tRNA. **KC 16.5** One possible concept map is shown below [**see Figure A16.1**].

Test Your Knowledge

1. c; **2.** c; **3.** d; **4.** c; **5.** c; **6.** d

Test Your Understanding

1. Basal transcription factors bind to promoter sequences in eukaryotic DNA and facilitate the positioning of RNA polymerase. As part of the RNA polymerase holoenzyme, sigma binds to a promoter sequence in bacterial DNA and initiates transcription by the core enzyme. **2.** If the wobble rules did not exist, it would take one tRNA for each amino-acid-specifying codon in the genetic code. This is inefficient. The wobble rules allow a single tRNA to match several mRNA codons, reducing the inefficiency caused by redundancy in the genetic code. **3.** Eukaryotic mRNAs contain introns which must be removed by splicing. Splicing takes place in the nucleus and is accomplished by a large complex called a spliceosome, composed of many snRNPs. **4.** After a peptide bond forms between the polypeptide and the amino acid held by the tRNA in the A site, the ribosome moves down the mRNA. As it does, an empty tRNA leaves the E site. The now-empty tRNA that was in the P site enters the E site; the tRNA holding the polypeptide chain moves from the A site to the P site, and a new aminoacyl tRNA enters the A site. **5.** The ribosome's active site is made up of RNA, not protein. **6.** The separation allows the aminoacyl tRNA to fit into the ribosome correctly, such that the anticodon binds to the mRNA and the amino acid fills the active site.

Applying Concepts to New Situations

1. Ribonucleases degrade mRNAs that are no longer needed by the cell. If an mRNA for a hormone that increased heart rate were never degraded, the hormone would be produced continuously and heart rate would stay elevated—a dangerous situation. **2.** Cordycepin triphosphate has a triphosphate group on the 5′ carbon, but no 3′-OH group. If RNA were built 3′ → 5′, then the incoming nucleoside triphosphates would have to have an OH group on the 3′ carbon in order for a phosphodiester bond to be formed. Since cordycepin lacks a 3′-OH, it could not be added to the 5′ end of an RNA chain growing in the 3′ → 5′ direction. **3.** The regions most crucial to the ribosome's function should be the most highly conserved: the active site, the E-, P-, and A-sites, and the site where mRNAs initially bind. Other regions should be more variable among species. **4.** The most likely locations are one of the grooves or channels where RNA, DNA, and ribonucleotides move through the enzyme—plugging one of them would prevent transcription.

CHAPTER 17

Check Your Understanding (CYU)

CYU p. 314 (1) The genes for metabolizing lactose should be expressed only when lactose is available. (2) [**See Figure A17.1**] **CYU p. 316** (1)

FIGURE A16.1

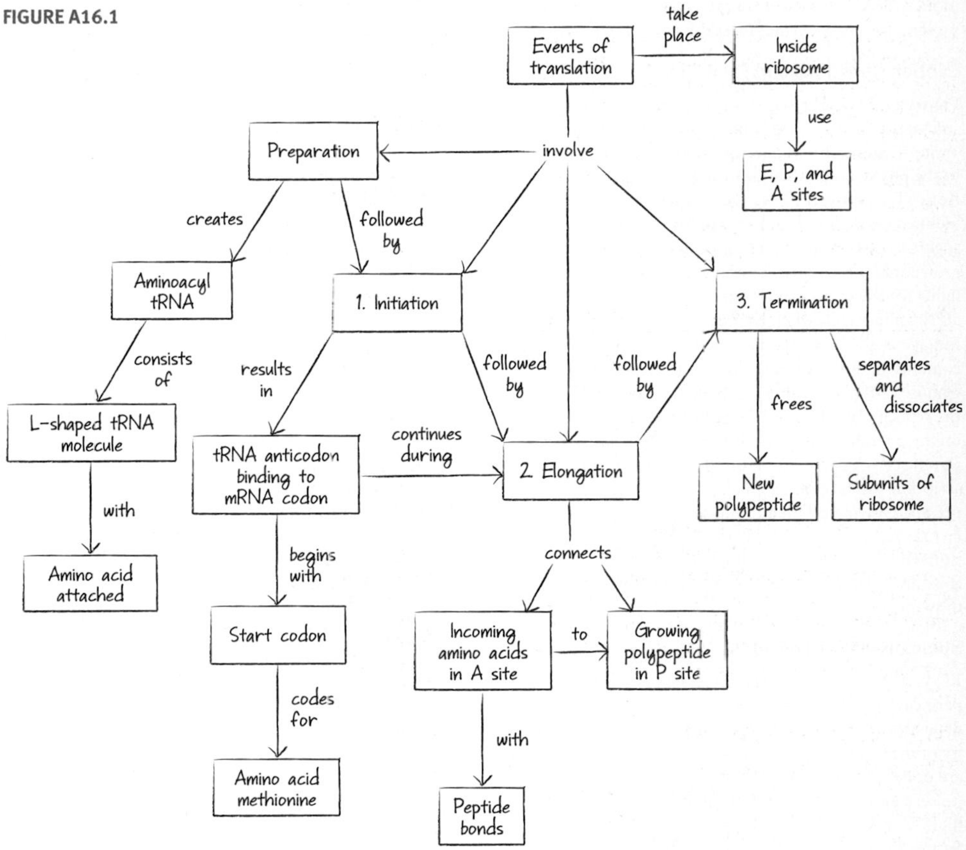

FIGURE A17.1

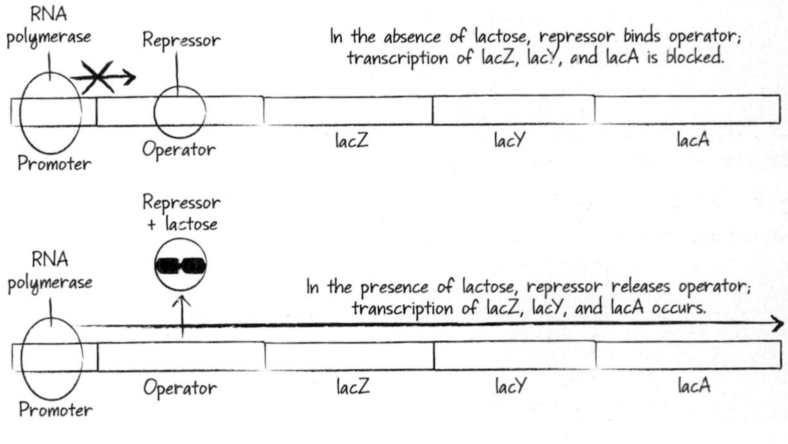

[See Figure A17.2] (2) Two conditions must be met for the *lac* operon to be expressed at a high rate. First, glucose levels must be low. Absence of glucose exerts positive control, because the cAMP-CAP complex binds to the CAP site and increases transcription rate. Second, lactose must be present. Lack of lactose exerts negative control, because the repressor remains attached to the operator and reduces transcription rate.

You Should Be Able To (YSBAT)

YSBAT p. 310 [See Table A17.1] **YSBAT p. 312** *lacZ* codes for the β-galactosidase enzyme, which breaks the disaccharide lactose into glucose and galactose. *lacY* codes for the lactose permease enzyme, which transports lactose into the bacterial cell. *lacI* codes for a protein that shuts down production of the other *lac* products. When lactose is absent, the *lacI* product prevents transcription. This is logical because there is no reason for the cell to make β-galactosidase and lactose permease if there is no lactose to metabolize. But when lactose is present, it interacts with *lacI* in some way so that *lacZ* and *lacY* are induced (their transcription can occur). When lactose is present, the enzymes that metabolize it are expressed.

Caption Questions and Exercises

Figure 17.1 Write "Slowest response, most efficient resource use" next to the transcriptional control label. Write "Fastest response, least efficient resource use" next to the post-translational control label. **Figure 17.2** Plates from all three treatments must be identical and contain identical growth medium, except for the presence of the sugars labeled in the figure. Also, all plates must be grown under the same physical conditions (temperature, light) for the same time. **Figure 17.3** Use a medium with all 20 amino acids when producing a master plate of mutagenized *E. coli* colonies, then use a replica plate that contains all of the amino acids except tryptophan. Choose cells from the master plate that did *not* grow on the replica plate. **Figure 17.8** Put the "Repressor protein" on the operator. No transcription will take place. Then put the "RNA polymerase" on the promoter. No transcription will take place. Finally, put "lactose" on the repressor protein and then remove the resulting lactose-repressor complex from the operon. Transcription will begin. **Figure 17.10** Write "Glucose low" above the drawing in part (a). Write "Glucose high" above the drawing in part (b).

Summary of Key Concepts

KC 17.1 Production of β-galactosidase and galactosidase permease are under transcriptional control—transcription depends on the action of regulatory proteins. The activity of the repressor and CAP—the regulatory proteins—are under post-translational control. **KC 17.2** After a dessert, glucose concentrations in your gut should be high, causing the *lac* operon in *E. coli* to shut down. After drinking milk, glucose concentrations should be low and lactose concentrations high, causing the *lac* operon in *E. coli* cells to be highly expressed. **KC 17.3** (1) The operator is the parking brake; the repressor locks it in place, and the inducer releases it. (2) The CAP-binding site is the gas pedal; the CAP-cAMP complex is a heavy foot on it.

Test Your Knowledge

1. b; **2.** d; **3.** b; **4.** c; **5.** a; **6.** c

Test Your Understanding

1. The glycolytic enzymes are always needed in the cell, because they are required to produce ATP, and ATP is always needed. **2.** Positive control means that a regulatory protein, when present, causes transcription to increase. Negative control means that a regulatory protein, when present, prevents transcription. **3.** The combination of positive and negative control allows *E. coli* to shut down the *lac* operon when glucose is present or lactose absent, but activate it when glucose is absent and lactose present. If only negative control occurred, the operon would be activated when glucose is present. If only positive control occurred, the operon would be activated when lactose is absent. **4.** cAMP levels rise when glucose is low. If glucose is low, the cell is in danger of starving unless it can switch to another food source. **5.** CAP acts like a receptor for cAMP—when cAMP binds to it, its activity changes. **6.** Unlike lactose, glucose can enter the glycolytic pathway directly—without being converted to another molecule. Less ATP and fewer enzymes are needed to acquire and use glucose as an energy source.

Applying Concepts to New Situations

1. Set up cultures with individuals that all come from the same colony of toluene-tolerating bacteria. Half of the cultures should have toluene as the only source of carbon; half should have glucose or another common source of carbon as well as toluene. Cells will grow in both cultures if they are able to use toluene as a source of carbon. **2.** Because toluene is not common in the environment, it is most likely that the toluene genes are under negative control. That is, the presence of toluene—by removing some brake on expression of the enzyme's genes—would trigger production of whatever enzymes are needed to break down toluene. **3.** The *LacI*^{S} mutants would not respond to the inducer, so the repressor would remain on the operator and block transcription even in the presence of the inducer, lactose. These mutants would soon die in an environment where lactose was the only sugar present. **4.** Cells with functioning β-galactosidase are blue; cells that are not producing normal β-galactosidase are white. The sequence of the β-galactosidase gene in structural mutants would differ from the sequence of the wild-type gene, whereas the sequence of the β-galactosidase gene in regulatory mutants would be the same as the sequence of the wild-type gene.

CHAPTER 18

Check Your Understanding (CYU)

CYU p. 323 (1) The basic double helical structure of DNA is the same, but the DNA in eukaryotic cells is complexed with histones to form chromatin, which is tightly packed into 30-nm fibers attached to scaffolding proteins. In bacteria, DNA is associated with histone-like proteins but is not organized into complex structures. (2) Addition of acetyl or methyl groups to histones can cause chromatin to open or close. Different patterns of acetylation or methylation will determine which genes in muscle cells versus brain cells can be transcribed and which are not available for transcription. **CYU p. 326** (1) Bacterial regulatory sequences are found close to the promoter; eukaryotic regulatory sequences can be close to the promoter or far from it. Bacterial regulatory proteins interact directly with RNA polymerase to initiate or prevent transcription; eukaryotic regulatory proteins influence transcription by altering chromatin structure or binding to the basal transcription complex through mediator proteins. (2) Certain regulatory proteins open

FIGURE A17.2

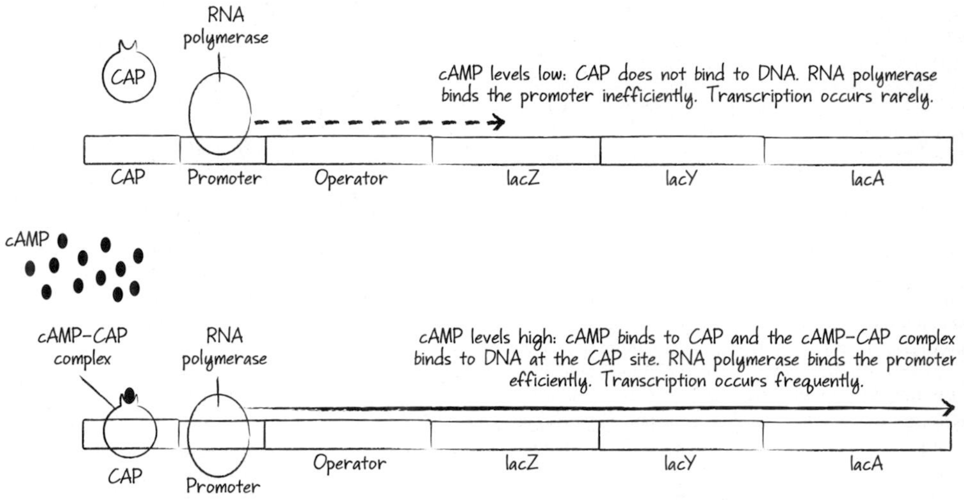
RNA polymerase
CAP
cAMP levels low: CAP does not bind to DNA. RNA polymerase binds the promoter inefficiently. Transcription occurs rarely.
CAP
Promoter
Operator
lacZ
lacY
lacA
cAMP
cAMP–CAP complex
RNA polymerase
cAMP levels high: cAMP binds to CAP and the cAMP–CAP complex binds to DNA at the CAP site. RNA polymerase binds the promoter efficiently. Transcription occurs frequently.
Operator
lacZ
lacY
lacA
CAP
Promoter

TABLE A17.1

Name	Function
lacZ	Gene for β-galactosidase
lacY	Gene for galactosidase permease
Operator	Binding site for repressor
Promoter	Binding site for RNA polymerase
CAP site	Binding site for CAP–cAMP
Repressor	Shuts down transcription
CAP	Increases transcription rate when bound to cAMP
cAMP	Produced when glucose is low; binds to CAP and activates it
Lactose	Binds to repressor and stimulates transcription (removes negative control)
Glucose	At low concentration, increases transcription (via increased production of cAMP)

chromatin at muscle- or brain-specific genes, and then activate or repress the transcription of cell-type-specific genes. Muscle-specific genes are expressed only if muscle-specific regulatory proteins are produced and activated. **CYU p. 330** (1) It became clear that a single gene can code for multiple products instead of a single one. (2) miRNAs interfere with mRNAs by targeting them for destruction or preventing them from being translated. **CYU p. 333** (1) Many different types of mutations can disrupt control of the cell cycle and initiate cancer. These mutations can affect any of the six levels of control over gene regulation outlined in Figure 18.1. (2) The p53 protein is responsible for shutting down the cell cycle in cells with damaged DNA. If the protein does not function, then cells with damaged DNA—and thus many mutations—continue to divide. If these cells have mutations in genes that regulate the cell cycle, then they may continue to divide in an uncontrolled fashion.

You Should Be Able To (YSBAT)

YSBAT p. 326a The basal transcription factors found in muscle and nerve cells are similar or identical; the regulatory transcription factors found in each cell type are different. **YSBAT p. 326b** DNA forms loops when distant regulatory regions, such as silencers and enhancers, are brought close to the promoter through binding of regulatory transcription factors to the mediator complex. **YSBAT p. 329** Alternative splicing does not occur in bacteria because bacterial genes do not contain introns—in bacteria, each gene codes for a single product. Alternative splicing is part of step 3 in Figure 18.1. **YSBAT p. 330** Step 4 and step 5. RNA interference either (1) decreases the life span of mRNAs or (2) inhibits translation.

Caption Questions and Exercises

Figure 18.4 Acetylation of histones decondenses chromatin and allows transcription to begin, so HATs are elements in positive control. Deacetylation condenses the chromatin and inactivates transcription, so HDACs are elements in negative control. **Figure 18.6** This treatment allowed them to measure the normal amount of antibody mRNA produced, for comparison with the other treatments. **Figure 18.7** A typical eukaryotic gene usually contains introns and is regulated by multiple enhancers. Bacterial operons lack introns and enhancers. The promoter-proximal element found in some eukaryotic genes is comparable to the CAP binding site in the *lac* operon of bacteria. Bacterial operons have a single promoter but code for more than one protein; eukaryotic genes code for a single product.

Summary of Key Concepts

KC 18.1 Because eukaryotic RNAs are not translated as soon as transcription occurs, it is possible for RNA processing to occur, which creates variation in the mRNAs produced from a primary RNA transcript and their life span. **KC 18.2** The default state of eukaryotic genes is "off," because the highly condensed state of the chromatin makes DNA unavailable to RNA polymerase. **KC 18.3** [See Figure A18.1] **KC 18.4** Alternative splicing makes it possible for a single gene to code for multiple products. **KC 18.5** Cancer is a disease that results from accumulated mutations. Radiation damages DNA and increases the rate of mutation. Mutations accumulate as cells undergo repeated divisions, so are more common in older people.

Test Your Knowledge

1. c; **2.** a; **3.** d; **4.** d; **5.** b; **6.** d

Test Your Understanding

1. (a) Enhancers and the CAP site are similar because both are sites in DNA where regulatory proteins bind. They are different because enhancers generally are located at great distances from the promoter, whereas the CAP site is located near the promoter. (b) Promoter-proximal elements and the *lac* operon operator are both regulatory sites in DNA located close to the promoter. (c) Basal transcription factors and sigma are proteins that must bind to the promoter before RNA polymerase can initiate transcription. They differ because sigma is part of the RNA polymerase holoenzyme, while the basal transcription complex recruits RNA polymerase to the promoter. **2.** If changes in the environment cause changes in how spliceosomes function, then the RNAs and proteins produced from a particular gene could change in a way that helps the cell cope with the new environmental conditions. **3.** (a) Enhancers and silencers are both regulatory sequences located at a distance from the promoter. Enhancers bind regulatory transcription factors that activate transcription; silencers bind regulatory proteins that shut down transcription. (b) Promoter-proximal elements and enhancers are both regulatory sequences that bind positive regulatory transcription factors. Promoter-proximal elements are located close to the promoter; enhancers are far from the promoter. (c) Transcription factors bind to regulatory sites in DNA; the mediator complex does not bind to DNA but instead forms a bridge between regulatory transcription factors and basal transcription factors. **4.** RNA interference happens when complementary base pairing occurs between a small single-stranded miRNA and a single-stranded mRNA. **5.** Certain chemical modifications of histones mark sections of chromatin for transcription activation or repression. These modifications are passed to daughter cells when cells divide. A certain array of chromatin modifications or "histone codes" is associated with the production of muscle- or nerve-specific proteins. **6.** Lack of functional p53 protein prevents damaged cells from being destroyed or shut down (which would prevent their continuing to divide). As a result, cells with damaged DNA and thus many mutations continue to divide, possibly initiating cancerous growth.

Applying Concepts to New Situations

1. In eukaryotic cells, histones are involved in the basic folding of DNA molecules in chromatin. Mutations in histones are likely to interfere with this folding, and thus the expression of large numbers of genes. Individuals with mutant forms of histones are likely to die as a result—meaning that histone structure should be highly conserved over time. **2.** Tissues that divide more often must replicate their DNA more often; thus they are more susceptible to mutation. Because the cell cycle is accelerated in these cell types, it is logical to expect that control over the cell cycle could easily be lost. **3.** You could treat a culture of DNA-damaged cells with a drug that stops transcription and then compare them with untreated DNA-damaged cells. If transcriptional control regulates p53 levels, then the p53 level would be lower in the treated cells versus control cells. If control of p53 levels is post-translational, then in both cultures the p53 level would be the same. *Other approaches:* add labeled NTPs to damaged cells and see if they are incorporated into mRNAs for p53; or add labeled amino acids to damaged cells and see if they are incorporated into completed p53 proteins; or add labeled phosphate groups and see if they are added to p53 proteins. **4.** The miRNA would bind to viral RNA inside infected cells and prevent the viral RNA from working, thus stopping the infection.

THE BIG PICTURE: GENETIC INFORMATION

Check Your Understanding (CYU), p. 336

1. Star = DNA, mRNA, proteins. **2.** RNA "is reverse transcribed by" reverse transcriptase "to form" DNA. **3.** E = splicing, etc.; E = meiosis and sexual reproduction (along with their links). **4.** Chromatin "makes up" chromosomes; independent assortment and recombination "contribute to" high genetic diversity.

CHAPTER 19

Check Your Understanding (CYU)

CYU p. 344 (1) When the endonuclease makes a staggered cut in a palindromic sequence and the strands separate, the single-stranded bases that are left will bind ("stick") to the single-stranded bases left where the endonuclease cut the same palindrome at a different location. (2) The word probe means to examine thoroughly. A DNA probe "examines" a large set of sequences thoroughly, and binds to one—the one that has a complementary base sequence. **CYU p. 346** (1) Denaturation makes DNA single stranded so the primer can bind to the sequence during the annealing step. Once the primer is in place, *Taq* polymerase can synthesize the rest of the strand during the extension step. It is a "chain reaction" because the products of each reaction cycle are used in the next reaction cycle—this is why the number of copies doubles in each cycle. (2) One of many possible

FIGURE A18.1

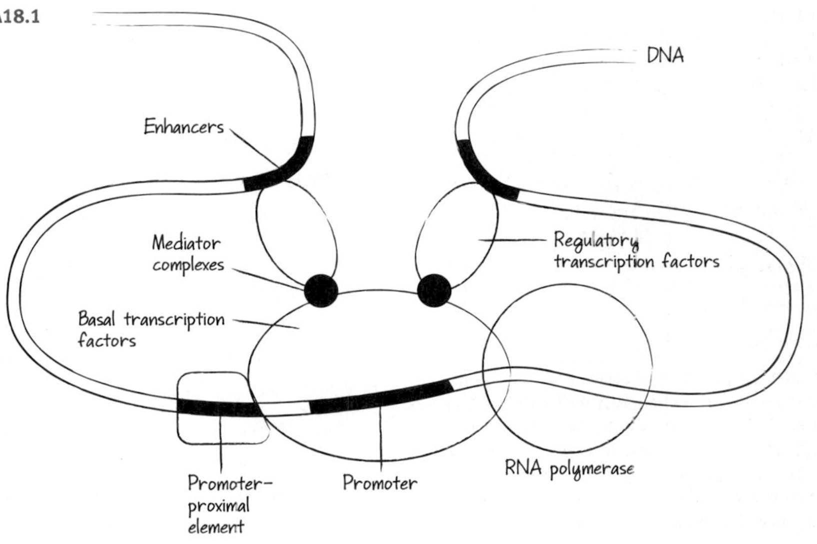

answers is shown below [see Figure A19.1]. **CYU p. 348** If ddNTPs were present at high concentration, they would almost always be incorporated—meaning that only fragments from the first complementary base in the sequence would be produced. **CYU p. 351** Start with a genetic map with as many polymorphic markers as possible. Get the genotype at these markers for a large number of individuals who have the same type of alcoholism—one that is thought to have a genetic component—as well as a large number of unaffected individuals. Look for particular versions of a marker that is almost always found in affected individuals. Genes that contribute to a predisposition to alcoholism will be near that marker.

You Should Be Able To (YSBAT)

YSBAT p. 342a Isolate the DNA and cut it into small fragments with EcoR1, which leaves sticky ends. Cut copies of a plasmid or other vector with EcoR1. Mix the fragments and plasmids under conditions that promote complementary base pairing by sticky ends of fragments and plasmids. Use DNA ligase to catalyze formation of phosphodiester bonds and seal the sequences. **YSBAT p. 342b** The probe must be single stranded so that it will bind by complementary base pairing to the target DNA, and it must be labeled so that it can be detected. The probe will only base pair with fragments that include a sequence complementary to the probe's sequence. A probe with the sequence 5′ AATCG 3′ will bind to the region of the target DNA that has the sequence 5′ AATCG 3′ as shown below:

```
     5′ AATCG 3′
3′ TCCGGTTAGCATTACCATTTT 5′
```

YSBAT p. 348 Using a map with many markers makes it more likely that there will be one very tightly linked to the gene of interest—meaning that a version of the marker will almost always be associated with the phenotype you are tracking.

Caption Questions and Exercises

Figure 19.4 No—many times, because there were many copies of each type of mRNA present in the pituitary cells and many pituitary cells were used to prepare the library. **Figure 19.7** The polymerase will begin at the 3′ end of each primer. On the top strand in part (b), it will move to the left; on the bottom strand it will move to the right. As always, synthesis is in the 5′ to 3′ direction. **Figure 19.10** The insertion will probably disrupt the gene and have serious consequences for the cell and potentially the individual.

Summary of Key Concepts

KC 19.1 When "sticky ends" anneal, the recombinant product would not be stable—it would tend to break apart because no phosphodiester bond would form between the DNA fragments. **KC 19.2** PCR advantages: It is relatively fast and easy, and can amplify a DNA sequence that is rare in the sample. PCR disadvantage: It requires knowledge of sequences on either side of the target gene, so primers can be designed. Plasmid advantages: No knowledge of the sequence is required. Plasmid disadvantages: It is slower and technically more difficult than PCR. **KC 19.3** The length of each fragment is dictated by where a ddNTP was incorporated into the growing strand, and each ddNTP corresponds to a base on the template strand. Thus, the sequence of fragment sizes corresponds to the sequence on the template DNA. **KC 19.4** Both affected and unaffected individuals have the same marker, so you have no way of knowing that it is close to the gene you are interested in. **KC 19.5** Genes can be inserted into the Ti plasmids, and the recombinant plasmids transferred to plant cells.

Test Your Knowledge

1. c; **2.** b; **3.** d; **4.** a; **5.** b; **6.** d

Test Your Understanding

1. When a restriction endonuclease cuts a "foreign gene" sequence and a plasmid, the same sticky ends are created on the excised foreign gene and the cut plasmid. After the sticky ends on the foreign gene and the plasmid anneal, DNA ligase catalyzes closure of the DNA backbone, sealing the foreign gene into the plasmid DNA. [See **Figure A19.2**] **2.** Vectors hold a piece of foreign DNA and carry it into an intact host cell. A "perfect" vector (1) has well-characterized restriction endonuclease sites, (2) can be inserted into host cells easily, (3) expresses recombinant genes in the host cell in a controlled way, and (4) doesn't damage the host cell. **3.** A cDNA library is a collection of complementary DNAs made from all the mRNAs present in a certain group of cells. A cDNA library from a human nerve cell would be different from one made from a human muscle cell, because nerve cells and muscle cells express many different, cell type–specific genes. **4.** Genetic markers are genes or other loci that have known locations in the genome. When these locations are diagrammed, they represent the physical relationships among landmarks—in other words, they form a map. **5.** The genes for making β-carotene need to be expressed in the endosperm. Adding an endosperm-specific promoter adjacent to the genes ensures that they are expressed in the endosperm. In general, including specific promoter and enhancer sequences with recombinant genes helps researchers control where and when the recombinant genes are expressed. **6.** PCR and cellular DNA synthesis are similar in the sense of producing copies of a template DNA. Both rely on primers and DNA polymerase. The major difference between the two is that PCR copies only a specific target sequence, while the entire genome is copied during cellular DNA synthesis.

Applying Concepts to New Situations

1. You could use a computer program to identify likely promoter sequences in the sequence data, and then look for sequences just downstream that have an AUG start codon and a stop codon about 1500 base pairs apart. **2.** Yes—you would transform cells that are the precursors to sperm or eggs, so that the recombinant DNA is found in mature sperm or eggs and passed on to offspring. **3.** In both techniques, researchers use an indicator to identify either a gene of interest or a colony of bacteria with a particular trait. The problem is the same—picking one particular thing (a certain gene or a cell with a particular mutation) out of a large collection. **4.** (a) Primer 1b binds to the top right strand and would allow DNA polymerase to synthesize the top strand across the target gene. Primer 1a, however, binds to the upper left strand and would allow DNA polymerase to synthesize the upper strand *away* from the target gene. Primer 2a binds to the bottom left strand and would allow DNA polymerase to synthesize the bottom strand across the target gene. Primer 2b, however, binds to the bottom right strand and would allow DNA polymerase to synthesize away from the target gene. (b) She could use primer 1b with primer 2a.

CHAPTER 20

Check Your Understanding (CYU)

CYU p. 364 (1) Parasites don't need genes that code for enzymes required to synthesize molecules they acquire from their hosts. (2) If two closely related species inherited the same gene from a common ancestor, the genes should be similar. But if one species acquired the same gene from a distantly related species via lateral gene transfer, then the genes should be much less similar. **CYU p. 370** (1) Unequal crossover and mistakes in DNA synthesis are so common in simple tandem repeats that new alleles—consisting of the same chromosome region with different repeat numbers—are created almost every generation, producing a molecular fingerprint. (2) Unequal crossover occurs when homologous chromosomes

FIGURE A19.1

```
5′ CATGACTATTACGTATCGGGTACTATGCTATCGATCTAGCTACGCTAGCT 3′
3′ GTACTGATAATGCATAGCCCATGATACGATAGCTAGATCGATGCGATCGA 5′
```

Probe #1, which will anneal to the 3′ end of the top strand:

```
5′ AGCTAGCGTAGCTAGATCGAT 3′
```

Probe #2, which will anneal to the 3′ end of the bottom strand:

```
5′ CATGACTATTACGTATCGGGG 3′
```

The primers will bind to the separated strands of the parent DNA sequence as follows:

```
5′ CATGACTATTACGTATCGGGTACTATGCTATCGATCTAGCTACGCTAGCT 3′
                                3′ TAGCTAGATCGATGCGATCGA 5′

5′ CATGACTATTACGTATCGGG 3′
3′ GTACTGATAATGCATAGCCCATGATACGATAGCTAGATCGATGCGATCGA 5′
```

FIGURE A19.2

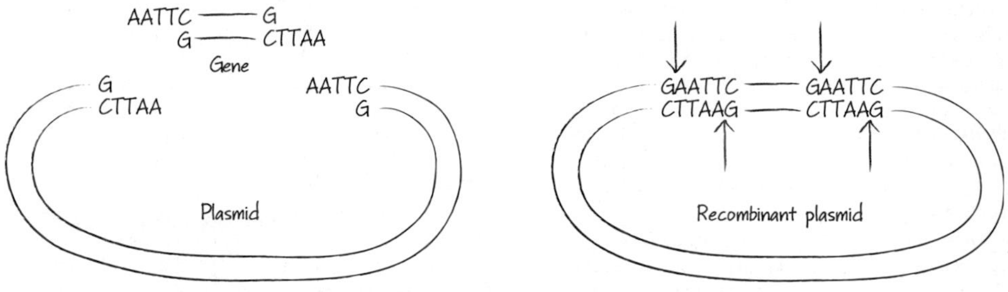

Sticky ends on cut plasmid and gene to be inserted are complementary and can base pair

DNA ligase forms four phosphodiester bonds between gene and plasmid DNA at the points indicated by arrows align incorrectly during synapsis. When crossing over occurs within a misaligned section of chromosome, one homolog ends up with less DNA and the other ends up with more DNA. The extra DNA is made up of sequences from the other chromosome that are duplicates of the DNA on the original strand.

You Should Be Able To (YSBAT)

YSBAT p. 360 If no overlap occurred, there would be no way of ordering the fragments correctly. You would only be able to put fragments in random order—not the correct order. **YSBAT p. 371** Start with a microarray containing exons from a large number of human genes. Isolate mRNAs from brain tissue and liver tissue, and make labeled cDNAs from each. Probe the microarray with each type of probe, and record where binding occurs. Binding events identify exons that are transcribed in each type of tissue. Compare the results to identify genes that are expressed in brain but not liver, or liver but not brain.

Caption Questions and Exercises

Figure 20.2 Shotgun sequencing is based on fragmenting the genome into many small pieces. **Figure 20.3** Because the mRNA codons that match up with each strand are oriented in the 5′ to 3′ direction. **Figure 20.8** Chromosome 1:
ψβ2-ε-Gγ-Aγ-ψβ1-δ-ψβ2-ε-Gγ-Aγ-ψβ1-δ-β
Chromosome 2:
β

Summary of Key Concepts

KC 20.1 If a search of human gene sequence databases revealed a gene that was similar in base sequence, and if follow-up work confirmed that the mouse and human genes were similar in their pattern of exons and introns and regulatory sequences, then the researchers could claim that they are homologous. **KC 20.2** DNA can move from one prokaryotic species to another packaged in plasmids or viruses, or by the direct uptake of DNA fragments from the environment (transformation). **KC 20.3** Junk DNA refers to noncoding regions of DNA—sequences that are not transcribed into RNA. The function of repeated sequences is not yet clear, but many noncoding sequences are interesting because they are relicts from transposable elements. **KC 20.4** Expression of the gene may be involved in the development of cancer.

Test Your Knowledge

1. c; **2.** a; **3.** d; **4.** b; **5.** d; **6.** d

Test Your Understanding

1. Computer programs are used to scan sequences in both directions to find ATG start codons, a gene-sized logical sequence with recognizable codons, and then a stop codon. One can also look for characteristic promoter, operator, and other regulatory sites. It is more difficult to identify open reading frames in eukaryotes because their genomes are so much larger and because of the presence of introns and repeated sequences. **2.** To have the potential to thrive in different environments, a cell requires a large array of enzymes and proteins—to use different sources of food, cope with different temperature or pH conditions and so on—and thus a large array of genes. **3.** They use the transcription and translation machinery of the host cell to copy themselves and insert themselves into the host's genome, and this may have a negative effect on the host cell. **4.** Each individual has a DNA fingerprint—a unique collection of microsatellite and minisatellite alleles (length variants). Because alleles are inherited, closely related individuals share more alleles than more distantly related individuals. **5.** A DNA microarray experiment identifies which genes are being expressed in a particular cell at a particular time. If a series of experiments shows that different genes are expressed in cells at different times or under different conditions, it implies that expression was turned on or off in response to changes in age or changes in conditions. **6.** Homology is a similarity among different species that is due to their inheritance from a common ancestor. If a newly sequenced gene is found to be homologous with a known gene of a different species, it is assumed that the gene products have similar function. Duplicated genes are homologous because they share a common ancestor gene.

Applying Concepts to New Situations

1. If "gene A" is not necessary for existence, it can be lost by an event like unequal crossing over (on the chromosome with deleted segments) with no ill effects on the organism. In fact, individuals who have lost unnecessary genes are probably at a competitive advantage, because they no longer have to spend time and energy copying and repairing unused genes. **2.** If the grave was authentic, it might include two very different parental patterns along with three children whose patterns each represented a mix between the two parents. The other unrelated individuals would have patterns not shared by anyone else in the grave. **3.** The genes may have similar DNA sequences because the viral gene was incorporated into the genome of a human host in the past, and subsequently passed on to future generations. **4.** You would expect that the livers and blood of chimps and humans would function similarly, but that strong differences occur in brain function. The microarray data support this prediction and suggest that even though the brain proteins might have similar sequences, chimp and human brains are different because certain genes are turned on or off at different times and expressed in different amounts.

CHAPTER 21

Check Your Understanding (CYU)

CYU p. 379 Biologists have been able to grow entire plants from a single differentiated cell taken from an adult. They have also succeeded at producing animals by transferring the nucleus of a fully differentiated cell to an egg whose nucleus has been removed. **CYU p. 384** (1) Bicoid is a transcription factor, so cells that experience a high versus medium versus low concentration transcribe different amounts of Bicoid-regulated genes. (2) Bicoid proteins control the expression of gap genes, which—in effect—tell cells which third of the embryo they are in along the anterior-posterior axis. Gap genes control the expression of pair-rule genes, which organize the embryo into individual segments. Pair-rule genes control expression of segment polarity genes, which establish an anterior-posterior polarity to each individual segment. Segment polarity genes turn on homeotic genes, which then trigger genes for producing segment-specific structures like wings or legs.

Caption Questions and Exercises

Figure 21.2 It would look like the embryo of the left side of part (b)—a normal phenotype. **Figure 21.5** In cells that are expressing a particular gene, the gene is transcribed into RNA. In situ hybridization locates these copies of RNA, identifying which cells are expressing the gene. **Figure 21.10** Genes are usually considered homologous if they have similar structure—meaning similar DNA sequence and exon-intron structure.

Summary of Key Concepts

KC 21.1 The seedling would grow very slowly and be small, but development should proceed normally otherwise. **KC 21.2** Parenchyma cells in the stem still contain a complete set of genes, so if they can be de-differentiated and received appropriate signals from other cells, they could start producing vascular-tissue-specific proteins. **KC 21.3** Different concentrations of transcription factors or cell-cell signals will cause differences in the amount of expression from certain genes—so a high concentration means something different than a low concentration. **KC 21.4** The idea is that a single molecule can set up the anterior-posterior or apical-basal axis of the embryo. Once the axis is established, a sequence of events follows in which cells get increasingly more precise information about where they are located along that axis—based on the original concentration of the master regulator. **KC 21.5** When the patterns of expression of two key genes involved in limb versus rib formation are compared in chick and snake embryos, they differ. In snakes, the "form ribs here" pattern occurs instead of the "form legs here" pattern.

Test Your Knowledge

1. b; **2.** d; **3.** a; **4.** c; **5.** a; **6.** a

Test Your Understanding

1. They occur in both plants and animals, and are responsible for the changes that occur as an embryo develops. **2.** If the transplanted nucleus has undergone some permanent change or loss of genetic information during development, it should not be able to direct the development of a viable adult. But if the transplanted nucleus is genetically equivalent to the nucleus of a fertilized egg, then it should be capable of directing the development of a new individual. **3.** The researchers exposed adult flies to treatments that induce mutations, then looked for embryos with defects in the anterior-posterior body axis or body segmentation. The embryos had mutations in genes required for body axis formation and segmentation. **4.** Development—specifically differentiation—depends on changes in gene expression. Changes in gene expression depend on differences in regulatory transcription factors. **5.** Differentiation is triggered by the presence or absence of external signals. These signals trigger the production of transcription factors, which induce other transcription factors, and so on—a sequence that constitutes a regulatory cascade—as development progresses. At each step in the cascade, a new subset of genes is activated—resulting in a step-by-step progression from undifferentiated to fully differentiated cells. **6.** *Hox* genes were first discovered in *Drosophila* and have been identified in nearly every animal. The number of *Hox* genes varies across species, but their arrangement on the chromosome and pattern of expression in the embryo are strikingly similar across animal species.

Applying Concepts to New Situations

1. There would be more Bicoid protein farther toward the posterior and less in the anterior—meaning that the anterior segments would be "less anterior" in their characteristics and the posterior segments would be "more anterior." **2.** The result means that the human gene can be transcribed in worms and that its protein product has performed the same function in worms as in humans. This is strong evidence that the genes are homologous, and that their function has been evolutionarily conserved. **3.** The stem cells would have to be stimulated with nerve-specific signals to trigger their differentiation into nerve cells. **4.** Homeotic genes—such as *Hox* genes—are responsible for triggering the production of structures like wings or legs in particular segments. One hypothesis to explain the variation is that changes in gene expression led to different numbers of segments that express leg-forming *Hox* genes.

CHAPTER 22

Check Your Understanding (CYU)

CYU p. 392 (1) If a mutation in either bindin or the egg-cell receptor for sperm prevented the proteins from binding to each other, then fusion of sperm and egg-cell membranes could not take place and fertilization would not occur. (2) The entry of a sperm triggers a wave of calcium ion release around the egg-cell membrane. Thus, calcium ions act as a signal that indicates "A sperm has entered the egg." In response to the increase in calcium ion concentration, cortical granules fuse with the egg-cell membrane and release their contents to the exterior, causing the fertilization envelope to develop. **CYU p. 394** If cytoplasmic determinants were distributed in equal concentration throughout the egg, all blastomeres would end up with the same types and concentrations of cytoplasmic determinants. **CYU p. 395** (1) Endoderm is in the interior, ectoderm is on the outside, mesoderm is in between. (2) These cells become endoderm that forms the gut lining. The direction of the gut helps define the anterior-posterior axis of the body—it connects mouth and anus.

You Should Be Able To (YSBAT)

YSBAT p. 391 If the bindin-like protein were blocked, it would not be able to bind to the egg-cell receptor for sperm. Fertilization would not take place.

Caption Questions and Exercises

Figure 22.4 The researchers didn't start out with a hypothesis about which egg-cell membrane protein bound to bindin. Instead, they tested every type of protein projecting from the egg-cell surface to see which one, if any, bound to bindin. **Figure 22.8** [See Figure A22.1] **Figure 22.9** Ectoderm forms most of the structures on the outside of the adult. Endoderm forms mostly interior structures—for example, the inner lining of the respiratory and digestive tracts. Mesoderm forms structures located between the ectoderm-derived and endoderm-derived structures, including bones and muscles and the bulk of most of the organ systems. **Figure 22.15** It made gene expression possible in non-muscle cells. A "general purpose" promoter can lead to transcription of an inserted cDNA in any type of cell, including fibroblasts.

Summary of Key Concepts

KC 22.1 Bindin and the egg-cell receptor for sperm bind to each other. For this to happen, their tertiary structures must fit together somewhat like the structures of a lock and key. **KC 22.2** If the master regulator were localized to a certain part of the egg cytoplasm or formed a Bicoid-like concentration gradient throughout the eggs, blastomeres would contain or not contain the regulator or contain a distinct concentration of it. **KC 22.3** Both cells would contain cell-cell signals and transcription factors specific to mesodermal cells, but the anterior cell would contain anterior-specific signals and regulatory transcription factors while the posterior cell would contain posterior-specific signals and regulators. **KC 22.4** Cell-cell signals from the notochord direct the formation of the neural tube; subsequently, cell-cell signals from the notochord, neural tube, and ectoderm direct the differentiation of somite cells. Notochord cells die later in development. Cells expand during neural tube formation. Somite cells move to new positions, proliferate, and differentiate.

Test Your Knowledge

1. d; **2.** c; **3.** c; **4.** d; **5.** c; **6.** a

Test Your Understanding

1. Eggs contain stores of nutrients. **2.** If more than one sperm nucleus entered the egg, the resulting cell would contain more than two copies of each chromosome. Mitosis could not occur correctly and the embryo would be deformed or die. **3.** They contain different types and/or concentrations of cytoplasmic determinants. **4.** They give rise to all of the tissue and organs of the adult (from the Latin *germen*, meaning shoot or sprout). **5.** When transplanted early in development, somite cells become the cell type associated with their new location. But when transplanted later in development, somite cells become the cell type associated with the original location. These observations indicate that the same cell is not committed to a particular fate until later in development. **6.** If the gene for MyoD is expressed in non-muscle cells, the cells begin producing muscle-specific proteins. It triggers differential gene expression typical of muscle cells.

Applying Concepts to New Situations

1. The interactions that allow proteins to bind to each other are extremely specific—certain amino acids have to line up in certain positions with specific charges and/or chemical groups exposed. Sperm-egg binding is species specific because each form of bindin binds to a different egg cell receptor for sperm. **2.** Translation is not needed for cleavage to occur normally—meaning that all the proteins needed for early cleavage are present in the egg. **3.** The yellow pigment is a cytoplasmic determinant that triggers a gene regulatory cascade leading to differentiation of muscle cells. **4.** The embryo cannot yet feed, so all of its activity is fueled by the nutrient stores in the egg. It is not taking in a lot of new molecules that would allow it to get bigger.

CHAPTER 23

Check Your Understanding (CYU)

CYU p. 404 (1) In the male reproductive organs of a mature flower, cells undergo meiosis to produce haploid cells. Each of these haploid cells divides by mitosis to produce a pollen grain containing haploid cells. One of the cells in the pollen grain divides again by mitosis to form two sperm cells. (2) They ensure that the plant is fertilized by a pollen grain from the same species. If any pollen could germinate on any stigma, pollen from a pine tree could grow on the stigma of an orchid. **CYU p. 406** (1) Cells located on the outside of the embryo become epidermal tissue; cells in the interior become vascular tissue; cells in between become ground tissue. (2) The *MONOPTEROS* gene product would be overactivated. The embryonic cells would get information indicating that they are in the shoot apex, where MONOPTEROS levels are high. As a result, the root would not develop normally. **CYU p. 409** [See Figure A23.1] If a new supply of fertilizer (cow dung) appeared near the plant's left side, the roots on that side would extend toward the nutrients. The shoot system should grow in response to the new nutrients but, all other things being equal (e.g., no shading of plant from sunlight), shoot growth should be symmetrical. **CYU p. 411** (1) The A protein might function as a transcriptional repressor by binding to a regulatory site near the *C* gene, thereby preventing transcription of the *C* gene. (2) Like *Hox* genes, *MADS-box* genes code for DNA-binding proteins that function as transcriptional regulators. Both Hox proteins and MADS-box proteins are involved in regulatory transcription cascades that lead to development of specific structures in specific locations. Although they perform similar functions, *Hox* genes and *MADS-box* genes evolved independently and differ in their base sequences. Thus, these genes and their protein products are not homologous.

You Should Be Able To (YSBAT)

YSBAT p. 410 If four genes coded for flower structure and each gene specified one of the four organs, then each mutant should lack an element in only one of the whorls. If each of three genes is expressed in two adjacent whorls, then four different gene expression patterns are possible—one for each of the four different whorls.

FIGURE A22.1

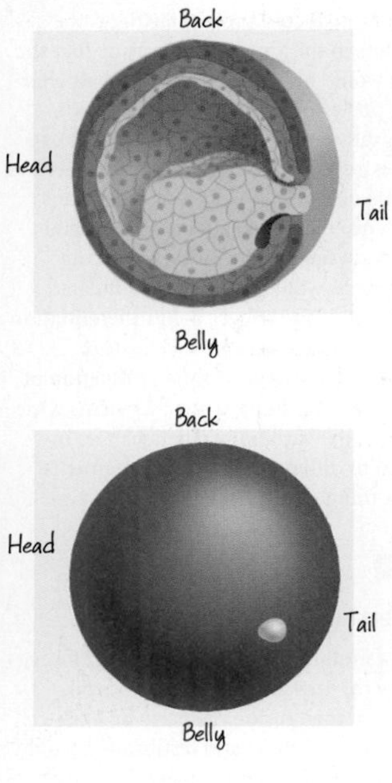

FIGURE A23.1

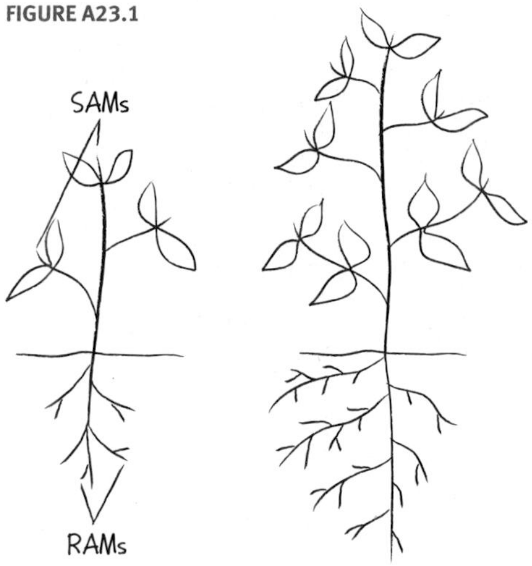

FIGURE A23.2

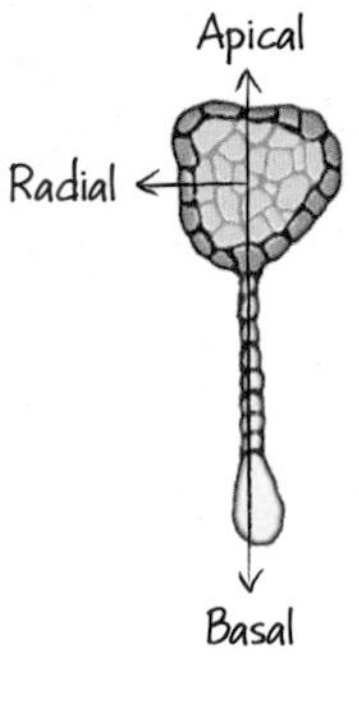

Caption Questions and Exercises

Figure 23.3 In both cases, protein-protein interactions between male and female structures in two different individuals prevent fertilization between members of different species. **Figure 23.5** [See Figure A23.2] **Figure 23.9** Many animals have three body axes analogous to the three axes of leaves: anterior-posterior (like the leaf's proximal-distal), dorsal-ventral (like the leaf's adaxial-abaxial), and a left-right axis (like the leaf's lateral axis). The main plant body, in contrast, has only two axes: apical-basal and radial.

Summary of Key Concepts

KC 23.1 Bicoid defines the anterior-posterior axis of the fly embryo; auxin defines the apical-basal axis of the plant embryo. Both work via concentration gradients. Bicoid is a regulatory transcription factor while auxin is a cell-cell signal. **KC 23.2** There are two fusion events between sperm and haploid cells in the female reproductive structure: one sperm fuses with the egg and leads to the development of an embryo; the other sperm fuses with two (or more) nuclei and leads to the development of endosperm. **KC 23.3** [See Figure A23.3] **KC 23.4** The presence of multiple SAMs and RAMs provides plants with flexibility in adapting to environmental changes and acquiring sunlight and nutrients, thereby increasing their ability to grow and reproduce. For example, multiple SAMs allow a plant to extend shoots (branches) in multiple directions to receive optimal sunlight, and multiple RAMs allow it to extend roots in multiple directions to acquire optimal water and nutrients. Also, a plant would be in trouble if its only SAM or RAM got eaten. **KC 23.5** Homeotic mutants are defined by a phenotype in which one body structure is replaced by another. *Arabidopsis* plants with a defective *B* gene have flowers in which petals are replaced by sepals and stamens are replaced by carpels.

Test Your Knowledge

1. a; **2.** b; **3.** d; **4.** d; **5.** c; **6.** a

Test Your Understanding

1. Proteins on the surface of a pollen grain interact with proteins on the surface of the stigma. Signals from the egg direct the growth of the pollen tube. **2.** Animal cells that lead to sperm or egg formation are "sequestered" early in development, so that they are involved only in reproduction and undergo few rounds of mitosis. Because mutations occur in every round of mitosis, plant eggs and sperm should contain many more mutations than animal eggs and sperm. **3.** Both meristems and stem cells can produce differentiated cells in an adult. Meristems are much more flexible than stem cells, as they can form all parts of the plant, and they increase the size of the body. Animal stem cells can usually develop into only a few differentiated cell types, and only replace lost or damaged cells. **4.** Just like the three tissue layers in plant embryos, the tissues produced in the SAMs and RAMs of a 300-year-old oak tree can differentiate into all of the specialized cell types found in a mature plant. **5.** The ABC model predicted that the *A*, *B*, and *C* genes would each be expressed in specific whorls in the developing flower. The experimental data showed that the prediction was correct. **6.** Cell proliferation is responsible for expansion of meristems and growth in size. Cell death occurs when leaves drop, and when three or four meiotic products die during gametogenesis in female tissues. Asymmetrical cell proliferation along with cell expansion is responsible for giving the adult plant a particular shape. Differentiation is responsible for generating functional tissues and organs in the adult plant. Cell-cell signals are responsible for setting up the apical-basal axis of the embryo and for guiding the pollen tube to the egg.

Applying Concepts to New Situations

1. If an individual is likely to die, it should throw all of its remaining resources into reproduction. **2.** Because vegetative growth is continuous and occurs in all directions (unrestricted), it could be considered indeterminate. However, reproductive growth produces mature reproductive organs and stops when those organs are complete. Because its duration is limited, it could be considered determinant. **3.** High light conditions could trigger changes in gene expression that reduced the rate of cell proliferation during leaf development. In contrast, low light conditions might trigger changes in gene expression that increase the rate of cell proliferation. **4.** [See Figure A23.4] In a young oak tree, SAMs would be evenly distributed in the upper parts of the plant and the RAMs would be evenly distributed in the tips of the roots. But after 50 years in the situation described above, the RAMs would be concentrated on the right side of the plant's roots, enabling the roots to grow into the area where water is available from the leaky pipe next to the building. The SAMs would be concentrated on the left side of the plant, allowing the upper parts of the tree to grow away from billboard, where light is available.

CHAPTER 24

Check Your Understanding (CYU)

CYU p. 429 (1) *Postulate 1:* Traits vary within a population. *Postulate 2:* Some of the trait variation is heritable. *Postulate 3:* There is variation in reproductive success (some individuals produce more offspring than others). *Postulate 4:* Individuals with certain heritable traits produce the most offspring. The first two postulates describe heritable variation; the second two describe differential reproductive success. (2) Beak size and shape and body size vary among individual finches, in part because of differences in their genotypes (some alleles lead to larger or narrower beaks, for example). When a drought hit, individuals with deep beaks survived better and produced more offspring than individuals with shallow beaks. **CYU p. 432** (1) When certain individuals are selected, their traits do not change—they simply produce more offspring than other individuals.
(2) Adaptations are not optimal solutions to challenges posed by a particular environment because they are compromised by (1) the necessity of meeting many challenges at the same time (an adaptive "solution" to one problem—such as flying faster—may make another problem worse—such as being maneuverable in flight); (2) lack of "optimal" alleles or presence of alleles that affect more than one trait; and (3) the necessity of selecting only preexisting variation in traits.

You Should Be Able To (YSBAT)

YSBAT p. 425 (1) Relapse occurred because the few bacteria remaining after drug therapy were not eliminated by the patient's weakened immune system and began to reproduce quickly. (2) No—almost all of the cells present at the start of the infection would have been resistant to the drug. **YSBAT p. 430** In biology, an adaptation is any heritable trait that increases an individual's ability to produce offspring in a particular environment. In everyday English, adaptation is often used to refer to an individual's nonheritable adjustment to meet an environmental challenge, a phenomenon that biologists call acclimation. The phenotypic changes resulting from acclimation are not passed on to offspring.

Caption Questions and Exercises

Figure 24.4 The theory of special creation would claim that the fossil and extant sloths were both created 6000 years ago, but that the fossil species became extinct during the flood in Noah's time, described in the Bible. It is not clear how the theory of special creation would explain transitional features observed in the fossil record. **Figure 24.5** If vestigial traits result from inheritance of acquired characteristics, some individuals must have lost the traits during their own lifetimes and passed the reduced traits on to their offspring. For example, a certain monkey's long tail might have been bitten off by a predator, or an ape's hair might have been pulled out of its skin by a rival during a fight. The new traits would then somehow have passed to the individuals' eggs and sperm, resulting in shorter-tailed and less-hairy offspring, until humans with a coccyx and goose bumps resulted. **Figure 24.16** "Prediction": Beak measurements were different before and after the drought. "Prediction of null hypothesis": No difference in beak measurements before and after the drought.

Summary of Key Concepts

KC 24.1 Under the theory of special creation, changes in *Mycobacterium* populations would be explained as individual creative events governed by an intelligent creator. Under the theory of evolution by inheritance of acquired characters, changes in *Mycobacterium* populations would be explained by the cells trying to transcribe genes in the presence of the drug, and their *rpoB* gene becoming altered as a result. **KC 24.2** In biology, fitness is the ability of an individual to produce offspring, relative to that ability in other individuals in the population.

FIGURE A23.3

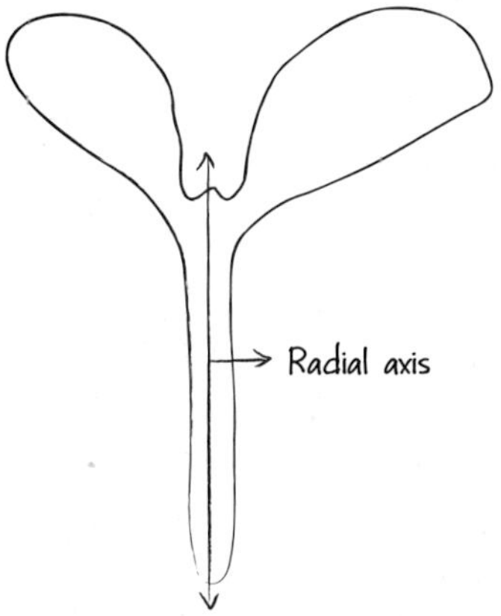

FIGURE A23.4

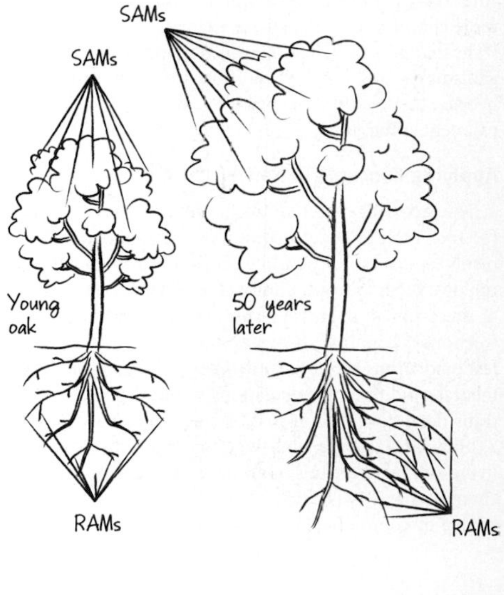

In everyday English, fitness is a physical attribute that is acquired as a result of practice or exercise. **KC 24.3** Brain size in *H. sapiens* might be constrained by the need for babies to pass through the mother's birth canal, by the energy required to maintain a large brain as an adult, or by lack of genetic variation for even larger brain size. Flying speed might be constrained by loss of maneuverability (and thus less success in hunting), the energy demands of extremely rapid flight, or lack of genetic variation for even faster flight.

Test Your Knowledge

1. b; **2.** d; **3.** a; **4.** d; **5.** a; **6.** c

Test Your Understanding

1. The theory of evolution by natural selection predicts that when a population's environment changes, individuals with certain traits will produce the most offspring, and the frequency of those traits will increase in the population. These changes should accumulate over long periods of time, eventually resulting in the formation of a new species that is geographically close to its ancestral species, and a fossil record should reflect changes in traits over time. The theory of special creation predicts that species do not change over time (or do so only due to divine intervention). Finally, the theory of acquired characters predicts that changes that occur in individuals in response to the environment are passed on to the next generation. **2.** Mutation produces new genetic variations, at random, without any forethought as to which variations might prove adaptive in the future. Individuals with mutations that are disadvantageous won't produce many offspring, but individuals with beneficial mutations will produce many offspring. **3.** The evidence for within-patient evolution is that DNA sequences at the start and end of treatment were identical except for a single nucleotide change in the *rpo* gene. If rifampin were banned, it is likely that *rpoB* mutant strains would have had lower fitness in the drug-free environment and would not continue to increase in frequency in *M. tuberculosis* populations. **4.** Typological thinking is based on the idea that species are unchanging and that any differences between individuals within a species are unimportant. Population thinking treats variation among individuals as critically important. That variation is what natural selection acts on, or selects. Population thinking was a radical break with typological thinking. **5.** Yes—the predictions made by a theory can be tested with indirect evidence in convincing ways. For example, if evolutionary theory predicts that certain transitional fossils should be found in rocks of a certain age, then researchers can search in those rocks and see if those fossils are indeed found. New groups of species found on island chains should be each others' closest relatives. And so on. **6.** Organisms do not necessarily become more complex over time; there are many examples of traits that have been lost. The biggest and strongest organisms do not necessarily produce the most offspring because they may not be the best adapted to the environment at that time.

Applying Concepts to New Situations

1. The theory of evolution fits the six criteria as follows: (1) and (2): It provides a common underlying mechanism responsible for puzzling observations such as homology, geographic proximity of similar species, the law of succession in the fossil record, vestigial traits, and extinctions. (3) and (4): It suggests new lines of research to test predictions about the outcome of changing environmental conditions in populations, about the presence of transitional forms in the fossil record, and so on. (5) It is a simple idea that explains the tremendous diversity of living and fossil organisms and why species continue to change today. (6) The realization that all organisms are related by common descent and that none are higher or lower than others was a surprise. **2.** It is well documented that people who are healthy and well fed grow taller than people who are sick and malnourished. In the absence of data indicating that alleles associated with increased height have increased in frequency recently, it is more logical to hypothesize that the observed change is due to changes in the environment—not evolutionary (genetic) changes. **3.** Compare the sequences of the 20 genes from many samples of preserved human tissue with the sequences of the same genes from a large sample of currently living humans. If evolution has occurred, the frequency of alleles correlated with "tallness" should be significantly greater in living humans than in those who lived a hundred years ago. **4.** The ability to tan is an adaptation because it is passed on from one generation to the next. The tan itself is not passed on, but the ability to tan is passed on; therefore it is heritable and can be classified as an adaptation.

CHAPTER 25

Check Your Understanding (CYU)

CYU p. 440 Given the observed genotype frequencies, the observed allele frequencies are freq(A_1) = 0.574 + ½(0.339) = 0.744; freq(A_2) = ½(0.339) + 0.087 = 0.256. Given these allele frequencies, the genotype frequencies expected under the Hardy-Weinberg principle are A_1A_1: $0.744^2 = 0.554$; A_1A_2: $2(0.744 \times 0.256) = 0.381$; A_2A_2: $0.256^2 = 0.066$. There are 4 percent too few heterozygotes observed, relative to the expected proportion. One of the assumptions of the Hardy-Weinberg principle is not met at this gene in this population, at this time. **CYU p. 446** (1) When allele frequencies fluctuate randomly up and down, sooner or later the frequency of an allele will hit 0. That allele thus is lost from the population, and the other allele at that locus is fixed. (2) In small populations, sampling error is large. For example, the accidental death of a few individuals would have a large impact on allele frequencies. **CYU p. 455** (1) Sperm are small and hence relatively cheap to produce, whereas eggs are large and require a large investment of resources to produce. (2) Sperm are inexpensive to produce, so reproductive success for males depends on their ability to find mates—not on their ability to find resources to produce sperm. The opposite pattern holds for females. Sexual selection is based on variation in ability to find mates, so is more intense in males—leading to more exaggerated traits.

You Should Be Able To (YSBAT)

YSBAT p. 444 If allele frequencies are changing due to drift, the populations in Table 25.1 would behave like the simulated populations in Figure 25.6—frequencies would drift up and down over time, and diverge. **YSBAT p. 451** The proportions of homozygotes should increase, and the proportions of heterozygotes should decrease. **YSBAT p. 452** (1) More recessive deleterious alleles are found in homozygotes and eliminated by natural selection. (2) If there are few or no deleterious recessives in a population, there is less inbreeding depression.

Caption Questions and Exercises

Table 25.1 The observed allele frequencies, calculated from the observed genotype frequencies, are 0.43 for *M* and 0.57 for *N*. The expected genotypes, calculated from the observed allele frequencies under the Hardy-Weinberg principle, are 0.185 for *MM*; 0.49 for *MN*; 0.325 for *NN*. **Figure 25.6** [See Figure A25.1] **Figure 25.8** Original population: freq(A_1) = (9 + 9 + 11) / 54 = 0.54; New population: freq(A_1) = (2 + 2 + 1) / 6 = 0.83. The frequency of A_1 has increased dramatically. **Figure 25.11** Yes—the frozen cells are traces of organisms that lived in the past. **Figure 25.13** The difference between number of flowers in individuals from the two types of matings represents inbreeding depression. This difference increases with age, which means that inbreeding depression increases with age. **Figure 25.16** An allele in males, because a successful male can have as many as 100 offspring—10 times more than a successful female.

Summary of Key Concepts

KC 25.1 There will be an excess of observed genotypes containing the favored allele compared to the proportion expected under Hardy-Weinberg proportions. **KC 25.2** Genetic drift will rapidly reduce genetic variation in small populations. Because captive individuals are usually housed under optimal conditions, selection will not be as intense as it would be in the wild. Alleles that allow individuals to thrive in captivity will increase; alleles that might lower fitness may not be eliminated. **KC 25.3** It increased homozygosity of recessive alleles associated with diseases such as hemophilia. As a result, the royal families were plagued by genetic diseases. **KC 25.4** In this case, reproductive success in females will depend on the number of male mates they can attract. It is reasonable to predict that female red-necked phalaropes are under intense sexual selection for traits that can help them attract mates and are more brightly colored than males.

Test Your Knowledge

1. b; **2.** a; **3.** a; **4.** b; **5.** d; **6.** c

Test Your Understanding

1. Selection may decrease genetic variation or maintain it (e.g., if heterozygotes are favored). Drift reduces it. Gene flow may increase or decrease it, depending on whether immigrants bring new alleles or emigrants remove alleles. Mutation increases it. **2.** Selective pressures often change over time or in different areas occupied by the same species. Even if selective pressures do not vary, mutation continually introduces new alleles. **3.** Sexual selection is most intense on the sex that makes the least investment in the offspring, so that sex tends to have exaggerated traits that make individuals successful in competition for mates. In this way, sexual dimorphism evolves. **4.** The prediction of a null hypothesis states what you should observe if the hypothesis you are testing is not correct. If you are testing the hypothesis that allele frequencies are changing or that mating is nonrandom with respect to a certain gene, the Hardy-Weinberg principle furnishes the appropriate null. **5.** Even if individuals in a small population mate at random or attempt to avoid inbreeding, over time all individuals in a small population are closely related. There are simply not enough nonrelatives to mate with. **6.** Alleles associated with extreme phenotypes are eliminated; alleles associated with intermediate phenotypes increase in frequency.

FIGURE A25.1

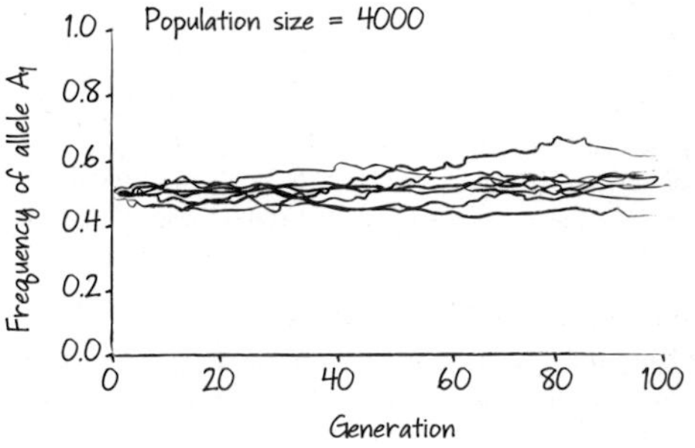

Applying Concepts to New Situations

1. The frequency of heterozygotes is $2p_1p_2$, or $2 \times 0.01 \times 0.99$, which is 0.0198. **2.** Inbreeding causes a decrease in heterozygotes, and a loss of fitness due to the increased numbers of homozygotes for deleterious alleles. Introducing new genetic variants through gene flow will increase allelic diversity in the population and result in more heterozygotes. **3.** If males never lift a finger to help females raise children, the fundamental asymmetry of sex is pronounced and sexual dimorphism should be high. If males invest a great deal in raising offspring, then the fundamental asymmetry of sex is small and sexual dimorphism should be low. **4.** Captive-bred young, transferred adults, and habitat corridors could be introduced to counteract the damaging effects of drift and inbreeding in the small, isolated populations. It is important, though, to introduce individuals only from similar habitats and connect patches of similar habitat with corridors—to avoid introducing alleles that lower fitness.

CHAPTER 26

Check Your Understanding (CYU)

CYU p. 462 (1) Biological species concept: Reproductive isolation means that no gene flow is occurring. Morphological species concept: If populations are evolving independently, they may have evolved morphological differences. Phylogenetic species concept: If populations are evolving independently, they should have synapomorphies that identify them as independent twigs on the tree of life. (2) Biological species concept: It cannot be used to evaluate fossils, species that reproduce asexually, or species that do not occur in the same area and therefore never have the opportunity to mate. Morphological species concept: It cannot identify species that differ in traits other than morphology and is subjective by nature—it can lead to differences of opinion that cannot be resolved by data. Phylogenetic species concept: Reliable phylogenetic information exists only for a small number of organisms. **CYU p. 468** Use one of the cases given in the chapter to illustrate colonization (Galápagos finches), vicariance (shrimp), or habitat specialization (apple maggot flies). In each case, drift will cause allele frequencies to change randomly. Differences in the habitats occupied by the isolated populations will cause allele frequencies to change under natural selection.

You Should Be Able To (YSBAT)

YSBAT p. 465 Selection will favor individuals that (1) breed on apples and have enzymes that are good at digesting apple fruit and that function well as larvae develop in warm temperatures and spend a relatively long time overwintering, and (2) breed on hawthorns and have enzymes that are good at digesting hawthorn fruit and that function well as larvae develop in cool temperatures and spend a relatively short time overwintering. **YSBAT p. 467a** A diploid individual experiences a defect in meiosis resulting in the formation of diploid gametes. The individual self-fertilizes, producing tetraploid offspring. The tetraploid individuals self-fertilize or mate with other tetraploid individuals, producing a tetraploid population. **YSBAT p. 467b** A tetraploid ($4n$) species gives rise to diploid ($2n$) gametes, and a diploid species gives rise to haploid (n) gametes. When a haploid gamete (n) fertilizes a diploid gamete ($2n$), a triploid offspring results. If this individual self-fertilizes, then a population of hexaploid wheat is formed.

Caption Questions and Exercises

Figure 26.3 Save one from the Atlantic Coast and one from the Gulf Coast. Their DNA sequences differ more than do subspecies within a geographic region, so you would be preserving more genetic diversity. **Figure 26.7** It creates genetic isolation—the precondition for divergence. **Figure 26.11** If the same processes were at work in the past as are at work in an experiment with living organisms, then the outcome of the experiment should be a valid replication of the outcome that occurred in the past.

Summary of Key Concepts

KC 26.1 (Many possibilities) Catch a group of fruit flies that are living in a lab, separate the individuals into different cages, and expose the two new populations to dramatically different environmental conditions—heat, lighting, food sources. In essence: create genetic isolation and then conditions for divergence due to drift and selection. **KC 26.2** Human populations would not be considered separate species under the biological concept because human populations can successfully interbreed. They would not be considered separate species under the morphological species concept because all human populations have the same basic morphology. (Although human races differ in minor superficial attributes such as skin color and hair texture, they have virtually identical anatomy and physiology in all other regards.) Nor would human populations be separate species under the phylogenetic species concept because they all arose from a very recent common ancestor. (DNA comparisons have revealed that human races are remarkably similar genetically and do not differ enough genetically to qualify even for subspecies status.) **KC 26.3** The sample experiment described above is an example of vicariance, because the experimenter fragmented the habitat into isolated cages. (If your own experiment differed from the one described above, then a different mechanism of speciation may have been involved.) **KC 26.4** Hybrid individuals will increase in frequency, and natural selection will favor parents that hybridize. The two parent populations may go extinct if hybrids live in the same environment. If the hybrids occupy a different environment, a new species may form.

Test Your Knowledge

1. a; **2.** b; **3.** d; **4.** a; **5.** a; **6.** a

Test Your Understanding

1. Direct observation: Galápagos finch colonization, apple maggot flies, *Tragopogon* allopolyploids, maidenhair ferns. Indirect evidence: snapping shrimp. **2.** On the basis of biological and morphospecies criteria (breeding range and morphological traits), six different species/subspecies of seaside sparrows were recognized. However, based on phylogenetic criteria (comparison of gene sequences), biologists concluded that seaside sparrows represent only two monophyletic groups. **3.** The colonizing population is typically small, and genetic drift has a large effect on smaller populations. If the newly colonized habitat differs from the original one, natural selection will favor individuals with alleles that increase fitness in the new environment. **4.** Some flies breed on apple fruits, others breed on hawthorn fruits. This reduces gene flow. Because apple fruits and hawthorn fruits have different scents and other characteristics, selection is causing the populations to diverge. **5.** Cells that go through many rounds of mitosis often accumulate errors, including mistakes that produce a tetraploid cell. If such a cell becomes part of a developing flower, and goes through meiosis, it will produce diploid gametes that can fuse to form polyploidy offspring. This is less likely to happen in animals because the future reproductive cells undergo fewer rounds of mitosis, so they have fewer chances to become polyploid. **6.** If the species have lived in the same area for a long time, there has been opportunity for selection to favor traits that prevent hybridization. This is reinforcement. If the species do not live in the same area, then there has been no opportunity for selection to favor traits that prevent hybridization.

Applying Concepts to New Situations

1. Decreasing. Gene flow tends to equalize allele frequencies among populations. **2.** These data cast doubt on the hypothesis. If founder events trigger speciation, then at least some speciation events due to introductions should be in progress. **3.** Two things: (1) Some flies happen to have alleles that allow them to respond to apple scents; others happen to have alleles that allow them to respond to hawthorn scents. There is no "need" involved. (2) The alleles for scent response exist; they are not acquired by spending time on the fruit. **4.** If the populations and habitat fragments are small enough, the species is likely to dwindle to extinction due to inbreeding and loss of genetic variation or catastrophes like a severe storm or a disease outbreak. If the populations survive, they are likely to diverge into new species because they are genetically isolated and because the habitats may differ.

CHAPTER 27

Check Your Understanding (CYU)

CYU p. 479 Hair and limb structures in humans and whales are examples of homology because they are traits that can be traced to a common ancestor. All mammals have hair and similar limb bone structure. However, extensive hair loss and advanced social behavior in whales and humans are examples of homoplasy. These traits are not common to all mammalian species and likely arose independently during the evolution of specific mammalian lineages. **CYU p. 488** (1) Doushantuo fossils are microscopic and include sponges and embryos from unknown species. Ediacaran fossils include sponges, jellyfish, and comb jellies along with traces of many other unidentified animals. Most were filter feeders. Burgess Shale fossils include every major animal lineage. The species present are larger, have much more complex morphology, and lived in a wide array of niches. (2) There were many resources available because no other animals (or other types of organisms) existed to exploit them. Also, the evolution of new species in new niches made new niches available for predators. Morphological innovations like limbs and complex mouthparts were important because they made it possible for animals to live in habitats other than the benthic area. **CYU p. 492** (1) Evidence includes the high content of iridium in sedimentary rock formed at the K-T boundary, shocked quartz and microtektite in rock layers that date 65 mya, and a large crater that was found off the coast of Mexico's Yucatán peninsula with microtektite in the crater's walls. Any of these three could be considered convincing, as they are known to form only at impact sites. Taken together, they are particularly convincing. (2) Mass extinctions are caused by such rapid and unusual environmental changes that any species is "lucky" to survive—it is not possible to adapt to such fast changes and such unusual environments.

You Should Be Able To (YSBAT)

YSBAT p. 479 Because whales and hippos share SINEs 4–7 and no other species has these four, they are synapomorphies that define them as a monophyletic group. The similarity is unlikely to be due to homoplasy because the chance of four SINEs inserting in exactly the same place in two different species, independently, is almost astronomically small. **YSBAT p. 485** In habitats on the mainland, there are more competitors and so species are already using the niches (resources) that are available to silverswords in Hawaii and *Anolis* lizards on Caribbean islands.

Caption Questions and Exercises

Figure 27.5 [See Figure A27.1] **Figure 27.16** The dinosaurs went extinct during the end-Cretaceous.

Summary of Key Concepts

KC 27.1 Humans are the only mammal that walks upright on two legs. **KC 27.2** They are the hardest structures in vertebrates and plants, so fossilize most readily. **KC 27.3** They could eat food that was available off the substrate. And once animals lived off the bottom, it created an opportunity for other animals who could eat them to evolve. **KC 27.4** Human-induced extinctions today are not due to poor adaptation to the normal environment. Humans are changing habitats rapidly and in unusual ways, so it is not possible for most species to adapt to these changes rapidly enough to avoid going extinct.

Test Your Knowledge

1. c; **2.** a; **3.** b; **4.** d; **5.** d; **6.** d

Test Your Understanding

1. The fossil record is biased because recent, abundant organisms with hard parts that live underground or in environments where sediments are being deposited are most likely to fossilize. Even so, it is the only data available on what organisms that lived in the past looked like, and where they lived. **2.** Diverse animal forms are found in the Cambrian that are not present in earlier strata. An enormous amount of diversity appeared in a time frame that was short compared to the sweep of earlier Earth history. **3.** Homoplasy is rare relative to homology, so phylogenetic trees that minimize the total change required are usually more accurate. In the case of the artiodactyl astralagus, parsimony was misleading because the astralagus was lost when whales evolved limblessness—creating two changes (a loss following a gain) instead of one (a gain). **4.** (Many answers are possible.) Adaptive radiation of *Anolis* lizards after they colonized new islands in the Caribbean. *Hypothesis:* After lizards arrived on each new island, where there were no predators or competitors, they rapidly diversified to occupy four distinct types of habitats on each island. (Many answers are possible.) Adaptive radiation during the Cambrian period following a morphological innovation. *Hypothesis:* Additional *Hox* genes made it possible to organize a large, complex body; the evolution of complex mouthparts and limbs made it possible for animals to move and find food in new ways. **5.** Environments all over the world deteriorated rapidly—large parts of the ocean lacked oxygen, many coastal habitats disappeared due to change in sea level, and little oxygen was available in the atmosphere. The underlying cause of these changes is not known. **6.** If a trait evolved in a common ancestor, then all of its descendants should share that trait, unless it is lost.

Applying Concepts to New Situations

1. Place the corpse in an environment in which decomposition is slow. One possibility is a swamp or bog; another is a beach or other coastal environment where mud or sand are being deposited. **2.** The fossil record and phylogenetic trees of extinct species indicate that whales evolved from a semiaquatic ancestor. Phylogenetic trees of living species indicate that hippos and whales share a recent common ancestor—meaning that one of the semiaquatic organisms that gave rise to today's whales also gave rise to the semiaquatic species that are today's hippos. **3.** It would be helpful to have (1) a large crater or other physical evidence from a major impact dated at 251 Mya, (2) shocked quartz and microtektites in rock layers dating to the end-Permian era, and (3) high levels of iridium or other elements common in space rocks and rare on Earth, dated to the time of the extinctions. **4.** Oxygen is an effective final electron acceptor in cellular respiration because of its high electronegativity. Organisms that use it as a final electron acceptor can produce more usable energy than organisms that do not use oxygen, but only if it is available. With more available energy, aerobic organisms can grow larger and move faster.

THE BIG PICTURE: EVOLUTION

Check Your Understanding (CYU), p. 494

1. Circle = inbreeding, sexual selection, natural selection, genetic drift, mutation, and gene flow. **2.** Adaptation "increases" fitness; synapomorphies "identify branches on" the tree of life. **3.** Several answers possible: e.g., the flower in angiosperms, the pharyngeal jaw in cichlid fishes, feathers and flight in birds. **4.** Genetic drift, mutation, and gene flow "are random with respect to" fitness.

CHAPTER 28

Check Your Understanding (CYU)

CYU p. 504 (1) Conditions should mimic a spill—sand or stones with a layer of crude oil or seawater with oil floating on top. Add samples, from sites contaminated with oil, that might contain cells capable of using molecules in oil as electron donors or electron acceptors. Other conditions (temperature, pH, etc.) should be realistic. (2) Use direct sequencing. Choose a well-studied gene, such as 16S RNA, with reliable PCR primers. After isolating DNA from a soil sample, do a PCR reaction and clone the resulting genes into plasmids grown in *E. coli* cells, so that each culture contains a different gene. Once many copies are available, sequence the genes and use the data to place the original organisms on the tree of life. **CYU p. 512** Eukaryotes can only (1) fix carbon via the Calvin-Benson pathway, (2) use aerobic respiration with organic compounds as electron donors, and (3) perform oxygenic photosynthesis. Among bacteria and archaea, there is much more diversity in pathways for carbon fixation, respiration, and photosynthesis, along with many more fermentation pathways.

You Should Be Able To (YSBAT)

YSBAT p. 506 cyanobacteria—photoautotrophs; *Clostridium aceticum*—chemoorganoautotroph; *Nitrosomonas* sp.—chemolithotrophs; heliobacteria—photoheterotrophs; *Escherichia coli*—chemoorganoheterotroph; *Beggitoa*—chemolithotrophic heterotrophs **YSBAT p. 513a, YSBAT p. 513b, YSBAT p. 514a, YSBAT p. 514b, YSBAT p. 515** [See Figure A28.1]

Caption Questions and Exercises

Figure 28.1 Prokaryotic—as it would require just one evolutionary change, the origin of the nuclear envelope in Eukarya. If it were eukaryotic, it would require that the nuclear envelope was lost in both Bacteria and Archaea—two changes and less parsimonious (see Chapter 27). **Table 28.1** [See Figure A28.2] **Figure 28.2** Improved nutrition made people better able to fight off disease, and improved sanitation lowered transmission of disease-causing bacteria. **Figure 28.4** Weak—different culture conditions may have revealed different species. **Figure 28.9** Table 28.5 contains the answers. For example, for organisms called sulfate reducers you would have H_2 as the electron donor, SO_4^{2-} as the electron acceptor, and H_2S as the reduced by-product. For humans the electron donor is glucose ($C_6H_{12}O_6$), the electron acceptor is O_2, and the reduced by-product is water (H_2O). **Table 28.5** Organotrophs use sugars, which are organic compounds, as their electron donor—getting energy by "feeding" on them. Sulfate reducers use sulfate ions as their electron acceptors, thus reducing the sulfate ion. Methanogens generate methane as a by-product. **Figure 28.11** Aerobic respiration. More free energy is released when oxygen is the final electron acceptor than when any other molecule is used, so more ATP can be produced and used for growth.

FIGURE A27.1

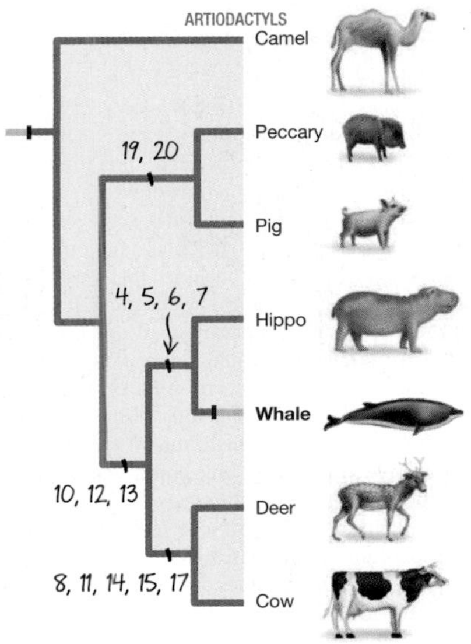

FIGURE A28.1

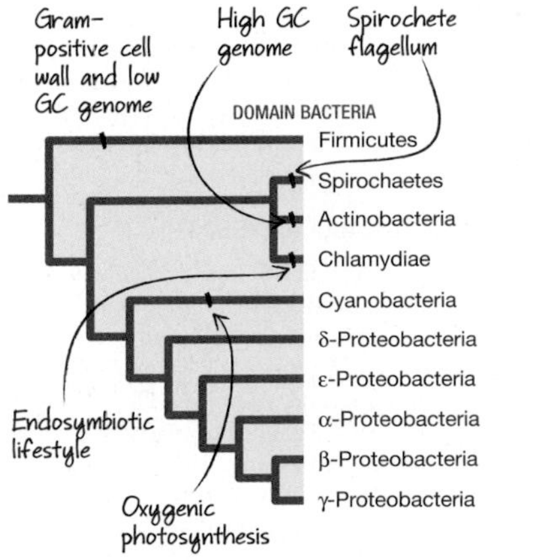

FIGURE A28.2

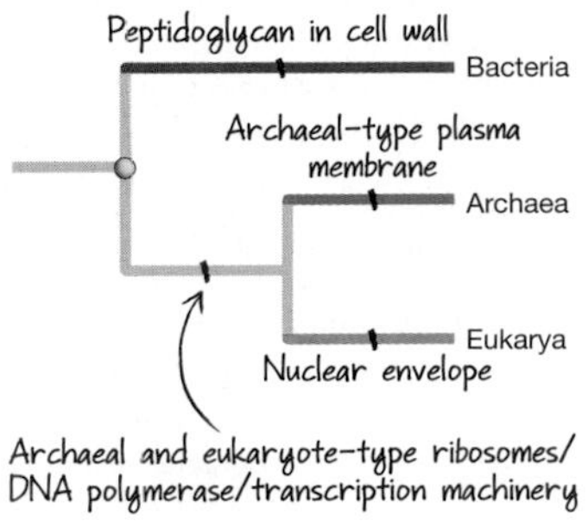

Summary of Key Concepts

KC 28.1 Eukaryotes have a nuclear envelope that encloses their chromosomes; bacteria and archaea do not. Bacteria have cell walls that contain peptidoglycan, and archaea have phospholipids containing isoprene subunits in their plasma membranes. Thus, the exteriors of a bacterium and archaeon are radically different. Archaea and eukaryotes also have similar machinery for processing genetic information. **KC 28.2** They are compatible, and thrive in each others' presence—one species' waste product is the other species' food. **KC 28.3** If bacteria and archaea did not exist, then (1) the atmosphere would have little or no oxygen, and (2) almost all nitrogen would exist in molecular form (the gas N_2).

Test Your Knowledge

1. c; **2.** b; **3.** d; **4.** d; **5.** b; **6.** c

Test Your Understanding

1. An electron donor provides the potential energy required to produce ATP. **2.** Yes. The array of substances that bacteria and archaea can use as electron donors, electron acceptors, and fermentation substrates, along with the diversity of ways that they can fix carbon and perform photosynthesis, allows them to live just about anywhere. **3.** You can sample and amplify genes—getting DNA sequence data that allow you to place unseen species on the tree of life. **4.** Large amounts of potential energy are released and ATP produced when oxygen is the electron acceptor, because oxygen is so electronegative. Large body size and high growth rates are not possible without large amounts of ATP. **5.** [See Figure A28.3] The following can serve as both electron donors and by-products: CH_4, NO_2^-, H_2S. The following can serve as both electron acceptors and by-products: SO_4^{2-}, CO_2, and NO_3^-. **6.** They are paraphyletic because the prokaryotes include some (bacteria and archaea) but not all (eukaryotes) groups derived from the common ancestor of all organisms living today.

Applying Concepts to New Situations

1. This result supports their hypothesis, because the drug poisons the enzymes of the electron transport chain and prevents electron transfer to Fe^{3+}, which is required to drive magnetite synthesis. If magnetite had still formed, another explanation would have been needed. **2.** Hypothesis: A high rate of tooth cavities in Western children is due to an excess of sucrose in the diet, which is absent from the diets of East African children. To test this hypothesis, have East African children switched to a diet that contains sucrose and monitor the presence of *S. mutans.* Have Western children switch to a diet lacking sucrose and monitor the presence of *S. mutans.* **3.** Look in waters or soil polluted with benzene-containing compounds. Put samples from these environments in culture tubes where benzene is the only source of carbon. Monitor the cultures and study the cells that grow efficiently. **4.** They should get energy from reduced organic compounds "stolen" from their hosts.

CHAPTER 29

Check Your Understanding (CYU)

CYU p. 523 (1) *Plasmodium* species are transmitted to humans by mosquitoes. If mosquitoes can be prevented from biting people, they cannot spread the disease. (2) Iron added → primary producers (photosynthetic protists and bacteria) bloom → more carbon dioxide taken up from atmosphere during photosynthesis → consumers bloom, eat primary producers → bodies of primary producers and consumers fall to bottom of ocean → large deposits of carbon-containing compounds form on ocean floor. **CYU p. 526** (1) Opisthokonts have a flagellum at the base or back of the cell; alveolate cells contain unique support structures called alveoli; stramenopiles have straw-like hairs on their flagella. (2) In direct sequencing, DNA is isolated directly from the environment and analyzed to place species on the tree of life. It is not necessary to actually see the species being studied.

You Should Be Able To (YSBAT)

YSBAT p. 526a [See Figure A29.1 on p. A:28] **YSBAT p. 526b** Membrane infoldings observed in bacterial species today support the hypothesis's plausibility—they confirm that the initial steps could have actually happened. The continuity of the nuclear envelope and ER are consistent with the hypothesis, which predicts that the two structures are derived from the same source (infolded membranes). **YSBAT p. 528** A photosynthetic bacterium (e.g., a cyanobacterium) could have been engulfed by a larger eukaryotic cell. If it was not digested, it could continue to photosynthesize and supply sugars to the host cell. **YSBAT p. 531** The chloroplast genes label should come off of the branch that leads to cyanobacteria. (If you had a phylogeny just of the cyanobacteria, the chloroplast branch would be located somewhere inside.) **YSBAT p. 532** The acquisition of the mitochondrion and the chloroplast represent the transfer of entire genomes, and not just single genes, to a new organism. **YSBAT p. 533** Yes—when food is scarce or population density is high, the environment is changing rapidly (deteriorating). Offspring that are genetically unlike their parents may be better able to cope with the new and challenging environment. **YSBAT p. 536a** (1) Alternation of generations refers to a life cycle in which there are multicellular haploid phases and multicellular diploid phases. A gametophyte is the multicellular haploid phase; the sporophyte is the multicellular diploid phase. A spore is a cell that grows into a multicellular individual, but is not produced by fusion of two cells. A zygote is a cell that grows into a multicellular individual, but *is* produced by fusion of two cells (gametes). Gametes are haploid cells that fuse to form a zygote. (2) [See Figure A29.2 on p. A:28] **YSBAT p. 536b, YSBAT p. 536c, YSBAT p. 536d, YSBAT p. 537a, YSBAT p. 537b, YSBAT p. 537c** [See Figure A29.1 on p. A:28] **YSBAT p. 537d** It is most likely that the two types of amoebae evolved independently. The alternative hypothesis is that the common ancestor of alveolates, stramenopiles, rhizarians, plants, opisthokonts, and amoebozoa were amoeboid, and that this growth form was lost many times. [See Figure A29.1 on p. A:28] **YSBAT p. 538a, YSBAT p. 538b** [See Figure A29.1 on p. A:28] **YSBAT p. 539a** If euglenids could take in food via phagocytosis (ingestive feeding), then it would have provided a mechanism by which a smaller photosynthetic protist could have been engulfed and incorporated into the cell via secondary endosymbiosis. **YSBAT p. 539b, YSBAT p. 540a, YSBAT p. 540b, YSBAT p. 541a, YSBAT p. 541b, YSBAT p. 542a, YSBAT p. 542b, YSBAT p. 543** [See Figure A29.1 on p. A:28]

Caption Questions and Exercises

Figure 29.9 Two—one derived from the original bacterium and one derived from the eukaryotic cell that engulfed the bacterium. **Table 29.3** Yes—green algae, euglenids, and chlorarachniophytes all have chlorophyll *a* and *b*, as predicted by the hypothesis that a green algal chloroplast was transferred to the ancestor of euglenids and of chlorarachniophytes. Red algae have chlorophyll *a*, and chromalveolates (which include the brown algae, diatoms, and dinoflagellates) have chlorophyll *a* and *c*. If the hypothesis is correct, chlorophyll *c* must have evolved in an ancestor of the chromalveolates independently of the acquisition of chlorophyll *a* from red algae.

Summary of Key Concepts

KC 29.1 Photosynthetic protists use CO_2 and light to produce sugars and other organic compounds, so they furnish the first or primary source of organic material in an ecosystem. **KC 29.2** (1) Outside membrane was from host eukaryote; inside from engulfed cyanobacterium. (2) From the outside in, the four membranes are derived from the eukaryote that engulfed a chloroplast-containing eukaryote, the plasma membrane of the eukaryote that was engulfed, the outer membrane of the engulfed cell's chloroplast, and the inner membrane of its chloroplast. **KC 29.3** Each set of pigments absorbs most strongly in a certain part of the electromagnetic spectrum. If different species have different pigments, they can live in close proximity and perform photosynthesis without competing for the same wavelengths of light. **KC 29.4** A gametophyte is haploid, and a sporophyte is diploid.

Test Your Knowledge

1. b; **2.** b; **3.** a; **4.** b; **5.** d; **6.** b

FIGURE A28.3

TABLE 28.5 **Some Electron Donors and Acceptors Used by Bacteria and Archaea**

		By-Products		
Electron Donor	**Electron Acceptor**	**From Electron Donor**	**From Electron Acceptor**	**Category***
Sugars	O_2	CO_2	H_2O	Organotrophs
H_2 or organic compounds	SO_4^{2-}	H_2O or CO	H_2S or S^{2-}	Sulfate reducers
H_2	CO_2	H_2O	CH_4	Methanogens
CH_4	O_2	CO_2	H_2O	Methanotrophs
S^{2-} or H_2S	O_2	SO_4^{2-}	H_2O	Sulfur bacteria
Organic compounds	Fe^{3+}	CO_2	Fe^{2+}	Iron reducers
NH_3	O_2	NO_2^-	H_2O	Nitrifiers
Organic compounds	NO_3^-	CO_2	N_2O, NO, or N_2	Denitrifiers (or nitrate reducers)
NO_2^-	O_2	NO_3^-	H_2O	Nitrosifiers

*The name biologists use to identify species that use a particular metabolic strategy.

Test Your Understanding

1. An extensive cytoskeleton is required to shape the cell membrane so that the pseudopod can form and surround the food item. A cell wall is not flexible enough to wrap around the food item as the plasma membrane does. **2.** Because all eukaryotes living today have cells with a nuclear envelope, it is valid to infer that their common ancestor also had a nuclear envelope. Because bacteria and archaea do not have a nuclear envelope, it is valid to infer that the trait arose in the common ancestor of eukaryotes. **3.** Meiosis is required for alternation of generations because it converts a diploid phase of the life cycle to a haploid phase. Without it, the generations wouldn't "alternate." **4.** The host cell provided a protected environment and carbon compounds for the endosymbiont; the endosymbiont provided increased ATP from the carbon compounds. **5.** It confirmed a fundamental prediction made by the hypothesis, and could not be explained by any alternative hypothesis. **6.** All alveolates have alveoli, which are unique structures that function in supporting the cell. Among alveolates there are species that are (1) ingestive feeders, photosynthetic, or parasitic, and (2) move using cilia, flagella, or a type of amoeboid movement.

Applying Concepts to New Situations

1. The observation suggests that eukaryotes did not acquire mitochondria until there was enough oxygen present to make aerobic respiration efficient. (Oxygen is also poisonous to cells at high concentration, so mitochondria gave the early eukaryotes a way to "detoxify" oxygen.) **2.** If the apicoplast that is found in *Plasmodium* (the organism that causes malaria) is genetically similar to chloroplasts, and if glyphosate poisons chloroplasts, it is reasonable to hypothesize that glyphosate will poison the apicoplast and potentially kill the *Plasmodium.* This would be a good treatment strategy for malaria because humans have no chloroplasts, provided that the glyphosate produces no other effects that would be detrimental to humans. **3.** Primary producers usually grow faster when CO_2 concentration increases, but to date, they have not grown fast enough to make CO_2 levels drop—CO_2 levels have been increasing steadily over decades. **4.** Given that lateral gene transfer can occur at different points in a phylogenetic history, specific genes can become part of a lineage by a different route from that taken by other genes of the organism. In the case of chlorophyll *a*, its history traces back to a bacterium being engulfed by a protist and forming a chloroplast.

CHAPTER 30

Check Your Understanding (CYU)

CYU p. 552 (1) Green algae and land plants share an array of morphological traits that are synapomorphies, including the chlorophylls they contain, (2) green algae appear before land plants in the fossil record, and (3) on phylogenetic trees estimated from DNA sequence data, green algae and land plants share a most recent common ancestor, with green algae being the initial groups to diverge and land plants diverging subsequently. **CYU p. 566** (1) Cuticle prevents water loss from the plant; vascular tissue moves water up from the soil and moves photosynthetic products down to the roots. (2) [See Figure A30.1]

FIGURE A29.1

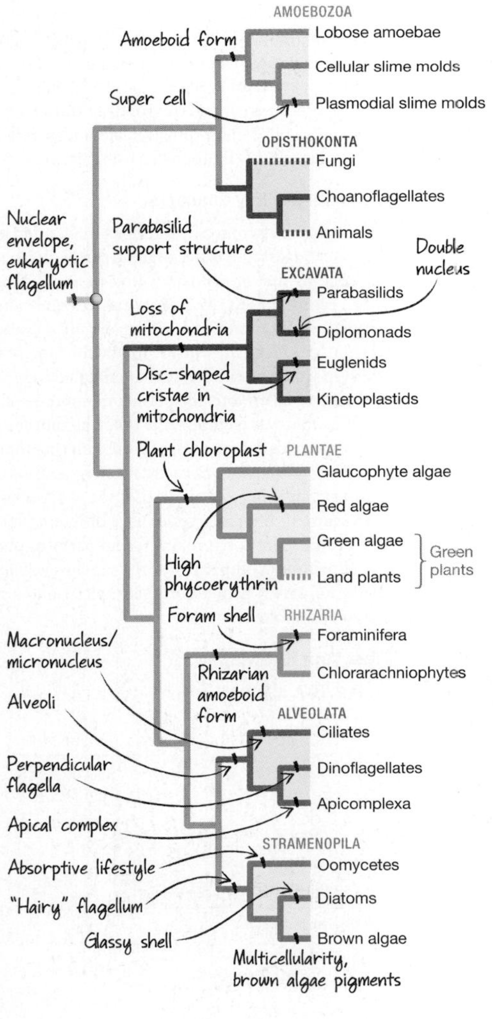

FIGURE A29.2

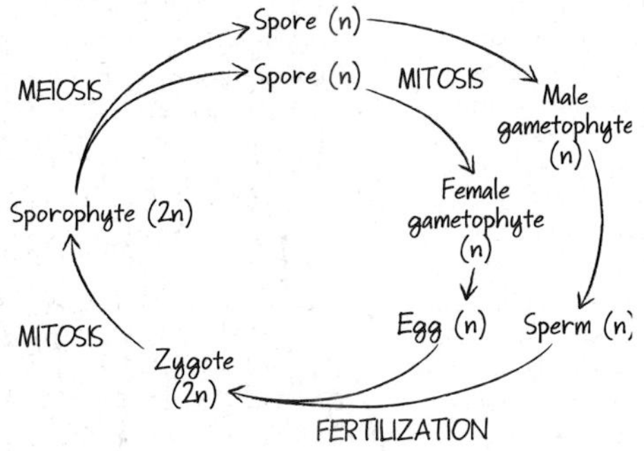

FIGURE A30.1

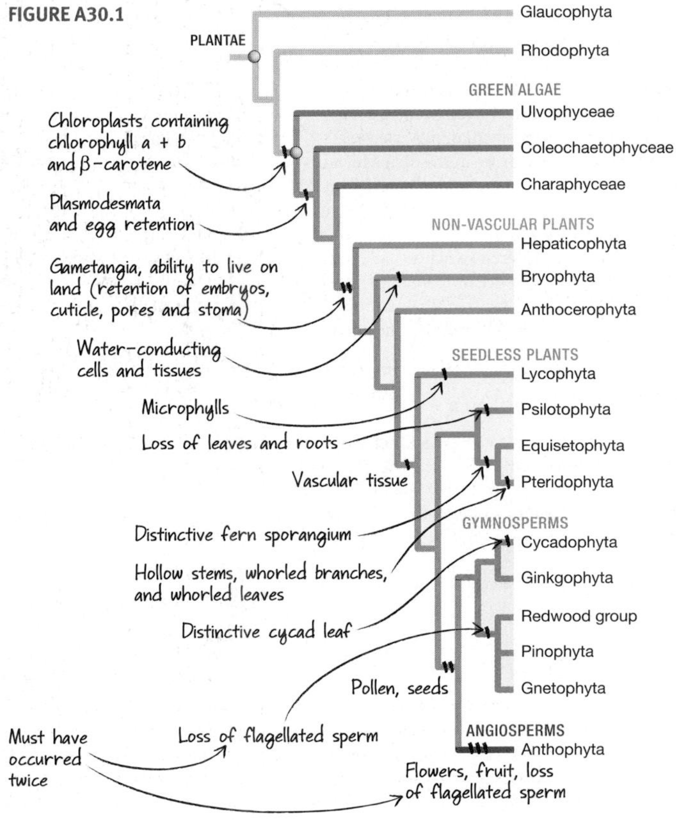

You Should Be Able To (YSBAT)

YSBAT p. 558 Alternation of generations occurs when there are multicellular haploid individuals and multicellular diploid individuals in a life cycle. A sporophyte is a multicellular diploid individual that produces spores by meiosis. A spore is a single cell that is not formed by fusion of two cells, and that grows by mitosis into a multicellular adult. **YSBAT p. 559** In the hornwort photo, the sporophyte is the spike-like green and brown structure; the gametophyte is the leafy-looking structure underneath. The horsetail gametophyte is the microscopic individual on the left; the sporophyte is the much larger individual on the right. **YSBAT p. 569a, YSBAT p. 569b, YSBAT p. 570, YSBAT p. 571, YSBAT p. 572a, YSBAT p. 572b, YSBAT, p. 573, YSBAT p. 574** [See Figure A30.1]

Caption Questions and Exercises

Figure 30.7 If you find the common ancestor of all the green algae (at the base of the Ulvophyceae), the lineages that are collectively called green algae don't include that common ancestor and all of its descendants—only some of its descendants. The same is true for non-vascular plants (the common ancestor here is at the base of the Hepaticophyta) and seedless vascular plants (the common ancestor here is at the base of the Lycophyta). **Figure 30.9** Water flows more easily through a short, wide pipe than through a long, skinny one because there is less resistance from the walls of the pipe. **Figure 30.10** To map an innovation on a phylogenetic tree, biologists determine the location(s) on the tree that is consistent with all descendants from that point having the innovation (unless, in rare cases, the innovation was lost in a few descendants). **Figure 30.15** There are multicellular haploid stages and multicellular diploid stages in these plants. **Figure 30.19** Gymnosperm gametophytes are microscopic, so they are even smaller than fern gametophytes. The gymnosperm gametophyte is completely dependent on the sporophyte for nutrition, while fern gametophytes are not. Fern gametophytes are photosynthetic and even supply nutrition to the young sporophytes. **Figure 30.20** Consistent—the fossil data suggest that gymnosperms evolved earlier, and gymnosperms have larger gametophytes than angiosperms. **Figure 30.22** It tests the hypothesis that the presence of yarn on the spur changes pollinator behavior and thus reproductive success. **Figure 30.24** [See Figure A30.1]

Summary of Key Concepts

KC 30.1 Rates of soil formation will drop; rates of soil loss will increase. **KC 30.2** Because so many adaptations are required—starting with cuticle and pores or stomata—it is unlikely that the transition from water to land would happen more than once. Stated another way, the aquatic and terrestrial environments are so different for plants that the transition was difficult and thus unlikely. **KC 30.3** Both spores and seeds have a tough, protective coat, so can survive while being dispersed to a new location. Seeds have the advantage of carrying a store of nutrients with them—when a spore germinates, it has to make its own food via photosynthesis right away.

Test Your Knowledge

1. c; **2.** b; **3.** d; **4.** c; **5.** c; **6.** a

Test Your Understanding

1. Plants build and hold soils required for human agriculture and forestry, and increase water supplies that humans can use for drinking, irrigation, or industrial use. Plants release oxygen that we breathe. **2.** Cuticle prevents water loss from leaves but also prevents entry of CO_2 required for photosynthesis. Stomata allow CO_2 to diffuse but can close to minimize water loss. Liverwort pores allow gas exchange but cannot be closed if conditions become dry. Liverworts that lack pores have a cuticle that is thin enough to allow some gas exchange. **3.** They provided the support needed for plants to grow upright and not fall over in response to wind or gravity. Erect growth allowed plants to compete for light. **4.** (1) Gametangia are found in all land plant groups except angiosperms; (2) transfer cells are found in all land plants; (3) pollen is found in gymnosperms and angiosperms; (4) seeds are found in gymnosperms and angiosperms, (5) fruit is found in angiosperms. **5.** In a gametophyte-dominant life cycle, the gametophyte is larger and longer lived than the sporophyte and produces most of the nutrition. In a sporophyte-dominant life cycle, the sporophyte generation is the larger, longer-lived, and photosynthetic phase of the life cycle. **6.** Homosporous plants produce a single type of spore that develops into a gametophyte that produces both egg and sperm. Heterosporous plants produce two different types of spores that develop into two different gametophytes that produce either egg or sperm. In a tulip, the microsporangium is found within the stamen, and the megasporangium is found within the ovule. Microspores divide by mitosis to form male gametophytes (pollen grains); megaspores divide by mitosis to form the female gametophyte.

Applying Concepts to New Situations

1. Homosporous plants produce a single type of spore that develops into a gametophyte that produces both egg and sperm. Heterosporous plants produce two different types of spores that develop into two different gametophytes that produce either egg or sperm. In a tulip, the microsporangium is found within the stamen, and the megasporangium is found within the ovule. Microspores divide by mitosis to form male gametophytes (pollen grains); megaspores divide by mitosis to form the female gametophyte. **2.** The combination represents a compromise between efficiency and safety. Tracheids can still transport water if vessels become blocked by air bubbles. **3.** A "reversion" to wind pollination might be favored by natural selection because it is costly to produce a flower that can attract animal pollinators. Because wind-pollinated species grow in dense clusters, they can maximize the chance that the wind will carry pollen from one individual to another (less likely if the individuals are far apart). Wind-pollinated deciduous trees flower in early spring before their developing leaves begin to block the wind. **4.** Alter one characteristic of a flower and present the flower to the normal pollinator. As a control, present the normal (unaltered) flower to the normal pollinator. Record the amount of time the pollinator spends in the flower, the amount of pollen removed, or some other measure of pollination success. Repeat for other altered characteristics. Analyze the data to determine which altered characteristic affects pollination success the most.

CHAPTER 31

Check Your Understanding (CYU)

CYU p. 586 (1) Because they are made up of a network of thin, branching hyphae, mycelia have a large surface area, which makes absorption efficient. (2) Swimming spores and gametes, zygosporangia, basidia, asci. **CYU p. 594** (1) When birch tree seedlings are grown in the presence and absence of EMF, individuals denied their normal EMF are not able to acquire sufficient nitrogen and phosphorus. Isotope-tracing experiments also show that sugars are transferred from host plant to EMF and that nitrogen and phosphorus are transferred from EMF to host plant. (2) Meiosis and production of haploid spores.

You Should Be Able To (YSBAT)

YSBAT p. 591 Human sperm and egg undergo plasmogamy followed by karyogamy during fertilization, but heterokaryosis does not occur in humans or other eukaryotes besides fungi. **YSBAT p. 594, YSBAT p. 596a, YSBAT p. 596b, YSBAT p. 596c, YSBAT p. 597, YSBAT p. 598a, YSBAT p. 598b** [See Figure A31.1]

Caption Questions and Exercises

Figure 31.10 Labeled-nutrient experiments identify which nutrients are exchanged and in which direction. They explain *why* plants do better in the presence of mycorrhizae, and why the fungus also benefits.

FIGURE A31.1

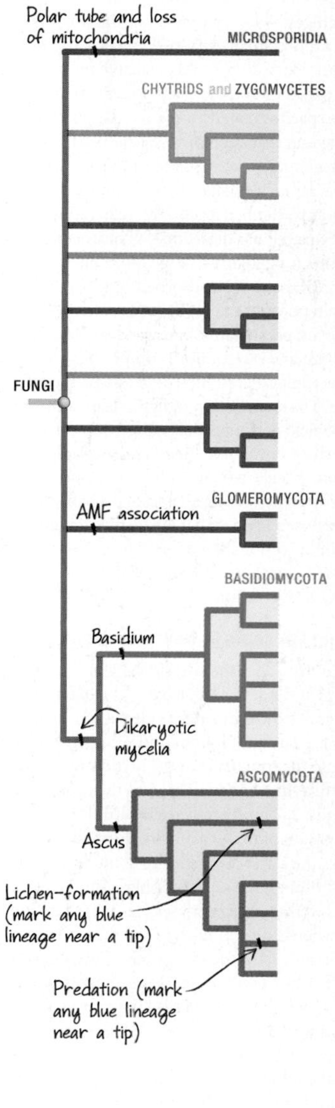

Figure 31.11 [See Figure A31.2] **Figure 31.13** The haploid mycelium. **Figure 31.14** Mycelia produced by asexual reproduction are genetically identical to their parent; mycelia produced by sexual reproduction are genetically different from both parents (each spore has a unique genotype).

Summary of Key Concepts

KC 31.1 Loss of mycorrhizal fungi would decrease nutrient delivery to plants and reduce their growth inside the experimental plots, compared to control plots with intact fungi. Also, lack of fungi would slow decay of dead plant material, causing a dramatic buildup of dead organic material. **KC 31.2** Absorption has to occur across a surface (usually via transport proteins in the plasma membrane). Other things being equal, more absorption will occur across a large surface area than a small surface area. **KC 31.3** Haploid hyphae fuse, forming a heterokaryotic mycelium. When karyogamy occurs, a diploid nucleus forms—just as when gametes fuse.

Test Your Knowledge

1. a; **2.** d; **3.** c; **4.** b; **5.** b; **6.** d

Test Your Understanding

1. Along with a few bacteria, fungi are the only organisms that can digest wood completely. If the wood is not digested, carbon remains trapped in wood. Without fungi, CO_2 would be tied up and unavailable for photosynthesis, and the presence of undecayed organic matter would reduce the space available for plants to grow. **2.** Fungi produce enzymes that degrade cellulose and lignin. **3.** Plant roots have much smaller surface area than EMF or AMF. Hyphae are much smaller than the smallest portions of plant roots so can penetrate dead material more efficiently. Extracellular digestion—which plant roots cannot do—allows fungi to break large molecules into small compounds that can be absorbed. **4.** DNA, RNA, and ATP all contain phosphorus. Without adequate amounts of phosphorus, plants cannot grow by synthesizing more DNA, RNA, and ATP. **5.** Both compounds are processed via extracellular digestion. Different enzymes are involved, however. The degradation of lignin is uncontrolled and does not yield useful products; digestion of cellulose is controlled and produces useful glucose molecules. **6.** Most fungi do not make gametes. Instead, the products of specific alleles identify mating types—probably to encourage fusion between hyphae that are dissimilar genetically, resulting in the production of genetically diverse offspring. It is possible for thousands of different mating-type alleles to exist in a population.

Applying Concepts to New Situations

1. (1) Confirm that the chytrid fungus is found only in sick frogs and not healthy frogs. (2) Isolate the chytrid fungus and grow it in a pure culture. (3) Expose healthy frogs to the cultured fungus and see if they become sick. (4) Isolate the fungus from the experimental frogs, grow it in culture, and test whether it is the same as the original fungus. **2.** The claim is reasonable because fungi have so many ways of making a living from plant tissues. For example, the same plant species could have several species of fungi that are endophytic, mycorrhizal, or parasitic, as well as an array of species that break down its tissues when it dies. **3.** Each of the different cellulase enzymes attacks cellulose in a different way, so producing all the enzymes together increases the efficiency of the fungus in breaking down cellulose completely. It is likely that lignin peroxidase is produced along with cellulases, so they can act in concert to degrade wood. To test this idea, you could harvest enzymes secreted from a mycelium before and after it contacts wood in a culture dish, and see if the cellulases and lignin peroxidase appear together once the mycelium begins growing on the wood. **4.** Collect a large array of colorful mushrooms that are poisonous and capture mushroom-eating animals, such as squirrels. Present a hungry squirrel with a choice of mushrooms that have been dyed or painted a drab color versus treated with a solution that is identical to the dye or paint used but uncolored. Record which mushrooms the squirrel eats. Repeat the test with many squirrels and many mushrooms.

CHAPTER 32

Check Your Understanding (CYU)

CYU p. 609 (1) Bilateral symmetry led to the evolution of long, slender body plans. Triploblasty gave rise to an inner tube (gut), an outer tube (body lining), and muscles and organs in between. The coelom functions as a hydrostatic skeleton to facilitate movement in soft-bodied animals that lack limbs. (2) When an animal moves through an environment in one direction, its ability to acquire food and perceive and respond to threats is greater if its feeding, sensing, and information-processing structures are at the leading end. **CYU p. 617** (1) The mouthparts of deposit feeders are relatively simple, as they simply gulp relatively soft material. The mouthparts of mass feeders are more complex, because they have to tear off and process chunks of relatively hard material. (2) Gametes that are shed into aquatic environments can float or swim. This cannot happen on land, so internal fertilization is more common.

You Should Be Able To (YSBAT)

YSBAT p. 606 If you do the exercise correctly, the balloon should wriggle. The wriggling movements would produce movement if the balloon (or animal) were surrounded by water or soil. **YSBAT p. 618** The origin of epithelial tissue should be marked on the same branch as multicellularity. **YSBAT p. 619** The origin of cnidocytes should be marked on the Cnidaria branch. **YSBAT p. 620** The origin of cilia-powered swimming should be marked on the Ctenophora branch.

Caption Questions and Exercises

Figure 32.6 The animal would get shorter and fatter. **Figure 32.9** The label and bar should go on the branch to the left of the clam shell and "Mollusk" label. **Figure 32.19** Stained *Dll* gene products would be located in the legs of the insect but would not be concentrated anywhere in the onychophoran and segmented worm. **Figure 32.22** No—there is no multicellular haploid form.

Summary of Key Concepts

KC 32.1 It should increase dramatically—animals would no longer be consuming plant material. **KC 32.2** Your drawing should show the tube-within-a-tube design of a worm, with one end labeled head and containing the mouth and brain. From the brain, one or two major nerve tracts should run the length of the body. The gut should go from the mouth to the other end, ending in the anus. In between the gut and outer body wall, there should be blocks or layers of muscle. **KC 32.3** During gastrulation, the initial pore becomes the mouth in protostomes, whereas in deuterostomes the pore becomes the anus. The protostome coelom is formed from blocks of mesodermal tissue that form cavities. The deuterostome coelom forms when mesodermal tissue around the gut pinches off. **KC 32.4** Natural selection has produced eye structures that function well in a particular species' habitat, and mollusks live in a wide array of habitats. Some mollusks live buried in sand and have no eyes—eyes would have no function in this habitat. Mollusks like squid and octopuses live in open water and hunt prey. They have complex eyes that form images and help them find food. **KC32.5** Juveniles look like miniature adults and feed on the same foods as adults. Larvae look much different from adults, live in different habitats, and eat different foods. The difference is important because larvae do not compete with adults for food.

FIGURE A31.2

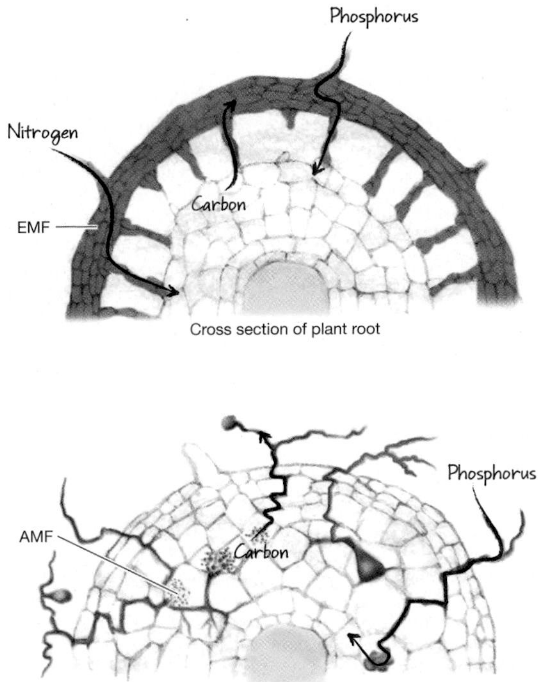

Test Your Knowledge

1. a; **2.** d; **3.** b; **4.** a; **5.** d; **6.** c

Test Your Understanding

1. Diploblasts have two types of embryonic tissue; triploblasts have three. Mesoderm made it possible for an enclosed, muscle-lined cavity to develop, creating a coelom. **2.** A hydrostatic skeleton is a pressurized, fluid-filled chamber (the coelom) that is surrounded by muscles. Contractions on either side of the body change the pressure of the fluid in the hydrostatic skeleton and thus the shape of the coelom. When muscle contractions are coordinated throughout the length of the animal, the shape of the hydrostatic skeleton changes and results in movement. **3.** There are many unicellular organisms that are heterotrophic, but they can consume only small packets of food. Animals are multicellular, so they are larger and can consume larger packets of food. **4.** A long, hollow structure that is sharp enough to pierce the wall of the stem; a simple, muscular mouth opening that can gulp soft, decaying material; sharp, stiff structures that can move in a scissoring action; a long, hollow structure that can be inserted into the flower. **5.** In oviparous species, the mother adds nutrient-rich yolk to the egg that nourishes the developing embryo. In viviparous species, the mother transfers nutrients directly from her body to the growing embryo. **6.** Radially symmetric organisms encounter the environment in many directions, so it is advantageous to have an equal number of neurons located throughout the body. Bilaterally symmetric organisms encounter the environment in one direction, so it is advantageous to have most neurons clustered in the head region.

Applying Concepts to New Situations

1. Yes—if the same gene is found in nematodes and humans, it was likely found in the common ancestor of protostomes and deuterostomes. If so, then fruit flies should also have this gene. **2.** Mosquitoes, because larval and adult mosquitoes feed on different food sources in different habitats and do not compete with one another. You could test this prediction by collecting data on the total number of mosquito versus tick species, their abundance, and geographic distribution. **3.** It should resemble a nerve net, because echinoderms need to take in and process information from multiple directions—not just one. **4.** Web-building spiders are some of the only terrestrial animals that suspension feed.

CHAPTER 33

Check Your Understanding (CYU)

CYU p. 630 (1) Arthropods have a tube-within-a-tube body plan with a drastically reduced coelom. They have a hemocoel body cavity that holds internal organs and body fluids. The body is segmented, with segments grouped into tagmata. They have jointed limbs and an exoskeleton made of chitin. Compared to unjointed limbs, jointed limbs allow animals to move faster and with more precision. (2) Supporting the body (and moving) without support from water. Preventing the body from drying out. Facilitating gas exchange.

You Should Be Able To (YSBAT)

YSBAT p. 631a, YSBAT p. 631b, YSBAT p. 632, YSBAT p. 633, YSBAT p. 637, YSBAT p. 638 [See Figure A33.1]

Caption Questions and Exercises

Figure 33.3 The tuft is the cluster of ciliated tentacles in part (a); the wheel is the ring of cilia in part (b). **Figure 33.6** Different phyla of worms have different feeding strategies, different mouthparts, and eat different foods, so they are not in direct competition for food. **Figure 33.22** [See Figure A33.2]

Summary of Key Concepts

KC 33.1 All of the phyla that are considered protostomes share a pattern of early development found in no other animals. Based on this observation, it is logical to claim that this pattern of development arose in the common ancestor of these phyla—meaning that it is a synapomorphy. **KC 33.2** Flatworms likely lost the adaptation of a coelom because as they evolved a thin, flattened body plan, the coelom was no longer needed for internal gas circulation or movement. Arthropods have a drastically reduced coelom because they have evolved limbs and complex musculature for movement and no longer require the coelom to provide a hydrostatic skeleton. **KC 33.3** (1) Because larvae and adults live in a different habitat, young and adults do not compete for food. (2) Because larvae can swim, they can disperse to new locations.

Test Your Knowledge

1. d; **2.** a; **3.** c; **4.** c; **5.** a; **6.** b

Test Your Understanding

1. Ecdysozoans grow by molting; lophotrochozoans grow by incremental additions to their bodies. In addition, some lophotrochozoan phyla have lophophores and/or trochophore larvae. Both groups are bilaterally symmetric triploblasts with the protostome pattern of development. **2.** Multiple times, because phylogenies show that segmented groups evolved from unsegmented ancestors in both lophotrochozoans (annelids) and ecdysozoans (arthropods). **3.** The ability to live on land offered the opportunity of exploiting new habitats and food sources, with minimal competition from other animals. **4.** An exoskeleton provided a stiff surface for muscle attachment—facilitating rapid, precise movement—as well as protection from enemies. **5.** The ability to fly allowed insects to disperse to new habitats and find new food sources efficiently. **6.** The leading hypothesis is variation in mouthpart structure and feeding strategies.

Applying Concepts to New Situations

1. [See Figure A33.3] **2.** If the ancestors of brachiopods and mollusks lived in similar habitats and experienced natural selection that favored similar traits, then they would have evolved to have similar forms and habitats. This is called convergent evolution (see Chapter 27). **3.** Aplacophora are probably basal—meaning that they would branch off between the ancestral form at the base of the tree and the rest of the groups. Because they have reduced forms of key molluscan synapomorphies, it is likely that they are the least-derived of the Mollusca. Because they lack a shell for muscle attachment and a well-developed foot, they probably move slowly via undulating movements and live in benthic habitats. **4.** Agree—it is likely that the common ancestor of arthropods was marine, and that terrestrial forms evolved from aquatic ancestors independently in crabs, isopods, and insects.

CHAPTER 34

Check Your Understanding (CYU)

CYU p. 650 (1) If it's an echinoderm, it should have five-part radial symmetry, a calcium carbonate endoskeleton just underneath the skin, and a water vascular system (e.g., visible podia). **CYU p. 659** (1) Jaws allow animals to capture food efficiently and process it by crushing or tearing. (2) The increased cushioning from enclosed fluids provided mechanical support, and the increased surface area made transport of gases and other materials more efficient.

FIGURE A33.1

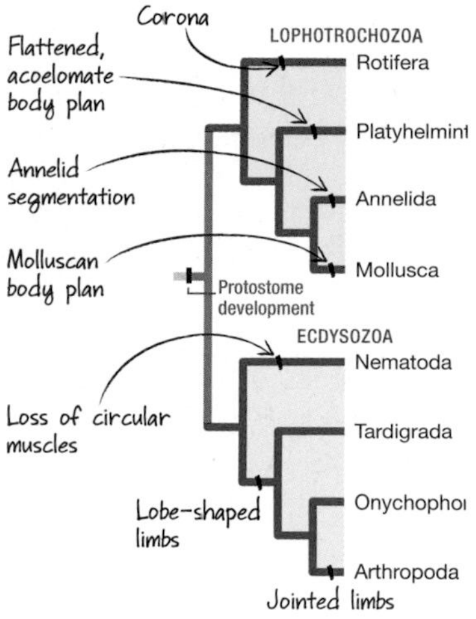

FIGURE A33.2

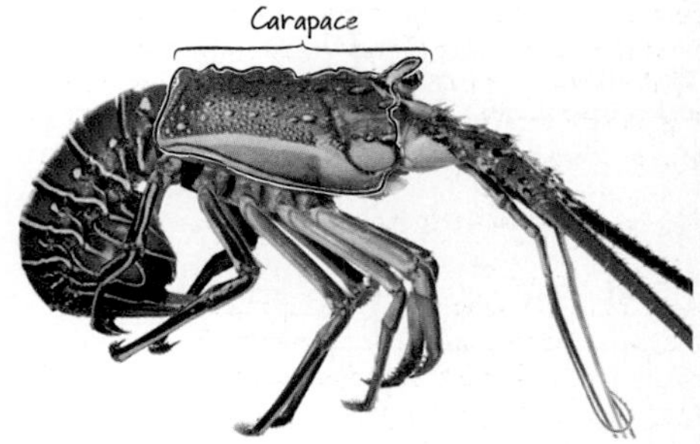

FIGURE A33.3

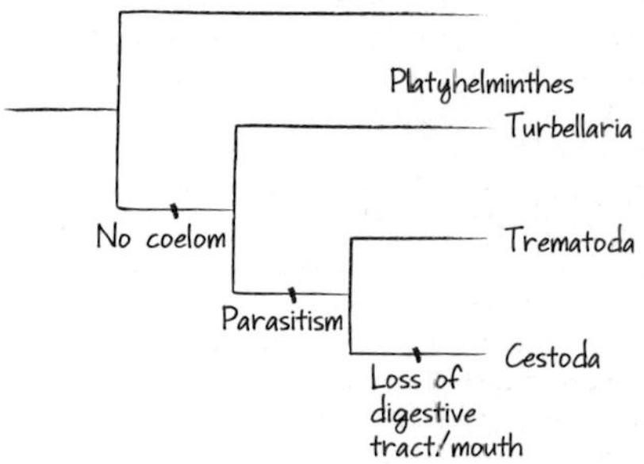

You Should Be Able To (YSBAT)

YSBAT p. 649a, YSBAT p. 649b, YSBAT p. 650 [See Figure A34.1] **YSBAT p. 655** [See Figure A34.2] **YSBAT p. 660** [See Figure A34.3] **YSBAT p. 662a, YSBAT p. 662b, YSBAT p. 663, YSBAT p. 664** [See Figure A34.4] **YSBAT p. 665, YSBAT p. 666a, YSBAT p. 666b, YSBAT p. 667a, YSBAT p. 667b, YSBAT p. 668** [See Figure A34.3]

Caption Questions and Exercises

Figure 34.1 Invertebrates are a paraphyletic group. The group does not include all the descendants of the common ancestor because vertebrates are excluded. **Figure 34.12** Mammals and birds are equally related to amphibians, because birds and mammals share a common ancestor, and this ancestor shares a common ancestor with amphibians. **Figure 34.14** Yes—even a rudimentary jaw could have been useful to better grasp prey or to change the size of the mouth. **Figure 34.16** If this phylogeny were estimated on the basis of limb traits, it would be based on the *assumption* that the limb evolved from fish fins in a series of steps. But because the phylogeny was estimated from other types of data, it is legitimate to use it to analyze the evolution of the limb without any assumptions about how limbs evolved. **Figure 34.20** They are smaller—their function in an amniotic egg has been taken over by the placenta. **Figure 34.37** Four hominin species existed 2.2 mya, five existed 1.8 mya, and four existed 100,000 years ago. **Figure 34.38** The forehead became much larger with the face becoming "flatter"; the brow ridges are less prominent in later skulls than in earlier skulls.

Summary of Key Concepts

KC 34.1 (1) Mussels and clams will increase dramatically; (2) kelp density will increase. **KC 34.2** Like today's lungfish, the earliest tetrapods could have used their limbs for pulling themselves along the substrate in shallow-water habitats. **KC 34.3** The earliest fossils of *H. sapiens* are found in Africa, and phylogenetic analyses of living human populations indicate that the most basal groups are all African.

Test Your Knowledge

1. a; **2.** a; **3.** d; **4.** a; **5.** d; **6.** b

Test Your Understanding

1. It is an enclosed, fluid-filled structure surrounded by muscle. Muscular contraction forces water into the tube feet, resulting in extension of the podia. **2.** Pharyngeal gill slits function in suspension feeding. The notochord furnishes a simple endoskeleton that stiffens the body; electrical signals that coordinate movement are carried by the dorsal hollow nerve cord to the muscles in the tail, which beats back and forth to make swimming possible. Cephalochordates and ascidians are chordates but they are not vertebrates. **3.** If jaws are derived forms of gill arches, then the same genes and the same cells should be involved in the development of the jaw and the gill arches. **4.** Homologous genes are involved in the formation of the fins of ray-finned fish and the limbs of tetrapods. This observation supports a prediction of the fins-to-limbs hypothesis. **5.** The hominins fulfill the criteria for an adaptive radiation: Over a short time interval, many species that occupy an array of foods and habitats evolved. Changes in tooth and jaw structure, tool use, and body size suggest that different hominin species exploited different types of food. **6.** Increased parental care allows offspring to be better developed, and thus have increased chances of survival, before they have to live on their own.

Applying Concepts to New Situations

1. If confirmed, it means that pharyngeal gill slits were present in the earliest echinoderms and lost later.
2. No—most of the feathered dinosaurs known from fossils did not fly. (The first feathers may have functioned in display or as insulation.) **3.** Xenoturbellidans either retain traits that were present in the common ancestor or all deuterostomes, or they have lost many complex morphological characteristics. **4.** Independently. Birds are a highly derived lineage of reptiles, and most reptiles are not endothermic, so it is logical to infer that the common ancestor of birds and mammals was also not endothermic.

FIGURE A34.1

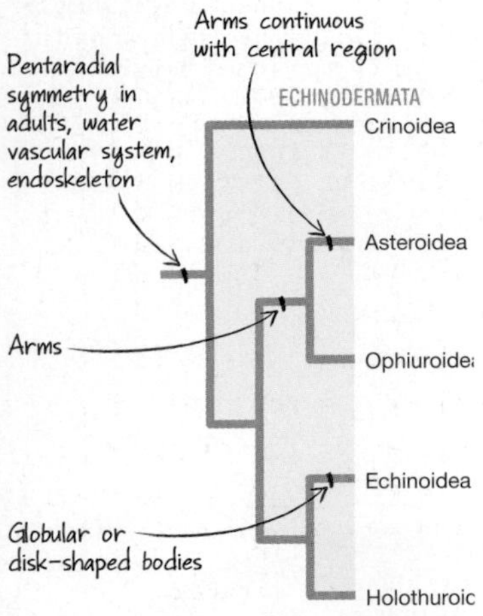

FIGURE A34.3

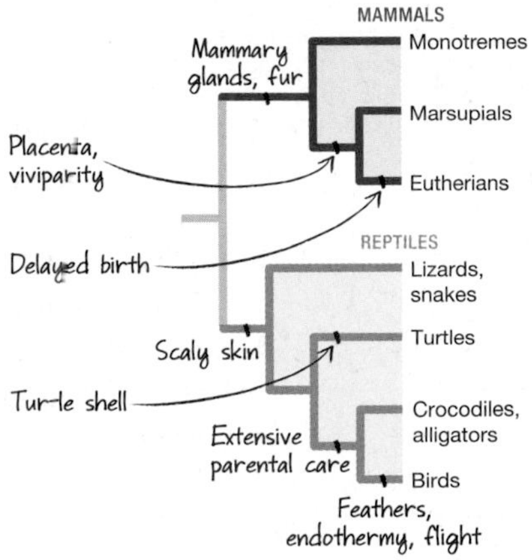

FIGURE A34.4

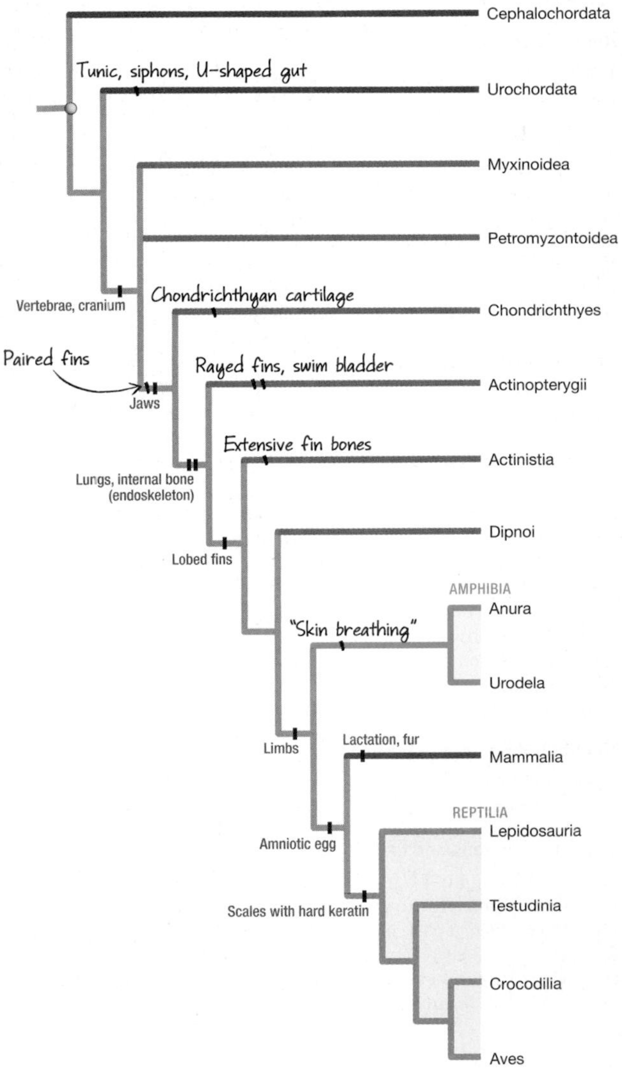

FIGURE A34.2

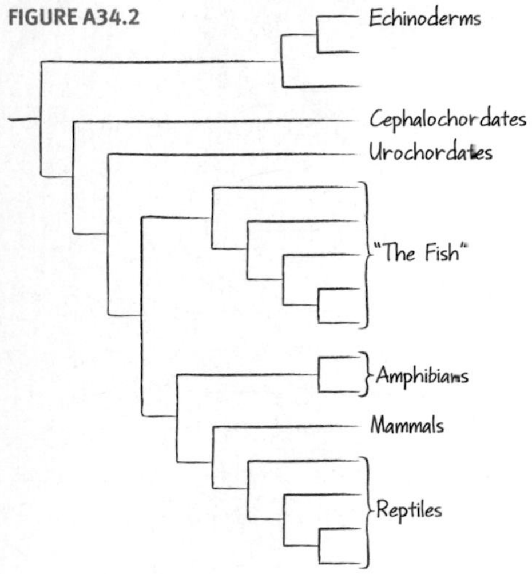

CHAPTER 35

Check Your Understanding (CYU)

CYU p. 686 HIV binds to CD4 and a co-receptor on the surface of T cells. This binding allows the viral envelope and the T-cell plasma membrane to fuse. The viral capsid then enters the cell. **CYU p. 689** (1) After the genome is copied to form positive-sense RNA, the transcript has to be copied to form a negative-sense, single-stranded genome. (2) Much more dangerous—with airborne transmission, a noninfected individual can become infected simply by inhaling virus particles released from an infected individual. This mode of transmission is much more efficient than transmission via body fluids.

You Should Be Able To (YSBAT)

YSBAT p. 680 In the lytic cycle, viral genes are transmitted to a new generation of virions. In lysogenic growth, viral genes are transmitted to daughter cells of the host cell. **YSBAT p. 686** [See Figure A35.1]

Caption Questions and Exercises

Figure 35.1 Many answers are possible: Herpes simplex 2: genitalia; sexually transmitted; HIV: immune system; sexually transmitted; Influenza virus: respiratory tract; coughing/sneezing (airborne); Chicken pox (varicella zoster virus): skin; direct contact. **Figure 35.8** Lysogenic bacteriophages are similar to transposable elements because they insert their DNA into the host cell chromosome and are passed on to daughter cells. They are unlike transposable elements because they can switch to a lytic cycle. **Figure 35.9** No—it only shows the CD4 is required. (Subsequent work showed that other proteins are involved as well.) **Figure 35.14** Budding viruses may disrupt the integrity of the host-cell plasma membrane enough to kill the cell. **Figure 35.15** You should have the following bars and labels: on branch to HIV-2 (sooty mangabey to human); on branch to HIV-1 strain O (chimp to human); on branch to HIV-1 strain N (chimp to human); on branch to HIV-1 strain M (chimp to human).

Summary of Key Concepts

KC 35.1 There is heritable variation among virions, due to random changes that occur as their genomes are copied. There is also differential reproductive success among virions in their ability to successfully infect host cells. This differential success is due to the presence of certain heritable traits. **KC 35.2** As the virus's envelope proteins are produced, they are inserted into the host cell plasma membrane. During budding, the capsid becomes surrounded by host cell membrane that includes the envelope proteins. **KC 35.3** (1) Fusion inhibitors block viral envelope proteins or host cell receptors; (2) protease inhibitors prevent processing/assembly of viral proteins; (3) reverse transcriptase inhibitors block reverse transcriptase, preventing replication of genome; (4) use of condoms or practice of monogamy prevents transmission. **KC 35.4** To replicate an ssDNA genome, a viral DNA polymerase copies the genome into a complementary strand, which then is used as a template to generate an ssDNA that is identical to the original viral genome.

Test Your Knowledge

1. b; **2.** c; **3.** d; **4.** b; **5.** b; **6.** d

Test Your Understanding

1. HIV has an envelope on its outer surface; adenovirus has only a capsid on its outer surface. A virus with an envelope exits host cells by budding. A virus that lacks an envelope exits host cells by lysis (or other mechanisms that don't involve budding). **2.** T4 bacteriophage should have a larger genome than HIV does, because T4 is much more complex morphologically—it should have genes that code for the proteins required for its head and tail regions. **3.** Growth rate is much higher during lytic growth, but viral DNA can increase during latent/lysogenic growth if the host cell is actively dividing and transmitting viral genomes to daughter cells. **4.** Viruses rely on host-cell enzymes to replicate, whereas bacteria do not. Therefore, many drugs designed to disrupt the virus life cycle cannot be used because they would kill host cells as well. Only viral-specific proteins are good targets for drug design. **5.** Each major hypothesis to explain the origin of viruses—from transposable-element-like sequences, from symbiotic bacteria, and from RNA genomes present early in Earth history—is associated with a different type of genome: single-stranded DNA or possibly single-stranded RNA, double-stranded DNA, and single- or double-stranded RNA, respectively. **6.** The phylogenies of SIVs and HIVs show that the two shared common ancestors, but that SIVs are ancestral to the HIVs. Also, there are plausible mechanisms for SIVs to be transmitted to humans through butchering or contact with pets, but fewer or no plausible mechanisms for HIVs to be transmitted to monkeys or chimps.

Applying Concepts to New Situations

1. Culture the *Staphylococcus* strain outside the human host and then add the virus to determine whether the virus kills the bacterium efficiently. Then test the virus on cultured human cells to determine whether the virus harms human cells. Then test the virus on monkeys or other animals to determine if it is safe. Finally, the virus could be tested on human volunteers. **2.** Prevention is currently the most cost-effective program but does not help people who are already infected. Treatment with effective drugs not only prolongs lives but also reduces virus loads in infected people, so that they have less chance of infecting others. **3.** Eukarya evolved from ancestors that did not contain genes encoding restriction endonucleases. Or, methylation of DNA performs other important functions in eukaryotic gene expression (see Chapter 18), so is not appropriate as a guard against endonuclease action. **4.** Viruses cannot be considered to be alive by the definition given in (a), because viruses are not capable of replicating by themselves. By the definition given in (b), it can be argued that viruses are alive because they store, maintain, replicate, and use genetic information—although they cannot perform all these tasks on their own.

FIGURE A35.1

$(+)\text{ssRNA} \xrightarrow{\text{translation}} \text{Viral proteins}$

$(-)\text{ssRNA} \xrightarrow[\text{replicase}]{\text{RNA}} \text{mRNA} \xrightarrow{\text{translation}} \text{Viral proteins}$

$\text{dsRNA} \xrightarrow{\text{transcription}} \text{mRNA} \xrightarrow{\text{translation}} \text{Viral proteins}$

$\text{dsDNA or ssDNA} \xrightarrow{\text{transcription}} \text{mRNA} \xrightarrow{\text{translation}} \text{Viral proteins}$

$(+)\text{ssRNA} \xrightarrow[\text{transcription}]{\text{reverse}} \text{dsDNA} \xrightarrow{\text{transcription}} \text{mRNA} \xrightarrow{\text{translation}} \text{Viral proteins}$

CHAPTER 36

Check Your Understanding (CYU)

CYU p. 704 (1) [See Figure A36.1] (2) The generalized body of a plant has a taproot with many lateral roots and broad leaves. Examples of deviations from this generalized body structure include: Modified roots: Unlike taproots, *fibrous roots* do not have one central root, and *adventitious roots* arise from stems. Modified stems: *Stolons* grow along the soil and grow roots and leaves at each node. *Rhizomes* grow horizontally underground. Both of these modified stems function in asexual reproduction. Modified leaves: The *needle-shaped leaves* of cacti lose less water to transpiration than do typical

FIGURE A36.1

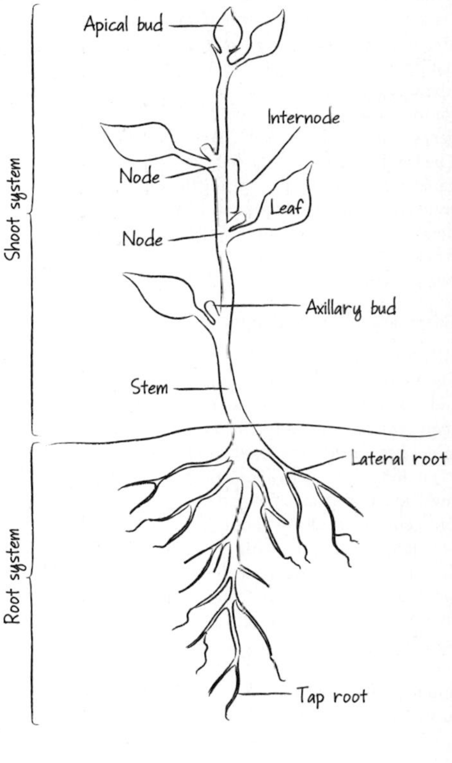

broad leaves, and they protect the plant from predation. *Tendrils* on climbing plants are modified leaves that do not photosynthesize but wrap around trees or other substrates to facilitate climbing. **CYU p. 712** (1) The apical meristem gives rise to the three primary meristems. The apical meristem is a single mass of cells localized at the tip of a root or shoot; the primary meristems are localized in distinctive locations behind the apical meristem. (2) Epidermal cells are flattened and lack chloroplasts; they secrete the cuticle (in shoots) or extend water and nutrient-absorbing root hairs (in roots) and protect the plant. Parenchyma cells are "workhorse cells" found throughout the plant body; they perform photosynthesis and synthesize and/or store materials. Tracheids are long, thin cells with pits in their secondary cell walls; they are found in xylem and are dead when mature; they conduct water and solutes up the plant. Sieve-tube elements are long, thin cells found in phloem; they are alive when mature and conduct sugars and other solutes up and down the plant. **CYU p. 715** (1) Primary growth increases the length of roots and shoots, and secondary growth increases their width. The function of primary growth is to extend the reach of the root and shoot system and thus increase a plant's ability to absorb light and acquire carbon dioxide, water, and nutrients. The function of secondary growth is to increase the amount of lateral conducting tissue and provide the structural support required for extending primary growth at the top of the plant. (2) The rings are small on the shaded side and large on the sunny side.

You Should Be Able To (YSBAT)

YSBAT p. 713a, YSBAT p. 713b, YSBAT p. 713c, YSBAT p. 714 [See Figure A36.2]

Caption Questions and Exercises

Figure 36.2 New branches and leaves should develop to the right (the plant will also lean that way); new lateral roots will develop to the left. **Figure 36.4** Lawn grasses are shallow rooted, so when the top part of soil dries out, the plants cannot grow. **Figure 36.5** For aerobic respiration, which generates the ATP needed to keep the cells alive. **Figure 36.7** The individuals are genetically identical—thus, any differences in size or shape in the different habitats are due to phenotypic plasticity and not genetic differences. **Figure 36.9** Large surface area provides more area to capture photons, but also more area from which to lose water and be exposed to potentially damaging tearing forces from wind. **Figure 36.20** Many possible answers; for example, in leaf cells, genes for forming chloroplasts would be expressed. These genes wouldn't be expressed in the root cells. **Figure 36.25** The leading hypothesis is that they lack many organelles so that the space inside the cell is available to transport nutrients. **Figure 36.27** You should have labeled a light band as early wood, and the dark band to its right as late wood; both bands together comprise one growth ring.

Summary of Key Concepts

KC 36.1 Phenotypic plasticity is more important in (1) environments where conditions vary because it gives individuals the ability to change the growth pattern of their roots, shoots, and stems to access sunlight, water, and other nutrients as the environment changes; and (2) in long-lived species because it gives individuals a mechanism to change their growth pattern as the environment changes throughout their lifetime. **KC 36.2** (1) Plants that live in warm, arid climates would have large root systems to increase their chances of accessing water, and small shoot systems because water is lost from surfaces exposed to air. (2) Plants that live in very wet climates would have small root systems because water is plentiful, and large shoot systems because the abundant water would support more aboveground growth especially in light-limiting environments such as forests. **KC 36.3** (1) In both cases, the structure would stop growing; (2) the shoot would lack epidermal cells and would die. **KC 36.4** (1) The plant would not produce secondary xylem and phloem, so its girth and transport ability would be reduced, though it would still produce bark. (2) The trunk would have much wider tree rings on one side than the other side.

Test Your Knowledge

1. b; **2.** b; **3.** a; **4.** d; **5.** c; **6.** a

Test Your Understanding

1. The general function of both systems is to acquire resources: The shoot system captures light and carbon dioxide; the root system absorbs water and nutrients. Vascular tissue is continuous throughout both the shoot and root systems. Diversity in roots and shoots enables plants of different species to live together in the same environment without directly competing for resources. **2.** Continuous growth enhances phenotypic plasticity because it allows plants to grow and respond to changes or challenges in their environment (such as changes in light and water availability). **3.** Cactus spines are modified leaves; thorns are modified stems. **4.** Cuticle reduces water loss; stomata facilitate gas exchange. Plants from wet habitats should have a relatively large number of stomata and thin cuticle. Plants living in dry habitats should have relatively few stomata and thick cuticle. **5.** Parenchyma cells in the ground tissue perform photosynthesis and/or synthesize and store materials. In vascular tissue, parenchyma cells in "rays" conduct water and solutes across the stem; other parenchyma cells differentiate into the sieve-tube members and companion cells in phloem. **6.** Cells produced to the inside of the vascular cambium differentiate into secondary xylem; cells produced to the outside of the vascular cambium differentiate into secondary phloem.

Applying Concepts to New Situations

1. Fewer sclerenchyma cells, because sclerenchyma cells have extremely tough walls and these populations have been selected to be tender. Humans select individuals with fewer sclerenchyma cells to be the parents of the next generation, so the frequency of sclerenchyma cells will decline over time. **2.** Asparagus—stem; Brussels sprouts—lateral bud; celery—petiole; spinach—leaf (petiole and blade); carrots—taproot; potato—modified stem. **3.** They grow continuously, so do not have alternating groups of large and small cells that form rings. **4.** Girdling disrupts transport of solutes in secondary phloem, and damages (but probably doesn't eliminate) the transport of water and solutes in secondary xylem. The tree starves.

CHAPTER 37

Check Your Understanding (CYU)

CYU p. 721 (1) Wet soils have almost no pressure potential, while dry soils have a highly negative pressure potential because the few water molecules present cling to soil particles. (2) Salty soils have extremely low solute potentials compared to typical soils, because the concentration of solutes is high. **CYU p. 727** (1) At the air-water interface under a stoma, hydrogen bonding is asymmetrical—the water molecules can bond only with water molecules below them. This pulls the water molecules down—creating tension at the surface. When transpiration occurs, there are few water molecules at the surface, which makes the asymmetry in bonding even more pronounced—increasing tension. (2) When a nearby stoma closes, the humidity inside the air space increases, transpiration decreases, and surface tension on menisci decreases. The same changes occur when a rain shower starts. When air outside the leaf dries, though, the opposite occurs: The humidity inside the air space decreases, transpiration increases, and surface tension on menisci increases. **CYU p. 734** (1) In early spring, the water potential of phloem sap near leaves is low because developing leaves are using sucrose much faster than they make it; root cells are releasing sucrose so the water potential of phloem cells in roots is high. (2) Midday in summer, the water potential of phloem sap near leaves is high because leaves are making much more sucrose than they consume. Root cells are storing sugar, however, so the water potential of phloem in roots is low.

You Should Be Able To (YSBAT)

YSBAT p. 719 (1) Add a solute (e.g., salt or sugar) to the left side of the U-tube, so that solute concentration is higher than on the right side. (2) Water would move from the right side of the U-tube into the left side. **YSBAT p. 720** Pull the plunger up. **YSBAT p. 723** The individual would not be able to exclude certain solutes from the xylem. If they were transported throughout the plant in high enough concentrations, they could damage tissues.

Caption Questions and Exercises

Figure 37.2 It increases (becomes less negative), which reduces the solute potential gradient between the two

FIGURE A36.2

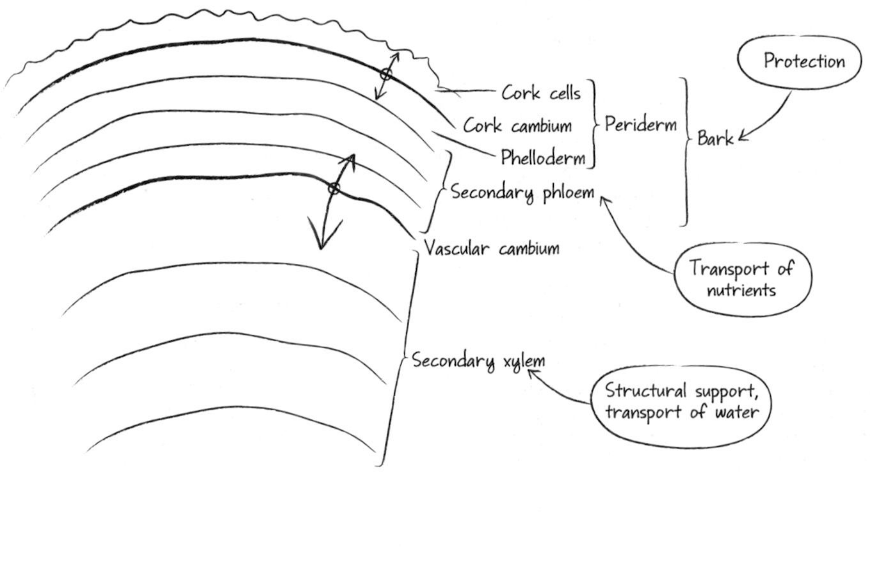

sides. **Figure 37.7** [See Figure A37.1] **Figure 37.10** Your labels should read as follows: Atmosphere: $\psi = -95.2$ MPa; Leaves: $\psi = -0.8$ MPa; Roots: $\psi = -0.6$ MPa; Soil: $\psi = -0.3$ MPa. **Figure 37.12** The line on the graph would go up, with large jumps occurring each time the light level was turned up, because increased transpiration rates would increase negative pressure (tension) at the leaf surface, and thus increase the water potential gradient. **Figure 37.17** [See Figure A37.2]

Summary of Key Concepts

KC 37.1 Because xylem cells are dead at maturity, there are no plasma membranes for water to cross and thus no solute potential. The only significant force acting on water in xylem is pressure. **KC 37.2** It drops, because transpiration rates from stomata decrease, reducing the water potential gradient between roots and leaves. **KC 37.3** To create high pressure in phloem near sources and low pressure near sinks, water has to move from xylem to phloem near sources and from phloem back to xylem near sinks. If xylem were not close to phloem, this water movement could not occur and the pressure gradient in phloem wouldn't exist.

Test Your Knowledge

1. d; **2.** c; **3.** b; **4.** a; **5.** b; **6.** c

Test Your Understanding

1. [See Figure A37.3] **2.** The force responsible for root pressure is the high solute potential of root cells at night, when they continue to accumulate ions from soil, with water following by osmosis. The mechanism is active, in the sense that ions are imported against their concentration gradient. Capillarity is passive; it is driven by adhesion of water molecules to the sides of xylem cells that creates a pull upwards, and cohesion with water molecules below. Cohesion-tension is also passive, because it is driven by the loss of water molecules from menisci in leaves, which creates a large negative pressure (tension). **3.** Water moves up xylem because of transpiration-induced tension created at the air-water interface under stomata, which is communicated down to the roots by the cohesion of water molecules. Sap flows in phloem because of a pressure gradient that exists between source and sink cells, driven by differences in sucrose concentration and flows of water into or out of nearby xylem. **4.** Aphid A is attacking a source, and Aphid B is attacking a sink. If sucrose concentrations are higher in phloem sap near sources, then aphids should be found primarily near mature leaves. (It's actually more common to find them near apical meristems, though—probably because the tissues are not as stiff and strong and difficult to pierce.) **5.** By pumping protons out, companion cells create a strong electrochemical gradient favoring entry of protons. A cotransport protein (a symporter) uses this proton gradient to import sucrose molecules *against* their concentration gradient. **6.** When it germinates—sucrose stored inside the seed is released to the growing embryo, which cannot yet make enough sucrose to feed itself.

Applying Concepts to New Situations

1. Plants have to gain CO_2 for photosynthesis to occur, which means that stomata have to be open. But transpiration occurs whenever stomata are open and the surrounding air is drier than the inside of the leaf. A similar "side effect" occurs when terrestrial animals breathe—they have to take dry air into their lungs, where water evaporates from the body and is lost. **2.** Plants would not grow as quickly or as tall, because the taller they grew, the more energy they would need to use to transport water and thus the less they would have available for growth. **3.** Because they cannot readily replace water that is lost to transpiration, they have to close stomata. During a heat wave, transpiration cannot cool the plant's tissues. They bake to death. **4.** Closing aquaporins will slow or stop movement of water from cells into the xylem and out of the plant via transpiration. Because water does not leave the cells, they are able to maintain turgor (normal solute potentials).

CHAPTER 38

Check Your Understanding (CYU)

CYU p. 748 (1) Proton pumps establish an electrical gradient across root hair membranes, with the inside of the membrane being much more negative than the outside.

FIGURE A37.1

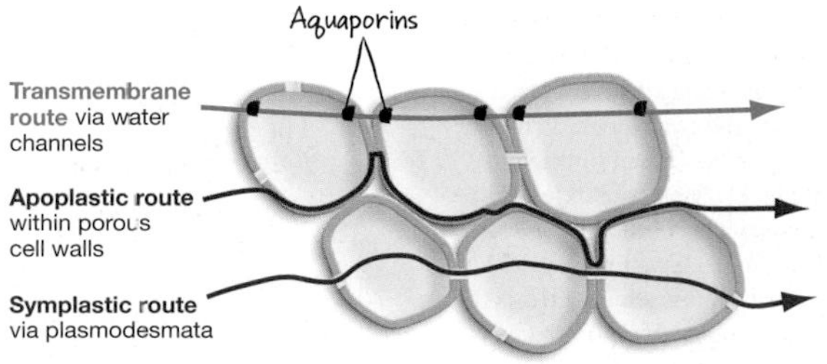

FIGURE A37.2

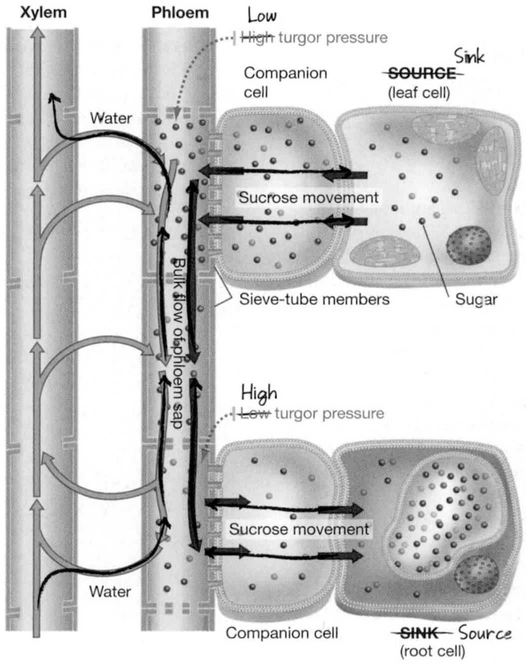

FIGURE A37.3

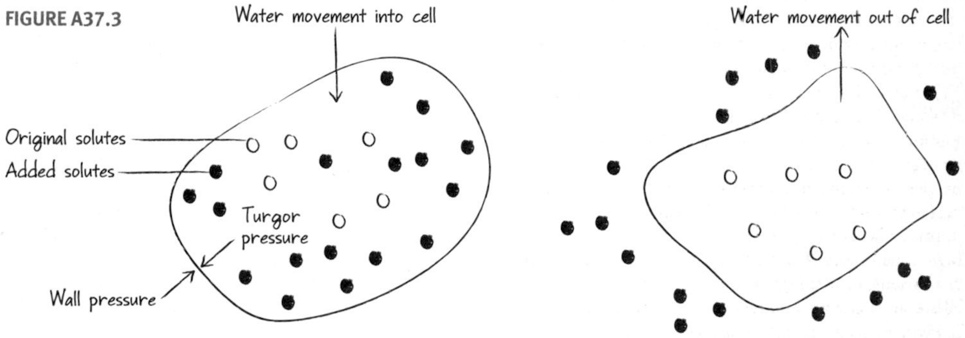

Cations follow this electrical gradient into the cell. (2) [See Figure A38.1] **CYU p. 750** (1) Many possible answers, including the following: If N-containing ions are abundant in the soil, the energetic cost of maintaining nitrogen-fixing bacteria may outweigh the benefits in terms of increased nitrogen availability. If mycorrhizal fungi provide a plant with nitrogen-containing ions, the energetic cost of maintaining nitrogen-fixing bacteria may outweigh the benefits in terms of increased nitrogen availability. Plants that grow near species with N-fixing bacteria might gain nitrogen by absorbing nitrogen-containing ions after root nodules die, or by "stealing it" (roots of different species will often grow together). There could be a genetic constraint (see Chapter 24): Plant species may simply lack the alleles required to manage the relationship. Note: To be considered correct, your hypothesis should be plausible (based on the underlying biology) and testable. (2) Many possible answers, including: Because the bacterial cell enters root cells, it is important for the plant to have reliable signals that it is not a parasitic bacterium. It is advantageous for the plant to be able to reject nitrogen-fixing bacteria if usable nitrogen is already abundant in the soil. The signals have no function.

You Should Be Able To (YSBAT)

YSBAT p. 745 (1) If there is no voltage across the root hair membrane, there is no electrical gradient favoring entry of cations, and absorption stops. (2) There is no route for cations to cross the root hair membrane, so absorption stops. **YSBAT p. 746** (1) If there is no proton gradient across the root hair membrane, there is no gradient favoring entry of protons, so symport of anions stops. (2) There is no route for anions to cross the root hair membrane, so absorption stops.

Caption Questions and Exercises

Figure 38.1 Point out that most of a plant's mass is due to carbon-containing molecules, but that water does not contain carbon. To test the hypothesis that most of a plant's mass comes from CO_2 in the atmosphere, design an experiment where plants are grown in the presence or absence of CO_2 and compare growth, and/or grow plants in the presence of labeled CO_2 and document the presence of labeled carbon in plant tissues. **Figure 38.2** Proteins and nucleic acids would be affected because they contain nitrogen. Most cell processes, including photosynthesis, would be affected because they depend on proteins. **Figure 38.3** Because soil is so complex, it would be difficult to defend the assumption that the soils with and without added copper are identical except for the difference in copper concentrations. Also, it would be better to compare treatments that had no copper versus normal amounts of copper, instead of comparing with and without added copper. **Figure 38.8** One cation (a proton) is being exchanged for another (calcium or magnesium). **Figure 38.10** The proton gradient arrow in (b) should begin above the membrane and cross the membrane pointing down. The electrical gradient arrow in (c) should begin above the membrane (marked positive) and cross the membrane pointing down (toward negative). **Figure 38.17** Mistletoe is parasitic (it extracts water and certain nutrients from host trees); bromeliads are not (they don't affect the fitness of their host plants).

Summary of Key Concepts

KC 38.1 Add nitrogen to an experimental plot and compare growth within the plot to growth of a similar plot nearby. Do many replicates of the comparison, to convince yourself and others that the results are not due to unusual circumstances in one or a few plots. **KC 38.2** A higher membrane voltage would allow cations in soil to cross root-hair membranes more readily through membrane channels and would allow anions to enter more readily via symporters. Nutrient absorption should increase. **KC 38.3** Grow pea plants in the presence of rhizobia, exposed to air with (1) N_2 containing the heavy isotope of nitrogen, and (2) radioactive carbon dioxide. Allow the plant to grow and then analyze the rhizobia and plant tissues. If the rhizobia-plant interaction is mutualistic, then ^{15}N-containing compounds and radioactive-carbon-containing compounds should be observed in both rhizobia and plant tissues. As a control, grow pea plants in the presence of the labeled compounds but without rhizobia. If the mutualism hypothesis is correct, the plant should contain labeled carbon but no heavy nitrogen.

Test Your Knowledge

1. b; **2.** d; **3.** d; **4.** c; **5.** b; **6.**d

Test Your Understanding

1. (1) Grow a large sample of corn plants with a solution containing all essential nutrients needed for normal growth and reproduction. Grow another set of genetically identical (or similar) corn plants in a solution containing all essential nutrients except iron. Compare growth, color, and seed production (and/or other attributes) in the two treatments. (2) Grow corn plants in solutions containing all essential nutrients in normal amounts, but with varying concentrations of iron. Determine the lowest iron concentration at which plants exhibit normal growth rates. **2.** If nutrients in soil are scarce, there is intense natural selection favoring alternative ways of obtaining nutrients—for example, by digesting insects or stealing nutrients. **3.** Higher in soils with both sand and clay. Clay fills spaces between sand grains and slows (1) passage of water through soil and away from roots, and (2) leaching of anions in soil water. Clay particles also provide negative charges that retain cations so they are available to plants. The presence of sand keeps soil loose enough to allow roots to penetrate. **4.** H^+-ATPases in the plasma membrane pump protons out of the cell—making the inside of the membrane less positive (more negative) and the outside of the membrane more positive. By convention, the voltage on a membrane is expressed as inside-relative-to-outside. **5.** Some metal ions are poisonous to plants, and high levels of other ions are toxic. Passive mechanisms of ion exclusion—such as a lack of ion channels that allow passage into root cells—do not require an expenditure of ATP. Active mechanisms of exclusion—such as production of metallothioneins—require an expenditure of ATP. **6.** (1) The amounts required and the mechanisms of absorption are different. Much larger amounts of C, H, and O are needed than mineral nutrients. CO_2 enters plants via stomata and water is absorbed passively along a water potential gradient; mineral nutrients enter via active uptake in root hairs or via mycorrhizae. (2) Macronutrients are required in relatively large quantities and usually function as components of macromolecules; micronutrients are required in relatively small quantities and usually function as enzyme cofactors.

Applying Concepts to New Situations

1. The water that he used may have contained many of the macro- and micronutrients that were incorporated into plant mass. To test this hypothesis, conduct an experiment of the same design but have one treatment with pure water and one treatment with water containing solutes ("hard water"). Compare the percentage of soil mass incorporated into the willow tree in both treatments. Willow may be unusual. To test this hypothesis, repeat the experiment with willow and several other species under identical conditions, and compare the percentage of soil mass incorporated into the plants. Van Helmont's measurements were inaccurate. Repeat the experiment. **2.** Acid rain inundates the soil with protons, which bind with the negatively charged clay parti-

FIGURE A38.1

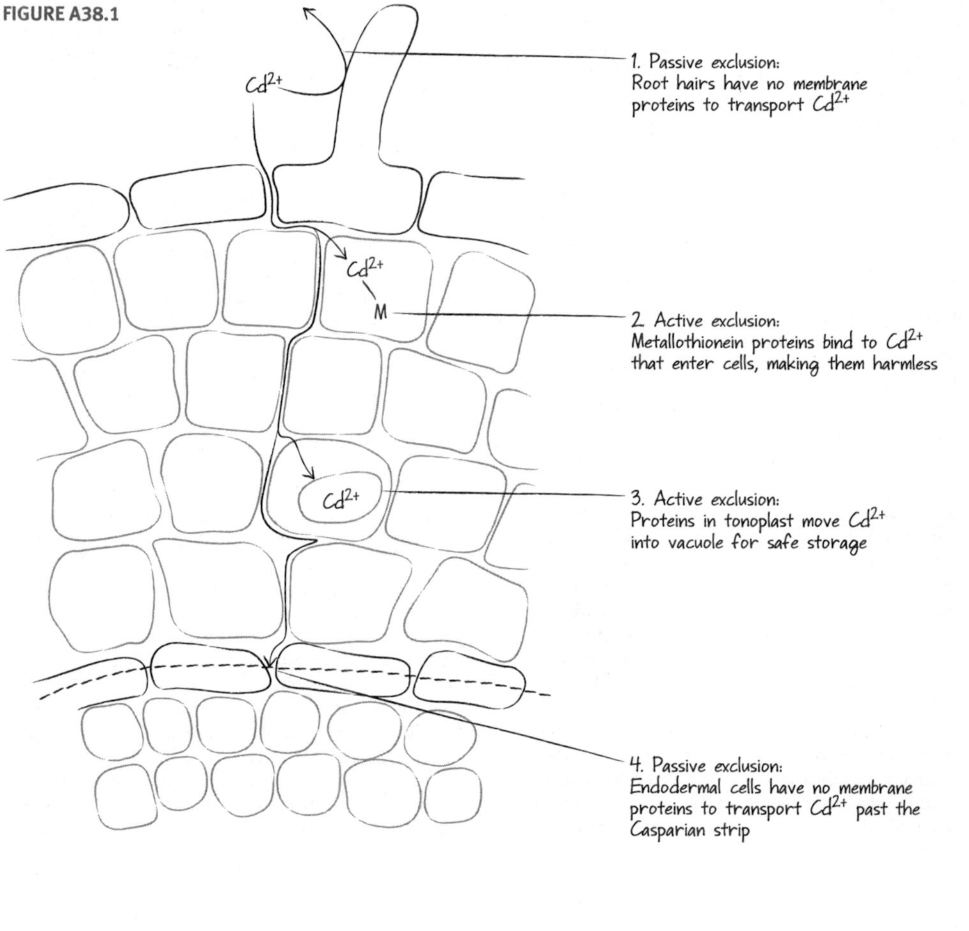

Cd^{2+}
Cd^{2+}
M
Cd^{2+}
1. Passive exclusion:
Root hairs have no membrane proteins to transport Cd^{2+}
2. Active exclusion:
Metallothionein proteins bind to Cd^{2+} that enter cells, making them harmless
3. Active exclusion:
Proteins in tonoplast move Cd^{2+} into vacuole for safe storage
4. Passive exclusion:
Endodermal cells have no membrane proteins to transport Cd^{2+} past the Casparian strip

cles—displacing the cations that are normally bound there. Cations may then be leached away. **3.** If the hypothesis is correct, vanadate should inhibit phosphorus uptake. To test this hypothesis, grow plants in solution with radioactively labeled phosphorus without vanadate and in solution with vanadate. Measure the amount of labeled phosphorus in root cells with and without vanadate. **4.** Alder decomposes much faster, because it provides more nitrogen to the fungi and bacteria that are responsible for decomposition—meaning that they can grow faster.

CHAPTER 39

Check Your Understanding (CYU)

CYU p. 757 (1) The only cells that can respond to a specific environmental signal or hormone are cells that have an appropriate receptor for that signal or hormone. (2) Once a receptor is activated, it triggers production of many second messengers or the activation of many proteins in a phosphorylation cascade. **CYU p. 762** (1) Using a syringe or other device, apply auxin to the west side of each stem, near the base. **CYU p. 764** (1) Lettuce grows best in full sunlight. If a seed receives red light, it indicates that the seed is in full sunlight—meaning that conditions for germination are good. But if a seed receives far-red light, it indicates that the seed is in shade—meaning that conditions for germination are poor. (2) Far-red light indicates shade, and stem-elongation gives plants a chance to grow above nearby plants into full sunlight. **CYU p. 766** (1) In both roots and shoots, auxin is redistributed asymmetrically in response to a signal. In phototropism, auxin is redistributed to the shaded side of the plant in response to blue light. In gravitropism, auxin is redistributed to the lower part of the root or shoot in response to gravity. (2) In root cells, auxin concentrations indicate the direction of gravity. The gravitropic response is triggered by changes in the distribution of auxin in root-tip cells. **CYU p. 776** (1) Short term, stomata should close and photosynthesis should stop. Long term, the plants will not grow well, compared to individuals that do not receive extra ABA, or die. (2) Their stems should elongate rapidly, compared to plants that do not receive extra GAs. **CYU p. 781** (1) Because specific *R* gene products recognize and match specific proteins produced by pathogens, having a wide array of *R* alleles allows individuals to recognize and respond to a wide array of pathogens. (2) If herbivores are not present, synthesizing large quantities of proteinase inhibitors wastes resources (ATP and substrates) that could be used for growth and reproduction.

You Should Be Able To (YSBAT)

YSBAT p. 777 Virtually all bananas have the same *R* alleles. If a pathogen evolves that is not identified by the existing *R* alleles, all bananas would be susceptible to the new disease.

Caption Questions and Exercises

Figure 39.4 These cells served as controls. If the radioactive band appeared in cells that did not have the *PHOT1* gene, it clearly could not be a *PHOT1* band. If the band appeared in cells that had not been exposed to blue light, then *PHOT1* would not be acting as a blue-light receptor. **Figure 39.5** For "Tip removed": cut the tip and replace it, to control for the hypothesis that the wound itself influences bending (not the loss of the tip). For "Tip covered": cover the tip with a transparent cover instead of opaque one, to control for the hypothesis that the cover itself affects bending (not excluding light). **Figure 39.7** Two possibilities: (1) none of the decapitated shoots would bend, or (2) the blocks from the completely divided tip would bend much less than the partially divided tip, because more cells were disrupted. **Table 39.1** The average germination rate of lettuce seeds last exposed to red light is about 99 percent. The average for seeds last exposed to far-red light is about 50 percent. Both of these values are much higher than the germination rate of buried seeds that receive no light at all; their germination rate is only 9 percent. **Figure 39.10** A signal in the form of a light wavelength is changed into a signal in the form of a protein's shape. In this way, a signal from outside the cell is changed to a signal inside the cell. **Figure 39.12** (One possibility) Direct pressure from a statolith changes the shape of the receptor protein. The shape change triggers a signal transduction cascade leading to a gravitropic cell response. **Figure 39.14** In nature, this response is most likely to occur in windy environments. A reduction in height might keep plants beneath the path of the wind, and stockier plants would be better able to resist bending when blown. **Figure 39.19** The dwarfed individuals can put more of their available energy into reproduction because they are using less energy for growth than taller individuals. **Figure 39.21** Without these data, a critic could argue that the stomata closed in response to water potentials in the leaf, not a signal from the roots. **Figure 39.24** The transported nutrients are conserved in the plant instead of being lost when leaves fall. Cessation of chlorophyll synthesis saves plants the energy and nutrients that would have gone into producing chlorophyll in leaves that soon will fall off. **Figure 39.28** The molecule travels throughout the shoot system (and root system).

Summary of Key Concepts

KC 39.1 They do not perform photosynthesis and therefore do not need to switch behavior based on being located in full sunlight versus shade. **KC 39.2** An external signal arrives at a receptor cell/protein (a person sees a barn on fire); the person calls (the receptor transduces the signal); the dispatcher rings a siren (a second messenger is produced or phosphorylation cascade occurs); the fire department members get to the pumper truck (gene expression or some other cell activity changes). **KC 39.3** Sun-adapted (high-light) plants are likely to have a significant positive phototropic response (i.e., growth toward light), whereas shade-adapted (low-light) plants are less likely to respond phototropically. **KC 39.4** They need to be transported throughout the plant (because cells throughout the body need to respond to the signal)—not in a single direction.

Test Your Knowledge

1. d; **2.** a; **3.** b; **4.** b; **5.** c; **6.** b

Test Your Understanding

1. The P_{fr}-P_r switch in phytochromes lets plants know whether they are in shade or sunlight, and triggers responses that allow sun-adapted plants to avoid shade. Phototropins are activated by blue light, which indicates full sunlight, and trigger responses that allow plants to photosynthesize at full capacity. **2.** If *PHOT1* were inserted into a plant, a critic could contend that it was being phosphorylated by a plant protein. But because the *PHOT1* was in an insect cell, this alternative hypothesis was not credible. **3.** *Transduce* means "to convert energy from one form to another." There are many examples: a touch may be converted to an electrochemical potential, light energy or pressure into a shape change in a protein, and so on. **4.** (Many possibilities) Cells or tissue involved: Auxin directs gravitropism in roots but phototropism in shoots. Developmental stage or age: GA breaks dormancy in seeds but extends stems in older individuals. Concentration: Large concentrations of auxin in stems promote cell elongation; small concentrations do not. Other hormones: The concentration of GA relative to ABA influences whether germination proceeds. **5.** Receptor cells sense changes in the availability of blue light received. They respond by changing the distribution of auxin. A change in auxin concentration causes target cells in the stem to grow, and results in the stem bending toward the source of blue light. In this way, plants can respond to changes in shading by growing toward areas where light is available. **6.** Ethylene promotes senescence. Fruits exposed to ethylene ripen faster than fruits that are not exposed to ethylene; increased ethylene sensitivity in leaves (combined with reduced auxin concentrations) causes abscission.

Applying Concepts to New Situations

1. Small-seeded plants need to perform photosynthesis early in seedling development or they will starve. Therefore, they need to germinate in direct sunlight. The food reserves in large-seeded plants can support seedling growth for a relatively long time without photosynthesis. Therefore, they do not need to germinate in direct sunlight. **2.** The experiment provides evidence that cytokinins stimulate cell division in lateral buds—suggesting that large amounts of cytokinins can overcome apical dominance. **3.** In these seeds, (a) little or no ABA is present, or (b) ABA is easily leached from the seeds, so its inhibitory effects are eliminated. Test hypothesis (a) by determining whether ABA is present in the seeds. Test hypothesis (b) by comparing the amount of ABA present before and after running water—enough to mimic the amount of rain that falls in a tropical rain forest over a few days or weeks—over the seeds. **4.** Yes—the experiments are consistent with the hypothesis that ABA must be on the surface of cells to trigger a response, meaning that the receptor must be located there.

CHAPTER 40

Check Your Understanding (CYU)

CYU p. 792 (1) Both are diploid cells that divide by meiosis to produce spores. The megasporocyte produces a megaspore (female spore); a microsporocyte produces a microspore (male spore). (2) Both are multicellular individuals that produce gametes by mitosis. The female gametophyte is larger than the male gametophyte and produces an egg; the male gametophyte produces sperm. **CYU p. 795** (1) Insects feed on nectar and/or pollen in flowers. Flowers provide food rewards to insects to encourage visitation. The individuals that attract the most pollinators produce the most offspring. (2) One product of double fertilization is the zygote, which will eventually grow by mitosis into a mature sporophyte. The other product is the endosperm nucleus, which will grow by mitosis to form a source of nutrients for the embryo.

You Should Be Able To (YSBAT)

YSBAT p. 786 (1) The microsporocyte is the male spore; the megasporocyte is the female spore. (2) In angiosperms the female gametophyte stays within the ovary at the base of the flower even after it matures and produces an egg cell. Fertilization and seed development take place in the same location. **YSBAT p. 795** The endosperm nucleus in the central cell is triploid; the zygote is diploid; the synergid and other cells remaining from the female gametophyte are haploid. **YSBAT p. 796** The protoderm gives rise to the epidermal tissue; the ground meristem gives rise to the ground tissue; the procambium gives rise to the vascular tissue.

Caption Questions and Exercises

Figure 40.5 An appropriate control treatment would be to graft a leaf from a plant that had never been exposed to a short-night photoperiod onto a new plant that also had never been exposed to the correct conditions for flowering. If the hypothesis is correct, this grafted leaf should not induce flowering. **Figure 40.9** A gametophyte is the multicellular individual that produces gametes by mitosis. The embryo sac conforms to this definition because it is a multicellular form that produces an egg by mitosis. **Figure 40.10** The pollen grain is a gametophyte because it is multicellular and produces sperm by mitosis.

Figure 40.17 Hypothesis: Fruit changes color when it ripens as a signal to fruit eaters. The color change is advantageous because seeds are not mature in unripe fruit, and fruit eaters disperse mature seeds in ripe fruit. **Figure 40.21** Beans—their cotyledons are aboveground.

Summary of Key Concepts

KC 40.1 The sporophyte is diploid, the gametophyte is haploid, spores and gametes are haploid, and zygotes are diploid. **KC 40.2** There would be four embryo sacs in each ovule, and thus four eggs. **KC 40.3** The carpel is the female portion of the flower and contains the ovary. The ovary contains ovules, which house the female gametophytes. Once fertilization has occurred, the ovary develops into the fruit and the ovules develop into seeds within the fruit.

Test Your Knowledge

1. b; **2.** c; **3.** a; **4.** d; **5.** d; **6.** c

Test Your Understanding

1. Asexual reproduction is a quick, efficient way for a plant to reproduce large numbers of offspring. The disadvantage is that offspring are genetically identical to the parent, and thus vulnerable to the same diseases and only able to thrive in habitats similar to those inhabited by the parent. **2.** Megasporocytes (female) and microsporocytes (male) undergo meiosis to generate megaspores and microspores, respectively. These divide mitotically to give rise to female and male gametophytes—the embryo sac and pollen grain, respectively. **3.** Sepals are the outermost structure in a flower and usually protect the structures within; petals are located just inside the sepals and usually function to advertise the flower to pollinators. In wind- and bee-pollinated flowers, sepals probably do not vary much in overall function or general structure. Petals are often lacking in wind-pollinated species; they tend to be broad and flat (to provide a landing site) and colored purple or blue or yellow in bee-pollinated species (often with ultraviolet markings). **4.** Outcrossing increases genetic diversity among offspring, making them more likely to thrive if environmental conditions change from the parental generation. However, outcrossing requires that cross-pollination is successful. Self-fertilization results in relatively low genetic diversity among offspring but ensures that pollination is successful. **5.** The female gametophyte develops inside the ovule; the ovary develops into the pericarp of the fruit after fertilization. A mature ovule contains both gametophyte and sporophyte tissue; the ovary and carpel are all sporophyte tissue. **6.** The endosperm of a corn seed and the cotyledons of a bean seed both contain nutrients needed for the seed to germinate. The bean cotyledons have absorbed the nutrients of the endosperm.

Applying Concepts to New Situations

1. Because the island is near the equator, photoperiod will not provide much information about when conditions are optimal for germination. Presumably, daily temperature fluctuations will also remain relatively constant. Based on the information provided, the most logical hypotheses are that flowering occurs at the onset of the rainy season and that dry, dormant seeds germinate when exposed to moisture—cues that indicate that conditions for growth are good. **2.** In response to "cheater" plants, insects should be under intense selection pressure to detect and avoid species that do not offer a food reward. Individuals would have to be able to distinguish scents, colors, or other traits that identify cheaters. In response to "cheater" insects, plants possessing unusually thick petals or sepals that prevented insects from bypassing the anthers on their way to the nectaries would be more successful at producing seed in the next generation. **3.** One possibility: Under identical growth conditions, pollinate many individuals of the same species with two pollen grains, and many individuals of this species with one pollen grain. Measure and compare the rates of pollen tube growth. **4.** Acorns have a large, edible mass and are probably animal dispersed (e.g., by squirrels that store them and forget some). Cherries have an edible fruit and are probably animal dispersed. Burrs stick to animals and are dispersed as the animals move around. Milkweed seeds float in wind. To estimate the distance that each type of seed is dispersed from the parent, (1) set up "seed traps" to capture seeds at various distances from the parent plant, (2) sample locations at various distances from the parent and analyze young individuals—using techniques introduced in Chapter 20 to determine if they are offspring from the parent being studied, (3) mark seeds, if possible, and re-find them after dispersal.

CHAPTER 41

Check Your Understanding (CYU)

CYU p. 810 *Connective tissues* consist of cells embedded in a matrix. The density of the matrix determines the rigidity of the connective tissue and its function in padding/protection, structural support, or transport. *Nervous tissue* is composed of neurons and support cells. Neurons have long projections that function as "biological wires" for transmitting electrical signals. *Muscle tissue* has cells that can contract and functions in movement. *Epithelial tissue* has tightly packed layers of cells with distinct apical and basal surfaces. They protect underlying tissues and regulate the entry and exit of materials into these tissues. **CYU p. 814** (1) As three-dimensional size increases, volume increases more quickly than does surface area, so the surface area/volume ratio decreases. (2) They are inversely proportional (negatively correlated). On a per gram basis, small animals have higher basal metabolic rates than large animals. **CYU p. 819** (1) Endotherms can remain active during the winter and at night and sustain high levels of aerobic activities such as running or flying, but require large amounts of food energy. Ectotherms need much less food and can devote a larger proportion of their food intake to reproduction, but have a hard time maintaining high activity levels at night or in cold weather. (2) [**See Figure A41.1**]

You Should Be Able To (YSBAT)

YSBAT p. 812 A newborn.

Caption Questions and Exercises

Figure 41.1 There would be no relationship between the variables. In all three cases, the graph would be a flat line. **Figure 41.3** Loose connective tissues have a soft matrix that is effective for cushioning and protecting organs. Dense connective tissues have a fiber-rich matrix that allows them to link adjacent structures. Structural connective tissues have a stiff or hard matrix that allows them to withstand bending or compressing forces. Fluid connective tissues are effective in transport because liquids move easily and carry solutes. **Figure 41.4** The projections allow electrical signals to be transmitted over long distances. **Figure 41.10** The dog has to eat more because it has a higher mass-specific metabolic rate than a human. **Figure 41.11** There should be no oxygen uptake on either side of the chamber.

Summary of Key Concepts

KC 41.1 Loosening tight junctions would compromise the barrier between the external and internal environments. It would allow water and other substances (including toxins) to pass more readily across the epithelial boundary. **KC 41.2** King Kong is endothermic and his huge mass would generate a great deal of heat. His relatively small surface area would not be able to dissipate the heat—especially if it were covered with fur. **KC 41.3** Individuals could acclimatize by normal homeostatic mechanisms (increased sweating, seeking shade, etc.). Individuals with alleles that allowed them to function better at higher temperatures would produce more offspring than individuals without those alleles. Over time, this would lead to adaptation via natural selection. **KC 41.4** They multiply the amount of heat or material exchanged between the two countercurrent flows, compared to a "con-current" system.

Test Your Knowledge

1. b; **2.** d; **3.** b; **4.** d; **5.** a; **6.** a

Test Your Understanding

1. It is where environmental changes are sensed first. As an effector, it has a large surface area for losing or gaining heat. **2.** A warm frog can move, breathe, and digest much faster than a cold frog, because enzymes work faster (rates of chemical reactions increase) at 35°C than at 5°C. A warm frog needs much more food, though, to support this high metabolic rate. **3.** *Absorptive regions* have numerous folds and projections, which increase their surface area for absorption via diffusion. *Capillaries* have a high surface area because they are thin and highly branched, making exchange of fluids and gases efficient. *Beaks of Galápagos finches* have sizes and shapes that correlate with the type of food. Large beaks are used to crack large seeds; long, thin beaks are used to pick insects off surfaces; etc. *Fish gills* are thin, flattened structures with a large surface area, which facilitates the efficient exchange of gases and wastes. **4.** A larger sphere has a relatively smaller surface area for its interior volume than does a smaller sphere, because surface area increases with size at a lower rate than volume does. Diffusion will be more efficient in the case of the smaller sphere. **5.** In the morning, an ant should move to a sunlit spot to gain heat by conduction from the ground and radiation striking its body directly. In midday, the ant should avoid overheating by losing heat from wet body surfaces, standing in breezy areas to lose heat by convection, or retreating to the cool burrow to lose heat by conduction and avoid gaining heat by radiation. **6.** The feedback causes a response that counters the current state of the system. It is negative in the sense of reducing the difference between the current state and the set point.

Applying Concepts to New Situations

1. Large individuals require more food, so are more susceptible to starvation if food sources can't be defended (e.g., when seeds are scattered around). They may also be slower and/or more visible to predators. **2.** You would have to show changes in the frequencies of alleles whose products are affected by temperature—for example, increases in the frequencies of alleles for enzymes that operate best at higher temperature.

FIGURE A41.1

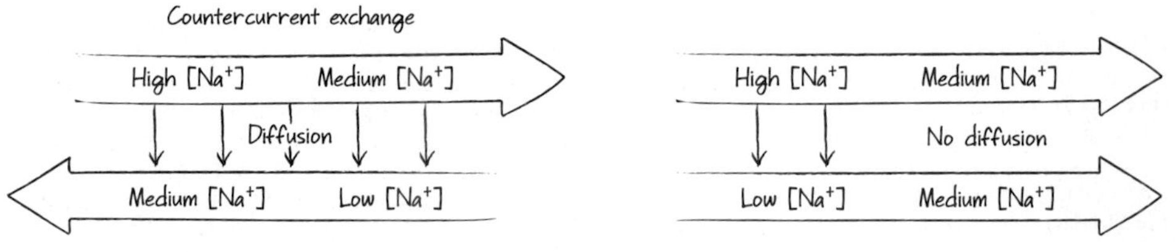

3. The heat-dissipating system should have a very high surface area/volume ratio. Based on biological structures, it should be flattened (thin) and highly folded, branched, and/or contain tubelike projections. Another approach inspired by biological systems would be to set up a countercurrent heat exchanger with fluid heated by the engine. **4.** Buy the turtle. Turtles are ectothermic, meaning they do not expend energy to regulate their body temperature. Mice are endothermic, meaning they do expend energy to regulate body temperature. Mice will therefore have a higher basal metabolic rate and consume more food.

CHAPTER 42

Check Your Understanding (CYU)

CYU p. 828 Without the "master gradient" established by the pump, sodium ions and chloride ions cannot move out of the epithelial cells into the surrounding seawater. Salt will build up in the shark's tissues. **CYU p. 831** (1) Uric acid is insoluble in water and can be excreted without much water loss. (2) When electrolytes are reabsorbed in the hindgut epithelia, water follows along an osmotic gradient. **CYU p. 838** (1) Aldosterone stimulates Na^+ reabsorption in pre-urine in the distal tubule. This increases the Na^+ level in the blood and leads to a more dilute (hypotonic) urine. (2) Water intake increases blood pressure and filtration rate in the renal corpuscle, and leads to lowered electrolyte concentration in the blood and filtrate and production of dilute urine. Eating large amounts of salt results in concentrated, hypertonic urine. Water deprivation triggers ADH release and the production of concentrated, hypertonic urine.

You Should Be Able To (YSBAT)

YSBAT p. 833 Blood contains cells and large molecules as well as electrolytes and wastes; the filtrate contains only electrolytes and wastes. **YSBAT p. 835** If sodium reabsorption is inhibited, then less water will be reabsorbed along an osmotic gradient and more urine will be produced. **YSBAT p. 837** (1) Less, because the osmotic gradient is not as steep. (2) Lower, because more water has been retained. (3) Less, because the concentration gradient is not as steep. **YSBAT p. 838** Ethanol consumption inhibits water reabsorption, leading to a larger volume of less concentrated urine. Nicotine consumption increases water reabsorption, leading to a smaller volume of more concentrated urine.

Caption Questions and Exercises

Figure 42.1 The purple and white molecules are moving down their concentration gradients, and the concentration of red molecules does not affect the concentration of purple or white molecules. **Figure 42.6** The sodium-potassium ATPase does active transport; the sodium-chloride-potassium cotransporter does secondary active transport; the chloride and potassium channels do passive transport. **Figure 42.8** The underside of the abdomen is shaded by the grasshopper's body. This location is relatively cool and should reduce the loss of water during respiration.

Summary of Key Concepts

KC 42.1 In the ocean, river otters must excrete the additional salt that is taken up from saltier marine foods and seawater. In freshwater, their bodies must conserve electrolytes and excrete the water taken in during drinking. **KC 42.2** A sodium gradient establishes an osmotic gradient that moves water. It also sets up an electrochemical gradient that can move ions against their electrochemical gradient by secondary transport through a cotransporter, or with their electrochemical gradient by passive diffusion through a channel.
KC 42.3 The desert locust—to save water, it has to actively reabsorb ions in the hindgut so that almost no water is lost during excretion. **KC 42.4** The longer the loop of Henle, the steeper the osmotic gradient in the medulla. Steep osmotic gradients allow a great deal of water to be reabsorbed as urine passes through the collecting duct.

Test Your Knowledge

1. c; **2.** a; **3.** d; **4.** d; **5.** a; **6.** c

Test Your Understanding

1. In salt water, a salmon gains NaCl by diffusion and loses water by osmosis. It replaces water by drinking and secretes ions through chloride cells in its gill epithelium. In freshwater, it gains water by osmosis and loses NaCl by diffusion. Epithelial cells in the gills take up ions, it stops drinking, and produces copious amounts of urine to rid itself of excess water. **2.** Mitochondria are needed to produce ATP. A key component of salt regulation in fish is the Na^+/K^+-ATPase, which requires ATP to function. Because ATP fuels establishment of the ion gradients necessary for the cotransport mechanisms utilized by chloride cells as well as other epithelial cells, an abundance of mitochondria would be expected. **3.** Microvilli increase the surface area available for pumps, cotransporters, and channels.
4. The wax layer should be thinner in tropical insects, because the animals are under less osmotic stress (less evaporation due to high humidity). You could test this prediction by measuring the thickness of the wax layer in two closely related species from each habitat, or in a wide array of species from each habitat. Insects that live in extremely humid habitats may lack the ability to close the openings to their respiratory passages; animals that live in humid habitats may excrete primarily ammonia or urea and have relatively short loops of Henle. In each case, you could test these predictions by comparing individuals of closely related species that live in dry versus humid habitats. **5.** In both cases, the initial filtrate is isotonic with blood and reabsorption of certain solutes depends on Na^+/K^+-ATPase activity and is energetically costly. One difference is that the initial filtrate forms by active pumping of ions followed by osmosis in insects versus by filtration in the renal corpuscle of mammals. Another difference is that mammals make use of a countercurrent exchanger in the loop of Henle versus direct pumping of ions in the insect hindgut. **6.** Ammonia is toxic and must be diluted with large amounts of water to be excreted safely. Urea and uric acid are safer and do not have to be excreted with large amounts of water, but are more expensive to produce in terms of energy expenditure. Fish excrete ammonia; mammals excrete urea; insects excrete uric acid. You would expect the embryos inside terrestrial eggs to excrete uric acid, as it is the least toxic and is insoluble in water.

Applying Concepts to New Situations

1. The countercurrent flow removes salt from the tissues around the ascending limb, raising the osmolarity of blood, and then water from the tissues around the descending limb, along an osmotic gradient. The countercurrent arrangement maintains a concentration and osmotic gradient all along the length of the nephron.
2. The observation that the Na^+ concentration decreased by 30 percent, even as volume also decreases, indicates that Na^+ ions were reabsorbed to a greater degree than water. The observation that the urea concentration increased by 50 percent indicates that urea was not reabsorbed. **3.** Increased salt concentrations in freshwater environments put organisms under a new kind of osmotic stress. They may not have chloride cells or other adaptations that allow them to rid themselves of additional salt. **4.** Their urine volume is much higher (actually 10 times).

CHAPTER 43

Check Your Understanding (CYU)

CYU p. 856 (1) *Mouth:* Food is taken in; teeth physically break down food into smaller particles; salivary amylase begins to break down carbohydrates; lingual lipase initiates the digestion of fats. *Esophagus:* Food is moved to the stomach via peristaltic contractions. *Stomach:* HCl denatures proteins; pepsin begins to digest them. *Small intestine:* Pancreatic enzymes complete the digestion of carbohydrates, proteins, lipids, and nucleic acids. Most of the water and all of the nutrients are absorbed here. *Large intestine:* Water is reabsorbed and feces are formed. *Anus:* Feces accumulate in the rectum and are expelled out the anus. (2) If the release of bile salts is inhibited, fats would not be digested and absorbed efficiently, and they would pass into the large intestine. The individual would likely produce fatty feces and lose weight over time. Trypsin inactivation would inhibit protein digestion and reduce absorption of amino acids, likely leading to weight loss and muscle atrophy due to protein deprivation. If the Na^+-glucose cotransporter were blocked, then glucose would not be absorbed into the bloodstream. The individual would likely become sluggish from lack of energy. **CYU p. 858** (1) Both lead to high urine volume due to reduced water reabsorption in the kidneys, but for different reasons. In diabetes mellitus, less water is reabsorbed because glucose concentrations in the filtrate from blood are high. In diabetes insipidus, less water is reabsorbed because of defects in the collecting ducts. (2) When the blood sugar of individuals with type I diabetes mellitus gets too high, they should inject themselves with insulin, which will trigger the absorption and storage of glucose by their cells. When their blood glucose gets too low, they should eat something with a high concentration of sugar (such as orange juice or a candy bar) to increase their blood glucose levels quickly.

You Should Be Able To (YSBAT)

YSBAT p. 851 Nutrient absorption occurs in the small intestine, but not in the esophagus or stomach, and the rate of absorption increases with available surface area. **YSBAT p. 854** (1) Decreased energy yield from foods due to reduced glucose absorption, (2) increased feces production due to the passing of unabsorbed glucose into the colon, (3) watery feces due to decreased water reabsorption in the large intestine (lower osmotic gradient). As an aside, also increased flatulence due to the metabolism of unabsorbed glucose by bacteria that produce methane as a waste product. **YSBAT p. 857** Individuals with type I diabetes have normal insulin receptors on their liver cells, but do not produce and release insulin from the pancreas. Individuals with type II diabetes produce and release insulin, but their insulin receptors are defective so liver cells cannot respond to insulin. In both types of diseased individuals, high glucose concentrations in the blood remain high. In individuals without disease, both insulin production and insulin receptors are normal, so liver cells respond to insulin by taking up glucose from the blood and storing it.

Caption Questions and Exercises

Figure 43.3 Your label should point to the rightmost two bones in the illustration. **Figure 43.14** Frog eggs don't normally make the sodium-glucose cotransporter protein, so the researchers could be confident that if the protein appeared, it was from the injected RNA. They would not be confident of this if the RNAs were injected into rabbit epithelial cells, where the protein was probably already present. **Figure 43.16** In type 2 diabetes mellitus, the top, black arrow from glucose (in the blood) to glycogen (inside liver and muscle cells) is disrupted. In type 1 diabetes mellitus, the green arrow on the top left, indicating insulin produced by the pancreas, is disrupted. **Figure 43.17** The leading hypothesis is an increased incidence of obesity, which is linked to development of type 2 diabetes mellitus.

Summary of Key Concepts

KC 43.1 Form two groups of mice selected from a common population, so that individuals have similar genetic characteristics and history. One group (the control group) would receive a regular diet of rat chow, and the other group (the experimental group) would receive a diet of rat chow that was deficient in magnesium. By comparing the two groups on a number of behavioral and physical measures, you could determine the effects of magnesium deficiency. **KC 43.2** Their mouth or tongue would be modified to make probing of flowers and sucking nectar efficient. They would not require teeth, and their digestive tract would be relatively simple. For example, the stomach would not have to pulverize food or digest large amounts of protein, and the large intestine would not have to process and store large amounts of bulky waste. **KC 43.3** By making the stomach smaller, gastric bypass surgery decreases the amount of food that can be comfortably contained in the stomach and digested. By bypassing a portion of the small intestine, the surgery decreases the absorption of food. (As an aside, people who have this surgery are at risk for serious nutrient deficiencies, chronic intestinal upset, ulcers, and flatulence.) **KC 43.4** The excess glucose is eventually eliminated in urine. (This takes much longer than sequestering the excess glucose in liver, muscle, or adipose cells.)

Test Your Knowledge

1. a; **2.** a; **3.** d; **4.** d; **5.** b; **6.** d

Test Your Understanding

1. The bird crop is an enlarged sac that can hold quickly ingested food; in leaf-eating species it is filled with symbiotic organisms and functions as a fermentation vessel. The cow rumen is an enlarged portion of the stomach; the elephant large intestine is enlarged relative to other species. Both structures are filled with symbiotic organisms and function as fermentation vessels. **2.** Digestive enzymes break down macromolecules (proteins, carbohydrates, nucleic acids, lipids). If they weren't produced in an inactive form, they would destroy the cells that produce and secrete them. **3.** Most biologists accept the hypothesis, for two reasons: (1) the structure is unique among fish in being effective at biting, and (2) its surface correlates with the type of food each species eats. **4.** In most terrestrial vertebrates, the primary function of the large intestine is water reabsorption; fish do not need to reabsorb water. **5.** When an individual ingests a solution of glucose and electrolytes, the solutes are absorbed. Water follows by osmosis, preventing dehydration. **6.** Insulin triggers negative feedback to high blood glucose concentrations by stimulating glucose uptake by several types of cells. Glucagon triggers negative feedback to low blood glucose concentrations by stimulating glucose release by several types of cells.

Applying Concepts to New Situations

1. Because they are feeding growing embryos and newborns, it is likely that female mammals during pregnancy and breastfeeding would require higher levels of almost every nutrient—particularly calcium, used by offspring for bone growth. **2.** Individuals with defects in pancreatic amylase would not be able to complete carbohydrate digestion and probably would be lethargic (low energy) and experience weight loss. Pepsin defects would reduce or eliminate protein digestion in the stomach, leading to severe amino acid deficiency. Defects in the fatty-acid binding protein would reduce or eliminate fatty-acid absorption in the small intestine, likely producing fatty feces, weight loss, and diarrhea. Aquaporin defects would prevent water reabsorption in the large intestine and lead to diarrhea and dehydration. **3.** The result is still valid because the injection was correlated with secretion—if lack of injection was correlated with lack of secretion. The criticism is also somewhat valid, however, because the researchers couldn't rule out the hypothesis that signaling from nerves plays some sort of role, too. **4.** Terrestrial animals are exposed to increased risk of water loss, and the large intestine is where water reabsorption occurs. In most cases fish do not need to reabsorb large amounts of water from their feces.

CHAPTER 44

Check Your Understanding (CYU)

CYU p. 864 (1) Oxygen partial pressure is high in mountain streams because the water is cold, mixes constantly, and has a high surface area (due to white water). Oxygen partial pressure is low at the ocean bottom because the area is far from the surface where gas exchange takes place and there is relatively little mixing. (2) *Warm-water species:* Large amount of air because the oxygen-carrying capacity of warm water is low. *Vigorous algal growth:* Small amount of air because algae contribute oxygen to the water through photosynthesis. *Sedentary animals:* Small amount of air because sedentary animals require relatively little oxygen. **CYU p. 870** (1) Common features include large surface area, short diffusion distance (a thin gas exchange membrane), and a mechanism that keeps fresh air or water moving over the gas-exchange surface. Only fish gills use a countercurrent exchange mechanism; only tracheae deliver oxygen directly to cells without using a circulatory system; only lungs contain "dead space"—areas that are not involved in gas exchange. (2) The anterior and posterior air sacs provide additional compartments for inhaled and exhaled air; they are arranged so that air moves through them and the lungs in one direction. **CYU p. 874** The saturation curve for Tibetans should be shifted to the left relative to people from sea level—meaning that their hemoglobin has a higher affinity for oxygen at all partial pressures. **CYU p. 883** (1) If the blood plasma that leaks out of the capillaries is not reabsorbed in capillaries, it enters lymphatic vessels that eventually merge with blood vessels. (2) [See **Figure A44.1**]

You Should Be Able To (YSBAT)

YSBAT p. 872 There would be an even larger change in the oxygen saturation of hemoglobin in response to an even smaller change in the partial pressure of oxygen.

Caption Questions and Exercises

Figure 44.4 External gills are ventilated passively and are efficient because they are in direct contact with water. They are exposed to predators and mechanical damage, however. Internal gills are protected but have to be ventilated by some type of active mechanism for water flow. **Figure 44.7** Using several to many different animals increases confidence that the results are true for most or all individuals in the population, and not due to one or a few unusual individuals or circumstances. **Figure 44.10** Sighing increases pressure in the lung cavity, forcing air out of the lungs more rapidly and completely than occurs during normal exhalation. When taking a deep breath, volume in the lung cavity is increased, resulting in lower pressure and causing more air to enter the lungs than during normal inhalation. **Figure 44.15** According to data in the figure, the oxygen saturation of hemoglobin is about 15 percent for blood at pH 7.2 and about 25 percent at pH 7.4. Therefore, about 85 percent of the oxygen is released from hemoglobin at pH 7.2, but only about 75 percent of the oxygen is released at pH 7.4. **Figure 44.17** In the lungs, a strong partial pressure gradient favors diffusion of dissolved CO_2 from blood into the alveoli. As the partial pressure of CO_2 in the blood declines, hydrogen ions leave hemoglobin and react with bicarbonate to form more CO_2, which then diffuses into the alveoli and is exhaled from the lungs. **Figure 44.23** Air from the alveoli mixes with air in the "dead space" in the bronchi and trachea on its way out of the body. This dead-space air is from the previous inhalation (P_{O_2} = 160 mm Hg; P_{CO_2} = 0.3), so when the alveolar air mixes with the dead-space air, the partial pressures in the exhaled air achieve levels intermediate between that of inhaled and alveolar air.

Summary of Key Concepts

KC 44.1 It is harder to extract oxygen from water than it is to extract it from air. This limits the metabolic rate of water breathers. **KC 44.2** Large animals have a relatively small body surface area relative to their volume. If they had to rely solely on gas exchange across their skin, they would not have enough skin surface area to exchange the volume of oxygen needed to meet their metabolic needs. **KC 44.3** Cold water carries more oxygen than warm water, so icefish blood can carry enough oxygen to supply the tissues with oxygen even in the absence of hemoglobin. The oxygen and carbon dioxide are simply dissolved in the blood. **KC 44.4** Their circulatory systems should have relatively high pressures—achieved by independent systemic and pulmonary circulations powered by a four-chambered heart—to maximize the delivery of oxygenated blood to metabolically active tissues.

Test Your Knowledge

1. b; **2.** c; **3.** d; **4.** d; **5.** c; **6.** a

Test Your Understanding

1. The insect tracheal system delivers O_2 directly to respiring cells, but in humans, O_2 is first taken up by the blood of the circulatory system, which then delivers it to the cells. The human respiratory system also has specialized muscles devoted to ventilating the lungs. The open circulatory system of the insect is a low-pressure system, which makes use of pumps (hearts) and body movements to circulate hemolymph. The closed circulatory system of humans is a high-pressure system, which can respond to rapid changes in O_2 demand by tissues. **2.** During exercise, P_{O_2} decreases in the tissues and P_{CO_2} increases. The increase in P_{CO_2} lowers tissue pH. The drop in P_{O_2} and pH causes more oxygen to be released from hemoglo-

FIGURE A44.1

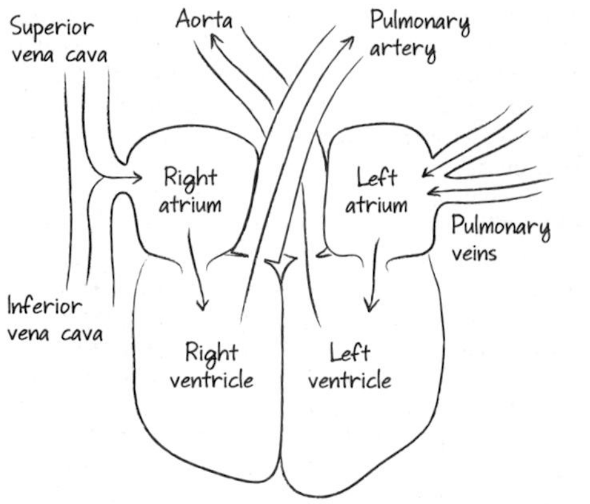

bin. **3.** In blood, CO_2 is converted to carbonic acid, which dissociates into a proton and a bicarbonate ion. Because the protons bind to deoxygenated hemoglobin, they do not cause a dramatic drop in blood pH (hemoglobin acts as a buffer). Also, small decreases in blood pH trigger a homeostatic response that increases breathing rate and expulsion of CO_2. **4.** Airflow through bird lungs is unidirectional, so it allows for continuous ventilation of gas-exchange surfaces with fresh, oxygenated air. The trachea and bronchi in a mammalian lung do not have a gas-exchange surface, and bidirectional airflow in these species means that "stale" air has to be expelled before "fresh" air can be inhaled. As a result, the alveoli are not ventilated continuously. **5.** Lungs increase the temperature of the air and are moist to allow greater solubility of gases (increasing k); alveoli present a large surface area (large A), the epithelium of alveoli is thin (small D), and constant delivery of deoxygenated blood to alveoli maintains a steep partial pressure gradient favoring diffusion of oxygen into the body ($P_2 - P_1$ is high). **6.** If the pulmonary circulation was under pressure as high as that found in the systemic circulation, large amounts of fluid would be forced out of capillaries in the lungs. There is a conflict between the thin surface required for efficient gas exchange and thick blood vessels required to withstand high pressure.

Applying Concepts to New Situations

1. In species that depend on rapid movement to chase down prey or perform other key actions, individuals with larger gill surfaces would absorb more O_2 from water and likely produce more offspring than individuals with smaller gill-surface areas. In slow-moving species, there would be no advantage to increasing the surface area in the gills, so such an adaptation would not be preferentially selected for. **2.** A shift to the right of the oxygen-hemoglobin dissociation curve represents a decrease in the affinity of hemoglobin for O_2. Thus DPG promotes the release of O_2 from hemoglobin into the tissues. **3.** Yes—the trait compensates for the small P_{O_2} gradient between stagnant water and the blood of the carp. **4.** Since O_2 cannot compete as well as CO for binding sites in hemoglobin, O_2 transport decreases. As oxygen levels in the blood drop, tissues (particularly the brain) become deprived of oxygen, and suffocation occurs.

CHAPTER 45

Check Your Understanding (CYU)

CYU p. 891 (1) The resting potential would fall (be much less negative) because K^+ could no longer leak out of the cell. (2) The size of an action potential from a particular neuron does not vary, so it cannot contain information. Only the frequency of action potentials from the same neuron varies. **CYU p. 895** (1) Once the threshold level of depolarization is attained, the probability that the voltage-gated sodium channels will open approaches 100 percent. But below threshold, the massive opening of Na^+ channels does not occur. This is why the action potential is "all or none." (2) Na^+ would continue to move into the cell as voltage-gated potassium channels opened and potassium began to diffuse out of the cell. As a result, repolarization would take much longer. **CYU p. 899** If neurons made direct electrical connections, action potentials would simply travel from one neuron to the next. The presence of synapses allows information from many different neurons to affect the activity of a postsynaptic neuron, through the process of summation. **CYU p. 904** (1) [See Figure A45.1] (2) 1. Study individuals with brain damage and correlate the location of the defect with a deficit in mental or physical function. 2. Directly stimulate brain areas in conscious patients during brain surgery and record the response.

You Should Be Able To (YSBAT)

YSBAT p. 890 The Na^+/K^+-ATPase makes the inside of the membrane less positive (more negative) than the outside, and generates a concentration gradient favoring movement of K^+ out of the cell, through the K^+ leak channel. As K^+ leaves, the resting potential becomes even more negative. **YSBAT p. 891** [See Figure A45.2] **YSBAT p. 892** (1) Na^+ channels open when the cell membrane depolarizes, and as more channels open, more Na^+ flows into the cell, further depolarizing the cell, causing more voltage-gated Na^+ channels to open. (2) After opening, the voltage-gated Na^+ channels close and temporarily cannot reopen. Also, sodium reaches its equilibrium potential, so no net force is available to drive Na^+ movement. (3) K^+ channels open in response to membrane depolarization. **YSBAT p. 897** The inside of the cell becomes more positive (depolarized). This shifts the membrane potential closer to the threshold, making it easier for the postsynaptic cell to fire an action potential.

Caption Questions and Exercises

Figure 45.3 No—as K^+ leaves the cell along its concentration gradient, the interior of the cell becomes more negative. As a result, an electrical gradient favoring movement of K^+ into the cell begins to counteract the concentration gradient favoring movement of K^+ out of the cell. Eventually, the two opposing forces balance out, and there is no net movement of K^+. **Figure 45.10** (1) Get a solution taken from the synapse between the heart muscle and the vagus nerve *without* the nerve being stimulated. Expose a second heart to this solution. There should be no change in heart rate. **Figure 45.16** In the "rest and digest" mode, pupils take in less light stimulation, the heartbeat slows to conserve energy, and liver conserves glucose and promotes digestion by stimulating release of gallbladder products. In the "fight or flight" mode, the pupils open to take in more light, the heartbeat increases to support muscle activity, and the liver releases glucose and inhibits digestion. **Figure 45.18** Point to your forehead and top of your head for the frontal lobe, the top and top rear of your head for the parietal lobe, the back of your head for the occipital lobe, and the sides of your head (just above your ear openings) for the temporal lobes. **Figure 45.19** No—for example, part (b) indicates that the size of the brain area devoted to the trunk is no bigger than the size of the brain area devoted to the thumb.

FIGURE A45.1

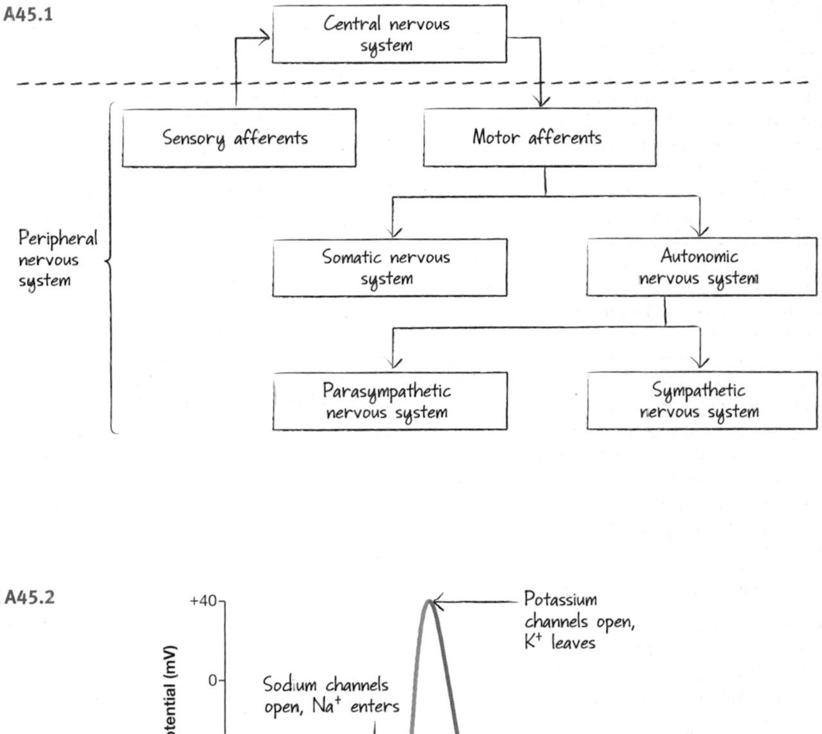

FIGURE A45.2

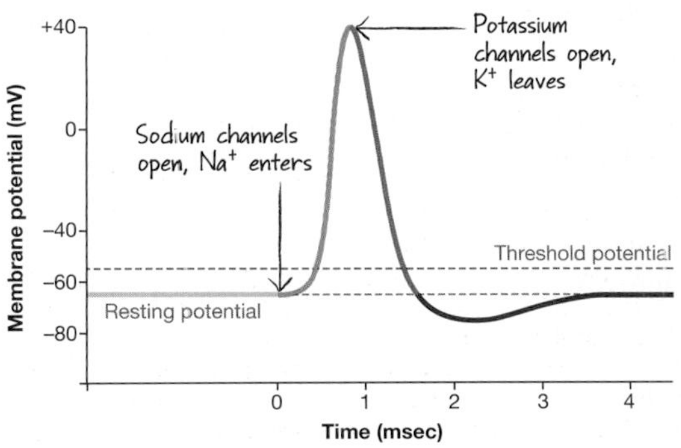

Summary of Key Concepts

KC 45.1 [See Figure A45.3] **KC 45.2** Because the behavior of voltage-gated Na^+ and K^+ channels does not change. Every time a membrane depolarizes past threshold and all of the voltage-gated Na^+ channels are open, the same amount of Na^+ will enter the cell and the cell will depolarize to the same extent. In addition, K^+ channels open on the same time course and allow the same amount of K^+ to leave the cell, meaning that it repolarizes to the same extent. **KC 45.3** Summation means that input from many neurons is involved in triggering a response from a postsynaptic cell. Thus, an action potential from a postsynaptic cell requires integration of input from many synapses. **KC 45.4** The animal's ability to learn—and perhaps remember—would be impaired or nonexistent.

Test Your Knowledge

1. b; **2.** d; **3.** c; **4.** a; **5.** c; **6.** a

Test Your Understanding

1. The Na^+/K^+-ATPase pumps 3 Na^+ out of the cell for every 2 K^+ it brings in. Since more positive charges leave the cell than enter it, there is a difference in charge on the two sides of the membrane and thus a voltage.
2. [See Figure A45.4] **3.** The ligand-gated channel opens or closes in response to binding by a small molecule (for example, a neurotransmitter); the voltage-gated channel opens or closes in response to changes in membrane potential. **4.** EPSPs and IPSPs are based on flows of ions, which change the membrane potential in the postsynaptic cell. If a flow of ions at one point depolarizes the cell but a nearby flow of ions hyperpolarizes the cell, then in combination the flows cancel each other out. But if two adjacent flows of ions depolarize the cell, then the total amount of ion movement—and thus the total change in membrane potential—sums. **5.** The somatic system responds to external stimuli and controls voluntary skeletal muscle activity, such as movement of arms and legs. The autonomic system responds to internal stimuli and controls internal involuntary activities, such as digestion, heart rate, and gland activities. **6.** Sympathetic nerves trigger responses in an array of organs and tissues that promote increased physical activity and heightened awareness. Parasympathetic nerves cause the opposite response in the same organs and tissues, promoting reduced physical and mental activity.

Applying Concepts to New Situations

1. The neurotransmitters will stay in the synaptic cleft longer so the amount of binding to ligand-gated channels will increase. Ion flows into the postsynaptic cell will increase dramatically, affecting its membrane potential and likelihood of firing action potentials. **2.** Diseased or damaged brains may not respond like healthy and undamaged brains. Also, the extent of lesions and the exact location of electrical stimulation may be difficult to determine, making correlations between the regions affected and the response imprecise. **3.** Because the current is reduced from normal levels, but not eliminated completely, there must be more than one type of potassium channel present. (The poison probably knocks out just one specific type of channel.) **4.** Record from the neuron while the individual is performing various tasks—feeding, moving, courting, etc.

CHAPTER 46

Check Your Understanding (CYU)

CYU p. 913 (1) The outer ear collects sound waves from the environment and directs them into the ear canal. The middle ear amplifies sound. The inner ear contains hair cells that transduce the information in sound waves to action potentials that are sent to the brain. (2) A punctured eardrum wouldn't vibrate correctly and would result in hearing loss at all frequencies in the affected ear. If the stereocilia are too short to come into contact with the tectorial membrane, vibration of the basilar membrane will not cause them to bend, and sound will not be detected. A loss in basilar membrane flexibility would result in the inability to hear lower-pitched sounds, such as human speech. **CYU p. 918** (1) Retinal acts like an on-off switch that indicates whether light has fallen on a rod cell. When retinal absorbs light, it changes shape. The shape change triggers events that result in a change in action potentials that signals that light has been absorbed. (2) A tear in the fovea would likely result in blurred vision in the affected eye. The mutation would produce blue-purple color blindness because this opsin responds to wavelengths in that region of the spectrum. A clouded lens would reduce the amount of light that reaches the retina, reducing visual sensitivity. **CYU p. 920** (1) Extremely hot food damages taste receptors, so proteins would no longer be able to respond to their chemical triggers. (2) Dogs have more than twice the number of odor receptors as humans. **CYU p. 926** (1) In a sarcomere, thick myosin filaments are sandwiched between thin actin filaments. When the heads on myosin contact actin and change conformation, they pull the actin filaments toward one another, shortening the whole sarcomere. (2) Increased ACh release would result in an increased rate of muscle cell contraction. Preventing conformational changes in troponin would prevent muscle contraction. Blocking the uptake of calcium ions into the sarcoplasmic reticulum would lead to sustained muscle contraction.

You Should Be Able To (YSBAT)

YSBAT p. 916 cGMP is a ligand that opens sodium channels and a second messenger in the sense that lack of it carries a signal from activated transducin. Transducin is like a G protein because it switches from "off" to "on" in response to a receptor protein, and activates a key pro-

FIGURE A45.3

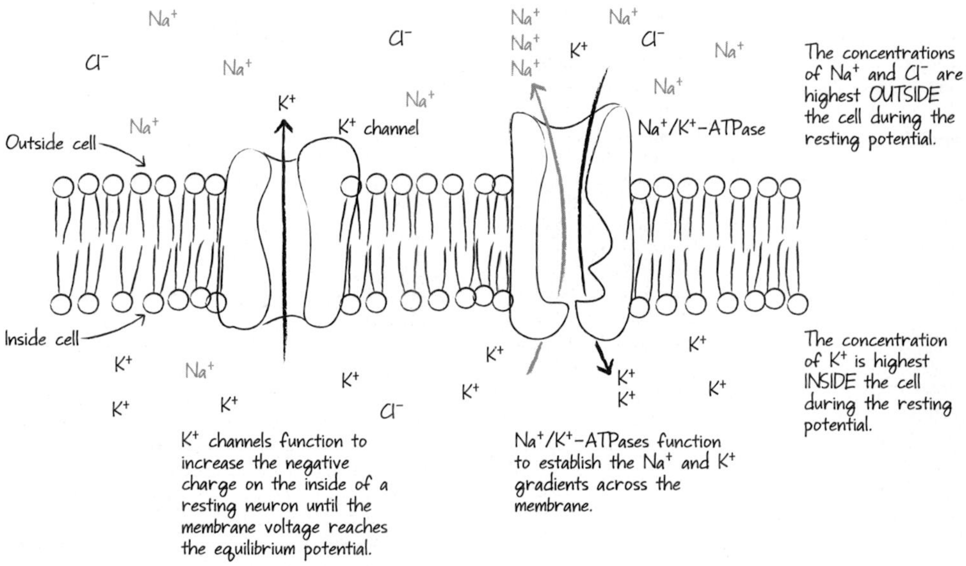

Outside cell
Inside cell
K+ channel
Na+/K+-ATPase
The concentrations of Na+ and Cl− are highest OUTSIDE the cell during the resting potential.
The concentration of K+ is highest INSIDE the cell during the resting potential.
K+ channels function to increase the negative charge on the inside of a resting neuron until the membrane voltage reaches the equilibrium potential.
Na+/K+-ATPases function to establish the Na+ and K+ gradients across the membrane.

FIGURE A45.4

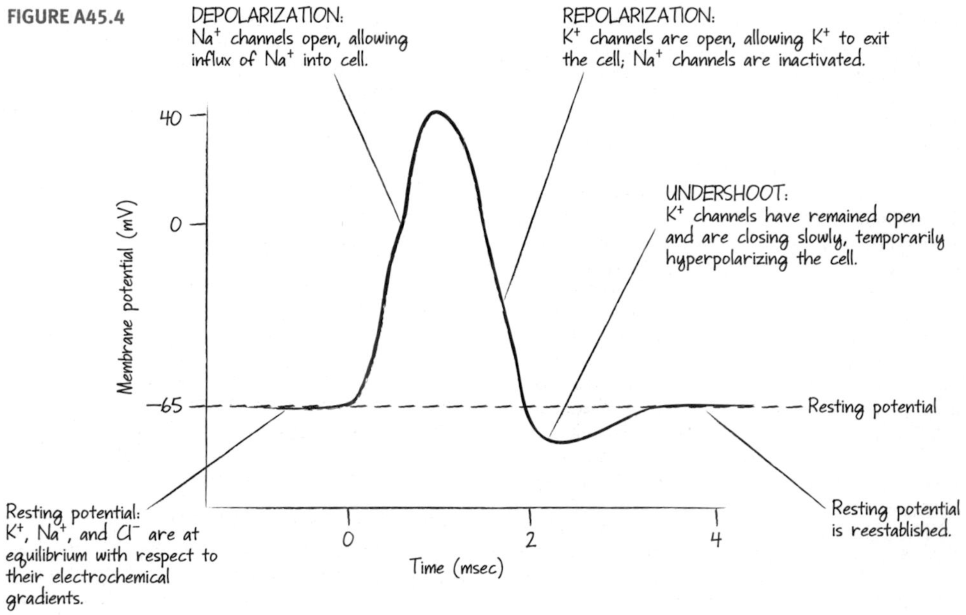

DEPOLARIZATION:
Na+ channels open, allowing influx of Na+ into cell.
REPOLARIZATION:
K+ channels are open, allowing K+ to exit the cell; Na+ channels are inactivated.
UNDERSHOOT:
K+ channels have remained open and are closing slowly, temporarily hyperpolarizing the cell.
Membrane potential (mV)
40
0
−65
Resting potential
Resting potential:
K+, Na+, and Cl− are at equilibrium with respect to their electrochemical gradients.
Resting potential is reestablished.
0
2
4
Time (msec)

tein in response. Closing Na^+ channels stops the entry of positive charges into the cell, making the inside of the cell more negative relative to the outside. **YSBAT p. 923** The trucks are the Z disks, the ropes are the thin filaments, and the burly weightlifters are the thick filaments.

Caption Questions and Exercises

Figure 46.9 The change in retinal's shape would induce a change in opsin's shape. A protein's function correlates with its shape, so opsin's function is likely to change as well. **Figure 46.10** This ion flow occurs when the photoreceptor cell is not receiving light—when the cell is in the dark. **Figure 46.12** The S, M, and L opsin proteins are each different in structure. These structural differences affect the ability of retinal to absorb specific frequencies of light and change shape. **Figure 46.18** The dark band includes thin filaments and a dense concentration of bulbous structures extending from the thick filament; the light band consists of just thin filaments.

Summary of Key Concepts

KC 46.1 The sensory neurons that used to serve that limb are cut, but may still fire action potentials that stimulate brain areas associated with awareness of pain in the missing limb. **KC 46.2** If dense, starch-filled amyloplasts exerted pressure on hair-cell-like receptors, the stereocilia would bend, causing a change in the membrane potential of the receptors and thus transducing the signal from gravity. **KC 46.3** Because different animal species have opsin molecules in their cone cells that respond to different wavelengths of light absorbed by retinal, they see different colors. **KC 46.4** Smells are part of the sensation of flavor, and people with nasal congestion cannot smell. In addition to smell being important, the small number of taste receptors can send many different frequencies of action potentials to the brain, and the taste receptors can be stimulated in many different combinations. **KC 46.5** Paralysis, because ACh has to bind to its receptors on the membrane of postsynaptic muscle fibers for action potentials to propagate in the muscle and cause contraction.

Test Your Knowledge

1. a; **2.** d; **3.** c; **4.** b; **5.** d; **6.** a

Test Your Understanding

1. The brain can distinguish sensory stimuli because the axons from different sensory neurons go to different and specific areas of the brain. For example, the axons of all the olfactory neurons that have the same receptor proteins go to the same location in the olfactory bulb, resulting in the sensation of smelling a particular odor. **2.** Many possibilities, including several blue opsins allowing coelacanths to distinguish the blue wavelengths present in their deep-sea habitat, infrasound hearing allowing elephants to hear over long distances, ultrasonic hearing allowing bats to hunt by echolocation and moths to avoid bat predators, red/yellow opsins allowing fruit-eating mammals to distinguish ripe from unripe fruit. **3.** It answered the question of why so many different smells can be detected and identified separately. Each type of odorant has its own receptor. **4.** Ion channels in hair cells are thought to open in response to physical distortion caused by the bending of stereocilia. Ion channels in taste receptors, in contrast, change shape and open when certain bitter-tasting molecules bind to them. **5.** Dalton's hypothesis was reasonable because blue fluid absorbs red light and would prevent it from reaching the retina, and could be tested by dissecting his eyes. Although the hypothesis was rejected, it inspired a rigorous test and required researchers to think of alternative explanations. **6.** The key observation was that the banding pattern of sarcomeres changed during contraction. Even though the entire unit became shorter, only some portions moved relative to each other. This observation suggested that some portions of the structure slid past other portions. Muscle fibers have to have many mitochondria because large amounts of ATP are needed to power myosin heads to move along actin filaments; large amounts of calcium stored in smooth ER are needed to initiate contraction by binding to troponin.

Applying Concepts to New Situations

1. If ACh receptors are covered, they cannot bind to ACh and become activated. Muscles will have trouble contracting. **2.** Block all but a few hundred ommatidia in the center of each compound eye of many dragonflies with an opaque material, and observe any differences that might occur in their ability to detect and pursue prey compared to individuals whose ommatidia had been covered by a transparent material. **3.** Ground-dwelling birds have large amounts of "white meat" in their breasts, used for powering short, rapid flights, but "dark meat" in their legs, used for endurance running. Birds that rely on long, sustained flights need "dark meat" in the breast muscles that power flight. **4.** When a person focuses directly on an object, the image lands on the fovea, which contains relatively few rods. But if the image lands beside the fovea, there are many more rod cells. Rod cells are specialized for detecting dim light.

CHAPTER 47

Check Your Understanding (CYU)

CYU p. 940 (1) By reducing immune system function, promoting release of fatty acids from storage cells and use of amino acids from muscles for energy production, and preventing release of glucose in response to signals from insulin, cortisol conserves glucose supplies for use by the brain. (2) Increased heart rate, glucose and fatty-acid release, and blood pressure all support rapid action in response to danger. If an animal were not able to respond quickly and at the maximum level, it might not escape predators or successfully challenge rivals. **CYU p. 943** (1) ACTH triggers the release of cortisol, but cortisol inhibits ACTH release by blocking the release of CRH from the hypothalamus and suppressing ACTH production. High levels of cortisol tend to lower cortisol levels in the future. (2) Processing centers in the brain are responsible for synthesizing a wide array of sensory input. To start a response to this sensory input, they stimulate neurosecretory cells in the hypothalamus. Secretions from these cells travel to the anterior pituitary, where they trigger the production and release of hormones that control hormones that stimulate the appropriate activity. **CYU p. 947** (1) The steroid-hormone receptor complex would probably fail to bind to the hormone-response element. If so, then gene expression would not change—the arrival of the hormone would have little or no effect on the target cell. (2) The mode of action would be similar—hormone binding would have to trigger a signal transduction event that resulted in production of a second messenger or a phosphorylation cascade.

You Should Be Able To (YSBAT)

YSBAT p. 932 Room temperature is analogous to the sensory input; the thermostat, to the CNS; the thermostat signal, to the cell-to-cell signal; and the furnace, to the effector. If feedback inhibition fails, hormone production will not stop and the response will continue indefinitely. **YSBAT p. 936** The cells could have (1) different receptors, (2) different signal transduction systems, or (3) different response systems (genes or proteins that can be activated).

Caption Questions and Exercises

Figure 47.4 Injecting the dog with a liquid extract from a different part of the body, or with a solution that is similar in chemical composition to the extract from the pancreas (e.g., similar pH or ions present) but lacking molecules produced by the pancreas. **Figure 47.7** The saline injection controlled for any stress induced by the injection procedure and for introducing additional fluids. **Figure 47.9** Same—a defect in both signal and receptor has the same effect—no response—as a defective signal alone or a defective receptor alone. **Figure 47.10** Inject ACTH and monitor cortisol levels. If cortisol does not increase, adrenal failure is likely. **Figure 47.14** The arrow that joins the signal and the receptor. The subsequent events also fail to occur.

Summary of Key Concepts

KC 47.1 The production and/or release of hormones are controlled, directly or indirectly, by electrical or chemical signals from the nervous system. A good example is the hypothalamic-pituitary axis, where neuroendocrine signals regulate endocrine signals. **KC 47.2** Electrical signals are short lived and localized. All of the responses listed in the question are long-term changes in the organism that require responses by tissues and organs throughout the body. **KC 47.3** The hypothalamus would continue to release CRH, stimulating the pituitary to release ACTH, which in turn would stimulate continued release of cortisol from the adrenal cortex. Sustained, high circulating levels of cortisol have negative effects. **47.4** A cell can respond to more than one hormone at a time if the receptors for the hormones are different and the cell produces receptors for each hormone.

Test Your Knowledge

1. c; **2.** a; **3.** d; **4.** a; **5.** c; **6.** a

Test Your Understanding

1. The posterior pituitary is an extension of the hypothalamus; the anterior pituitary is independent—it communicates with the hypothalamus via chemical signals in blood vessels. The posterior pituitary is a storage area for hypothalamic hormones; the anterior pituitary synthesizes and releases an array of hormones in response to releasing hormones from the hypothalamus.
2. Steroid hormones act directly, and alter gene expression. They bind to receptors inside the cell, forming a complex that binds to DNA and activates transcription. Nonsteroid hormones act indirectly, and activate proteins. They bind to receptors on the cell surface and trigger production of a second messenger or a phosphorylation cascade, ending in activation of proteins already present in the cell. **3.** The pituitary produces hormones that regulate hormone production in other glands. **4.** This is one of the reasons that the same hormone can trigger different effects in different tissues. For example, epinephrine binds to four different types of receptors in different tissues—eliciting a different response from each. **5.** The hormonal signal is amplified through production of many copies of a second messenger, the activation of many proteins through a phosphorylation cascade, or the transcription and translation of many mRNAs. **6.** When fat stores are high, leptin release acts on the brain to reduce feeding. In this way, it helps maintain weight and energy balance—preventing obesity in mice.

Applying Concepts to New Situations

1. There are two basic strategies: (1) remove the structure from some individuals, and compare their behavior and condition to untreated individuals in the same environment, or (2) make a liquid extract from the structure, inject it into some individuals, and compare their behavior and condition to individuals in the same environment who were injected with water or a saline solution.
2. If sustained high cortisone levels suppress wound healing, individuals receiving large doses might become more susceptible to disease or heal slowly. **3.** The patient will be unable to produce oxytocin and ADH. Although the patient could manage without oxytocin, artificial

administration of ADH would be required to maintain water balance. You would need to monitor the individual's total water intake and the water content of the urine, along with symptoms of dehydration or water retention (such as swelling or high blood pressure) and increase or decrease ADH dosage accordingly. **4.** Label the hormone and introduce it into an experimental animal. Analyze fat cells by isolating cell components, testing for the presence of the label, and eventually isolating just the labeled hormone-receptor complex. To test the second messenger hypothesis, treat fat cells growing in culture with the hormone and monitor levels of cAMP, Ca^{2+}, and other known second messengers.

CHAPTER 48

Check Your Understanding (CYU)

CYU p. 957 (1) Oviparity usually requires less energy input from the mother after egg laying, and mothers do not have to carry eggs around as long—meaning that they can lay more eggs and be more mobile. Mothers have to produce all the nutrition required by the embryo prior to egg laying, however, and eggs may not be well protected after laying. Viviparity usually increases the likelihood that the developing offspring will survive until birth, but limits the number of young that can be produced to the space available in the mother's reproductive tract. If viviparous young can be nourished longer than oviparous young, then they may be larger and more capable of fending for themselves. (2) Divide a population of sperm into two groups. Subject one group of sperm to spermkillerene at concentrations observed in the female reproductive tract (the experimental group) but not the other group (the control). Document the number of sperm that are still alive over time in both groups. **CYU p. 966** (1) FSH triggers maturation of an ovarian follicle. It rises at the end of a cycle because it is no longer inhibited by progesterone, which stops being produced at high levels when the corpus luteum degenerates. (2) The drug would keep FSH levels low, meaning that follicles would not mature and would not begin producing estradiol and progesterone. The uterine lining would not thicken.

You Should Be Able To (YSBAT)

YSBAT p. 952 Asexual reproduction would be expected in environments where conditions change little over time. **YSBAT p. 965** (1) There should be an LH spike early in the cycle, followed by early ovulation. (2) LH levels should remain low—no mid-cycle spike and no ovulation. **YSBAT p. 968** (1) The uterine lining may not be maintained adequately—if it degenerates, a miscarriage is likely. (2) After the first trimester, the placenta produces enough progesterone to maintain the pregnancy, so supplementation is no longer needed.

Caption Questions and Exercises

Figure 48.3 Isolate and identify the molecules found in crowded water. Test each molecule by adding it, at the same concentration found in "crowded water," to clean water occupied by a single *Daphnia* and recording whether the female produces a male-containing brood. Repeat with many test females, and for each molecule identified. As a control, record the number of male-containing broods produced in clean water. **Figure 48.5** No—the data are consistent with the displacement hypothesis, but there is no direct evidence for it. There are other plausible explanations for the data. **Figure 48.11** (Note: There is more than one possible answer—this is an example.) Having two gonads is an "insurance policy" against loss or damage. To test this idea, surgically remove one gonad from a large number of male and female rats. Do a similar operation on a large number of similar male and female rats but do not remove either gonad. Once the animals have recovered, place them in a barn or other "natural" setting and let them breed. Compare reproductive success of individuals with paired vs. unpaired gonads. **Figure 48.12** It is similar. In both cases, the hypothalamus produces a releasing factor (GnRH or CRH) that acts on the pituitary. The releasing factor stimulates release of regulatory hormones from the anterior pituitary (LH and FSH or ACTH). These hormones travel via the bloodstream and act on the gonads or adrenals to induce the release of hormones from these glands. In both cases, the hormones are involved in negative feedback control of the regulatory hormones from the pituitary. **Figure 48.20** The birth will be more difficult, as limbs or other body parts can get caught. **Figure 48.21** A mother's chance of dying in childbirth in 1760 was a little over 1000 in 100,000 live births, or close to 1.0 percent. If she gave birth 10 times, she would have a 10 percent chance of dying in childbirth.

Summary of Key Concepts

KC 48.1 Parental care demands resources (time, nutrients) that cannot be used to produce more eggs. Sperm competition is not likely when external fertilization occurs, and most fishes use external fertilization. **KC 48.2** A surgeon can ligate (tie off) the fallopian tubes in a woman to stop the delivery of the egg to the uterus and can cut the vas deferens in a man to stop the delivery of sperm into the semen. **KC 48.3** Progesterone suppresses the release of GnRH and FSH through negative feedback. When progesterone levels are high, an LH spike does not occur, ovulation is prevented, and no egg is available for fertilization.

Test Your Knowledge

1. a; **2.** b; **3.** a; **4.** c; **5.** d; **6.** d

Test Your Understanding

1. Every offspring that is produced sexually is genetically unique; every offspring that is produced asexually is genetically identical to its parent. **2.** *Daphnia* females only produced males when they were exposed to short day lengths, *and* water from crowded populations, *and* low food levels. These conditions are likely to occur in the fall. Sexual reproduction could be adaptive in fall if genetically variable offspring are better able to thrive in conditions that occur the following spring. **3.** Spermatogenesis generates four haploid sperm cells from each primary spermatocyte; oogenesis produces only one haploid egg cell from each primary oocyte. Egg cells are much larger than sperm cells because they contain more cytoplasm. In males, the second meiotic division occurs right after the first meiotic division, but in females it is delayed until fertilization. **4.** LH triggers release of estradiol, but at low levels estradiol inhibits further release of LH. LH and FSH trigger release of progesterone, but progesterone inhibits further release of LH and FSH. High levels of estrogen trigger release of more LH. The follicle can produce high levels of estrogen only if it has grown and matured—meaning that it is ready for ovulation to occur. **5.** Ethanol can cross the placenta, enter the fetus, and adversely affect development. Ethanol abuse by pregnant women is correlated with fetal alcohol syndrome (FAS), a condition caused by loss of neurons in developing embryos. **6.** Females should be choosier about their second mate because the sperm of the second male has an advantage in sperm competition.

Applying Concepts to New Situations

1. If all sheep were genetically identical, it is less likely that any individuals could survive a major adverse event (e.g., a disease outbreak or other environmental challenge) than a population composed of genetically diverse individuals, which are likely to vary in their ability to cope with the new conditions. **2.** Compare the number of gametes produced by species that have similar body sizes, but undergo external versus internal fertilization. By comparing similar-sized animals, you would be able to rule out the possibility that differences in gamete production are due to differences in body size. **3.** Oviparous populations should produce larger eggs than viviparous populations. Because oviparous species deposit eggs in the environment, the eggs must contain all the nutrients and water required for the entire period of embryonic development. But because embryos in viviparous species receive nourishment directly from the mother, it is likely that their eggs are smaller. **4.** The shell membrane is a vestigial trait (see Chapter 24)—specifically, an "evolutionary holdover" from an ancestor of today's marsupials that laid eggs.

CHAPTER 49

Check Your Understanding (CYU)

CYU p. 977 Applying direct pressure constricts blood vessels, mimicking the effect of histamine released from mast cells. Applying bandages impregnated with platelet-recruiting compounds mimics the effect of chemokines released from injured tissues and macrophages, which attract circulating leukocytes and platelets to facilitate blood clotting and wound repair. **CYU p. 984** (1) They are identical, except that a B-cell receptor has a transmembrane domain that allows it to be located in the B-cell plasma membrane. (2) Lymphocytes that respond to self molecules would circulate in the blood, and trigger immune system responses to self cells—leading to their destruction. The mutation would probably be fatal. **CYU p. 987** Individuals who are heterozygous for MHC genes have greater variability in their MHC proteins than do individuals who are homozygous for these genes. The increased variability in MHC proteins allows a greater variety of antigens to be presented to T cells, which thus would be able to recognize, attack, and eliminate a wider range of pathogens compared with T cells in homozygotes. As a result, heterozygous individuals would likely suffer fewer infections than homozygous individuals. **CYU p. 990** A vaccination is a "false alarm" that triggers a primary response from the immune system. The memory cells that result produce the secondary response when and if an authentic infection occurs.

You Should Be Able To (YSBAT)

YSBAT p. 982 Have the lottery numbers contain a large number of digits—say 5, or even 10. To generate a number, pick a number from 0–9 for each of the 5 or 10 places. If there are 5 places, there would be $10 \times 10 \times 10 \times 10 \times 10 = 10{,}000$ different numbers. If there were 10 places, there would be 10^{10} (10 billion) different numbers. **YSBAT p. 985** Rapidly dividing B and T cells may be destroyed by chemotherapy, along with cancer cells. If clonal expansion cannot occur, antibody production and other aspects of the immune response will be dampened and possibly ineffective. **YSBAT p. 986** (1) All nucleated cells in the body express Class I MHC proteins, whereas only B cells and some other leukocytes express Class II MHC proteins. (2) Cytotoxic T cells interact only with cells that display antigens presented on Class I MHC proteins. Helper T cells interact only with antigens presented on Class II MHC proteins. (3) An MHC-peptide displayed on an infected cell means "kill me"; displayed on a dendritic cell it means "activate killer T cells"; displayed on a B cell it means "helper T cells should activate me."

Caption Questions and Exercises

Figure 49.1 Pathogens stick to mucus; contaminated mucus is then swept to a disposal point by cilia. **Figure 49.3** Blood vessels near the wound would be constricted, which restricts blood loss. Blood vessels slightly farther from the wound would be dilated, which increases blood flow and the delivery rate of platelets, neutrophils, and macrophages to the surrounding area. **Figure 49.18** [See Figure A49.1]

Summary of Key Concepts

KC 49.1 Leukocytes could not move to the damaged tissue efficiently to clean up debris, remove pathogens, and begin the repair process. **KC 49.2** The drug would inhibit the activation of T and B cells by that antigen, because it would prevent an interaction between the epitope and the appropriate TCRs and BCRs. The cell-mediated response would also fail if the drug blocked the interaction between cytotoxic T cells and infected cells displaying the antigen. **KC 49.3** Cowpox is similar enough to smallpox to be antigenic, but cannot grow effectively enough in humans to cause disease.

Test Your Knowledge

1. c; **2.** a; **3.** b; **4.** d; **5.** b; **6.** a

Test Your Understanding

1. Rubor (reddening) is due to increased blood flow to the infected or injured area. Mast cells trigger dilation of vessels by releasing histamine and other signaling molecules. Calor (heat) is the result of fever, which is activated by the release of cytokines from macrophages that are in the area of infection. Dolor (pain) occurs when tissue damage stimulates pain receptors. Tumor (swelling) occurs due to dilation of blood vessels triggered by histamines and other compounds released by macrophages. **2.** Both the BCR and TCR interact with antigens via binding sites located in the variable regions of their polypeptide chains. A TCR is like one "arm" of a BCR. [See Figure A49.2] **3.** Vaccines have to contain an antigen that can stimulate an appropriate primary immune response. Vaccines have not worked for HIV because the antigens on this virus are constantly changed through mutation, rendering it unrecognizable to memory cells generated following vaccination. **4.** "Clonal" refers to the cloning—producing many exact copies—of cells that are "selected" by the binding of their receptor to an antigen. **5.** Pattern-recognition receptors on leukocytes bind to surface molecules that are present on many pathogens. BCRs and TCRs bind to particular epitopes of antigens. Pattern-recognition-receptor binding is nonspecific; BCR and TCR binding is specific. **6.** By mixing and matching different combinations of gene segments from the variable and joining regions of the light-chain immunoglobulin gene, along with diversity regions of the heavy-chain gene, lymphocytes end up with a unique sequence for both chains in the BCR and TCR.

Applying Concepts to New Situations

1. Irrigating the wound removes dirt and debris that may contain pathogens. Scrubbing with soapy water removes more pathogens and may also kill them (soap may disrupt plasma membranes enough to be fatal). Antibiotics are toxic to pathogenic bacteria. **2.** If the antibody had a fluorescent molecule or other type of label attached, you could treat various types of cells with it, examine them under the microscope, and determine the location of the antibody-pump complexes. **3.** Natural selection favors individuals that can create a large array of antibodies, because the high mutation rates and rapid evolution observed in pathogens means that they will constantly present the immune system with new antigens. If these antigens were not recognized by the immune system, the pathogens would multiply freely and kill the individual. **4.** Tissue from the patient's own body is marked with the major histocompatibility (MHC) proteins that the patient's immune system recognizes as self. This results in the preservation, rather than the destruction, of the transplanted tissue. Tissue from a different person is marked with MHC proteins that are unique to that individual. The body of the transplant recipient will recognize the MHC proteins (and other molecules) of the donor as foreign, resulting in a full immune response and rejection of the grafted tissue.

CHAPTER 50

Check Your Understanding (CYU)

CYU p. 1001 (1) They are shallow, so a relatively large proportion of the water receives enough sunlight to support photosynthesis. In addition, nutrients are available from rivers that contribute nutrients from inland and upwellings that bring nutrients up from the ocean bottom. (2) It sinks, and the lower-density water below it rises, bringing nutrients to the surface. **CYU p. 1008** Tropical dry forests are probably less productive than tropical wet forests because they have less water available to support photosynthesis during some periods of the year. **CYU p. 1012** Your answer will depend on where you live. For example, if global warming continues at the current projected rates, several effects can be expected. If you live in a coastal area, you can expect water levels to rise. In general, plant communities will change because average temperature and moisture and variation in temperature and moisture will change.

You Should Be Able To (YSBAT)

YSBAT p. 997 The littoral zone, because light is abundant and nutrients are available from the substrate. **YSBAT p. 998** Bogs are nitrogen-poor, so plants that are able to capture and digest insects have a large advantage. This advantage does not exist in marshes and swamps, where nutrients are more readily available. **YSBAT p. 999** Cold water contains more oxygen than warm water, so it can support much more cellular respiration in fish. **YSBAT p. 1000** Species that live in estuaries must be able to tolerate variable salinity; marsh species do not. The abiotic environment (salt concentration) is so different that few species grow well in both habitats. **YSBAT p. 1001** No—the aphotic zone is lightless, so natural selection favors individuals that do not invest energy in developing and maintaining eyes. **YSBAT p. 1003** Vines and epiphytes increase productivity because they are photosynthetic organisms that fill space between small trees and large trees—they capture light and use nutrients that might not be used in a forest that lacked vines and epiphytes. **YSBAT p. 1004** Most leaves have a large surface area to capture light. Light is abundant in deserts, however, and a leaf with a large surface area would be susceptible to high water loss and/or overheating. **YSBAT p. 1005** Vegetation is continuous in a grassland, so a fire carries better. In a desert, the fire is likely to run out of fuel. **YSBAT p. 1006** There is a continuous grass cover and scattered trees. (A biome like this is called a savannah.) **YSBAT p. 1007** Boreal forests should move north. **YSBAT p. 1008** High elevations present an abiotic environment (precipitation and temperature) that is similar to the conditions in arctic tundras. As a result, the plant species present will have similar adaptations to cope with these physical conditions. **YSBAT p. 1010** [See Figure A50.1]

Caption Questions and Exercises

Figure 50.4 Freshwater, because light penetrates much deeper. **Figure 50.23** Dry, because dry air is descending here. This is consistent with the very low rainfall amounts in Barrow, Alaska, shown in Figure 50.21. **Figure 50.27** Visible wavelengths enter the chamber

FIGURE A49.1

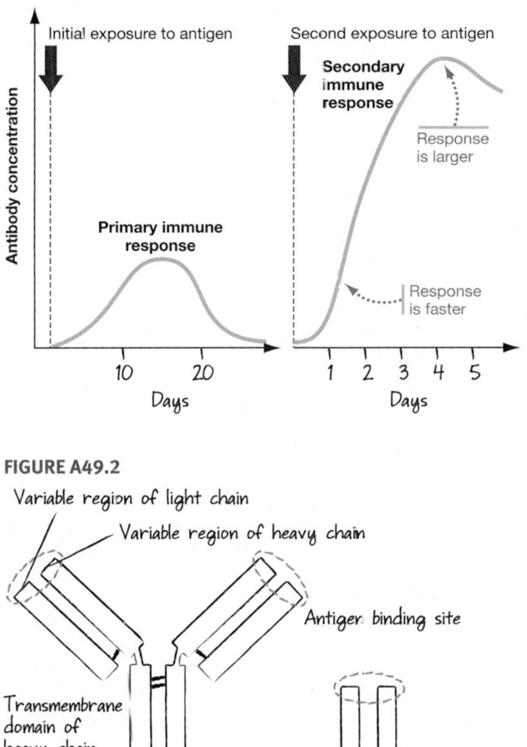

FIGURE A49.2

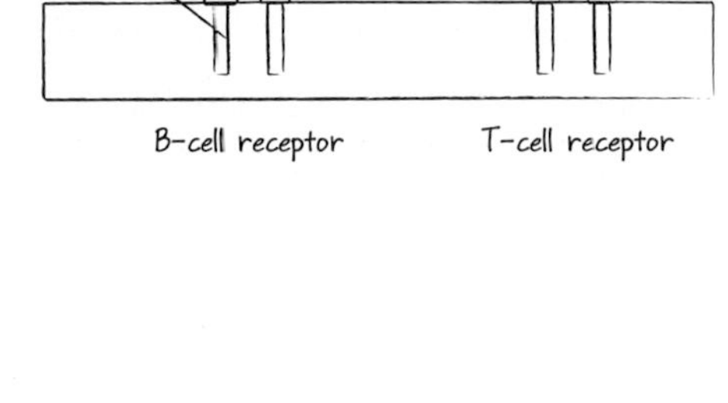

FIGURE A50.1

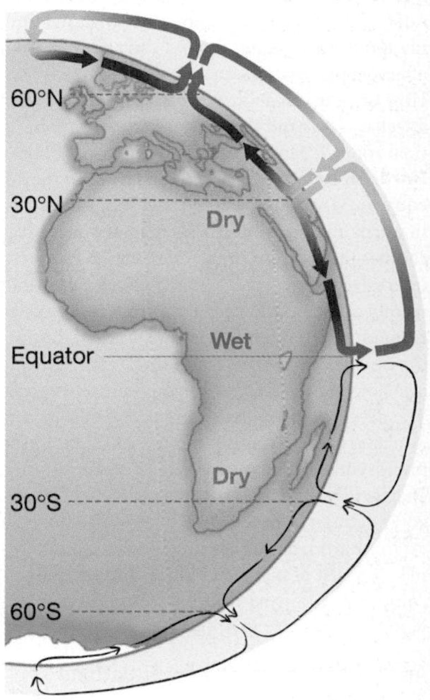

from the top and through the glass, warming the plants and ground inside the chamber. Some of this heat energy is retained within the mini-greenhouse. **Figure 50.28** They address the question posed because they capture important aspects of a plant community. NPP indicates how much biomass is available for other organisms to use; species diversity indicates how many different species are present. **Figure 50.34** (Several possible answers; one possibility is provided.) If water is short, there is not enough available to support a continuous cover of photosynthetic tissue aboveground. Roots of bunchgrasses can spread out and harvest water below spaces that are "bare" aboveground.

Summary of Key Concepts

KC 50.1 Populations are made up of individual organisms; communities are made up of populations of different species; ecosystems are made up of groups of communities (along with the abiotic environment). **KC 50.2** Increased evaporation from the oceans should increase precipitation on land—at least in some places. If precipitation does not increase, then increased transpiration rates will put plants under water stress and reduce productivity. **KC 50.3** All latitudes would get equal amounts of sunlight all year round. There would be no seasons and no changes in climate with latitude. **KC 50.4** It would increase, because cattle would no longer succumb to the disease carried by tsetse flies. (In Africa, cattle are limited by a biotic condition, not abiotic conditions.)

Test Your Knowledge

1. b; **2.** a; **3.** d; **4.** b; **5.** a; **6.** c

Test Your Understanding

1. Organismal, population, community, and ecosystem. Examples of possible questions: *Organismal:* How do humans cope with extremely hot (or cold) weather conditions? *Population:* How large will the total human population be in 50 years? *Community:* How are humans affecting prey species such as cod and tuna? *Ecosystem:* How will human-induced changes in global temperature affect sea level and thus organisms that live in the intertidal zone? **2.** More—in June the Sun's rays strike the Northern Hemisphere more directly than they do the equator. **3.** A Hadley cell just south of the equator results in warm, dry air falling to Earth at about 30 degrees south—the location of the Outback. **4.** Productivity in intertidal and neritic zones is high because sunlight is readily available and because nutrients are available from estuaries and deep-ocean currents. Productivity in the oceanic zone is extremely low—even though light is available at the surface—because nutrients are scarce. The deepest part of the oceanic zone may have nutrients available from the substrate but lacks light to support photosynthesis. **5.** As you increase in elevation, biomes change in a similar way as increasing in latitude (e.g., you might go from temperate forest to a boreal-type forest to tundra). **6.** The combination of low oxygen, low pH, and lack of nitrogen reduces productivity in bogs compared with what would be found in marshes or lakes with similar solar radiation.

Applying Concepts to New Situations

1. Yes—as Mars orbits the Sun, the part of the planet that is tilted toward the Sun is warmest because it receives the most direct sunlight. But as the planet orbits, that portion of the planet begins to tilt away from the Sun. Its temperature drops as a result, producing something like Earth's cycle of winter and summer. **2.** Mountaintops are cold because as air rises, it expands; this expansion cools the air, making the mountaintop cold. Also, there is less land surface nearby to absorb solar radiation and heat up. **3.** The hypothesis is that winds are usually from the northeast, and that higher elevations in the middle of the islands cause a rain shadow in the southwest corner. **4.** Edinburgh is surrounded by water. Because water has a high specific heat, it keeps temperatures moderate. Moscow, in contrast, is surrounded by a large landmass that does not have the same moderating effects on temperature as the ocean does.

CHAPTER 51

Check Your Understanding (CYU)

CYU p. 1022 If a sitter-like allele in bee-eaters was favored when population density is low, then it should be at high frequency in small colonies. If a rover-like allele was favored when population density is high, then it should be at high frequency in large colonies. **CYU p. 1025** (1) In the experiment, females came into breeding condition slowly if they were exposed only to springlike light conditions, but much more quickly if they were exposed to springlike light conditions *and* displaying males. Both are required for maximum effect. (2) Many possible answers, but based on the experiments with *Anolis*, it would be legitimate to hypothesize that they have to be exposed to springlike light and temperature conditions as well as courtship displays from males. **CYU p. 1031** (1) Auditory communication allows individuals to communicate over long distances but can be heard by predators. Olfactory communication is effective in the dark and scents can continue to carry information long after the signaler has left, but scents do not carry long distances. Visual communication is effective during the day but can be seen by predators. (2) The ability to detect deceit protects the individual from fitness costs (e.g., being eaten, having a mate's eggs fertilized by someone else); avoiding or punishing "liars" should also lower the frequency of deceit (because it becomes less successful). Alleles associated with detecting and avoiding or punishing "liars" should be favored by natural selection and increase in frequency. **CYU p. 1034** (1) Altruism is likely when B is high, as when a predator is about to pounce; r is high, as when a close relative is threatened; C is low, as when the caller is far from the predator. Altruism is unlikely under the opposite conditions. (2) For reciprocal altruism to work, individuals have to interact repeatedly and remember who has helped whom in the past.

You Should Be Able To (YSBAT)

YSBAT p. 1033 Humans often live near kin, are good at recognizing kin, and in many cases can give kin resources or protection that increase fitness. It is not known whether kin selection occurs in plants. In many cases, limited seed dispersal means that close relatives live together—making kin selection more likely. But it is not known whether plants can recognize kin and confer benefits preferentially to kin.

Caption Questions and Exercises

Figure 51.1 Several legitimate hypotheses are possible. Examples at the proximate level: The increased size of the antennae provide space for more sensory neurons; or large antennae allow individuals to sense a larger area. Examples at the ultimate level: large antennae help individuals navigate better and thus escape predation; or large antennae allow individuals to forage more successfully in the dark. **Figure 51.5** Bring females into the lab and give them the same food and housing conditions as the treatment groups, but expose them to artificial lighting that simulates the short daylight conditions of winter. **Figure 51.7** Because individuals were chosen for the different treatments at random, the investigator could claim that on average, the only thing that differed among individuals in the different groups was average tail length. **Figure 51.10** [See Figure A51.1] **Figure 51.14** Between first cousins, $r = \frac{1}{2} \times \frac{1}{2} \times \frac{1}{2} = \frac{1}{8}$. **Figure 51.15** The control would be to drag an object of similar size through the colony, at similar times of day. This would test the hypothesis that prairie dogs are reacting to the presence of the experimenter and the disturbance—not a predator.

Summary of Key Concepts

KC 51.1 At the proximate level, different alleles of the *for* gene affect the probability that an individual will move or stay after feeding. At the ultimate level, roving is favored when population density is high; sitting is favored when population density is low. **KC 51.2** Individuals have the capacity to act altruistically or non-altruistically toward kin. Hamilton's rule specifies the conditions—in terms of the costs and benefits of the act and the degree of relationship between the participants—under which altruism is favored. **KC 51.3** Roving is expensive energetically and could make individuals more susceptible to predation, but could allow individuals to find unexploited food sources. Sitting is inexpensive energetically, but may restrict individuals to areas where food supplies have been depleted. **KC 51.4** If only males that are well fed and free of parasites and disease are able to produce a brightly colored dewlap and do the bobbing display vigorously, then these traits would be a reliable indication of their condition. **KC 51.5** If migration does not occur when extra food is supplied, it suggests that the "decision" to migrate depends on conditions. **KC 51.6** Deceitful communication should decrease. If large males are gone, there are few nests available and more female-mimic males would compete at them. Most female-mimic males would have no (or fewer) eggs to fertilize and no male to take care of the eggs they did fertilize. There would be less natural selection pressure favoring deceitful communication. **KC 51.7** (1) Long-lived, so that extensive interactions with kin and nonrelatives are possible; (2) good memory, to record reciprocal interactions; (3) kin nearby, to make inclusive fitness gains possible; (4) ability to share fitness benefits (e.g., food) between individuals.

Test Your Knowledge

1. c; **2.** d; **3.** a; **4.** c; **5.** d; **6.** c

Test Your Understanding

1. At the ultimate level, homing to a safe den increases survival. The proximate cause for this behavior is still not fully known, but appears to involve the ability to sense changes in Earth's magnetic field. **2.** When optimal foraging occurs, an individual maximizes the benefit of foraging in terms of energy gain and minimizes the cost of foraging in terms of time, energy expended, and risk of predation. **3.** Only males that are in good shape (well-nourished and free of disease) have the resources required to produce a trait like a long tail. **4.** Longer day lengths indicate the arrival of spring, when renewed plant growth and insect activity make more food available. Courtship displays from males indicate the presence of males that can fertilize eggs. **5.** A map provides information about the spatial relationships of places on

FIGURE A51.1

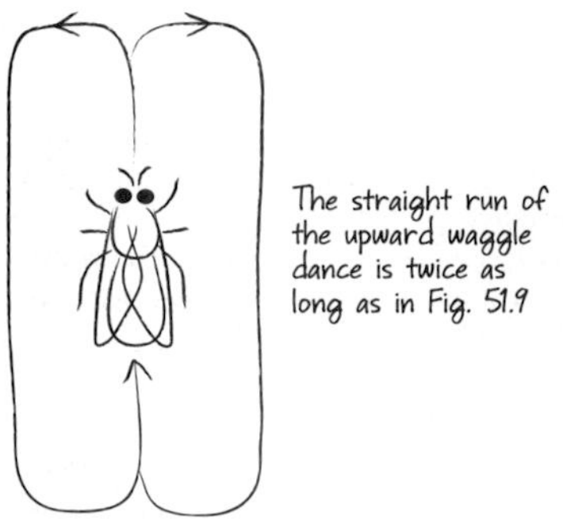

The straight run of the upward waggle dance is twice as long as in Fig. 51.9

a landscape; a compass allows you to orient the map—so that it aligns with the actual landscape. Animals have been shown to use the Sun's position, the position of the North star, and information from Earth's magnetic field as a compass (a way to find North). **6.** Kin selection can only act on kin; reciprocal altruism can occur between nonrelatives. Kin selection should be more common, because it potentially can occur any time that kin interact. Reciprocal altruism should be relatively rare because it is only possible when unrelated individuals interact repeatedly, have resources to share, and can remember the outcomes of past interactions.

Applying Concepts to New Situations

1. One possibility would be to set up a testing arena with food sources at various distances from a food source where adult flies are originally released. Test adults with a rover versus sitter genotype individually, and record how far each flies over the course of a set time interval. If the rover and sitter alleles affect foraging movements in adults, then rovers should be more likely to move and move farther than sitters. **2.** Simulate the onset of rains after a dry period, by using a hose to sprinkle water on the habitat where lizards are being maintained. **3.** Individuals have an r of 1/2 with full siblings and an r of 1/8 with first cousins. If the biologist will lose his life, he needs to save two siblings to keep the "lost copy" of his altruism alleles in the population but eight first cousins to maintain a copy. **4.** People would be expected to donate blood under two conditions: (1) They either received blood before or expect to have a transfusion in the future, and/or (2) they receive some other benefit in return, such as a good reputation among people who might be able to help them in other ways.

CHAPTER 52

Check Your Understanding (CYU)

CYU p. 1046 One possibility is competition for food. To test this idea, compare carrying capacity in identical vials that differ only in the amount of food added. (You could also test a hypothesis of oxygen limitation by adding more oxygen to one set of vials, or test the hypothesis of space limitation by doubling the volume but keeping the amount of food and oxygen the same.) **CYU p. 1053** (1) In developed nations, there are roughly equal numbers of individuals in each age class because the fertility rate has been constant and survivorship high for many years. In developing countries there are many more children and young people than older people, because the fertility rate and survivorship have been increasing. (2) Because survivorship is high in most human populations, changes in overall growth rate depend almost entirely on fertility rates.

You Should Be Able To (YSBAT)

YSBAT p. 1041 (1) Compared to northern populations, fecundity should be high and survivorship low. (2) Fecundity can be much higher if females lay eggs instead of retaining them in their bodies and giving birth to live young. **YSBAT p. 1055** The population appears to be increasing over time: from the original 1000 females, there are 2308 females in the third year.

Caption Questions and Exercises

Figure 52.2 [See Figure A52.1] **Figure 52.4** [See Table A52.1] **Figure 52.8** At high population density, competition for food limits the amount of energy available to female song sparrows for egg production. **Figure 52.12** Large-scale field experiments like this are extremely expensive and difficult to implement—it was not practical to replicate all treatments. **Table 52.2** From 1999 to 2009, $6800 = 6000\,\lambda^{10}$ (see Eq. 52.5 in Box 52.2). Solving, $\lambda = 1.013$. If $e^r = 1.013$, then $r = 0.0125$. Now use the expression $N_t = N_0e^{rt}$ (Eq. 52.7 in Box 52.2) to solve the problems.

Number of years required to add 1 billion people to 2009 level:

$$7800 = 6800e^{0.0125t}$$
$$ln(1.15) = 0.0125t$$
$$t = \sim 11 \text{ years}$$

Number of years required to double population from 2009 level:

$$13{,}600 = 6800e^{0.0125t}$$
$$ln(2) = 0.0125t$$
$$t = \sim 5.5 \text{ years}$$

Figure 52.17 [See Figure A52.2 on p. A:48]

Summary of Key Concepts

KC 52.1 A life table represents a snapshot of how a particular population is growing. As conditions change, survivorship and fecundity may change. In ancient Rome, for example, survivorship was probably low and fecundity high—women started reproducing at a young age and did not live long, on average. In Rome today, the population has high survivorship and low fecundity. **KC 52.2** [See Figure A52.3 on p. A:48] **KC 52.3** Fewer children are being born per female, but there are many more females of reproductive age due to high fecundity rates in the previous generation. **KC 52.4** Small, isolated populations are likely to be wiped out by bad weather, a disease outbreak, or changes in the habitat. For example, primrose populations may die out when a gap is shaded. But in a metapopulation, migration between the individual small populations helps reestablish subpopulations and maintain the overall size of the metapopulation.

Test Your Knowledge

1. b; **2.** d; **3.** a; **4.** a; **5.** d; **6.** c

Test Your Understanding

1. Equation 52.4: After one breeding interval, the population size is equal to the original population size times the discrete growth rate. Equation 52.5: After t breeding intervals, the population size is equal to the original population size times the discrete growth rate multiplied by itself t times (i.e., raised to the tth power). **2.** Species with type I survivorship curves have high survivorship until old age, so population growth is based on the number of offspring produced—not on how many of the

FIGURE A52.1

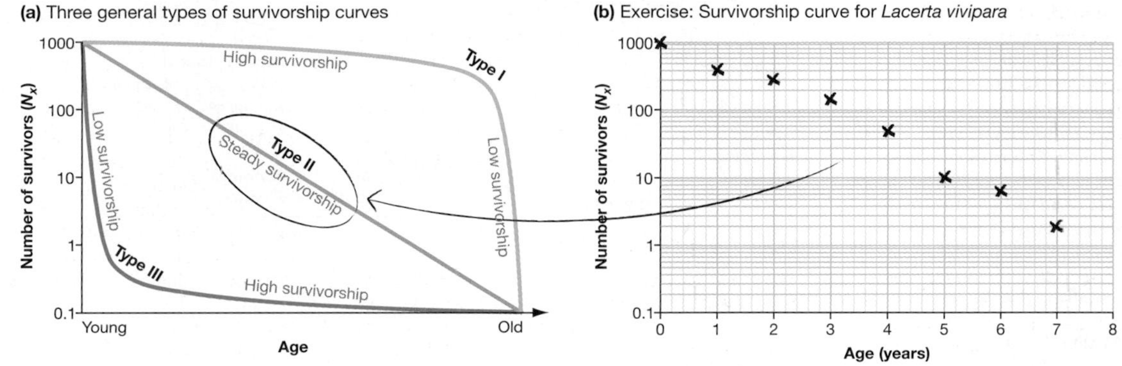

TABLE A52.1

Trait	Life-History Continuum		
	Left	Middle	Right
Growth habit	Herbaceous	Shrub	Tree
Disease- and predator-fighting ability	Low	Medium	High
Seed size	Small	Medium	Large
Seed number	Many	Moderate	Few

offspring survive. Species with type III survivorship curves exhibit very high juvenile mortality, but high survivorship of those few individuals able to reach adulthood. Because only the very few adults can reproduce, any changes to survivorship at the adult level will have a large effect on the population as a whole. **3.** The population has undergone near-exponential growth recently because advances in nutrition, sanitation, and medicine have allowed humans to live at high density without suffering from decreased survivorship and fecundity. Eventually, however, growth rates must slow as density-dependent effects increase death rates and lower birth rates. Experiments on snowshoe hares have shown that food availability and predation are density-dependent factors that influence population growth. **4.** (a) Corridors allow individuals to move between subpopulations, increasing gene flow and making it possible to recolonize habitats where populations have been lost.
(b) Maintaining unoccupied habitat makes it possible for the habitat to be recolonized. **5.** As a sexually transmitted disease, AIDS will reduce the number of sexually active adults. If the epidemic continues unabated, the numbers of both reproductive-age adults and children will decline, causing a top-heavy age distribution dominated by older adults and the elderly. [See Figure A52.4] **6.** Lambda is simple to calculate but only describes growth over a discrete time interval. Determining R_0 requires knowledge of all age-specific birth and death rates. r indicates the growth rate at any instant, so is independent of generation time and can be applied to any population. r_{max} gives the population growth rate in the absence of density-dependent limitation; r is the actual growth rate, which is usually affected by density-dependent factors.

Applying Concepts to New Situations

1. Fewer older individuals will be left in the population; there will be relatively more young individuals. If too many older individuals are taken, growth rate may decline sharply as reproduction stops or slows. (But if relatively few older individuals are taken, more resources are available to younger individuals and their survivorship and fecundity, and the population's overall growth rate, may increase.) **2.** The sunflowers and beetles are both metapopulations. To preserve them, you must preserve as many of the sunflower patches as possible (or plant more) and maintain corridors (which may be smaller sunflower patches) along which the beetles can migrate between the patches. **3.** Large-scale immigration into this population is projected. **4.** The growth rate of any population is female dependent, because females can produce a limited number of offspring at a time, regardless of how many males are in the population. This means that there are "extra" men in the Chinese population—they will not participate in reproduction. Growth rate should decrease, as a result.

CHAPTER 53

Check Your Understanding (CYU)

CYU p. 1070 (1) The individuals do not choose or try to have traits that reduce competition—they simply have those traits (or not). Resource partitioning just happens, because individuals with traits that allow them to exploit different resources produce more offspring, which also have those traits. (Recall from Chapter 24 that natural selection occurs on individuals, but adaptive responses such as resource partitioning are properties of populations.) (2) When species interact via consumption, a trait that gives one species an advantage will exert natural selection on individuals of the other species who have traits that reduce that advantage. This reciprocal adaptation will continue indefinitely. An example is the interaction of *Plasmodium* with the human immune system: The human immune system has evolved the

FIGURE A52.2

All the females are dead at the END of year 4 ...

... but they've produced 3 offspring each that year before they died.

(a) Life table

Age (x)	Survivorship (l_x)	Fecundity (m_x)
0 (birth)	----	0.0
1	0.33	3.0
2	0.2	4.0
3	0.04	5.0
4	0.00	3.0

(b) Fate of first-generation females

Year	0 (newborns)	1-year-olds	2-year-olds	3-year-olds	Total population size (N)
1st	1000 (just introduced)				1000
2nd	990 (= 330 × 3.0)	330 (= 1000 × 0.33)			1320 (= 990 + 330)
3rd	800 (= 200 × 4.0)		200 (= 1000 × 0.20)		800 + 200 = 1000
4th	200 (= 40 × 5.0)			40 (= 1000 × 0.04)	200 + 40 = 240
5th	(40 × 3.0 =) 120			0	120 + 0 = 120

(reproduced before they died)

(c) Fate of first- and second-generation females

Year	0 (newborns)	1-year-olds	2-year-olds	3-year-olds	Total population size (sum across all rows)
1st	1000				1000
2nd	990	330			1320
3rd	800 + 981 (981 = 327 × 3.0)	327 (= 990 × 0.33)	200		2308 (= 800 + 981 + 327 + 200)
4th	200 + 792 (792 = 198 × 4.0)		198 (= 990 × 0.20)	40	200 + 792 + 198 + 40 = 1230
5th	120 + 195 (195 = 39 × 5.0)			39 (= 990 × 0.04)	120 + 195 + 39 + 0 = 354

FIGURE A52.3

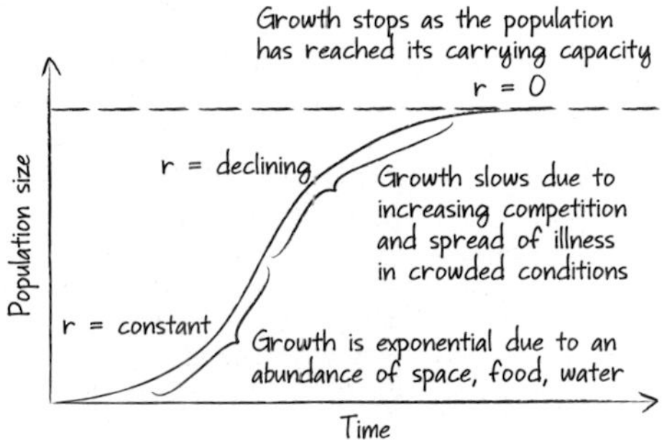

Growth stops as the population has reached its carrying capacity
r = 0
r = declining
Growth slows due to increasing competition and spread of illness in crowded conditions
r = constant
Growth is exponential due to an abundance of space, food, water
Population size
Time

FIGURE A52.4

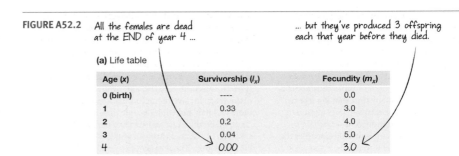

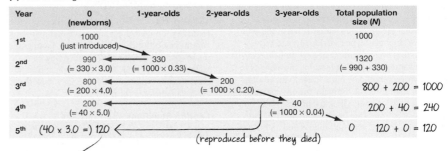

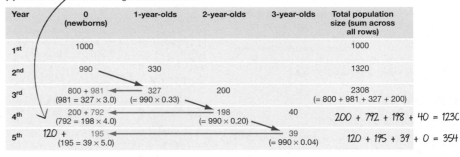

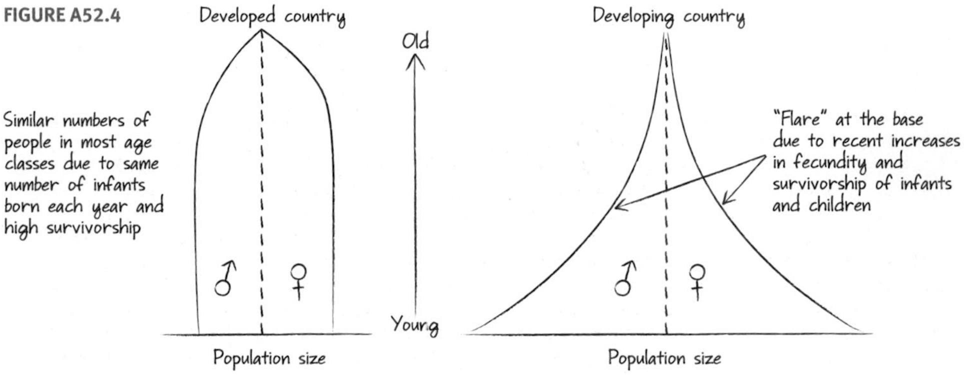

Developed country
Similar numbers of people in most age classes due to same number of infants born each year and high survivorship
♂
♀
Population size
Old
Young
Developing country
"Flare" at the base due to recent increases in fecundity and survivorship of infants and children
♂
♀
Population size

ability to detect proteins from the *Plasmodium* and kill infected cells; in response, *Plasmodium* has evolved different proteins that the immune system does not detect. **CYU p. 1077** (1) The shade provided by early successional species increases humidity, and decomposition of their tissues adds nutrients and organic material to the soil. These conditions favor growth by later successional species, which can outcompete the early successional species. (2) The presence or absence of a plant species where nitrogen-fixation occurs would dramatically alter nutrient conditions, and thus the speed of succession and the types of species that could become established. For example, species that require high nitrogen would be favored on sites where alder grew, while species that can tolerate low nitrogen would thrive on sites where alder is absent.

You Should Be Able To (YSBAT)

YSBAT p. 1079a The calculations for Community 1 are as follows:

Species A	Species B
$p_A = 10/18 = 0.55$	$p_B = 1/18 = 0.055$
$\ln 0.55 = -0.598$	$\ln 0.055 = -2.9$
$-0.598 \times 0.55 = -0.329$	$-2.9 \times 0.055 = -0.16$

Species C	Species D
$p_C = 1/18 = 0.055$	$p_D = 3/18 = 0.167$
$\ln 0.055 = -2.9$	$\ln 0.167 = -1.78$
$-2.9 \times 0.055 = -0.16$	$-1.78 \times 0.167 = -0.299$

Species E	Species F
$p_E = 2/18 = 0.11$	$p_F = 1/18 = 0.055$
$\ln 0.11 = -2.21$	$\ln 0.055 = -2.9$
$-2.2 \times 0.11 = -0.243$	$-2.9 \times 0.055 = -0.16$

Summing the values of $p \times \ln p$ for each species and multiplying by -1 gives the Shannon index of species diversity for Community 1:

$$(-1)[(-0.329) + (-0.16) + (-0.16) + (-0.299) + (-0.243) + (-0.16)] = 1.351$$

Similar calculations would give species diversity values of 1.79 for Community 2 and 1.61 for Community 3.
YSBAT p. 1079b
Community 1: Species richness = 12; Diversity = 2.04
Community 2: Species richness = 12; Diversity = 2.49
Community 3: Species richness = 10; Diversity = 2.3
YSBAT p. 1079c Species richness doubled in each community because the number of species doubled. Although the species diversity values differ from those in Figure 53.25, the communities with more equal numbers of individuals in each species still have the higher diversity values.

Caption Questions and Exercises

Figure 53.6 If he had done the treatments on different rocks, a critic could argue that differences in survival were due to differences in the nature of the rocks—not differences in competition. **Figure 53.7** The species partition, or divide up, resources. In this way, evolution leads to a reduction or elimination of competition. **Figure 53.12** This experimental design tested the hypothesis that mussels can sense the presence of crabs. If the crabs had been fed mussels, a critic could argue that the mussels detect the presence of damaged mussels—not crabs. (As it turns out, the mussels can do both. The experimenters did the same experiment with broken mussel shells in the chamber instead of a crab-fed fish.) **Figure 53.13** Beavers must represent the greater overall threat to cottonwoods, because the trees have evolved a compound that deters further beaver attack but does not deter beetle larvae. Getting the trunk cut down probably reduces fitness more than getting leaves eaten. **Figure 53.15** Put equal numbers of infected and uninfected ants in a pen that includes a bird predator. Count how many of each type are eaten over time. **Figure 53.16** Many possible answers are reasonable hypotheses. For example, acacia trees may expend energy in producing the large bulbs in which the ants live and the food they eat. The *Crematogaster* ants may expend energy and risk injury or death in repelling herbivores. The cleaner shrimp may occasionally get injured or eaten by the fish they are cleaning, or their diets may be somewhat restricted by the association. The host fish may miss meals or be at greater risk of predation when they are undergoing cleaning by the shrimp. In both examples of mutualism, however, the overall associations are positive for both parties. **Figure 53.17** These steps controlled for the alternative hypothesis that differences in treehopper survival were due to differences in the plants they occupied—not the presence or absence of ants. **Figure 53.18** Predictable, at least to a degree. **Figure 53.24** The effect would be to make the island more remote, which would lower the rate of immigration and move the whole immigration curve downward. The rate of extinction would increase, shifting the curve upward. The overall effect would be to decrease the number of species because the island would have effectively become more remote.

Summary of Key Concepts

KC 53.1 A mutualistic relationship becomes a parasitic one if one of the species stops receiving a benefit. The treehopper-ant mutualism becomes parasitic in years when spiders are rare, because the treehoppers no longer derive a benefit but pay a fitness cost (producing honeydew that the ants eat). **KC 53.2** All species at Glacier Bay must be able to survive in a cold climate with the local amount of precipitation. The species earliest in the succession must also be able to grow on rock exposed as the glacier melts. But chance, historical differences in seed sources, and presence or absence of alder (and nitrogen-fixation) created differences in the species present.
KC 53.3 Lakes are abundant in northern latitudes but rare in the tropics. The presence of much greater amounts of habitat supports higher species richness for breeding shorebirds at high latitudes.

Test Your Knowledge

1. d; **2.** a; **3.** b; **4.** d; **5.** c; **6.** d

Test Your Understanding

1. Yes—the treehopper-ant mutualism is parasitic or mutualistic depending on conditions; competition can evolve into no competition over time if niche differentiation occurs; arms races mean that the outcome of host-parasite interactions can change over time, and so forth (many other examples). **2.** *Top-down control:* Herbivore populations are kept at relatively low levels by predation and disease. *Bottom-up control:* Plant matter is low in nitrogen or defended by chemicals that are toxic to herbivores. These hypotheses are certainly not mutually exclusive. For example, resprouted cottonwoods produce compounds that make them less edible for beavers, which is a plant defense. In addition, these cottonwoods might not provide enough nitrogen for beavers to grow quickly, and species that prey on beavers or make them sick could be keeping their population under control. **3.** (a) If community composition is predictable, then the species present should not change over time. But if composition is not predictable, then there should be significant changes in the species present over time. (b) If community composition is predictable, then two sites with identical abiotic factors should develop identical communities. If community composition is not predictable, then sites with identical abiotic factors should develop variable communities. In most tests, the data best match the predictions of the "not predictable" hypothesis, though communities show elements of both. **4.** Disturbance is an event that removes biomass. Compared to low-frequency fires, high-frequency fires would tend to be less severe (less fuel builds up) and would tend to exert more intense natural selection for adaptations to resist the effects of fire. Compared to low-severity fires, high-severity fires would open up more space for pioneering species and would tend to exert more intense natural selection for adaptations to resist the effects of fire. **5.** Early successional species are adapted to disperse to new environments (small seeds) and grow and reproduce quickly (reproduce at an early age, grow quickly). They can tolerate severe abiotic conditions (high temperature, low humidity, low nutrient availability) but have little competitive ability. These species are able to enter a new environment (with no competitors) and thrive. **6.** The idea is that high productivity will lead to high population density of consumers, leading to competition and intense natural selection favoring niche differentiation that leads to speciation.

Applying Concepts to New Situations

1. Natural selection will favor orchid individuals that have traits that resist bee attack: thicker flower walls, nectar storage in a different position, a toxin in the flower walls, or other. Individuals could also be favored if their anthers were in a position that accomplished pollination, even if bees eat through the walls of the nectar-storage structure. **2.** Set fires or let natural fires burn, to match the time interval between fires recorded prior to the arrival of European settlers. Note that because fires have been suppressed for so long, the initial fires will have to be carefully controlled or fuel will have to be removed manually—for example by selective logging. The logic is that species in this habitat are adapted to a disturbance regime of frequent, low-intensity fires. For them to thrive, these conditions have to be re-created. **3.** The exact answer will depend on the location of the campus. The first species to appear must possess good dispersal ability, rapid growth, quick reproductive periods, and tolerance for very harsh and severe conditions. The two-acre plot is likely to be colonized first by pioneer species that have very "weedy" characteristics. But once colonization is under way, the course of succession will depend more on how the various species interact with each other. The presence of one species can inhibit or facilitate the arrival and establishment of another. For example, an early-arriving species might provide the shade and nutrients required by a late-arriving species. The site's history and nearby ecosystems may influence which species appear at each stage; for instance, an undisturbed ecosystem nearby could be a source for native species. The pattern and rate of this succession is also influenced by the overall environmental conditions affecting it. Only species with traits appropriate to the local climate are likely to colonize the site. **4.** One reasonable experiment would involve constructing artificial ponds and introducing different numbers of plankton species to different ponds, but the same total number of individuals. (Any natural immigration to the ponds would have to be prevented.) After a period of time, remove all of the plankton and measure the biomass present. Make a graph with number of species on the x-axis and total biomass on the y-axis. If the hypothesis is correct, the line of best fit through the data should have a positive slope.

CHAPTER 54

Check Your Understanding (CYU)

CYU p. 1092 (1) At each trophic level, most of the energy that is consumed is lost to heat, metabolism, or other maintenance activities, which leaves only a small percentage of energy for biomass production (growth and reproduction). (2) More food will be available to primary consumers, resulting in increases in biomass at each trophic level. An increase in biomass entering decomposer food chains will also increase decomposer populations. **CYU p. 1098** Perhaps the most direct impacts concern rates of groundwater replenishment. Converting biomes into farms or suburbs decreases groundwater recharge and increases runoff. Irrigation pumps groundwater to the surface, where much of it runs off into streams.

You Should Be Able To (YSBAT)

YSBAT p. 1086 To grow a kilogram of beef, you have to first grow 10 kilograms of grain or grass and feed it to the cow. Only 10 percent of this 10 kg will be used for growth and reproduction—the other 9 kg is used for maintenance or lost as heat. **YSBAT p. 1087a** Crustaceans and fish are ectothermic, so they are much more efficient at converting primary production into the biomass in their bodies than endothermic birds and mammals. **YSBAT p. 1087b** Bigger: elk to wolf, mule deer to wolf, aspen/cottonwood/willows to beaver, mice to rough-legged hawk. Smaller: mice to coyote, aspen/cottonwood/willows to elk and mule deer. **YSBAT p. 1094** The uptake arrow was removed—there were no plants to take up nutrients from the soil nutrient pool.

Caption Questions and Exercises

Figure 54.3 The percentage of energy used for maintenance would be much lower because they are ectothermic and sedentary. The percentage excreted would also be much lower because their diet contains little indigestible material. Thus the percentage of energy available for growth and reproduction would be much higher. **Figure 54.10** The diatoms' bodies sink, removing nutrients from the surface ocean and feeding organisms in the deeper ocean. **Figure 54.13** One logical hypothesis is that the total amount exported increases as tree roots and other belowground organic material decay and begin to wash into the stream; the amount exported should begin to decline as nitrate reserves become exhausted—eventually, there is no more nitrate to wash away. **Figure 54.14** Much more water should evaporate. One possibility is that more water vapor will be blown over land and increase precipitation over land. **Figure 54.21** Photosynthesis increases in summer, resulting in removal of CO_2 from the atmosphere.

Summary of Key Concepts

KC 54.1 [See Figure A54.1] **KC 54.2** A food chain consisting of only primary producers and primary decomposers would minimize energy loss. If organisms at higher trophic levels were present, the efficiency of energy transfer would be highest if they are ectothermic, move little, and have extremely efficient digestion. **KC 54.3** Decomposition is slowest in cold, wet temperatures on land and in anoxic conditions in the ocean (or in freshwater). Stagnant water often becomes anoxic as decomposers use up available oxygen and it is not replenished by diffusion from the atmosphere. **KC 54.4** Many possible answers, including: (1) reducing carbon dioxide emissions by finding alternatives to fossil fuels should decrease the amount of carbon dioxide released into the atmosphere, (2) recycling programs decrease carbon dioxide emissions from manufacturing plants and power plants, (3) reforestation can tie up carbon dioxide in wood; (4) iron fertilization could increase NPP in the open ocean—meaning that more CO_2 would be used.

Test Your Knowledge

1. a; **2.** d; **3.** a; **4.** c; **5.** c; **6.** b

Test Your Understanding

1. A food chain shows just a single connection between organisms at different trophic levels; a food web shows many or most of the connections that exist. The arrows represent the movement of chemical energy, obtained through feeding. **2.** If predation by cougars succeeds in reducing the deer population, then trees and low-growing plant species should increase. More (uneaten) biomass will remain in primary producers, meaning that more will be available for other primary consumers besides deer. **3.** Warmer, wetter climates speed decomposition; cool temperatures slow it. Lower amounts of nitrogen and oxygen also slow decomposition by limiting decomposer growth. Wood is slow to decompose because it contains lignin. Decomposition regulates nutrient availability because it releases nutrients from detritus and allows them to reenter the food web. **4.** Bleaching occurs when corals release symbiotic algae that perform photosynthesis; acidification is occurring as increased CO_2 in the atmosphere leads to increased carbonic acid in the ocean and a lower pH. Corals can starve if bleaching is extensive; acidification slows the growth of corals because it makes formation of calcium carbonate skeletons more difficult. **5.** Positive feedbacks include increased forest fires (due to drier, hotter summers) and increased rates of decomposition in tundras (due to warming); negative feedbacks include increased NPP due to warmer temperatures, increased precipitation, and/or increased access to CO_2 for photosynthesis. **6.** The open ocean has almost no nutrient input from the land and has little upwelling to supply nutrients from the deep ocean. In contrast, intertidal and coastal areas receive large inputs of nutrients from rivers as well as from upwellings from ocean depths.

Applying Concepts to New Situations

1. One possibility is to add radioactive phosphorus to the water, then follow this isotope through the primary producers, primary consumers, and secondary consumers in the system by measuring the amount of radioisotope in the tissues of organisms at each trophic level. **2.** If large amounts of organic material were produced in the open ocean and then moved to nearshore areas by currents, then decomposition of the material might use up all available oxygen and lead to creation of anoxic dead zones. **3.** Without herbivores, there is no link in the nitrogen cycle between primary producers and secondary consumers. All of the plant nitrogen would go to the primary decomposers and back into the soil. If decomposition is rapid enough, nitrogen would cycle quickly between primary producers, decomposers, the soil, and back to primary producers. **4.** Atmospheric oxygen would increase due to extensive photosynthesis, but carbon dioxide levels would decrease because little decomposition was occurring. The temperature would drop because fewer greenhouse gases would be trapping heat reflected from the Earth's surface.

CHAPTER 55

Check Your Understanding (CYU)

CYU p. 1109 Do an all-taxon survey: organize experts and volunteers to collect, examine, and identify all species present. This could include direct sequencing studies (see Chapter 29) to document the bacteria and archaea present in different habitats on campus. **CYU p. 1117** (1) Fragmentation reduces habitat quality by creating edges that are susceptible to invasion and loss of species, due to changed abiotic conditions. Genetic problems occur inside fragments as species become inbred and/or lose genetic diversity via genetic drift. (2) Using the equation $S = cA^z$, then $S = 19(100{,}000^{0.20}) = 190$ prior to habitat destruction and $S = 19(10{,}000^{0.20}) = 120$ afterwards, for a total loss of 70 species. Using the same equation for a z of 0.25 suggests that 338 species existed prior to extinction and 190 exist afterwards, a difference of 148. Twice as much diversity was lost due to the increased z. **CYU p. 1120** Establish a large number of study plots on the same hillside. Randomly assign the plots to contain 0, 1, 2, 4, 16, 32, or 64 species (or some other combination, to test a range of species richness), with several replicates for each level of species richness. Prior to planting, document soil depth and soil nutrient levels. Over time, maintain each plot by weeding to keep the original species richness intact. After several years, document soil depth and soil nutrient levels. Compare values as a function of species richness.

Caption Questions and Exercises

Figure 55.6 Nearly all endangered species are threatened by more than one factor and therefore appear on the graph more than once, leading to species totals greater than 100 percent. **Figure 55.10** Because the treatments and study areas were assigned at random, there is no bias in terms of picking certain areas that are unusual in terms of their biomass or species diversity. They should represent a random sample of biomass and species diversity in fragments of various sizes versus intact forest. **Figure 55.11** Draw a horizontal line at lambda of 1.0. **Figure 55.14** Null hypothesis: NPP is not affected by species richness or functional diversity of species. Prediction: NPP will be greater in plots with more species and more functional groups. Prediction of null: There will be no difference in NPP based on species richness or number of functional groups. **Figure 55.15** This line represents no change in biomass, which would mean the

FIGURE A54.1

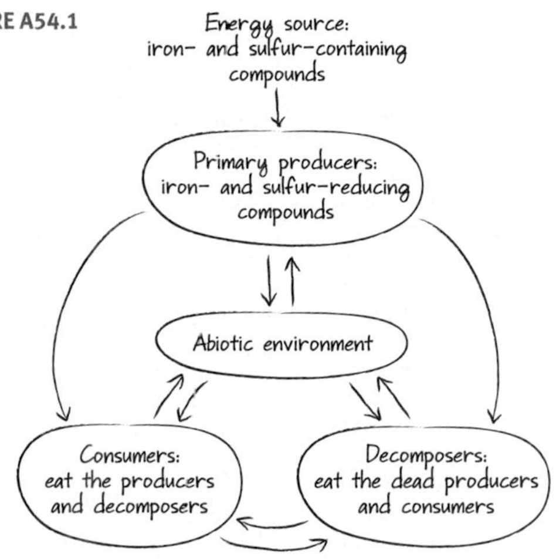
Energy source:
iron- and sulfur-containing compounds
Primary producers:
iron- and sulfur-reducing compounds
Abiotic environment
Consumers:
eat the producers and decomposers
Decomposers:
eat the dead producers and consumers

community was completely resistant to drought—that is, unaffected. **Figure 55.16** [See Figure A55.1]

Summary of Key Concepts

KC 55.1 By sequencing the genes present in a sample from a habitat, you get a broad view of the allelic diversity present—even if you don't know which species are there. **KC 55.2** Small, geographically isolated populations have low genetic diversity due to lack of gene flow and genetic drift and can suffer from inbreeding depression. Low genetic diversity makes them vulnerable to changes in their habitat, and small population size makes them susceptible to catastrophic events such as storms, disease outbreaks, or fires. **KC 55.3** If many species are present, and each uses the available resources in a unique way, then a larger proportion of all available resources should be used—leading to higher biomass production. **KC 55.4** 11percent is already protected, but a mass extinction event is still going on. This is because 11 percent is not sufficient to save most species and because many or most protected areas are not located in biodiversity or conservation hotspots.

Test Your Knowledge

1. d; **2.** a; **3.** b; **4.** d; **5.** c; **6.** a

Test Your Understanding

1. The threats have increased, and instead of most endangered species occurring on islands and being threatened by introduced species and overexploitation, now most threatened species are on continents (where habitat destruction is reducing habitats to small islands) and are in trouble due to habitat destruction and climate change. **2.** Level of resource use—because resources are only used at the rate at which they are replenished. **3.** By comparing the number of species estimated to reside in the park before and after the survey, biologists will have an idea of how much actual species diversity is being underestimated in other parts of the world—where research was at the level of Great Smoky Mountains National Park before the survey. The limitation is that it may be difficult to extrapolate from the data—no one knows if the situation at this park is typical of other habitats. **4.** Species–area curves specify how many species are expected to occur in habitats of various size. After projecting amounts of habitat loss, biologists can read a species–area curve to estimate how many species will be left in the amount of area that remains. **5.** In experiments with species native to North American grasslands, study plots that have more species produce more biomass, change less during a disturbance (drought), and recover from a disturbance faster than study plots with fewer species. **6.** Wildlife corridors facilitate the movement of individuals. Corridors allow areas to be recolonized if a species is lost and the introduction of new alleles that can counteract genetic drift and inbreeding in small isolated populations.

Applying Concepts to New Situations

1. Ecosystem services are beneficial effects that ecosystems have on the abiotic environment (soil, air, water quality, climate) for humans. No one owns or pays for these services, so no one has a vested interest in maintaining them. This is one of the primary reasons that ecosystems are destroyed, even though the services they offer are valuable. **2.** Catalog existing biodiversity at the genetic, species, and ecosystem levels. Using these data, find areas that have the highest concentration of biodiversity at each level. Protect as many of these areas as possible, and connect them with corridors of habitat. (In essence, protect and connect the biodiversity hotspots.) **3.** No "correct" answer. One argument is that conservationists could lobby officials in Brazil and Indonesia to learn from the mistakes made in developed nations, and preserve enough forests to maintain biodiversity and ecosystem services intact—avoiding the expenditures that developed countries had to make to clean up pollution and restore ecosystems and endangered species. **4.** Species with specialized food or habitat requirements, large size (and thus large requirements for land area), small population size, and low reproductive rate are vulnerable to extinction. A good example is the koala bear. Species that are particularly resistant to pressure from humans possess the opposite of many of these traits. A good example of a resistant species is the Norway rat.

THE BIG PICTURE: ECOLOGY

Check Your Understanding (CYU), p. 1126

1. Habitat loss "may eliminate" species; elimination of species (1) may disrupt community structure, (2) may reduces species richness, and (3) may reduce primary productivity. **2.** CO_2 "increases/contributes to" global warming; global warming (1) may disrupt community structure, (2) may reduce species richness, and (3) may reduce primary productivity. **3.** Populations "make up" species; ecosystems "are based on" primary productivity; water temperature, etc. "dictate species that can be found in certain" aquatic ecosystems; CO_2 "concentrations affect" climate.

APPENDIX A: BIOSKILLS

BIOSKILLS 1 (p. B:2) 1. 3.1 miles **2.** 36.96 yards **3.** 37°C **4.** Multiply your weight in pounds by 1/2.2 (0.45). **5.** 4 **BIOSKILLS 2 (p. B:4) 1.** about 18% **2.** a dramatic drop (almost 10%) **3.** No—the order of presentation in a bar chart does not matter (though it's convenient to arrange the bars in a way that reinforces the overall message). **4.** about 11 **5.** 68 inches **BIOSKILLS 3 Figure B3.1** [See Figure AB.1] **Figure B3.2** [See Figure AB.2]

FIGURE A55.1

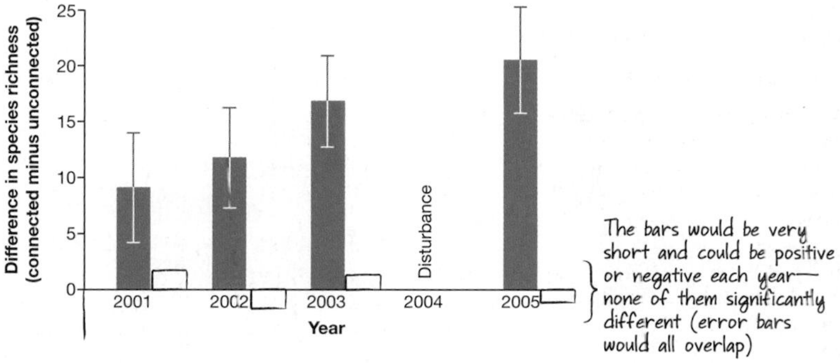

FIGURE AB.1

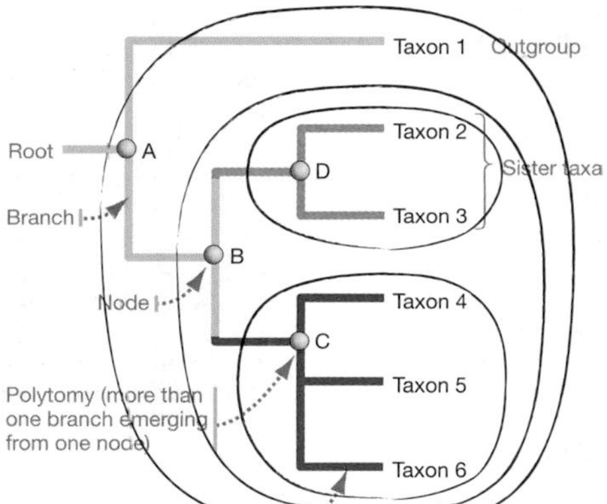

FIGURE AB.2

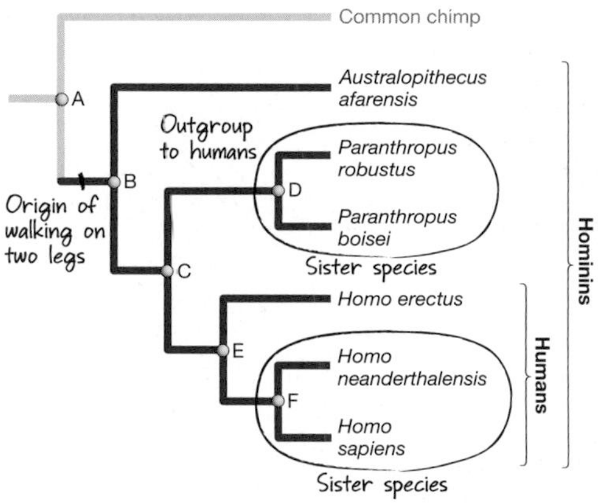

Figure B3.3 d **BIOSKILLS 4 (p. B:6) 1.** "different-yoked-together" **2.** "sugary-loosened" **3.** "study-of-form" **4.** "three-bodies" **BIOSKILLS 5 (p. B:7)** The estimate based on the larger sample—the more replicates or observations you have, the more precise your estimate of the average should be. **BIOSKILLS 6 Figure B6.1** [See Figure AB.3] **BIOSKILLS 7 (p. B:9) 1.** exponential **2.** $\ln N_t = \ln N_0 + rt$ **BIOSKILLS 8 (p. B:10) 1.** [See Figure AB.4] **2.** [See Figure AB.4] **BIOSKILLS 9 Figure B9.1** DNA and RNA are acids that tend to drop a proton in solution, giving them a negative charge. **(p. B:13)** The lane with no band comes from a sample where RNA X is not present. The same size RNA X is present in the next two lanes, but the light band has very few copies while the dark band has many. In the fourth lane, the band is formed by a smaller version of RNA X, with relatively few copies present. **BIOSKILLS 10 (p. B:16) 1.** No—it's just that no mitochondria happened to be present in this section sliced through the cell. **2.** Understanding a molecule's structure is often critical to understanding how it functions in cells. **BIOSKILLS 11 (p. B:17) 1.** size and/or density **2.** The use of high speeds generates the much higher forces needed to separate molecules that are similar in size and mass. **BIOSKILLS 12 (p. B:18) 1.** It may not be clear that the results are relevant to noncancerous cells that are not growing in cell culture—that is, that the artificial conditions mimic natural conditions. **2.** It may not be clear that the results are relevant to individuals that developed normally, from an embryo—that is, that the artificial conditions mimic natural conditions. **BIOSKILLS 13 (p. B:19) 1.** $\frac{1}{2} \times \frac{1}{2} \times \frac{1}{2} \times \frac{1}{2} = \frac{1}{16}$ **2.** $\frac{1}{6} + \frac{1}{6} + \frac{1}{6} = \frac{1}{2}$ **BIOSKILLS 14 Figure B14.1** This is human body temperature—the natural habitat of *E. coli.* **(p. B:22) 1.** *Caenorhabditis elegans* would be a good possibility, because the cells that normally die have already been identified. You could find mutant individuals that lacked normal cell death; you could compare the resulting embryos to normal embryos and be able to identify exactly which cells change as a result. **2.** Any of the multicellular organisms in the list would be a candidate, but *Dictyostelium discoideum* might be particularly interesting because cells only stick to each other during certain points in the life cycle. **3.** *Mus musculus*—as the only mammal in the list, it is the organism most likely to have a gene similar to the one you want to study.

FIGURE AB.3

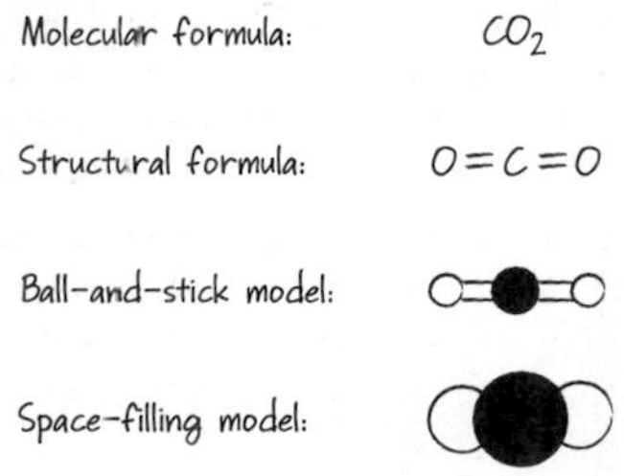

FIGURE AB.4

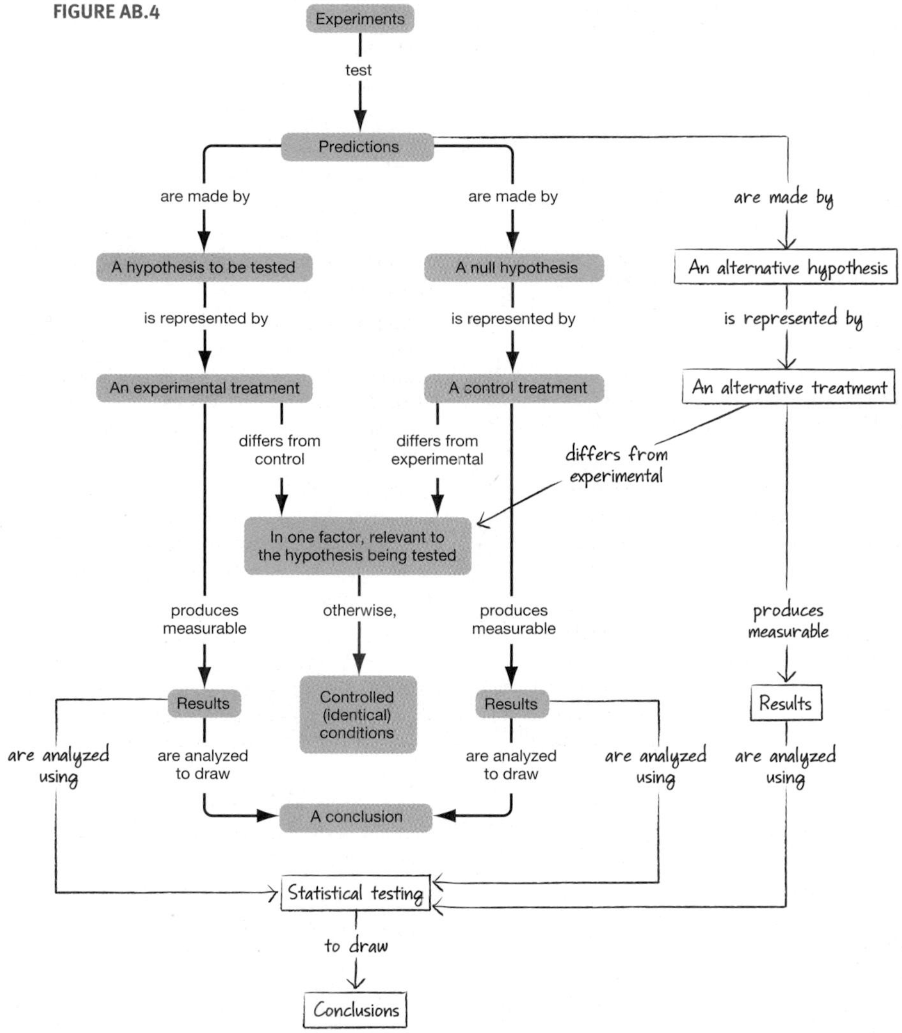

Glossary

5′ cap A chemical grouping, consisting of 7-methylguanylate and three phosphate groups, that is added to the 5′ end of newly transcribed messenger RNA molecules.

20-hydroxyecdysone See **ecdysone**.

abdomen A region of the body; in insects, one of the three prominent body regions called tagmata.

abiotic Not alive (e.g., air, water, and soil). Compare with **biotic**.

aboveground biomass The total mass of living plants in an area, excluding roots.

abscisic acid (ABA) A plant hormone that inhibits cell elongation and stimulates leaf shedding and dormancy.

abscission In plants, the normal (often seasonal) shedding of leaves, fruits, or flowers.

abscission zone The region at the base of a petiole that thins and breaks during dropping of leaves.

absorption In animals, the uptake of ions and small molecules derived from food across the lining of the intestine and into the bloodstream.

absorption spectrum The amount of light of different wavelengths absorbed by a pigment. Usually depicted as a graph of light absorbed versus wavelength. Compare with **action spectrum**.

acclimation, acclimatization Gradual physiological adjustment of an organism to new environmental conditions that occur naturally or as part of a laboratory experiment.

acetyl CoA A molecule produced by oxidation of pyruvate (the final product of glycolysis) in a reaction catalyzed by pyruvate dehydrogenase. Can enter the citric acid cycle and also is used as a carbon source in the synthesis of fatty acids, steroids, and other compounds.

acetylation Addition of an acetyl group (CH_3COO^-) to a molecule.

acetylcholine (ACh) A neurotransmitter, released by nerve cells at neuromuscular junctions, that triggers contraction of muscle cells. Also used as a neurotransmitter between neurons.

acid Any compound that gives up protons or accepts electrons during a chemical reaction or that releases hydrogen ions when dissolved in water.

acid-growth hypothesis The hypothesis that auxin triggers elongation of plant cells by inducing the synthesis of proton pumps whose activity makes the cell wall more acidic, leading to expansion of the cell wall and an influx of water.

acoelomate An animal that lacks an internal body cavity (coelom). Compare with **coelomate** and **pseudocoelomate**.

acquired immune deficiency syndrome (AIDS) A human disease characterized by death of immune system cells (in particular helper T cells and macrophages) and subsequent vulnerability to other infections. Caused by the human immunodeficiency virus (HIV).

acquired immunity Immunity to a particular pathogen or other antigen conferred by antibodies and activated B and T cells following exposure to the antigen. Is characterized by specificity, diversity, memory, and self-nonself recognition. Also called *acquired immune response*. Compare with **innate immunity**.

acrosome A caplike structure, located on the head of a sperm cell, that contains enzymes capable of dissolving the outer coverings of an egg.

ACTH See **adrenocorticotropic hormone**.

actin A globular protein that can be polymerized to form filaments. Actin filaments are part of the cytoskeleton and constitute the thin filaments in skeletal muscle cells.

actin filament A long fiber, about 7 nm in diameter, composed of two intertwined strands of polymerized actin protein; one of the three types of cytoskeletal fibers. Involved in cell movement. Also called a *microfilament*. Compare with **intermediate filament** and **microtubule**.

action potential A rapid, temporary change in electrical potential across a membrane, from negative to positive and back to negative. Occurs in cells, such as neurons and muscle cells, that have an excitable membrane.

action spectrum The relative effectiveness of different wavelengths of light in driving a light-dependent process such as photosynthesis. Usually depicted as a graph of some measure of the process versus wavelength. Compare with **absorption spectrum**.

activation energy The amount of energy required to initiate a chemical reaction; specifically, the energy required to reach the transition state.

active site The portion of an enzyme molecule where substrates (reactant molecules) bind and react.

active transport The movement of ions or molecules across a plasma membrane or organelle membrane against an electrochemical gradient. Requires energy (e.g., from hydrolysis of ATP) and assistance of a transport protein (e.g., pump).

adaptation Any heritable trait that increases the fitness of an individual with that trait, compared with individuals without that trait, in a particular environment.

adaptive immunity Immunity to a particular pathogen or other antigen conferred by antibodies and activated B and T cells following exposure to the antigen. Is characterized by specificity, diversity, memory, and self-nonself recognition. Also called *adaptive immune response*. Compare with **innate immunity.**

adaptive radiation Rapid evolutionary diversification within one lineage, producing numerous descendant species with a wide range of adaptive forms.

adenosine diphosphate (ADP) A molecule consisting of adenine, a sugar, and two phosphate groups. Addition of a third phosphate group produces adenosine triphosphate (ATP).

adenosine triphosphate (ATP) A molecule consisting of adenine, a sugar, and three phosphate groups that can be hydrolyzed to release energy. Universally used by cells to store and transfer energy.

adenylyl cyclase An enzyme that can catalyze the formation of cyclic AMP (cAMP) from ATP. Involved in controlling transcription of various operons in prokaryotes and in some eukaryotic signal-transduction pathways.

adhesion The tendency of certain dissimilar molecules to cling together due to attractive forces. Compare with **cohesion**.

adipocyte A fat cell.

adipose tissue A type of connective tissue whose cells store fats.

ADP See **adenosine diphosphate**.

adrenal glands Two small endocrine glands that sit above each kidney. The outer portion (cortex) secretes several steroid hormones; the inner portion (medulla) secretes epinephrine and norepinephrine.

adrenaline See **epinephrine**.

adrenocorticotropic hormone (ACTH) A peptide hormone, produced and secreted by the anterior pituitary, that stimulates release of steroid hormones (e.g., cortisol, aldosterone) from the adrenal cortex.

adult A sexually mature individual.

adventitious root A root that develops from a plant's shoot system instead of from the plant's root system.

aerobic Referring to any metabolic process, cell, or organism that uses oxygen as an electron acceptor. Compare with **anaerobic**.

afferent division The part of the nervous system, consisting mainly of sensory neurons, that transmits information about the internal and external environment to the central nervous system. Compare with **efferent division**.

age class All the individuals of a specific age in a population.

age structure The proportion of individuals in a population that are of each possible age.

age-specific fecundity The average number of female offspring produced by a female in a certain age class.

agglutination Clumping together of cells, typically caused by antibodies.

aggregate fruit A fruit (e.g., raspberry) that develops from a single flower that has many separate carpels. Compare with **multiple** and **simple fruit**.

AIDS See **acquired immune deficiency syndrome**.

albumen A solution of water and protein (particularly albumins), found in amniotic eggs, that nourishes the growing embryo. Also called *egg white*.

albumin A class of large proteins found in plants and animals, particularly in the albumen of eggs and in blood plasma.

alcohol fermentation Catabolic pathway in which pyruvate produced by glycolysis is converted to ethanol in the absence of oxygen.

aldosterone A hormone produced in the adrenal cortex that stimulates the kidney to conserve salt and water and promotes retention of sodium.

alimentary canal See **digestive tract**.

allele A particular version of a gene.

allergen A molecule (antigen) that triggers an allergic response (an allergy).

allergy An abnormal response to an antigen, usually characterized by dilation of blood vessels, contraction of smooth muscle cells, and increased activity by mucous-secreting cells.

allopatric speciation The divergence of populations into different species by physical isolation of populations in different geographic areas. Compare with **sympatric speciation**.

allopatry Condition in which two or more populations live in different geographic areas. Compare with **sympatry**.

allopolyploidy (adjective: allopolyploid) The state of having more than two full sets of chromosomes (polyploidy) due to hybridization between different species. Compare with **autopolyploidy**.

allosteric regulation Regulation of a protein's function by binding of a regulatory molecule, usually to a specific site distinct from the active site, causing a change in the protein's shape.

α-amylase See **amylase**.

α-helix (alpha-helix) A protein secondary structure in which the polypeptide backbone coils into a spiral shape stabilized by hydrogen bonds between atoms.

alternation of generations A life cycle involving alternation of a multicellular haploid stage (gametophyte) with a multicellular diploid stage (sporophyte). Occurs in most plants and some protists.

alternative splicing In eukaryotes, the splicing of primary RNA transcripts from a single gene in different ways to produce different mature mRNAs and thus different polypeptides.

altruism Any behavior that has a cost to the individual (such as lowered survival or reproduction) and a benefit to the recipient. See **reciprocal altruism**.

alveolus (plural: alveoli) One of the tiny air-filled sacs of a mammalian lung.

ambisense virus A virus whose genome contains both positive-sense and negative-sense sequences.

amino acid A small organic molecule with a central carbon atom bonded to an amino group ($-NH_3$), a carboxyl group ($-COOH$), a hydrogen atom, and a side group. Proteins are polymers of 20 common amino acids.

aminoacyl tRNA A transfer RNA molecule that is covalently bound to an amino acid.

aminoacyl tRNA synthetase An enzyme that catalyzes the addition of a particular amino acid to its corresponding tRNA molecule.

ammonia (NH_3) A small molecule, produced by the breakdown of proteins and nucleic acids, that is very toxic to cells. Is a strong base that gains a proton to form the ammonium ion (NH_4^+).

amnion The membrane within an amniotic egg that surrounds the embryo and encloses it in a protective pool of fluid (amniotic fluid).

amniotes A major lineage of vertebrates (Amniota) that reproduce with amniotic eggs. Includes all reptiles (including birds) and mammals.

amniotic egg An egg that has a watertight shell or case enclosing a membrane-bound water supply (the amnion), food supply (yolk sac), and waste sac (allantois).

amoeboid motion See **cell crawling**.

amphibians A lineage of vertebrates many of whom breathe through their skin and feed on land but lay their eggs in water. Represent the earliest tetrapods; include frogs, salamanders, and caecilians.

amphipathic Containing hydrophilic and hydrophobic elements.

amylase Any enzyme that can break down starch by catalyzing hydrolysis of the glycosidic linkages between the glucose residues.

amyloplasts Dense, starch-storing organelles that settle to the bottom of plant cells and that may be used as gravity detectors.

anabolic pathway Any set of chemical reactions that synthesizes larger molecules from smaller ones. Generally requires an input of energy. Compare with **catabolic pathway**.

anadromous Having a life cycle in which adults live in the ocean (or large lakes) but migrate up freshwater streams to breed and lay eggs.

anaerobic Referring to any metabolic process, cell, or organism that uses an electron acceptor other than oxygen, such as nitrate or sulfate. Compare with **aerobic**.

anaphase A stage in mitosis or meiosis during which chromosomes are moved to opposite ends of the cell.

anatomy The study of the physical structure of organisms.

ancestral trait A trait found in ancestors.

aneuploidy (adjective: aneuploid) The state of having an abnormal number of copies of a certain chromosome.

angiosperm A flowering vascular plant that produces seeds within mature ovaries (fruits). The angiosperms form a single lineage. Compare with **gymnosperm**.

animal A member of a major lineage of eukaryotes (Animalia) whose members typically have a complex, large, multicellular body, eat other organisms, and are mobile.

animal model Any disease that occurs in a nonhuman animal and has many parallels to a similar disease of humans. Studied by medical researchers in hope that findings may apply to human disease.

anion A negatively charged ion.

annelids Members of the phylum Annelida (segmented worms). Distinguished by a segmented body and a coelom that functions as a hydrostatic skeleton. Annelids belong to the lophotrochozoan branch of the protostomes.

annual Referring to a plant whose life cycle normally lasts only one growing season—less than one year. Compare with **perennial**.

anoxygenic Referring to any process or reaction that does not produce oxygen. Photosynthesis in purple sulfur and purple nonsulfur bacteria, which does not involve photosystem II, is anoxygenic. Compare with **oxygenic**.

antenna (plural: antennae) A long appendage that is used to touch or smell.

antenna complex Part of a photosystem, containing an array of chlorophyll molecules and accessory pigments, that receives energy from light and directs the energy to a central reaction center during photosynthesis.

anterior Toward an animal's head and away from its tail. The opposite of posterior.

anterior pituitary The part of the pituitary gland containing endocrine cells that produce and release a variety of peptide hormones in response to other hormones from the hypothalamus. Compare with **posterior pituitary**.

anther The pollen-producing structure at the end of a stamen in flowering plants (angiosperms).

antheridium (plural: antheridia) The sperm-producing structure in most land plants except angiosperms.

anthropoids One of the two major lineages of primates, including apes, humans, and all monkeys. Compare with **prosimians**.

antibiotic Any substance, such as penicillin, that can kill or inhibit the growth of bacteria.

antibody An immunoglobulin protein, produced by B cells, that can bind to a specific part of an antigen, tagging it for attack by the immune system. All antibody molecules have a similar Y-shaped structure and, in their monomer form, consist of two identical light chains and two identical heavy chains.

anticodon The sequence of three bases (triplet) in a transfer RNA molecule that can bind to a mRNA codon with a complementary sequence.

antidiuretic hormone (ADH) A peptide hormone, secreted from the posterior pituitary gland, that stimulates water retention by the kidney. Also called *vasopressin*.

antigen Any foreign molecule, often a protein, that can stimulate a specific response by the immune system.

antigen presentation Process by which small peptides, derived from ingested particulate antigens (e.g., bacteria) or intracellular antigens (e.g., viruses in infected cell) are complexed with MHC proteins and transported to the cell surface where they are displayed and can be recognized by T cells.

antiparallel Describing the opposite orientation of the strands in a DNA double helix with one strand running in the $5' \rightarrow 3'$ direction and the other in the $3' \rightarrow 5'$ direction.

antiporter A carrier protein that allows an ion to diffuse down an electrochemical gradient, using the energy of that process to transport a different substance in the opposite direction *against* its concentration gradient. Compare with **symporter**.

antiviral Any drug or other agent that can kill or inhibit the transmission or replication of viruses.

anus In a multicellular animal, the end of the digestive tract where wastes are expelled.

aorta In terrestrial vertebrates, the major artery carrying oxygenated blood away from the heart.

aphotic zone Deep water receiving no sunlight. Compare with **photic zone**.

apical Toward the top. In plants, at the tip of a branch. In animals, on the side of an epithelial layer that faces the environment and not other body tissues. Compare with **basal**.

apical bud A bud at the tip of a stem, where growth occurs to lengthen the stem.

apical dominance Inhibition of lateral bud growth by the apical meristem at the tip of a plant branch.

apical meristem A group of undifferentiated plant cells, at the tip of a stem or root, that is responsible for primary growth. Compare with **lateral meristem**.

apical–basal axis The long or vertical, shoot-to-root axis of a plant.

apomixis The formation of mature seeds without fertilization occurring; a type of asexual reproduction.

apoplast In plant roots, a continuous pathway through which water can flow, consisting of the porous cell walls of adjacent cells and the intervening extracellular space. Compare with **symplast**.

apoptosis Series of genetically controlled changes that lead to death of a cell. Occurs frequently during embryological development and later may occur in response to infections or cell damage. Also called *programmed cell death*.

appendix A blind sac (having only one opening) that extends from the colon in some mammals.

aquaporin A type of channel protein through which water can move by osmosis across a plasma membrane.

arbuscular mycorrhizal fungi (AMF) Fungi whose hyphae enter the root cells of their host plants. Also called *endomycorrhizal fungi*.

Archaea One of the three taxonomic domains of life consisting of unicellular prokaryotes distinguished by cell walls made of certain polysaccharides not found in bacterial or eukaryotic cell walls, plasma membranes composed of unique isoprene-containing phospholipids, and ribosomes and RNA polymerase similar to those of eukaryotes. Compare with **Bacteria** and **Eukarya**.

archegonium (plural: archegonia) The egg-producing structure in most land plants except angiosperms.

arteriole One of the many tiny vessels that carry blood from arteries to capillaries.

artery Any thick-walled blood vessel that carries blood (oxygenated or not) under relatively high pressure away from the heart to organs throughout the body. Compare with **vein**.

arthropods Members of the phylum Arthropoda. Distinguished by a segmented body; a hard, jointed exoskeleton; paired appendages; and an extensive body cavity called a hemocoel. Arthropods belong to the ecdysozoan branch of the protostomes.

articulation A movable point of contact between two bones of a skeleton. See **joint**.

artificial selection Deliberate manipulation by humans, as in animal and plant breeding, of the genetic composition of a population by allowing only individuals with desirable traits to reproduce.

ascocarp A large, cup-shaped reproductive structure produced by some ascomycete fungi. Contains many microscopic asci, which produce spores.

ascus (plural: asci) Specialized spore-producing cell found at the ends of hyphae in sac fungi (ascomycetes).

asexual reproduction Any form of reproduction resulting in offspring that are genetically identical to the parent. Includes binary fission, budding, and parthenogenesis. Compare with **sexual reproduction**.

asymmetric competition Ecological competition between two species in which one species suffers a much greater fitness decline than the other. Compare with **symmetric competition**.

atomic number The number of protons in the nucleus of an atom, giving the atom its identity as a particular chemical element.

ATP See **adenosine triphosphate**.

ATP synthase A large membrane-bound protein complex in chloroplasts, mitochondria, and some bacteria that uses the energy of protons flowing through it to synthesize ATP. Also called F_oF_1 *complex*.

atrial natriuretic hormone An animal hormone that stimulates excretion of salt from the kidneys.

atrioventricular (AV) node A region of the heart between the right atrium and right ventricle where electrical signals from the atrium are slowed briefly before spreading to the ventricle. This delay allows the ventricle to fill with blood before contracting. Compare with **sinoatrial (SA) node**.

atrium (plural: atria) A thin-walled chamber of the heart that receives blood from veins and pumps it to a neighboring chamber (the ventricle).

autocrine Relating to a chemical signal that affects the same cell that produced and released it.

autoimmunity A pathological condition in which the immune system attacks self cells or tissues of an individual's own body.

autonomic nervous system The part of the peripheral nervous system that controls internal organs and involuntary processes, such as stomach contraction, hormone release, and heart rate. Includes parasympathetic and sympathetic nerves. Compare with **somatic nervous system**.

autophagy The process by which damaged organelles are surrounded by a membrane and delivered to a lysosome to be destroyed.

autopolyploidy (adjective: autopolyploid) The state of having more than two full sets of chromosomes (polyploidy) due to a mutation that doubled the chromosome number.

autoradiography A technique for detecting radioactively labeled molecules separated by gel electrophoresis by placing an unexposed film over the gel. A black dot appears on the film wherever a radioactive atom is present in the gel. Also can be used to locate labeled molecules in fixed tissue samples.

autosomal inheritance The inheritance patterns that occur when genes are located on autosomes rather than on sex chromosomes.

autosome Any chromosome that does not carry genes involved in determining the sex of an individual.

autotroph Any organism that can synthesize reduced organic compounds from simple inorganic sources such as CO_2 or CH_4. Most plants and some bacteria and archaea are autotrophs. Also called *primary producer*. Compare with **heterotroph**.

auxin Indoleacetic acid, a plant hormone that stimulates phototropism and some other responses.

avirulence (*avr*) genes Genes in pathogens encoding proteins that trigger a defense response in plants. Compare with **resistance (*R*) genes**.

axillary bud A bud that forms in the angle between a leaf and a stem and may develop into a lateral (side) branch. Also called *lateral bud*.

axon A long projection of a neuron that can propagate an action potential and transmit it to another neuron.

axon hillock The site in a neuron where an axon joins the cell body and where action potentials are first triggered.

axoneme A structure found in eukaryotic cilia and flagella and responsible for their motion; composed of two central microtubules surrounded by nine doublet microtubules (9 + 2 arrangement).

B cell A type of leukocyte that matures in the bone marrow and, with T cells, is responsible for adaptive immunity. Produces antibodies and also functions in antigen presentation. Also called *B lymphocyte*.

BAC library A collection of all the sequences found in the genome of a species, inserted into bacterial artificial chromosomes (BACs).

background extinction The average rate of low-level extinction that has occurred continuously throughout much of evolutionary history. Compare with **mass extinction**.

Bacteria One of the three taxonomic domains of life consisting of unicellular prokaryotes distinguished by cell walls composed largely of peptidoglycan, plasma membranes similar to those of eukaryotic cells, and ribosomes and RNA polymerase that differ from those in archaeans or eukaryotes. Compare with **Archaea** and **Eukarya**.

bacterial artificial chromosome (BAC) An artificial version of a bacterial chromosome that can be used as a cloning vector to produce many copies of large DNA fragments.

bacteriophage Any virus that infects bacteria.

baculum A bone inside the penis usually present in mammals with a penis that lacks erectile tissue.

balancing selection A pattern of natural selection in which no single allele is favored in all populations of a species at all times. Instead, there is a balance among alleles in terms of fitness and frequency.

ball-and-stick model A representation of a molecule where atoms are shown as balls—colored and scaled to indicate the atom's identity—and covalent bonds are shown as rods or sticks connecting the balls in the correct geometry.

bar coding The use of well-characterized gene sequences to identify species.

bark The protective outer layer of woody plants, composed of cork cells, cork cambium, and secondary phloem.

baroreceptors Specialized nerve cells in the walls of the heart and certain major arteries that detect changes in blood pressure and trigger appropriate responses by the brain.

basal Toward the base. In plants, at the base of a branch where it joins the stem. In animals, on the side of an epithelial layer that abuts underlying body tissues. Compare with **apical**.

basal body A structure of nine pairs of microtubules arranged in a circle at the base of eukaryotic cilia and flagella where they attach to the cell. Structurally similar to a centriole.

basal lamina A thick, collagen-rich extracellular matrix that underlies most epithelial tissues (e.g., skin) in animals.

basal metabolic rate (BMR) The total energy consumption by an organism at rest in a comfortable environment. For aerobes, often measured as the amount of oxygen consumed per hour.

basal transcription complex A large multi-protein structure that assembles near the promoter of eukaryotic genes and initiates transcription. Composed of basal transcription factors, TATA-binding protein, coactivators, and RNA polymerase.

basal transcription factor General term for proteins, present in all cell types, that bind to eukaryotic promoters and help initiate transcription. Compare with **regulatory transcription factor**.

base Any compound that acquires protons or gives up electrons during a chemical reaction or accepts hydrogen ions when dissolved in water.

basidium (plural: basidia) Specialized spore-producing cell at the ends of hyphae in club fungi.

basilar membrane The membrane in the vertebrate cochlea on which the bottom portion of hair cells sit.

basolateral Toward the bottom and sides. In animals, the side of an epithelial layer that faces other body tissues and not the environment.

Batesian mimicry A type of mimicry in which a harmless or palatable species resembles a dangerous or poisonous species. Compare with **Müllerian mimicry**.

B-cell receptor (BCR) An immunoglobulin protein (antibody) embedded in the plasma membrane of mature B cells and to which antigens bind.

beak A structure that exerts biting forces and is associated with the mouth; found in birds, cephalopods, and some insects.

behavior Any action by an organism.

beneficial In genetics, referring to any mutation, allele, or trait that increases an individual's fitness.

benign tumor A mass of abnormal tissue that grows slowly or not at all, does not disrupt surrounding tissues, and does not spread to other organs. Benign tumors are not cancers. Compare with **malignant tumor**.

benthic Living at the bottom of an aquatic environment.

benthic zone The area along the bottom of an aquatic environment.

β-pleated sheet (beta-pleated sheet) A protein secondary structure in which the polypeptide backbone folds into a sheetlike shape stabilized by hydrogen bonding.

bilateral symmetry An animal body pattern in which there is one plane of symmetry dividing the body into a left side and a right side. Typically, the body is long and narrow, with a distinct head end and tail end. Compare with **radial symmetry**.

bilaterian A member of a major lineage of animals (Bilateria) that are bilaterally symmetrical at some point in their life cycle, have three embryonic germ layers, and have a coelom. All protostomes and deuterostomes are bilaterians.

bile A complex solution produced by the liver, stored in the gall bladder, and secreted into the intestine. Contains steroid derivatives called bile salts that are responsible for emulsification of fats during digestion.

biodiversity The diversity of life considered at three levels: genetic diversity (variety of alleles in a population, species, or group of species); species diversity (variety and relative abundance of species present in a certain area); and ecosystem diversity (variety of communities and abiotic components in a region).

biodiversity hot spot A region that is extraordinarily rich in species.

biofilm A hard, polysaccharide-rich layer secreted by bacterial cells, which allows them to attach to a surface.

biogeochemical cycle The pattern of circulation of an element or molecule among living organisms and the environment.

biogeography The study of how species and populations are distributed geographically.

bioinformatics The field of study concerned with managing, analyzing, and interpreting biological information, particularly DNA sequences.

biological fitness See **fitness**.

biological species concept The definition of a species as a population or group of populations that are reproductively isolated from other groups. Members of a species have the potential to interbreed in nature to produce viable, fertile offspring but cannot produce viable, fertile hybrid offspring with members of other species. Compare with **morphospecies** and **phylogenetic species concept**.

bioluminescence The emission of light by a living organism.

biomagnification In animal tissues, an increase in the concentration of particular molecules that may occur as those molecules are passed up a food chain.

biomass The total mass of all organisms in a given population or geographical area; usually expressed as total dry weight.

biome A large terrestrial ecosystem characterized by a distinct type of vegetation and climate.

bioprospecting The effort to find commercially useful compounds by studying organisms—especially species that are poorly studied to date.

bioremediation The use of living organisms, usually bacteria or archaea, to degrade environmental pollutants.

biotechnology The application of biological techniques and discoveries to medicine, industry, and agriculture.

biotic Living, or produced by a living organism. Compare with **abiotic**.

bipedal Walking primarily on two legs.

bipolar cell A cell in the vertebrate retina that receives information from one or more photoreceptors and passes it to other bipolar cells or ganglion cells.

bivalves A lineage of mollusks that have two shells, such as clams and mussels.

bladder A mammalian organ that holds urine until it can be excreted.

blade The wide, flat part of a plant leaf.

blastocoel Fluid-filled cavity in the blastula of many animal species.

blastocyst Specialized type of blastula in mammals. A spherical structure composed of trophoblast cells on the exterior and a cluster of cells (the inner cell mass), which fills part of the interior space.

blastomere A small cell created by cleavage divisions in early animal embryos.

blastopore A small opening (pore) in the surface of an early vertebrate embryo, through which cells move during gastrulation.

blastula In vertebrate development, a hollow ball of cells (blastomere cells) that is formed by cleavage of a zygote and immediately undergoes gastrulation. See **blastocyst**.

blood A type of connective tissue consisting of red blood cells and leukocytes suspended in a fluid portion called plasma.

blood pressure See **diastolic blood pressure** and **systolic blood pressure**.

body mass index (BMI) A mathematical relationship of weight and height used to assess obesity in humans. Calculated as weight (kg) divided by the square of height (m^2).

body plan The basic architecture of an animal's body, including the number and arrangement of limbs, body segments, and major tissue layers.

bog A wetland that has no or almost no water flow, resulting in very low oxygen levels and acidic conditions.

Bohr shift The rightward shift of the oxygen-hemoglobin dissociation curve that occurs with decreasing pH. Results in hemoglobin being more likely to release oxygen in the acidic environment of exercising muscle.

bone A type of vertebrate connective tissue consisting of living cells and blood vessels within a hard extracellular matrix composed of calcium phosphate ($CaPO_4$) and small amounts of calcium carbonate ($CaCO_3$) and protein fibers.

bone marrow The soft tissue filling the inside of long bones containing stem cells that develop into red blood cells and leukocytes throughout life.

Bowman's capsule The hollow, double-walled cup-shaped portion of a nephron that surrounds a glomerulus in the vertebrate kidney.

brain A large mass of neurons located in the head region of an animal, that is involved in information processing; may also be called the cerebral ganglion.

brain stem The most posterior portion of the vertebrate brain, connecting to the spinal cord and responsible for autonomic body functions such as heart rate, respiration, and digestion.

braincase See **cranium**.

branch (1) A part of a phylogenetic tree that represents populations through time. (2) Any extension of a plant's shoot system.

brassinosteroids A family of steroid hormones found in plants.

bronchiole One of the small tubes in mammalian lungs that carry air from the bronchi to the alveoli.

bronchus (plural: bronchi) In mammals, one of a pair of large tubes that lead from the trachea to each lung.

bryophytes Members of several phyla of green plants that lack vascular tissue including liverworts, hornworts, and mosses. Also called *non-vascular plants.*

budding Asexual reproduction via outgrowth from the parent that eventually breaks free as an independent individual; occurs in yeasts and some invertebrates.

buffer A substance that, in solution, acts to minimize changes in the pH of that solution when acid or base is added.

bulbourethral glands In male mammals, small paired glands at the base of the urethra that secrete an alkaline mucus (part of semen), which lubricates the tip of the penis and neutralizes acids in the urethra during copulation. In humans, also called *Cowper's glands.*

bulk flow The directional movement of a substantial volume of fluid due to pressure differences, such as movement of water through plant phloem and movement of blood in animals.

bundle-sheath cell A type of cell found around the vascular tissue (veins) of plant leaves.

C_3 photosynthesis The most common form of photosynthesis in which atmospheric CO_2 is used to form 3-phosphoglycerate, a three-carbon sugar.

C_4 photosynthesis A variant type of photosynthesis in which atmospheric CO_2 is first fixed into four-carbon sugars, rather than the three-carbon sugars of classic C_3 photosynthesis. Enhances photosynthetic efficiency in hot, dry environments, by reducing loss of oxygen due to photorespiration.

cadherin Any of a class of cell-surface proteins involved in cell adhesion and important for coordinating movements of cells during embryological development.

callus In plants, a mass of undifferentiated cells that can generate roots and other tissues necessary to create a mature plant.

Calvin cycle In photosynthesis, the set of light-independent reactions that use NADPH and ATP formed in the light-dependent reactions to drive the fixation of atmospheric CO_2 and reduction of the fixed carbon, ultimately producing sugars. Also called *carbon fixation* and *light-independent reactions.*

calyx All of the sepals of a flower.

CAM See **crassulacean acid metabolism**.

cambium (plural: cambia) See **lateral meristem**.

Cambrian explosion The rapid diversification of animal body types and lineages that occured between the species present in the Ediacaran faunas (565–542 mya) and the Cambrian faunas (525–515 mya).

camera eye The type of eye in vertebrates and cephalopods, consisting of a hollow chamber with a hole at one end (through which light enters) and a sheet of light-sensitive cells against the opposite wall.

cAMP See **cyclic AMP**.

cancer General term for any tumor whose cells grow in an uncontrolled fashion, invade nearby tissues, and spread to other sites in the body.

canopy The uppermost layers of branches in a forest (i.e., those fully exposed to the Sun).

CAP binding site A DNA sequence upstream of certain prokaryotic operons to which catabolite activator protein can bind, increasing gene transcription.

capillarity The tendency of water to move up a narrow tube due to surface tension, adhesion, and cohesion.

capillary One of the numerous small, thin-walled blood vessels that permeate all tissues and organs, and allow exchange of gases and other molecules between blood and body cells.

capillary bed A thick network of capillaries.

capsid A shell of protein enclosing the genome of a virus particle.

carapace In crustaceans, a large platelike section of the exoskeleton that covers and protects the cephalothorax (e.g., a crab's "shell").

carbohydrate Any of a class of molecules that contain a carbonyl group, several hydroxyl groups, and several to many carbon-hydrogen bonds. See **monosaccharide** and **polysaccharide**.

carbon cycle, global The worldwide movement of carbon among terrestrial ecosystems, the oceans, and the atmosphere.

carbon fixation See **Calvin cycle**.

carbonic anhydrase An enzyme that catalyzes the formation of carbonic acid (H_2CO_3) from carbon dioxide and water.

carboxylic acids Organic acids with the form R-COOH (a carboxyl group).

cardiac cycle One complete heartbeat cycle, including systole and diastole.

cardiac muscle The muscle tissue of the vertebrate heart. Consists of long branched fibers that are electrically connected and that initiate their own contractions; not under voluntary control. Compare with **skeletal** and **smooth muscle**.

carnivore (adjective: carnivorous) An animal whose diet consists predominantly of meat. Most members of the mammalian taxon Carnivora are carnivores. Some plants are carnivorous, trapping and killing small animals, then absorbing nutrients from the prey's body. Compare with **herbivore** and **omnivore**.

carotenoid Any of a class of accessory pigments, found in chloroplasts, that absorb wavelengths of light not absorbed by chlorophyll; typically appear yellow, orange, or red. Includes carotenes and xanthophylls.

carpel The female reproductive organ in a flower. Consists of the stigma, to which pollen grains adhere; the style, through which pollen grains move; and the ovary, which houses the ovule. Compare with **stamen**.

carrier A heterozygous individual carrying a normal allele and a recessive allele for an inherited trait; does not display the phenotype of the trait but can pass the recessive gene to offspring.

carrier protein A membrane protein that facilitates diffusion of a small molecule (e.g., glucose) across the plasma membrane by a process involving a reversible change in the shape of the protein. Also called *carrier* or *transporter.*

carrying capacity (*K*) The maximum population size of a certain species that a given habitat can support.

cartilage A type of vertebrate connective tissue that consists of relatively few cells scattered in a stiff matrix of polysaccharides and protein fibers.

Casparian strip In plant roots, a waxy layer containing suberin, a water-repellent substance, that prevents movement of water through the walls of endodermal cells, thus blocking the apoplastic pathway.

cast A type of fossil, formed when the decay of a body part leaves a void that is then filled with minerals that later harden.

catabolic pathway Any set of chemical reactions that breaks down larger, complex molecules into smaller ones, releasing energy in the process. Compare with **anabolic pathway**.

catabolite activator protein (CAP) A protein that can bind to the CAP binding site upstream of certain prokaryotic operons, facilitating binding of RNA polymerase and stimulating gene expression.

catabolite repression A type of positive transcriptional control in which the end product of a catabolic pathway inhibits further transcription of the gene encoding an enzyme early in the pathway.

catalysis (verb: catalyze) Acceleration of the rate of a chemical reaction due to a decrease in the free energy of the transition state, called the activation energy.

catalyst Any substance that increases the rate of a chemical reaction without itself undergoing any permanent chemical change.

catecholamines A class of small compounds, derived from the amino acid tyrosine, that are used as hormones or neurotransmitters. Include epinephrine, norepinephrine, and dopamine.

cation A positively charged ion.

cation exchange In botany, the release (displacement) of cations, such as magnesium and calcium from soil particles, by protons in acidic soil water. The released cations are available for uptake by plants.

CD4 A membrane protein on the surface of some T cells in humans. $CD4^+$ T cells can give rise to helper T cells.

CD8 A membrane protein on the surface of some T cells in humans. $CD8^+$ T cells can give rise to cytotoxic T cells.

Cdk See **cyclin-dependent kinase**.

cDNA See **complementary DNA**.

cDNA library A set of cDNAs from a particular cell type or stage of development. Each cDNA is carried by a plasmid or other cloning vector and can be separated from other cDNAs. Compare with **genomic library**.

cecum A blind sac between the small intestine and the colon. Is enlarged in some species (e.g., rabbits) that use it as a fermentation vat for digestion of cellulose.

cell A highly organized compartment bounded by a thin, flexible structure (plasma membrane) and containing concentrated chemicals in an aqueous (watery) solution. The basic structural and functional unit of all organisms.

cell body The part of a neuron that contains the nucleus and where incoming signals are integrated. Also called the *soma*.

cell crawling A form of cellular movement involving actin filaments in which the cell produces bulges (pseudopodia) that stick to the substrate and pull the cell forward. Also called *amoeboid motion*.

cell culture See **culture**.

cell cycle Ordered sequence of events in which a eukaryotic cell replicates its chromosomes, evenly partitions the chromosomes to two daughter cells, and then undergoes division of the cytoplasm.

cell-cycle checkpoint Any of several points in the cell cycle at which progression of a cell through the cycle can be regulated.

cell division Creation of new cells by division of pre-existing cells.

cell-mediated (immune) response The type of immune response that involves generation of cytotoxic T cells from $CD8^+$ T cells. Defends against pathogen-infected cells, cancer cells, and transplanted cells. Compare with **humoral (immune) response**.

cell membrane See **plasma membrane**.

cell plate A double layer of new plasma membrane that appears in the middle of a dividing plant cell; ultimately divides the cytoplasm into two separate cells.

cell sap An aqueous solution found in the vacuoles of plant cells.

cell theory The theory that all organisms are made of cells and that all cells come from preexisting cells.

cell wall A fibrous layer found outside the plasma membrane of most bacteria and archaea and many eukaryotes.

cellular respiration A common pathway for production of ATP, involving transfer of electrons from compounds with high potential energy (often NADH and $FADH_2$) to an electron transport chain and ultimately to an electron acceptor (often oxygen).

cellulose A structural polysaccharide composed of β-glucose monomers joined by β-1,4-glycosidic linkages. Found in the cell wall of algae, plants, bacteria, fungi, and some other groups.

Cenozoic era The most recent interval of geologic time, beginning 65.5 million years ago. during which mammals became the dominant vertebrates and angiosperms became the dominant plants.

central dogma The long-accepted hypothesis that information in cells flows in one direction: DNA codes for RNA, which codes for proteins. Exceptions are now known (e.g., retroviruses).

central nervous system (CNS) The brain and spinal cord of vertebrate animals. Compare with **peripheral nervous system (PNS)**.

centriole One of two small cylindrical structures, structurally similar to a basal body, found together within the centrosome near the nucleus of a eukaryotic cell.

centromere Constricted region of a replicated chromosome where the two sister chromatids are joined and the kinetochore is located.

centrosome Structure in animal and fungal cells, containing two centrioles, that serves as a microtubule-organizing center for the cell's cytoskeleton and for the spindle apparatus during cell division.

cephalization The formation of a distinct anterior region (the head) where sense organs and a mouth are clustered.

cephalochordates One of the three major chordate lineages (Cephalochordata), comprising small, mobile organisms that live in marine sands; also called *lancelets* or *amphioxi*. Compare with **urochordates** and **vertebrates**.

cephalopods A lineage of mollusks including the squid, octopuses, and nautiluses. Distinguished by large brains, excellent vision, tentacles, and a reduced or absent shell.

cerebellum Posterior section of the vertebrate brain that is involved in coordination of complex muscle movements, such as those required for locomotion and maintaining balance.

cerebrum The most anterior section of the vertebrate brain. Divided into left and right hemispheres and four lobes: parietal lobe, involved in complex decision making (in humans); occipital lobe, receives and interprets visual information; parietal lobe, involved in integrating sensory and motor functions; and temporal lobe, functions in memory, speech (in humans), and interpreting auditory information.

cervix The narrow passageway between the vagina and the uterus of female mammals.

chaetae (singular: chaeta) Bristle-like extensions found in some annelids.

channel A protein that forms a pore in a cell membrane. The structure of most channels allows them to admit just one or a few types of ions or molecules.

character displacement The tendency for the traits of similar species that occupy overlapping ranges to change in a way that reduces interspecific competition.

chelicerae A pair of clawlike appendages found around the mouth of certain arthropods called chelicerates (spiders, mites, and allies).

chemical bond An attractive force binding two atoms together. Covalent bonds, ionic bonds, and hydrogen bonds are types of chemical bonds.

chemical carcinogen Any chemical that can cause cancer.

chemical energy The potential energy stored in covalent bonds between atoms.

chemical equilibrium A dynamic but stable state of a reversible chemical reaction in which the forward reaction and reverse reactions proceed at the same rate, so that the concentrations of reactants and products remain constant.

chemical evolution The theory that simple chemical compounds in the ancient atmosphere and ocean combined via spontaneous chemical reactions to form larger, more complex substances, eventually leading to the origin of life and the start of biological evolution.

chemical reaction Any process in which one compound or element is combined with others or is broken down; involves the making and/or breaking of chemical bonds.

chemiosmosis An energetic coupling mechanism whereby energy stored in an electrochemical proton gradient (proton-motive force) is used to drive an energy-requiring process such as production of ATP.

chemokine Any of several chemical signals that attract leukocytes to a site of tissue injury or infection.

chemolithotroph An organism that produces ATP by oxidizing inorganic molecules with high potential energy such as ammonia (NH_3) or methane (CH_4). Also called *lithotroph*. Compare with **chemoorganotroph**.

chemoorganotroph An organism that produces ATP by oxidizing organic molecules with high potential energy such as sugars. Also called *organotroph*. Compare with **chemolithotroph**.

chemoreceptor A sensory cell or organ specialized for detection of specific molecules or classes of molecules.

chiasma (plural: chiasmata) The X-shaped structure formed during meiosis by crossing over between non-sister chromatids in a pair of homologous chromosomes.

chitin A structural polysaccharide composed of *N*-acetylglucosamine monomers joined end to end by β-1,4-glycosidic linkages. Found in cell walls of fungi and many algae, and in external skeletons of insects and crustaceans.

chitons A lineage of marine mollusks that have a protective shell formed of eight calcium carbonate plates.

chlorophyll Any of several closely related green pigments, found in chloroplasts and photosynthetic protists, that absorb light during photosynthesis.

chloroplast A chlorophyll-containing organelle, bounded by a double membrane, in which photosynthesis occurs; found in plants and photosynthetic protists. Also the location of amino acid, fatty acid, purine, and pyrimidine synthesis.

choanocyte A specialized flagellated feeding cell found in choanoflagellates (protists that are the closest living relatives of animals) and sponges (the oldest animal phylum).

cholecystokinin A peptide hormone secreted by cells in the lining of the small intestine. Stimulates the secretion of digestive enzymes from the pancreas and of bile from the liver and gallbladder.

chordates Members of the phylum Chordata, deuterostomes distinguished by a dorsal hollow nerve cord, pharyngeal gill slits, a notochord, a dorsal hollow nerve cord, and a post-anal tail. Include vertebrates, cephalochordata, and urochordata.

chromatid One of the two identical strands composing a replicated chromosome that is connected at the centromere to the other strand.

chromatin The complex of DNA and proteins, mainly histones, that compose eukaryotic chromosomes. Can be highly compact (heterochromatin) or loosely coiled (euchromatin).

chromatin remodeling The process by which the DNA in chromatin is unwound from its associated proteins to allow transcription or replication. May involve chemical modification of histone proteins or reshaping of the chromatin by large multi-protein complexes in an ATP-requiring process.

chromosome Gene-carrying structure consisting of a single long molecule of DNA and associated proteins (e.g., histones). Most prokaryotic cells contain a single, circular chromosome; eukaryotic cells contain multiple noncircular (linear) chromosomes located in the nucleus.

chromosome theory of inheritance The principle that genes are located on chromosomes and that patterns of inheritance are determined by the behavior of chromosomes during meiosis.

chylomicron A ball of protein-coated lipids used to transport the lipids through the bloodstream.

cilium (plural: cilia) One of many short, filamentous projections of some eukaryotic cells containing a core of microtubules. Used to move the cell and/or to move fluid or particles along a stationary cell. See **axoneme**.

circadian clock An internal mechanism found in most organisms that regulates many body processes (sleep-wake cycles, hormonal patterns, etc.) in a roughly 24-hour cycle.

circulatory system The system in animals responsible for moving oxygen, carbon dioxide, and other materials (hormones, nutrients, wastes) around the body.

cisternae (singular: cisterna) Flattened, membrane-bound compartments that make up the Golgi apparatus.

citric acid cycle A series of eight chemical reactions that starts with citrate (citric acid, when protonated) and ends with oxaloacetate, which reacts with acetyl CoA to form citrate—forming a cycle that is part of the pathway that oxidizes glucose to CO_2. Also known as the *Krebs cycle, tricarboxylic acid cycle*, and *TCA cycle*.

clade See **monophyletic group**.

cladistic approach A method for constructing a phylogenetic tree that is based on identifying the unique traits of each monophyletic group. Compare with **phenetic approach**.

Class I MHC protein An MHC protein that is present on the plasma membrane of virtually all nucleated cells and functions in presenting antigen to $CD8^+$ T cells.

Class II MHC protein An MHC protein that is present only on the plasma membrane of dendritic cells, macrophages, and B cells and functions in presenting antigen to $CD4^+$ T cells.

cleavage In animal development, the series of rapid mitotic cell divisions, with little cell growth, that produces successively smaller cells and transforms a zygote into a multicellular blastula, or blastocyst in mammals.

cleavage furrow A pinching-in of the plasma membrane that occurs as cytokinesis begins in animal cells and deepens until the cytoplasm is divided.

climate The prevailing long-term weather conditions in a particular region.

climax community The stable, final community that develops from ecological succession.

clitoris A small rod of erectile tissue in the external genitalia of female mammals. Is formed from the same embryonic tissue as the male penis and has a similar function in sexual arousal.

cloaca An opening to the outside used by the excretory and reproductive systems in a few mammals and many nonmammalian vertebrates.

clonal-selection theory The dominant explanation of the development of adaptive immunity in vertebrates. According to the theory, the immune system retains a vast pool of inactive lymphocytes, each with a unique receptor for a unique antigen. Lymphocytes that encounter their antigens are stimulated to divide (selected and cloned), producing daughter cells that combat infection and confer immunity.

clone (1) An individual that is genetically identical to another individual. (2) A lineage of genetically identical individuals or cells. (3) As a verb, to make one or more genetic replicas of a cell or individual.

cloning vector A plasmid or other agent used to transfer recombinant genes into cultured host cells. Also called simply *vector*.

closed circulatory system A circulatory system in which the circulating fluid (blood) is confined to blood vessels and flows in a continuous circuit. Compare with **open circulatory system**.

club mosses Common name for species in the land plant lineage Lycophyta.

cnidocyte A specialized stinging cell found in cnidarians (e.g., jellyfish, corals, and anemones) that is used in capturing prey.

cochlea The organ of hearing in the inner ear of mammals, birds, and crocodilians. A coiled, fluid-filled tube containing specialized pressure-sensing neurons (hair cells) that detect sounds of different pitches.

coding strand See **non-template strand**.

codominance An inheritance pattern in which heterozygotes exhibit both of the traits seen in either kind of homozygous individual.

codon A sequence of three nucleotides in DNA or RNA that codes for a certain amino acid or that initiates or terminates protein synthesis.

coefficient of relatedness (*r*) A measure of how closely two individuals are related. Calculated as the probability that an allele in two individuals is inherited from the same ancestor.

coelom An internal, usually fluid-filled, body cavity that is lined with mesoderm.

coelomate An animal that has a true coelom. Compare with **acoelomate** and **pseudocoelomate**.

coenocytic Containing many nuclei and a continuous cytoplasm through a filamentous body, without the body being divided into distinct cells. Some fungi are coenocytic.

coenzyme A small organic molecule that is a required cofactor for an enzyme-catalyzed reaction. Often donates or receives electrons or functional groups during the reaction.

coenzyme A (CoA) A nonprotein molecule that is required for many cellular reactions involving transfer of acetyl groups (($-COCH_3$).

coenzyme Q A nonprotein molecule that shuttles electrons between membrane-bound complexes in the mitochondrial electron transport chain. Also called *ubiquinone* or *Q*.

coevolution A pattern of evolution in which two interacting species reciprocally influence each other's adaptations over time.

coevolutionary arms race A series of adaptations and counter-adaptations observed in species that interact closely over time and affect each other's fitness.

cofactor A metal ion or small organic compound that is required for an enzyme to function normally. May be bound tightly to an enzyme or associate with it transiently during catalysis.

cohesion The tendency of certain like molecules (e.g., water molecules) to cling together due to attractive forces. Compare with **adhesion**.

cohesion-tension theory The theory that water movement upward through plant vascular tissues is due to loss of water from leaves (transpiration), which pulls a cohesive column of water upward.

cohort A group of individuals that are the same age and can be followed through time.

coleoptile A modified leaf that covers and protects the stems and leaves of young grasses.

collagen A fibrous, pliable, cable-like glycoprotein that is a major component of the extracellular matrix of animal cells. Various subtypes differ in their tissue distribution.

collecting duct In the vertebrate kidney, a large straight tube that receives filtrate from the distal tubules of several nephrons. Involved in the regulated reabsorption of water.

collenchyma cell In plants, an elongated cell with cell walls thickened at the corners that provides support to growing plant parts; usually found in strands along leaf veins and stalks. Compare with **parenchyma cell** and **sclerenchyma cell**.

colon The portion of the large intestine where feces are formed by compaction of wastes and reabsorption of water.

colony An assemblage of individuals. May refer to an assemblage of semi-independent cells or to a breeding population of multicellular organisms.

commensalism (adjective: commensal) A symbiotic relationship in which one organism (the commensal) benefits and the other (the host) is not harmed. Compare with **mutualism** and **parasitism**.

communication In ecology, any process in which a signal from one individual modifies the behavior of another individual.

community All of the species that interact with each other in a certain area.

companion cell In plants, a cell in the phloem that is connected via numerous plasmodesmata to adjacent sieve-tube members. Companion cells provide materials to maintain sieve-tube members and function in the loading and unloading of sugars into sieve-tube members.

compass orientation A type of navigation in which movement occurs in a specific direction.

competition In ecology, the interaction of two species or two individuals trying to use the same limited resource (e.g., water, food, living space). May occur between individuals of the same species (intraspecific competition) or different species (interspecific competition).

competitive exclusion principle The principle that two species cannot coexist in the same ecological niche in the same area because one species will out-compete the other.

competitive inhibition Inhibition of an enzyme's ability to catalyze a chemical reaction via a nonreactant molecule that competes with the substrate(s) for access to the active site.

complement system A set of proteins that circulate in the bloodstream and can form holes in the plasma membrane of bacteria, leading to their destruction.

complementary base pairing The association between specific nitrogenous bases of nucleic acids stabilized by hydrogen bonding. Adenine pairs only with thymine (in DNA) or uracil (in RNA), and guanine pairs only with cytosine.

complementary DNA (cDNA) DNA produced in the laboratory using an RNA transcript as a template and reverse transcriptase; corresponds to a gene but lacks introns. Also produced naturally by retroviruses.

complementary strand A newly synthesized strand of RNA or DNA that has a base sequence complementary to that of the template strand.

complete digestive tract A digestive tract with two openings, usually called a mouth and an anus.

complete metamorphosis See **holometabolous metamorphosis**.

compound eye An eye formed of many independent light-sensing columns (ommatidia); occurs in arthropods. Compare with **simple eye**.

concentration gradient Difference across space (e.g., across a membrane) in the concentration of a dissolved substance.

condensation reaction A chemical reaction in which two molecules are joined covalently with the removal of an —OH from one and an —H from another to form water. Also called a *dehydration reaction*. Compare with **hydrolysis**.

conduction (1) Direct transfer of heat between two objects that are in physical contact. Compare with **convection**. (2) Transmission of an electrical impulse along the axon of a nerve cell.

cone cell A photoreceptor cell with a cone-shaped outer portion that is particularly sensitive to bright light of a certain color. Also called simply *cone*. Compare with **rod cell**.

connective tissue An animal tissue consisting of scattered cells in a liquid, jellylike, or solid extracellular matrix. Includes bone, cartilage, tendons, ligaments, and blood.

conservation biology The effort to study, preserve, and restore threatened populations, communities, and ecosystems.

constant (C) region The portion of an antibody's light chains or heavy chains that has the same amino acid sequence in the antibodies produced by every B cell of an individual. Compare with **variable (V) region**.

constitutive Always occurring; always present. Commonly used to describe enzymes and other proteins that are synthesized continuously or mutants in which one or more genetic loci are constantly expressed due to defects in gene control.

constitutive defense A defensive trait that is always manifested even in the absence of a predator or pathogen. Also called *standing defense*. Compare with **inducible defense**.

constitutive mutant An abnormal (mutated) gene that produces a product at all times, instead of under certain conditions only.

consumer See **heterotroph**.

consumption Predation or herbivory.

continental shelf The portion of a geologic plate that extends from a continent to under seawater.

continuous strand See **leading strand**.

control In a scientific experiment, a group of organisms or samples that do not receive the experimental treatment but are otherwise identical to the group that does.

convection Transfer of heat by movement of large volumes of a gas or liquid. Compare with **conduction**.

convergent evolution The independent evolution of analogous traits in distantly related organisms due to adaptation to similar environments and a similar way of life.

cooperative binding The tendency of the protein subunits of hemoglobin to affect each other's oxygen binding such that each bound oxygen molecule increases the likelihood of further oxygen binding.

coprophagy The eating of feces.

copulation The act of transferring sperm from a male directly into a female's reproductive tract.

coral reef A large assemblage of colonial marine corals that usually serves as shallow-water, sunlit habitat for many other species as well.

core enzyme The enzyme responsible for catalysis in a multi-part holoenzyme.

co-receptor Any membrane protein that acts with some other membrane protein in a cell interaction or cell response.

cork cambium One of two types of lateral meristem, consisting of a ring of undifferentiated plant cells found just under the cork layer of woody plants; produces new cork cells on its outer side. Compare with **vascular cambium**.

cork cell A waxy cell in the protective outermost layer of a woody plant.

corm A rounded, thick underground stem that can produce new plants via asexual reproduction.

cornea The transparent sheet of connective tissue at the very front of the eye in vertebrates and some other animals. Protects the eye and helps focus light.

corolla All of the petals of a flower.

corona The cluster of cilia at the anterior end of a rotifer.

corpus callosum A thick band of neurons that connects the two hemispheres of the cerebrum in the mammalian brain.

corpus luteum A yellowish structure in an ovary that secretes progesterone. Is formed from a follicle that has recently ovulated.

cortex (1) The outermost region of an organ, such as the kidney or adrenal gland. (2) In plants, a layer of ground tissue found outside the vascular bundles and pith of a plant stem.

cortical granules Small enzyme-filled vesicles in the cortex of an egg cell. Involved in formation of the fertilization envelope after fertilization.

corticotropin-releasing hormone (CRH) A peptide hormone, produced and secreted by the hypothalamus, that stimulates the anterior pituitary to release ACTH.

cortisol A steroid hormone, produced and secreted by the adrenal cortex, that increases blood glucose and prepares the body for stress. The major glucocorticoid hormone in some mammals. Also called *hydrocortisone*.

cost-benefit analysis Decisions or analyses that weigh the fitness costs and benefits of a particular action.

cotransport Transport of an ion or molecule against its electrochemical gradient, in company with an ion or molecule being transported with its electrochemical gradient. Also called *secondary active transport.*

cotransporter A transmembrane protein that facilitates diffusion of an ion down its previously established electrochemical gradient and uses the energy of that process to transport some other substance, in the same or opposite direction, *against* its concentration gradient. Also called *secondary active transporter.* See **antiporter** and **symporter**.

cotyledon The first leaf, or seed leaf, of a plant embryo. Used for storing and digesting nutrients and/or for early photosynthesis.

countercurrent exchanger In animals, any anatomical arrangement that allows the maximum transfer of heat or a soluble substance from one fluid to another. The two fluids must be flowing in opposite directions and have a heat or concentration gradient between them.

covalent bond A type of chemical bond in which two atoms share one or more pairs of electrons. Compare with **hydrogen bond** and **ionic bond**.

cranium A bony, cartilaginous, or fibrous case that encloses and protects the brain of vertebrates. Forms part of the skull. Also called *braincase.*

crassulacean acid metabolism (CAM) A variant type of photosynthesis in which CO_2 is stored in organic acids at night when stomata are open and then released to feed the Calvin cycle during the day when stomata are closed. Helps reduce water loss and oxygen loss by photorespiration in hot, dry environments.

cristae (singular: crista) Sac-like invaginations of the inner membrane of a mitochondrion. Location of the electron transport chain and ATP synthase.

Cro-Magnon A prehistoric European population of modern humans (*Homo sapiens*) known from fossils, paintings, sculptures, and other artifacts.

crop A storage organ in the digestive systems of certain vertebrates.

cross-fertilization A mating that combines gametes from different individuals, as opposed to combining gametes from the same individual (self-fertilization).

crossing over The exchange of segments of non-sister chromatids between a pair of homologous chromosomes that occurs during meiosis I.

cross-pollination Pollination of a flower by pollen from another individual, rather than by self-fertilization. Also called *crossing.*

cross-talk Interactions among signaling pathways that modify a cellular response.

crustaceans A lineage of arthropods that includes shrimp, lobster, and crabs. Many have a carapace (a platelike portion of the exoskeleton) and mandibles for biting or chewing.

cryptic species A species that cannot be distinguished from other species by easily identifiable morphological traits.

culture In cell biology, a collection of cells or a tissue growing under controlled conditions, usually in suspension or on the surface of a dish on solid growth medium.

cup fungus See **sac fungus**.

Cushing's disease A human endocrine disorder caused by loss of feedback inhibition of cortisol on ACTH secretion. Characterized by high ACTH and cortisol levels and wasting of body protein reserves.

cuticle A protective coating secreted by the outermost layer of cells of an animal or a plant.

cyanobacteria A lineage of photosynthetic bacteria formerly known as blue-green algae. Likely the first life-forms to carry out oxygenic photosynthesis.

cyclic AMP (cAMP) Cyclic adenosine monophosphate; a small molecule, derived from ATP, that is widely used by cells in signal transduction and transcriptional control.

cyclic photophosphorylation Path of electron flow during the light-dependent reactions of photosynthesis in which photosystem I transfers excited electrons back to the electron transport chain of photosystem II, rather than to $NADP^+$. Also called *cyclic electron flow.* Compare with **Z scheme**.

cyclin One of several regulatory proteins whose concentrations fluctuate cyclically throughout the cell cycle.

cyclin-dependent kinase (Cdk) Any of several related protein kinases that are active only when bound to a cyclin. Involved in control of the cell cycle.

cytochrome *c* (cyt *c*) A soluble iron-containing protein that shuttles electrons between membrane-bound complexes in the mitochondrial electron transport chain.

cytokines A diverse group of autocrine signaling proteins, secreted largely by cells of the immune system, whose effects include stimulating leukocyte production, tissue repair, and fever. Generally function to regulate the intensity and duration of an immune response.

cytokinesis Division of the cytoplasm to form two daughter cells. Typically occurs immediately after division of the nucleus by mitosis or meiosis.

cytokinins A class of plant hormones that stimulate cell division and retard aging.

cytoplasm All of the contents of a cell, excluding the nucleus, bounded by the plasma membrane.

cytoplasmic determinant A regulatory transcription factor or signaling molecule that is distributed unevenly in the cytoplasm of the egg cells of many animals and that directs early pattern formation in an embryo.

cytoplasmic streaming The directed flow of cytosol and organelles that facilitates distribution of materials within some large plant and fungal cells. Occurs along actin filaments and is powered by myosin.

cytoskeleton In eukaryotic cells, a network of protein fibers in the cytoplasm that are involved in cell shape, support, locomotion, and transport of materials within the cell. Prokaryotic cells have a similar but much less extensive network of fibers.

cytosol The fluid portion of the cytoplasm.

cytotoxic T cell An effector T cell that destroys infected cells and cancer cells. Is descended from an activated $CD8^+$ T cell that has interacted with antigen on an infected cell or cancer cell. Also called *cytotoxic T lymphocyte (CTL)* and *killer T cell.* Compare with **helper T cell**.

dalton (Da) A unit of mass equal to 1/12 the mass of one carbon-12 atom; about the mass of 1 proton or 1 neutron.

day-neutral plant A plant whose flowering time is not affected by the relative length of day and night (the photoperiod). Compare with **long-day** and **short-day plant**.

dead space Portions of the air passages that are not involved in gas exchange with the blood, such as the trachea and bronchi.

deciduous Describing a plant that sheds leaves or other structures at regular intervals (e.g., each fall)

decomposer See **detritivore**.

decomposer food chain An ecological network of detritus, decomposers that eat detritus, and predators and parasites of the decomposers.

definitive host The host species in which a parasite reproduces sexually. Compare with **intermediate host**.

dehydration reaction See **condensation reaction**.

deleterious In genetics, referring to any mutation, allele, or trait that reduces an individual's fitness.

demography The study of factors that determine the size and structure of populations through time.

denaturation (verb: denature) For a macromolecule, loss of its three-dimensional structure and biological activity due to breakage of hydrogen bonds and disulfide bonds, usually caused by treatment with excess heat or extreme pH conditions.

dendrite A short extension from a neuron's cell body that receives signals from other neurons.

dendritic cell A type of leukocyte that ingests and digests foreign antigens, moves to a lymph node, and presents the antigens displayed on its membrane to $CD4^+$ T cells.

dense connective tissue A type of connective tissue, distinguished by having an extracellular matrix dominated by collagen fibers.

density dependent In population ecology, referring to any characteristic that varies depending on population density.

density independent In population ecology, referring to any characteristic that does not vary with population density.

deoxyribonucleic acid (DNA) A nucleic acid composed of deoxyribonucleotides that carries the genetic information of a cell. Generally occurs as two intertwined strands, but these can be separated. See **double helix**.

deoxyribonucleoside triphosphate (dNTP) A monomer that can be polymerized to form DNA. Consists of deoxyribose, a base (A, T, G, or C), and three phosphate groups; similar to a nucleotide, but with two more phosphate groups.

deoxyribonucleotide See **nucleotide**.

depolarization Change in membrane potential from its resting negative state to a less negative or a positive state; a normal phase in an action potential. Compare with **hyperpolarization**.

deposit feeder An animal that eats its way through a food-containing substrate.

derived trait A trait that is clearly homologous with a trait found in an ancestor, but which has a new form.

dermal tissue system The tissue forming the outer layer of an organism. In plants, also called *epidermis;* in animals, forms two distinct layers: *dermis* and *epidermis.*

descent with modification The phrase used by Darwin to describe his hypothesis of evolution by natural selection.

desmosome A type of cell-cell attachment structure, consisting of cadherin proteins, that binds the cytoskeletons of adjacent animal cells together. Found where cells are strongly attached to each other. Compare with **gap junction** and **tight junction**.

detergent A type of small amphipathic molecule used to solubilize hydrophobic molecules in aqueous solution.

determination In embryogenesis, progressive changes in a cell that commit it to a particular cell fate. Once a cell is fully determined, it can differentiate only into a particular cell type (e.g., liver cell, brain cell).

detritivore An organism whose diet consists mainly of dead organic matter (detritus). Various bacteria, fungi, and protists are detritivores. Also called *decomposer.*

detritus A layer of dead organic matter that accumulates at ground level or on seafloors and lake bottoms.

deuterostomes A major lineage of animals that share a pattern of embryological development, including formation of the anus earlier than the mouth, and formation of the coelom by pinching off of layers of mesoderm from the gut. Includes echinoderms and chordates. Compare with **protostomes**.

developmental homology A similarity in embryonic form, or in the fate of embryonic tissues, that is due to inheritance from a common ancestor.

diabetes insipidus A human disease caused by defects in the kidney's system for conserving water. Characterized by production of large amounts of dilute urine.

diabetes mellitus A human disease caused by defects in insulin production (type I) or the response of cells to insulin (type II). Characterized by abnormally high blood glucose levels and large amounts of glucose-containing urine.

diaphragm An elastic, sheetlike structure. In mammals, the muscular sheet of tissue that separates the chest and abdominal cavities. Contracts and moves downward during inhalation, expanding the chest cavity.

diastole The portion of the heartbeat cycle during which the atria or ventricles of the heart are relaxed. Compare with **systole**.

diastolic blood pressure The force exerted by blood against artery walls during relaxation of the heart's left ventricle. Compare with **systolic blood pressure**.

dicot Any plant that has two cotyledons (embryonic leaves) upon germination. The dicots do not form a monophyletic group. Also called *dicotyledonous plant.* Compare with **eudicot** and **monocot**.

dideoxy sequencing A laboratory technique for determining the exact nucleotide sequence of DNA. Relies on the use of dideoxynucleotide triphosphates (ddNTPs), which terminate DNA replication.

diencephalon The part of the mammalian brain that relays sensory information to the cerebellum and functions in maintaining homeostasis.

differential centrifugation Procedure for separating cellular components according to their size and density by spinning a cell homogenate in a series of centrifuge runs. After each run, the supernatant is removed from the deposited material (pellet) and spun again at progressively higher speeds.

differential gene expression Expression of different sets of genes in cells with the same genome. Responsible for creating different cell types.

differentiation The process by which a relatively unspecialized cell becomes a distinct specialized cell type (e.g., liver cell, brain cell) usually by changes in gene expression. Also called *cell differentiation.*

diffusion Spontaneous movement of a substance from a region of high concentration to one of low concentration (i.e., down a concentration gradient).

digestion The physical and chemical breakdown of food into molecules that can be absorbed into the body of an animal.

digestive tract The long tube that begins at the mouth and ends at the anus. Also called *alimentary canal, gastrointestinal (GI) tract,* or the *gut.*

dihybrid cross A mating between two parents that are heterozygous for both of the two genes being studied.

dikaryotic Having two nuclei.

dimer An association of two molecules, which may be identical (homodimer) or different (heterodimer).

dioecious Describing an angiosperm species that has male and female reproductive structures on separate plants. Compare with **monoecious**.

diploblast (adjective: diploblastic) An animal whose body develops from two basic embryonic cell layers—ectoderm and endoderm. Compare with **triploblast**.

diploid (1) Having two sets of chromosomes ($2n$). (2) A cell or an individual organism with two sets of chromosomes, one set inherited from the maternal parent and one set from the paternal parent. Compare with **haploid**.

direct sequencing A technique for identifying and studying microorganisms that cannot be grown in culture. Involves detecting and amplifying copies of certain specific genes in their DNA, sequencing these genes, and then comparing the sequences with the known sequences from other organisms.

directional selection A pattern of natural selection that favors one extreme phenotype with the result that the average phenotype of a population changes in one direction. Generally reduces overall genetic variation in a population.

disaccharide A carbohydrate consisting of two monosaccharides (sugar residues) linked together.

discontinuous strand See **lagging strand**.

discrete trait An inherited trait that exhibits distinct phenotypic forms rather than the continuous variation characteristic of a quantitative trait such as body height.

dispersal The movement of individuals from their place of origin (birth, hatching) to a new location.

disruptive selection A pattern of natural selection that favors extreme phenotypes at both ends of the range of phenotypic variation. Maintains overall genetic variation in a population. Compare with **stabilizing selection**.

distal tubule In the vertebrate kidney, the convoluted portion of a nephron into which filtrate moves from the loop of Henle. Involved in the regulated reabsorption of sodium and water. Compare with **proximal tubule**.

disturbance In ecology, any event that disrupts a community, usually causing loss of some individuals or biomass from it.

disturbance regime The characteristic disturbances that affect a given ecological community.

disulfide bond A covalent bond between two sulfur atoms, typically in the side groups of some amino acids (e.g., cysteine). Often contributes to tertiary structure of proteins.

DNA See **deoxyribonucleic acid**.

DNA cloning Any of several techniques for producing many identical copies of a particular gene or other DNA sequence.

DNA fingerprinting Any of several methods for identifying individuals by unique features of their genomes. Commonly involves using PCR to produce many copies of certain simple sequence repeats (microsatellites) and then analyzing their lengths.

DNA library See **cDNA library** and **genomic library**.

DNA ligase An enzyme that joins pieces of DNA by catalyzing formation of a phosphodiester bond between the pieces.

DNA microarray A set of single-stranded DNA fragments, representing thousands of different genes, that are permanently fixed to a small glass slide. Can be used to determine which genes are expressed in different cell types, under different conditions, or at different developmental stages.

DNA polymerase Any enzyme that catalyzes synthesis of DNA from deoxyribonucleotides.

domain (1) A section of a protein that has a distinctive tertiary structure and function. (2) A taxonomic category, based on similarities in basic cellular biochemistry, above the kingdom level. The three recognized domains are Bacteria, Archaea, and Eukarya.

dominant Referring to an allele that determines the phenotype of a heterozygous individual. Compare with **recessive**.

dopamine A catecholamine neurotransmitter that functions mainly in a part of the mammalian brain involved with muscle control. Also functions as a hypothalamic inhibitory hormone that inhibits release of prolactin from the interior pituitary; also called *prolactin-inhibiting hormone (PIH).*

dormancy A temporary state of greatly reduced, or no, metabolic activity and growth in plants or plant parts (e.g., seeds, spores, bulbs, and buds).

dorsal Toward an animal's back and away from its belly. The opposite of ventral.

double fertilization An unusual form of reproduction seen in flowering plants, in which one sperm nucleus fuses with an egg to form a zygote and the other sperm nucleus fuses with two polar nuclei to form the triploid endosperm.

double helix The secondary structure of DNA, consisting of two antiparallel DNA strands wound around each other.

Down syndrome A human developmental disorder caused by trisomy of chromosome 21.

downstream In genetics, the direction in which RNA polymerase moves along a DNA strand. Compare with **upstream**.

dynein Any one of a class of motor proteins that use the chemical energy of ATP to "walk" along an adjacent microtubule. Dyneins are responsible for bending of cilia and flagella, play a role in chromosome movement during mitosis, and can transport certain organelles.

early endosome A small membrane-bound vesicle, formed by endocytosis, that is an early stage in the formation of a lysosome.

ecdysone An insect hormone that triggers either molting (to a larger larval form) or metamorphosis (to the adult form), depending on the level of juvenile hormone.

ecdysozoans A major lineage of protostomes (Ecdysozoa) that grow by shedding their external skeletons (molting) and expanding their bodies. Includes arthropods, insects, crustaceans, nematodes, and centipedes. Compare with **lophotrochozoans**.

echinoderms A major lineage of deuterostomes (Echinodermata) distinguished by adult bodies with five-sided radial symmetry, a water vascular system, and tube feet. Includes sea urchins, sand dollars, and sea stars.

echolocation The use of echoes from vocalizations to obtain information about locations of objects in the environment.

ecology The study of how organisms interact with each other and with their surrounding environment.

ecosystem All the organisms that live in a geographic area, together with the nonliving (abiotic) components that affect or exchange materials with the organisms; a community and its physical environment.

ecosystem diversity The variety of biotic components in a region along with abiotic components, such as soil, water, and nutrients.

ecosystem services Alterations of the physical components of an ecosystem by living organisms, especially beneficial changes in the quality of the atmosphere, soil, water, etc.

ecotourism Tourism that is based on observing wildlife or experiencing other aspects of natural areas.

ectoderm The outermost of the three basic cell layers in most animal embryos; gives rise to the outer covering and nervous system. Compare with **endoderm** and **mesoderm**.

ectomycorrhizal fungi (EMF) Fungi whose hyphae form a dense network that covers their host plant's roots but do not enter the root cells.

ectoparasite A parasite that lives on the outer surface of the host's body.

ectotherm An animal that does not use internally generated heat to regulate its body temperature. Compare with **endotherm**.

effector Any cell, organ, or structure with which an animal can respond to external or internal stimuli. Usually functions, along with a sensor and integrator, as part of a homeostatic system.

efferent division The part of the nervous system, consisting primarily of motor neurons, that carries commands from the central nervous system to the body.

egg A mature female gamete and any associated external layers (such as a shell). Larger and less mobile than the male gamete. In animals, also called *ovum.*

ejaculation The release of semen from the copulatory organ of a male animal.

ejaculatory duct A short duct connecting the vas deferens to the urethra, through which sperm move during ejaculation.

elastic Referring to a structure (e.g., lungs) with the ability to stretch and then spring back to its original shape.

electric current A flow of electrical charge past a point. Also called *current.*

electrical potential Potential energy created by a separation of electric charges between two points. Also called *voltage.*

electrocardiogram (EKG) A recording of the electrical activity of the heart, as measured through electrodes on the skin.

electrochemical gradient The combined effect of an ion's concentration gradient and electrical (charge) gradient across a membrane that affects the diffusion of ions across the membrane.

electrolyte Any compound that dissociates into ions when dissolved in water. In nutrition, refers to the major ions necessary for normal cell function.

electromagnetic spectrum The entire range of wavelengths of radiation extending from short wavelengths (high energy) to long wavelengths (low energy). Includes gamma rays, X-rays, ultraviolet, visible light, infrared, microwaves, and radio waves (from short to long wavelengths).

electron acceptor A reactant that gains an electron and is reduced in a reduction-oxidation reaction.

electron carrier Any molecule that readily accepts electrons from and donates electrons to other molecules.

electron donor A reactant that loses an electron and is oxidized in a reduction-oxidation reaction.

electron microscope See **scanning electron microscope** and **transmission electron microscope**.

electron shell A group of orbitals of electrons with similar energies. Electron shells are arranged in roughly concentric layers around the nucleus of an atom, with electrons in outer shells having more energy than those in inner shells. Electrons in the outermost shell, the valence shell, often are involved in chemical bonding.

electron transport chain (ETC) Any set of membrane-bound protein complexes and smaller soluble electron carriers involved in a coordinated series of redox reactions in which the potential energy of electrons transferred from reduced donors is successively decreased and used to pump protons from one side of a membrane to the other.

electronegativity A measure of the ability of an atom to attract electrons toward itself from an atom to which it is bonded.

electroreceptor A sensory cell or organ specialized to detect electric fields.

element A substance, consisting of atoms with a specific number of protons, that cannot be separated into or broken down to any other substance. Elements preserve their identity in chemical reactions.

elongation (1) The process by which messenger RNA lengthens during transcription. (2) The process by which a polypeptide chain lengthens during translation.

elongation factors Proteins involved in the elongation phase of translation, assisting ribosomes in the synthesis of the growing peptide chain.

embryo A young developing organism; the stage after fertilization and zygote formation.

embryo sac The female gametophyte in flowering plants that exhibit alternation of generations.

embryogenesis The process by which a single-celled zygote becomes a multicellular embryo.

Embryophyta An increasing popular name for the lineage called land plants.

embryophyte A plant that nourishes its embryos inside its own body. All land plants are embryophytes.

emergent vegetation Any plants in an aquatic habitat that extend above the surface of the water.

emerging disease Any infectious disease, often a viral disease, that suddenly afflicts significant numbers of humans for the first time; often due to changes in the host species for a pathogen or host population movements.

emigration The migration of individuals away from one population to other populations. Compare with **immigration**.

emulsification (verb: emulsify) The dispersion of fat into an aqueous solution. Usually requires the aid of an amphipathic substance such as a detergent or bile salts, which can break large fat globules into microscopic fat droplets.

endangered species A species whose numbers have decreased so much that it is in danger of extinction throughout all or part of its range.

endemic species A species that lives in one geographic area and nowhere else.

endergonic Referring to a chemical reaction that requires an input of energy to occur and for which the Gibbs free-energy change (ΔG) > 0. Compare with **exergonic**.

endocrine Relating to a chemical signal (hormone) that is released into the bloodstream by a producing cell and acts on a distant target cell.

endocrine system All of the glands and tissues that produce and secrete hormones into the bloodstream.

endocrine gland A gland that secretes hormones directly into the bloodstream or interstitial fluid instead of into ducts. Compare with **exocrine gland**.

endocytosis General term for any pinching off of the plasma membrane that results in the uptake of material from outside the cell. Includes phagocytosis, pinocytosis, and receptor-mediated endocytosis. Compare with **exocytosis**.

endoderm The innermost of the three basic cell layers in most animal embryos; gives rise to the digestive tract and organs that connect to it (liver, lungs, etc.). Compare with **ectoderm** and **mesoderm**.

endodermis In plant roots, a cylindrical layer of cells that separates the cortex from the vascular tissue.

endomembrane system A system of organelles in eukaryotic cells that performs most protein and lipid synthesis. Includes the endoplasmic reticulum (ER), Golgi apparatus, and lysosomes.

endomycorrhizal fungi See **arbuscular mycorrhizal fungi (AMF)**.

endoparasite A parasite that lives inside the host's body.

endophyte (adjective: endophytic) A fungus that lives inside the aboveground parts of a plant in a symbiotic relationship. Compare with **epiphyte**.

endoplasmic reticulum (ER) A network of interconnected membranous sacs and tubules found inside eukaryotic cells. See **rough** and **smooth endoplasmic reticulum**.

endoskeleton Bony and/or cartilaginous structures within the body that provide support. Examples are the spicules of sponges, the plates in echinoderms, and the bony skeleton of vertebrates. Compare with **exoskeleton**.

endosome See **early** and **late endosome**.

endosperm A triploid ($3n$) tissue in the seed of a flowering plant (angiosperm) that serves as food for the plant embryo. Functionally analogous to the yolk in some animal eggs.

endosymbiont An organism that lives in a symbiotic relationship inside the body of its host.

endosymbiosis An association between species in which one lives inside the cell or cells of the other.

endosymbiosis theory The theory that mitochondria and chloroplasts evolved from prokaryotes that were engulfed by host cells and took up a symbiotic existence within those cells, a process termed primary endosymbiosis. In some eukaryotes, chloroplasts originated by secondary endosymbiosis, that is, by engulfing a chloroplast-containing protist and retaining its chloroplasts.

endotherm An animal whose primary source of body heat is internally generated heat. Compare with **ectotherm**.

endothermic Referring to a chemical reaction that absorbs heat. Compare with **exothermic**.

energetic coupling In cellular metabolism, the mechanism by which energy released from an exergonic reaction (commonly, hydrolysis of ATP) is used to drive an endergonic reaction.

energy The capacity to do work or to supply heat. May be stored (potential energy) or available in the form of motion (kinetic energy).

enhancer A regulatory sequence in eukaryotic DNA that may be located far from the gene it controls or within introns of the gene. Binding of specific proteins to an enhancer enhances the transcription of certain genes.

enrichment culture A method of detecting and obtaining cells with specific characteristics by placing a sample, containing many types of cells, under a specific set of conditions (e.g., temperature, salt concentration, available nutrients) and isolating those cells that grow rapidly in response.

entropy (*S*) A quantitative measure of the amount of disorder of any system, such as a group of molecules.

envelope, viral A membrane-like covering that encloses some viruses and their capsid coats, shielding them from attack by the host's immune system.

environmental sequencing The inventory of all the genes in a community or ecosystem by sequencing, analyzing, and comparing the genomes of the component organisms.

enzyme A protein catalyst used by living organisms to speed up and control biological reactions.

epicotyl In some embryonic plants, a portion of the embryonic stem that extends above the cotyledons.

epidemic The spread of an infectious disease throughout a population in a short time period. Compare with **pandemic**.

epidermis The outermost layer of cells of any multicellular organism.

epididymis A coiled tube wrapped around the testis in reptiles, birds, and mammals. The site of the final stages of sperm maturation and storage.

epigenetic inheritance Pattern of inheritance involving differences in phenotype that are not due to changes in the nucleotide sequence of genes.

epinephrine A catecholamine hormone, produced and secreted by the adrenal medulla, that triggers rapid responses relating to the fight-or-flight response. Also called *adrenaline*.

epiphyte (adjective: epiphytic) A nonparasitic plant that grows on trees or other solid objects and is not rooted in soil.

epithelium (plural: epithelia) An animal tissue consisting of sheet-like layers of tightly packed cells that lines an organ, a duct, or a body surface. Also called *epithelial tissue*.

epitope A small region of a particular antigen to which an antibody, B-cell receptor, or T-cell receptor binds.

equilibrium potential The membrane potential at which there is no net movement of a particular ion into or out of a cell.

ER signal sequence A short amino acid sequence that marks a polypeptide for transport to the endoplasmic reticulum where synthesis of the polypeptide chain is completed and the signal sequence removed. See **signal recognition particle**.

erythropoietin (EPO) A peptide hormone, released by the kidney in response to low blood oxygen levels, that stimulates the bone marrow to produce more red blood cells.

esophagus The muscular tube that connects the mouth to the stomach.

essential amino acid An amino acid that an animal cannot synthesize and must obtain from the diet. May refer specifically to one of the eight essential amino acids of adult humans: isoleucine, leucine, lysine, methionine, phenylalanine, threonine, tryptophan, and valine.

essential nutrient Any chemical element, ion, or compound that is required for normal growth, reproduction, and maintenance of a living organism and that cannot be synthesized by the organism.

ester linkage The covalent bond formed by a condensation reaction between a carboxyl group (—COOH) and a hydroxyl group (—OH). Ester linkages join fatty acids to glycerol to form a fat or phospholipid.

estradiol The major estrogen produced by the ovaries of female mammals. Stimulates development of the female reproductive tract, growth of ovarian follicles, and growth of breast tissue.

estrogens A class of steroid hormones, including estradiol, estrone, and estriol, that generally promote female-like traits. Secreted by the gonads, fat tissue, and some other organs.

estrous cycle A female reproductive cycle, seen in all mammals except Old World monkeys and apes (including humans), in which the uterine lining is reabsorbed rather than shed in the absence of pregnancy, and the female is sexually receptive only briefly during mid-cycle (estrus). Compare with **menstrual cycle**.

estuary An environment of brackish (partly salty) water where a river meets the ocean.

ethylene A gaseous plant hormone that induces fruits to ripen, flowers to fade, and leaves to drop.

eudicot A member of a monophyletic group (lineage) of angiosperms that includes complex flowering plants and trees (e.g., roses, daisies, maples). All eudicots have two cotyledons, but not all dicots are members of this lineage. Compare with **dicot** and **monocot**.

Eukarya One of the three taxonomic domains of life consisting of unicellular organisms (most protists, yeast) and multicellular organisms (fungi, plants, animals) distinguished by a membrane-bound cell nucleus, numerous organelles, and an extensive cytoskeleton. Compare with **Archaea** and **Bacteria**.

eukaryote A member of the domain Eukarya; an organism whose cells contain a nucleus, numerous membrane-bound organelles, and an extensive cytoskeleton. May be unicellular or multicellular. Compare with **prokaryote**.

eutherians A lineage of mammals (Eutheria) whose young develop in the uterus and are not housed in an abdominal pouch. Also called *placental mammals.*

evaporation The energy-absorbing phase change from a liquid state to a gaseous state. Many organisms evaporate water as a means of heat loss.

evo-devo Research field focused on how changes in developmentally important genes have led to the evolution of new phenotypes.

evolution (1) The theory that all organisms on Earth are related by common ancestry and that they have changed over time, predominantly via natural selection. (2) Any change in the genetic characteristics of a population over time, especially, a change in allele frequencies.

ex situ conservation Preserving species outside of natural areas; e.g., in zoos, aquaria, or botanic gardens.

excitable membrane A plasma membrane that is capable of generating an action potential. Neurons, muscle cells, and some other cells have excitable membranes.

excitatory postsynaptic potential (EPSP) A change in membrane potential, usually depolarization, at a neuron dendrite that makes an action potential more likely.

exergonic Referring to a chemical reaction that can occur spontaneously, releasing heat and/or increasing entropy, and for which the Gibbs free-energy change (ΔG) < 0. Compare with **endergonic**.

exocrine gland A gland that secretes some substance through a duct into a space other than the circulatory system, such as the digestive tract or the skin sufrace. Compare with **endocrine gland**.

exocytosis Secretion of intracellular molecules (e.g., hormones, collagen), contained within membrane-bounded vesicles, to the outside of the cell by fusion of vesicles to the plasma membrane. Compare with **endocytosis**.

exon A region of a eukaryotic gene that is translated into a peptide or protein. Compare with **intron**.

exoskeleton A hard covering secreted on the outside of the body, used for body support, protection, and muscle attachment. Examples are the shell of mollusks and the outer covering (cuticle) of arthropods. Compare with **endoskeleton**.

exothermic Referring to a chemical reaction that releases heat. Compare with **endothermic**.

exotic species A nonnative species that is introduced into a new area. Exotic species often are competitors, pathogens, or predators of native species.

expansins A class of plant proteins that actively increase the length of the cell wall when the pH of the wall falls below 4.5.

exponential population growth The accelerating increase in the size of a population that occurs when the growth rate is constant and density independent. Compare with **logistic population growth**.

expressed sequence tag (EST) A portion of a transcribed gene (synthesized from an mRNA in a cell), used to find the physical location of that gene in the genome.

extant species A species that is living today.

extensor A muscle that pulls two bones farther apart from each other, as in the extension of a limb or the spine. Compare with **flexor**.

extinct Said of a species that has died out.

extracellular digestion Digestion that takes place outside of an organism, as occurs in many fungi that make and secrete digestive enzymes.

extracellular matrix (ECM) A complex meshwork of proteins (e.g., collagen, fibronectin) and polysaccharides secreted by animal cells and in which they are embedded.

extremophile A bacterium or archaean that thrives in an "extreme" environment (e.g., high-salt, high-temperature, low-temperature, or low-pressure).

F_1 generation First filial generation. The first generation of offspring produced from a mating (i.e., the offspring of the parental generation).

facilitated diffusion Movement of a substance across a plasma membrane down its concentration gradient with the assistance of transmembrane carrier proteins or channel proteins.

facilitation In ecological succession, the phenomenon in which early-arriving species make conditions more favorable for later-arriving species. Compare with **inhibition** and **tolerance**.

facultative aerobe Any organism that can perform aerobic respiration when oxygen is available to serve as an electron acceptor but can switch to fermentation when it is not.

FAD/$FADH_2$ Oxidized and reduced forms, respectively, of flavin adenine dinucleotide. A nonprotein electron carrier that functions in the citric acid cycle and oxidative phosphorylation.

fallopian tube A narrow tube connecting the uterus to the ovary in humans, through which the egg travels after ovulation. Site of fertilization and cleavage. In nonhuman animals, called *oviduct.*

fat A lipid consisting of three fatty acid molecules joined by ester linkages to a glycerol molecule. Also called *triacylglycerol* or *triglyceride.*

fatty acid A lipid consisting of a hydrocarbon chain bonded to a carboxyl group (—COOH) at one end. Used by many organisms to store chemical energy; a major component of animal and plant fats.

fatty-acid binding protein Proteins that bind to fatty acids and enable them to be transported into cells.

fauna All the animals characteristic of a particular region, period, or environment.

feather A specialized skin outgrowth, composed of β-keratin, present in all birds and only in birds. Used for flight, insulation, display, and other purposes.

feces The waste products of digestion.

fecundity The average number of female offspring produced by a single female in the course of her lifetime.

feedback inhibition A type of metabolic control in which high concentrations of the product of a metabolic pathway inhibit one of the enzymes early in the pathway. A form of negative feedback.

fermentation Any of several metabolic pathways that make ATP by transferring electrons from a reduced compound such as glucose to a final electron acceptor other than oxygen. Allows glycolysis to proceed in the absence of oxygen.

ferredoxin In photosynthetic organisms, an iron- and sulfur-containing protein in the electron transport chain of photosystem I. Can transfer electrons to the enzyme $NADP^+$ reductase, which catalyzes formation of NADPH.

fertilization envelope A physical barrier that forms around a fertilized egg in amphibians and some other animals. Formed by an influx of water under the vitelline membrane.

fertilization Fusion of the nuclei of two haploid gametes to form a zygote with a diploid nucleus.

fetal alcohol syndrome A condition, marked by hyperactivity, severe learning disabilities, and depression, thought to be caused by exposure of an individual to high blood alcohol concentrations during embryonic development.

fetus In live-bearing animals, the unborn offspring after the embryonic stage, which usually are developed sufficiently to be recognizable as belonging to a certain species. In humans, from 9 weeks after fertilization until birth.

fiber In plants, a type of elongated sclerenchyma cell that provides support to vascular tissue. Compare with **sclereid**.

fibronectin An abundant protein in the extracellular matrix that binds to other ECM components and to integrins in plasma membranes; helps anchor cells in place. Numerous subtypes are found in different tissues.

Fick's law of diffusion A mathematical relationship that describes the rates of gas exchange in animal respiratory systems.

fight-or-flight response Rapid physiological changes that prepare the body for emergencies. Includes increased heart rate, increased blood pressure, and decreased digestion.

filament Any thin, threadlike structure, particularly (1) the threadlike extensions of a fish's gills or (2) the slender stalk that bears the anthers in a flower.

filter feeder See **suspension feeder**.

filtrate Any fluid produced by filtration, in particular the fluid ("pre-urine") in the nephrons of vertebrate kidneys.

filtration A process of removing large components from a fluid by forcing it through a filter. Occurs in a renal corpuscle of the vertebrate kidney, allowing water and small solutes to pass from the blood into the nephron.

finite rate of increase (λ) The rate of increase of a population over a given period of time. Calculated as the ending population size divided by the starting population size. Compare with **intrinsic rate of increase**.

first law of thermodynamics The principle of physics that energy is conserved in any process. Energy can be transferred and converted into different forms, but it cannot be created or destroyed.

fission (1) A form of asexual reproduction in which a prokaryotic cell divides to produce two genetically similar daughter cells by a process similar to mitosis of eukaryotic cells. Also called *binary fission.* (2) A form of asexual reproduction in which an animal splits into two or more individuals of approximately equal size; common among invertebrates.

fitness The ability of an individual to produce viable offspring relative to others of the same species.

fitness trade-off See **trade-off**.

fixed action pattern (FAP) Highly stereotyped behavior pattern that occurs in a certain invariant way in a certain species. A form of innate behavior.

flaccid Limp as a result of low internal pressure (e.g., a wilted plant leaf). Compare with **turgid**.

flagellum (plural: flagella) A long, cellular projection that undulates (in eukaryotes) or rotates (in prokaryotes) to move the cell through an aqueous environment. See **axoneme**.

flatworms Members of the phylum Platyhelminthes. Distinguished by a broad, flat, unsegmented body that lacks a coelom. Flatworms belong to the lophotrochozoan branch of the protostomes.

flavin adenine dinucleotide See **FAD/FADH$_2$**.

flexor A muscle that pulls two bones closer together, as in the flexing of a limb or the spine. Compare with **extensor**.

floral meristem A group of undifferentiated plant cells that can give rise to the four organs making up a flower.

florigen In plants, a protein hormone that is synthesized in leaves and transported to the shoot apical meristem where it stimulates flowering.

flower In angiosperms, the part of a plant that contains reproductive structures. Typically includes a calyx, a corolla, and one or more stamens and/or carpels. See **perfect** and **imperfect flower**.

fluid connective tissue A type of connective tissue, distinguished by having a liquid extracellular matrix.

fluid feeder An animal that feeds by sucking or mopping up liquids such as nectar, plant sap, or blood.

fluid-mosaic model The widely accepted hypothesis that the plasma membrane and organelle membranes consist of proteins embedded in a fluid phospholipid bilayer.

fluorescence The spontaneous emission of light from an excited electron falling back to its normal (ground) state.

follicle An egg cell and its surrounding ring of supportive cells in a mammalian ovary.

follicle-stimulating hormone (FSH) A peptide hormone, produced and secreted by the anterior pituitary, that stimulates (in females) growth of eggs and follicles in the ovaries or (in males) sperm production in the testes.

follicular phase The first major phase of a menstrual cycle during which follicles grow and estrogen levels increase; ends with ovulation.

food Any nutrient-containing material that can be consumed and digested by animals.

food chain A relatively simple pathway of energy flow through a few species, each at a different trophic level, in an ecosystem. Might include, for example, a primary producer, a primary consumer, a secondary consumer, and a decomposer. Compare with **food web**.

food web Any complex pathway along which energy moves among many species at different trophic levels of an ecosystem.

foot One of the three main parts of the mollusk body; a muscular appendage, used for movement and/or burrowing into sediment.

foraging Searching for food.

forebrain One of the three main regions of the vertebrate brain; includes the cerebrum, thalamus, and hypothalamus. Compare with **hindbrain** and **midbrain**.

fossil Any trace of an organism that existed in the past. Includes tracks, burrows, fossilized bones, casts, etc.

fossil record All of the fossils that have been found anywhere on Earth and that have been formally described in the scientific literature.

founder effect A change in allele frequencies that often occurs when a new population is established from a small group of individuals (founder event) due to sampling error (i.e., the small group is not a representative sample of the source population).

fovea In the vertebrate eye, a portion of the retina where incoming light is focused; contains a high proportion of cone cells.

free energy See **Gibbs free-energy change**.

free radical Any substance containing one or more atoms with an unpaired electron. Unstable and highly reactive.

frequency The number of wave crests per second traveling past a stationary point. Determines the pitch of sound and the color of light.

frequency-dependent selection A pattern of selection in which certain alleles are favored only when they are rare; a form of balancing selection.

fronds The large leaves of ferns.

frontal lobe In the vertebrate brain, one of the four major areas in the cerebrum.

fruit In flowering plants (angiosperms), a mature, ripened plant ovary (or group of ovaries), along with the seeds it contains and any adjacent fused parts. See **aggregate, multiple,** and **simple fruit**.

fruiting body A structure formed in some prokaryotes, fungi, and protists for spore dispersal; usually consists of a base, a stalk, and a mass of spores at the top.

functional genomics The study of how a genome works, that is, when and where specific genes are expressed and how their products interact to produce a functional organism.

functional group A small group of atoms bonded together in a precise configuration and exhibiting particular chemical properties that it imparts to any organic molecule in which it occurs.

fundamental niche The ecological space that a species occupies in its habitat in the absence of competitors. Compare with **realized niche**.

fungi A lineage of eukaryotes that typically have a filamentous body (mycelium) and obtain nutrients by absorption.

fungicide Any substance that can kill fungi or slow their growth.

G protein Any of various peripheral membrane proteins that bind GTP and function in signal transduction. Binding of a signal to its receptor triggers activation of the G protein, leading to production of a second messenger or initiation of a phosphorylation cascade.

G_1 phase The phase of the cell cycle that constitutes the first part of interphase before DNA synthesis (S phase).

G_2 phase The phase of the cell cycle between synthesis of DNA (S phase) and mitosis (M phase); the last part of interphase.

gallbladder A small pouch that stores bile from the liver and releases it as needed into the small intestine during digestion of fats.

gametangium (plural: gametangia) (1) The gamete-forming structure found in all land plants except angiosperms. Contains a sperm-producing antheridium and an egg-producing archegonium. (2) The gamete-forming structure of some chytrid fungi.

gamete A haploid reproductive cell that can fuse with another haploid cell to form a zygote. Most multicellular eukaryotes have two distinct forms of gametes: egg cells (ova) and sperm cells.

gametogenesis The production of gametes (eggs or sperm).

gametophyte In organisms undergoing alternation of generations, the multicellular haploid form that arises from a single haploid spore and produces gametes. A female gametophyte is commonly called an *embryo sac;* a male gametophyte, a *pollen grain.* Compare with **sporophyte**.

ganglia A mass of neurons in a centralized nervous system.

ganglion cell A neuron in the vertebrate retina that collects visual information from one or several bipolar cells and sends it to the brain via the optic nerve.

gap junction A type of cell-cell attachment structure that directly connects the cytoplasms of adjacent animal cells, allowing passage of water, ions, and small molecules between the cells. Compare with **desmosome** and **tight junction**.

gastrin A hormone produced by cells in the stomach lining in response to the arrival of food or to a neural signal from the brain. Stimulates other stomach cells to release hydrochloric acid.

gastrointestinal (GI) tract See **digestive tract**.

gastropods A lineage of mollusk distinguished by a large muscular foot and a unique feeding structure, the radula. Include slugs and snails.

gastrulation The process by which some cells on the outside of a young embryo move to the interior of the embryo, resulting in the three distinct germ layers (endoderm, mesoderm, and ectoderm).

gated channel A channel protein that opens and closes in response to a specific stimulus, such as the binding of a particular molecule or a change in the electrical charge on the outside of the membrane.

gel electrophoresis A technique for separating molecules on the basis of size and electric charge, which affect their differing rates of movement through a gelatinous substance in an electric field.

gemma (plural: gemmae) A small reproductive structure that is produced in some liverworts during the gametophyte phase and can grow into mature gametophyte.

gene A section of DNA (or RNA, for some viruses) that encodes information for building one or more related polypeptides or functional RNA molecules along with the regulatory sequences required for its transcription.

gene duplication The formation of an additional copy of a gene, typically by misalignment of chromosomes during crossing over. Thought to be an important evolutionary process in creating new genes.

gene expression Overall process by which the information encoded in genes is converted into an active product, most commonly a protein. Includes transcription and translation of a gene and in some cases protein activation.

gene family A set of genetic loci whose DNA sequences are extremely similar. Thought to have arisen by duplication of a single ancestral gene and subsequent mutations in the duplicated sequences.

gene flow The movement of alleles between populations; occurs when individuals leave one population, join another, and breed.

gene pool All of the alleles of all of the genes in a certain population.

gene therapy The treatment of an inherited disease by introducing normal alleles.

gene-for-gene hypothesis The hypothesis that there is a one-to-one correspondence between the resistance (*R*) loci of plants and the avirulence (*avr*) loci of pathogenic fungi; particularly, that *R* genes produce receptors and *avr* genes produce molecules that bind to those receptors.

generation The average time between a mother's first offspring and her daughter's first offspring.

genetic bottleneck A reduction in allelic diversity resulting from a sudden reduction in the size of a large population (population bottleneck) due to a random event.

genetic code The set of all 64 codons and the particular amino acids that each specifies.

genetic correlation A type of evolutionary constraint in which selection on one trait causes a change in another trait as well; may occur when the same gene(s) affect both traits.

genetic diversity The diversity of alleles in a population, species, or group of species.

genetic drift Any change in allele frequencies due to random events. Causes allele frequencies to drift up and down randomly over time, and eventually can lead to the fixation or loss of alleles.

genetic homology Similarities in DNA sequences or amino acid sequences that are due to inheritance from a common ancestor.

genetic map An ordered list of genes on a chromosome that indicates their relative distances from each other. Also called a *linkage map* or *meiotic map*. Compare with **physical map**.

genetic marker A genetic locus that can be identified and traced in populations by laboratory techniques or by a distinctive visible phenotype.

genetic model A set of hypotheses that explain how a certain trait is inherited.

genetic recombination A change in the combination of genes or alleles on a given chromosome or in a given individual. Also called *recombination*.

genetic screen Any of several techniques for identifying individuals with a particular type of mutation. Also called a *screen*.

genetic variation (1) The number and relative frequency of alleles present in a particular population. (2) The proportion of phenotypic variation in a trait that is due to genetic rather than environmental influences in a certain population in a certain environment.

genetics The field of study concerned with the inheritance of traits.

genitalia External copulatory organs.

genome All of the hereditary information in an organism, including not only genes but also other non-gene stretches of DNA.

genomic library A set of DNA segments representing the entire genome of a particular organism. Each segment is carried by a plasmid or other cloning vector and can be separated from other segments. Compare with **cDNA library**.

genomics The field of study concerned with sequencing, interpreting, and comparing whole genomes from different organisms.

genotype All of the alleles of every gene present in a given individual. May refer specifically to the alleles of a particular set of genes under study. Compare with **phenotype**.

genus (plural: genera) In Linnaeus' system, a taxonomic category of closely related species. Always italicized and capitalized to indicate that it is a recognized scientific genus.

geologic time scale The sequence of eons, eras, and periods used to describe the geologic history of Earth.

germ cell In animals, any cell that can potentially give rise to gametes. Also called *germ-line cells*.

germ layer In animals, one of the three basic types of tissue formed during gastrulation; gives rise to all other tissues. See **endoderm, mesoderm,** and **ectoderm**.

germ theory of disease The theory that infectious diseases are caused by bacteria, viruses, and other microorganisms.

germination The process by which a seed becomes a young plant.

gestation The duration of embryonic development from fertilization to birth in those species that have live birth.

gibberellins A class of hormones found in plants and fungi that stimulate growth. Gibberellic acid is one of the major gibberellins.

Gibbs free-energy change (ΔG) A measure of the change in potential energy and entropy that occurs in a given chemical reaction. $\Delta G < 0$ for spontaneous reactions and > 0 for nonspontaneous reactions.

gill Any organ in aquatic animals that exchanges gases and other dissolved substances between the blood and the surrounding water. Typically, a filamentous outgrowth of a body surface.

gill arch In aquatic vertebrates, curved region of tissue between the gills. Gills are suspended from the gill arches.

gill filament In fish, one of the many long, thin structures that extend from gill arches into the water and across which gas exchange occurs.

gill lamella (plural: gill lamellae) One of hundreds to thousands of sheetlike structures, each containing a capillary bed, that makes up a gill filament.

gland An organ whose primary function is to secrete some substance, either into the blood (endocrine gland) or into some other space such as the gut or skin (exocrine gland).

glia Collective term for several types of cells in nervous tissue that are not neurons and do not conduct electrical signals but provide support, nourishment, and electrical insulation and perform other functions. Also called *glial cells*.

global carbon cycle See **carbon cycle, global**.

global nitrogen cycle See **nitrogen cycle, global**.

global warming A sustained increase in Earth's average surface temperature.

global water cycle See **water cycle, global**.

glomalin A glycoprotein that is abundant in the hyphae of arbuscular mycorrhizal fungi; when hyphae decay it an important component of soil.

glomerulus (plural: glomeruli) (1) In the vertebrate kidney, a ball-like cluster of capillaries, surrounded by Bowman's capsule, at the beginning of a nephron. (2) In the brain, a ball-shaped cluster of neurons in the olfactory bulb.

glucagon A peptide hormone produced by the pancreas in response to low blood glucose. Raises blood glucose by triggering breakdown of glycogen and stimulating gluconeogenesis. Compare with **insulin**.

glucocorticoids A class of steroid hormones, produced and secreted by the adrenal cortex, that increase blood glucose and prepare the body for stress. Include cortisol and corticosterone. Compare with **mineralocorticoids**.

gluconeogenesis Synthesis of glucose from non-carbohydrate sources (e.g., proteins and fatty acids). Occurs in the liver in response to low insulin levels and high glucagon levels.

glucose Six-carbon monosaccharide whose oxidation in cellular respiration is the major source of ATP in animal cells.

glyceraldehyde-3-phosphate (G3P) The phosphorylated three-carbon compound formed as the result of carbon fixation in the first step of the Calvin cycle.

glycerol A three-carbon molecule that forms the "backbone" of phospholipids and most fats.

glycogen A highly branched storage polysaccharide composed of α-glucose monomers joined by 1,4- and 1,6-glycosidic linkages. The major form of stored carbohydrate in animals.

glycolipid Any lipid molecule that is covalently bonded to a carbohydrate group.

glycolysis A series of 10 chemical reactions that oxidize glucose to produce pyruvate and ATP. Used by all organisms as part of fermentation or cellular respiration.

glycoprotein Any protein with one or more covalently bonded carbohydrate groups.

glycosidic linkage The covalent bond formed by a condensation reaction between two sugar monomers; joins the residues of a polysaccharide.

glycosylation Addition of a carbohydrate group to a molecule.

glyoxisome Specialized type of peroxisome found in plant cells and packed with enzymes for processing the products of photosynthesis.

gnathostomes Animals with jaws.

Golgi apparatus A eukaryotic organelle, consisting of stacks of flattened membranous sacs (cisternae), that functions in processing and sorting proteins and lipids destined to be secreted or directed to other organelles. Also called *Golgi complex*.

gonad An organ that produces reproductive cells, such as a testis or an ovary.

gonadotropin-releasing hormone (GnRH) A peptide hormone, produced and secreted by the hypothalamus, that stimulates release of FSH and LH from the anterior pituitary.

grade In taxonomy, a group of species that share a position in an inferred evolutionary sequence of lineages but that are not a monophyletic group. Also called a *paraphyletic group*.

Gram stain A dye that distinguishes the two general types of cell walls found in bacteria. Used to routinely classify bacteria as Gram-negative or Gram-positive.

Gram-negative Describing bacteria that look pink when treated with a Gram stain. These bacteria have a cell wall composed of a thin layer of peptidoglycan and an outer phospholipid layer.

Gram-positive Describing bacteria that look purple when treated with a Gram stain. These bacteria have cell walls composed of a thick layer of peptidoglycan.

granum (plural: grana) In chloroplasts, a stack of flattened, membrane-bound vesicles (thylakoids) where the light reactions of photosynthesis occur.

gravitropism The growth or movement of a plant in a particular direction in response to gravity.

grazing food chain The ecological network of herbivores and the predators and parasites that consume them.

great apes See **hominids**.

green algae A paraphyletic group of photosynthetic organisms that contain chloroplasts similar to those in green plants. Often classified as protists, green algae are the closest living relatives of land plants and form a monophyletic group with them.

greenhouse gas An atmospheric gas that absorbs and reflects infrared radiation, so that heat radiated from Earth is retained in the atmosphere instead of being lost to space.

gross photosynthetic productivity The efficiency with which all the plants in a given area use the light energy available to them to produce sugars.

gross primary productivity In an ecosystem, the total amount of carbon fixed by photosynthesis, including that used for cellular respiration, over a given time period. Compare with **net primary productivity**.

ground meristem The middle layer of a young plant embryo. Gives rise to the ground tissue system.

ground tissue An embryonic tissue layer that gives rise to parenchyma, collenchyma, and sclerenchyma—tissues other than the epidermis and vascular tissue.

groundwater Any water below the land surface.

growth factor Any of a large number of signaling molecules that are secreted by certain cells and that stimulate other cells to divide or to differentiate.

growth hormone (GH) A peptide hormone, produced and secreted by the mammalian anterior pituitary, that promotes lengthening of the long bones in children and muscle growth, tissue repair, and lactation in adults. Also called *somatotropin*.

GTP See **guanosine triphosphate**.

guanosine triphosphate (GTP) A molecule consisting of guanine, a sugar, and three phosphate groups. Can be hydrolyzed to release free energy. Commonly used in many cellular reactions; also functions in signal transduction in association with G proteins.

guard cell One of two specialized, crescent-shaped cells forming the border of a plant stoma. Guard cells can change shape to open or close the stoma. See also **stoma**.

gustation The perception of taste.

guttation Excretion of water droplets from plant leaves in the early morning, caused by root pressure.

gymnosperm A vascular plant that makes seeds but does not produce flowers. The gymnosperms include four lineages of green plants (cycads, ginkgoes, conifers, and gnetophytes). Compare with **angiosperm**.

H^+-ATPase See **proton pump**.

habitat destruction Human-caused destruction of a natural habitat with replacement by an urban, suburban, or agricultural landscape.

habitat fragmentation The breakup of a large region of a habitat into many smaller regions, separated from others by a different type of habitat.

Hadley cell An atmospheric cycle of large-scale air movement in which warm equatorial air rises, moves north or south, and then descends at approximately 30° N or 30° S latitude.

hair cell A pressure-detecting sensory cell, found in the cochlea, that has tiny "hairs" (stereocilia) jutting from its surface.

hairpin A secondary structure in RNA consisting of a stable loop formed by hydrogen bonding between purine and pyrimidine bases on the same strand.

halophile A bacterium or archaean that thrives in high-salt environments.

Hamilton's rule The proposition that an allele for altruistic behavior will be favored by natural selection only if $Br > C$, where B = the fitness benefit to the recipient, C = the fitness cost to the actor, and r = the coefficient of relatedness between recipient and actor.

haploid (1) Having one set of chromosomes ($1n$). (2) A cell or an individual organism with one set of chromosomes. Compare with **diploid**.

haploid number The number of distinct chromosome sets in a cell. Symbolized as n.

Hardy-Weinberg principle A principle of population genetics stating that genotype frequencies in a large population do not change from generation to generation in the absence of evolutionary processes (e.g., mutation, migration, genetic drift, random mating, and selection).

heart A muscular pump that circulates blood throughout the body.

heart murmur A distinctive sound caused by backflow of blood through a defective heart valve.

heartwood The older xylem in the center of an older stem or root, containing protective compounds and no longer functioning in water transport.

heat Thermal energy that is transferred from an object at higher temperature to one at lower temperature.

heat of vaporization The energy required to vaporize 1 gram of a liquid into a gas.

heat-shock proteins Proteins that facilitate refolding of proteins that have been denatured by heat or other agents.

heavy chain The larger of the two types of polypeptide chains in an antibody molecule; composed of a variable (V) region, which contributes to the antigen-binding site, and a constant (C) region. Differences in heavy-chain constant regions determine the different classes of immunoglobulins (IgA, IgE, etc.). Compare with **light chain**.

helicase An enzyme that catalyzes the breaking of hydrogen bonds between nucleotides of DNA, "unzipping" a double-stranded DNA molecule.

helper T cell An effector T cell that secretes cytokines and in other ways promotes the activation of other lymphocytes. Is descended from an activated $CD4^+$ T cell that has interacted with antigen presented by dendritic cells, macrophages, or B cells.

heme A small molecule that binds to four polypeptides to form hemoglobin; contains an iron atom that can bind oxygen.

hemimetabolous metamorphosis A type of metamorphosis in which the animal increases in size from one stage to the next, but does not dramatically change its body form. Also called *incomplete metamorphosis*.

hemocoel A body cavity, present in arthropods and some mollusks, containing a pool of circulatory fluid (hemolymph) bathing the internal organs.

hemoglobin An oxygen-binding protein consisting of four polypeptide subunits, each containing an oxygen-binding heme group. The major oxygen carrier in mammalian blood.

hemolymph The circulatory fluid of animals with open circulatory systems (e.g., insects) in which the fluid is not confined to blood vessels.

hemophilia A human disease, caused by an X-linked recessive allele, that is characterized by defects in the blood-clotting system.

herbaceous Referring to a plant that is not woody.

herbivore (adjective: herbivorous) An animal that eats primarily plants and rarely or never eats meat. Compare with **carnivore** and **omnivore**.

herbivory The practice of eating plant tissues.

heredity The transmission of traits from parents to offspring via genetic information.

heritable Referring to traits that can be transmitted from one generation to the next.

hermaphrodite An organism that produces both male and female gametes.

heterokaryotic Describing a cell or fungal mycelium containing two or more nuclei that are genetically distinct.

heterospory (adjective: heterosporous) In seed plants, the production of two distinct types of spore-producing structures and thus two distinct types of spores: microspores, which become the male gametophyte, and megaspores, which become the female gametophyte. Compare with **homospory**.

heterotherm An animal whose body temperature varies markedly with environmental conditions. Compare with **homeotherm**.

heterotroph Any organism that cannot synthesize reduced organic compounds from inorganic sources and that must obtain them by eating other organisms. Some bacteria, some archaea, and virtually all fungi and animals are heterotrophs. Also called *consumer*. Compare with **autotroph**.

heterozygote advantage A pattern of natural selection that favors heterozygous individuals compared with homozygotes. Tends to maintain genetic variation in a population. Also called *heterozygote superiority*.

heterozygous Having two different alleles of a certain gene.

hexose A monosaccharide (simple sugar) containing six carbon atoms.

hibernation An energy-conserving physiological state, marked by a decrease in metabolic rate, body temperature, and activity, that lasts for a prolonged period (weeks to months). Occurs in some animals in response to winter cold and scarcity of food. Compare with **torpor**.

hindbrain One of the three main regions of the vertebrate brain; includes the cerebellum and medulla oblongata. Compare with **forebrain** and **midbrain**.

histamine A molecule released from mast cells during an inflammatory response that causes blood vessels to dilate and become more permeable.

histone One of several positively charged (basic) proteins associated with DNA in the chromatin of eukaryotic cells.

histone acetyl transferase (HAT) In eukaryotes, one of a class of enzymes that loosen chromatin structure by adding acetyl groups to histone proteins.

histone code The hypothesis that specific combinations of chemical modifications of histone proteins contain information that influences gene expression.

histone deacetylase (HDAC) In eukaryotes, one of a class of enzymes that recondense chromatin by removing acetyl groups from histone proteins.

HIV See human immunodeficiency virus (HIV).

holoenzyme A multipart enzyme consisting of a core enzyme (containing the active site for catalysis) along with other required proteins.

holometabolous metamorphosis A type of metamorphosis in which the animal completely changes its form. Also called *complete metamorphosis*.

homeosis Replacement of one body part by another normally found elsewhere in the body as the result of mutation in certain developmentally important genes (homeotic genes).

homeostasis (adjective: homeostatic) The array of relatively stable chemical and physical conditions in an animal's cells, tissues, and organs. May be achieved by the body's passively matching the conditions of a stable external environment (conformational homeostasis) or by active physiological processes (regulatory homeostasis) triggered by variations in the external or internal environment.

homeotherm An animal that has a constant or relatively constant body temperature. Compare with **heterotherm**.

homeotic gene Any gene that specifies a particular location within an embryo, leading to the development of structures appropriate for that location. Mutations in homeotic genes cause the development of extra body parts or body parts in the wrong places.

hominids Members of the family Hominidae, which includes humans and extinct related forms; chimpanzees, gorillas, and orangutans. Distinguished by large body size, no tail, and an exceptionally large brain. Also called *great apes*.

hominins Humans and extinct related forms; species in the lineage that branched off from chimpanzees and eventually led to humans.

homologous chromosomes In a diploid organism, chromosomes that are similar in size, shape, and gene content. Also called *homologs*.

homology (adjective: homologous) Similarity among organisms of different species due to their inheritance from a common ancestor. Features that exhibit such similarity (e.g., DNA sequences, proteins, body parts) are said to be homologous. Compare with **homoplasy**.

homoplasy Similarity among organisms of different species due to convergent evolution. Compare with **homology**.

homospory (adjective: homosporous) In seedless vascular plants, the production of just one type of spore. Compare with **heterospory**.

homozygous Having two identical alleles of a certain gene.

hormone Any of numerous different signaling molecules that circulate throughout the body in blood or other body fluids and can trigger characteristic responses in distant target cells at very low concentrations.

hormone-response element A specific sequence in DNA to which a steroid hormone-receptor complex can bind and affect gene transcription.

host cell A cell that has been invaded by an organism such as a parasite or a virus.

***Hox* genes** A class of homeotic genes found in several animal phyla, including vertebrates, that are expressed in a distinctive pattern along the anterior-posterior axis in early embryos and control formation of segment-specific structures.

human Any member of the genus *Homo*, which includes modern humans (*Homo sapiens*) and several extinct species.

human chorionic gonadotropin (hCG) A glycoprotein hormone produced by the human placenta from about week 3 to week 14 of pregnancy. Maintains the corpus luteum, which produces hormones that preserve the uterine lining.

Human Genome Project The multinational research project that sequenced the human genome.

human immunodeficiency virus (HIV) A retrovirus that causes AIDS (acquired immune deficiency syndrome) in humans.

humoral (immune) response The type of immune response that involves generation of antibody-secreting plasma cells from activated B cells. Defends against extracellular pathogens. Compare with **cell-mediated (immune) response**.

humus The completely decayed organic matter in soils.

Huntington's disease A degenerative brain disease of humans caused by an autosomal dominant allele.

hybrid The offspring of parents from two different strains, populations, or species.

hybrid zone A geographic area where interbreeding occurs between two species, sometimes producing fertile hybrid offspring.

hydrocarbon An organic molecule that contains only hydrogen and carbon atoms.

hydrogen bond A weak interaction between two molecules or different parts of the same molecule resulting from the attraction between a hydrogen atom with a partial positive charge and another atom (usually O or N) with a partial negative charge. Compare with **covalent bond** and **ionic bond**.

hydrogen ion (H^+) A single proton with a charge of 1+; typically, one that is dissolved in solution or that is being transferred from one atom to another in a chemical reaction.

hydrolysis A chemical reaction in which a molecule is split into smaller molecules by reacting with water. In biology, most hydrolysis reactions involve the splitting of polymers into monomers. Compare with **condensation reaction**.

hydrophilic Interacting readily with water. Hydrophilic compounds are typically polar compounds containing charged or electronegative atoms. Compare with **hydrophobic**.

hydrophobic Not interacting readily with water. Hydrophobic compounds are typically nonpolar compounds that lack charged or electronegative atoms and often contain many C—C and C—H bonds. Compare with **hydrophilic**.

hydroponic growth Growth of plants in liquid cultures instead of soil.

hydrostatic skeleton A system of body support involving fluid-filled compartments that can change in shape but cannot easily be compressed.

hydroxide ion (OH^-) An oxygen atom and a hydrogen atom joined by a single covalent bond and carrying a negative charge; formed by dissociation of water.

hyperpolarization Change in membrane potential from its resting negative state to an even more negative state; a normal phase in an action potential. Compare with **depolarization**.

hypersensitive reaction An intense allergic response by cells that have been sensitized by previous exposure to an allergen.

hypersensitive response In plants, the rapid death of a cell that has been infected by a pathogen, thereby reducing the potential for infection to spread throughout a plant. Compare with **systemic acquired resistance**.

hypertension Abnormally high blood pressure.

hypertonic Comparative term designating a solution that has a greater solute concentration, and therefore a lower water concentration, than another solution. Compare with **hypotonic** and **isotonic**.

hypha (plural: hyphae) One of the strands of a fungal mycelium (the meshlike body of a fungus). Also found in some protists.

hypocotyl The stem of a very young plant; the region between the cotyledon (embryonic leaf) and the radicle (embryonic root).

hypothalamic-pituitary axis The functional interaction of the hypothalamus and the pituitary gland, which are anatomically distinct but work together to regulate most of the other endocrine glands in the body.

hypothalamus A part of the brain that functions in maintaining the body's internal physiological state by regulating the autonomic nervous system, endocrine system, body temperature, water balance, and appetite.

hypothesis A proposed explanation for a phenomenon or for a set of observations.

hypotonic Comparative term designating a solution that has a lower solute concentration, and therefore a higher water concentration, than another solution. Compare with **hypertonic** and **isotonic**.

immigration The migration of individuals into a particular population from other populations. Compare with **emigration**.

immune system In vertebrates, the system whose primary function is to defend the body against pathogens. Includes several types of cells (e.g., lymphocytes and macrophages) and several organs where they develop or reside (e.g., lymph nodes and thymus).

immunity (adjective: immune) State of being protected against infection by disease-causing pathogens either by relatively nonspecific mechanisms (innate immunity) or by specific mechanisms triggered by exposure to a particular antigen (adaptive immunity).

immunization The conferring of immunity to a particular disease by artificial means.

immunoglobulin (Ig) Any of the class of proteins that function as antibodies.

immunological memory The ability of the immune system to "remember" an antigen and mount a rapid, effective response to a pathogen encountered years or decades earlier.

impact hypothesis The hypothesis that a collision between the Earth and an asteroid caused the mass extinction at the K-P boundary, 65 million years ago.

imperfect flower A flower that contains male parts (stamens) *or* female parts (carpels) but not both. Compare with **perfect flower**.

implantation The process by which an embryo buries itself in the uterine wall and forms a placenta. Occurs in mammals and a few other vertebrates.

in situ hybridization A technique for detecting specific DNAs and mRNAs in cells and tissues by use of labeled probes. Can be used to determine where and when particular genes are expressed in embryos.

inbreeding Mating between closely related individuals. Increases homozygosity of a population and often leads to a decline in the average fitness (inbreeding depression).

inbreeding depression In inbred offspring, fitness declines due to deleterious recessive alleles that are homozygous.

inclusive fitness The combination of (1) direct production of offspring (direct fitness) and (2) extra production of offspring by relatives in response to help provided by the individual in question (indirect fitness).

incomplete digestive tract A digestive tract that has just one opening.

incomplete dominance An inheritance pattern in which the heterozygote phenotype is a blend or combination of both homozygote phenotypes.

incomplete metamorphosis See **hemimetabolous metamorphosis**.

independent assortment, principle of The concept that each pair of hereditary elements (alleles of the same gene) behaves independently of other genes during meiosis. One of Mendel's two principles of genetics.

indeterminate growth A pattern of growth in which an individual continues to increase its overall body size throughout its life.

indicator plate A laboratory technique for detecting mutant cells by growing them on agar plates containing a compound that when metabolized by wild-type cells yields a colored product.

induced fit Change in the shape of the active site of an enzyme, as the result of the initial weak binding of a substrate, so that it binds substrate more tightly.

inducer A small molecule that triggers transcription of a specific gene, often by binding to and inactivating a repressor protein.

inducible defense A defensive trait that is manifested only in response to the presence of a consumer (predator or herbivore) or pathogen. Compare with **constitutive defense**.

infection thread An invagination of the membrane of a root hair through which beneficial nitrogen-fixing bacteria enter the roots of their host plants (legumes).

inflammatory response An aspect of the innate immune response, seen in most cases of infection or tissue injury, in which the affected tissue becomes swollen, red, warm, and painful.

inhibition In ecological succession, the phenomenon in which early-arriving species make conditions less favorable for the establishment of certain later-arriving species. Compare with **facilitation** and **tolerance**.

inhibitory postsynaptic potential (IPSP) A change in membrane potential, usually hyperpolarization, at a neuron dendrite that makes an action potential less likely.

initiation (1) In an enzyme-catalyzed reaction, the stage during which enzymes orient reactants precisely as they bind at specific locations within the enzyme's active site. (2) In DNA transcription, the stage during which RNA polymerase and other proteins assemble at the promoter sequence. (3) In RNA translation, the stage during which a complex consisting of a ribosome, a mRNA molecule, and an aminoacyl tRNA corresponding to the start codon is formed.

initiation factors A class of proteins that assist ribosomes in binding to a messenger RNA molecule to begin translation.

innate behavior Behavior that is inherited genetically, does not have to be learned, and is typical of a species.

innate immune response See **innate immunity**.

innate immunity A set of nonspecific defenses against pathogens that exist before exposure to an antigen and involves mast cells, neutrophils, and macrophages; typically results in an inflammatory response. Compare with **acquired immunity**.

inner cell mass (ICM) A cluster of cells in the interior of a mammalian blastocyst that undergo gastrulation and eventually develop into the embryo.

inner ear The innermost portion of the mammalian ear, consisting of a fluid-filled system of tubes that includes the cochlea (which receives sound vibrations from the middle ear) and the semicircular canals (which function in balance).

insulin A peptide hormone produced by the pancreas in response to high levels of glucose (or amino acids) in blood. Enables cells to absorb glucose and coordinates synthesis of fats, proteins, and glycogen. Compare with **glucagon**.

integral membrane protein Any membrane protein that spans the entire lipid bilayer. Also called *transmembrane protein*. Compare with **peripheral membrane protein**.

integrated pest management In agriculture or forestry, systems for managing insects or other pests that include carefully controlled applications of toxins, introduction of species that prey on pests, planting schemes that reduce the chance of a severe pest outbreak, and other techniques.

integrator A component of an animal's nervous system that functions as part of a homeostatic system by evaluating sensory information and triggering appropriate responses. See **effector** and **sensor**.

integrin Any of a class of cell-surface proteins that bind to fibronectins and other proteins in the extracellular matrix, thus holding cells in place.

intercalated disc A specialized junction between adjacent heart muscle cells that contains gap junctions, allowing electrical signals to pass between the cells.

intermediate disturbance hypothesis The hypothesis that moderate ecological disturbance is associated with higher species diversity than either low or high disturbance.

intermediate filament A long fiber, about 10 nm in diameter, composed of one of various proteins (e.g., keratins, lamins); one of the three types of cytoskeletal fibers. Form networks that help maintain cell shape and hold the nucleus in place. Compare with **actin filament** and **microtubule**.

intermediate host The host species in which a parasite reproduces asexually. Compare with **definitive host**.

interneuron A neuron that passes signals from one neuron to another. Compare with **motor neuron** and **sensory neuron**.

internode The section of a plant stem between two nodes (sites where leaves attach).

interphase The portion of the cell cycle between one mitotic (M) phase and the next. Includes the G_1 phase, S phase, and G_2 phase.

interspecific competition Competition between members of different species for the same limited resource. Compare with **intraspecific competition**.

interstitial fluid The plasma-like fluid found in the region (interstitial space) between cells.

intertidal zone The region between the low-tide and high-tide marks on a seashore.

intraspecific competition Competition between members of the same species for the same limited resource. Compare with **interspecific competition**.

intrinsic rate of increase (r_{max}) The rate at which a population will grow under optimal conditions (i.e., when birthrates are as high as possible and death rates are as low as possible). Compare with **finite rate of increase**.

intron A region of a eukaryotic gene that is transcribed into RNA but is later removed, so it is not translated into a peptide or protein. Compare with **exon**.

invasive species An exotic (nonnative) species that, upon introduction to a new area, spreads rapidly and competes successfully with native species.

inversion A mutation in which a segment of a chromosome breaks from the rest of the chromosome, flips, and rejoins with the opposite orientation as before.

invertebrates A paraphyletic group composed of animals without a backbone; includes about 95 percent of all animal species. Compare with **vertebrates**.

involuntary muscle Muscle that cannot respond to conscious thought.

ion An atom or a molecule that has lost or gained electrons and thus carries an electric charge, either positive (cation) or negative (anion), respectively.

ion channel A type of channel protein that allows certain ions to diffuse across a plasma membrane down an electochemical gradient.

ionic bond A chemical bond that is formed when an electron is completely transferred from one atom to another so that the atoms remain associated due to their opposite electric charges. Compare with **covalent bond** and **hydrogen bond**.

iris A ring of pigmented muscle just below the cornea in the vertebrate eye that contracts or expands to control the amount of light entering the eye through the pupil.

isotonic Comparative term designating a solution that has the same solute concentration and water concentration than another solution. Compare with **hypertonic** and **hypotonic**.

isotope Any of several forms of an element that have the same number of protons but differ in the number of neutrons.

joint A place where two components (bones, cartilages, etc.) of a skeleton meet. May be movable (an articulated joint) or immovable (e.g., skull sutures).

juvenile An individual that has adult-like morphology but is not sexually mature.

juvenile hormone An insect hormone that prevents larvae from metamorphosing into adults.

karyogamy Fusion of two haploid nuclei to form a diploid nucleus. Occurs in many fungi, and in animals and plants during fertilization of gametes.

karyotype The distinctive appearance of all of the chromosomes in an individual, including the number of chromosomes and their length and banding patterns (after staining with dyes).

keystone species A species that has an exceptionally great impact on the other species in its ecosystem relative to its abundance.

kidney In terrestrial vertebrates, one of a paired organ situated at the back of the abdominal cavity that filters the blood, produces urine, and secretes several hormones.

kilocalorie (kcal) A unit of energy often used to measure the energy content of food. A kcal of energy raises 1 g of water 1°C.

kin selection A form of natural selection that favors traits that increase survival or reproduction of an individual's kin at the expense of the individual.

kinesin Any one of a class of motor proteins that use the chemical energy of ATP to transport vesicles, particles, or chromosomes along microtubules.

kinetic energy The energy of motion. Compare with **potential energy**.

kinetochore A protein structure at the centromere where kinetochore microtubules attach to the sister chromatids of a replicated chromosome. Contains motor proteins that move a chromosome along a microtubule.

kinetochore microtubules Microtubules that form during mitosis and meiosis, and which extend from a spindle apparatus to an attachment point—the kinetochore—on a chromosome.

kinocilium (plural: kinocilia) A single cilium that juts from the surface of many hair cells and functions in detection of sound or pressure.

Klinefelter syndrome A syndrome seen in humans who have an XXY karyotype. People with this syndrome have male sex organs, may have some female traits, and are sterile.

knock-out allele A mutant allele that does not function at all, or an organism homozygous for such a mutation. Also called *null allele* or *loss-of-function allele*.

Koch's postulates Four criteria used to determine whether a suspected infectious agent causes a particular disease.

labia major (plural: labium majus) One of two outer folds of skin that protect the labia minora, clitoris, and vaginal opening of female mammals.

labia minora (plural: labium minus) One of two inner folds of skin that protect the opening of the urethra and vagina.

labor The strong muscular contractions of the uterus that expel the fetus during birth.

lactation (verb: lactate) Production of milk from mammary glands of mammals.

lacteal A small lymphatic vessel extending into the center of a villus in the small intestine. Receives chylomicrons containing fat absorbed from food.

lactic acid fermentation Catabolic pathway in which pyruvate produced by glycolysis is converted to lactic acid in the absence of oxygen.

lagging strand In DNA replication, the strand of new DNA that is synthesized discontinuously in a series of short pieces that are later joined together. Also called *discontinuous strand*. Compare with **leading strand**.

large intestine The distal portion of the digestive tract consisting of the cecum, colon, and rectum. Its primary function is to compact the wastes delivered from the small intestine and absorb enough water to form feces.

larva (plural: larvae) An immature stage of a species in which the immature and adult stages have different body forms.

late endosome A membrane-bound vesicle that arises from an early endosome and develops into a lysosome.

latency In viruses that infect animals, the ability to exist in a quiescence state without producing new virions.

lateral bud A bud that forms in the angle between a leaf and a stem and may develop into a lateral (side) branch. Also called *axillary bud*.

lateral gene transfer Transfer of DNA between two different species, especially distantly related species. Commonly occurs among bacteria and archaea via plasmid exchange; also can occur in eukaryotes via viruses and some other mechanisms.

lateral line system A pressure-sensitive sensory organ found in many aquatic vertebrates.

lateral meristem A layer of undifferentiated plant cells found in older stems and roots that is responsible for secondary growth. Also called *cambium* or *secondary meristem*. Compare with **apical meristem**.

lateral root A plant root extending from another, older root.

leaching Loss of nutrients from soil via percolating water.

leading strand In DNA replication, the strand of new DNA that is synthesized in one continuous piece, with nucleotides added to the 3′ end of the growing molecule. Also called *continuous strand.* Compare with **lagging strand**.

leaf The main photosynthetic organ of vascular plants.

leak channel Potassium channel that allows potassium ions to leak out of a neuron in its resting state.

learning An enduring change in an individual's behavior that results from specific experience(s).

leghemoglobin An iron-containing protein similar to hemoglobin. Found in root nodules of legume plants where it binds oxygen, preventing it from poisoning a bacterial enzyme needed for nitrogen fixation.

legumes Members of the pea plant family that form symbiotic associations with nitrogen-fixing bacteria in their roots.

lens A transparent, crystalline structure that focuses incoming light onto a retina or other light-sensing apparatus of an eye.

lenticels Spongy segments in bark that allow gas exchange between cells in a woody stem and the atmosphere.

leptin A hormone produced and secreted by fat cells (adipocytes) that acts to stabilize fat tissue mass in part by inhibiting appetite and increasing energy expenditure.

leukocytes Several types of blood cells, including neutrophils, macrophages, and lymphocytes, that circulate in blood and lymph and function in defense against pathogens. Also called *white blood cells.*

lichen A symbiotic association of a fungus and a photosynthetic alga.

life cycle The sequence of developmental events and phases that occurs during the life span of an organism, from fertilization to offspring production.

life history The sequence of events in an individual's life from birth to reproduction to death, including how an individual allocates resources to growth, reproduction, and activities or structures that are related to survival.

life table A data set that summarizes the probability that an individual in a certain population will survive and reproduce in any given year over the course of its lifetime.

ligand Any molecule that binds to a specific site on a receptor molecule.

ligand-gated channel An ion channel that opens or closes in response to binding by a certain molecule. Compare with **voltage-gated channel**.

light chain The smaller of the two types of polypeptide chains in an antibody molecule; composed of a variable (V) region, which contributes to the antigen-binding site, and a constant (C) region. Compare with **heavy chain**.

lignin A substance found in the secondary cell walls of some plants that is exceptionally stiff and strong. Most abundant in woody plant parts.

limiting nutrient Any essential nutrient whose scarcity in the environment significantly reduces growth and reproduction of organisms.

limnetic zone Open water (not near shore) that receives enough light to support photosynthesis.

lineage See **monophyletic group**.

LINEs (long interspersed nuclear elements) The most abundant class of transposable elements in human genomes; can create copies of itself and insert them elsewhere in the genome. Compare with **SINEs**.

linkage In genetics, a physical association between two genes because they are on the same chromosome; the inheritance patterns resulting from this association.

linkage map See **genetic map**.

lipase Any enzyme that can break down fat molecules into fatty acids and monoglycerides.

lipid Any organic subtance that does not dissolve in water, but dissolves well in nonpolar organic solvents. Lipids include fats, oils, phospholipids, and waxes.

lipid bilayer The basic structural element of all cellular membranes consisting of a two-layer sheet of phospholipid molecules whose hydrophobic tails are oriented toward the inside and hydrophilic heads, toward the outside. Also called *phospholipid bilayer.*

littoral zone Shallow water near shore that receives enough sunlight to support photosynthesis. May be marine or freshwater; often flowering plants are present.

liver A large, complex organ of vertebrates that performs many functions including storage of glycogen, processing and conversion of food and wastes, and production of bile.

lobe-finned fish Fish with fins supported by bony elements that extend down the length of the structure.

locomotion Movement of an organism under its own power.

locus (plural: loci) A gene's physical location on a chromosome.

logistic population growth The density-dependent decrease in growth rate as population size approaches the carrying capacity. Compare with **exponential population growth**.

long interspersed nuclear elements See **LINEs**.

long-day plant A plant that blooms in response to short nights (usually in late spring or early summer in the northern hemisphere). Compare with **day-neutral** and **short-day plant**.

loop of Henle In the vertebrate kidney, a long U-shaped loop in a nephron that extends into the medulla. Functions as a countercurrent exchanger to set up an osmotic gradient that allows reabsorption of water from a subsequent portion of the nephron.

loose connective tissue A type of connective tissue consisting of fibrous proteins in a soft matrix. Often functions as padding for organs.

lophophore A specialized feeding structure found in some lophotrochozoans and used in filter feeding.

lophotrochozoans A major lineage of protostomes (Lophotrochozoa) that grow by extending the size of their skeletons rather than by molting. Many phyla have a specialized feeding structure (lophophore) and/or ciliated larvae (trochophore). Includes rotifers, flatworms, segmented worms, and mollusks. Compare with **ecdysozoans**.

loss-of-function allele See **knock-out allele**.

lumen The interior space of any hollow structure (e.g., the rough ER) or organ (e.g., the stomach).

lung Any respiratory organ used for gas exchange between blood and air.

luteal phase The second major phase of a menstrual cycle, after ovulation, when the progesterone levels are high and the body is preparing for a possible pregnancy.

luteinizing hormone (LH) A peptide hormone, produced and secreted by the anterior pituitary, that stimulates estrogen production, ovulation, and formation of the corpus luteum in females and testosterone production in males.

lymph The mixture of fluid and white blood cells that circulates through the ducts and lymph nodes of the lymphatic system in vertebrates.

lymph node One of numerous small oval structures through which lymph moves in the lymphatic system. Filter the lymph and screen it for pathogens and other antigens. Major sites of lymphocyte activation.

lymphatic system In vertebrates, a body-wide network of thin-walled ducts (or vessels) and lymph nodes, separate from the circulatory system. Collects excess fluid from body tissues and returns it to the blood; also functions as part of the immune system.

lymphocytes Two types of leukocyte—B cells and T cells—that circulate through the bloodstream and lymphatic system and that are responsible for the development of acquired immunity.

lysogenic cycle A type of viral replication in which a viral genome enters a host cell, is inserted into the host's chromosome, and is replicated whenever the host cell divides. When activated, the viral DNA enters the lytic cycle, leading to production of new virus particles. Also called *lysogeny* or *latent growth.* Compare with **lytic cycle**.

lysosome A small organelle in an animal cell containing acids and enzymes that catalyze hydrolysis reactions and can digest large molecules. Compare with **vacuole**.

lysozyme An enzyme that functions in innate immunity by digesting bacterial cell walls. Occurs in saliva, tears, mucus, and egg white.

lytic cycle A type of viral replication in which a viral genome enters a host cell, new virus particles (virions) are made using host enzymes and eventually burst out of the cell, killing it. Also called *replicative growth.* Compare with **lysogenic cycle**.

macromolecule Any very large organic molecule, usually made up of smaller molecules (monomers) joined together into a polymer. The main biological macromolecules are proteins, nucleic acids, and polysaccharides.

macronutrient Any element (e.g., carbon, oxygen, nitrogen) that is required in large quantities for normal growth, reproduction, and maintenance of a living organism. Compare with **micronutrient**.

MADS box A DNA sequence that codes for a DNA-binding motif in proteins; present in floral organ identity genes in plants. Functionally similar sequences are found in some fungal and animal genes.

major histocompatibility protein See **MHC protein**.

maladaptive A trait that lowers fitness.

malaria A human disease caused by four species of the protist *Plasmodium* and passed to humans by mosquitoes.

malignant tumor A tumor that is actively growing and disrupting local tissues and/or is spreading to other organs. Cancer consists of one or more malignant tumors. Compare with **benign tumor**.

Malpighian tubules A major excretory organ of insects, consisting of blind-ended tubes that extend from the gut into the hemocoel. Filter hemolymph to form "pre-urine" and then send it to the hindgut for further processing.

mammals One of the two lineages of amniotes (vertebrates that produce amniotic eggs) distinguished by hair (or fur) and mammary glands. Includes the monotremes (platypuses), marsupials, and eutherians (placental mammals).

mammary glands Specialized exocrine glands that produce and secrete milk for nursing offspring. A diagnostic feature of mammals.

mandibles Any mouthpart used in chewing. In vertebrates, the lower jaw. In insects, crustaceans, and myriapods, the first pair of mouthparts.

mantle One of the three main parts of the mollusk body; the thick outer tissue that protects the visceral mass and may secrete a calcium carbonate shell.

Marfan syndrome A human syndrome involving increased height, long limbs and fingers, an abnormally shaped chest, and heart disorders. Probably caused by mutation in one pleiotropic gene.

marsh A wetland that lacks trees and usually has a slow but steady rate of water flow.

marsupials A lineage of mammals (Marsupiala) that nourish their young in an abdominal pouch after a very short period of development in the uterus.

mass extinction The extinction of a large number of diverse evolutionary groups during a relatively short period of geologic time (about 1 million years). May occur due to sudden and extraordinary environmental changes. Compare with **background extinction**.

mass feeder An animal that takes chunks of food into its mouth.

mass number The total number of protons and neutrons in an atom.

mast cell A type of leukocyte that is stationary (embedded in tissue) and helps trigger the inflammatory response to infection or injury, including secretion of histamine. Particularly important in allergic responses and defense against parasites.

maternal chromosome A chromosome inherited from the mother.

mechanoreceptor A sensory cell or organ specialized for detecting distortions caused by touch or pressure. One example is hair cells in the cochlea.

mediator complex regulatory proteins that form a physical link between regulatory transcription factors that are bound to DNA and the basal transcription complex

medium A liquid or solid in which cells can grow in vitro. Also called *growth medium.*

medulla The innermost part of an organ (e.g., kidney or adrenal gland).

medulla oblongata In vertebrates, a region of the brain stem that along with the cerebellum forms the hindbrain.

medusa (plural: medusae) The free-floating stage in the life cycle of some cnidarians (e.g., jellyfish). Compare with **polyp**.

megapascal (MPa) A unit of pressure (force per unit area), equivalent to 1 million pascals (Pa).

megasporangium (plural: megasporangia) In heterosporous species of plants, a spore-producing structure that produces megaspores, which go on to develop into female gametophytes.

megaspore In seed plants, a haploid (n) spore that is produced in a megasporangium by meiosis of a diploid ($2n$) megasporocyte; develops into a female gametophyte. Compare with **microspore**.

meiosis In sexually reproducing organisms, a special two-stage type of cell division in which one diploid ($2n$) parent cell produces four haploid (n) reproductive cells (gametes); results in halving of the chromosome number. Also called *reduction division.*

meiosis I The first cell division of meiosis, in which synapsis and crossing over occur, and homologous chromosomes are separated from each other, producing daughter cells with half as many chromosomes (each composed of two sister chromatids) as the parent cell.

meiosis II The second cell division of meiosis, in which sister chromatids are separated from each other. Similar to mitosis.

meiotic map See **genetic map**.

membrane potential A difference in electric charge across a cell membrane; a form of potential energy. Also called *membrane voltage.*

memory Retention of learned information.

memory cells A type of lymphocyte responsible for maintenance of immunity for years or decades after an infection. Descended from a B cell or T cell activated during a previous infection.

meniscus (plural: menisci) The concave boundary layer formed at most air-water interfaces due to surface tension.

menstrual cycle A female reproductive cycle seen in Old World monkeys and apes (including humans) in which the uterine lining is shed (menstruation) if no pregnancy occurs. Compare with **estrous cycle**.

menstruation The periodic shedding of the uterine lining through the vagina that occurs in females of Old World monkeys and apes, including humans.

meristem (adjective: meristematic) In plants, a group of undifferentiated cells that can develop into various adult tissues throughout the life of a plant.

mesoderm The middle of the three basic cell layers in most animal embryos; gives rise to muscles, bones, blood, and some internal organs (kidney, spleen, etc.). Compare with **ectoderm** and **endoderm**.

mesoglea A gelatinous material, containing scattered ectodermal cells, that is located between the ectoderm and endoderm of cnidarians (e.g., jellyfish, corals, and anemones).

mesophyll cell A type of cell, found near the surfaces of plant leaves, that is specialized for the light-dependent reactions of photosynthesis.

Mesozoic era The interval of geologic time, from 251 million to 65.5 million years ago, during which gymnosperms were the dominant plants and dinosaurs the dominant vertebrates. Ended with extinction of the dinosaurs.

messenger RNA (mRNA) An RNA molecule that carries encoded information, transcribed from DNA, that specifies the amino acid sequence of a polypeptide.

meta-analysis A comparative analysis of the results of many smaller, previously published studies.

metabolic pathway An ordered series of chemical reactions that build up or break down a particular molecule. Often, each reaction is catalyzed by a different enzyme.

metabolic rate The total energy use by all the cells of an individual. For aerobic organisms, often measured as the amount of oxygen consumed per hour.

metabolic water The water that is produced as a by-product of cellular respiration.

metabolism All the chemical reactions occurring in a living cell or organism.

metagenomics Sequencing of all or most of the genes present in an environment directly (also called environmental sequencing).

metallothioneins Small plant proteins that bind to and prevent excess metal ions from acting as toxins.

metamorphosis Transition from one developmental stage to another, such as from the larval to the adult form of an animal.

metaphase A stage in mitosis or meiosis during which chromosomes line up in the middle of the cell.

metaphase plate The plane along which chromosomes line up during metaphase of mitosis or meiosis; not an actual structure.

metapopulation A population made up of many small, physically isolated populations.

metastasis The spread of cancerous cells from their site of origin to distant sites in the body where they may establish additional tumors.

methanogen A prokaryote that produces methane (CH_4) as a by-product of cellular respiration.

methanotroph An organism that uses methane (CH_4) as its primary electron donor and source of carbon.

methyl salicylate (MeSA) A molecule that is hypothesized to function as a signal, transported among tissues, that triggers systematic acquired resistance in plants—a response to pathogen attack.

methylation The addition of a methyl ($-CH_3$) group to a molecule.

MHC protein One of a large set of mammalian cell-surface glycoproteins involved in marking cells as self and in antigen presentation to T cells. Also called *MHC molecule*. See **Class I** and **Class II MHC protein**.

microbe Any microscopic organism, including bacteria, archaea, and various tiny eukaryotes.

microbiology The field of study concerned with microscopic organisms.

microfilament See **actin filament**.

micrograph A photograph of an image produced by a microscope.

micronutrient Any element (e.g., iron, molybdenum, magnesium) that is required in very small quantities for normal growth, reproduction, and maintenance of a living organism. Compare with **macronutrient**.

micropyle The tiny pore in a plant ovule through which the pollen tube reaches the embryo sac.

microRNA (miRNA) A small, single-stranded RNA associated with proteins in an RNA-induced silencing complex. Can bind to complementary sequences in mRNA molecules, allowing the associated proteins to degrade the bound mRNA or inhibit its translation. See **RNA interference**.

microsatellite A noncoding stretch of eukaryotic DNA consisting of a repeating sequence 1- to 5-base pair long. Also called *simple sequence repeat.*

microsporangim (plural: microsporangia) In heterosporous species of plants, a spore-producing structure that produces microspores, which go on to develop into male gametophytes.

microspore In seed plants, a haploid (n) spore that is produced in a microsporangium by meiosis of a diploid ($2n$) microsporocyte; develops into a male gametophyte. Compare with **megaspore**.

microtubule A long, tubular fiber, about 25 nm in diameter, formed by polymerization of tubulin protein dimers; one of the three types of cytoskeletal fibers. Involved in cell movement and transport of materials within the cell. Compare with **actin filament** and **intermediate filament**.

microtubule-organizing center (MTOC) General term for any structure (e.g., centrosome and basal body) that organizes microtubules in cells.

microvilli (singular: microvillus) Tiny protrusions from the surface of an epithelial cell that increase the surface area for absorption of substances.

midbrain One of the three main regions of the vertebrate brain; includes sensory integrating and relay centers. Compare with **forebrain** and **hindbrain**.

middle ear The air-filled middle portion of the mammalian ear, which contains three small bones (ossicles) that transmit and amplify sound from the tympanic membrane to the inner ear. Is connected to the throat via the eustachian tube.

migration (1) In ecology, a cyclical movement of large numbers of organisms from one geographic location or habitat to another. (2) In population genetics, movement of individuals from one population to another.

millivolt (mV) A unit of voltage equal to 1/1000 of a volt.

mimicry A phenomenon in which one species has evolved (or learns) to look or sound like another species. See **Batesian mimicry** and **Müllerian mimicry**.

mineralocorticoids A class of steroid hormones, produced and secreted by the adrenal cortex, that regulate electrolyte levels and the overall volume of body fluids. Aldosterone is the principal one in humans. Compare with **glucocorticoids**.

minisatellite A noncoding stretch of eukaryotic DNA consisting of a repeating sequence that is 6 to 500 base pairs long. Also called *variable number tandem repeat (VNTR).*

mismatch repair The process by which mismatched base pairs in DNA are fixed.

missense mutation A point mutation (change in a single base pair) that causes a change in the amino acid sequence of a protein. Also called *replacement mutation.*

mitochondrial matrix Central compartment of a mitochondrion, which is lined by the inner membrane; contains the enzymes and substrates of the citric acid cycle and mitochondrial DNA.

mitochondrion (plural: mitochondria) A eukaryotic organelle that is bounded by a double membrane and is the site of aerobic respiration.

mitosis In eukaryotic cells, the process of nuclear division that results in two daughter nuclei genetically identical to the parent nucleus. Subsequent cytokinesis (division of the cytoplasm) yields two daughter cells.

mitosis-promoting factor (MPF) A complex of a cyclin and cyclin-dependent kinase that phosphorylates a number of specific proteins needed to initiate mitosis in eukaryotic cells.

mitotic (M) phase The phase of the cell cycle during which cell division occurs. Includes mitosis and cytokinesis.

model organism An organism selected for intensive scientific study based on features that make it easy to work with (e.g., body size, life span), in the hope that findings will apply to other species.

molarity A common unit of solute concentration equal to the number of moles of a dissolved solute in 1 liter of solution.

mole The amount of a substance that contains 6.022 $\times 10^{23}$ of its elemental entities (e.g., atoms, ions, or molecules). This number of molecules of a compound will have a mass equal to the molecular weight of that compound expressed in grams.

molecular chaperone A protein that facilitates the three-dimensional folding of newly synthesized proteins, usually by an ATP-dependent mechanism.

molecular formula A notation that indicates only the numbers and types of atoms in a molecule, such as H_2O for the water molecule. Compare with **structural formula**.

molecular weight The sum of the mass numbers of all of the atoms in a molecule; roughly, the total number of protons and neutrons in the molecule.

molecule A combination of two or more atoms held together by covalent bonds.

mollusks Members of the phylum Mollusca. Distinguished by a body plan with three main parts: a muscular foot, a visceral mass, and a mantle. Include bivalves (clams, oysters), gastropods (snails, slugs), chitons, and cephalopods (squid, octopuses). Mollusks belong to the lophotrochozoan branch of the protostomes.

molting A method of body growth, used by ecdysozoans, that involves the shedding of an external protective cuticle or skeleton, expansion of the soft body, and growth of a new external layer.

monocot Any plant that has a single cotyledon (embryonic leaf) upon germination. Monocots form a monophyletic group. Also called a monocotyledonous plant. Compare with **dicot**.

monoecious Describing an angiosperm species that has both male and female reproductive structures on each plant. Compare with **dioecious**.

monohybrid cross A mating between two parents that are both heterozygous for a given gene.

monomer A small molecule that can covalently bind to other similar molecules to form a larger macromolecule. Compare with **polymer**.

monophyletic group An evolutionary unit that includes an ancestral population and all of its descendants but no others. Also called a *clade* or *lineage.* Compare with **paraphyletic group**.

monosaccharide A small carbohydrate, such as glucose, that has the molecular formula $(CH_2O)_n$ and cannot be hydrolyzed to form any smaller carbohydrates. Also called *simple sugar.* Compare with **disaccharide** and **polysaccharide**.

monosomy Having only one copy of a particular type of chromosome.

monotremes A lineage of mammals (Monotremata) that lay eggs and then nourish the young with milk. Includes just three living species: the platypus and two species of echidna.

morphogen A molecule that exists in a concentration gradient and provides spatial information to embryonic cells.

morphospecies concept The definition of a species as a population or group of populations that have measurably different anatomical features from other groups. Also called *morphological species concept.* Compare with **biological** and **phylogenetic species concept**.

morphology The shape and appearance of an organism's body and its component parts.

motor neuron A nerve cell that carries signals from the central nervous system (brain and spinal cord) to an effector, such as a muscle or gland. Compare with **interneuron** and **sensory neuron**.

motor protein A class of proteins whose major function is to convert the chemical energy of ATP into motion. Includes dynein, kinesin, and myosin.

MPF See **mitosis-promoting factor**.

mRNA See **messenger RNA**.

mucigel A slimy substance secreted by plant root caps that eases passage of the growing root through the soil.

mucosal-associated lymphoid tissue (MALT) Collective term for lymphocytes and other leukocytes associated with skin cells and with mucus-secreting epithelial tissues in the gut and respiratory tract. Plays important role in preventing entry of pathogens into the body.

mucous cell A type of cell found in the epithelial layer of the stomach; responsible for secreting mucus into the stomach.

mucus (adjective: mucous) A slimy mixture of glycoproteins (called mucins) and water that is secreted in many animal organs for lubrication.

Müllerian inhibitory substance A peptide hormone secreted by the embryonic testis that causes regression (withering away) of the female reproductive ducts.

Müllerian mimicry A type of mimicry in which two (or more) harmful species resemble each other. Compare with **Batesian mimicry**.

multicellularity The state of being composed of many cells that adhere to each other and do not all express the same genes with the result that some cells have specialized functions.

multienzyme complex A group of enzymes that are physically attached to each other, even though each of the enzymes catalyzes a separate—but usually related—chemical reaction.

multiple allelism In a population, the existence of more than two alleles of the same gene.

multiple fruit A fruit (e.g., pineapple) that develops from many separate flowers and thus many carpels. Compare with **aggregate** and **simple fruit**.

multiple sclerosis (MS) A human autoimmune disease caused by the immune system attacking the myelin sheaths that insulate nerve axons.

muscle fiber A single muscle cell.

muscle tissue An animal tissue consisting of bundles of long, thin contractile cells (muscle fibers).

mutagen Any physical or chemical agent that increases the rate of mutation.

mutant An individual that carries a mutation, particularly a new or rare mutation.

mutation Any change in the hereditary material of an organism (DNA in most organisms, RNA in some viruses).

mutualism (adjective: mutualistic) A symbiotic relationship between two organisms (mutualists) that benefits both. Compare with **commensalism** and **parasitism**.

mycelium (plural: mycelia) A mass of underground filaments (hyphae) that form the body of a fungus. Also found in some protists and bacteria.

mycorrhiza (plural: mycorrhizae) A mutualistic association between certain fungi and most vascular plants, sometimes visible as nodules or nets in or around plant roots.

myelin sheath Multiple layers of myelin, derived from the cell membranes of certain glial cells, that is wrapped around the axon of a neuron, providing electrical insulation.

myoD A transcription factor that is critical for differentiation of muscle cells (short for "*myo*blast *d*etermination").

myofibril Long, slender structure composed of contractile proteins organized into repeating units (sarcomeres) in vertebrate heart muscle and skeletal muscle.

myosin Any one of a class of motor proteins that use the chemical energy of ATP to move along actin filaments in muscle contraction, cytokinesis, and vesicle transport.

myriapods A lineage of arthropods with long segmented trunks, each segment bearing one or two pairs of legs. Includes millipedes and centipedes.

NAD^+/NADH Oxidized and reduced forms, respectively, of nicotinamide adenine dinucleotide. A nonprotein electron carrier that functions in many of the redox reactions of metabolism.

$NADP^+$/NADPH Oxidized and reduced forms, respectively, of nicotinamide adenine dinucleotide phosphate. A nonprotein electron carrier that is reduced during the light-dependent reactions in photosynthesis and extensively used in biosynthetic reactions.

natural experiment A situation in which groups to be compared are created by an unplanned, natural change in conditions rather than by manipulation of conditions by researchers.

natural selection The process by which individuals with certain heritable traits tend to produce more surviving offspring than do individuals without those traits, often leading to a change in the genetic makeup of the population. A major mechanism of evolution.

nauplius A distinct planktonic larval stage seen in many crustaceans.

Neanderthal A recently extinct European species of hominid, *Homo neanderthalensis*, closely related to but distinct from modern humans.

nectar The sugary fluid produced by flowers to attract and reward pollinating animals.

nectary A nectar-producing structure in a flower.

negative control Of genes, when a regulatory protein shuts down expression by binding to DNA on or near the gene

negative feedback A self-limiting, corrective response in which a deviation in some variable (e.g., body temperature, blood pH, concentration of some compound) triggers responses aimed at returning the variable to normal. Compare with **positive feedback**.

negative pressure ventilation Ventilation of the lungs that is accomplished by "pulling" air into the lungs by expansion of the rib cage. Compare with **positive pressure ventilation**.

negative-sense virus A virus whose genome contains sequences complementary to those in the mRNA required to produce viral proteins. Compare with **ambisense virus** and **positive-sense virus**.

nematodes See **roundworms**.

nephron One of the tiny tubes within the vertebrate kidney that filter blood and concentrate salts to produce urine. Also called *renal tubule*.

neritic zone Shallow marine waters beyond the intertidal zone, extending down to about 200 meters, where the continental shelf ends.

nerve A long, tough strand of nervous tissue typically containing thousands of axons wrapped in connective tissue; carries impulses between the central nervous system and some other part of the body.

nerve cord A bundle of nerves extending from the brain along the dorsal (back) side of a chordate animal, with cerebrospinal fluid inside a hollow central channel. One of the defining features of chordates.

nerve net A nervous system in which neurons are diffuse instead of being clustered into large masses or tracts.

nervous tissue An animal tissue consisting of nerve cells (neurons) and various supporting cells.

net primary productivity (NPP) In an ecosystem, the total amount of carbon fixed by photosynthesis over a given time period minus the amount oxidized during cellular respiration. Compare with **gross primary productivity**.

net reproductive rate (R_0) The growth rate of a population per generation; equivalent to the average number of female offspring that each female produces over her lifetime.

neural Relating to nerve cells (neurons) and the nervous system.

neural tube A folded tube of ectoderm that forms along the dorsal side of a young vertebrate embryo and that will give rise to the brain and spinal cord.

neuroendocrine Referring to nerve cells (neurons) that release hormones into the blood or to such hormones themselves.

neuron A cell that is specialized for the transmission of nerve impulses. Typically has dendrites, a cell body, and a long axon that forms synapses with other neurons. Also called *nerve cell*.

neurosecretory cell A nerve cell (neuron) that produces and secretes hormones into the bloodstream. Principally found in the hypothalamus. Also called *neuroendocrine cell*.

neurotoxin Any substance that specifically destroys or blocks the normal functioning of neurons.

neurotransmitter A molecule that transmits electrical signals from one neuron to another or from a neuron to a muscle or gland. Examples are acetylcholine, dopamine, serotonin, and norepinephrine.

neutral In genetics, referring to any mutation or mutant allele that has no effect on an individual's fitness.

neutrophil A type of leukocyte, capable of moving through body tissues, that engulfs and digests pathogens and other foreign particles; also secretes various compounds that attack bacteria and fungi.

niche The particular set of habitat requirements of a certain species and the role that species plays in its ecosystem.

niche differentiation The change in resource use by competing species that occurs as the result of character displacement.

nicotinamide adenine dinucleotide See **NAD^+/NADH**.

nicotinamide adenine dinucleotide phosphate See **NADP+/NADPH**.

nitrogen cycle, global The movement of nitrogen among terrestrial ecosystems, the oceans, and the atmosphere.

nitrogen fixation The incorporation of atmospheric nitrogen (N_2) into forms such as ammonia (NH_3) or nitrate (NO_3^-), which can be used to make many organic compounds. Occurs in only a few lineages of bacteria and archaea.

nociceptor A sensory cell or organ specialized to detect tissue damage, usually producing the sensation of pain.

Nod factors Molecules produced by nitrogen-fixing bacteria that help them recognize and bind to roots of legumes.

node (1) In animals, any small thickening (e.g., a lymph node). (2) In plants, the part of a stem where leaves or leaf buds are attached. (3) In a phylogenetic tree, the point where two branches diverge, representing the point in time when an ancestral group split into two or more descendant groups. Also called *fork*.

node of Ranvier One of the periodic unmyelinated sections of a neuron's axon at which an action potential can be regenerated.

nodule Lumplike structure on roots of legume plants that contain symbiotic nitrogen-fixing bacteria.

noncyclic electron flow See **Z scheme**.

nondisjunction An error that can occur during meiosis or mitosis in which one daughter cell receives two copies of a particular chromosome, and the other daughter cell receives none.

nonpolar covalent bond A covalent bond in which electrons are equally shared between two atoms of the same or similar electronegativity. Compare with **polar covalent bond**.

non-sister chromatids The chromatids of a particular type of chromosome (after replication) with respect to the chromatids of its homologous chromosome. Crossing over occurs between non-sister chromatids. Compare with **sister chromatids**.

non-template strand The strand of DNA that is not transcribed during synthesis of RNA. Its sequence corresponds to that of the mRNA produced from the other strand. Also called *coding strand*.

non-vascular plants See **bryophytes**.

norepinephrine A catecholamine used as a neurotransmitter in the sympathetic nervous system. Also is produced by the adrenal medulla and functions as a hormone that triggers rapid responses relating to the fight-or-flight response.

notochord A long, gelatinous, supportive rod down the back of a chordate embryo, below the developing spinal cord. Replaced by vertebrae in most adult vertebrates. A defining feature of chordates.

nuclear envelope The double-layered membrane enclosing the nucleus of a eukaryotic cell.

nuclear lamina A lattice-like sheet of fibrous nuclear lamins, which are one type of intermediate filaments. Lines the inner membrane of the nuclear envelope, stiffening the envelope and helping organize the chromosomes.

nuclear lamins Intermediate filaments that make up the nuclear lamina layer—a lattice-like layer inside the nuclear envelope that stiffens the structure.

nuclear localization signal (NLS) A short amino acid sequence that marks a protein for delivery to the nucleus.

nuclear pore An opening in the nuclear envelope that connects the inside of the nucleus with the cytoplasm and through which molecules such as mRNA and some proteins can pass.

nuclear pore complex A large complex of dozens of proteins lining a nuclear pore, defining its shape and transporting substances through the pore.

nuclease Any enzyme that can break down RNA or DNA molecules.

nucleic acid A macromolecule composed of nucleotide monomers. Generally used by cells to store or transmit hereditary information. Includes ribonucleic acid and deoxyribonucleic acid.

nucleoid In prokaryotic cells, a dense, centrally located region that contains DNA but is not surrounded by a membrane.

nucleolus In eukaryotic cells, specialized structure in the nucleus where ribosomal RNA processing occurs and ribosomal subunits are assembled.

nucleosome A repeating, bead-like unit of eukaryotic chromatin, consisting of about 200 nucleotides of DNA wrapped twice around eight histone proteins.

nucleotide A molecule consisting of a five-carbon sugar (ribose or deoxyribose), a phosphate group, and one of several nitrogen-containing bases. DNA and RNA are polymers of nucleotides containing deoxyribose (deoxyribonucleotides) and ribose (ribonucleotides), respectively. Equivalent to a nucleoside plus one phosphate group.

nucleotide excision repair The process of removing a damaged region in one strand of DNA and correctly replacing it using the undamaged strand as a template.

nucleus (1) The center of an atom, containing protons and neutrons. (2) In eukaryotic cells, the large organelle containing the chromosomes and surrounded by a double membrane. (3) A discrete clump of neuron cell bodies in the brain, usually sharing a distinct function.

null allele See **knock-out allele**.

null hypothesis A hypothesis that specifies what the results of an experiment will be if the main hypothesis being tested is wrong. Often states that there will be no difference between experimental groups.

nutrient A substance that an organism requires for normal growth, maintenance, or reproduction.

occipital lobe In the vertebrate brain, one of the four major areas in the cerebrum.

oceanic zone The waters of the open ocean beyond the continental shelf.

oil A fat that is liquid at room temperature.

Okazaki fragment Short segment of DNA produced during replication of the 5′ to 3′ template strand. Many Okazaki fragments make up the lagging strand in newly synthesized DNA.

olfaction The perception of odors.

olfactory bulb A bulb-shaped projection of the brain just above the nose. Receives and interprets odor information from the nose.

oligodendrocyte A type of glial cell that wraps around axons of some neurons in the central nervous system, forming a myelin sheath that provides electrical insulation. Compare with **Schwann cell**.

oligopeptide A chain composed of fewer than 50 amino acids linked together by peptide bonds. Often referred to simply as *peptide*.

ommatidium (plural: ommatidia) A light-sensing column in an arthropod's compound eye.

omnivore (adjective: omnivorous) An animal whose diet regularly includes both meat and plants. Compare with **carnivore** and **herbivore**.

oncogene Any gene whose protein product stimulates cell division at all times and thus promotes cancer development. Often is a mutated form of a gene involved in regulating the cell cycle. See **proto-oncogene**.

one-gene, one-enzyme hypothesis The hypothesis that each gene is responsible for making one (and only one) protein, in most cases an enzyme that catalyzes a specific reaction. Many exceptions to this hypothesis are now known.

oocyte A cell in the ovary that can undergoes meiosis to produce an ovum.

oogenesis The production of egg cells (ova).

oogonia (singular: oogonia) The diploid cells in an ovary that can divide by mitosis to create more oogonia and primary oocytes, which can undergo meiosis.

open circulatory system A circulatory system in which the circulating fluid (hemolymph) is not confined to blood vessels. Compare with **closed circulatory system**.

open reading frame (ORF) Any DNA sequence, ranging in length from several hundred to thousands of base pairs long, that is flanked by a start codon and a stop codon. ORFs identified by computer analysis of DNA may be functional genes, especially if they have other features characteristic of genes (e.g., promoter sequence).

operator In prokaryotic DNA, a binding site for a repressor protein; located near the start of an operon.

operculum The stiff flap of tissue that covers the gills of teleost fishes.

operon A region of prokaryotic DNA that codes for a series of functionally related genes and is transcribed from a single promoter into a polycistronic mRNA.

opsin A transmembrane protein that is covalently linked to retinal, the light-detecting pigment in rod and cone cells.

optic nerve A bundle of neurons that runs from the eye to the brain.

optimal foraging The concept that animals forage in a way that maximizes the amount of usable energy they take in, given the costs of finding and ingesting their food and the risk of being eaten while they're at it.

orbital The spherical region around an atomic nucleus in which an electron is present most of the time.

ORF See **open reading frame**.

organ A group of tissues organized into a functional and structural unit.

organ system Groups of tissues and organs that work together to perform a function.

organelle Any discrete, membrane-bound structure within a cell (e.g., mitochondrion) that has a characteristic structure and functions.

organic For a compound, containing carbon and hydrogen and usually containing carbon-carbon bonds. Organic compounds are widely used by living organisms.

organism Any living entity that contains one or more cells.

organogenesis A stage of embryonic development, just after gastrulation in vertebrate embryos, during which major organs develop from the three embryonic germ layers.

origin of replication The site on a chromosome at which DNA replication begins.

osmoconformer An animal that does not actively regulate the osmolarity of its tissues but conforms to the osmolarity of the surrounding environment.

osmolarity The concentration of dissolved substances in a solution, measured in moles per liter.

osmoregulation The process by which a living organism controls the concentration of water and salts in its body.

osmoregulator An animal that actively regulates the osmolarity of its tissues.

osmosis Diffusion of water across a selectively permeable membrane from a region of high water concentration (low solute concentration) to a region of low water concentration (high solute concentration).

osmotic potential See **solute potential**.

ossicles, ear In mammals, three bones found in the middle ear that function in transferring and amplifying sound from the outer ear to the inner ear.

ouabain A plant toxin that poisons the sodium-potassium pumps of animals.

outcrossing Reproduction by fusion of the gametes of different individuals, rather than self-fertilization. Typically refers to plants.

outer ear The outermost portion of the mammalian ear, consisting of the pinna (ear flap) and the ear canal. Funnels sound to the tympanic membrane.

outgroup A taxon that is closely related to a particular monophyletic group but is not part of it.

out-of-Africa hypothesis The hypothesis that modern humans (*Homo sapiens*) evolved in Africa and spread to other continents, replacing other *Homo* species without interbreeding with them.

oval window A membrane separating the fluid-filled cochlea from the air-filled middle ear through which sound vibrations pass from the middle ear to the inner ear in mammals.

ovary The egg-producing organ of a female animal, or the seed-producing structure in the female part of a flower.

oviduct See **fallopian tube**.

oviparous Producing eggs that are laid outside the body where they develop and hatch. Compare with **ovoviviparous** and **viviparous**.

ovoviviparous Producing eggs that are retained inside the body until they are ready to hatch. Compare with **oviparous** and **viviparous**.

ovulation The release of an ovum from an ovary of a female vertebrate. In humans, an ovarian follicle releases an egg at the end of the follicular phase of the menstrual cycle.

ovule In flowering plants, the structure inside an ovary that contains the female gametophyte and eventually (if fertilized) becomes a seed.

ovum (plural: ova) See **egg**.

oxidation The loss of electrons from an atom during a redox reaction, either by donation of an electron to another atom or by the shared electrons in covalent bonds moving farther from the atomic nucleus.

oxidative phosphorylation Production of ATP molecules from the redox reactions of an electron transport chain.

oxygen-hemoglobin equilibrium curve The graphical depiction of the percentage of hemoglobin in the blood that will bind to oxygen at various partial pressures of oxygen.

oxygenic Referring to any process or reaction that produces oxygen. Photosynthesis in plants, algae, and cyanobacteria, which involves photosystem II, is oxygenic. Compare with **anoxygenic**.

oxytocin A peptide hormone, secreted by the posterior pituitary, that triggers labor and milk production in females and that stimulates pair bonding, parental care, and affiliative behavior in both sexes.

p53 A tumor-suppressor protein (molecular weight of 53 kilodaltons) that responds to DNA damage by stopping the cell cycle and/or triggering apoptosis. Encoded by the *p53* gene.

pacemaker cell A specialized cardiac muscle cell in the sinoatrial (SA) node of the vertebrate heart that has an inherent rhythm and can generate an electrical impulse that spreads to other heart cells.

paleontologists Scientists who study the fossil record and the history of life.

Paleozoic era The interval of geologic time, from 542 million to 251 million years ago, during which fungi, land plants, and animals first appeared and diversified. Began with the Cambrian explosion and ended with the extinction of many invertebrates and vertebrates.

pancreas A large gland in vertebrates that has both exocrine and endocrine functions. Secretes digestive enzymes into a duct connected to the intestine and several hormones (notably, insulin and glucagon) into the bloodstream.

pancreatic lipase An enzyme that is produced in the pancreas and acts in the small intestine to break bonds in complex fats, releasing small lipids.

pandemic The spread of an infectious disease in a short time period over a wide geographic area and affecting a very high proportion of the population. Compare with **epidemic**.

parabiosis An experimental technique for determining whether a certain physiological phenomenon is regulated by a hormone, by surgically uniting two individuals so that hormones can pass between them.

paracrine Relating to a chemical signal that is released by one cell and affects neighboring cells.

paraphyletic group An evolutionary unit that includes an ancestral population and *some* but not all of its descendants. Paraphyletic groups are not meaningful units in evolution. Compare with **monophyletic group**.

parapodia (singular: parapodium) Appendages found in some annelids from which bristle-like structures (chaetae) extend.

parasite An organism that lives on or in a host species and that damages its host.

parasitism (adjective: parasitic) A symbiotic relationship between two organisms that is beneficial to one organism (the parasite) but detrimental to the other (the host). Compare with **commensalism** and **mutualism**.

parasitoid An organism that has a parasitic larval stage and a free-living adult stage. Most parasitoids are insects that lay eggs in the bodies of other insects.

parasympathetic nervous system The part of the autonomic nervous system that stimulates functions for conserving or restoring energy, such as reduced heart rate and increased digestion. Compare with **sympathetic nervous system**.

parathyroid glands Four small glands, located near or embedded in the thyroid gland of vertebrates, that secrete parathyroid hormone (PTH), a peptide hormone that increases blood calcium.

parenchyma cell In plants, a general type of cell with a relatively thin primary cell wall. These cells, found in leaves, the centers of stems and roots, and fruits, are involved in photosynthesis, starch storage, and new growth. Compare with **collenchyma cell** and **sclerenchyma cell**.

parental care Any action by which an animal expends energy or assumes risks to benefit its offspring (e.g., nest-building, feeding of young, defense).

parental generation The adult organisms used in the first experimental cross in a formal breeding experiment.

parietal cell A cell in the stomach lining that secretes hydrochloric acid.

parietal lobe In the vertebrate brain, one of the four major areas in the cerebrum.

parsimony The logical principle that the most likely explanation of a phenomenon is the most economical or simplest. When applied to comparison of alternative phylogenetic trees, it suggests that the one requiring the fewest evolutionary changes is most likely to be correct.

parthenogenesis Development of offspring from unfertilized eggs; a type of asexual reproduction.

partial pressure The pressure of one particular gas in a mixture; the contribution of that gas to the overall pressure.

particulate inheritance The observation that genes from two parents do not blend together to form a new physical entity in offspring, but instead remain separate or particle-like.

pascal (Pa) A unit of pressure (force per unit area).

passive transport Diffusion of a substance across a plasma membrane or organelle membrane. When this occurs with the assistance of membrane proteins, it is called facilitated diffusion.

patch clamping A technique for studying the electrical currents that flow through individual ion channels by sucking a tiny patch of membrane to the hollow tip of a microelectrode.

paternal chromosome A chromosome inherited from the father.

pathogen (adjective: pathogenic) Any entity capable of causing disease, such as a microbe, virus, or prion.

pattern formation The series of events that determines the spatial organization of an embryo, including alignment of the major body axes and orientation of the limbs.

pattern-recognition receptor One of a class of membrane proteins on leukocytes that bind to molecules on the surface of many bacteria. Part of the innate immune response.

PCR See **polymerase chain reaction.**

peat Semidecayed organic matter that accumulates in moist, low-oxygen environments such as bogs.

pectin A gelatinous polysaccharide found in the primary cell wall of plant cells. Attracts and holds water, forming a gel that helps keep the cell wall moist.

pedigree A family tree of parents and offspring, showing inheritance of particular traits of interest.

penis The copulatory organ of male mammals, used to insert sperm into a female.

pentose A monosaccharide (simple sugar) containing five carbon atoms.

PEP carboxylase An enzyme that catalyzes addition of CO_2 to phosphoenol pyruvate, a three-carbon compound, forming a four-carbon organic acid. Found in mesophyll cells of plants that perform C_4 photosynthesis.

pepsin A protein-digesting enzyme present in the stomach.

pepsinogen The precursor of the digestive enzyme pepsin. Is secreted from cells in the stomach lining and converted to pepsin by the acidic environment of the stomach lumen.

peptide See **oligopeptide.**

peptide bond The covalent bond (C—N) formed by a condensation reaction between two amino acids; links the residues in peptides and proteins.

peptidoglycan A complex structural polysaccharide found in bacterial cell walls.

perennial Describing a plant whose life cycle normally lasts for more than one year. Compare with **annual.**

perfect flower A flower that contains both male parts (stamens) and female parts (carpels). Compare with **imperfect flower.**

perforation In plants, a small hole in the primary and secondary cell walls of vessel elements that allow passage of water.

pericarp The part of a fruit, formed from the ovary wall, that surrounds the seeds and protects them. Corresponds to the flesh of most edible fruits and the hard shells of most nuts.

pericycle In plant roots, a layer of cells that give rise to lateral roots.

peripheral membrane protein Any membrane protein that does not span the entire lipid bilayer and associates with only one side of the bilayer. Compare with **integral membrane protein.**

peripheral nervous system (PNS) All the components of the nervous system that are outside the central nervous system (the brain and spinal cord). Includes the somatic nervous system and the autonomic nervous system.

peristalsis Rhythmic waves of muscular contraction that push food along the digestive tract.

permafrost A permanently frozen layer of icy soil found in most tundra and some taiga.

permeability The tendency of a structure, such as a membrane, to allow a given substance to diffuse across it.

peroxisome An organelle found in most eukaryotic cells that contains enzymes for oxidizing fatty acids and other compounds including many toxins, rendering them harmless. See **glyoxisome.**

petal One of the leaflike organs arranged around the reproductive organs of a flower. Often colored and scented to attract pollinators.

petiole The stalk of a leaf.

pH A measure of the concentration of protons in a solution and thus of acidity or alkalinity. Defined as the negative of the base-10 logarithm of the proton concentration: $pH = -\log[H^+]$.

phagocytosis Uptake by a cell of small particles or cells by pinching off the plasma membrane to form small membrane-bound vesicles; one type of endocytosis.

pharyngeal gill slits A set of parallel openings from the throat through the neck to the outside. A diagnostic trait of chordates.

pharyngeal jaw A secondary jaw in the back of the mouth, found in some fishes. Derived from modified gill arches.

phelloderm A component of bark; specifically, a tissue layer produced to the inside of the cork cambium in plants with secondary growth.

phelloderm In the stems of woody plants, a thin layer of cells located between the outer cork cells and inner cork cambium.

phenetic approach A method for constructing a phylogenetic tree by computing a statistic that summarizes the overall similarity among populations, based on the available data. Compare with **cladistic approach.**

phenology The timing of events during the year, in environments where seasonal changes occur.

phenotype The detectable physical and physiological traits of an individual, which are determined its genetic makeup. Also the specific trait associated with a particular allele. Compare with **genotype.**

phenotypic plasticity Within-species variation in phenotype that is due to differences in environmental conditions. Occurs more commonly in plants than animals.

phenylketonuria (PKU) A disease caused by the inability to process the amino acid phenylalanine.

pheophytin In photosystem II, a molecule that accepts excited electrons from a reaction center chlorophyll and passes them to an electron transport chain.

pheromone A chemical signal, released by one individual into the external environment, that can trigger responses in a different individual.

phloem A plant vascular tissue that conducts sugars; contains sieve-tube members and companion cells. Primary phloem develops from the procambium of apical meristems; secondary phloem, from the vascular cambium of lateral meristems. Compare with **xylem.**

phosphodiester linkage Chemical linkage between adjacent **nucleotide residues** in DNA and RNA. Forms when the phosphate group of one nucleotide condenses with the hydroxyl group on the sugar of another nucleotide. Also known as *phosphodiester bond.*

phosphofructokinase The enzyme that catalyzes synthesis of fructose-1,6-bisphosphate from fructose-6-phosphate, a key reaction (step 3) in glycolysis.

phospholipid A class of lipid having a hydrophilic head (a phosphate group) and a hydrophobic tail (one or more fatty acids). Major components of the plasma membrane and organelle membranes.

phosphorylase An enzyme that breaks down glycogen by catalyzing hydrolysis of the α-glycosidic linkages between the glucose residues.

phosphorylation (verb: phosphorylate) The addition of a phosphate group to a molecule.

phosphorylation cascade A series of enzyme-catalyzed phosphorylation reactions commonly used in signal transduction pathways to amplify and convey a signal inward from the plasma membrane.

photic zone In an aquatic habitat, water that is shallow enough to receive some sunlight (whether or not it is enough to support photosynthesis). Compare with **aphotic zone.**

photon A discrete packet of light energy; a particle of light.

photoperiodism Any response by an organism to the relative lengths of day and night (i.e., photoperiod).

photophosporylation Production of ATP molecules using the energy released as light-excited electrons flow through an electron transport chain during photosynthesis. Involves generation of a proton-motive force during electron transport and its use to drive ATP synthesis.

photoreceptor A molecule, a cell, or an organ that is specialized to detect light.

photorespiration A series of light-driven chemical reactions that consumes oxygen and releases carbon dioxide, basically reversing photosynthesis. Usually occurs when there are high O_2 and low CO_2 concentrations inside plant cells, often in bright, hot, dry environments when stomata must be kept closed.

photoreversibility A change in conformation that occurs in certain plant pigments when they are exposed to their preferred wavelengths of light and that triggers responses by the plant.

photosynthesis The complex biological process that converts the energy of light into chemical energy stored in glucose and other organic molecules. Occurs in plants, algae, and some bacteria.

photosystem One of two types of units, consisting of a central reaction center surrounded by antenna complexes, that is responsible for the light-dependent reactions of photosynthesis.

photosystem I A photosystem that contains a pair of P700 chlorophyll molecules and uses absorbed light energy to produce NADPH.

photosystem II A photosystem that contains a pair of P680 chlorophyll molecules and uses absorbed light energy to split water into protons and oxygen and to produce ATP.

phototroph An organism that produces ATP through photosynthesis.

phototropins A class of plant photoreceptors that detect blue light and initiate phototropic responses.

phototropism Growth or movement of an organism in a particular direction in response to light.

phylogenetic species concept The definition of a species as the smallest monophyletic group in a phylogenetic tree. Compare with **biological** and **morphospecies concept**.

phylogenetic tree A diagram that depicts the evolutionary history of a group of species and the relationships among them.

phylogeny The evolutionary history of a group of organisms.

phylum (plural: phyla) In Linnaeus' system, a taxonomic category above the class level and below the kingdom level. In plants, sometimes called a *division*.

physical map A map of a chromosome that shows the number of base pairs between various genetic markers. Compare with **genetic map**.

physiology The study of how an organism's body functions.

phytoalexin Any small compound produced by a plant to combat an infection (usually a fungal infection).

phytochrome A specialized plant photoreceptor that exists in two shapes depending on the ratio of red to far-red light and is involved in the timing certain physiological processes, such as flowering, stem elongation, and germination.

pigment Any molecule that absorbs certain wavelengths of visible light and reflects or transmits other wavelengths.

piloting A type of navigation in which animals use familiar landmarks to find their way.

pinocytosis Uptake by a cell of extracellular fluid by pinching off the plasma membrane to form small membrane-bound vesicles; one type of endocytosis.

pioneering species Those species that appear first in recently disturbed areas.

pit In plants, a small hole in the secondary cell walls of tracheids that allow passage of water.

pitch The sensation produced by a particular frequency of sound. Low frequencies are perceived as low pitches; high frequencies, as high pitches.

pith In the shoot systems of plants, ground tissue located to the inside of the vascular bundles.

pituitary gland A small gland directly under the brain that is physically and functionally connected to the hypothalamus. Produces and secretes an array of hormones that affect many other glands and organs.

placenta A structure that forms in the pregnant uterus from maternal and fetal tissues. Exchanges nutrients and wastes between mother and fetus, anchors the fetus to the uterine wall, and produces some hormones. Occurs in most mammals and in a few other vertebrates.

placental mammals See **eutherians**.

plankton Any small organism that drifts near the surface of oceans or lakes and swims little if at all.

Plantae The monophyletic group that includes red, green, and glaucophyte algae and land plants.

plantlet A small plant, particularly one that forms on a parent plant via asexual reproduction and drops, becoming an independent individual.

plasma The non-cellular portion of blood.

plasma cell An effector B cell, which produces large quantities of antibodies. Is descended from an activated B cell that has interacted with antigen.

plasma membrane A membrane that surrounds a cell, separating it from the external environment and selectively regulating passage of molecules and ions into and out of the cell. Also called *cell membrane*.

plasmid A small, usually circular, supercoiled DNA molecule independent of the cell's main chromosome(s) in prokaryotes and some eukaryotes.

plasmodesmata (singular: plasmodesma) Physical connection between two plant cells, consisting of gaps in the cell walls through which the two cells' plasma membranes, cytoplasm, and smooth ER can connect directly. Functionally similar to gap junctions in animal cells.

plasmogamy Fusion of the cytoplasm of two individuals. Occurs in many fungi.

plastocyanin A small protein that shuttles electrons from photosystem II to photosystem I during photosynthesis.

plastoquinone (PQ) A nonprotein electron carrier in the chloroplast electron transport chain. Receives excited electrons from pheophytin and passes them to more electronegative molecules in the chain. Also carries protons to the lumen side of the thylakoid membrane, generating a proton-motive force.

platelet A small membrane-bound cell fragment in vertebrate blood that functions in blood clotting. Derived from large cells in the bone marrow.

pleiotropy (adjective: pleiotropic) The ability of a single gene to affect more than one phenotypic trait.

ploidy The number of complete chromosome sets present. *Haploid* refers to a ploidy of 1; *diploid*, a ploidy of 2; *triploid*, a ploidy of 3; and *tetraploid*, a ploidy of 4.

podium (plural: podia) See **tube foot**.

point mutation A mutation that results in a change in a single nucleotide pair in a DNA molecule.

polar (1) Asymmetrical or unidirectional. (2) Carrying a partial positive charge on one side of a molecule and a partial negative charge on the other. Polar molecules are generally hydrophilic.

polar bodies The tiny, nonfunctional cells produced during meiosis of a primary oocyte, due to most of the cytoplasm going to the ovum.

polar covalent bond A covalent bond in which electrons are shared unequally between atoms differing in electronegativity, resulting in the more electronegative atom having a partial negative charge and the other atom, a partial positive charge. Compare with **nonpolar covalent bond**.

polar microtubules Microtubules that form during mitosis and meiosis, and which extend from a spindle apparatus and overlap with each other in the middle of the cell.

polar nuclei In flowering plants, the nuclei in the female gametophyte that fuse with one sperm nucleus to produce the endosperm. Most species have two.

pollen grain In seed plants, a male gametophyte enclosed within a protective coat.

pollen tube In flowering plants, a structure that grows out of a pollen grain after it reaches the stigma, extends down the style, and through which two sperm cells are delivered to the ovule.

pollination The process by which pollen reaches the carpel of a flower (in flowering plants) or reaches the ovule directly (in conifers and their relatives).

poly(A) tail In eukaryotes, a sequence of 100–250 adenine nucleotides added to the 3′ end of newly transcribed messenger RNA molecules.

polygenic inheritance The inheritance patterns that result when many genes influence one trait.

polymer Any long molecule composed of small repeating units (monomers) bonded together. The main biological polymers are proteins, nucleic acids, and polysaccharides.

polymerase chain reaction (PCR) A laboratory technique for rapidly generating millions of identical copies of a specific stretch of DNA by incubating the original DNA sequence of interest with primers, nucleotides, and DNA polymerase.

polymerization (verb: polymerize) The process by which many identical or similar small molecules (monomers) are covalently bonded to form a large molecule (polymer).

polymorphism (adjective: polymorphic) (1) The occurrence of more than one allele at a certain genetic locus in a population. (2) The occurrence of more than two distinct phenotypes of a trait in a population.

polyp The immotile (sessile) stage in the life cycle of some cnidarians (e.g., jellyfish). Compare with **medusa**.

polypeptide A chain of 50 or more amino acids linked together by peptide bonds. Compare with **oligopeptide** and **protein**.

polyploidy (adjective: polyploid) The state of having more than two full sets of chromosomes.

polyribosome A structure consisting of one messenger RNA molecule along with many attached ribosomes and their growing peptide strands.

polysaccharide A linear or branched polymer consisting of many monosaccharides joined by glycosidic linkages. Carbohydrate polymers with relatively few residues often are called *oligosaccharides*.

polyspermy Fertilization of an egg by multiple sperm.

population A group of individuals of the same species living in the same geographic area at the same time.

population density The number of individuals of a population per unit area.

population dynamics Changes in the size and other characteristics of populations through time.

population ecology The study of how and why the number of individuals in a population changes over time.

population thinking The ability to analyze trait frequencies, event probabilities, and other attributes of populations of molecules, cells, or organisms.

pore In land plants, an opening in the epithelium that allows gas exchange.

positive control Of genes, when a regulatory protein triggers expression by binding to DNA on or near the gene.

positive feedback A physiological mechanism in which a change in some variable stimulates a response that increases the change. Relatively rare in organisms but is important in generation of the action potential. Compare with **negative feedback**.

positive pressure ventilation Ventilation of the lungs that is accomplished by "pushing" air into the lungs by positive pressure in the mouth. Compare with **negative pressure ventilation**.

positive-sense virus A virus whose genome contains the same sequences as the mRNA required to produce viral proteins. Compare with **ambisense virus** and **negative-sense virus**.

posterior Toward an animal's tail and away from its head. The opposite of anterior.

posterior pituitary The part of the pituitary gland that contains the ends of hypothalamic neurosecretory cells and from which oxytocin and antidiuretic hormone are secreted. Compare with **anterior pituitary**.

postsynaptic neuron A neuron that receives signals, usually via neurotransmitters, from another neuron at a synapse. Muscle and gland cells also may receive signals from presynaptic neurons.

post-translational control Regulation of gene expression by modification of proteins (e.g., addition of a phosphate group or sugar residues) after translation.

postzygotic isolation Reproductive isolation resulting from mechanisms that operate after mating of individuals of two different species occurs. The most common mechanisms are the death of hybrid embryos or reduced fitness of hybrids.

potential energy Energy stored in matter as a result of its position or molecular arrangement. Compare with **kinetic energy**.

prebiotic soup A hypothetical solution of sugars, amino acids, nitrogenous bases, and other building blocks of larger molecules that may have formed in shallow waters or deep-ocean vents of ancient Earth and given rise to larger biological molecules.

Precambrian The interval between the formation of the Earth, about 4.6 billion years ago, and the appearance of most animal groups about 542 million years ago. Unicellular organisms were dominant for most of this era, and oxygen was virtually absent for the first 2 billion years.

predation The killing and eating of one organism (the prey) by another (the predator).

predator Any organism that kills other organisms for food.

prediction A measurable or observable result of an experiment based on a particular hypothesis. A correct prediction provides support for the hypothesis being tested.

pressure potential (ψ_P) A component of the potential energy of water caused by physical pressures on a solution. In plant cells, it equals the wall pressure plus turgor pressure. Compare with **solute potential (ψ_S)**.

pressure-flow hypothesis The hypothesis that sugar movement through phloem tissue is due to differences in the turgor pressure of phloem sap.

presynaptic neuron A neuron that transmits signals, usually by releasing neurotransmitters, to another neuron or to an effector cell at a synapse.

prezygotic isolation Reproductive isolation resulting from any one of several mechanisms that prevent individuals of two different species from mating.

primary cell wall The outermost layer of a plant cell wall, made of cellulose fibers and gelatinous polysaccharides, that defines the shape of the cell and withstands the turgor pressure of the plasma membrane.

primary consumer An herbivore; an organism that eats plants, algae, or other primary producers. Compare with **secondary consumer**.

primary decomposer A decomposer (detritivore) that consumes detritus from plants.

primary growth In plants, an increase in the length of stems and roots due to the activity of apical meristems. Compare with **secondary growth**.

primary immune response An acquired immune response to a pathogen that the immune system has not encountered before. Compare with **secondary immune response**.

primary oocyte The large diploid cell in an ovarian follicle that can initiate meiosis to produce a haploid ovum.

primary producer Any organism that creates its own food by photosynthesis or from reduced inorganic compounds and that is a food source for other species in its ecosystem. Also called *autotroph*.

primary RNA transcript In eukaryotes, a newly transcribed messenger RNA molecule that has not yet been processed (i.e., it has not received a 5′ cap or poly(A) tail, and still contains introns). Also called *pre-mRNA*.

primary spermatocyte A diploid cell in the testis that can initiate meiosis I to produce two secondary spermatocytes.

primary structure The sequence of amino acids in a peptide or protein; also the sequence of nucleotides in a nucleic acid. Compare with **secondary, tertiary,** and **quaternary structure**.

primary succession The gradual colonization of a habitat of bare rock or gravel, usually after an environmental disturbance that removes all soil and previous organisms. Compare with **secondary succession**.

primase An enzyme that synthesizes a short stretch of RNA to use as a primer during DNA replication.

primates The lineage of mammals that includes prosimians (lemurs, lorises, etc.), monkeys, and great apes (including humans).

primer A short, single-stranded RNA molecule that base pairs with the 5′ end of a DNA template strand and is elongated by DNA polymerase during DNA replication.

prion An infectious form of a protein that is thought to cause disease by inducing the normal form to assume an abnormal three-dimensional structure. Cause of spongiform encephalopathies, such as mad cow disease.

probe A radioactively or chemically labeled single-stranded fragment of a known DNA or RNA sequence that can bind to and thus detect its complementary sequence in a sample containing many different sequences.

proboscis A long, narrow feeding appendage through which food can be obtained.

procambium A group of cells in the center of a young plant embryo that gives rise to the vascular tissue.

product Any of the final materials formed in a chemical reaction.

productivity The total amount of carbon fixed by photosynthesis per unit area per year.

progesterone A steroid hormone produced in the ovaries and secreted by the corpus luteum after ovulation; causes the uterine lining to thicken.

programmed cell death See **apoptosis**.

prokaryote A member of the domain Bacteria or Archaea; a unicellular organism lacking a nucleus and containing relatively few organelles or cytoskeletal components. Compare with **eukaryote**.

prolactin A peptide hormone, produced and secreted by the anterior pituitary, that promotes milk production in female mammals and has a variety of effects on parental behavior and seasonal reproduction in other vertebrates.

prometaphase A stage in mitosis or meiosis during which the nuclear envelope breaks down and kinetochore microtubles attach to chromatids.

promoter A short nucleotide sequence in DNA that binds RNA polymerase, enabling transcription to begin. In prokaryotic DNA, a single promoter often is associated with several contiguous genes. In eukaryotic DNA, each gene generally has its own promoter.

promoter-proximal elements In eukaryotes, regulatory sequences in DNA that are close to a promoter and that can bind regulatory transcription factors.

proofreading The process by which a DNA polymerase recognizes and removes a wrong base added during DNA replication and then continues synthesis.

prophase The first stage in mitosis or meiosis during which chromosomes become visible and the spindle apparatus forms. Synapsis and crossing over occur during prophase of meiosis I.

prosimians One of the two major lineages of primates, including lemurs, tarsiers, pottos, and lorises. Compare with **anthropoids**.

prostate gland A gland in male mammals that surrounds the base of the urethra and secretes a fluid that is a component of semen.

protease An enzyme that can degrade proteins by cleaving the peptide bonds between amino acid residues.

proteasome A multi-molecular machine that destroys proteins that have been bound to ubiquitin.

protein A macromolecule consisting of one or more polypeptide chains composed of 50 or more amino acids linked together. Each protein has a unique sequence of amino acids and, in its native state, a characteristic three-dimensional shape.

protein kinase An enzyme that catalyzes the addition of a phosphate group to another protein, typically activating or inactivating the substrate protein.

proteinase inhibitors Defense compounds, produced by plants, that induce illness in herbivores by inhibiting digestive enzymes.

proteome The complete set of proteins produced by a particular cell type.

proteomics The systematic study of the interactions, localization, functions, regulation, and other features of the full protein set (proteome) in a particular cell type.

protist Any eukaryote that is not a green plant, animal, or fungus. Protists are a diverse paraphyletic group. Most are unicellular, but some are multicellular or form aggregations called colonies.

protoderm The exterior layer of a young plant embryo that gives rise to the epidermis.

proton pump A membrane protein that can hydrolyze ATP to power active transport of protons (H^+ ions) across a plasma membrane against an electrochemical gradient. Also called *H^+-ATPase.*

proton-motive force The combined effect of a proton gradient and an electric potential gradient across a membrane, which can drive protons across the membrane. Used by mitochondria and chloroplasts to power ATP synthesis via the mechanism of chemiosmosis.

proto-oncogene Any gene that normally encourages cell division in a regulated manner, typically by triggering specific phases in the cell cycle. Mutation may convert it into an oncogene.

protostomes A major lineage of animals that share a pattern of embryological development, including formation of the mouth earlier than the anus, and formation of the coelom by splitting of a block of mesoderm. Includes arthropods, mollusks, and annelids. Compare with **deuterostomes**.

proximal tubule In the vertebrate kidney, the convoluted section of a nephron into which filtrate moves from Bowman's capsule. Involved in the largely unregulated reabsorption of electrolytes, nutrients, and water. Compare with **distal tubule**.

proximate causation In biology, the immediate, mechanistic cause of a phenomenon (how it happens), as opposed to why it evolved. Also called *proximate explanation*. Compare with **ultimate causation**.

pseudogene A DNA sequence that closely resembles a functional gene but is not transcribed. Thought to have arisen by duplication of the functional gene followed by inactivation due to a mutation.

pseudopodium (plural: pseudopodia) A temporary bulge-like extension of certain cells used in cell crawling and ingestion of food.

puberty The various physical and emotional changes that an immature animal undergoes leading to reproductive maturity. Also the period when such changes occur.

pulmonary artery A short, thick-walled artery that carries oxygen-poor blood from the heart to the lungs.

pulmonary circulation The part of the circulatory system that sends oxygen-poor blood to the lungs. Is separate from the rest of the circulatory system (the systemic circulation) in mammals and birds.

pulmonary vein A short, thin-walled vein that carries oxygen-rich blood from the lungs to the heart.

pulse-chase experiment A type of experiment in which a population of cells or molecules at a particular moment in time is marked by means of a labeled molecule and then their fate is followed over time.

pump Any membrane protein that can hydrolyze ATP to power active transport of a specific ion or small molecule across a plasma membrane against its electrochemical gradient. See **proton pump**.

Punnett square A diagram that depicts the genotypes and phenotypes that should appear in offspring of a certain cross.

pupa (plural: pupae) A metamorphosing insect that is enclosed in a protective case.

pupil The hole in the center of the iris through which light enters a vertebrate or cephalopod eye.

pure line In animal or plant breeding, a strain of individuals that produce offspring identical to themselves when self-pollinated or crossed to another member of the same population. Pure lines are homozygous for most, if not all, genetic loci.

purifying selection Selection that lowers the frequency or even eliminates deleterious alleles.

purines A class of small, nitrogen-containing, double-ringed bases (guanine, adenine) found in nucleotides. Compare with **pyrimidines**.

pyrimidines A class of small, nitrogen-containing, single-ringed bases (cytosine, uracil, thymine) found in nucleotides. Compare with **purines**.

pyruvate dehydrogenase A large enzyme complex, located in the inner mitochondrial membrane, that is responsible for conversion of pyruvate to acetyl CoA during cellular respiration.

quantitative trait A heritable feature that exhibits phenotypic variation along a smooth, continuous scale of measurement (e.g., human height), rather than the distinct forms characteristic of discrete traits.

quaternary structure The overall three-dimensional shape of a protein containing two or more polypeptide chains (subunits); determined by the number, relative positions, and interactions of the subunits. Compare with **primary, secondary,** and **tertiary structure**.

quorum sensing Cell-cell signaling in bacteria, in which cells of the same species communicate via chemical signals. It is often observed that cell activity changes dramatically when the population reaches a threshold size, or quorum.

radial symmetry An animal body pattern in which there are least two planes of symmetry. Typically, the body is in the form of a cylinder or disk, with body parts radiating from a central hub. Compare with **bilateral symmetry**.

radiation Transfer of heat between two bodies that are not in direct physical contact. More generally, the emission of electromagnetic energy of any wavelength.

radicle The root of a plant embryo.

radula A rasping feeding appendage in gastropods (snails, slugs).

rain shadow The dry region on the side of a mountain range away from the prevailing wind.

range The geographic distribution of a species.

Ras protein A type of G protein that is activated by binding of signaling molecules to receptor tyrosine kinases and then initiates a phosphorylation cascade, culminating in a cell response.

rays In plant shoot systems with secondary growth, a lateral array of parenchyma cells produced by vascular cambium.

Rb protein A tumor-suppressor protein that helps regulate progression of a cell from the G_1 phase to the S phase of the cell cycle. Defects in Rb protein are found in many types of cancer.

reactant Any of the starting materials in a chemical reaction.

reaction center Centrally located component of a photosystem containing proteins and a pair of specialized chlorophyll molecules. Is surrounded by antenna complexes and receives excited electrons from them.

reactive oxygen intermediates (ROIs) Highly reactive oxygen-containing compounds that are used in plant and animal cells to kill infected cells and for other purposes.

reading frame The division of a sequence of DNA or RNA into a particular series of three-nucleotide codons. There are three possible reading frames for any sequence.

realized niche The ecological niche that a species occupies in the presence of competitors. Compare with **fundamental niche**.

receptor tyrosine kinase (RTK) Any of a class of cell-surface signal receptors that undergo phosphorylation after binding a signaling molecule. The activated, phosphorylated receptor then triggers a signal-transduction pathway inside the cell.

receptor-mediated endocytosis Uptake by a cell of certain extracellular macromolecules, bound to specific receptors in the plasma membrane, by pinching off the membrane to form small membrane-bound vesicles.

recessive Referring to an allele whose phenotypic effect is observed only in homozygous individuals. Compare with **dominant**.

reciprocal altruism Altruistic behavior that is exchanged between a pair of individuals at different points in time (i.e., sometimes individual A helps individual B, and sometimes B helps A).

reciprocal cross A breeding experiment in which the mother's and father's phenotypes are the reverse of that examined in a previous breeding experiment.

recombinant DNA technology A variety of techniques for isolating specific DNA fragments and introducing them into different regions of DNA and/or a different host organism.

recombinant Possessing a new combination of alleles. May refer to a single chromosome or DNA molecule, or to an entire organism.

recombination See **genetic recombination**.

rectal gland A salt-excreting gland in the digestive system of sharks, skates, and rays.

rectum The last portion of the digestive tract where feces are held until they are expelled.

red blood cell A hemoglobin-containing cell that circulates in the blood and delivers oxygen from the lungs to the tissues.

redox reaction Any chemical reaction that involves the transfer of one or more electrons from one reactant to another. Also called *reduction-oxidation reaction.*

reduction An atom's gain of electrons during a redox reaction, either by acceptance of an electron from another atom or by the electrons in covalent bonds moving closer to the atomic nucleus.

reduction-oxidation reaction See **redox reaction**.

reflex An involuntary response to environmental stimulation. May involve the brain (e.g., conditioned reflex) or not (e.g., spinal reflex).

refractory No longer responding to stimuli that previously elicited a response. For example, the tendency of voltage-gated sodium channels to remain closed immediately after an action potential.

regulatory genes DNA sequences that code for regulatory proteins—products that alter gene expression.

regulatory sequence, DNA Any segment of DNA that is involved in controlling transcription of a specific gene by binding certain proteins.

regulatory site A site on an enzyme to which a regulatory molecule can bind and affect the enzyme's activity; separate from the active site where catalysis occurs.

regulatory transcription factor General term for proteins that bind to DNA regulatory sequences (eukaryotic enhancers, silencers, and promoter-proximal elements), but not to the promoter itself, leading to an increase or decrease in transcription of specific genes. Compare with **basal transcription factor**.

reinforcement In evolutionary biology, the natural selection for traits that prevent interbreeding between recently diverged species.

release factors Proteins that can trigger termination of RNA translation when a ribosome reaches a stop codon.

renal corpuscle In the vertebrate kidney, the ball-like structure at the beginning of a nephron, consisting of a glomerulus and the surrounding Bowman's capsule. Acts as a filtration device.

replacement mutation See **missense mutation**.

replacement rate The number of offspring each female must produce over her entire life to "replace" herself and her mate, resulting in zero population growth. The actual number is slightly more than 2 because some offspring die before reproducing.

replica plating A method of identifying bacterial colonies that have certain mutations by transferring cells from each colony on a master plate to a second (replica) plate and observing their growth when exposed to different conditions.

replication fork The Y-shaped site at which a double-stranded molecule of DNA is separated into two single strands for replication.

replisome The multi-molecular machine that copies DNA; includes DNA polymerase, helicase, primase, and other enzymes.

repolarization Return to a normal membrane potential after it has changed; a normal phase in an action potential.

repressor Any regulatory protein that inhibits transcription.

reptiles One of the two lineages of amniotes (vertebrates that produce amniotic eggs) distinguished by adaptations for reproduction on land. Includes turtles, snakes and lizards, crocodiles and alligators, and birds. Except for birds, all are ectotherms.

resilience, community A measure of how quickly a community recovers following a disturbance.

resistance, community A measure of how much a community is affected by a disturbance.

resistance (*R*) genes Genes in plants encoding proteins involved in sensing the presence of pathogens and mounting a defensive response. Compare with **avirulence (*avr*) genes**.

respiratory system The collection of cells, tissues, and organs responsible for gas exchange between an animal and its environment.

resting potential The membrane potential of a cell in its resting, or normal, state.

restriction endonucleases Bacterial enzymes that cut DNA at a specific base-pair sequence (restriction site). Also called *restriction enzymes.*

retina A thin layer of light-sensitive cells (rods and cones) and neurons at the back of a camera-type eye, such as that of cephalopods and vertebrates.

retinal A light-absorbing pigment, derived from vitamin A, that is linked to the protein opsin in rods and cones of the vertebrate eye.

retrovirus A virus with an RNA genome that reproduces by transcribing its RNA into a DNA sequence and then inserting that DNA into the host's genome for replication.

reverse transcriptase A enzyme of retroviruses (RNA viruses) that can synthesize double-stranded DNA from a single-stranded RNA template.

rhizobia (singular: rhizobium) Members of the bacterial genus *Rhizobia;* nitrogen-fixing bacteria that live in root nodules of members of the pea family (legumes).

rhizoid The hairlike structure that anchors a bryophyte (non-vascular plant) to the substrate.

rhizome A modified stem that runs horizontally underground and produces new plants at the nodes (a form of asexual reproduction). Compare with **stolon**.

rhodopsin A transmembrane complex that is instrumental in detection of light by rods and cones of the vertebrate eye. Is composed of the transmembrane protein opsin covalently linked to retinal, a light-absorbing pigment.

ribonucleic acid (RNA) A nucleic acid composed of ribonucleotides that usually is single stranded and functions as structural components of ribosomes (rRNA), transporters of amino acids (tRNA), and translators of the message of the DNA code (mRNA).

ribonucleotide See **nucleotide**.

ribosomal RNA (rRNA) A RNA molecule that forms part of the structure of a ribosome.

ribosome A large complex structure that synthesizes proteins by using the genetic information encoded in messenger RNA strands. Consists of two subunits, each composed of ribosomal RNA and proteins.

ribosome binding site In a bacterial mRNA molecule, the sequence just upstream of the start codon to which a ribosome binds to initiate translation. Also called the *Shine-Dalgarno sequence.*

ribozyme Any RNA molecule that can act as a catalyst, that is, speed up a chemical reaction.

ribulose bisphosphate (RuBP) A five-carbon compound that combines with CO_2 in the first step of the Calvin cycle during photosynthesis.

RNA interference (RNAi) Degradation of an mRNA molecule or inhibition of its translation following its binding by a short RNA (microRNA) whose sequence is complementary to a portion of the mRNA.

RNA polymerase One of a class of enzymes that catalyze synthesis of RNA from ribonucleotides using a DNA template. Also called *RNA pol.*

RNA processing In eukaryotes, the changes that a primary RNA transcript undergoes in the nucleus to become a mature mRNA molecule, which is exported to the cytoplasm. Includes the addition of a

5′ cap and poly(A) tail and splicing to remove introns.

RNA replicase A viral enzyme that can synthesize RNA from an RNA template.

RNA See **ribonucleic acid**.

rod cell A photoreceptor cell with a rod-shaped outer portion that is particularly sensitive to dim light, but not used to distinguish colors. Also called simply *rod*. Compare with **cone cell**.

root (1) An underground part of a plant that anchors the plant and absorbs water and nutrients. (2) In a phylogenetic tree, the bottom, most ancient node.

root apical meristem (RAM) A group of undifferentiated plant cells at the tip of a plant root that can differentiate into mature root tissue.

root cap A small group of cells that covers and protects the tip of a plant root. Senses gravity and determines the direction of root growth.

root hair A long, thin outgrowth of the epidermal cells of plant roots, providing increased surface area for absorption of water and nutrients.

root pressure Positive (upward) pressure of xylem sap in the vascular tissue of roots. Is generated during the night as a result of the accumulation of ions from the soil and subsequent osmotic movement of water into the xylem.

root system The belowground part of a plant.

rough endoplasmic reticulum (rough ER) The portion of the endoplasmic reticulum that is dotted with ribosomes. Involved in synthesis of plasma membrane proteins, secreted proteins, and proteins localized to the ER, Golgi apparatus, and lysosomes. Compare with **smooth endoplasmic reticulum**.

roundworms Members of the phylum Nematoda. Distinguished by an unsegmented body with a pseudocoelom and no appendages. Roundworms belong to the ecdysozoan branch of the protostomes. Also called *nematodes*.

rRNA See **ribosomal RNA**.

rubisco The enzyme that catalyzes the first step of the Calvin cycle during photosynthesis: the addition of a molecule of CO_2 to ribulose bisphosphate.

ruminants A group of hoofed mammals (e.g., cattle, sheep, deer) that have a four-chambered stomach specialized for digestion of plant cellulose. Ruminants regurgitate the cud, a mixture of partially digested food and cellulose-digesting bacteria, from the largest chamber (the rumen) for further chewing.

salivary glands Vertebrate glands that secrete saliva (a mixture of water, mucus-forming glycoproteins, and digestive enzymes) into the mouth.

sampling error The accidental selection of a nonrepresentative sample from some larger population, due to chance.

saprophyte An organism that feeds primarily on dead plant material.

sapwood The younger xylem in the outer layer of wood of a stem or root, functioning primarily in water transport.

sarcomere The repeating contractile unit of a skeletal muscle cell; the portion of a myofibril located between adjacent Z disks.

sarcoplasmic reticulum Sheets of smooth endoplasmic reticulum in a muscle cell. Contains high concentrations of calcium, which can be released into the cytoplasm to trigger contraction.

saturated Referring to fats and fatty acids in which all the carbon-carbon bonds are single bonds. Such fats have relatively high melting points. Compare with **unsaturated**.

scanning electron microscope (SEM) A microscope that produces images of the surfaces of objects by reflecting electrons from a specimen coated with a layer of metal atoms. Compare with **transmission electron microscope**.

scarify To scrape, rasp, cut, or otherwise damage the coat of a seed. Necessary in some species to trigger germination.

Schwann cell A type of glial cell that wraps around axons of some neurons outside the brain and spinal cord, forming a myelin sheath that provides electrical insulation. Compare with **oligodendrocyte**.

scientific name The unique, two-part name given to each species, with a genus name followed by a species name—as in *Homo sapiens*. Scientific names are always italicized, and are also known as Latin names.

sclereid In plants, a type of sclerenchyma cell that usually functions in protection, such as in seed coats and nutshells. Compare with **fiber**.

sclerenchyma cell In plants, a cell that has a thick secondary cell wall and provides support; typically contains the tough structural polymer lignin and usually is dead at maturity. Includes fibers and sclereids. Compare with **collenchyma cell** and **parenchyma cell**.

screen See **genetic screen**.

scrotum A sac of skin, containing the testes, suspended just outside the abdominal body cavity of many male mammals.

second law of thermodynamics The principle of physics that the entropy of the universe or any closed system increases during any spontaneous process.

second messenger A nonprotein signaling molecule produced or activated inside a cell in response to stimulation at the cell surface. Commonly used to relay the message of a hormone or other extracellular signaling molecule.

secondary active transport Transport of an ion or molecule against its electrochemical gradient, in company with an ion or molecule being transported with its electrochemical gradient. Also called *cotransport*.

secondary active transporter A transmembrane protein that facilitates diffusion of an ion down its previously established electrochemical gradient and uses the energy of that process to transport some other substance, in the same or opposite direction, *against* its concentration gradient. Also called *cotransporter*. See also **antiporter** and **symporter**.

secondary cell wall The inner layer of a plant cell wall formed by certain cells as they mature. Provides support or protection.

secondary consumer A carnivore; an organism that eats herbivores. Compare with **primary consumer**.

secondary growth In plants, an increase in the width of stems and roots due to the activity of lateral meristems. Compare with **primary growth**.

secondary immune response The acquired immune response to a pathogen that the immune system has encountered before. Compare with **primary immune response**.

secondary metabolites Molecules that are closely related to compounds in key synthetic pathways, and that often function in defense.

secondary spermatocyte A cell produced by meiosis I of a primary spermatocyte in the testis. Can undergo meiosis II to produce spermatids.

secondary structure In proteins, localized folding of a polypeptide chain into regular structures (e.g., α-helix and β-pleated sheet) stabilized by hydrogen bonding between atoms of the backbone. In nucleic acids, elements of structure (e.g., helices and hairpins) stabilized by hydrogen bonding and other interactions between complementary bases. Compare with **primary, tertiary,** and **quaternary structure**.

secondary succession Gradual colonization of a habitat after an environmental disturbance (e.g., fire, windstorm, logging) that removes some or all previous organisms but leaves the soil intact. Compare with **primary succession**.

second-male advantage The reproductive advantage of a male who mates with a female last, after other males have mated with her.

secretin A peptide hormone produced by cells in the small intestine in response to the arrival of food from the stomach. Stimulates secretion of bicarbonate (HCO_3^-) from the pancreas.

sedimentary rock A type of rock formed by gradual accumulation of sediment, as in riverbeds and on the ocean floor. Most fossils are found in sedimentary rocks.

seed A plant reproductive structure consisting of an embryo, associated nutritive tissue (endosperm), and an outer protective layer (seed coat). In angiosperms, develops from the fertilized ovule of a flower.

seed bank A repository where seeds, representing many different varieties of domestic crops or other species, are preserved.

seed coat A protective layer around a seed that encases both the embryo and the endosperm.

segment A well-defined region of the body along the anterior-posterior body axis, containing similar structures as other, nearby segments.

segmentation Division of the body or a part of it into a series of similar structures; exemplified by the body segments of insects and worms and by the somites of vertebrates.

segmentation genes A group of genes that affect body segmentation in embryonic development. Includes gap genes, pair-rule genes, and segment polarity genes.

segregation, principle of The concept that each pair of hereditary elements (alleles of the same gene) separate from each other during the formation of offspring (i.e., during meiosis). One of Mendel's two principles of genetics.

selective adhesion The tendency of cells of one tissue type to adhere to other cells of the same type.

selective permeability The property of a membrane that allows some substances to diffuse across it much more readily than other substances.

selectively permeable membrane Any membrane across which some solutes can move more readily than others.

self Property of a molecule or cell such that immune system cells do not attack it, due to certain molecular similarities to other body cells.

self molecule A molecule that is synthesized by an organism and is a normal part of its cells and/or body; as opposed to non-self or foreign molecules.

self-fertilization In plants, the fusion of two gametes from the same individual to form a diploid offspring. Also called *selfing*.

semen The combination of sperm and accessory fluids that is released by male mammals and reptiles during ejaculation.

semiconservative replication The mechanism of replication used by cells to copy DNA. Results in each daughter DNA molecule containing one old strand and one new strand.

seminal vesicles In male mammals, paired reproductive glands that secrete a sugar-containing fluid into semen, which provides energy for sperm movement. In other vertebrates and invertebrates, often stores sperm.

senescence The process of aging.

sensor Any cell, organ, or structure with which an animal can sense some aspect of the external or internal environment. Usually functions, along with an integrator and effector, as part of a homeostatic system.

sensory neuron A nerve cell that carries signals from sensory receptors to the central nervous system. Compare with **interneuron** and **motor neuron**.

sepal One of the protective leaflike organs enclosing a flower bud and later supporting the blooming flower.

septum (plural: septa) Any wall-like structure. In fungi, septa divide the filaments (hyphae) of mycelia into cell-like compartments.

serotonin A neurotransmitter involved in many brain functions, including sleep, pleasure, and mood.

serum The liquid that remains when cells and clot material are removed from clotted blood. Contains water, dissolved gases, growth factors, nutrients, and other soluble substances. Compare with **plasma**.

sessile Permanently attached to a substrate; not capable of moving to another location.

set point A normal or target value for a regulated internal variable, such as body heat or blood pH.

severe combined immunodeficiency disease (SCID) A human disease characterized by an extremely high vulnerability to infectious disease. Caused by a genetic defect in the immune system.

sex chromosome Any chromosome carrying genes involved in determining the sex of an individual. Compare with **autosome**.

sex-linked inheritance Inheritance patterns observed in genes carried on sex chromosomes, so females and males have different numbers of alleles of a gene and may pass its trait only to one sex of offspring. Also called *sex-linkage*.

sexual dimorphism Any trait that differs between males and females.

sexual reproduction Any form of reproduction in which genes from two parents are combined via fusion of gametes, producing offspring that are genetically distinct from both parents. Compare with **asexual reproduction**.

sexual selection A pattern of natural selection that favors individuals with traits that increase their ability to obtain mates. Acts more strongly on males than females.

shell A hard protective outer structure.

Shine-Dalgarno sequence See **ribosome binding sequence**.

shoot apical meristem (SAM) A group of undifferentiated plant cells at the tip of a plant stem that can differentiate into mature shoot tissues.

shoot The combination of hypocotyl and cotyledons in a plant embryo, which will become the aboveground portions of the body.

shoot system The aboveground part of a plant comprising stems, leaves, and flowers (in angiosperms).

short interspersed nuclear elements See **SINEs**.

short tandem repeats (STRs) Relatively short DNA sequences that are repeated, one after another, down the length of a chromosome. The two major types are microsatellites and minisatellites.

short-day plant A plant that blooms in response to long nights (usually in late summer or fall in the northern hemisphere). Compare with **day-neutral** and **long-day plant**.

shotgun sequencing A method of sequencing genomes that is based on breaking the genome into small pieces, sequencing each piece separately, and then figuring out how the pieces are connected.

sieve plate In plants, a pore-containing structure at one end of a sieve-tube member in phloem.

sieve-tube member In plants, an elongated sugar-conducting cell in phloem that has sieve plates at both ends, allowing sap to flow to adjacent cells.

sign stimulus A simple stimulus that elicits an invariant, stereotyped behavioral response (fixed action pattern) from an animal. Also called a *releaser*.

signal In behavioral ecology, any information-containing behavior.

signal receptor Any cellular protein that binds to a particular signaling molecule (e.g., a hormone or neurotransmitter) and triggers a response by the cell. Receptors for water-soluble signals are transmembrane proteins in the plasma membrane; those for many lipid-soluble signals (e.g., steroid hormones) are located inside the cell.

signal recognition particle (SRP) A RNA-protein complex that binds to the ER signal sequence in a polypeptide as it emerges from a ribosome and transports the ribosome-polypeptide complex to the ER membrane where synthesis of the polypeptide is completed.

signal transduction cascade See **phosphorylation cascade**.

signal transduction The process by which a stimulus (e.g., a hormone, a neurotransmitter, or sensory information) outside a cell is amplified and converted into a response by the cell. Usually involves a specific sequence of molecular events, or signal transduction pathway.

silencer A regulatory sequence in eukaryotic DNA to which repressor proteins can bind, inhibiting transcription of certain genes.

silent mutation A mutation that does not detectably affect the phenotype of the organism.

simple eye An eye with only one light-collecting apparatus (e.g., one lens), as in vertebrates. Compare with **compound eye**.

simple fruit A fruit (e.g., apricot) that develops from a single flower that has a single carpel or several fused carpels. Compare with **aggregate** and **multiple fruit**.

simple sequence repeat See **microsatellite**.

SINEs (short interspersed nuclear elements) The second most abundant class of transposable elements in human genomes; can create copies of itself and insert them elsewhere in the genome. Compare with **LINEs**.

single nucleotide polymorphism (SNP) A site on a chromosome where individuals in a population have different nucleotides. Can be used as a genetic marker to help track the inheritance of nearby genes.

single-strand DNA-binding proteins (SSBPs) A class of proteins that attach to separated strands of DNA during replication or transcription, preventing them from re-forming a double helix.

sink Any tissue, site, or location where an element or a molecule is consumed or taken out of circulation (e.g., in plants, a tissue where sugar exits the phloem). Compare with **source**.

sinoatrial (SA) node A cluster of cardiac muscle cells, in the right atrium of the vertebrate heart, that initiates the heartbeat and determines the heart rate. Compare with **atrioventricular (AV) node**.

siphon A tubelike appendage of many mollusks, that is often used for feeding or propulsion.

sister chromatids The paired strands of a recently replicated chromosome, which are connected at the centromere and eventually separate during anaphase of mitosis and meiosis II. Compare with **non-sister chromatids**.

sister species Closely related species, which occupy adjacent branches in a phylogenetic tree.

skeletal muscle The muscle tissue attached to the bones of the vertebrate skeleton. Consists of long, unbranched muscle fibers with a characteristic striped (striated) appearance; controlled voluntarily. Also called *striated muscle*. Compare with **cardiac** and **smooth muscle**.

sliding-filament model The hypothesis that thin (actin) filaments and thick (myosin) filaments slide past each other, thereby shortening the sarcomere. Shortening of all the sarcomeres in a myofibril results in contraction of the entire myofibril.

small intestine The portion of the digestive tract between the stomach and the large intestine. The site of the final stages of digestion and of most nutrient absorption.

small nuclear ribonucleoproteins See **snRNPs**.

smooth endoplasmic reticulum (smooth ER) The portion of the endoplasmic reticulum that does not have ribosomes attached to it. Involved in synthesis and secretion of lipids. Compare with **rough endoplasmic reticulum**.

smooth muscle The unstriated muscle tissue that lines the intestine, blood vessels, and some other organs. Consists of tapered, unbranched cells that can sustain long contractions. Not voluntarily controlled. Compare with **cardiac** and **skeletal muscle**.

snRNPs (small nuclear ribonucleoproteins) Complexes of proteins and small RNA molecules that function in splicing (removal of introns from primary RNA transcripts) as components of spliceosomes.

sodium-potassium pump A transmembrane protein that uses the energy of ATP to move sodium ions out of the cell and potassium ions in. Also called *Na^+/K^+-ATPase.*

soil organic matter Organic (carbon-containing) compounds found in soil.

solute Any substance that is dissolved in a liquid.

solute potential (ψ_S) A component of the potential energy of water caused by a difference in solute concentrations at two locations. Also called *osmotic potential.* Compare with **pressure potential (ψ_P)**.

solution A liquid containing one or more dissolved solids or gases in a homogeneous mixture.

solvent Any liquid in which one or more solids or gases can dissolve.

soma See **cell body**.

somatic cell Any type of cell in a multicellular organism except eggs, sperm, and their precursor cells. Also called *body cells.*

somatic hypermutation Mutations that occur in the immunoglobulin genes of the immune system's memory cells, resulting in novel variation in the receptors that bind to antigens.

somatic nervous system The part of the peripheral nervous system (outside the brain and spinal cord) that controls skeletal muscles and is under voluntary control. Compare with **autonomic nervous system**.

somatostatin A hormone secreted by the pancreas and hypothalamus that inhibits the release of several other hormones.

somites Paired blocks of mesoderm on both sides of the developing spinal cord in a vertebrate embryo. Give rise to muscle tissue, vertebrae, ribs, limbs, etc.

sori In ferns, a cluster of spore-producing structures (sporangia).

source Any tissue, site, or location where a substance is produced or enters circulation (e.g., in plants, the tissue where sugar enters the phloem). Compare with **sink**.

space-filling model A representation of a molecule where atoms are shown as balls—color-coded and scaled to indicate the atom's identify—attached to each other in the correct geometry.

speciation The evolution of two or more distinct species from a single ancestral species.

species A distinct, identifiable group of populations that is thought to be evolutionarily independent of other populations and whose members can interbreed. Generally distinct from other species in appearance, behavior, habitat, ecology, genetic characteristics, etc.

species diversity The variety and relative abundance of the species present in a given ecological community.

species richness The number of species present in a given ecological community.

species–area relationship The mathematical relationship between the area of a certain habitat and the number of species that it can support.

specific heat The amount of energy required to raise the temperature of 1 gram of a substance by 1°C; a measure of the capacity of a substance to absorb energy.

sperm A mature male gamete; smaller and more mobile than the female gamete.

sperm competition Competition to fertilize eggs between the sperm of different males, inside the same female.

spermatid An immature sperm cell.

spermatogenesis The production of sperm. Occurs continuously in a testis.

spermatogonia (singular: spermatogonium) The diploid cells in a testis that can give rise to primary spermatocytes.

spermatophore A gelatinous package of sperm cells that is produced by males of species that have internal fertilization without copulation.

sphincter A muscular valve that can close off a tube, as in a blood vessel or a part of the digestive tract.

spicule Stiff spike of silica or calcium carbonate found in the body of many sponges.

spindle apparatus The array of microtubules responsible for contacting and moving chromosomes during mitosis and meiosis; includes kinetochore microtubules and polar microtubules.

spines In plants, modified leaves that are stiff and sharp and that function in defense.

spiracle In insects, a small opening that connects air-filled tracheae to the external environment, allowing for gas exchange.

spleen A dark red organ, found near the stomach of most vertebrates, that filters blood, stores extra red blood cells in case of emergency, and plays a role in immunity.

spliceosome In eukaryotes, a large, complex assembly of snRNPs (small nuclear ribonucleoproteins) that catalyzes removal of introns from primary RNA transcripts.

splicing The process by which introns are removed from primary RNA transcripts and the remaining exons are connected together.

sporangium (plural: sporangia) A spore-producing structure found in seed plants, some protists, and some fungi (e.g., chytrids).

spore (1) In bacteria, a dormant form that generally is resistant to extreme conditions. (2) In eukaryotes, a single cell produced by mitosis or meiosis (not by fusion of gametes) that is capable of developing into an adult organism.

sporophyte In organisms undergoing alternation of generations, the multicellular diploid form that arises from two fused gametes and produces haploid spores. Compare with **gametophyte**.

sporopollenin A watertight material that encases spores and pollen of modern land plants.

stabilizing selection A pattern of natural selection that favors phenotypes near the middle of the range of phenotypic variation. Reduces overall genetic variation in a population. Compare with **disruptive selection**.

stamen The male reproductive structure of a flower. Consists of an anther, in which pollen grains are produced, and a filament, which supports the anther. Compare with **carpel**.

standing defense See **constitutive defense**.

stapes The last of three small bones (ossicles) in the middle ear of vertebrates. Receives vibrations from the tympanic membrane and by vibrating against the oval window passes them to the cochlea.

starch A mixture of two storage polysaccharides, amylose and amylopectin, both formed from α-glucose monomers. Amylopectin is branched, and amylose is unbranched. The major form of stored carbohydrate in plants.

start codon The AUG triplet in mRNA at which protein synthesis begins; codes for the amino acid methionine.

statocyst A sensory organ of many arthropods that detects the animal's orientation in space (i.e., whether the animal is flipped upside down).

statolith A tiny stone or dense particle found in specialized gravity-sensing organs in some animals such as lobsters.

statolith hypothesis The hypothesis that amyloplasts (dense, starch-storing plant organelles) serve as statoliths in gravity detection by plants.

STATs See **signal transducers and activators of transcription**.

stem cell Any relatively undifferentiated cell that can divide to produce daughter cells identical to itself or more specialized daughter cells, which differentiate further into specific cell types.

stems Vertical, aboveground structures that make up the shoot system of plants.

stereocilium (plural: stereocilia) One of many stiff outgrowths from the surface of a hair cell that are involved in detection of sound by terrestrial vertebrates or of waterborne vibrations by fishes.

steroid A class of lipid with a characteristic four-ring structure.

steroid-hormone receptor One of a family of intracellular receptors that bind to various steroid hormones, forming a hormone-receptor complex that acts as a regulatory transcription factor and activates transcription of specific target genes.

sticky end The short, single-stranded ends of a DNA molecule cut by a restriction endonuclease.

Tend to form hydrogen bonds with other sticky ends that have complementary sequences.

stigma The moist tip at the end of a flower carpel to which pollen grains adhere.

stolon A modified stem that runs horizontally over the soil surface and produces new plants at the nodes (a form of asexual reproduction). Compare with **rhizome**.

stoma (plural: stomata) Generally, a pore or opening. In plants, a microscopic pore on the surface of a leaf or stem through which gas exchange occurs.

stomach A tough, muscular pouch in the vertebrate digestive tract between the esophagus and small intestine. Physically breaks up food and begins digestion of proteins.

stop codon One of three mRNA triplets (UAG, UGA, or UAA) that cause termination of protein synthesis. Also called a *termination codon.*

strain A population of genetically similar or identical individuals.

stream A body of water that moves constantly in one direction.

striated muscle See **skeletal muscle**.

stroma The fluid matrix of a chloroplast in which the thylakoids are embedded. Site where the Calvin cycle reactions occur.

structural formula A two-dimensional notation in which the chemical symbols for the constituent atoms are joined by straight lines representing single (—), double (=), or triple (≡) covalent bonds. Compare with **molecular formula**.

structural gene A stretch of DNA that codes for a functional protein or functional RNA molecule, not including any regulatory sequences (e.g., a promoter, enhancer).

structural homology Similarities in organismal structures (e.g., limbs, shells, flowers) that are due to inheritance from a common ancestor.

style The slender stalk of a flower carpel connecting the stigma and the ovary.

subspecies A population that has distinctive traits and some genetic differences relative to other populations of the same species but that is not distinct enough to be classified as a separate species.

substrate (1) A reactant that interacts with an enzyme in a chemical reaction. (2) A surface on which a cell or organism sits.

substrate-level phosphorylation Production of ATP by transfer of a phosphate group from an intermediate substrate directly to ADP. Occurs in glycolysis and in the citric acid cycle.

succession In ecology, the gradual colonization of a habitat after an environmental disturbance (e.g., fire, flood), usually by a series of species. See **primary** and **secondary succession**.

sucrose A disaccharide formed from glucose and fructose. One of the two main products of photosynthesis.

sugar Synonymous with carbohydrate, though usually used in an informal sense to refer to small carbohydrates (monosaccharides and disaccharides).

sulfate reducer A prokaryote that produces hydrogen sulfide (H_2S) as a by-product of cellular respiration.

summation The additive effect of different postsynaptic potentials at a nerve or muscle cell, such that several subthreshold stimulations can cause an action potential.

supporting connective tissue A type of connective tissue, distinguished by having a firm extracellular matrix.

surface tension The cohesive force that causes molecules at the surface of a liquid to stick together, thereby resisting deformation of the liquid's surface and minimizing its surface area.

surfactant A mixture of phospholipids and proteins produced by lung cells that reduces surface tension, allowing the lungs to expand more.

survivorship On average, the proportion of offspring that survive to a particular age.

survivorship curve A graph depicting the percentage of a population that survives to different ages.

suspension feeder Any organism that obtains food by filtering small particles or small organisms out of water or air. Also called *filter feeder.*

sustainability The planned use of environmental resources at a rate no faster than the rate at which they are naturally replaced.

sustainable agriculture Agricultural techniques that are designed to maintain long-term soil quality and productivity.

swamp A wetland that has a steady rate of water flow and is dominated by trees and shrubs.

swim bladder A gas-filled organ of many ray-finned fishes that regulates buoyancy.

symbiosis (adjective: symbiotic) Any close and prolonged physical relationship between individuals of two different species. See **commensalism, mutualism,** and **parasitism**.

symmetric competition Ecological competition between two species in which both suffer similar declines in fitness. Compare with **asymmetric competition**.

sympathetic nervous system The part of the autonomic nervous system that stimulates fight-or-flight responses, such as increased heart rate, increased blood pressure, and decreased digestion. Compare with **parasympathetic nervous system**.

sympatric speciation The divergence of populations living within the same geographic area into different species as the result of their genetic (not physical) isolation. Compare with **allopatric speciation**.

sympatry Condition in which two or more populations live in the same geographic area, or close enough to permit interbreeding. Compare with **allopatry**.

symplast In plant roots, a continuous pathway through which water can flow through the cytoplasm of adjacent cells that are connected by plasmodesmata. Compare with **apoplast**.

symporter A carrier protein that allows an ion to diffuse down an electrochemical gradient, using the energy of that process to transport a different substance in the same direction *against* its concentration gradient. Compare with **antiporter**.

synapomorphy A shared, derived trait found in two or more taxa that is present in their most recent common ancestor but is missing in more distant ancestors. Useful for inferring evolutionary relationships.

synapse The interface between two neurons or between a neuron and an effector cell.

synapsis The physical pairing of two homologous chromosomes during prophase I of meiosis. Crossing over occurs during synapsis.

synaptic cleft The space between two communicating nerve cells (or between a neuron and effector cell) at a synapse, across which neurotransmitters diffuse.

synaptic plasticity Long-term changes in the responsiveness or physical structure of a synapse that can occur after particular stimulation patterns. Thought to be the basis of learning and memory.

synaptic vesicle A small neurotransmitter-containing vesicle at the end of an axon that releases neurotransmitter into the synaptic cleft by exocytosis.

synaptonemal complex A network of proteins that holds non-sister chromatids together during synapsis in meiosis I.

synthesis (S) phase The phase of the cell cycle during which DNA is synthesized and chromosomes are replicated.

systemic acquired resistance (SAR) A slow, widespread response of plants to a localized infection that protects healthy tissue from invasion by pathogens. Compare with **hypersensitive response**.

systemic circulation The part of the circulatory system that sends oxygen-rich blood from the lungs out to the rest of the body. Is separate from the pulmonary circulation in mammals and birds.

systemin A peptide hormone, produced by plant cells damaged by herbivores, that initiates a protective response in undamaged cells.

systole The portion of the heartbeat cycle during which the heart muscles are contracting. Compare with **diastole**.

systolic blood pressure The force exerted by blood against artery walls during contraction of the heart's left ventricle. Compare with **diastolic blood pressure**.

T cell A type of leukocyte that matures in the thymus and, with B cells, is responsible for acquired immunity. Involved in activation of B cells ($CD4^+$ helper T cells) and destruction of infected cells ($CD8^+$ cytotoxic T cells). Also called *T lymphocytes.*

T tubules Membranous tubes that extend into the interior of muscle cells. Propagate action potentials throughout a muscle cell and trigger release of calcium from the sarcoplasmic reticulum.

taiga A vast forest biome throughout subarctic regions, consisting primarily of short conifer trees. Characterized by intensely cold winters, short summers, and high annual variation in temperature.

taproot A large vertical main root of a plant.

taste bud Sensory structure, found chiefly in the mammalian tongue, containing spindle-shaped cells that respond to chemical stimuli.

TATA box A short DNA sequence in many eukaryotic promoters about 30 base pairs upstream from the transcription start site.

TATA-binding protein (TBP) A protein that binds to the TATA box in eukaryotic promoters and is a component of the basal transcription complex.

taxon (plural: taxa) Any named group of organisms at any level of a classification system.

taxonomy The branch of biology concerned with the classification and naming of organisms.

TBP See **TATA-binding protein**.

T-cell receptor (TCR) A transmembrane protein found on T cells that can bind to antigens displayed on the surfaces of other cells. Composed of two polypeptides called the alpha chain and beta chain. See **antigen presentation**.

tectorial membrane A membrane in the vertebrate cochlea that takes part in the transduction of sound by bending the stereocilia of hair cells in response to sonic vibrations.

telomerase An enzyme that replicates the ends of chromosome (telomeres) by catalyzing DNA synthesis from an RNA template that is part of the enzyme.

telomere The region at the end of a linear chromosome.

telophase The final stage in mitosis or meiosis during which sister chromatids (replicated chromosomes in meiosis I) separate and new nuclear envelopes begin to form around each set of daughter chromosomes.

temperate Having a climate with pronounced annual fluctuations in temperature (i.e., warm summers and cold winters) but typically neither as hot as the tropics nor as cold as the poles.

temperature A measurement of thermal energy present in an object or substance, reflecting how much the constituent molecules are moving.

template strand (1) The strand of DNA that is transcribed by RNA polymerase to create RNA. (2) An original strand of RNA used to make a complementary strand of RNA.

temporal lobe In the vertebrate brain, one of the four major areas in the cerebrum.

tendon A band of tough, fibrous connective tissue that connects a muscle to a bone.

tentacle A long, thin, muscular appendage of gastropod mollusks.

termination (1) In enzyme-catalyzed reactions, the final stage in which the enzyme returns to its original conformation and products are released. (2) In DNA transcription, the dissociation of RNA polymerase from DNA when it reaches a termination signal sequence. (3) In RNA translation, the dissociation of a ribosome from mRNA when it reaches a stop codon.

territory An area that is actively defended by an animal from others of its species.

tertiary consumers In a food chain or food web, organisms that feed on secondary consumers. Compare with **primary consumer** and **secondary consumer**.

tertiary structure The overall three-dimensional shape of a single polypeptide chain, resulting from multiple interactions among the amino acid side chains and the peptide backbone. Compare with **primary, secondary,** and **quaternary structure**.

testcross The breeding of an individual of unknown genotype with an individual having only recessive alleles for the traits of interest in order to infer the unknown genotype from the phenotypic ratios seen in offspring.

testis (plural: testes) The sperm-producing organ of a male animal.

testosterone A steroid hormone, produced and secreted by the testes, that stimulates sperm production and various male traits and reproductive behaviors.

tetrad The structure formed by synapsed homologous chromosomes during prophase of meiosis I.

tetrapod Any member of the taxon Tetrapoda, which includes all vertebrates with two pairs of limbs (amphibians, mammals, birds, and other reptiles).

texture A quality of soil, resulting from the relative abundance of different-sized particles.

theory A proposed explanation for a broad class of phenomena or observations.

thermal energy The kinetic energy of molecular motion.

thermocline A gradient (cline) in environmental temperature across a large geographic area.

thermophile A bacterium or archaean that thrives in very hot environments.

thermoreceptor A sensory cell or an organ specialized for detection of changes in temperature.

thermoregulation Regulation of body temperature.

thick filament A filament composed of bundles of the motor protein myosin; anchored to the center of the sarcomere. Compare with **thin filament**.

thigmotropism Growth or movement of an organism in response to contact with a solid object.

thin filament A filament composed of two coiled chains of actin and associated regulatory proteins; anchored at the Z disk of the sarcomere. Compare with **thick filament**.

thorax A region of the body; in insects, one of the three prominent body regions called tagmata.

thorn A modified plant stem shaped as a sharp protective structure. Helps protect a plant against feeding by herbivores.

threshold potential The membrane potential that will trigger an action potential in a neuron or other excitable cell. Also called simply *threshold.*

thylakoid A flattened, membrane-bound vesicle inside a plant chloroplast that functions in converting light energy to chemical energy. A stack of thylakoids is a granum.

thymus An organ, located in the anterior chest or neck of vertebrates, in which immature T cells generated in the bone marrow undergo maturation.

thyroid gland A gland in the neck that releases thyroid hormone (which increases metabolic rate) and calcitonin (which lowers blood calcium).

thyroid-stimulating hormone (TSH) A peptide hormone, produced and secreted by the anterior pituitary, that stimulates release of thyroid hormones from the thyroid gland.

thyroxine (T_4) A peptide hormone containing four iodine atoms that is produced and secreted by the thyroid gland. Acts primarily to increase cellular metabolism. In mammals, T_4 is converted to the more active hormone triiodothyronine (T_3) in the liver.

Ti plasmid A plasmid carried by *Agrobacterium* (a bacterium that infects plants) that can integrate into a plant cell's chromosomes and induce formation of a gall.

tight junction A type of cell-cell attachment structure that links the plasma membranes of adjacent animal cells, forming a barrier that restricts movement of substances in the space between the cells. Most abundant in epithelia (e.g., the intestinal lining). Compare with **desmosome** and **gap junction**.

tip The end of a branch on a phylogenetic tree. Represents a specific species or larger taxon that has not (yet) produced descendants—either a group living today or a group that ended in extinction. Also called *terminal node.*

tissue A group of similar cells that function as a unit, such as muscle tissue or epithelial tissue.

tolerance In ecological succession, the phenomenon in which early-arriving species do not affect the probability that subsequent species will become established. Compare with **facilitation** and **inhibition**.

tonoplast The membrane surrounding a plant vacuole.

top-down control The hypothesis that population size is limited by predators or herbivores (consumers).

topoisomerase An enzyme that cuts and rejoins DNA downstream of the replication fork, to ease the twisting that would otherwise occur as the DNA "unzips."

torpor An energy-conserving physiological state, marked by a decrease in metabolic rate, body temperature, and activity, that lasts for a short period (overnight to a few days or weeks). Occurs in some small mammals when the ambient temperature drops significantly. Compare with **hibernation**.

totipotent Capable of dividing and developing to form a complete, mature organism.

trachea (plural: tracheae) (1) In insects, one of the small air-filled tubes that extend throughout the body and function in gas exchange. (2) In terrestrial vertebrates, the airway connecting the larynx to the bronchi. Also called *windpipe.*

tracheid In vascular plants, a long, thin water-conducting cell that has gaps in its secondary cell wall, allowing water movement between adjacent cells. Compare with **vessel element**.

trade-off In evolutionary biology, an inescapable compromise between two traits that cannot be optimized simultaneously. Also called *fitness trade-off.*

trait Any heritable characteristic of an individual.

transcription The process by which RNA is made from a DNA template.

transcriptional control Regulation of gene expression by various mechanisms that change the rate at which genes are transcribed to form messenger RNA. In negative control, binding of a regulatory protein to DNA represses transcription; in

positive control binding of a regulatory protein to DNA promotes transcription.

transcriptome The complete set of genes transcribed in a particular cell.

transduction Conversion of information from one mode to another. For example, the process by which a stimulus outside a cell is converted into a response by the cell.

transfer cell In land plants, a cell that transfers nutrients from a parent plant to a developing plant seed.

transfer RNA (tRNA) One of a class of RNA molecules that have an anticodon at one end and an amino acid binding site at the other. Each tRNA picks up a specific amino acid and binds to the corresponding codon in messenger RNA during translation.

transformation (1) Incorporation of external DNA into the genome. Occurs naturally in some bacteria; can be induced in the laboratory by certain processes. (2) Conversion of a normal cell to a cancerous one.

transgenic Referring to an individual plant or animal whose genome contains DNA introduced from another individual, either from the same or a different species.

transition state A high-energy intermediate state of the reactants during a chemical reaction that must be achieved for the reaction to proceed. Compare with **activation energy**.

transitional feature A trait that is intermediate between a condition observed in ancestral species and the condition observed in more derived species.

translation The process by which proteins and peptides are synthesized from messenger RNA.

translational control Regulation of gene expression by various mechanisms that alter the life span of messenger RNA or the efficiency of translation.

translocation (1) In plants, the movement of sugars and other organic nutrients through the phloem by bulk flow. (2) A type of mutation in which a piece of a chromosome moves to a nonhomologous chromosome. (3) The process by which a ribosome moves down a messenger RNA molecule during translation.

transmembrane protein Any membrane protein that spans the entire lipid bilayer. Also called *integral membrane protein.*

transmission The passage or transfer (1) of a disease from one individual to another or (2) of electrical impulses from one neuron to another.

transmission electron microscope (TEM) A microscope that forms an image from electrons that pass through a specimen. Compare with **scanning electron microscope**.

transpiration Water loss from aboveground plant parts. Occurs primarily through stomata.

transport protein Collective term for any membrane protein that enables a specific ion or small molecule to cross a plasma membrane. Includes carrier proteins and channel proteins, which carry out passive transport (facilitated diffusion), and pumps, which carry out active transport.

transporter See **carrier protein**.

transposable elements Any of several kinds of DNA sequences that are capable of moving themselves, or copies of themselves, to other locations in the genome. Include LINEs and SINEs.

tree of life A diagram depicting the genealogical relationships of all living organisms on Earth, with a single ancestral species at the base.

trichome A hairlike appendage that grows from epidermal cells of some plants. Trichomes exhibit a variety of shapes, sizes, and functions depending on species.

triglyceride See **fat**.

triiodothyronine (T_3) A peptide hormone containing three iodine atoms that is produced and secreted by the thyroid gland. Acts primarily to increase cellular metabolism. In mammals, T_3 has a stronger effect than does the related hormone thyroxine (T_4).

triose A monosaccharide (simple sugar) containing three carbon atoms.

triplet code A code in which a "word" of three letters encodes one piece of information. The genetic code is a triplet code because a codon is three nucleotides long and encodes one amino acid.

triploblast (adjective: triploblastic) An animal whose body develops from three basic embryonic cell layers: ectoderm, mesoderm, and endoderm. Compare with **diploblast**.

trisomy The state of having three copies of one particular type of chromosome.

tRNA See **transfer RNA**.

trochophore A larva with a ring of cilia around its middle that is found in some lophotrochozoans.

trophic cascade A series of changes in the abundance of species in a food web, usually caused by the addition or removal of a key predator.

trophic level A feeding level in an ecosystem.

trophoblast The exterior of a blastocyst (the structure that results from cleavage in embryonic development of mammals).

tropomyosin A regulatory protein present in thin (actin) filaments that blocks the myosin-binding sites on these filaments, thereby preventing muscle contraction.

troponin A regulatory protein, present in thin (actin) filaments, that can move tropomyosin off the myosin-binding sites on these filaments, thereby triggering muscle contraction. Activated by high intracellular calcium.

true navigation The type of navigation by which an animal can reach a specific point on Earth's surface.

trypsin A protein-digesting enzyme present in the small intestine that activates several other protein-digesting enzymes.

trypsinogen The precursor of protein-digesting enzyme trypsin. Secreted by the pancreas and activated by the intestinal enzyme enterokinase.

tube foot One of the many small, mobile, fluid-filled extensions of the water vascular system of echinoderms; the part extending outside the body is called a podium. Used in locomotion and feeding.

tuber A modified plant rhizome that functions in storage of carbohydrates.

tuberculosis A disease of the lungs caused by infection with the bacterium *Mycobacterium tuberculosis.*

tumor A mass of cells formed by uncontrolled cell division. Can be benign or malignant.

tumor suppressor A gene (e.g., *p53* and *Rb*) or the protein it encodes that prevents cell division, particularly when the cell has DNA damage. Mutated forms are associated with cancer.

tundra The treeless biome in polar and alpine regions, characterized by short, slow-growing vegetation, permafrost, and a climate of long, intensely cold winters and very short summers.

turgid Swollen and firm as a result of high internal pressure (e.g., a plant cell containing enough water for the cytoplasm to press against the cell wall). Compare with **flaccid**.

turgor pressure The outward pressure exerted by the fluid contents of a plant cell against its cell wall.

Turner syndrome A human genetic disorder caused by the presence of only one X chromosome and no Y chromosome ("XO"). Individuals with this condition are female but sterile.

turnover In lake ecology, the complete mixing of upper and lower layers of water that occurs each spring and fall in temperate-zone lakes.

tympanic membrane The membrane separating the middle ear from the outer ear in terrestrial vertebrates, or similar structures in insects. Also called the *eardrum.*

ubiquinone See **coenzyme Q**.

ulcer A hole in an epithelial layer that damages the underlying basement membrane and tissues.

ultimate causation In biology, the reason that a trait or phenomenon is thought to have evolved; the adaptive advantage of that trait. Also called *ultimate explanation.* Compare with **proximate causation**.

umami The taste of glutamate, responsible for the "meaty" taste of most proteins and of monosodium glutamate.

umbilical cord The cord that connects a developing mammalian embryo or fetus to the placenta and through which the embryo or fetus receives oxygen and nutrients.

unequal crossover An error in crossing over during meiosis I in which the two non-sister chromatids match up at different sites. Results in gene duplication in one chromatid and gene loss in the other.

universal tree The phylogenetic tree that includes all organisms.

unsaturated Referring to fats and fatty acids in which at least one carbon-carbon bond is a double bond. Double bonds produce kinks in the fatty acid chains and decrease the compound's melting point. Compare with **saturated**.

upstream In genetics, opposite to the direction in which RNA polymerase moves along a DNA strand. Compare with **downstream**.

ureter In vertebrates, a tube that transports urine from one kidney to the bladder.

urethra The tube that drains urine from the bladder to the outside environment. In male vertebrates, also used for passage of sperm during ejaculation.

uric acid A whitish excretory product of birds, reptiles, and terrestrial arthropods. Used to remove from the body excess nitrogen derived from the breakdown of amino acids. Compare with **urea**.

urochordates One of the three major chordate lineages (Urochordata), comprising sessile, filter-feeding animals that have a polysaccharide exoskeleton (tunic) and two siphons through which water enters and leaves; also called tunicates or sea squirts. Compare with **cephalochordates** and **vertebrates**.

uterus The organ in which developing embryos are housed in those vertebrates that give live birth. Common in most mammals and in some lizards, sharks, and other vertebrates.

vaccination The introduction into an individual of weakened, killed, or altered pathogens to stimulate development of acquired immunity against those pathogens.

vaccine A preparation designed to stimulate an immune response against a particular pathogen without causing illness. Vaccines consist of inactivated (killed) pathogens, live but weakened (attenuated) pathogens, or portions of a viral capsid (subunit vaccine).

vacuole A large organelle in plant and fungal cells that usually is used for bulk storage of water, pigments, oils, or other substances. Some vacuoles contain enzymes and have a digestive function similar to lysosomes in animal cells.

vagina The birth canal of female mammals; a muscular tube that extends from the uterus through the pelvis to the exterior.

valence electron An electron in the outermost electron shell, the valence shell, of an atom. Valence electrons tend to be involved in chemical bonding.

valence The number of unpaired electrons in the outermost electron shell of an atom; determines how many covalent bonds the atom can form.

valves In circulatory systems, flaps of tissue that prevent backward flow of blood, particularly in veins and between the chambers of the heart.

van der Waals interactions A weak electrical attraction between two hydrophobic side chains. Often contributes to tertiary structure in proteins.

variable (V) region The portion of an antibody's light chains or heavy chains that has a highly variable amino acid sequence and forms part of the antigen-binding site. Compare with **constant (C) region**.

variable number tandem repeat See **minisatellite**.

vas deferens (plural: vasa deferentia) A muscular tube that stores and transports semen from the epididymis to the ejaculatory duct. In nonhuman animals, called the *ductus deferens*.

vasa recta In the vertebrate kidney, a network of blood vessels that runs alongside the loop of Henle of a nephron. Functions in reabsorption of water and solutes from the filtrate.

vascular bundle A cluster of xylem and phloem strands in a plant stem.

vascular cambium One of two types of lateral meristem, consisting of a ring of undifferentiated plant cells inside the cork cambium of woody plants; produces secondary xylem (wood) and secondary phloem. Compare with **cork cambium**.

vascular tissue In plants, tissue that transports water, nutrients, and sugars. Made up of the complex tissues xylem and phloem, each of which contains several cell types.

vector A biting insect or other organism that transfers pathogens from one species to another. See also **cloning vector**.

vegetative organs The nonreproductive parts of a plant including roots, leaves, and stems.

vein Any blood vessel that carries blood (oxygenated or not) under relatively low pressure from the tissues toward the heart. Compare with **artery**.

veliger A distinctive type of larva, found in mollusks.

vena cava (plural: vena cavae) A large vein that returns oxygen-poor blood to the heart.

ventral Toward an animal's belly and away from its back. The opposite of dorsal.

ventricle (1) A thick-walled chamber of the heart that receives blood from an atrium and pumps it to the body or to the lungs. (2) One of several small fluid-filled chambers in the vertebrate brain.

venules Small veins (blood vessels that return blood to the heart).

vertebra (plural: vertebrae) One of the cartilaginous or bony elements that form the spine of vertebrate animals.

vertebrates One of the three major chordate lineages (Vertebrata), comprising animals with a dorsal column of cartilaginous or bony structures (vertebrae) and a skull enclosing the brain. Includes fishes, amphibians, mammals, reptiles, and birds. Compare with **cephalochordates** and **urochordates**.

vessel element In vascular plants, a short, wide water-conducting cell that has gaps through both the primary and secondary cell walls, allowing unimpeded passage of water between adjacent cells. Compare with **tracheid**.

vestigial trait Any rudimentary structure of unknown or minimal function that is homologous to functioning structures in other species. Vestigial traits are thought to reflect evolutionary history.

vicariance The physical splitting of a population into smaller, isolated populations by a geographic barrier.

villi (singular: villus) Small, fingerlike projections (1) of the lining of the small intestine or (2) of the fetal portion of the placenta adjacent to maternal arteries. Function to increase the surface area available for absorption of nutrients and gas exchange (in the placenta).

virion A single mature virus particle.

virulence The ability of a pathogen or parasite to cause disease and death.

virulent Referring to pathogens that can cause severe disease in susceptible hosts.

virus A tiny intracellular parasite that uses host cell enzymes to replicate; consists of a DNA or RNA genome enclosed within a protein shell (capsid). In enveloped viruses, the capsid is surrounded by a phospholipid bilayer derived from the host cell plasma membrane, whereas nonenveloped viruses lack this protective covering.

visceral mass One of the three main parts of the mollusk body; contains most of the internal organs and external gill.

visible light The range of wavelengths of electromagnetic radiation that humans can see, from about 400 to 700 nanometers.

vitamin An organic micronutrient that usually functions as a coenzyme.

vitelline envelope A fibrous sheet of glycoproteins that surrounds mature egg cells in many vertebrates. Surrounded by a thick gelatinous matrix (the jelly layer) in some species. In mammals, called the *zona pellucida*.

viviparous Producing live young (instead of eggs) that develop within the body of the mother before birth. Compare with **oviparous** and **ovoviviparous**.

volt (V) A unit of electrical potential (voltage).

voltage Potential energy created by a separation of electric charges between two points. Also called *electrical potential*.

voltage clamping A technique for imposing a constant membrane potential on a cell. Widely used to investigate ion channels.

voltage-gated channel An ion channel that opens or closes in response to changes in membrane voltage. Compare with **ligand-gated channel**.

voluntary muscle Muscle tissue that can respond to conscious thought.

wall pressure The inward pressure exerted by a cell wall against the fluid contents of a plant cell.

Wallace line A line that demarcates two areas in the Indonesian region, each of which is characterized by a distinct set of animal species.

water cycle, global The movement of water among terrestrial ecosystems, the oceans, and the atmosphere.

water potential (ψ) The potential energy of water in a certain environment compared with the potential energy of pure water at room temperature and atmospheric pressure. In living organisms, ψ equals the solute potential (ψ_S) plus the pressure potential (ψ_P).

water potential gradient A difference in water potential in one region compared with that in another region. Determines the direction that water moves, always from regions of higher water potential to regions of lower water potential.

water table The upper limit of the underground layer of soil that is saturated with water.

water vascular system In echinoderms, a system of fluid-filled tubes and chambers that functions as a hydrostatic skeleton.

watershed The area drained by a single stream or river.

Watson-Crick pairing See **complementary base-pairing**.

wavelength The distance between two successive crests in any regular wave, such as light waves, sound waves, or waves in water.

wax A class of lipid with extremely long hydrocarbon tails, usually combinations of long-chain alcohols with fatty acids. Harder and less greasy than fats.

weather The specific short-term atmospheric conditions of temperature, moisture, sunlight, and wind in a certain area.

weathering The gradual wearing down of large rocks by rain, running water, and wind; one of the processes that transform rocks into soil.

weed Any plant that is adapted for growth in disturbed soils.

wetland A shallow-water habitat where the soil is saturated with water for at least part of the year.

white blood cells See **leukocytes**.

wild type The most common phenotype seen in a population; especially the most common phenotype in wild populations compared with inbred strains of the same species.

wildlife corridor Strips of wildlife habitat connecting populations that otherwise would be isolated by man-made development.

wilt To lose turgor pressure in a plant tissue.

wobble hypothesis The hypothesis that some tRNA molecules can pair with more than one mRNA codon, tolerating some variation in the third base, as long as the first and second bases are correctly matched.

wood Xylem resulting from secondary growth. Also called *secondary xylem.*

xeroderma pigmentosum A human disease characterized by extreme sensitivity to ultraviolet light. Caused by an autosomal recessive allele that results in a defective DNA repair system.

X-linked inheritance Inheritance patterns for genes located on the mammalian X chromosome. Also called *X-linkage.*

X-ray crystallography A technique for determining the three-dimensional structure of large molecules, including proteins and nucleic acids, by analysis of the diffraction patterns produced by X-rays beamed at crystals of the molecule.

xylem A plant vascular tissue that conducts water and ions; contains tracheids and/or vessel elements. Primary xylem develops from the procambium of apical meristems; secondary xylem, or wood, from the vascular cambium of lateral meristems. Compare with **phloem**.

xylem sap The watery fluid found in the xylem of plants.

yeast Any fungus growing as a single-celled form. Also, a specific lineage of ascomycetes.

Y-linked inheritance Inheritance patterns for genes located on the mammalian Y chromosome. Also called *Y-linkage.*

yolk The nutrient-rich cytoplasm inside an egg cell; used as food for the growing embryo.

Z disk The structure that forms each end of a sarcomere. Contains a protein that binds tightly to actin, thereby anchoring thin filaments.

Z scheme Path of electron flow in which electrons pass from photosystem II to photosystem I and ultimately to $NADP^+$ during the light-dependent reactions of photosynthesis. Also called *noncyclic electron flow.*

zero population growth (ZPG) A state of stable population size due to fertility staying at the replacement rate for at least one generation.

zona pellucida The gelatinous layer around a mammalian egg cell. In other vertebrates, called the *vitelline envelope.*

zone of (cellular) division In plant roots, a group of apical meristematic cells just behind the root cap where cells are actively dividing.

zone of (cellular) elongation In plant roots, a group of young cells, located behind the apical meristem, that are increasing in length.

zone of (cellular) maturation In plant roots, a group of plant cells, located several centimeters behind the root cap, that are differentiating into mature tissues.

zygosporangium (plural: zygosporangia) The spore-producing structure in fungi that are members of the Zygomycota.

zygote The diploid cell formed by the union of two haploid gametes; a fertilized egg. Capable of undergoing embryological development to form an adult.

Credits

Frontmatter

p. ix Oscar Miller/SPL/Photo Researchers **p. xi** Photo by Jason Rick, supplied by Loren Rieseberg **p. xiv** Jeff Rotman/NPL/Minden Pictures **p. xvi** Frans Lanting/Minden Pictures **p. xxT** Natalie B. Fobes Photography **p. xxB** David Quillin

Chapter 1

Opener Marjorie Kibby **1.1** 'Animalcules' observed by Anton van Leeuwenhoek, c1795, HIP/Art Resource, NY **1.6a** Samuel F. Conti and Thomas D. Brock **1.6b** Kwangshin Kim/Photo Researchers **1.7b** Michael Hughes/laif/Aurora Photos **1.8R** From "The ant odometer: stepping on stilts and stumps." Wittlinger, M.; Wehner, R.; and Wolf, H. Science. 2006 June 30; 312(5782): 1965–7

Chapter 2

Opener Frans Lanting/Corbis **2.1b** Dragan Trifunovic/Shutterstock **2.6c** Photos. com **2.11** Robert and Beth Plowes, ProteaPIX **2.14b** Dietmar Nill/Picture Press/Photolibrary **2.15c** Jan Vermeer/Minden Pictures **2.20** Shutterstock **2.20 inset** PhotoSpin/Alamy

Chapter 3

Opener Fritz Wilhelm (heisingart.com) **3.10a** Microworks/Phototake **3.10b** Walter Reinhart/Phototake

Chapter 4

Opener SSPL/The Image Works **4.5** National Cancer Institute

Chapter 5

Opener Peter Arnold/Alamy

Chapter 6

6.5L James J. Cheetham, Carleton University **6.9a** iStockphoto **6.9b** iStockphoto **6.9c** Shutterstock **6.18** Harold Edwards/Visuals Unlimited **6.24T** David M. Phillips/Photo Researchers **6.24M** David M. Phillips/Photo Researchers **6.24B** Joseph F. Hoffman/Yale University School of Medicine

Chapter 7

Opener Torsten Wittmann/Photo Researchers **7.1** T. J. Beveridge/Visuals Unlimited **7.2** Gopal Murti/Visuals Unlimited **7.3** Wanner/Eye of Science/Photo Researchers **7.4** Stanley C. Holt/Biological Photo Service **7.5** From: Machinery of Life, Goodsell, David S., 2nd ed., 2009. Copyright 2009, Springer-Verlag,. Reprinted with permission. **7.7** Don W. Fawcett/Photo Researchers **7.8** Don W. Fawcett/Photo Researchers **7.9** Don W. Fawcett/Photo Researchers **7.10** Biophoto Associates/Photo Researchers **7.11** Omikron/Photo Researchers **7.12** Fawcett & Friend/Photo Researchers **7.13** Gopal Murti/Visuals Unlimited **7.16** E. H. Newcomb & W. P. Wergin/Biological Photo Service **7.17** Kanaseki & Donald Fawcett/Visuals Unlimited **7.18** E. H. Newcomb & W. F. Wergin/Biological Photo Service **7.19** E. H. Newcomb & S. E. Frederick/Biological Photo Service **7.20** From: Biochemistry. Moran, L.A. and Scrimgeour K. G. (1994). Neil Patterson Publishers /Prentice Hall, North Carolina. © 1994 Neil Patterson Publishers. Reprinted with permission from Pearson Science. **7.21a** Don Fawcett & S. Ito & A. Like/Photo Researchers **7.21b** Don W. Fawcett/Photo Resarchers **7.21c** Biophoto Associates/Photo Researchers **7.21d** Dennis Kunkel/Visuals Unlimited **7.22** Don W. Fawcett/Photo Researchers **7.25a** James D. Jamieson, Yale University School of Medicine **7.25b** From: Intracellular transport of secretory proteins in the pancreatic exocrine cell: II. Transport to condensing vacuoles and zymogen granules. Jamieson, J. D. and Palade, G. E. (1967). Reproduced from The Journal of Cell Biology, 1967, v. 34 , pp. 597–615 by permission of the Rockefeller University Press. **7.25c** From: Intracellular transport of secretory proteins in the pancreatic exocrine cell: II. Transport to condensing vacuoles and zymogen granules. Jamieson, J. D. and Palade, G. E. (1967). Reproduced from The Journal of Cell Biology, 1967, v. 34 , pp. 597–615 by permission of the Rockefeller University Press. **7.30** Conly L. Rieder, Wadsworth Center, Albany, New York **7.31a** From: Molecular Biology of the Cell 9(12), December 1998, cover. Copyright © 1999 by the American Society for Cell Biology. Image courtesy of Bruce J. Schnapp, Oregon Health Sciences University. Reproduced by permission of the American Society for Cell Biology. **7.31b** From: Molecular Biology of the Cell 9(12), December 1998, cover. Copyright © 1999 by the American Society for Cell Biology. Image courtesy of Bruce J. Schnapp, Oregon Health Sciences University. Reproduced by permission of the American Society for Cell Biology. **7.32a** John E. Heuser, Washington University School of Medicine, St. Louis, Missouri **7.33L** SPL/Photo Researchers **7.33R** Visuals Unlimited/Corbis **7.34a** Don W. Fawcett/Photo Researchers

Chapter 8

Opener Caroline Weight, Molecular Immunology, Institute of Food Research, Norwich, UK **8.1** William James Warren/Science Faction **8.2** Biophoto Associates/Photo Researchers **8.3b** Barry King, University of California, Davis/Biological Photo Service **8.5** SPL/Photo Researchers **8.6** Photo Researchers **8.7aL** Don W. Fawcett/Photo Researchers **8.7aR** Don W. Fawcett/Photo Researchers **8.8a** Don W. Fawcett & Gida Matoltsy/Photo Researchers **8.11a** E. H. Newcomb & W. P. Wergin/Biological Photo Service **8.11b** Don W. Fawcett/Photo Researchers **8.18** CDC/Rodney M. Donlan & Janice Carr

Chapter 9

Opener Richard Megna/Fundamental Photographs **9.7T** Shutterstock **9.7B** Shutterstock **9.13** Terry Frey **9.22a** Science VU/B. Bhatnagar/Visuals Unlimited

Chapter 10

Opener Visuals Unlimited/Corbis **10.2T** John Durham/SPL/Photo Researchers **10.2M** Electron micrograph by Wm. P. Wergin, courtesy of E. H. Newcomb, University of Wisconsin **10.2B** Electron micrograph by Wm. P. Wergin, courtesy of E. H. Newcomb, University of Wisconsin **10.4b** Sinclair Stammers/SPL/Photo Researchers **10.17L** James A. Bassham, Lawrence Berkeley Laboratory, UCB (retired) **10.17R** James A. Bassham, Lawrence Berkeley Laboratory, UCB (retired) **10.20a** Jeremy Burgess/SPL/Photo Researchers

Chapter 11

Opener CNRI/Phototake **11.1a** Originally published in Walter Flemming, Zellsubstanz, Kern, und Zelltheilung. Leipzig: Verlag von F. C. W. Vogel, 1882. Image courtesy of Conly Rieder, Wadsworth Center, New York State Department of Health. **11.1b** Photo Researchers **11.2T** Gopal Murti/Photo Researchers **11.2B** Biophoto Associates/Photo Researchers **11.5 1.** Micrographs by Conly L. Rieder, Division of Molecular Medicine, Wadsworth Center, Albany, New York **11.5 2.** Micrographs by Conly L. Rieder, Division of Molecular Medicine, Wadsworth Center, Albany, New York **11.5 3.** Micrographs by Conly L. Rieder, Division of Molecular Medicine, Wadsworth Center, Albany, New York **11.5 4.** Micrographs by Conly L. Rieder, Division of Molecular Medicine, Wadsworth Center, Albany, New York **11.5 5.** Micrographs by Conly L. Rieder, Division of Molecular Medicine, Wadsworth Center, Albany, New York **11.5 6.** Micrographs by Conly L. Rieder, Division of Molecular Medicine, Wadsworth Center, Albany, New York **11.5 7.** Micrographs by Conly L. Rieder, Division of Molecular Medicine, Wadsworth Center, Albany, New York **11.5 8.** Micrographs by Conly L. Rieder, Division of Molecular Medicine, Wadsworth Center, Albany, New York **11.6a** Ed Reschke/Peter Arnold **11.6b** Visuals Unlimited/Getty Images

Chapter 12

Opener David Phillips/The Population Council/Photo Researchers **12.6** Photos .com**12.7 1.** Edward Novitski, courtesy of Charles Novitski **12.7 2.** Edward Novitski, courtesy of Charles Novitski **12.7 3.** Edward Novitski, courtesy of Charles Novitski **12.7 4.** Edward Novitski, courtesy of Charles Novitski **12.7 5.** Edward Novitski, courtesy of Charles Novitski **12.7 6.** Edward Novitski, courtesy of Charles Novitski **12.77.** Edward Novitski, courtesy of Charles Novitski **12.7 8.** Edward Novitski, courtesy of Charles Novitski **12.79.** Edward Novitski, courtesy of Charles Novitski

Chapter 13

Opener Brian Johnston **13.9a** Benjamin Prud'homme/Nicolas Gompel **13.9bL** From: "Learning to Fly: Phenotypic Markers in Drosophila." A poster of common phenotypic markers used in Drosophila genetics. Jennifer Childress, Richard Behringer, and Georg Halder. 2005. Genesis 43(1). Cover illustration. **13.9bR** From: "Learning to Fly: Phenotypic Markers in Drosophila." A poster of common phenotypic markers used in Drosophila genetics. Jennifer Childress, Richard Behringer, and Georg Halder. 2005. Genesis 43(1). Cover illustration. **13.15a** Robert Calentine/Visuals Unlimited

Chapter 14

Opener Gopal Murti/SPL/Photo Researchers **14.1b** Eye of Science/Photo Researchers **14.5** From: "The replication of DNA in *Escherichia coli.*" M. Meselson and F.W. Stahl. Proceedings of the National Academy of Sciences 44(7):671–682 (July 1958) Reproduced by permission of Matthew S. Meselson, Harvard University. **14.7a** Gopal Murti/SPL/Photo Researchers

Chapter 15

Opener Alfred Pasieka/Photo Researchers **15.4a** Rod Williams/Nature Picture Library **15.4b** The Peromyscus Genetic Stock Center at the University of South Carolina, Photograph by Clint Cook. **15.9a** Peter Duesberg, UC Berkeley **15.9b** Peter Duesberg, UC Berkeley

Chapter 16

Opener Oscar Miller/SPL/Photo Researchers **16.5a** Bert W. O'Malley, Baylor College of Medicine **16.8a** From Hamkalo, B. A. and Miller, O. L. Jr., "Electronmicroscopy of genetic material." Annual Review of Biochemistry, 1973; 42: 379–96. Reproduced with permission. **16.9b** E. V. Kiseleva and Donald Fawcett/Visuals Unlimited

Chapter 17

Opener Stephanie Schuller/Photo Researchers **UN17.1** EDVOTEK, The Biotechnology Education Company (www.edvotek.com)

Chapter 18

18.2a Ada Olins & Don Fawcett/Photo Researchers **18.4T** From: Molecular Biology of the Cell 4/e fig. 4-23 by Bruce Alberts et al. Copyright © 2002. Reproduced by permission of Victoria E. Foe, and Garland Science/Taylor & Francis Books, Inc. **18.4B** From: Molecular Biology of the Cell 4/e fig. 4-23 by Bruce Alberts et al. Copyright © 2002. Reproduced by Permission of Victoria E. Foe, and Garland Science/Taylor & Francis Books, Inc.

Chapter 19

Opener AJ/IRRI/Corbis **19.1b** Brady-Handy Photograph Collection (Library of Congress). Reproduction number: LC-DIG-cwpbh-02977. **19.12** Baylor College of Medicine/Peter Arnold **19.16bL** Golden Rice Humanitarian Board (www.goldenrice.org) **19.16bR** Golden Rice Humanitarian Board (www.goldenrice.org)

Chapter 20

Opener Sanger Institute/Wellcome Photo Library **20.7bL** Test results provided by GENDIA (www.gendia.net) **20.7bR** Test results provided by GENDIA (www.gendia.net) **20.10** Agilent Technologies **20.11** Camilla M. Kao and Patrick O. Brown, Stanford University

Chapter 21

Opener Anthony Bannister/Gallo Images/Corbis **21.1a** From: The BMP antagonist Gremlin regulates outgrowth, chondrogenesis and programmed cell death in the developing limb." R. Merino, J. Rodriguez-Leon, D. Macias, Y. Gañan, A. N. Economides and J. M Hurle. Development 126: 5515–5522 (1999), Fig 6. **21.1b** From: Reduced apoptosis and cytochrome c-mediated caspase activation in mice lacking caspase 9." K. Kuida et al. Cell, 94:325–337 (1998), figs. 2E and 2F. Copyright 1998 by Elsevier Science Ltd. Reproduced by permission of Elsevier Science. Image courtesy of Keisuke Kuida, Vertex Pharmaceuticals. **21.2** Roslin Institute/Phototake **21.3a** Adapted from: "Embryonic and neonatal phenotyping of genetically engineered mice." Kulandavelu, S.; Qu, D.; Sunn, N.; Mu, J.; Rennie, M. Y.; Whiteley, K. J.; Walls, J. R.; Bock, N. A.; Sun, J. C.; Covelli, A.; Sled, J. G.; Adamson, S. L. ILAR Journal. 2006; 47(2): 103–17; Fig. 12A. **21.3b** From: Embryonic and Neonatal Phenotyping of Gentically Engineered Mice. Adamson and Walls. The ILAR Journal, 47 (2); Fig. 12A. Optical projection tomographhy (OPT) image of a day 12.5 mouse embryo. (The rotation of the embryo and the art overlays do not appear in original paper.) **21.4a** F. Rudolf Turner, Indiana University **21.4b** Christiane Nusslein-Volhard, Max Planck Institute of Developmental Biology, Tubingen, Germany **21.5** Prof. Dr. Christiane Nusslein-Volhard **21.6** Wolfgang Driever, University of Freiburg, Germany **21.7a** Jim Langeland, Stephen Paddock, and Sean Carroll, University of Wisconsin-Madison **21.7b** Stephen J. Small, New York University **21.7c** Jim Langeland, Stephen Paddock, and Sean Carroll, University of Wisconsin-Madison **21.8L** Visuals Unlimited/Getty Images **21.8M** David Scharf/Science Faction/Corbis **21.8R** Eye of Science/Photo Researchers **21.10a** F. Rudolf Turner, Indiana University **21.11** Anthony Bannister/NHPA **21.12a** Reprinted from "The Origin of Form" by Sean B. Carroll, Natural History, November 2005. Burke, A.C. 2000, "Hox" Genes and the Global Patterning of the Somitic Mesoderm. In Somitogenesis. C. Ordahl (ed.) "Current Topics in Developmental Biology", Vo. 47. Academic Press. **21.12b** Reprinted from "The Origin of Form" by Sean B. Carroll, Natural History, November 2005. Burke, A.C. 2000, "Hox" Genes and the Global Patterning of the Somitic Mesoderm. In Somitogenesis. C. Ordahl (ed.) "Current Topics in Developmental Biology", Vo. 47. Academic Press

Chapter 22

Opener Yorgos Nikas/Photo Researchers **22.2** Holger Jastrow **22.5a1** Michael Whitaker/SPL/Photo Researchers **22.5a2** Michael Whitaker/SPL/Photo Researchers **22.5a3** Michael Whitaker/SPL/Photo Researchers **22.5b1** Victor D. Vacquier, Scripps Institution of Oceanography, University of California at San Diego **22.5b2** Victor D. Vacquier, Scripps Institution of Oceanography, University of California at San Diego **22.5b3** Victor D. Vacquier, Scripps Institution of Oceanography, University of California at San Diego **22.11a** Kathryn W. Tosney, University of Michigan **22.11b** Kathryn Tosney, University of Miami

Chapter 23

Opener John Runions/Oxford Brookes University **23.3b** Cabisco/Visuals Unlimited **23.4b** Michael Clayton, Department of Botany, University of Wisconsin **23.8b** Ken Wagner/Phototake **23.10a** From: Kim et al. Nature. Vol 424, July 24, 2003, pg 439. Copyright 2004. Macmillan Publishers Ltd. Reprinted by permission. Photograph supplied by Dr. Neelima Sinah. **23.10bT** From: Kim et al. Nature. Vol 424, July 24, 2003, pg 439. Copyright 2004. Macmillan Publishers Ltd. Reprinted by permission. Photograph supplied by Dr. Neelima Sinah. **23.10bB** From: Kim et al. Nature. Vol 424, July 24, 2003, pg 439. Copyright 2004. Macmillan Publishers Ltd. Reprinted by permission. Photograph supplied by Dr. Neelima Sinah. **23.12L** John L. Bowman, University of California at Davis **23.12ML** John L. Bowman, University of California at Davis **23.12MR** John L. Bowman, University of California at Davis **23.12R** John L. Bowman, University of California at Davis

Chapter 24

Opener Art Wolfe/Stone/Getty Images **24.2a** Sinclair Stammers/Photo Researchers **24.2b** From: Ichnotaxonomy of Bird-Like Footprints: An Example from the Late Triassic-Early Jurassic of Northwest Argentina. Silvina De Valais and Ricardo N. Melchor. Journal of Vertebrate Paleontology 28(1):145–159, March 2008, Figure 5c. © 2008 by the Society of Vertebrate Paleontology **24.2c** The Natural History Museum, London **24.3L** Gerd Weitbrecht for Landesbildungsserver Baden-Württemberg (www.schule-bw.de) **24.3R** iStockphoto **24.5aL** CMCD/PhotoDisc/Getty Images **24.5aR** Vincent Zuber/Custom Medical Stock Photo **24.5bL** Mary Beth Angelo/Photo Researchers **24.5bR** Custom Medical Stock Photo **24.6a1** age fotostock/SuperStock **24.6a2** George D. Lepp/Photo Researchers **24.6a3** Tui De Roy/Minden Pictures **24.6a4** Stefan Huwiler/age fotostock **24.8L** From: Anatomy and Embryology 305, Fig. 7. Copyright © Springer-Verlag GmbH & Co KG, Heidelberg, Germany. Reproduced by permission. Photo by Michael K. Richardson. **24.8M** From: Anatomy and Embryology 305, Fig. 8. Copyright © Springer-Verlag GmbH & Co KG, Heidelberg, Germany. Reproduced by permission. Photo by Ronan O'Rahilly. **24.8R** From: Richardson, M. K., et al. Science. Vol 280: Pg 983c, Issue # 5366, May 15, 1998. Embryo from Professor R.

O'Rahilly. National Museum of Health and Medicine/Armed Forces Institute of Pathology **24.10** Walter J. Gehring **24.15aL** G. Gerra & S. Sommazzi /www .justbirds.it **24.15aR** Tui De Roy/Minden Pictures **24.15b** Huw Lewis **24.18T** From: "Bmp4 and Morphological Variation of Beaks in Darwin's Finches". Abzhonov et al 2004 Science 305:1462–1465, Fig 1C. **24.18B** From: A phylogeny of Darwin's finches based on microsatellite DNA length variation. K. Petren et al. Proceedings of the Royal Society of London, Series B, 266:321–329 (1999), p. 327, fig. 3. Copyright © 1999 Royal Society of London. Reproduced by permission. Image courtesy of Kenneth Petren, University of Cincinnati. **24.20** Erlend Haarberg/NHPA/Photoshot

Chapter 25

Opener Phil Savoie/Minden Pictures **25.7a** André Karwath/Wikimedia Commons. Licensed under the Creative Commons Attribution ShareAlike 2.5 **25.14a** Cyril Laubscher/Dorling Kindersley **25.15** Otorohanga Kiwi House, New Zealand **25.16a** Robert Rattner **25.17TL** Robert & Linda Mitchell Photography **25.17TR** Robert & Linda Mitchell Photography **25.17BL** Fotolia **25.17BR** Fotolia

Chapter 26

Opener Photo by Jason Rick, supplied by Loren Rieseberg **26.9bL** Michael Lustbader/Photo Researchers **26.9bR** From: Recent and recurrent polyploidy in Tragopogon (Asteraceae): genetic, genomic, and cytogenetic comparisons. Soltis, D. E., P. S. Soltis, J. C. Pires, A. Kovarik, and J. Tate. Biological Journal of the Linnaean Society 82:485–501, 2004. Photo D. Soltis, P. Soltis, and A Doust. **26.10aT** H. Douglas Pratt/National Geographic Stock **26.10aM** H. Douglas Pratt/National Geographic Stock **26.10aB** H. Douglas Pratt/National Geographic Stock

Chapter 27

Opener Merv Feick (www.indiana9fossils.com) **27.7a** Stefan Piasecki, GEUS **27.7b** iStockphoto **27.7c** From The Virtual Petrified Wood Museum (petrifiedwoodmuseum.org/). by Mike Viney **27.7d** From The Virtual Petrified Wood Museum (petrifiedwoodmuseum.org/) by Mike Viney **27.11LT** Gerald D. Carr **27.11LB** Forest & Kim Starr **27.11R** Forest & Kim Starr **27.12a** Eladio Fernandez/Caribbean Nature Photography **27.12b** Jonathan B. Losos and Kevin de Quieroz, Washington University in St. Louis **27.13a** Colin Keates/Dorling Kindersley, Courtesy of the Natural History Museum, London **27.13b** Tim Fitzharris/Minden Pictures **27.13c** Don P. Northup, www.africancichlidphotos. com**27.13d** James L. Amos/Corbis **27.14bTL** Chip Clark, National Museum of Natural History, Smithsonian Institution **27.14bTR** Ed Reschke/Peter Arnold **27.14bML** Simon Conway Morris, University of Cambridge, Cambridge, United Kingdom **27.14bMR** J. Gehling, South Australian Museum **27.14bBL** Shuhai Xiao, Virginia Tech **27.14bBM** Shuhai Xiao, Virginia Tech **27.14bBR** Shuhai Xiao, Virginia Tech **27.17bL** Glen A. Izett, U.S. Geological Survey, Denver **27.17bR** Glen A. Izett, U.S. Geological Survey, Denver **27.17c** Peter H. Schultz, Brown University; Steven D'Hondt, University of Rhode Island Graduate School of Oceanography

Chapter 28

Opener John Hammond **28.3** Science VU/Visuals Unlimited **28.4** Reprinted with permission of the American Association for the Advancement of Science from S. V. Liu et al., Thermophilic Fe(III)-reducing bacteria from the deep subsurface. Science 277:1106–1109 (1997), fig. 2. Copyright © 1997 American Association for the Advancement of Science. Image courtesy of Yul Roh, Oak Ridge National Laboratory. Permission from AAAS is required for all other uses. **28.7aT** Daniel Quilten/Visuals Unlimited **28.7aB** Reprinted with permission from H.N. Schulz, et al., Dense populations of a giant sulfur bacterium in Namibian shelf sediments. Science 284:493–495, Fig. 1b. August 4, 1999. Copyright 1999 American Association for the Advancement of Science. Image courtesy of Dr. Heide Schulz, Max-Planck-Institute for Marine Microbiology, Bremen, Germany. **28.7bT** SciMAT/Photo Researchers **28.7bB** SciMAT/Photo Researchers **28.7cT** K.S. Kim/Peter Arnold **28.7cB** Ralf Wagner **28.8a** R. Hubert, Microbiology Program, Dept. of Animal Science, Iowa State University **28.10** Bruce J. Russell/BioMEDIA Associates **28.12** Wally Eberhart/Visuals Unlimited/Getty Images **28.15** Andrew Syred/Photo Researchers **28.16** Bill Schwartz, Public Health Image Library (PHIL), CDC **28.17** S. Amano, S. Miyadoh and T. Shomura/The Society for Actinomycetes Japan **28.18** From the Cover: Cytoplasmic lipid droplets are translocated into the lumen of the Chlamydia trachomatis parasitophorous vacuole. Jordan L. Cocchiaro, Yadunanda Kumar, Elizabeth R. Fischer, Ted Hackstadt, and Raphael H. Valdivia Proc Natl Acad Sci U.S.A. 2008 July 8; 105(27): 9379–9384. Copyright 2008 National Academy of Sciences, U.S.A. **28.19** Peter Siver/Visuals Unlimited **28.20a** Yves V. Brun **28.20b** George Barron, University of Guelph, Canada **28.21** Kenneth M. Stedman/Portland State University **28.21 inset** Eye of Science/Photo Researchers **28.22** Doc Searls **28.22 inset** Priya DasSarma, Center of Marine Biotechnology, University of Maryland Biotechnology Institute

Chapter 29

Opener Wim van Egmond/Visuals Unlimited/Getty Images **29.2L** D. P. Wilson/FLPA/Minden Pictures **29.2M** Mark Conlin/V&W/imagequestmarine .com**29.2R** E. R. Degginger/Animals Animals-Earth Scenes **29.4** Pete Atkinson/Photographer's Choice/Getty Images **29.4 inset** D. Anderson, WHOI **29.6** Biophoto Associates/Photo Researchers **29.11a** Steve Gschmeissner/Photo Researchers **29.11b** Biophoto Associates/Photo Researchers **29.11c** Andrew Syred/Photo Researchers **29.12a** Biophoto Associates/Photo Researchers **29.12b** Bruce Coleman/Photoshot **29.15** Wim van Egmond/Visuals Unlimited/Getty Images **29.19** Scott Camazine/Photo Researchers **29.20** Dennis Kunkel/Visuals Unlimited **29.21** Tai-Soon Yong (www.atlas.or.kr) **29.22** Image by D. Patterson, provided by the microscope web project. **29.23** Francis Abbott/NPL/Minden Pictures **29.24** Peter Parks/imagequestmarine.com **29.25** Image by D. Patterson, provided by the micro*scope web project. **29.26** Maurice LOIR (www.diatomloir.eu) **29.27** Visuals Unlimited/Corbis **29.28** Andrew Syred/SPL/Photo Researchers **29.29** Wim van Egmond/Visuals Unlimited **29.30** Joyce Photographics/Photo Researchers **T29.31** Peter Siver/Visuals Unlimited/Getty Images **T29.32** Visuals Unlimited/Corbis **T29.33** Jeff Rotman/Stone/Getty Images **T29.34** iStockphoto **T29.35** Yuuji Tsukii, Protist Information Server (protist.i.hosei.ac.jp) **T29.36** Dr. Kawachi of National Institue for Environmental Studies, Japan

Chapter 30

Opener Jean-Paul Ferrero/Auscape/Minden Pictures **30.1** Lynn Betts/USDA Natural Resources Conservation Service **30.2b** Hugh Iltis/The Doebley Lab, University of Wisconsin-Madison (teosinte.wisc.edu) **30.4L** Linda Graham, University of Wisconsin-Madison **30.4R** Lee W. Wilcox **30.5L** Lee W. Wilcox **30.5M** iStockphoto **30.5R** Rob Whitworth/GAP Photos/Getty Images **30.11a** Lee W. Wilcox **30.11b** Lee W. Wilcox **30.12** Biodisc/Visuals Unlimited/Alamy **30.16a** Lee W. Wilcox **30.16bL** Lee W. Wilcox **30.16bR** Lee W. Wilcox **30.18** Carolina Biological Supply Company/ Phototake **30.21a** Larry Deack/ Wikimedia Commons **30.21b** Shutterstock **30.21c** Andreas Lander/dpa/Corbis **30.23a** Nigel Cattlin/Holt Studios/Photo Researchers **30.23b** Shutterstock **30.25T1** Jerome Wexler/Visuals Unlimited/Getty Images **30.25T2** Keith Wheeler/SPL/Photo Researchers **30.25T3** iStockphoto**30.25T4** Noodle snacks (www.noodlesnacks.com/) via Wikipedia. This file is licensed under the Creative Commons Attribution ShareAlike 3.0 and GNU Free Documentation License. **30.25B1** Shutterstock **30.25B2** ISM/Phototake **30.25B3** iStockphoto **30.25B4** Shutterstock **30.27a** Doug Allan/NPL/Minden Pictures **30.27c** Lee W. Wilcox **30.27b** Wim van Egmond/Visuals Unlimited/Getty Images **30.28** Peter Verhoog/Foto Natura/Minden Pictures **30.29** Wim van Egmond **30.30** CAIP IMAGES Copyright © 2006 University of Florida (plants.ifas.ufl.edu/slidecol. html) **30.31** Adrian Davies/NPL/Minden Pictures **30.32a** Wildlife GmbH/Alamy **30.32b** Colin McPherson/Corbis **30.33** Steven P. Lynch, Lynch Images **30.34** Lee W. Wilcox **30.35** Frans Lanting/Corbis **30.36** Peter Arnold/Alamy **30.37L** David T. Webb, Botany, University of Hawaii **30.37R** Jami Tarris/Workbook Stock/Getty Images **30.38** Lee W. Wilcox **30.39** iStockphoto **30.40a** Martin Land/SPL/Photo Researchers **30.40b** iStockphoto **30.41** Peter Chadwick/Dorling Kindersley **30.42a** Lee W. Wilcox **30.42b** Lee W. Wilcox **30.43** Michael & Patricia Fogden/Minden Pictures **30.44a** Brian Johnston **30.44bL** Lee W. Wilcox **30.44bR** Lee W. Wilcox

Chapter 31

Opener Richard Shiell/Animals Animals-Earth Scenes **31.1a** David Reed, Department of Animal and Plant Sciences, University of Sheffield **31.1b** Mycorrhizal Applications Inc. (www.mycorrhizae.com) **31.2** John Cang Photography **31.3a** Visuals Unlimited/Corbis **31.3b** iStockphoto **31.4a** Eye of Science/Photo Researchers **31.4b** Tony Brain/Photo Researchers **31.5bL** George Barron, University of Guelph, Canada **31.5bR** Biophoto Associates/Photo Researchers **31.6a** Melvin Fuller/ The Mycological Society of America

31.6b James Richardson/Visuals Unlimited **31.6c** Biophoto Associates/Photo Researchers **31.6d** Nino Santamaria **31.10a** Mark Brundrett (http://mycorrhizas.info) **31.10b** Mark Brundrett (http://mycorrhizas.info) **31.13** J. I. Ronny Larsson, Professor of Zoology **31.14** Thomas J. Volk, TomVolkFungi.net **31.15a** Gregory G. Dimijian/Photo Researchers **31.15b** David M. Phillips/Photo Researchers **31.16** Jim Deacon **31.17** Michel Poinsignon/NPL/Minden Pictures **31.18b** Daniel Mosquin **31.19** N. Allen and G. Barron, University of Guelph, Guelph, Ontario **31.20** Michael Fogden/OSF/Photolibrary

Chapter 32

Opener Bill Curtsinger/National Geographic Stock **32.1a** Satoshi Kuribayashi/Nature Production/Minden Pictures **32.1b** Mark Moffett/Minden Pictures **32.2** Walker England/Photo Researchers **32.3a** Roberto Rinaldi/NPL/Minden Pictures **32.3b** Ingo Arndt/Minden Pictures **32.8bT** Robert Steene/imagequestmarine.com **32.8bB** Russ Hopcroft/University of Alaska Fairbanks **32.12a** J.P. Ferrero/Jacana/Photo Researchers **32.12b** Todd Husman (www.soundaquatic.com) **32.13a** Pete Oxford/Minden Pictures **32.13b** Heidi & Hans-Jurgen Koch/Minden Pictures **32.14a overlay** Tim Ridley/Dorling Kindersley, Courtesy of the Royal Veterinary College, University of London **32.14a background** Shutterstock **32.14b** Adrian Costea/Fotolia **32.14b inset** Josef Ramsauer and Prof. Dr. Robert Patzner, Salzburg University, Austria **32.15a** Satoshi Kuribayashi/OSF/Photolibrary **32.15b** Radius Images/Corbis **32.16a** Andrew Syred/Photo Researchers **32.16b** Eye of Science/Photo Researchers **32.17a** Roland Seitre/Peter Arnold **32.17b** Jeff Foott/NPL/Minden Pictures **32.18a** D. P. Wilson/FLPA/Minden Pictures **32.18b** Sue Scott **32.19L** Reproduced by permission from Grace Panganiban et al., The origin and evolution of animal appendages. Proceedings of the National Academy of Sciences 94:5162–5166 (May 1997), p. 5163, Fig. 1c. Copyright © 1997 National Academy of Sciences, U.S.A. **32.19M** Reproduced by permission from Grace Panganiban et al., The origin and evolution of animal appendages. Proceedings of the National Academy of Sciences 94:5162-5166 (May 1997), p. 5163, Fig. 1c. Copyright © 1997 National Academy of Sciences, U.S.A. **32.19R** Reproduced by permission from Grace Panganiban et al., The origin and evolution of animal appendages. Proceedings of the National Academy of Sciences 94:5162–5166 (May 1997), p. 5163, Fig. 1f. Copyright © 1997 National Academy of Sciences, U.S.A. **32.20a** Shutterstock **32.20b** D. Parer & E. Parer-Cook/Auscape/Minden Pictures **32.21a** Nigel Cattlin/FLPA/Minden Pictures **32.21b inset** Frank Greenaway/Dorling Kindersley **32.21b** Kim Taylor/NPL/Minden Pictures **32.23** Roberto Rinaldi/NPL/Minden Pictures **32.24** Gary Bell/Taxi/Getty Images **32.25** Gregory G. Dimijian/Photo Researchers **32.26** Matthew D. Hooge, University of Maine

Chapter 33

Opener Barbara Strnadova/Photo Researchers **33.3a** Peter Parks/imagequestmarine.com **33.3b** Carmel McDougall, University of Oxford **33.4a** Dave King/Dorling Kindersley **33.4b** Breck P. Kent/Animals Animals–Earth Scenes **33.8a** Amazon-Images/Alamy **33.8b** Wim van Egmond/Visuals Unlimited **33.8c** Wim van Egmond/Visuals Unlimited/Corbis **33.9a** Pankaj Bajpai **33.9b** Bates Littlehales/Animals Animals–Earth Scenes **33.9c** Jane Burton/NPL/Minden Pictures **33.11** Antonio Guillén Oterino **33.12a** Doug Anderson (doug.deep via Flickr) **33.12b** Manfred Kage/Peter Arnold **33.12c** Oliver Meckes/Photo Researchers **33.13a** J. W.Alker/Imagebroker/Alamy **33.13b** Jasper Nance (www.flickr.com/photos/nebarnix) **33.13c** Robert and Linda Mitchell **33.14a** Marevision/age fotostock **33.15a** Roger Steen/imagequestmarine.com **33.15b** Daniela Wolf (www.taucher.li) **33.16** Kjell B. Sandved/Photo Researchers **33.17** Mark Conlin/VWPICS/age fotostock **33.18a** Shutterstock **33.18b** Andrew Syred/Photo Researchers **33.19a** Eye of Science/Photo Researchers **33.19b** C.Hough/Custom Medical Stock Photo **33.20** Mark Smith/Photo Researchers **33.21a** Holger Mette/Fotolia **33.21a inset** Holger Mette/Fotolia **33.21b** Manfred Kage/Peter Arnold **33.22a** iStockphoto **33.22b** National Geographic/Getty Images **33.23** Colin Milkins/OSF/Photolibrary **T33.11** Meul/ARCO/NPL/Minden Pictures **T33.12** iStockphoto **T33.13** iStockphoto **T33.14** Wikimedia Commons **T33.15** J. Gall/Photo Researchers **T33.16** iStockphoto **T33.17** J. K. Lindsey **T33.18** Brand X Pictures/Jupiter Images

Chapter 34

34.2a D. P. Wilson/FLPA/Minden Pictures **34.2b** Juri Vinokurov/Fotolia **34.3b** Jeff Rotman/NPL/Minden Pictures **34.4a** Sue Scott/OSF/Photolibrary **34.4b** Gary Bell/OceanwideImages.com **34.6** Alan James/NPL/Minden Pictures **34.7a** Audra Reese Kirkland **34.7b** Jeffrey L. Rotman/CORBIS **34.9** Heather Angel/Natural Visions **34.10a** Gary Bell/Getty Images **34.10b** Sinclair Stammers/NPL/Minden Pictures **34.15** Rudie Kuiter/OceanwideImages.com **34.17** Pavel Riha **34.21a** Hiroya Minakuchi/Minden Pictures **34.21b** Shutterstock **34.23a** Tom & Theresa Stack/NHPA/Photoshot **34.23b** James L. Amos/National Geographic Stock **34.24a** Fred McConnaughey/Photo Researchers **34.24b** Paul Souders/Corbis **34.25** Oceans Image/Photoshot **34.26** Arnaz Mehta **34.27a** Michael & Patricia Fogden/Minden Pictures **34.27b** Michael & Patricia Fogden/Minden Pictures **34.28a** Tom McHugh/Photo Researchers **34.28b** Reg Morrison/Auscape/Minden Pictures **34.29** Frank Lukasseck/Photographer's Choice/Getty Images **34.30** Anup Shah/NPL/Minden Pictures **34.31** George Grall/National Geographic Stock **34.32** iStockphoto **34.33** iStockphoto **34.34a** Tui De Roy/Minden Pictures **34.34b** Ingo Arndt/Minden Pictures **34.35a** Cyril Ruoso/J. H. Editorial/Minden Pictures **34.35b** Ingo Arndt/Minden Pictures

Chapter 35

Opener Jed Fuhrman, University of Southern California **35.2** SPL/Photo Researchers **35.5** David M. Phillips/Photo Researchers **35.6a** Harold Fisher/Visuals Unlimited/Getty **35.6b** Biophoto Associates/Photo Researchers **35.6c** Gary Gaugler/Visuals Unlimited/Getty Images **35.6d** Oliver Meckes/E.O.S./Max-Planck-Institut-Tubingen/Photo Researchers **35.12L** Abbott Laboratories **35.12R** Abbott Laboratories **35.14a** Hans R. Gelderblom/Eye of Science **35.14b** James L. Van Etten, University of Nebraska. Originally published in R. H. Meints, J. L. Van Etten, D. Kuczmarski, K. Lee, and B. Ang, Viral infection of the symbiotic Chlorella-like alga present in Hydra viridis. Virology 113:698–703 (1981), Fig. C. **35.16** Jean Roy/CDC **35.17** Philip Leder, Dept. of Genetics, Harvard Medical School **35.18** Nigel Cattlin/Holt Studios/Photo Researchers **35.19** Lowell Georgia/Science Source/Photo Researchers **35.20** David Parker/SPL/Photo Researchers

Chapter 36

Opener Thomas Marent/Minden Pictures **36.5a** Matt Meadows/Peter Arnold **36.5bL** Alain Mafart-Renodier/Biosphoto/Peter Arnold **36.5bR** Rukhsana Photography/Fotolia **36.8a** Geoff Dann/Dorling Kindersley **36.8b** Dorling Kindersley **36.8c** Lee W. Wilcox **36.8d** Nigel Cattlin/FLPA/Minden Pictures **36.8e** Will Cook/Charles W. Cook **36.9a** Lee W. Wilcox **36.9b** Lee W. Wilcox **36.9c** David T. Webb **36.9d** David T. Webb **36.10a** Lee W. Wilcox **36.10b** Lee W. Wilcox **36.10c** Lee W. Wilcox **36.10d** Victoria Firmston/GAP Photos/Getty Images **36.11** RDF/Visuals Unlimited **36.12a** Lee W. Wilcox **36.12b** Ivan Kmit/Alamy **36.12c** Doug Wechsler/Animals Animals-Earth Scenes **36.12d** Lee W. Wilcox **36.12e** Lee W. Wilcox **36.13a** Ed Reschke/Peter Arnold **36.13b** Lee W. Wilcox **36.16a** Keith Wheeler/SPL/Photo Researchers **36.16b** ISM/Phototake **36.18L** Ed Reschke/Peter Arnold **36.18R** Ed Reschke/Peter Arnold **36.19** Andrew Syred/SPL/Photo Researchers **36.20a** John Durham/Photo Researchers **36.20b** Lee W. Wilcox **36.21L** Lee W. Wilcox **36.21R** Lee W. Wilcox **36.22a** Lee W. Wilcox **36.22b** Lee W. Wilcox **36.22c** Lee W. Wilcox **36.23a** Paul Schulte, University of Nevada **36.23b** Jack Bostrack/Visuals Unlimited **36.24a** John D. Cunningham/Visuals Unlimited/Getty Images **36.24b** Richard Kessel & Gene Shih/Visuals Unlimited/Getty Images **36.24c** Richard Kessel & Gene Shih/Visuals Unlimited/Getty Images **36.25L** Lee W. Wilcox **36.25M** Lee W. Wilcox **36.25R** Lee W. Wilcox **36.26L** Biodisc/Visuals Unlimited/Alamy **36.26R** Michael Clayton **36.27a** Lee W. Wilcox **36.27b** Biodisc/Visuals Unlimited/Alamy **36.27c** Adam Hart-Davis/SPL/Photo Researchers

Chapter 37

Opener Creatas/Photolibrary **37.4L** David Cook/blueshiftstudios/Alamy **37.4R** David Cook/blueshiftstudios/Alamy **37.6L** Lee W. Wilcox **37.6R** Lee W. Wilcox **37.8** Bruce Peters **37.10T** Ken Wagner/Phototake NYC **37.10M** G. Shih and R. Kessel/Visuals Unlimited **37.10B** Lee W. Wilcox **37.13** Lee W. Wilcox **37.18** Martin H. Zimmerman/Harvard Forest **37.22** Michael R. Sussman/American Society of Plant Biologists

Chapter 38

Opener Angelo Cavalli/Getty Images **38.2a** Photo Researchers **38.2b** Nigel Cattlin/Photo Researchers **38.2c** Photo Researchers **38.3** Emanuel Epstein **38.6a** The Institute of Texan Cultures **38.6b** Steve Ringman/The Seattle Times **38.9** Ed Reschke/Peter Arnold **38.13b** Eduardo Blumwald **38.14T** Hugh Spencer/Photo Researchers **38.14BL** Andrew Syred/SPL/Photo Researchers

38.14BR E.H. Newcomb & S. R. Tandon/Biological Photo Service **38.16a** Ed Reschke/Peter Arnold **38.16b** Ed Reschke/Peter Arnold **38.17L** Frank Greenaway/Dorling Kindersley Media Library **38.17R** Frank Greenaway/Dorling Kindersley Media Library **38.18** Noah Elhardt

Chapter 39

Opener Lee W. Wilcox **39.3** Malcolm B. Wilkins **39.4bL** John M. Christie/ American Association for the Advancement of Science **39.4bR** John M. Christie/ American Association for the Advancement of Science **39.9T** Malcolm B. Wilkins **39.9B** Malcolm B. Wilkins **39.11a** American Society of Plant Biologists **39.14** Thomas Bjorkman **39.15** Donald Specker/Animals Animals-Earth Scenes **39.16L** Richard Shiell/Animals Animals-Earth Scenes **39.16R** blickwinkel/Alamy **39.17a** Lee W. Wilcox **39.17b** Lee W. Wilcox **39.19** Malcolm B. Wilkins **39.23** Adel Kader **39.29** Nigel Cattlin/Holt Studios International/Photo Researchers

Chapter 40

Opener Walter Siegmund/Wikimedia Commons. GNU Free Documentation License **40.3a** Dan Suzio/Photo Researchers **40.3b** Jerome Wexler/Photo Researchers **40.3c** Jerome Wexler/Photo Researchers **40.6bT** berniekasper.com **40.6bM** Rod Planck/Photo Researchers **40.6bB** Tom & Therisa Stack/Tom Stack & Associates **40.7a** Leonard Lessin/Photo Researchers **40.7b** Leonard Lessin/ Photo Researchers **40.8a** iStockphoto **40.8bT** Lee W. Wilcox **40.8bB** Lee W. Wilcox **40.11L** Frans Lanting/Corbis **40.11R** Nick Garbutt/NPL/Minden Pictures **40.17aL** Brian Johnston **40.17aR** Pablo Galán Cela/age fotostock **40.17bL** Kerstin Layer/Mauritius/Photolibrary **40.17bR** Laurie Campbell/ NHPA/Photoshot **40.17cL** Nicholas and Sherry Lu Aldridge/FLPA/Minden Pictures **40.17cR** James Hardy/PhotoAlto Agency RF Collections/Getty Images **40.18** Lee W. Wilcox **40.19L** Scott Camazine/Alamy **40.19M** Jonathan Watts/OSF/Photolibrary **40.19R** Daniel Heuclin/NHPA/Photoshot

Chapter 41

Opener Joe McDonald/Corbis **41.2** The Royal Society of London **41.3aL** Educational Images/Custom Medical Stock Photo **41.3aR** Nina Zanetti **41.3b** Peter Arnold/Alamy **41.3cL** Nina Zanetti **41.3cR** Nina Zanetti **41.3d** Carolina Biological Supply Company/Phototake **41.4a** Deco Images II/Alamy **41.4b** Dennis Kunkel Microscopy/Phototake **41.5a** ISM/Phototake **41.5b** Manfred Kage/Peter Arnold **41.5c** Biophoto Associates/Photo Researchers **41.6** Ed Reschke/Peter Arnold **41.8** Tom Stewart/CORBIS **41.12a** G. Kruitwagen, H.P.M. Geurts and E.S. Pierson/Radboud University Nijmegen (vcbio.science.ru .nl/en) **41.12b** Innerspace Imaging/SPL/Photo Researchers **41.12c** P.M. Motta, A. Caggiati, G. Macchiarelli/SPL/Photo Researchers **41.14** Derrick Hamrick/ imagebroker.net /Photolibrary **41.15** Nutscode/T Service/Photo Researchers **41.17** From: Thermoregulation in the Mouths of Feeding Gray Whales. John E. Heyning, James G. Mead Science 7 November 1997: Vol. 278. no. 5340, pp. 1138–1140, Figure 1. (B). Image provided by the Natural History Museum of Los Angeles, County, From specimen: LACM 92044.

Chapter 42

Opener Frans Lanting/Minden Pictures **42.13a** Image taken by H. Wartenberg © H. Jastrow from Dr. Jastrow's Electron Microscopic Atlas (www.drjastrow.de)

Chapter 43

Opener Jonathan Blair/Corbis **43.2a** Wallace63/Wikimedia. Creative Commons Attribution ShareAlike 3.0 License. **43.2b** Kim Taylor/NPL/Minden Pictures **43.4L** Karel F. Liem **43.4M** Karel F. Liem **43.4R** Karel F. Liem **43.12** Biophoto Associates/Photo Researchers

Chapter 44

Opener Tony Freeman/PhotoEdit **44.4a** Walter E. Harvey/Photo Researchers **44.4b** John D. Cunningham/Visuals Unlimited **44.19a** Lennart Nilsson, Albert Bonniers Forlag AB **44.19b** Kessel & Kardon/Tissues & Organs/Visuals Unlimited/Getty Images

Chapter 45

Opener Marcus E. Raichle **45.9a** C. Raines/Visuals Unlimited **45.11** Dennis Kunkel/Visuals Unlimited **45.20a** Genny Anderson **45.22a** Elsevier Science Ltd.

Chapter 46

Opener Neil Hardwick/Alamy **46.1** Dmitry Smirnov **46.3a** Carole M. Hackney **46.7a** David Scharf/Peter Arnold **46.9a** Don Fawcett/T. Kwwabara/Photo Researchers **46.11L** David Quillin **46.11R** David Quillin **46.17T** James E. Dennis/Phototake NYC **46.17B** James E. Dennis/Phototake NYC **T46.1L** ISM/Phototake **T46.1M** Manfred Kage/Peter Arnold **T46.1R** Biophoto Associates/Photo Researchers

Chapter 47

Opener Ralph A. Clevenger/Corbis **47.5L** Stephen Dalton/Photo Researchers **47.5M** Duncan McEwan/Nature Picture Library **47.5R** Bernard Castelein/Nature Picture Library **47.6L** BIOS/Heras Joël/Peter Arnold **47.6M** Hans Pfletschinger/Peter Arnold **47.6R** Perennou Nuridsany/Photo Researchers **47.8** The Jackson Laboratory **47.9L** BIOS Heras Joll/Peter Arnold **47.9M** Hans Pfletschinger/Peter Arnold **47.9R** Perennou Nuridsany/Photo Researchers

Chapter 48

Opener Jurgen & Christine Sohns/FLPA/Minden Pictures **48.1a** NHPA/ Photoshot **48.1b** Andrew J. Martinez/Photo Researchers **48.1c** Charles J. Cole **48.2** Bruce J. Russell/BioMEDIA Associates **48.5T** Reprinted by permission from Catherine S.C.Price et al., Sperm competition between Drosophila males involves both development and incapacitation. Nature 400:449–452 (1999), figs. 2 and 3. Copyright © 1999 Macmillan Magazines Limited. Image courtesy of Jerry A. Coyne, University of Chicago/Nature Magazine **48.5B** Reprinted by permission from Catherine S.C.Price et al., Sperm competition between Drosophila males involves both development and incapacitation. Nature 400:449–452 (1999), figs. 2 and 3. Copyright © 1999 Macmillan Magazines Limited. Image courtesy of JerryA. Coyne, University of Chicago/Nature Magazine **48.6** From "A stimulatory phalloid organ in a weaverbird" M. Winterbottom, T. Burke & T. R. Birkhead/Nature/6 May 1999, 399: 28 Fig 1b. Photo courtesy T. R. Birkhead. **48.7a** Suzanne L. Collins/Photo Researchers **48.8a** Johanna Liljestrand Rönn **48.8a inset** Fleur Champion de Crespigny **48.16a** Geoff Shaw, Zoology, University of Melbourne, Australia/Wikipedia. GNU Free Documentation License. **48.16b** Mitsuaki Iwago/Minden Pictures **48.16c** Mitsuaki Iwago/Minden Pictures **48.17a** Claude Edelmann/Petit Format/Photo Researchers **48.17b** Lennart Nilsson, Albert Bonniers Forlag AB **48.17c** Petit Format/Nestle/Science Source/Photo Researchers **48.19aL** American Association for the Advancement of Science **48.19aR** American Association for the Advancement of Science

Chapter 49

Opener Eye of Science/Photo Researchers **49.1** Susumu Nishinaga/Photo Researchers **49.2a** Don W. Fawcett/Photo Researchers **49.2b** Steve Gschmeissner/SPL/Photo Researchers **49.2c** Steve Gschmeissner/Photo Researchers **49.5a** David M. Phillips/Visuals Unlimited **49.5b** Steve Gschmeissner/SPL/Photo Researchers **49.5c** David M. Phillips/Visuals Unlimited

Chapter 50

Opener Jeff Rotman/NPL/Minden Pictures **50.1a** Paul Nicklen/National Geographic/Getty Images **50.1b** Sergey Gorshkov/Minden Pictures **50.1c** Werner & Kerstin Layer/Photolibrary **50.1d** Lynn M. Stone/NPL/Minden Pictures **50.6a** Gerry Ellis/Minden Pictures **50.6b** Gerry Ellis/Minden Pictures **50.6c** Tim Fitzharris/Minden Pictures **50.7T** George Gerster/Photo Researchers **50.7B** Rich Wheater/Aurora/Getty Images **50.8** David Zimmerman/UpperCut Images/Getty Images **50.12** Gerry Ellis/Minden Pictures **50.14** Michael & Patricia Fogden/Minden Pictures **50.16** D. Hartnett, Konza Prairie Biological Station **50.18L** Alex Hyde/Minden Pictures **50.18R** Alex Hyde/Minden Pictures **50.20L** Tim Fitzharris/Minden Pictures **50.20R** Stephen J. Krasemann/NHPA/ Photoshot **50.22** Pixtal Images/Photolibrary **50.27** Fairfax Photos/Sandy Scheltema/Courtesy of the Age **50.29T** Purestock/Getty Images **50.29B** Comstock/Getty Images **50.30** Bruce Coleman/Alamy **50.31** Michael & Patricia Fogden/Minden Pictures **50.32 inset** Steven Mihok (www.nzitrap.com) **50.33L** John M. Randall **50.33R** John M. Randall **50.34L** Shutterstock **50.34R** USGS Canyonlands Research Station (www.soilcrust.org)

Chapter 51

Opener Konrad Wothe/Minden Pictures **51.1** Peter Arnold/Alamy **51.3aL** Peet van Schalkwyk (www.flickr.com/photos/peetvs) **51.3aR** David Dean Peterson **51.4a** iStockphoto **51.6** Gary Woodburn (pbase.com/woody) **51.8** Operation Migration (www.operationmigration.org) **51.11a** Shinji Kusano/Nature Production/Minden Pictures **51.11b** James E. Lloyd/Animals Animals-Earth Scenes **51.12** Bryan D. Neff **51.13** Tom Vezo/Peter Arnold

Chapter 52

Opener Annie Griffiths Belt/NGS Image Collection/Getty Images **52.1** David Kjaer/NPL/Minden Pictures **52.6** Mark Trabue/USDA/NRCS **52.9L** JC Schou/Biopix.dk (www.Biopix.dk) **52.9R** Marko Nieminen, University of Helsinki, Helsinki, Finland **52.13a** John Downer/NPL/Minden Pictures

Chapter 53

Opener Scott Tuason/imagequestmarine.com **53.1a** PREMAPHOTOS/Nature Picture Library **53.1b** David Tipling/Alamy **53.8L** From: Evolution of Character Displacement in Darwin's Finches Peter R. Grant and B. Rosemary Grant Science 313, 224 (2006); Figure 1. **53.8R** From: Evolution of Character Displacement in Darwin's Finches Peter R. Grant and B. Rosemary Grant Science 313, 224 (2006); Figure 1. **53.9L** blickwinkel/Alamy **53.9M** Chris Newbert/Minden Pictures **53.9R** J. Sneesby/B. Wilkins/Stone/Getty Images **53.11L** Rainer Zenz/Wikimedia Commons.GNU Free Documentation License and Creative Commons Attribution ShareAlike. **53.11R** CSIRO Marine and Atmospheric Research (www.scienceimage.csiro.au) **53.13a** Thomas G. Whitham/Northern Arizona University **53.15a** Stephen P. Yanoviak **53.15b** Stephen P. Yanoviak **53.15c** Stephen P. Yanoviak **53.16a** Mark Moffett/Minden Pictures **53.16b** David B. Fleetham/OSF/Photolibrary **53.19a** Nancy Sefton/Photo Researchers **53.19b** Thomas Kitchin/Tom Stack & Associates **53.20** Raymond Gehman/Corbis **53.21a** Tony C. Caprio **53.22T** G. Carleton Ray/Photo Researchers **53.22MT** Michael P. Gadomski/Animals Animals-Earth Scenes **53.22MB** James P. Jackson/Photo Researchers **53.22B** Michael P. Gadomski/Photo Researchers **53.23L** Glacier Bay National Park **53.23M** Christopher L. Fastie, Middlebury College **53.23R** Glacier Bay National Park

Chapter 54

Opener Jan Tove Johansson/The Image Bank/Getty Images **54.8** MODIS/NASA **54.12a** Richard Hartnup/Wikipedia Commons **54.12b** Randall J. Schaetzl, Michigan State University **54.13** John Campbell, U.S. Forest Service, Northern Research Station **54.15** Pornchai Kittiwongsakul/AFP/Getty Images

Chapter 55

Opener Ben Simmons/Photolibrary **55.1a** DLILLC/Corbis **55.1b** Xiaoqiang Wang **55.2** Edward S. Ross, California Academy of Sciences **55.5** Theo Allofs/The Image Bank/Getty Images **55.7** Karl Ammann (KarlAmmann.com) **55.8** Lester Lefkowitz/Corbis **55.9T** NASA/USGS (earthobservatory.nasa.gov) **55.9B** NASA/USGS (earthobservatory.nasa.gov) **55.12a** F. Mercay, BIOS/Peter Arnold **55.16** Ellen Damschen **55.17** Photo by Ariel Poster, provided by Marlboro Productions, photographed while filming TAKING ROOT: The Vision of Wangari Maathai, A Film by Lisa Merton & Alan Dater (www.takingrootfilm.com), © 2008. **55.18a** Daniel H. Janzen, University of Pennsylvania **55.18b** Daniel H. Janzen, University of Pennsylvania **55.19a** Charles Curtin **55.19b** Charles Curtin **55.19c** Charles Curtin

Appendix A

BS9.3 Reproduced by permission from J.P. Ferris et al., Synthesis of long prebiotic oligomers on mineral surfaces. Nature 381:59–61 (1996), Fig. 2. Copyright © 1996 Macmillan Magazines Limited. Image courtesy of James P. Ferris, Rensselaer Polytechnic Institute **BS10.1a** Biology Media/Photo Researchers **BS10.1b** Janice Carr/CDC **BS10.2a** Michael W. Davidson/Florida State University/Molecular Expressions **BS10.2b** Michael W. Davidson/Florida State University/Molecular Expressions **BS10.3** Rosalind Franklin/Photo Researchers **BS12.1aL** National Cancer Institute **BS12.1aR** E.S. Anderson/Photo Researchers **BS12.1b** Sinclair Stammers/Photo Researchers **BS14.1a** Kwangshin Kim/Photo Researchers **BS14.1b** Mark J. Grimson & Richard L. Blanton **BS14.1c** Holt Studios International/Photo Researchers **BS14.1d** Custom Medical Stock Photo **BS14.1e** Graphic Science/Alamy **BS14.1f** Sinclair Stammers/Photo Researchers **BS14.1g** iStockphoto

Illustration Credits

Chapter 1

1.3 Data from S. P. Moose, J. W. Dudley, and T. R. Rocheford. 2004. Maize selection passes the century mark: A unique resource for 21st century genomics. *Trends in Plant Science* 9 (7): 358–364, Fig. 1a. **1.8** After M. Wittlinger, R. Wehner, and H. Wolf. 2006. The ant odometer: Stepping on stilts and stumps. *Science* 312: 1965–1967, Figs.1, 2, 3.

Chapter 3

3 Opener PDB ID: 2DN2. S. Y. Park et al. 2006. 1.25 A resolution crystal structures of human haemoglobin in the oxy, deoxy and carbonmonoxy forms. *J Mol Biol* 360: 690–701. **3.9a** PDB ID: 1YTB. Y. Kim et al. 1993. Crystal structure of a yeast TBP/TATA-box complex. *Nature* 365: 512–520. **3.9b** PDB ID: 2F1C. G. V. Subbarao and B. van den Berg. 2006. Crystal structure of the monomeric porin OmpG. *J Mol Biol* 360: 750–759. **3.9c** PDB ID: 2PTN. J. Walter et al. 1982. On the disordered activation domain in trypsinogen. chemical labelling and low-temperature crystallography. *Acta Crystallogr, Sect B* 38: 1462. **3.9d** PDB ID: 1CLG. J. M. Chen 1991. An energetic evaluation of a "Smith" collagen microfibril model. *J Protein Chem* 10: 535–552. **3.12bL** PDB ID: 2MHR. S. Sheriff et al. 1987. Structure of myohemerythrin in the azidomet state at 1.7/1.3 Å resolution. *J Mol Biol* 197: 273–296. **3.12bM** PDB ID: 1FTP. N. H. Haunerland et al. 1994. Three-dimensional structure of the muscle fatty-acid-binding protein isolated from the desert locust *Schistocerca gregaria. Biochemistry* 33: 12378–12385. **3.12bR** PDB ID: 1IXA. M. Baron et al. 1992. The three-dimensional structure of the first EGF-like module of human factor IX: Comparison with EGF and TGF-alpha. *Protein Sci* **1:** 81–90. **3.23a, b** Data from: T. Hansen, B. Schlichting, and P. Schönheit. 2002. Glucose-6-phosphate dehydrogenase from the hyperthermophilic bacterium *Thermotoga maritima*: Expression of the *g6pd* gene and characterization of an extremely thermophilic enzyme. Elsevier Science, Federation of European Microbiological Societies (FEMS), *Microbiology Letters*, 216: 249–253, Fig. 1; N. N. Nawani and B. P. Kapadnis. 2001. One-step purification of chitinase from *Serratia marcescens* NK1, a soil isolate. *Journal of Applied Microbiology* 90: 803–808, Fig. 3; N. N. Nawani et al. 2002. Purification and characterization of a thermophilic and acidophilic chitinase from *Microbispora* sp.V2. The Society for Applied Microbiology, *Journal of Applied Microbiology* 93: 965–975, Fig. 7. **3.13a** PDB ID: 1D1L. P. B. Rupert et al. 2000. The structural basis for enhanced stability and reduced DNA binding seen in engineered second-generation Cro monomers and dimers. *J Mol Biol* 296: 1079–1090. **3.13b** PDB ID: 2DN2. S. Y. Park et al. 2006. 1.25 Å resolution crystal structures of human haemoglobin in the oxy, deoxy and carbonmonoxy forms. *J Mol Biol* 360: 690–701. **3.19** PDB IDs: 1Q18, 1SZ2. V. V. Lunin et al. 2004. Crystal structures of *Escherichia coli* ATP-dependent glucokinase and its complex with glucose. *J Bacteriol* 186: 6915–6927.

Chapter 4

4.11 PDB ID: 1GRZ. B. L. Golden et al. 1998. A preorganized active site in the crystal structure of the *Tetrahymena* ribozyme. *Science* 282: 259–264.

Chapter 6

6.11 Data from J. de Gier, J. G. Mandersloot, and L. L. M. Van Deenen. 1968. Lipid composition and permeability of liposomes. *Biochimica et Biophysica Acta* 150: 666–675. **6.22** Data from C. E. Bear et al. 1992. Purification and functional reconstitution of the cystic fibrosis transmembrane conductance regulator (CFTR). *Cell* 68: 809–818. **6.25a** PDB ID: 1J4N. H. Sui et al. Structural basis of water-specific transport through the AQP1 water channel. 2001. *Nature* 414:872–878. **6.25b** PDB IDs: 1ORS, 1ORQ. Y. Jiang et al. 2003. X-ray structure of a voltage-dependent K^+ channel. *Nature* 423: 33–41.

Chapter 8

8.7b After B. Alberts et al.2002. *Molecular Biology of the Cell.* 4th ed. Garland Science/Taylor & Francis, LLC, Fig. 19.5.

Chapter 9

9.2L PDB ID: 1IRK. S. R. Hubbard et al. 1994. Crystal structure of the tyrosine kinase domain of the human insulin receptor. *Nature* 372: 746–754. **9.2R** PDB ID: 1IR3. S. R. Hubbard 1997. Crystal structure of the activated insulin receptor tyrosine kinase in complex with peptide substrate and ATP analog. *EMBO J* 16: 5572–5581. **9.12** PDB ID: 4PFK. P. R. Evans and P. J. Hudson. 1981. Phosphofructokinase: structure and control. *Philos Trans R Soc Lond B Biol Sci* 293: 53–62.

Chapter 10

10.10 Data from Emerson, R. and W. Arnold. 1932. The photochemical reaction in photosynthesis. *The Journal of General Physiology* 16: 191–205.

Chapter 11

11.13a American Cancer Society. *Cancer Facts & Figures 2009*. Age-adjusted Cancer Death Rates, Males by Site, US, 1930–2005 (www.cancer.org). Reprinted with permission.**11.13b** American Cancer Society. *Cancer Facts & Figures 2009.* Age-adjusted Cancer Death Rates, Females by Site, US, 1930–2005 (www.cancer.org). Reprinted with permission.

Chapter 12

12.13 Data from C. Lively and J. Jokela. 2002. Temporal and spatial distributions of parasites and sex in a freshwater snail. *Evolutionary Ecology Research* 4: 219–226.

Chapter 13

13.13 Data from T. H. Morgan. 1911. An attempt to analyze the constitution of the chromosomes on the basis of sex-limited inheritance in *Drosophila. Journal of Experimental Zoology* 11: 365–414.

Chapter 14

14.5 Data from M. Meselson and F. W. Stahl. 1958. The replication of DNA in *Escherichia coli. PNAS* 44: 671–682. **14.17a, b** Data from J. E. Cleaver. 1968. Defective repair replication of DNA in Xeroderma pigmentosum. *Nature* 218: 652–656, Fig. 5. **14-UN02** Data from P. Howard-Flanders and R. P. Boyce. 1966. DNA repair and genetic recombination: Studies on mutants of *Escherichia coli* defective in these processes. Radiation Research Society, *Radiation Research Supplement* 6: 156–184, Fig. 8.

Chapter 15

15.2 Data from Srb, A. M. and N. H. Horowitz. 1944. The ornithine cycle in *Neurospora* and its genetic control. *The Journal of Biolcgical Chemistry* 154:129–139.

Chapter 16

16.2 PDB ID: 3IYD. B. P. Hudson et al. 2009. Three-dimensional EM structure of an intact activator-dependent transcription initiation complex. *PNAS* 106:19830–19835. **16.11** PDB ID: 1ZJW. I. Gruic-Sovulj et al. 2005. tRNA-dependent aminoacyl-adenylate hydrolysis by a nonediting class I aminoacyl-tRNA synthetase. *J Biol Chem* 280: 23978–23986. **16.12** Data from M. B. Hoagland et al. 1957. A soluble ribonucleic acid intermediate in protein synthesis. *The Journal of Biological Chemistry* 231: 241–257, Fig. 6. **16.14b** PDB IDs: 3FIK, 3FIH. E. Villa et al. 2009. Ribosome-induced changes in elongation factor Tu conformation control GTP hydrolysis. *PNAS* 106: 1063–1068.

Chapter 17

17.2 Data from A. B. Pardee, F. Jacob, and J. Monod. 1959. The genetic control and cytoplasmic expression of "inducibility" in the synthesis of β-glactosidase by *E. coli. Journal of Molecular Biology* 1: 165–178.

Chapter 18

18 Opener PDB ID: 1ZBB. T. Schalch et al. 2005. X-ray structure of a tetranucleosome and its implications for the chromatin fibre. *Nature* 436: 138–141. **18.6** Data from Hozumi, N. and S. Tonegawa. 1976. Evidence for somatic rearrangement of immunoglobulin genes coding for variable and constant regions. *PNAS* 73: 3628–3632.

Chapter 20

20.4 Data from S. J. Giovannoni et al. 2005. Genome streamlining in a cosmopolitan oceanic bacterium. *Science* 309: 1242–1245, Fig. 1. **20.9** Data from P. D. Thomas et al. 2003. PANTHER: a library of protein families and subfamilies indexed by function. *Genome Res* 13: 2129–2141 (www.pantherdb.org).

Chapter 21

21.10a, b After S. B. Carroll. 1995. Homeotic genes and the evolution of arthropods and chordates. *Nature* 376: 479–485, Fig. 1.

Chapter 24

24.4 After: E. B. Daeschler, N. H. Shubin, and F. A. Jenkins, Jr. 2006. A Devonian tetrapod-like fish and the evolution of the tetrapod body plan. *Nature* 440: 757–763, Fig. 6; P. E. Ahlberg and J. A. Clack. 2006. A firm step from water to land. *Nature* 440: 747–749, Fig. 1; N. H Shubin, E. B. Daeschler, and F. A. Jenkins, Jr. 2006. The pectoral fin of *Tiktaalik roseae* and the origin of the tetrapod limb. *Nature* 440: 764–771, Fig. 4; G. E. Goslow and M. Hildebrand. 2001. *Analysis of Vertebrate Structures.* 5th ed. John Wiley and Sons, Inc., Fig. 4. **24.11** After: P. D. Gingerich et al. 2001. *Science,* New Series 293 (5538): 2239–2242, Fig. 3; J. G. M. Thewissen et al. 2007. Whales originated from aquatic artiodactyls in the Eocene epoch of India.*Nature* 450: 1190–1194, Fig. 5; P. D. Gingerich et al. 2009. New protocetid whale from the middle Eocene of Pakistan: Birth on land, precocial development, and sexual dimorphism. *PLoS ONE* 4 (2): 1–20, Fig. 1. **24.14** Data from D. P. Genereux and C. T. Bergstrom. "Chapter 13: Evolution in Action: Understanding Antibiotic Resistance," Fig. 3. In *Evolutionary Science and Society: Educating a New Generation,* edited by J. Cracraft and R. W. Bybee, 145–153. Washington, DC: American Institute of Biological Sciences, 2005. **24.16** Data from P. T. Boag and P. R. Grant. 1981. Intense natural selection in a population of Darwin's finches (*Geospizinae*) in the Galápagos. *Science* New Series 214 (4516): 82–85, Table 1. **24.17** Data from P. R. Grant and R. Grant. 2006. Evolution of character displacement in Darwin's finches. *Science* 313: 224–226, Fig. 2; P. R. Grant and B. R. Grant. 2002. Unpredictable evolution in a 30-year study of Darwin's finches. *Science* 296: 707–711, Fig. 1.

Chapter 25

25.3b Data from C. R. Brown and M. B. Brown. 1998. Intense natural selection on body size and wing and tail asymmetry in cliff swallows during severe weather. *Evolution* 52: 1461–1475. **25.4b** Data from M. N. Karn et al. 1951. Birth weight, gestation time and survival in sibs. *Annals of Eugenics* 15: 306–322. **25.5b** Data from T. B. Smith. 1987. Bill size polymorphism and intraspecific niche utilization in an African finch. *Nature* 329: 717–719. **25.10b** E. Postma and A. J. van Noordwijk. 2005. Gene flow maintains a large genetic difference in clutch size at a small spatial scale. *Nature* 433: 65–68, Fig. 5.**25.11** Data from S. F. Elena, V. S. Cooper, and R. E. Lenski. 1996. Punctuated evolution caused by selection of rare beneficial mutations. *Science* New Series 272: 1802–1804. **25.13** Data from M. Johnston. 1992. Effects of cross and self-fertilization on progeny fitness in *Lobelia cardinalis* and *L. siphilitica. Evolution* 46: 688–702, Fig. 1. **25.14b** Data from J. D. Blount et al. 2003. Carotenoid modulation of immune function and sexual attractiveness in zebra finches. *Science* 300: 125–127, Fig. 2. **25.16b, c** Data from B. J. Le Boeuf and R. S. Peterson. 1969. Social status and mating activity in elephant seals. *Science* 163: 91–93.

Chapter 26

26.3a, b AfterJ. C. Avise and W. S. Nelson. 1989. Molecular genetic relationships of the extinct dusky seaside sparrow. *Science* 243 (4891): 646–648, Figs. 1, 2, 3. **26.5b** Data from N. Knowlton et al. 1993. Divergence in proteins, mitochondrial DNA, and reproductive compatibility across the Isthmus of Panama. *Science* 260 (5114):1629–1632, Fig. 1. **26.7** H. R. Dambroski et al. 2005. The genetic basis for fruit odor discrimination in rhagoletis flies and its significance for sympatric host shifts. *The Society for the Study of Evolution* 59 (9):1953–1964. **26.10b** Data from S. E. Rohwer et al. 2001. Plumage and mitochondrial DNA haplotype variation across a moving hybrid zone. *Evolution* 55: 405–422. **26.11** Data from L. H. Rieseberg et al. 1996. Role of gene interactions in hybrid speciation: Evidence from ancient and experimental hybrids. *Science* 272 (5262): 741–745.

Chapter 27

27.5 a, b, c After M. A. Nikaido, P. Rooney, and N. Okada. 1999. Phylogenetic relationships among cetartiodactyls based on insertions of short and long interspersed elements: Hippopotamuses are the closest extant relatives of whales. *PNAS* 96: 10261–10266, Figs. 1, 7. **27.11** B. G. Baldwin, D. W. Kyhos, and J. Dvorak. 1990. Chloroplast DNA evolution and adaptive radiation in the Hawaiian Silversword Alliance (Asteraceae-Madiinae). *Annals of the Missouri Botanical Garden* 77 (1): 96–109, Fig. 2. **27.12c** J. B Losos et al. 1997. Adaptive differentiation following experimental island colonization in *Anolis* lizards. *Nature* 387: 70–73. **27.15** Data from: J. W. Valentine, D. H. Erwin, and D. Jablonski. 1996. Developmental evolution of metazoan body plans: The fossil evidence. *Developmental Biology* 173: 373–381, Fig. 3; R. de Rosa et al. 1999. *Hox* genes in brachiopods and priapulids and protostome evolution. *Nature* 399: 772–776; D. Chourrout et al. 2006. Minimal ProtoHox cluster inferred from bilaterian and cnidarian *Hox* complements. *Nature* 442: 684–687. **27.16** Data from M. J. Benton. 1995. Diversification and extinction in the history of life. *Science* 268: 52–58. **27.17a** Data from W. Alvarez, F. Asaro, and A. Montanari. 1990. Iridium profile for 10 million years across the Cretaceous-Tertiary boundary at Gubbio (Italy). *Science* 250: 1700–1702, Fig. 1.

Chapter 28

28.2 Data from G. L. Armstrong, L. A. Conn, and R. W. Pinner. 1999. Trends in infectious disease mortality in the United States during the 20th century. *JAMA* 281: 61–66, Fig. 1. **28.6** After: J. G. Elkins et al. 2008. A korarchaeal genome reveals insights into the evolution of the Archaea. Procedings of the National Academy of Sciences USA 105: 8102–8107, Fig. 2; F. D. Ciccarelli et al. 2006. Toward automatic reconstruction of a highly resolved tree of life. *Science* 311:1283–1287, Fig. 2. **28.11** Data from Madigan M. T. and J. M. Martinko. 2006. *Brock Biology of Microorganisms.* 11th ed. Fig. 5.9.

Chapter 29

29.1, 29.7 S. M. Adl et al. 2005. The new higher level classification of eukaryotes with emphasis on the taxonomy of protists. International Society of Protistologists 52(5): 399–45; N. Arisue, M. Hasegawa, and T. Hashimoto. 2005. Root of the eukaryota tree as inferred from combined maximum likelihood analyses of multiple molecular sequence data. Society for Molecular Biology and Evolution, *Molecular Biology and Evolution* 22 (3); V. Hampl et al. 2009. Phylogenomic analyses support the monophyly of excavata and resolve relationships among eukaryotic "supergroups." *PNAS* 106 (10): 3859–3864, Figs. 1, 2, 3; J. D. Hackett et al. 2007. Phylogenomic analysis supports the monophyly of cryptophytes and haptophytes and the association of rhizaria with chromalveolates. Society for Molecular Biology and Evolution, Oxford University Press: 1702–1713, Fig. 1; P. Schaap et al. 2006. Molecular phylogeny and evolution of morphology in the social amoebas. *Science* 314: 661–663.

Chapter 30

30.3 Data from Owen et al. Natural Resource Conservation, 7th ed., 509. **30.7, 30.10, 30.24** Y. Qiua et al. 2006. The deepest divergences in land plants inferred from phylogenomic evidence. *PNAS* 103 (42):15511–15516, Fig. 1; K. S. Renzaglia et al. 2007. Bryophyte phylogeny: Advancing the molecular and morphological frontiers. American Bryological and Lichenological Society, Inc, *The Bryologist* 110 (2): 179–213; J. Pombert et al. 2005. The chloroplast genome sequence of the green alga *Pseudendoclonium akinetum* (ulvophyceae) reveals unusual structural features and new insights into the branching order of chlorophyte lineages. Society for Molecular Biology and Evolution, Oxford University Press: 1903–1918. **30.22** Data from Johnson, S. D. and K. E. Steiner. 1997. Long-tongued fly pollination and evolution of floral spur length in the *Disa draconis* complex (Orchidaceae). *Evolution* 51: 45–53. **30.26** P. S. Soltis and D. E. Soltis. 2004. The origin and diversification of angiosperms. *American Journal of Botany* 91 (10): 1614–1626, Figs. 1, 2, 3.

Chapter 31

31.7 S. M. Adl et al. 2005. The new higher level classification of eukaryotes with emphasis on the taxonomy of protists. International Society of Protistologists 52(5): 399–45; N. Arisue, M. Hasegawa, and T. Hashimoto. 2005. Root of the eukaryota tree as inferred from combined maximum likelihood analyses of multiple molecular sequence data. Society for Molecular Biology and Evolution, *Molecular Biology and Evolution* 22 (3); V. Hampl et al. 2009. Phylogenomic analyses support the monophyly of excavata and resolve relationships among eukaryotic "supergroups." *PNAS* 106 (10): 3859–3864, Figs. 1, 2, 3; J. D. Hackett et al. 2007. Phylogenomic analysis supports the monophyly of cryptophytes and haptophytes and the association of rhizaria with chromalveolates. Society for Molecular Biology and Evolution, Oxford University Press: 1702–1713, Fig. 1; P. Schaap et al. 2006. Molecular phylogeny and evolution of morphology in the social amoebas. *Science* 314: 661–663. **31.8** T. Y. James et. al. 2006. Reconstructing the early evolution of fungi using a six-gene phylogeny. *Nature* 443: 818–822, Fig. 1.

Chapter 32

32.9 A. M. A. Aguinaldo et al. 1997. Evidence for a clade of nematodes, arthropods, and other moulting animals. *Nature* 387: Figs. 1, 2, 3; C. W. Dunn et al. 2008. Broad phylogenomic sampling improves resolution of the animal tree of life. *Nature* 452: 745–750, Figs. 1, 2.

Chapter 33

33.2 A. M. A. Aguinaldo et al. 1997. Evidence for a clade of nematodes, arthropods, and other moulting animals. *Nature* 387: Figs. 1, 2, 3; C. W. Dunn et al. 2008. Broad phylogenomic sampling improves resolution of the animal tree of life. *Nature* 452: 745–750, Figs. 1, 2. **33.7** T. H. Struck et al. 2007. Annelid phylogeny and the status of Sipuncula and Echiura. BioMed Central Ltd., *Evolutionary Biology* 7: 571–11.

Chapter 34

34.1 After C. W. Dunn et al. 2008. Broad phylogenomic sampling improves resolution of the animal tree of life. *Nature* 452: 745–750, Figs. 1, 2; F. Delsuc et al. 2006. Tunicates and not cephalochordates are the closest living relatives of vertebrates. *Nature* 439: 965–968, Fig. 1. **34.12** J. E. Blair and S. B. Hedges.2005. Molecular phylogeny and divergence times of deuterostome animals. Society for Molecular Biology and Evolution, Oxford University Press 22 (11): 2275–2284, Figs. 1, 3, 4. **34.16** After: E. B. Daeschler et al. 2006. A Devonian tetrapod-like fish and the evolution of the tetrapod body plan. *Nature* 440:757–763, Fig. 6; N. H. Shubin et al. 2006. The pectoral fin of Tiktaalik roseae and the origin of the tetrapod limb. *Nature* 440:764–771, Fig. 4. **34.22** J. E. Blair and S. B. Hedges.2005. Molecular phylogeny and divergence times of deuterostome animals. Society for Molecular Biology and Evolution, Oxford University Press 22 (11): 2275–2284, Figs. 1, 3, 4; F. R. Liu et al. 2001. Molecular and morphological supertrees for eutherian (placental) mammals. *Science* 291: 1786–1789; A. B. Prasad et al. 2008. Confirming the phylogeny of mammals by use of large comparative sequence data sets. Oxford University Press, *Molecular Biology and Evolution* 25 (9):1795–1808. **34.36** A. B. Prasad et al. 2008. Confirming the phylogeny of mammals by use of large comparative sequence data sets. Oxford University Press, *Molecular Biology and Evolution* 25 (9):1795–1808, Figs. 1, 2, 3. **34.39** J. Z. Li et al. 2008. Worldwide human relationships inferred from genome-wide patterns of variation. *Science* 319: 1100–1104, Fig. 1. **34.40** L. L. Cavalli-Sforza and M. W. Feldman. 2003. The application of molecular genetic approaches to the study of human evolution. *Nature Genetics Supplement* 33: 266–275, Fig 3.

Chapter 35

35.3 Data from A. Fauci et al. 1996. Immunopathogenic mechanisms of HIV infection. *Annals of Internal Medicine* 124 (7): 654–663. **35.15** F. Gao et al. 1999. Origin of HIV-1 in the chimpanzee Pan troglodytes troglodytes. *Nature* 397: 436–441, Fig. 2.

Chapter 37

37.3 Data from R. G. Cline and W. S. Campbell. 1976. Seasonal and diurnal water relations of selected forest species. *Ecology* 57: 367–373, Fig. 1. **37.12** Data from C. Wei, M. T. Tyree, and E. Steudle. 1999. Direct measurement of xylem pressure in leaves of intact maize plants; A test of the cohesion-tension theory taking hydraulic architecture into consideration. *Plant Physiology* 121: 1191–1205. **37.22** Data from N. D. DeWitt and M. R. Sussman. 1995. Immunocytological localization of an epitope-tagged plasma membrane proton pump (H^+-ATPAse) in phloem companion cells. *Plant Cell* 7: 2053–2067, Fig. 7.

Chapter 39

39.7 Data from T. I. Baskin, M. Iino, P. G. Green, and W. R. Briggs. 1985. High-resolution measurements of growth during first positive phototropism in maize. *Plant Cell and Environment* 8: 595–603.

Chapter 40

40.12 Y. Qiua et al. 2006. The deepest divergences in land plants inferred from phylogenomic evidence. *PNAS* 103 (42):15511–15516, Fig. 1; K. S. Renzaglia et al. 2007. Bryophyte phylogeny: Advancing the molecular and morphological frontiers. American Bryological and Lichenological Society, Inc, *The Bryologist* 110 (2): 179–213; J. Pombert et al. 2005. The chloroplast genome sequence of the green alga *Pseudendoclonium akinetum* (ulvophyceae) reveals unusual structural features and new insights into the branching order of chlorophyte lineages. Society for Molecular Biology and Evolution, Oxford University Press: 1903–1918.

Chapter 41

41.1 Data from B. P. Sinervo, Doughty, R. B. Huey, and K. Zamudio. 1992. Allometric engineering: A causal analysis of natural selection on offspring size. *Science* 258: 1927–1930. **41.10** Data from K. Schmidt-Nielsen. 1984. Scaling: Why is animal size so important? Cambridge University Press. **41.11** Data from Wells, P. R. and A. W. Pinder. 1996. The respiratory development of Atlantic salmon. *Journal of Experimental Biology* 199: 2737–2744.

Chapter 42

42.14a, b Data from K. J. Ullrich, K. Kramer, and J. W. Boyer. 1961. Present knowledge of the counter-current system in the mammalian kidney. Progress in Cardiovascular Diseases 3 (5): 395–431, Figs. 5, 7.

Chapter 43

43.17a, b Data fromL. O. Schultz et al. 2006. Effects of traditional and western environments on prevalence of type 2 diabetes in Pima Indians in Mexico and the U.S. American Diabetes Association, *Diabetes Care* 29: 1866–1871, Fig. 1, Table 2.

Chapter 44

44.7 Data from Komai, Y. 1998. Augmented respiration in a flying insect. *Journal of Experimental Biology* 201: 2359–2366. **44.11** W. Bretz and K. Schmidt Nielsen. 1971. Bird respiration: Flow patterns in the duck lung. *Journal of Experimental Biology* 54:103–118. **44.21** J. E. Blair and S. B. Hedges. 2005. Molecular phylogeny and divergence times of deuterostome animals. Society for Molecular Biology and Evolution, Oxford University Press 22 (11): 2275–2284, Figs. 1, 3, 4. **44.20** Data from E. Starling.1896. On the absorption of fluids from the connective tissue spaces. *Journal of Physiology* 19: 312–326.

Chapter 45

45.21 Data from K. C. Martin et al.1997. Synapse-specific, long-term facilitation of aplysia sensory to motor synapses: A function for local protein synthesis in memory storage. *Cell* 91: 927 938. **45.22b** Data from K. C. Martin et al. 1997. Synapse-specific, long-term facilitation of aplysia sensory to motor synapses: A function for local protein synthesis in memory storage. *Cell* 91: 927–938.

Chapter 46

46.2b Data from J. E. Rose et al. 1971. Some effects of stimulus intensity on response of auditory nerve fibers in the squirrel monkey. Laboratory of Neurophysiology, University of Wisconsin, Madison: 685–699. **46.12** Data from D. M. Hunt et al. 1995. The chemistry of John Dalton's color blindness. *Science* 267: 984–988, Fig 3. **46.19** PDB ID: 1KWO. D. M. Himmel et al. 2002. Crystallographic findings on the internally uncoupled and near-rigor states of myosin: further insights into the mechanics of the motor. *PNAS* 99: 12645–12650.

Chapter 47

47.7 Data from B. D. King, L. Sokoloff, and R. L. Wechsler. 1952. The effects of *l*-epinephrineand *l*-nor-epinephrine upon cerebral circulation and metabolism in man. *Journal of Clinical Investigation* 31: 273–279; Mueller, P. S. and D. Horwitz. 1962. Plasma free fatty acid and blood glucose responses to analogues of norepinephrine in man. *Lipid Research* 3: 251–255. **47.9** Data from Coleman, D. L. 1973. Effects of parabiosis of obese with diabetes and normal mice. *Diabetologia* 9: 294–298. **47.10a, b** Data from G. V. Upton et al. 1973. Evidence for the internal feedback phenomenon in human subjects: Effects of ACTH on plasma CRF. *Acta Endocrinologica* 73: 437–443. **47.13** Data from D. Toft, and J. Gorski. 1966. A receptor molecule for estrogens: Isolation from the rat uterus and preliminary characterization. *PNAS* 55: 1574–1581. **47.15b** Data from T. W. Rall et al. 1956. The relationship of epinephrine and glucagon to liver phosphorylase. *The Journal of Biological Chemistry,* 224: 463–475. **47.16** Data from E. W. Sutherland. 1972. Studies on the mechanism of hormone action. *Science* 177: 401–408, Fig. 6.

Chapter 48

48.3a Data from R. G. Stross and J. C. Hill. 1965. Diapause induction in Daphnia requires two stimuli. Science 150: 1462–1464, Fig.3. **48.3b** Data from O. T. Kleiven, Petter Larsson, and Anders Hobæk. 1992. Sexual reproduction in Daphnia magna requires three stimuli. *Oikos* 65: 197–206, Table 4. **48.5** Data from C. S. C. Price, K. A. Dyer, and J. A. Coyne. 1999. Sperm competition between *Drosophila* males involves both displacement and incapacitation. *Nature* 400: 449–452. **48.7b** R. Shine and M. S. Y. Lee. 1999. A reanalysis of the evolution of viviparity and egg-guarding in squamate reptiles. The Herpetologists' League, Inc., *Herpetologica* 55 (4): 538–549, Figs. 1, 2, 3. **48.8b** Data from C. Hotzyand and G. Arnqvist. 2009. Sperm competition favors harmful males in seed beetles. *Current Biology* 19: 404–407. **48.19b** Data from C. Ikonomidou et al. 2000. Ethanol-induced apoptotic neurodegeneration and fetal alcohol syndrome. *Science* 287: 1056–1060, Fig. 4. **48.21** Data from U. Högberg and I. Joelsson. 1985. The decline in maternal mortality in Sweden, 1931–1980. *Acta Obstetrics Gynecology Scandinavica* 64: 583–592, Fig. 1.

Chapter 49

49.6a PDB ID: 1IGT. L. J. Harris et al. 1997. Refined structure of an intact IgG2a monoclonal antibody. *Biochemistry* 36: 1581–1597. **49.6b** PDB ID: 1TCR. K. C. Garcia et al. 1996. An alphabeta T cell receptor structure at 2.5 Å and its orientation in the TCR-MHC complex. *Science* 274: 209–219.

Chapter 50

50.4a Data from M. B. Saffo. 1987. New light on seaweeds. American Institute of Biological Sciences, *BioScience* 37 (9): 654–664, Fig. 1. **50.28** Data from A. K. Knapp et al. 2002. Rainfall variability, carbon cycling, and plant species diversity in a mesic grassland. *Science* 298: 2202–2205.

Chapter 51

51.3b Data from R. E. Hegner, S. T. Emlen, and N. J. Demong. 1982. Spatial organization of the white-fronted bee-eater. *Nature* 298: 264–266. **51.4b** Data from D. Crews. 1975. Psychobiology of reptilian reproduction. *Science* 189: 1059–1065. **51.5** Data from D. Crews. 1975. Psychobiology of reptilian reproduction. *Science* 189: 1059–1065. **51.7** Data from Møller, A. P. 1988. Female choice selects for male sexual tail ornaments in the monogamous swallow. *Nature* 332: 640–642. **51.15** Data from J. L. Hoogland. 1983. *Animal Behavior* 31: 472–479. **51.16** Data from M. Daly and M. Wilson. 1988. Evolutionary social psychology and family homicide. *Science* 242 (4878): 519–524, Fig. 1.

Chapter 52

52.3 Data from Ghalambor and Martin. 2001. *Science* 292: 494–496, Fig. 1c. **52.7b** Data from G. F. Gause. 1934. *The Struggle for Existence.* Baltimore: Williams & Wilkins. **52.8a** Data from G. E. Forrester. 1995. Strong density-dependant survival and recruitment regulate the abundance of a coral reef fish. Springer, International Association for Ecology, *Oecologia* 103: 275–282, Fig. 2. **52.8b** Data from P. Arcese and J. N. M. Smith. 1988. Effects of population density and supplemental food on reproduction in song sparrows. British Ecological Society, *Journal of Animal Ecology* 57: 119–136, Fig. 2. **52.12** Data from C. J. Krebs et al. 1995. Impact of food and predation on the snowshow hare cycle. *Science* 269: 1112–1115, Fig. 3. **52.13b** Data from T. Valverde and J. Silvertown. 1998. Variation in the demography of a woodland understorey herb (*Primula vulgaris*) along the forest regeneration cycle: Projection matrix analysis. *Journal of Ecology* 86: 545–562. **52.14a, b** Data from U.S. Census Bureau, International Data Base (IDB) (www.census.gov).

Chapter 53

53.2a, b, 53.3, 53.4a, b G. F. Gause. 1934. *The Struggle for Existence.* Baltimore: Williams & Wilkins. **53.6** Data from Connell, J. H. 1961. The influence of interspecific competition and otherfactors on the distribution of the barnacle *Chthamalus stellatus. Ecology* 42: 710–723. **53.8** Data from P. R. Grant and B. R. Grant. 2006. Evolution of character displacement in Darwin's finches. *Science* 313: 224–226, Fig. 2. **53.11** Data from G. H. Leonard, M. D. Bertness, and P. O. Yund. 1999. Crab predation, waterborne clues, and inducible defenses in the blue mussel, *Mytolus edulis. Ecology* 80: 1–14. **53.12** Data from G. H. Leonard, M. D. Bertness, and P. O. Yund. 1999. Crab predation, waterborne clues, and inducible defenses in the blue mussel, *Mytolus edulis. Ecology* 80: 1–14. **53.13b, c** Data from G. D. Martinsen, E. M. Driebe, and T. G. Whitham. 1998. Indirect interactions mediated by changing plant chemistry: Beaver browsing benefits beetles. *Ecology* 79: 192–200. **53.17** Data from Cushman, J. H. and T. G. Whitham. 1989. Conditional mutualism in a membracid-ant association: Temporal, age-specific, and density-dependant effects. *Ecology* 70: 1040–1047. **53.18** Data from Jenkins, D. G. and A. L. Buikema. 1998. Do similar communities develop in similar sites? A test with zooplankton structure and function. *Ecological Monographs* 68: 421–443. **53.19c** Data from R. T. Paine. 1966. Food web complexity and species diversity. *The American Naturalist* 100: 65–75; R. T. Paine. 1974. Intertidal community structure. *Oecologia* 15: 93–120. **53.21b** Data from T. W. Swetnam. 1993. Fire history and climate change in giant sequoia groves. *Science* 262: 885–889. **53.24a, b, c** After R. MacAurthor and E. O. Wilson. 1967. The theory of island biogeography. Princeton, University Press. **53.26** Data from W. V. Reid and K. R. Miller. 1989. Keeping options alive: The scientific basis for conserving biodiversity. Washington, DC: World Resources Institute.

Chapter 54

54.3 Data from J. R. Gosz et al. 1978. The flow of energy in a forest ecosystem. *Scientific American* 238: 93–102. **54.7** Data from C. A. Mackenzie, A. Lockridge, and M. Keith. 2005. Declining sex ratio in a first nation community. *Environmental Health Perspectives* 113 (10): 12–96. **54.9a, b, c** Data from R. H. Whittaker and G. E. Likens. 1973. Primary production: The biosphere and man. Springer, *Human Ecology,* 1 (4): 357–369, Table 1; H. Lieth.1973. Primary production: Marine ecosystems. Springer, *Human Ecology* 1 (4): 303–332, Tables 2, 3; J. S. Bunt. 1973. Primary production: Marine ecosystems. Springer, *Human Ecology* 1 (4): 333–345, Table 1; G. E. Likens. 1973. Primary Production: Freshwater Ecosystems. Springer, *Human Ecology* 1 (4): 347–356, Table 3. **54.10** Data from A. Tsuda et al. 2003. A mesoscale iron enrichment in the western subarctic Pacific induces a large centric diatom bloom. *Science* 300: 958–961. **54.13** Data from G. E. Likens et al. 1970. Effects of cutting and herbicide treatment on nutrient budgets in the Hubbard Brook watershed-ecosystem. *Ecological Monographs* 40: 23–47. **54.16** Data from W. H. Schlesinger. 1997. Biogeochemistry: An Analysis of Global Change. 2nd Ed. Academic Press, San Diego. **54.17** Data from P. M. Vitousek et al. 1997. Human alteration of the global nitrogen cycle: Sources and consequences. Ecological Society of America, *Ecological Applications* 7 (3): 737–750. **54.20** Data from BP Statistical Review of World Energy 2007, 6–21. **54.21** Data from NOAA/ESRL Global Monitoring Division. "Trends in Atmospheric Carbon Dioxide—Mauna Loa" (www.esrl.noaa.gov/gmd/ccgg/trends/). **54.22a** Data fromG. Beaugrand et al. 2002. Reorganization of North Atlantic marine copepod biodiversity and climate. *Science* 296: 1692–1694. **54.22b** Data fromN. L. Bradley et al. 1999. Phenological changes reflect climate change in Wisconsin. *PNAS* 96: 9701–9704. **54.23a** Data from R. R. Nemani et al. 2003. Climate-driven increases in global terrestrial net primary production from 1982 to 1999. *Science* 300: 1560–1563, Fig 2. **54.23b** Data from M. J. Behrenfeld. 2006. Climate-driven trends in contemporary ocean productivity. *Nature* 444: 752–755.

Chapter 55

55.1 F. R. Liu et al. 2001. Molecular and morphological supertrees for eutherian (placental) mammals. *Science* 291: 1786–1789, Fig. 1; A. B. Prasad et al. 2008. Confirming the phylogeny of mammals by use of large comparative sequence data sets. Oxford University Press, *Molecular Biology and Evolution* 25(9): 1795–1808; U. Arnason, A. Gullberg, and A. Janke. 2004. Mitogenomic analyses provide new insights into cetacean origin and evolution. Elsevier, *Gene* 333: 27–34, Fig. 1. **55.3a, b** Data from C. D. L. Orme et al. 2005. Global hotspots of species richness are not congruent with endemism or threat. *Nature* 436: 1016–1019; N. Myers et al. 2000. Biodiversity hotspots for conservation priorities. *Nature* 403: 853–858. **55.4** Data from Conservation International, *Annual Report 2008*, 12–13 (www.conservation.org). **55.6** Data from O. Venter et al. 2006. Threats to endangered species in Canada. *BioScience* 56: 903–910, Fig 2. **55.10** Data from W. F. Laurance et al. 2000. Rainforest fragmentation kills big trees. *Nature* 404: 836. **55.11** Data from J. T. Hogg et al. 2006. Genetic rescue of an insular population of large mammals. *PNAS* 273: 1491–1499, Fig 1. **55.12b** Data from D. B. Lindenmayer and R. C. Lacy. 1995. Metapopulation viability of Leadbeater's possum, *Gymnobelideus leadbeateri*, in fragmented old-growth forests. *Ecological Applications* 5: 164–182. **55.13** Data from J. M. Diamond and R. M. May. 1981. *Theoretical Ecology.* 2nd ed. Blackwell: Oxford, 228–252.**55.14** Data from D. Tilman et al. 1997. The influence of functional diversity and composition on ecosystem processes. *Science* 277: 1300–1302. **55.15** Data from D. Tilman and J. A. Downing. 1994. Biodiversity and stability in grasslands. *Nature* 367: 363–365. **55.16** Data from E. I. Damschen et al. 2006. Corridors increase plant species richness at large scales. *Science* 313: 1284–1286. **55-UN01** Data from A. P. Dobson. 1996. *Conservation and Biodiversity.* Scientific American Library, W. H. Freeman and Company, New York, Fig.1; D. S. Wilcove et al. 1986. *M. E. Soulé Conservation Biology.* Sinauer Associates; M. Williams. 1989. *Americans and their forests.* Cambridge University Press.

Index

Boldface page numbers indicate a glossary entry; page numbers followed by *f* indicate a figure; page numbers followed by *t* indicate a table.

Boldface page numbers indicate a glossary entry; page numbers followed by an *f* indicate a figure; page numbers followed by *t* indicate a table.

Boldface page numbers indicate a glossary entry; page numbers followed by an *f* indicate a figure; page numbers followed by *t* indicate a table.

Boldface page numbers indicate a glossary entry; page numbers followed by an *f* indicate a figure; page numbers followed by *t* indicate a table.

D

Boldface page numbers indicate a glossary entry; page numbers followed by an *f* indicate a figure; page numbers followed by *t* indicate a table.

F

Boldface page numbers indicate a glossary entry; page numbers followed by an *f* indicate a figure; page numbers followed by *t* indicate a table.

Boldface page numbers indicate a glossary entry; page numbers followed by an *f* indicate a figure; page numbers followed by *t* indicate a table.

H

I

Boldface page numbers indicate a glossary entry; page numbers followed by an *f* indicate a figure; page numbers followed by *t* indicate a table.

M

Boldface page numbers indicate a glossary entry; page numbers followed by an *f* indicate a figure; page numbers followed by *t* indicate a table.

N

O

Boldface page numbers indicate a glossary entry; page numbers followed by an *f* indicate a figure; page numbers followed by *t* indicate a table.

Boldface page numbers indicate a glossary entry; page numbers followed by an *f* indicate a figure; page numbers followed by *t* indicate a table.

Boldface page numbers indicate a glossary entry; page numbers followed by an *f* indicate a figure; page numbers followed by *t* indicate a table.

S

Boldface page numbers indicate a glossary entry; page numbers followed by an *f* indicate a figure; page numbers followed by *t* indicate a table.

Boldface page numbers indicate a glossary entry; page numbers followed by an *f* indicate a figure; page numbers followed by *t* indicate a table.